A Comprehensive Review of Lubricant Chemistry, Technology, Selection, and Design

Syed Q. A. Rizvi, Ph.D.

ASTM Stock Number: MNL59

ASTM International
100 Barr Harbor Drive
PO Box C700
West Conshohocken, PA 19428–2959

Printed in the U.S.A.

Library of Congress Cataloging-in-Publication Data

Rizvi, Syed Q. A., 1945–

A comprehensive review of lubricant chemistry, technology, selection, and design / Syed Q. A. Rizvi.

"ASTM stock number: MNL59."
ISBN 978-0-8031-7000-1
1. Lubricating oils. I. Title.
TJ1077.R48 2008
621.8'9—dc22 2008003579

Printed in Baltimore, MD
March, 2009

Contents

Preface

This book provides an overview of the technology of lubricants and its use in devising new ones. During my research, I found lubricant literature, both academic and commercial, to be rich with respect to information, with the Worldwide Web being an excellent source. The monographs pertaining to lubricant technology and tribology that I found most useful are listed below.

1. **Motor Oils and Engine Lubrication** by Alphonse Schilling, published by Scientific Publications, Great Britain, 1968.
2. **Lubricants and Related Properties**—Synthesis, Properties, Applications, International Standards, authored by Professor Dieter Klamann, published by Verlag Chemie, Hamburg, 1984.
3. **Chemistry and Technology of Lubricants**, Edited by R. M. Mortier and S. T. Orszulik, published by VCH Publishers, Inc. (Blackie), New York, 1992.
4. **Automotive Lubricants Reference Handbook** by Arthur Caines and Roger Haycock, published by SAE International, 1996.
5. **Lubricant Additives: Chemistry and Applications**, Leslie R. Rudnick, Ed., Marcel Dekker, Inc., New York, 2003.
6. **Fuels and Lubricants Handbook: Technology, Properties, and Testing**, George E. Totten, Ed., ASTM International, Philadelphia, 2003.
7. **CRC Handbook of Lubrication, Vol. I, Theory and Practice of Tribology: Applications and Maintenance**, Richard E. Booser, Ed., CRC Press, Boca Raton, 1983; **CRC Handbook of Lubrication, Theory and Practice in Tribology, Vol. II, Theory and Design**, Richard E. Booser Ed., CRC Press, Boca Raton, 1988; **CRC Handbook of Lubrication, Theory and Practice in Tribology, Vol. III, Monitoring, Materials, Synthetic Lubricants, and Applications**, Richard E. Booser, Ed., CRC Press, Boca Raton, 1993.
8. **Handbook of Lubrication and Tribology, Vol. I, Applications and Maintenance**, 2nd ed., George E. Totten, Ed., CRC Press, Boca Raton, 2006.
9. **Fundamentals of Tribology**, N. P. Suh and N. Saka, Eds., MIT Press, Cambridge, Massachusetts, 1980.
10. **Friction, Wear, Lubrication: A Text Book of Tribology**, Kenneth C. Ludema, Ed., CRC Press, Boca Raton, 1996.
11. **Introduction to Tribology**, Bharat Bhushan, John Wiley & Sons, Inc., 2002.

The present manuscript differs from these in that it is written both from a chemist's perspective and a lubrication engineer's perspective and therefore places equal emphasis on the chemistry and the formulation of lubricants. In addition, it includes additive technology that is used in petroleum exploration, refining, and fuels.

The finalizing of this manuscript would not have been possible without the help of many of my colleagues, at Lubrizol and elsewhere, my wife Marie, and my children Sophia and Ali. My colleagues provided me the technical advice and my family provided the needed moral support. In addition, I am indebted to Marie and Sophia for typing portions of the manuscript and reviewing the document for content, legibility, and grammar. I also wish to express my gratitude to the ASTM publications staff including Monica Siperko, most of who worked behind the scenes to make this publication possible. Above all, I am indebted to Kathy Dernoga of ASTM International who faciliated the completion of this book by providing continuous support.

Syed Q. Rizvi
January 2009

1

Lubrication Fundamentals

THIS CHAPTER DEALS WITH THE FUNDAMENTALS of lubrication. It covers the lubricant functions, nature and composition of the lubricants market, concepts of friction, lubrication, viscosity, and wear, lubricant types, and lubricant selection, performance specifications, and composition. A brief description of the lubricant classes and additives is also provided. The concepts addressed in this chapter are invaluable in understanding the modern lubricant technology and the material covered in the subsequent chapters.

Almost all modern machines require the use of a lubricant. Power generation in such equipment is achieved by the use of engines that mostly comprise metal parts that move against one another. In many cases, there is metal-to-metal contact that leads to the generation of friction and heat, which results in wear. The extent of wear in equipment depends upon the degree of the metal-to-metal contact, either due to the equipment design or the nature of the operation. For example, the equipment that is designed to experience minimal metal-to-metal contact, as is the case in most parts of an internal combustion engine, there is little friction and wear. However, the parts that are designed to have intimate metal-to-metal contact, such as gears and bearings, wear due to friction is extensive. With respect to the effect of equipment operation on wear, high-speed, low-load operation leads to lower wear than slow-speed, high-load operation. This is because in the former case there is minimal metal-to-metal contact. A lubricant can be a solid, liquid, or gas, and lubrication is its primary function. The usual objective of the lubrication is to lubricate surfaces to minimize direct metal-to-metal contact and, hence, reduce friction and wear. The term lubricant is also loosely applied to many other fluids that do not specifically perform this function. Examples include power and heat transmission fluids, hydraulic fluids, dielectric fluids, process oils, and the others. Incidentally, in this book the term "lubricant" pertains to a finished lubricant, that is, it comprises base fluid and additives.

A lubricant performs many diverse functions, which help protect and prolong the life of the equipment [1]. These include the following:

1. *Lubrication* (reduce friction and wear)—Lubricant helps reduce friction and wear by introducing a lubricating film between mechanical moving parts, such as gears and bearings. Essentially, the presence of a lubricating film minimizes the metal-to-metal contact and reduces the force necessary to move one surface against the other, thereby reducing wear and saving energy.
2. *Cooling* (heat transfer)—Lubricant acts as a heat sink and dissipates the heat away from the critical moving parts of the equipment, thereby decreasing the possibility of the machine component deformation and wear. The heat is either frictional heat that results from the metal surfaces rubbing against one another, such as in gears, or is conducted and radiated heat, which is due to the close proximity of the parts to a combustion source, such as the combustion chamber in an automobile engine.
3. *Cleaning and Suspending*—Lubricant facilitates smooth operation of the equipment by removing and suspending potentially harmful products, such as carbon, sludge, and varnish, and the other materials, such as dirt and wear debris. This lubricant function is important in operations that involve high operating temperatures, as in the case of an internal combustion engine or a transmission. This is because in these applications the lubricant gets oxidized to form deposit precursors that can separate on hot surfaces and get converted into deposits.
4. *Protection*—Lubricant prevents metal damage due to oxidation products, corrosion, and wear. It achieves this by forming a physical film on metal surfaces that is impervious to oxygen, water, and acids, or by forming physical and chemical films by additives, such as rust and corrosion inhibitors, extreme-pressure (EP) additives, and anti-wear agents, that are present in the lubricant.
5. *Transfer Power*—Lubricant is used as a power transfer medium in some applications, for example, in hydraulic systems. The lubricant performs this function in addition to its normal function of lubrication. Examples of equipment that use hydraulics technology include transmissions, circulating systems, lifts used in automotive service stations, log splitters, fork lifts, dump trucks, and underground continuous mining equipment such as drills, loaders, and miners.

Lubricant Market

The world lubricant market for the year 2006 is estimated at around 38.5 million metric tons (85 billion pounds), with the United States consuming about 9.5 million metric tons (21 billion pounds), or almost 25 % of the total world use [2]. Asia Pacific and the Near/Middle East have the next highest share and are the regions with the fastest growth in lubricant consumption. They are followed by Central and Eastern Europe, Western Europe, and the others. The 2005 world demand breakdown by geographic areas is presented in Table 1.1 and in Fig. 1.1 [2]. The demand for lubricants is projected to increase at an average rate of ~2 % over the next decade,

TABLE 1.1—2005 world lubricant consumption by region (in million metric tons) [2].

Region	Estimated Use
North America	9.48
Asia Pacific, Near/Middle East, Africa	14.02
Central and Eastern Europe	4.80
Russia	1.67
Western Europe	4.69
Latin America	3.00
Others	0.24
Total	37.9

bringing the world lubricant consumption to 41.8 million tons in the year 2010 [2]. This estimate takes into consideration the expected growth rates of 5 % for China, 3.5 % for India, and 2.5 to 3.3 % for Malaysia, Indonesia, and Thailand [2].

Engine oils, that is, the internal combustion engine lubricants, accounted for approximately 57 % of the 2005 lubricant use, more than half being in commercial fleet and off-road vehicles (29 %) and the rest in passenger cars and vans (28 %). The balance of 43 % is used in nonautomotive (industrial) applications. World lubricant use by product type is provided in Fig. 1.2 [2]. It is important to note that the use ratio between the two types of applications varies across regions. For example, automotive use in Western Europe is 47 %, and in the relatively less industrialized nations of the Near/Middle East it is as high as 74 %.

The U.S. lubricant growth in recent years has been slow, on the order of around 1 %. The slow growth of the automotive lubricants in the United States can be ascribed to smaller crankcases in newer cars, decrease in the distance driven per year, reduction in new U.S. made car purchases, and tighter modern engines which consume less oil per traveled mile. The sluggish growth in industrial lubricants reflects the general trend of the United States away from being a heavy industry-based economy. Historical data show that the overall global lubricant market has not changed much over the last decade. This is because the increased demand in Asia and Latin America is offset by a drop in demand in Western Europe and North America. The additional factors that are either affecting the present lubricant demand or will impact future consumption of the lubricants, hence the additives, include the following:

- New engine designs to achieve more efficient combustion.
- Continuously variable transmissions (CVTs).

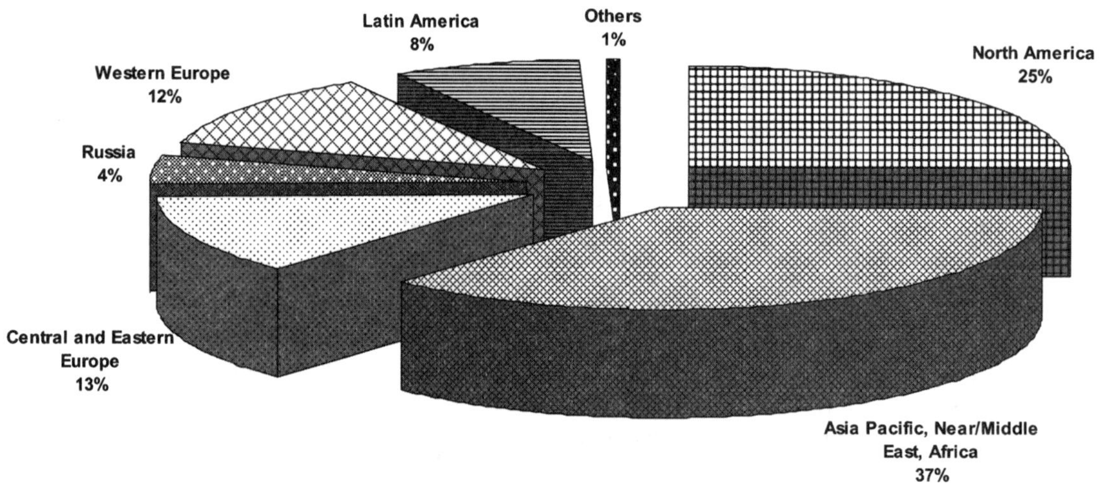

Fig. 1.1—2005 world lubricant consumption by region.

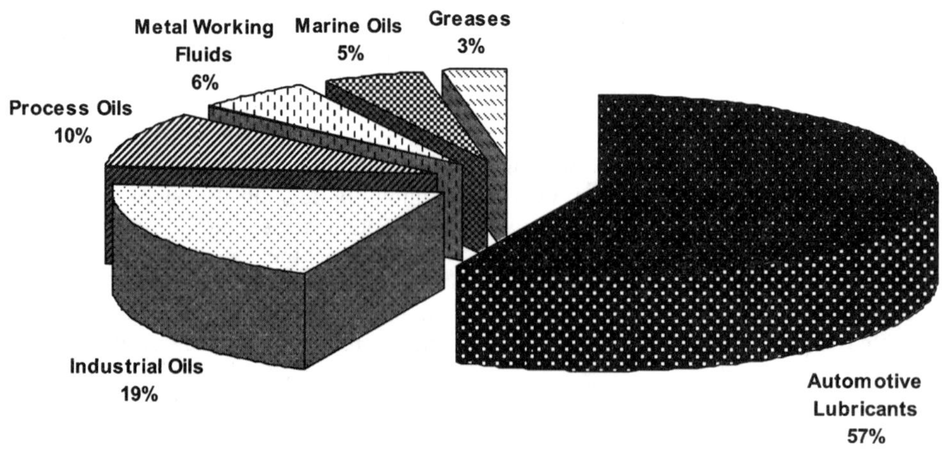

Fig. 1.2—World lubricant use by product type [2].

- On-board oil monitoring.
- New performance classifications or specifications.
- Government regulations pertaining to the stringent emissions standards and lowering of the certain chemical elements, such as phosphorus, chlorine, and some metals, in lubricants due to their real or perceived adverse effects on the environment.
- Extended service intervals.
- Filled-for-life, sealed-for-life drive lines.
- Fuel cell technology.
- Development and or growth of electric cars and hybrid cars.

Friction and Lubrication

Tribology and tribo-technology are two terms that are often used in relation to lubrication and wear. Tribology is the science and technology of friction, wear, and lubrication, derived from the Greek word tribo, meaning "to rub." The term was first introduced in a British study in 1966 (The Jost Report), which pointed out the magnitude of the annual monetary loss in the U.K. ascribed to the consequences of friction, wear, and corrosion. Since then the term has gained worldwide use in lubrication and mechanical engineering. More precisely, tribology is the science and technology of the interacting surfaces in relative motion, irrespective of whether or not they involve mechanics [3]. Tribology is an interdisciplinary approach that involves a scientific basis to understand surfaces in contact and the lubrication needs of a given tribological system. The term friction has its origin from the Latin word "fricare" that means "to rub." Applied tribology, or tribo-technology, primarily deals with the maintenance of machines and the minimization of wear and energy losses due to friction.

Friction

Friction is the force that hinders or resists the relative motion of the two contacting bodies and, depending on the application, high friction may either be desirable or undesirable. For example, in tire traction on pavement and braking, high friction is desirable. However, in applications such as the operation of engines or of equipment with bearings and gears, high friction is undesirable. This is because friction causes wear and generates heat which can lead to premature failure of the functioning machine parts. Frictional heat not only causes wear through welding but it is also considered a waste since it does not result in useful work. Friction originates from complex molecular and mechanical interactions between the contacting surfaces. Two bodies in direct contact with each other experience dry or solid friction. When they are separated by a solid, liquid, or gaseous medium, they experience fluid friction. Between these two extremes is the situation where some parts of the two bodies are in direct contact while the others are separated by a fluid film. This is called mixed friction. Friction between two solids is dependent upon the size of the contact zone, surface roughness (asperities), and load or pressure, on surfaces, but is independent of the materials. The friction may even involve a single body, in which case it is related to the dissipation of the internal energy within the body and is called internal friction [4].

Friction is commonly represented by the friction coefficient, signified by the symbol μ. The coefficient of friction (μ) is a unit-less ratio that equals F_M/F_N, where F_M represents the frictional force experienced by the two contacting bodies in motion and F_N represents the normal force pressing the same two bodies together. The value of the coefficient of friction typically ranges from 0 to 1; the higher the value, the higher the frictional force or the resistance of the contacting bodies towards motion. Under boundary lubrication conditions, μ usually approaches 1.

Minimizing friction is one of the fundamental functions of a lubricant. If friction is not controlled it can lead to wear and surface damage, and ultimately to catastrophic failure of the equipment. Because of a generally direct correlation between friction and wear [5,6], proper lubrication of the equipment is important if its integrity is to be preserved over its designated lifetime. However, it is important to note that the correlation between friction and wear is a function of the system and is not always direct [7].

In lubricant-related applications, we are concerned with all three types of friction, that is, solid friction, fluid friction, and the internal friction. The major function of a lubricant is to minimize solid friction which it achieves by forming a fluid film between the two contacting (metal) surfaces. Usually, a fluid's internal friction is not of any major consequence except at very low temperatures. At these temperatures the lubricant gains viscosity which can interfere with the smooth operation of the equipment. Internal friction is important while dealing with a lubricant's intrinsic properties, such as viscosity and pour point.

All metal surfaces, irrespective of their finish, contain ridges, valleys, asperities, and depressions. When two metal surfaces come in contact, solid friction, sometimes called static or adhesive friction, ensues and the surfaces undergo adhesion and cold welding. The strength of such an association depends upon the hardness of the materials, the cleanliness of the surfaces, and the electronic structure of the metals as related to their tendency to form metal-metal solutions, or alloys [8,9]. As soon as the surfaces start to move, kinetic friction comes into play. Kinetic friction results from plowing of the asperities of the one surface across the other surface, plastic deformation or elastic hysteresis, and wear debris getting lodged between the moving surfaces [9].

Friction is also related to the type of motion of the two contacting bodies. Sliding motion, for instance, leads to higher friction than rolling motion, and hence results in more wear. Rolling friction and sliding friction are two general cases of friction. For example, when force is applied to slide a steel block sitting on a steel table, both will experience sliding friction. If a weight, or load, is placed on top of the metal block, the force necessary to cause sliding will increases significantly. When a metal cylinder is made to roll on the surface of metal table, the cylinder will experience rolling friction, or the rolling resistance. Experience shows that in general less force is required to roll an object than to slide or drag it. For lubricated surfaces, friction is governed by different laws than those for dry surfaces [10]. Table 1.2 shows the relationship between wear and different types of friction [4].

The data in the table show the following:

- A positive relationship exists between the coefficient of

TABLE 1.2—Relationship between friction and wear [4].

Friction Regime	Coefficient of Friction (μ)	Wear
Dry friction (sliding)	0.300[a]	High
Dry friction (rolling)	0.005	Very low
Mixed friction (rolling)	0.005–0.300[a]	Noticeable
Fluid friction	0.005–0.100	Practically zero

[a]Some plastics show significantly lower values.

friction (μ) and wear; that is, the higher the coefficient of friction, the higher the wear.

- Solid or dry friction is more severe than the mixed friction, which in turn is more severe than the fluid friction.
- Sliding friction is higher than the rolling friction, which is primarily a consequence of the larger contact zone of the sliding surfaces.

Lubrication

Applications that encounter metal-to-metal contact involve either no lubrication (dry), solid lubrication, or liquid lubrication. Wear resistance in equipment designed to operate without lubrication is introduced by the use of low-wear metals or surface treatment, or both, such as hardening or coating. Agricultural plows and certain parts of ore handling machines are examples of such equipment. Solid lubrication is common where liquid lubrication is unwanted or is difficult because of the equipment design or extremely high operating temperatures. Solid lubricants, exemplified by graphite and molybdenum disulfide, have multi-layered structures with low shear strength in some directions. Movement in these directions is therefore facilitated. These lubricants are applied to equipment in a number of ways, such as bonded dry films, sputtered films, and loose flakes.

When we talk about lubrication, we usually imply liquid lubrication, that is by the use of lubricating oil, which is normally a blend of oil and additives that perform various functions. Lubrication efficiency of an oil depends not only upon its properties, such as composition, consistency, flow properties, and surface activity, but also on the needs of the tribological system. Lubrication environments, often called lubrication regimes, are primarily defined by considering these needs. The factors that are considered include gross geometry of the contacting surfaces, their texture and roughness, the nature of the contact (rolling versus sliding), the contacting load, ambient pressure and temperature, the environmental conditions, material composition, and the properties of the surface layers. Lubrication effectiveness is measured by the film thickness, the ability to handle pressure (load-carrying capacity), and the coefficient of friction.

In extreme environmental conditions, such as ambient temperatures above 500°C (930°F) or the vacuum environment, conventional liquid lubricants often become less effective. This is because either their viscosity is too low or they rapidly oxidize, decompose, or under high vacuum vaporize away from the surfaces. Although many synthetic base fluids with high viscosity-temperature (VT) properties, good thermo-oxidative stability, and low volatility have been developed, in certain cases the use of the solid lubricants is imperative. Over the past 50 years, the use of these lubricants has grown extensively.

Lubrication Regimes

As mentioned earlier, the primary functions of a lubricant are to minimize friction between the surfaces in contact, prevent wear, and remove frictional heat. Tribological parameters that usually define a lubrication environment are friction, lubricant viscosity, and the equipment speed and load. The relationship of the coefficient of friction (μ) and the oil film thickness to lubricant viscosity (Z), equipment speed (N), and equipment load, or pressure (P), are graphically presented by the Stribeck curve [11] in Fig. 1.3. The ratio of ZN/P is related directly to the oil film thickness but inversely to the coefficient of friction (μ). This implies that high lubricant viscosity (Z), high equipment speed (N), and low equipment load (P) will allow the formation of a thick lubricant film, and hence the equipment will encounter little or no friction. Conversely, low lubricant viscosity, low equipment speed, and high equipment load will create a situation where the film thickness will be inappropriate and the equipment will encounter high friction, as indicated in the figure.

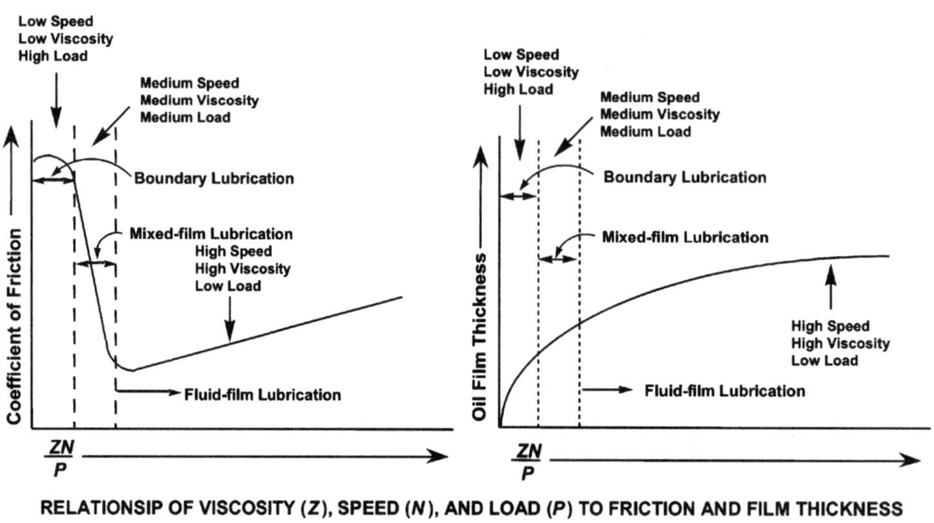

Fig. 1.3—Types of lubrication [11,12].

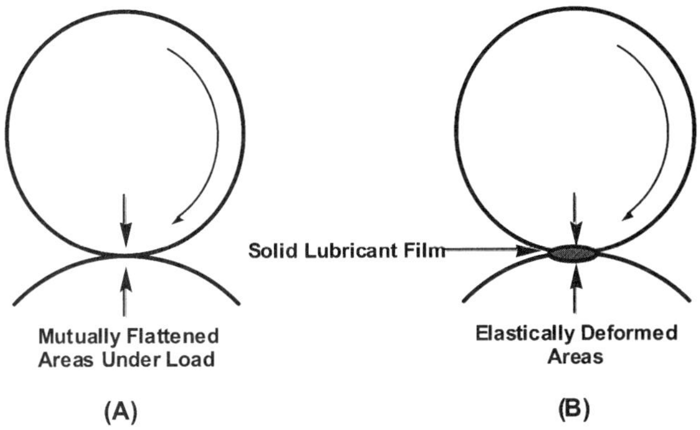

Fig. 1.4—Elasto-hydrodynamic lubrication.

Incidentally, the observed initial drop in the coefficient of friction while moving from fluid-film to mixed-film lubrication reflects a decrease in viscous drag due to a decrease in lubricant viscosity. Depending upon the lubricating environment, lubrication regimes can be divided into fluid-film, boundary, mixed-film, and hydrostatic types. The first three of these states of lubrication are depicted in Fig. 1.3 [12].

Fluid-Film Lubrication

Fluid-film lubrication, also known as hydrodynamic lubrication, is the most desirable type. This type of lubrication depends upon the viscosity of the lubricant and is effective only when the load in the contact zone is low. Under these circumstances, the sliding or the rolling surfaces are separated by a lubricant film several times the thickness of the surface roughness (asperities). The film thickness in this lubrication regime is estimated to be 2–100 μm. Lubrication of the thrust bearings, journal bearings, and most of the internal combustion engine parts experience fluid-film lubrication.

Another type of hydrodynamic lubrication, referred to as elasto-hydrodynamic lubrication, or EHD [13–15], commonly occurs in roller element bearings (ball and roller types), cams, and gears. In this type of lubrication, the lubricant is exposed to high contact pressures and undergoes a large viscosity increase. This results in an extremely rigid lubricant film (0.01–5.0-μm thick), which causes elastic deformation of the surfaces in the lubricating zone. Elasto-hydrodynamic lubrication in rolling contacts is depicted in Fig. 1.4.

Boundary Lubrication

Boundary lubrication represents the opposite extreme of the lubrication environment spectrum. Under this kind of lubrication, high loads and very slow speeds produce extreme pressures that can lead to the lack of effective lubrication. The film thickness in this regime is in the order of 0.0–2.0 μm only, and hence maximum metal-to-metal contact occurs. If not controlled, the resulting dry metallic friction will cause catastrophic wear, and ultimately will lead to total seizure. Reactive chemicals called anti-wear and extreme pressure agents provide protection in this kind of lubrication environment. Examples of equipment that rely exclusively on boundary lubrication include reciprocating parts of an engine and compressor pistons, slow-moving equipment such as turbine wicket gates, and gears. It is important to note that the anti-wear agents are effective only up to a maximum temperature of about 250°C (480°F), above which they essentially become ineffective. Typically, heavy loading causes the oil temperature to increase beyond the effective range of the anti-wear agents. This is because the degree of contact between the surface asperities further increases due to flattening and the consequence being greater friction, hence higher temperatures. When the load exceeds the equipment's recommended limit, the asperities, instead of sliding, experience shearing and removing the lubricant and the oxide layer. The result is catastrophic failure.

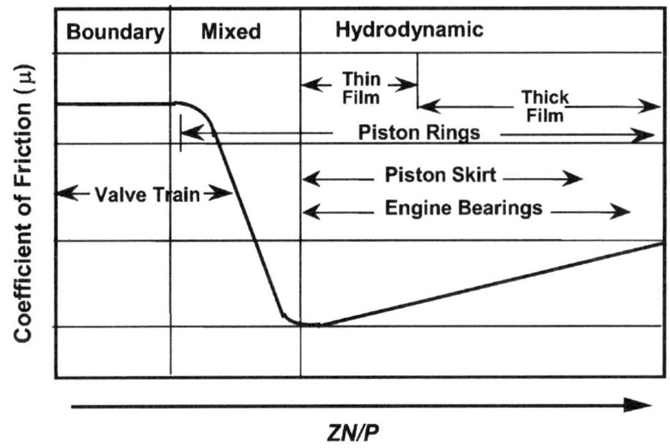

Fig. 1.5—Stribeck diagram showing the type of lubrication encountered by various engine parts.

Mixed-film Lubrication

Mixed-film lubrication falls between the two extremes mentioned above and contains characteristics of both the fluid-film and the boundary lubrication. There are regions of no metal-to-metal contact and of extensive metal-to-metal contact.

Hydrostatic Lubrication

Unlike the other types of lubrication discussed above, hydrostatic lubrication has the advantage of not depending upon the motion of the surfaces. Hence, this type of lubrication is invaluable in applications that involve little or no surface movement. This lubrication regime is characterized by the lack of wear, low friction, high load capacity, and the ability to dampen vibration. Examples of hydrostatic lubrication include lubrication of some type of bearings, such as hydrostatic bearings, and certain metal-forming equipment involving simple pressure [16]. This type of lubrication allows complete separation of the surfaces by the static film of the lubricant.

Although boundary lubrication is encountered in certain parts of the engine, such as valve train, cylinder bores, and piston rings, most of the lubrication in an engine is hydrodynamic in nature [17,18], see Fig. 1.5. Boundary lubrication is more common in rear axles, gears, and bearings. The surfaces in these parts are designed to mesh closely so as to efficiently transfer power generated by the power source to parts that work. Figure 1.6 shows the lubrication regimes encountered in various automotive applications.

Lubrication Methods

In most applications, the lubricant is delivered from a reservoir to various parts of the equipment that require lubrication and cooling. As stated earlier, lubrication is the process in which the lubricant reduces friction by forming a film be-

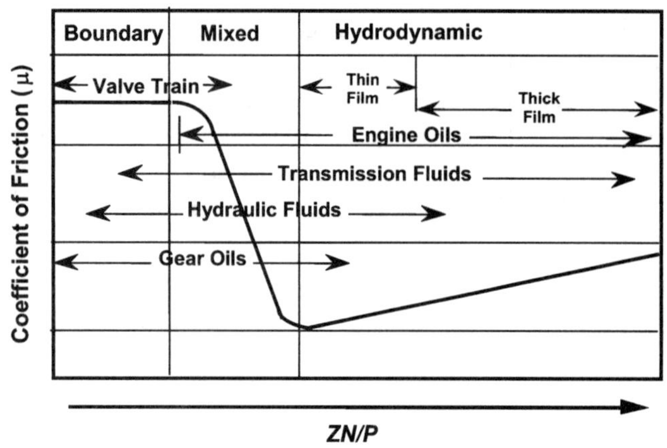

Fig. 1.6—Stribeck diagram showing the types of lubrication encountered in various automotive applications.

tween two contacting surfaces in motion. The strength and the durability of this film are related to lubricant viscosity and the speed and load experienced by the moving surfaces. The lubrication effectiveness depends upon both the quality and the quantity of the lubricant and its delivery to parts that need lubrication. A number of methods are used to lubricate various mechanical devices [19]. These include the following:

- Manual lubrication
- Bath or splash lubrication
- Waste-type lubrication
- Air-oil mist lubrication
- Centralized lubricating systems
- Drop-feed lubrication
- Ring, chain, or collar lubrication
- Positive force-feed lubrication
- Pressure circulating systems
- Built-in lubrication

Manual Lubrication
Manual lubrication involves human action of some sort. While a number of devices are available for this purpose, an oil squirt can, a sprayer, or a brush are commonly used. Manual lubrication suffers from low reliability; hence, its use is primarily limited to small machine elements and slow-speed bearings.

Drop-feed Lubrication
Drop-feed lubrication depends upon the flow of a lubricant under the influence of gravity. The lubricant is delivered to the machine elements one drop at a time. This method of lubrication is commonly used for parts that are easily accessible. Drop-feed lubrication is used for journal and roller bearings, gears, chains, engine guides, pumps, and compressors.

Bath Or Splash Lubrication
Bath or splash lubrication is used for machinery that contains high-speed moving parts. This lubrication method depends upon a rotating or reciprocating mechanism that partly sits in the lubricant. In bath lubrication, the rotating or reciprocating part transfers lubricant to the neighboring parts through contact. In splash lubrication, the rotating part physically throws the lubricant at the neighboring parts. The excess lubricant cycles back to the reservoir. This method, although expensive, allows continuous and dependable lubrication of the parts that are hard to lubricate by the other methods. Bath or splash lubrication is used for different parts of internal combustion engines, chain drives, and the enclosed gear sets.

Ring, Chain, Or Collar Lubrication
This lubrication method is used for horizontal rotating shafts. These devices, partially submersed in the lubricant, lubricate the shaft by carrying the lubricant from the reservoir through rotation. The ring and chain turn freely around the shaft, but the collar is fixed to the shaft. This method is commonly used for electric motors, fans, blowers, compressors, and line shaft bearings.

Pad and Waste Lubrication
Pad and waste lubrication depends upon the oil-retaining characteristics of the felt pads and waste packing. The pad or the packing extends into the lubricant reservoir, and the lubricant is transferred from the reservoir to the machine element by the capillary action of the pad material. This type of lubrication is used for railroad and traction motor bearings.

Positive Force-feed Lubrication
This type of lubrication uses one or more plunger-type adjustable-stroke pumps to transfer the lubricant from a reservoir to the parts requiring lubrication. By connecting the rotating shaft that drives the pump to the moving elements of the machine to be lubricated, the whole operation can be made automatic. This type of lubrication is used for steam cylinders, bearings for gasoline and diesel-fueled engines, oil drilling rigs, and metal press bearings.

Air-oil Mist Lubrication
Air-oil mist lubrication is based on the use of the oil atomized by compressed air or steam. This type allows lubrication across distance and is used for high-speed bearings, enclosed gears, chains, slides, and guides.

Pressure Circulating Systems
This method involves the use of pressure for lubrication. An appropriate amount of pressure is generated either through gravity or with pumps. The pressure ensures uniform and continuous delivery of the lubricant to parts that require lubrication. These systems are devised to lubricate a number of parts simultaneously. Pressure circulating systems are used for internal combustion engines, gears, and bearings.

Centralized Lubrication Systems
These lubrication systems employ a centrally located lubricant reservoir and a pump. Such systems are usually automatic in that they start and stop with the machinery being lubricated. A centralized system is ideally suited to equipment that has multiple lubrication points.

Built-in Lubrication
This type pertains to components that do not need any external lubrication. This may be due to the inherent nature of the material used to fabricate the part or because the part design makes external lubrication difficult. Built-in lubrication is used for sleeve bearings, gears, and slide ways.

Each of these methods has its advantages and limitations. Depending upon the cost constraints and the lubricating needs of the equipment, some methods may be more appropriate than the others.

Lubricant Selection
Selecting a lubricant that matches the performance requirements (specifications) of the intended application is the first step. The performance requirements are defined by a num-

TABLE 1.3—Lubricant properties of high importance for various applications.

Application	Dispersancy	Friction and Wear Control	Oxidation Resistance	Corrosion Control	Low-temperature Fluidity	Foam Control	Viscosity Index	Viscosity/ Pressure Relationship
Engine Oils								
Gasoline	✓	✓	✓	✓	✓	✓	✓	
Diesel	✓	✓	✓	✓	✓	✓	✓	
Stationary Gas	✓	✓	✓	✓	✓		✓	
Aviation	✓	✓	✓	✓	✓		✓	
Two-stroke Cycle	✓	✓	✓	✓				
Transmission and Hydraulic Fluids								
Automatic Transmission	✓	✓	✓	✓	✓	✓	✓	✓
Tractor Hydraulic	✓	✓	✓	✓	✓	✓	✓	✓
Industrial Hydraulic	✓	✓	✓	✓	✓	✓		✓
Gear Oils								
Automotive		✓	✓	✓	✓	✓	✓	✓
Industrial		✓	✓	✓	✓	✓	✓	✓
Miscellaneous Lubricants								
Metalworking Fluids	✓	✓	✓	✓		✓		✓
Industrial Oils		✓	✓	✓		✓	✓	
Turbine Oils		✓	✓	✓		✓	✓	
Greases		✓	✓	✓				✓

ber of national and international organizations and the end-users. Specifications are established by considering many equipment-related parameters. These include system design, operating conditions, lubrication needs of the equipment, duration of use, safety, health, and environmental considerations, and cost. Typically, lubricants designed for one application are not suitable for use in another application, without deterioration in performance. Original equipment manufacturers (OEMs) play a predominant role in recommending viscosity grades and lubricant quality, which are based upon their system needs. A user must follow manufacturer recommendations to ensure warranty protection. Lubricant properties that are commonly considered for assessing the suitability of a lubricant for a particular application include the following. Incidentally, lubrication-related terms, acronyms, and organizations are included in the appendix.

1. Viscosity
2. Fluidity Range
3. Viscosity-temperature Relationship (Viscosity Index)
4. Low-temperature Fluidity
5. Oxidation Stability (Inhibited)
6. Hydrolytic Stability
7. Thermal Stability
8. Mineral Oil Compatibility
9. Additive Solvency
10. Volatility
11. Rust Control (Inhibited)
12. Boundary Lubrication
13. Fire Resistance
14. Elastomer Compatibility, Especially with Buna Rubber
15. Relative Cost

In addition, there are other properties that are important in some end-use applications. These include color, density, API gravity, volatility, bulk modulus, viscosity-pressure relationship, shear stability, acidity and alkalinity, detergency, dispersancy, and foaming and air release tendency. Lubricant properties of high importance for various applications are listed in Table 1.3.

A formulated lubricant comprises a base fluid and a performance package, and in the case of multi-grade oils, an additional viscosity modifier. The amount of each of these components varies based upon the application and the desired service. Figure 1.7 provides the approximate ranges of each in the lubricant and Table 1.4 [20] provides the typical ranges of additives and the viscosity modifier used in automotive lubricants. The performance package contains a number of additives, the quality and quantity of which depend upon the quality of the base fluid and the lubricant's intended use. The performance package can make up to 20 %, and sometimes even higher, of the total lubricant composition, depending upon the desired performance level and the severity of the end-use requirements. In general, the base fluids of inferior quality need better additives and in larger amounts than the base fluids of good quality. Likewise, the applications, such as combustion engines and automotive gears, which place a higher demand on the lubricant, require superior additives than the less-demanding applications, such as some industrial and metalworking operations. Since the base fluid is the

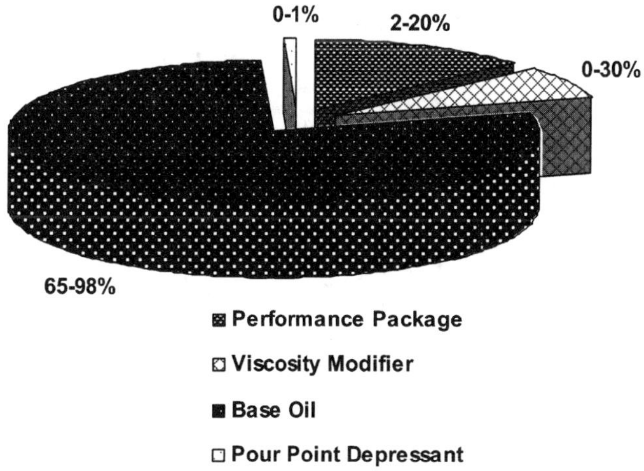

Fig. 1.7—Lubricant composition.

TABLE 1.4—Typical lubricant composition [20].

Application	SAE Grade	Additive[a] (%)	Viscosity Modifier[b] (%)
Passenger Car Motor Oil	Single grade	6-14	–
	Multi-grade	6-14	2-9
Heavy-duty Diesel Oil	Single grade	12-18	–
	Multi-grade	12-19	6-12
Automatic Transmission Fluids	Multi-grade	6-12	3-14
Automotive Gear and Transmission Oils	Single grade	5-12	–
	Multi-grade	5-12	25-35
Tractor Hydraulic Fluids	Single grade	6-9	–
	Multi-grade	6-9	4-8
Industrial Hydraulic Fluids	Single grade	0.5-1.5	–
	Multi-grade	0.5-1.5	2-12
Industrial Turbine Oils	Single grade	0.5-1.5	–
Greases	–	2-6	Contains a thickener[c]

[a] Level depends upon the performance desired.
[b] Level depends upon the viscosity of the base oil and additive package, the viscosity grade, and the nature and molecular weight of the polymer.
[c] Also contains a polymer to improve tack and water sheddability.

largest component in the lubricant, its properties primarily determine the properties of the lubricant. These are considered in detail in the sections that follow.

As far as the automotive applications are concerned, we are primarily interested in liquid lubrication, that is, to separate the contacting surfaces by introducing a liquid film. Figure 1.8 depicts the general arrangement of a cam-in-block

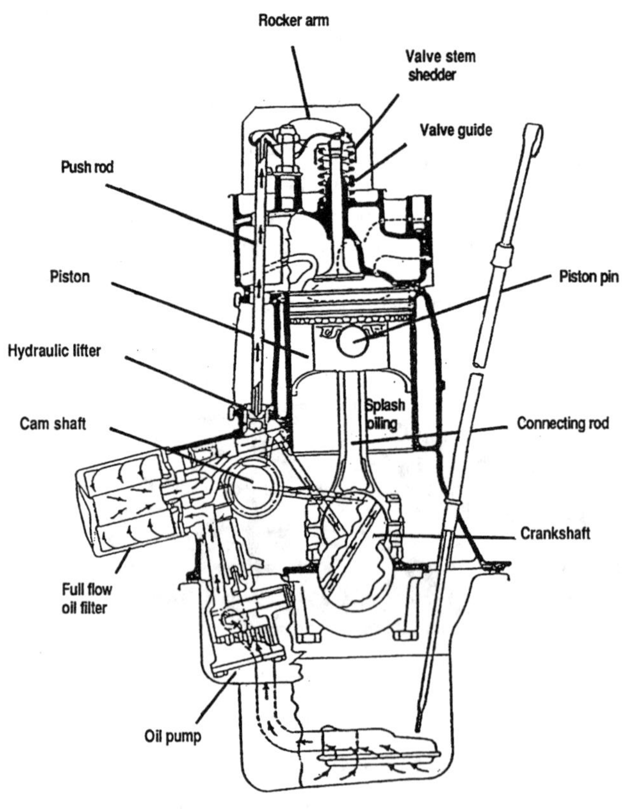

Fig. 1.8—Lubrication system of a cam-in-block overhead valve engine [21b].

overhead valve engine's lubrication system [21b]. While such a system primarily employs full-pressure and force-feed lubrication, other methods, such as splash and air-oil mist lubrication, are also used for certain parts [22]. Large engines and most automotive engines use a full-pressure system. Small engines use a splash or modified splash system [23].

In the full-pressure system, the oil is pumped from the oil sump to the main bearing and connecting rods and up the connecting rods to the piston pin. In overhead valve engines, a portion of the pumped oil travels through push rods (in some cases), over rocker arms, past valve stems, and down the valve guides. In many engines, the cylinder walls and the piston pins depend upon splash lubrication by the oil that is thrown off of the main bearing. Air-oil mist lubrication is used to lubricate the rocker boxes.

The choice of a lubrication method for equipment other than the combustion engines depends upon its nature and use. As mentioned previously, for transmissions, axles, and enclosed gears, forced feed and splash lubrication are the methods of choice. For open gears, drop feed or air-oil mist lubrication may be appropriate.

Viscosity and Wear

Viscosity

The role of viscosity in forming effective lubricating films makes it one of the most important properties of the lubricant [24]. Viscosity is defined as a fluid's resistance to flow and, as mentioned earlier, is primarily a consequence of the

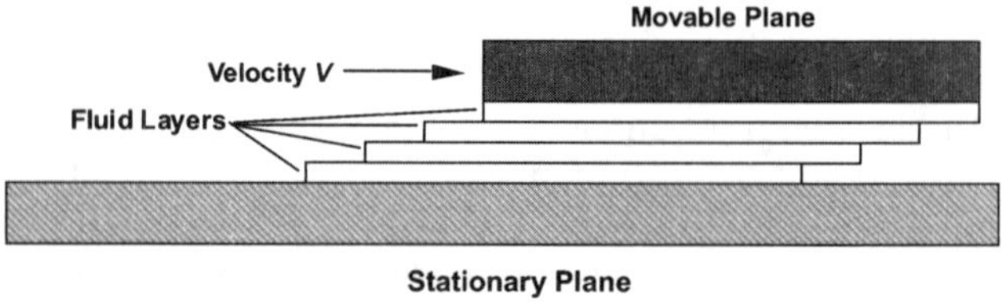

Fig. 1.9—Shearing planes in laminar flow [23].

TABLE 1.5—Divergence of kinematic viscosity (ν) and absolute viscosity (η) data with increasing temperature [4].

Paraffinic Base Oil, d_4^{15}=0.871 g/mL				Naphthenic Base Oil, d_4^{15}=0.925 g/mL			
Temp., °C	ν, mm²/s	η, mPa·s	Δ, %	Temp., °C	ν, mm²/s	η, mPa·s	Δ, %
0	287	253	13.4	0	1330	1245	6.8
20	78.4	68	15.3	20	218	201.0	8.5
40	30.2	25.8	17.1	40	60.5	55.0	10.0
60	14.7	12.33	19.2	60	23.6	21.2	11.3
80	8.33	6.91	20.5	80	11.6	16.2	13.7
100	5.3	4.32	22.7	100	6.66	5.80	14.8
120	3.65	2.93	24.6	120	4.27	3.66	16.7
150	2.33	1.83	27.3	150	2.53	2.12	19.3

internal friction of the fluid [4]. A simple model to help explain the concept of viscosity is presented in Fig. 1.9 [23].

The model shows the fluid in the form of parallel layers of certain molecular thickness between a stationary plane and a movable plane. When a tangential force F is applied, the top plane moves at a constant velocity V. The force that tends to cause the plane of an area to slide on adjacent planes is called shear. Because the fluid usually wets or adheres to surfaces, the layer of the fluid next to the moving plane will move at velocity V, the same velocity as that of the moving plane. This movement is transmitted through successive layers of the fluid in a dissipating manner until the fluid velocity approaches zero near the stationary plane. The slowdown in movement occurs due to friction between the fluid layers, each of which drags the layer above it and the layer below it. The overall effect is the fluid's resistance to free flow, termed viscosity. Viscosity without the influence of gravity is called absolute or dynamic viscosity and that under the influence of gravity is called kinematic viscosity.

Absolute or dynamic viscosity (η) equals τ/s, where τ represents the shear stress and s represents the shear rate. Shear stress (τ), the force applied per unit area, is denoted by F/A, where F represents the force in the shearing direction and A represents the area. Shear rate (s), the velocity gradient, is denoted by V/Y, where V is the velocity of the moving plane and Y is the fluid film thickness.

$$\eta = \frac{\tau}{s} = \frac{F/A}{V/Y}$$

The viscosity of certain fluids is independent of the shear rate. Such fluids are called Newtonian, named after Sir Isaac Newton, who first made this observation [24]. Conversely, fluids whose viscosities vary with the shear rate are called non-Newtonian. The addition of chemicals with polymeric structures, for example, viscosity improvers and dispersants, to Newtonian fluids makes them non-Newtonian. Hence, mineral base oils are Newtonian, but the finished lubricants are non-Newtonian.

Absolute viscosity is commonly expressed in poise (P) units. The SI (Système International d'Unites) unit of absolute viscosity is Pascal-second (Pa·s), which is equal to 10^3 centipoise (cP). Certain methods, such as capillary or effluent viscometers, measure viscosity under the accelerating influence of gravity. This type of viscosity, termed kinematic viscosity (ν), equals absolute viscosity (η) divided by fluid density (ρ), or η/ρ. Kinematic viscosity is expressed in SI units of m^2/s (square metres per second). One m^2/s is equal to 10^6 centistokes (cSt) and 1 cSt equals 1 mm^2/s. For an exact inter conversion of the two viscosities, the density of the liquid at the temperature of measurement must be known. This is because the drop in the viscosity of an oil with an increase in temperature is related to a decrease in its density [4], which is not the same across all temperature ranges. The measured viscosity-temperature data for a paraffinic oil and a naphthenic oil, oils of different densities, are presented in Table 1.5 [4]. While in both cases there is a decrease in viscosity with increasing temperature, the decrease at each temperature for the lower density paraffinic oil is larger in magnitude than that for the higher density naphthenic oil. In the absence of the knowledge of the oil density at a particular temperature, a temperature coefficient of density 0.00065 K^{-1} is often used for mineral-based lubricants (see DIN 51 757 or ASTM D1250 conversion tables). Dynamic viscosity is required for comprehending the lubrication processes in bearings, gears, etc. However, determining this viscosity is not straight forward. On the other hand, the kinematic viscosity is easy to measure and with greater precision. It is therefore the preferred viscosity for production control and the characterization of the lubricants [4].

The popularly used viscosity units, Poise and Stoke, are named after the scientists Dr. J. L. M. Poiseuille and Sir George Stokes in honor of their work on viscosity and viscous fluids [24]. Previously, Saybolt Seconds Universal (SSU or SUS) and Saybolt Furol Seconds (SFS) were also used to express the kinematic viscosity. However, these units are now obsolete. The methodology to convert these units into Stokes is described in the ASTM Standard D2161.

A number of factors can affect viscosity. These include temperature, pressure, time, and the structures of the organic compounds present in the fluid and their response to the shear forces [4]. Viscosity decreases with increasing temperature, increases with increasing pressure, and generally decreases with increased shear. Viscosity also depends upon chemical structures that make up the lubricant and their molecular size and shape. Within the same structural type, it increases with the molecular size [25,26]. Aliphatic structures (paraffinics) are less sensitive to temperature than cycloaliphatic and aromatic structures (naphthenics and bright stocks). In the case of the mineral oil base stocks, which contain molecules that are derived from all three types of structures, the viscosity-temperature (VT) relationship depends upon the ratio of the number of carbon atoms

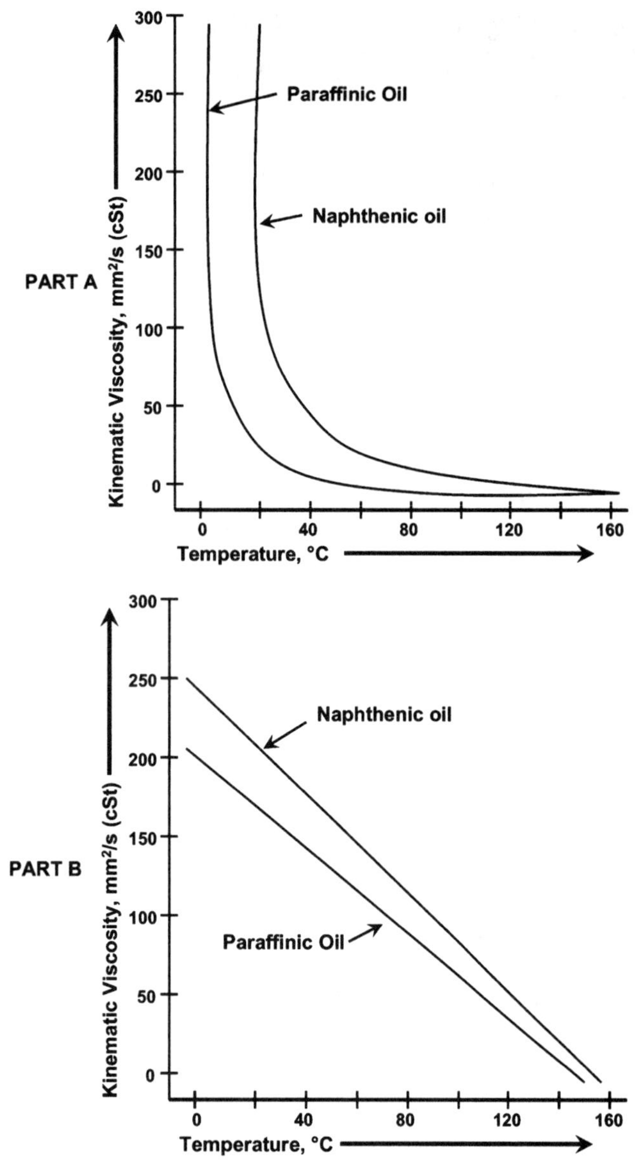

Fig. 1.10—Viscosity-temperature relationship for mineral base oils [4].

in the hydrocarbon chains and the number of carbon atoms in the ring structures. The effect of pressure on the viscosity of the aliphatic hydrocarbons is a lot lower than on that of the naphthenics and aromatics. Branching of hydrocarbon chains increases viscosity since it increases the molecular size. Oxygen-containing structures, such as those present in the synthetic ether- and polysiloxane base stocks, reduce rigidity, hence they reduce viscosity. Polar groups such as halogens and hydroxyls have the opposite effect [4]. The viscosity of the fluids with weight average molecular weights of less than 20,000 g/mol is not greatly affected by shear, but those with a higher molecular weight experience a profound drop in viscosity with increased shear. For a discussion on molecular weight averages, refer to the polymeric additives in Chapter 4.

Viscosity-temperature (VT) Relationship

Mineral oils and synthetic fluids exhibit an inverse viscosity-temperature relationship. That is, their viscosity decreases

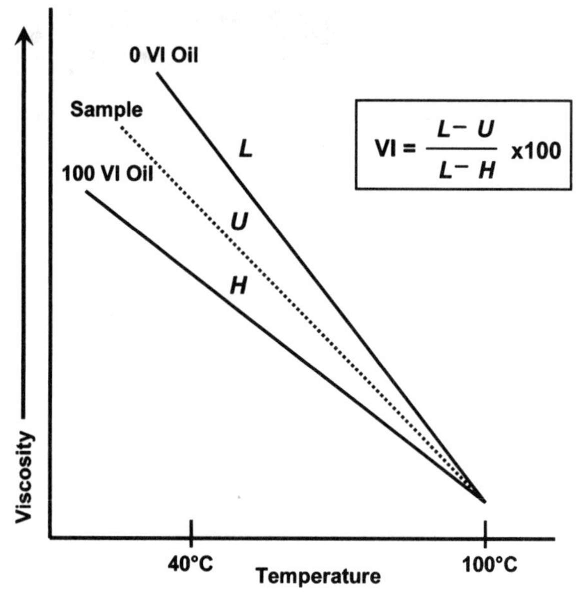

Fig. 1.11—Viscosity-temperature relationship for calculating viscosity index (VI) [24].

with an increase in temperature. This is shown in Part A of Fig. 1.10 where the kinematic viscosities of a paraffinic oil and a naphthenic oil as a function of temperature are plotted [4]. Lubricants are usually formulated by the use of the base fluids with a good VT relationship. This is because such lubricants maintain adequate viscosity at high temperatures to provide effective lubrication. Knowledge of the VT function of the oils is of great importance in practice since it helps in judging the operating range of the lubricants. There are three common ways to determine VT characteristics of a fluid. These are viscosity index (VI), viscosity-temperature constant (VTC), and the ASTM charts (ASTM D341). Viscosity index (VI) (ISO 2909, ASTM D2270) is the most common method to indicate the VT characteristics. Viscosity index, an arbitrary scale from 0 to 100, is based upon kinematic viscosity and is quite useful in comparing the VT characteristics of the different oils. The oils whose viscosities have a *high sensitivity to temperature* have a low VI and those whose viscosities have *low sensitivity to temperature* have a high VI. High VI oils are generally preferred for use in most lubricants. The viscosity index of an oil is determined by comparing its 40°C and 100°C kinematic viscosities with the viscosities of 0 and 100 VI oils [24]. The reference point for determining VI is the 100°C viscosity. To be useful in VI measurement, the 0 and 100 VI oils must have the same 100°C viscosity as the oil of interest. Viscosity index of the new oil can be calculated from its 40°C viscosity by using the following relationship, where VI is the viscosity index; L is the 40°C viscosity of the 0 VI oil, U is the kinematic viscosity (cSt) of the oil of interest at 40°C, and H is the 40°C viscosity of the 100 VI oil.

$$\mathrm{VI} = \frac{L - U}{L - H} \times 100$$

The viscosity relationship for calculating VI is pictorially presented in Fig. 1.11 [24]. The need to find oils that have the same 100°C viscosity as the new oil is eliminated by the use

TABLE 1.6—Basic *L* and *H* values for kinematic viscosity in 40–100°C system, ASTM D2270 [28].

Kinematic Viscosity at 100°C mm^2/s (cSt)	*L*	*H*	Kinematic Viscosity at 100°C mm^2/s (cSt)	*L*	*H*	Kinematic Viscosity at 100°C mm^2/s (cSt)	*L*	*H*	Kinematic Viscosity at 100°C mm^2/s (cSt)	*L*	*H*	Kinematic Viscosity at 100°C mm^2/s (cSt)	*L*	*H*	Kinematic Viscosity at 100°C mm^2/s (cSt)	*L*	*H*
2.00	7.994	6.394	7.00	78.00	48.57	12.0	201.9	108.0	17.0	369.4	180.2	24.0	683.9	301.8	42.5	1935	714.9
2.10	8.640	6.894	7.10	80.25	49.61	12.1	204.8	109.4	17.1	373.3	181.7	24.2	694.5	305.6	43.0	1978	728.2
2.20	9.309	7.410	7.20	82.39	50.69	12.2	207.8	110.7	17.2	377.1	183.3	24.4	704.2	309.4	43.5	2021	741.3
2.30	10.00	7.944	7.30	84.53	51.78	12.3	210.7	112.0	17.3	381.0	184.9	24.6	714.9	313.0	44.0	2064	754.4
2.40	10.71	8.496	7.40	86.66	52.88	12.4	213.6	113.3	17.4	384.9	186.5	24.8	725.7	317.0	44.5	2108	767.6
2.50	11.45	9.063	7.50	88.85	53.98	12.5	216.6	114.7	17.5	388.9	188.1	25.0	736.5	320.9	45.0	2151	780.9
2.60	12.21	9.647	7.60	91.04	55.09	12.6	219.6	116.0	17.6	392.7	189.7	25.2	742.2	324.9	45.5	2197	794.5
2.70	13.00	10.25	7.70	93.20	56.20	12.7	222.6	117.4	17.7	396.7	191.3	25.4	758.2	328.8	46.0	2243	808.2
2.80	13.80	10.87	7.80	95.43	57.31	12.8	225.7	118.7	17.8	400.7	192.9	25.6	769.3	332.7	46.5	2268	821.9
2.90	14.63	11.50	7.90	97.72	58.45	12.9	228.8	120.1	17.9	404.6	194.6	25.8	779.7	336.7	47.0	2333	835.5
3.00	15.49	12.15	8.00	100.0	59.60	13.0	231.9	121.5	18.0	408.6	196.2	26.0	790.4	340.5	47.5	2380	849.2
3.10	16.36	12.82	8.10	102.3	60.74	13.1	235.0	122.9	18.1	412.6	197.8	26.2	801.6	344.4	48.0	2426	863.0
3.20	17.26	13.51	8.20	104.6	61.89	13.2	238.1	124.2	18.2	416.7	199.4	26.4	812.8	348.4	48.5	2473	876.9
3.30	18.18	14.21	8.30	106.9	63.05	13.3	241.2	125.6	18.3	420.7	201.0	26.6	824.1	352.3	49.0	2521	890.9
3.40	19.12	14.93	8.40	109.2	64.18	13.4	244.3	127.0	18.4	424.9	202.6	26.8	835.5	356.4	49.5	2570	905.3
3.50	20.09	15.66	8.50	111.5	65.32	13.5	247.4	128.4	18.5	429.0	204.3	27.0	847.0	360.5	50.0	2618	919.6
3.60	21.08	16.42	8.60	113.9	66.48	13.6	250.6	129.8	18.6	433.2	205.9	27.2	857.5	364.6	50.5	2667	933.6
3.70	22.09	17.19	8.70	116.2	67.64	13.7	253.8	131.2	18.7	437.3	207.6	27.4	869.0	368.3	51.0	2717	948.2
3.80	23.13	17.97	8.80	118.5	68.79	13.8	257.0	132.6	18.8	441.5	209.3	27.6	880.6	372.3	51.5	2767	962.9
3.90	24.19	18.77	8.90	120.9	69.94	13.9	260.1	134.0	18.9	445.7	211.0	27.8	892.3	376.4	52.0	2817	977.5
4.00	25.32	19.56	9.00	123.3	71.10	14.0	263.3	135.4	19.0	449.9	212.7	28.0	904.1	380.6	52.5	2867	992.1
4.10	26.50	20.37	9.10	125.7	72.27	14.1	266.6	136.8	19.1	454.2	214.4	28.2	915.8	384.6	53.0	2918	1007
4.20	27.75	21.21	9.20	128.0	73.42	14.2	269.8	138.2	19.2	458.4	216.1	28.4	927.6	388.8	53.5	2969	1021
4.30	29.07	22.05	9.30	130.4	74.57	14.3	273.0	139.6	19.3	462.7	217.7	28.6	938.6	393.0	54.0	3020	1036
4.40	30.48	22.92	9.40	132.8	75.73	14.4	276.3	141.0	19.4	467.0	219.4	28.8	951.2	396.6	54.5	3073	1051
4.50	31.96	23.81	9.50	135.3	76.91	14.5	279.6	142.4	19.5	471.3	221.1	29.0	963.4	401.1	55.0	3126	1066
4.60	33.52	24.71	9.60	137.7	78.08	14.6	283.0	143.9	19.6	475.7	222.8	29.2	975.4	405.3	55.5	3180	1082
4.70	35.13	25.63	9.70	140.1	79.27	14.7	286.4	145.3	19.7	479.7	224.5	29.4	987.1	409.5	56.0	3233	1097
4.80	36.79	26.57	9.80	142.7	80.46	14.8	289.7	146.8	19.8	483.9	226.2	29.6	998.9	413.5	56.5	3286	1112
4.90	38.50	27.53	9.90	145.2	81.67	14.9	293.0	148.2	19.9	488.6	227.7	29.8	1011	417.6	57.0	3340	1127
5.00	40.23	28.49	10.0	147.7	82.87	15.0	296.5	149.7	20.0	493.2	229.5	30.0	1023	421.7	57.5	3396	1143
5.10	41.99	29.46	10.1	150.3	84.08	15.1	300.0	151.2	20.2	501.5	233.0	30.5	1055	432.4	58.0	3452	1159
5.20	43.76	30.43	10.2	152.9	85.30	15.2	303.4	152.6	20.4	510.8	236.4	31.0	1086	443.2	58.5	3507	1175
5.30	45.53	31.40	10.3	155.4	86.51	15.3	306.9	154.1	20.6	519.9	240.1	31.5	1119	454.0	59.0	3563	1190
5.40	47.31	32.37	10.4	158.0	87.72	15.4	310.3	155.6	20.8	528.8	243.5	32.0	1151	464.9	59.5	3619	1206
5.50	49.09	33.34	10.5	160.6	88.95	15.5	313.9	157.0	21.0	538.4	247.1	32.5	1184	475.9	60.0	3676	1222
5.60	50.87	34.32	10.6	163.2	90.19	15.6	317.5	158.6	21.2	547.5	250.7	33.0	1217	487.0	60.5	3734	1238
5.70	52.64	35.29	10.7	165.8	91.40	15.7	321.1	160.1	21.4	556.7	254.2	33.5	1251	498.1	61.0	3792	1254
5.80	54.42	36.26	10.8	168.5	92.65	15.8	324.6	161.6	21.6	566.4	257.8	34.0	1286	509.6	61.5	3850	1270
5.90	56.20	37.23	10.9	171.2	93.92	15.9	328.3	163.1	21.8	575.6	261.5	34.5	1321	521.1	62.0	3908	1286
6.00	57.97	38.19	11.0	173.9	95.19	16.0	331.9	164.6	22.0	585.2	264.9	35.0	1356	532.5	62.5	3966	1303
6.10	59.74	39.17	11.1	176.6	96.45	16.1	335.5	166.1	22.2	595.0	268.6	35.5	1391	544.0	63.0	4026	1319
6.20	61.52	40.15	11.2	179.4	97.71	16.2	339.2	167.7	22.4	604.3	272.3	36.0	1427	555.6	63.5	4087	1336
6.30	63.32	41.13	11.3	182.1	98.97	16.3	342.9	169.2	22.6	614.2	275.8	36.5	1464	567.1	64.0	4147	1352
6.40	65.18	42.14	11.4	184.9	100.2	16.4	346.6	170.7	22.8	624.1	279.6	37.0	1501	579.3	64.5	4207	1369
6.50	67.12	43.18	11.5	187.6	101.5	16.5	350.3	172.3	23.0	633.6	283.3	37.5	1538	591.3	65.0	4268	1386
6.60	69.16	44.24	11.6	190.4	102.8	16.6	354.1	173.8	23.2	643.4	286.8	38.0	1575	603.1	65.5	4329	1402
6.70	71.29	45.33	11.7	193.3	104.1	16.7	358.0	175.4	23.4	653.8	290.5	38.5	1613	615.0	66.0	4392	1419
6.80	73.48	46.44	11.8	196.2	105.4	16.8	361.7	177.0	23.6	663.3	294.4	39.0	1651	627.1	66.5	4455	1436
6.90	75.72	47.51	11.9	199.0	106.7	16.9	365.6	178.6	23.8	673.7	297.9	39.5	1691	639.2	67.0	4517	1454
												40.0	1730	651.8	67.5	4580	1471
												40.5	1770	664.2	68.0	4645	1488
												41.0	1810	676.6	68.5	4709	1506
												41.5	1851	689.1	69.0	4773	1523
												42.0	1892	701.9	69.5	4839	1541
															70.0	4906	1558

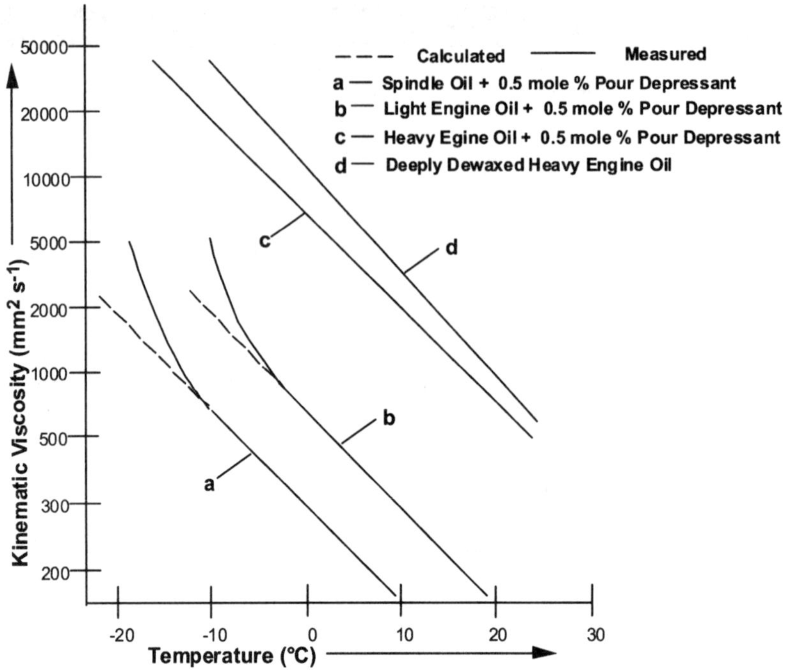

Fig. 1.12—Anomalous viscosities of base oils at low temperatures [4].

of the ASTM Standards D2270 or D39B [27]. These standards contain VI data for many hypothetical reference oils. All one has to do is to measure the 40°C and 100°C kinematic viscosities of the new oil, choose the oil from the table with the matching 100°C viscosity, and read the VI based on its 40°C viscosity. Table 1.6 provides *L* and *H* values for oils with different 100°C viscosities [28].

The next VT measure that is proposed f'or differentiating oils with a low VT relationship is viscosity-temperature constant (VTC) [29,30]. While the procedure works well for the high VI oils, it does not numerically differentiate oils with moderate or poor VT characteristics [4]. For additional information, refer to Chapter 3 on Synthetic Base Fluids. Petroleum oils exhibit a change in viscosity with temperature which is near logarithmic within a limited temperature range. ASTM International has produced a chart which is somewhat better than an ordinary logarithmic chart, with a near linear viscosity-temperature function. By using this chart, the slope of the viscosity-temperature lines of the various lubricants can be compared. Since viscosity has an inverse relationship to temperature, the smaller values for the slope are better. This implies that the drop in viscosity is lower with increasing temperature. Standard viscosity-temperature charts for liquid petroleum products in six ranges are provided in ASTM Standard D341. A unique advantage of these charts is that it is convenient to predict kinematic viscosity of a petroleum oil, or a liquid hydrocarbon, at any temperature within a limited range, if the kinematic viscosities of the oil at two temperatures are known. These charts also have the advantages of needing only two data points, magnifying small differences that exist between oils, and providing the ability to read viscosities at other temperatures from the VT line by interpolation. The data in Fig. 1.10, Part A, are replotted in Part B, using a logarithmic scale to achieve linearity and because of this the viscosity value at a particular temperature can be easily obtained from the VT line. It is important to note that the values near or below the cloud point of the oil or at very high temperatures may deviate from the straight-line relationship. This is because at these temperatures phase transitions, such as precipitation or thermal degradation, are possible. Such transitions can lead to unexpected changes in viscosity, thereby leading to erroneous results. This is aptly demonstrated by the data depicted in Fig. 1.12 [4]. As one can see, the measured viscosities below 0°C, represented by the solid line, are much

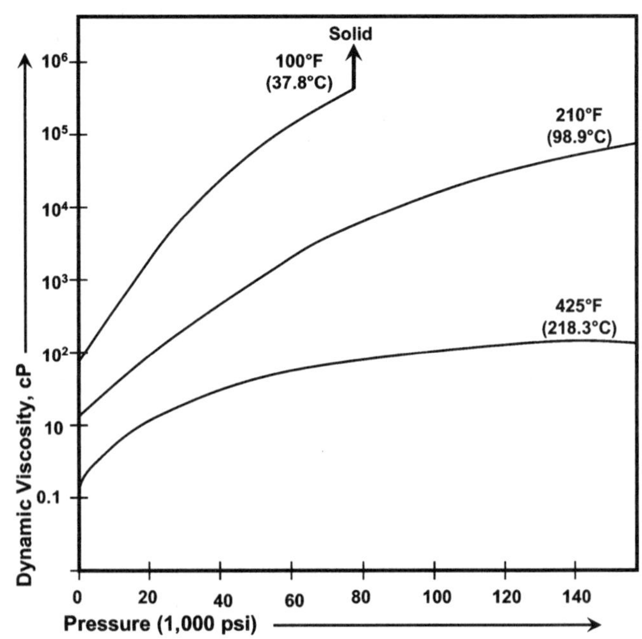

Fig. 1.13—Viscosity-pressure relationship.

higher than those predicted through interpolation, represented by the dotted line, because of a phase transition at −2°C and −10°C.

Viscosity-pressure (VP) Relationship

Unlike temperature which has an inverse relationship to viscosity, pressure bears a direct relationship. This means that as the pressure increases, the viscosity increases as well. However, the increase, which is a function of the chemical structure of the fluid, is not as dramatic as in the case of temperature [24]. As a result, the effect is observable only at relatively high pressures (Fig. 1.13). The dynamic viscosity (η_p) at pressure P is equal to $\eta_0 e^{\alpha P}$, where η_0 is the dynamic viscosity at atmospheric pressure and α is the constant that depends on the temperature and the structural characteristics of the oil [4]. A detailed discussion on the significance of α is provided in Chapter 3, the Synthetic Base Stocks chapter.

Viscosity-shear Rate (VS) Relationship

Hydrocarbon molecules or molecular aggregates that make up the oil experience shear forces while in a container or during flow. These forces can cause a degradation of these molecules, thereby leading to a viscosity loss. Newtonian fluids, such as pure mineral oils, are not affected by such shear forces and retain their viscosity. However, shear forces can significantly affect the viscosity of the non-Newtonian fluids, such as the finished lubricants. The finished lubricants generally contain additives that lead to structural viscosity, termed as apparent viscosity, which progressively decreases under the influence of shear [4]. For example, mineral oils and synthetic fluids of similar molecular weights can accommodate shear rates of over 10^9 s^{-1} that occur in gears without an adverse effect on viscosity. However, engine oils and industrial oils, which contain high-molecular weight VI improvers or dispersants, or both, have viscosities that are highly affected by shear. Under the worst situation, the viscosity can drop down to that of the base oil and is a consequence of the complete breakdown of the polymeric structures of these additives. Engine oils also experience an increase in apparent viscosity due to the presence of soot, resulting from inefficient combustion, which enters the oil. While such oils also undergo shear-related viscosity decrease, their viscosity loss is a lot lower in magnitude and their viscosity is never reduced to that of the base oil.

Shear-related viscosity loss in a finished lubricant may be temporary or permanent, depending upon the magnitude of the shear force or forces. It is termed temporary, if the viscosity reverts to its original value, or close to it, after the shear forces are removed. Temporary viscosity loss is more common in fluids such as mineral oils and synthetic base fluids that are devoid of the structure-modifying additives. Permanent viscosity loss occurs when the fluid does not regain its original viscosity after the removal of the shear forces. This type of loss is typical in finished lubricants that contain polymeric additives, such as dispersants and viscosity modifiers. For further details on the shear-related viscosity loss, refer to the section on viscosity modifiers in Chapter 4 which deals with additives.

Viscosity, especially the low-temperature viscosity, of a formulated oil can also change as a function of time, depending upon the components it contains. For example, polymer-treated oils experience a significant increase in viscosity at

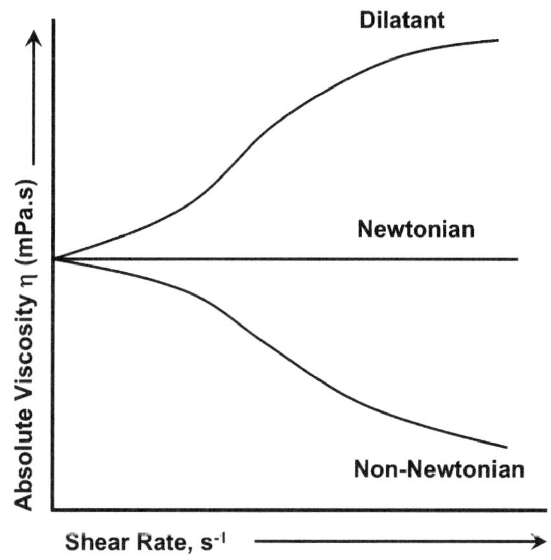

Fig. 1.14—Flow characteristics of a fluid as a function of shear rate [4].

low temperatures. The knowledge of the time effect on viscosity is important; otherwise low-temperature starting problems will be encountered.

Dilatant fluids are a special type of non-Newtonian fluids. Their behavior under shear is opposite to that shown by the additive-treated lubricants. Their viscosity increases on shearing instead of decreasing. Figure 1.14 depicts the viscosity-shear rate relationship for Newtonian, non-Newtonian, and the dilatant fluids [4].

Viscosity Classifications

Because the lubricant viscosity plays a predominant role in minimizing friction and wear, most lubricants must meet viscosity requirements that are established by a number of organizations. These include the SAE (Society of Automotive Engineers) viscosity grades and the ISO (International Standardization Organization) viscosity grades. The SAE viscosity classification systems primarily apply to oils for use in automotive applications. The SAE viscosity system for engine oils contains twelve viscosity grades and for gear oils contains six viscosity grades. ISO viscosity classification system contains 18 viscosity grades, which are based on kinematic viscosity of the oil/fluid at 40°C. The ISO system is primarily used for industrial oils, such as turbine oils, compressor oils, and others. Unlike the ISO system, the SAE system is not based on a specific viscosity at a particular temperature, but it instead specifies the low-temperature and the high-temperature dynamic viscosities and the kinematic viscosities at 40 and 100°C. In both systems, the higher the number, the higher is the viscosity. It is important to note that although the viscosity grade numbers of the ISO classification are the same as those of ASTM International and BSI (British Standards Institution) viscosity systems, the viscosities for the ISO grades are measured at 40°C, while those for the ASTM and BSI grades are measured at 100°F (37.8°C). This makes the measured viscosities in the latter case somewhat higher than those of the equivalent ISO viscosity grades.

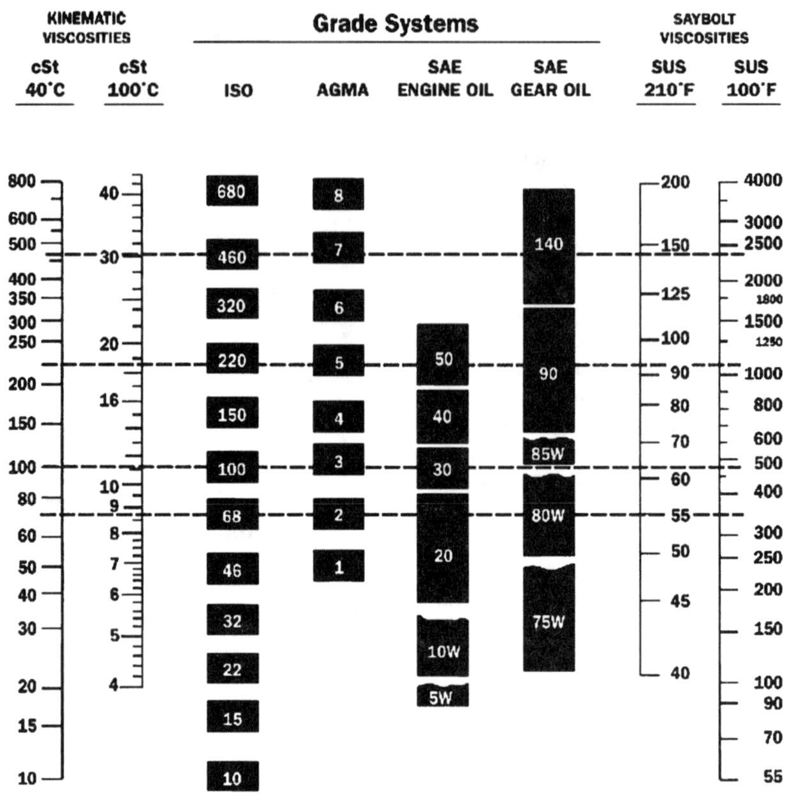

Viscosities can be related horizontally only. For example, the following oils have similar viscosities: ISO 460, AGMA 7 and SAE GEAR OIL 140.

The viscosity/temperature relationships are based on 95 VI oils and are usable only for mono grade engine oils, gear oils and other 95 VI oils.

Crankcase oils and gear oils are based on 100°C viscosity. The "W" grades are classified on low temperature properties. ISO oils and AGMA grades are based on 40°C viscosity.

Fig. 1.15—Viscosity classification systems comparison.

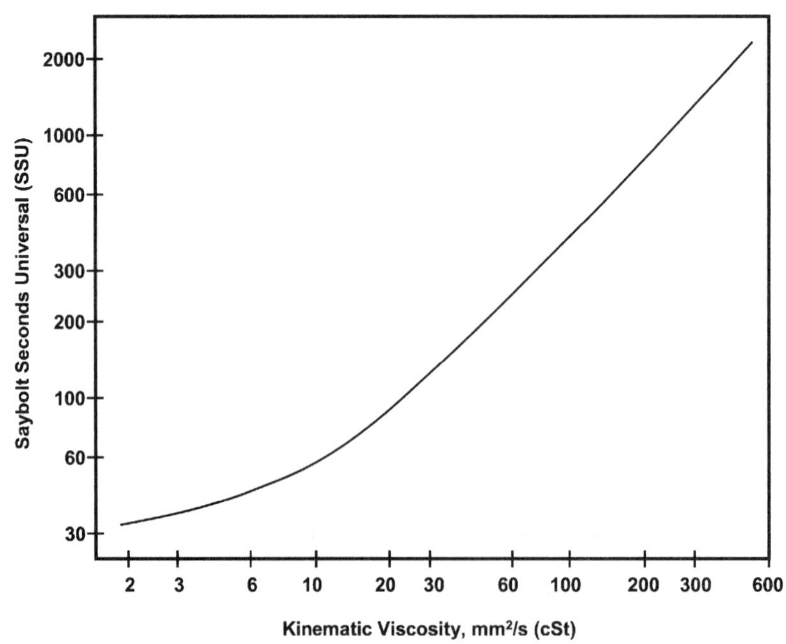

Fig. 1.16—Relationship between Saybolt Seconds Universal (SSU) and kinematic viscosity (cSt) scales [28].

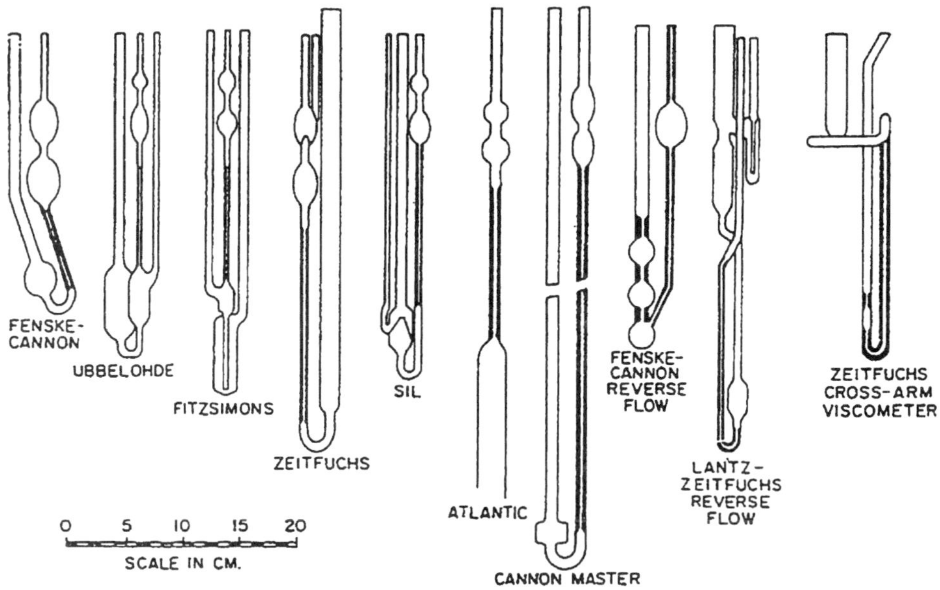

Fig. 1.17—Types of capillary viscometers [24].

To summarize:

1. ASTM/BSI Viscosity Classification System for industrial fluids/lubricants contains minimum and maximum viscosities, expressed in the most commonly used units. ASTM International previously used Saybolt Seconds Universal, SSU or SUS, but now it uses centistokes (cSt), the same as the BSI. For both systems, the viscosity measuring temperature is 100°F (37.8°C).
2. International Organization for Standardization (ISO) Viscosity Classification System for industrial fluids/lubricants provides minimum and maximum viscosity ranges in centistokes at 40°C, for its grade numbers.
3. Axle and Manual Transmission Lubricants Viscosity Classification, SAE J306, provides a range of SAE viscosity grades based on maximum temperature in °C for a viscosity of 150,000 cP and a minimum and maximum viscosity (cSt) at 100°C.
4. Engine Oil Viscosity Classification, SAE J300, provides oil viscosity grades based on low temperature cranking and pumping viscosities as well as low and high shear viscosities.

Fig. 1.18—Viscosity measurement using Canon-Fenske viscometers [318]. Reprinted with permission from the Lubrizol Corporation.

A viscosity grade comparison chart is provided in Fig. 1.15. The relationship between Saybolt Seconds Universal (SSU or SUS) and centistokes (cSt) viscosity scale is shown in Fig. 1.16 [28]. Detailed viscosity grades for each lubricant type will be discussed in detail in the appropriate chapters.

Viscosity Measurement

Instruments called viscometers are used to measure the viscosity of lubricants. Viscometers are of three general types: capillary, rotary, and miscellaneous others [24,28].

Capillary Viscometers

Capillary viscometers are used to measure kinematic viscosity. The lubricant is allowed to flow down a capillary at the prescribed temperature and the viscosity is calculated by taking into account the flow rate, the length and radius of the bore, the pressure drop between inlet and outlet, and the fluid density. Common types of capillary viscometers are shown in Fig. 1.17 [24] and the actual setup is shown in Fig. 1.18.

The ASTM D445 procedure is used to measure the kinematic viscosity of a lubricant. In this procedure, viscosity is measured by allowing the fluid to flow through the viscometer under the normal force of gravity. Shear rate for this measurement is less than 10 s^{-1}, which is fairly low. The ASTM D4624 procedure is used to measure apparent viscosity of a fluid at high temperature (150°C) and high shear rates (10^6 s^{-1}). In this procedure, a fixed volume of the fluid is forced through the capillary viscometer using gas pressure. The procedure simulates the viscosity of engine oils in operating crankshaft bearings. The ASTM D1092 procedure, which is used to measure the viscosity of lubricating greases, also uses pressure [24]. The type of viscometer used for this purpose is shown in Fig. 1.19 [24].

Rotary Viscometers—Rotary viscometers use torque on a rotating shaft to measure absolute or dynamic viscosity and include cold cranking simulator (CCS), mini-rotary vis-

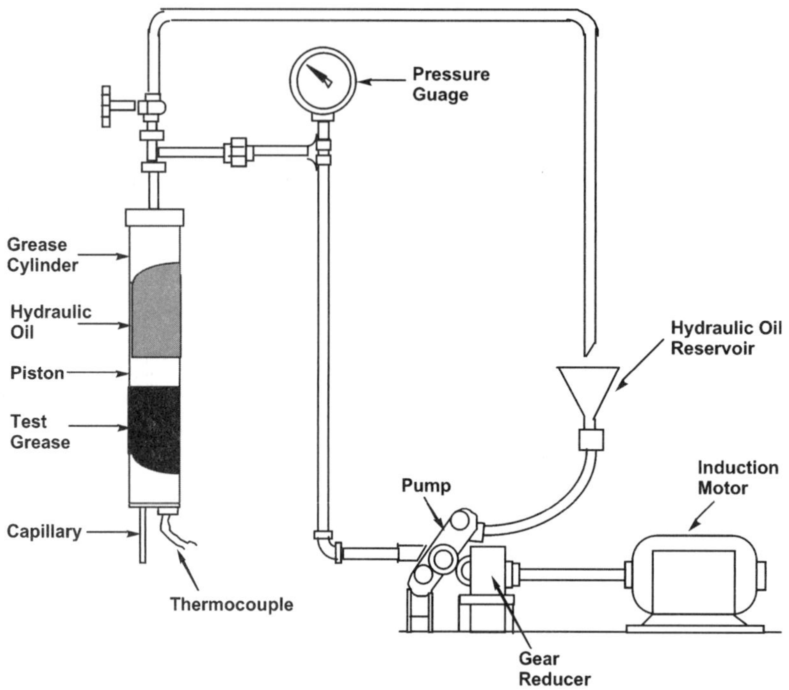

Fig. 1.19—Pressure viscometer [24].

cometer (MRV), Brookfield viscometer, and tapered bearing simulator.

Cold cranking simulator, shown in Fig. 1.20, is used to measure the low-temperature viscosity of the engine oils, a requirement specified in the SAE Viscosity Classification J300 (ASTM D2602) [24,27]. The instrument measures the apparent viscosity of a lubricant to reflect its cold cranking resistance. The viscosity measured is between 500 and 10,000 cP at operating temperatures of between −5 and −30°C. The viscometric cell of the apparatus, shown in the lower half of Fig. 1.20, consists of a stainless steel rotor inside a closely fitted copper stator. The clearance between the rotor and stator is very small, in the order of 0.01 mm, and simulates the high shear rate (10^5–10^4 s^{-1}) in the engine bearings.

The sample is introduced into the apparatus through the fill tube and is rapidly cooled to the prescribed temperature by cold methanol. The rotor is then turned using a motor whose speed has an inverse relationship to viscosity, that is, the higher the viscosity, the lower the speed, and vice versa. The viscosity of the test oil provides the load for the drive motor, a situation that parallels that in an automobile engine. With the motor speed known, the viscosity can be obtained by using speed-viscosity calibration charts. These charts are developed by plotting the motor speed as a function of viscosity for various reference oils at the prescribed temperatures.

Mini-rotary viscometers are used to measure the low-temperature, low-shear (0.4 to 15 s^{-1}) viscosity. Prior to the measurement, the samples are slowly cooled to the specified temperature using a prescribed procedure. The cooling time is 10 h for the ASTM D3829 procedure and 40 h for the ASTM D4864 procedure. The ASTM D3829 procedure is used to predict borderline pumping temperature (BPT) at which the viscosity exceeds 30,000 cP and the ASTM D4684

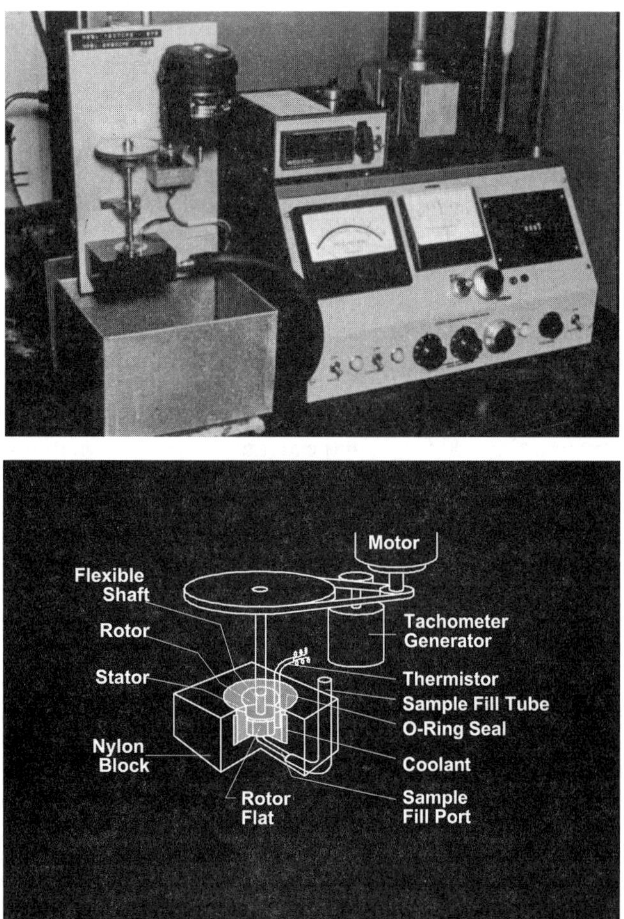

Fig. 1.20—Cold cranking simulator [318]. Reprinted with permission from the Lubrizol Corporation.

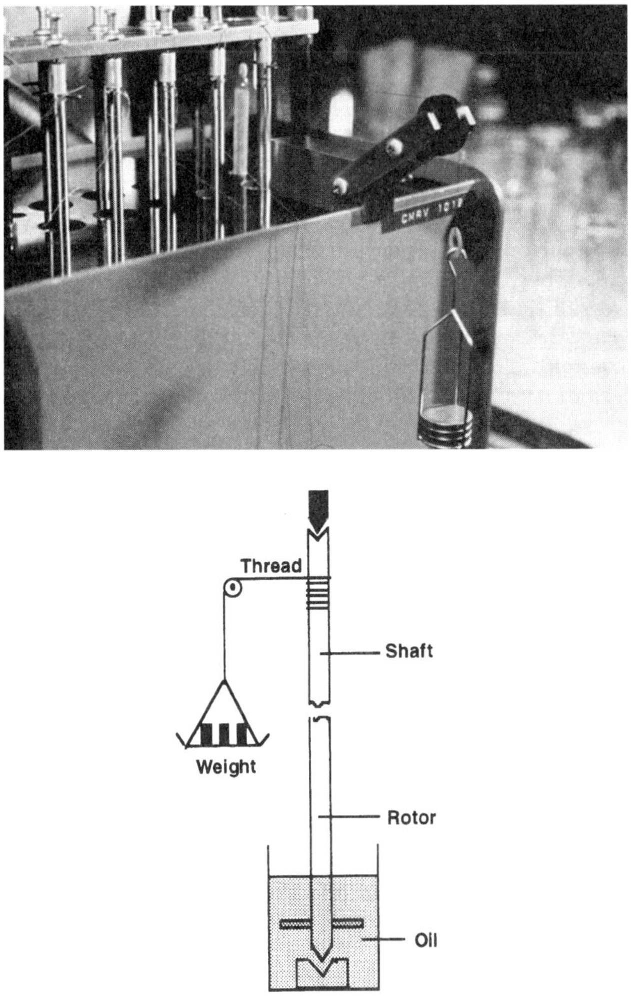

Fig. 1.21—Mini-rotary viscometer [318]. Reprinted with permission from the Lubrizol Corporation.

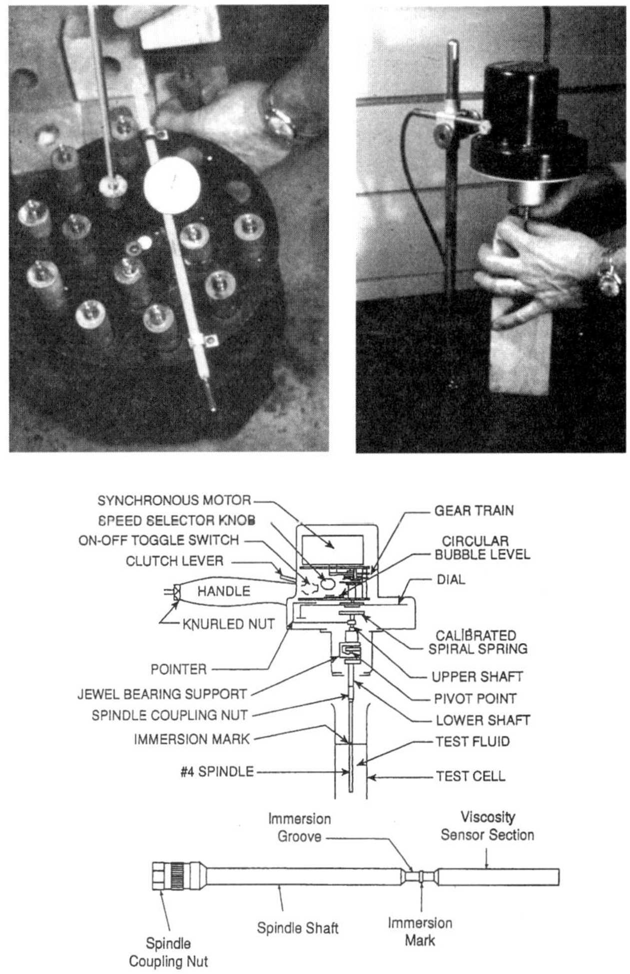

Fig. 1.22—Brookfield viscometer [318]. Reprinted with permission from the Lubrizol Corporation.

procedure is used to measure the apparent low-temperature viscosity.

Such viscometers, one shown in Fig. 1.21, contain a number of small viscometric cells in an aluminum block equipped with a thermostat. The samples are poured into the outer cylinder and the rotors with a length of thread wound around their shafts are inserted. The samples, covered to minimize condensation and frost formation, are cooled to the desired temperature. Measurements are made by removing the protective cover and placing a pulley assembly on the top cover of the aluminum block in front of the viscometric cells. The thread is drawn over the pulley and a small platform weight holder (10 g) is attached. If the rotor does not turn, additional 10 g weights are added until there is movement. The load needed to cause rotation is then used to compute the yield stress, that is, the minimum stress needed for the oil to flow. Following the yield stress measurement, a 150 g load is attached to the thread and the time to complete three full revolutions of the rotor is recorded. This and the yield stress values are used to calculate the sample viscosity. The details for calculating the borderline pumping temperature (BPT) and the apparent viscosity are provided in the above-mentioned ASTM standards.

The Brookfield viscometer, shown in Fig. 1.22, is also used for the low-temperature, low-shear (1.7 s^{-1}) viscosity measurement. The rotor is driven by a synchronous motor, via a multi-speed gear box and spring. A scale which is attached to the output shaft of the gearbox rotates with it. The viscous drag of the fluid on the rotor causes an angular displacement between the rotor and the scale at the spring, which is registered on the scale. The scale reading for each speed can be converted into viscosity either through calibration with Newtonian fluids of known viscosity or from the calibration constants provided by the manufacturer. Two procedures, ASTM D2983 and D5133, employ the Brookfield viscometer. The ASTM D2983 procedure is primarily used to determine the low temperature (−5 to −40°C) viscosity of the driveline lubricants and the industrial fluids. ASTM D5133, the Scanning Brookfield procedure, the testing component of the viscometer shown in Fig. 1.23, measures viscosity of the fluid during cooling at a rate of 1°C per hour. The torque required to shear the fluid is continuously recorded and the record is converted into critical pumpability viscosity. Temperature, where the viscosity increase is greater than desired, usually 30,000 cP, is reported as the critical pumpability temperature.

Table 1.7 compares various low-temperature viscosity

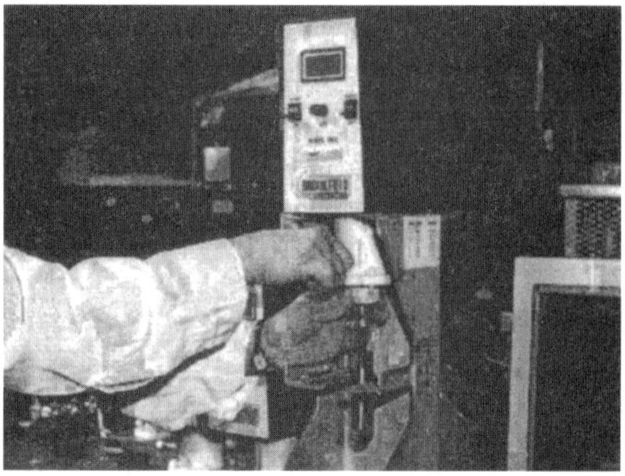

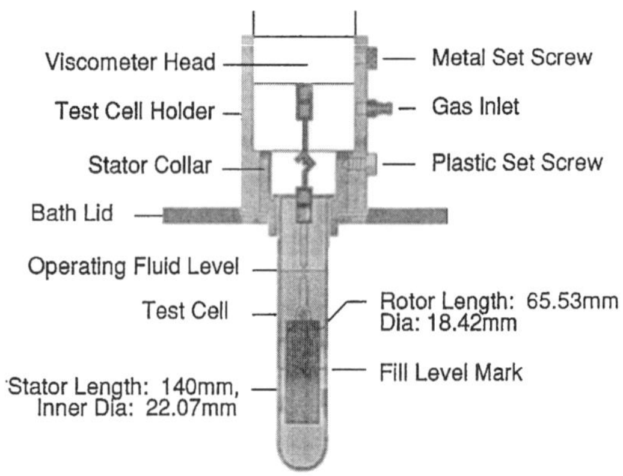

Fig. 1.23—Scanning Brookfield Viscometer [318]. Reprinted with permission from the Lubrizol Corporation.

measurement methods in terms of shear rates and their application.

Wear

When the lubricant fails to provide an adequate protection by forming an effective film, wear initiates, which if not protected against can lead to equipment seizure. Wear can simply be defined as the damage to a surface due to a progressive loss of material. The lost material becomes a part of the lubricant directly or indirectly, that is, via transfer to the other surface. Such material is referred to as *wear metals*, if it is present in the liquid phase and is in a nonoxidized form. As mentioned earlier, the wear is a consequence of friction. Four major types of wear that occur in machinery are abrasive, corrosive, fatigue, and adhesive or sliding wear.

Abrasive Wear

This is the most common type of wear that occurs in mechanical equipment. It primarily results from the hard particles, such as foreign matter, and hard protuberances or asperities that are either forced against or move between the surfaces when they mesh. All metal surfaces, irrespective of the quality of the surface finish, are rough on a microscopic scale. They consist of protuberances, or asperities, and depressions. Figure 1.24 depicts the mechanism for this kind of wear. Abrasive wear can be controlled by the use of filters, strainers, magnetic drain plugs to remove iron particles, and periodic draining and refilling of the equipment with clean oil.

Corrosive Wear

This type of wear results from the attack of the lubricant or the corrosive contaminants, such as salts, water, and acids, on the metal surfaces. These materials cause chemical or electrochemical reaction with the metal. See the section on corrosion inhibitors in Chapter 4, the Additives section. The surfaces that experience metal-to-metal contact show a great deal more damage than those that do not. This is primarily due to the generation of the frictional heat and the fresh metal. The mechanism for this type of wear is shown in Fig. 1.25. Corrosive wear, which causes pitting or polishing of the metal surfaces, can be minimized by using acid-neutralizing and film-forming agents. Acid-neutralizing agents, such as basic detergents, remove acids. Film-forming agents adsorb on or chemically react with the surfaces to form protective films, thereby keeping acids, salts, and water away from the metal.

Fatigue Wear

Fatigue wear results from repeated stressing of the metal surfaces. Such stressing arises from continuous temperature changes due to frictional heat, periodic impacts, repeated contacts with abrasive asperities, or the rubbing cycles. This kind of wear usually starts as a surface or subsurface crack which progressively grows until a piece of metal is removed from the surface. This damage is called pitting; when pits grow larger the damage is called spalling. Fatigue failure commonly occurs in pinion gears and bearings that experience a large amount of stress during service. Figure 1.26 illustrates the mechanism of the fatigue wear. This type of wear is not easy to control. However, fatigue life of the equipment can be improved by using a good lubricant, by se-

TABLE 1.7—Comparison of low-temperature viscosity measurement methods.

Test	Temperature	Shear Rate (s^{-1})	Application
Cold Cranking Simulator (ASTM D5293)	-5 to -30°C	10^5-10^4	Starting, Cranking
Mini-rotary Viscometer (ASTM D4684)	-15 to -35°C	0.4 - 15	Pumpability
Mini-rotary Viscometer (ASTM D3829)	0 to -40°C	—	Pumpability
Scanning Brookfield Viscometer (ASTM D5133)	-10 to -35°C	1.7	Pumpability

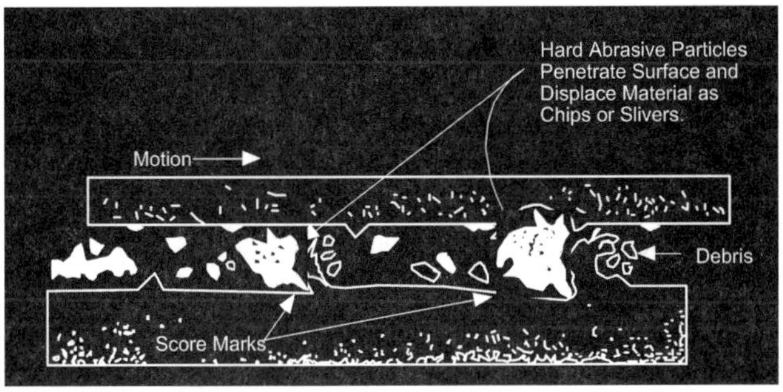

Fig. 1.24—Mechanism of abrasive wear [318]. Reprinted with permission from the Lubrizol Corporation.

lecting proper design and material, and by controlling its operating conditions.

Adhesive Wear

Adhesive wear results from frictional heat as a consequence of the surface-to-surface contact. Heat causes both surface and subsurface damage. The most common types of damage include adhesion, welding, scoring (scuffing), pitting, cracking, and plastic deformation. Adhesive wear essentially involves adhesion and pulling away of the adhered sections of one sliding surface by the other. Figure 1.27 graphically presents the mechanism for this kind of wear. This type of wear can be classified as *mild or severe*, depending upon the rate of wear and the size of the wear debris.

Adhesive wear can be minimized by using a lubricant with good extreme pressure and film-forming properties. Physical and chemical film-forming agents, called the friction modifiers and the extreme pressure (EP) agents are added to the lubricant to achieve this goal. Adhesive wear, which is more prevalent in gears and axles, will be discussed in more detail in Chapter 8 on Gear Lubricants. It is important to note that the success in minimizing friction and wear requires an understanding of the tribo-system and the elements involved [31].

Types of Lubricants

The lubricant industry uses a number of ways to classify lubricants. Some of these are as follows:

1. Based upon viscosity—viscosity grades
2. Base stock source—mineral versus synthetic
3. Type of additives—R&O oils, EP lubricants
4. Use application—engine oil, gear oil, hydraulic fluid

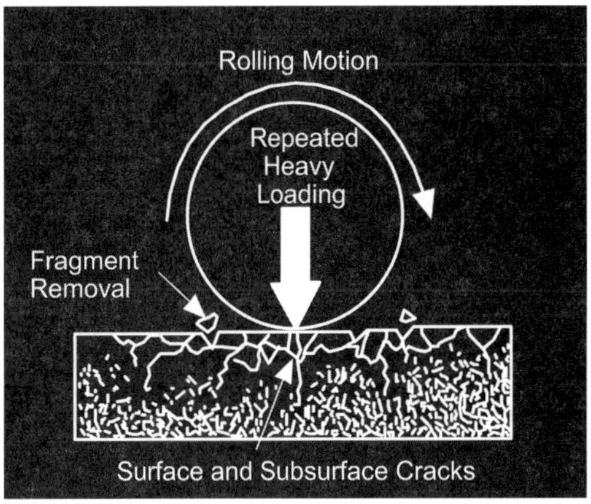

Fig. 1.26—Mechanism of fatigue wear [318]. Reprinted with permission from the Lubrizol Corporation.

5. By sector—automotive (passenger car, heavy-duty diesel) versus industrial
6. By function—crankcase oils versus driveline lubricants
7. A combination of these—industrial gear oil, synthetic hydraulic fluid, automotive grease

In this book, we broadly classify lubricants into engine lubricants and nonengine lubricants. Engine lubricants are those that are used to lubricate components in an internal combustion engine. Nonengine lubricants are those that are used to lubricate parts and mechanisms that help transfer power from the power source, such as an engine, to parts that perform the actual work. Transmissions, hydraulic systems, and gears are examples of such mechanisms. Consideration behind this classification is that the environment for the two types of applications is vastly different. Engine lubricants perform in an environment that is open to the atmosphere, is highly oxidative, and involves contaminants from combustion. Nonengine lubricants, on the other hand, perform in an environment that does not involve combustion-derived contaminants and is somewhat closed, and hence is less oxidative in nature. In order to facilitate discussion, we will further subdivide the two groups into smaller groups based upon application or the end use. Various end-use classes are listed below in the lubricant classifications section.

Since all lubricants must possess proper viscometrics to perform effectively, they will be included in the discussion of specific lubricants.

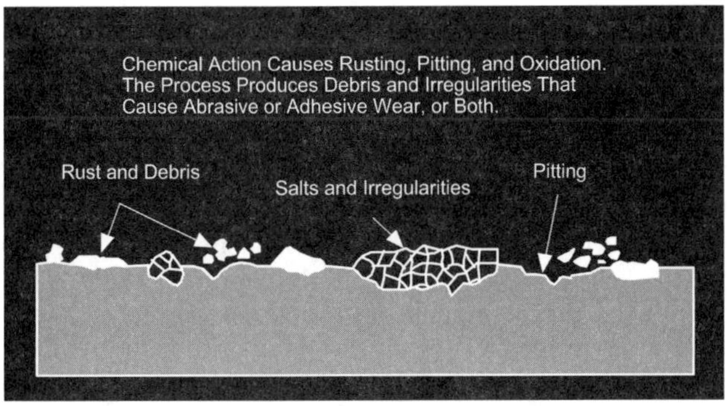

Fig. 1.25—Mechanism of corrosive wear [318]. Reprinted with permission from the Lubrizol Corporation.

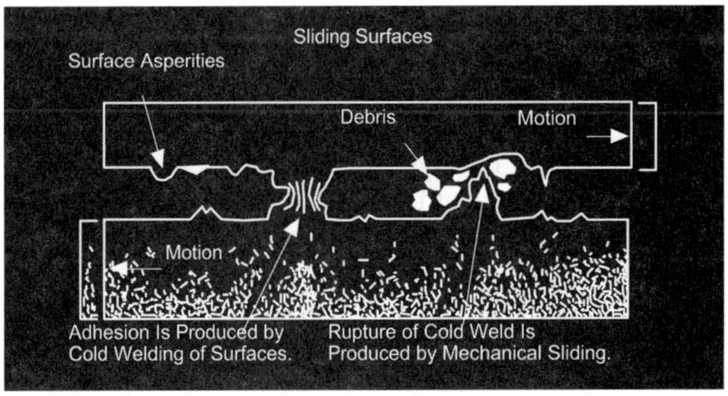

Fig. 1.27—Mechanism of adhesive wear [318]. Reprinted with permission from the Lubrizol Corporation.

Lubricant Selection and Specifications

Lubricant Selection Criteria

As mentioned earlier, a lubricant must possess the appropriate properties necessary to perform in a particular application. Desirable properties include the following:

1. Proper low and high temperature viscometrics
2. Good lubricity
3. Low volatility, low flash point, and nonflammability
4. Good thermo-oxidative and chemical stability
5. Neutralizing and suspending ability
6. Low corrosivity
7. Low foaming tendency
8. Elastomer compatibility
9. High biodegradability and low toxicity
10. Low cost

Proper viscometrics and good lubricity are important to ensure proper lubrication across the application's operating temperature range. Low volatility, low flash point, and nonflammability are important to minimize the loss of lubricant through evaporation at high temperatures and to lower the fire hazard during transport and use. Good thermo-oxidative and chemical stability are important so that the lubricant does not lose its integrity or form corrosive products and deposits during service, or both. Oxidation is also undesirable because in certain cases it can increase lubricant viscosity, hence its ability to be pumped to parts needing lubrication. Neutralizing and suspending ability are important, especially in engine oils, where the lubricant must neutralize and suspend the potentially harmful oxidation derived acidic products and the deposit precursors, thereby keeping them away from surfaces. Lubricants derived from synthetic base fluids can be acid sensitive or base sensitive and can lose their structural integrity. Low corrosivity is important for the lubricant or its components so as not to attack metals used in the forging of the equipment. A low-foaming tendency, that is, good air release properties are important because in some machine elements air gets entrained into lubricants to form foam. This reduces effective heat transfer, interferes with the lubricant flow, causes lubricant loss through vents, and accelerates oxidation. Elastomer compatibility is important since seals used in some parts are made of elastomers. An aggressive lubricant can migrate into or remove the plasticizer out of the seals and impair their function by damaging them. Low toxicity and high biodegradability are important so as not to harm personnel or the environment. A number of tests are used to evaluate these attributes of a lubricant. An exhaustive list of tests is provided in Chapter 12 on Testing.

Lubricant Classifications

The main objective of lubricant selection is to match the lubrication needs of the equipment with the properties of the lubricant that will meet these needs. Equipment manufacturers and lubricant producers attempt to make the lubricant selection process simple for the consumers. They accomplish this by recommending a quality lubricant for use in their equipment. Lubricant quality is defined by the use of the lubricant qualifications or specifications. A number of professional societies and organizations are involved in establishing these standards. Such organizations include ASTM International, CEC (Coordinating European Council), DIN (Deutsches Institut für Normung), SAE (Society of Automotive Engineers), API (American Petroleum Institute), ACEA (the Association des Constructeurs Europeens D' Automobiles), AGMA (American Gear Manufacturers Association), ISO (International Standards Organization), NLGI (National Lubricating Grease Institute), NMMA (National Marine Manufacturers Association), the U.S. military, and the U.S., European, and Japanese OEMs (original equipment builders). The U.S. military uses existing nongovernmental specifications to establish requirements for products it uses or identifies commercial products that must be modified to include the military-unique requirements. If a nongovernmental standard exists that contains the basic technical requirements for a product or a process, it is referenced in the military specification. The military specification itself contains only the additional requirements needed by the Department of Defense which are listed in the Department of Defense Index of Specifications.

TABLE 1.8—Lubricant classification.

Engine Oils	Nonengine Lubricants
Gasoline engine oils	***Transmission fluids***
Diesel engine oils	1. Automatic transmission fluids
1. Automotive diesel oils	2. Manual transmission fluids
2. Stationary diesel oils	3. Power transmission fluids
3. Railroad diesel oils	***Gear oils***
4. Marine diesel oils	1. Automotive gear oils
Stationary gas engine oils	2. Industrial gear oils
Aviation engine oils	***Hydraulic fluids***
Two-stroke cycle engine oils	1. Tractor hydraulic fluids
	2. Industrial hydraulic fluids
	Turbine oils
	Miscellaneous industrial oils
	Metalworking fluids
	Greases

Different societies are involved in establishing specifications for different applications. For example, the API is active in engine oil specifications, but the OEMs dominate in establishing performance of the automatic transmission fluids; and AGMA plays a leading role in devising specifications for the industrial gear oils. Viscosity classification is the domain of the SAE and ISO, and ASTM, DIN, and AFNOR are the primary organizations that establish the test methods. While one will expect OEMs to have the greatest knowledge of the lubrication needs of the equipment that they design, their recommendations are not necessarily always the best. This is because the definition of "the best" is a matter of one's opinion. Nonetheless, their opinion must be taken into consideration while designing a lubricant since most manufacturers will not honor equipment warranty if their recommendations are not followed. If the recommended lubricant fails to perform satisfactorily, the manufacturer should be consulted for alternative recommendations.

The lubricant producer is another source of advice during the selection of a suitable lubricant, especially if the manufacturer recommends a lubricant based upon specifications or properties. Many lubricant producers employ product specialists to advise users in selecting lubricants that meet the relevant specifications and to answer technical questions. It is essential for the user to select the lubricant that meets the OEM's performance criteria, irrespective of the identity of the lubricant manufacturer. Additional factors that need to be considered while selecting a lubricant are described in Ref. [32]. In the absence of the manufacturer's specifications, one must consider lubricants that are being used in the intended application and determine their adequacy prior to selecting one. A list of such lubricants can be obtained from the various lubricant producers.

We stated earlier in the chapter that in this book we will classify lubricants into engine lubricants and nonengine lubricants, primarily because of the differences in their operating environments. Table 1.8 contains a list of lubricants belonging to the two classes. Almost all lubricants must meet the SAE, ISO, and ASTM established viscosity requirements and the performance requirements established by various organizations, which for each lubricant type are listed below:

1. Engine Oils
   a. Passenger Car—ILSAC, API, ACEA, JASO
   b. Heavy-duty Diesel—API, OEMs, U.S. Military
2. Transmission Fluids
   a. Automatic—OEMs (GM, Ford, Chrysler)
   b. Power—OEMs
3. Automotive Gear Oils—API, U.S. Military
4. Tractor Fluids—OEMs
5. Industrial Lubricants
   a. Anti-wear Hydraulic Fluids—OEMs, Government Agencies, Standards Organizations
   b. Industrial Gear Oils—United States Steel (USS), AGMA, Cincinnati Machine
   c. R&O Turbine Oils—OEMs, U.S. Military, Technical Societies
6. Metalworking Fluids—Some Standardization Tests
7. Greases—NLGI

Lubricant Composition

Almost all commercial lubricants are formulated oils, that is, they comprise base stock(s) and performance additives. When present in the proper concentration, these components impart the formulated lubricant properties necessary to perform effectively in the intended application. In addition to performing the principle functions of lubrication, cooling, containment/suspension, corrosion protection, and power transfer, the lubricant must also fulfill additional functions that are unique to the application. Lubrication is the base oil's or the base fluid's exclusive domain but for achieving the other functions the additives play a predominant role. Additives belong to two general classes: those that affect or impart to the physical properties of the base oil and those that improve the chemical properties of the base oil. While in-depth discussions on additives are presented in Chapter 4, the functions of some of these are briefly described below.

Detergents

These additives perform two major functions. They neutralize the acidic by-products of combustion and lubricant oxidation and keep the deposit precursors and contaminants, which have marginal oil solubility, in oil. This minimizes deposit formation on engine or transmission parts. It is their base reserve, or the TBN, that help neutralize acids and the soap content that help suspend the polar products in oil. These materials are alkali metal or alkaline earth metal salts of organic acids, with or without the reserve base. Common acids include alkylbenzenesulfonic acids, alkylphenols, and fatty carboxylic acids.

Dispersants

These additives perform the same function as the soap component of the detergents. That is, they suspend polar contaminants of low oil solubility in the bulk lubricant. They do so by associating with these species via their polar ester or imide functionalities and keeping them dissolved in oil by associating with it via their nonpolar hydrocarbon chains. Dispersants are more effective in performing this function than detergents because of their higher molecular weight, that is, the higher hydrocarbon content. The suspended harmful products are removed when the oil is changed. Common dispersants are polyamine and polyhydric alcohol-derived polyisobutylene derivatives.

Oxidation Inhibitors

These additives control the oxygen-initiated degradation of the lubricant. They belong to three general classes: hydroperoxide decomposers, free radical scavengers, and metal deactivators. *Hydroperoxide decomposers* promote the decomposition of the hydroperoxides either to innocuous materials or to free radicals. Common additives of this class include organo-sulfur and organo-phosphorus compounds. *Free radical scavengers* remove the free radicals that are primarily responsible for the oxidation chain reaction. Common additives of this class are zinc dialkyl dithiophosphates (ZDTPs), hindered phenols, and alkylated arylamines. *Metal deactivators* complex with metallic cations, which are oxidation catalysts, and make them inactive. Poly-functional (polydentate) compounds, such as ethylenediaminetetraacetic acid (EDTA) and salicylaldoxime, are useful in controlling oxidation by this mechanism.

Rust and Corrosion Inhibitors

These additives protect metal surfaces against the attack of oxygen, water, acids, bases, and salts. They achieve this by physically adsorbing on the metal surfaces via their polar functional group and by maintaining a resilient protective film on the surfaces by associating with the lubricant. Rust inhibitors are additives that protect ferrous metals and corrosion inhibitors are additives that protect nonferrous metals. Both types perform by coating the surfaces of the metal parts and forming a barrier between them and the environment.

Friction Modifiers

These are additives that usually reduce friction. The mechanism of their performance is similar to that of the rust and corrosion inhibitors in that they form durable low-resistance lubricant films via adsorption on surfaces and via association with the oil. Common materials that are used for this purpose include long-chain fatty acids, their derivatives, and the molybdenum compounds. In addition to reducing friction, the friction modifiers also reduce wear, especially at low temperatures where the anti-wear agents are inactive, and they improve fuel efficiency.

Anti-wear Agents and Extreme-pressure Additives

These additives form extremely durable protective films by thermo-chemically reacting with the metal surfaces. This film can withstand extreme temperatures and mechanical pressures and minimizes direct contact between surfaces, thereby protecting them from scoring and seizing. Typically, anti-wear (AW) agents have a lower activation temperature than the extreme-pressure (EP) agents. The latter are also referred to as anti-seize and anti-scuffing additives. Organosulfur and organo-phosphorus compounds, such as organic polysulfides, phosphates, dithiophosphates, and dithiocarbamates are the most commonly used AW and EP agents. For further details of the film-forming mechanism, please refer to the section on film-forming agents in Chapter 4.

Foam Inhibitors

Formation of foam in most lubrication applications is undesirable since it impedes lubrication, promotes lubricant oxidation, obstructs narrow passages, and reduces a lubricant's cooling ability. Foam inhibitors are additives that reduce the foam-forming tendency of the lubricant. Common additives used to accomplish foaming control include polysiloxanes and styrene ester polymers. These materials have borderline oil solubility and perform by lowering the surface tension of the foam bubbles.

Viscosity Modifiers

The viscosity of liquids decreases with an increase in temperature. An oil's viscosity at high temperatures can drop to a low level that will make it lose its ability to maintain a lubricating film on surfaces. Viscosity modifiers are polymers that help a lubricant maintain its lubricating ability at high temperatures. They do so by increasing their molecular size, hence increasing association with the oil so that it does not flow away from the surfaces. These additives increase both the low-temperature viscosity and the high-temperature viscosity of the oil, but to a varying degree. They are often used to make multi-grade oils. Common polymers that are used in this capacity include polymethacrylates, olefin copolymers (OCPs), styrene-diene copolymers, and styrene-ester copolymers.

Pour Point Depressants

Petroleum base oil-derived fluids contain waxes, which at low temperatures start to crystallize to form network structures. These structures absorb oil and impede its flow. Pour point depressants (PPDs) prevent crystalline network formation and permit oil flow at low temperatures. Common PPDs include wax-alkylated naphthalenes and phenols, polymethacrylates, and styrene-ester copolymers

In addition to the types of additives described here, there are additional types. They will be discussed in the Additives chapter, Chapter 4. Also, please note that not all lubricants contain all types of additives and in the same amounts. Typically, automotive lubricants contain more classes of additives and in greater amounts than industrial lubricants.

2

Mineral Base Oils

IN THIS CHAPTER WE DESCRIBE PETROLEUM composition and the oil field and refinery chemicals that are used to facilitate petroleum drilling to extract crude petroleum from beneath the earth's surface and refine it to yield value-added products, such as fuels, lubricant base stocks, and petrochemicals. The chapter also describes many of the refinery processes in some detail to explain the manner in which the hydrocarbon cuts from petroleum with suitable properties for use as lubricant base stocks are obtained. Commonly referred to as mineral oils, they are the cheapest and the most abundant base stocks available and therefore are often used to formulate lubricants. Discussion also includes the desirable properties of the mineral oils that are critical to formulating a quality lubricant.

Untreated or nonformulated lubricants (mineral base oils and synthetic base stocks) do not possess the necessary properties to perform effectively in today's demanding lubricating environments. To function properly in such environments, base fluids need the help of chemicals, called additives. Additives improve the lubricating ability of the base oils either by enhancing the desirable properties already present or by adding new properties. Most of today's lubricants are formulated lubricants, and additives are their integral part [33].

The world consumption of lubricant additives has increased from 2.6 million metric tons (5.7 billion lb) in 1997 to about 3 million metric tons (~6.6 billion lb) in 2006. The consumption is expected either to plateau or grow slowly in North America and Western Europe, which consume the largest share of the total. The developing economies of Asia, such as India and China, and of Latin America, such as Brazil, Chile, and Argentina, are expected to see a faster growth.

A formulated lubricant comprises a base fluid and a performance package, and in the case of multi-grade oils, an additional viscosity modifier. The amount of the base fluid in a lubricant can be anywhere from 70 % to greater than 99 %, based on the desired performance level and the severity of the end-use requirements. Base fluid is derived from three sources: petroleum, synthetic, and biological, i.e., plant or animal in origin.

Petroleum Composition

Petroleum, or the crude oil, is the main source of a number of products that are essential to modernization. These include fuels for household, industrial, and transportation use, lubricants, and chemicals that are used as raw materials to manufacture a variety of synthetic products. Crude oils are classified as *light, medium, and heavy*, based on density, *sweet or sour*, based on sulfur level, *low quality* or *high quality*, based on wax content, or by geographical region.

Petroleum, or crude oil, as it is recovered from the ground is primarily composed of organics and residuum, which are mixed with metals and the metal salts as contaminants. The organic portion of petroleum consists of saturated hydrocarbons, unsaturates, aromatics, asphaltenes, high molecular weight resins, and hetero-organic compounds containing sulfur, nitrogen, and oxygen atoms in their structures. Hydrocarbons are the major constituent of the crude petroleum. These compounds contain carbon and hydrogen atoms only and are classified into alkanes, alkenes, alicyclics, and aromatics. *Alkanes*, also known as *paraffins*, are compounds with saturated linear or branched structures. The latter are also called iso-paraffins. They do not contain cyclic structures, or rings; hence they are sometimes referred to as *acyclics*. *Alkenes*, also known as *olefins*, are unsaturated molecules (contain double bonds) that do not occur to a great degree in crude petroleum but result from cracking or dehydrogenation reaction during certain refining processes. *Naphthenes*, also called *alicyclics*, are saturated compounds that contain five- or six-membered cyclic rings. *Aromatics* also contain cyclic rings, but these rings are aromatic; that is, they contain conjugated double bonds (alternating single and double bonds). When present in the crude oil, they are primarily based on the six-membered benzene ring. Average aromatics content of the most crude oils is around 50 %, but it can range from 25 % in the light paraffinic crudes to 75 % in the heavy crudes. In the crude petroleum most compounds have composite structures; that is, they contain linear or branched hydrocarbon chains and rings in the same structure. Paraffins to naphthenes ratio varies widely among the crude oils from different sources and is used to classify them as *paraffinic* or *naphthenic*. Other components in the crude oils are undesired and must be removed during the manufacture of fuels and base oils.

In addition to the simple hydrocarbon molecules described so far, crude petroleum contains compounds that have elements other than carbon and hydrogen, either in the side chain or in a ring. When nitrogen, oxygen, sulfur, or other elements, commonly found in petroleum, are present in cyclic structures, the compounds are called *heterocyclics*. Oxygen-containing compounds are usually noncyclic, such as carboxylic acids, and therefore are not classified as heterocyclics. Sulfur- and nitrogen-containing compounds, on the other hand, are usually cyclic. Asphaltenes are very high molecular weight compounds with heterocyclic and aromatic structures.

Figure 2.1 depicts the structures of these classes of compounds. The size of the molecules that make up petroleum ranges from very simple gaseous molecules, such as meth-

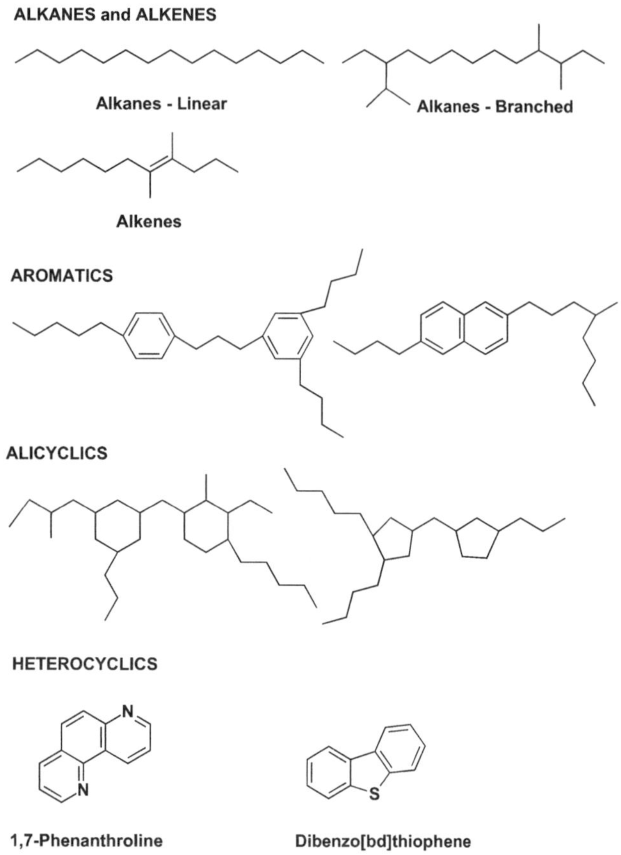

Fig. 2.1—Typical classes of compounds present in crude petroleum.

ane, to very complex high molecular weight asphaltic components [34]. The boiling ranges of the commercial products [35] that are isolated from petroleum are provided in Table 2.1. The approximate amount of each of these components in the crude oil depends upon its source. This is shown in Table 2.2.

Oil Field and Refinery Chemicals

Oil field and refinery chemicals are specialty chemicals that facilitate oil drilling and the refining operations. These include the following types.

- Paraffin Control Agents
- Drilling Fluid Components

TABLE 2.1—Products from petroleum refining [35].

Product	Boiling Range °C	Boiling Range °F
Liquefied Petroleum Gas	−40 to 0	−40 to 32
Motor Gasoline	30 to 200	90 to 400
Kerosene, Jet Fuel	170 to 270	340 to 520
Diesel Fuel	180 to 340	360 to 650
Furnace Oil	180 to 340	360 to 650
Base Oils for Lubricants	340 to 540	650 to 1010
Residual Fuel	340 to 650	650 to 1210
Asphalt	540+	1000+
Petroleum Coke	Solid	...

TABLE 2.2—Typical composition of common crude oils.

Product	U.S. (Texas)	Britain	Saudi Arabia	Nigeria
Liquefied Petroleum Gas (LPG)	1 %	2 %	1 %	1 %
Naphtha	29 %	18 %	18 %	13 %
Kerosene/Middle Distillate	42 %	37 %	33 %	47 %
Residue	29 %	43 %	48 %	39 %

- Refinery Process Chemicals
  —Wetting Agents
  —Anti-foulant
  —Corrosion Inhibitors
  —Foam Inhibitors
  —Demulsifiers
  —Combustion Improvers
  —Catalyst Presulfiding Agents

Paraffin Control Agents

These chemicals are used to improve the flow characteristics of petroleum so as to facilitate its recovery and processing. As mentioned earlier, petroleum and its products contain paraffins. The paraffins with high molecular weight and largely linear structures can separate at low temperatures as waxes. The resulting crystalline networks, or crystal lattices, have the tendency to capture the fluid components of petroleum, making them immobile. Classes of compounds that are used to overcome this problem are called pour point depressants, which are described in Chapter 4, the Additives chapter.

Drilling Fluids and Their Components

Most crude oil and gas reserves are either underground (on shore) or submerged (off shore). Well known on-shore locations include West Texas, Rocky Mountains Region, North Slope of Alaska, and the Middle East. Off-shore locations include the Gulf of Mexico, North Sea, and the China Sea. The drilling systems used for the crude oil recovery are of three types.

- *Rotary Drilling*—This is the traditional method for both on-shore and off-shore drilling. With this type of drilling, the drill bit is rigidly attached to the end of a drill pipe while the entire drill string is rotated at the rig floor by a rotary table.
- *Down-hole Motor*—This type of drilling utilizes a conventional rotary rig with a motor attached to the end of the drill string. The drilling fluid acts as the hydraulic fluid and the power turbines within the motor turn the bit. Down-hole motor drilling is primarily used while drilling horizontal and extended reach wells.
- *Slim-hole or Coiled Tubing*—This type of drilling uses a roll of tubing fed from a spool to deliver the drilling fluid to the bit. The continuous length of small diameter tubing serves the same function as the larger diameter drill pipe used on the conventional rigs. A down-hole motor is attached to the end of the tubing to power the drill bit.

Drilling systems employ fluids that perform a number of important functions, which include the following:

1. Lubricate and cool the bit.
2. Clean and transport cuttings to the surface.
3. Help stabilize the well bore.

4. Contain formation pressures to reduce the risk of a blowout.

Drilling fluids, sometimes called drilling muds because of their consistency and appearance, are usually composed of a liquid phase (aqueous or organic), soluble inorganic and organic additives, and suspended solids. For hard rock drilling, the gas-based muds derived from dry gas or natural gas are often used. Based on the liquid phase, the drilling fluids are classified into three types: water-based fluids, oil-based fluids, and synthetic oil-based fluids.

Water-based Drilling Fluids

These drilling fluids are the most common. They employ fresh water, seawater, or the common saltwater mixture. Saltwater-based drilling fluids are used when drilling through salt beds or domes. Over 85 % of the wells drilled in the United States use these fluids because they are more environmentally acceptable than the other types. In most cases, both the drilling fluid and the cuttings can be discharged overboard from off-shore rigs. Water-based fluids can be of dispersed or nondispersed types. The dispersed systems use dispersants, such as lignosulfonates, to help suspend clay particles and fines in the mud. This is the mud type of choice in the United States, both for on-shore and off-shore drilling, primarily because of its relatively low cost. Nondispersed systems, such as polymer muds, do not use dispersants because they do not depend on clays to impart viscosity. They use water-compatible polymers instead [36]. Potassium chloride-based saltwater polymer mud, a mud of the nondispersed type belonging to the class of shale-inhibitive muds, is invaluable for drilling highly reactive shales in the North Sea.

Oil-based Drilling Fluids

Oil-based drilling fluids are used for drilling high angle or extended reach wells, drilling through reactive shales, and when high temperatures are involved. These fluids are water-in-oil emulsions with brine as the dispersed phase. Oil-based fluids usually contain mineral oil or diesel oil as the continuous phase. Therefore, the used fluid and the drilled debris cannot be discharged overboard and must be brought back to the shore for disposal [37].

Synthetic Oil-based Drilling Fluids

These fluids are developed as environmentally friendly alternatives to the oil-based drilling fluids and are water-in-oil emulsions, with brine as the dispersed phase. However, they are formulated using synthetic fluids, such as esters, ethers, polyalphaolefins, and olefin isomers. Their characteristics are similar to those of the oil-based drilling fluids. For example, they can be used for drilling high angle and extended reach wells, and highly reactive shales. They have inherent lubricity and therefore require fewer lubricity additives than the water-based drilling fluids. However, for certain applications, such as extended reach wells, the lubricity additives may still be required. Because of the low toxicity and good biodegradability of these fluids, in many cases the drilled debris can be discharged directly overboard while the drilling fluid can be saved for re-use [37].

A number of synthetic oil-based fluids unrelated to drilling applications are also used in oil recovery. These include completion fluids and fracture-acidizing fluids. *Completion fluids* are used to finish work on wells after the drilling is complete. They control formation pressure while minimizing formation damage. Such fluids include water-based muds, oil-based muds, saltwater fluids, and brines. One such fluid is an emulsion (oil-in-water) composed of water, oil, and a blend of zinc and calcium bromides. *Fracture acidizing fluids* are used prior to mechanical fracturing of the formation. They etch heterogeneous carbonates, thereby increasing porosity and wall productivity. These are used as acid-in-oil emulsions to avoid direct contact between the acid and the carbonates in the rock.

Drilling Fluid Properties

For drilling fluids to be effective they must possess a number of properties. These include suitable density, viscosity, flow properties, wall-building properties, filtration properties, and stability, both during manufacture and use [36–38]. Density relates to a fluid's ability to withstand the hydrostatic pressures. Powdered high density solids or salts are used to build up the fluid density. High-density fluids can withstand the formation pressures better than the low-density fluids, as well as prevent the collapse of weak formations into the bore hole. Most drilling fluids have a fluid density above that of water, that is, from 1000 kg/m^3 (8.33 lb/gal) to over 2500 kg/m^3 (20.8 lb/gal).

Fluid viscosity and annular flow viscosity are the flow characteristics of interest in the drilling fluids. The fluid viscosity relates to a fluid's ability to be pumped down the drill and to suspend the resulting debris. The annular flow viscosity reflects its ability to carry the suspended material up the annulus to the surface for separation and disposal. This can be achieved for both low-viscosity and high-viscosity fluids, by simply adjusting the circulation flow rate. While developing a fluid of proper flow characteristics it is important to consider the drill speed because it can influence the effective viscosity due to shear. Non-Newtonian fluids are the best in meeting the desired flow characteristics. They have lower viscosity under the influence of shear and higher viscosity in the absence of shear.

Wall-building properties of the drilling fluids are also important. Since drilling fluids are introduced under pressure, they can enter the pores and crevices of the formation. However, if the pores are too small to allow the suspended particles to enter, the solids stay at the surface as a cake and the liquid passes into the formation. This is called leak-off or the filtration loss. High liquid loss is undesired because it can lead to drilling difficulties. These arise from an increase in fluid viscosity and an increase in the thickness of the wall cake, both of which hinder the fluid circulation. A similar situation results when the formation hydrates and swells because it contains clay or shale and the fluid contains water. The amount of filtration loss depends on the porosity of the formation and the permeability of the cake. The more porous and fractured the formation, the higher the loss, and the more permeable the cake, the higher the loss. On the other hand, an impermeable cake can hinder and even stop circulation by preventing the fluid's approach to the formation. To control the loss of circulation, additives are used that help form a low permeability cake on the surface of the formation. This type of cake maintains reasonable circulation and low fluid loss. The additives, either colloidal solids or organic polymers, can hinder the drilling process. This is countered by the use of other additives that control excessive fluid loss and cake buildup.

Water is an essential ingredient in all but the gaseous drilling fluids. It is present as a mist or in an emulsified form in oil-based fluids, and as a continuous phase in water-based fluids. The quality of water can therefore affect the stability of the dispersed phase in these fluids. The water's low salt content and low *p*H can destabilize the colloidal dispersion and destroy the fluid's structural integrity. For good mud stability, a salt content of 20 % or greater and a *p*H of greater than 8 are most suitable [37,38]. Hence, it is important to monitor the quality of water and its effect on the drilling mud stability.

Drilling Fluid Composition

Drilling fluids contain many organic and inorganic components, each of which performs a very specialized function [36,37,38].

Density Enhancers

Mud density is controlled by the use of the weighting material, barite ($BaSO_4$) being the most common. Other materials that are used to build the mud density include hematite (Fe_2O_3), magnetite (Fe_3O_4), ilmenite ($TiO_2 \cdot FeO$), siderite ($FeCO_3$), dolomite ($CaCO_3 \cdot MgCO_3$), calcite ($CaCO_3$), and sodium chloride (NaCl). While the drilled solids can also be used for this purpose, their small size may lead to the formation of the impervious cakes, thus hindering the fluid circulation. For solids-free fluids that are used for work-over and completion operations, water soluble salts, such as alkali and alkaline earth metal halides and formates are often used.

Viscosity Control Agents

The cuttings removal efficiency of a fluid largely depends upon its viscosity, with the high-viscosity fluids being superior. For water-based fluids, a variety of clays are used to obtain fluids of good viscosity, with bentonite, attapulgite, and sepiolite being the most important. Bentonite is usually used for fresh water muds and attapulgite and sepiolite are used for muds based on salt water. High performance grade bentonite contains montmorillonite clay in sodium and calcium forms that have the ability to react with high molecular weight polyacrylamides and polyacrylates. The viscosity-building ability of these materials is double that of the untreated bentonite. Low solids, nondispersed muds use these chemicals to extend the viscosity-building properties of bentonite. Magnesium-aluminum and calcium-magnesium-aluminum mixed metal hydroxides are also effective in extending the viscosity of bentonite and attapulgite.

To build the viscosity of the mineral oil-based and synthetic oil-based muds, oil-dispersible or organophillic clays are used. These clays are C_{12} or higher amine salts of bentonite, hectorite, and attapulgite. The amino group is believed to displace sodium and calcium originally present in these clays. A number of organic polymers are also used in drilling fluids, not only to build viscosity but also to control filtration loss. These include natural polysaccharides, such as starch, guar gum, and xanthan gum; modified natural polymers, such as cellulosics (modified cellulose), lignosulfonate, and lignite; and synthetic polymers and copolymers of acrylic acid, acrylonitrile, acrylamide, and 2-acrylamido-2-methylpropanesulfonic acid. Of these, cellulosics, xanthan gum, and polyacrylamides are used most often.

Filtration Control Agents

Proper flow does not always require an increase in viscosity. Sometimes, a decrease in viscosity is desired. Because the drilling muds have a high solids content and thixotropic (gelling) behavior over time, they can be treated as Bingham fluids. Hence, their flow can be described in terms of their yield point and plastic flow. Yield point and thixotropy reflect the tendency of the clay sheets to interlock (structure) and plastic flow relates to the force necessary to move them past one another. All three flow properties, that is, plastic flow, yield point, and thixotropy, are believed to involve charge interactions. These properties can be improved by lowering the mud viscosity. This can be achieved by decreasing the solids content or by the use of viscosity-reducing agents, or thinners. These chemicals function by minimizing the clay's tendency to form structure (presumably by neutralizing charges), thereby facilitating flow. In the water-based muds, polyphosphates, tannins, lignites, lignosulfates, and low molecular weight polyacrylates are used for this purpose.

Alkalinity Aids

Most muds have a *p*H in the range of 6–13. For the water-based muds, alkalinity agents are used to maintain the *p*H on the basic side in order to control corrosion. The bases used include sodium hydroxide, potassium hydroxide, calcium oxide (lime), and magnesium oxide. Sometimes there is a need to lower the *p*H, which is achieved by the use of the organic acids, such as acetic acid, citric acid, and oxalic acid, or mild inorganic acidic materials, such as sodium bicarbonate and sodium acid pyrophosphate.

Contaminant Removal Chemicals

Drilled solids, if present in a large amount, can impair proper functioning of the drilling fluid. These are removed by the use of screens, hydro-cyclones, centrifuges, or by the use of chemicals, called flocculants. Soluble contaminants encountered during drilling, such as sodium chloride, are taken care of by dilution and calcium salts are taken care of by precipitation as a phosphate, a carbonate, or an oxalate.

Formation Stabilizing Agents

As mentioned earlier, formations that contain shale are water sensitive and swell in the presence of water. The presence of too much trapped water in some formations leads to plastic flow into the borehole leading to operational difficulties. While a number of methods are used to stabilize such formations, the key strategy is to minimize contact between water in the mud and the formation. The use of the oil-based muds can help achieve this to some degree. The problem is that in such fluids the oil phase forms a surface barrier and does not allow water to escape from the formation. This is overcome by the use of the saltwater muds that remove water from the formation by osmosis.

Because the oil-based muds are higher in cost and difficult to dispose of, water-based muds with additives are the alternative option. Such additives include salts, polymers, and other organic materials. Saltwater muds are very effective as formation-stabilizing agents. The higher the concentration of salt in the mud, the lower is its formation-wetting ability. Hence, saturated saltwater muds are better than the seawater muds. Calcium compounds, such as hydrated calcium sulfate (gypsum) and calcium oxide (lime), are also effective. They function by converting the more water sensitive sodium clay into the less hydrophilic calcium clay. This lowers the flow of water from the drilling fluid to the formation by osmosis, thereby leading to its stabilization. Potassium, ammonium, and magnesium salts also convert sodium clays into less sensitive potassium, ammonium, and magnesium clays. However, the ammonium mud is not always preferred because of its highly alkaline nature.

A number of organic polymers are used to perform the same function. They can be added to a fresh water mud or a mud that contains one or more of the salts mentioned above. These polymers form a barrier against water on the formation surface. Nonionic polymers do this simply through physical association. Cationic and anionic polymers, on the other hand, are believed to associate with the clay through electrostatic charge interactions, with opposite charges on the clay surface. The formation-stabilizing activity of these polymers is hard to determine because they also behave as filtration control agents. Commonly used polymers include modified starches, cellulosics, gums such as guar and xanthan, quaternary ammonium salts, high molecular weight polyacrylamides, cationic polyacrylamides, and poly(vinyl alcohol).

Surfactants

Surfactants perform a number of important functions in drilling fluids. They can help emulsify oil and organic additives in water, water in oil, and act as foaming and defoaming agents. However, their most prevalent use is as emulsifiers. A wide variety of natural and synthetic materials have potential use. However, lignites, lignosulfonates, alkylarylsulfonates and sulfates, and polyalkylated acids and esters are among those most commonly used.

Lost Circulation Additives

As stated earlier, circulation can be lost if the fluid encounters highly permeable and or fractured formation because of the leak-off. If drilling is to continue, circulation must be regenerated. This can be achieved by the use of a wide variety of materials. Lost circulation additives, based on their texture, can be classified as flaky, fibrous, and granular. Flaky materials include cellophane, paper, mica, and rice and cotton seed hulls; fibrous materials include cellulose, saw dust, sugar cane residue (bagasse) after the juice is removed; and the granular materials include ground rubber, nylon, plastics, asphalt, and ground nut shells. Basically, any imaginable material can be used. The prerequisite is its particle size which must be large enough to plug up the pores and crevices to avoid the fluid loss.

Flocculating Agents

Solids that result from drilling are removed from the mud either mechanically, through dilution, or by flocculation. The suspended solids slow down the drilling operation. For separation, all drilled solids must be removed so that the drilling fluid is clean. Polymers, such as high molecular weight polyacrylamides (with various degrees of hydrolysis) and vinyl acetate-maleic anhydride copolymers, are used to remove solids by flocculation.

Lubricants and Spotting Fluids

Drill string may experience resistance during raising and lowering if the well is straight, during angular entry if the hole changes directions, and during rotation when against the formation. In order to overcome this problem, muds are treated with lubricants. Modern lubricants are basically blends of anionic or nonionic surfactants, glycols, glycerols, fatty acid esters, synthetic hydrocarbons, and vegetable oil derivatives. Sometimes solid lubricants, such as graphite and plastic or glass beads, are also used to minimize pipe sticking, a situation that arises when the drill pipe stops against the formation wall due to a physical obstruction. Another type of sticking, called differential sticking, results when the mud and the formation pressure differential holds the pipe against the formation wall. In such instances, water- or oil-based spotting fluids are introduced into the interface to break the pressure lock. Water-based spotting fluids have a composition similar to that of the lubricants.

Corrosion Inhibitors

Corrosion is a major problem in drilling and is caused by oxygen or acidic gases. Oxygen corrosion is more prevalent at pH of <11, hence maintaining a higher pH rectifies this problem. Acidic gases, such as carbon dioxide and hydrogen sulfide cause chemical corrosion and again a high pH takes care of it. Zinc and iron compounds, such as oxides, hydroxides, and carbonates, are used to scavenge sulfide, and sulfites are used to scavenge oxygen.

Drilling Rate Enhancers

Drilling debris from some formations tends to stick to the drill bit which can seriously impair drilling. The use of oil- or synthetic oil-based muds can minimize this. For water-based muds, with a higher tendency to experience this problem, water-soluble glycols and some terpenes are somewhat helpful.

Environmental Aspects

North American Off-shore Drilling

The Environmental Protection Agency (EPA) has established a list of water-based drilling fluid formulations that are used in the field. The "eight generic muds," as they are called, are standardized fluids that are used to test additives. The 96-h LC_{50} Test, or the "shrimp test," is used to determine toxicity. The minimum LC_{50} value allowable for discharge is 30,000 ppm SPP (Suspended Particulate Phase), which correlates to the toxicity of Generic Mud #1, the most toxic of the eight generic muds. Generic Mud #7 is the preferred drilling fluid for this determination. Most major mud companies prefer LC_{50} values greater than 30,000 ppm. Because the regulations for off-shore drilling are more stringent than those for on-shore drilling, the additives restricted for off-shore use, in some instances, can be used for on-shore drilling fluids. In the United States, the LC_{50} test is not generally required for testing additives in the oil-based drilling fluids.

North Sea Drilling

In this region, drilling is regulated by the MAFF (Ministry of Agriculture Fisheries and Food) of the U.K., the State Supervision of Mines of the Netherlands, and the SFT (State Pollution Control Authority) of Norway. The toxicity tests to evaluate additives include a 48-h LC_{50} Acartia Test and a 72-h EC_{50} Algae Test. These tests are used for both the water-based and the oil-based drilling fluids. For use in Europe, the additives are tested independent of the drilling fluid. Other important tests include a ten-day LC_{50} Corophium Test, OECD 306 Biodegradation Test, and OECD 117 Octanol/Water Partitioning Test.

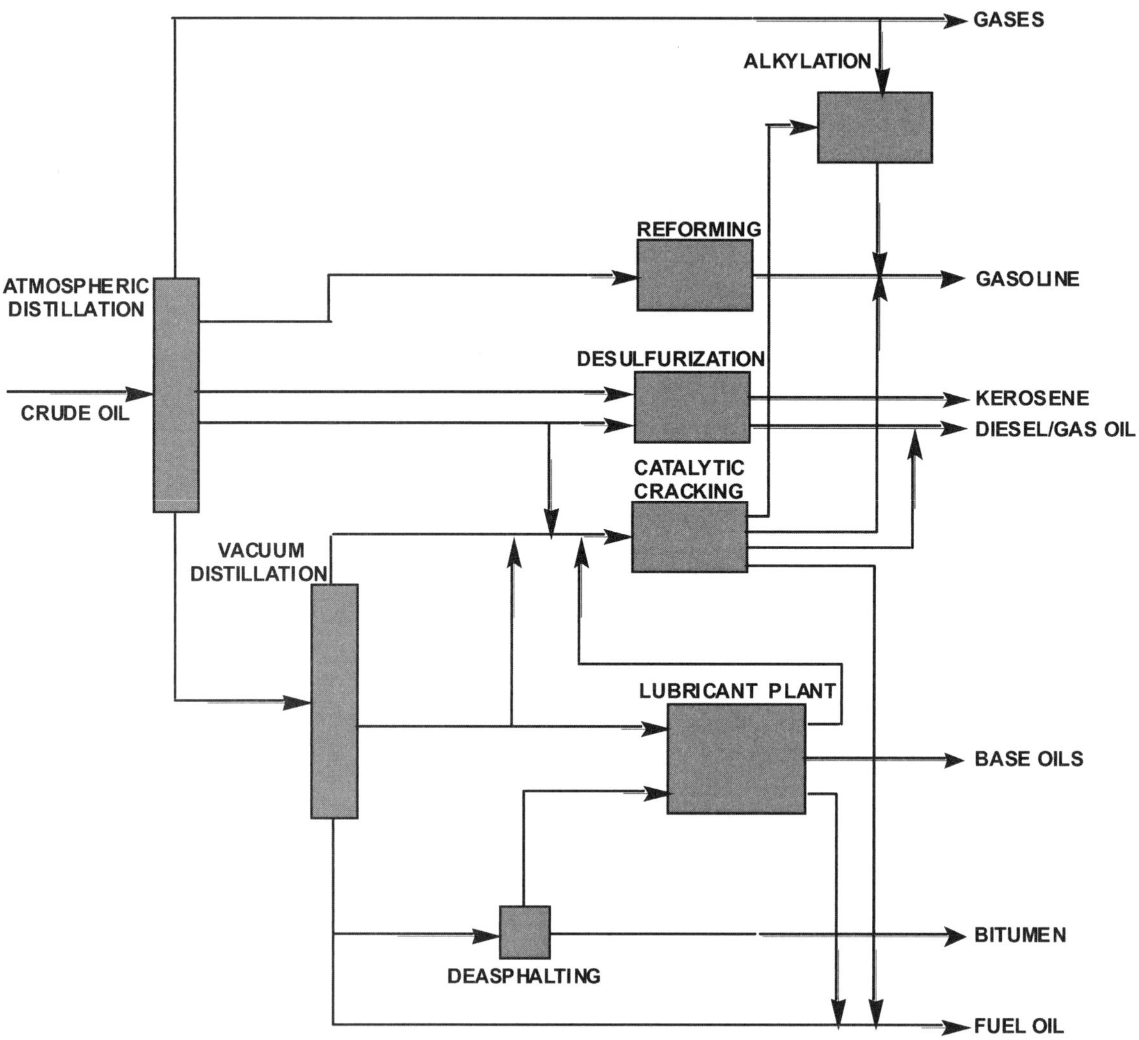

Fig. 2.2—A simplified refinery flow scheme [41].

Petroleum Refining

Refinery Processes

The petroleum refining process is used to manufacture fuels and base oils for use in formulating lubricants [39,40]. The amount and the quality of the fuels obtained from the crude petroleum depend on its composition.

Isolation of different fuels from the crude oil is carried out in a refinery and involves separation based on boiling ranges or distillation. Refineries have the capability to change product composition based on the market demand. For example, they can produce more gasoline from the higher boiling naphtha through cracking (breaking down) than that naturally present in the crude oil. Conversely, it can produce higher boiling fractions suitable as diesel fuel or base stocks/oils by the reaction of the lower boiling fractions. A simplified refinery scheme is shown in Fig. 2.2 [41], which shows some of the refinery operations listed below.

- Desalting
- Distillation
- Hydrogen Processing (Hydroprocessing)
- Hydrogen Cracking (Hydrocracking)
- Catalytic Cracking
- Coking
- Vis-breaking
- Solvent Deasphalting
- Steam Cracking
- Catalytic Reforming
- Alkylation
- Isomerization
- Extraction
- Gasoline Blending and Treating
- Lubricant Manufacture

Desalting

This operation is used to remove salts, primarily sodium, magnesium, and calcium chlorides, clay, or the other suspended solids from the crude oil, prior to distillation. This minimizes potential problems due to sedimentation. Salt, if not removed, can hydrolyze during distillation to release hydrogen chloride that can either cause corrosion of the equipment or lead to deposits in the presence of ammonia. The desalting process involves washing of the crude oil with 9–10 % water at 120–140°C. Emulsifiers or demulsifiers may be

TABLE 2.3—Effect of various refining processes on yield.

Process	Conversion, % w	Objective
Hydrotreating	0	Cleanup without conversion
Hydrorefining	As little as possible	Cleanup of higher boiling streams with minimum conversion
Hydrocracking	25+	Cleanup+conversion

needed to promote mixing of the crude oil and water, or the separation of the two at the end of the washing process.

Distillation

Distillation is the process of isolating products based on boiling point. Five fractions are typically isolated. These are gas and light fractions, naphtha (used for motor gasoline), kerosene, middle distillate (used for diesel fuel, jet fuel, and heating oil), and the bottoms fraction that comprises higher boiling materials. Distillation is carried out at atmospheric pressure using a fractionation plate tower with the bottom temperature of 370–400°C. During this step, about 50 % of the crude oil is converted into the indicated feed streams, ranging from the low boiling gases to the gas oil. The final boiling point of the atmospheric gas oil can range from as low as 340°C to as high as 410°C. Isolation of the additional product, called the vacuum gas oil, requires the use of vacuum, typically in the range of 7–25 mm of Hg. Vacuum gas oil has the boiling range of 500–575°C and its yield is up to 30 %. The residual fraction after the removal of these components, referred at as the vacuum residue, is either used as asphalt or as heavy fuel oil. It can be processed further, if absolutely necessary.

Kerosene fraction is used as jet fuel and to blend in lighter fuel oils and diesel fuels. Atmospheric gas oil is heavier than kerosene but can be distilled and hydrotreated to produce low-sulfur diesel. Vacuum gas oils, both light and heavy, are the additional source of the jet and diesel fuel, and for the production of the lubricating oil base stocks.

Hydrogen Processing (Hydroprocessing)

Hydroprocessing is used either to improve the quality of the isolated products or crack the heavy carbonaceous materials to lower boiling, more valuable products. Hydroprocessing is the main technology for removing sulfur- and nitrogen-containing compounds and for reducing the aromatics and the olefins from various fractions. These materials in the refinery liquid streams are not wanted, because they either impart color, odor, or thermo-oxidative instability [42,43]. Hydroprocessing is a generic term used for a process that involves treating various refinery streams with hydrogen. It includes hydrotreating, hydrorefining, and hydrocracking, depending upon the rigor of the operation, product conversion, and the product cleanliness achieved. Product conversion is defined as 100 minus the amount of the product in the same boiling range as the feed. Higher conversion implies that the product/s obtained have significantly different boiling points, hence have different composition/s. Table 2.3 lists the differences between the three processes. Based upon the criteria provided in the table, one can conclude that with respect to conversion, hydrotreating is at one end of the scale and hydrocracking is at the other end of the scale, and hydrorefining falls in the middle.

There are a number of other refining terms used, based on the objective met. For example, the term de-sulfurization implies removal or reduction in the amount of sulfur and its compounds; denitrogenation implies removal of the nitrogen-containing materials; and dearomatization implies removal of the aromatics or their conversion into nonaromatic compounds, primarily naphthenes.

Hydroprocessing involves exposing the refinery stream to hydrogen in the presence of a catalyst, and at appropriate temperature and pressure. In general, higher temperatures and pressures are required, if the boiling range of the fraction to be cleaned is high, dearomatization is intended, or hydrocracking is the objective. Hydroprocessing entails hydrogen gas circulation, at a high rate, through or over the refinery stream. If desulfurization is the goal, sulfur and sulfur-containing materials will be converted into hydrogen sulfide, which is removed by scrubbing with an amine solution. Mild treatment involves low hydrogen pressures (200–300 psi, 1379–2069 kPa) and severe treatment involves high hydrogen pressures (1000–3000 psi, 6895 –20,685 kPa). The temperature for both treatments is in the 285–400°C range.

In hydroprocessing, a variety of catalysts are employed. These include combinations of cobalt and molybdenum or nickel and molybdenum promoters on an alumina base. The first catalyst system is commonly used for sulfur removal and the second catalyst system is used for nitrogen removal and for some aromatics to remove unsaturation. Other catalysts include nickel/tungsten, noble metals, such as platinum, and zeolites, especially for a high degree of dearomatization. Cobalt/molybdenum and nickel/molybdenum on alumina catalysts contain 3–25 % promoter metals, depending upon the quality of the feed and the intended hydroprocessing severity.

Hydrocracking (Hydrogen Cracking)

Hydrocracking is used to convert high molecular weight (higher boiling) materials into low boiling materials, that is, to convert naphtha fraction into premium-octane gasoline or aromatic petrochemical materials. This process, like *hydrotreating*, uses hydrogen, except that it involves higher temperatures and longer contact times. Catalysts for the two processes are also similar.

Catalytic Cracking

This process allows the conversion of heavy distillate into gasoline and middle distillate, by fragmenting the high molecular weight hydrocarbons to low molecular weight hydrocarbons. The process involves suspending the finely divided catalyst (particle size of ~20–200 μm) in gaseous heavy distillate and circulating it through pipes and valves between the reaction and the regeneration vessels. Cracking temperature is around 525–550°C and the process takes only a few seconds to complete.

Coking

Coking is high-severity thermal cracking or destructive distillation, which is used to convert the heavy low-value residuum into the low boiling value-added products, such as gasoline, naphtha, and gas oil. *Delayed coking* involves holding large drums of hot oil at 450 °C and 5–10 psi and collecting the cracked vapors that form in the head space. *Fluid coking*, on the other hand, involves spraying the hot residuum into a fluidized bed of hot coke particles that are held at 22–36 psi and 51 °C, or higher. The vaporized and cracked feed forms a liquid film on the particle surface, which is ultimately collected.

Vis-breaking

The purpose of *vis-breaking* is to decrease the viscosity of the heavy fuel oil to permit its handling at lower temperatures. This, like coking, is also a thermal cracking process using only heat to crack the residuum to lower boiling products. The process involves heating of heavy oil in a furnace at ~480 °C for a sufficient time to obtain gas, naphtha, and the distillate gas oil.

Solvent Deasphalting

This process is used to remove asphaltenes from the residuum, so that its useful components can be processed in a cracking unit. The removal of the asphaltenes is warranted since these are highly aromatic entities containing nitrogen and sulfur that are difficult to crack. The process involves dissolving the vacuum residue in a hydrocarbon, such as propane, butane, or pentane. The insoluble asphaltenes are removed and after removing the solvent from the soluble phase, the residue is sent to the cracking unit.

Steam Cracking

Steam cracking is used to convert ethane, butane, and naphtha into olefins that are used as raw materials for the manufacture of the petrochemicals. The process involves thermal cracking at ~800–850 °C at slightly above the atmospheric pressure.

Catalytic Reforming

This process is used to convert the straight-run naphtha (octane number 30-50) into motor gasoline (octane number 94-103). The feed is usually a C_6-C_{10} cut. The process involves passing the naphtha fraction of boiling range 80–230 °C over a catalyst held at 430–530 °C and 145–870 psi. Reforming reactions improve octane by converting linear molecules into cyclic molecules (naphthenes), dehydrogenating naphthenes into aromatics, and by converting high-boiling large hydrocarbons into low-boiling lighter compounds. Originally, an alumina base promoted with 0.25–0.75 % platinum was used as the reforming catalyst. However, today's producers are using proprietary technology and the catalyst may be platinum alone, or bi- or tri-metallic systems based on additional metals, such as rhenium and tin. Some producers may even be using zeolites.

This process is a major source of industrial raw materials, such as benzene, toluene, and ortho- and para-xylenes. These materials are obtained by extracting reformate with a solvent, such as sulfolane; or di, tri, and tetra(ethylene glycol)s. The aromatics become part of the extract, from which the solvent is distilled off to yield the aromatic products. The component consisting of paraffins and naphthenes, which is not soluble in these solvents, is the raffinate. This is marketed as a solvent for various end uses or is used as a feed to ethylene plants running on naphtha. The aromatics fraction is separated into its components by passing it through a series of distillation columns. It is not easy to separate meta-xylene from para-xylene through distillation since both have close boiling points. They are separated either via crystallization or by the use of the molecular sieves.

Alkylation

Alkylation reaction involves the reaction of a low molecular weight olefin (containing 3, 4, or 5 carbon atoms) with an alkane, alkene, or an aromatic compound. The reaction of isobutylene with isobutane yields 2,2,4-trimethylpentane (isooctane)—a high octane product. The reaction is carried out in the presence of sulfuric acid or hydrofluoric acid catalyst and at a temperature of 4–15 °C and slightly higher than atmospheric pressure. Similarly, benzene is alkylated with ethylene to form ethylbenzene, which is ultimately converted into styrene, and with propylene to ultimately produce alkylphenol. Both these materials are important petrochemicals that are used in a variety of chemical transformations.

Alkylation of an alkane with an olefin requires tertiary alkanes; primary and secondary alkanes do not undergo alkylation. This is because tertiary alkanes have a carbon that has single hydrogen attached to it. This hydrogen is more reactive than the hydrogens on a secondary or a primary carbon. Alkylation of benzene and phenol are straight forward reactions. Again, 98–99 % sulfuric acid (H_2SO_4), or concentrated hydrofluoric acid (HF) are used as catalysts. Sulfuric acid catalyzed alkylations are carried out at low temperatures (5–10 °C) to minimize olefin polymerization. Hydrofluoric acid (HF) alkylation can be carried out at 25–50 °C, which eliminates the need for refrigeration, making it a lower cost process than H_2SO_4 catalyzed alkylations. The use of the solid catalysts is presently being explored.

Isomerization

This process is used mainly to convert normal paraffins into isoparaffins. The technology is employed to increase the octane quality of the light straight run gasoline/naphtha by converting *n*-pentane and *n*-hexane into mixed C_5–C_6 isoparaffins. The process is also employed to convert *n*-butane into isobutane, either to provide feed for the alkylation process or for use to make methyl *tert*-butyl ether (MTBE) octane improver, when isobutylene is not available. The process involves reacting paraffins feed and hydrogen at a temperature of 180–260 °C in the presence of an isomerization catalyst.

Gasoline Blending and Treating

These processes are used to obtain different grades of gasoline by blending separate components. Blending involves adding alcohol as an anti-icing aid and treatment involves removing sulfur and gum by the use of caustic, clay, and other chemicals. Hydrotreating and hydrofinishing are also used to take care of these problems.

Lubricant Base Stock Manufacture

The base stock manufacture involves vacuum distillation/fractionation of the bottoms fraction (the vacuum gas oil), or the residuum, that is left after the removal of the low boiling fractions used to manufacture fuels. There are many critical requirements for a finished lubricant, as we will discuss later. Base oil quality is primarily determined by its viscosity index, low-temperature properties (pour point), and volatility (flash point). In terms of the chemical composition, the

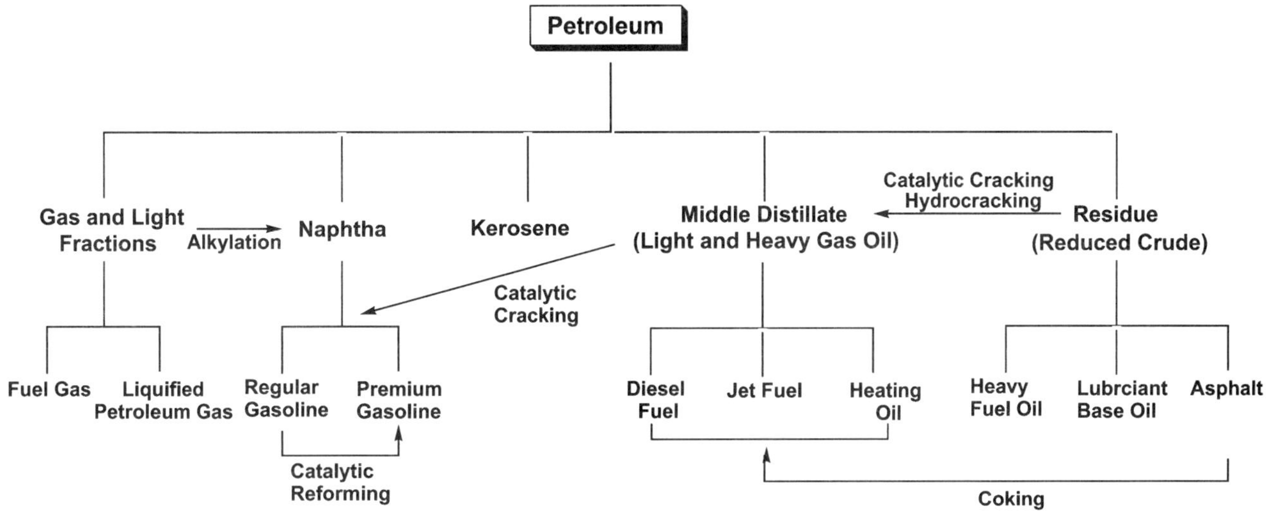

Fig. 2.3—Common refinery operations.

aromatics in the base stocks are usually undesired and a relatively lower amount of linear, or normal, paraffins is preferred. This is because aromatics have a low viscosity index (VI) and impart oxidative instability to the base stock [42]. On the other hand, normal paraffins have good VIs, but they have high pour points because of the tendency to crystallize at low temperatures to form network structures. Incidentally, branched paraffins also have good VIs and reasonable pour points, but they are somewhat more volatile than their unbranched counterparts. Hence, the base oil manufacture has the primary objectives of lowering the amount of the aromatics and improving the pour point. These objectives are achieved by either removing the aromatics via solvent extraction or by converting them into naphthenes through hydrotreating. Normal paraffins are removed through dewaxing. A new process, called iso-dewaxing, is gaining popularity since it can attain both objectives in a single step. The process is essentially severe hydrocracking in which the aromatics are converted into naphthenes and a significant amount of the normal paraffins are converted into isoparaffins. Details of these processes will be described under the base stock manufacture.

Figure 2.3 shows the products that are isolated from simple distillation and those that are manufactured through other common refinery processes.

Refinery Process Chemicals

Chemical Wetting Agents

Crude oil contains many impurities that must be removed prior to processing. These include solids (dirt), naphthenic acids, salts (chlorides and sulfides), and metals, such as nickel and vanadium, that can "poison" the hydrotreating catalysts. Dirt (solids) and most of the salts in the field are removed by water washing and in the refinery by electric desalting. Their extraction into water is greatly enhanced by the use of the chemical wetting agents that possess surfactant properties. Despite the extensive effort, all solids and salts are not removed and some remain.

Anti-foulants

The solids and salts that escape removal can deposit in areas of low velocity in heat exchangers and other process equipment and lead to fouling. *Fouling* is unwanted deposition of materials on the heat transfer surfaces and other process equipment. This results in a loss of heat transfer, constriction of the heat transfer equipment, lower conversion rates, and many other problems. The materials that form such deposits are called foulants. Besides inorganic solids and salts, some organic materials also lead to fouling, asphaltenes being the most prominent organic foulant. Asphaltenes are a class of poly-nuclear hydrocarbons that contain nitrogen and sulfur as heterocyclic moieties. A hypothetical structure of an asphaltene molecule is provided in Fig. 2.4. Inorganic compounds that are responsible for fouling include salts, dirt, and compounds resulting from corrosion.

Three types of fouling mechanisms have been proposed.

1. Thermal—Involves coking of the organics at high temperatures.
2. Chemical—Involves oxidation and polymerization resulting in tacky oxygenated hydrocarbons that can coke easily at high temperatures.
3. Physical—Involves deposition of the inorganics, salts, and asphaltenes; the latter, being organic, can also coke.

Anti-foulants are materials that are used in refinery op-

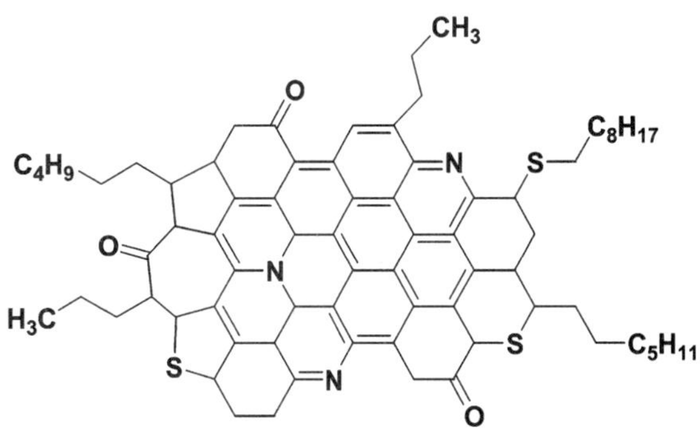

Fig. 2.4—A hypothetical asphaltene molecule.

erations to overcome fouling. The processes that use anti-foulants include distillation, hydroprocessing, catalytic cracking, delayed coking, vis-breaking, and catalytic reforming. Many classes of compounds have potential as anti-foulants: dispersants, detergents, and oxidation and corrosion inhibitors are among those used most often. Oxidation inhibitors are useful when fouling results from thermo-oxidative degradation of organics, such as asphaltenes. However, once the deposits form, conventional detergents and dispersants can help in removing them. For inorganic deposits, dispersants are deemed to be more effective. Since fouling is likely to involve both the inorganic salts and the organic-derived thermo-oxidative decomposition products, a combination strategy using oxidation inhibitors, detergents, and dispersants has a higher degree of success.

A number of tests are used to determine the effectiveness of the anti-foulants. These include the Static Thermal/Gum Polymerization Test, Static Rod/Filament Deposition Test, and Hot Liquid Process Simulator (HLPS) Test. The HLPS Test is used most widely in the industry and consists of a resistance heated tube-in-shell exchanger with dynamic once-through or a recycle flow. The apparatus is pressurized with nitrogen to reduce volatility. The lines, pump, and the vessel are heated to ensure flow of the test sample. A sample of untreated refinery feedstock is passed through the simulator to determine its fouling tendency. The fouling tendency of an anti-foulant treated feedstock is then determined in a similar manner. Comparison of the two results helps in determining the additive's ability to prevent fouling.

Corrosion and Foam Inhibitors

Acidic components and salts, present in petroleum, can lead to chemical and electrochemical corrosion. Corrosion can also contribute towards fouling because it produces metal oxides and salts. Corrosion inhibitors are used to control this type of metal damage. These inhibitors are of two types: film-forming agents and acid-neutralizing agents; see Chapter 4 on Additives. Some protect against corrosion by both mechanisms. Acid-neutralizing agents guard against chemical attack and film-forming agents guard against both chemical and electrochemical attack. Amines, commonly used to control *p*H, are examples of the acid-neutralizing agents and organic soaps, such as neutral metal sulfonates, are examples of the film-forming agents. Basic detergents protect against both fouling (deposition) and corrosion. This is because of their ability to form protective surface films due to their soap content and their acid-neutralizing ability because of the base reserve. Foaming in coking units is a prevalent problem. It is controlled by the use of silicone-type foam inhibitors.

Miscellaneous Additives

A variety of additives that perform diverse functions are also used. These include scaling and biological growth inhibitors for cooling towers, poly-electrolytes and demulsifiers for effluent treatment, and combustion improvers for furnaces [43]. Another class of additives, called the catalyst presulfiding agents, is used to produce new hydrogen treatment catalysts or to regenerate them. Sulfide is the active form of the catalyst. During hydrotreating and hydrocracking, sulfur and its compounds can generate hydrogen sulfide that can convert the metal oxide catalysts into metal sulfides. However, this process involves two steps. During the first step the oxides are reduced to the free metals and during the second step the metals are converted into their sulfides. The problems the refiners encounter are that the first step generates excessive exotherms and that it takes time to convert the metal oxides into sulfides during which the feed stream goes untreated. Refineries overcome both these problems through presulfiding. During use, the metal sulfide catalysts get reduced to free metals and must be regenerated to their active sulfide form. The chemicals used for this purpose are organic polysulfides of high sulfur content.

Trace amounts of metal ions, such as those of nickel (Ni), vanadium (V), iron (Fe), and sodium (Na), are known to lower the catalyst activity. This problem can be suppressed by the use of the metal passivators that perform through various mechanisms, including chelation. Typical nickel passivators for fluid catalytic cracking catalysts are antimony (Sb), bismuth (Bi), and cerium (Ce) compounds; typical vanadium passivators are tin (Sn) compounds; and typical sodium passivators are aluminum (Al) compounds.

Lubricant Base Stocks

The term "lubricant" is used to describe materials, usually liquid, that are employed to perform a number of diverse functions in machinery to prolong its useful life. These functions were commented upon in Chapter 1 while discussing the Fundamentals of Lubrication. A modern lubricant comprises a base fluid and chemical additives.

A vast majority of lubricants are based on mineral oils, which is because of their low cost, ready availability, and overall adequate performance. Since the base fluid makes up the bulk of the lubricant, its properties greatly impact the properties of the lubricant. The properties of a mineral base stock depend on its source, viscosity, and the degree of refining. As stated before, mineral base stocks or oils are classified on the basis of the predominant hydrocarbon type present in their composition, that is, as *paraffinic, naphthenic*, and *aromatic*, or bright stocks. Unfortunately, the hydrocarbon-based classification is not clear cut but is vague because these base stocks are manufactured based on viscosity and not on structural composition. Hence, neither base stock is purely paraffinic, naphthenic, or aromatic, but is a blend of all three types of hydrocarbons.

While each type of hydrocarbons has its own advantages and disadvantages, to formulate lubricants for most applications, the paraffinic base stocks are often preferred. However, they do suffer from poor low-temperature performance because of the wax formation, which incidentally is corrected by the use of additives, called pour point depressants. Naphthenic base stocks are the next best in desirable overall properties. While these base stocks have mediocre viscosity indices and oxidative stability, they have good viscosity-pressure relationship and good low-temperature properties. Aromatics are the least suited to formulate most modern lubricants, but they do have good solvency, thermal stability, and viscosity-pressure properties. Their oxidation stability and VI characteristics are the poorest among the three groups. As a general rule, the base oils of different origin may be similar in many respects, but they subtly differ in some properties, such as oxidation stability, which are determined by their overall structural composition. Desirable base oil

TABLE 2.4—Physical properties of hydrofinished HVI stocks [44]. Reprinted with permission from the Lubrizol Corporation.

	Specific Gravity at 60°F	Sulfur (% wt)	Viscosity Index	Kinematic Viscosity (cSt) at 40°C	Kinematic Viscosity (cSt) at 100°C	Pour Point (°C)	COC Flash (°C)
Source 1							
100 Neutral	0.860	0.065	101	20.39	4.11	−13	192
200 Neutral	0.872	0.096	99	40.74	6.23	−20	226
350 Neutral	0.877	0.126	97	65.59	8.39	−18	252
650 Neutral	0.882	0.155	96	117.90	12.43	−18	272
150 Bright Stock	0.895	0.263	95	438.00	29.46	−18	302
Source 2							
150 Neutral	0.861	0.036	98	24.38	4.55	−23	210
250 Neutral	0.872	0.055	96	48.96	6.94	−21	238
600 Neutral	0.878	0.099	95	108.00	11.64	−23	262
150 Bright Stock	0.892	0.147	95	473.00	30.90	−15	294
Source 3 (Hydrotreated)							
100 Neutral	0.868	0.018	100	25.18	4.66	−20	200
200 Neutral	0.869	0.012	101	39.78	6.19	−21	216
500 Neutral	0.869	0.015	105	89.37	10.78	−21	254
Source 4							
100 Neutral	0.862	0.278	107	21.26	4.28	−16	186
200 Neutral	0.877	0.571	103	30.53	5.26	−13	194
500 Neutral	0.888	0.729	98	95.48	10.89	−10	244
600 Neutral	0.891	0.760	96	111.80	11.99	−13	258
150 Bright Stock	0.903	0.843	96	477.80	30.99	−13	290

properties include good oxidation resistance, high viscosity index, good low-temperature performance, and low volatility. The refining process has the flexibility to produce base stocks where these properties are optimized. However, the challenge is to achieve this cost effectively. Good quality crudes, such as those from the Middle East, help accomplish this manufacturing goal. Base stock production results in a number of products based upon viscosity. Properties of some of these products are provided in Table 2.4 [44].

These base stocks are blended to make base oils of the desired viscometrics. The resulting base oils do not always possess the optimal properties for use in a specific application. Hence, they need performance-enhancing additives. Some additives enhance physical properties of the base oils while others impart or improve their chemical properties and behavior.

Comparison between Naphthenic and Paraffinic Base Oils

The base oils/stocks are classified the same as the crude oils into paraffinic, naphthenic, or aromatic, depending upon the predominant proportion of the hydrocarbon molecules they contain. Mixed-base crude oils have the varying amounts of each type of hydrocarbons. The refinery crude base stocks usually consist of mixtures of two or more different crude oils; hence they will affect the composition of the obtained base stocks. Three methods that are commonly used to classify the crude oils are API Gravity, Universal Oil Products' characterization factor (UOP "K" factor), and the United States Bureau of Mining classification.

API (American Petroleum Institute) gravity is based upon specific gravity, or relative weight, at 60°F (15.6°C). Specific gravity (SG) is the density of a substance divided by the density of water at a specific temperature. The oil industry uses API Gravity, which is related to specific gravity as provided in the equation below.

$$\text{API Gravity} = \frac{141.5}{\text{Specific Gravity at } 60\,^\circ\text{F}} - 131.5$$

Thus, a heavy oil with a specific gravity of 1.0, the same as of water, would have an API Gravity of (141.5/1.0) − 131.5 = 10.0°API. API Gravity is an arbitrary scale which was devised jointly by the American Petroleum Institute and the United States National Bureau of Standards, for expressing the gravity, or the density, of the liquid oil products. The API Gravity scale was designed so that most values fell between 10 and 70°API. Sixty degrees Fahrenheit is used as the normal value for the measurements and further tables give adjustments for temperature, see the ASTM D1298 Standard [27]. Crude oils are classified as light, medium, or heavy, according to the API Gravity. Light crude oils have an API Gravity of higher than 31.1°API, medium oils have API Gravity between 22.3 and 31.1°API, and heavy oils have an API Gravity of less than 22.3°API. Oils that do not flow at normal temperatures or without dilution, normally called bitumen, have an API Gravity of less than 10°API. API Gravity values (ASTM D287, D1298) are sensitive to the relative amounts of paraffins, cycloparaffins, and aromatics in crude oil; a higher paraffinic content increases the API Gravity values.

Universal Oil Products' characterization factor (UOP K factor) is based upon gravity and boiling point or the viscosity data. The UOP K factor is indicative of the origin and the nature of the petroleum stock. Values of 12.5 and higher indicate a material predominantly paraffinic in nature. Highly aromatic materials have characterization factors of 10.0 or less. The value between 10 and 12.5 indicates a crude oil of high naphthenic content. Typical API Gravity values of the mineral base oil fractions are provided in Table 2.5 [45].

TABLE 2.5—Typical values for gravity of mineral base oil fractions [45].

Oil Type	API Gravity	Specific Gravity 60/60°F
Aromatic	10.0	1.000
Naphthenic	20.0	0.934
Paraffinic	30.0	0.876

There are other approaches that are used to classify the crude oils. For example, one approach considers the chemical composition of the crude. It considers the crude to be paraffinic, if paraffins and naphthenes together make up greater than 50 % of the composition and the paraffins are greater than 40 %. It considers the crude to be naphthenic, if the paraffinic and naphthenic content are greater than 50 % and the naphthenes are greater than 40 %. The aromatic crudes are those that have an aromatic content greater than 50 %.

Typically, base stocks obtained from vacuum distillation of the paraffinic crudes have a high paraffinic content and those obtained from naphthenic crudes have a high naphthenic content. Each type of base stock has attributes and deficiencies. These will be commented upon later in this chapter. In general, the paraffinic oils have good high-temperature properties and the naphthenic oils have good low-temperature properties. Naphthenic base stocks also have the disadvantage of a higher aromatics content, which limits their use in applications where oxidative stability and toxicity are important considerations. A properly formulated lubricant, in addition to performing functions listed earlier, conserves energy by lowering the fuel consumption, minimizes damage to the equipment by reducing wear, and lowers the maintenance cost by extending the service interval.

Petroleum is the main source of the lubricant base fluids and as stated earlier, price, availability, and performance are the primary reasons for their prevalent use. However, the performance requirements of some applications, such as aviation, cannot be satisfied by the mineral oil-based lubricants because they are not very effective at extreme temperatures, both high and low. In these cases synthetic fluids are used by necessity. Many synthetics have much higher thermal and oxidative stability, low volatility, and excellent low-temperature fluidity. Nonpetroleum base fluids are also used in applications where special properties are desired, petroleum base oils are in short supply, or substitution by natural products is easily feasible or more desirable.

The structures of compounds that are present in petroleum are shown in Fig. 2.1. Since all these structures are not appropriate for use as lubricant base fluids, a series of physical refining steps are needed to isolate those components that are suitable. In general, only true hydrocarbon molecules are useful in the lubricating compositions. Hydrocarbon molecules, which contain oxygen, sulfur, or nitrogen in their structure, are usually not preferred in lubricants. This is because such compounds are highly susceptible to oxidation or degradation, or both. However, some organo-sulfur compounds are desirable since they tend to provide oxidative stability.

Mineral Base Oil Manufacture

Mineral base oils are manufactured by petroleum refining, which was described earlier. Base oils are obtained from the stocks isolated from atmospheric distillation residue, after the removal of the low-boiling gases and the liquid fractions that are used to obtain fuels. The process to produce mineral oils involves the following steps.

1. Vacuum distillation to separate aliphatic components from the aromatic components.
2. Acid refining of aliphatic components to improve their aging tendency and the viscosity-temperature characteristics.
3. De-asphalting of the vacuum distillates to remove aromatic impurities and of the distillation residue to obtain bright stocks.
4. Solvent refining or catalytic dewaxing to improve low-temperature properties.
5. Hydrogen finishing, or clay treatment, to remove trace quantities of the hetero-aromatics, olefins, or other compounds that can lead to instability of the base stock/oil [46,47].
6. Blending of the different base stocks to obtain mineral oils.

The purpose of these steps is to isolate materials that have boiling points and physical and chemical properties suitable for use in lubricants. Tha major processes that are used to obtain mineral base stocks from petroleum include distillation, de-asphalting, solvent extraction, solvent dewaxing, and finishing.

Fractional distillation by boiling point is the fundamental operation in refining. Various products that are isolated from the crude oil, along with their approximate boiling ranges, were provided in Table 2.1. As mentioned earlier, the amount of each isolated product depends on the quality of the crude and the extent of refining. After the removal of the fuel components, the remaining heavy black residuum (atmospheric residue) is drawn to the base of the column so as to convert it into the lubricant base stocks. In the case of good quality crude, such as the Middle Eastern crude, the amount of the atmospheric residue may be as much as 50 % of the total. Atmospheric residue is next passed through a vacuum distillation unit to separate various grades of the petroleum base stocks. Vacuum is needed in this step for two reasons. First, the fractions constituting the base stocks have high boiling points. Second, the residuum has low thermal stability and, therefore, has a tendency to decompose above 300°C.

Vacuum distillation produces hydrocarbon fractions which must be treated further to yield base stocks of the desired quality. These processes include deasphalting, extraction, dewaxing, and hydrogen finishing. The base oil manufacturing process produces a large quantity of by-products, which even in the case of a good quality crude, such as Middle East crude, can be as high as 75 % [41]. This translates into the base oil or base stock yield of only 25 %. Gas oil, light fuels, bitumen or fuel oil, aromatics. and wax make up the rest. Obviously, the yield of the base oils from the poor quality crudes is even lower. Considering that the atmospheric residue represents ~50 % of the crude oil, the yield of the base oils starting from the crude is only about 12 %. This means that the quality of the crude is critical to the cost-effective manufacture of the lubricant base stocks.

Poor quality crudes impose a penalty on the refinery by producing a larger percentage of products unsuitable for use

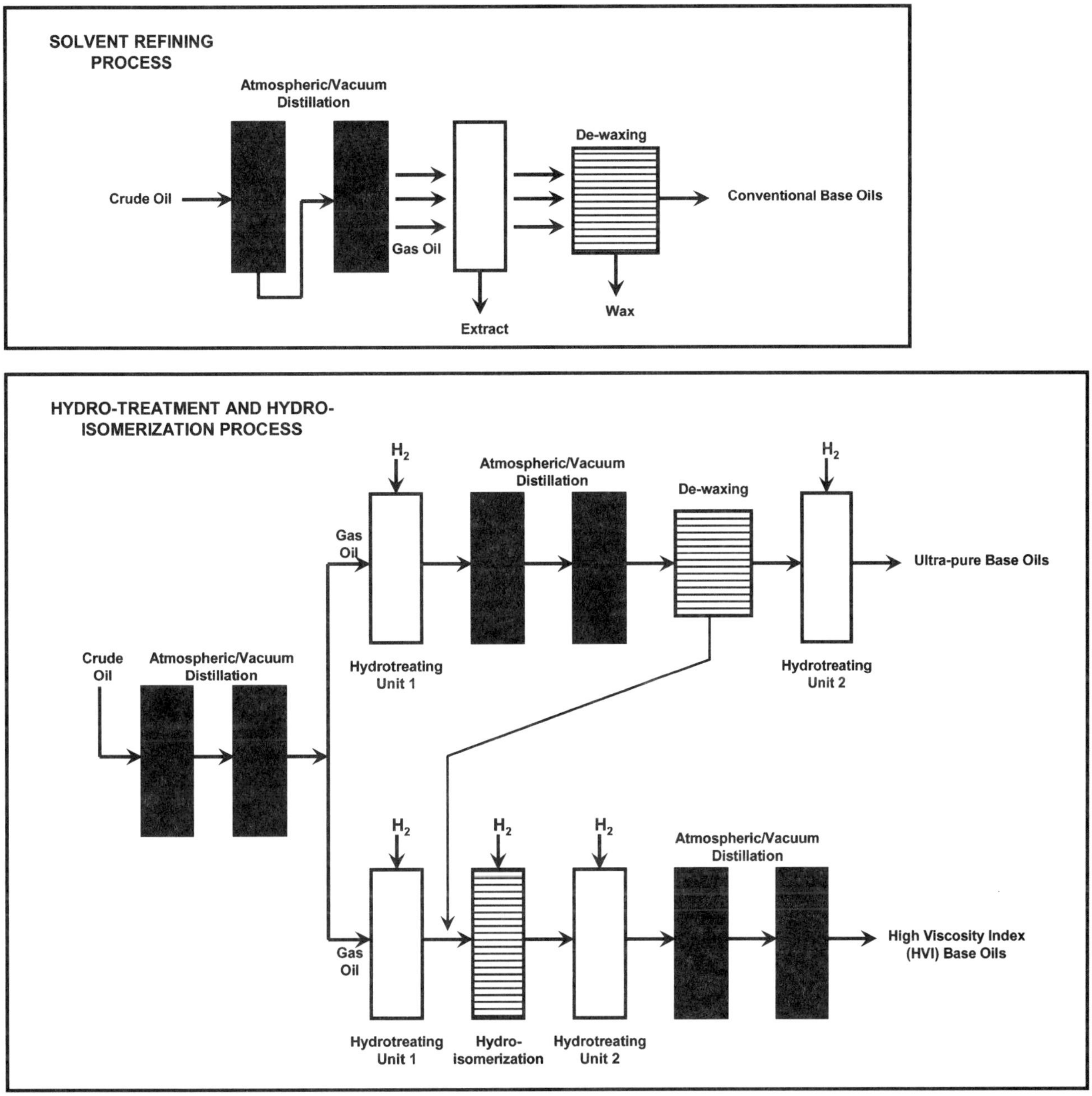

Fig. 2.5—Solvent refining process versus hydro-treatment and hydro-isomerization processes [514].

to formulate lubricants. The refining processes achieve two goals. They help isolate components with desirable properties and they help minimize gross variations in the base stocks that exist in crude oils from different sources. These processes are described below [41].

Distillation is the primary process for separating fractions based on boiling point. Atmospheric distillation of the crude yields components that boil below 350°C, such as gases, naphtha, kerosene, and gas oil. Atmospheric residue is the source of the lubricant base stocks. Since hydrocarbons that form such base stocks generally have boiling points above 350°C at atmospheric pressure, vacuum distillation is necessary to isolate them. Typically, steam is injected into the atmospheric residue feedstock, and the mixture preheated in a furnace before entering it into the distillation unit. Vacuum is provided at the top of the unit. The pressure at the top of the column is 60–80 mm of Hg and at the bottom is 100–140 mm of Hg. The temperature at the top is about 140°C and at the bottom is about 360°C. The vaporizing steam-feedstock mixture separates into different boiling range fractions, usually three, which are collected using various devices. The residue, which typically has a boiling point above 550°C, is drawn from the base of the tower.

The major goals of distillation are to isolate products based on viscosity and to improve flash point. Viscosity of a base stock depends upon the boiling range of the components present. Since this component parameter is hard to control, distillation is used to obtain a variety of distillation cuts, including base stocks, which are later blended to obtain products with the desired viscometrics (Table 2.4). Flash point of a fraction is largely due to the presence of the small amounts of the low-boiling components, which are removed during the distillation process.

Base stock refining involves the removal of certain materials from the vacuum distillate, which if not removed, will cause darkening of the base oils and hence the lubricants, their viscosity increase, and the formation of the oil-insoluble matter, called sludge. These materials include olefins, acids such as naphthenic acid, certain sulfur compounds, some polyaromatics, nitrogen compounds, and asphaltenes and resins. The paraffin wax present in the vacuum distillate must also be removed to obtain base oils with better low-temperature properties. Refining consists of acid refining, solvent refining, and hydrotreating.

The goal of the acid refining is to remove olefins and undesirable aromatic compounds. The process involves treatment of the base stock distillates with sulfuric acid or oleum (fuming sulfuric acid). During acid treatment, olefins and undesirable aromatics are converted into sulfates and sulfonates, which because of high polarity separate out from the raffinate as a dark liquid. Since this process generates a significant amount of acid waste, it is being replaced by less waste-generating processes. Acid washing not only improves color and aging stability, it also improves other desirable properties, such as viscosity, viscosity index, and oxidative stability. All these properties relate to the removal of aromatics, which have higher density, low oxidative stability, and a very low VI. The major disadvantages of acid refining include a large amount of waste generation, loss of viscosity, and an increase in the pour point of the raffinates. The latter occurs due to the removal of materials, such as alkylaromatics, which have the ability to act as pour point depressants.

Because of the waste issue, acid refining is only used if the objective is to manufacture white oil and produce sulfonic acids. Sulfonic acids are used in a variety of industries as raw materials and for a variety of uses. The major uses include lubricant additive manufacture, as catalysts (cure agents) in paints, coatings, and inks, etching agents to produce electronic circuit boards, and polymerization catalysts. The oil after the treatment with sulfuric acid contains acidic resins and traces of sulfuric acid, which are removed by treating with calcium hydroxide and an adsorbent, usually a bleaching clay. After these treatments, the resulting raffinate is neutral and is light in color. The temperature in this step is kept low so that the reactions are slow and easier to control. This is especially true for the low viscosity oils. For high viscosity oils, higher reaction temperatures are needed to permit better mixing. In general, aromatics react with oleum (SO_3) at a faster rate than olefins. The treatment time is usually on the order of 20 to 40 minutes. A significant amount of material is lost during the acid treatment. The quantity losses essentially depend upon the quality of the crude oil and the processing conditions. Acid refining can be carried out as a batch process or as a continuous process, although most modern refineries prefer continuous process because of the higher throughput and better economy. The continuous sulfuric acid refining process has the additional advantages of reduced acid consumption due to better mixing, and more efficient removal of the acid tar from the base oil. The latter factor translates into the lower amount of clay necessary for bleaching and, hence, smaller amounts of oil losses.

Deasphalting is the removal of asphaltic and resinous components, present in the vacuum distillation residue, by the use of a solvent. This process is analogous to that used to prepare the residuum for the cracking unit. The process involves mixing the distillation residue, a viscous black liquid, with a hydrocarbon liquid, such as propane, butane, pentane, or hexane. The hydrocarbon insoluble phase, which settles to the bottom, contains asphaltic and resinous components. The hydrocarbon phase, which is removed from the top, contains the deasphalted oil dissolved in the solvent. The oil is recovered by removing the solvent, which is recycled. Propane deasphalting is a low yield process, but product quality is good. These oils are used as lubricant bright stocks, which are useful in certain applications. Butane, pentane, and hexane deasphalted oils contain contaminants that make them less suitable for use as base stocks.

Besides solvent, other processing variables that affect deasphalted oil quality include extraction temperature, extraction pressure, and the solvent to oil ratio. At constant pressure and the solvent to oil ratio, the lower extraction temperature increases the oil yield but decreases the quality, and vice versa. This is because high extraction temperatures reduce the solubility of the heavier feedstock components, which get removed with asphaltenes. However, the extraction temperature is limited by the critical temperature of the solvent. This is the temperature at which none of the feedstock components dissolve, therefore separation is not possible. The solubility of the oil in propane increases with increasing pressure. This is because higher pressures ensure solvent-oil mixture to stay liquid, thereby improving contact and improving extraction efficiency. A higher solvent to oil ratio improves the deasphalted oil quality by facilitating the separation of the various components.

The viscosity index of the solvent-neutral oils from the paraffinic crudes is usually between 90 and 100, and the residual aromatics content is between 5 and 25 % [42]. Solvent deasphalting does not remove these molecules because they have low aromatic character due to the presence of the long alkyl chains. For this reason and because of their relatively low amount, they have little or no negative effect on the VI behavior and the aging characteristics of the base oil. In addition to improving the aging stability, solvent deasphalting lowers the density and the viscosity of the raffinates. Viscosity index improves because of the removal of a significant amount of aromatics, which have a very low VI. Asphaltene removal has little effect on the flash point. The pour point of the raffinates is higher than that of the feed distillates because many alkylated polyaromatics that are removed have pour point depressing properties. Sulfur content is greatly reduced, by as much as 50 %, and the carbon residue, color, and color stability are also improved.

Despite the removal of the aromatics and some olefins, raffinates still contain traces of olefins and unstable compounds. In this respect, solvent deasphalting is less efficient than acid refining, for removing the undesirables. These contaminants must be removed to make the raffinates suitable for use as base stocks. The inability to completely remove olefins is due to their solubility in the hydrocarbon solvents that are used in deasphalting. In terms of solubility, they lie between the saturated hydrocarbons and the aromatics;

hence, they persist in raffinates. Previously, the sulfuric acid treatment was used to remove them, but today they are removed by hydrogen finishing.

Solvent refining, or solvent extraction, improves the oxidative stability and the viscosity-temperature (VT) characteristics of the base oils through preferential removal of the undesirable polar materials and aromatics. A variety of materials are used for this purpose. They include sulfur dioxide, furfural, phenol, and N-methyl-2-pyrrolidone which is also called N-methyl-2 pyrrolidinone. Among these, furfural and N-methylpyrrolidone are by far the most commonly used solvents [47,48]. Since this step lowers the base stock yield and can remove some beneficial lubricant components, over-extraction is avoided.

Sulfur dioxide is one of the first materials used on a large scale. Its use has dwindled over the years because of its low selectivity and the formation of the corrosive sulfurous acid in the presence of water. Furfural has better selectivity than that of sulfur dioxide and its selectivity decreases less with increasing temperature than that of the other extractants. Furfural is especially useful for refining paraffinic oils that have higher pour points, hence require higher extraction temperatures. Furfural is a good solvent for materials present in the distillates that are highly prone to oxidation and form sludge and coke. It is also effective in removing colored constituents, resins, and sulfur from the raffinates. It has the further advantages of being noncorrosive and of forming an azeotropic mixture with water that has a boiling point of 97°C. With respect to properties, phenol is very similar to furfural. It forms an azeotrope with water that has a boiling point of 99.5°C. N-Methyl-2-pyrrolidone (NMP) is a highly effective, nontoxic solvent which has gained popularity during the past few years. Its selectivity can be further improved by the addition of water, the same as in the case of phenol. An NMP-water system can be used to extract the lubricating oil distillates of low quality and the bright stocks to make high quality base stocks. A heterogeneous solvent system comprising propane, phenol, and cresols has been found effective in removing the undesirable aromatics, naphthenes, asphaltenes, resins, and unsaturated and highly colored constituents. The propane portion removes asphaltenes and resins and phenol and cresols remove the other components. This solvent system essentially carries out the functions of deasphalting and solvent extraction [4]. Despite the effectiveness and the use of various extractants, nontoxicity, good selectivity, and adaptability to process both paraffinic and naphthenic oils make NMP an attractive candidate for use.

Hydrogen finishing or hydrofinishing is mild hydrotreating, which is applied after solvent refining to remove remnants of the undesirable compounds that remain. The process basically involves the removal of the trace amounts of olefins and undesirable oxygen, nitrogen, and sulfur-containing impurities, through selective catalytic hydrogenation, or hydrotreating. Depending upon the process conditions, aromatics, if present in the oil, are partially or fully hydrogenated. During hydrotreating, there is little material loss. The process involves heating the raffinate at temperatures of 150–420°C and pressures of 14–170 bars with hydrogen in the presence of a catalyst. Typically, the reaction is carried out at 250–350°C and 20–60 bars, using nickel and molybdenum catalysts supported on alumina. Hydrofinishing improves color, oxidative and thermal stability, demulsification properties, and electrical insulating properties. The removal of the organo-nitrogen compounds improves the oil color and stability.

Refining by the use of adsorbents is an alternative to hydrotreating. Adsorbents, such as bleaching clay, have been used for many years to remove the trace amounts of impurities from liquids, such as edible oils. The same adsorbents have been used to remove color compounds such as asphaltenes and resins from the petroleum oils. But the refiners prefer hydrotreating because of the lower amount of the material loss and the difficulty in disposing of spent absorbents. However, in some specialty products, clay refining is still used.

Dewaxing—Petroleum distillates have high pour points because of the tendency to precipitate wax on cooling. As such this makes them unsuitable for use in formulating lubricants. The wax content in these oils is reduced in three ways: solvent dewaxing, urea dewaxing, and isodewaxing. The first method involves dissolving the oil in a solvent, such as methyl ethyl ketone, and cooling the solution to precipitate wax. The second method involves the use of urea, which forms an inclusion complex with wax that is removed from the oil through filtration [4]. Isodewaxing, or catalytic dewaxing, is essentially hydrocracking which isomerizes linear chain hydrocarbons (wax) into branched hydrocarbons. The latter compounds have better low-temperature properties than their unbranched analogues.

In solvent dewaxing, a variety of solvents can be used. These include ketones, such as dimethyl ketone, methyl ethyl ketone (MEK), and methyl isobutyl ketone; chlorinated hydrocarbons, such as dichloromethane and dichloroethane; propane, toluene; and blends of these compounds. The dewaxed oil yield depends upon the wax content of the feed and the required pour point: the higher the wax content and the lower the pour point desired, the lower the yield, and vice versa.

Ketone dewaxing process is the classical method. In this method, the oil is mixed with the ketone and the solution cooled to $<10°C$. The precipitated wax is filtered off and the solvent removed from the filtrate to obtain the dewaxed oil. The solvent is then recycled [4]. The use of dichloroethane–dichloromethane (methylene chloride) mixtures, sometimes called DiMe solvent, is used to produce low pour point lubricating oils. This solvent system removes both hard and soft waxes from the oil. A newer process, called the DILCHILL (Dilution Chilling) process, involves adding a large quantity of the chilled solvent to the raffinate, causing the wax to precipitate. This dewaxing technology is superior to the other solvent dewaxing technologies because of the following reasons.

1. Crystal size is larger, which facilitates filtration.
2. Pour points of the dewaxed oil are lower (by as much as −35°C).
3. Yields are higher.
4. Operating and maintenance costs are lower.

Urea dewaxing is based upon the fact that urea forms an inclusion complex (adduct) with straight chain hydrocarbons, such as waxes. An inclusion complex forms between the host (urea), which forms a cavity in which the guest

(n-paraffins or wax) resides. In this process, the waxy distillate, diluent (methylene chloride), and saturated aqueous urea solution are mixed in a reaction vessel. The urea–wax adduct that forms as a solid is separated from the liquid via filtration. The solvent is removed from the filtrate through distillation to yield the dewaxed base oil. The solid urea–wax adduct is subjected to water at high temperature, which results in two layers: the top layer is the wax and the bottom layer is the aqueous urea solution. To this urea solution at 70 °C, more urea is added to make a saturated solution for re-use.

Hydrogen cracking or hydrocracking solvent refining is very effective in removing the undesirable polar materials and aromatics from the raffinate. However, a substantial amount of the material loss occurs, which is worse in distillates derived from the poor quality crude oils that contain high concentrations of polyaromatic compounds. Refiners minimize or eliminate this material loss by the use of hydrocracking. The process involves treatment of the distillate with hydrogen at high pressures (1500–4000+psi) in the presence of a catalyst. This converts the aromatic molecules into naphthenes and cracks larger molecules to compounds of molecular weight that are suitable for use as base oils [47]. The conversion of the low VI aromatics into the medium VI naphthenes improves the VI of the oil. In addition, the olefins are converted into saturates, the normal paraffins are isomerized to branched alkanes which improves the pour point, and the sulfur and nitrogen containing compounds are removed.

Hydrocracking under certain conditions provides fractions that are useful as very high VI base stocks and at a very reasonable cost. Conventional base oils with 100 °C viscosity of about 12 mm^2/s (12 cSt) have a VI of between 95 and 100, but the hydrocracked oils have a VI of 120 or higher. For example, hydrocracking of slack waxes from the dewaxing plants, followed by distillation and dewaxing yields oils of a viscosity index of up to 150. These oils are referred to as very high viscosity index (VHVI) oils.

When the objective of hydrocracking is primarily to minimize wax in the distillate by isomerizing linear hydrocarbons into branched hydrocarbons, the process is called iso-dewaxing. The details of this process will be discussed later.

Each refinery produces base stocks of different viscosities and chemical properties for use in different applications. More highly refined oils have little or no organic sulfur and, hence, are more prone to oxidation. The oxidation resistance of oils, which are more susceptible to oxidation due to the presence of unsaturation, can be greatly improved through hydrogenation. Most automotive and industrial lubricants and greases use solvent-extracted base oils. Solvent-extracted and hydrotreated base oils are mainly used for premium products such as turbine fluids, hydraulic fluids, and circulating oils [35].

As stated earlier, mineral oil base stocks typically consist of hydrocarbon molecules that contain 20–70 or more carbon atoms and are characterized as *paraffinic, naphthenic,* or *aromatic*. *Paraffinic base oils* are also called solvent neutral (SN) oils. They possess good viscosity-temperature (VT) characteristics (VI), adequate low-temperature properties, and good oxidative stability. They are called solvent neutral because they are solvent refined and they have a pH close to 7, which is the pH of neutral liquids and solutions. A high VI oil is extremely desirable since it will experience a relatively smaller drop in viscosity with increasing temperature than a lower VI oil, thereby making it a better choice in formulating high temperature lubricants. The viscosity index of oil can be determined by measuring its kinematic viscosity at 40 °C and 100 °C [27,28] and using the ASTM tables [49], see Table 1.6. High viscosity base oils, also known as high viscosity index (HVI) oils, owe their good viscosity-temperature characteristics to the presence of higher amounts of paraffinic structures. Their low temperature properties are a result of solvent refining (low-temperature dewaxing), and their good oxidative stability is a consequence of the catalytic hydrogenation. Paraffinic base oils make excellent general purpose lubricants and are used extensively to formulate engine oils, transmission fluids, journal oils, gear oils, hydraulic oils, paper machine oils, metalworking fluids, and greases [46].

TABLE 2.6—Oil classification according to ASTM D2226.

Oil Type	Asphaltenes	Polar Compounds	Saturates
Aromatic			
101	0.75 % max	25 % max	20 % max
102	0.5 % max	12 % max	20.1–35.1 % max
Naphthenic			
103	0.3 % max	6 % max	35.1–65 % max
Paraffinic			
104	0.1 % max	1 % max	65 % min

Naphthenic base oils, manufactured in smaller quantities than paraffinic base oils, have naturally low pour points because they are wax-free and have excellent solvency. However, their VT characteristics are inferior to those of the paraffinics; that is, they have low to medium VI. These base stocks are preferred only in those applications that will benefit from their lower wax content and lower pour points, and where the low VI is acceptable. Since such oils are devoid of wax, no dewaxing is needed during their manufacture. Nonetheless, solvent extraction and hydrotreatment are carried out to reduce the aromatics content, especially the polycyclic aromatics which are potential carcinogens. Some base oil producers are using hydrocracking technology, which hydrogenates aromatics to naphthenics, to achieve this goal. The same as paraffinics, naphthenic base oils are also used to make good general purpose lubricants, especially for low-temperature applications. They are used to formulate turbine oils, hydraulic oils, metalworking lubricants, rubber process oils, cylinder lubricants, and greases.

Bright stocks are manufactured from the residue left after the isolation of the paraffinic and the naphthenic base stocks, as discussed in the previous sections. The residue is subjected to propane deasphalting, solvent extraction, and solvent or catalytic dewaxing to obtain these oils. These stocks have a high aromatic content and are the least preferred because of their poor oxidation resistance, tendency to form black sludge at high engine operating temperatures, low VI character, and suspect carcinogenic nature. They are usually used as high viscosity blending stocks to formulate engine oils, gear oils, hydraulic oils, and greases.

Generally, the paraffinic oils have light color and good

TABLE 2.7—Relationship between hydrocarbon structure and physical properties in mineral base oils [50].

Hydrocarbon type	Structure	Main properties
Straight Chain Paraffin		1. Viscosity varies little with temperature. 2. Good oxidation resistance. 3. High pour point.
Branched Chain Paraffin	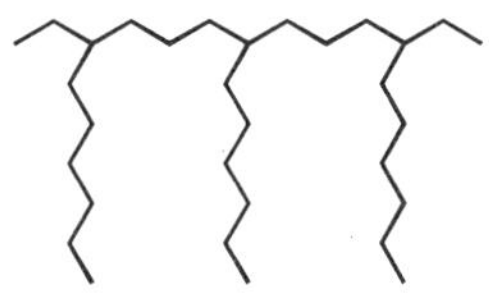	1. Viscosity varies little with temperature. 2. Good oxidation resistance. 3. May have low pour point.
Naphthenic Rings with Short Paraffinic Side Chains	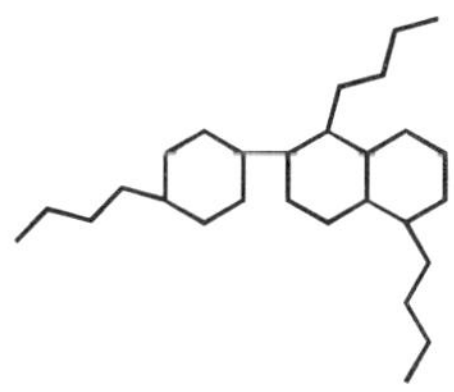	1. Viscosity varies greatly with temperature. 2. Becomes pseudo-plastic at low temperatures. 3. Good oxidation resistance. 4. Low pour point.
Aromatic Rings with Short Paraffinic Side Chains	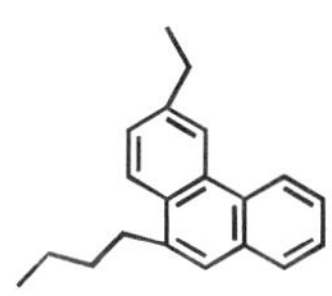	1. Viscosity varies greatly with temperature. 2. Poor oxidation resistance. 3. Good thermal stability. 4. Pour point varies according to structure.
Naphthenic Rings with Long Paraffinic Side Chains	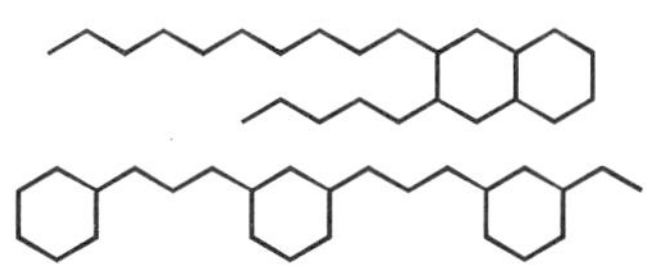	1. Viscosity varies little with temperature. 2. Good oxidation resistance. 3. May have low pour point.
Aromatic Rings with Long Paraffinic Side Chains	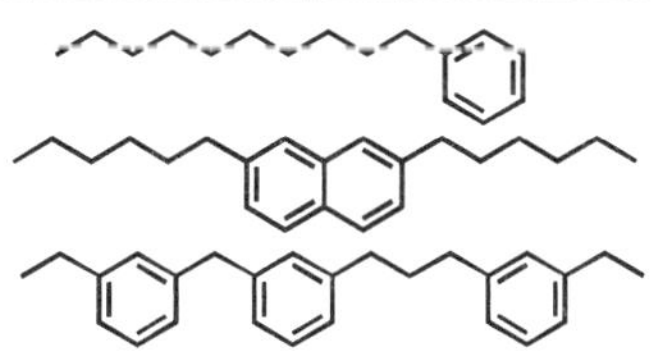	1. Viscosity varies little with temperature. 2. May have good oxidation resistance if the structure does not contain too many rings. 3. May have low pour point.

thermo-oxidative stability, but have a high cost. Aromatic oils, or bright stocks, are darker in color, have poor thermo-oxidative stability, possess good solvency towards polar compounds, and have lower cost. Naphthenic oils fall in between the two. Table 2.6 provides the composition of three classes of oils, as described in ASTM Standard D2226 [27]. Table 2.7 presents the relationship between various lubricant properties and chemical structures [50].

In addition to these categories, the base oil manufacturers produce a variety of other oils, such as cylinder oils, electrical oils, process oils, agricultural spray oils, refrigeration oils, transformer oils, turbine oils, and white oils. These oils, used in specialty applications, usually require more severe and extensive processing methods in their manufacture.

Recently developed catalyst technology by Chevron Oil Company has made it possible for Chevron to manufacture base stocks with higher VI and oxidative stability than conventional petroleum-derived base stocks. These stocks, popularly known as RLOP stocks, named after the place of first manufacture—Richmond Lubricating Oil Plant, are manufactured via hydrocracking. Chevron has the flexibility to produce consistent products from any crude, irrespective of its source. While it produces base stocks that meet the present market specifications, the Chevron catalyst is versatile enough to make base stocks of diverse properties, through more extensive cracking or cracking in the presence of waxes; thereby making the base stocks of even a higher VI accessible. RLOP stocks are primarily cycloparaffinic and are more oxidatively stable than the solvent refined base stocks and are consistent in quality. The consistency aspect is greatly preferred by the OEMs, such as GM and Ford, which produce passenger cars. The oxidative stability of these stocks is attributed to the absence of unsaturation and nitrogen. However, the same attributes become drawbacks as far as their deposit-suspending ability, or the solvency, is concerned. Many other refiners, such as Petro Canada, Pennzoil-Conoco, and Texaco are licensing Chevron catalyst technology and are either already producing or plan to produce similar base stocks for the North American Market. The process to make these base stocks, along with that used to

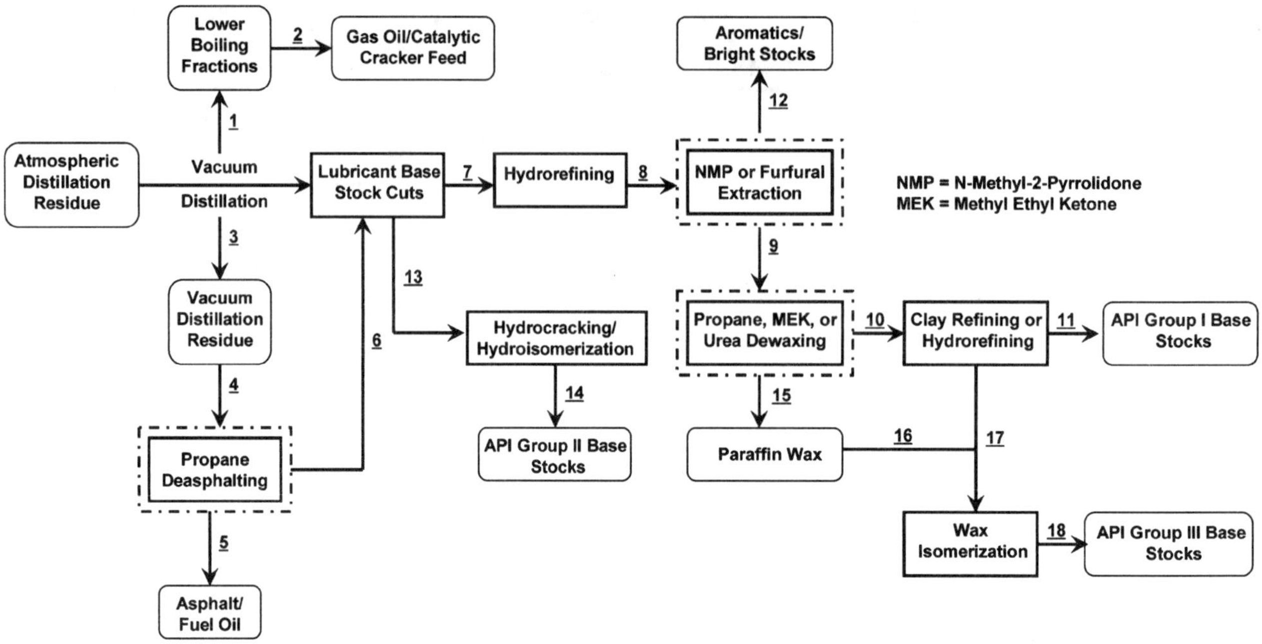

Fig. 2.6—Consolidated base oil manufacturing process.

manufacture conventional base stocks, is provided in Fig. 2.5. Figure 2.6 shows consolidated refining scheme to produce mineral oil base stocks. As shown in the figure, the manufacture of Group I base oils involves steps 7–12, the manufacture of Group II base oils involves steps 13 and 14, and the manufacture of Group III base oils involves steps 7–10 and 15–17.

Processes used in refining can be divided into two groups: those that involve separation and those that involve conversion. The first group includes chemical treatment to remove impurities and the undesirables by the use of the chemicals, for example acid refining, and physical separation, such as distillation or cooling. The second group includes processes, such as hydrorefining and hydrocracking, which convert various components of the raffinate into different components. Table 2.8 lists these processes and their effect on the yield of the base stocks [52].

One of the primary requirements for the organic compounds that comprise base oils is that they must be fluid at least at ambient temperature and may be below and have a reasonably low volatility and high oxidative stability during service. In a number of applications, fluidity at lower temperatures (down to a temperature of −40°C) is also required. Fluidity is related to the molecular weight and the type of hydrocarbons present in the base oil. *Alkanes* (paraffins), *alicyclics* (naphthenes), and *aromatics* of the same molecular weight have different physical and chemical characteristics.

Alkanes have lower density and viscosity than alicyclics (naphthenics), which in turn have lower density and viscosity than aromatics. Viscosity-temperature characteristics

TABLE 2.8—Lube refining processes and their effect on yield [52].

Process	Charge	Yield Range (Vol. %)
Separation		
Chemical-Acid Treatment (H_2SO_4, $AlCl_3$)	Distillation stock	20–95
Physical		
1. Distillation	Crudes, treated stocks	100
2. Solvent Extraction (SO_2, phenol, cresol, propane, solvent mixtures)	Distillation and residual stocks	40–80
3. Clay Treatment	Treated stocks	95–100
4. Dewaxing	Raffinates	75–95
Conversion (Hydro-treating)		
Mild—Hydrorefining to remove sulfur and nitrogen	Treated stocks	100
Moderate to severe—Hydrocracking to decrease molecular size or to isomerize	Distillate extracts, treated stocks	80–100

TABLE 2.9—Refining method comparison.

Base Oil—SSU Viscosity	% Saturates	% Aromatics
1. Solvent Refined—100 Neutral	70–90	10–30
2. Solvent Refined—320 Neutral (Medium Viscosity)	65–75	25–35
3. Solvent Refined—850 Neutral (High Viscosity)	60–70	30–40
4. Bright Stock	50–60	40–50
5. Hydrocracked—100	90–100	0–10
6. Hydroisomerized—100	95–100	0–5
7. Solvent Refined Naphthenic Crude	60–70	30–40

(VI character) and oxidative stability, however, are the best for alkanes but the worst for aromatics. *Alicyclics* fall in between the two extremes with respect to these properties. Based upon these properties, alkanes or paraffins are highly desired as base oils. However, their high melting points can hinder their use in low-temperature lubricating applications. This problem is partially overcome by the presence of the branched alkanes, which have somewhat lower melting points than the linear or unbranched alkanes. However, higher branching has a negative effect on the VI, which is another desired property. The higher the branching, the lower the melting point and the VI.

As mentioned earlier, a modern lubricant is composed of base oil and additives. As mentioned earlier, a modern lubricant is composed of base oil and additives with the base oil making up 75–99 % of the volume of the lubricant. Hence, its physical and chemical properties profoundly influence the properties of the lubricant and the properties of the additives. Because of the competitiveness of the lubricant market, the cost of the base oil is also an important consideration. Worldwide use of mineral base oils and waxes is estimated to be well in excess of 10 billion gallons per year, or 37,854,118 kiloliters [47]. The U.S. production is over 3.5 billion gallons, or 13,248,941 kiloliters per year [48]. Of this volume, about 20 % is naphthenic and 80 % is paraffinic.

Despite the fact that there is a growing trend towards the use of the synthetic base fluids, primarily because of their superior low- and high-temperature properties, mineral base oils are still used in the largest amount. This is because of their abundant availability, good to reasonable performance in most applications, and their lower cost. New technologies are constantly being developed to overcome their deficiencies. These technologies modify the hydrocarbon structure and the composition of the base stocks. API Base Oil Groups II and III are the result of the use of these technologies. Mineral base oil demand in North America is growing at a rate of about 3 % per year, but its demand in the developing South and Central American and Asian countries is estimated to be greater than 5 % per year [48]. While at present the API Group II and III oils are trailing behind in terms of use, the use of these oils is steadily increasing because of their superior properties [53]. The attributes of the oils belonging to the various base oil groups will be discussed later.

TABLE 2.10—Base oil categories for API interchange guidelines.

Base Oil Category	Sulfur (%)	Saturates (%)	Viscosity Index
Group I	>0.03	and or <90	80 to 120
Group II	≤0.03	and ≥90	80 to 120
Group III	≤0.03	and ≥90	≥120
Group IV	All Polyalphaolefins		...
Group V	All others not included in the above groups		...

TABLE 2.11—API base stock group composition for 100N oil [42].

Component	API Group I	API Group II	API Group III
Paraffins	24 %	22 %	70 %
Naphthenes	55 %	76 %	24 %
Aromatics	21 %	2 %	6 %
Total	100 %	100 %	100 %

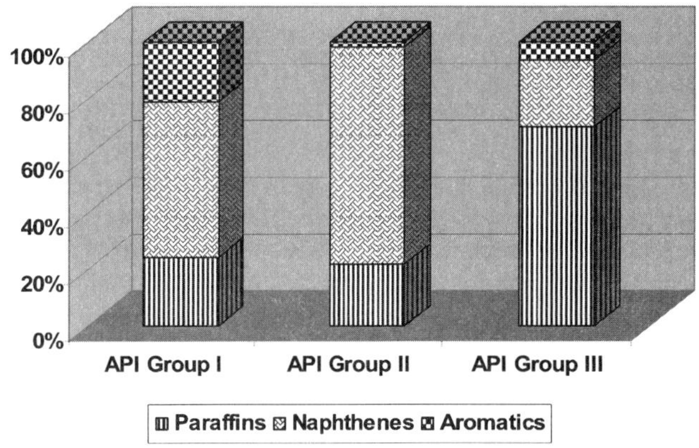

Fig. 2.7—Typical composition of the API group mineral oils.

Hydrocarbon Analysis

Chemical composition of the base oils depends upon the crude oil source, the refining process (solvent refining or hydroprocessing), the degree of refining, and the effectiveness of finishing. Table 2.9 compares the composition of the oils obtained by the use of the different refining methods; especially compare the solvent refined oil 1 with the hydrocracked oil 5 and the hydroisomerized oil 6. Please note that the viscosity of all three oils is very similar. The composition data underscore the effectiveness of hydroprocessing over solvent refining. There are a number of ways to determine the type and quantity of the hydrocarbon molecules present in the base oil. Many of these methods are covered in depth in the *ASTM Fuels and Lubricants Handbook* [42,54] and in Chapter 12 of this book on Lubricant Testing. While the cited references provide detailed hydrocarbon analysis of several mineral base oils, we are presenting the composition differently to facilitate the discussion that follows. Table 2.10 lists the API Base Oil Groups and Table 2.11 shows the composition of the 100N API Group I, II, and III mineral oils with respect to three components, *viz.* paraffinics, naphthenics, and aromatics [42]. These components are commonly used to judge the overall properties of the mineral oils since they have the most profound effect on such properties, hence those of the lubricant. The data in Table 2.11 is depicted in Figure 2.7 to facilitate the discussion that follows.

Base Oil Properties

Oxidation Properties

Good oxidation properties in a base oil and a lubricant are a pre-requisite to their use in most applications, especially so for applications such as engine oils that are exposed to high operating temperatures. Oxidation is initiated by the reaction of oxygen with the hydrocarbon materials to initially form hydroperoxides and peroxides, which thermally decompose or rearrange to form precursors to harmful products. See Chapter 4 on additives and Ref. [55] for the detailed oxidation mechanism. The effects of excessive lubricant oxidation are noticeable by oil thickening and the formation of the insolubles, such as sludge, varnish, lacquer, and hard carbonaceous deposits on hot surfaces. Some oxidation products are acidic and therefore cause corrosion of the metal surfaces. Hydrocarbon oil oxidation is initiated by the attack of oxygen on the weak carbon hydrogen bonds. These commonly occur in the vicinity of a double bond (allylic hydrogens) or are next to an aromatic ring (benzylic hydrogens). Because of this, the presence of unsaturation or the aromatic rings in the base oils is not desired. While the presence of the unsaturated compounds in hydrocarbon base oils is uncommon, they do form during base oil processing, as a result of the cracking reactions. Just the same, they are removed during the hydrofinishing or the clay treatment step [47]. On the other hand, aromatics occur abundantly in the mineral base oils. Even the API Group II and Group III oils, which are hydro-processed, contain a significant amount of aromatic content, see data in Tables 2.9 and 2.11. In saturated hydrocarbons, carbon hydrogen bond strength of the tertiary hydrogen is weaker than that of the secondary hydrogens, which in turn is weaker than that of the primary hydrogens. Hence, the hydrocarbons with tertiary and secondary hydrogens are also highly susceptible to the oxygen attack. This suggests that paraffinics with less branched structures or the cyclic rings will have better oxidation properties than naphthenics that have more cyclic rings, hence more secondary and tertiary hydrogens. For further details, please refer to the Oxidation Inhibitors section of Chapter 4. Lubricant oxidation is controlled by the use of the oxidation inhibitors. Common oxidation inhibitors are alkylphenols, arylamines, organic sulfides and polysulfides, and metal and amine salts of dialkyl dithiocarbamic and dialkyl dithiophosphoric acids. As a general rule, oils rich in aromatics require a higher amount of oxidation inhibitors than those that contain higher amounts of naphthenic and paraffinic hydrocarbons.

Effects of Sulfur and Nitrogen Compounds

For many years, the presence of sulfur in oil was equated with quality, primarily because such oils had good oxidation stability. The sulfur in oil exists in the form of organo-sulfur compounds and in domestic base oils its amount can range from <0.05 to 0.4 % by weight for light base stocks and up to 0.8 to 1.0 % for heavy base stocks. Sulfur levels in the base oils used in other countries is even higher. Modern base stocks are required to have low sulfur levels (see Table 2.10), which is achieved by the use of the better refining techniques. These were discussed in the base oil manufacture section. However, the sulfur-containing additives are often used in these oils to formulate lubricants in order to improve their oxidative stability and the antiwear performance.

The presence of the nitrogen compounds in base oils promotes oxidation and decreases useful life of the derived lubricants. This is because most nitrogen compounds present in the base oil, such as pyridines, are basic in character. They react with acids, which have oxidation-inhibiting ability, to form salts.

Other Properties

In addition to the properties discussed so far, the base oil composition affects a number of other physical properties. These include rheology, volatility, solvency, and elastomer compatibility.

Rheology, or the flow properties, of a lubricant depends upon the base oil composition used in its formulation. Flow properties include viscosity index (VI), low-temperature fluidity, and the pour point. Viscosity index is a measure of the viscosity-temperature (VT) relationship. All liquids lose viscosity with an increase in temperature. It is the rate of loss in viscosity which is of concern in lubricants; a lower decrease rate is preferred in most applications. With respect to this property, paraffinics are the best, which are followed by naphthenics and then aromatics, if the molecular weight of the components is the same. Hence, to formulate lubricants for use in high temperature applications, base oils with a high paraffinic content are chosen. With respect to the low-temperature fluidity, oils that contain a large amount of linear paraffins (paraffinics) have the tendency to form wax at low temperatures. This is not desired since waxes formed have a network structure which act like a sponge and adsorb most of the oil and hinder its flow. Fortunately, they respond well to additives, called the pour point depressants. This makes it possible to formulate lubricants from paraffinic base oils that have good low- and high-temperature properties. Naphthenic and aromatic base oils have good low-temperature properties, because they do not form waxes at low temperatures. However, their VI is too low for use in some applications. Figure 2.8 depicts the viscosity-temperature behavior of oils of various VIs. The top portion of the figure shows their behavior across the full temperature range of −40°C to 150°C and the bottom half of the figure shows behavior of these oils in the 40°C to 100°C range, which is used to define the viscosity index of an oil. As expected, high VI oils have a shallower slope than the low VI oils, in the 40°C to 100°C range. Slopes across the whole range are presented in Table 2.12. Data show that while the magnitude of the slopes outside this range is different, the oils maintain their viscosity-temperature behavior in the same order as indicated in the VI temperature range.

Viscosity-pressure (VP) relationship is another important property of the base oil that needs to be addressed. Many lubricant applications involve compression of the lubricant, such as hydraulics, roller element bearings, metalworking, and continuously variable transmissions (CVTs). Unlike the VT relationship where an increase in temperature causes a decrease in viscosity, in the VP relationship an increase in pressure leads to an increase in viscosity. This behavior is represented by the following equation [56–58].

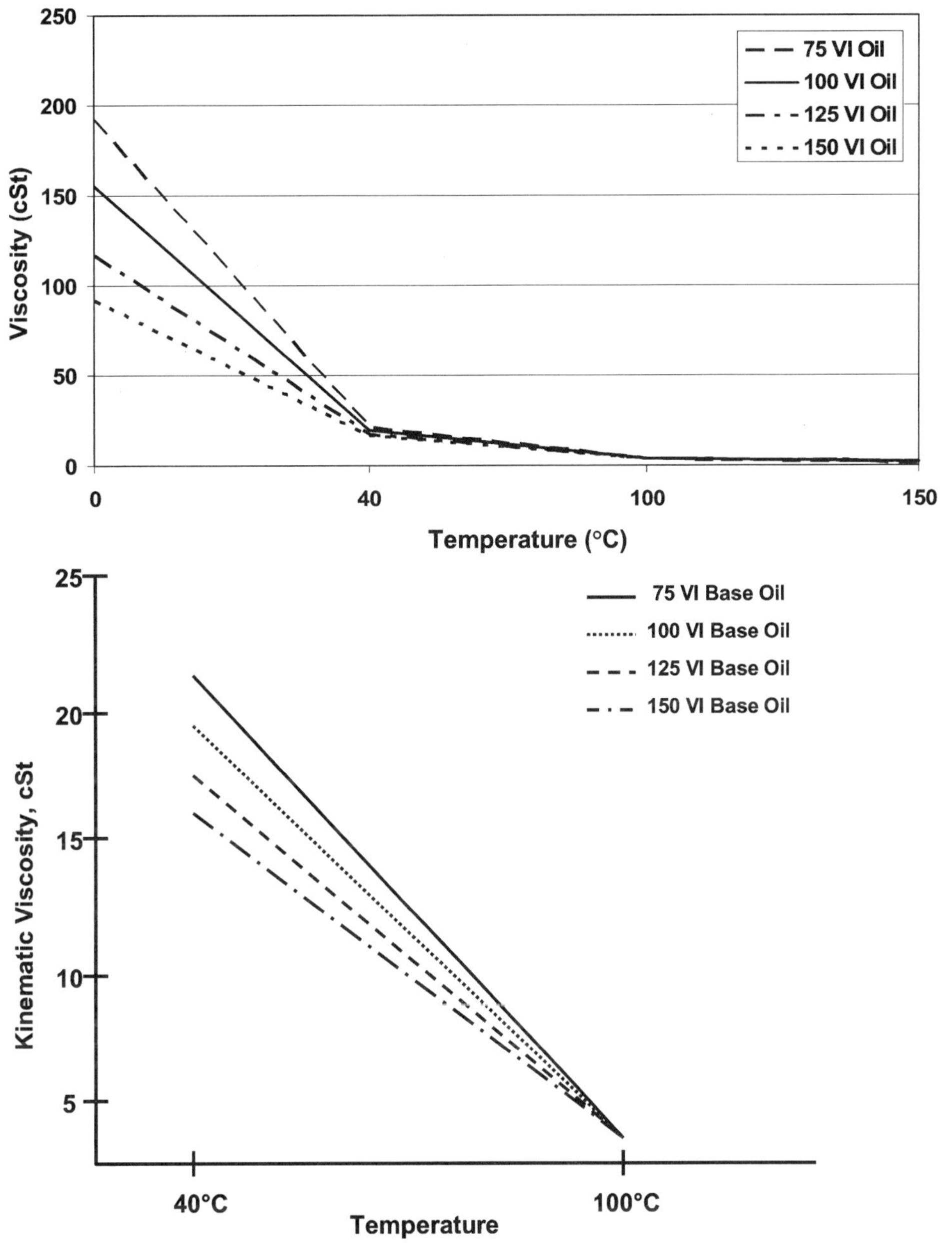

Fig. 2.8—Viscosity-temperature behavior of various mineral oils.

$$\eta_p = \eta_{atm} e^{\alpha p}$$

where η_p is viscosity at pressure p, N/mm^2, η_{atm} is viscosity at atmospheric pressure, N/mm^2, and α is pressure viscosity constant, N/mm^2 [58].

In general, the naphthenics have an excellent viscosity-pressure relationship and reasonable oxidation stability. Therefore, they are the base oils of choice in formulating lubricants for high pressure applications. Aromatics also have excellent viscosity-pressure relationship, but their oxidation stability is not as good. They are also used in such applications, but in conjunction with a hefty dose of proper oxidation inhibitors. Figure 2.9 [4] and 2.10 show the viscosity increase of base oils as a function of pressure [59]. Figure 2.10 shows the effect of pressure on the viscosity of an oil at various temperatures. As the figure shows, an increase in pressure increases the oil's viscosity at all temperatures. However, the viscosity increase at lower temperatures is more dramatic than at higher temperatures. In Fig. 2.10, viscosity-pressure behavior of various hydrocarbon oils is compared with that of the conventional naphthenic and paraffinic oils. It is obvious from the figure that naphthenes experience the highest viscosity increase and paraffinic polyolefins, designated as API Group IV oil, experience the lowest viscosity increase. This is not surprising since all hydrocarbon oils, except PAOs, have either naphthenes or aromatics in their composition, see Table 2.11.

Solvency is an oil's ability to dissolve polar additives. In this respect, the aromatic oils are the best, paraffinic oils are the worst, and the naphthenic oils fall in between the two. Since most modern lubricants contain additives, many of which are polar, reasonable solvency in a base oil is desired.

TABLE 2.12—VT relationship of various oils.

	Temperature			Temperature			Temperature		
Oil	0°C	40°C	Slope (m)	40°C	100°C	Slope (m)	100°C	150°C	Slope (m)
75 VI Oil	192.0	21.0	−4.2750	21.0	4.0	−0.4250	4.0	1.87	−0.0426
100 VI Oil	155.0	19.5	−3.3875	19.5	4.0	−0.2583	4.0	1.91	−0.0418
125 VI Oil	117.0	17.7	−2.4825	17.7	4.0	−0.2283	4.0	1.97	−0.0406
150 VI Oil	92.0	16.3	−1.8925	16.3	4.0	−0.2050	4.0	2.03	−0.0394

In addition, in some applications, such as engine oils, gear oils, and transmission fluids, the lubricant is required to dissolve or suspend the polar contaminants resulting from oxidation or the additive degradation. The measure of an oil solvency is its aniline point. Aniline point is determined by mixing aniline, a polar compound, with the base oil (ASTM D611). Aniline point is essentially the temperature at which the two phase aniline-base oil mixture becomes a single phase. Typically, base oils of high paraffinic content have aniline points of >230°F (110°C), solvent-refined base oils have aniline points of 200–215°F (93–102°C), and naphthenic base oils have an aniline point of <150°F (65°C), see the data presented in Table 2.13 [45]. In cases where aniline point is too high, that is, the base oil has a low solvency; the additives may not dissolve, creating a problem to formulate a lubricant. In such cases, the problem can be overcome by the addition of a small amount of the polar materials, such as polyol ester or an alkylbenzene. These are sometimes called base oil extenders.

Volatility relates to base oil or its components to evaporate under the influence of high temperatures. High volatility in modern lubricants is not acceptable because of its negative effect on the environment and the possibility of ignition. When a lubricant is used in applications that are open to the environment, such as some metalworking fluids, it can pose a health hazard to workers. Flash Point is used as a measure of an oil's volatility. In general, the paraffinic oils have the lowest volatility, the aromatic oils have the highest volatility, and the naphthenic oils fall in between the two extremes. Table 2.14 compares the flash points of the different types of oils of the same viscosity [45].

Elastomer compatibility of the base oil is important because during service a lubricant comes in contact with elastomer seals, gaskets, O-rings, and similar components. While some base oils are innocuous to elastomers, others are not. Base oil components can interact with the seals in many ways, some of which can lead to structural damage. For example, they can migrate into the elastomer matrix causing it to swell, or remove the plasticizer. Naphthenic and aromatic oils with high solvency (low aniline points) lead to elastomer swelling via migration. This causes a loss of proper fit and may impair proper functioning of the part. On the other hand, paraffinic oils and PAOs that have high aniline points remove the plasticizer, thereby causing elastomer parts to crack and a loss of lubricant. Additives, called the seal swell agents can help overcome this problem.

Biocompatibility of the lubricant with the environment is becoming increasingly important because of the health concerns. Base oils can contain components, such as polycyclic aromatic hydrocarbons, that are known to have toxicity, carcinogenicity, and mutagenic activity. Hence, it is important that either the amount of these components in the oil is minimized or the worker exposure to such oils is eliminated. The ASTM D3239 method can be used to determine the polycyclic aromatic content of the light and medium viscosity base oils, that is, those that are 70 to 150 Neutral or have a 100°C viscosity of 3–5 cSt. However, the method does not work well for medium to heavy grade oils, that is, those that are 320–850 Neutral. The method underestimates their polycyclic content [42]. Biodegradability is another desired property that is gaining importance. It is the ability of a lubricant to degrade naturally in the environment. Biodegradability depends upon the structure of the hydrocarbon materials that are present in the base oil. A number of standards are used to determine the biodegradability of the lubricants; ASTM D5864, D6066, and the CEC-L-33-T-82 being among them.

In the year 2003, the API has published a summary of the toxicological and ecotoxicological data on the lubricating oil base stocks. Biodegradability data summary from various studies is provided in Section 3E of the report. While the data are confusing because of the various methods used, most studies indicate paraffinic base stocks to be more biodegradable than naphthenic base stocks and bright stocks [54]. Of all the test methods used, the CEC method provides the highest biodegradability values. It is important to note that the base oils are expected to have higher biodegradability than the finished lubricants. This is because all of the additives used to formulate lubricants may not be biodegrad-

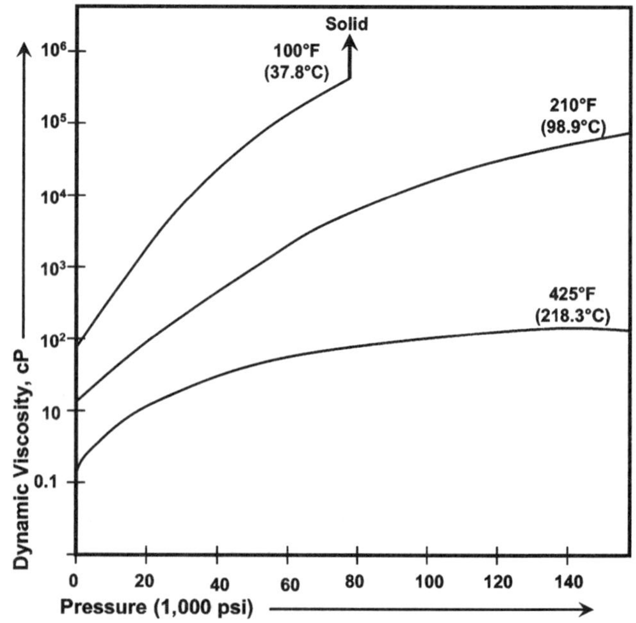

Fig. 2.9—Viscosity-pressure relationship [4].

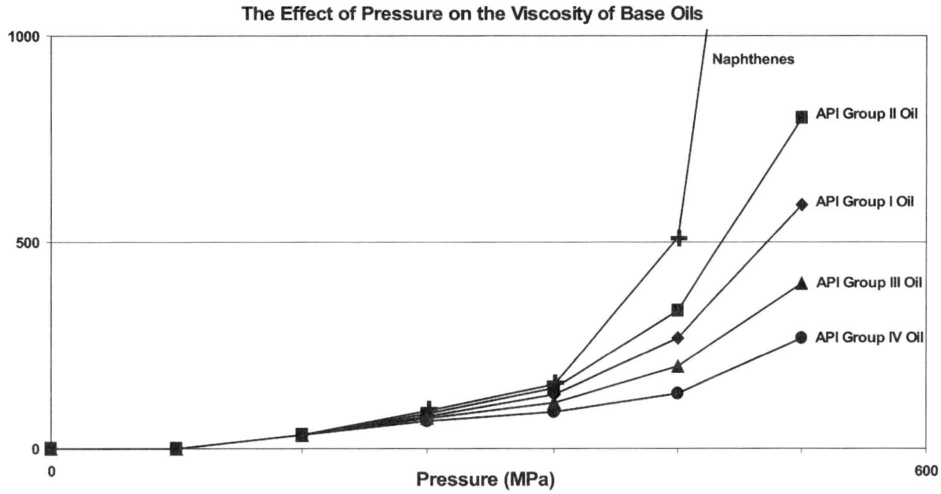

Fig. 2.10—Viscosity-pressure behavior of various mineral oils [59].

able or they may even lower the biodegradability rating because of being toxic to some of the microorganisms used in testing. It is therefore not too surprising that the hydraulic fluids based on mineral oils degraded to a level of only 40 %, while the average biodegradability of the paraffinic base stocks is 66 %. The above report provides a biodegradability range of 51–75 % for the paraffinic base stocks.

Table 2.15 summarizes the effect of the base oil composition on various base oil properties. A number of research articles correlating base oil composition with real engine performance have been published [59–64]. Similar studies pertaining to industrial lubricants have also been reported [60,63,66,67]. Testing standards and test methods used to assess various properties of the base stocks are provide elsewhere [42].

Gas to Liquid Technology

While discussing mineral base oils, we concentrated on petroleum refining and alteration of the base oil composition by the use of hydrorefining, hydrocracking, and hydroisomerization technologies. Another technology that has generated recent interest in obtaining mineral base stocks is gas to liquid technology [68].

Gas to liquids (GTL) technology involves converting natural gas into synthetic oil, which can be processed into fuels and other hydrocarbon-based products. The process essentially involves combining smaller natural gas molecules, such ac methane, into larger molecules that are suitable as diesel fuel, naphtha, and other liquid petroleum products. The molecules that contain over 20 carbon atoms are suitable as base stocks for lubricants. Because of the way these hydrocarbons are manufactured, they are devoid of the undesirable compounds, such as aromatics. These base stocks, dubbed as isoparaffins, are good substitutes for the blends of the API Group III and IV base stocks because of similar properties. Such mixtures are in great demand for use in automotive engine oils because of the increased emphasis on reduced emissions and energy efficiency, and shortages are expected.

The process to manufacture GTL base stocks involves first converting the natural gas, primarily methane, into syn-

TABLE 2.13—Aniline points of mineral oil fractions [45].

Oil Type	Aniline Point Range °F	Aniline Point Range °C
Paraffinic	200–250	93–121
Relative Naphthenic	190–225	88–107
Naphthenic	130–195	54–91
Aromatic	80–130	27–54

TABLE 2.14—Typical values for flash points [45].

Oil Type	Viscosity cSt	Flash Point °F	Flash Point °C
Aromatic	3000	430	221
Naphthenic	3000	450	232
Paraffinic	3000	575	302

TABLE 2.15—Effect of composition on various base oil properties.

Base Oil Property	Paraffins	Naphthenes	Aromatics
Viscosity Index (VI)	Excellent	Poor-Good	Poor
Volatility	Low	Medium	High
Low-temperature Fluidity[a]	Poor	Good	Good
Low-temperature Fluidity[b]	Excellent	Good	Good
Pour Point	Poor	Good	Excellent
Viscosity-Pressure Relationship	Poor	Good	Excellent
Oxidation/Thermal Stability	Excellent	Poor-Good	Poor
Solvency Towards Additives	Poor	Good	Excellent
Toxicity	None	Low	High
Biodegradability	High	Medium	Low

[a]Without pour point depressant.
[b]With pour point depressant.

TABLE 2.16—Physical properties comparison—Gas-to-liquid (GTL) technology base stocks versus API Group III base stocks [68].

Property	GTL-5 Base Oil	API Group III Base Oil (100N)	Industry Range
Viscosity at 100°C, cSt (ASTM D445)	4.5	4.28	4.0–5.0
Viscosity Index (ASTM D2270)	144	125	120–141
Pour Point, °C (ASTM D97)	−21	−16	−24 to−12
Cold Cranking Simulator Viscosity at −25°C, cP (ASTM D5293)	816	1500	729–2239
NOACK, % Wt Loss (ASTM D5800)	7.8	14	10.4–14.8
Composition, Mass %			
Paraffins (linear and iso)	100 (iso)	70	47–81
Mono-cycloparaffins	0	12	19–30
Poly-cycloparaffins	0	12	5–22
Aromatics	0	6	0–1

thesis gas, a mixture of hydrogen and carbon monoxide, via partial oxidation in the presence of a proprietary catalyst. The synthesis gas is then converted into the synthetic crude oil, which can be processed to yield hydrocarbon fractions suitable as lubricant base stocks. The synthesis is based upon Fischer-Tropsch technology. For further details, see the section on Hydrocarbon Polymers in Chapter 3. Sasol and Mossgas of South Africa and Shell are the present producers of the GTL base stocks and a number of major oil companies, such as BP/Amoco, Conoco, and ExxonMobil are considering commercializing the GTL based products in the near future. Isoparaffins possess structural resemblance to PAOs and have extremely high viscosity indices. In addition, these branched paraffinics exhibit excellent low- and high-temperature viscometrics, high oxidation resistance, low pour points, and low volatility, which are at par with or superior to the commercially available API Group III base stocks, see data in Table 2.16. The performance in several ILSAC GF-3 tests conducted on lubricants derived from these base stocks indicates these to perform well [68].

3

Synthetic and Biological (Natural) Base Stocks

THIS CHAPTER COVERS NON-PETROLEUM lubricant base stocks. These are either man-made (synthetic) or biological in origin. These base stocks possess certain advantages over mineral base stocks, which make their use more suitable in lubricants that are employed in applications that experience temperature extremes or operating conditions, or both. Interest in the base stocks of biological origin is primarily due to their superior environmental compatibility and availability from renewable sources. In this chapter, we discuss their chemistry, synthesis or isolation, and the advantages.

As stated previously, the base stocks used for formulating lubricants are derived from three sources: petroleum, raw materials derived from petroleum, and plants and animal (natural). Here we will discuss materials that are either synthesized from petroleum-derived raw materials or are isolated from seeds and fruits. Animal fats, except for fish oils, because of their nonfluid nature at room temperature are unsuitable for use as liquid base stocks. At one time fish oils were abundantly used either as base fluids or starting materials to manufacture additives, such as sulfurized sperm oil, for use as EP agents. However, fish oils are no longer readily available because of the conservation interest against the slaughter of the oil bearing large fish.

Synthetic Base Stocks

Synthetics have a number of advantages or disadvantages over the mineral oils which pertain to the following properties:

1. Thermal stability
2. Oxidation stability
3. Viscosity-temperature behavior
4. Hydrolytic stability
5. Tribological properties (lubricity and load-carrying properties)
6. Corrosion resistance
7. Volatility
8. Biodegradability
9. Toxicity
10. Compatibility with other materials
11. Miscibility with mineral oils
12. Solvency (additive solubility)
13. General availability and only in certain viscosity grades
14. Ionizing radiation stability
15. Cost

It is important to note that no one base stock has all of the listed attributes. Two properties that are common to most synthetic fluids are good thermo-oxidative stability and chemical stability. Synthetic fluids belong to two broad groups: synthetic hydrocarbons and other fluids. Synthetic hydrocarbons comprise polyalphaolefins, polybutenes (polyisobutylenes), and dialkylbenzenes and alkylnaphthalenes. Other fluids include dicarboxylate esters (di-esters), polyol esters, poly(alkylene glycol)s, phosphate esters, polysiloxanes (silicones), poly(phenyl ether)s, and perfluoroalkyl and chlorofluoroalkyl ethers. Depending upon the advantages and the disadvantages some synthetic fluids are suitable for use in some applications but not in others.

Unlike petroleum-derived base oils (mineral oils) that occur in nature, synthetic base fluids are man-made products. These are manufactured from the petroleum-derived low-molecular weight raw materials by chemical reactions such as alkylation, polymerization, and esterification. These materials have a well-defined structure or structures with predictable properties. This is in contrast to mineral oils which are complex mixtures because of which they have less-defined properties. Previously, the term synthetic was limited to base stocks that were made outside the refining unit. However, the distinction is becoming less clear since some base stocks, such as Chevron's RLOP type, discussed in Chapter 2, which are produced in the refinery, are being marketed as synthetic base stocks. We consider the first definition of synthetic as proper. Based on this definition, the following classes of synthetic base fluids are commercially available [69–74]:

1. Synthetic Hydrocarbon Polymers
   a. Polyalphaolefins
   b. Polybutenes
   c. Alkylated Aromatics
2. Carboxylate Esters
   a. Aliphatic Esters
   b. Polyol Esters
3. Phosphate Esters
4. Poly(alkylene glycol)s
5. Silicon Compounds
   a. Silicones
   b. Silicate Esters
6. Poly (phenyl ether)s
7. Halogenated Hydrocarbons

Synthetic Hydrocarbon (SHC) Polymers

Polyalphaolefins

Most olefins can be polymerized to yield hydrocarbon molecules of the size that can be used as lubricant base stocks. Common olefins which are used to make these polymers include ethylene, propylene, isobutylene, α-olefins (alphaole-

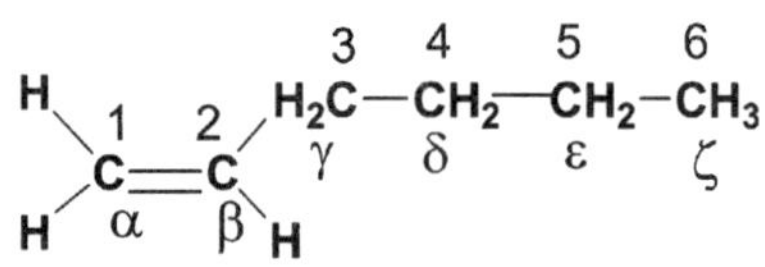

Fig. 3.1—1-Hexene—A typical linear α-olefin (alphaolefin).

fins), and internal olefins. The resulting polyolefins have viscosity indices of 110 or higher. For example, polymerization of ethylene provides materials of VIs of up to 125 and low pour points. Polymerization of propylene yields polymers of VIs of about 110, but with poor thermal stability. The equimolar mixture of the two provides copolymers that have VIs of up to 140, have low pour points, and have good oxidation resistance and thermal stability. Butene (butylene) polymerization, described later, leads to materials of 120 VI that have low pour points, but have only moderate resistance to oxidation. Of all the polyolefins, polyalphaolefins are among those used most often for formulating lubricants.

Polyalphaolefins, or PAOs, are saturated hydrocarbons that are synthesized by the polymerization of alphaolefins, followed by hydrogenation [74]. Linear alphaolefins, or normal alphaolefins, are olefins or alkenes with a general chemical formula C_nH_{2n}. These are distinguished from the other mono-olefins of the same molecular formula by linearity of the hydrocarbon chain and the position of the double bond being at the primary or the alpha position. Structure of 1-hexene, one of the common α-olefins, is shown in Fig. 3.1. Linear alphaolefins include a range of industrially important alphaolefins, such as 1-butene, 1-hexene, 1-octene, 1-decene, 1-dodecene, 1-tetradecene, 1-hexadecene, 1-octadecene, and the C_{20}–C_{24}, C_{24}–C_{30}, and C_{20}–C_{30} range higher olefin blends.

The term α-olefin is used for olefins that have a terminal double bond. One of the common olefins that is used to make PAOs is α-decene, also called decene-1, to indicate that the double bond is at carbon 1. This is in contrast to polyolefins that are made by the polymerization of the olefins that may or may not have a terminal double bond. Hydrogenated polyinternalolefins, or PIOs, described later, have recently gained use in Europe as a base oil, unofficially designated as the API Group V oil.

The starting material for PAOs is usually an even number olefin that is obtained by the oligomerization (low degree of polymerization) of ethylene. Oligomers typically contain between 30–100 monomer units. Petroleum distillates contain significant amounts of olefins, but they are difficult to isolate. Hence, olefins used for making olefin oligomers are obtained by synthetic methods, which include ethylene oligomerization, paraffin wax cracking, dehydrogenation of paraffins, and chlorination/dehydrochlorination of paraffins. Ethylene oligomerization provides C_6 to C_{20} linear α-olefins in high purity. Steam cracking of paraffin wax is carried out in the vapor phase at 510 to 540°C in the presence of 10 % steam. This results in the formation of up to 90 %, primarily odd and even carbon numbered, linear α-olefins of the 130 to 300°C boiling point range. Dehydrogenation and chlorination/dehydrochlorination of paraffins primarily results in linear internal olefins [75].

Linear alphaolefins are manufactured by two main processes: Fischer-Tropsch synthesis [76] and the oligomerization of ethylene, followed by purification. A smaller scale route employs the dehydration of alcohols. Prior to the use of these methods, various other methods, such as thermal cracking of waxes and chlorination/dehydrochlorination of linear paraffins, were used.

$$\underset{\text{Methane}}{CH_4} + 1/2\,O_2 \longrightarrow \underset{\text{Synthesis Gas}}{2\,H_2 + CO}$$

$$\underset{\text{Hydrocarbon Source}}{\geq CH} + O_2 \longrightarrow 1/2\,H_2 + CO$$

$$\underset{\text{Synthesis Gas}}{(2n+1)\,H_2 + n\,CO} \longrightarrow \underset{\text{Paraffins}}{C_n H_{2n+2}} + n\,H_2O$$

For n = 4 example, the above equation can be written as follows.

$$9\,H_2 + 4\,CO \longrightarrow \underset{\text{Butane}}{C_4 H_{10}} + 4\,H_2O$$

Fig. 3.2—Fischer-Tropsch technology [77].

The Fischer-Tropsch synthesis product consists of a complex multi-component mixture of linear and branched hydrocarbons and oxygenated products. However, the main products are linear paraffins and α-olefins. The original Fischer-Tropsch process can be described by the chemical equation shown in Fig. 3.2 [77]. The initial reactants in the reaction, that is, CO and H_2, can be produced by reactions such as the partial combustion of methane, as in the case of gas to liquids technology, gasification of coal, or biomass.

The Fischer-Tropsch synthesis can be carried out in the presence of iron, cobalt, or the ruthenium-based catalysts. Iron-based catalysts are much less expensive than the cobalt-based catalysts. The cost is an important economical consideration while choosing a catalyst since many times the catalyst must be replaced due to deactivation. However, when iron catalysts are used, there is a competing reaction between the carbon monoxide and the hydrogen to form water. The Fischer-Tropsch reaction may be considered a methylene polymerization reaction where the monomer unit (CH_2), although not initially present, forms in situ from the hydrogenation of CO [78]. Sasol Ltd., a South African oil, gas, and petrochemical company is the only company that commercially employs Fischer-Tropsch synthesis to make fuels from synthesis gas derived from coal and recovers 1-hexene from these fuel streams. Sasol employs iron and cobalt catalysts at pressures of 10 to 60 bars and at temperatures of 200–300°C. The reaction occurs on the surface of the catalyst, as is graphically shown in Fig. 3.3.

Ethylene oligomerization to alphaolefins also involves

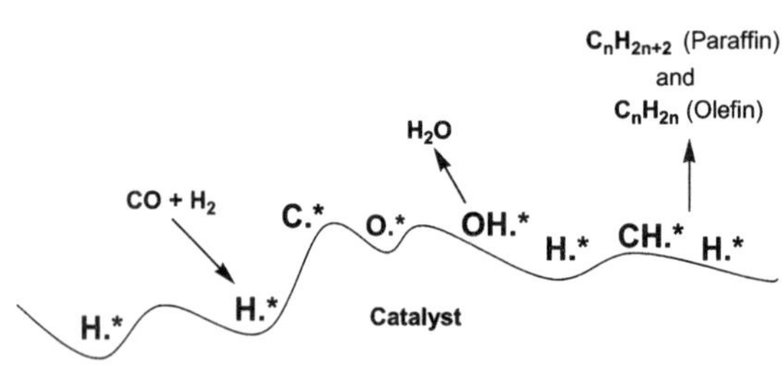

Fig. 3.3—Graphic presentation of the mechanism of Fischer-Tropsch reaction.

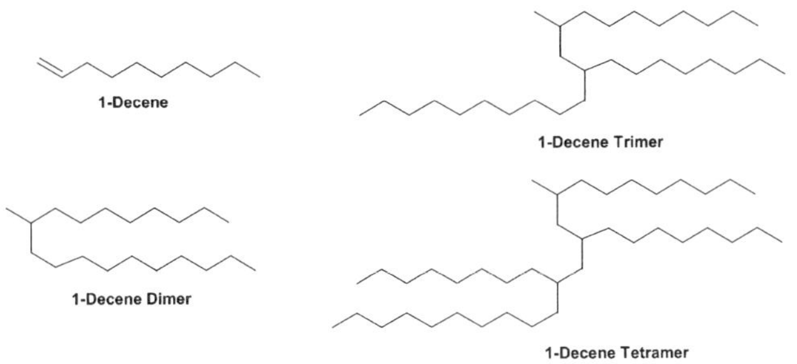

Fig. 3.4—Structures of 1-decene and its hydrogenated oligomers (PAOs) [79].

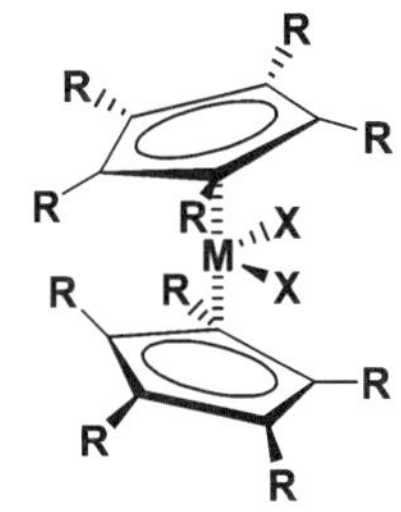

Fig. 3.5—General structure of metallocene.

the use of a catalyst. Chevron and Ethyl employ a Ziegler catalyst (triethylaluminum), while Shell uses a proprietary nickel coordinated complex system. Saudi Arabian Basic Industries Company (SABIC) has developed a new low pressure ethylene oligomerization process that is conducted over a heterogeneous catalyst in a slurry bed [77].

Dehydration of alcohols to linear alpha olefins by passing the alcohols in the vapor phase over acidic alumina catalyst has been practiced periodically by Ethyl Corporation (later BP, now Ineos), Chevron Phillips, Sasol (formerly Vista Chemical), and Godredj Industries, an Indian petrochemical and specialty chemical company. Normally, the process is not economical as the linear fatty alcohols are more valuable than the corresponding linear alphaolefins. However, the process has been applied whenever the value of the fatty alcohols dipped below that of the linear olefins because of the market dynamics or the regional supply-demand issues [76].

Polyalphaolefins, or PAOs, are synthetic, saturated hydrocarbons that are made from linear alpha olefins by a two-step process. The first step involves oligomerization of the olefin to yield an unsaturated olefin oligomer, called linear unsaturated alphaolefins (LAOs). The second step involves their hydrogenation to PAOs, slightly branched alkanes, for use as synthetic lubricant base fluids. Hydrogenation is carried out in the presence of either nickel catalyst or palladium catalyst. This step is usually followed by distillation to remove unreacted starting materials and separate oligomers based upon viscosity. In some cases the distillation step precedes hydrogenation [79]. Although 1-decene is the common α-olefin used, higher or lower carbon number homologues may also be used [79]. Typical 100°C viscosities of the common PAOs range from 2–10 cSt. The structures of some of the hydrogenated tridecene oligomers (PAOs) are provided in Fig. 3.4.

Polyalphaolefins can be synthesized either by the use of a free radical process, Ziegler catalysis, or the cationic Friedel-Crafts catalysis.

Free Radical Process

While olefins can be oligomerized thermally, the products are of poor quality and the yield is poor. Product quality improves if free radical initiators, such as benzoyl peroxide and di-*t*-butyl peroxide are used. However, the yield still stays low.

Ziegler Catalysis

A Ziegler catalyst system, such as triethylaluminum-titanium tetrachloride, can be used to synthesize polyalphaolefins. The process requires a solvent, and there are difficulties in reclaiming the catalyst. Both these factors make the process costly and inconvenient. Also, the products obtained are of a broad molecular weight range and hence of widely variable physical properties. The mechanism by which these catalysts affect polymerization is believed to simultaneously involve a number of reaction sites on the catalyst. This is primarily the reason for the broad molecular weight distribution since the polymer growth at different sites proceeds at different rates. The invention of the new metallocene catalysts helps overcome this problem [80].

Metallocenes are positively charged metal ions, most commonly titanium or zirconium, sandwiched between two negatively charged cyclopentadienyl rings, see structure in Fig. 3.5. These catalysts have the advantage of catalyzing the reaction of olefins via only one reactive site; hence they result in polymers of narrow molecular weights, which have more predictable and desirable properties [81].

Another group of catalysts that is presently being explored for polymerizing olefins consists of the late transition metal compounds. These catalysts are based upon Group VI and the higher periodic group metals and have the additional advantage of polymerizing polar olefins. The most commercially advanced catalysts of this type are the Brookhart catalysts [81], which are diimine complexes of palladium or nickel, see the structure in Fig. 3.6.

Friedel-Crafts Catalysis

While many Lewis acid catalysts, such as $AlCl_3$, $SiCl_4$, $TiCl_4$, $SnCl_4$, $BiCl_3$, and mixtures of $AlCl_3$ and $FeCl_3$ can be used, aluminum chloride is the most commonly used Lewis acid for preparing olefin oligomers. Aluminum chloride initiates oligomerization by protonating the olefin to form a carbocation, which being the active species adds to olefin molecules until the reaction is terminated [74]. The mechanism for the oligomerization reaction is shown in Fig. 3.7. The proton source used to terminate the reaction is either water or alcohol. Polyalphaolefins contain many products that result from the cationic rearrangement. The formation of such products in the case of polypropylene is hypothesized in Fig. 3.8. In the scheme, $(+H_2)$ indicates hydrogenation of the resulting olefin, (LAO).

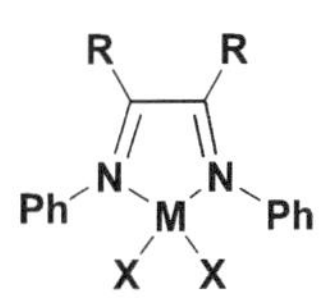

Fig. 3.6—Brookhart catalyst [81].

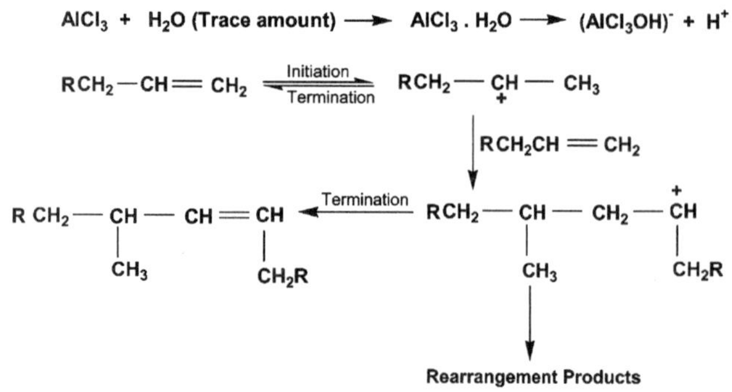

Fig. 3.7—Mechanism of Friedel-Crafts oligomerization reaction.

The boron trifluoride-alcohol (BF_3-ROH) catalyst system is another Friedel-Crafts type catalyst system that can be used to synthesize PAOs. Under proper conditions, this system yields an oligomer mixture that is rich in the trimer. Other co-catalysts that also lead to trimer-rich PAOs are water and carboxylic acids [84]. As mentioned earlier, 1-decene is one of the most common olefins used. After the oligomerization is complete, water is added to the crude reaction mixture to quench the catalyst and the obtained product mixture hydrogenated and fractionated to yield the desired polyalphaolefin fraction. The number of monomer units in the polyolefin depends upon the starting olefin, the type of catalyst, the reaction temperature, and the reaction time and pressure. For example, if one uses BF_3/n-butanol at 30°C, the product primarily contains tridecene (C_{30}), with only minor amounts of the other oligomers. Tridecene does not comprise a single structure but is a mixture of as much as 30 isomers, resulting from various carbon size pendent groups. The formation of the isomers occurs due to the rearrangement of the intermediates via a methide (CH_3^-) or a hydride (H^-) shift, as is shown in Fig. 3.8. Since the polyalphaolefins are classified based on their 100°C viscosity; in general the higher the viscosity, the longer the average pendant group chain length.

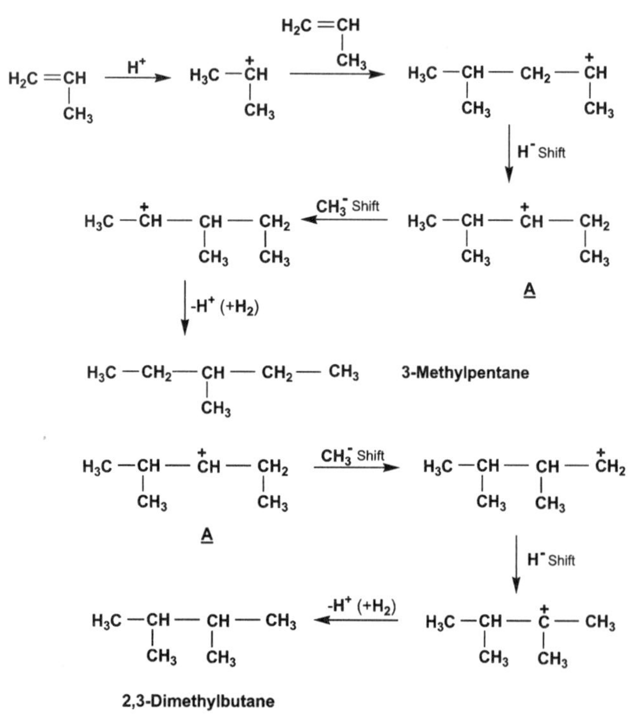

Fig. 3.8—Polypropylene oligomerization and rearrangement reactions.

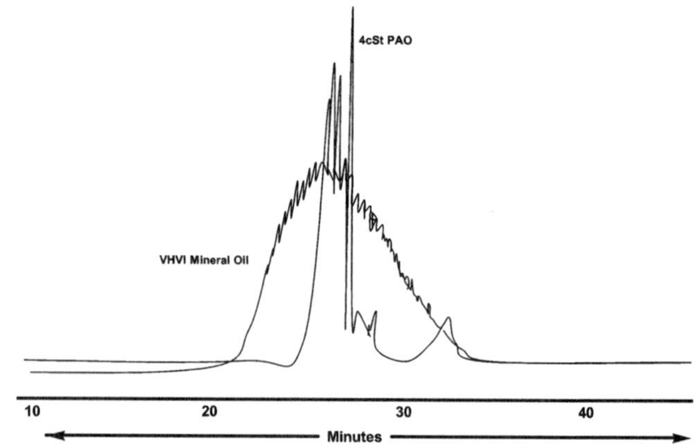

Fig. 3.9—Compositional comparison of polyalphaolefins and mineral oil [87].

In order to synthesize 2–10 cSt PAO, BF_3-ROH [85] and BF_3–H_2O [86] are often used, but for synthesizing higher viscosity PAOs, it is necessary to use Ziegler-Natta catalyst systems [86]. Even in the latter case, the resulting polyolefin oligomer contains a double bond, which must be removed by hydrogenation. Besides α-decene, other olefins may also be used, either as such or as mixtures with dodecene-1. The use of the olefin mixtures provides polyalphaolefins with uniquely different properties. This is because the polymer structure has different carbon size branches and different isomers.

One of the reasons for the popularity of PAOs is their similarity to the hydrocarbon base oils, but without the presence of naphthenics and aromatics that negatively impacts properties, such as VI, volatility, and oxidation stability. As stated earlier, commercial PAOs are classified according to their approximate kinematic viscosity at 100°C. For example, the PAO of viscosity 1.80 is classified as PAO 2 and the PAO of viscosity 9.87 is classified as PAO 10. Common PAOs are of 2, 4, 6, 8, 40, and 100 cSt viscosities. PAOs have superior low and high temperature properties relative to those of the similar viscosity mineral base oils. These properties include low pour point, high viscosity index, and high flash and fire points. Because of these, the PAOs are excellent base stocks for use in applications that experience broad temperature ranges. Good low-temperature properties are due to the presence of the long branches because of which at low temperatures these materials do not form crystalline networks. Good volatility is because of their uniform or close to uniform composition. In mineral oils, it is the presence of the low molecular weight low boiling components that causes volatility problems. PAOs are essentially free of such components, hence they have better volatility. Figure 3.9 compares the compositional difference between PAO and a mineral oil of similar viscosity, as determined by the gas chromatography (GC) [87]. As one can see, the PAO has a narrower composition than the mineral oil. PAOs are also thermally more stable than the mineral base oils, as indicated by

TABLE 3.1—Properties of polyalphaolefins [87].

Parameter	PAO 2	PAO 4	PAO 6	PAO 10	PAO 40	PAO 100
Kinematic Viscosity at 100°C cSt	1.80	3.90	5.90	9.60	40.0	100
Kinematic Viscosity at 40°C cSt	5.5	16.8	31.0	45.8	395	1250
Kinematic Viscosity at −40°C cSt	310	2540	7800	19000	...	...
Viscosity Index	...	122	137	134	150	170
Pour Point, °C	−63	−69	−63	−54	−34	−20
Flash Point, °C	>155	215	225	264	280	290
Noack Volatility, % Loss	99	12.0	6.7	2.0	0.8	0.6

the Panel Coker Test. This is primarily because of the PAOs being free of the aromatics that are the reason for poor performance of mineral oil in the test.

Table 3.1 summarizes the physical properties of the polyalphaolefins [87]. As one can see, PAOs 2 to10 have excellent low-temperature properties; they flow at −40°C and have a pour point of less than −54°C. Therefore, they are especially suitable in formulating lubricants for use in cold climates, such as that of North America and Northern Europe, without the need of a pour point depressant. Their viscosity indices are well over 122, making them suitable to formulate lubricants for the warm climates as well. The VI increases with the increasing molecular weight of the starting olefin. The main advantage of the good viscosity index is that the base oil does not need a viscosity improver to maintain its viscosity at high temperatures. Flash points of the PAOs are almost equivalent to those of the mineral oils of similar viscosity.

Higher viscosity grades, PAO 40 and PAO 100, are also listed in Table 3.1. These are similar to the low viscosity grade PAOs in having good viscometrics and therefore have possible use in broad temperature applications. New PAOs with medium viscosity grades 5, 7, and 9; and very high viscosity grades, of up to 3000 cSt, have also been developed for use in specialty applications [88]. The volatility of these base oils is somewhat superior to that of the mineral oils, which will be discussed later. The lower volatility of the PAOs can be ascribed to their structural homogeneity, that is, they contain well defined hydrocarbon structures. There are no light ends to evaporate, which is the case in the mineral oil base stocks. The volatility of the PAO 2 is too high to use it in lubricants for high temperature applications. Low volatility in a base oil is desired to eliminate the need to replenish the lost oil, an increase in viscosity during use, and the negative impact of the evaporated oil on the environment. In addition, it lowers the flash and the fire points. While the PAOs 4-10 are suitable for use in lubricants for broad temperature applications, the PAO 40 and 100 are only useful in formulating products for use in high temperature applications. This is because their pour points are somewhat higher. Some of these properties among PAOs are compared in Table 3.1 [79,87]. With respect to the resistance to oxidation, PAOs are at least equivalent to mineral oils that have low or no aromatics. However, they are superior to mineral oils with significant aromatics content. Their response to oxidation inhibitors is similar to that of the mineral oils. PAOs are essentially nontoxic and have a reasonable biodegradability. These properties make them well suited for use in insulating oils, cable impregnating oils etc. ASTM methods to determine these properties are described in the *ASTM Annual Book of Standards* [27].

Prior to the use of ethylene-derived α-olefins, cracked olefins were used to make the polyolefin oligomers. The polymerization process involved the use of $AlCl_3$ at 20 to 100°C and the obtained polyolefins had good low- and high-temperature properties [4].

The major application area for the PAOs is automotive engine oils that require low pour point, low volatility, and good thermo-oxidative stability. In automotive lubricants, PAOs have better oxidation and deposit control and fuel economy than those that are mineral oil derived [70]. Hence, PAOs are the most frequently used synthetic fluids. This is detailed in the section that compares petroleum and synthetic base stocks. Other automotive applications include two-stroke cycle engine oils, automatic transmission fluids, multi-grade gear oils, and greases. PAOs are also used in industrial applications, such as hydraulic fluids, compressor oils, heat transfer fluids, food grade oils, and greases. Engine oils based on the PAOs are claimed to have extended drain intervals, improved fuel economy, enhanced wear protection, and broad temperature performance. These and other beneficial properties of PAOs are described in Ref [87].

While PAOs have a number of very desirable properties, they do not reflect a high degree of biodegradability relative to that of the synthetic and natural esters. The biodegradability of the PAOs decreases as their molecular weight and the viscosity increases [87], see Fig. 3.10. In general, biodegradability performance of the decene-based PAOs with viscosities greater than 2 cSt is poor, even in the CEC-L33-A93

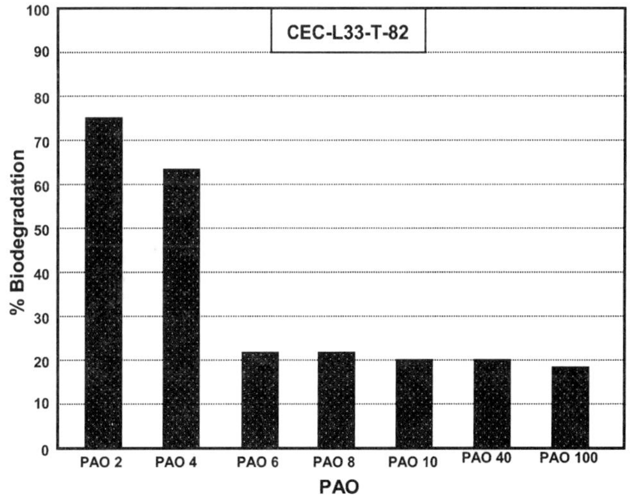

Fig. 3.10—PAO biodegradability after 21 days [87].

TABLE 3.2—Biodegradability of C_{10} to C_{14} derived PAOs [88].

	% Biodegradability via CEC L33 A93		
Test Compound	Test Lab A	Test Lab B	Average
2 cSt C_{10} PAO	74.0	88.5	81.3
6 cSt C_{10} PAO	...	28.4	28.4
3 cSt C_{12}/C_{14} PAO	71.0	72.8	71.9
6 cSt C_{12}/C_{14} PAO	56.0	59.3	57.7
7 cSt C_{12}/C_{14} PAO	...	42.30	42.3

test that does not measure complete biodegradation to CO_2 and H_2O. This precludes the use of the PAOs in hydraulic fluids, especially in off-road applications, which is a large market for biodegradable fluids [88]. However, it was found that the use of a combination of C_{12} and C_{14} α-olefins yields products which have superior biodegradability than that of the C_{10} (1-decene)-based products [89]; for comparative data, see Table 3.2. Incidentally, CEC-L33-A93 (formerly L-33-T-82) test procedure is designed to determine the persistence of the two-stroke cycle outboard engine oils in aquatic environments. Hence, good results in this test do not make the fluids derived from the mixed C_{12}/C_{14} PAOs *readily biodegradable* since this test is not appropriate to determine the ready biodegradability of a substance.

Despite this, there are reasons to use these higher biodegradability PAOs to design lubricants for use in environmentally sensitive applications, such as farm equipment, two-stroke cycle engines, forestry and marine equipment, and earth-moving equipment. Another advantage of these olefins is that they provide PAOs of higher viscosity indices than those derived from 1-decene, thereby requiring a lower amount of viscosity improver to attain the same viscosity. Hydraulic formulations provided in Table 3.3 demonstrate this advantage [88]. The reference also provides performance data on the other lubricants based on these PAOs.

A relatively new class of polyolefins, called polyinternalolefins (PIOs), has come to the market. There is a lot of discussion in the United States whether the API should create a new base oil category, API Group VI, to accommodate this new class of base fluids. These oils, which at present have limited availability, have properties that are somewhat similar to those of the PAOs. In the interim, Association Technique de L'Industrie Europeene Des Lubrifiants [The Technical Association of the European Lubricants Industry (ATIEL)] has included these base stocks as Group VI in its Guidelines on Base Oil Quality Assurance and Base Oil Interchange [90,91]. The guidelines allow interchange of PIOs with each other and with PAOs without the need for additional engine tests. ATIEL reviewed a substantial amount of engine test data, provided by Sasol—the present manufacturer of PIOs, prior to including this base stock in its interchange guidelines. ATIEL maintains definitions of the base stock categories in Europe which, except for the new Group VI category, are identical to those maintained by the American Petroleum Institute for North America. So far, the API has not been approached about establishing an API Group VI for PIOs. While both PIOs and PAOs are similar to each other in being synthetic and largely linear, they are not the same. This is because PAOs primarily use C_{10} α-olefin as the starting material and PIOs use internal C_{15} and C_{16} olefins. The two also differ in some properties. For example, PIO 4 has a 100°C viscosity of 4.3 cSt, pour point of −51°C, NOACK volatility of 15.3 %, and a VI of 127. PAO 4, on the other hand, has a 100°C viscosity of 3.8 cSt, pour point of −64°C, NOACK volatility of 13.5 %, and a VI of 137. There is uncertainty as to the significance of these differences in affecting the final lubricant properties.

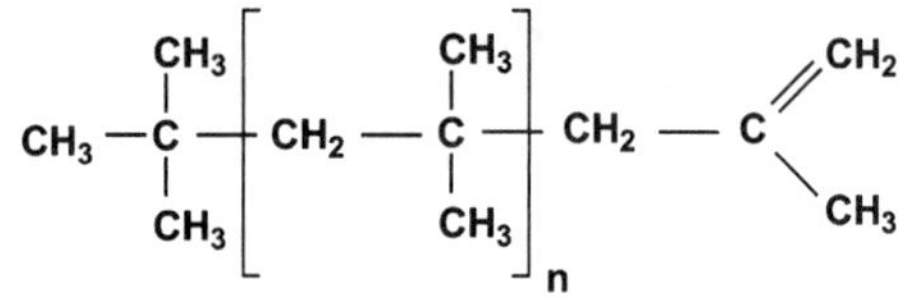

Fig. 3.11—Polyisobutylene oligomer.

Polybutenes [74]

Polyisobutylene (PIB), a major component in polybutenes, is also an olefin oligomer. However, its properties differ radically from those of the polyalphaolefins of the analogous molecular weight. The primary reason for the difference is their structural dissimilarity. For the same molecular size, the PIBs contain smaller and more numerous branches than the PAOs that contain larger and fewer branches. The structure of PIB shown in Fig. 3.11 is for the vinylidene isomer of polyisobutylene. Many other isomers make up the actual composition. The amount of each isomer present depends upon many factors, including the nature of the catalyst, the temperature, and the presence of the solvent during the polymerization reaction.

Low molecular weight PIBs, the molecular weight of 400–1300 g/mol, are used as lubricants in high-pressure and high-temperature applications, such as compressors used in polyethylene manufacture, ovens, dryers, furnaces, and metalworking. In the high-temperature applications, the main reason for their use is their ability to depolymerize and burn off cleanly. They are also used as starting materials

TABLE 3.3—Hydraulic fluids based on C_{10} to C_{14} PAOs [88].

ISO Grade	ISO 15	ISO 15	ISO 32	ISO 32	ISO 46	ISO 46
Hydraulic Additive Package	1.00	1.00	1.00	1.00	1.00	1.00
2 cSt C_{10} PAO	81.00	...	74.00	...	70.50	...
3 cSt C_{12}/C_{14} PAO	...	86.00	...	79.00	...	76.00
Viscosity Modifier	8.00	3.00	15.00	10.00	18.50	13.00
Ester	10.00	10.00	10.00	10.00	10.00	10.00
Kinematic Viscosity at 40°C	15.25	15.01	33.15	32.17	46.97	44.48
% Biodegradability (CEC L-33 T-82)	78.90	79.07	76.00	75.40	78.00	70.70
Flash Pt (COC), °C	164	196	164	188	162	198

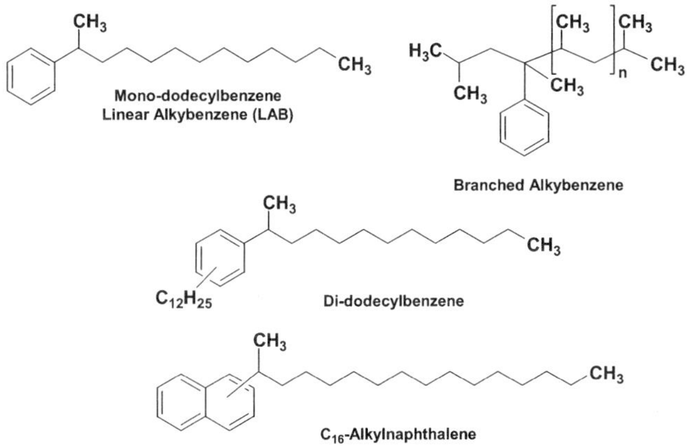

Fig. 3.12—Common structures in manufactured alkylated aromatics.

in the manufacture of the dispersant additives. High molecular weight PIBs, the molecular weight of over 10,000 g/mol, are used more extensively, but as additives and not as lubricants. They are used as shear-stable viscosity modifiers for gear oils and as additives in two-stroke cycle engine oils, to guard against scuffing and wear.

Alkylated Aromatics [74]

Alkylbenzenes and alkylnaphthalenes are the alkylated aromatics used most often as synthetic lubricants. They are prepared by Friedel-Crafts type alkylation of benzene and naphthalene with an aliphatic alcohol, alkyl halide, α-olefin, or an olefin oligomer. Typical structures for alkylated aromatics are presented in Fig. 3.12, where n equals 3. Over the years, many catalysts have been used to synthesize alkylaromatics. Prior to the 1960s, proton acids (mineral acids), such as hydrofluoric acid (HF), sulfuric acid (H_2SO_4), and silica supported phosphoric acid; and Lewis acids, such as boron trifluoride (BF_3) and aluminum chloride ($AlCl_3$), were used as alkylation catalysts [92–95]. In most cases, fractionation needs to be employed to separate the mono-alkylate from the di-alkylate. These catalysts suffered from a number of deficiencies. They were corrosive, were not usable more than once, and yielded a mixture of mono-alkylated, polyalkylated, and isomeric products. Also in most cases, they were difficult to handle and the product separation from the spent catalyst was difficult. To overcome these deficiencies, Zeolites catalysts, such as ZSM-5, were discovered in the 1980s. These were noncorrosive, regeneratable, selective, and easy to handle. In the 1990s, the next generation of Zeolites, such as USY, Beta, and MCM-22, with even higher selectivity, greater stability, and longer life were introduced [96]. More recently, a new generation of alkylation catalysts based upon WOX/ZrO_2 is being explored [51,97]. The alkylation mechanism is shown in Fig. 3.13. The characteristics of mono- and di-alkylbenzenes are described in a number of patents [98–100].

Alkylaromatics are similar in chemical structure to the aromatic base oils (bright stocks) obtained from petroleum. While some of their properties are identical with those of the mineral oils, the others radically differ. These materials have excellent low-temperature fluidity (low pour points), high flash points, good electrical insulating properties, refrigerant compatibility, and good corrosion properties. Their viscosity indices are generally higher than those of the high viscosity index mineral oils. They have good oxidation resistance (the use of oxidation inhibitors is generally required) and are thermally and hydrolytically quite stable.

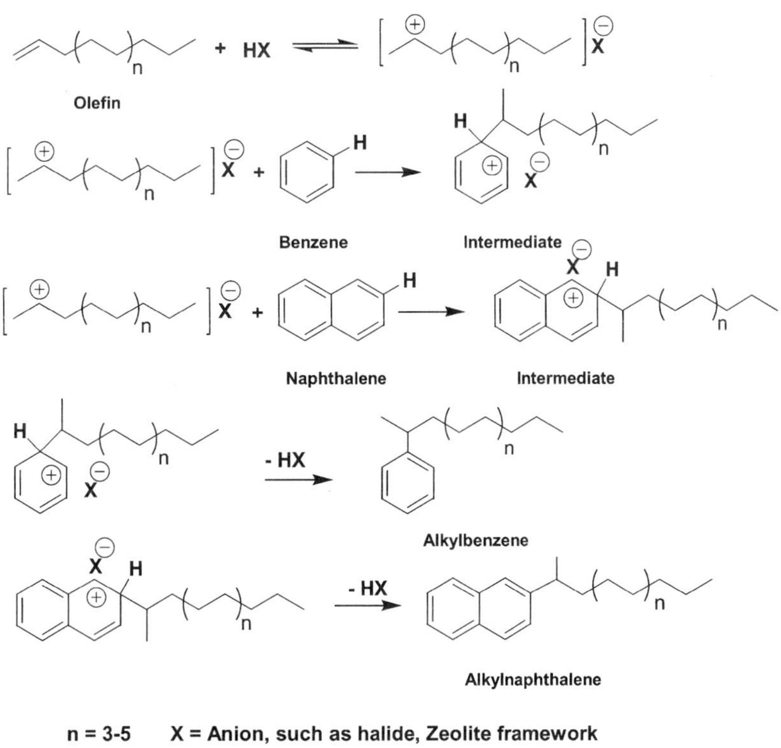

Fig. 3.13—Alkylated aromatic synthesis.

These synthetic hydrocarbons are compatible with mineral oils and a variety of synthetic fluids. Therefore, they are used as the base oil, or the base oil extenders, to formulate a variety of lubricants for diverse applications. The applications include engine oils, gear oils, hydraulic fluids, air compressor and gas turbine fluids, heat transfer fluids, and greases.

Linear dialkylbenzenes (LABs) are used to produce engine oils with good low-temperature properties. These are also used to formulate manual transmission lubricants, hydraulic oils, industrial gear oils, greases, and lubricants for chlorofluorocarbon type air conditioning and refrigeration compressors. However, they are not suitable for use in compressors that use HFC-134a (1,1,1,2-tetrafluoroethane) because of immiscibility [101]. The same is true of the alkylated naphthalenes. Branched alkylbenzenes, especially the mono-alkyl derivatives, are used as raw materials for making synthetic sulfonates for use in a variety of applications.

Alkylated naphthalenes are often used in formulating automotive and industrial lubricants. This is because of their superior thermal and oxidative stability than the other types of base stocks, hydrolytic stability, elastomer compatibility, and solvency. They are used as blend stocks for PAOs to improve their compatibility with the high polarity additives, such as rust and corrosion inhibitors and anti-wear/EP agents. The lubricants that use such blends include engine oils, hydraulic fluids, compressor oils, high-temperature gear and bearing industrial lubricants, paper machine oils, and high-temperature greases. A proprietary additive treated blend of alkylated naphthalene, polyalphaolefins, and synthetic ester is marketed as Mobil 1 for use in diesel and gasoline engines [102]. Alkylnaphthalene (Paraflow® 149) was originally developed for use as a pour point depressant for paraffinic oils. It was made by the reaction of an alkylating agent, such as chlorowax, with naphthalene in the presence of a suitable catalyst [103–105].

Carboxylate Esters [74]

Esters are the reaction products of acids or their derivatives with alcohols. Carboxylate esters are manufactured by the chemical reaction of carboxylic acids with alcohols, usually in the presence of a catalyst. The by-product of the esterification reaction is water, which is generally removed by an azeotroping agent, such as toluene. A variety of catalysts are used in the esterification process. These include sulfuric acid, *p*-toluenesulfonic acid, tetra-alkyl titanate, anhydrous sodium hydrogen sulfate, phosphorus oxides, and stannous octanoate. Typically, the reaction is run at 300°C and 50–760 mm of Hg pressure. Esterification is a versatile reaction in the sense that careful selection of the raw materials allows the synthesis of a base stock with very specific physical properties [106]. In some cases, the esters are made by a trans-esterification reaction, where a low molecular weight alcohol derived ester is reacted with a high molecular weight alcohol. In this case, the by-product is the low molecular weight alcohol, which is usually removed by distillation.

Esters are more polar than the mineral base oils, which is due to the presence of the alkoxycarbonyl (ester) functional group. Therefore, they have lower volatility and higher flash points. Other properties that are affected by the presence of the ester functional group are thermal stability, hydrolytic stability, solvency, lubricity, and biodegradability [106]. Figure 3.14 shows the influence of the chain variations on ester viscosity, viscosity index, and pour point [107].

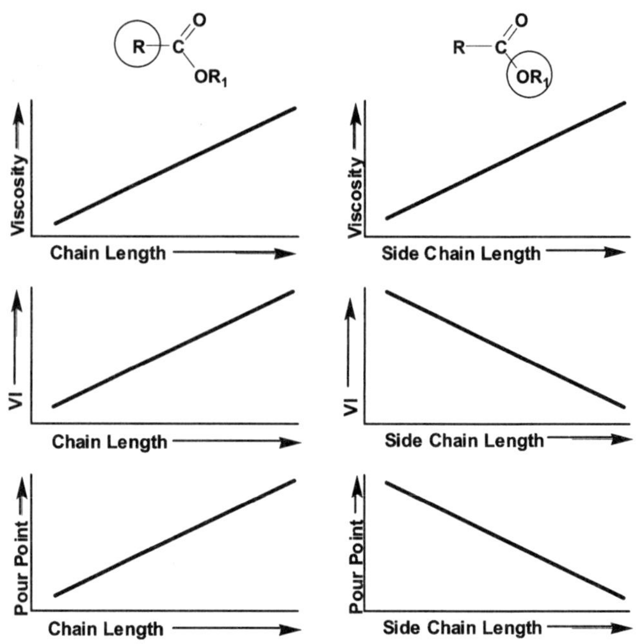

Fig. 3.14—The influence of chain variations on ester properties [107].

The ester molecule is composed of a carbon chain from the acid and a carbon chain from the alcohol. Viscosity increases as the carbon chain length of either the acid or the alcohol increases. Viscosity index and pour point, on the other hand, increase only with an increase in the acid chain length but decrease with an increase in the alcohol chain length. Chain branching has additional impact on these properties.

Esters fall under two general classes: esters derived from monohydric alcohols and esters derived from polyhydric alcohols. Monohydric alcohols are those that contain a single hydroxyl group and polyhydric alcohols are those that contain more than one hydroxyl group. Butanol, 2-ethylhexanol, and octanol are the examples of the first group and ethylene glycol, neopentyl glycol, glycerol, trimethylolpropane, and pentaerythritol are the examples of the second group. The carboxylic portion can either be aliphatic or aromatic; those derived from aliphatic carboxylic acids, such as heptanoic acid, are called aliphatic esters and those derived from aromatic carboxylic acids, such as phthalic acid and trimellitic acid, are called aromatic esters.

Aliphatic esters are synthesized by the reaction of monohydric alcohols with aliphatic carboxylic acids. *Diesters* are a subgroup of the aliphatic esters where the product results from the reaction of an aliphatic dicarboxylic acid with a monohydric alcohol. *Polyol esters*, on the other hand, are prepared by the reaction of a polyhydric alcohol, such as neopentyl glycol, trimethylolpropane, or pentaerythritol, with an aliphatic mono-carboxylic acid. Some esters are made by the reaction of the polyhydric alcohols with a dicarboxylic acid or a polycarboxylic acid. The result is the formation of a polymeric molecule. The size of such a molecule needs to be controlled, otherwise the polymerization will continue and the resulting material will not be fluid. The

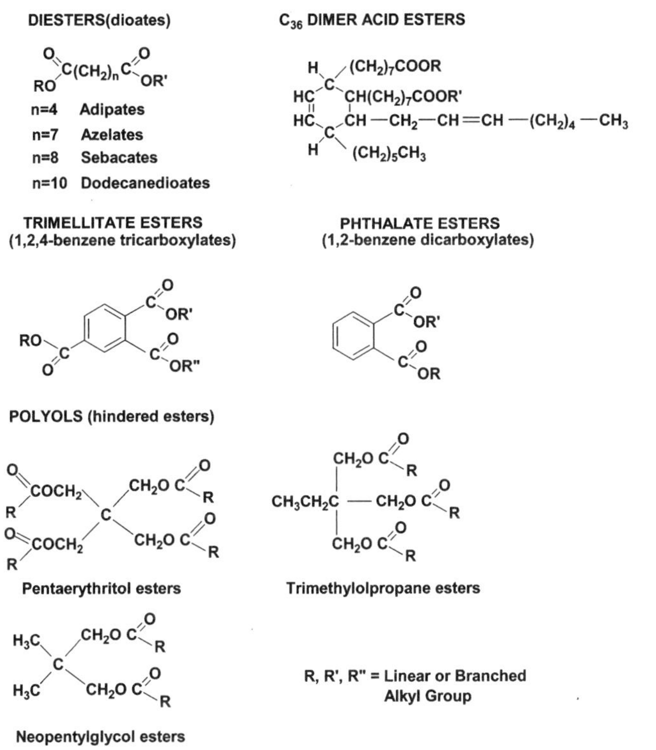

Fig. 3.15—Synthetic esters [106].

control is usually achieved by terminating the polymerization by the use of a monohydric alcohol or a monocarboxylic acid. Such oligomeric materials, called the complex esters, have high molecular weights, hence they have higher viscosities. However, these esters have a higher degree of polydispersity, that is, they contain different molecular weight components, because of which they have poor volatility properties and flash points [4]. Natural fats and vegetable oils, commonly called triglycerides, fall under the class of polyol esters. These will be discussed separately since they are not synthetic in origin. Figure 3.15 shows the structures of the esters that are commonly used as synthetic fluids [106]. Figure 3.16 shows the structures of two complex esters. One is derived from the reaction of 2,4-dimethyladipic acid (dicarboxylic acid), triethylene glycol (dihydric alcohol), and n-octanol (monohydric alcohol). The other is derived from the n-nonanoic acid (mono-carboxylic acid), poly(alkylene glycol), and adipic acid (dicarboxylic acid). Esters that contain unsaturation (multiple bonds) can be fully hydrogenated to improve their oxidative stability.

The invention of gas turbine and jet engines created a need for lubricants that had greater thermal stability and better low-temperature properties than the petroleum-derived oils. This is because turbines have extremely high operating temperatures and modern aircrafts generally fly at high altitudes where the ambient temperature is quite low. Initially, diesters were used as base fluids to formulate lubricants for use in aircraft engines. However, as aviation technology evolved, the operating temperatures increased further and a need for base fluids with even higher thermal stability was realized. This resulted in the use of the polyol ester-derived lubricants that are thermally more stable than those based on the diesters. Besides possessing superb broad-temperature properties, polyol esters have good lubricity, high VI, low volatility, and are compatible with additives and the most other base fluids. Because of the availability of a large number of carboxylic acids and alcohols and the simplicity of the esterification reaction, a wide variety of products with diverse properties are made for use in many applications.

Biodegradability is another desirable property in modern lubricants. This can be imparted in esters by the use of linear alcohols and carboxylic acids, which are more biodegradable than materials with branched structures. Because of these properties, the use of esters is extended beyond turbine and aviation applications. Today, they are commonly used in automotive engine oils, gear oils, hydraulic and transmission fluids, metalworking fluids, and compressor lubricants.

As mentioned earlier, esters are the reaction products of carboxylic acids and alcohols. The properties of the esters depend upon the carboxylic acid and the alcohol used, the number of ester groups per molecule, and the degree of branching in the acid and the alcohol portion. Carboxylic acids and alcohols used to make synthetic esters include the following:

1. Monocarboxylic acids—C_8–C_{16} saturated and C_{18}–C_{22} unsaturated acids.
2. Dicarboxylic and polycarboxylic acids—adipic, azelaic, sebacic, dodecanedioic, and C_{36} dimer acids.

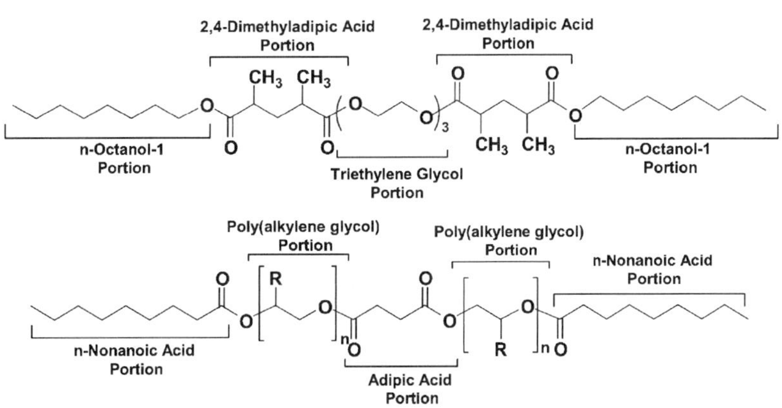

Fig. 3.16—Structures of complex esters.

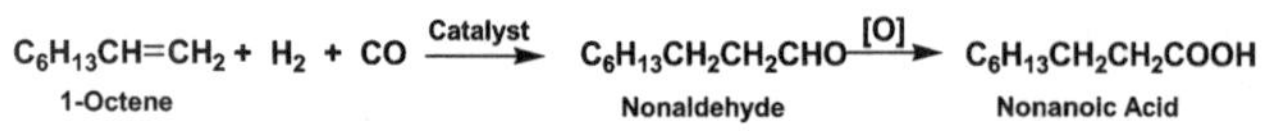

Fig. 3.17—Synthesis of mono-carboxylic acids by oxidative carbonylation reaction.

3. Aromatic acids and anhydrides—phthalic acid or anhydride and trimellitic acid.
4. Monohydric alcohols—C_8–C_{10} alcohols and 2-ethylhexanol.
5. Polyhydric alcohols—neopentyl glycol, trimethylolpropane, and pentaerythritol.

Saturated monocarboxylic acids are made either from petroleum-derived raw materials or are isolated from natural products. Alcohols can be synthetic or natural. A petrochemical route provides linear saturated fatty alcohols of C_6 to C_{20} chain length via the Ziegler process based on aluminum, hydrogen, and ethylene. Similarly, linear oxo-alcohols can be manufactured from linear olefins that are obtained from dehydrogenation of paraffins present in crude oil [108]. Semilinear fatty alcohols, such as Neodols® are produced via selective hydroformylation of α-olefins. The resulting products contain approximately 80 % linear and saturated primary alcohols (modified oxo-alcohol) and approximately 20 % 2-alkyl branched alcohols. Multiply branched alcohols are manufactured by hydroformylation of the oligomers of propene or butene, or both. Typical chain length in these alcohols is C_6 to C_{15}. Examples of such alcohols include *iso*-nonanol, *iso*-decanol, and *iso*-tridecanol. A new class of fatty alcohols is available via hydroformylation of the Fischer-Tropsch olefins. These alcohols show some unique structural features, but their primary alcohol content is only 50 % and the branching is not in the 2-position. 2-Alkyl branched alcohols, such as 2-ethylhexanol and 2-propylheptanol, can be manufactured via the Guerbet reaction, which involves aldol condensation of the aldehydes and hydrogenation of the resulting 2-alkyl branched unsaturated aldehydes. Any of the alcohols obtained by the use of the above listed processes can be oxidized to carboxylic acids.

The natural source of octanoic (C_8), decanoic (C_{10}), and dodecanoic (C_{12}) acids is coconut oil. Common names of these acids are caprylic, capric, and lauric acids. The natural source for tetradecanoic (C_{14}) and hexadecanoic (C_{16}) acids is palm kernel oil and that for octadecanoic acid (C_{18}) are animal fats. Common names of these acids are myristic, palmitic, and stearic acids. Unsaturated C_{18} to C_{22} acids include 9-octadecenoic acid (oleic acid), 9-eicosenoic acid (gadoleic acid), and 13-docosenoic acid (erucic acid). They are obtained from vegetable oils or animal fats. Natural fatty acids are usually unbranched (linear) and contain an even number of carbons. Petroleum-derived fatty acids can be unbranched or branched, depending upon the method of their manufacture. Oxidative carbonylation of the linear olefins results in the unbranched acids and with an odd number of carbon atoms. Olefins are ethylene oligomers, hence they contain an even number of carbons; it is the addition of another carbon group (from CO) that introduces the extra carbon in the molecule. The reaction is shown in Fig. 3.17. Aromatic acids or anhydrides are manufactured by the oxidation of the corresponding hydrocarbons, such as naphthalene, xylenes, and pseudo-cumene [109,110]. See Fig. 3.18 for the synthetic scheme.

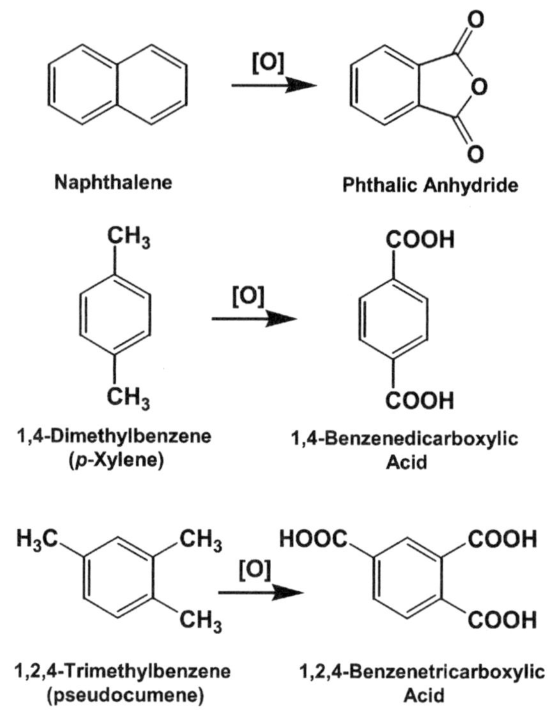

Fig. 3.18—Synthesis of aromatic carboxylic acids.

Alpha, omega (α,ω) diacids, such as adipic acid (1,6-hexanedioic acid), azelaic acid (1,9-nonanedioic acid), sebacic acid (1,10-decanedioic acid), and 1,12-dodecanedioic acid are also produced through oxidation [111]. Figure 3.19 shows the formation of the adipic acid from cyclohexanol or cyclohexanone. The most important dicarboxylic acids to make diester base stocks are sebacic, adipic, and azelaic acid. Pelargonic acid (nonanoic acid) is a mono-carboxylic acid that is obtained together with azelaic acid by the oxidation of the oleic acid. Sebacic acid is obtained by the alkaline oxidative cleavage of the ricinoleic acid (12-hydroxy-9-

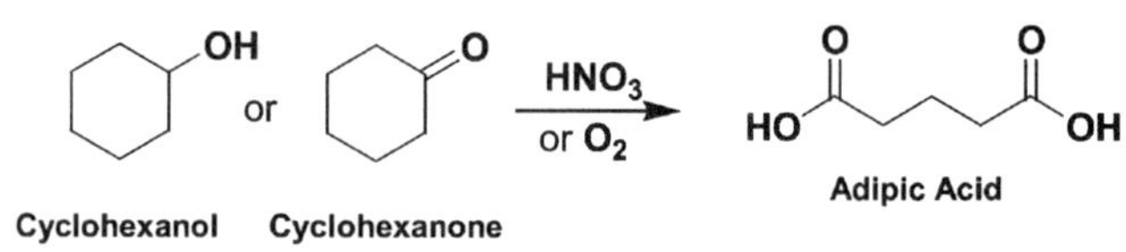

Fig. 3.19—Synthesis of adipic acid.

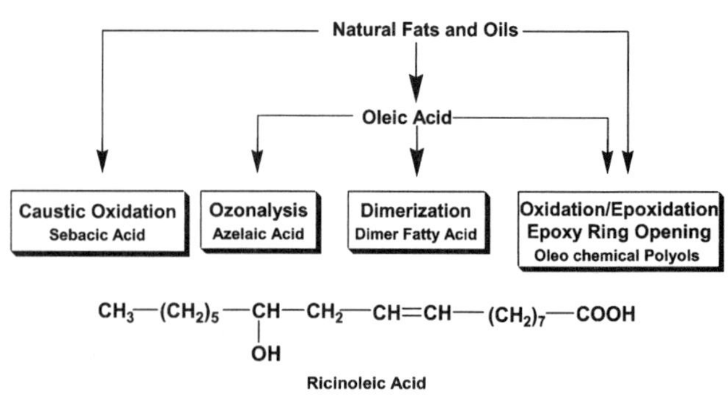

Fig. 3.20—Natural fat-derived oleo chemicals.

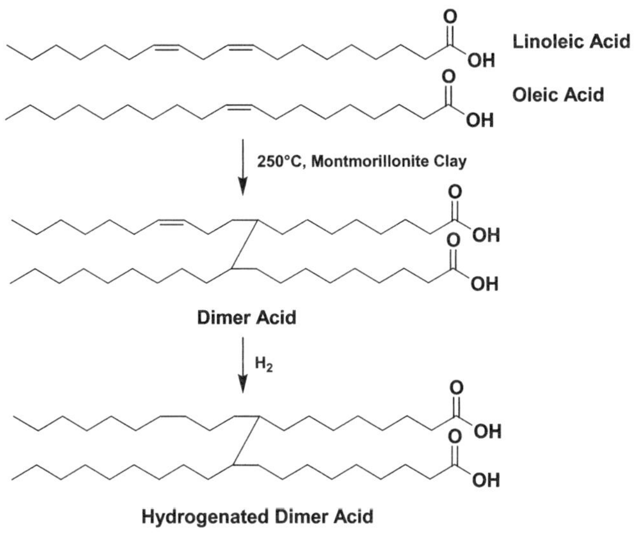

Fig. 3.21—Manufacture of dimer acids.

octadecenoic acid), which is obtained from castor bean oil. See Fig. 3.20 for oleo chemicals that are used to produce synthetic esters. C_{36}-Dimer acid is made via a thermal reaction between two molecules of an unsaturated acid, such as oleic acid, in the presence of clay, as shown in Fig. 3.21.

Linear saturated fatty alcohols of chain length C_8 up to C_{22} are sometimes made by the hydrogenation of the fatty carboxylic acids or their esters. Monohydric alcohols are also manufactured by the acid-catalyzed hydration of the olefins. As stated earlier, some branched alcohols are made by the aldol condensation of an aldehyde. In the case of 2-ethylhexanol, the aldehyde is butanal (butyraldehyde), which is a product of propene and syngas (CO/H_2). Polyhydric alcohols, or polyols, that are used in the manufacture of the synthetic ester base stocks include neopentyl glycol (NPG), trimethylolpropane (TMP), and pentaerythritol (PE). All three are derivatives of formaldehyde with other aldehydes. Neopentyl glycol is obtained by the hydrogenation of hydroxypivaldehyde, which in turn is a product of formaldehyde and iso-butyraldehyde [112]. Pentaerythritol is prepared from the reaction of acetaldehyde with formaldehyde in the presence of a base, such as sodium hydroxide, and acidifying the resulting product [113]. Trimethylolpropane is made by the hydrogenation of 2,2-dimethylolbutanal. This aldehyde forms when *n*-butyraldehyde is reacted with formaldehyde in the presence of catalytic amounts of tertiary amine [114]. Structures of these polyhydric alcohols are provided in Figs. 3.22 and 3.23, along with the methods to manufacture them.

Synthetic esters are made by a high-temperature reaction of the acid and the alcohol in the presence of a catalyst. Common catalysts that are used include sodium hydrogen sulfate ($NaHSO_4$), phosphoric acid and its salts, trialkyl or triaryl phosphates, *p*-toluenesulfonic acid, and various mineral acids. Water, the by-product of reaction, needs to be removed to shift the equilibrium towards the formation of the product. While high temperatures increase the rate of esterification, too high a temperature may result in a dark-colored product, which is undesired. The reaction temperature between 200 to 250°C is considered most suitable. While making esters derived from the monohydric alcohols, an excess amount of alcohol is employed. This is done to bring faster completion to the esterification reaction. At the end of the reaction, the excess alcohol is removed by distillation. It is important that the acidity of the product is very low; otherwise reverse reaction will be promoted in the presence of mois-

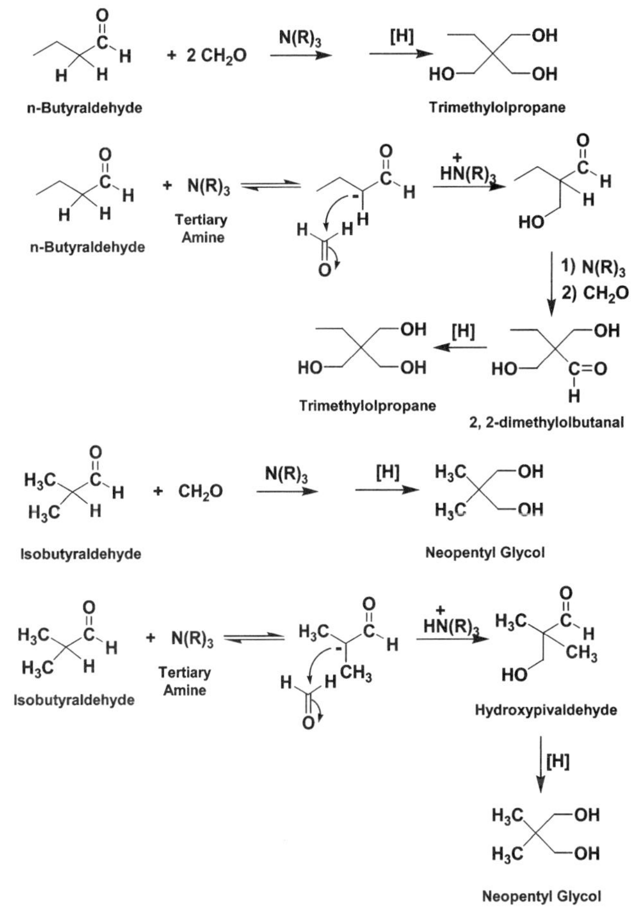

Fig. 3.22—Synthesis of trimethylolpropane and neopentyl glycol.

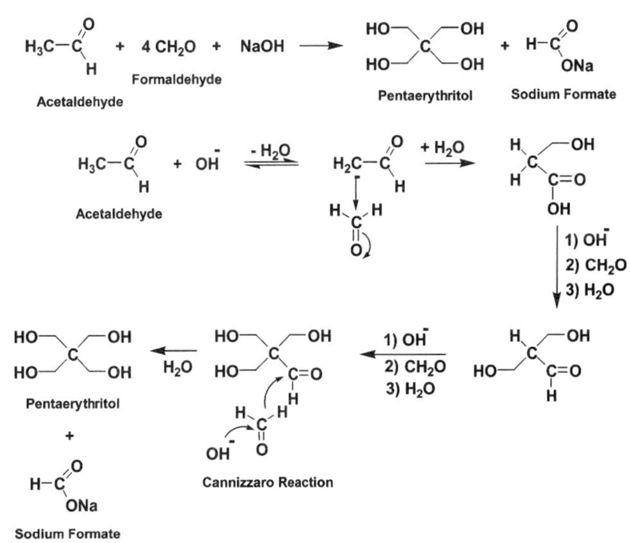

Fig. 3.23—Pentaerythritol synthesis.

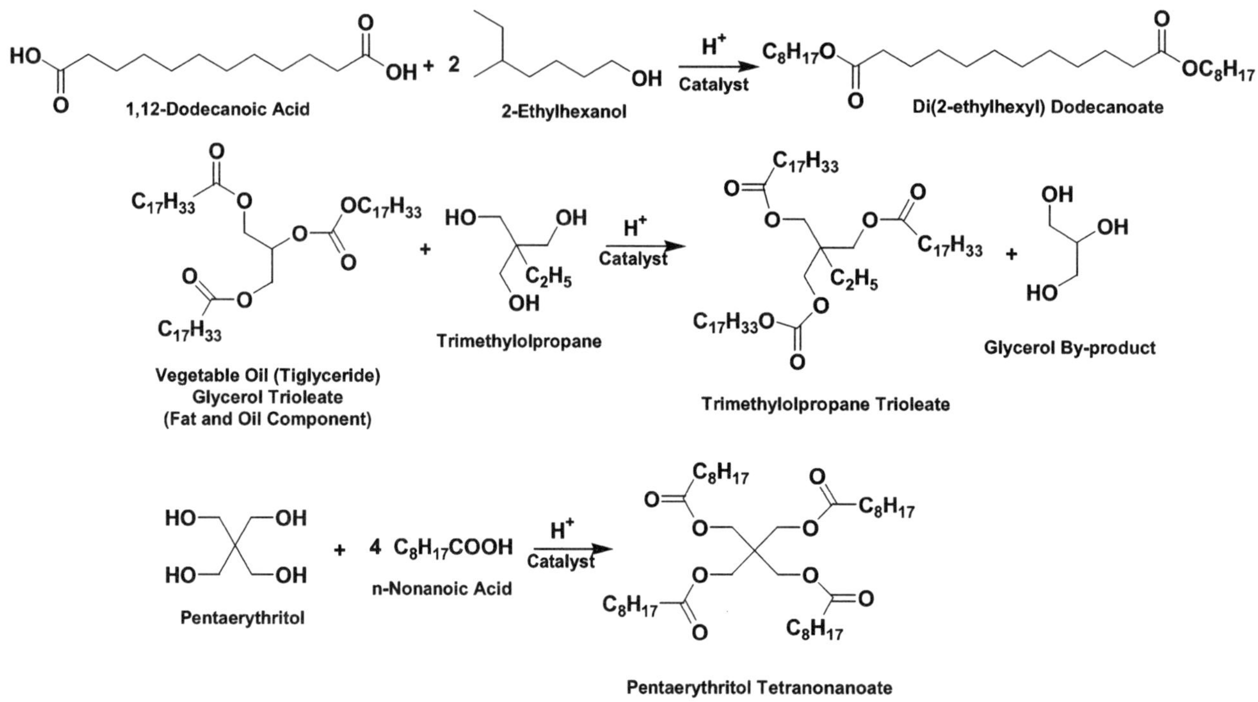

Fig. 3.24—Esterification reaction.

ture. In the case of the synthetic esters derived from the high boiling polyhydric alcohols, the carboxylic acid is used in excess, which is removed at the end of the esterification reaction. Alternatively, such esters can be produced via the transesterification reaction between a polyhydric alcohol and a low molecular weight alcohol carboxylate. Synthesis of esters via esterification reaction is depicted in Fig. 3.24.

Acid number (ASTM D974) and hydroxyl number (ASTM E326) are two of the parameters used to make certain that the unreacted starting materials have been completely removed from the product. Common catalysts used in the esterification reaction include *p*-toluenesulfonic acid and organic complexes of titanium and tin. Heterogeneous catalysts, such as ion exchange resins (Amberlysts®) may also be used. Homogeneous catalysts need to be removed prior to purification by adding water and a base. If it is not done, the product will revert back to the starting materials in the presence of water. Distillation is the common process used to purify esters.

Properties of Synthetic Esters

Ester properties that need to be considered are the same as those discussed in the PAO section. These include viscometrics, low-temperature properties, and volatility. Table 3.4 compares various physical properties of the different types of esters [4,106]. Complex esters, on account of being oligomeric, have higher molecular weights than those of the simple esters. Hence, they have higher base viscosities, but their VI, pour point, and flash points are not much different from those of the simple esters [4]. Chemical properties of interest in the synthetic esters include thermal stability, hydrolytic stability, and oxidative stability.

Viscometrics

The viscosity of the synthetic esters is easily controlled by selecting the proper starting materials. Synthetic esters with 40°C viscosities of 5 cSt for simple diesters and up to >1000 cSt for complex esters are therefore commercially available [79]. Viscosity index, or the VI, a measure of a loss of viscosity with increasing temperature, is a function of the degree and the type of branching in the molecule. In general, the more linear the structure, the greater is the VI. Again, esters of a good VI can be obtained by the proper selection of the starting materials, that is, those that have linear structures [79]. However, such materials form esters that have the tendency to wax out at low temperatures. Hence, there is an

TABLE 3.4—Comparison of ester properties [4,106].

Property	Diesters	Phthalates	Trimellitates	C_{36}-Dimer Esters	Polyol Esters	Poly-oleates
Viscosity at 40°C (cSt)	6 to 46	29 to 94	47 to 366	90 to 185	14 to 35	8 to 95
Viscosity at 100°C (cSt)	2 to 8	4 to 9	7 to 22	13 to 20	3 to 6	10 to 15
Viscosity Index	90 to 170	40 to 90	60 to 120	120 to 150	120 to 130	130 to 180
Pour Point (°C)	−70 to −40	−50 to −30	−55 to −25	−50 to −15	−60 to −9	−40 to −5
Flash Point (°C)	200 to 260	200 to 270	270 to 300	240 to 310	250 to 310	220 to 280
Thermal Stability	Good	Very Good	Very Good	Very Good	Excellent	Fair
Conradson Carbon (%)	0.01–0.06	0.01–0.03	0.01–0.40	0.20–0.70	0.01–0.10	...
Biodegradability (%)	75–100	46–88	0–69	18–78	90–100	80–100
Cost (PAO=1)	0.9–2.5	0.5–1.0	1.5–2.0	1.2–2.8	2.0–2.5	0.6–1.5

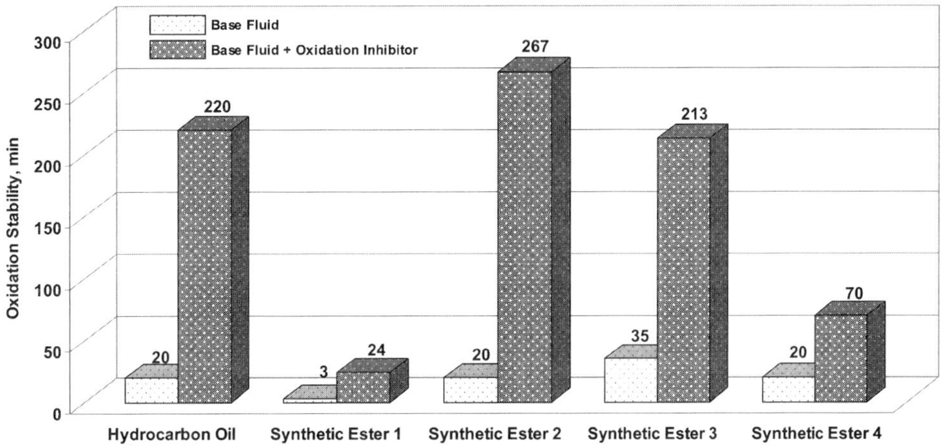

Fig. 3.25—Oxidation stability of various esters (ASTM D2272) [115].

optimal chain length in both the acid and the alcohol portion of the ester to yield esters with good VI and low pour point. The size of the optimal chain length is around eight carbon atoms. Since branching in the hydrocarbon chain lowers the pour point without significantly affecting the VI, the synthetic esters for use in the low-temperature applications are made by using a mixture of linear and branched starting materials.

The presence of the ester functional group has no effect on the viscosity due to the hydrocarbon chain of the synthetic ester molecule. Hence the synthetic ester viscosity is comparable to that of the hydrocarbons of similar molecular weight and branching. One way to explain the similarity of most physical properties between synthetic esters and hydrocarbon oils is by considering the ester functional group an equivalent of a methylene group. Properties that differ between the two as a consequence of the ester functional group include VI, boiling points, and pour points. All three parameters are lower in esters than in hydrocarbons of the same molecular size.

Since synthetic esters contain hydrocarbon chain in the alcohol- and the acid-derived portion, their properties vary depending upon the structures of these moieties. Detailed viscometrics, VI, pour point, flash point, and vapor pressure data on synthetic esters made from the same acid but different alcohols, and those made from the same alcohol and different acids are reported elsewhere [4]. These are visually summarized in Fig. 3.14. The graphs on the left side of the figure depict the effect of the acid-derived hydrocarbon portion and the graphs on the right side depict the effect of the alcohol-derived hydrocarbon portion. As the figure shows, an increase in the chain length of the acid increases viscosity, viscosity index, and pour point of a synthetic ester. Conversely, an increase in the chain length of the alcohol only increases viscosity but decreases viscosity index and pour point.

Volatility

The carbon oxygen bond in ester functional group is polar due to the electronegativity difference between carbon and oxygen atoms. This causes ester molecules to interact strongly with one another than the hydrocarbon molecules of similar carbon number and branching. Such forces are called London Forces, or van der Waals Forces. The greater intermolecular interaction translates into esters having lower volatility, and hence higher flash points.

Stability

Unlike hydrocarbon materials where oxidation stability is of primary concern, in the case of the synthetic esters, hydrolytic stability and thermal stability also need to be considered. As stated earlier, all hydrocarbon materials are susceptible to oxidation and their rate of oxidation depends upon the type of the hydrocarbon structures present and the ambient temperature. Since the esters contain hydrocarbon chains in both the carboxylic acid and the alcohol portions of their structure, the oxidation stability in esters is also of concern. Esters differ in their rates of oxidation, depending upon the structure and the branching of the hydrocarbon chains. This is shown in Fig. 3.25. Data for the figure were taken from Ref [115]. Like other hydrocarbon materials, the oxidation rate of esters is higher at higher temperatures. Figure 3.26 aptly demonstrates this [115]. Primary oxidation products are hydroperoxides and free radicals, the formation of which must be controlled; otherwise an increase in oil viscosity and the formation of sludge and deposits on hot surfaces will occur. See the section on oxidation in Chapter 4, the Additives chapter. Hence, ester-derived lubricants

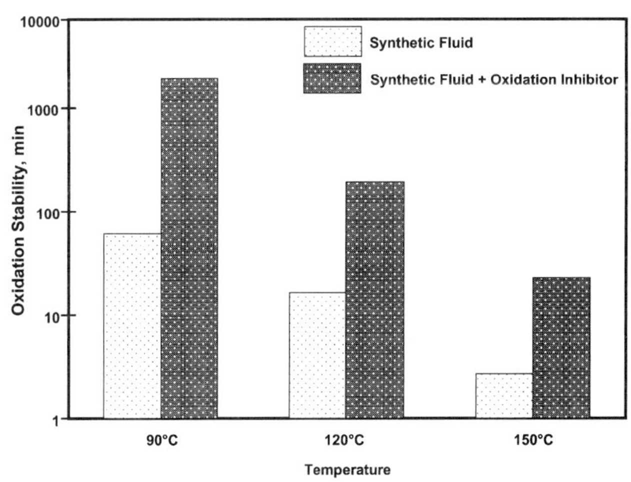

Fig. 3.26—Effect of temperature on oxidation stability of ester-type hydraulic fluids (ASTM D2272) [115].

must be treated with oxidation inhibitors, the same as mineral oil lubricants.

Hydrolytic stability is not a concern in purely hydrocarbon materials. But it is a concern for many other types of synthetic base fluids, including esters. Hydrolytic instability in esters refers to the reversal of the ester to the starting carboxylic acid and alcohol, in the presence of water. Hydrolysis rate is slow at low temperatures, but in applications where the lubricant is exposed to water and high temperatures, it is quite fast. Under these conditions, hydrolysis will occur and the corrosive acids will be produced. The rate of hydrolysis depends upon a number of factors, which include structure and purity of the ester, reaction conditions, and the nature of the additives present.

Hydrolysis reaction is *p*H and temperature dependent. *p*H is defined as the minus log of the hydrogen ion concentration ($-\log_{10} \cdot \alpha H^+$) and is a measure of acidity. Acidity in esters can be due to the presence of the residual acid catalyst, or a result of the thermal decomposition of additives, such as EP/AW agents, under frictional heat. Once the hydrolysis starts, the reaction is autocatalytic since it results in the formation of more acid. Typically, the more hindered the ester functional group, as is the case in polyol esters, the higher is the hydrolytic stability. This is because the water molecule cannot easily access the carbonyl group to cause hydrolysis. Similarly, the more hydrophobic the ester, one with less affinity towards water, the higher is its hydrolytic stability. This is true of esters that contain long hydrocarbon chains, such as the dimer acid esters.

Synthetic esters differ from one another in their thermal stability. Some are more stable than the others. For example, thermal stability of the polyol esters and the aromatic esters, such as phthalates and trimellitates, is quite good but of the diesters and the dioleates is not quite as good (see Table 3.4). The reason is that the diesters and dioleates contain β hydrogens that can lead to the β-elimination reaction. This results in the formation of a carboxylic acid and an olefin. Esters made from the secondary alcohols decompose around 190–260 °C and those made from primary alcohols decompose around 260–315 °C.

The possibility of decomposition of polyol and aromatic esters via β-elimination reaction does not exist because their structures lack β-hydrogens. However, when the temperature is high enough, even the polyol esters and aromatic esters decompose. The process involves a free radical mechanism. The thermal decomposition mechanism of diesters and polyol esters is shown in Fig. 3.27.

Because of the superior thermal and oxidation stability, polyol esters are favored over other types of synthetic esters in most high temperature applications. And among polyol esters, neopentyl glycol esters are preferred over the other types, such as butanediol and glycerol esters. A number of applications employ glycerol esters, the so-called mid-chain triglycerides (MCTs). These are produced by the chemical reaction of the short chain natural fatty acids with glycerol. These differ from the natural triglycerides based on long chain fatty acids that contain a higher degree of unsaturation. Oxidation stability of the MCTs falls in between the natural esters and the polyol esters. The reasons for the use of these esters as synthetic base oils are their higher degree

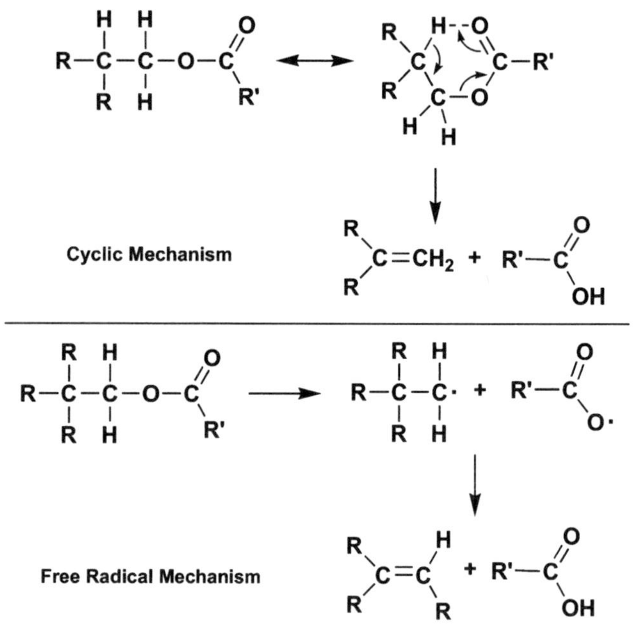

Fig. 3.27—Thermal degradation of esters.

of biodegradability and a drive to use replenishable raw materials.

Biodegradability

Carboxylate esters are more biodegradable than the hydrocarbon base stocks, because of their ready hydrolyzability in the presence of lipase, an enzyme produced by microorganisms. The resulting products, water-soluble acid and alcohol have the ability to further degrade via enzyme-catalyzed oxidation reactions. Water solubility is a prerequisite to biodegradability and so is the linearity of the hydrocarbon portions of the ester. As a general rule, materials that contain a linear hydrocarbon structure biodegrade much more rapidly than those that contain branched structures. Trimethylolpropyl oleate is an example of such a material. It has the added advantage of being based upon oleic acid, a natural material possessing good biodegradability because of the olefinic double bond. ASTM D5864 and OECD 301B are the most common test methods used to determine biodegradability of the organic substances.

Use in Lubricants

Data indicate that despite class differences, most esters possess properties that make them ideal base stocks for formulating lubricants. They, in addition, possess good solvency towards additives, most of which are polar, and have good film-forming ability. Both these properties are due to the presence of the highly polar ester functional group.

Synthetic esters are compatible with mineral base oils and most other types of synthetic fluids. Compatibility with mineral oils is of special significance since in many applications a blend of synthetic esters and mineral oils, often called semisynthetics, are used. Such blends provide good performance and at a lower cost than if a pure synthetic ester base stock is used. Somewhat higher polarity of the synthetic esters imparts good solvency, hence polar lubricant additives are easier to dissolve. Because of this, the esters way may be used to improve additive compatibility of the base stocks that have poor solvency. Examples of such base stocks in-

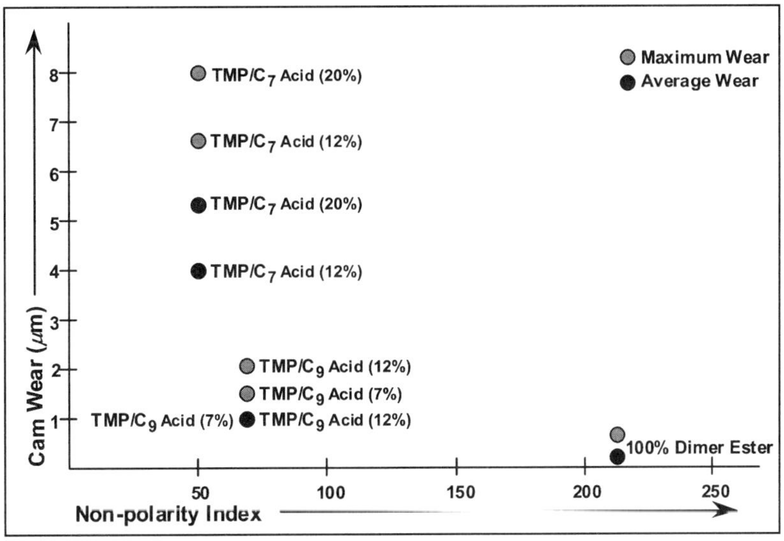

Fig. 3.28—Nonpolarity index versus engine wear [107].

clude PAOs and some API Group III oils, which have little or no aromatic components that impart solvency. Because of the compatibility of esters with both polar and nonpolar base stocks, they are sometimes used for material compatibility, such as poly(alkylene glycol)s and mineral oils, which are ordinarily immiscible.

Higher polarity of these base stocks has its own repercussions. It makes esters more surface active, because of which they compete with the surface-active additives, such as EP/antiwear agents and rust and corrosion inhibitors, for the surface [107]. These additives are designed to separate on and react with the metal surfaces to form protective chemical films. Ester lubricants overwhelm the surface, thereby preventing the absorption of these additives. This interferes with the EP mechanism and the result is increased wear. This problem can be solved by increasing the amount of such additives in the formulation, but of course at an increased cost. The additive treat rate in an ester-derived lubricant may need to be higher than in a mineral oil formulation in order to ensure an effective surface concentration of the additive.

Whether an ester lubricant will associate strongly or weakly with the surface depends upon its overall affinity for the metal, which is a function of its polarity. Polarity can be increased or decreased by varying the size of the hydrocarbon chains of the alcohol and the acid. The less polar (or more nonpolar) the base fluid, the lower is the interference in the EP mechanism; hence, the lower the wear. G. van der Waal used the *nonpolarity index* to differentiate between different ester lubricants [107]. He defines the index as:

$$\text{Nonpolarity index} = \frac{\text{Total number of C atoms} \times \text{Molecular weight}}{\text{Number of carboxylic acid groups} \times 100}$$

The nonpolarity index is high for nonpolar materials and low for polar materials. Average and maximum wear data for esters of different polarity at various concentrations in engine oils are presented in Fig. 3.28 [107].

TMP/C_7 is a trimethylolpropane [2-ethyl-2-(hydroxymethyl)-1,3-propanediol] ester of heptanoic acid, TMP/C_9 is a trimethylolpropane ester of nonanoic acid, and the dimer ester is di-2-ethylhexyl ester of the dimer (C_{36}) acid. The figure shows a direct correlation between polarity and wear, and between concentration and wear. The higher the polarity, the lower the nonpolarity index, the higher is the wear; and the higher the concentration, the higher the wear. TMP/C_7, with a nonpolarity index of 42 shows higher wear than TMP/C_9 with a nonpolarity index of 61. The dimer ester baseline with a nonpolarity index of over 200 shows minimum wear, even when present alone (at 100 %). These results indicate that increasing the molecular weight decreases the surface activity, or improves the lubricity.

We commented earlier on the oxidation stability of the synthetic esters. We stated that they are prone to oxidation, the same as any other hydrocarbon material. Oxidation stability of esters is slightly better than that of the mineral oils and is comparable to that of the synthetic hydrocarbons, such as PAOs. In order to use them in high-temperature applications with oxygen exposure, they need to be supplemented with oxidation inhbitors. Their response to these additives is excellent [79].

Esters are commonly used as synthetic base oils to formulate engine oils, two-stroke cycle oils, compressor oils, hydraulic fluids, greases, and aviation oils. Synthetic engine oils often contain ester-PAO blends for improved performance. Table 3.5 predicts the oxidation life of various fluids and their blends, with and without additives, by the use of high pressure differential scanning calorimetry (PDSC) [79]. Oxidation onset temperature (OOT) is the temperature at which the oxidation begins. In the instrument this is indicated by a deflection in the baseline resulting in a signal. The method used to measure OOT is called the nonisothermal process, that is, the temperature of the measuring cell that contains the test oil is raised by an increment of 10 °C while maintaining the oxygen pressure. Higher OOT suggests better oxidative stability of the oil. OOT suffers from the disadvantages of the large measurement error and poor discrimination, which makes drawing meaningful conclusions from

TABLE 3.5—Oxidative stability comparison of various oils by PDSC [79].

Oil	Onset Temperature (°C)
Mineral Oil-Based Lubricant	187
Synthetic Lubricants	
Polyalphaolefins	187
Polyol Esters	210
Diesters	198
Blends of Synthetic Lubricants	
Polyalphaolefins/Polyol Ester (80:20) Blend	196
Polyalphaolefin/Diester (80:20) Blend	196
Base Oil/Mixtures with Additives	
Mineral Oil-based lubricant	254
Mineral Oil/Polyalphaolefins (55:22) Blend	260
Mineral Oil-based Diesel Engine Lubricant	262
Polyalphaolefins/Synthetic Ester (80:20) Blend	274

the data difficult [116]. The OOT data in Table 3.5, while in some cases close, suggest the following:

1. Polyolefins have oxidative stability similar to that of the other additive-free oils.
2. Polyol ester possesses the highest oxidation stability.
3. Additives improve the oxidative stability of all mixtures, but to a varying degree. The 80:20 mixture of the polyalphaolefins and synthetic ester possesses the maximum oxidation stability.

Four stroke cycle engine oils must have good low- and high-temperature viscometrics, low volatility, and good thermal and oxidative stability. Synthetic esters are ideal base stocks with respect to these properties. The main disadvantage is their cost, which is often lowered by using their blends with polyolefins. Both diesters and polyol esters are used in such blends. As mentioned earlier, one of the functions polyesters perform is improving the solvency of the PAOs. Other advantages of using synthetic esters are to improve seal-swell properties and thermal stability of the derived lubricants. Two-stroke cycle oils for marine and nonmarine use require biodegradability as well, in the case of an oil spill, and low smoking tendency. With respect to these properties, diesters derived from very linear C_{36} dimer acid and polyol esters are the most suitable. As mentioned earlier, diesters made from secondary alcohols decompose around 190–260°C and those made from primary alcohols decompose around 260–315°C. Aviation engine oils primarily require high thermal stability; hence polyol esters that have higher thermal stability are used to formulate these lubricants. While previously mineral oils fortified with natural oils to improve lubricity were used, the extreme-temperature operation of the modern jet engines precludes the use of such lubricants. This is because at high operating temperatures, natural oils oxidize to form resins that can cause engine failure. In addition to having excellent thermal stability, polyol esters have the additional advantages of possessing good lubricity which is due to the presence of the polar ester functional group, low volatility (high flash points), and low pour points. Temperature profile of different parts of a supersonic jet engine as a function of speed is provided in Table 3.6 [4]. As one can see, the temperatures that some parts experience are quite high and they increase further with an increase in speed. This obviates the importance of the thermal stability in aviation lubricants.

In addition to the attributes listed above, aviation lubricants have the additional requirements of being nontoxic and being compatible with rubber, metals, plastics, and paints. In terms of viscometrics, low viscosity is important to facilitate starting and low pour point and low volatility are important for high-altitude operation, where temperatures and pressures are low. Good VI and oxidative stability are important because of the high operating temperatures.

Synthetic esters are gaining use in lubricating greases, especially in lithium soap greases since they have good compatibility with the soaps. Greases made by the use of synthetic esters have excellent low temperature and bearing wear performance and those made from diester-silicone mixtures have superb volatilities and bleeding characteristics. They can be used in applications that experience temperatures as low as −75°C.

Hydraulic fluids typically use polyol esters or diesters, especially if good oxidative stability and biodegradability are desired. Polyol esters are also used as compressor lubricants that are hydrofluorocarbons (HFC) compatible. Unlike the previously used hydrochlorofluorocarbon (HCFC) refrigerants that were miscible with the mineral oils, HFCs are not; hence there is a need to use polyol esters.

Poly(Alkylene Glycol)s [74]

Poly(alkylene glycol)s, or PAGs, are products that are obtained from the polymerization of one or more alkylene oxides. Alternative names for PAGs are polyethers, poly(alkylene glycol) ethers, and polyglycols. A variety of PAGs are commercially available, each with diverse physical and chemical properties. Their essential structural feature is the repeating ether linkage. Usually they are extensively

TABLE 3.6—Temperature conditions in supersonic-speed jet engines [4].

Part	Speed Range		
	Mach <1-2 (750-1500mph)	Mach 3 (2250mph)	Mach 4 (3000mph)
Bearings Temperatures	205°C	315°C	425°C
Gear Temperatures	120°C	205°C	315°C
Hot Parts Temperatures	315 to 425°C	424 to 540°C	540 to 650°C
Oil Temperature, inlet	135°C	205°C	315°C
Oil Temperature, outlet	190°C	260°C	370°C

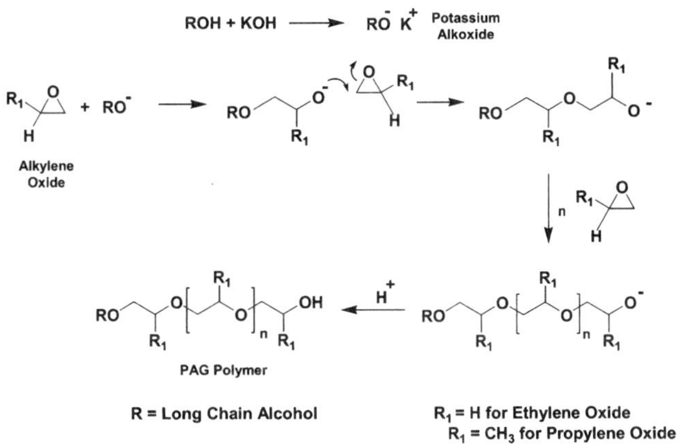

Fig. 3.29—Mechanism of poly(alkylene glycol) formation through base catalysis.

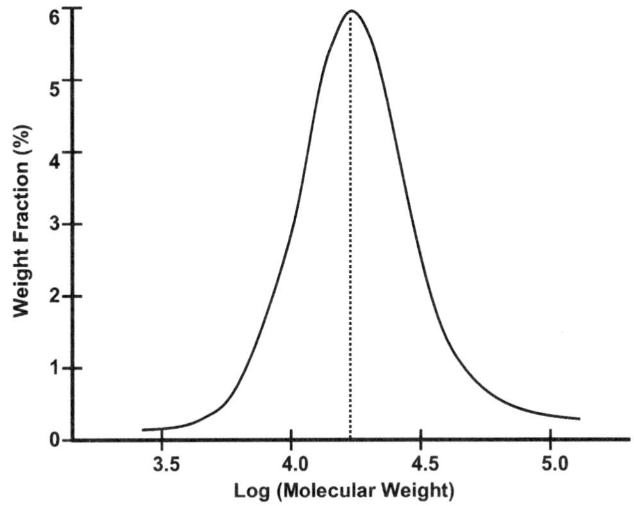

Fig. 3.30—Typical molecular weight distribution of poly(alkylene glycol) [79].

branched, which is a consequence of the alkylene oxide monomer used. Commonly used alkylene oxides are ethylene oxide (C_2H_4O) and propylene oxide (C_3H_6O). However, higher alkylene oxides can also be used. In certain instances a mixture containing two or more alkylene oxides are employed.

A poly(alkylene glycol) molecule consists of two parts: an initiator, which is an active hydrogen compound, such as an alcohol, and a polyether portion, which is derived from an alkylene oxide. The number of the monomer-derived groups in the polyether portion may vary. Therefore, a large number of products with different properties are possible from the same starting materials. The reaction is usually carried out in the presence of a base, which acts as a catalyst. However, an acid catalyst may also be used. Common catalysts include sodium hydroxide, potassium hydroxide or alkoxide, and tertiary amines. The reaction sequence that leads to the formation of PAG by a base catalysis is shown in Fig. 3.29. First the initiator is converted into an alkoxide, which is achieved by the reaction of the initiator with a strong base, such as potassium hydroxide (KOH) or a potassium *t*-butylate (*t*-BuOK). Alkylene oxide is then added to the alkoxide at a temperature of 80–150°C in a pressure reactor, to minimize the loss of the volatile alkylene oxide. Generally, a pressure of around 10 bars, or less, is maintained. Once all of the alkylene oxide has reacted, the mixture is cooled, and the resulting PAG precursor is acidified to obtain the free PAG. Typically, the higher alkylene oxides require a stronger base catalyst.

When ethylene oxide is used as the alkylene oxide, the resulting product is a primary alcohol. When propylene oxide is used, the product is a secondary alcohol. Depending upon the carbon chain length of the starting alcohol and the number of polymerized ethylene oxide or propylene oxide molecules, these polymers can be water soluble or oil soluble. PAG is a polymer composition that contains different sized polymers, that is, it has a high polydispersity index. Typical PAG distribution on a logarithmic scale is shown in Fig. 3.30 [79]. The vertical line running through the center of the distribution indicates the $\log_{10}$ of the molecular weight of the component expected from the polymerization of the $n+2$ alkylene oxide units. As the number of the monomer units (n) increases, so does the viscosity. While alcohols are the common initiators, other active hydrogen compounds, such as water [H_2O (HOH)], carboxylic acids (RCOOH), mercaptans (RSH), and amines may also be used. The amine can be aliphatic or aromatic, primary ($RNH_2, ArNH_2$), secondary (R_2NH, Ar_2NH), or alkanolamines type [$HO(CH_2)_xNH_2$]. When the amine is the initiator, the resulting polymers will be basic or alkaline in nature. Typically, the lower amounts of catalyst yield polymers of higher molecular weight and the higher amounts of catalyst yield polymers of lower molecular weight. This is because in the latter case many more polymer chains get started than in the former case. The average chain length in PAG can be controlled by the manner of addition of the alkylene oxide. Lewis acid catalysts, such as $AlCl_3$, $FeCl_3$, and BF_3, result in even broader molecular weight distributions than those that obtained by the use of the base catalysts. The reaction scheme in Fig. 3.31 shows PAG formation through Lewis acid catalysis. The reaction mechanism involves an oxonium ion inter-

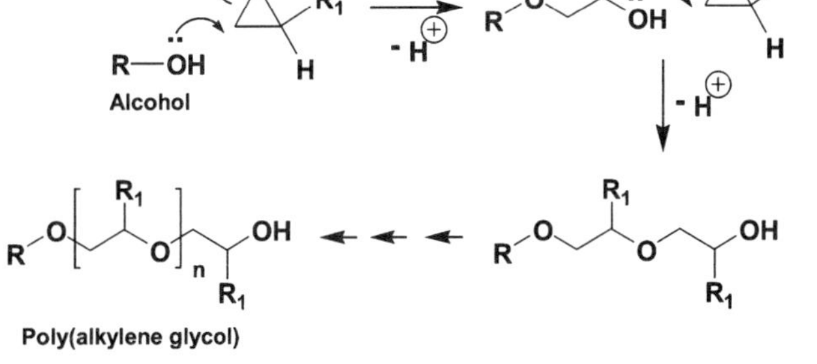

Fig. 3.31—Mechanism of poly(alkylene glycol) formation through Lewis acid catalysis.

mediate. The reaction is exothermic and the reaction temperature reaches 80–150°C, and the pressure reaches up to nine Atmospheres (9 bars). The relatively poor oxidation stability of the poly(alkylene glycol)s requires the reaction to be carried out in an inert gas atmosphere, usually nitrogen.

Alcohols used as initiators may be monohydric, that is, they contain one hydroxyl (OH) group, or polyhydric—they contain two or more OH groups. The effect of more than one hydroxyl group is that there is more than one site for the polymer chain growth. This leads to PAGs of a higher molecular weight than if the alcohol used is monohydric. Also, polyhydric alcohols result in PAGs of higher polarity since they have a relatively lower hydrocarbon content. For most lubricant applications, lower polarity or higher hydrocarbon content is desired so that the resulting PAG is compatible with other organic base stocks, if used in combination, and the additives of lower polarity, such as the friction modifiers.

The polyether section in the PAG can be derived from one type of alkylene oxide or two or more types of alkylene oxides. The PAGs made from a single alkylene oxide are called homo polymers and are used most often. Those derived from two or more alkylene oxides are called copolymers. They can be block copolymers, random copolymers, or alternating copolymers. They somewhat differ from one another in their properties. These structural types will be discussed in detail in the section on viscosity modifiers in Chapter 4 on Additives. These are prepared by altering the mode of addition of the alkylene oxide/s to the initiator. The PAGs derived from ethylene oxide have greater water solubility, or less hydrocarbon solubility, than those made from propylene oxide. Propylene oxide-derived PAGs have lower polarity, hence better compatibility with the organic materials, and superior low-temperature properties. Better low-temperature properties are due to the presence of the methyl group side chains, which decrease the tendency of the poly(alkylene glycol)s to crystallize at low temperatures. Hydrocarbon compatibility of the PAGs can be further improved by the reaction of the terminal OH group with another alcohol to form an ether, or a carboxylic acid to form an ester.

PAGs can be divided into water soluble and water insoluble types. Water solubility or insolubility depends upon the ethylene oxide content in the polymer. The PAGs solely made from ethylene oxide have high water solubility and those made from propylene and higher olefin oxides have

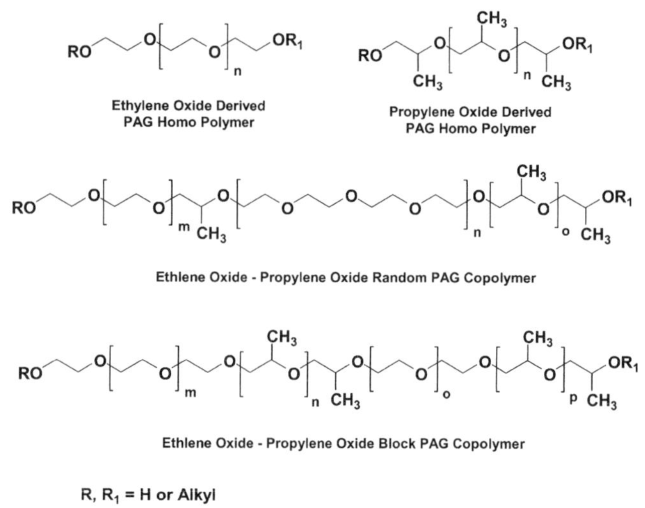

Fig. 3.32—Various types of PAG structures.

low water solubility or a high solubility in organic materials. Those made from the mixtures of the two types of olefin oxides have intermediate properties. The structure, hence properties, of these products can be further modified by altering the ethylene oxide to a higher olefin oxide ratio, using various blocking groups, and their molecular weights. The properties also vary depending upon if the resulting polymer is a block copolymer or a random copolymer. When a mixture of olefin oxides is used without controlling the sequence of addition, ethylene oxide being more reactive will be incorporated first and the higher olefin oxide will be incorporated last. Hence the polymer will have higher olefin oxide units at the end of the molecule. If one wants to control the properties of the resulting PAG, block polymerization is the correct approach. Block polymers are made by adding olefin oxide in a sequential fashion. Various types of polymer structures that are of interest in PAGs are shown in Fig. 3.32.

Poly(ethylene glycol)s of molecular weight 200 to 20,000 g/mol and poly(propylene glycol)s of molecular weight 400 to 4000 g/mol are liquids or waxy materials. They have specific gravities of 1, or higher. They do not exhibit a direct viscosity temperature relationship; their viscosities are higher at low temperatures and lower at high temperatures, lower than those expected. High molecular

TABLE 3.7—Physical properties of commercial poly(alkylene glycol)s [79].

Initiator Alcohol	EtO/PrO Ratio	Viscosity (cSt) 40°C	Viscosity (cSt) 100°C	VI	Mol. Wt[a]	Density at 20°C (g/mL)	Cloud Point (°C)	Pour Point (°C)	Flash Point[b] (°C)
1. Monohydric	0:1	11	3	103	350	0.9573	Insoluble	−53	80
2. Monohydric (a)	0:1	126	22.5	204	1900	0.9940	Insoluble	−36	225
3. Dihydric (b)	0:1	142	22.2	184	2000	1.0035	Insoluble	−36	230
4. Dihydric	0:1	387	65	242	2600	1. 0031	Insoluble	−23	232
5. Trihydric (c)	3:1	127	18	157	1200	1.0951	>100	−28	254
6. Monohydric (d)	1:1	132	25	225	1650	1.0564	59	−42	230
7. Monohydric	1:1	1050	180	287	4500	1.0574	53	−28	230
8. Dihydric	3:1	19500	2400	408	12500	1.0908	81	4	240
9. Trihydric	3:1	45000	6500	489	25000	1.0905	76	7	240

[a]Molecular weight calculation: Mol. wt.=Functionality of the initiator * 56 100/OH value.
[b]Cleveland Open Cup.

weight poly(ethylene glycol)s can have viscosity index (VI) values of 400 or more.

A number of physical properties of the commercial PAGs are listed in Table 3.7 [79]. These include viscosities at 40°C and 100°C (ASTM D445), viscosity index (ASTM D2270), molecular weight, density (ASTM D70), cloud point (ASTM D2024), pour point (ASTM D97), and flash point (ASTM D92). Items 1–4 employ only propylene oxide and Items 5–9 employ mixtures of ethylene oxide and propylene oxide. Molecular weight of the polymer is calculated by multiplying the number of OH groups in the initiator with 56,100 and dividing it with the initiator's OH value. The cloud point of the PAGs by the ASTM D2024 method is determined as an aqueous solution. Examination of the data in the table suggests the following:

1. The viscosity of the PAGs spans a wide range and it correlates well with the molecular weight of the resulting polymer. The molecular weight in turn depends upon the number of OH groups in the initiator and the alkylene oxide/s used.
2. Increasing the number of OH groups in the initiator from one to two causes a drop in the VI. Compare Items 2 and 3. The reasons for selecting these items to draw this conclusion are that they use the same ethylene oxide/propylene oxide ratio (solely uses propylene oxide) and have similar 40°C and 100°C viscosities.
3. Increasing the ethylene oxide/propylene oxide ratio leads to an increase in the VI. Compare Item 2 with Item 5.
4. The water solubility of the polymers increases with an increase in ethylene oxide content of the alkylene oxides mixture. Compare the cloud point data for Items 1-4 with Items 5–9.
5. Most PAGs have excellent pour points and flash points.
6. An increase in molecular weight leads to an increase in the VI, pour point, and flash point.

Poly(alkylene glycol)s exhibit inverse solubility in water; that is, they become less soluble as the solution temperature increases. This phenomenon is believed to occur due to a decrease in hydrogen bonding with water at elevated temperatures. This is reflected by the high cloud point of these materials. The cloud point is the temperature at which the reversal of water solubility occurs. The higher the ethylene oxide content, the higher the cloud point. Below the cloud point, the normal solubility is observed. The PAGs, on account of being polar, are compatible with other polar materials, for example, hydrofluorocarbon (HFC) refrigerants, such as R134a or 1,1,1,2-tetrafluoroethane. This makes PAGs an ideal replacement for mineral oil lubricants that are not miscible with HFCs and hence cannot be used [117]. In addition, the PAGs are compatible with most system components and have excellent thermal stability.

All PAGs are hygroscopic, which is primarily due to the presence of the hydroxyl end groups, so is their water solubility. Both of these properties decrease with an increase in the molecular weight and the number of ether linkages. PAGs derived from ethylene oxide are readily water soluble, regardless. PAGs derived from propylene oxide are largely water insoluble but are readily soluble in organic substances, if their molecular weight is above 900 g/mol. These properties are difficult to predict if the PAGs are made by using a mixture of olefin oxides. In some applications, such as fire-resistant hydraulic fluids, metalworking fluids, and textile lubricants, their water miscibility is beneficial since the fluid can be washed off the part or the equipment by the use of water. The water miscibility of the PAGs is also an advantage in the food, pharmaceutical, tobacco, and cosmetics industries.

Poly(alkylene glycol)s possess a number of unique properties, which facilitate their use in industrial gear oils, greases, brake fluids, compressor lubricants, metalworking fluids, aqueous quenching fluids, and fire-resistant hydraulic fluids. These include good thermal and shear stability, good VT properties, high flash points, low toxicity, and low flammability of their aqueous solutions.

PAGs possess good thermal and hydrolytic stability, but poor oxidative stability. Their rate of decomposition is a function of the temperature and the presence of oxygen. In the absence of oxygen, PAGs are thermally stable up to 250°C. However, in the presence of oxygen, their stability drops to around 180°C. Decomposition products are acids which result from the oxidation of the aldehyde depolymerization products and their condensates. Since bases catalyze this decomposition reaction, it is important to completely remove the basic catalysts used in the manufacture of PAGs. Oxidative stability of the PAGs can be improved by the use of the arylamine-type oxidation inhibitors, such as phenyl α-naphthylamine, or PANA. In the absence of oxygen, thermal decomposition results in the formation of alcohols, ethers, hydrocarbons, carbonyl compounds, and lower molecular weight polymeric fragments. The decomposition mechanism involves the formation of the oxygen-initiated free radicals, which lead to the fragmentation of the polymer. The use of oxidation inhibitors does improve oxidation properties of the PAGs. Interestingly, PAGs do not form carbon residue on combustion. The carbon-forming tendency of a fluid is determined by the use of two tests: Conradson carbon and Ramsbottom carbon (ASTM D189 and D524). Typical values for PAGs are less than 0.01 %. Residue-free decomposition is an advantage in some applications, such as high-temperature chain oils and compressor fluids.

PAGs have low toxicity and are highly biodegradable, especially those that have a high ethylene oxide content. This is partly due to their hydrophilic nature. Biodegradability is high in PAGs that are largely based upon ethylene oxide, for example around 80 % for PAGs of 80 % ethylene oxide content. This is because such polymers have a linear structure, which usually provides favorable results in the OECD 301 biodegradability test.

The higher polarity of these polymers makes them highly surface-active. As a consequence, they form durable lubricating films, which minimize metal-to-metal contact, hence wear. These lubricating films are persistent even at high temperatures and heavy loads. High polarity has a number of disadvantages as well. These include PAGs' aggressiveness towards coatings and some elastomers, and incompatibility with lubricant additives of low polarity, such as olefin copolymers (OCPs) and hydrogenated styrene-butadiene polymers. Such polymers are often used as viscosity improvers. In addition, just like the synthetic esters, PAGs compete with surface-active additives, such as EP/antiwear agents and rust and corrosion inhibitors, for the surface, thereby interfering in their performance. Increasing the propylene oxide content can improve the situation but only to a limited degree. The rust situation gets further complicated because PAGs tend to retain water because of their hydrophilic nature.

Applications

PAGs possess a number of desirable properties due to which they are used as base fluids in a variety of applications. These properties include lubricity, water miscibility, high viscosity index, high solvency, clean decomposition, thermal and hydrolytic stability, low pour point, high boiling point and flash point, thermal conductivity, and elastomer compatibility. Lubricity implies low coefficient of friction, which means improved antiwear properties. Water miscibility is related to the ethylene oxide content and organic compatibility is related to the propylene oxide or higher olefin oxide content. The compatibility with the hydrocarbon materials can be further improved by capping or blocking the hydroxyl end group with a long chain alcohol or a carboxylic acid, via ether formation or esterification. High solvency of the PAGs makes them dissolve many types of surface deposits, thereby keeping metal surfaces clean. Ash-free decomposition allows their use in applications involving high temperatures. Their clean decomposition is contrary to other synthetic fluids, except polybutenes, which leave a residue on decomposition. High thermal and hydrolytic stability allows their use in lubricants with long service life, especially after supplementing them with oxidation inhibitors to improve their oxidation stability as well. High boiling points and flash points make PAGs safe to use. Good high temperature properties, excellent thermal stability, and low pour points make PAGs suitable in applications that involve a broad temperature range. They also have good thermal conductivity, which in combination with water solubility makes poly(ethylene glycol)-based lubricants ideal for use in polyethylene extruders. Their compatibility with most elastomer materials is good to excellent, except with the Buna S elastomer, where it is fair. High viscosity PAGs are somewhat less aggressive toward elastomers than the low viscosity poly(alkylene glycol)s; hence they can be used without any problems as hydraulic brake fluids for brake systems based upon natural and synthetic rubbers, for example Buna S and N, butyl, neoprene, and silicone rubbers.

PAGs are used as base fluids to formulate compressor lubricants, gear oils, engine oils, lubricating greases, brake fluids, metalworking fluids, fire resistant hydraulic fluids, industrial lubricants, and refrigeration lubricants. Very high viscosity water-soluble PAGs are used in glycol-water based fire resistant hydraulic fluids as shear-stable thickening agents and as pour point depressants. Industrial lubricants benefit from good lubricity, low pour points, clean thermal degradation, high VIs and shear stability of the PAGs. These characteristics allow PAG-based industrial lubricants to be used year-round and over a broad temperature range.

PAGs are ideal for use in water-based metalworking fluids, such as metal removal fluids, metal forming fluids, and quenching fluids. Water is the best coolant and its presence extends tool life. Ethylene oxide-based PAGs have a number of advantages, which include water miscibility in all proportions, hydrolytic stability, little affect by water quality, good lubricity, wetting and penetrating ability, low foaming tendency, low toxicity, resistance to bacteria, noncorrosivity to most metals, and safety of use. As mentioned earlier, PAGs, if they are water soluble, lose their water solubility at high temperatures and come out of solution to coat the surfaces. That is how they provide lubricity that facilitates the metalworking process. Their heat transfer properties are superior to those of the hydrocarbon fluids. This, along with their high flash points, fire points, and clean thermal decomposition, makes them ideal for use as heat transfer fluids. PAGs with a variety of end groups, such as hydroxyl, alkoxy, and alkylamino, are commercially available for use in metalworking applications.

PAGs are excellent as base fluids to suspend solids, such as graphite and molybdenum disulfide (MoS_2), commonly used in lubricating greases for high-temperature applications, such as refractory kiln car bearings, which encounter temperatures of 2000°F (1093°C), and oven chain drives and gears. In these applications, the sludge-free decomposition of the PAGs is an added advantage. On decomposition, polyglycol decomposition products flash off or burn clean, leaving the solid lubricant without any tar deposits.

The use of the PAGs in compressor lubricants is due to their excellent lubricity, high temperature stability, and clean thermal decomposition. Gas compression leads to entrainment of the gas into the lubricant, which causes the viscosity of the lubricant to drop. As a consequence, the lubricant loses its ability to form an effective lubricating film between the metal surfaces. PAG base fluids are ideally suited for this application because of their high polarity since gases, being of low polarity, are not soluble in them. Hence, no drop in the lubricating ability of oil is observed. Polypropylene glycol works extremely well in propane gas refrigeration compressors. Poly(ethylene glycol) lubricant of ISO viscosity grade 150 is highly resistant to thinning due to condensable hydrocarbons with four or more carbons. In applications, such as well gas that involves both the condensable hydrocarbons and the water vapor, a poly(ethylene glycol)/poly(propylene glycol) co-polymer lubricant is suitable. While the poly(alkylene glycol) lubricants can be designed to resist dilution, the hydrocarbons will condense at high operating pressures. If such a lubricant enters the compressor, a bearing failure will occur.

As stated earlier, PAGs are good base fluids for refrigeration applications, which is because of their compatibility with new chlorine-free hydrofluorocarbon refrigerants. The impetus to replace chlorofluoro-hydrocarbons (CFCs) with hydrofluorocarbons (HFCs) is to protect against chlorine-initiated damage to the ozone layer. Compatibility of the refrigerant with the lubricant is important because in mobile air conditioning units, for example, those used in automobiles, the lubricant travels through the system together with the refrigerant. PAGs, because of their superior lubricity, help lubricate the critical parts of the refrigeration system.

PAGs are also used to formulate lubricants for heavy-duty worm gears, plain and rolling contact bearings, and other industrial gearing. Worm gears are used in conveyers, escalators, material handling equipment, press drives, packaging machinery, ski lifts, and agitators and mixers. Other gear and bearing applications are in the cement, metalworking, plastics, food, rubber, paper, and textile industries. In these applications, PAG's low coefficient of friction, high VI,

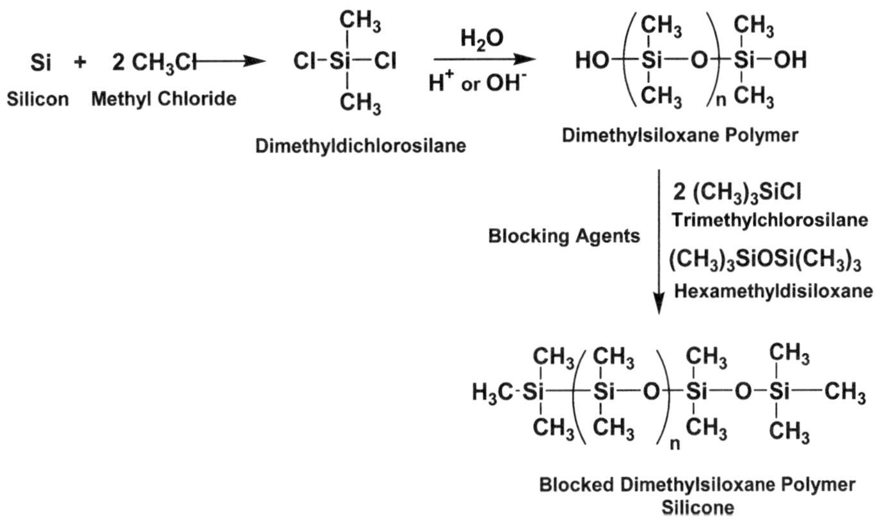

Fig. 3.33—Polysiloxane synthesis.

and superior load-carrying capacity under boundary lubrication conditions are beneficial. Worm gears experience a high degree of sliding motion which results in extensive metal-to-metal contact. The result is the high surface temperatures due to friction. PAGs provide lubricity and cooling because of their high heat capacity. Heat capacity is defined as the amount of heat needed to raise the temperature of a material by one degree. Liquids that exhibit hydrogen bonding, such as water and alcohols, have higher heat capacities than those that do not have hydrogen bonding, such as hydrocarbons. That is, they require greater amounts of heat to increase their temperatures than do substances with low heat capacities. The reason for this is that part of the energy supplied to such liquids is consumed to break the hydrogen bonds that are present in their structure. Despite their superior load-carrying capacity relative to the mineral oils [4], PAGs need to be supplemented with rust-inhibiting and extreme-pressure additives. However, such additives for use in PAGs need to have better water tolerance and different solubility characteristics than those used to formulate hydrocarbon lubricants. This is again due to their competing with the additives for the surface, the same as in the case of synthetic esters.

Polyglycols have been tested extensively for use in automobile engines but have not gained much use, primarily because of their corrosion tendency that cannot be controlled in the presence of the acidic fuel combustion products. However, the diesters made by the fatty carboxylic acid blocking of the poly(ethylene glycol)s and poly(propylene glycol)s, made from the glycol initiators, are used in gasoline. These can be mixed with the mineral oil and used in two-stroke cycle applications. Such blends have the advantages of providing proper lubrication without causing extensive exhaust port deposits [4].

Other Synthetic Base Stocks

There are many specialized applications which require a unique set of properties in a lubricant that are not present in the base stocks described so far. By taking advantage of the flexibility of the synthetic methods, a number of materials have been developed that meet the requirements of such applications. Of course, the cost of producing them is largely irrelevant since no other material/s are good substitutes. This section covers these specialized fluids.

Silicon Compounds [74]

Silicones and silicate esters are two types of silicon compounds that are used as synthetic lubricants. Of these, silicones are the most popular and are used in high-temperature lubricants. The term silicone is used for polysiloxanes. Polysiloxanes are linear polymers that contain Si-O-Si linkages with pendent hydrocarbon groups: the methyl group being the most common alkyl pendent group and phenyl being the most common aryl pendent group. Other alkyl or aryl groups can also be used if the improvement of certain properties is the objective. Polysiloxanes are products of hydrolysis of the dialkyldichlorosilane. This intermediate is made by the Rochow synthesis, which involves direct reaction of an alkyl halide, such as methyl chloride, with silicon metal. One obtains a mixture of alkylchlorosilanes that contains 70–90 % dimethyldichlorosilane, $(CH_3)_2SiCl_2$. This silane is isolated from the others by distillation and reacted with water to yield the dimethylsiloxane polymer—$[Me_2SiO]_n$. Since dichlorosilanes are bifunctional, polymerization continues until it is terminated by the use of a mono-functional blocking agent, such as trialkylchlorosilane or hexamethyldisiloxane. The reaction scheme to make these materials is provided in Fig. 3.33.

Molecular weight and many of the other properties, such as viscosity, are determined by the number of the monomer units in the polymer and the nature of the pendent group. Silicone fluids have a viscosity range of between 1 cSt and 500,000 cSt. Their desirable properties include high compressibility, low pour points, low surface tension, low volatility, good fire resistance, excellent thermal and chemical resistance, good dielectric properties, and a good VT relationship.

Silicone fluids have relatively high boiling points, hence have low volatility (low vapor pressure) and high flash points; depending upon the molecular weight and the viscosity. Silicones with viscosities above 50 mm^2/s (50 cSt) at 25°C do not reach their boiling points, even under vacuum.

Volatility is a function of the molecular size of the polymer chain, hence substituting a phenyl group in place of a methyl group has little or no effect on volatility. Pour points and setting points (solidification temperatures) increase with viscosity, but they can be lowered by partially replacing the methyl groups with higher carbon number groups, such as ethyl or phenyl. Density also increases with viscosity, but within a given class. Silicones, especially dimethylsiloxanes, have excellent VT behavior relative to that of the mineral oil. However, an increase in the phenyl content of the polymer leads to slight to mediocre deterioration of the VT properties [4]. Silicone fluids have good shear stability. However, higher molecular weight polysiloxanes lose their apparent viscosity with increased shear more rapidly than their lower molecular weight counterparts. This loss is temporary and the polymer viscosity reverts once the shear forces are removed.

Silicones are good lubricants for bearings and gears with rolling friction. However, when the mechanism involves high sliding friction, their lubricating properties depend on the metal pairs. In the case of steel-on-steel, the performance of silicones is not very good, presumably because of their low surface tension and the weakness of the lubricating film, due to low affinity for the metal surface. Such a film is easy to remove because of high shear in sliding contacts. When the metals involved are dissimilar, such as steel and the nonferrous metals, the lubricating properties of these polymers parallel those of the aliphatic hydrocarbons. The use of the conventional EP/AW agents causes only limited change in performance. Silicones are excellent lubricants for bearings that are made from polymeric materials, such as polyamides and polystyrene [4].

Silicones are thermally stable up to approximately 315°C, beyond which they decompose. The reason is the presence of water, catalysts (bases and acids), and other ionic materials that catalyze the decomposition of polysiloxanes to dimethylcyclosiloxanes, carbon monoxide, carbon dioxide, silicon dioxide, and formaldehyde. The stability of silicones can be increased to 425–450°C by incorporating phenyl groups in the polysiloxane structure. Despite the fact that silicones dissolve oxygen, they are oxidatively stable up to 200°C, which is much higher than the additive-free hydrocarbons, esters, and poly(alkylene glycol)s. Above this temperature, their oxidation results in siloxyl and silyl free radicals, which cross link to form structures with gel-like properties. Since the carbon side chains are the first point of attack, phenyl substitution improves the oxidation stability. Oxidation can be controlled by the use of the aromatic amines and a variety of other types of oxidation inhibitors.

Silicones are also susceptible to hydrolysis at high temperatures. The result is silica (SiO_2) and silicic acid gel formation. Ionizing radiation stability of these polymers is similar to their oxidation stability, that is, dimethyl derivatives are more likely to decompose to form gel-like products than those that contain the phenyl groups.

Silicones are used as lubricants both in industrial and military applications. Dialkylsilicones are used to lubricate equipment that uses dissimilar metal pairs, such as bronze/brass on aluminum and copper and zinc. Examples of such equipment are precision mechanisms, speedometers, synchronous motors, and those containing plastic and rubber parts. Silicones are also used as switch and transformer oils, for the impregnation of the porous bronze bearings, and as VI improvers for silicate esters. Silicones of high aromatic content are used to make lubricants for turbines, ball bearings, watches, electric shavers, and various other devices. Since they have excellent thermal stability and high flash points, they are used in a variety of high-temperature applications. These properties, combined with the low pour points and the imperviousness to water, makes silicones ideal for use in hydraulic, brake, and heat transfer fluids, and refrigerator oils.

Silicones are also used in textile applications as fiber, thread, and yarn lubricants, which is primarily due to their low surface tension. Silicones are not compatible with mineral oils and other synthetic base fluids, which makes them unsuitable for use as lubricant base stocks. Their use as foam inhibitors is primarily due to their low surface tension and incompatibility with other fluids. However, their use in metalworking fluids to suppress foam is not preferred because they leave a film on metal parts, which precludes adhesion of paint. These fluids are nonbiodegradable, because they linger in the environment for an extremely long period. Their excellent thermal stability is beneficial to their use as heat transfer oils and their good dielectric properties justify their use as dielectric fluids in transformers. Their load-carrying capacity is extremely poor; hence they are not normally used as base fluids, except in specialty greases. Such greases are made by using both organic and inorganic thickeners. Silicone-based lithium soap greases can be used at temperatures up to 200°C. Specialty silicone greases based on carbon black, phthalocyanins, polytetrafluoroethylene, Indanthrene Blue, and arylureas can be used in applications that experience temperatures up to 370°C. While the rheological properties of these greases are similar to those of the mineral oil-based greases, these have better chemical stability, compatibility with plastics and rubber materials, and ability to function over a broad range of temperatures. These characteristics make them suitable for use in filled-for-life lubrication. Silicone greases are also well suited for lubricating rolling element bearings operating at elevated temperatures. The low vapor pressure of these greases makes them suitable for use in equipment involving high vacuum, such as diffusion pumps and aerospace applications. Silicone-based greases are also used in aviation and automotive industries to lubricate linkages, bearings and bushings, and instrument components [79,118]. Silicones are also used to formulate hydraulic fluids and brake fluids.

Unlike silicones that contain silicon-carbon bonds, silicate esters contain none [4]. They contain only silicon-oxygen bonds. They are organic esters of orthosilicic acid, H_4SiO_4. Materials that are commonly used as base fluids include hexa-alkoxy and hexa-aryloxy di- and tri-siloxanes. Alkyl and aryl groups may or may not contain substituent groups, such as chloro, fluoro, nitro, alkoxy, and thioalkoxy. These esters are prepared by the hydrolysis of the di- and or tri-alkoxysilicon chloride, prepared from the reaction of silicon tetrachloride with an alcohol. Representative structures of the silicate esters are provided in Fig. 3.34. The properties of these compounds radically differ from those of the silicones. Although low volatility and excellent viscosity-temperature characteristics make them suitable for use as lubricant base stocks, their applications are limited due to

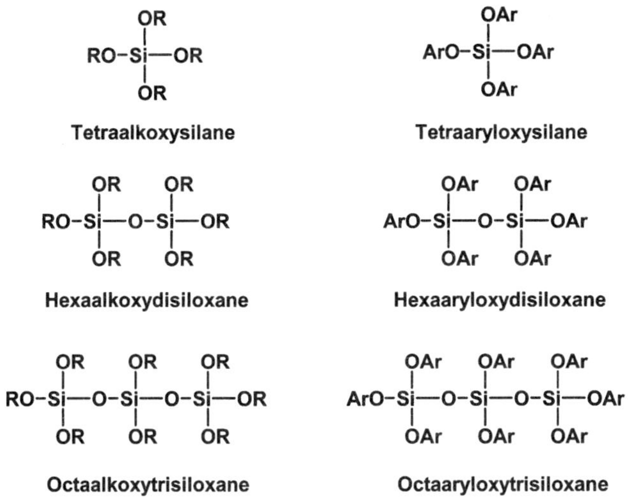

Fig. 3.34—Representative structures of silicate esters.

their poor hydrolytic stability, resulting in the formation of gels in the presence of moisture. This also limits their use in closed lubrication systems, such as those found in low-temperature refrigeration compressors.

Physical and chemical properties of these esters depend upon the presence of alkyl or aryl substituent, molecular weight, and the structural symmetry. They possess excellent low- and high-temperature properties (low pour point, high boiling point, and high flash and auto-ignition points), excellent dielectric properties, good oxidative stability, lubricity, and radiation resistance, noncorrosivity to most metals, and good to reasonable compatibility with most elastomers and plastics. The viscosity and VT characteristics of the silicate esters are superb, especially in the case of the alkyl derivatives. Their viscosity indices are estimated to be between 140 and 230. Their boiling points are in 100 to 200°C range at a reduced pressure of 1.3 mbar and they have pour points of less than −65°C. This makes their operating range from −50°C to 200°C. Because of the high boiling points, their low dielectric constants, and excellent thermal stability,

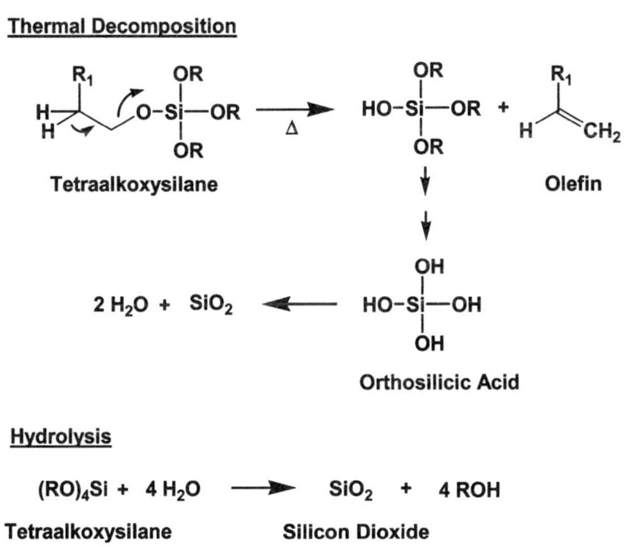

Fig. 3.35—Thermal and hydrolytic decomposition of silicate esters.

$C_3F_7O[CF(CF_3)CF_2O]_xC_2F_5$
PFPE-1
Krytox®

$CF_3O(CF(CF_3)CF_2O)_x(CF_2O)_yCF_3$
PFPE-2

$C_3F_7O(CF_2CF_2CF_2O)_xC_2F_5$
PFPE-4
Demnum®

$CF_3O(CF_2CF_2O)_x(CF_2O)_yCF_3$
PFPE-3
Fomblin® Z

$C_3F_7O[CF(CF_3)CF_2O]_x(CF_2O)_yC_2F_5$
Fomblin® Y

Fig. 3.36—Structures of perfluoropolyethers [120].

these materials are used as dielectric heat transfer fluids. Thermal stability of the alkyl silicate esters is similar to that of the synthetic esters and so is the mechanism of decomposition. The products of thermal decomposition are olefins and silicic acid, which further dehydrates to silica. The resulting water initiates their hydrolytic decomposition to alcohol and SiO_2. Figure 3.35 shows the decomposition mechanism of these esters. Aryl silicates are thermally more stable (~450°C) than alkyl silicates since they do not have the β-hydrogens that facilitate the decomposition of the alkyl esters. Oxidation stability of these esters parallels that of the hydrocarbons and it can be improved by the use of the oxidation inhibitors. They themselves are noncorrosive, but they need corrosion inhibitors to protect metals against corrosion. Because of the presence of oxygen, they have polarity, hence surface affinity, making them good lubricants [4]. Their toxicity is mild to none.

The use of silicones as base fluids is primarily confined to hydraulic and heat exchange fluids, lubricants for some military applications, cooling fluids, refrigeration lubricants, and lubricating greases. Their use in other lubricants is limited because of their poor hydrolytic stability. A number of silicate ester products are marketed under the trade name Coolanol® for use as dielectric heat transfer fluids. The product brochure summarizes various properties of these silicate esters along with their recommended uses. One of the brochures can be accessed via Ref [119].

Fluorine-derived Fluids

Perfluoropolyethers [120]

Perfluoropolyethers (PFPEs) are poly(alkylene glycol) analogues, with the difference that all the hydrogens are replaced by fluorine. While originally developed for use in aerospace applications, they are gaining use in applications that require oxidation and the chemical resistance. Their oxidation resistance can be ascribed to the lack of the carbon hydrogen bonds in their structures and chemical resistance is due to the strong electron withdrawing nature of the fluorine. Essentially, the fluoro-alkyl group withdraws electrons away from the ether linkage to make these materials very much hydrocarbon-like. They are made by the polymerization of the perfluorinated monomers. Four classes of PFPEs are commercially available. Krytox® [121], or PFPE-K for short, is made by cesium fluoride catalyzed polymerization of hexafluoropropylene oxide, yielding a branched series of polymers. Fomblin® Y, or PFPE-Y for short, is made by the UV catalyzed oxidation of hexafluoropropene [122]. This polymer is similar to PFPE-K in that it also contains pendent groups. Fomblin® Z, or PFPE-Z for short, is made by the same method as PFPE-Y, except that it utilizes tetrafluoroethylene instead, which gives it a linear structure [123].

TABLE 3.8—Physical properties of perfluoropolyethers [120].

Lubricant	Average Molecular Weight	Viscosity at 20°C, cSt	Viscosity Index	Pour Point, °C	Vapor Pressure, Torr 20°C	100°C
Fomblin® Z-25	9500	255	355	−66	2.9×10^{-12}	1×10^{-8}
Krytox® 143AB	3700	230	113	−40	1.5×10^{-6}	3×10^{-4}
Krytox® 143AC	6250	800	134	−35	2×10^{-8}	8×10^{-6}
Demnum® S-200	8400	500	210	−53	1×10^{-10}	1×10^{-7}

Demnum®, PFPE-D for short, is produced by polymerizing hexafluorooxetane, followed by fluorination to replace the remaining C-H bonds with C-F bonds [124]. Krytox® is a Dupont trade name, Fomblin® is an Ausimont trade name, and Demnum® is a Diakin trade name. All four products are polydisperse polymers, that is, they contain a distribution of different size molecules, or the different molecular weight components. They are subjected to fractional distillation to obtain components with the desired physical properties. Despite the structural differences between the different types, they have similar gross properties. A structure of each type is shown in Fig. 3.36 and their properties are summarized in Table 3.8 [120].

For aerospace applications, a lubricant must have vacuum stability, low creeping tendency, high VI (broad temperature operating range), resistance to radiation and oxidation, optical transparency, and good boundary and EHD film-forming ability [120]. Creep is the tendency of a liquid lubricant to migrate from one environment into another. It is inversely proportional to the surface tension. PFPEs have extremely low surface tensions (17 to 25 dynes/cm at 20°C); hence their creeping tendency is expected to be high. Although these fluids are contained by using low surface energy fluorocarbon barrier films [125], PFPEs can dissolve such films on prolonged contact [126].

PFPE products are available with 40°C viscosities ranging from about 5 to about 500 cSt and have high viscosity indices. Among the four types of PFPEs, the K and Y types have lower VI than the D and Z types. Space applications typically involve temperatures of between −20°C to 60°C, hence high VI in fluids is not a requirement, but the pour point is. These materials have extremely low pour points. Their VI is related to the carbon to oxygen ratio in the polymer repeating unit. The presence of the pendent group, as in the case of Krytox® and Fomblin® Y, lowers the VI. Demnum® is the next best, and Fomblin® Z is the best. High temperature volatility is generally low [127]. This is a very desirable feature since in space mechanisms the lubricant loss over the long term (7–30 years) is a concern [120]. In this regard, Fomblin® Z is the best, which is followed by Krytox® 143AC, Apiezon® C (mineral oil), synthetic ester, and polyolefins. Vapor pressure data in Table 3.8 shows the comparative volatilities of the four PFPEs. Fomblin® Z has the lowest volatility and hence is the best, which is followed by Demnum® S-200, Krytox® 143AC, and Krytox® 143AB.

The concern for optical properties of a lubricant is related to the sensitive optical components, such as mirrors that are used in the measurement devices. Incidental contamination of the component surfaces by a lubricant of inappropriate quality can hinder precise data collection. Hence, optically transparent lubricants are preferred. Since PFPEs are devoid of hydrogens, they are transparent over most of the infrared region. They have an absorbance below 1400 cm^{-1}.

PFPEs have poor radiation resistance. They can degrade from low-energy and high-energy electrons, and ion beams. Demnum® is the most stable of the three examined. The other two are Fomblin® Z-25 and Krytox® 16256 [120]. Oxidation resistance of these fluids is only a concern because in the low earth orbit neutral atomic oxygen exists. Experimental data indicate these fluids to be almost completely impervious to oxygen, which is not too surprising since they are free of the carbon hydrogen bonds [128]. In addition to being resistant to oxidation, PFPEs are resistant to most corrosive chemicals, such as acids and alkalis. Because of these attributes, they are suitable for use in many fill-for-life applications.

Film-forming ability of a lubricant is another consideration. Elasto-hydrodynamic (EHD) properties are important in continuously rotating medium- to high-speed bearings. Lubricant properties that determine the quality of the EHD film are its absolute viscosity (μ) and its viscosity-pressure coefficient (α). Viscosity of a fluid in turn depends upon its molecular weight and structure. For low molecular weight fluids, such as PFPEs, viscosity-pressure coefficient (α) solely determines the EHD film-forming ability of the fluid. The α values for various fluids including three PFPEs are presented in Table 3.9 [74,129]. Typical bearings temperatures for most space applications are 0 to 40°C. Hence, from the EHD lubrication viewpoint, Krytox® 143AB with the highest α value is the preferred fluid. In the boundary lubrication regime, where the surface-to-surface contact is high, protection against adhesive wear is usually provided by the thermally labile EP/AW agents, which react with metal surfaces to form chemical protective films. Since PFPEs are pure, additive-free fluids, they must perform the boundary

TABLE 3.9—Pressure-viscosity coefficient (α^a, $Pa^{-1}\times10^8$) for various base fluids [129].

Fluid	Temperature (°C) 38	99	149
Ester	1.3	1.0	0.85
Synthetic Paraffin	1.8	1.5	1.1
Fomblin® Z-25	1.8	1.5	1.3[a]
Naphthenic Mineral Oil	2.5	1.5	1.3
Traction Fluid	3.1	1.7	0.94
Krytox® 143AB	4.2	3.2	3.0

[a] Extrapolated Value

$CF_3O(CF_2O)_xCF_3$	$CF_3O(CF_2CF_2O)_xCF_3$
Perfluoropolymethylene oxide (PFPMO)	**Perfluoropolyethylene oxide (PFPEO)**
$CF_3O(CF_2CF_2OCF_2O)_xCF_3$	$CF_3O[(CF_2CF_2O)_2CF_2O]_xCF_3$
Perfluoropolydioxolane (PFP-DIOX)	**Perfluoropolytrioxocane (PFP-TRIOX)**

Fig. 3.37—Developmental perfluoropolyethers [134].

lubrication function as well. Interestingly, these fluids do decompose under boundary lubrication conditions to produce corrosive gases that react with the metal oxide surface film to form fluorides, particularly those of aluminum and iron. However, they are strong Lewis acids and they catalyze the break down of the PFPE structure. This generates more of these species and the cycle continues. The same effect can lead to oxidative corrosion of some metals by PFPEs in air. Both thermal decomposition and oxidative corrosion are inhibited in the absence of oxygen [130]. However, the ultimate result is the bearings failure. Unfortunately, the compatibility of the PFPEs with the conventional antiwear agents is low [79,127] and such chemicals for use in PFPE are not presently available. The research in this area is in progress [131]. The challenge for such additives is not to undermine the strengths of the PFPEs because they are used in certain applications.

A boundary lubrication study of the three commercial fluids in air and vacuum was recently reported. In terms of wear in the presence of air, data indicated Krytox® and Fomblin® Z to be almost equivalent, and both are better than Demnum®. In the absence of air, Krytox® and Demnum® were almost equivalent and both were significantly better than Fomblin® Z. Experience with respect to the use of Demnum® in space applications is lacking, but Krytox® and Fomblin® Z have been successfully used. A number of new PFPE fluids are being developed by the use of direct fluorination technology [132–134]. Some of these structures are shown in Fig. 3.37.

As mentioned earlier, of the commercial PFPE fluids, only Krytox® and Fomblin® Z have been extensively used as liquid lubricants and in greases for space applications [135]. Recently, the use of a Demnum® fluid has also been tested. The extremely high cost of PFPEs is the major deterrent for their use in applications other than those pertaining to space. The largest current application is in specialty vacuum pump oils for use in contact with reactive chemicals in electronics manufacture. PFPEs are also used in the formulation of greases for some sealed for life bearings.

Fluoroester Fluids [4]

Despite the fact that fluorocarbons have excellent oxidation stability, low coefficient of friction, and good fire resistance, they have low VIs, high pour points, and high volatility. Because of these limitations they have not gained use as synthetic base fluids for lubricants. Ester functional group in a hydrocarbon has the tendency to lower the pour point and increase the boiling point; hence efforts have been made to synthesize fluoroesters and to explore their use as base fluids in specialty applications. Perfluoro mono- and di-carboxylic acids can be obtained by the electrofluorination of the aliphatic acid chlorides or fluorides or by telomerization with tetrafluoroethylene and oxidation of the resulting products. Fluorinated alcohols are prepared by the reduction of the fluoro acids or by telomerization. The carbon atoms of these alcohols, which carry the OH groups, are substituted by H atoms or CH_3 groups [4]. Strong acids, such as sulfuric acid, are used as esterification catalysts. Fluoroester synthesis is shown in Fig. 3.38.

Fluorinated esters are better than their nonfluorinated analogs with respect to thermal and oxidative stability, flash point (~20°C higher), and the wear performance. However, they are more susceptible to facile hydrolysis. Their viscosities at 38°C (100°F) are higher than those of the nonfluorinated materials. However, their viscosity indices and pour points are much lower. While fluoroesters are soluble in po-

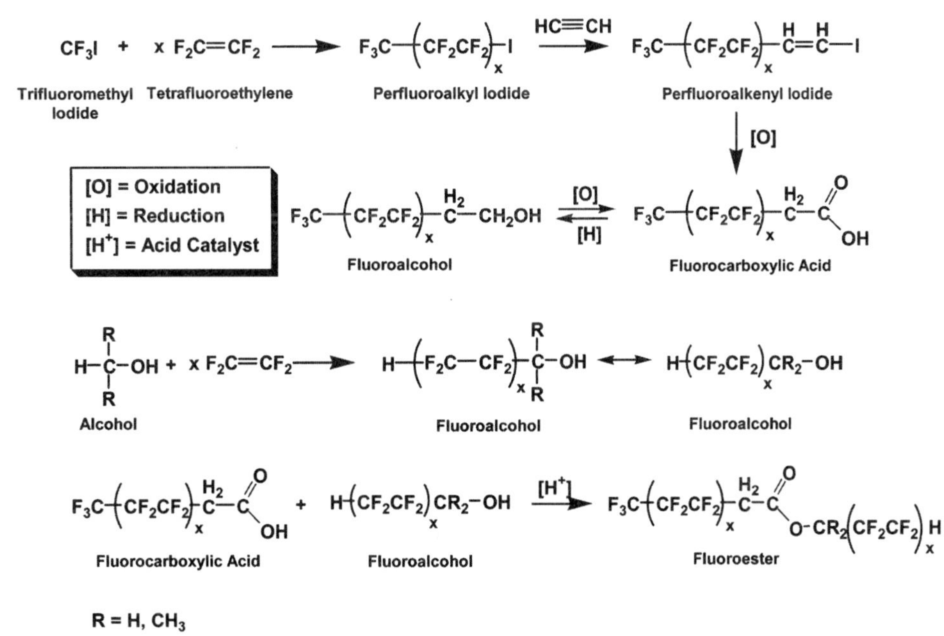

Fig. 3.38—Fluoroester synthesis.

TABLE 3.10—U.S. Military Specification MIL-H-19457 D (SH) for fire-resistant hydraulic fluids [137].

Description	Requirement	Typical Result	Test Method
Kinematic Viscosity, cSt			ASTM D445
40 °C, minimum	38.5 - 45.5 cSt	40.0 cSt	
100 °C, minimum	4.8 cSt	4.2 cSt	
Neutrality (qualitative)	Neutral	Neutral	Method 5101
Evaporation loss, max at 100 °C	0.3	0.2	ASTM D972
Refractive Index, Nd at 20°C	0.005	0.002	ASTM D1218
Foaming Tendency at 24°C, max	65	50	ASTM D892
Foam Stability, ml after 10min	Complete collapse	Complete collapse	ASTM D892
Fire Point, °C minimum	246°C	250°C	ASTM D92
Total acid number, max.	0.10 mg KOH/gm	0.05 mg KOH/gm	ASTM D974
Water content, % maximum	None	None	ASTM D95
Hydrolytic Stability			MIL TEST
Copper weight loss, max	0.3	0.2	
Appearance	No corrosion	No corrosion	
Acid number increase, max	0.2	0.1	
Acidity of water layer, max	5.0	4.5	
Insolubles, %, max	0.5	0.2	

lar solvents, such as alcohols, esters, ethers, aromatic hydrocarbons, poly(alkylene glycol)s, and triaryl phosphates, they are incompatible with aliphatic hydrocarbons, mineral oils, poly(phenyl ether)s, silicones, and silicate esters. They are also incompatible with some seal materials since they remove the plasticizer. Because of their high thermal stability, high dielectric constants, and low solvency, they are suitable for use in high-temperature electrical applications. Greases made from these fluids and soap and nonsoap thickeners (copper phthalocyanins) are sometimes used in electrical machinery. The high density of these esters allows their use in submarines [4].

Phosphate Esters [74]

These esters can be expressed by the general formula $(RO)_3PO$ where R can be aliphatic or aromatic. Phosphate esters are usually prepared by the reaction of an alcohol or a phenol with phosphoryl chloride (phosphorus oxychloride) in the presence of a metal catalyst. For making aryl phosphates, phenol, or mixtures of alkylphenols, for example, isobutylphenol and a mixture of *t*-butylphenols, are used as the starting materials. The equation below generically represents the esterification reaction to form the phosphate esters.

$$3ROH + POCl_3 \rightarrow (RO)_3PO + 3HCl$$

Although aromatic esters have excellent lubricating properties, they have a low VI and suffer from poor hydrolytic and thermal stability. The result is the formation of acids, which can attack and corrode vital metal components. This precludes their widespread use in synthetic lubricants. They also possess high solvency and, hence, are incompatible with certain seal materials. Fluid degradation through hydrolysis and oxidation can be minimized by lowering the water level to less than 1000 ppm, keeping the acid number at 0.2 or less, and controlling air entrainment. Vacuum is sometimes used to remove the water and the entrained air. Poor thermal stability of these materials is advantageous in their use as additives in applications such as gear oils, transmission fluids, and hydraulic fluids that require EP/AW performance. In such applications, these materials thermally decompose to phosphoric acid, which can react with the metal surfaces to form a protective metal phosphate film. Inhibited phosphate esters possess excellent oxidation stability and good antiwear properties under critical loading conditions.

Phosphate esters are primarily used in fire-resistant synthetic fluids and aircraft hydraulic fluids. These are tertiary esters of *ortho* phosphoric acid, $O=P(OH)_3$, and may be triaryl, trialkyl, and alkyl/aryl. At present, triaryl phosphates are the most significant commercial fire-resistant products [74]. All three organic groups in the phosphate ester may be the same, as in tricresyl or trixylenyl phosphate, or may be different, as in *iso*-propylphenyl diphenyl phosphate or cresyl diphenyl phosphate. Of the trialkyl phosphate esters, tri-butyl phosphate is the most important of the synthetic base stocks. Most are used in aircraft hydraulic fluids. Dibutyl phenyl phosphate, also used in an aircraft hydraulic fluid, is the most important of the alkyl/aryl phosphate esters. Products may be either mixtures of the phosphate esters resulting directly from the manufacturing process or are obtained by post-blending with additives. The key properties of these materials include excellent fire resistance, outstanding antiwear properties, excellent oxidation stability, high resistance to shear-related viscosity loss, and low foaming tendency.

For aviation use, tert-butylated triphenyl phosphate is the ester of choice. For most applications, water-based flame retardant fluids are used. However, these cannot be used in systems where fluid temperature exceeds 60°C, or the pres-

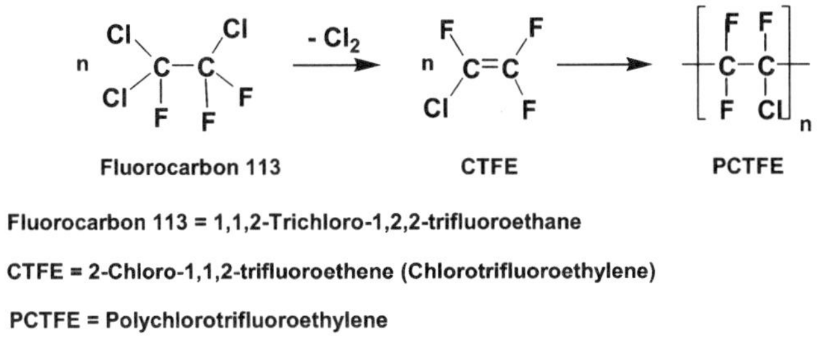

Fig. 3.39—Polychlorotrifluoroethylene synthesis.

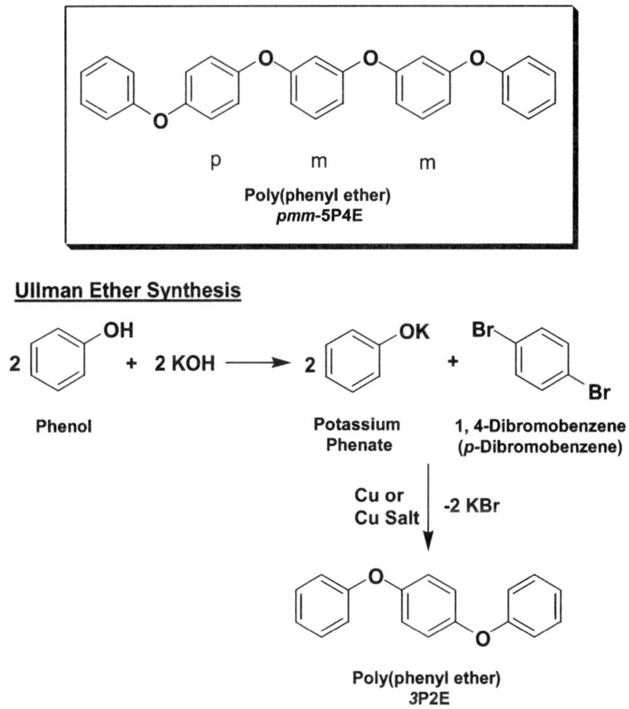

Fig. 3.40—Poly(phenyl ether) structure.

sure of the system or other aspects of its design require a fluid with superior lubricating properties as well. That is where phosphate esters are used. Phosphate esters are also used in circulating oils, turbine oils, metalworking fluids, and industrial lubricants. The reason for using phosphate esters in these applications is because of their low flammability [74]. These fluids *will* burn if sufficient amount of heat and flame are applied. Their autoignition temperature is over 1000°F (537.5°C), but they do not support combustion. Hence, when the ignition source is removed, they self extinguish. However, when they burn, the result is the formation of toxic gases which pose a serious health hazard. The ignition-related risk is the most common in plants where metals are smelted or hot-worked, for example, those involved in aluminum die casting, steel production, and metal quenching [136]. Phosphate esters have a high bulk modulus, giving extremely fast response in electrohydraulic servo systems. These systems are commonly used where precise control is required, such as turbine speed control.

The use of triaryl phosphates in aircraft applications is covered by the military specification MIL-H-19457, which is provided in Table 3.10 [137]. Lubricating properties, corrosion resistance, viscosity control, and the stability of such fluids is proven to be comparable to those of the petroleum-derived fluids. Their major drawback is their aggressiveness to paints and adhesives, insulation in electric cables, and gasket and seal materials. Hence, it is important to use materials that are resistant to attack by the phosphate esters. The U.S. Military specifies paints for the exterior surfaces of the hydraulic systems, components on ship structure and decks, and gasket and seal materials that can come in contact with these fluids without harm. A number of commercial phosphate materials that do not conform to the MIL-H-19457 specification are also being marketed.

Phosphate esters may contain components that are toxic; hence proper precautions must be taken during handling and use. Tri-*ortho*-cresyl phosphate (TOCP) was previously identified as one of the neurotoxins of concern. Hence, today's triaryl phosphates consist of 98 % *meta*- and *para*-isomers and are free of TOCP.

Because of the listed drawbacks, these fluids are being replaced in some applications with silicon-based fire-resistant fluids. While silicone fluids do not possess the detrimental properties of the phosphate ester fluids and provide excellent fire protection, they lack the ability to adequately protect against corrosion and provide lubrication.

Halogenated Hydrocarbons [74]

Chlorotrifluoroethylene oligomers are the most commonly used halogenated hydrocarbon lubricants. These are made from chlorotrifluoroethylene via free radical-initiated polymerization to yield a homo polymer of structure shown in Fig. 3.39. Chlorotrifluoroethylene is produced commercially by dechlorination of 1,1,2-trichloro-1,2,2-trifluoroethane (Fluorocarbon 113). This material has been used as a cold degreasing agent, dry cleaning solvent, refrigerant, blowing agent, chemical intermediate, fire extinguisher fluid, and the drying agent. Because of the negative impact of the chlorofluorocarbons (CFCCs) on the earth's ozone layer, its use is being minimized. Fluorocarbon 113 is dechlorinated to form chlorotrifluoroethylene (CTFE) by using zinc in methanol or by gas phase dechlorination by the use of the aluminum fluoride-nickel phosphate catalyst [138].

Polychlorotrifluoroethylene (PCTFE) is nonflammable and is the basis of the nonflammable hydraulic fluid with the military designation MIL-H-53119. While excellent as a fire retarding fluid, it suffers from high cost, aggressiveness to copper and other nonferrous alloys and incompatibility with certain seal materials which are used in some hydraulic systems. PCTFE-derived fluids have high specific gravity (1.87–1.96 at 38°C, or 100°F) and low vapor pressure (0.07–2.2 mm Hg at 93°C or 200°F). They are water insoluble, have boiling points of over 260°C, or 500°F, and are thermally stable. Above this temperature, PCTFE depolymerizes and decomposes to yield toxic gaseous products, which include hydrogen fluoride, hydrogen chloride, chlorotrifluoroethylene, and other toxic organic fluorine compounds. It is important to avoid friction or galling in contact with the aluminum or magnesium parts. This is because local heat and pressure may cause detonation, accompanied by extensive decomposition. PCTFE has limited solubility in polyalphaolefins; the blends show excessive foaming and other property changes [139]. PCTFE has an LD_{50} oral (rat) of >9200 mg/kg and LD_{50} dermal (rabbit: >3700 mg/kg),

TABLE 3.11—Physical properties of poly(phenyl ether)s [143].

Poly(phenyl ether)	Appearance	Pour Point °F (°C)	Thermal Stability °F (°C)	Viscosity (cSt) 100°F (38°C)	Viscosity (cSt) 210°F (99°C)
6-Ring 6P5E	Clear Liquid	50 (10)	836 (447)	2000	25
5-Ring 5P4E	Clear Liquid	40 (4.5)	847 (453)	360	13
4-Ring 4P3E	Clear Liquid	10 (−12)	825 (441)	70	6
3- and 4-Ring Oxythio	Hazy Liquid	−20 (−29)	693 (367)	25	4
3-Ring 3P2E	Solid	···	800 (427)	12	3
2-Ring 2P1E	Solid	···	>600 (316)	2.4	1.6

making them practically nontoxic (toxicity rating of 5). Toxicity of PCTFE—oil-based hydraulic fluid, with three parts PCTFE and one part oil, is believed to be due to the conversion of the neutral chlorotrifluoroethylene (CTFE) oligomers into the corresponding halogenated fatty acids.

Because of their resistance to chemicals, nonflammability, high thermal stability, good lubricity, and low compressibility, these materials are used to lubricate equipment in industries, such as aerospace, nuclear, and chemical, where traditional lubricants will not suffice. PCTFE is used in hydraulic fluids for advanced aircraft and armored vehicle turret-and-gun control system. CTFE-based polymers and copolymers are also used in the synthesis of high performance lubricants, plastics, and elastomers, which are marketed by Allied Chemical under the trade name of Halar® [140]. Synthetic chlorotrifluoroethylene derived oils, which can be used in rotary piston and rotary vane pumps, are marketed under the trade name of Halovac™ fluids made by the Halocarbon Corporation [141]. These are sold in three viscosity grades, Halovac™ 100, 125, 190, as the cost effective alternative to PFPE fluids. Their listed attributes include sealing and lubricity required by the mechanical vacuum pumps, low vapor pressure, and chemical stability up to 230–260°C, or 450–500°F. CTFEs are also used in designing health-related chemicals, such as anesthetics, and are used in telomerization with carbon tetrachloride or chloroform for use as inert fluids, hydraulic fluids, or lubricants [142].

TABLE 3.12—ASTM slope values for poly(phenyl ether)s [4].

Poly(phenyl ether)	ASTM Slope 38–99°C	99–205°C	205–315°C	315–370°C
pp-4P3E	···	0.78	0.85	0.96
mp-4P3E	0.86	0.81	0.89	1.01
mm-4P3E	0.84	0.82	0.92	1.06
op-4P3E	0.88	0.87	0.94	1.11
mo-4P3E	0.95	0.88	0.96	1.13
oo-4P3E	···	0.95	0.98	1.22
m-3P2E		0.90	1.09	1.25
mm-4P3E	0.84	0.82	0.92	1.06
mmm-5P4E	0.90	0.77	0.84	0.94
mmmm-6P5E	0.93	0.75	0.80	0.89
mmmmm-7P6E	0.98	0.73	0.76	0.86

Poly(phenyl ether)s [4]

Proper name for phenyl ether polymers is poly(phenyl ether), or polyphenyl polyether but the name polyphenyl ether, (PPE) is widely accepted. Poly(phenyl ether)s contain benzene rings that are joined through ether linkages, usually in the *meta* orientation. These materials are obtained by the reaction of an alkali metal phenate with a halogenated ben-

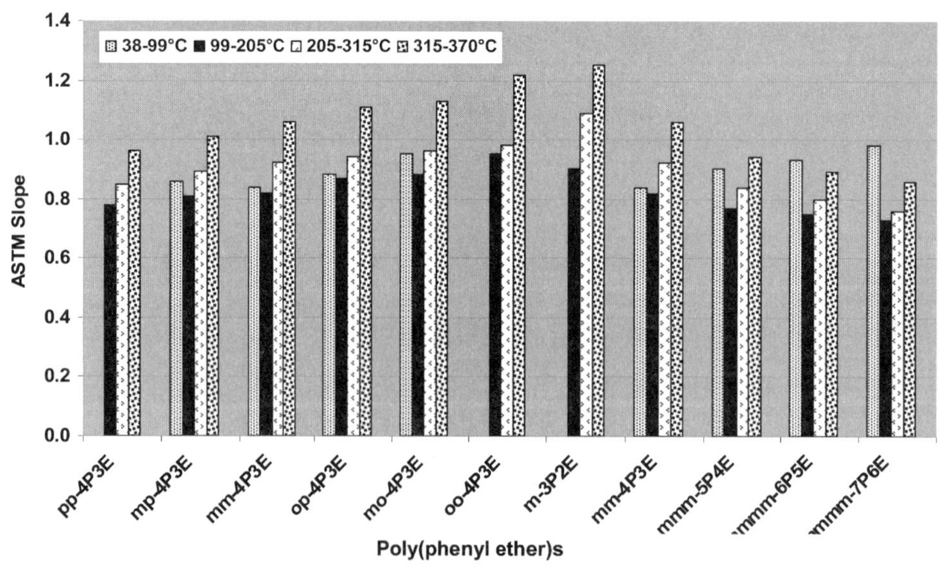

Fig. 3.41—ASTM slope data for poly(phenyl ether)s as a function of temperature.

TABLE 3.13—Fluid stability test results [145].

Parameter	Base Fluid 600°F (316°C)	Base Fluid+Inhibitor 600°F (316°C)	Base Fluid 650°F (343°C)	Base Fluid+Inhibitor 600°F (316°C)
% Change in Viscosity at 212°F (100°C)	32	17	1809	22

zene in the presence of copper catalysts (Ullman Ether Synthesis). Ullman Synthesis for a 3-ring PPE is shown in the lower half of Fig. 3.40. PPEs of up to six phenyl rings are commercially available. They are characterized by indicating the substitution pattern of each ring, followed by the number of the phenyl rings and the number of the ether linkages. Thus, the structure in Fig. 3.40 is identified as *pmm*-5P4E, indicating a total of five rings and four ether linkages, the di-substituted rings having *para*, *meta*, and *meta* orientation. Longer chain analogues with up to ten benzene rings are also available. As stated earlier, *meta* substitution of the aryl rings in these materials is the most common.

Typical physical properties of poly(phenyl ether)s are provided in Table 3.11 [143]. The important attributes of these materials include their thermal and oxidative stability and the stability against ionizing radiation. However, they have the disadvantage of having high pour points; those containing two and three benzene rings are actually solids at room temperatures. Their melting points are lowered if they contain more *m*-phenylene rings, alkyl groups, or are mixtures of isomers. PPEs that contain *o*- and *p*-substituted rings have the highest melting points.

These fluids have excellent high temperature properties and good oxidation and radiation stability. While the presence of the aromatic rings lowers their VT properties somewhat, they are still reasonable. Within the class, *p*-ethers are the best, *m*-ethers are in the middle, and *o*-ethers are the worst, with respect to the VT properties, as shown by the ASTM slopes [4]. These data are provided in Table 3.12 and plotted in Fig. 3.41. The ASTM slope is used to correlate viscosity and temperature. High values indicate large viscosity changes with temperature and low values indicate small viscosity changes with temperature. The ASTM slope essentially indicates the same parameter as the VI, but for temperature ranges other than for 40 and 100°C, typically used to define the VI. Lower slope values are better than the higher slope values.

Data in the table and on the left side of the figure show the ASTM slopes to increase with increasing molecular weight; their value for the mixed ring substitutions falling between those with uniform ring substitutions. For example, ASTM slope values for *mp*-4P3E fall between those of the *pp*-4P3E and *mm*-4P3E and the values for *mo*-3P4E fall between those for *mm*-4P3E and *oo*-4P3E. For *m*-substituted phenolic polyethers, shown on the right side of the figure, the ASTM slope values decrease as the number of phenolic groups increase. With respect to volatilities, *p*-derivatives have the lowest volatilities and *o*-derivatives have the highest volatilities. The opposite is true for the flash points and fire points. Spontaneous ignition temperatures of the poly(phenyl ether)s lie between 550 and 595°C, alkyl substitution reduces this value by ~50°C. PPEs swell the common seal materials [4].

Oxidation stability of the unsubstituted derivatives is quite good, partly because they lack easily oxidizable carbon-hydrogen bonds. However, alkylated diphenyl oxides are highly susceptible to oxidation. Their decomposition does not generate any residue or corrosive materials; therefore their viscosity increase due to oxidation is minimal. Oxidative stability of the alkylated derivatives can be improved by the use of the oxidation inhibitors. See Table 3.13 for the stability data [144]. Thermal decomposition temperature, as measured by the isoteniscope procedure is between 440 and 465°C. This procedure, described as ANSI/ASTM Method

TABLE 3.14—Poly(phenyl ether) film strength test results [145].

Shell Four-Ball Wear Test 5-Ring PPE, 400°F (204°C), 600 r/min, 1 h Duration		Ryder Gear Scuff Test PPE Lubricants	
Load	Scar Diameter	Temperature	Rating (psi)
10 kg	0.80 mm	160°F (71°C)	2450±300
30 kg	0.89 mm	400°F (204°C)	1000±150
50 kg	1.13 mm		

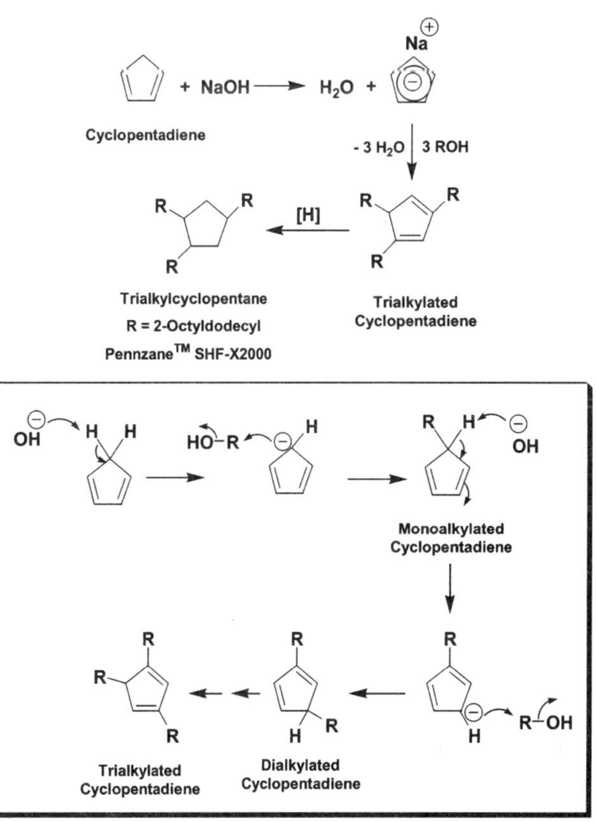

Fig. 3.42—Synthesis mechanism of polyalkylated cyclopentanes.

D2879, measures vapor pressures of liquids with vapor pressures of 0.1 to 100 kilopascals (kPa) (0.75 to 750 torr). Methyl, ethyl, and isopropyl substituents reduce the decomposition temperatures to about 380 °C and n-propyl and higher alkyl groups drop the decomposition temperature to 330–350 °C [4]. PPEs have high radiation resistance. Ionizing radiation, the same as in other cases, increases viscosity, acidity, evaporation loss, corrosivity, and coke formation. It also lowers the PPEs' flash and ignition points. Of all classes of synthetic lubricants, poly(phenyl ether)s are the most radiation resistant.

High-temperature lubricating properties of the poly(phenyl ether)s, measured in the laboratory at 200–300 °C, are comparable to those of the mineral oils and synthetic esters, see Table 3.14 [144]. Alkylsubstituted poly(phenyl ether)s are somewhat better [4]. However, in real life applications, their load-carrying capacity and wear resistance is inferior to those of the mineral oils [4]. Poly(phenyl ether)s are used as base oils for use in aviation gas turbine lubricants, ultra-high vacuum diffusion pump fluids, electronics lubricants, high-temperature and radiation-resistant hydraulic fluids, and lubricating greases [143–145].

Alkylated Cyclopentanes [79]

Polyalkylated or multiply-alkylated cyclopentanes, or MACs, were developed recently for use in lubricants. They are made by the reaction of cyclopentadiene, in the form of dicyclopentadiene, with an alcohol or an alkyl halide in the presence of a strong base [146]. The resulting alkylated cyclopentadiene is catalytically hydrogenated to yield a mixture containing di-, tri-, tetra-, and penta-alkylated cyclopentanes. Fractionation of the mixture results in the isolation of the components suitable for use as base oils [147–151]. The reaction scheme for the synthesis of these hydrocarbons is provided in Fig. 3.42, along with the possible mechanism. It is important to note that while we stopped the mechanism at tri-alkylated cyclopentadiene, theoretically five alkyl substituents on cyclopentadiene ring are possible.

A number of MACs have been developed, out of which three have been commercialized under the trade name of Pannzane®. Pennzane® X-1000 has a viscosity of ~8 cSt at 100 °C, Pennzane® 2000 has 100 °C viscosity of 14.5 cSt, and Pennzane® X-3000 has 100 °C viscosity of 220 cSt. [150]. Pennzane® X-1000 remains fluid between -50 and -60 °C, compared to Pennzane® X-2000's low temperature rating of -45 °C. While chemically Pennzane® X-2000 is a tri-2-octyldodecyl substituted cyclopentane [151], Pennzane® 1000 is a dialkylated cyclopentane.

MACs are synthetic hydrocarbons that combine excellent wear protection and low vapor pressure, two of the most critical properties needed for lubricants to be used in space and other vacuum environments [152]. Of the three, Pennzane® 2000 is the oldest product, which has been extensively used in satellites, space vehicles, and space suits [153]. It has a vapor pressure of about 1×10^{-12} torr at 25 °C. It is also used in the manufacture of the semiconductors, precision instruments, and other clean-room applications. Low vapor pressure in vacuum applications is important so that the loss of lubricant due to out-gassing and possible contamination of the nearby critical components, such as semiconductor wafers, sensors, and optics, is minimized. Pennzane® X-1000, because of its excellent low-temperature viscometrics is used to lubricate mechanisms with very low starting torque, even at very low temperatures; however, its higher volatility precludes its use in many high vacuum applications. Pennzane® X-3000, the highest viscosity product, offers superior wear protection, especially in applications that experience higher loads. Its high viscosity makes it suitable for use as a damping fluid to reduce noise and undesired motion. Nye Lubricants, Inc. markets these products under the name of Nye Synthetic oils [150]. Pennzane® hydrocarbons have a functional temperature range of -50 °C to 125 °C. Viscometric properties are typical for hydrocarbons and similar to those of the PAOs. The main distinguishing characteristic of MACs is that they are essentially mono-disperse and their volatility is therefore substantially lower than that of the mineral oil or the PAO of similar viscosity. Viscosity and viscosity index can be adjusted by the appropriate selection of the chain length and the degree of branching of the alcohol raw material, and by the degree of substitution obtained. MACs are fully compatible with conventional additives and many other types of base stocks. These materials have a high surface tension, approximately 32 dynes/cm, which makes them good lubricants since they make more durable surface films. Their properties can be considerably improved by the use of the conventional additives, such as extreme pressure agents, antiwear agents, and oxidation inhibitors. Comparative properties of these lubricants with respect to their use in space applications are discussed in a NASA report [153]. Pennzoil also markets Pennzane®-based synthetic engine oils. The claimed benefits of such oils include extra engine protection against deposits and wear resulting from extreme driving conditions, good broad-temperature performance, and low oil consumption.

Cyclohexane Derivatives [79]

Unlike alkylated cyclopentanes, which were devised primarily for use in high-temperature, high-vacuum applications, cyclohexane derivatives were developed for use in automotive and industrial transmission applications, requiring high traction. Previously, a number of products with this chemistry were available, but today only 2,4-dicyclohexyl-2-methylcyclopentane is commercially available [74]. It is made by the catalytic hydrogenation of α-methylstyrene dimer.

Most of the properties of this hydrocarbon are similar to those of the other hydrocarbon materials of analogous molecular weight, except that it exhibits high traction coefficients under high pressures that occur during EHD contact.

Petroleum Base Stocks Versus Synthetic Base Fluids

Typically, synthetic fluids are superior to petroleum-refined base oils with respect to thermal stability, oxidation resistance, viscosity index, low-temperature fluidity—their pour points are -40 °C (-50 °F) or lower, coefficient of friction, and volatility. The advantages of synthetics are mostly realized at extreme temperatures. For example, low pour point gains importance at very low temperatures since the oil stays fluid at these temperatures and provides adequate lubrication. High viscosity index, low volatility, and high thermal and oxidation resistance are important at high tempera-

TABLE 3.15—Comparison of synthetic fluid properties.

Fluid Property	Mineral Oil	Di-ester	Phosphate Ester	Inhibited Esters	Methylsilicone	Phenylmethyl Silicone	Chlorinated Phenylmethyl Silicone
Maximum temp. (°C) – no air	200	250	300	110	220	320	305
Maximum temp. (°C) - air	150	210	240	110	180	250	230
Maximum temp. (°C) due to decrease in viscosity	200	150	180	100	200	250	280
Maximum temp. (°C) due to increase in viscosity	0 to -50	-35	-65	-55	-50	-30	-65
Density (g/ml)	0.88	0.91	1.01	1.12	0.97	1.06	1.04
Viscosity index	0 to 140	145	140	0	200	175	195
Flash point (°C)	150 to 200	230	255	200	310	290	270
Spontaneous ignition temp.	Low	Low	Medium	Very high	High	High	Very high
Thermal conductivity (W/M °C)	0.13	0.15	0.14	0.13	0.16	0.15	0.15
Thermal capacity (J/kg °C)	2,000	2,000	1,700	1,600	1,550	1,550	1,550
Bulk modulus	Fairly high	Medium	Medium	Medium	Very low	Low	Low
Boundary lubrication	Good	Good	Good	Very good	Fair (Poor for steel on steel)	Fair (Poor for steel on steel)	Good
Toxicity	Slight	Slight	Slight	Some Toxicity	Nontoxic	Nontoxic	Nontoxic
Suitable rubbers	Nitrile	Nitrile, silicone	Silicone	Butyl, EPR	Neoprene, Viton	Neoprene, Viton	Viton, fluoro-silicone
Effect on plastics	Generally slight	May act as plasticizers	Powerful solvent	Slight - may leach out Plasticizers	Slight - may leach out Plasticizers	Slight - may leach out Plasticizers	
Resistance to attack by water	Excellent	Good	Good	Fair	Very good	Very good	Good
Resistance to chemicals	Very resistant	Attacked by alkali	Attacked by alkali	Attacked by many chemicals	Attacked by strong alkali	Attacked by strong alkali	Attacked by alkali
Effect on metals	Non-corrosive when pure	Slightly Corrosive to non-ferrous metals	Corrosive to non-ferrous metals when hot	Enhanced corrosion in presence of water	Non-corrosive	Non-corrosive	Corrosive in presence of water to ferrous metals
Cost (relative to mineral oil)	1	4	6	6	15	25	40

tures. Oil maintains reasonable viscosity, to provide effective lubrication, its volatility-related losses are low, and there is minimum thermo-oxidative breakdown. An additional but important advantage of the synthetic fluids is the flexibility to create structures that provide optimal performance in an application. A number of other properties are also considered while selecting a suitable fluid. These include demulsibility, lubricity, rust and corrosion protection, antiwear properties, elastomer and other materials compatibility (seals and paints). The biggest drawback of synthetics is their high cost of production. As a result, synthetic-mineral oil blends are often used as base fluids to formulate lubricants for some applications. In this case, the idea is to take advantage of the superior properties of the synthetics but at a reasonable cost. Of course, in such cases compatibility between the fluids being mixed becomes an important issue since not all synthetics are compatible with one another or with the mineral oils.

Modern mechanical equipment places increased demand on the lubricant because of the extreme operating environments. In some such applications, for example those pertaining to aviation, mineral oil-based lubricants are incapable of delivering the necessary performance, but synthetic base stocks do. This is because synthetics possess a number of properties that are better than those of the mineral oils, or are unique. Such properties include:

- Better low- and high-temperature properties, such as viscosity, pour point, viscosity index, and volatility.
- Good thermal and oxidative stability.
- Low flammability or nonflammability.
- Ionizing radiation stability.
- Improved lubricity, usually at very high and very low temperatures.
- Biocompatibility, i.e., reduced environmental impact, biodegradability, and reduced toxicity.

Base stocks are used either directly or as blends to formulate lubricants and their properties largely determine the properties of the lubricant. A number of properties for some of the commonly used synthetic base stocks are listed in Table 3.15, along with the relative cost. The examination of the data suggests that while some base stocks are better than the others in some aspects, none is perfect. Data from Table 3.15 are consolidated with additional data and presented in Tables 3.15 and 3.16. Table 3.16 visually compares the overall attributes of the synthetics among one another [4]. Table 3.17 presents the physical properties of the common base stocks, both mineral and synthetic, which are a primary consideration while selecting a base stock for formulating a lubricant for a particular application. Information provided in these tables should facilitate the base stock selection process [44,318]. In terms of the cost and the overall performance, mineral oil base stocks are the most attractive; however, in

TABLE 3_15a—Comparison of synthetic fluid properties (Continued)

Fluid Property	Mineral Oil	Polyglycol Inhibited	Perfluorinated Polyester	Chlorinated Diphenyl	Silicate Ester or Disiloxane	Poly(phenyl ether)	Fluorocarbon
Maximum Temp. (°C) – no air	200	260	370	315	300	450	300
Maximum Temp. (°C) - air	150	200	310	145	200	320	300
Maximum temperature (°C) due to decrease in viscosity	200	200	300	100	240	150	140
Maximum temperature (°C) due to increase in viscosity	0 to -50	-20	-60	-10	-60	0	0 -50
Density (g/ml)	0.88	1.02	1.88	1.42	1.02	1.19	1.95
Viscosity index	0 to 140	160	100-300	-200 to +25	150	-60	-25
Flash point (°C)	150 to 200	180		180	170	275	None
Spontaneous ignition Temp.	Low	Medium	Very high	Very high	Medium	High	Very high
Thermal conductivity (W/M °C)	0.13	0.15		0.12	0.15	0.14	0.13
Thermal capacity (J/kg °C)	2,000	2,000		1,200	1,700	1,750	1,350
Bulk modulus	Fairly high	Medium	Low	Medium	Low	Medium	Low
Boundary lubrication	Good	Very good	Poor	Very good	Fair	Fair	Very good
Toxicity	Slight	Believed to be low	Low	Irritant vapor when hot	Slight	Believed to be low	Nontoxic unless overheated
Suitable rubbers	Nitrile	Nitrile	Many	Viton	Viton, nitrile, fluoro-silicone	(None for very high temperatures	Silicone
Resistance to attack by water	Excellent	Good	Very good	Excellent	Poor	Very good	Excellent
Resistance to chemicals	Very resistant	Attacked by oxidants	Very good	Very resistant	Generally poor	Resistant	Resistant but attacked by alkali and amines
Effect on metals	Non-corrosive when pure	Non-corrosive	Removes oxide films at high temperatures	Some corrosion of copper alloys	Non-corrosive	Non-corrosive (unsafe with aluminum and magnesium)	Non-corrosive when pure
Cost (relative to mineral oil)	1	4	500	10	8	100	300

certain cases the cost is irrelevant and there is merit in using synthetic base stocks. Some of these applications are listed in Table 3.18. Incidentally, these tables contain data on some fluids that we did not include in our discussion either due to their use in very specialized applications or their low-volume manufacture. In addition to the desirable base stock properties discussed above, there are other base oil properties that must be considered when designing a lubricant for specialty applications.

Viscosity-temperature and Viscosity-pressure Behavior

Proper viscosity of a lubricant is critical to its ability to perform the lubrication function. Typically, the lubricant viscosities at 40°C (104°F) and 100°C (212°F) are reported. Historically, the temperatures of 100°F (37.8°C) and 210°F (99°C) were of interest because the kinematic viscosities at these temperatures were previously used to calculate the viscosity index. These temperatures were rounded up to 40°C and 100°C during transition to the metric system. Incidentally, ISO grades for industrial lubricants have always been based upon lubricant viscosity at 40°C. Previously, the oil's VI range of between 0 and 100 was considered relevant, but the VIs of today's oils can be less than 0 and as high as 300, see Table 3.15. It is important to note that while the VI is a good criterion for an oil's viscosity-temperature relationship, it may or may not predict viscosity outside this range. Ultimately, it is the viscosity of the oil at the operating tem-

TABLE 3.16—Lubricant base stock properties [4].

Property	Mineral Oil	PAO[a]	Alkylated Aromatics	Aliphatic Di-ester	Polyol Ester	Phosphate Ester	PAGs[b]	Silicones	Silicate esters	Fluoro-carbons	Poly(phenyl ether)s
Fluidity Range	0	+	+	++	++	0	+	+++	...	...	...
Viscosity Temperature (Viscosity Index)	0	+	0	+++	++	...	+	+++	+++	0	...
Low Temperature Fluidity	...	+	+	+	+	0	+	+	0	+	...
Oxidation Stability (Inhibited)	0	++	0	++	0	+	...	++	++	+++	++
Hydrolytic Stability	+++	+++	+++	0	0	0	+	+	...	++	+++
Thermal Stability	0	0	0	+	+	0	+	++	+	++	+++
Mineral Oil Compatibility	...	+++	+++	+	0	0	...	...	0	...	+
Additive Solvency	+++	+	+++	++	++	+	0	...	...	...	...
Volatility	0	+	+	+++	+++	+	+	+	+	0	+
Rust Control (Inhibited)	+++	+++	+++	0	0	0	+	+	...	...	...
Boundary Lubrication	+	+	+	++	++	+++	+	0	0	+++	+++
Fire Resistance	...	...	...	0	0	+++	0	0	0	+++	0
Elastomer Swell (Buna)	Low	None	Low	Medium	High	High	Low	Low	Low	Medium	Low
Relative Cost	$	$$	$$	$$	$$	$$	$$	$$$	$$$	$$$$	$$$

Note: − Poor + Good 0 Moderate ++ Very Good +++ Excellent
[a]Polyalphaolefin.
[b]Poly(alkylene glycol).

TABLE 3.17—Physical properties of lubricant base stocks [44,318]. Reprinted with permission from Lubrizol Corporation.

Type	Kinematic Viscosity, cSt		Dynamic Viscosity, at −40°C (cP)	Viscosity Index	Pour Point (°C)	COC Flash Point (°C)
	40°C	100°C				
(A) Mineral Oils						
90 Neutral	17.40	3.68	Solid	92	−15	190
100 Neutral	20.39	4.11	Solid	101	−13	192
200 Neutral	40.74	6.23	Solid	99	−20	226
350 Neutral	65.59	8.39	Solid	97	−18	252
650 Neutral	117.90	12.43	Solid	96	−18	272
150 Bright Stock	438.00	29.46	Solid	95	−18	302
(B) Synthetic Base Stocks						
Alkylated aromatics	26.84	4.99	9047	119	−18.3	224
Olefin oligomers	16.77	3.87	2371	126	−43.0	221
Dibasic acid esters (Dioctyl sebacate)	18.2	...	3450	176	−15.6	232
Trimethylolpropane esters (C_7 Acid)	13.94	3.4	2360	...	−18.3	232
Poly(alkylene glycol)s	45.69	...	...	150	...	177

TABLE 3.18—Suitable uses of synthetic base stocks.

Polyalphaolefins	Polyisobutylenes	Poly(alkylene glycol)s	Synthetic Esters
♦ Engine oils ♦ Compressor oils ♦ Hydraulic fluids ♦ Gear oils ♦ Greases	♦ Two-stroke cycle engine oils ♦ Ethylene gas compressor lubricants ♦ Metalworking fluids ♦ Greases ♦ Wire rope lubricants	♦ Worm gear lubricants ♦ Fire-resistant hydraulic fluids	♦ Aviation turbine oils ♦ Engine oils ♦ Compressor lubricants ♦ Gear and hydraulic oils ♦ Refrigeration oils ♦ Fast biodegradable lubricants
Phosphate Esters	**Perfluoroalkyl Ethers**	**Polysiloxanes**	**Poly(phenyl ether)s**
♦ Fire resistant hydraulic fluids (hydraulic fluids and bearing lubricants for use in power stations) ♦ Gas turbine oils	♦ Extreme fire-resistant hydraulic fluids ♦ Nuclear reactor lubricants ♦ Specialty greases ♦ Space applications	♦ High temperature hydraulic fluids ♦ Greases ♦ Specialty lubricants for use in chemical and electric environments	♦ Special high temperature lubricants ♦ Nuclear reactor lubricants

TABLE 3.19—Viscosity-temperature parameters of industrial fluids.

Industrial Oil ISO VG 32	Viscosity at 40°C (cSt)	Viscosity at 100°C (cSt)	Viscosity Index	Viscosity-temperature Coefficient
Machine Oil	30.4	4.8	58	0.842
Turbine Fluid	32.0	5.4	102	0.831
Hydraulic Fluid	30.4	6.1	154	0.799

TABLE 3.20—Extrapolated viscosities at elevated temperatures [154].

	Measured Viscosity, cSt			% Deviation from Extrapolated Viscosity[a]	
Test Fluid	40°C	254°C	375°C	254°C	375°C
Methyl, phenyl silicone	44.4	3.25	1.46-1.37	−17	−36
Chlorinated silicone	59.9	4.52	2.2-2.0	−14	−25
Tetra-2-ethylhexyl silicate	6.79	0.59	...	−10	...
Pentaerythritol ester	22.4	0.83	0.42-0.59	−15	−27
Naphthenic bright stock	201	1.16	0.57-0.66	−4	−8
Naphthenic neutral	44.6	0.87	0.45	−11	−18
Chlorinated aromatic hydrocarbon	42.5	0.55	...	+15	...
Tricresyl phosphate	38.3	0.76	...	+3	...

[a]Based on 40 and 100°C viscosities.

TABLE 3.21—Extrapolated viscosities at low temperatures [154].

Test fluid	Cloud Point (°C)	Pour Point (°C)	Measured Viscosity, cSt 40°C	-18°C	-40°C	% Deviation from Extrapolation -18°C	-40°C
Naphthenic Oil	<- 23	-23	201	156,000	—	-2	—
Paraffinic Oil+ Pour Point Depressant	- 4	-23	40.4	9,550	—	-300	—
De-waxed Paraffinic Oil	- 14	-18	45.1	10,800	—	-210	—
Oil + Pour Point Depressant	- 14	-29	46.3	4,010	—	+38	—
Naphthenic Oil + Polymethacrylate Viscosity Modifier	<- 54	—	33.0	655	7,880	+61	+250
Naphthenic Oil + Polybutene Viscosity Modifier	<- 54	—	33.3	817	10,700	+49	+170
Spec. MIL-H-5606 Hydraulic fluid	<- 54	—	14.2	100	183	+29	+110
Di-capryl phthalate	<-40	—	26.8	—	31,800	—	-56
Pentaerythritol ester	<-54	—	22.4	—	5,310	—	-21
Tri-*n*-butyl Phosphate	<-46	—	2.66	—	46.9	—	-42
Di-*n*-hexyl carbonate	<-40	—	2.93	—	59.1	—	-26
Tri-2-ethylhexyl borate	<-40	—	6.28	—	433	—	-12
Poly-α-olefin	<-54	—	20.9	—	44,000[b]	—	45[b]
Super-refined mineral oil	<-54	—	14.4	—	21,000[b]	—	53[b]

[a] Based on 40 and 100°C viscosities.
[b] Values at -54°C.

TABLE 3.22—Viscosity indices and pressure coefficients of synthetic base fluids [79].

Base Fluid	Viscosity Index	VPC-α(GPa^{-1})	VPC-α(10^3 bar^{-1})
Naphthenic Mineral	0–80	18.0–36.0	1.8–3.6
Faraffinic Mineral	80–120	18.0–23.0	1.8–2.3
Polyalphaolefin (PAO)	120–150	10.5–12.6	1.05–1.26
Di- and Tri-ester	50–150	5.3–19.9	0.53–1.99
Polyol Ester	50–170	5.3–21.1	0.53–2.11
Poly(alkylene glycol) (PAG)	150–280	7.7–19.1	0.77–1.91
Alkylbenzene	<0–110	11.8–33.4	1.18–3.34

perature that is of interest to a formulator, or a user. At high temperatures, lubricant or base oil decomposition may occur, thereby unexpectedly increasing or decreasing viscosity. Also, it is important to note that the increase or decrease in the operating temperature in equipment may not be consistent or it may fluctuate, thereby making it difficult to predict the viscometric requirements of the equipment. The same problem can be encountered if the system being lubricated dissipates heat at a different rate with a change in its operating parameters.

Most industrial mineral lubricating oils have a VI between 55 and 100, but VI can be up to 175 for the high VI oils. Viscosity Indices of various industrial oils are shown in Table 3.19. A less arbitrary indication of the change in viscosity with temperature is the viscosity temperature coefficient (VTC). For 40 to 100°C range, it can be calculated by the following expression.

$$\mathrm{VTC} = \frac{\text{Viscosity at } 40\,^\circ\mathrm{C} - \text{Viscosity at } 100\,^\circ\mathrm{C}}{\text{Viscosity at } 40\,^\circ\mathrm{C}}$$

$$= 1 - \frac{\text{Viscosity at } 100\,^\circ\mathrm{C}}{\text{Viscosity at } 40\,^\circ\mathrm{C}}$$

The lower the value of the coefficient, the higher is the VI. The coefficient for mineral oils can vary by a factor of 10 depending on the temperatures.

In Chapter 1 on Lubrication Fundamentals, we stated that extrapolation of the viscosity-temperature curve outside of the 40°C and 100°C range is not advisable because of the unexpected transitions that may occur. This is because viscosity index, as measured by ASTM D2270, is based upon Walther plots (ASTM D341), which allow extrapolation of viscosities based on two measured values. The extrapolation range for mineral oils is -7°C to 177°C (20 to 350°F). Incidentally, the lower temperature limit of -7°C is around the cloud point of napthenic oil −14°C for paraffinic oil and −23°C for the napthenic oil, as shown in Table 3.21. Synthetics also follow the Walther relationship but their extrapolation range is -18°C to 175°C (0°F to 347°F) [154]. Since the operating range of the synthetics is well beyond these temperatures, extrapolation outside this range will yield erroneous values. Table 3.20 provides extrapolation of viscosities at high temperatures and Table 3.21 provides extrapolation at low temperatures. Data in the tables indicate the following:

1. Deviations from the measured values increase as the extrapolation temperature is further away from the extrapolation temperature range limit of -18 to 175°C.
2. At elevated temperatures, the discrepancy is small for mineral oil stocks.
3. At low temperatures, the discrepancy is high for paraffinic mineral oils.
4. Deviation for mineral oils at low temperatures is positive but for synthetic oils it is negative. This implies that for the mineral oils, the Walther equation underpredicts viscosities and for the synthetics, it overpredicts them. Incidentally, silicones and poly(glycol ether)s, not included in the table, like mineral oils, have measured viscosities that are higher than predicted.

While formulating a lubricant, the influence of pressure on viscosity is usually neglected. This is reasonable for low pressures, but for pressures greater than 1000 psi (68 atmospheres, 6895 kPa), the absolute viscosity of a fluid may be a strong function of pressure. Viscosity-pressure coefficient (VPC) measures the rate of viscosity-increase with increased load (pressure). Viscosity increases with pressure because the molecules of the fluid come closer together, which increases their intermolecular van der Waals type interactions.

TABLE 3.23—Viscosity-pressure coefficients for synthetic base fluids and expected increase in viscosity at 2000 bar (200 mPa) [4].

Base Fluid	VPC-$\alpha_{25^\circ C}\cdot 10^3$ bar^{-1}	Viscosity at 25°C	Viscosity at 80°C
Paraffinic Mineral	1.5–2.4	15–100	10–30
Naphthenic Mineral	2.5–3.5	150–800	40–70
Aromatic Solvent Extracts (Bright Stocks)	4.0–8.0	1000–200,000	100–1000
Polyalphaolefin (PAO)	1.5–2.0	10–50	8–20
Polyester (Branched Diesters)	1.5–2.5	20–50	12–20
Poly(alkylene glycol) (PAG)	1.1–1.7	9–30	7–13
Silicone (Aliphatic)	1.4–1.8	9–16	7–9
Silicone (Aromatic)	3.0–5.0	600	...
Chlorinated Paraffin (Depends upon Halogen Content)	0.7–5.0	5–2000	...

TABLE 3.24—Comparison of low-temperature properties.

Fluid	Viscosity at 100°C (cSt)	Viscosity at −18°C (cP)	Pour Point (°C)	VI	Flash Point (°C)
Mineral Oil[a]	4.02	610[b]	−18	98	192
Dibasic Acid Ester	3.22	<500	−71	147	230
Polyol Ester	3.40	<550	<−57	130	232
Polyalphaolefin	3.93	<550	<−54	127	221
Mineral Oil[c]	6.12	2200[b]	−15	95	226
Alkylated Aromatic	4.95	750	−57	107	224
Polyalphaolefin	5.83	700	−57	138	243
Dibasic Acid Ester	5.41	1040	−54	134	232

[a]100 Neutral solvent-extracted mid-continent mineral oil.
[b]Contains 0.5 % pour point depressant.
[c]200 Neutral solvent-extracted mid-continent mineral oil.

Viscosity at high pressures is measured by flow through pressurized capillary tubes, or a ball falling down a pressurized tube. Viscosity-pressure coefficient, α, is the slope of the line obtained by plotting the log of viscosity against pressure. The higher the temperature, the smaller is the viscosity increase due to pressure (see Fig. 3.11). The coefficient can be represented by the following equation.

$$\alpha = \frac{1}{\eta_p} \cdot \left(\frac{d\eta_p}{dp}\right)_T$$

The unit of pressure-viscosity coefficient, α, is the reciprocal of pressure. The SI units are 1/Pa or $m^2\ N^{-1}$. A variety of units are used to describe pressure. Those that are commonly used are interrelated by the following relationship.

Units of pressure: 1 Atmosphere = 760 Torr = 101.325 Kilo Pascals (kPa) = 101,325 Newton per square metre (N/m^2) = 14.696 psi = 1.01325 bar. For conversion between different units of pressure, use the link in Ref [155] or search under PRESSURE using wikipedia.org.

The viscosity-pressure coefficient is specific for a given material because it depends upon its molecular structure or the composition. For base fluids that predominantly contain paraffinic structures, the pressure-viscosity curve levels off with increasing pressure. In the case of fluids that contain cyclic structures, such as naphthenic oils or alkylaromatics, the slope of the curve continues to increase with an increase in pressure [4]. Viscosity-pressure coefficient increases with viscosity, but decreases with temperature [156]. Table 3.22 provides viscosity indices and viscosity-pressure coefficients of various mineral and synthetic base fluids [79]. Two α values are provided in the table. They differ from each other by a factor of 10. The reason for doing so is to facilitate comparison with the data in Table 3.23 [4]. Data in Table 3.22 show naphthenic mineral oil and alkylbenzene to have high α values. This implies that these fluids are good lubricants for high-pressure applications, such as bearings and gears. Data in Table 3.23 clearly support this conclusion. Fluids that contain aromatic or cyclic aliphatic groups (naphthenes) experience the highest viscosity increase with pressure. The data in the table also indicate a drop in the pressure-related viscosity increase with an increase in temperature. Compare data at 25°C with those at 80°C.

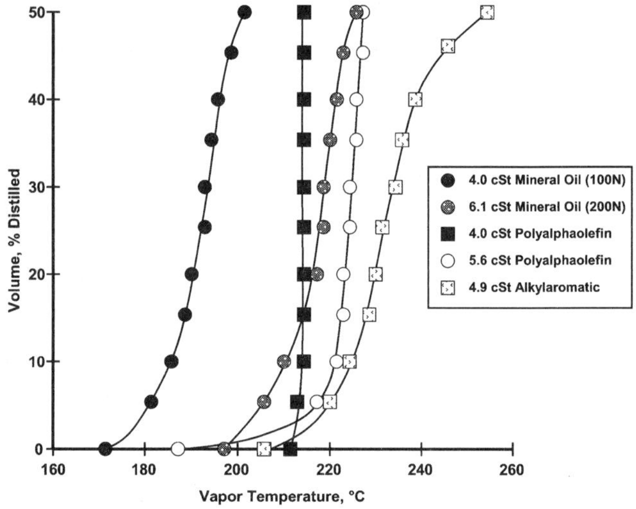

Fig. 3.43—Volatility—Synthetic hydrocarbons versus mineral oils.

In compressor applications, the pressurized air or gas gets entrained in the lubricant. This leads to a drop in lubricant viscosity, thereby diminishing its ability to lubricate. This often leads to excessive wear of the parts. While the problem can be minimized by the use of a higher viscosity fluid, such fluids frequently have poor low-temperature fluidity and hence are difficult to pump at low temperatures. Some synthetic fluids, such as poly(alkylene glycol)s (PAGs), are resistant to dilution-related decrease in viscosity. This is a consequence of their solvency, due to which they dissolve such gases without a loss in viscosity. Typically, polar base stocks, such as esters and PAGs, are suitable for use in applications that involve polar gases, such as chlorofluorocarbons, fluorocarbons, and carbon dioxide. One exception is ammonia where PAGS are suitable but esters are not. This is because esters can react with ammonia to form amides, thereby losing their structural integrity. Conversely, mineral oils and PAOs are effective only for largely nonpolar gases, such as hydrocarbons. Incidentally, all base oil classes are suitable for use in combination with air.

TABLE 3.25—Low-temperature engine oil requirements.

Lubricant base oil	Viscosity at 100°C, cSt	Cold-cranking Viscosity at −25°C, cP (mPa-s)
Mineral SN 100	3.8	1300
Polyalphaolefin	3.9	500
Polyol Ester	4.5	550

Low-temperature Properties

Pour point and cloud point are two commonly used measures of the low-temperature behavior of an oil. Pour point is the lowest temperature at which an oil will flow. This property is important for oils that involve the low-temperature operation. The general guideline for selecting an oil is that it has a pour point of at least 10°C or 20°F lower than the lowest anticipated ambient temperature. Cloud point is the temperature at which the dissolved solids in oil, such as the paraffin wax, begin to form crystals and separate from the oil.

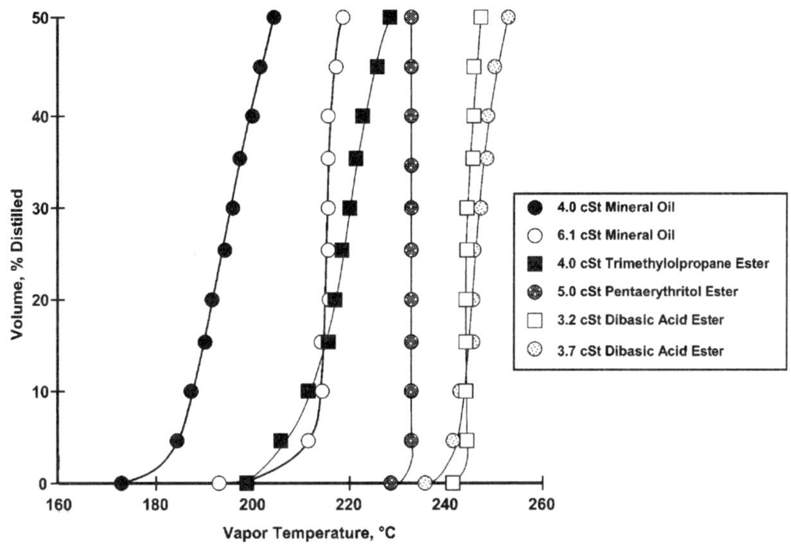

Fig. 3.44—Volatility—Synthetic esters versus mineral oils.

Certain oils must be maintained at temperatures above the cloud point to prevent plugging of filters and small orifices. Cloud point is usually higher than the pour point, that is, on cooling, visible crystal formation occurs first and the oil ceases flow afterwards. Table 3.24 provides pour points of the commonly used synthetic fluids. For automotive lubricants, three additional low-temperature measures are used. For engine oils, they are low-temperature cranking and pumping viscosities at prescribed temperatures and for transmission fluids and gear oils it is the Brookfield viscosity. Cold-cranking engine oil viscosity requirements are 6200 cP (−35°C) for 0W and 13 000 cP (−10°C) for 25W. Complete physical requirements are provided in the SAE Standard J300 and in Chapter 5, the Combustion Engine Lubricants chapter. For DEXRON®, MERCON®, and Mopar® type automatic transmission fluids, Brookfield viscosity of <20,000 cP at −40°C and for MERCON® V type transmission fluids, Brookfield viscosity of <13,000 cP at −40°C is specified. Incidentally, MERCON® specification was retired in the year 2007 and all Ford transmissions will be serviced by the MERCON® V quality fluids. The GM, Ford, and Chrysler ATF specifications are provided in the Hydraulic and Transmission Fluids chapter, Chapter 7. For gear oils, temperature at which the oil attains Brookfield viscosity of 150 000 cP needs to be reported. For details, see SAE Standard J306 and consult Chapter 8 on Gear Lubricants. Synthetic fluids with low pour points and a lower tendency to thicken at low temperatures perform better than their mineral oil counterparts. This is demonstrated by the data in Table 3.25 for PAO and polyol ester with similar 100°C viscosities as 100 solvent neutral mineral oil.

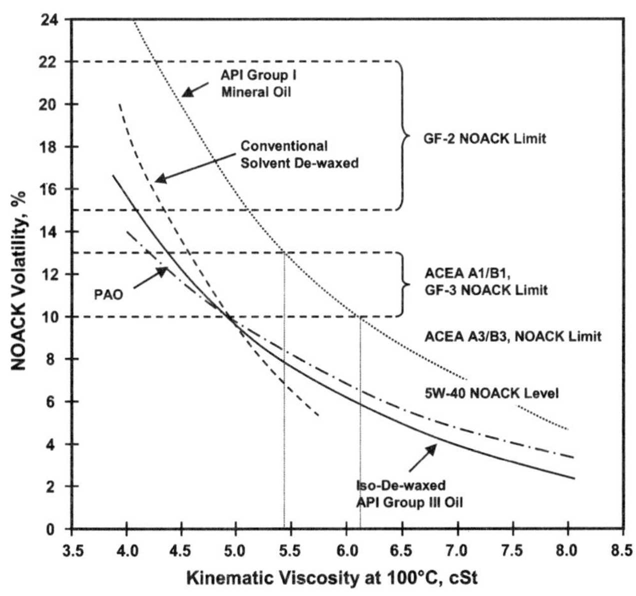

Fig. 3.45—Typical Noack volatilities for base oil types versus specifications [157].

Volatility

Base fluid volatility is also an important parameter since it will contribute to the volatility of the lubricant. High volatility is undesired because evaporation of the base fluid, especially in high-temperature applications, will not only require constant topping up and replenishment but will also pose a fire hazard. In engine oils, there is an additional concern regarding an increase in the undesired emissions. There are two methods to measure volatility of the base oils: the ASTM D1160 method and the NOACK Volatility Test. Similar to the better low-temperature performance, the volatility of the synthetics is in general better than that of the mineral oils. This is shown by the ASTM distillation curves at 1 mm of Hg (Figs. 3.43 and 3.44). Figure 3.43 compares the volatility of the polyolefins and the alkylaromatic base stocks with that of the mineral oils and Fig. 3.44 compares the volatility of the ester base stocks with that of the mineral oils. Despite the fact that the synthetic base stocks and the mineral oils under consideration differ in viscosities, one can still draw conclusions about the superior volatility of the synthetic base stocks. A consideration of the distillation profiles suggests that in terms of volatility:

1. Synthetics are better than mineral base stocks of a comparable or higher viscosity.

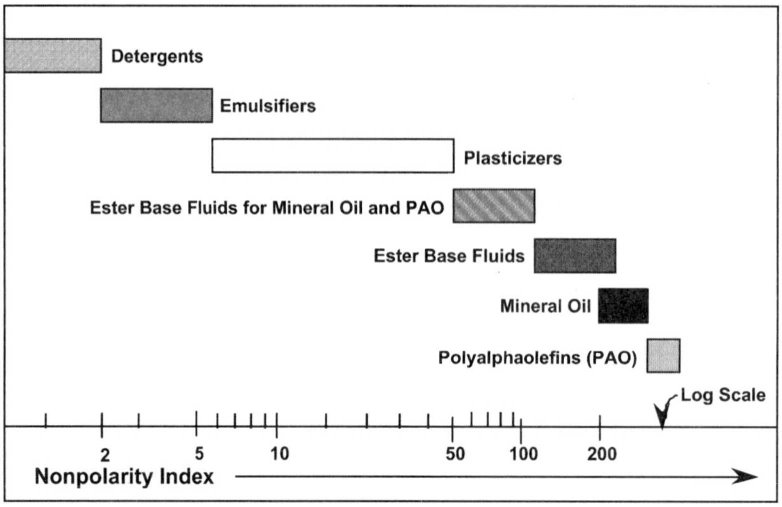

Fig. 3.46—Nonpolarity index of base oils and additives [107].

2. Alkylaromatics are superior to polyolefins.
3. Polyol esters are better than the diesters.

Volatility of lubricants in general and of engine oils in particular is under scrutiny because it contributes to undesirable emissions. Engine oil volatility limits are being decreased progressively. For passenger car engine oils, GF-2 NOACK volatility limit was 22 % but for GF-4, it is lowered to 15 %. Further lowering of this limit is expected in the upcoming lubricant specifications. Part of the problem is the growing trend to use low viscosity oils, which have good low-temperature fluidity that contributes towards fuel economy. As the data in Fig. 3.45 show, the mineral oils of 100°C viscosity of less than 5.5 cSt do not meet the ACEA A1/B1 and ILSAC GF-3 NOACK volatility limits [157].

Fluid Compatibility and Additive Solubility

While intercompatibility of the mineral oil-derived lubricants is not a problem, the intercompatibility of the synthetics is and so is their compatibility with the mineral oils. This is because not all synthetics are compatible with mineral oils or with one another. This is because of the polarity difference among them, as was stated earlier.

Besides compatibility, the polarity of the base oil plays an important role in many other applications. For example, blends of polar esters with nonpolar polyalphaolefins or paraffinic base oils have better viscometrics, as well as good lubricating ability. This is because in the hydrodynamic lubrication regime the nonpolar component is effective and in the mixed or boundary lubrication regime, the polar ester component forms a durable lubricating film. Of course, the greater surface affinity of the oil is one of the many properties that makes it an effective lubricant under boundary lubrication conditions. Despite this advantage, polar PAGs and synthetic esters can overwhelm the surfaces, making it difficult for the surface-active additives, such as extreme-pressure/antiwear additives and rust and corrosion inhibitors, to adsorb on the surface to perform their function. The result is greater wear and increased corrosion when these base fluids are used without properly compensating for their affinity for the metal surfaces. We explained the concept of nonpolarity index under synthetic ester base stocks. G. van der Waal also rated certain base oils and some additives with respect to the nonpolarity index [107]. The data, graphically presented in Fig. 3.46, suggest that base stocks, such as mineral oils and polyalphaolefins, with a higher nonpolarity index (>100), are less likely to interfere with the action of the EP agents than additives, such as detergents, that have a lower nonpolarity index (<2). Esters with the nonpolarity index of 50–100 fall in the middle [107].

The ability to dissolve additives, or solvency, of a base stock is a related problem. Most additives, except some viscosity improvers or some pour point depressants, contain hetero atoms, nitrogen, oxygen, sulfur, and phosphorus. Their presence makes these additives highly polar in a manner similar to that of the esters and PAGs, therefore raising the question of compatibility. While polar base stocks are quite compatible with the most classes of additives, PAOs sometimes have difficulty in keeping the additives dissolved. This is overcome by co-blending with either a mineral oil, an ester, and or an alkylaromatic; the latter is believed to be the case for the engine oil sold under the trade name of Mobil 1.

Thermo-oxidative and Chemical Stability

Synthetic fluids can lose their structural integrity via thermal, oxidative, and chemical degradation. And of course, this susceptibility relates to their chemical structure as well as composition. Chemical stability of a base fluid is one of the key properties that determines the performance of a lubricant. While for hydrocarbon fluids, such as mineral oils and polyolefins, stability largely implies oxidative stability; for organic fluids with high oxygen content, such as ester fluids, thermal and hydrolytic stability are also important.

A number of factors can affect oxidation testing of organic materials. These include exposure time, availability of oxygen (air), and the temperature. With respect to oxidation, base fluids that have low hydrogen content, such as perfluoropolyethers, fluoroesters, chlorofluorocarbons, poly(phenyl ether)s and arylsilicones, have better stability than those that have high hydrogen content. This is because these are the aliphatic carbon hydrogen bonds that react with the oxygen first, to form hydroperoxides, which are the oxidation initiation species. See Chapter 4 on Additives and the section

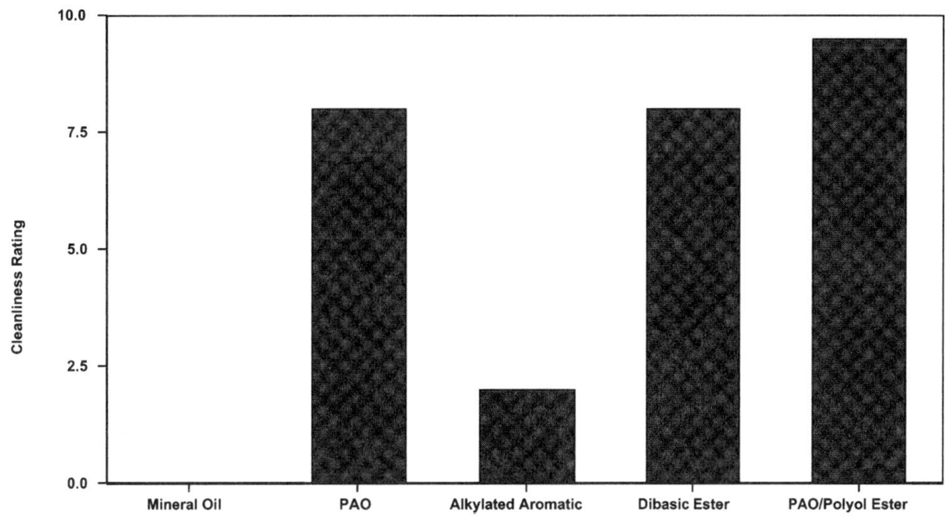

Fig. 3.47—Thermal stability panel coker test; Panel temperature=310°C; Sump temperature=121°C; operation=6 min splash/1.5 min bake; Clean rating=10 [71].

on Oxidation Inhibitors. Many oxidation tests are used to determine the susceptibility of a base fluid towards oxidation. In fact, each application has its own specific tests that try to simulate the operating conditions in that application, which is to observe their effect on lubricant oxidation. Pressure Differential Scanning Calorimetry (PDSC) is a useful technique for discriminating fluids with good oxidation stability [42, 116]. The data in Table 3.5 give the oxidation onset temperature (OOT) of the hydrogen-rich oils. According to these data, untreated PAOs and mineral oils are equivalent but are highly susceptible to oxidation. Diesters are better and polyol esters are the best with respect to oxidation resistance. The data also show that blending polyalphaolefins with ester base stocks increases their oxidation resistance significantly. This is one of the strategies that is often used to lower the cost of synthetics while achieving a reasonable oxidation performance. Further support for this is obvious from the OOT data for fully formulated oils; again the PAO/synthetic ester blends are better than the lubricants based on mineral oils and hydrocarbon mixtures.

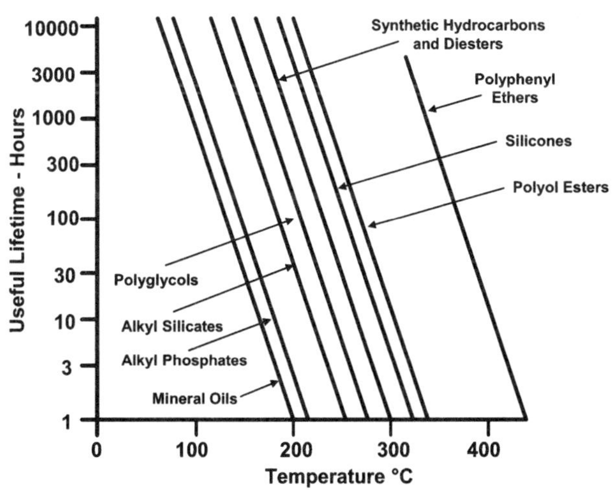

Fig. 3.48—Useful lifetime versus operating temperature—Synthetic lubricants [79,159].

A number of other tests are used to evaluate the oxidation performance of the engine oils. These include ASTM Rotating Pressure Vessel Oxidation Test (Rotary Bomb Test, ASTM D2272), Turbine Oil Oxidation Stability Test (ASTM D943), Cigre Oxidation Test (IP 280), Panel Coker (FTM 3462) and Wolf strip tests (UK 359), Conradson Carbon Residue (ASTM D189), Coke Residue (DIN 51 551), Pneurop Oxidation Test (DIN 51352-2), NOACK Volatility (ASTM D5800), and thermo-gravimetric analysis (TGA).

Performance comparison in oxidation screen tests indicates synthetics to have better overall performance than mineral oils [71]. However, some synthetics, such as esters, are better than others. It is possible that some esters, already being highly oxidized, have less affinity towards oxygen and are therefore less likely to undergo oxidative degradation. Alternatively, the oxidation screen tests measure thermo-oxidative stability and not oxidative stability. The results of the panel coker thermal stability test, depicted in Fig. 3.47 [71], are consistent with this rationale. It is important to point out that these tests were performed on fluids that were devoid of additives. Fully formulated oils *may* reflect different performance. With regard to the mechanical tests, the PAO-based engine oils were equivalent in performance to those that were mineral oil-based.

Oxidative stability measurement by MIL-L-9236A Gas

TABLE 3.26—High-temperature volatility and deposit-forming tendency of lubricants [79].

Lubricant Base Stock	Volatility at 250°C	Deposit Formation at 250°C
Mineral Oil 100N	Poor	Poor
Polyalphaolefins 6 (PAO 6)	Good	Fair
Alkylbenzene 150	Fair	Fair
Esters	Very Good	Excellent
Poly(alkylene glycol)s (PAGs)	Poor	Very Good
Polyisobutenes (PIBs)	Very Poor	Good
Silicones	Excellent	Very Good
Fluorocarbons	Good to Excellent	Very Good

TABLE 3.27—Approximate operating temperature limits for oils [160].

	Long Term[a]		Short Term[b]	
Lubricant	°C	°F	°C	°F
Petroleum oils	95–120	200–250	135–150	275–300
Super–refined petroleum oils	175–230	350–450	315–35	600–60
Synthetic hydrocarbons	175–230	350–450	350–345	600–650
Organic esters	175–190	350–375	220–230	425–450
Phosphate esters	95–175	200–350	135–230	275–450
Polyglycols	160–175	325–350	205–220	400–425
Poly(phenyl ether)s	315–370	600–700	425–480	800–900
Silicate esters	190–220	375–425	260–290	500–550
Silicones	220–275	425–525	315–345	600–650
EP oils	60–80	140–175	65–85	150–185

[a]Long term ≈ hundreds of hours.
[b]Short term ≈ few hours.

Turbine Test, a real life test, that uses a temperature of 260°C and air blow rate of 5 L/h, led to a large viscosity increase and acid build-up, showing esters to possess only poor to mediocre stability. However, lowering the temperature to 200°C prolonged their oxidative life considerably [72]. These findings may appear contradictory to those noted above but it is important to state that petroleum-derived and nonester base stocks do even more poorly in this test and that most turbine lubricants are of the ester type and contain a substantial amount of an oxidation inhibitor package.

Hydrolysis, molecular degradation due to water, is only a problem in esters and not in mineral oils and PAOs. It is usually catalyzed by strong acids and does not occur in their absence. However, the presence of the residual acidity after manufacture or exposure to acidic materials and environment during use can lead to hydrolytic cleavage of the ester functional group. Usually esters with steric crowding, such as those derived from neo-alcohols or geminal dibranched acids, hydrolyze much more slowly than the unbranched alcohol or mono-carboxylic acid derived esters. The life of an oil at a particular temperature depends upon the amount and the type of degradation that is acceptable. This in turn depends upon how much performance can be allowed to deteriorate during use.

Greater thermal and oxidative stability of the synthetics are two of the primary reasons to choose synthetic base stocks over mineral oils. With respect to thermal stability, esters are more stable than pure hydrocarbon base stocks, such as mineral oil and PAOs (Table 3.16). Among esters, polyol esters are three times more stable than monohydric alcohol derived esters. The stability relates to the number of β-hydrogens present in the alcohol portion of the molecule [72,158]. The higher the number of β-hydrogens, the lower is the thermal stability. The absence of the β-hydrogens in polyol esters greatly enhances their thermal stability. The presence of the β-hydrogens, as is the case in alkyl esters, facilitates thermal degradation via a six-membered transition state. When the β-hydrogens are absent, esters decompose

TABLE 3.28—Material—Synthetic base fluid compatibility (at room temperature) [161].

Material	Synthetic Hydrocarbons	Esters and Polyglycols	Silicones (All Types)	Fluorinated Ethers
Plastics				
Acetals	A	A	A	A
Polyamides	A	A	A	A
Phenolics	A	A	A	A
Terephthalates	A	A	A	A
Polycarbonates	A	C	A	A
ABS resins	A	C	A	A
Poly(phenylene oxide)	A	C	A	A
Polysulfones	A	C	A	A
Polyethylenes	B	B	A	A
Elastomers				
Natural Rubbers	C	C	A	A
Buna S	C	C	A	A
Butyl	C	C	A	A
Ethylene propylene	C	B	A	A
Nitrile (Buna N)	A	B	A	A
Neoprene	A	C	A	A
Silicone	B	B	C	A
Fluoroelastomers	A	C	A	A

Note: A = Usually OK; B = Be Careful; C = Causes Problems.

TABLE 3.29—Viscosity change in petroleum oils on irradiation [164].

Gamma Dose at 24°C, 10^8 Rads in air	*0*	*0.7–1.1*	*4–6*
Fluid	**Viscosity at 38°C cSt**	**% Viscosity Increase**	
25 VI Naphthenic (pale)	101	10	72
85 VI Paraffinic (bright stock)	939	11	75
85 VI Paraffinic (neutral)	112	8	316
95 VI Paraffinic (white)	76	46	757

via a free radical mechanism that requires much higher energy [158]. That is why polyol esters have higher decomposition temperatures (>300°C) than alkyl esters (~200°C). The mechanisms for the pyrolysis of esters are depicted in Fig. 3.27.

Figure 3.48 gives an indication of the useful life versus temperature for the formulated lubricants based on synthetic fluids. As expected, an increase in temperature decreases a lubricant's useful life. Useful life presented here is based on field experience in a range of applications [79,159].

Table 3.26 subjectively provides volatility and deposit forming tendencies of the various lubricants [79]. Materials with relatively higher polarity have a lower tendency to form deposits than the low polarity mineral oil, PAOs, and alkylaromatics. This is partly due to their higher solvency, which keeps the polar deposit precursors in solution instead of allowing them to separate out on the surfaces. These precursors on hot surfaces result in the formation of deposits. As mentioned earlier, volatility is a function of the molecular weight; the low molecular weight materials have higher volatility. A number of factors may be responsible for the poor volatility characteristics of a base stock, the presence of the low boiling components being one. Both mineral oils and polymeric synthetic materials, such as unpurified polyolefins or poly(alkylene glycol)s, with poly-dispersity index of higher than 1, contain such components. However, all synthetic fluids undergo a rigorous purification step; hence, their poor volatility is related to their thermal depolymerization or oxidative degradation to the lower boiling species. This may be the primary reason for the higher volatility of the PIBs and PAGs above their threshold temperature, where rapid degradation initiates. It is important to note that a temperature of 250°C is quite high and in most cases is outside the recommended operating range of the most synthetic fluids. See the data in Table 3.27 [160].

Hydrolytic Stability

Hydrolytic stability is an important consideration if the base stock is to be used either in combination with water, such as in the case of water-based metalworking fluids, or has a chance to be exposed to water, such as in combustion engine lubricants. Under these circumstances, the oil must keep its integrity and must not hydrolyze. Hydrolysis is a primary concern for silicate esters and phosphate esters and only a minor concern for carboxylate esters. High temperatures and a trace amount of acidic impurities catalyze the hydrolytic breakdown of these base stocks.

Materials Compatibility

Many applications employ plastics and elastomers. Hence compatibility of the synthetic fluids with these materials is important. While compatibility of the synthetic base oils with plastics is not a problem, except for esters and polyglycols, elastomer compatibility problems are more serious. The problems occur because of the fluid migrating into or plasticizer moving out of the elastomer seal material. The former type of fluids, exemplified by esters and PAGs, will cause excessive swelling of the seals and the latter types, exemplified by the PAOs and the alkylaromatics, will cause shrinking and cracking of the seals. Material compatibility of the synthetic base fluids is provided in Table 3.28 [161].

Compressibility/Bulk Modulus

As stated earlier, bulk modulus expresses the resistance of a fluid to compression, which is the reciprocal of compressibility. This property of the fluid is significant for use in hydraulic and servo systems. Of all synthetics, silicones and perfluoropolyethers show a relatively low bulk modulus (high compressibility) based on a density correlation. Bulk modulus is a physical property of the base fluid which cannot be changed significantly by additives.

Heat Transfer

A variety of industries employ heat transfer fluids. Those worth mentioning within the context of lubricants include food processing, oil and gas industry (natural gas compressor engines and high-speed industrial stationary engines, drilling equipment), petroleum and petroleum refining, metalworking and metal processing, and aerospace and military.

Lubricating advantage of a particular base fluid depends on a number of factors. It depends upon its viscosity-temperature (VT) and viscosity-pressure (VP) behavior, its affinity for the surface (polarity), and the lubrication regime (hydrodynamic, elasto-hydrodynamic, mixed, or boundary). When used as lubricants in the vicinity of the high-temperature environments, such as internal combustion engines, or where the machine elements generate a significant amount of frictional heat, such as gears, the VT properties of the fluids are of primary importance. This is because a substantial viscosity loss due to heat will diminish their ability to lubricate effectively. In addition, because of the high temperatures oxidative degradation of the lubricant and metal corrosion due to acidic oxidation products will also increase. This will result in shorter service life. In such situations, lubricants/base fluids with good thermal capacity, or heat transferring capability to dissipate heat, are useful. Heat transfer directly affects equipment performance, fuel efficiency, materials selection, emissions, and engine or equipment life.

Heat capacity of a liquid/oil is the amount of heat it can absorb or store before a change in its overall temperature occurs. Thermal capacity of a fluid is defined as the amount of energy required to raise the temperature of a sample by 1°C. The value of this parameter can be calculated from the specific heat capacity at constant pressure, thermal conductivity, density, viscosity, and the heat transfer coefficient of the

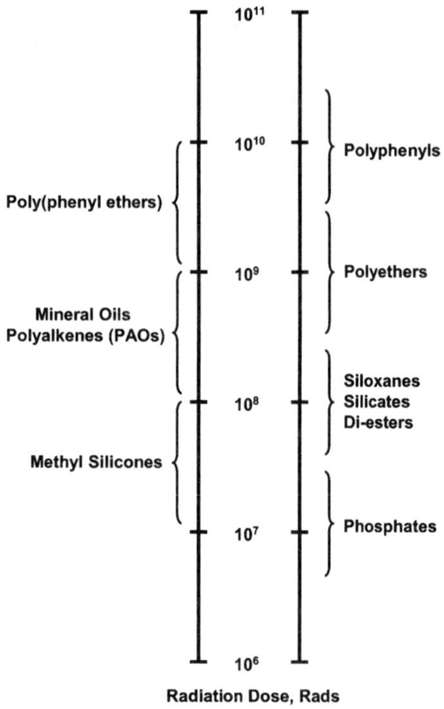

Fig. 3.49—Radiation resistance limits of various base fluids [165].

lubricant components [79]. Specific heat capacity of a substance is the amount of energy required to raise the temperature of one kilogram of the substance by 1 °C (1 °K). Heat capacity can be measured by using calorimetry. The measuring unit of heat capacity is joule per kilogram Kelvin, $J \cdot kg^{-1} \cdot K^{-1}$ or J/(kg·K), or BTU/lb °F. Higher values indicate higher heat capacity. The specific heat can also be interpreted as a measure of how well a substance maintains its temperature, i.e., "stores" heat. Thermal conductivity is a measure of the ability of a material to conduct heat. Again, the higher values are better. The oil must be able to absorb heat, store it, and release it without ever compromising its own stability. The rate at which this absorption, storage, and release occurs is known as the heat transfer coefficient; higher values of this parameter are highly desirable. In general, lubricants with good heat transferring ability have good heat capacity, high thermal conductivity, high density, and low working viscosity.

An ideal heat transfer fluid must have superior thermal conductivity, exceptional thermal stability (resistance to coking and oxidation up to 425 °C), low pour point (≤0 °C), low vapor pressure at high temperatures, high boiling point, long operating life, and the ability to be used over a broad temperature range. In addition, it must be nonflammable, nontoxic, and inexpensive. Commercial heat transfer fluids (HTFs) vary in performance. Polar materials have generally higher heat capacity than nonpolar materials. A rough estimate of the heat storage ability of the polar synthetics is around 10 % over that of the pure hydrocarbon fluids. Hydrocarbon fluids, especially petroleum oils, in addition, are susceptible to accelerated oxidation at temperatures of around 135 °C (275 °F). They also suffer from the disadvantage of high volatility and can evaporate at elevated temperatures. Synthetic fluids used to make heat transfer lubricants include organic esters and diesters, poly(alkylene glycol)s and water-glycol mixtures, and silicones. Current commercial HTFs are derived either from hydrocarbons (paraffinics or alkylaromatics), poly(alkylene glycol)s, silicones, or CFCs (only for closed systems). Some of these may even contain water. Properties of some commercial heat transfer fluids are provided in Ref [162]. While many of these meet the above listed criteria for the ideal heat transfer fluid, the search for better heat transfer fluids continues [163].

Radiation Stability

Radiation affects all organic compounds, causing a change in their properties. This is because radiation attacks the covalent bonds that are most prevalent in organic compounds.

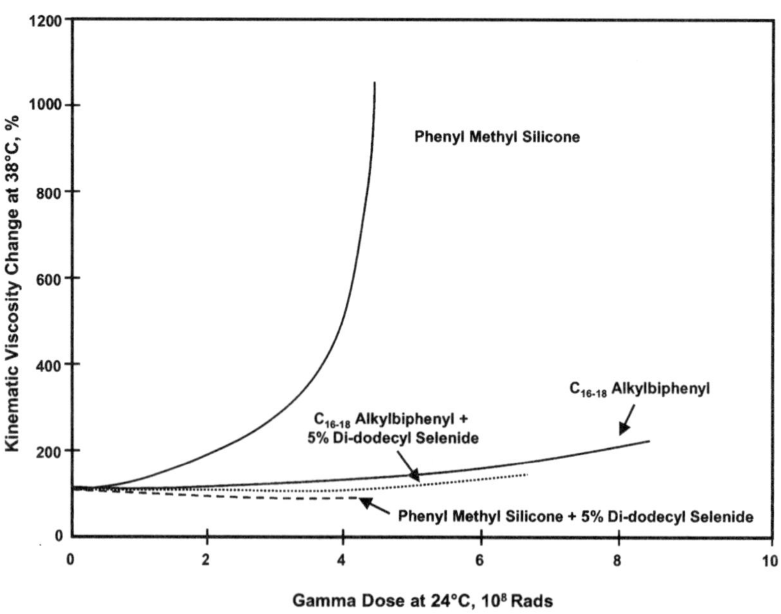

Fig. 3.50—Influence of di-dodecyl selenide on radiolytic viscosity change [166].

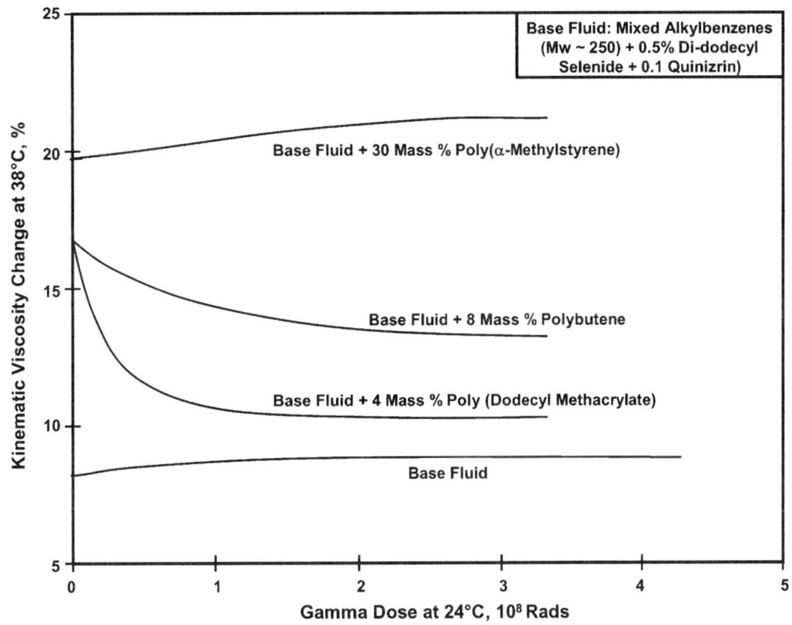

Fig. 3.51—Effect of radiation on polymer-treated fluids [166].

Since the strength of these bonds differs among organic compounds, so does their stability under radiation, especially short-wavelength, highly energetic ionizing radiation, such as gamma rays, X rays, and subatomic charged particles. One result of ionization is that the organic molecules disproportionate to form smaller hydrocarbon molecules as well as larger hydrocarbons molecules. Smaller molecules result from fragmentation and larger molecules result from their recombination, or polymerization. Other chemical reactions caused by radiation include oxidation and isomerization.

Lubricants as well as base fluids are primarily organic molecules, hence their radiation stability is a concern when used in applications with a possibility of radiation exposure. Examples of such an environment include nuclear power plants, space satellites, nuclear submarines, food sterilization equipment, radiation chemistry systems, and medical imaging systems. Radiation stability of a lubricant is typically assessed by an increase in its viscosity. Those lubricants/base fluids that exhibit the lowest increase in viscosity under radiation are the most desirable in the above-listed applications. Lubricants that contain a large number of ionizable carbon-carbon and carbon-hydrogen bonds, such as petroleum base stocks (mineral oils), have a lower stability than those that contain fewer such bonds. The examples of the latter type are poly(phenyl ether)s [PPE] and perfluoropolyethers (PFPEs). Table 3.29 shows the effect of radiation on the viscosity of mineral oils. In one study, the performance of the PPE under the influence of 10^{11} ergs/gram of radiation at 99°C (210°F) was compared with that of the aliphatic ester, aliphatic hydrocarbon, methylphenyl silicone, and chlorophenyl silicone [164]. PPE showed a viscosity increase of only 35 %, while all other fluids showed a viscosity increase of 1700 % and gelled. In some lubricants, radiation also creates free radicals that can cause corrosion. Radiation stability of the ortho-phosphoric acid esters, not part of this study, is also reported as poor [4]. PPE-derived lubricants are therefore widely used in radiation environments. Radiation resistance of various base fluids is shown in Fig. 3.49 [165,166]. From the figure it appears that the materials that are primarily aromatic in structure, such as PPEs and polyphenyls, are the most radiation stable. Their greater stability can be ascribed to the presence of the difficulty to ionize aromatic carbon hydrogen bonds. Radiation properties of the fluids can be improved by the use of additives. Among those tried, dialkyl selenide was found to be the most effective [165]. This is shown in Fig. 3.50. Without the selenide, phenylmethyl silicone shows a large viscosity increase at 4×10^8 Rads, but the addition of dodecyl selenide helps the fluid maintain its viscosity at a reasonable level, even at higher radiation dosage. The effect in C_{16-18} alkylbiphenyl, which maintains its viscosity well without the additive, the effect is a lot less dramatic. Other base fluids, such as esters, also respond well to such additives [165].

TABLE 3.30—Other synthetic oils and radiolysis effects.

Class	Radiolysis Effects
Poly(alkylene glycol)s	Large evolution and dilution; low viscosity increases; temperature sensitivity high
Poly(phenyl ether)s	Low gas evolution; good at high temperatures
Silicates and phosphates	Splitting to form acids
Silicones	Good oxidative and thermal stability; tendency to gel
Halogenated organics	Give off halogen acids; corrode metals

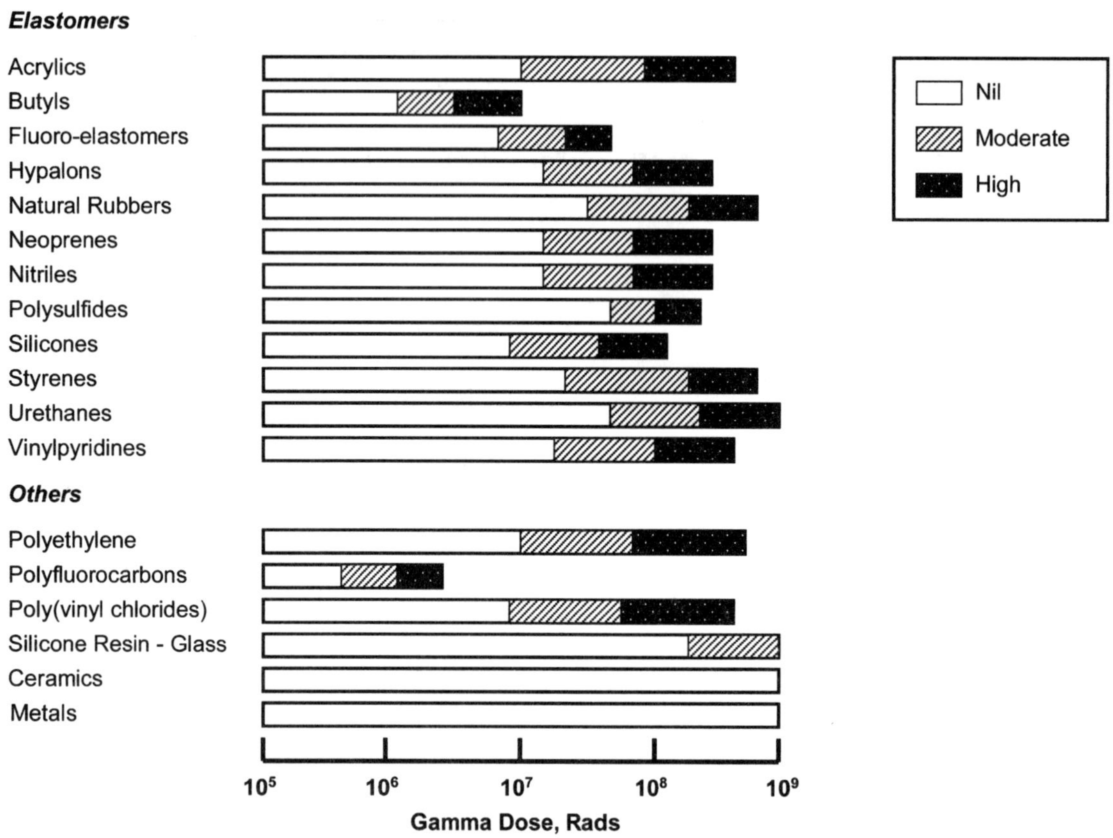

Fig. 3.52—Effect of radiation on various materials [165].

Polymers used in finished lubricants as viscosity modifiers respond differently to radiation, as shown in Fig. 3.51. Poly-(alkyl acrylate) and polybutene lose their viscosity but polystyrene somewhat gains viscosity on irradiation. The ideal fluid must do neither and by this criterion, finished fluid that contains poly (α-methylstyrene) is a good one. Please note that all fluids depicted in the figure contain radiation stability additives. Viscosity increase is not the only consequence of radiation exposure. Other radiological effects that are observed in synthetic fluids are listed in Table 3.30. In addition to the base fluids and additives, there are other materials in modern machines that are also affected by radiation. Prime examples include elastomers that are used to make seals and polymeric coatings. Radiation effects on these materials are provided in Fig. 3.52 [165–167]. Again, different materials have a varying degree of resistance to radiation exposure.

Lubrication

As mentioned in the earlier part of the discussion, the lubricating advantage of a particular base fluid depends upon a combination of factors. These include superior viscosity-temperature (VT) and viscosity-pressure (VP) behavior and high affinity towards surfaces, also termed polarity. Of these, a lubricant's surface affinity has the most direct influence on lubrication, the primary objectives of which are temperature control and minimizing metal-to-metal contact to avoid wear. The temperature control is critical since it can cause a reduction in lubricant viscosity, facilitate its oxidative degradation, and promote metal corrosion due to the resulting acidic oxidation products. The temperature-related viscosity loss can interfere in the lubricant's second critical function, which is to form a durable lubricating film between surfaces to prevent metal-to-metal contact. Situations that explicitly require effective lubrication primarily occur during the start-up, slow down to stop, overheating, and overloading. In such situations, the lubricating film is either not present or is ineffective. Lubricants derived from the polar base stocks, such as esters, have a greater affinity towards metal surfaces than those derived from the mineral oil; hence they form more durable surface films under all situations. The film formation capability in elasto-hydrodynamic contacts has been studied. The results indicate that the polar polyesters form thicker lubricating films that are maintained in highly loaded, high slip contacts. This was ascribed to their high surface affinity.

Low-temperature viscosity of a lubricant is also an important consideration. This is because during cold starts the main cause of engine wear is the inability of the lubricant to reach parts requiring lubrication. The lubricants that have poor low-temperature viscometrics take too long to reach such parts. Lubricants, especially those for automotive use, have low-temperature viscosity specifications. These are provided in the section on low-temperature properties and for the engine oils in Chapter 5 on Combustion Engine Lubricants. As stated earlier, cranking and pumping viscosities are two of the most critical requirements for engine oils. Engine lubricants derived from the PAOs and polyol esters have much better cold-cranking viscosity than those derived from the mineral oils. Part of the advantage here is the absence of

TABLE 3.31—Biodegradability of various fluids [171].

Base Fluid	CEC-L-33-A-94	CEC-L-33-T-82[a]	OEC 301B	EPA 560/6-82-003
Mineral Oil	10 to 45 %	15 to 30 %	10 to 40 %	29 to 49 %
White Oil	...	25 to 45 %	...	...
Polyalphaolefins	20 to 80 %	5 to 30 %	5 to 60 %	...
Natural Oils	...	70 to 100 %	...	72 to 80 %
Diesters	75 to 100 %	55 to 100 %	25 to 80 %	55 to 84 %
Aromatic Esters	0 to 95 %	5 to 80 %	5 to 45 %	...
Polyol Esters	0 to 100 %	55 to 100 %	0 to 80 %	55 to 84 %
Alkylbenzenes	5 to 20 %	...	0 to 20 %	...
Poly(alkylene glycol)s	5 to 70 %	0 to 25 %	5 to 80 %	6 to 38 %
Polybutenes	5 to 20 %	0 to 25 %	0 to 20 %	...

[a]Lubriztol's biodegradability review of current situation. Published in 1993 by Dr. Stephanie Harold

the high molecular weight linear hydrocarbons, the paraffin wax. Incidentally, lubrication characteristics of all base stocks can be improved by the use of the friction modifiers and antiwear and extreme-pressure agents. Friction modifiers adsorb on metal surfaces and associate with the oil to keep it on the surface and extreme-pressure/antiwear agents form more durable chemical films on metal surfaces via thermal reaction.

Environmental Compatibility

Governmental regulatory agencies, such as the Environmental Protection Agency (EPA), National Marine Fisheries Service (NMFS), U.S. Department of Agriculture (USDA), and European Environment Agency (EEA), and many private organizations are established to monitor and control the release of harmful substances into air, water, and soil. Various countries have initiated the use of environmental labels, or eco-labels. The objective of the scheme is to promote products with a reduced environmental impact, compared to other products in the same product group. Not all product groups are amenable to eco-labeling. Those that are must meet the following criteria [168]:

- They must represent a significant volume of sales and trade in the domestic market.
- They must have a significant environmental impact.
- They must present a significant potential for affecting environmental improvements through consumer choice.
- A significant part of the sales volume must be sold for final consumption or use.

The reason for choosing these criteria is to convey the environmental characteristics of the product to the consumer. The hope is that the consumers will choose environmentally friendly products, due to a concern for the environment. This in turn will increase the demand for such products, which will motivate manufacturers to produce products that are environmentally compatible. Lubricants, new or used, not only meet the above listed criteria but their release into the environment is also a concern because of the accidental release of its precursor, petroleum, into the environment and its well publicized negative impact. Eco-labeling of lubricants in Europe uses four measures to select lubricants; these are biodegradability, aquatic toxicity, health hazardness, and renewability. As of the year 2003, no eco-labels have been used for automotive lubricants, which make up ~47 % of the total lubricant use, compressor oils, drilling fluids, textile lubricants, and marine and aviation lubricants [169].

Biodegradability of the oils plays a predominant role in assessing the environmental compatibility of the lubricating oil, both in Europe and in the United States (ASTM D6046). There are a number of methods that are used to measure biodegradability [170]. These include CEC-L-33-T-82, CEC-L-33-A-94, OECD 301B, and ASTM D5864. The biodegradability results obtained by the use of different methods are compared in Table 3.31 [171]. As one can see, not only that the extent of biodegradability differs across methods but it also differs within the same class, as indicated by the biodegradability range. Results also differ if the sludge used in the test is obtained from a different source. Typically, linear hydrocarbon materials and natural oils exhibit the highest degree of biodegradability. ASTM D6006 is a guide for assessing the biodegradability of the hydraulic fluids and ASTM D6384 describes the terminology relating to the biodegradability and eco toxicity of the lubricants. For biodegradation mechanism, see Ref [172]. From the data in Table 3.31 one can see that the ester fluids, such as diesters and vegetable oils, are highly biodegradable. Please consult Chapter 13 for further discussion on this topic.

When assessing toxicity, one considers environmental fate (photodegradation, stability in water, biodegradation, fungicidal action), health effects (acute toxicity, repeated-dose toxicity, genetic toxicity, and reproductive/developmental toxicity), and ecological effects (harm to fish, invertebrates, and algae). As one can see, biodegradability is one of the many criteria that define environmental compatibility. A fluid may be highly biodegradable but still may be harmful to life or the environment. Numerous toxicity test methods have been available for years, and some of these have been used to evaluate lubricants. Some of the test methods can be applied as such to determine lubricant toxicity but others need extensive modification. Some others may prove totally inadequate. This is because the traditional toxicity test methods were devised for assessing toxicity of aqueous specimens and not nonaqueous specimens, such as lubricants. Most test methods attempt to determine the lethal dose (LD_{50}) of an animal species being tested. An LD_{50} value is the amount of a solid or liquid material that it takes to kill 50 % of the test animals in a single dose. This is closely related to the LD_{Lo} value which is the lowest dosage reported to have killed animals or humans. The species of animals

TABLE 3.32—Relative cost of vegetable and synthetic oils.

Fluid	Cost	Cost	Cost
Mineral Oil	1	1	1
Vegetable Oils	2–3	...	...
Synthetic Fluids			
Polybutenes	2	2–4	...
Polyalphaolefins	3	2–5	...
Dialkylbenzene	5	2–3	*
Poly(alkylene glycol)s	3–5	2–6	4
Polyol Esters	3–5	2–10	...
Diesters	2–6	2–10	4
Phosphate Esters	4–7	...	6
Cycloaliphatics	9–15	...	...
Silicone Fluids	12–14	10–15	15–40
Silicate Esters	33–45	...	8
Halogenated Hydrocarbons	100–450	75–300	300
Poly(phenyl ether)s	625–700	50–250	100

studied is virtually limitless and can range from water fleas to sea creatures to mammals.

The term environmentally friendly is commonly used for lubricants but the term's precise meaning is not certain. Not surprisingly, U.S. environmental regulations do not recognize the term and treats an environmentally friendly lubricant (EFL) spill as being that of toxic material. While the industry considers EFLs to be nontoxic and expects them to biodegrade in a relatively short time, adequate test methods to unequivocally determine these parameters are lacking. We can observe the nonuniformity of the biodegradability results across the methods described above. In addition, there is the nagging question of which test organisms will provide toxicity data to draw meaningful conclusions. At present, ASTM committees are in the process of developing standards to resolve these issues. A number of environmentally friendly fluids based on natural oils, such as rapeseed oil, are commercially available. So are the products based on the synthetic esters and PAGs derived from such natural products. They have good lubricating properties, sometimes superior to those of the mineral oils, and are used to formulate hydraulic fluids, other lubricating oils, and greases. Their essential feature is that they occur in nature, hence are expected to have low toxicity and high biodegradability. The problem is that once such base oils are formulated with additives to make lubricants, their biodegradability and especially toxicity benefits may be lost. Many of the lubricant additives are not as biodegradable or as nontoxic as these base fluids.

TABLE 3.33—Fastest-growing synthetic lubricants and functional fluids (demand in million dollars).

Product	1996	2001	2006	2001–2006 Annual Growth
Engine Oils	283	428	632	8.1 %
Hydraulic and Transmission Fluids	132	199	311	9.3 %
Metalworking Fluids	141	186	237	5.0 %
Dielectric Fluids	81	106	140	5.7 %
Others	192	254	345	6.3 %

Toxicity data on finished lubricants are abundant but on base fluids alone it is sparse. However, a year 2000 study, published by the U.S. Department of the Interior, Minerals Management Service, Gulf of Mexico OCS Region, on "Environmental Impacts of Synthetic Based Drilling Fluids" suggests normal (linear) paraffins (LPs), linear-α-olefins (LAOs), poly-α-olefins (PAOs), polyinternal olefins (PIOs), and esters to be largely nontoxic [172]. The study also provides the biodegradation mechanism of alkanes and alkenes. Please note that food grade lubricants need not meet the biodegradability requirements but only toxicity requirements with respect to human consumption. Toxicity towards other land or aquatic life forms is not a major consideration.

Cost of Synthetics

As mentioned in the earlier part of the discussion, the cost of the base oil is an important consideration if more than one base stock can be used to formulate lubricants within the desired performance criteria. Raw materials, which are usually petroleum-derived, are expensive and the processes to manufacture base stocks usually add to the cost. A number of cost estimates for synthetic base stocks relative to the cost of mineral oil are available in the literature. Three of them are provided in Table 3.32. As one can see, the estimates of the relative cost radically differ from one another. This may be related to the amounts manufactured and the fluctuation in demand and raw material prices. Despite differences across estimates, relative costs within each estimate follow the same order. Despite the higher cost, the market for synthetics is growing while that of the mineral base stocks is stagnating. This is because of the superior properties of the synthetics relative to the mineral oils which make them more suitable for the more demanding modern applications. Table 3.33 shows the rate of growth of synthetics in various lubricant segments for the 1996 to 2006 period. The data presented here are the Fredonia Group data. Table 3.34 lists the desirable properties of the synthetic base fluids, along with the recommended operating range [174].

One way to manage the cost of lubricants derived from synthetic fluids is to use their blends with mineral oils or the relatively cheaper synthetic fluids, such as polyalphaolefins (PAOs). Besides PAOs, polyesters are the most commonly used synthetics, and they often are used as blends with mineral oil, PAO, and or alkylaromatics.

Biological (Natural) Base Stocks

There has been a great deal of renewed interest in the use of oils from sources that are easy to replenish, such as animal fats and vegetable oils, to formulate lubricants. In addition to the advantage of continuous supply far into the future, these oils possess other properties that are important in their own right. A high degree of biodegradability is one. Others include governmental support, emerging technology, and ongoing uncertainty over the crude oil supply and its price. Lubricants from the mineral base oils are not always compatible with the environment; hence the disposal of mineral oil-based used lubricants poses a problem. Used lubricant is therefore recycled after removing the harmful products; a topic that is discussed in Chapter 13 on Lubricants and the Environment. However, the recycling process is costly, some question the quality of the recycled or the re-

TABLE 3.34—Synthetic Base fluids and their properties [174].

Synthetic Base Fluids	Properties
Synthetic Hydrocarbons *Temperature Range: - 60 to 125°C*	▪ Excellent thermal stability ▪ Good friction reduction and lubricity ▪ Wide range of viscosities ▪ Low-temperature serviceability ▪ Good plastic and elastomer compatibility ▪ Long and growing list of applications in many industries
Poly(alkylene glycol)s *Temperature Range: - 40 to 100°C*	▪ Non-carbonizing, no residue ▪ Good lubricity and film strength ▪ Wide range of viscosities ▪ Unusually good elastomer compatibility ▪ Good load-carrying capacity ▪ Only synthetic oils which include water-soluble versions ▪ Good high-temperature stability with proper antioxidant ▪ Commonly used in arcing switches, and particularly effective in large worm and planetary gears
Synthetic Esters — Includes diesters and polyol esters) *Temperature Range: - 65 to 150°C*	▪ Excellent oxidative and thermal stability ▪ Low volatility ▪ Excellent anti-wear properties ▪ Outstanding lubricity ▪ Good low-temperature properties ▪ Minimal viscosity change with temperature ▪ Excellent load-carrying ability for bearing applications
Silicones — Includes dimethyl, phenyl, halogenated *Temperature Range: - 70 to 200°C*	▪ Excellent oxidative and thermal stability ▪ Low volatility ▪ Wide range of viscosities ▪ Minimal viscosity change with temperature ▪ Excellent plastic and elastomer compatibility ▪ Good wetting capability ▪ Commonly used with plastic and elastomer components, including gears, control cables, and seals. Higher viscosities provide mechanical damping.
Perfluoropolyethers *Temperature Range: - 90 to 250°C*	▪ Excellent oxidative and thermal stability ▪ Low volatility and vapor pressure ▪ Nonflammable and chemically inert ▪ Excellent plastic and elastomer compatibility ▪ Resistant to aggressive chemicals and solvents ▪ Commonly used in extreme-temperature environments and applications which require chemical, fuel, or solvent resistance
Poly(phenyl ether)s *Temperature Range: - 0 to 250°C*	▪ Highest thermal and oxidative stability of all oils ▪ Excellent radiation, chemical, and acid resistance ▪ Excellent lubricity ▪ Excellent high-temperature stability ▪ Non-spreading even in thin film ▪ Traditional lubricant for noble metal connector applications; also used for high-temperature, specialty bearings
Multiply-alkylated Cyclopentane (Pennzane® SHF-X2000) *Temp. range: - 45 to 125°C*	▪ Proprietary fluid, manufactured by Pennzoil® and distributed exclusively by Nye, that combines the low vapor pressure of a PFPE with the lubricity and film strength of a synthetic hydrocarbon

claimed oils, and only about 24 % is truly recycled. The rest is either burned or disposed of illegally.

Animal fats and vegetable oils occur in nature and are therefore biodegradable [106]. As a result, the disposal of lubricants containing them is of less concern. Interest in biodegradable lubricants, such as those derived from the natural oils, has historically been in response to the environmental concerns, particularly relating to the effects of the oil entering the soil and the fresh water. Therefore, the lubricants used in agriculture, forestry, and marine applications are formulated by the use of biodegradable and nontoxic synthetic and natural ester base stocks. Such lubricants include hydraulic oils, gear oils, and greases. Biologically compatible or acceptable lubricants are particularly popular in the northern European countries. Incidentally, the U.S. EPA treats all oil spills alike and does not distinguish them based on whether they are petroleum or biological in origin. Interestingly, a significant driver for the industrial use of the vegetable oils is "green chemistry" initiatives by the U.S. Government, under the auspices of the 2002 Farm Bill which promote the development of the new bio-based products and processes. The objective of the bill is to promote cooperation between agriculture and other industries to exploit homegrown biochemical resources. A variety of terms are used to designate products derived from the natural oils, many of which are used in adhesives, personal care products, and fuels and lubricants. Common terms for such products include renewable, bio-based, green, environmentally sound, environmentally acceptable, natural, and so on, portraying an environmentally conscious image. Vegetable oil-based products, such as those produced from canola, soybean, palm, and sunflower seed, are referred to as natural esters since they are used only after ex-

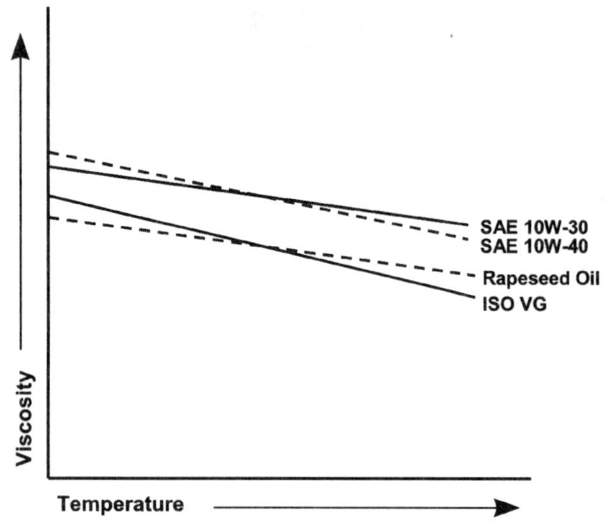

Fig. 3.53—Comparison of viscosity-temperature characteristics of different oils.

traction and purification that does not involve chemical changes to the oil at all.

Besides natural oils themselves, there are synthetic esters that are made by the use of the natural acids, such as oleic acids, or alcohols, such as oleyl alcohol. While these products have a certain amount of biological content, they are not usually claimed to be bio-based. They essentially combine the superior properties of the natural oils, such as biodegradability, with excellent low-temperature fluidity of the synthetic esters.

Beside biodegradability, natural oils have a number of other advantages. These include excellent lubricity, viscosity-temperature (VT) relationship (VI character), low volatility, safety to the environment, nontoxicity, and non-carcinogenicity. Their high lubricity is extremely useful for use as friction-reducing additives in many applications, for example, in tractor transmissions and hydraulic pumps (ASTM D2882 and D2271). Their viscosity indices are much higher than those of the mineral oils. For example, soybean oil has a viscosity index of 223 relative to 90 to 100 for most conventional petroleum oils, see Fig. 3.53. In the figure, the viscosity index of the rapeseed oil is compared with that of the other viscosity-modified oils. As one can see the rapeseed oil has a viscosity index that is better than even 10W-40 oil, an oil that contains a viscosity modifier, as indicated by the flatter slope. This implies that such oils maintain their viscosity well at high temperatures. Vegetable oils have flash and fire points of around 610°F (326°C) compared to a flash point of approximately 392°F (200°C) for the mineral oils. This makes them highly suitable for applications where a fire hazard may exist.

Vegetable oils suffer from a number of disadvantages which the user must either accept or must correct prior to their use in formulating high-performance lubricants. Performance-related deficiencies are their less than desired oxidative stability and poor low-temperature properties. Because of the advantages listed in the previous paragraph, the lower thermo-oxidative stability of the vegetable oils compared to the mineral oils must not discourage a formulator from using these oils in lubricants. It is also important to note that the thermo-oxidative decomposition products from the vegetable oils are much less harmful than those from the mineral oils. On oxidation, mineral oils form short chain acids, which react with metals to form oil-insoluble materials. Vegetable oils, on the other hand, form fatty acids, which are not only chemically less reactive but can also act as corrosion inhibitors and friction modifiers [175]. In addition, the viscosity increase due to oxidation is slow and continuous, which can be beneficial in some applications, such as hydraulic fluids and shock absorber lubricants.

Some vegetable oils, such as rapeseed oil, possess superior tribological and environmental properties, good lubrication characteristics, and are easily miscible with mineral oils and most synthetic fluids. In addition, they have high shear resistance, which makes them useful in formulating industrial lubricants for applications that involve high-shear hydraulic pumps [176]. It is important to note that because of the highly polar nature of the vegetable oils, the additive chemistry for these oils is somewhat different from that used for the mineral base oils.

With the exception of the thermo-oxidative stability, the rapeseed oil compares well in performance with most synthetic base stocks. Hence, it is being used to formulate certain specialty lubricants for the European market [176]. Although refined rapeseed oil is more expensive than refined mineral oil, it is lower in cost relative to synthetics and also provides better environmental compatibility.

Melting Point/Pour Point

At low temperatures vegetable oils solidify much more readily than mineral base oils. The typical melting range for fats and oils is between 11°C for cottonseed oil and 50°C for beef tallow. Other oils have melting points between these two extremes [175]. Typical liquefaction temperatures (pour points) for *mineral base oils* range from −23°C to −10°C. The petroleum fractions that are used to obtain these oils have pour point ranges between less than −12°C for the light gas oil to 49°C for the residual oil. The purification process improves their low-temperature fluidity characteristics. These fractions have higher pour points than the finished

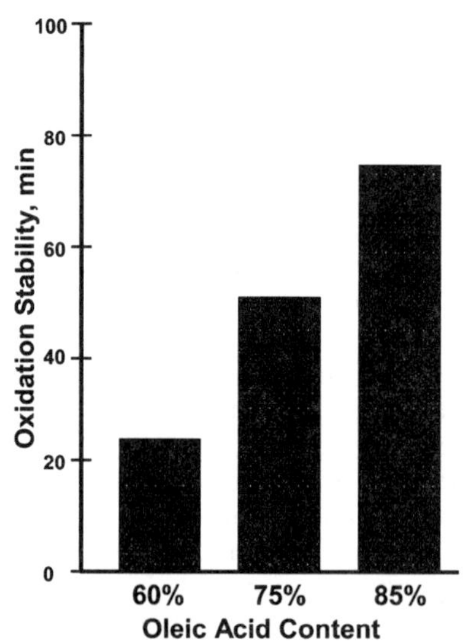

Fig. 3.54—Oxidation stability versus oleic acid content [178].

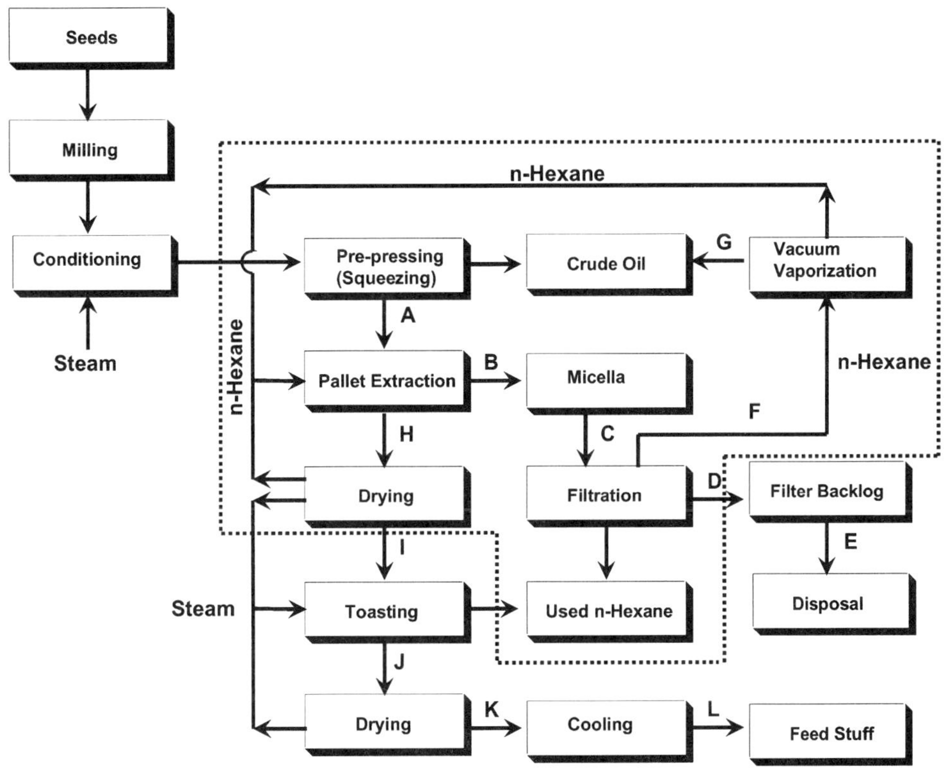

Fig. 3.55—Rapeseed oil extraction [115].

base oils because of the presence of wax that is removed during refining. The melting points of the *oils of biological origin*, on the other hand, are not due to the presence of the removable impurities but are related to the fundamental structure of the organic compounds that make up their bulk. Since these oils do not respond well to pour point depressants that are commonly used to improve the low-temperature properties of the mineral oils, blending with fluids that have good low-temperature fluidity is an option. Blending with some synthetic esters is quite effective and hydraulic fluids with pour points of −32.8°F (−36°C) have been developed [177]. If a high degree of biodegradability is to be maintained, biodegradable synthetic esters must be used as diluents.

Oxidative Stability

All hydrocarbon materials, irrespective of their origin, are prone to oxidation. Autoxidation of organics is discussed in detail in the oxidation inhibitors section of Chapter 4 on Additives. Atmospheric oxygen attacks the weaker portion of the molecule to form hydroperoxides. These materials can fall apart to form free radicals that start the oxidation chain mechanism. Allylic and benzylic hydrogens, if present in a molecule, are among those that are highly susceptible to oxidative attack. Most natural oils are primarily glycerol esters of unsaturated carboxylic acids, such as oleic acid, linoleic acid, and linolenic acid. These compounds contain allylic hydrogens that oxidize readily and are the reason for the poor oxidative stability of the natural oils. Mineral oils, on the other hand, either do not contain such components or if they do, their oxidative susceptibility can be reduced through hydrogenation. There is no straight forward way to achieve this for the natural oils.

Two ways to improve oxidation stability of these oils are the use of the traditional oxidation inhibitors, the same as for mineral oils, and the structural modifications. Unfortunately, this further exacerbates their cost disadvantage relative to the mineral oils. One approach that has been used to improve oxidation stability is to lower their overall olefinic content (unsaturation), either through selective breeding of the seed or the fruit-bearing plants by genetic manipulation or by chemical means. Olefinic sites in the natural oil structure are among those that are the most susceptible to oxidation. Selective breeding can result in genetic strains that can produce oils of up to 98 % oleic (mono-unsaturate) content. Oils of high mono-unsaturate (low poly-unsaturate) content have superior oxidation stability than those of the low mono-unsaturate content. This is demonstrated in Fig. 3.54 [178]. One can see that as the mono-unsaturate content increases, so does the oxidative stability. High oleic soybean, canola, rapeseed, and sunflower oils are now becoming standard base oils for formulating biodegradable lubricants and greases.

Chemical modification involves partial hydrogenation of the vegetable oil and the redistribution of the fatty acids. While hydrogenating, it is important to stop at a point where an oil of reasonable oxidative stability is obtained, without the loss of its fluidity. The low-temperature properties of the vegetable oil are not great to start with and too much hydrogenation can convert the oil into a product that is semi-solid, with grease-like consistency even at room temperature. Most genetically manipulated and chemically modified oils retain high biodegradability in tests established by ASTM and OECD. Most vegetable oils show a biodegradability of better than 70 % in 28 days. Petroleum oils normally show a

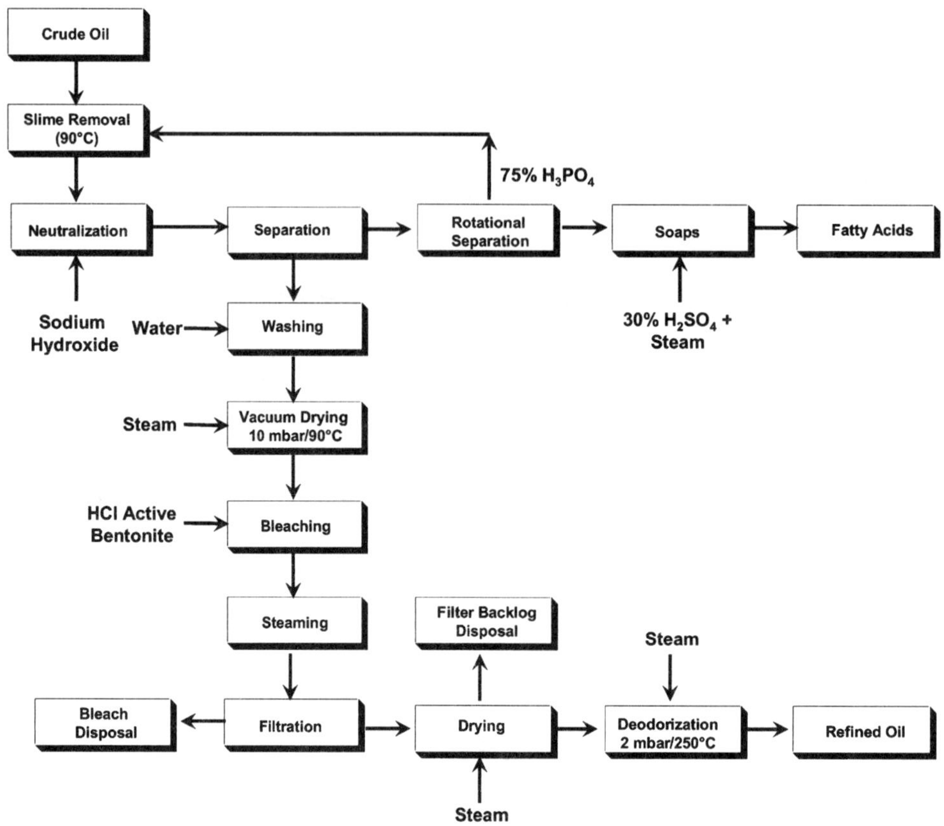

Fig. 3.56—Rapeseed oil purification [115].

biodegradability of only 15 to 35 %. Greater than 60 % biodegradability is the limit for *readily biodegradable* materials. Vegetable oils are fairly benign to fish, daphnia, and other organisms that are used to test toxicity. However, the toxicity increases when fully formulated lubricants are used. Hence, it is absolutely critical to choose highly biodegradable and nontoxic additives to maintain the environmental benefits of the vegetable oils.

Manufacture and Processing

Commercial sources of oils and fats are seeds, fruits, animals, and fish. The oil is primarily stored in the seed kernel, the fruit pulp, and the animal tissue. Soybean, sunflower, peanut, cottonseed, coconut, rapeseed, linseed, palm kernel, corn, castor seed, and safflower are the common seeds, and palm and olive are the two primary fruits that are used to extract oil. Because of the sheer volume and their more fluid

TABLE 3.35—Fatty acid composition of commonly used fats and oils in lubricant-related applications [179].

	Saturated					Mono-unsaturated	Poly-unsaturated	
Oil or fat	Capric Acid (C10:0)	Lauric Acid (C12:0)	Myristic Acid (C14:0)	Palmitic Acid (C16:0)	Stearic Acid (C18:0)	Oleic Acid (C18:1)	Linoleic Acid (ω6) (C18:2)	α-Linolenic Acid (ω3) (C18:3)
Beef Tallow	...	...	3 %	24 %	19 %	43 %	3 %	1 %
Canola Oil	...	...	...	4 %	2 %	62 %	22 %	10 %
Cocoa Butter	...	...	...	25 %	38 %	32 %	3 %	...
Coconut Oil	6 %	47 %	18 %	9 %	3 %	6 %	2 %	...
Corn Oil (Maize Oil)	...	...	...	11 %	2 %	28 %	58 %	1 %
Cottonseed Oil	...	...	1 %	22 %	3 %	19 %	54 %	1 %
Olive Oil	...	...	...	13 %	3 %	71 %	10 %	1 %
Palm Oil	...	...	1 %	45 %	4 %	40 %	10 %	...
Palm Kernel Oil	4 %	48 %	16 %	8 %	3 %	15 %	2 %	...
Peanut Oil	...	...	...	11 %	2 %	48 %	32 %	...
Safflower Oil[a]	...	...	...	7 %	2 %	13 %	78 %	...
Soybean Oil	...	...	...	11 %	4 %	24 %	54 %	7 %
Sunflower Oil[a]	...	...	...	7 %	5 %	19 %	68 %	1 %

[a]Not high-oleic variety. Percentages may not add to 100 % due to rounding and other constituents not listed. Where percentages vary, average values are used.

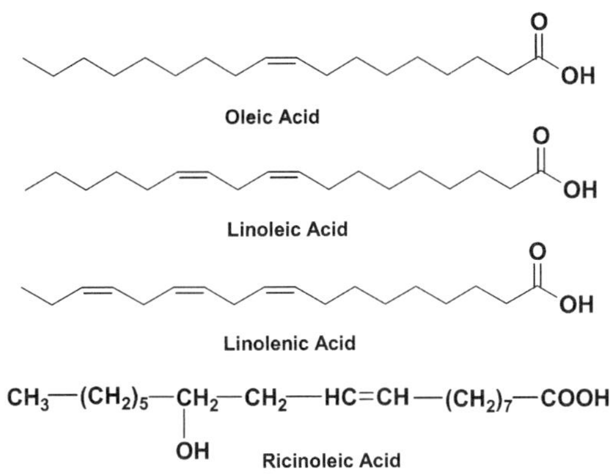

Fig. 3.57—Fatty carboxylic acids.

nature, the discussion in this book will be limited to the vegetable oils. The manufacture of the base stocks from these sources involves the following steps.

The dried, cleaned, and dehulled oilseed kernels are milled, cooked (conditioned) with steam, and prepressed (squeezed) in screw presses using 200 bar pressure to remove 40 to 60 % of the oil present in the seed. The use of steam (80–90 °C) makes the oil thinner that facilitates the oil separation. The press cake, called pallet (oil content reduced to 10–15 %), is flaked using steam rollers and the flakes are solvent extracted, primarily using hexane. The reason for prepressing is that most solvent extraction units cannot handle high-fat seeds. The solvent extract containing the oil is filtered by the use of micella to remove fine solids, which are dried to remove the residual hexane, prior to disposal. This portion of the process follows paths A, B, C, D, and E in Fig. 3.55 [115]. The solvent extract from this step is sent to a stripper to remove hexane by vacuum evaporation and yield the remaining oil left in the seed after prepressing, see paths F and G. The oil obtained from this step is combined with that obtained from the prepressing step to yield the crude vegetable oil, which is subjected to further purification to convert it into the lubricant base stock. The flakes coming out of the extraction process (oil content <1 %) go to the solvent stripper to remove all traces of solvent before being processed as a meal. The process follows paths H, I, J, K, and L in Fig. 3.55. Incidentally, the box in the figure shows the *n*-hexane use and recycling loop. The same extraction procedure applies to the recovery of the oil from the fruit pulp.

A number of purification steps are necessary to purify crude vegetable oils for use as lubricant base stocks [115]. These include precleaning to remove phosphatides (slime removal or degumming), neutralization to remove acidic impurities (refining), and decolorizing (bleaching) by the use of absorbents, e.g., bleaching clays. *Degumming* prior to refining is usually not necessary, except in the case of the soybean oil. Degumming involves stirring the crude oil with water at 70 °C, followed by high-speed centrifugation to remove lecithin. *Refining* is carried out to remove the free fatty acids, gums, and pigments. A number of methods are used for this purpose. These include alkali refining, micella refining, steam refining, and bleaching and decolorization. Alkali refining involves treating the oil with a small amount (0.1 %) of 12 % caustic solution (aqueous NaOH), mixing, and holding the mixture to hydrate gums. This is followed by heating to 75 °C to break the emulsion and high-speed centrifugation. No additional refining is carried out if the oil is to be used as a soap stock. If intended for other uses, the oil is washed with 10–20 % water using a water-wash centrifuge, vacuum dried, and cooled. Micella refining is similar to alkali refining, except that the oil is dissolved in a solvent, such as hexane. The hexane-oil mixture is transferred from the refining unit to the extraction unit where it is treated with gum conditioners and alkali, heated, mixed, and centrifuged before bleaching and removing the solvent. *Bleaching and decolorization* is carried out for most edible and industrial applications. The oil from the refining step has a dark color due to the presence of the oxidation products, carotenoids, and chlorophyll. Absorptive bleaching is commonly used to lighten the color. In this process, oils are mixed with 0.2–2.5 % neutral or acid-activated earths, such as bentonite clays, and are heated to 80–110 °C for about 15 minutes with mixing, preferably under vacuum, and filtered. Easily bleached oils are treated with neutral earths. Those containing more difficult to remove pigments require acid-activated earths. Chlorophyll colors often require the additional use of up to 10 % activated charcoal. Since these absorbents retain some oil, minimum amounts are used to lower the loss due to absorption. All purification steps are depicted in Fig. 3.56.

Composition of Natural Oils and Structural Modifications

A variety of vegetable and fruit oils can be used to formulate bio-lubricants. These are listed in Table 3.35, along with the composition [179]. These oils are mixtures of saturates, mono-unsaturates, and poly-unsaturates. While saturates are triglycerides of a variety of saturated acids, unsaturates are primarily triglycerides of oleic and linoleic acids. Another fatty acid, ricinoleic acid (12-hydroxyoleic acid), which is not present in common seeds and fruits, is isolated from castor oil. See Fig. 3.57 for the structures of the unsaturated acids that make up the structures of these oils. Ricinoleic acid is hydrogenated to 12-hydroxystearic acid which is used to make soap for the manufacture of the lubricating greases. Of the oils that are liquids at room temperature, canola (rapeseed) and soybean oils are among the most commonly used vegetable oils in lubricants. Both oleic and linoleic acids that make up the fat and the vegetable oil composition contain points of unsaturation, or carbon-carbon double bonds. The presence of these sites is responsible for the poor oxidation stability of the natural oils. As mentioned earlier, hydrogenation can correct the problem of oxidation to some degree, but at the expense of raising the oil's pour point. One strategy that is used to facilitate their use as lubricant base stocks is by blending them with synthetic esters. Such blends have improved oxidation stability and lower pour points than the vegetable oils alone and are lower in cost than that of the pure ester. In addition, if the synthetic ester is biodegradable, biodegradability is maintained in the final blend. Improved oxidation stability and good low-temperature properties are due to the dilution effect. The second approach is to modify the vegetable oil structure by changing the alcohol (glycerol) portion of the triglyceride structure. Although such oils can no longer be called natural, they have good oxidation stability, pour points, and biodegradability. Changing the alcohol portion can be achieved ei-

TABLE 3.36—Currently available base fluids and their properties [179].

Fluid	Kinematic Viscosity, cSt		VI	Pour Point	NOACK Loss, %	Biodegradability %	Source
	100°C	40°C					
Vegetable Oils							
Rape Seed	9.1	40.3	217	-18	–	75 - 100	Biological
Sunflower	7.7	31.6	226	–	–	75 - 100	Biological
Soybean	7.6	31	227	-9	–	75 - 100	Biological
Trimethylolpropane (TMP) Esters							
n-Heptanoic Acid	3.4	13.9	120	-60	11.8	100	Mineral + Biological
n-Nonanoic Acid	4.6	21	139	-51	2.3	100	Mineral
Oleic Acid	9.4	46.8	191	-39	–	100	Mineral + Biological
Iso-nonyl	7.2	51.7	98	-32	6.7	7	Mineral
Neopentyl Glycol (NPG) Esters							
n-Heptanoic Acid	1.9	5.6	–	-64	–	100	Mineral
n-Nonanoic Acid	2.6	8.6	145	-55	31.2	97	Mineral
Pentaerythritol (PE) Esters							
n-Heptanoic Acid	6.1	32.2	140	-7	0.9	100	Mineral
n-Nonanoic Acid	11.6	129.2	70	-22	–	8???	Mineral
Synthetic Di-esters							
Di(2-ethylhexyl) Adipate	2.4	8	124	-68	44.3	97	Mineral
Di(2-ethylhexyl) Azelate	3	10.7	137	-64	29	99	Mineral + Biological
Di(2-ethylhexyl) Sebacate	3.1	11.8	126	-60	18.3	96	Mineral + Biological
Di(2-ethylhexyl) Dodecanedioate	3.8	14.3	168	-57	–	–	Mineral
Di(iso-decyl) Adipate	3.6	15.2	121	-62	15.5	84	Mineral
Di(iso-decyl) Azelate	4.3	18.1	151	-65	9.8	86	Mineral + Biological
Di(iso-decyl) Sebacate	4.8	20.2	169	-60	6.2	100	Mineral + Biological
Di(iso-decyl) Dodecanedioate	5.2	23.4	162	-41	4.3	93	Mineral
Iso-tridecyl	Min						
Di(iso-tridecyl) Adipate	5.4	27	139	-51	4.8	92	Mineral
Di(iso-tridecyl) Sebacate	6.7	36.7	141	-52	3.7	80	Mineral + Biological
Di(iso-tridecyl) Dodecanedioate	7.6	40.7	156	-50	2.9	76	Mineral
Polyalphaolefins							
PAO 2	1.8	5.5	-	-65	99	75-92	Mineral
PAO 4	3.9	16.8	129	-70	12	52 - 72	Mineral
PAO 6	5.9	31	138	-68	7	20 - 23	Mineral
PAO 8	7.8	46	140	-63	3	20 - 22	Mineral
PAO 10	9.6	63	134	-53	2	16 - 21	Mineral
PAO 40	40	395	151	-34	0.8	–	Mineral

ther by direct transesterification reaction or by first saponifying the oil and then converting the isolated acid into an ester by reacting it with a mono- or polyhydric alcohol. These reactions are depicted in Fig. 3.24, along with other esters that possess biodegradability as well. Various properties of the parent vegetable oils and these modified esters are provided in Table 3.36 [180]. As one can see, the modified esters have excellent low- and high-temperature viscometrics, improved oxidation stability, and biodegradability. Although their biodegradability is somewhat lower, they still meet the biodegradability requirements to be labeled as biodegradable. There are a few on the list, such as higher grades of PAOs, which have low biodegradability. As mentioned earlier, additives can negatively impact biodegradability and nontoxicity of these base oils. Unfortunately, at this time, a full array of nontoxic additives is not yet available to the formulators.

So far the biological base stocks have been used to formulate lubricants for hydraulics and other devices that are utilized in natural settings. Industries that employ bio-based lubricants include farming, forestry, construction, off-shore drilling, and maritime; and a substantial amount of data supporting the benefits of these lubricants are available. However, data regarding their use in highly demanding automotive applications, such as engine oils, are either sparse or nonexistent. Interestingly, an article published in a recent issue of Lubes and Greases magazine provides an insight into the expected performance of the bio-based engine oils in current engine tests [180b]. It appears that the performance deficiencies that exist may be overcome by the use of suitable additives.

Based on the discussion in this section, it is clear that all lubricant base stocks are not alike. In practice, many lubricant base oils are manufactured according to the market specifications, by blending base stocks of various physical and chemical properties. Despite this effort to narrow their structural and performance differences, lubricants developed from the same additive package may have radically different performance. To minimize this problem, the API has classified base oils into five categories, Group I to Group V, and has published guidelines relating to interchange of base oils.

4

Lubricant Additives

IN THIS CHAPTER WE DESCRIBE THE CHEMISTRY, manufacture, and properties of the chemicals that are used in lubricants as additives. The function of these chemicals is either to enhance the already existing properties of the base fluids or to impart new properties that they lack. Discussion also includes the mechanism by which each additive type performs its functions, multifunctional nature of some additives, and their impact on the environment. The need and the development of a new additive and the process by which it is approved for use in a particular application are also explained.

All mechanical equipment must be lubricated. The primary purpose is to reduce friction and wear. If not controlled, these can lead to inefficiencies, damage, and ultimately to equipment seizure.

Pictorial records depicting the use of lubricants date as far back as 1650 B.C. Analysis of the residue from chariot axle hubs suggests the use of animal fat as a lubricant, as early as 1400 B.C. This practice continued until 1859 when petroleum-based lubricants became available. While the claimed use of additives in lubricants dates far back into the 1860s, the modern history of lubricant additives began in the early 20th century, with the use of fatty oils and sulfur in mineral oils to improve lubrication under high loads. The use of additives became common only after the 1930s when more compact and faster engines were developed and the OEMs started to specify oils for use in their equipment [181,182]. However, these were the military needs during and after the Second World War that were the major driving force behind the development of the lubricant additives as we know them today. This is because the fast-running engines had a complex design, which placed a heavy demand on the lubricants. At that time, besides the SAE viscosity classification system, there were no established performance criteria and the end-user had to depend on the lubricant supplier's claims regarding the suitability of the lubricants [182]. Additives are not only critical to the manufacture and use of automotive fuels and lubricants; they are also extensively used in petroleum recovery as drilling additives and refining as refinery chemicals [183]. This was discussed in Chapter 2 on Mineral Base Stocks. Some of the highlights of the additives development are as follows.

1. Sodium soap greases were developed in the 1930s. They did not suffer from water sensitivity and had better resistance to high temperatures than the lime or calcium soap (calcium carboxylate) greases, the use of which dates back to the 1880s.
2. In 1937, the lubricants containing a lead soap-active sulfur additive combination, known as an extreme-pressure or EP additive system, was found to be effective. When GM in its 1937 models introduced hypoid axles, it tested and recommended the use of a number of such lubricants.
3. Complex calcium soap greases were invented in 1940, lithium soap greases were invented in 1942/1943, and barium greases were invented in the 1950s. These greases retain their semi-solid morphology at much higher temperatures than their sodium and calcium analogues. They contain sodium nitrite to inhibit corrosion and oxidation.
4. Until 1947, the performance of the engine oils was defined solely by their viscosity grades, without due consideration to the engine design, its operating environments, and the fuel type and quality. In 1947, API introduced three performance categories, *regular, premium,* and *heavy-duty,* based on the severity of service. *Regular* oils were straight mineral oils with or without viscosity modifiers and corresponded to oils defined previously by the viscosity grades. These oils were for both gasoline and diesel engines operating under mild to moderate service conditions. *Premium* oils, designed for somewhat more severe operating conditions, generally contained oxidation and corrosion inhibitors, and, in some cases, mild detergents. *Heavy-duty* oils possessed better oxidation and corrosion resistance and detergency than the *premium* oils and were designed to withstand the most severe service.
5. In the early 1950s, multi-grade oils were an outcome of the development of the viscosity modifiers. These were made by dissolving polymeric materials to low viscosity, or thin, oils.

All of today's lubricants contain additives, which impart to the lubricants the necessary performance. The amount of additives in a lubricant can range between 1 to 25 %, and more. See Table 1.4 for typical additive concentrations for some commonly used lubricants. The largest use of additives is in automotive lubricants, such as engine oils (for gasoline and diesel engines), transmission fluids, gear oils, and greases. The primary role of the lubricants is to provide protection for the metal surfaces against wear and corrosion. Lubricant additives for automotive lubricants are generally supplied as a performance package, which is blended in base stocks or base oils to yield formulated lubricants. These lubricants must meet the performance requirements established by organizations such as the SAE, API, U.S. Military, OEMs, and the end-users. The performance package contains many classes and types of additives [184,185]. While there are a number of ways to classify additives, we have grouped them into stabilizers/

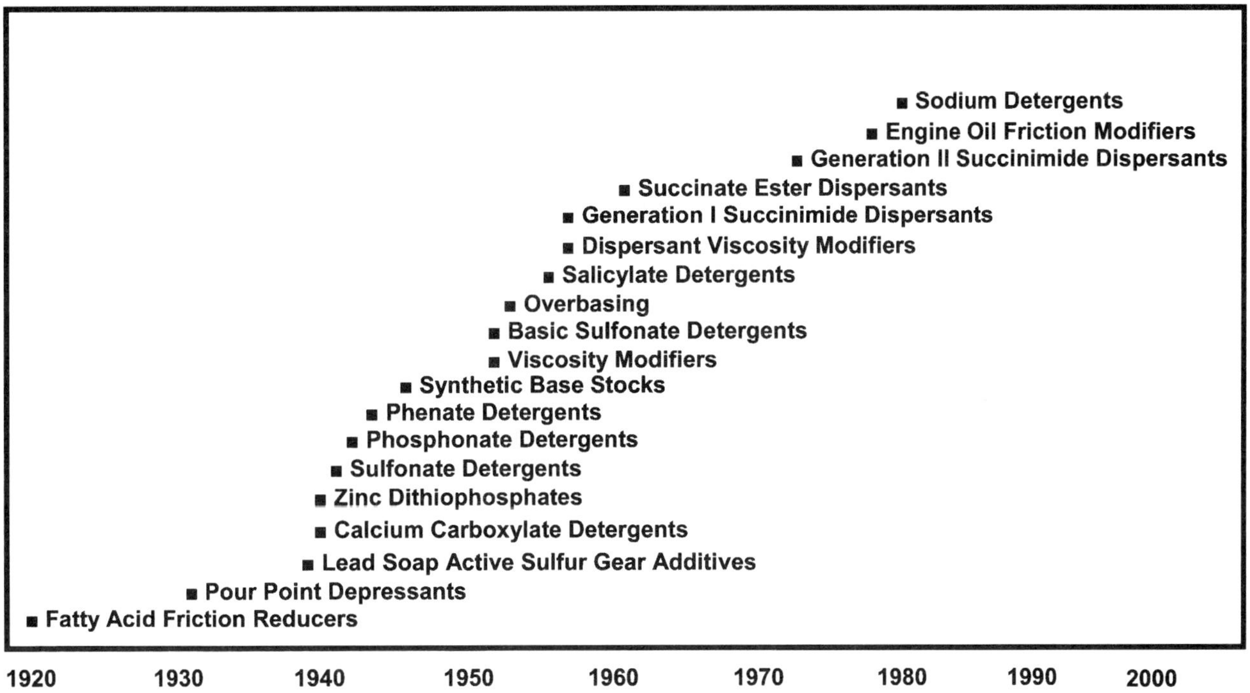

Fig. 4.1—Classes of additives—Chronology of their development.

deposit control agents, film-forming agents, polymeric additives, and miscellaneous others, based either upon their function or their structural characteristics. The primary functions of each type are briefly described below. A more detailed discussion on the chemistry and mode of their action will be provided in the subsequent discussion.

- Stabilizers/Deposit Control Agents—minimize the amount of deposit formation.
  Oxidation inhibitors and metal deactivators control oxidative decomposition of the lubricant and the additives.
  Dispersants keep normally insoluble contaminants dispersed in the lubricant.
  Detergents prevent metal attack by acidic byproducts of combustion and oxidation by neutralizing and suspending them, thereby keeping metal surfaces free of deposits.
- Film-forming Agents—either increase the durability of the lubricant film or form chemical films on metal surfaces.
  Friction modifiers generally lower the coefficient of friction, thereby leading to improved fuel economy.
  Antiwear and extreme pressure agents protect metal surfaces against wear and equipment seizure.
  Rust and corrosion inhibitors prevent corrosion and rusting of the metal parts that come in contact with the lubricant.
- Polymeric Additives—alter physical properties, such as viscosity and pour point, of the lubricants.
  Viscosity modifiers minimize the rate of viscosity decrease with an increase in temperature.
  Pour point depressants enable a lubricant to flow at low temperatures.
  Emulsifiers promote mixing of water and oil to form an emulsion. These additives are used to make water-based hydraulic fluids and metalworking lubricants, which employ such emulsions.
  Demulsifiers enhance water separation from the oil contaminated with water.
  Foam inhibitors prevent lubricant from forming persistent foam.
- Miscellaneous Additives—perform miscellaneous other functions.
  Seal swell agents swell elastomer seals.
  Dyes are used to color code lubricants and fuels.
  Biocides prevent degradation of the high water-based lubricants due to microbial attack.
  Couplers help stabilize water-organic micro-emulsions.

The evolution of the major classes of additives is provided in Fig. 4.1. The world lubricant additives demand for the year 2006 is estimated at 3 million metric tons (~6.6 billion lb), which is expected to grow at an average rate of less than 1 % per year, until the year 2010 [186]. The additive growth rate in North America over the next five years is predicted to remain flat; Europe, Middle East, and Africa will see a decline; Asia Pacific and Latin America will see an increase of 5 % and 2 %, respectively [186]. The U.S. consumption in 2006 was estimated to be about 862,000 metric tons (1.9 billion lb), almost 29 % of the world total. This volume reflects an average annual growth rate of around 2 % since 2002, which is primarily due to changing diesel engine oil specifications to meet stringent environmetal standards and the demand for high-performance industrial lubricants that meet the occupational safety and health requirements, established by Occupational Safety and Health Administration (OSHA). The increased use of synthetics, vegetable oils, and their blends, which require higher additive treat rate levels, have also contributed towards this growth.

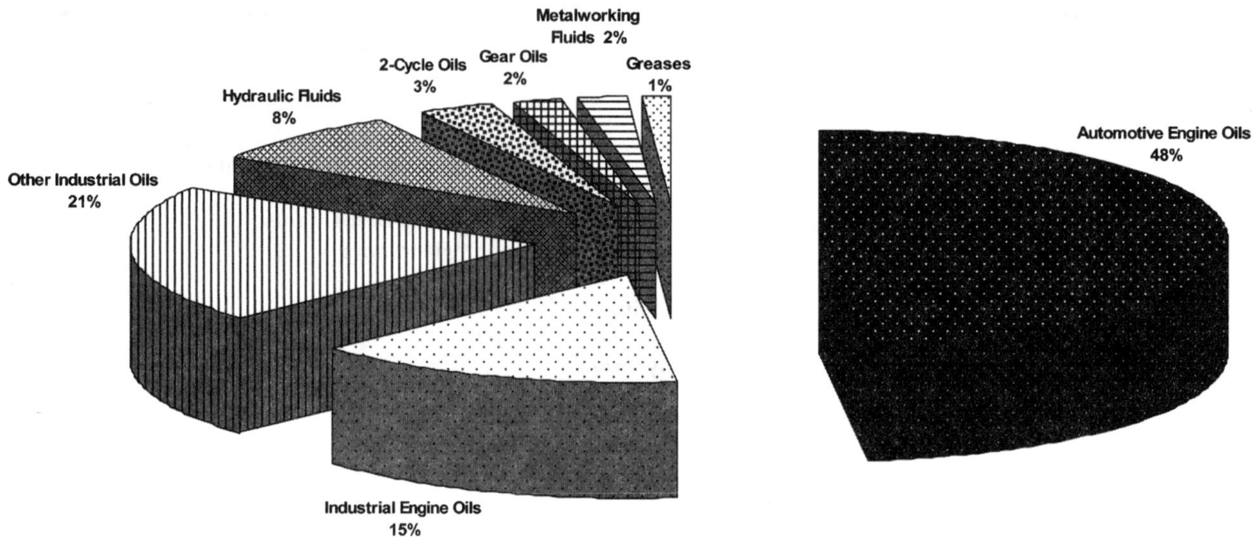

Fig. 4.2—Estimated additive use for 2006 by application, North America.

Estimated North American use for the year 2006 by application is depicted in Fig. 4.2 and by additive type is provided in Fig. 4.3. As one can see, engine oils account for 63 % of the total additive use. Detergents and dispersants, that are classified as deposit control additives, and viscosity modifiers account for 55 % of the total additive consumption. The engine lubricants are exposed to combustion products in the form of blow-by that travel past the piston rings into the lubricant. In addition to containing unburned and partially burned fuel, the blow-by contains reactive intermediates from fuel oxidation, such as peroxides and peroxy-free radicals, nitrogen oxides, resulting from the high-temperature reaction of nitrogen and oxygen from the air, soot, sulfur oxides, carbon monoxide, carbon dioxide, and water. Peroxides and peroxy-free radicals initiate lubricant oxidation to form deposit precursors and acidic products. The formation and or separation of these products on hot surfaces must be minimized; otherwise deposit formation and acid-related corrosion damage will occur. Hence, engine lubricants are formulated with oxidation inhibitors, which control the formation of these species, and dispersants and detergents that keep them away from the surfaces by neutralizing them or suspending them in the bulk lubricant. In addition, dispersants have the ability to largely contain undesired particulate emissions (PM) in the bulk oil. Hence, the large volume use of these classes of additives is consistent with the greater use of the engine lubricants. As will be discussed later, the use of detergents is expected to drop because of the incompatibility of the sulfonate detergents with particulate filtering devices (DPFs), discussed in the Chapter 6, the Emissions Chapter. As mentioned above, from 2006 to 2010, the growth of the lubricant additives is expected to plateau. This is because of the maturity of the North American lubricants market and a variety of other factors, such as the growing consumer interest in hybrid cars, extended service (drain) in-

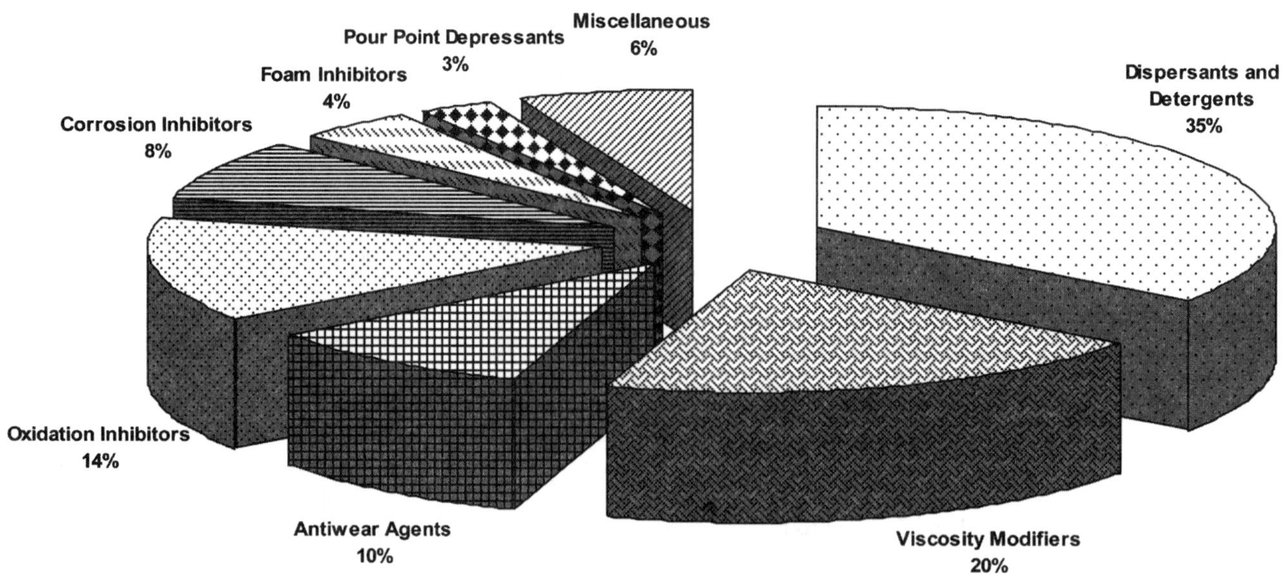

Fig. 4.3—Estimated additive use for 2006 by additive type, North America.

TABLE 4.1—U.S. Lubricant Additives Demand (million lb)—Source: Fredonia Group.

Item	Year				% Annual Growth	
	1989	1999	2004	2009	1989–1999	1999–2004
Finished Lubricants Demand (million gal)	2668	2850	2943	3066	0.7	0.6
Additives Use—lb/gal	0.57	0.66	0.65	0.65	...	...
Lubricant Additives Demand	1529	1870	1920	1995	2.0	0.5
Deposit Control Additives	760	905	903	917	1.8	0.0
Viscosity Modifiers	268	329	340	355	2.1	0.7
Antiwear and Extreme-pressure Additives	165	197	204	214	1.8	0.7
Corrosion Inhibitors	93	119	130	139	2.5	1.8
Oxidation Inhibitors	87	111	118	125	2.5	1.2
Foam Inhibitors	50	67	74	80	3.0	2.0
Pour Point Depressants	27	36	37	39	2.9	0.5
Miscellaneous Additives	79	106	114	126	3.0	1.5
Price—$/lb	0.77	0.84	0.86	0.88	0.9	0.5
Total Value ($ million)	1170	1570	1655	1765	3.0	1.1

tervals, low oil consumption engines, sealed-for-life components, and the use of highly refined base oils that require lower additive treatment. Additive use data from the Fredonia group for the 1989–2009 period, presented in Table 4.1, aptly support this prediction. Examination of the data reveals the following.

1. The growth of the additives for the 1989–99 period outpaced the growth of the lubricants, but for the 1999–2009 period, the two are almost equal.
2. The additive use per gallon increased during the first decade, but has leveled off in the second decade.
3. The strongest sustained growth occurred and will occur in oxidation inhibitors, corrosion inhibitors, and foam inhibitors.

Desirable Lubricant Properties

For lubricants to perform effectively, they must possess certain specific properties, which include suitable viscosity, slipperiness, high film strength, low corrosivity, low pour point, good cleansing and dispersing ability, nontoxicity, and low flammability and volatility. In addition, the lubricant must not foam. It must also be capable of getting rid of air (oxygen) to minimize oxidation and maintaining its lubricating characteristics. As mentioned in the prior discussion, the lubricants are formulated by blending base oils with additives to meet a series of performance specifications. These specifications relate to the physical and chemical properties of the lubricant, when it is new and during use, and its ability to protect the equipment against damage during service. Many of these properties were discussed under the lubricant selection criteria in Chapter 1. Previously, we also stated that the base fluid is the largest component in a lubricant and its properties are likely to have the greatest effect on lubricant properties. Hence, it is important to choose a base oil that by and large has the properties that are desired in a lubricant.

Criteria For Suitable Base Stocks

Although in a lubricant many of the base fluid's properties are modified or enhanced by the use of the additives, the knowledge of such properties is critical to the formulator. These properties deal with the base oil's density, viscosity, both at low and high temperatures, foaming characteristics, seal compatibility, oxidation resistance, corrosivity, the viscosity-temperature relationship (VI), low-temperature properties, such as cloud point and pour point, and high-temperature properties, such as volatility and flash point [4]. Other important properties include thermal expansion, bulk modulus or compressibility, thermal conductivity, thermal capacity, electrical conductivity, and surface tension. The knowledge of these properties is especially useful in designing lubricants for specialty applications.

The viscosity and viscosity-temperature relationship were discussed in the viscosity section of Chapter 1 on Lubrication Fundamentals. The interested reader may like to refer back to that chapter. The high-temperature properties of the base oil are governed by its distillation temperature, or its boiling range. *Volatility* is important because it is an indication of the oil's tendency to be lost during service due to vaporization, for example, in a hot engine. The methods used to determine volatility include distillation curves (see Chapter 3 on Synthetic Base Fluids), thermo-gravimetric analysis, and NOACK volatility. *Flash point* is the lowest temperature at which auto-ignition of the vapor above the heated sample occurs. Flash point of an oil is important from a safety point of view and is used to classify flammable liquids into hazard grades. There are two methods to determine the flash point: the closed cup method (Pensky-Martens Closed Cup or PMCC) and the open cup method (Cleveland Open Cup or COC) [4].

Density is important because oils may be formulated by weight but measured by volume. Density is also used to identify an oil, or its fractions, and in calculating kinematic viscosity from absolute or dynamic viscosity. Demulsibility is the ability of an oil to separate water. Foaming characteristics determine the tendency of the oil to form foam and the stability of the foam once it is formed. Seal compatibility of the oil is also important because the oil in most applications comes in contact with elastomer seals. If seals are damaged, the equipment will be left unprotected since the lubricant will be lost.

Oxidation resistance of the base oil depends largely upon the structure of the hydrocarbons present. As mentioned earlier, lubricants that do not contain aromatic structures or structures with unsaturation have better oxidative stability than those that do. Oxidation of the base oil results in the formation of the polar compounds that are either corrosive or lead to the formation of sludge and resin, both of which can impair proper functioning of the equipment. Previously, the sulfur content of the base oils was used as an indicator of their natural resistance to oxidation. This is because many naturally-occurring organo-sulfur compounds in crude oils are moderately effective in destroying the organic peroxides and breaking the oxidation chain mechanism. However, modern refining processes, which are used to enhance the other desirable properties of the base oils, result in the removal of these beneficial compounds.

Thermal expansion predicts an increase in volume of a given mass of oil with an increase in temperature and is accompanied by a decrease in density. The degree of expansion is expressed as the coefficient of thermal expansion. The knowledge of the oil's thermal expansion is useful for determining the size of the container needed when the oil is heated. Its knowledge is also useful in assessing the lubrication needs of the bearings since in elasto-hydrodynamic lubrication (EHD) regime the thermal expansion of the oil results in an increase in the hydraulic pressure.

As mentioned in Chapter 3, the bulk modulus expresses the resistance of a fluid to a decrease in volume due to compression and is accompanied by an increase in density. Compressibility is the reciprocal of the bulk modulus, or the tendency to be compressed. In high-pressure hydraulic systems, a high bulk modulus, or low compressibility, is required to transmit power efficiently and dynamically. Mineral oils have low compressibility, but low-viscosity polysiloxane fluids have high compressibility, or a low bulk modulus.

Thermal conductivity is the rate of dissipation of heat through a material. A fluid with high thermal conductivity is expected to lower the temperature of a bearing more quickly than a fluid of low thermal conductivity. Most mineral oils and hydrocarbon fluids, such as the polyperfluoroalkyl ethers and PAOs, have very low thermal conductivity of 0.13 W/m K, or W m^{-1} K^{-1}, relative to that of water and ethylene glycol at 0.6 and 0.42 W/m K, respectively. However, polyglycols and silicones with a little higher thermal conductivities of 0.15 and 0.14 W/m K are slightly better than the hydrocarbon oils. The quick transfer of heat from a hot spot in a bearing to the lubricant can control overheating of the bearing and minimize damage.

Thermal capacity, or heat capacity, is the heat required to raise the temperature of a body by one unit of temperature, i.e., 1 °K. Specific heat is the ratio of the thermal capacity of the substance to that of water at 15 °C. Specific heat is a function of the fluid structure and the density. Higher values are better since they imply that the fluid can contain larger amounts of heat. With respect to this parameter, hydrocarbon oils and polyglycols are better than silicone oils. In hydrodynamic lubrication, specific heat is used to calculate heat transfer, temperature rise, and other thermal factors in an oil film.

Electrical conductivity is a measure of a material's ability to conduct an electric current. This property is important for insulating oils, where low conductivity base stocks are needed. A variety of base fluids are used for this application, including hydrotreated heavy paraffinics, silicones, perfluoropolyalkylethers, and fluorinated esters; all of which have good electrical properties. At one time polychlorobiphenyls (PCBs) were extensively used as insulating oils in transformers. This was because of their low water solubility, excellent thermal stability, low volatility, and virtual inflammability. However, since the 1970s their use has declined almost to nothing because of their being persistent organic pollutants. It is important to note that while clean, dry base oil has a low electrical conductivity, the used, wet, and contaminated oil can have a high conductivity. Hence, quality control of such oils is imperative.

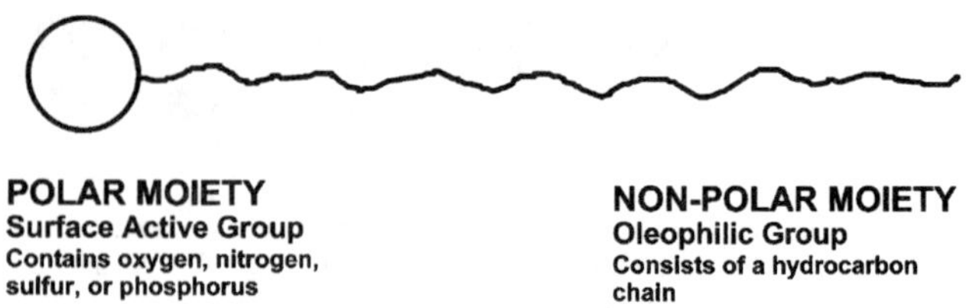

Fig. 4.4—A general representation of a typical additive molecule [50].

Surface tension is the surface energy between a liquid and its vapor, or air, or a metal surface. Surface tension is considered a factor in the ability of an oil to "wet" a surface, stability of its emulsions, and the stability of its dispersions that contain solids. Of these, the first two phenomena are most relevant to the present discussion. This is because they relate to surface affinity, a concept important in film-forming (surface-active) agents, and foaming tendency of the oil. High surface tension fluids form more persistent foams. One way to correct this problem is by the use of silicones, which reduce the surface tension of the bubbles, hence help break foams.

The base oil must not contain components that promote corrosion. Corrosion tests usually involve bringing the oil sample in contact with a metal, such as copper or silver, under controlled conditions. Discoloration of the metal, changes in its surface condition, or weight loss reflects the corrosive tendency of the oil.

Performance Additives

It is always implied that the lubrication involves the use of lubricating oil, usually formulated by blending the appropriate quality lubricant base stock(s) and additives. Incidentally, for some applications, such as automotive engine oils and transmission fluids, the additives manufacturer supplies additives as a package that meets certain industry specified testing criteria. In other cases, such as metalworking fluids and greases, the additives supplier furnishes individual additives, which the end-user adds to the base stock.

Most lubricant additives, except perhaps some viscosity modifiers and pour point depressants comprise an oleophilic (lipophilic) hydrocarbon group and a polar functional group, as shown in Figure 4.4 [50]. The polar functional group typically contains oxygen, nitrogen, sulfur, and or phosphorus. With respect to the polar group strength, the oxygen-based additives are the most polar, which are followed by the nitrogen-based additives, sulfur-based additives, and phosphorus-based additives. The polarity of the various polar functional groups follows Pauling's electronegativity scale—the electronegativities of oxygen, nitrogen,

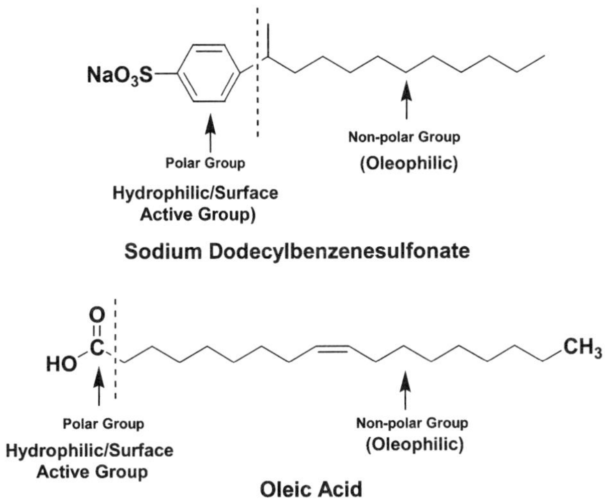

Fig. 4.5—Polar and non-polar groups in additive molecules.

TABLE 4.3—Additives and their nature.

Chemically Active/Reactive	Chemically Inert
Deposit Control Agents	*Viscosity Control Agents*
1. Oxidation Inhibitor	2. Viscosity Modifier
2. Dispersant	2. Pour Point Depressant
3. Detergent	
Film-forming Agents	*Miscellaneous Additives*
1. Friction Modifier	1. Foam Inhibitor
2. Antiwear/Extreme Pressure Agent	2. Demulsifier
3. Rust and Corrosion Inhibitor	3. Seal Swell Agent
	4. Emulsifier
	5. Dye
	6. Odor Mask

sulfur, and phosphorus are 3.44, 3.04, 2.58, and 2.19, respectively. Electronegativity is a measure of an atom's tendency to attract the bonding pair of electrons towards itself; the greater the attraction, the higher the electronegativity. Figure 4.5 shows the polar and the nonpolar portions in the structures of sodium dodecylbenzenesulfonate and oleic acid. Since initially all additives must be oil soluble, the presence of a hydrocarbon group of sufficient carbon chain length is essential.

Lubricant additives perform their function either in the bulk lubricant, such as detergents and dispersants, or on surfaces via adsorption and chemical reaction, or both, such as rust inhibitors and extreme pressure (EP) agents. We can design additives of the desired polarity either by altering the strength of the polar functional group or by changing the size of the hydrocarbon chain. Changing the strength of the polar functional group alone is difficult and has its limitations. Changing the size of the hydrocarbon chain, on the other hand, is much easier. In practice, both strategies are used. Whether an additive performs its function on the surface or in the bulk lubricant depends on its polar to nonpolar ratio. With the strength of the polar moiety constant, additives with small hydrocarbon groups have a higher polar to nonpolar ratio than those with large hydrocarbon groups. As a consequence, EP agents and rust inhibitors that require more surface activity have small hydrocarbon groups; and dispersants and detergents that require a higher solubility in oil contain large hydrocarbon groups [187,188]. Except in a very few cases, a connecting group or a link is necessary to tie the two functionalities together. The importance of such a group is described in detail in the dispersants section. Table 4.2 shows the relationship between the polar to nonpolar ratio and the additive's oil solubility and surface affinity. The table also identifies additives based on polar to non-polar ratio.

Besides the function, additives can also be classified based upon whether they are chemically active/reactive or chemically inert. Typically, chemically active/reactive additives are those that either impart or improve the chemical properties of the base fluid and chemically inert additives are those that enhance the inherent physical properties of the base fluid. Table 4.3 lists additive groups based upon these two criteria.

Stabilizers/Deposit Control Agents

Major causes of engine malfunction due to lubricant quality are deposit formation, lubricant contamination, oil thickening, oil consumption, ring sticking, corrosion, and wear. The terms that are commonly used to describe problems in automotive equipment [189] are defined below:

- Ash is the residue that results from combustion. It can be white, gray, or brown in color and is sometimes covered with a bright carbonaceous layer.
- Carbon is a firm black deposit that is readily definable by its thickness. This form of deposit is without luster, ex-

TABLE 4.2—Polar—non-polar ratio versus additive properties.

Polar Group[a]	Non-polar Group	Polar/Non-polar Ratio	Oil Solubility	Surface Affinity	Additives
Weak/Small	Large	Low	High	Low	Oxidation Inhibitor Detergent; Dispersant Viscosity Modifier Pour Point Depressant
Strong/Large	Large	Medium	Medium	Medium	Foam Inhibitor Demulsifier Friction Modifier
Strong/Large	Small	High	Low	High	Antiwear (AW) Agent Extreme Pressure (EP) Agent Corrosion Inhibitor

[a]Polar Group Strength: Oxygen-based additives > Nitrogen-based additives > Sulfur-based additives > Phosphorus-based additives.

cept when polished by various engine components. It is usually unaffected by rinsing with conventional solvents.

- Cavitation in solids refers to erosive wear that results when a solid and a fluid are in relative motion and bubbles are formed. This kind of damage appears as irregular cavities on the metal surface.
- Corrosion is the removal of metal through chemical attack on the metal surface.
- Cutting is the removal of piston or liner material that results from abrasion due to carbonaceous deposits on the mating surfaces.
- Emulsion is a cream-like substance that is often aerated and contains entrained water. It can be easily removed by wiping or rinsing with conventional solvents.
- Erosion is the mechanical removal of material due to a high velocity fluid with or without entrained particles.
- Glazing is a skin or a coating that forms chemically on the liner surface. It results from the interaction of the metal with iron oxide, graphite, or fuel and the lubricant decomposition products.
- Lacquer is a hard, dry, generally lustrous, oil-insoluble deposit. It cannot be removed through wiping. It can be gray, brown, amber, or black in color.
- Lead paint is a gray deposit that results primarily from the use of leaded fuel. It consists of lead salts that appear either as thin sludge or as baked-on deposit on parts such as piston skirts, that experience high temperatures. This is no longer a problem because of the worldwide phase out of the leaded gasoline.
- Pitting is metal damage resulting from the removal of small pieces of metal, leaving small irregular cavities.
- Polishing, commonly experienced in cylinder bores, is a bright mirror-like finish that results from the local mechanical wear of the surface.
- Rust is the oxidation of iron surfaces by an electrochemical process involving oxygen, water, and sometimes chemicals.
- Scoring is the metal damage due to metal-to-metal contact or metal contact with foreign matter. The metal is removed either through cutting or plastic deformation and the surface appears grooved or ridged.
- Scratching is metal damage due to fine line cutting in the direction of motion resulting from light metal-to-metal contact or the metal contact with small abrasive particles.
- Scuffing is surface roughening due to either progressive removal or the transfer of material resulting from localized welding and subsequent breaking.
- Sludge is primarily composed of oil and combustion products. It does not drain from the surfaces but can be easily wiped off.
- Soot is a loose powdery carbon deposit that leaves black marks on finger tips when touched. It does not adhere to any surface to any extent.

Deposit Formation

The term deposit refers to any material that has the tendency to separate out of the fuel or the lubricant onto the metal surface. Deposit formation occurs in many parts of the vehicle, including the engine, the transmission, and the gears. However, the amount of deposits in the vicinity of an engine is by far the largest. In an internal combustion engine, deposits result from two sources, the fuel and the lubricant. Fuel-related deposits occur either on the intake system, that is, intake valve deposits (IVDs), or in the combustion chamber, the combustion chamber deposits (CCDs). The nature, the mechanism, and the control of these deposits are discussed in Chapter 6 on Emissions in an Internal Combustion Engine. The lubricant-related deposit formation initiates when a lubricant is contaminated by chemically active or reactive species. Two main sources of lubricant contamination are the blow-by from the combustion chamber and the gases and the volatiles from the crankcase. Both are vented into the intake manifold as an anti-pollution measure [190]. The blow-by contains by-products of combustion and is a mixture of nitrogen oxides (NO_X), sulfur compounds (SO_2, SO_3, and H_2SO_4), carbonyl compounds, hydrocarbons (unburned fuel), fuel fragments, peroxides and free radicals, air, water, carbon monoxide, and carbon dioxide. Free radicals result when the high combustion temperatures fragment the fuel hydrocarbon molecules by exciting their bonding electrons and cleaving their bonds homolytically. High combustion temperatures are also accompanied by the electromagnetic radiation, which is the reason for the blue/violet color of the flame. Incidentally, the free radicals are the most reactive of the combustion-derived products. These species are blown past the piston rings and down the valve guides into the crankcase, where they initiate the lubricant decomposition. This suggests that the quality of the fuel, thermal and oxidative stability of the lubricant, and the efficiency of combustion all play an important role in the formation of harmful products. Such products include oxygenates, soot, carbon, lacquer, varnish, and sludge [190]. This sequence of reactions is represented by equations 1–4 in Fig. 4.6.

$$\text{Sulfur} \xrightarrow[O_2]{\text{Fuel}} H_2SO_4 \qquad (1)$$

$$\text{ROOH} + H_2SO_4 \longrightarrow R_2C{=}O \qquad (2)$$

$$R_2C{=}O \longrightarrow \text{Resins} \qquad (3)$$

$$\text{Resins} + \text{Soot} \longrightarrow \text{Carbon Deposits and Lacquer} \qquad (4)$$

Fig. 4.6—Formation of deposit precursors.

Resin is the key ingredient in the deposit formation. It consists of highly oxygenated hydrocarbon materials that result from the polymerization of the lubricant oxidation products. If the polar to nonpolar ratio of these products is low, they stay dissolved in oil, which leads to oil thickening, or a viscosity increase. On the other hand, if their polar to nonpolar ratio is high, they have low oil solubility and hence fall out of solution, to form a sticky film on metal surfaces. There, they either capture other species or chemically undergo further transformation to form deposits, see Fig. 4.7.

Soot and Carbon

Soot is an important particulate contaminant in the crankcase lubricants. It can be derived both from inefficient combustion of the fuel and burning of the lubricating oil, which travels past the piston rings into the combustion chamber. Fuel-derived soot, more commonly encountered in the diesel engines, is due to the diesel fuel's broad boiling range and

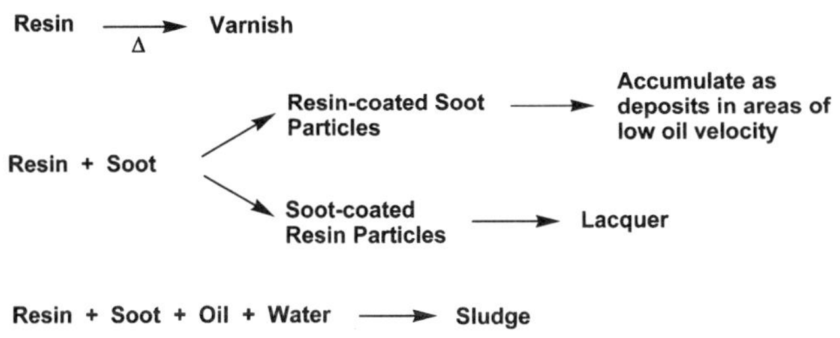

Fig. 4.7—Interaction between deposit-forming species.

the nonuniformity of its combustion mixture, which under certain operating conditions does not burn completely.

Soot is not pure carbon, but contains an appreciable amount of hydrogen, oxygen, and sulfur in a combined form. Soot particles are basically hydrocarbon fragments with hydrogen atoms partly stripped off. These particles are charged and are strongly attracted to one another and to polar compounds in the oil. When attracted to one another they form aggregates (soot deposits) and when they become part of the oil, they lead to oil thickening, or the viscosity increase. Soot deposits are soft and flaky in texture.

Carbon deposits, more prevalent in diesel engine operation, are hard and result from the carbonization of the liquid lubricating oil and the fuel on hot surfaces. These deposits have lower carbon content than soot and, in most cases, contain oily material and ash.

Soot is commonly found in the combustion chamber and the carbon deposits are commonly found on piston top lands and crowns, in piston ring grooves, and on valve stems.

Lacquer and Varnish

Unburned air-fuel mixture and oxidized, or partially oxidized, reactive intermediates in the blow-by promote lubricant oxidation. This results in the formation of a variety of oxygenated products, which when exposed to high temperatures polymerize to produce lacquer and varnish [191,192]. See Fig. 4.7 for the mechanism of their formation.

The term lacquer is usually used to describe deposits in the diesel engines and varnish to describe such deposits in the gasoline engines. Lacquer is often derived from the lubricant and is generally water-soluble. Varnish, on the other hand, is fuel-derived and is acetone-soluble.

Lacquer is commonly found on pistons, cylinder walls, and in the combustion chamber. Varnish occurs on the valve lifters, piston rings, piston skirts, valve covers, and the positive crankcase ventilation (PCV) valves.

Sludge

Lubricant oxidation, oxidation and combustion products in the blow-by, and the accumulation of combustion water and solids are the three major causes of the sludge formation. Sludge can vary in consistency from that of mayonnaise to that of a baked deposit. Heat can drive off water, thereby causing a change in sludge consistency. Low-temperature sludge, more prevalent in gasoline engines, is watery in appearance and forms below 95°C. High-temperature sludge, more common in diesel engines, forms above 120°C and is hard in consistency. Sludge is usually found in areas of low oil velocity, such as crankcase bottoms and the rocker boxes.

Mechanism of Deposit Formation

The volume of deposits in an engine depends upon the fuel quality, engine's operating conditions (speed, load, and temperature), the quality of combustion and the blow-by (its oxygen content, presence of sulfur and nitrogen compounds, etc.), and the integrity of the seal between the combustion chamber and the crankcase.

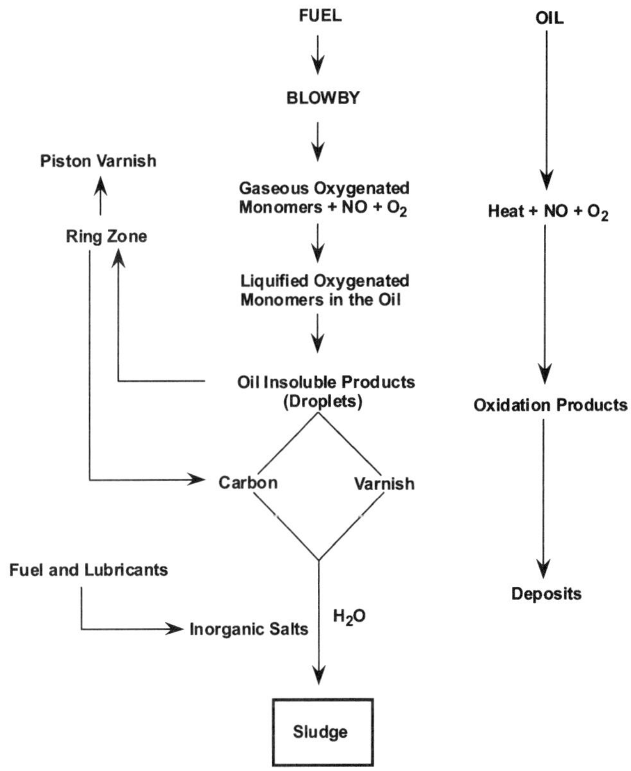

Fig. 4.8—Mechanism of deposit formation in gasoline engines [193].

Low-temperature deposits (soot, varnish, and low-temperature sludge) are usually encountered in gasoline engines with intermittent operation, that is, stop-and-go driving. This is because this type of operation does not allow the engine to achieve the optimum temperature necessary to drive off the contaminants [193]. High-temperature deposits include carbon, lacquer, and high-temperature sludge. These are typically found in gasoline engines with long continuous operation and in diesel engines [191,192]. These deposits result from thermal and oxidative degradation of the lubricant and the additives. Increased emphasis on cleaner exhaust emissions by assuring efficient combustion of the fuel at high temperatures, the use of the power accessories, such as air conditioning, and the body style changes, which affect cooling, additionally stress the lubricant.

Figure 4.8 shows the mechanism of deposit formation in gasoline engines [193]. As depicted in the figure, both the lubricant and the fuel contribute to deposit formation. Unreacted oxygen and NO_X, a consequence of combustion, are present in the blow-by. At high temperatures, these cause oxidative and thermal degradation of the lubricant. The result is the formation of the oxygenated products which act as precursors to deposit-forming species. This is depicted in Fig. 4.9. The chemical sequences involved in the formation of the deposit precursors are hypothesized in Figs. 4.10–4.12. The mechanism involves the formation of the free radicals and their disproportionation or rearrangement to

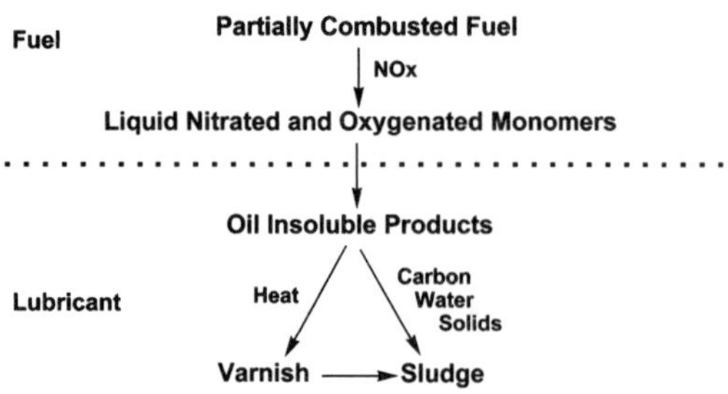

Fig. 4.9—Mechanism of deposit formation.

olefinic and carbonyl-containing reactive materials, which undergo further reactions to form the deposit precursors. Sometimes the rearrangement is facilitated by the combustion-derived acids. The precursors have approximately 15 to 50 carbon atoms and contain hydroxyl and carboxyl functional groups. Because of being poly-functional molecules, they have the ability to thermally polymerize to higher molecular weight products [192,193].

If the oxygen content of the precursors is low and the polymer product is of low molecular weight and of good oil solubility, only oil thickening is observed. However, if the oxygen content of the precursors is high and the polymerization results in products of low lubricant solubility, resin and varnish are formed. Resin forms when the polymerization of the oil-insoluble products occurs in the bulk lubricant and varnish forms when it occurs on hot metal surfaces. Oil-insoluble products can also interact with carbon, water, and solids to form sludge.

Table 4.4 shows typical operating temperatures in different parts of an internal combustion engine [21,23]. As can be seen, exhaust valves, combustion chamber, and piston crown experience extreme temperatures relative to the other parts. The upper parts of the piston, such as the crown, top land, ring grooves, and rings, being closer to the flame, are exposed to higher temperatures than the lower parts, such as the piston skirt. This is also true for the pistons of a two-stroke cycle engine. As a consequence, the deposit formation is more prevalent in the upper areas of the piston.

Resin, due to low solubility in the lubricant, tends to

Radical Formation

$$RH + O_2 \longrightarrow \underset{\text{Hydroperoxides}}{ROOH} \quad (5)$$

$$ROOH \longrightarrow RO\cdot + \cdot OH \quad (6)$$

$$2\,ROOH \longrightarrow RO\cdot + ROO\cdot + H_2O \quad (7)$$

$$ROOH + NO \longrightarrow ROH + NO_2 \quad (8)$$

$$2\,ROH + \underbrace{NO + NO_2}_{NO_x} \longrightarrow H_2O + \underbrace{RONO + RONO_2}_{\text{Nitrite and nitrate esters}} \quad (9)$$

$$RONO + RONO_2 \longrightarrow 2\,RO\cdot + NO + NO_2 \quad (10)$$

Decomposition and Rearrangement

$$ROOH + SO_2 \longrightarrow ROH + SO_3 \quad (11)$$

$$ROOH + H_2SO_4 \longrightarrow \text{Carbonyl compounds} \quad (12)$$

$$RO\cdot \text{ and } ROO\cdot \longrightarrow \text{Oxidation products} \quad (13)$$

Fig. 4.10—Formation of deposit precursors.

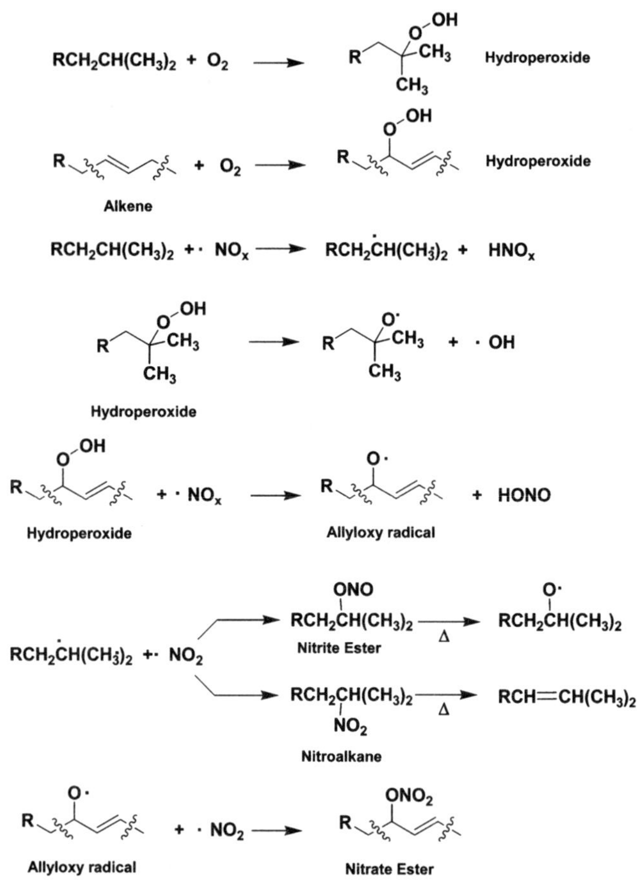

Fig. 4.11—Formation and decomposition of nitrite and nitrate esters.

separate out on the metal surfaces as lacquer. Lacquer can be amber or black in color. At low soot levels, the interaction of resin and soot results in the resin-coated soot particles, which separate out on piston surfaces as black lacquer. At high soot levels, extensive interaction between the soot and the resin results in soot-coated resin particles, which have little or no ability to adhere to the metal surfaces. Instead, they accumulate as deposits in areas of slow oil flow, such as grooves behind the piston rings [192]. Sludge results from the interaction of oxygenates with soot in the presence of oil and water. Deposit levels depend upon the fuel quality. In the case of the sulfur-rich fuels, the oxidation of sulfur results in the formation of sulfur acids which catalyze the rearrangement of hydroperoxides to carbonyl compounds (see Fig. 4.10, Eq. 12, and Fig. 4.12) and their subsequent polymerization to resin. The addition of the basic detergents, which are used in diesel engine oils (especially for marine and railroad use) to counter the adverse effects of sulfur, helps alleviate this problem.

Oil Thickening

Oil thickening can result from the combination of the oxidative degradation of the lubricant and the accumulation of the insolubles [194]. Autoxidation of the lubricant, accelerated near the oxidation inhibitor depletion stage, can lead to oxygenated products which, through polymerization, can cause a viscosity increase [192]. Contaminant-related thickening

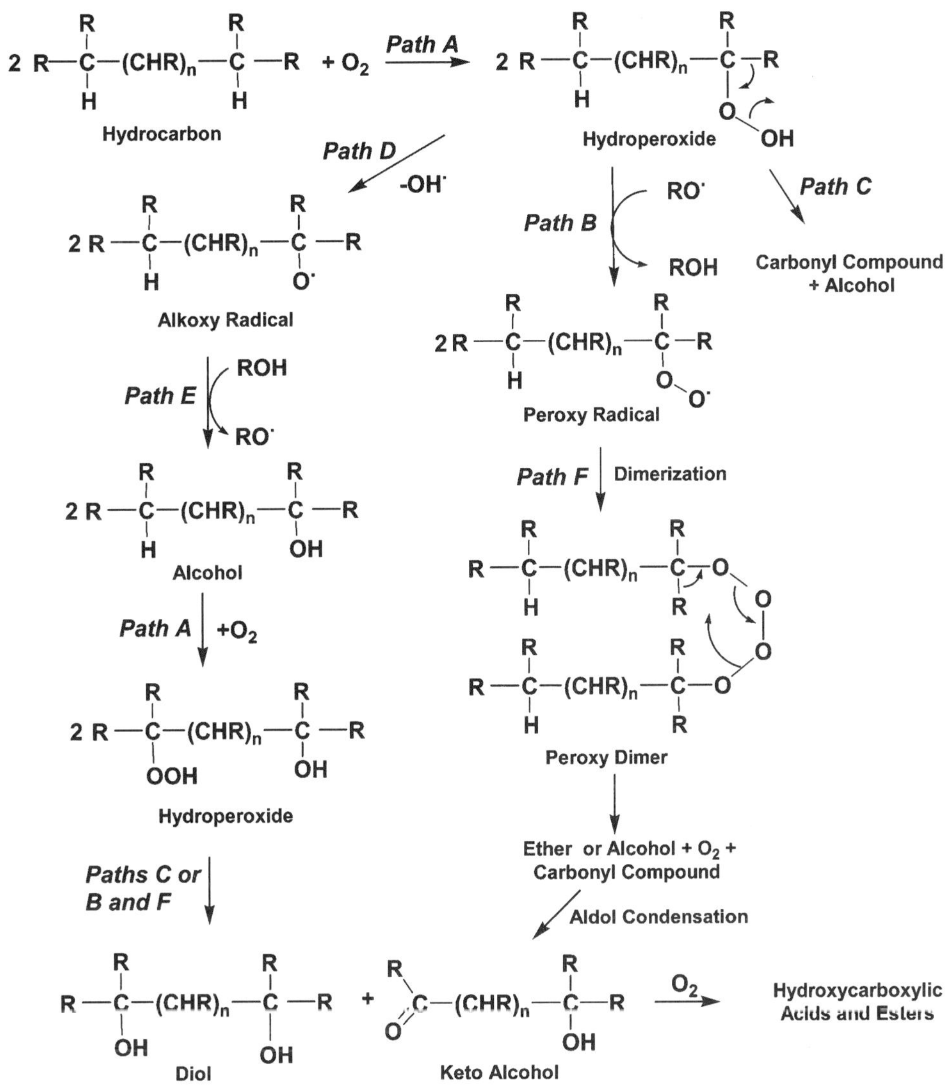

Fig. 4.12—Chemical transformations leading to the formation of deposit precursors.

arises from the suspension of the fuel-derived insolubles, such as soot, in the bulk lubricant.

TABLE 4.4—Approximate temperatures of internal surfaces in an internal combustion engine [21,23].

Area of the Engine	Temperature Range (°C)
Exhaust Valve Head	650–730
Exhaust Valve Stem	635–675
Combustion Chamber Gases	2300–2500
Combustion Chamber Wall	204–260
Piston Crown	204–426
Piston Rings	149–315
Piston Pin	120–230
Piston Skirt	93–204
Top Cylinder Wall	93–371
Bottom Cylinder Wall	Up to 149
Wrist Pin	121–232
Main Bearings	Up to 177
Connecting Rod Bearings	93–204

Oil Consumption

Oil consumption is related mainly to the lubricant that travels past the piston rings and the valves and burns in the combustion chamber. Burning of the lubricant, along with inefficient fuel combustion, leads to soot and carbon deposits on the inside of the combustion chamber, piston top lands, ring grooves, etc. [190]. The extent of the oil consumption depends upon a number of lubricant-related and equipment design-related factors and in this regard, the viscosity, the volatility, and the sealing characteristics of the lubricant play an important role.

A certain minimum amount of oil is necessary for proper lubrication of the cylinder walls and the pistons. High oil consumption, however, indicates problems in the pistons and the cylinders, such as increased wear of the cylinder, bore polishing, stuck piston rings, or out of square grooves [17]. Under these circumstances, the blow-by gases have an increased chance of entering the crankcase which is likely to complicate the situation further.

There is some evidence that the oil consumption contributes to particulates in diesel emissions. Due to the envi-

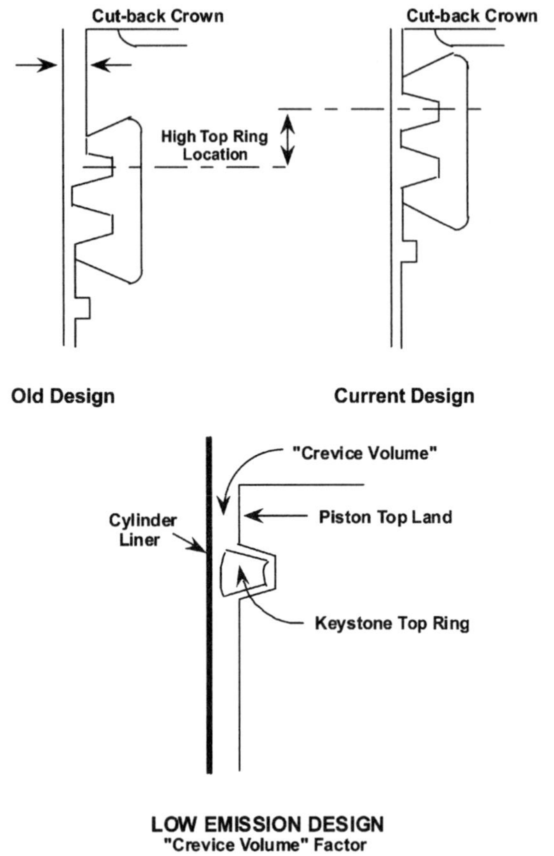

Fig. 4.13—U.S. diesel piston designs—old versus current.

ronmental (emissions) concerns, there is a consolidated effort by the OEMs and the lubricant marketers in the United States, Europe, and Japan to minimize the formation and release of the particulates into the environment. One strategy to achieve this goal is by the use of a lower thickness oil film. This strategy is quite effective except that beyond a certain limit, the oil film thickness is ineffective for proper lubrication and the ring/liner scuffing problem initiates. Lubricant volatility is another important factor responsible for the increased oil consumption. Lighter base oils, used in formulating multi-grade diesel oils, not only contribute to the formation of a less effective lubricating film but can also readily leak past the piston rings and burn [195]. The current piston design has been developed to minimize the formation of the lubricant-related particulates in the diesel engine emissions. Top piston ring in these pistons drags less oil into the combustion chamber. This is achieved by decreasing the "crevice volume" between the piston and the cylinder liner. These pistons have less cut back crowns and smaller first lands. The older and current piston designs are shown in Fig. 4.13.

Ring Sticking

The major cause of ring sticking is the formation of deposits in the piston grooves. The consequence is a loss of oil seal which not only favors blow-by but also leads to poor transfer of heat from the piston to the externally cooled cylinder wall. This is quite serious because this can cause nonuniform thermal expansion of the pistons, hence a loss of compression and ultimately to engine seizure.

Corrosion and Wear

Diesel fuel with high-sulfur content causes piston ring and cylinder wear, especially in large slow speed marine diesel engines. Corrosive wear, more commonly associated with combustion and oxidation products, results from the attack of the sulfur acids or the organic acids on the iron surfaces. This happens when the engine operating temperatures are below the dew point of these acids. Organic acids may originate from thermo-oxidative decomposition of the lubricant or the additive system. This kind of wear is controlled by the use of lubricants with a base reserve. Such additives control deposit formation by inhibiting the oxidative breakdown of the lubricant to deposit precursors and by suspending those already formed in the bulk lubricant.

Oxidation inhibitors intercept the oxidation mechanism and the dispersants and the detergents do the suspending part.

Oxidation Inhibitors

Since both the base fluid and the additives that comprise the lubricant are organic in nature, they are susceptible to oxidation. There are base fluids that either do not contain any carbon-hydrogen bonds, such as perfluoropolyethers, or contain carbon-hydrogen bonds that are not easily oxidizable as in the case of poly(phenyl ether)s. They are used in specialty applications involving extreme temperatures. See Synthetic Fluids, Chapter 3, for the pertinent discussion. It is obvious that with respect to oxidation not all base fluids are alike. As stated previously, base fluids are mineral, synthetic, or biological in origin. Each type has a stable threshold, beyond which the rate of oxidation increases rapidly and stabilizers, or oxidation inhibitors, are needed to retard oxidation. With respect to oxidative stability, synthetic oils are the most stable and vegetable oils are the least stable, and mineral oils fall between the two.

Most lubrication applications expose lubricants to oxygen in some manner, the exposure being more intensive when used to lubricate an internal combustion engine. This is because the engine's lubrication environment contains all the key elements that increase the rate of oxidation. These include the following:

1. The presence of the ample amount of oxygen since the engine is open to the atmosphere, via the combustion chamber.
2. High ambient temperatures because of the close proximity to the flame.
3. The presence of NO_X.
4. The presence of metals and their ions.

These factors not only facilitate the attack of oxygen on the hydrocarbon molecules that make up the lubricant, but they also exponentially increase the reaction rate. The reaction sites in hydrocarbon molecules, in order of decreasing ease of attack, are benzylic, allylic, tertiary alkyl, secondary alkyl, and primary alkyl hydrogens. The result is the formation of peroxy or other free radicals. The mechanism of oxidation is illustrated in Figs. 4.10–4.12 and Fig. 4.14. Oxidation proceeds in three stages: the initiation stage, the propagation stage, and the termination stage [196]. During the initiation stage, oxygen reacts with the fuel and the lubricant to form alkyl free radicals, see Fig. 4.14, Eqs. 14 and 15. During the propagation stage, these free radicals react with

Initiation:

$$RH \xrightarrow{\text{activation}} R^{\cdot} + H^{\cdot} \qquad (14)$$
$$RH + O_2 + RH \longrightarrow R^{\cdot} + H_2O + R^{\cdot} \qquad (15)$$

Propagation:

$$R^{\cdot} + O_2 \longrightarrow RO_2^{\cdot} \qquad (16)$$
$$RO_2^{\cdot} + RH \longrightarrow ROOH + R^{\cdot} \qquad (17)$$

Decomposition of Peroxide:

$$ROOH \longrightarrow RO^{\cdot} + OH^{\cdot} \qquad (18)$$
$$2\,ROOH \longrightarrow RO^{\cdot} + RO_2^{\cdot} \qquad (19)$$
$$RO^{\cdot} + ROOH \longrightarrow \text{Various Products} \qquad (20)$$
$$\left.\begin{array}{r} RO_2^{\cdot} \\ ROOH \end{array}\right] \longrightarrow \text{Nonradical Products} \qquad (21)$$

Induced Decomposition of Peroxide:

$$X + ROOH \longrightarrow \text{Free Radicals} \qquad (22)$$
$$Y + ROOH \longrightarrow ROH + YO \qquad (23)$$
$$Z + ROOH \longrightarrow \text{Inactive Prodcts} + Z \qquad (24)$$
$$M + ROOH \longrightarrow \text{Free Radicals} \qquad (25)$$

Self Termination:

$$RO_2^{\cdot} + RO_2^{\cdot} \longrightarrow \text{Inactive Products} \qquad (26)$$

Chain-Breaking Termination:

$$RO_2^{\cdot} + IH \longrightarrow ROOH + I^{\cdot} \qquad (27)$$
$$RO^{\cdot} + IH \longrightarrow ROH + I^{\cdot} \qquad (28)$$
$$R^{\cdot} + IH \longrightarrow RH + I^{\cdot} \qquad (29)$$

RH = Organic substrate; $RO_2^{\cdot}$ = Peroxy radical; ROOH = Hydroperoxide
$RO^{\cdot}$ = Alkoxy radical; X, Y, Z = Decomposers of Radicals
$R^{\cdot}$ = Alkyl radical; M = Metal; $I^{\cdot}$ = Inhibitor radical
IH = Chain-breaking inhibitor or inhibitor of free radicals

Fig. 4.14—Mechanism of oxidation [196].

oxygen and the lubricant hydrocarbons to form the peroxy free radicals and hydroperoxides, see Fig. 4.14, Eqs. 16 and 17. During the termination stage, free radicals get removed from the reaction sequence, thereby slowing down the oxidation reaction.

The initiation stage primarily starts in the combustion chamber and generates free radicals. Free radical formation purely due to heat is unlikely and is probably aided by the short to medium wavelength radiation in the ultraviolet to infrared region, the release of which accompanies combustion. See Fig. 4.15 for the electromagnetic radiation spectrum [197]. The energy required to homolytically cleave the covalent bonds, that are present in hydrocarbons, to free radicals is called homolytic bond dissociation energy, designated by the symbol DH°. Its magnitude is affected by the structure of the molecule as a whole, not just the two atoms that make the bond. In general, the lower the bond dissociation energy, the easier it is to cleave it. With respect to the hydrocarbon structures, the ease of carbon-hydrogen bond cleavage follows the order: benzylic hydrogens > allylic hydrogens > tertiary alkyl hydrogens > secondary alkyl hydrogens > primary alkyl hydrogens. Table 4.5 shows the relationship between the various hydrocarbon structures and the carbon-hydrogen bond strengths. Please note that the structure provided below the table was taken from Ref. [213]. The structures in the table with smaller DH° form free radicals easily. We believe that the free radical formation according to Eq. 15 in Fig. 4.14 is more likely since the energetics of this reaction are more favorable than that depicted in Eq. 14. During the propagation phase, the peroxy free radicals and hydroperoxides accumulate during the induction period, after which the oxidation auto-accelerates [198]. The auto-acceleration is indicated in Fig. 4.16 by the oxygen uptake, which drops after reaching a maximum at which time oxidation takes off [198].

During the propagation phase, hydroperoxides, either thermally or in the presence of metal, decompose either to additional free radicals or form nonradical oxygen compounds. The newly formed free radicals keep on supporting the oxidation chain mechanism. The nonradical oxygen compounds include alcohols, aldehydes, ketones, and carboxylic acids that form according Eqs. 18–25 in Fig. 4.14. The detailed mechanism of the thermal decomposition of the hydroperoxides and their rearrangement to form these species is shown in Figs. 4.11, 4.12, and 4.17. Aldehydes and ketones are highly reactive and can form polymers in the presence of acids, such as nitric acid and sulfuric acid, which are present in the blow-by. These acids result from the interaction of the nitrogen oxides and sulfur oxides, the products of combustion, with water. Carboxylic acids attack the iron metal and copper and lead bearings to form metal carboxylates which further increase the rate of oxidation. Figure 4.18 depicts the catalytic effect of metals and metal coatings [115] and Fig. 4.19 shows the effect of iron carboxylate metal salt on hydroperoxide formation [198].

An increase in temperature affects the oxidation process profoundly, with its rate approximately doubling with every ten degree rise in temperature. Wear metals can also enhance the rate of oxidation, especially after they get converted into salts by reacting with acids [199]. If the oxidation is not controlled, the lubricant decomposition will lead to oil thickening, sludge formation, and the formation of varnish, resin, and corrosive acids [191,192]. Oil thickening occurs mainly due to the polymerization or association of certain oxidation products in the bulk lubricant. A model showing oxidative and thermal degradation of lubricants [198] is shown in Fig. 4.20.

During the termination stage, the free radicals either self-terminate, as shown in Eq. 26 of Fig. 4.14, or terminate by reacting with the oxidation inhibitors [196]. The mechanism involving self-termination leading to the formation of the harmful carbonyl compounds is shown in Fig. 4.17. Oxidation inhibitors circumvent the free radical chain mechanism of the oxidation process as shown in Eqs. 23, 24, and 27–29 of Fig. 4.14 by converting the hydroperoxides and free radicals into innocuous species, thereby retarding oxidation. Some refined base oils contain sulfur and nitrogen-derived "natural" inhibitors, which under mild conditions are adequate in protecting the lubricants against oxidation. However, most modern uses require supplemental inhibitors to protect lubricants under increasingly demanding operating conditions. In addition, the new methods of refining remove these beneficial compounds from the base oils, warranting the use of more effective oxidation inhibitors just the same.

Oxidation inhibitors are classified into three groups: hydroperoxide decomposers, free radical scavengers, and metal deactivators, depending upon the mode of their con-

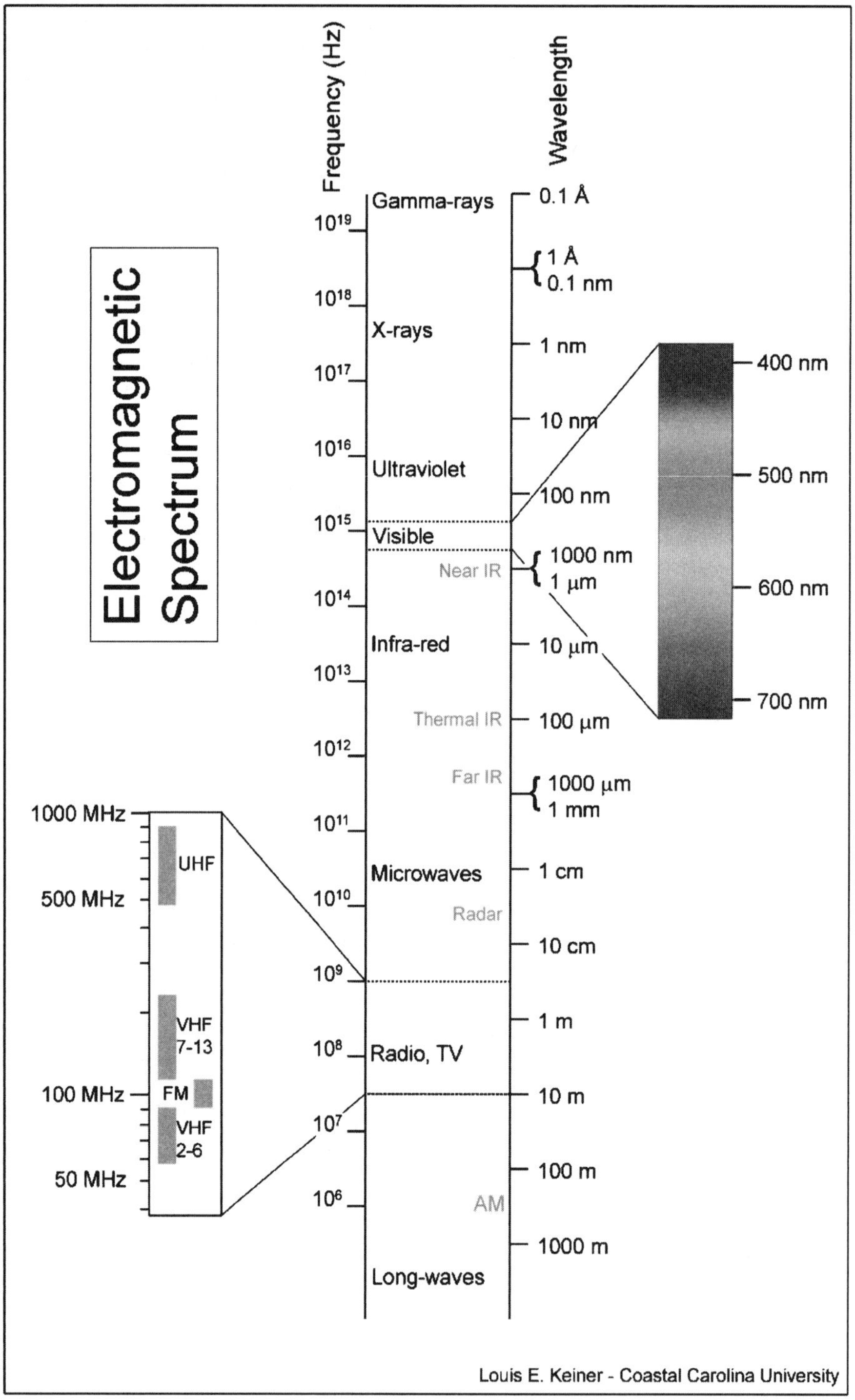

Fig. 4.15—Electromagnetic radiation spectrum [197].

trolling action. The structures of the commonly used oxidation inhibitors are shown in Fig. 4.21. Common hydroperoxide decomposers include sulfur and phosphorus-containing compounds, such as sulfides, dithiocarbamates, phosphites, and dithiophosphates. The mechanism by which the hydroperoxide decomposers perform is shown in Fig. 4.22. Hydroperoxide decomposers convert the chain-propagating hydroperoxides into alcohols while themselves getting oxidized to the higher oxidation levels. Sulfur compounds, represented by alkyl sulfides and alkyl polysulfides, react with hydroperoxides and are converted into alkyl sulfoxides, alkyl thiosulfinates, and alkyl sulfones, as shown in Part 1 of Fig. 4.22. Sulfoxides and thiosulfenates may disproportionate or arrange thermo-oxidatively to form other sulfur-containing compounds, such as sulfonic and sulfuric acids, which are also hydroperoxide decomposers. Figure 4.23 shows the mechanism of formation of these products from di-*t*-butyl sulfides. Organic sulfides used as oxidation inhibi-

TABLE 4.5—Bond dissociation energies of various carbon-hydrogen bonds.

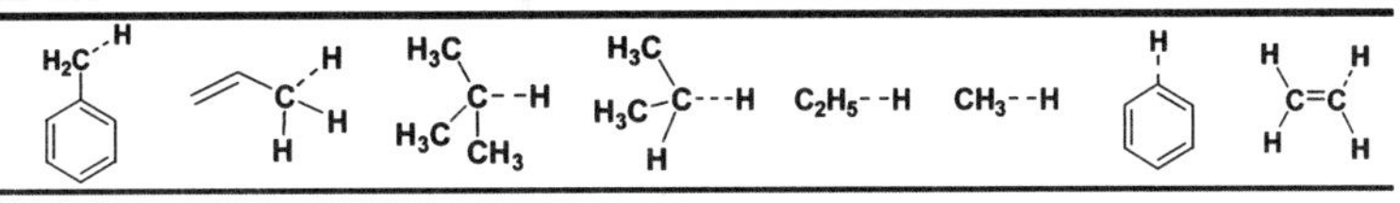

Structure								
Hydrogen Atom Attachment Group	Benzyl	Allyl	tert-Butyl	sec-Propyl	Ethyl	Methyl	Phenyl	Vinyl
Bond Dissociation Energy, DH° (kcal/mole)	85	88	93	95	98	103	110	112

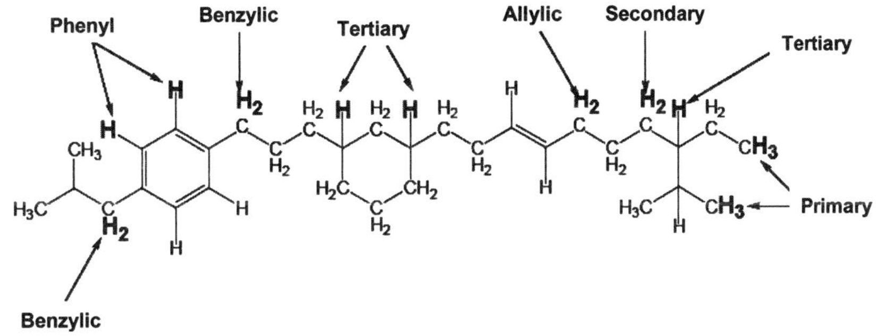

tors include phenol sulfides, sulfurized and phosphosulfurized fats, and alkyl dithiadiazoles. Many of these compounds are discussed in the subsequent sections. Organic sulfides are usually made by the reaction of an organic halide with sodium sulfide and removing the resulting sodium halide salt. Organic polysulfides can be made either by the reaction of a sodium polysulfide with an organic halide or by sulfurizing an olefin at high temperature with elemental sulfur. The synthesis of these compounds will be covered in the latter part of the chapter while discussing antiwear and extreme-pressure agents. Dilauryl selenide is another hydroperoxide decomposer that is related to the organic sulfides on account of the selenium belonging to the same group in the periodic table as sulfur. Its mechanism to inhibit oxidation is also via hydroperoxide decomposition. It is made by the reaction of lauryl chloride with dimethyl selenide. It is used as an antioxidant in greases and some synthetic fluids [165]. Molybdenum compounds, such as molybdenum sulfide, molybdenum-amine complexes, and molybdenum dithiocarbamates are considered multipurpose additives since they provide both the oxidation control and the antiwear/extreme-pressure properties. While these compounds are synergistic with alkylated diphenylamines, they are expensive and can cause copper and lead corrosion, under certain circumstances.

Phenol and nonhindered alkylphenols act as hydroperoxide decomposers as well and in the process they get con-

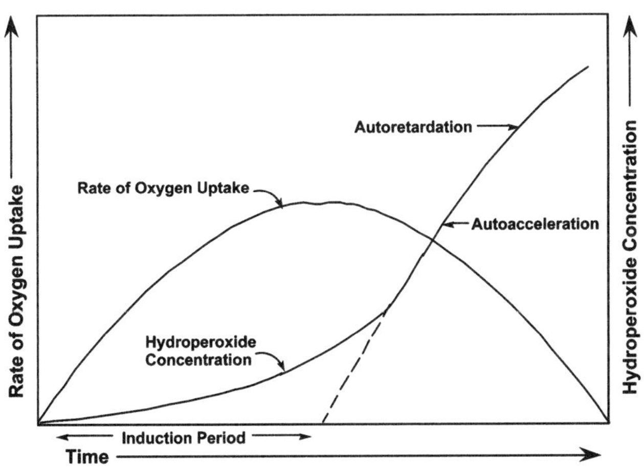

Fig. 4.16—Effect of hydroperoxide concentration on the rate of oxidation [198].

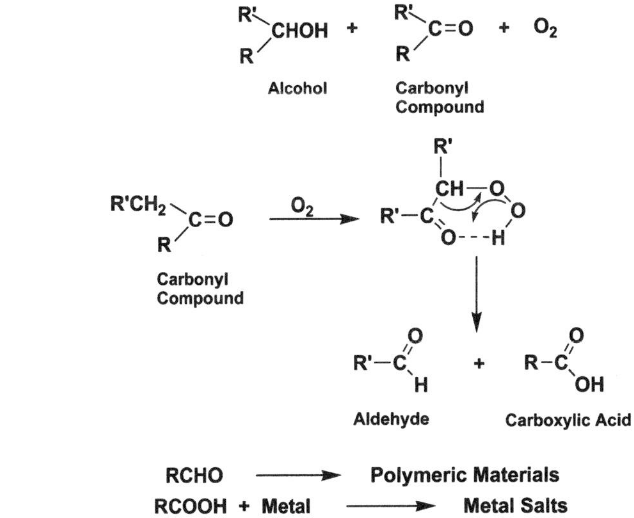

Fig. 4.17—The mechanism of hydroperoxide decomposition.

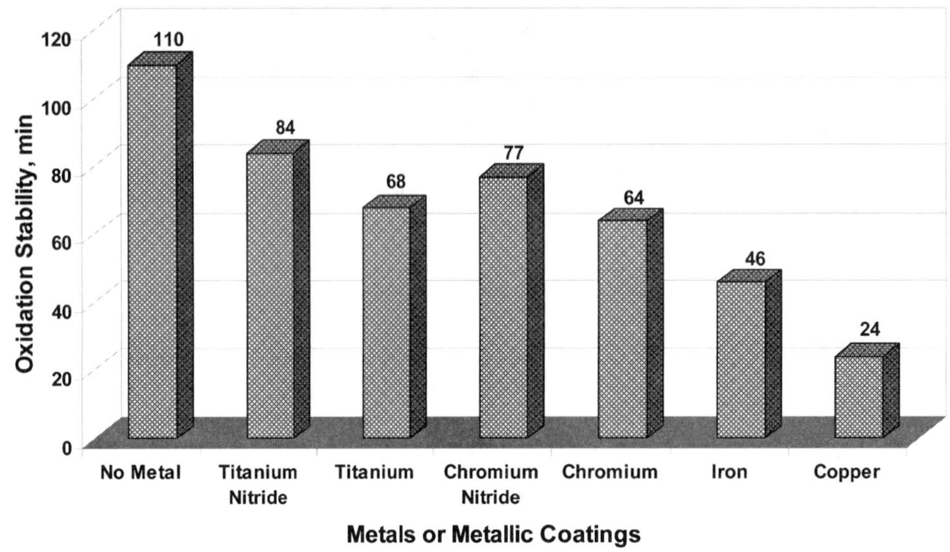

Fig. 4.18—Effect of metals and metal coatings on oxidation stability of oxidation-inhibited synthetic ester-based hydraulic fluid (ASTM D2272) [115].

verted into poly-hydroxy compounds. The manner in which they take part in making hydroperoxides innocuous is shown in Part 2 of the Fig. 4.22. The hydroperoxide decomposing action of the phosphorus compounds is depicted in Part 3 of the Fig. 4.22. Phosphines, rarely used as inhibitors because of their toxicity, are not as effective as other phosphorus derivatives. This is because they stoichiometrically react with hydroperoxides to form phosphine oxides, which lack further oxidation-inhibiting ability. Alkyl phosphites are better inhibitors because during the process of decomposing hydroperoxides they are converted into phosphates, which continue inhibiting oxidation by forming acidic materials, via thermal degradation or hydrolysis. Alkyl phosphites are made either by the reaction of alcohol with phosphorus trichloride or by the trans-esterification reaction involving an alcohol and a lower alkyl phosphite or hydrogen phosphite. Alkyl phosphates are analogously made by the reaction of an alcohol with phosphorus oxychloride in the presence of a base or by the oxidation of trialkyl or triaryl phosphites. Again, these methods will be discussed in some detail in the latter part of the chapter.

Dialkyl dithiophosphoric acid derivatives are the most potent of the phosphorus-based inhibitors, primarily because they inhibit oxidation both by hydroperoxide decomposition and free radical scavenging mechanisms. They decompose hydroperoxides catalytically as well as by reacting with them to form compounds that are peroxide decomposers in their own right. Various processes involved in the oxi-

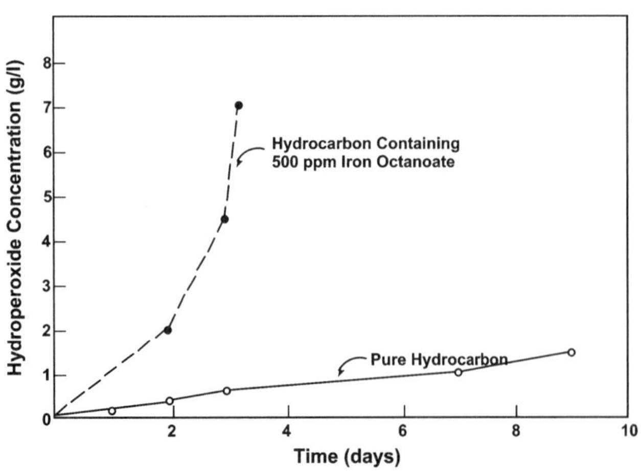

Fig. 4.19—Catalytic effect of iron carboxylate on the rate of oxidation [198].

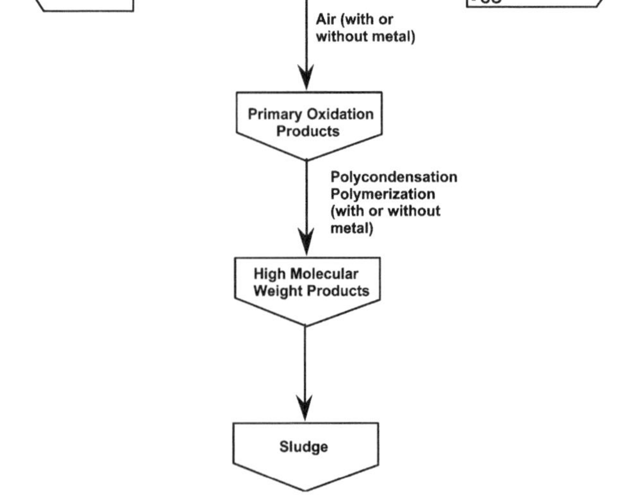

Fig. 4.20—High-temperature lubricant degradation model [198].

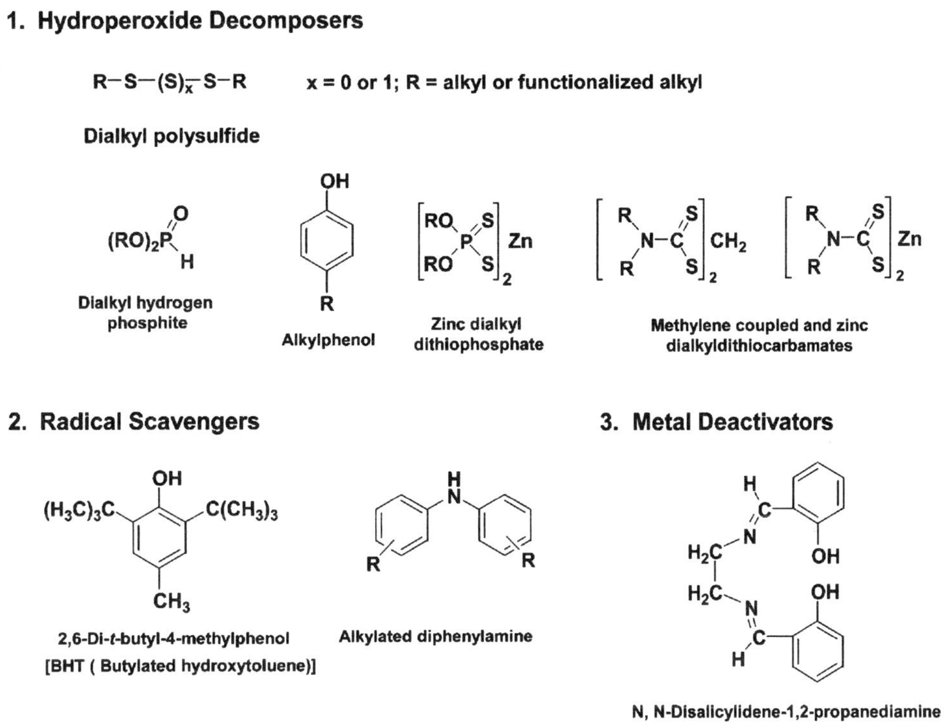

Fig. 4.21—Commonly used oxidation inhibitors.

dation performance of these additives are shown in Figs. 4.24 and 4.25. Their hydroperoxide decomposing mechanism is provided in Fig. 4.24 and the free radical scavenging mechanism is shown in Fig. 4.25. Free radical scavenging mechanism initially involves the conversion of these additives into dialkyl dithiophosphoryl disulfide, Structure 1 in Fig. 4.25, which plays a major role in the oxidation-inhibiting process. The inhibiting sequence, shown as Path A, is very similar to that encountered for alkyl sulfides. That is, it involves the oxidation of the disulfide group to thiosulfinate 2, followed by sulfur-sulfur bond cleavage to form the sulfenyl free radical 3. This free radical reacts with alkyl free radicals to form intermediates that, through decomposition

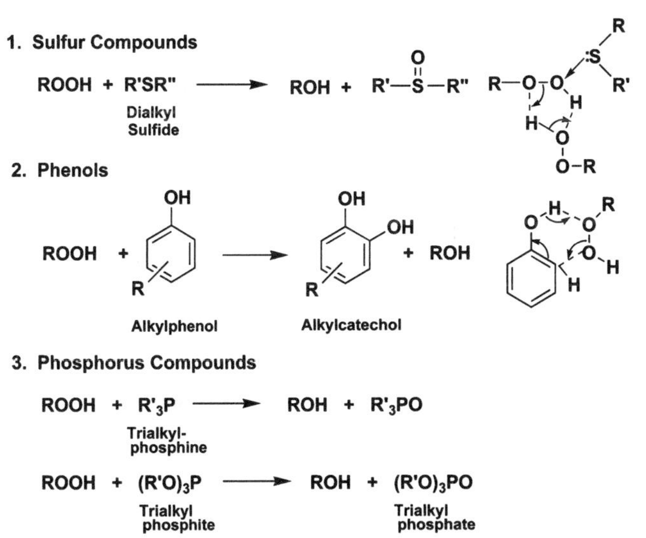

Fig. 4.22—Hydroperoxide decomposers—Mode of their action.

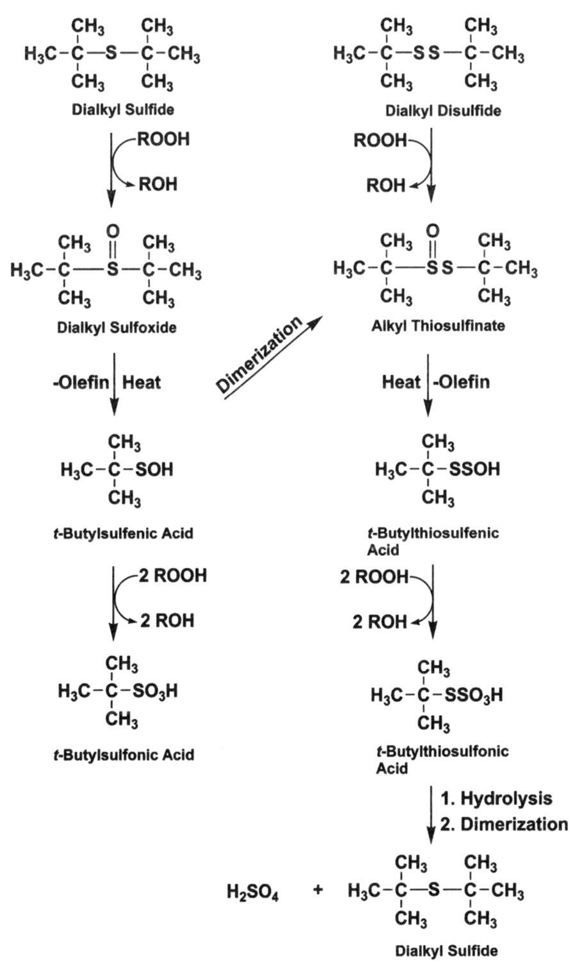

Fig. 4.23—Hydroperoxide decomposition by alkyl sulfides.

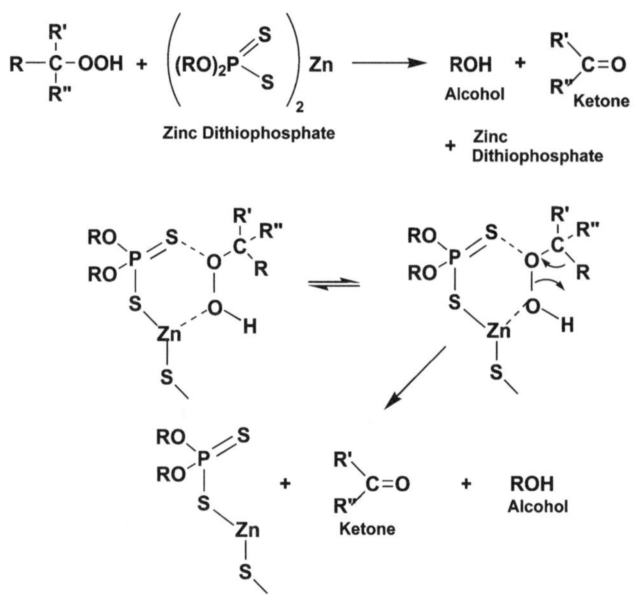

Fig. 4.24—Hydroperoxide decomposition by zinc dialkyl dithiophosphates.

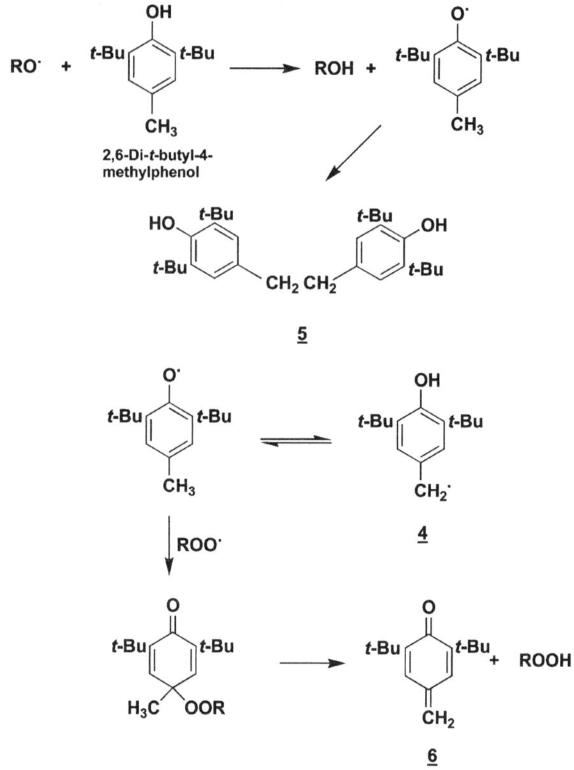

Fig. 4.26—Radical scavenging by a hindered phenol.

and reaction with hydroperoxides, result in strongly acidic species that are also responsible for oxidation inhibition. Dithiocarbamic acid derivatives, a related class of oxidation inhibitors, also act as both hydroperoxide decomposers and free radical scavengers and presumably via similar mechanisms [200]. Dithiophosphoric acids are made by the reaction of a hydroxy compound, such as an alcohol or a phenol, with phosphorus pentasulfide, which are neutralized to form an amine or a metal salt. Dithiocarbamates are made by the reaction of a secondary amine and carbon disulfide, followed by neutralization. The details of these reactions are provided under antiwear agents. Another class of potential inhibitors, which has not seen much use includes xanthates; presumably because of somewhat difficult process to make them and their lower hydrolytic stability. These are alcohol-derived analogues of dithiocarbamates, which are made by the reaction of an alkoxide with carbon disulfide, followed by the neutralization of the resulting xanthic acid [201].

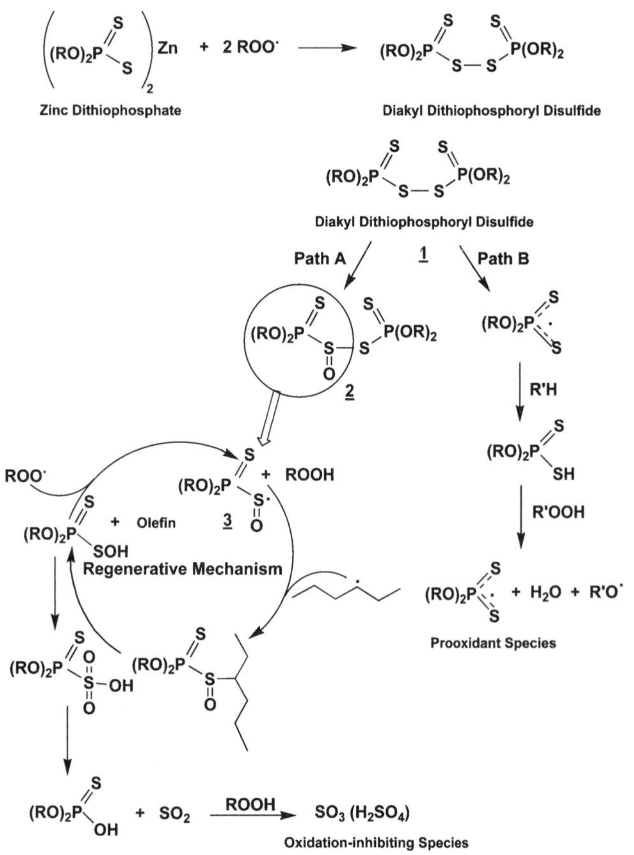

Fig. 4.25—Radical scavenging by zinc dialkyl dithiophosphates.

Free radical scavengers are inhibitors that render free radicals innocuous, either by transferring a hydrogen atom to them or by an oxidation-reduction mechanism. Hydrogen transfer from the inhibitor to the free radical generates a new inhibitor-derived free radical. However, unlike the free radicals from the oxidation process, the newly formed free radicals are incapable of propagating oxidation. This is because the new free radicals are either too sterically hindered or are too delocalized to take part in the oxidation process. The oxidation-reduction mechanism involves electron transfer to or from the peroxy free radical, thereby converting it into an ion and removing it from the oxidation process. The formation of dialkyl dithiophosphoryl disulfide from zinc dialkyl dithiophosphate is an example of such a mechanism, see Fig. 4.25.

Nitrogen and oxygen containing inhibitors, such as arylamines and hindered alkylphenols, act as free radical scavengers [196,200,202]. Both types are prominent examples of inhibitors that act as free radical scavengers via hydrogen

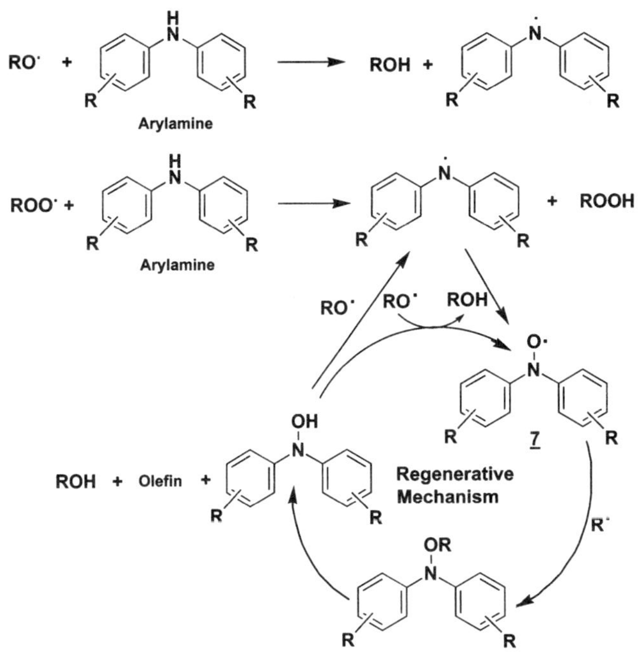

Fig. 4.27—Radical scavengers and mode of their action.

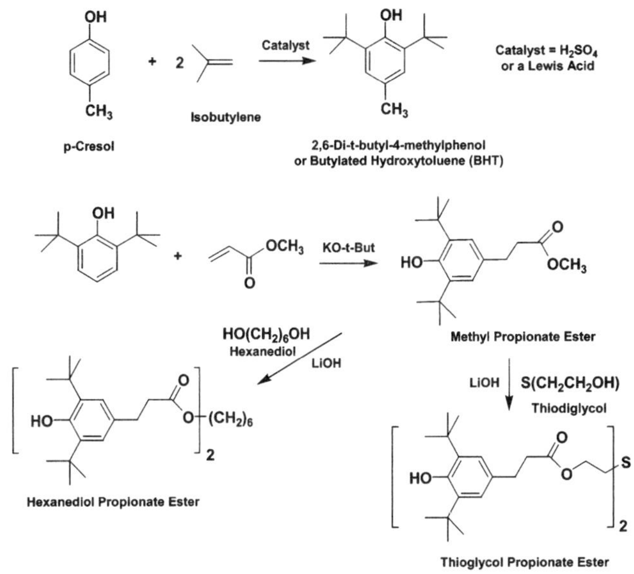

Fig. 4.28—Hindered phenol-derived oxidation inhibitors.

transfer. Figures 4.26 and 4.27 illustrate the mechanism of their performance. Hindered phenol, represented by 2,6-di-*t*-butyl-4-methylphenol in Fig. 4.26 transfers its phenolic hydrogen to the alkoxy free radical and converts it into an alcohol. The phenoxy-free radical that results from this process is incapable of propagating oxidation because of being sterically hindered. However, this free radical either rearranges to form the benzylic free radical 4, which dimerizes to form the diphenoxyethane 5, or reacts with an alkylperoxy free radical to form the quinone methide 6. The diphenoxyethane 5 still contains the hindered phenol functional group and the benzylic hydrogens. Therefore, it can continue to act as an oxidation inhibitor. 2,6-Di-*t*-butyl-4-methylphenol, also called butylated hydroxytoluene (BHT), is made by the reaction of *p*-cresol and isobutylene in the presence of a Lewis or a Brönsted acid [203]. The hindered phenol substituted methyl propionate ester is prepared by the base-catalyzed reaction of 2,6-di-*t*-butylphenol with methyl acrylate. This can be converted into other esters by the base-catalyzed trans-esterification reaction [204,205]. The synthesis of these materials is shown in Fig. 4.28. Thioglycol propionate ester in the figure contains both the hindered phenol moiety and the sulfide functional group. This means that it can act both as a hydroperoxide decomposer and a free radical scavenger and there are claims to its superior oxidation performance in engine oils [206]. Other additives that perform via free radical scavenging are shown in Fig. 4.29.

The mechanism of oxidation inhibition by arylamines, which inhibit oxidation by free radical scavenging, is presented in Fig. 4.27. The newly formed free radicals are resonance-stabilized and hence cannot start the new oxidation chains. However, they do react with hydroperoxides and peroxy free radicals to form the nitroxy free radical 7, which is also a potent inhibitor. This is because it has the ability to terminate a large number of oxidation chains through a catalytic action, which is represented by the cyclic sequence in the figure.

The oxidation-controlling action of an aromatic amine is shown in Fig. 4.30 [207]. Despite being from the *obsolete* D version of the ASTM Sequence III Test, the data are still useful in demonstrating the effectiveness of an arylamine in controlling the oxidation-related viscosity increase. The 10W-30 base formulation containing 8 % shear-stable polymethacrylate viscosity modifier and 5 % performance package shows a several hundred percent viscosity increase after 30 hours of the 64-hour test. However, when 0.5 % of the arylamine oxidation inhibitor was added to the formulation, the viscosity increase was only 90 %, which is well within the specified 375 % limit for the test [207]. Alkylated arylamines are synthesized by the reaction of an amine, such as diphenylamine, with an olefin, in the presence of a Lewis acid—aluminum chloride being a common one. Alkylated phenothiazines are a related class of multifunctional oxidation inhibitors. In addition to being an arylamine, they contain a sulfide group, which imparts them the ability to inhibit oxidation via hydroperoxide decomposition as well. However,

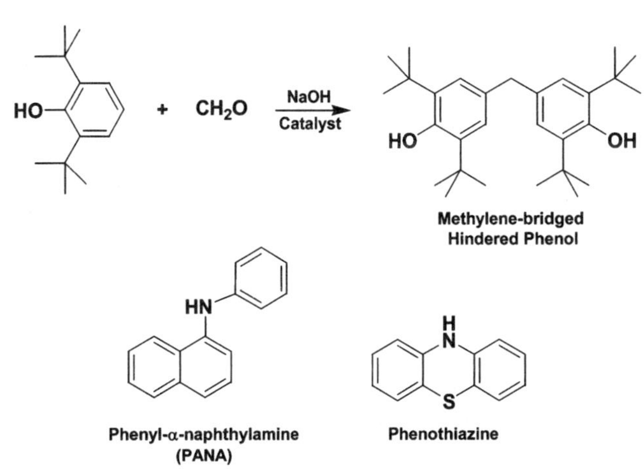

Fig. 4.29—Other commonly used oxidation inhibitors.

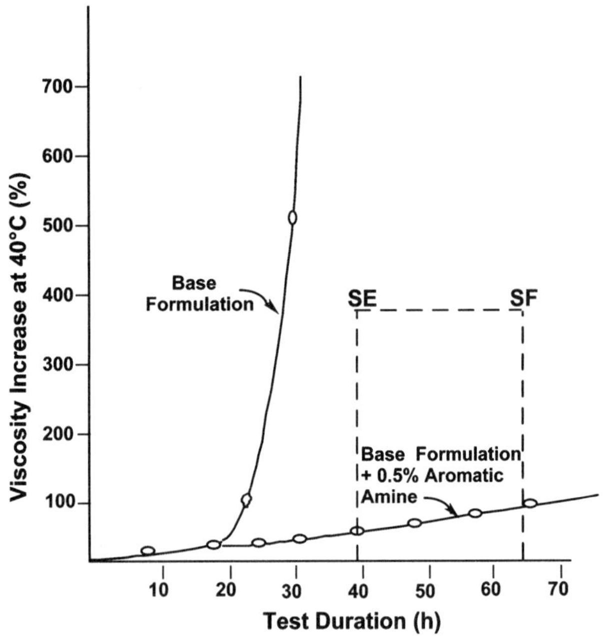

Base Formulation 10W-30

87.0% 10W Base Oil + 8% Shear-stable Polymethacrylate-based Dispersant VI + 5.0% Additive Package to Meet SD/CC Level Performance

Fig. 4.30—Oxidative viscosity control by an arylamine in ASTM sequence IIID Test [207].

their hydroperoxide decomposing action is limited since the sulfide group is aromatic and after reacting with two moles of hydroperoxides and getting converted into sulfone, its activity subsides. This is because the aromatic sulfides and sulfones lack the β-hydrogen atoms, which enhance the activity of the aliphatic sulfides via β-elimination, to form the further hydroperoxide decomposing sulfinic and sulfenic compounds. This is shown in Fig. 4.23. Alkyl phenothiazines are made by the reaction of diphenylamine with sulfur, followed by alkylation.

Transition metals can act both as oxidation initiators (promoters) and oxidation inhibitors, depending upon their oxidation state [196]. They act as promoters if they facilitate the formation of the free radicals, and they act as inhibitors, if they remove the free radicals from the oxidation process [208]. For example, heavy metals, such as iron and lead, and their salts are well known as oxidation promoters [199,209]; see Eq. 25 in Fig. 4.14. This is appropriately depicted in Fig. 4.31 [4]. The presence of copper, iron, and even calcium ions greatly increases the rate of the oxygen uptake to a varying degree. Compare the slopes of the Curves C, D, and E, which are for lubricants that contain metals, with the slope of the Curves B and A, which are for metal-free oils. The difference between Oil A and Oil B is the degree of refining. The use of an efficient oxidation inhibitor, such as zinc dialkyl dithiophosphate, does help control the oxygen uptake. Compare the slopes of the Curves G, H, and F, for oils that contain inhibitors, with those of the Curves C, D, and E, for oils that contain only the metal ions. The figure also depicts the effect of the degree of refining on the oxidation of the oil and its response to oxidation inhibitors. Highly refined oil (Curve A) takes up oxygen at a somewhat slower rate than the normally

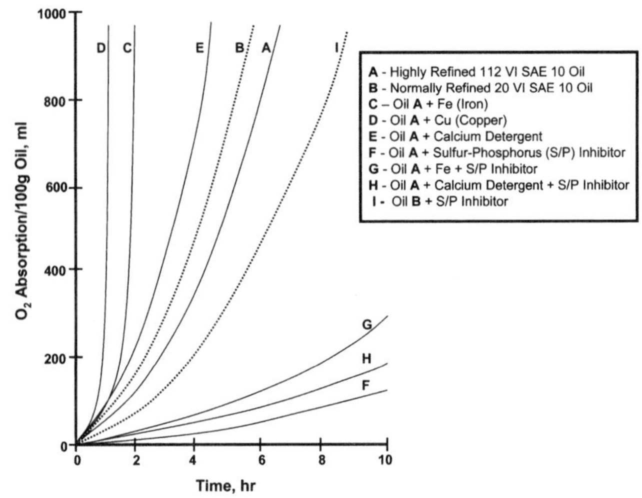

Fig. 4.31—High-temperature (175°C) oxidation inhibition of mineral oils [4].

refined oil (Curve B), but observe its dramatic response to the sulfur-phosphorus inhibitor as indicated by the slopes of the Curves F and I. Metal deactivators, another class of oxidation inhibitors, are used to control oxidation in the presence of the metal ions. These inhibitors, commonly used in fuels, perform by forming complexes with metal ions and taking them out of the chain reaction. Ethylenediaminetetraacetic acid derivatives and N,N-disalicylidene-1,2-propanediamine represent the most popular members of this class. Their structures are provided in Fig. 4.32. Other derivatives that are used in this application include lecithin, heterocycles, such as thiadiazole, imidazole, and pyrazole, and citric and gluconic acid derivatives [4].

Copper ions promote oxidation just like the other transition metal ions. They do this by forming free radicals both directly and by delivering the molecular oxygen in a more reactive state. They in addition catalyze the decomposition of hydroperoxides to free radicals. These reactions are shown in Eqs. 30–33 of Fig. 4.33. Copper ions also exhibit excellent oxidation-inhibiting ability [208]. They remove free radicals by converting them into ions which do not have the ability to take part in the oxidation process. The reactions leading to the free radical removal are represented in Eqs. 34–36 of Fig. 4.33. The positive effect of copper stearate in diminishing the rate of oxidation of tetralin catalyzed by iron stearate has been reported in Ref. [198]. Tetralin (tetrahydronaphthalene) is a bicyclic hydrocarbon that contains an aromatic

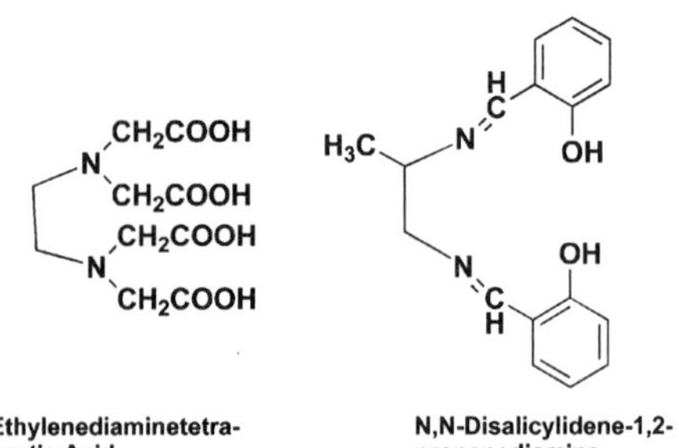

Ethylenediaminetetra-acetic Acid **N,N-Disalicylidene-1,2-propanediamine**

Fig. 4.32—Metal deactivators.

$$RH + Cu^{++} \longrightarrow Cu^{+} + R^{\cdot} + H^{+} \quad (30)$$
$$Cu^{+} + O_2 \longrightarrow Cu^{++}(O_2)^{-} + RH \longrightarrow Cu^{++} + RO^{\cdot} + OH^{-} \quad (31)$$
$$Cu^{++} + ROOH \longrightarrow Cu^{+} + RO_2^{\cdot} + H^{+} \quad (32)$$
$$Cu^{+} + ROOH \longrightarrow Cu^{++} + RO^{\cdot} + OH^{-} \quad (33)$$

$$RO_2^{\cdot} + Cu^{+} \longrightarrow RO_2^{-} + Cu^{++} \quad (34)$$
$$RO_2^{\cdot} + Cu^{++} \longrightarrow R^{+} + O_2 + Cu^{+} \quad (35)$$
$$RO^{\cdot} + Cu^{+} \longrightarrow RO^{-} + Cu^{++} \quad (36)$$

Fig. 4.33—Effect of copper ions on oxidation.

ring and a cyclic nonaromatic ring fused together; the latter ring can be considered a naphthenic ring.

Copper salts have proven to be quite effective in controlling the oxidation-related viscosity increase in a finished lubricant [207]. However, since wear and viscosity are inversely-related, as shown in Fig. 4.34 [210], the drop in viscosity beyond a certain threshold limit contributes to wear, causing it to exceed that permitted in the test. Acceptable test results both for wear and viscosity increase can be obtained by using 90–120 ppm copper in the formulation [207].

In some instances, compounds that intercept oxidation by different mechanisms reflect synergism when present together. Synergism is an effect greater than the additive effect of two or more compounds [211]. A combination of a sulfur compound with an arylamine or a hindered phenol is a common way to benefit from this phenomenon. The superior ability of a synergistic combination of a phenol and an aromatic amine to control oxidation in turbine oils is demonstrated in Figs. 4.35 and 4.36 [212]. Figure 4.35 shows inhibitor synergy in Rotary Pressurized Vessel Oxidation Test (RPVOT, ASTM D2272). Longer oxidation induction time (OIT) is desired since it is indicative of the greater oxidative stability of the lubricant. As shown in Fig. 4.35, alkylated diphenylamine at equal weight is better in controlling oxidation than the hindered phenol, but a combination of the two is vastly superior to even the alkylated diphenylamine. The same is indicated by the plotted data for the ASTM D943 Test (TOST) in Fig. 4.36 [212]. In this test, the oxidation stability of the oil is assessed by the number of hours it takes to reach the total acid number (TAN) of 2; the greater the number of

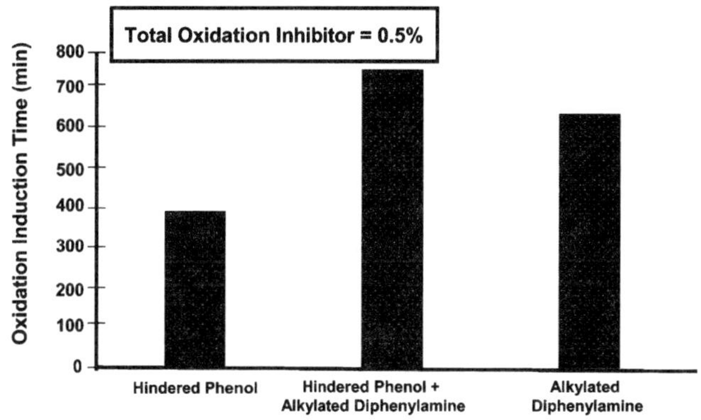

Fig. 4.35—Oxidation inhibitor synergy in turbine oils (RPVOT, ASTM D2272) [212].

hours the better. In this test, the hindered phenol is better than the alkylated diphenylamine. However, the combination of the two is again better than either. An excellent summary on oxidation inhibitors and their use in turbine oils is provided in Ref. [213].

Typically, arylamines are considered good oxidation inhibitors for high temperature applications, those involving temperatures above 120°C, and hindered phenols are considered good for low-temperature applications, those involving temperatures below 120°C. This implies that there is an advantage in combining the two types to obtain broad-temperature performance. It is also important to note that the base stock/s used to formulate lubricants respond differently to different classes of inhibitors. Overall, the API Group IV oils show the best inhibitor response and are followed by the API Group III oils, the API Group II oils, and the API Group I oils. This explains the trend away from the use of the API Group I oils in developing premium lubricants. Data depicted in Fig. 4.37 show the greater stability of the API Group II and higher group oils and the greater effectiveness of the arylamine type of oxidation inhibitors. Data in Fig. 4.38 also support the superior response of the API Group II oil over Group I oil in terms of the longer time to reach the total acid number of 2.0 in the ASTM D943 Test (Part A) and a lower amount of sludge formation in the ASTM D4310 Test

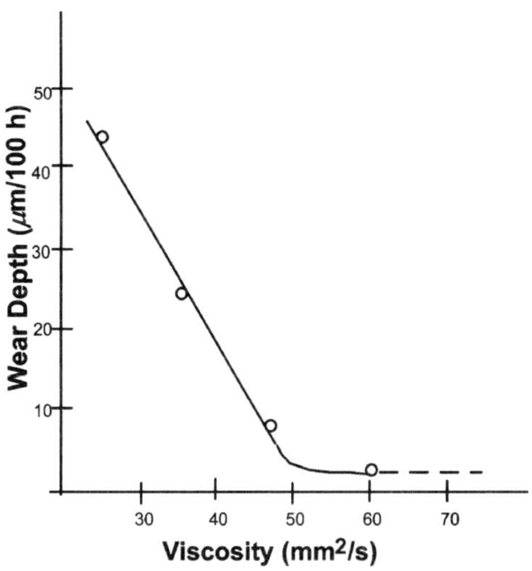

Fig. 4.34—Viscosity versus wear [210].

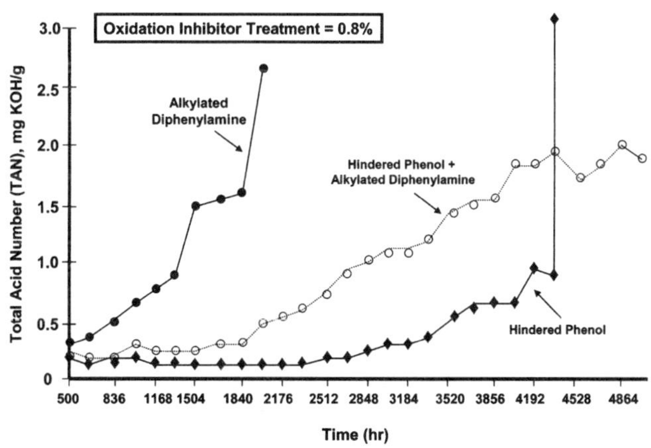

Fig. 4.36—Oxidation control synergy between hindered phenol and arylamine in turbine oil oxidation test (TOST, ASTM D943) [212].

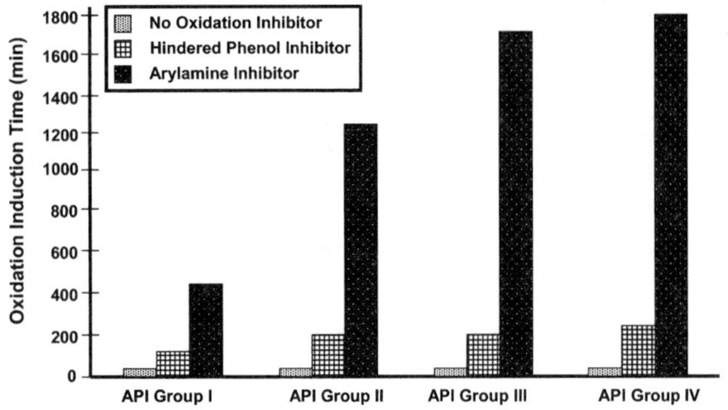

Fig. 4.37—Base oil response to oxidation inhibitors (RPVOT, ASTM D2272) [212].

(Part B). Please note that the data plotted in this figure were taken from Ref. [214]. In addition, the data demonstrate the greater effectiveness of the bifunctional inhibitor (Inhibitor II) than 2,6-di-*t*-butyl-4-methylphenol (Inhibitor I) in both these tests. Inhibitor II probably is a thioglycol ester of the hindered phenol derived propionic acid (Ciba's Irganox® 1035), shown in Fig. 4.28.

Oxidation inhibitors are used in almost all lubricants, with gasoline and diesel engine oils, and automatic transmission fluids accounting for ~60 % of the total use. High-temperature and high-air exposure applications require a higher level of oxidation protection. Zinc dialkyl dithiophosphates are the primary inhibitor type, followed by the aromatic amines, sulfurized olefins, and hindered phenols.

A number of tests are used to assess a lubricant's oxidation stability under accelerated oxidation conditions, both in bench tests and engine tests. The API required tests include the following:

1. Thermo-oxidation Engine Oil Simulation Test (TEOST, ASTM D6335)/TEOST (MHT4) for deposits.
2. Sequence IIIF/IIIG for viscosity increase and deposits.
3. Sequence IIIGA (ASTM D4684) for aged oil low-temperature viscosity.
4. Sequence VG for sludge and varnish.
5. Sequence VIII (ASTM D6709) for viscosity change.
6. Caterpillar 1K/1M-PC/1N/1R tests for deposits

ACEA in its engine oil specifications include the following tests.

1. CEC-L-88-T-02 (TU5JP-L4) for deposits, ring sticking, and oil thickening.
2. CEC-L-53-T-95 (M111) for black sludge.
3. CEC-L-46-T-93 (VW 1.6 TC D) for ring sticking and piston cleanliness.
4. CEC-L-51-A-98 (OM602A) for viscosity stability.
5. CEC-L-78-T-99 (VW DI) for diesel piston cleanliness and ring sticking.

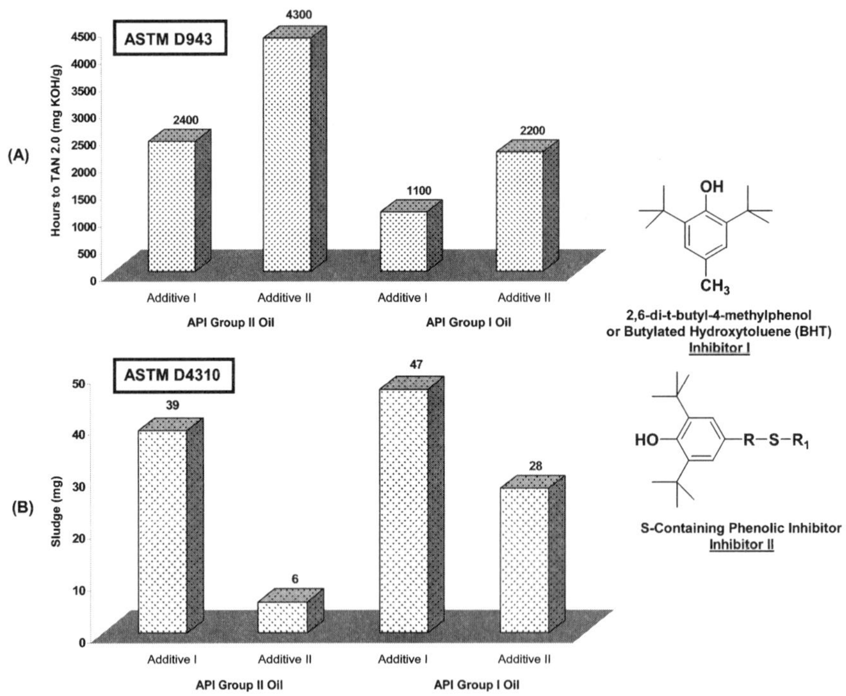

Viscosity Grade	ISO VG 32[1]		ISO VG 32[1]	
Test Parameter	Solvent Neutral (API Group I)[2]		Hydrotreated (API Group II)[3]	
Oxidation Inhibitor	I	II	I	II
Hrs (To Acid Number 2.0) – ASTM D943	2400	4300	1100	2200
Sludge (mg) – ASTM D4310	39	6	47	28

[1] The test oil contains 0.25% Oxidation Inhibitor and 0.05% Corrosion Inhibitor.
[2] The test oil contains 6.5% aromatics and 0.54% sulfur.
[3] The test oil contains 0% aromatics and 0% sulfur.

Fig. 4.38—Base oil response to oxidation inhibitors.

6. CEC-L-85-T-99 (PDSC) for oxidation.
7. CEC-L-52-T-97 (OM441LA) and CEC-L-42-T-99 (OM364LA) for piston cleanliness.

It is important to note that many of these tests do not measure solely the oxidation properties of the lubricant but they also measure its ability to disperse deposit precursors and the already formed deposits. The CRC L-60 test is used for gear oils and ASTM D943 and ASTM D2272 tests are commonly used for turbine oils.

Dispersants

Dispersants are metal-free additives that suspend oil-insoluble resinous oxidation products and particulate contaminants in the bulk oil. By doing so, they minimize sludge formation, particulate-related abrasive wear, lubricant viscosity increase, and oxidation-related deposit formation. Dispersants perform these functions by doing the following [187]:

- They include polar contaminants in their micelles.
- They associate with colloidal particles to prevent aggregation to larger particles so as to prevent them from separating out of oil.
- They suspend aggregates, if they form, in the bulk lubricant.
- They modify soot to minimize its aggregation and hence prevent soot-related oil thickening.
- They lower the surface/interfacial energy of the polar products to decrease their tendency to adhere to surfaces.

The undesirable polar materials, generically described as dirt, are a consequence of the oxidative degradation of the lubricant and or thermal decomposition of the thermally labile lubricant additives, such as extreme-pressure/antiwear agents; and the reaction of the resulting chemically reactive species, such as carboxylic acids, with the metal surfaces or wear debris in the engine. The lubricant consists of three components: the base fluid, the additives, and a viscosity modifier, in the case of a multi-grade lubricant. On account of being organic, all three are susceptible to attack by oxygen, resulting in the formation of the highly oxygenated polar materials.

In gasoline-fueled engines, the formation of these materials is catalyzed by nitrogen oxides, or NO_X. NO_X results when nitrogen and oxygen present in the air-fuel mixture react at high temperatures (1370°C). NO_X can react with hydrocarbons of the fuel and the lubricant to form nitrate esters [192,215], shown in Fig. 4.11, which along with the hydroperoxides generated from the direct oxidation of the lubricant thermally decompose to form carbonyl compounds, such as aldehydes and ketones. In the presence of bases or acids, these compounds undergo aldol-type condensation to form oligomeric or polymeric materials, which further oxidize to form oxygenates. These high molecular weight oxygenates are of sticky consistency and the term resin is often used to describe them.

In diesel-fueled engines, soot from the combustion chamber is the key component of the carbon and lacquer deposits, which occur on pistons, and sludge. As stated in an earlier part of this chapter, soot combines with resin to form lacquer and carbon deposits. In general, lacquer is rich in resin, and carbon is rich in soot. Sludge results when soot

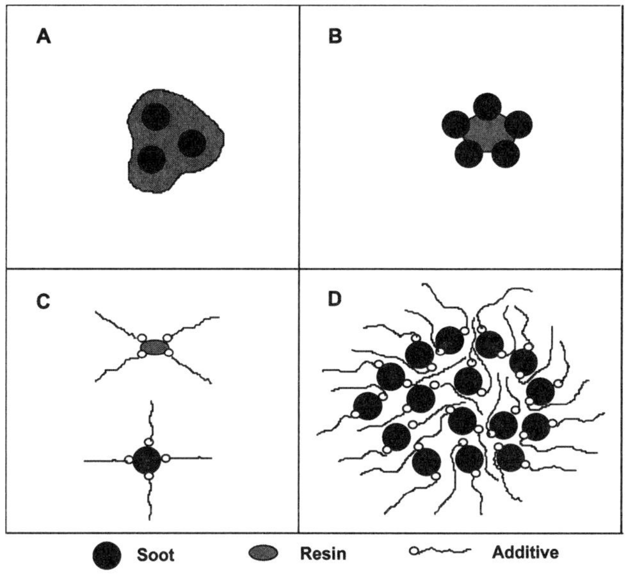

Fig. 4.39—Mechanism of soot-resin-additive interaction [191].

combines with the oxygenated species, oil, and water [191]. The local piston temperatures and the lubricant's ash-producing tendency have a profound effect on the composition of the carbon deposits. High temperatures and the lubricants with high metals content primarily produce deposits with high residue and low organic content [216–219]. Metals are the main source of ash, which is a part of some deposits. Basic detergents contain metals and hence are considered to make a contribution. Zinc dialkyl dithiophosphates also contribute towards deposits, but only slightly, because their amount in lubricants is much smaller than that of the detergents.

Because of the low oil-solubility, resin tends to separate out as amber lacquer on hot piston surfaces. If oil contains soot, soot separates with resin to form *"resin-coated soot particles,"* which appear as black lacquer. As the soot level increases, more and more soot associates with the resin to form *"soot-coated resin particles."* These events are shown in Parts A and B of Fig. 4.39. The size and the composition of these particles do not allow them to adhere to metal surfaces, but they have the propensity to collect in the areas of the low oil flow, such as piston grooves, as deposits.

Deposit Control by Dispersants

As stated before, resin and soot are of low lubricant (hydrocarbon) solubility, with a propensity to separate on surfaces and form varnish, lacquer, and carbon deposits. The separation tendency of these materials is a consequence of their particle size. Small particles are more likely to stay in oil than large particles. Therefore, resin and soot particles, which are the two essential components of all the deposit forming species, must grow in size via agglomeration, prior to separation. Growth occurs either because of the dipolar interactions, as is the case in resin molecules, or due to the adsorbed polar impurities, such as water and oxygen, as is the case in soot particles. Alternatively, soot particles get caught in the sticky resin, which is shown in Parts A and B of Fig. 4.39.

Dispersants suppress the agglomeration of the resin and

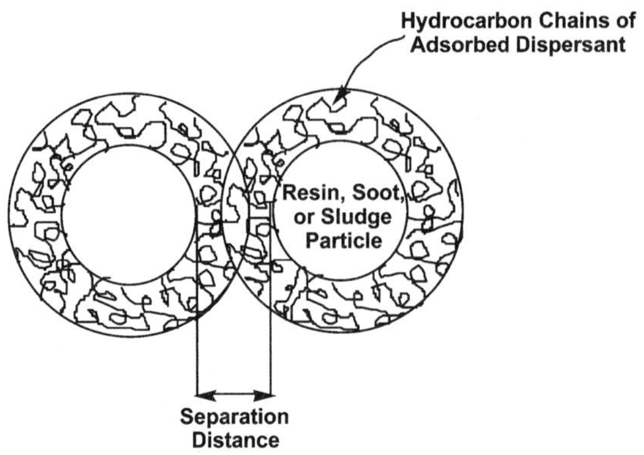

Fig. 4.40—Mechanism of steric stabilization.

the soot particles to a larger size and by minimizing their association with each other. As stated earlier, all additive types, except some viscosity modifiers and pour-point depressants, contain a polar functional group and a nonpolar oleophilic hydrocarbon moiety. Dispersants manage to keep resin and soot particles apart by preferentially associating with the individual particles via their polar end and keeping them suspended in the bulk lubricant by associating with the lubricant via their oleophilic hydrocarbon functional group [191]. The polar association between the dispersant and the resin and the soot particles is facilitated because they are also polar in character, either by their very nature or due to the adsorbed polar impurities. The inter-association between the two is depicted in Parts C and D of the Fig. 4.39. Incidentally, neutral detergents, or soaps, that are discussed in the next section operate by an analogous mechanism.

The inability of the particles with associated dispersant molecules to coalesce can be explained by invoking the concepts of steric stabilization and electrostatic stabilization [218,219]. According to steric stabilization, once the dispersant molecules attach themselves to the resin or the soot particles, their long hydrocarbon chains prevent agglomeration of the particles by keeping them distant [192]. This mechanism is depicted in Fig. 4.40.

Electrostatic stabilization is based upon a double-layer model [218]. According to this model, the resin and soot particles are charged and are surrounded by a layer of counter-ions. The region where the counter-ions exist is called the diffuse double layer. Since for low-dielectric liquids, such as lubricants, the double-layer thickness, represented as $1/\kappa$, is much larger than the particle radius, the double layer extends well into the bulk lubricant. However, beyond the double layer, the amount of the counter-ions diminishes greatly. At three times the thickness of the double layer ($3/\kappa$), the electrical potential due to the counter-ions is only 2 % of the original value. The double layer model is graphically shown in Fig. 4.41. Ψ is the electrical potential and Ψ_D is the potential at the surface of the particle that contains the adsorbed ions. The surface potential Ψ_D decays exponentially and can be represented by the equation

$$\Psi = \Psi_D e^{-\kappa x}$$

The term x is the distance between the particle surface and the solvent (bulk lubricant), and κ is a mathematical construct that relates the decay of the potential to the physical properties of the system.

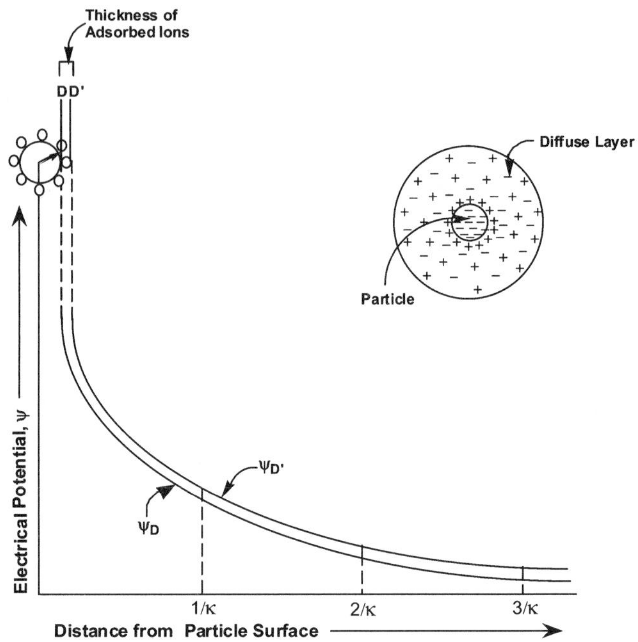

Fig. 4.41—The double-layer model.

$$\kappa = \sqrt{2n_{i0}z_i e/ee_0 kT}$$

In this equation, the value of κ depends upon the permittivity of the space or vacuum (ε_0), permittivity of the liquid medium (ε), concentration of ions (n_i) charge per ion (z_i), charge of an electron (e), Boltzman constant (k), and temperature (T).

The development of the charged particles in a low-dielectric medium, such as oil, can be explained in two ways. According to the first explanation, the dispersant is already ionized and through adsorption imparts a charge to the particle, as depicted by Mechanism A in Fig. 4.42. According to the second explanation, the nonionized dispersant adsorbs onto the surface of the particle, an acid-base reaction involving the transfer of a proton or another ion occurs, and the counter-ion desorbs resulting in the formation of a charged particle [218]. Incidentally, neutral and high soap detergents, discussed in next section, are believed to also perform via this mechanism. Particles with like charges repel each other, thereby preventing agglomeration. This is shown in Part B of Fig. 4.42. It is important to note that for certain classes of dispersants, both mechanisms might be operating simultaneously. Interestingly, empirical data suggest that the ionic mechanism, although seems unlikely due to the organic nature of the lubricant medium, does operate.

Desirable Dispersant Properties

While dispersing soot, deposit precursors, and deposits is clearly the primary function of a dispersant. Dispersants, in addition, need other attributes to perform effectively. These include good thermal and oxidative stability, improved shear stability, good low-temperature properties, being chlorine-free, noncorrosiveness to nonferrous metals such as copper and lead, elastomer seal compatibility, and reasonable cost.

If a dispersant suffers from poor thermal stability, it will

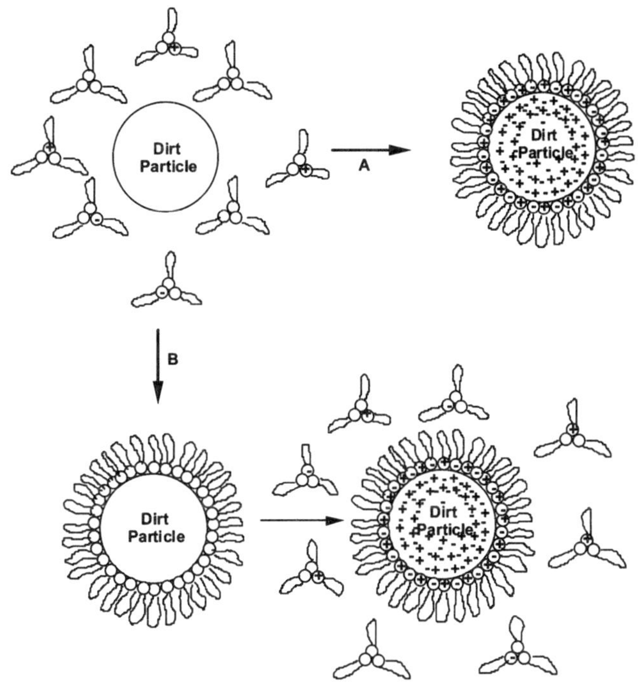

Fig. 4.42—The mechanism of electrostatic stabilization. (A) by ionized dispersant; (B) by un-ionized dispersant.

breakdown, thereby losing its ability to associate with and suspend potentially harmful products. Poor oxidative stability translates into the dispersant molecule contributing towards deposit formation. Improved shear stability is important, especially in the case of high molecular weight dispersant polymers, also called dispersant viscosity modifiers. Improved low-temperature properties of a lubricant are desired for many reasons. These include the ease of cold cranking, good lubricant circulation, and fuel economy. Base oil suppliers have developed a number of ways to achieve these properties. The methods they use include isomerization of the base stock hydrocarbons via hydrocracking and the use of special synthetic oils as additives. Dispersants must have no or low residual chlorine because of its role in dioxin formation, which has a negative effect on public welfare. Noncorrosiveness to copper and lead is important because of the use of these metals in bushings, bearings, cam followers, and oil coolers in diesel engines. Standard test methods ASTM D6594 and ASTM D5968 are used to determine diesel engine lubricant's tendency to corrode alloys of lead and copper. Elastomer seal compatibility, especially those made from Viton®, is desired. Since dispersant is one of the major components in the engine oil formulations, its presence can adversely affect these properties, which must be preserved. And, of course, reasonable cost is always a desirable objective.

Dispersant Structure

A dispersant molecule consists of three distinct structural features: a *hydrocarbon group*, a *polar group*, and a connecting group, or a *link*. These are graphically depicted in Fig. 4.43. The *hydrocarbon group* is polymeric in nature and depending upon its molecular weight, dispersants can be classified into *polymeric dispersants* and *dispersant polymers*. Polymeric dispersants are of lower molecular weight than the dispersant polymers. The molecular weight of polymeric dispersants ranges between 3000 and 7000 g/mol as compared to the dispersant polymers that have a molecular weight of 25,000 g/mol and higher. While a variety of polyolefins, such as polyisobutylene, polypropylene, polyalphaolefins, and mixtures thereof, can be used to make polymeric dispersants, polyisobutylene-derived dispersants are the most common. The primary reasons for polyisobutylene to be the olefin of choice are its thermal stability, large volume availability in the proper molecular weight range, and its relatively low cost. The desired number-average molecular weight (Mn) of polyisobutylene to make polymeric dispersants ranges between 500 and 3000 g/mol, with an Mn of 1000–2000 g/mol being typical [220]. In addition to the Mn, other polyisobutylene parameters, such as molecular weight distribution, the length and the degree of branching, and the reactivity are also important in determining the overall effectiveness of a dispersant. However, despite its strengths, the polyisobutylene use in dispersants has its limitations. For example, achieving high-temperature viscosity without adversely affecting the cold-cranking viscosity with polyisobutylene is a minor challenge. The other concern that is driving the new dispersant research, including the use of the new olefins, is to develop technology that is chlorine free, has a lower cost, and meets or exceeds the ever-changing performance specifications for the future engine oils. With respect to the use of the new olefins, any olefins that have the proper molecular weight range and reasonable reactivity are at present being explored. Incidentally, in the ensuing discussion the term dispersant pertains to polymeric dispersants.

Substances that are obtained via a polymerization reaction, especially those made by the use of an acid catalyst or a free radical initiator, often contain molecules of different sizes. Molecular weight distribution, or the polydispersity index, is commonly used to assess the heterogeneity in the molecular size. Polydispersity index is the ratio of the

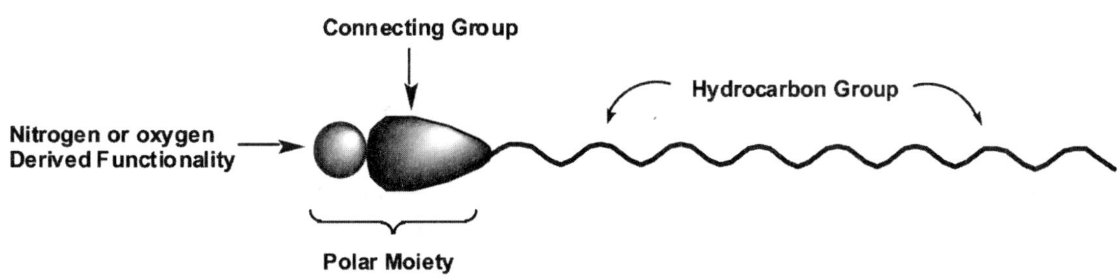

Fig. 4.43—Graphic representation of a dispersant molecule.

weight-average molecular weight (Mw) and the number-average molecular weight (Mn), or Mw/Mn, in a polymer [221–224]. These molecular weights are determined by subjecting the polymer to Gel Permeation Chromatography (GPC). The method separates molecules based on size [223,224]. The larger molecules come out first, followed by the next size. When the molecules are of the same size, Mw/Mn equals 1 and the polymer is called a mono-disperse polymer. The polymers with a polydispersity index of greater than 1 are called poly-disperse polymers. For most applications, mono-dispersity is desired. Polyisobutylene, on account of being derived from the acid-catalyzed polymerization reaction, typically has a polydispersity index of between 2 and 3. This impacts many of the dispersant properties that are described below.

Dispersant polymers, also called dispersant viscosity modifiers (DVMs) and dispersant viscosity index improvers (DVIIs), are derived from hydrocarbon polymers of molecular weights between 25,000 and 500,000 g/mol. Polymer substrates used to make DVMs include high molecular weight olefin copolymers (OCPs), such as ethylene-propylene rubbers (EPRs), ethylene-propylene-diene copolymers (EPDMs), polymethacrylates (PMAs), styrene-diene rubbers (SDRs) of both linear and star configurations, and styrene-ester polymers (SEs).

The *polar group* in dispersants is usually nitrogen or oxygen-derived. Nitrogen-based groups are derived from amines and are usually basic in character. Oxygen-based groups are alcohol-derived and are neutral. The amines that are commonly used to synthesize dispersants are polyalkylenepolyamines, such as diethylenetriamine and triethylenetetramine. In polymeric dispersants, the basic functional group needs to be attached to the hydrocarbon portion by way of a connecting group, which is either a carboxylic acid or an anhydride, such as succinic acid or succinic anhydride, or a phenol. The dispersants are called succinimide dispersants, if the polar group was introduced by the reaction of a polyamine, and succinate dispersants, if the polar group was introduced by the reaction of a polyhydric alcohol [226,227]. However, in the case of the dispersant viscosity modifiers, or the dispersant polymers, there are the options of either directly introducing the polar group by grafting or by copolymerization; or indirectly, in a manner similar to that of the polymeric dispersants. That is, by introducing a connecting link and then functionalizing it. The compounds that are suitable for direct grafting or copolymerization include monomers, such as 2- and 4-vinylpyridine, N-vinylpyrrolidinone, and N, N-dialkylaminoalkyl acrylate. The connecting groups that are used to introduce the dispersant moiety indirectly include unsaturated anhydrides and acids, such as maleic anhydride, acrylic acid, and glyoxylic acid; which are then functionalized by a reaction with an amine. The details of these reactions are described in the latter part of this chapter that deals with the dispersant synthesis.

Besides the amine-derived (nitrogen or imide) and alcohol-derived (oxygen or ester) dispersants, at one time the oxygen derived phosphonate ester dispersants were also used. However, their use in engine oils is now restrained because of the phosphorus limit, which was established because of its tendency to poison noble metal catalysts that are used in catalytic converters in automobiles. In addition, formulators prefer to use the allowed amount of phosphorus as zinc dialkyl dithiophosphates, which are excellent oxidation inhibitors and antiwear agents. In the case of the amine dispersants, it is customary to leave some of the amino groups unreacted to impart basicity to the dispersant. The reasons for this are described later.

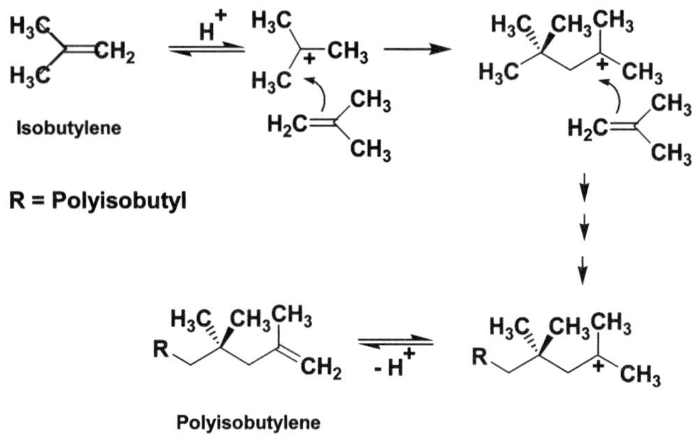

Fig. 4.44—Acid-catalyzed polymerization of isobutylene.

Dispersant Synthesis

Since it is not easy to attach the polar group directly to the hydrocarbon group, except in the case of olefins that are used to make dispersant viscosity modifiers, the need for a connecting group or a link arises. While many such groups can be used, the two common ones are phenol and succinic anhydride. Olefin, such as polyisobutylene, is reacted either with phenol to form an alkylphenol or with maleic anhydride to form an alkenylsuccinic anhydride. These substrates are then reacted with the appropriate amines or the alcohols to introduce the polar functional group.

Hydrocarbon Group

As mentioned earlier, polyisobutylene is the most common olefin used to introduce the hydrocarbon group in polymeric dispersants. Polyisobutylene is manufactured via acid-catalyzed polymerization of isobutylene [106,228]. The mechanism of its formation is depicted in Fig. 4.44. While in the figure, polyisobutylene is shown as a terminal olefin, in reality it is a mixture of a variety of isomers. Those that predominate include geminally di-substituted (vinylidene), tri-substituted, and tetra-substituted olefins. Figure 4.45 shows some of these structures and the possible mechanism of their formation. Polyisobutylenes of Structures I and II result from the loss of a proton from Carbon 1 and Carbon 3 of the intermediate of structure V. Polyisobutylenes of Structures III and IV result from the rearrangement of the initially formed carbocation, as shown in Fig. 4.45. The reactivity of these olefins towards phenol and maleic anhydride varies. In general, the more substituted the olefin, the lower the reactivity, which is a consequence of the steric factors. Similarly, the larger the size of the polyisobutyl pendent group, that is, the higher the molecular weight, the lower is the reactivity. This is due to the dilution effect, which results from the low olefin to hydrocarbon ratio. One of the reasons for the preferred use of polyisobutylene is its extensive branching. This makes the derived dispersants possess excellent oil solubility, both in the nonassociated form and in the associated

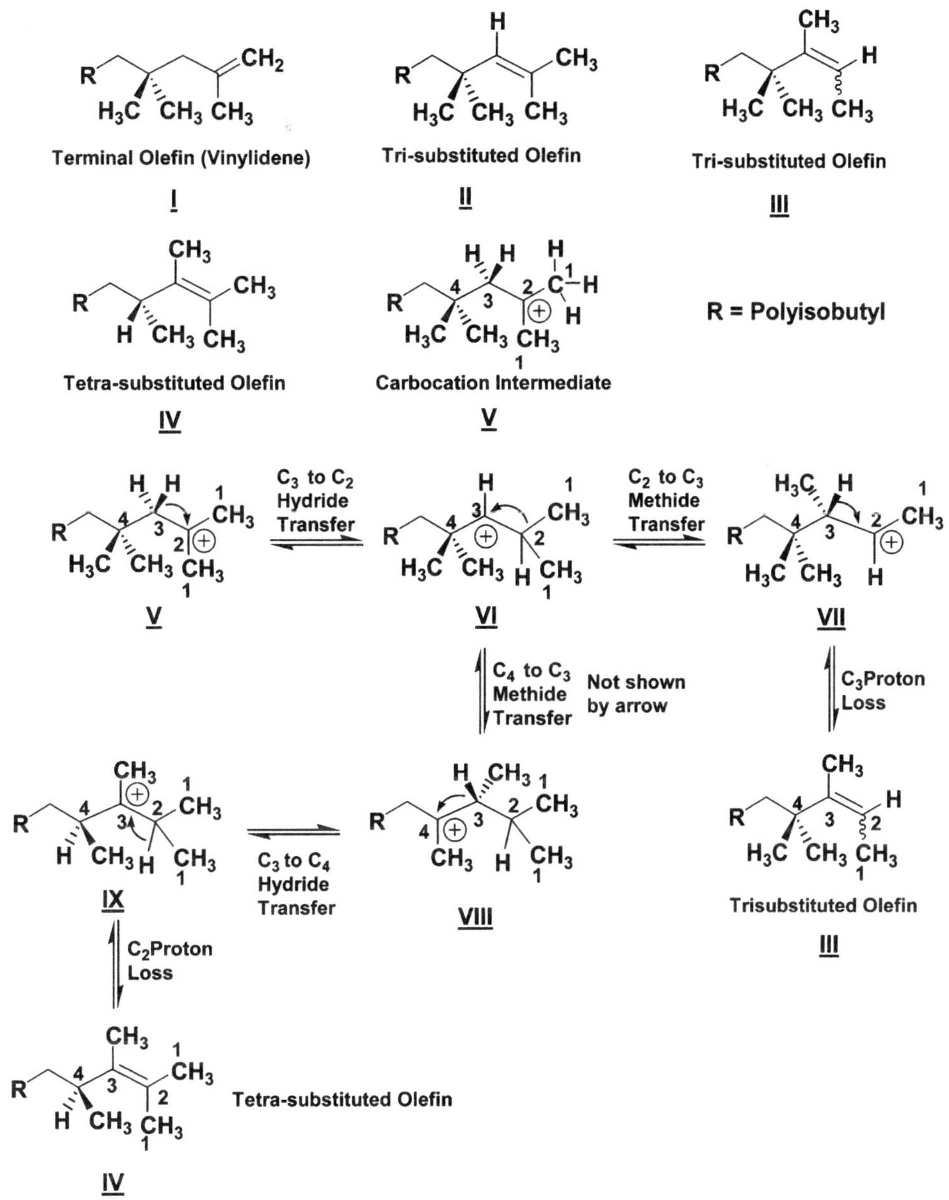

Fig. 4.45—Polyisobutylene structures and the mode of their formation.

form. However, if the hydrocarbon chain in the dispersant is too small, the lubricant solubility greatly suffers. Because of this, the low molecular weight components in polyisobutylene are not desired. This is despite their higher reactivity. These must be removed, which is achieved by distillation. Alternatively, one can minimize the formation of these components by decreasing the amount of the catalyst or by lowering the polymerization reaction temperature, or both.

A new class of dispersants derived from ethylene/α-olefin copolymer with a number-average molecular weight (Mn) of 300 to 20,000 g/mol has also been reported, primarily by the Exxon scientists [229,230]. Such dispersants are claimed to have superior low and high-temperature viscometrics to those of the polyisobutylene-derived materials. Polyalphaolefin derived dispersants are also being explored, presumably because of their superior dispersancy, good low-temperature performance, and better conversion [231,232]. Trilene® is another hydrocarbon the use of which has been reported to introduce the hydrocarbon group in dispersants [233,234]. Trilene® is a liquid ethylene-propylene-diene rubber (EPDM) that is available from Uniroyal Chemical Co.

As mentioned earlier, dispersant polymers are derived from ethylene-propylene copolymers, styrene-butadiene copolymers, polyacrylates, polymethacrylates, and styrene-esters. Ethylene-propylene rubbers are synthesized by Ziegler-Natta catalysis [235]. Styrene-butadiene rubbers are synthesized via anionic polymerization. Polyacrylates and polymethacrylates are synthesized via polymerization of the monomers using the free radical initiators. Styrene esters are made by the reaction of styrene-maleic anhydride copolymer or styrene-maleic anhydride-alkyl acrylate terpolymer with alcohols, usually in the presence of a proton acid, such as sulfuric acid or methanesulfonic acid, catalyst. Since

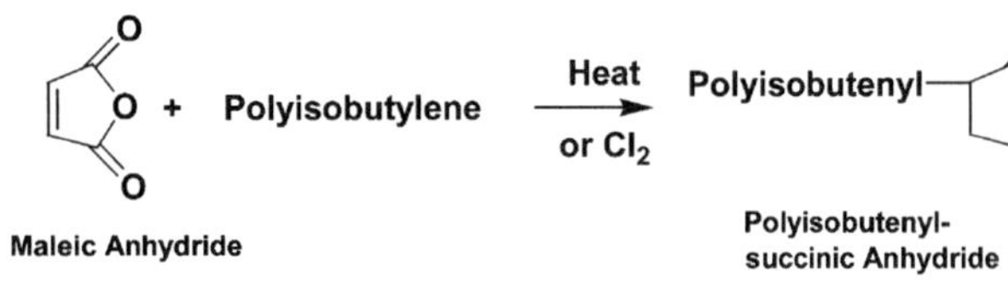

Fig. 4.46—Alkenylsuccinic anhydride formation.

complete esterification of the anhydride is hard to achieve, the neutralization of the residual carboxylic acid anhydride is carried out by alternative means [236–238]. The synthetic details of these materials are provided in the section on polymeric additives.

Connecting Group

As stated earlier, succinimide, phenol, and phosphonate are the common connecting groups that are used to make dispersants. Of these, succinimide and phenol are the most prevalent [239]. Succinimide group results when an alkenylsuccinic anhydride, a cyclic dicarboxylic acid anhydride, is reacted with a primary amino group. Alkenylsuccinic anhydride, the precursor for introducing the succinimide connecting group in dispersants, is synthesized by the reaction of an olefin, such as polyisobutylene, with maleic anhydride [239], as shown in Fig. 4.46. The reaction is carried out either thermally [220,240,241] or in the presence of chlorine [242]. The *thermal process* involves heating the two reactants together, usually above 200°C. The *chlorine-mediated reaction*, on the other hand, is carried out by introducing chlorine into the reaction containing polyisobutylene and maleic anhydride [242–246]. Depending upon the manner in which the chlorine is added, the procedure is either a *One-step Procedure* or a *Two-step Procedure* [243]. If the chlorine is first reacted with polyisobutylene, prior to adding maleic anhydride, the procedure is considered a two-step procedure. If the chlorine is added to a mixture of polyisobutylene and maleic anhydride, it is considered a one-step procedure. The one-step procedure is generally preferred.

The chlorine-mediated process has several advantages, which include low reaction temperature, faster reaction rate, and that it works well with internalized, or highly substituted, olefins. The low reaction temperature minimizes the chances of thermal breakdown of the polyisobutylene and saves energy. The major drawback of the chlorine-mediated process is that the resulting dispersants contain residual chlorine as organic chlorides. The presence of chlorine in the environment is becoming a concern because it takes part in the formation of the carcinogenic dioxins. A number of strategies are reported in the literature to decrease the chlorine content of the dispersants [247–252]. The thermal process, on the other hand, does not suffer from the presence of chlorine. However, the thermal process involves direct alkylation, which is less energy efficient since the reaction requires a high temperature. It also requires the use of the predominantly terminal olefin, that is, polyisobutylene of high vinylidene content. It does not work well if the double bond in polyisobutylene is highly substituted.

The mechanism by which the two processes proceed is also different [245,246,248–250]. *Thermal process* is postulated to occur via an Alder ENE reaction. *Chlorine-mediated reaction*, on the other hand, is postulated to proceed via a Diels-Alder reaction. The mechanism by which polyisobutylene gets converted into alkylbutadiene is suggested in Fig. 4.47. The chlorine first reacts with polyisobutylene I to form the allylic chloride II. This by the loss of the chloride radical yields the intermediate III, which via C_4 to C_3 methyl radical transfer is converted into intermediate IV. A C_3 to C_4 hydrogen shift in the intermediate results in the formation of the radical V. This radical can lose hydrogen either from C_4 to yield the diene VI or from C_5 to result in the diene VII. The resulting dienes then react with the maleic anhydride via a 4:2 addition reaction, commonly called a Diels-Alder reaction [253], to form alkenyltetrahydrophthalic anhydrides [248,250]. These reactions are shown in Fig. 4.48.

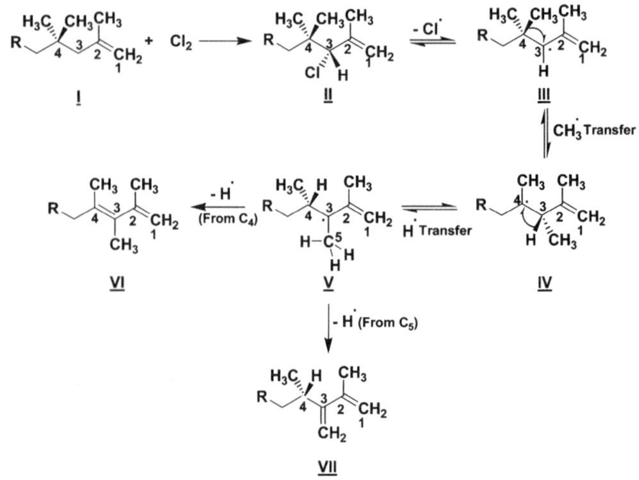

Fig. 4.47—Mechanism of chlorine-assisted diene formation.

These anhydrides can be converted into phthalic anhydrides via dehydrogenation by the use of sulfur [248–250]. These compounds can then be transformed into dispersants by their reaction with polyamines and polyhydric alcohols. During the *thermal reaction* of the polyisobutylene with maleic anhydride (the ENE reaction) the vinylidene double bond moves down the chain to the next carbon. Since thermal reaction requires a terminal olefin, further reaction of the new olefin with another mole of maleic anhydride will not occur if the double bond internalizes and the reaction will stop at this stage. This is shown in Eq. 3 of Fig. 4.48. If, on the other hand, the new double bond is external, the reaction with another molecule of maleic anhydride is possible [244]. This is shown in Eq. 4 of Fig. 4.48.

Of the new carboxylate functionalities that are being explored as connecting groups to synthesize new dispersants, two are worth mentioning. One is based upon methyl glycolate methyl hemiacetal (GMHA) and the other is based upon α,β-unsaturated polycarboxylate. See Fig. 4.49 for structures and the synthesis of the latter. Their use as dispersant intermediates or the actual dispersants made from them have been reported in the current patent literature [233,234,254–260]. However, at present there appear to be no commercial products based on this chemistry.

For phenol-derived dispersants, i.e., the Mannich dispersants, polyisobutylphenol is the alkylphenol of choice. It is synthesized by the reaction of polyisobutylene with phenol in the presence of an acid catalyst [261,262]. Lewis acid catalysts, such as aluminum chloride and boron trifluoride, are often employed. Boron trifluoride is preferred over alu-

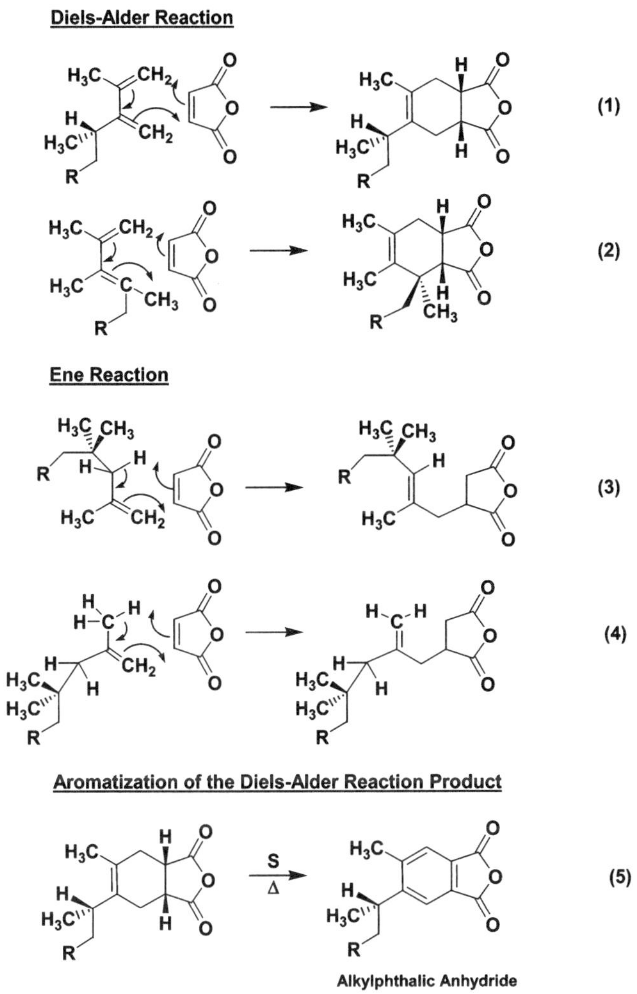

Fig. 4.48—Mechanism of alkenylsuccinic anhydride formation.

minum chloride because the reaction can be carried out at low temperatures, which minimizes acid-mediated breakdown of the polyisobutylene [262]. This is desired because the dispersants derived from the low molecular weight phenols are not very effective. Other catalysts, such as sulfuric acid, methanesulfonic acid, and porous acid catalysts of the Amberlyst® type, can also be used to make alkylphenols [263,264]. Polyisobutylene also reacts with phosphorus pentasulfide via an ENE reaction, as described later in the section on detergents. The resulting adduct is hydrolyzed by the

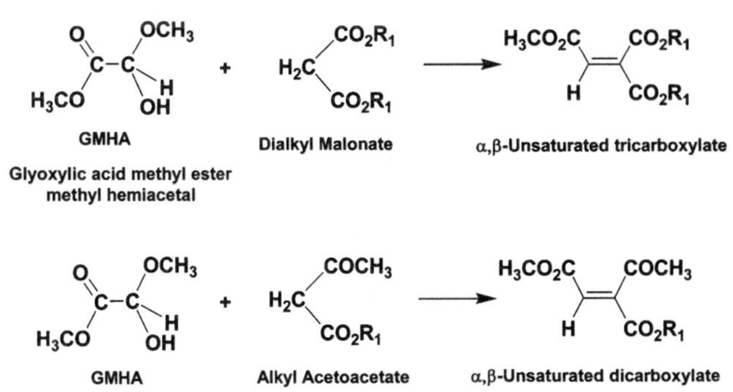

Fig. 4.49—New acylating groups for dispersants.

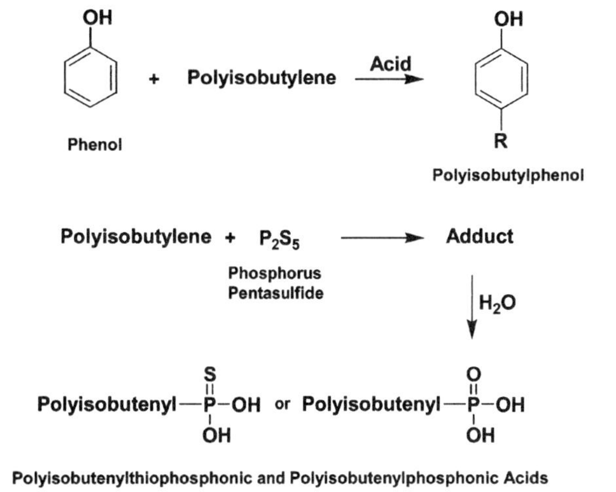

Fig. 4.50—Synthesis of alkylphenols and alkenylphosphonic acids.

use of steam to alkenylphosphonic and alkenylthiophosphonic acids [239]. The methods to synthesize alkylphenols and alkenylphosphonic acids are summarized in Fig. 4.50.

Polar Moiety

As mentioned above, the two common polar moieties in dispersants are based upon polyamines and polyhydric alcohols. The structures of the common amines and the alcohols that are used to make dispersants are shown in Fig. 4.51.

The polyamines are manufactured from ethylene via chlorination, followed by the reaction with ammonia [265]. The reaction scheme is given in Fig. 4.52. As shown, the

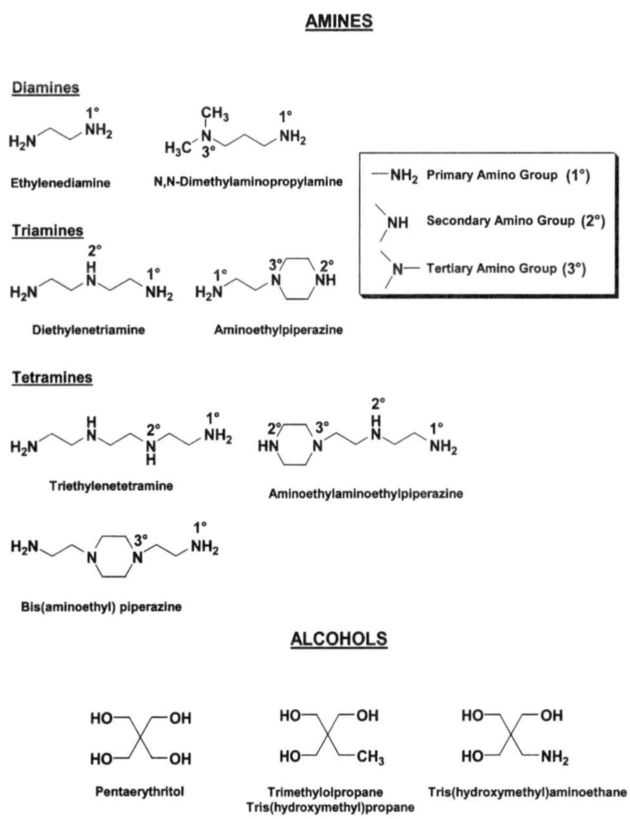

Fig. 4.51—Amines and alcohols used to synthesize dispersants.

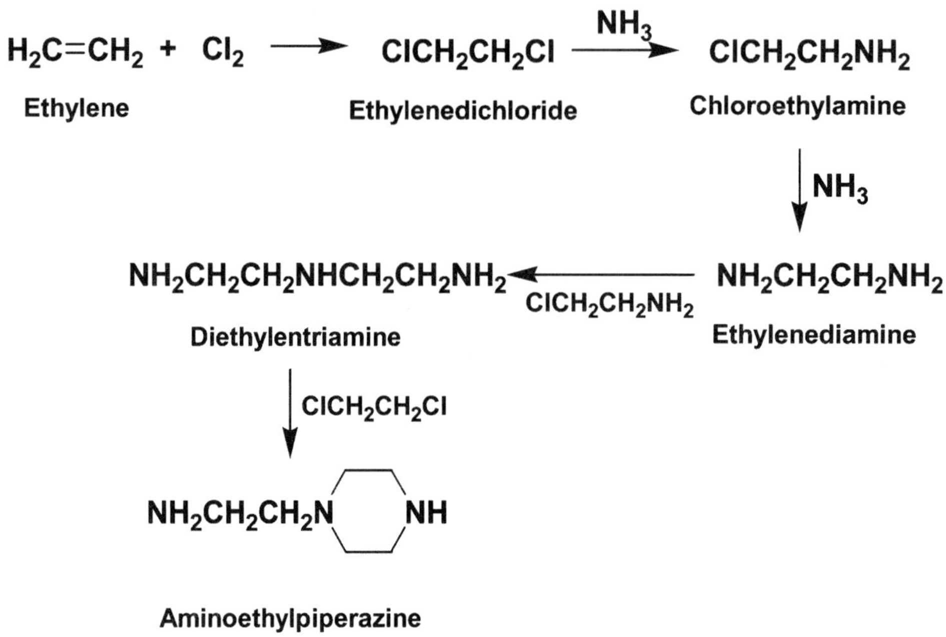

Fig. 4.52—Manufacture of polyamines.

polyamines contain piperazines as a by-product. If one examines the structures of the various amines, one can see that they contain primary amino groups, secondary amino groups, and tertiary amino groups. Each type of amino group has different reactivity towards alkenylsuccinic anhydride. Primary amino group reacts with the anhydride to form a cyclic imide, the secondary amino group reacts with the anhydride to form an amide/carboxylic acid, and the tertiary amine does not react with the anhydride at all [266].

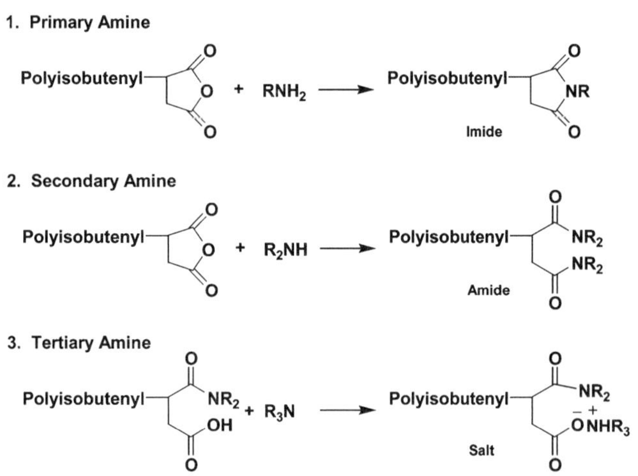

Fig. 4.53—Amine-anhydride reaction products.

However, the tertiary amine can make a salt if there is free carboxylic acid functional group present in the molecule, as is the case in amide/carboxylic acid. These reactions are shown in Figure 4.53. New high molecular weight amines derived from phosphoric acid catalyzed condensation of the polyhydroxy compounds, such as pentaerythritol, with polyalkylene polyamines, such as triethylenetetramine, are known [267]. These amines are claimed to form high TBN (total base number) dispersants with low free amine content and better engine test performance than dispersants made from the conventional polyamines.

Polyhydric alcohols that are commonly used to make dispersants include trimethylolpropane, tris(hydroxymethyl)aminomethane, and pentaerythritol, all of which are base catalyzed reaction products of formaldehyde, either with another aldehyde or with nitromethane. Reaction mechanisms for the formation of trimethylolpropane and pentaerythritol were provided in Figs. 3.22 and 3.23 in the Synthetic Esters section of the Synthetic and Biological Base Stocks chapter. Hence, here we present only the synthetic sequence for tris(hydroxymethyl)aminomethane (THAM), see Fig. 4.54. The reaction mechanism in this case is analogous to that of the trimethylolpropane formation. That is, it involves the base-catalyzed addition of three moles of formaldehyde to nitromethane. The resulting nitro derivative is then reduced to the amino compound [268].

Imide and ester dispersants are made by the reaction of the polyamines and the polyhydric alcohols with alkenylsuc-

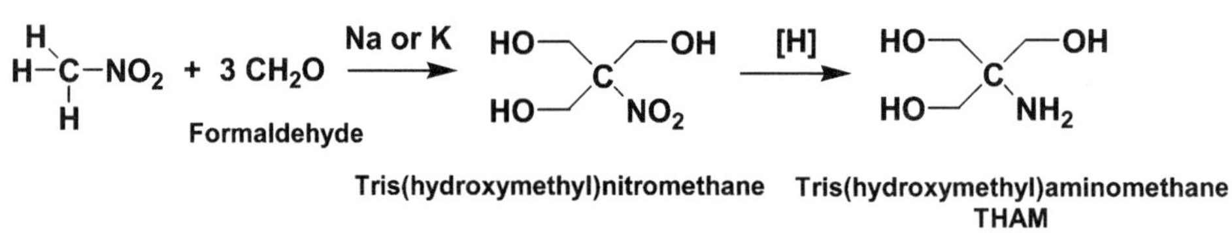

Fig. 4.54—Synthesis of tris(hydroxymethyl)aminomethane (THAM).

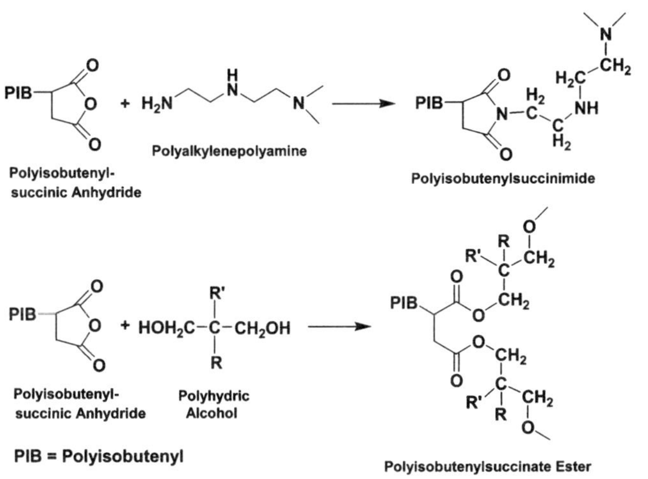

Fig. 4.55—Synthesis of alkenylsuccinimides and alkenylsuccinates.

cinic anhydrides. The reaction typically requires a reaction temperature of between 130 and 200°C, to remove the resulting water and complete the reaction [243]. As mentioned earlier, imide dispersants are made by the use of polyalkylene-polyamines and ester dispersants are made by the use of polyhydric alcohols; and that the imide dispersants are basic and the ester dispersants are neutral. When one uses an amino alcohol, such as tris(hydroxymethyl)aminomethane, one obtains an ester dispersant with basicity. The reaction schemes to make succinimide and succinate dispersants are depicted in Fig. 4.55.

Alkylphenol-derived dispersants are made by the reaction of an alkylphenol, such as polyisobutylphenol, with formaldehyde and a polyamine [262,269]. The result is the formation of 2-aminomethyl-4-polyisobutylphenol. The reaction of ammonia or an amine, formaldehyde, and a compound with active hydrogen/s, such as a phenol, is called the Mannich reaction [270,271]. Hence, such dispersants are called Mannich dispersants. The use of glyoxylic acid derivatives to make phenolic dispersants is described in references [254–260]. For making phosphonate dispersants, the common method is to react the free acid with an olefin epoxide, such as propylene oxide or butylene oxide, or an amine [272–274]. These reactions are summarized in Fig. 4.56. The salts derived from the direct reaction of the amine and the metal bases with olefin-phosphorus pentasulfide adduct are also known [274,275]. It is important to note that the structures in the figures are idealized structures. The actual structures will depend upon the substrate (alkylphenol and alkenylsuccinic anhydride) to the reactant (formaldehyde and polyamines) ratio.

Because of the polyfunctionality of the succinic anhydride group and of the amines and the polyhydric alcohols, a variety of dispersants can be made by altering the anhydride to amine or anhydride to alcohol ratios. These dispersants not only differ in their molecular weight but also in their properties. Polyfunctionality of the two reactants leads to dispersants, which have molecular weights that are three to seven times higher than expected, if the two reactants were monofunctional. The bridged structure of a dispersant is conceptually presented in Fig. 4.57, and for illustration purposes, two structures that can result from the same starting

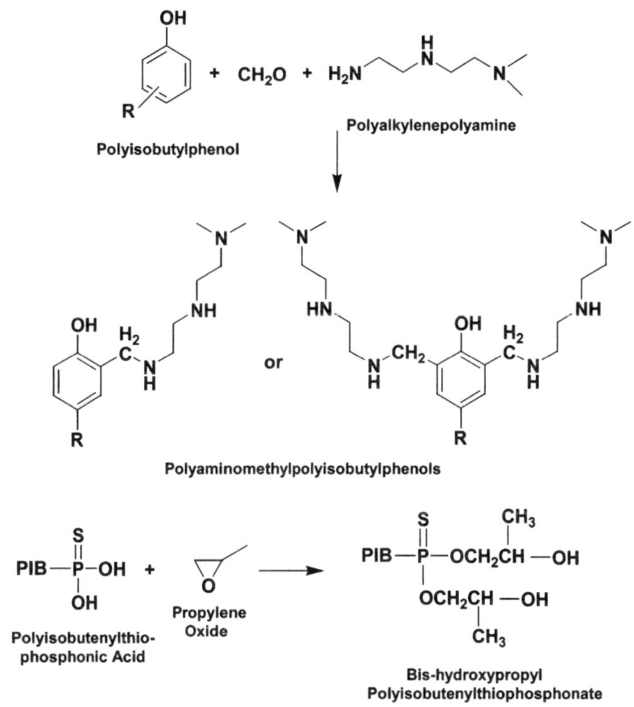

Fig. 4.56—Synthesis of Mannich and phosphonic acid dispersants.

materials, viz., a polyisobutenylsuccinic anhydride and a polyamine, but reacted in different ratios are depicted in Fig. 4.58. It is important to note that the dispersant of Structure 1, because of having more amino groups unreacted will be more basic than that of Structure 2.

In addition to the high molecular weight dispersants discussed so far, somewhat lower molecular weight materi als derived from oleic acid, iso-stearic acid, or naphthenic acid and polyamines are also used in some applications. Such materials, referred to as imidazolines, have five-membered heterocyclic rings. These are used as detergents and inhibitors in two-stroke cycle engine oils, industrial oils, and friction modifiers in engine oils. Possible structure of one such imidazoline is presented in Fig. 4.59.

The methods to make dispersant viscosity modifiers are shown in Figs. 4.60–4.62. These are synthesized by the following methods.

- Grafting or reacting of a dispersancy-imparting monomer on an already formed polymer, as in the case of EPRs and SDRs [276–285].
- Including such a monomer during the polymerization process, as in the case of polyacrylates and polymethacrylates [285–287].
- Introducing a reactive functional group in the polymer that can be reacted with a reagent to impart dispersancy, as in the case of styrene-maleic anhydride copolymers [238,259,288–294].

While most of the examples in the figures pertain to the introduction of the basic nitrogen-containing moieties, dispersant viscosity modifiers that are neutral are also known in the literature. These are made by using nonbasic reactants, such as N-vinyl-pyrrolidinone, alcohols, or polyether derived methacrylate ester [279,286,295]. Recently, dispersant viscosity improving additives with built-in oxidation inhibit-

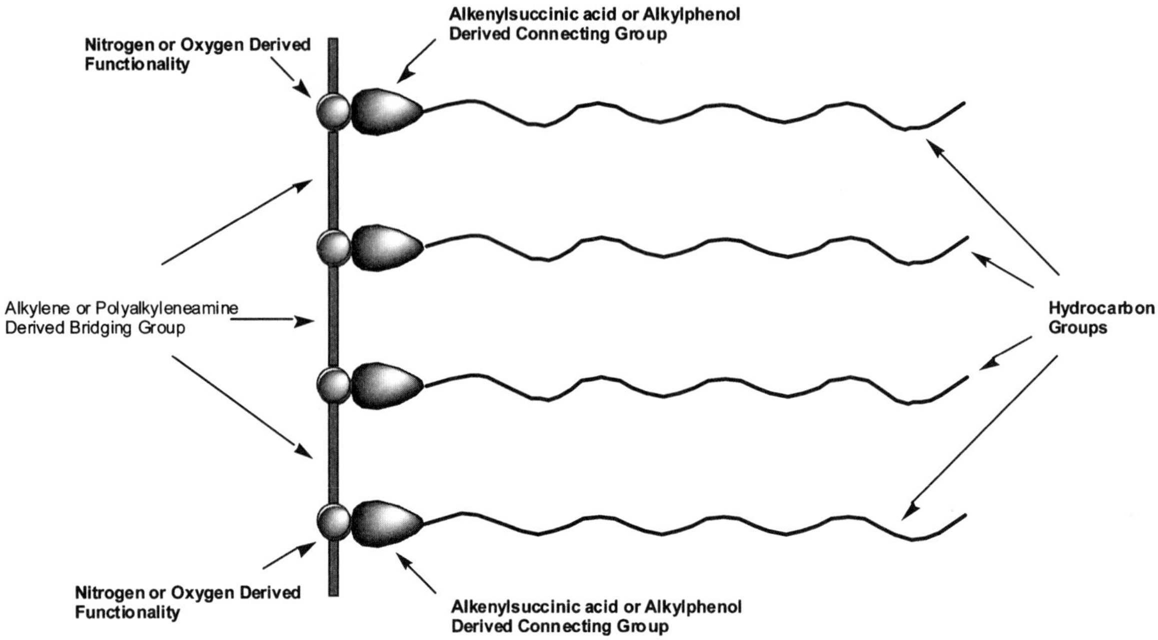

Fig. 4.57—A representation of the bridged structure of a dispersant.

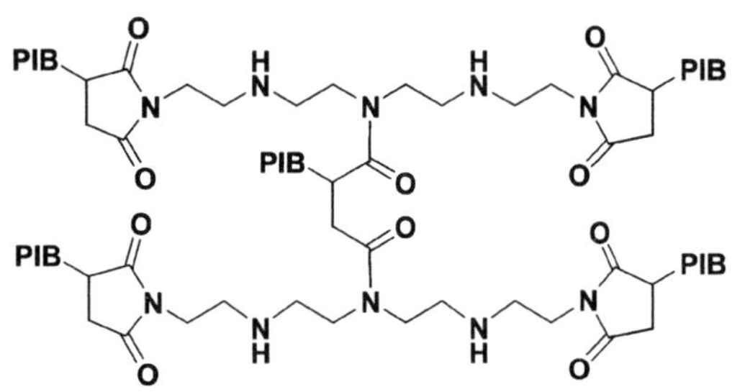

PIB/TEPA (1COOH:1N)

Dispersant very basic since 4 out of 10 nitogens are basic

1

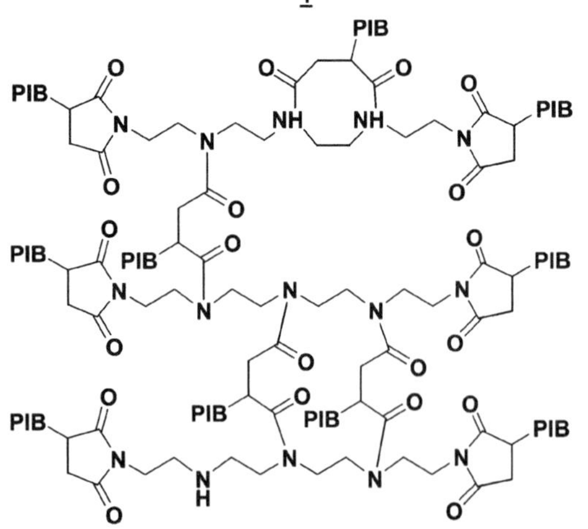

PIB/TEPA (4COOH:3N)

Dispersant not very basic since 1 out of 15 nitogens is basic

2

Fig. 4.58—Bridged structure of dispersants.

ing and antiwear moieties have been reported in the patent literature [277,296,297]. Dispersant polymers containing oxidation-inhibiting moieties are commercially available from Texaco, now part of Ethyl Petroleum Additives Company (Afton). As the examples show, grafting usually allows the introduction of the connecting group in the dispersant polymers at the same time as the polar moiety. The structures of some common monomers that are used in making dispersant viscosity modifiers are provided in Fig. 4.63.

Dispersant Properties

Dispersant properties that define dispersant performance include dispersancy, thermo-oxidative stability, viscosity characteristics, and elastomer seal compatibility. These properties in turn are a function of a dispersant's structural features—the hydrocarbon chain, the connecting group, and the polar functional group; individually and in combination. Please note that our primary interest is in assessing performance of the dispersant in engine oils where they find major use.

Dispersancy

Dispersancy pertains to a dispersant's ability to suspend by-products of combustion, such as soot, and of lubricant degradation, such as resin, varnish, lacquer, and carbon deposits. Dispersancy of a dispersant depends upon all three of its

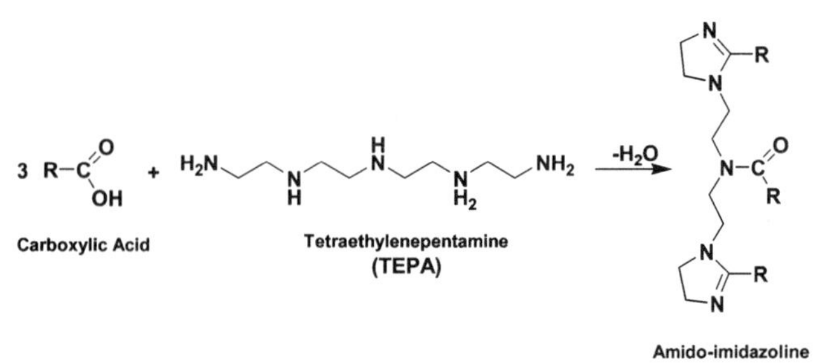

Fig. 4.59—Imidazoline dispersant.

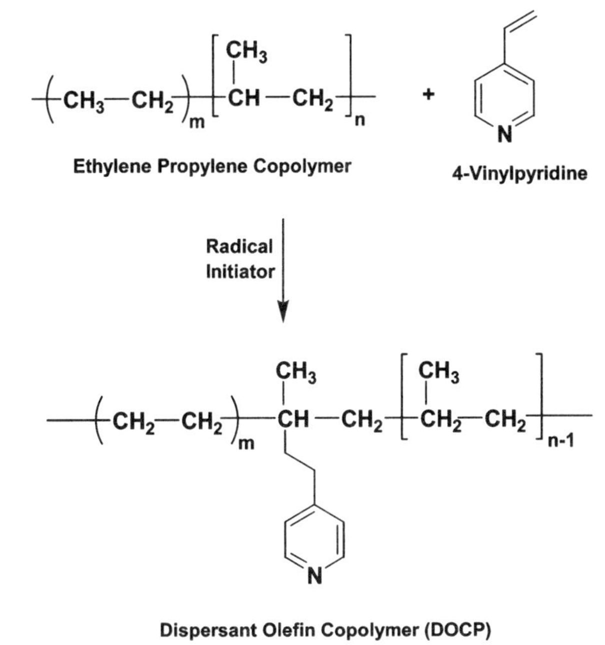

Fig. 4.60—Dispersant viscosity modifier synthesis via grafting.

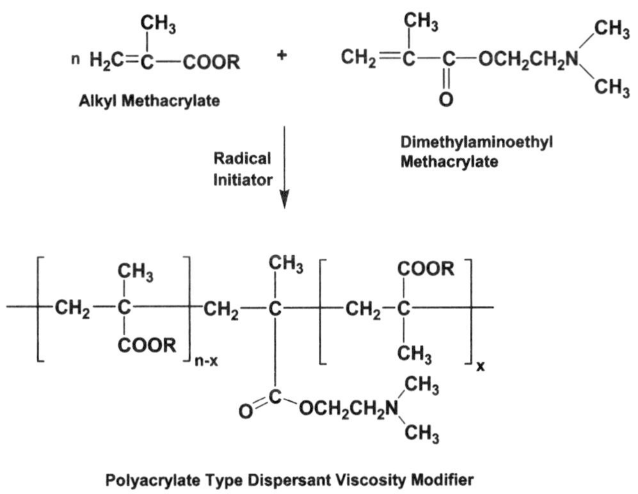

Fig. 4.61—Dispersant viscosity modifier synthesis via copolymerization.

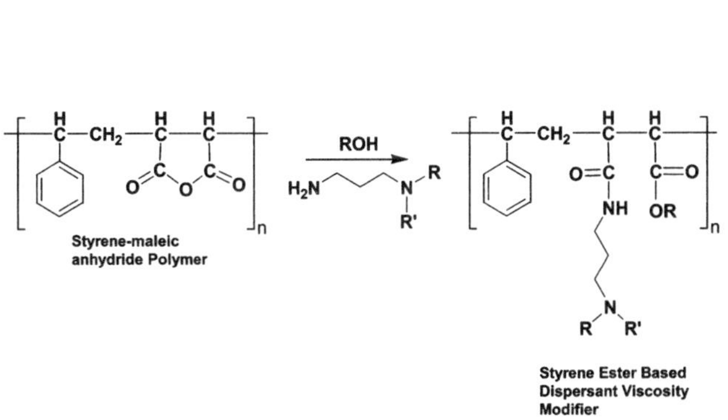

Fig. 4.62—Dispersant viscosity modifier synthesis via chemical reaction.

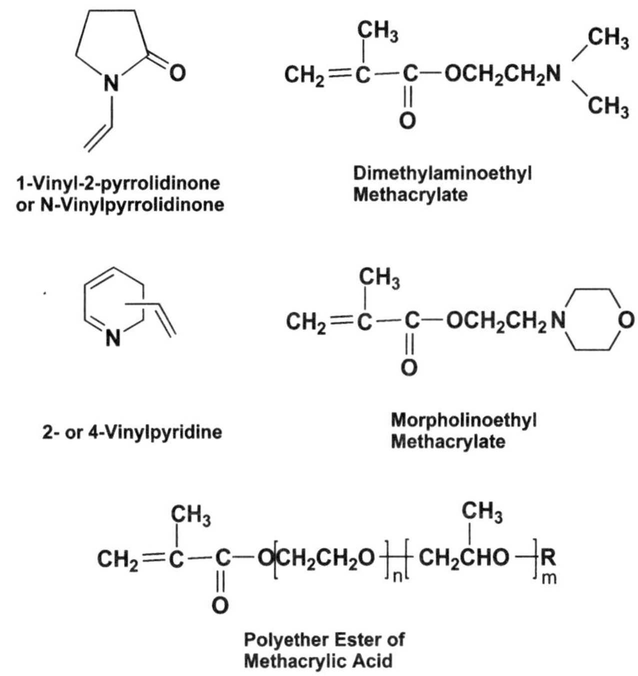

Fig. 4.63—Monomers used in synthesizing dispersant viscosity modifiers.

structural features. The molecular weight of the hydrocarbon group in a dispersant determines its ability to associate with the undesirable polar species and suspend them in the bulk lubricant. For dispersants that have the same connecting group and the polar moiety, the lower the molecular weight of the hydrocarbon group, the higher the ability of the dispersant to associate with the polar materials but the lower the ability to suspend them in the lubricant. Because of this trade off between the associating ability and the suspending ability, the hydrocarbon chain must be of the correct size and branching.

The size affects a dispersant's affinity towards polar materials and the branching affects its solubility, both before association and after association with the species that a dispersant is designed to suspend in oil. Experience has demonstrated that hydrocarbon groups containing 70 to 200 carbon atoms and extensive branching, as in the case of polyisobutylenes, are extremely suitable to devise dispersants with good dispersancy. The hydrocarbon chains of a larger size, even if the branching is similar, lead to dispersants with low affinity towards polar materials. That is why dispersant polymers possess lower dispersancy than polymeric dispersants. However, since the dispersant polymers have additional attributes, such as good thickening efficiency and in some cases good thermal and oxidative stability, their use in lubricants is advantageous. They usually replace additives, called the viscosity modifiers, in the package. Since dispersant polymers impart some dispersancy because of their structure, the amount of the polymeric dispersant in engine oil formulations is decreased somewhat [279,298].

The connecting group and the polar moiety are also important to the dispersancy of the dispersant molecule. However, they must be considered together since both contribute towards polarity. In Mannich dispersants, the phenol functional group, and in imide and ester dispersants, the succinimide, succinate, and phosphonate functional groups, are polar, the same as the amine and the alcohol-derived portion of the molecule. The polarity is a consequence of the electronegativity difference between carbon, oxygen, nitrogen, and phosphorus atoms. The greater the electronegativity difference, the higher is the polarity. This implies that groups that contain phosphorus oxygen bonds are more polar than those containing carbon oxygen bonds, carbon nitrogen bonds, and the carbon phosphorus bonds. The electronegativity difference for such bonds is 1.4, 1.0, 0.5, and 0.4, respectively [299]. However, since the dispersants have many bonds with various combinations of atoms, the overall polarity in a dispersant and its ability to associate with the polar materials is not easy to predict. Because some of the materials with which the dispersant associates are acidic, such as carboxylic acids derived from lubricant oxidation, the presence of an amine nitrogen is an advantage because of its basic character. Therefore, in certain gasoline engine tests, the nitrogen dispersants are superior to the ester dispersants. Ester dispersants, on the other hand, are usually superior in diesel engine tests because of their higher thermo-oxidative stability. Mannich dispersants are good low temperature dispersants; hence they are typically used in gasoline engine oils.

As mentioned in the earlier part of the chapter, commercial polyisobutylenes have a molecular weight distribution. This leads to dispersant structures of varying sizes, hence varying molecular weights. An optimum ratio between the molecular weight of the hydrocarbon chain and that of the polar functional group (nonpolar to polar ratio) is a prerequisite for good dispersancy. If a dispersant composition has an excessive amount of components with low molecular weight (short) hydrocarbon chains, its associating ability with the polar species increases but its oil solubility suffers. This is likely to deteriorate its dispersancy, especially after associating with the polar impurities, which will increase its polar to non-polar ratio. Such structures in dispersants are therefore undesired. These can be minimized either by using polyolefins of low polydispersity index [300], controlling the formation of the low molecular weight components, removing such components via distillation, or post reacting with another reagent, preferably with a large hydrocarbon group. Polyolefins of low polydispersity index ($\leqslant$2.0) are available from British Petroleum (BP) and Exxon Chemical Company. Controlling the formation of the low molecular weight components is exemplified by the use of boron trifluoride catalyst for making alkylphenols instead of aluminum chloride that has the tendency to fragment polyisobutylene. Removing the lower molecular weight components, although not easy, is possible at the precursor stage, which is prior to reacting with the alcohol or the amine. A number of reagents can be used for the post reaction [300]. Hydrocarbon post-treatment agents include polyepoxides [302], polycarboxylic acid [303], alkylbenzenesulfonic acids [304], and alkenylnitriles [305]. Whenever post reacted dispersants are used in engine oils, improved dispersancy, viscosity index credit, improved fluorocarbon elastomer compatibility, hydrolytic stability, and shear stability are often claimed.

Thermal and Oxidative Stability

All three components of the dispersant structure also determine its thermal and oxidative stability, the same as dispersancy. The *hydrocarbon group* can oxidize in the same manner as the lubricant hydrocarbons to form oxidation products that can contribute towards the deposit forming species [55,196]. This was described earlier on the nature of deposits and the mode of their formation. While the rate of oxidation of largely paraffinic hydrocarbon groups, such as polyisobutyl group, is quite slow, those that contain multiple bonds, such as polyisobutenyl, and the benzylic groups, it is quite high. The benzylic functional group is present in styrene-butadiene and styrene-ester derived dispersant polymers, and in Mannich dispersants. Purely paraffinic hydrocarbon groups that contain tertiary hydrogen atoms, such as ethylene-propylene copolymers, oxidize at a faster rate than those that contain only primary and secondary hydrogen atoms. Styrene-isoprene derived materials contain both benzylic and tertiary hydrogen atoms. This implies that highly branched alkyl groups, such as polyisobutyl and polyisobutenyl, have a higher susceptibility towards oxidation than linear or unbranched alkyl groups. Dispersant polymers with built-in oxidation-inhibiting moieties are known in the literature [277,296]. The polar moiety in a dispersant that is amine-derived is also likely to oxidize at a faster rate than the oxygen-derived moiety. This is because of the facile formation of the amine oxide functional group on oxidation. Such groups are known to thermally undergo β-elimination [306], called the Cope Reaction, to form an olefin. These olefins can polymerize to form polymeric deposit precursors.

From a thermal stability perspective, the hydrocarbon group in the case of high molecular weight dispersant polymers, such as those derived from OCPs, is more likely to fragment than the hydrocarbon groups present in the low molecular weight polymers. Dispersants based upon 1000 to 2000 g/mol molecular weight polyisobutylenes are relatively stable, except at very high temperatures that are experienced in some engine parts, such as near the top of the piston [21,23]. Thermal breakdown of the oxidized amine polar group was already mentioned in the previous paragraph.

Chemical reactivity of certain dispersants towards water and other reactive chemicals present in the lubricant formulation or generated during use is an additional concern. The most likely reaction site is the connecting group. The common connecting groups are amide and imide in amine-derived dispersants and ester in the alcohol-derived dispersants. All three can hydrolyze in the presence of water [307], but at different rates. Esters are easier to hydrolyze than amides and imides. The hydrolysis is facilitated by the presence of bases and acids. Basic detergents are the source of the metal carbonate and metal hydroxide bases, which at high temperatures can cause saponification reaction or catalyze hydrolysis reaction. Additives, such as zinc dialkyl dithiophosphates, are a source of strong acids that result when these additives hydrolyze, thermally decompose, or oxidize. The fate of the ester, amide, and imide type dispersant polymers, such as those derived from polyacrylates, polymethacrylates, and styrene ester substrates, is the same. Some OCP-derived dispersant polymers, such as those obtained by grafting of the monomers, 2- or 4-vinylpyridine and 1-vinyl-2-pyrrolidinone [276,280], do not suffer from this problem since they do not contain easily hydrolyzable groups. Reactivity towards other chemicals present in the formulation is again prevalent in the case of the ester-derived dispersants. Reaction with metal containing additives, such as detergents and zinc dialkyl dithiophosphates, can occur after hydrolysis to form the metal salts. This can destroy the polymeric structure of the dispersant and hence its effectiveness. Some formulations contain amines or their salts as corrosion inhibitors or friction modifiers. Depending upon the molecular weight and the ambient temperature, these can displace the polyol or sometimes the polyamine, thereby again altering the dispersant structure, hence the dispersant properties.

Viscosity Characteristics

The amount of dispersant in automotive engine oils typically ranges between 3–7 % by weight [279], making it the highest among additives. In addition, the dispersant is the highest molecular weight component except the viscosity improver [308]. Both these factors can alter some physical properties, such as viscosity, of the lubricant. A boost in the viscosity of a lubricant at high temperatures is desired but at low temperatures it is a disadvantage. At high temperatures, the lubricant loses some of its viscosity [4], hence its film-forming ability, resulting in poor lubrication. Maintaining good high-temperature viscosity in a lubricant is therefore imperative in order to minimize wear damage. This is usually achieved by the use of the polymeric viscosity modifiers [50,226]. Some dispersants, especially those that are based on high molecular weight polyolefins which have been oversuccinated also partly fulfill this need [243]. Hence, in this case the amount of the polymeric viscosity modifier necessary to achieve specific high-temperature viscosity is reduced. Unfortunately, dispersants that provide a viscosity advantage lead to a viscosity increase at low temperatures also. The low-temperature viscosity requirements for engine oils have two components: cranking viscosity and pumping viscosity [309]. Cranking viscosity is an indication of how easily the engine will turn over in extreme cold weather conditions. Pumping viscosity is the ability of the lubricant to be pumped at low temperatures to reach the various parts of an engine. For cold weather operation, low to moderate cranking and pumping viscosities are highly desirable. While pumping viscosity and the pour point can be lowered by the use of additives, called the pour point depressants [21,226], lowering the cranking viscosity is not easy. The use of the pour point depressants is only effective in paraffinic base stocks, where these additives help suspend wax in a finely crystalline form. Base oil manufacturers therefore use a number of other strategies to develop base oils with good low-temperature properties. These include carefully blending the selected base oils, isomerization via hydrocracking, and the use of the special synthetic oils as additives. An ideal polymeric dispersant must provide high-temperature viscosity advantage, without adversely affecting the cold cranking viscosity of the lubricant. Dispersant polymers have the same requirement. Good high-temperature viscosity to cranking viscosity ratio in polymeric dispersants can be achieved by taking the following actions.

- Carefully balancing the type and the molecular weight of the hydrocarbon chain [310].
- Choosing the optimum olefin to maleic anhydride molar ratio [311].
- Selecting the type and the amount of the polyamine used.

In dispersant polymers, the viscosity balance can be achieved by selecting a polymer of the correct molecular weight and branching, and a suitable pendent group. Dispersant polymers derived from medium molecular weight, highly branched structures and ester type pendent groups are best suited for use as additives. Examples include polyacrylate, polymethacrylate, and styrene ester-derived dispersants. These additives not only act as viscosity modifiers and dispersants; they also act as pour point depressants, thereby improving the low-temperature properties of the lubricant.

A number of patents pertaining to dispersants with balanced high-temperature viscosity and low-temperature properties have been reported in the patent literature [312–315]. A Mannich (alkylphenol) dispersant, derived from ethylene/1-butene polymers of Mn 1500–7500, has been claimed to possess both improved dispersancy and the pour point [312]. Another patent claiming the synthesis of a dispersant with superior dispersancy and pour point depressing properties has also been issued [313]. The dispersant in this patent is based upon the reaction of the maleic anhydride/lauryl methacrylate/stearyl methacrylate terpolymer with dimethylaminopropylamine and a Mannich base, obtained by reacting N-aminoethylpiperazine, paraformaldehyde, and 2,6-di-*t*-butyl phenol. A number of patents describe the use of ethylene/α-olefin/diene inter-polymers to make dispersants [314–316]. These dispersants are claimed to possess excellent high-temperature and low-temperature viscosities, as defined by VR^1/VR ratio. Here VR^1 pertains to the dispersant and VR pertains to the precursor, such as alkylphenol or alkenylsuccinic anhydride. VR^1 is the ratio of the $-20°C$ cold cranking simulator (CCS) viscosity (cP) of a 2 % solution of the dispersant in a reference oil and the $100°C$ kinematic viscosity (cSt) of the dispersant. VR is the ratio of the $-20°C$ cold cranking simulator (CCS) viscosity (cP) of a 2 % solution of the precursor in the reference oil and the $100°C$ kinematic viscosity (cSt) of the precursor. The values of 2.0 to 3.9 for VR and VR^1 and of less than 1.11 for the VR^1/VR ratio are considered suitable for balanced low- and high-temperature viscosities.

Seal Performance

Seals in automotive equipment are used for many purposes, the most prominent of which are to have an easy access to the malfunctioning parts to perform repair and to minimize contamination and the loss of lubricant. A variety of polymeric materials is used to make seals. These include fluoroelastomers, nitrile rubber, polyacrylates, and polysiloxanes (silicones). Maintaining the integrity of the seals is critical, otherwise the lubricant will be lost and wear damage and equipment failure will occur. Seals fail in a number of ways. They can shrink, elongate, or become brittle and thus deteriorate. The damage to elastomer seals is assessed by soaking the elastomers in the lubricant for a prolonged period of time and examining the change in volume, hardness, and tensile strength; and the tendency to elongate and rupture [317]. Two primary mechanisms by which the seal damage can occur include abrasion due to particulate matter in the lubricant and the attack of the various lubricant components on the seals. The lubricant-related damage can occur when some of its components diffuse into the seals. This will either cause a change in the seal's hardness, thereby leading to swelling and or elongation, or extract the plasticizer, an agent used to impart flexibility and strength to the polymeric materials.

Abrasive damage is not common since most equipment has an installed lubricant filtration system. The lubricant-related damage, however, is of primary interest to us. The lubricant is a blend of base stocks and an additive package. Certain base stocks, such as those of high aromatics content or those that are of ester type, have the tendency to extract the plasticizer because of their high polarity. Additives, on the other hand, have the ability to diffuse into the seal material and alter its properties as well as remove the plasticizer. Among additives, dispersants are those that are most implicated in causing the seal damage, especially to fluoroelastomer (Viton®) seals. While in many cases, seal failure can be corrected by the use of additives, called the seal-swell agents, it is wise to eliminate such damage via prevention. Elastomer compatibility requirements are a part of ACEA's (Association des Consructeurs Européens de l'Automobile) 2007 standard for engine oils and worldwide automotive transmission and tractor hydraulic fluid specifications [318]. Damage to seals is prevalent in the case of nitrogen dispersants. In general, the higher the nitrogen content, the greater the seal problems [317]. Rationally, these problems occur due to the presence of the low molecular weight molecules in the dispersant. These include free amine either as such or in a labile form, such as an alkylammonium salt, or low molecular weight succinimides and succinamides. These molecules, because of their high polarity and smaller size, are more likely to diffuse into the seal material and alter its physical and mechanical properties [319]. It is believed that in the case of the Viton® seals, the loss of the fluoride ions is responsible for seal deterioration. Removal of the free amine and of low molecular weight succinimides will improve seal performance. Alternatively, one can post-treat dispersants with reagents, such as boric acid and epoxides, which will either make such species innocuous or hinder their diffusion into the seal material. Many chemical treatments of dispersants cited above claim to improve the seal performance of the dispersants and engine lubricants that use them. These reagents react with the seal damaging amines and low molecular weight succinimides to make them harmless. Strategies other than listed above are also reported in the patent literature [320–324].

Performance Testing

Engine oils account for almost 80 % of the total dispersant use. It includes use in gasoline engine oils, heavy-duty and railroad diesel engine oils, natural gas engine oils, and aviation piston engine oils. Other applications that use these additives include automatic transmission fluids, power steering fluids, gear lubricants, hydraulic fluids, and in refinery processes as antifoulants. The dispersants of a relatively lower molecular weight are also used in fuels to control injector and combustion chamber deposits [325,326]. Such dispersants usually contain the polyether functional group [327].

Succinimide and succinate ester type polymeric dispersants are used both in gasoline and diesel engine oils, but the use of the alkylphenol-derived dispersants, that is, of the Mannich type, is limited to gasoline engine oils. Dispersant polymers derived from ethylene-propylene rubbers, styrene-diene copolymers, and polymethacrylates are also used in both gasoline and diesel engine oils. Polymethacrylate and styrene ester-derived dispersant polymers are primarily used in automatic transmission fluids, power steering fluids, and, to a limited extent, in gear oils.

As mentioned earlier, dispersant polymers lack sufficient dispersancy to be used alone and hence they are used in combination with the polymeric dispersants. In gasoline and diesel engine oils, the effectiveness of a dispersant is determined by its ability to disperse lamp black or used engine oil sludge. Laboratory screen tests, the ASTM Sequence VE/VG, and the various diesel engine tests are used for this purpose. Figure 4.64 shows the effectiveness of a dispersant in controlling sludge in the Sequence VE Test [318]. The exhibit portrays the effect of the dispersant treat level at different stages of the test. The treat level increases on going from left to right and the stages are determined by the number of hours into the test. If one examines the degree of agglomeration (growth) of the dirt particles (indicated by dark areas), it is clear that the particle growth is higher when the dispersant treat level is low, compare the left most plates in the figure with the right most plates, and in the latter parts of the test, compare the first row plates with the last row plates. The most desirable performance is depicted by the right most plate at the end of the test (288 h), which signifies the dispersant treat of 5 %. The data in the figure clearly demonstrate the importance of the dispersant quality and quantity in providing sustainable performance.

Additives manufacturers use a variety of laboratory screen tests and engine tests to evaluate a dispersant's effectiveness. Many of the screen tests are proprietary but all are developed around evaluating performance in terms of a dispersant's ability to disperse lamp black or used engine oil sludge. The laboratory engine tests are industry required tests and include both the gasoline engine and the diesel engine tests. These are listed in the ILSAC (International Lubricant Standardization and Approval Committee), API (American Petroleum Institute), ACEA 2007, JASO (Japanese Automobile Standards Organization), and BIS (Bureau of Indian Standards) standards. Some of these standards are listed in Chapter 5 on Combustion Engine Lubricants. It is important to note that the U.S. Military and the original equipment manufacturers (OEMs) have their own performance requirements, which are over and above those of the API. While the details of the various tests are available in these standards and elsewhere [318], the important engine tests that possibly evaluate a dispersant's performance are listed in Tables 4.6–4.8.

As mentioned earlier in the chapter, the soot-related viscosity increase and deposit-related factors are the primary criteria for evaluating a dispersant's performance. Moreover, it is important to note that the neutral detergents (soaps) also help control deposits, such as varnish, lacquer, sludge, and carbon. Therefore, besides controlling the soot-related viscosity increase, which is the sole domain of the dispersants, the deposit control is the result of a joint performance of the oxidation inhibitor, the detergent, and the dispersant. How-

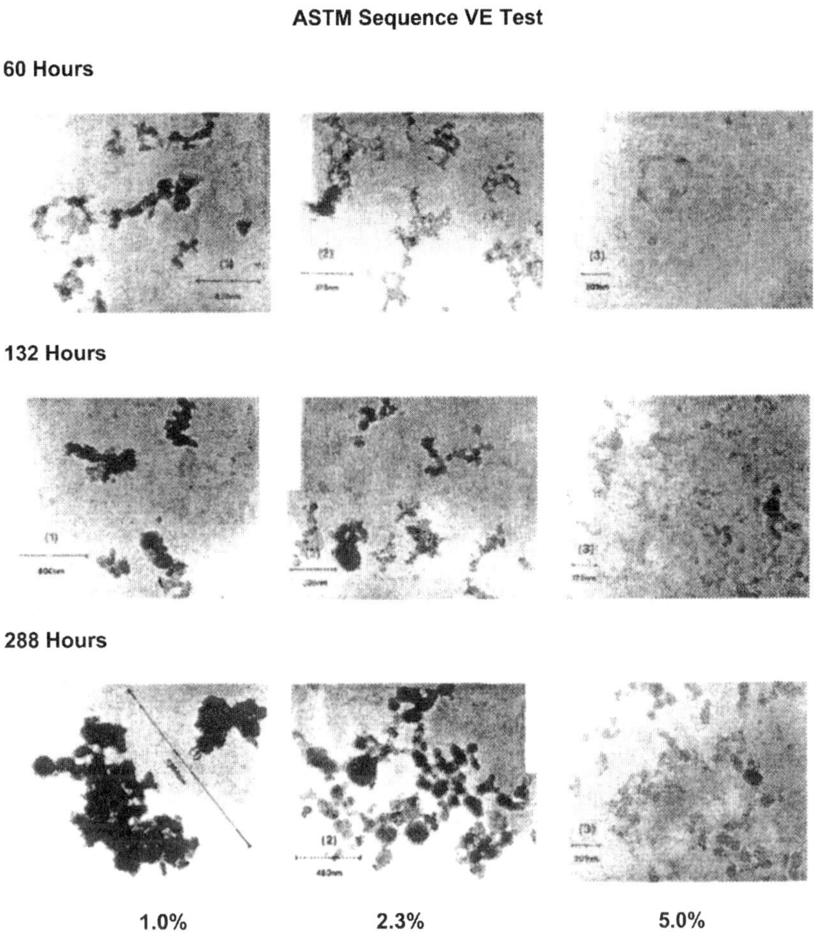

Fig. 4.64—The effectiveness of a dispersant in the ASTM sequence VE test [318b]. Reprinted with permission from the Lubrizol Corporation.

ever, in this regard the oxidation inhibitor and the dispersant play a more prominent role.

Besides engine oils, transmission fluids are the primary users of the dispersants. Certain parts of the transmission see very high temperatures, which lead to extensive lubricant oxidation. The oxidation products, such as sludge and varnish, appear on parts, for instance, clutch housing, clutch piston, control valve body, and oil screen components. These can impair the functioning of these parts. Turbohydramatic Oxidation Test (THOT) is used to determine a transmission fluid's oxidative stability. Polymeric dispersants are useful in controlling the sludge build-up [328]. When friction modification of the transmission fluid is the goal, either the dispersants or their precursors, such as alkenylsuccinic acids or anhydrides, are used in combination with metal sulfonates [329–333]. In many such formulations, either the borated dispersant or the borated detergent (metal sulfonate), or both, are used.

Dispersants are used in gear oils to also improve their properties. Gear oils usually contain thermally labile ex-

TABLE 4.6—U.S. gasoline tests that determine dispersancy/detergency.

Engine Test	Engine Type	ASTM Test	Evaluation Criteria
CRC L-38	CLR single cylinder engine	D5119	Bearing corrosion, viscosity increase
Sequence IIIF	A 1996/1997 GM V-6 fuel-injected engine	...	Viscosity increase, varnish, deposits, and wear
Sequence IIIG	1996/1997 Series II General Motors V-6 engine	...	Viscosity increase, deposits, wear, and oil consumption
Sequence VE	Ford 2.3-liter 4 cylinder engine	D5302	Sludge, varnish, and wear
Sequence VG	2000 Ford 4.6-liter V-8 engine	D6593	Sludge, varnish, and wear
Sequence VIII	CLR Single Cylinder engine	...	Bearing weight loss, viscosity change
TEOST, TEOST (MHT4)	Bench test	D6335	Thermal and oxidative stability, deposits

TABLE 4.7—U.S. diesel engine tests.

Engine Test	Engine Type	Evaluation Criteria
Caterpillar 1K	Caterpillar 1Y540 single cylinder direct injection engine (Aluminum)	Piston deposits and oil consumption
Caterpillar 1M-PC	Caterpillar 1Y73 single cylinder indirect injection engine	Piston, rings, and liner distress
Caterpillar 1N	Caterpillar 1Y540 single-cylinder direct-injection engine (Aluminum)	Piston deposits and oil consumption
Caterpillar 1P	Caterpillar 1Y3700 single-cylinder direct-injection engine	Piston, rings, and liner distress
Caterpillar 1R	High-speed four-stroke cycle Caterpillar 1Y3700 single cylinder engine (Steel)	Piston, rings, and liner distress; and oil consumption
Caterpillar C-13	Six cylinder turbocharged engine	Steel piston deposits and oil consumption
Mack T-8	Mack E7-350, in-line six cylinder 11GBA77623, direct injection engine	Soot-related viscosity increase and oil consumption
Mack T-11 (ASTM D 7156)	Mack E-TECH V-MAC III, electronically-controlled fuel injection engine	Soot-related viscosity increase and oil consumption
Mack T-11A	–	Soot-related low-temperature viscosity increase
Cummins M-11 EGR	Cummins 1SM 425 engine	Crosshead weight loss
Cummins 1SM	–	Valve train wear, filter plugging, Δ P and sludge at 6% soot

TABLE 4.8—European engine tests.

Engine Test	Engine Type	Test Method	Evaluation Criteria
Peugeot DV4TD Medium Temperature Test	Peugeot DV4 TD/L4 four-cylinder common rail diesel engine	(CEC L-93-04)	Piston deposits and ring sticking.
Peugeot TU5 JP-L4 High-Temperature Test	Peugeot TU5JP-L4, 1.5 liter, 4-cylinder engine	(CEC L-88-T-02)	Piston deposits and ring sticking, and oil Thickening
Peugeot XUD11BTE	Peugeot 4 cylinder diesel engine	(CEC-L-56-T-98)	Piston deposits and viscosity increase
M-B M111 Black Sludge Test	Mercedes-Benz M111 E20 four cylinder gasoline engine	(CEC L-53-T-95)	Engine sludge and cam wear
M-B OM 602A	Mercedes-Benz OM 602 A in line five cylinder engine	(CEC-L-51-A-98)	Engine wear and cleanliness
Nissan TD25 Detergency Test	Nissan 2.5L four-cylinder, inline diesel engine	(JASO M336:1998)	Piston deposits and ring sticking
Volkswagen Turbocharged DI Diesel Test	Volkswagen 1.9-liter 4-cylinder, turbocharged, DI diesel engine (VW TDi)	(CEC L-78-T-99)	Pistons deposits and ring sticking.
Volkswagen T4 Test	Volkswagen 2.0L four-cylinder gasoline engine	(VW PV 1449)	Extended drain capability
VW 1.6TC Diesel Intercooler	VW 4 cylinder turbocharged inter-cooled IDI Diesel engine	(CEC L-46-T-93)	Piston deposits, varnish, and ring sticking
M-B OM 364A/LA	Four cylinder direct injection engine	CEC-L-42-T-99	Piston deposits, varnish, sludge, wear, and oil consumption

treme pressure additives. Their decomposition products are highly polar and the dispersants are used to contain them, in order to avoid corrosion and deposit formation [334,335]. Polymeric dispersants are used in hydraulic fluids to overcome wet filtration (AFNOR) problems. Wet filtration is often required for HF-0 type fluids [336]. Filtration problems occur due to the interaction of water with the metal sulfonate detergent and zinc dialkyl dithiophosphate, which are often used as additives in hydraulic fluid formulations. Fouling is a common problem in many processes, including the refinery processes. Fouling refers to the deposition of the various inorganic and organic materials, such as salt, dirt, and asphaltenes, on heat transfer surfaces and other processing equipment. This results in a poor heat transfer, which is one of the many problems caused by fouling. Antifoulants are chemicals that are used in refinery operations to overcome fouling. Detergents and dispersants are often used for this purpose [337–339].

Detergents

Detergents are the third member of the deposit control agents group. Detergents are not only similar to dispersants in performing the suspending function, but they also have the added advantage of being able to neutralize acidic combustion and oxidation products [188]. This is due to their usually having a metal hydroxide or a metal carbonate base reserve. Hence, they control rust, corrosion, and the resinous build-up in the engine. It is important to note though that the detergents have a molecular weight typically in the 500–1500 g/mol range, compared to that of the dispersants which is between 3000 and 7000 g/mol. Also, unlike dispersants that have the great ability to suspend polar materials, the detergents are much less effective. This is because in the case of detergents the moiety that performs the suspending function has a molecular weight of around 600 g/mol, or less. The balance of the molecular weight is due to the encapsulated metal hydroxide or metal carbonate, which has no polars suspending ability, only the acid-neutralizing ability.

The organic portion of the detergent (soap) structure is remarkably similar to that of the dispersant; that is, it contains a surface-active polar functional group, which is connected to an oleophilic moiety through a connecting group. The only exception to this is the carboxylate detergents that lack the connecting group and the polar carboxylate group is directly attached to the hydrocarbon moiety. The connecting group in the case of most detergents is an aromatic ring, either a phenyl group or a phenol group. In a way, the phenyl ring can be treated as a part of the hydrocarbon group, instead of treating it separately. The oleophilic group in detergents is an alkyl group that typically comprises 18 to 30 carbon atoms and is normally derived from an olefin or its oligomer. It is important for the alkyl group to have an appropriate number of carbon atoms to render the derived detergent oil or fuel soluble. The common polar groups that are connected to the alkyl group, via the aromatic ring, are sulfonate, phenate, carboxylate, and salicylate.

Detergent Substrates

Detergents are the metal salts of organic acids, which include alkylbenzenesulfonic acids; alkylnaphthalenesulfonic acids [340–343], alkylphenols [344–348], long chain aliphatic carboxylic acids such as natural fatty carboxylic acids, naphthenic acids, and petroleum oxidates [349–352], and alkylphosphonic and alkylthiophosphonic acids [353–356]. A mixture of two or more of these acids may also be employed [333]. Detergents are identified as sulfonates, phenates, carboxylates, salicylates, and phosphonates depending upon the acid precursor. These acids are called detergent substrates [188].

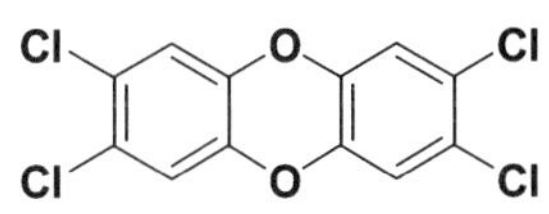

2,3,7,8-Tetrachlorodibenzo-*p*-dioxin

Fig. 4.65—A dioxin structure.

Alkylbenzene and alkylnaphthalene sulfonic acids are made by the sulfonation reaction of alkylbenzenes and alkylnaphthalenes, which in turn are obtained from the alkylation of benzene and naphthalene [357–361]. These detergent starting materials are called *synthetic sulfonic* acids because of their intended synthesis. This is in contrast to *natural sulfonic acids,* which are isolated from petroleum during the manufacture of white oil. The alkylating agent used to make alkylaromatics can be an alkyl halide or an olefin. Olefin is the preferred alkylating agent because it does not produce any waste and there is no chance of chlorine incorporation into the product. The latter is a serious concern because of it taking part in the formation of dioxin, which are confirmed human carcinogens. *Dioxin* is the popular name for a family of chlorinated organic compounds consisting of polychlorinated dibenzofurans (PCDFs) and polychlorinated dibenzodioxins (PCDDs). Of these, 2,3,7,8-tetrachlorodibenzo-*p*-dioxin, the structure in Fig. 4.65, is the most toxic. PCDDs result from the reaction of the chlorine donors, such as chlorinated aromatics, with organics at high incineration temperatures of approximately 700°C.

The olefins used for alkylating aromatics may be α-olefins, internal olefins, or olefin oligomers, such as polypropylene and polyisobutylene. Some of the structures, which are made by the oligomerization of ethylene, propylene, or isobutylene, are provided in Fig. 4.66. A number of catalysts are employed in the oligomerization reaction. Ethylene oligomerization is carried out by using homogeneous as well as heterogeneous catalysts. However, the product selectivity and the molecular weight vary, depending upon the type of catalyst and the reaction conditions. While Ziegler-

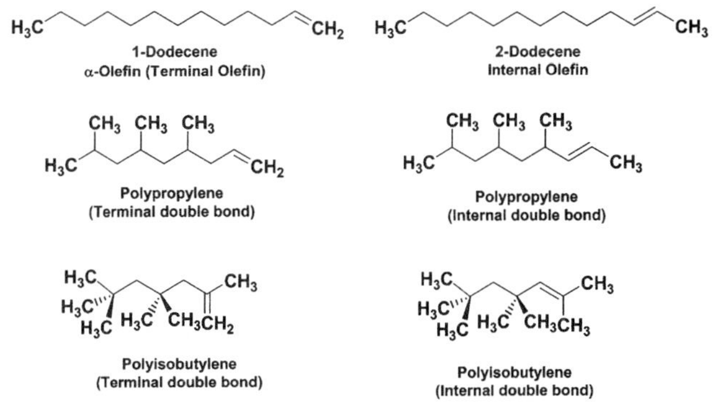

Fig. 4.66—Common olefins used in detergents.

Natta catalysts and metallocenes are used to produce polyethylene, the degree of polymerization is hard to control to obtain oligomers containing less than 20 units, or 40 carbon atoms. Such oligomers are commonly used as synthetic base stocks and olefin starting materials to make alkylates for use in detergent synthesis. The correct molecular weight α-olefins are obtained by the use of the transition metal complexes, such as those of zirconium, nickel, and iron [362–366]. In many cases, alkylaluminum is used as a co-catalyst. Polypropylene is also made by the use of Ziegler-Natta and metallocene catalysts. Polyisobutylene is made from the monomer isobutylene, by Lewis acid catalyzed vinyl polymerization. However, because of the steric crowding, the reaction rate of the polyisobutylene thus obtained with certain substrates, such as maleic anhydride (see section on dispersants), is very slow. Hence, polyisobutylene with a terminal double bond (vinylidene) is greatly desired. Such polyisobutylenes are obtained by diminishing the reactivity of the BF_3 catalyst by the use of oxygenates, such as alcohols and methyl *tertiary*-butyl ether (MTBE). This helps in preventing the rearrangement of the vinylidene enriched polyisobutylene to polyisobutylene with the internalized double bond [367,368].

The alkylation of aromatics requires the use of an acid catalyst. There are many acids to choose from and they include mineral acids, such as sulfuric acid and phosphoric acid, Lewis acids, such as aluminum chloride and boron trifluoride, organic acids, such as methanesulfonic acid, and the mixtures thereof [360,361]. Some of the inorganic acids, such as sulfuric acid, are also available on a solid support, such as Fuller's earth or silica. Unlike other catalysts that require neutralization at the end of the reaction, these catalysts just require filtration to remove them. Zeolites and related mixed metal oxides also enjoy the same advantage as the solid alkylation catalysts [358,359]. Another class of catalysts, exemplified by Amberlysts®, is aromatic polymer derived sulfonic acids [359,369]. While they have the advantages of being insoluble, hence easier to remove, and of multiple uses, they have the disadvantage of being expensive. It is important to note that not all catalysts have equal effectiveness in all alkylations.

Detergents made from *synthetic* sulfonic acids are called *synthetic* sulfonates and those made from *natural* sulfonic acids are called *natural* or *petroleum* sulfonates. The steps involved in making *synthetic* sulfonic acids are shown in Fig. 4.67. The degree of branching in alkylbenzenes and alkylnaphthalenes, commonly called the *"alkylate,"* increases as we go from α-olefins to internal olefins to olefin oligomers. The branching also increases when the inherently more branched olefins, such as polypropylene and polyisobutylene, are used. More branching in the alkylate implies somewhat less efficient sulfonation, but better oil solubility of the final sulfonate detergent. The common reagents that are used to sulfonate alkylaromatics are sulfur trioxide, fuming sulfuric acid or oleum, and chlorosulfonic acid [370]. Of these, sulfur trioxide is the best sulfonating agent since it produces no waste and results in little tar, which comprises poly-sulfonated aromatics. Oleum is the next best. It is 15–30 % sulfur trioxide dissolved in concentrated sulfuric acid. The major disadvantage of this reagent is that after the sulfonation, the so-called spent sulfuric acid is a potential waste,

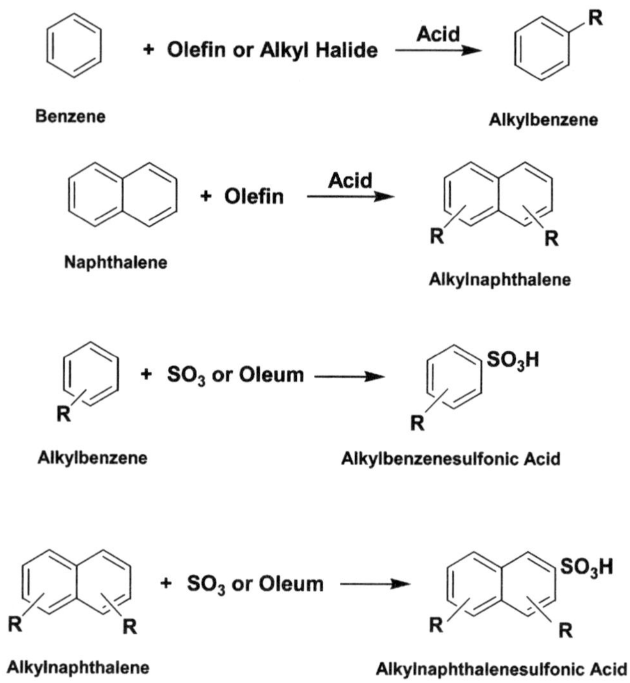

Fig. 4.67—Synthesis of sulfonic acid substrates.

which must either be disposed off or returned to the oleum suppliers for reuse. Some oleum suppliers do not accept spent sulfuric acid because of the presence of the organic contaminants. Chlorosulfonic acid again suffers from the disadvantage of generating acid waste, hydrogen chloride in this case. In all cases, the sulfonation is affected by reacting the alkylate, dissolved in a hydrocarbon solvent such as hexane or heptane, with the sulfonating reagent. Depending upon the reactants ratio, sulfonating agent residence time, and the reaction temperature, one obtains a mixture of the normally desired mono-sulfonic acid and the undesired di and higher sulfonic acids. The latter must be removed because of the high polarity of their metal salts, which have potentially lower oil solubility. Also, when the metal used to make a detergent is polyvalent, such as calcium, magnesium, barium, and aluminum, which is often the case, polysulfonated aromatics give polymeric salts that have even lower solubility in organics. The undesired poly-sulfonic acids can be easily removed by water washing the sulfonated product.

Not all components of the alkylate are sulfonatable. In the case of alkylbenzenes, the species that do not sulfonate easily include polyalkylated benzenes, such as trialkylbenzene, or highly branched dialkylbenzenes. Their sulfonation difficulty is primarily a consequence of the steric crowding of the sulfonatable positions. Mono-alkylbenzenes, on the other hand, do not suffer from this drawback and hence sulfonate easily. In general, the sulfonation of the branched alkylbenzenes is slower than linear alkylbenzenes, primarily because of the steric reasons. In the case of alkylnaphthalenes, however, steric factors are not as important, because of its bicyclic nature. The alkyl groups are likely to be attached to different aryl rings, except for very highly alkylated naphthalenes. Commercial Aristonate® sulfonates are alkylbenzene based and NA-SUL® products are based upon alkyl-

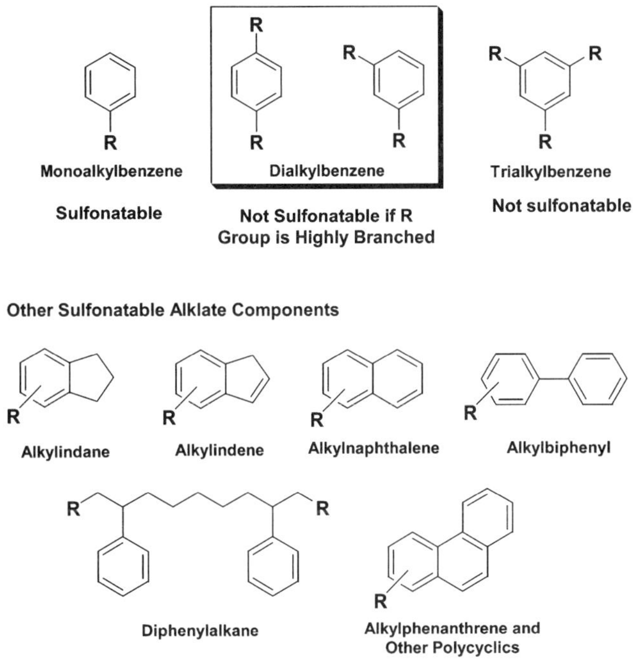

Fig. 4.68—Alkylbenzene structures.

naphthalene chemistry. In the case of alkylbenzenes, the structures that are sulfonatable and those that are not are shown in Fig. 4.68. The figure also shows other sulfonatable materials that are present in the commercial alkylates. Some of these, such as alkylated phenanthrene and other fused polycyclics, are not desired because of their carcinogenicity.

Natural sulfonic acids are obtained during petroleum refining when the crude mineral oil is washed with a sulfonating agent, such as sulfur trioxide or oleum [371]. Crude mineral oil contains reactive unsaturated compounds containing multiple bonds and alkylaromatics. These react with sulfur trioxide to form sulfonic acids. This is a desirable step because the oils containing unsaturates and aromatics have a greater susceptibility towards oxidative breakdown, which can lead to the formation of the increased deposits. If this occurs, it is likely to lead to equipment malfunction [191,192,196]. An analogous process is used to manufacture medicinal quality white oil from petroleum. The sulfonic acid fraction thus obtained in the subsequent reaction is reacted with sodium hydroxide to convert the acids into the sodium salts. These salts are washed with water to extract the *green acid soaps*, which are used in many consumer products. The residual water-insoluble material is then extracted with alcohol. This results in the isolation of the *mahogany acid soaps*, which are useful in making detergent additives. The process is summarized in Fig. 4.69. So far we covered only the aromatic sulfonic acids. There are a number of routes to obtain the aliphatic sulfonates. These include sodium bisulfite addition to olefins and chlorosulfonation and sulfoxidation of paraffins [372,373]. However, these aliphatic sulfonates are primarily used in household cleaners and detergents and as surfactants. See Fig. 4.70 for chemistry.

Alkylphenols are made in a manner analogous to that of the alkylbenzenes; that is, by alkylating phenol with an olefin in the presence of an acid catalyst. The preferred catalysts

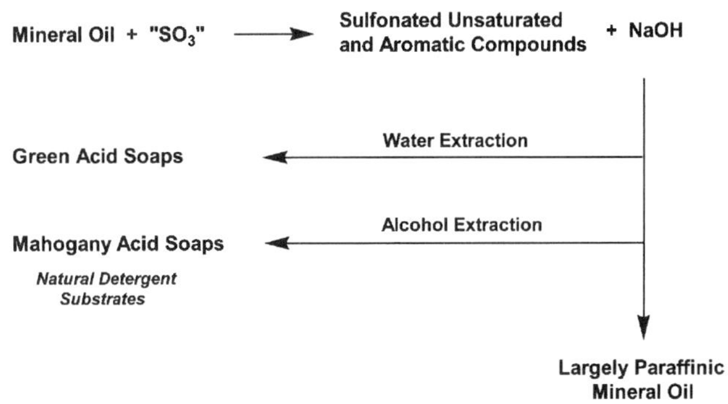

Fig. 4.69—Isolation of natural sodium sulfonates.

are sulfuric acid, aluminum chloride, and boron trifluoride [261,263,264], although sulfuric acid supported on Fuller's earth and Amberlyst® can also be used. If one uses an α-olefin, the alkylated mixture contains only 33 % of the desired *para*-alkylphenol, the rest being the *ortho* isomer. However, when a branched olefin, such as polypropylene, is used, the product is almost exclusively the desired *para*-alkylphenol. This can be explained in terms of the steric factors because of which the more branched polypropylene cannot approach the ortho-position. The alkylphenols can either be converted directly into their neutral or basic salts, or can be further reacted with sulfur or sulfur dichloride to form sulfur-bridged alkylphenols (alkylphenol sulfides) and with formaldehyde to form methylene-bridged alkylphenols. Figure 4.71 shows the generic structures of these materials. Figure 4.72 shows the detailed mechanism of formation of the methylene-bridged alkylphenols. The probable intermediate is the quinone methide that results from the dehydration of the hydroxymethylalkylphenol. Once the mono-methylene bridged alkylphenol forms, it can undergo further formylation and coupling with another molecule of

1. Bisulfite Addition

$$R{-}HC{=}CH_2 + NaHSO_3 \xrightarrow{\text{Free Radical Initator}} R{-}CH_2{-}CH_2SO_3Na$$

2. S ulfochlorination (Reed Process)

$$\text{Paraffin} + SO_2 + Cl_2 \xrightarrow{\text{UV Radiation}} RSO_2Cl \xrightarrow{H_2O} RSO_3H$$

2. S ulfoxidation

$$\text{Paraffin} + SO_2 + [O] \xrightarrow{\text{UV Radiation}} RSO_3H$$

Fig. 4.70—Routes to aliphatic sulfonic acids/sulfonates.

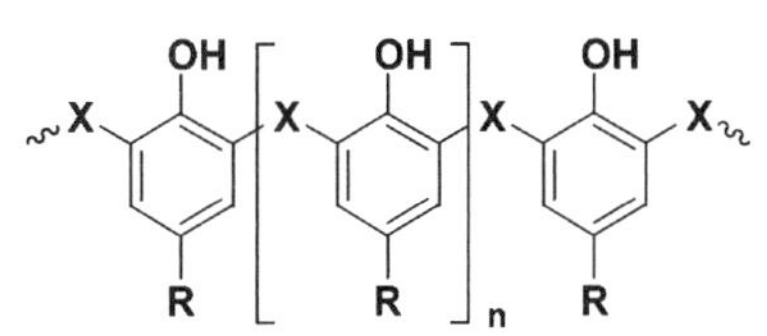

Bridged Alkylphenol Oligomer

R = C_6 to C_{12}; X = S or CH_2; n = 0-2

Fig. 4.71—Sulfide and methylene bridged alkylphenol structure.

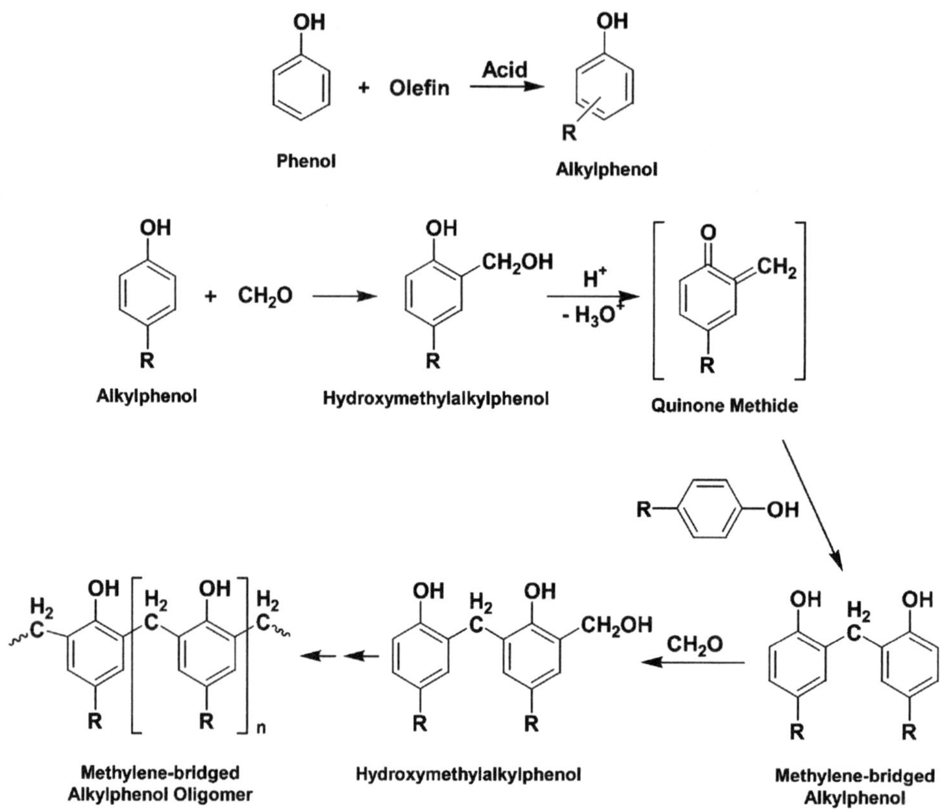

Fig. 4.72—Methylene bridged alkylphenol.

alkylphenol to form another bridge, and so on. One side product that is undesirable is the hyform, which is polymeric material with low oil solubility. It is shown in Fig. 4.72 as methylene bridged alkylphenol oligomer. Figure 4.73 shows the formation of alkylphenol sulfides by the use of sulfur dichloride and Fig. 4.74 shows the formation of alkylphenol sulfides by the use of elemental sulfur. While sulfur dichloride has the advantage of stopping the reaction at a single sulfide bridge stage, it results in chlorination of the alkylphenol by chlorine, resulting from the disproportionation of sulfur dichloride, as shown in the boxed portion of Fig. 4.73. As mentioned earlier, dioxins result from the reaction of chlorine/chlorine donors with organics in the environment and are a concern. The use of the elemental sulfur to make phenol sulfides does not "suffer" from this disadvantage. However, one obtains di- and tri-sulfide bridged alkylphenols. They result from sulfur acting as an oxidizing agent, as shown in Fig. 4.74. The tri-sulfide bridged materials are not desirable since they can separate reactive sulfur, which is

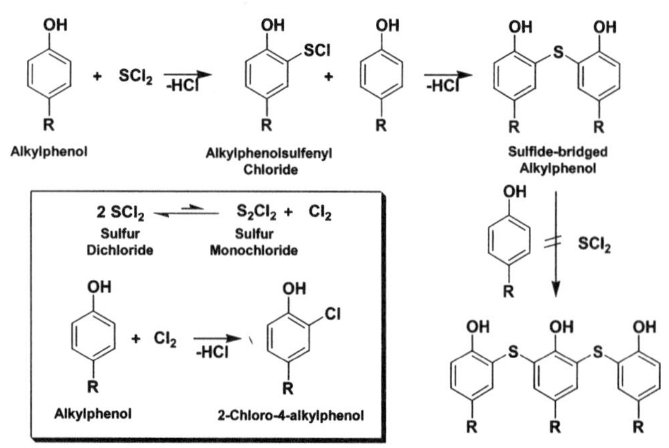

Fig. 4.73—Alkylphenol sulfide formation by the use of sulfur dichloride.

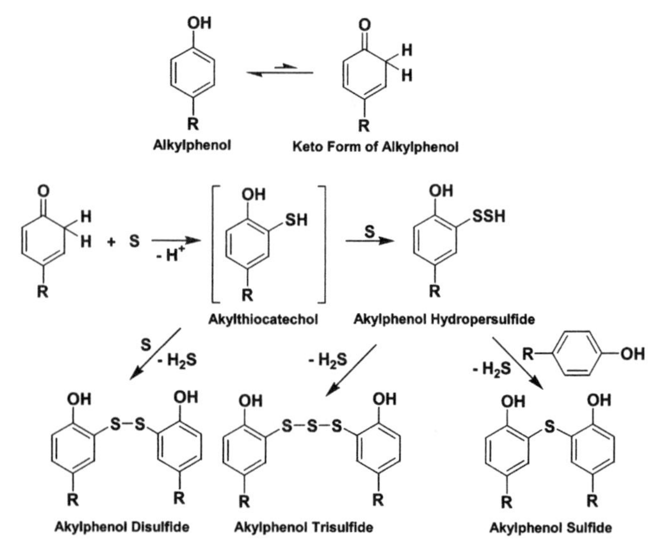

Fig. 4.74—Possible mechanism for alkylphenol sulfide formation by the use of elemental sulfur.

corrosive to yellow metals, such as copper and bronze. Here we show only one isomer of the bridged products. There are many other isomers that form, although those shown in the figure predominate.

Common carboxylic acids, which are used to make detergents, include natural fatty acids, branched long-chain acids, naphthenic acids, and alkylsalicylic acids. Fatty acids are the products of hydrolysis of the natural fats and oils, which are triglycerides. Fatty acids from natural sources contain 12 to 18 carbon atoms and can have zero to three double bonds. Acids that have no double bonds are called saturated and are exemplified by dodecanoic acid (lauric acid), hexadecanoic acid (palmitic acid), and octadecanoic acid (stearic acid). These acids have melting points of 44.2, 63.1, and 69.6 °C respectively, and are solids at room temperature. Because of this, they are not good substrates for making detergents for use in lubricants since their metal salts are also solids, with low oil solubility/miscibility. Carboxylic acids that contain double bonds (unsaturated) are liquids at room temperature; hence are ideal for synthesizing detergents. Oleic acid (*cis* 9-octadecenoic acid), which is one of the common fatty acids used for this purpose, has a melting point of 16.2 °C [374]. While C_{18} carboxylic acids with two or three double bonds, *viz.* 9,12-octadecadienoic (linoleic) acid and 9,12,15-octadecatrienoic (α-linolenic) acid have even lower melting points of −5 °C and −11 °C, they are not suitable substrates to make detergents. This is because as the number of the double bonds in the structure increases, the oxidation susceptibility increases as well.

Two other types of carboxylic acids that are worth noting as substrates for making detergents are branched and cyclic carboxylic acids. These include iso-stearic acid, neodecanoic acid, and naphthenic acids. Iso-stearic acid, or 16-methyl heptadecanoic acid, is liquid around 20 °C. It is obtained as a by-product during high-temperature oligomerization of the unsaturated fatty acids to monomers, dimers, and higher oligomers. This acid has the advantage of yielding additives, such as detergents, with superior oxidation stability than their oleic acid-derived counterparts, primarily because of the lack of the double bond. Despite being a saturated acid, its products have good oil solubility, a consequence of branching. The detergents derived from this acid have the superior oxidation resistance of those derived from stearic acid and the oil solubility of those derived from oleic acid. Naphthenic acids consist of acids that contain five- and six-membered cyclic rings either single or fused, often called naphthenes in the petroleum industry, in their structures. Naphthenic acids are liquids with low freezing points and a wide boiling point range (250–350 °C). They are good acids to use in detergents since they are completely soluble in organic solvents and behave like typical carboxylic acids and have acid strengths similar to those of the higher fatty acids. Compared to the fatty acid derivatives, naphthenic acid derivatives offer many advantages, particularly high oxidation stability and good solubility in hydrocarbon materials. The structures of these carboxylic acids are provided in Fig. 4.75. A number of other carboxylic acids have also been used to make detergents [375–377]. The structures of these acids are also included in Fig. 4.75.

Alkylsalicylic acids are prepared from alkylphenols by the reaction of an alkali metal, especially potassium, to form

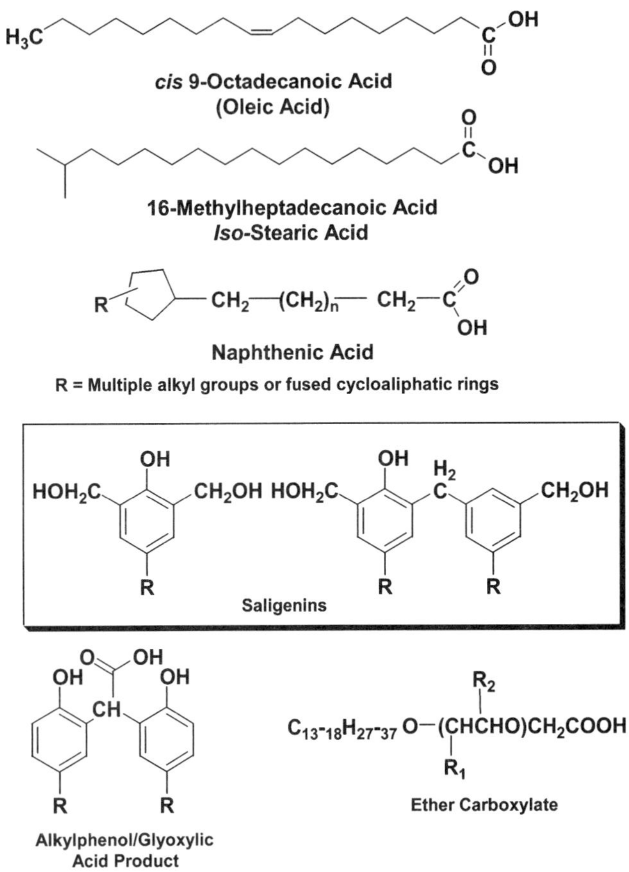

Fig. 4.75—Carboxylic acid and phenolic substrates for detergents.

a phenate, which is then reacted with carbon dioxide to form an aromatic carbonate. This rearranges to potassium salicylate under the influence of heat, which can then be treated with an acid to release the free alkylsalicylic acid. The reaction is known as the Kolbe-Schmitt Reaction [378]. Like the *natural* sulfonate process, this process yields alkali metal salts. There are few suppliers of this chemistry because of the potential waste that is produced, which we will comment upon during the conversion of the potassium salt into detergents. The reaction scheme for the synthesis of the alkylsalicylic acid is provided in Fig. 4.76. An overbased metal salt of an alkylsalicylamide with good engine oil performance has also been reported in the patent literature [379].

The alkylphosphonic and alkylthiophosphonic acids are prepared by the reaction of polyisobutylene of varying molecular weights with phosphorus pentasulfide, and the subsequent hydrolysis of the resulting adduct [272,274]. This adduct is believed to result via an ENE type addition of the polyolefin to phosphorus pentasulfide. This type of addition does not result in the loss of the double bond but shifts the double bond down the carbon chain. Unless steric factors hinder further reaction, at least theoretically, the ENE product can react with another molecule of the phosphorus pentasulfide. This process can extend even further. The polyolefin-phosphorus pentasulfide adduct is hydrolyzed by the use of steam to obtain a complex mixture of acids that include fully hydrolyzed (sulfur-free) alkenylphosphonic acids and partially hydrolyzed (contain residual sulfur) alkenylthiophosphonic acids. The reaction schemes to synthe-

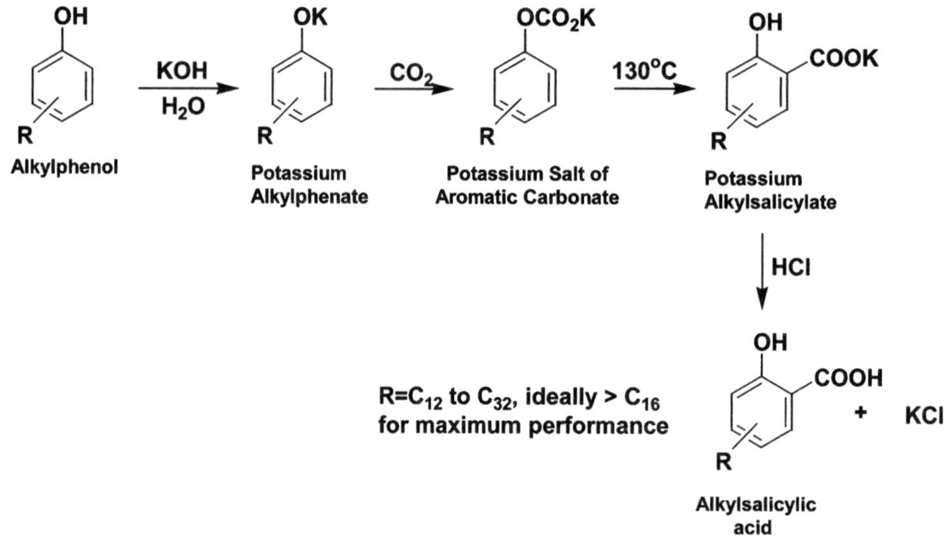

Fig. 4.76—Synthesis of alkylsalicylic acid.

size alkenylphosphonic and thiophosphonic acids are shown in Fig. 4.77. Please note that in this scheme and in the subsequent discussion, we represent phosphorus pentasulfide as P_2S_5, although in reality the solid phosphorus pentasulfide exists as a dimer; that is, as P_4S_{10} with an adamantane type structure [380], as is shown in Fig. 4.78.

In addition to the acidic detergent substrates discussed so far, nonacidic substrates have also been used to make detergents [381,382]. These include tallow- and oleyl-diaminopropane, sulfurized C_{16-18} α-olefins, Propomeen® T/12, *t*-dodecyl hydroxypropyl thioether, trioctylamine, Tergitol® 26L-5, and didodecyl ether of ethylene glycol pentamer [382]. The structures of some of these are shown in Fig. 4.79. Alkaline earth metal salts of cyclic alkenylsuccinimide with good engine test performance have also been reported. Magnesium salt was made by the reaction of alkenylsuccinimide with magnesium methoxide [240].

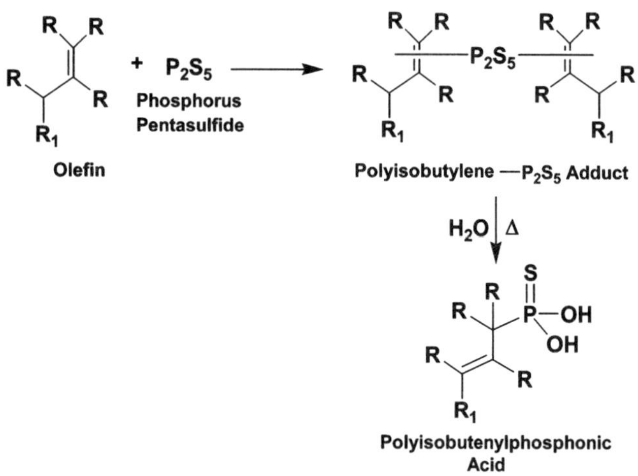

Fig. 4.77—Polyisobutenyl-phosphonic acid substrates.

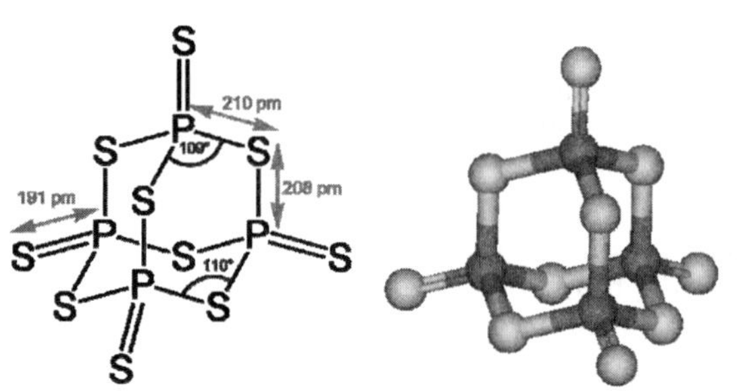

Fig. 4.78—Phosphorus pentasulfide crystal structure [380].

Neutral and Basic Detergent Synthesis

The acidic detergent substrates are reacted with a metal base, such as a metal oxide, hydroxide, or carbonate, to form a neutral salt or a basic or overbased material. Basic materials are made by the overbasing process that allows entrainment of a large amount of inorganic base in a detergent in a soluble form. Neutral salts by analogy with traditional neutral carboxylates are called soaps. The base that is normally used is derived from an alkali metal, an alkaline earth metal, or aluminum. In general, the reaction between the organic acid and the inorganic base is not easy, because of the poor contact between the two reactants. This is due to the base being highly polar and most times being a solid, and the substrates being relatively nonpolar. To facilitate contact and hence the reaction between the two, a number of compounds, called promoters, are used both in salt formation as well as during carbonation or a related reaction. Common promoters include ammonium hydroxide, low molecular

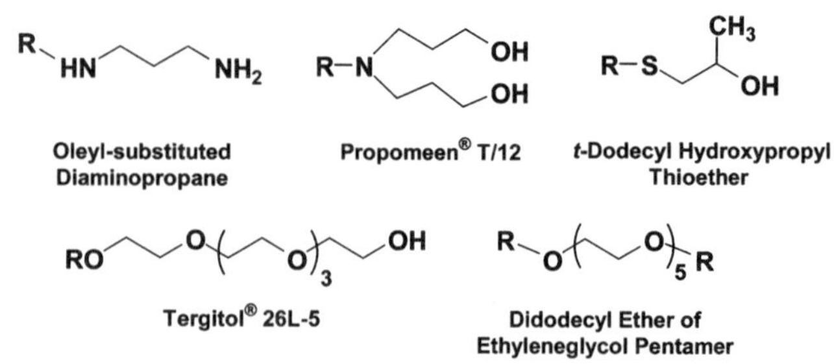

Fig. 4.79—Structures of some nonconventional detergent substrates.

weight carboxylic acids, such as formic acid and acetic acid, low molecular weight alkylphenols, and other polar compounds, such as nitroalkanes and imidazolines. A comprehensive list of such agents is provided elsewhere [383]. Most of these reagents are used in combination with water, except for the high temperature overbasing reactions, where water will not stay in the reaction. In such cases, alcohols, such as 2-ethylhexanol or iso-octyl alcohol, and alkylphenols that have high boiling points are used. When water is present as part of the promoter system, it is either added or results from the neutralization reaction. The promoters are surfactants, that is, they contain a hydrophilic moiety, such as a hydroxyl group or a carboxylic acid functional group, and a reasonable size alkyl group to impart somewhat of a hydrophobic character to the molecule. Not all promoters are effective for all overbasing reactions and one must experiment to select the right promoter system. For overbasing that is carried out at 100°C or less, alcohol-water mixtures are commonly used. For the high-temperature overbasing, which involves temperatures over 100°C, low molecular weight alkylphenols, such as heptylphenol, are used. The structure of the final detergent from the two processes is perceived to be different because there is evidence of different performance in certain tests. The role of promoters in the overbasing reaction is not well understood. One explanation regarding their role is based upon their preferential reaction with the base to form an alkoxide or a phenoxide. This species then transfers the metal to the substrate, thereby facilitating salt formation and or overbasing. The other explanation is based upon their acting as a surfactant and a wetting agent. This improves the contact between the base and the substrate, thereby assisting the reaction to occur. The second explanation is definitely more plausible than the first. However, in high-temperature overbasing reactions, which are usually carried out under anhydrous conditions, the first explanation may have merit.

In order to make detergents, one can either use the quantity of the metal base that is equal to or in excess of the precise amount (stoichiometric amount) necessary to completely neutralize the acid function. When the metal is present in the stoichiometric amount, the detergents are called "neutral." When it is present in excess, they are called "basic, overbased, or superbased." While many metals can be used to make neutral salts (soaps), fewer metals have the ability to result in oil-soluble basic or overbased detergents. The common metals that can be used for this purpose include lithium, sodium, and potassium in Group I (alkali metals) and magnesium, calcium, strontium, and barium in Group II (alkaline-earth metals) of the periodic table. Aluminum is the only metal in Group III that can be used to make the overbased detergents. Basic transition metal salts of zinc, copper, cadmium, molybdenum, manganese, cobalt, nickel and iron of sulfonic acids, alkylphenols, and naphthenic acids are also reported in the patent literature [343,351,384,385]. The ability to overbase relates to a metal's base strength: the higher the basic character, the easier it is to overbase. For Group I metals, where the basic character increases while going from lithium to sodium to potassium, it is easier to overbase potassium than it is to overbase lithium. In Group II metals, the basic character increases while going from magnesium to calcium to strontium to

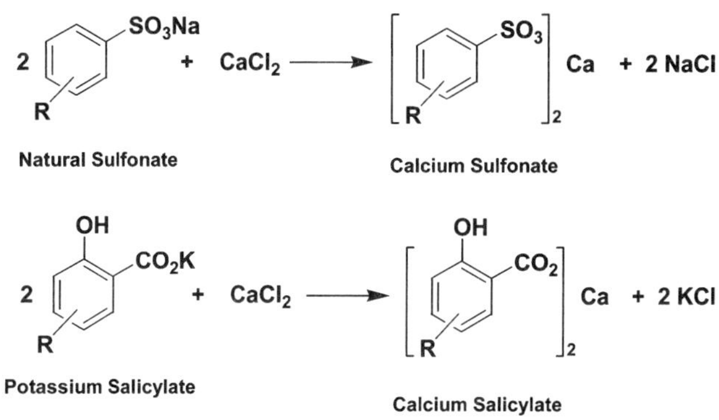

Fig. 4.80—Double decomposition reaction.

barium; hence it is easier to overbase barium than it is to overbase magnesium.

A variety of detergents based upon metal anions other than hydroxide and carbonate are also reported in the patent literature. The anions include sulfites, sulfates, thiosulfates, borates, and phosphates [386–389]. These detergents are obtained either from the carbonate detergent by displacing the carbonate anion with the alternative anion, or by using the desired anion precursor during overbasing. For example, one can obtain metal sulfite overbased detergents by either blowing sulfur dioxide during overbasing or by displacing carbon dioxide in a carbonate detergent with sulfur dioxide. The resulting metal sulfite detergent may be oxidized to a metal sulfate detergent by using an oxygen source, such as oxygen gas or a peroxide, or to a thiosulfate detergent by reacting it with elemental sulfur [383,386,387]. The borate and phosphate overbased compositions can be made by using boric acid or phosphoric acid during the reaction [388,389].

Common commercial detergents are derived from calcium, magnesium, sodium, and barium. The metals are listed in order of preference. In order to make calcium and magnesium salts from *natural* sulfonic acids and alkylsalicylic acids, one must convert the commercially available alkali metal (sodium and potassium) salts into the free acids by reaction with a mineral acid (see Fig. 4.76) and then reacting the free acids with magnesium oxide or calcium hydroxide. Alternatively, alkali metal salts can be converted directly into magnesium and calcium salts via a double decomposition reaction with a metal halide, such as calcium chloride or magnesium chloride, as is shown in Fig. 4.80. This reaction will convert the sodium sulfonate soap into calcium sulfonate, which can be overbased, if desired. The major drawback of either approach for *natural* sulfonates and salicylates is the generation of the sodium chloride waste. In addition, for alkylsalicylates, it is not advisable to neutralize the salt to the free acid until one is ready to use it. This is because the free acid on long-term storage has the tendency to slowly decompose to alkylphenol by the loss of carbon dioxide. Obviously, if one wants to make soaps or overbased materials for use as detergents, double decomposition is the best option. If, on the other hand, one is interested in chemistry other than the metal salt formation, freeing the acid may be necessary. Because of the extensive branching, petroleum derived sulfonates have better oil solubility than synthetic sulfonates of similar molecular weight. The idealized structures for neutral detergents are presented in Fig. 4.81.

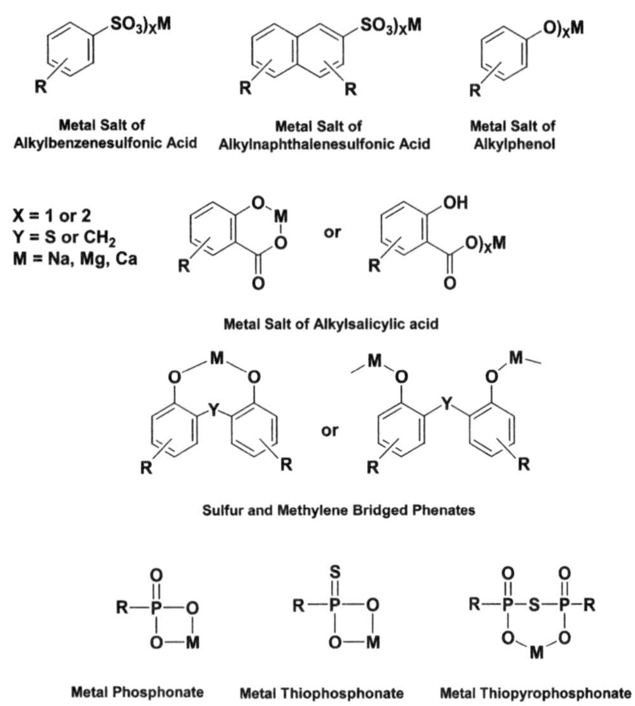

Fig. 4.81—Idealized Structures of neutral salts (soaps)

In order to synthesize the overbased detergents, one can either use a two-step process or a one-step process. Generally, the one-step process is preferred over the two-step process. In the two-step process, the neutral salt or the soap is made first, which is subsequently overbased. In the one-step process, the excess metal base is charged to the reaction and once the neutral salt formation is complete, carbon dioxide blowing (carbonation) of the reaction is initiated. When carbon dioxide uptake stops, the reaction is considered complete and is worked up to isolate the product. The two processes are summarized in Fig. 4.82. For making overbased *natural* sulfonates and alkylsalicylates, one can double decompose the alkali metal salts in situ by reacting with a metal halide and overbasing. The alkali metal halide byproduct need not be removed until the overbasing is complete. It comes out during the final filtration, which is employed to

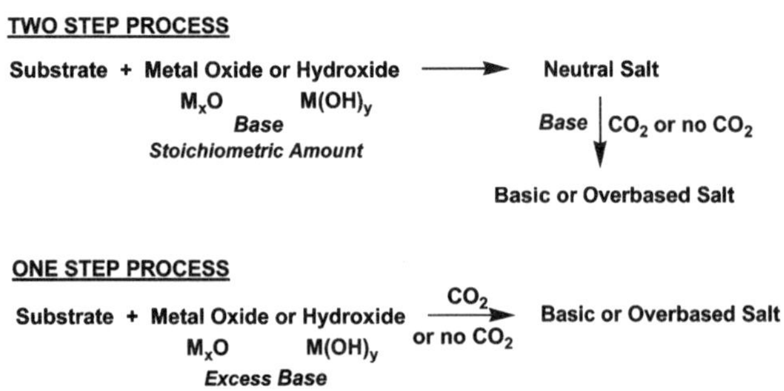

Basic Salts = Neutral Salts · M_xO, $M(OH)_y$, or M_xCO_3
M = Na, Mg, Ca, or **Ba**
x = 1 and **y = 2** for **Mg, Ca,** and **Ba** (divalent metals)
x = 2 and **y = 1** for **Na** (monovalent metal)

Fig. 4.82—Processes to make basic detergents.

$(RSO_3)_a M{\cdot}xM_bCO_3{\cdot}yM(OH)_c$
Basic Sulfonate

$(RPhO)_a M{\cdot}xM_bCO_3{\cdot}yM(OH)_c$
Basic Phenate

$(RCOO)_a M{\cdot}xM_bCO_3{\cdot}yM(OH)_c$
Basic Carboxylate

a and **c** = 1 and **b** = 2, if the metal **M** is mono-valent; **a** and **c** = 2 and **b** = 1, if the metal **M** is divalent.

Fig. 4.83—General structures for overbased detergents [227].

remove any unreacted excess base and other particulate materials.

Calcium- and magnesium-derived detergents find most extensive use as lubricant additives, with a preference for calcium due to its lower cost. The use of the barium-derived detergents is decreasing due to a concern for barium's toxicity. While technically one can use metal oxides, hydroxides, and carbonates to manufacture neutral (nonoverbased) detergents; for overbased detergents, oxides and hydroxides are the preferred bases. For sodium, calcium, and barium detergents, sodium hydroxide, calcium hydroxide, and barium hydroxide are often used. For magnesium detergents, however, magnesium oxide is the preferred base.

During the synthesis of the calcium detergents, overbasing is usually stopped before all the metal base is converted into the calcium carbonate. As a result, the excess base is present as a mixture of calcium hydroxide and calcium carbonate, with calcium carbonate predominating. This is because if the reaction is overblown with carbon dioxide, the amorphous calcium carbonate, which is desired, gets converted into crystalline calcium carbonate. This, being of low solubility in the overbased system, falls out of solution and one obtains an oil insoluble gel-like product. While such products are of little use as lubricant additives, they are useful as rheology control agents in coatings. The challenge is to make them on a consistent basis. Lubrizol supplies such products, derived from alkylbenzenesulfonic acids, as their Ircogel® product line. Gelled carboxylates and solid calcium micellar complexes have also been reported in the patent literature [390–392]. The general structures of the overbased detergents are shown in Fig. 4.83 [227]. For neutral detergents, x and y in the formulas are zero. On the other hand, for low overbased detergents, such as those with a base number of about 50 or less, x may be zero and y may be a low number, or both x and y may be low numbers. This implies that the detergents that are only slightly overbased are either carbonate free or contain a mixture of both the hydroxide and the carbonate. Highly overbased detergents invariably have a large amount of carbonate as the reserve base. That is, in their case, y is low and x is very high. In some cases x can be as high as 20, or more. In summary, the excess base per equivalent of acid in the metal hydroxide-containing detergents is generally lower than that in the metal carbonate-containing detergents.

Some detergents are marketed as neutral or nonoverbased. However, most of them have a small amount of reserve base present. In other words, they are overbased to a slight degree. This implies that there was no effort made to overbase them and their reserve base is because of the presence of the unreacted base used to make them. For example, commercially available *neutral* sulfonates have a TBN of 30, or less, and the base is commonly present as a hydroxide, such as calcium hydroxide. Conversely, "basic" or overbased

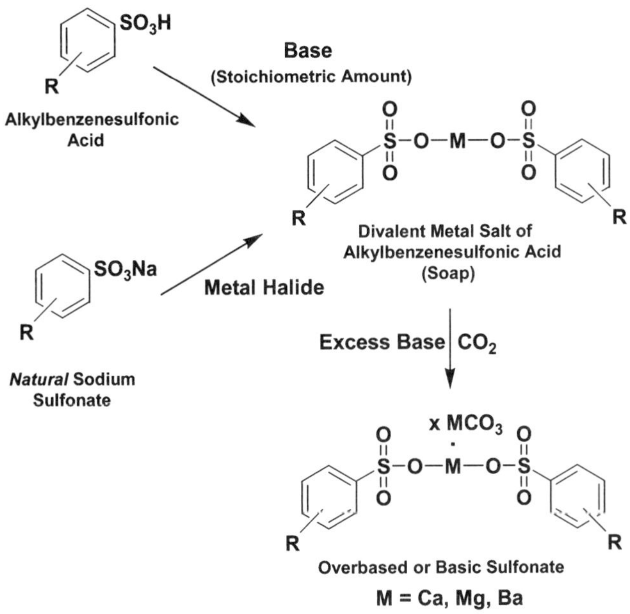

Fig. 4.84—Synthesis of neutral and basic metal sulfonates.

detergents have a much higher base reserve. They typically have a TBN of 200 to 500 and the base is commonly present as a metal carbonate.

The calcium-based phenate detergents are easier to make than the magnesium-based phenate detergents. This is because the alkylphenols are weak acids and their reaction with magnesium oxide, a weak base, is not facile. In order to make the neutral salt, the alkylphenol must be reacted with a strong base, such as magnesium alkoxide. This reagent can be prepared by the reaction of magnesium metal with an excess of highly reactive alcohol, such as methanol. However, this method is hazardous, because of the hydrogen gas by-product, and costly, because of the price of the magnesium metal. Once the neutral salt or soap formation occurs, the excess alcohol is exchanged with an inert solvent, such as toluene or mineral oil, prior to overbasing. Alternatively, one can use the high-temperature overbasing procedure using a low molecular alkylphenol as a promoter [393]. In the case of methylene or sulfur bridged phenols that are more acidic than the regular alkylphenols, the reactivity towards magnesium oxide is not a problem. And, these compounds form neutral and overbased magnesium salts without difficulty. Neutral and basic calcium phenates, bridged or unbridged, are easy to make because calcium hydroxide is a strong base; hence it readily reacts with alkylphenols. The other acids, i.e., alkylsalicylic acids, fatty carboxylic acids, and alkenylphosphonic acids react with calcium and magnesium bases without any problems.

The synthetic sequences that are used to make common neutral and carbonate overbased detergents are outlined in Figs. 4.84–4.89. Sulfonate, salicylate, and carboxylate detergents are commercially available as calcium and magnesium salts and phosphonates are available as calcium salts. Some specialty sulfonates, for example NA-SUL® BSB, are also available as barium salts. Phenate detergents are commonly available as calcium salts and phosphonate detergents are available as both calcium and barium salts.

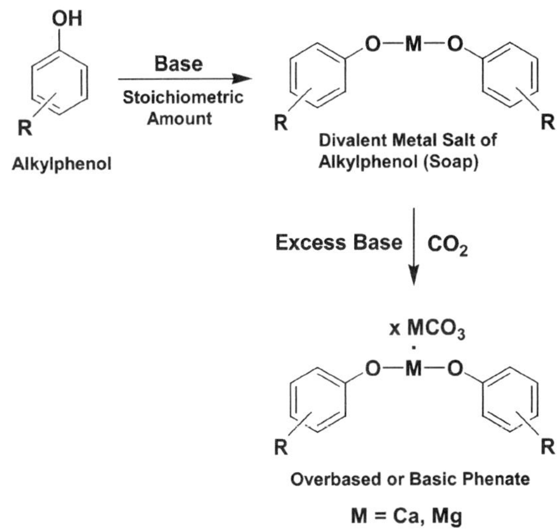

Fig. 4.85—Synthesis of neutral and basic metal phenates

Detergent Parameters

Chemically, detergents are described by the use of a number of parameters. These include metal ratio, percent sulfated ash, degree of overbasing or conversion, soap content, and total base number (TBN) [184].

The *metal ratio* is defined as the total equivalents of metal per equivalent of acid. The *percent sulfated ash* is the ash produced when the detergent is treated with sulfuric acid and burned. All organic material in the detergent burns, leaving behind the metal sulfate ash. Sulfate ash results from the reaction of the metal compounds with sulfuric acid either directly, as in the case of metal hydroxide and metal carbonate, or through oxidative degradation of the metal sulfonate. Detergents are not the only additives that produce sulfated ash due to the reaction with sulfuric acid. Other metal containing additives in the lubricant also contribute towards it. Such additives include metal carboxylates and metal dialkyl dithiophosphates, such as zinc dialkyl dithio-

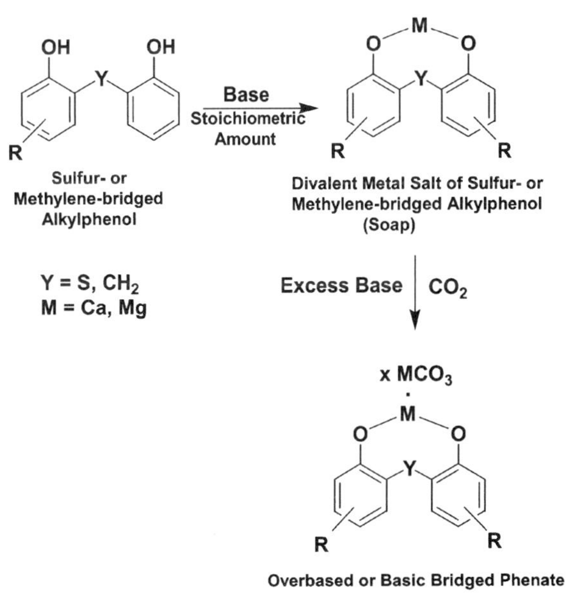

Fig. 4.86—Synthesis of neutral and basic bridged metal phenates.

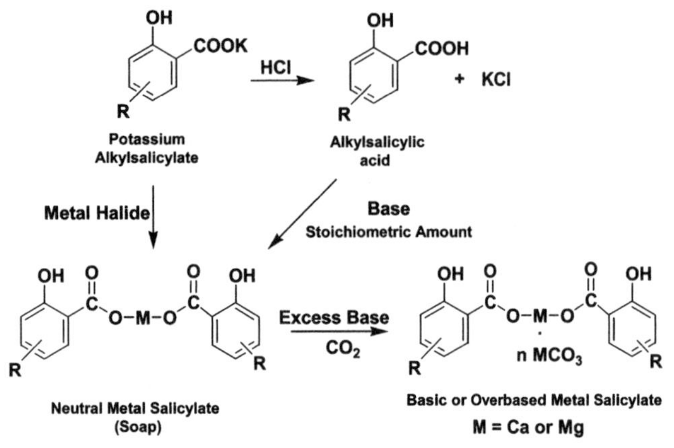

Fig. 4.87—Synthesis of neutral and basic metal salicylates.

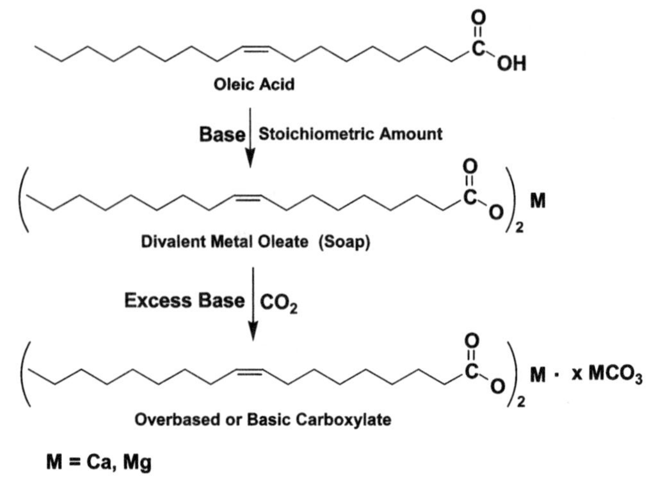

Fig. 4.89—Synthesis of neutral and basic fatty carboxylates.

phosphate. The former compounds are sometimes used as friction modifiers and corrosion inhibitors, and the latter compounds are commonly used as oxidation inhibitors and antiwear agents. Because the metal compounds can lead to the formation of the inorganic material (ash) on combustion, a formulator must know the metal content of the formulation to offset any problems that might occur. This is because when a lubricant travels past piston rings into areas that either have flame or experience high temperatures, such as the top land and the groove behind the top ring, it burns/decomposes to produce ash. Ash is undesired because it is believed to initiate deposit formation. Sulfated ash is one of the methods used to assess the metal content of a lubricant and the methods to determine this are described in the ASTM Standards D482 and D874 [27].

The *degree of overbasing* is the number of equivalents of the metal base per equivalent of the acid substrate. This is usually expressed as conversion. *Conversion* indicates the amount of the inorganic material relative to that of the organic material, and is expressed as the number of equivalents of base per equivalent of acid, multiplied by 100 [184]. The *soap content* is the amount of the neutral salt as a percent of the detergent composition.

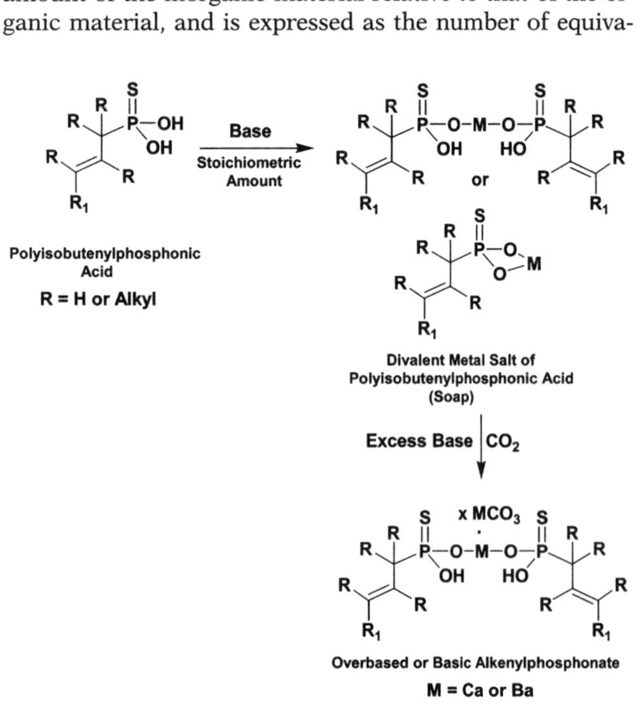

Fig. 4.88—Synthesis of neutral and basic phosphonates.

The *total base number*, or *TBN*, of the detergent reflects its ability to neutralize acids. In the case of the basic sulfonate and phosphonate detergents, only the overbased portion of the detergent, i.e., the carbonate and the hydroxide, possess this capability. The neutral metal sulfonates and phosphonates, or the soaps, lack this ability. However, in the case of the basic carboxylates, salicylates, and phenates, the soaps also possess the acid neutralizing ability. This is because unlike the sulfonates and phosphonates that are strong acid—strong base salts, metal carboxylates, metal salicylates, and metal phenates are strong base-weak acid salts. This makes them Lewis bases, hence the acid-neutralizing ability.

As an example, let us try and calculate the above described parameters for a detergent of a hypothetical molecular formula $(RSO_3)_2Ca{\cdot}xCaCO_3{\cdot}yCa(OH)_2$. We will rewrite the formula as $(RSO_3)_vCa_w(CO_3)_x(OH)_y$ to facilitate understanding of the calculations that follow. In this formula, v, w, x, and y denote the number of sulfonate groups, the number of calcium atoms, the number of carbonate groups, and the number of hydroxyl groups, respectively. The *metal ratio*, the total equivalents of metal per equivalent of acid, for such a detergent equals $2w/v$. The coefficient 2 signifies the divalent nature of calcium. For metals such as sodium and potassium that are monovalent, the metal ratio equals w/v. The *degree of overbasing or conversion*, which is the metal ratio times 100, is $(w \times 100)/v$ for monovalent metals and $(2w \times 100)v$ for divalent metals. Neutral detergents, or soaps, have a conversion of 100 because the ratio of the equivalents of the base to the equivalents of the acid is 1. *Soap content* for such a detergent can be calculated by the use of the following formula.

$$\%\text{Soap} = \frac{\text{Formula Weight}[(RSO_3)_2Ca] \times 100}{\text{Effective Formula Weight}}$$

The effective formula weight is the weight of all the atoms that make up the formula $(RSO_3)_vCa_w(CO_3)_x(OH)_y$, plus the diluent, if present. The diluent can either be the incidental alkylate that did not get sulfonated or the diluent oil that is

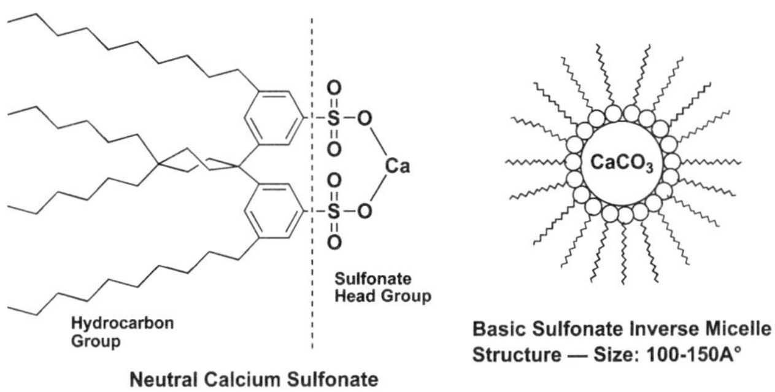

Fig. 4.90—Micelle structure of basic detergents.

intentionally added. If one must add oil, most of it is added to reactants at the beginning of the reaction, especially during the manufacture of the basic detergents. The presence of the diluent is believed to facilitate micelle formation, thereby making the process more efficient. Adding oil after the reaction is complete is not as effective.

The total base number, or *TBN*, indicates a detergent's ability to neutralize acids. In additives and formulated lubricants, the *TBN* is expressed as mg KOH/g of additive [27]. The method to determine base numbers is described in the ASTM Standard D974 [27]. For sulfonate and phosphonate detergents, it can be calculated by using the number of equivalents of the excess metal after salting the acid, that is ($2w$-v), according to the following equation.

$$\text{TBN (mg KOH/g)} = \frac{(2w - v) \times 56100}{\text{Effective Formula Weight}}$$

To calculate the base number of monovalent metal (lithium, sodium, and potassium) derived sulfonates, one must use only (w-v) in the above equation. For divalent metal (magnesium, calcium, and barium) derived carboxylate, salicylate, and phenate detergents, the equation to be used is as follows.

$$\text{TBN (mg KOH/g)} = \frac{(2w) \times 56\,100}{\text{Effective Formula Weight}}$$

For monovalent metal salts of this type, the nominator will be $w \times 56\,100$. As mentioned earlier, the *percent sulfated ash* is the quantity of the solid metal sulfate that results when the detergent is treated with sulfuric acid and the mixture ignited. Theory sulfated ash for divalent and monovalent metals can be calculated by using the following equations.

For Divalent Metals

$$\%\ \text{Sulfated Ash} = \frac{w \times \text{Molecular Weight of } MSO_4 \times 100}{\text{Effective Formula Weight}}$$

For Monovalent Metals

$$\%\ \text{Sulfated Ash} = \frac{0.5w \times \text{Molecular Weight of } M_2SO_4 \times 100}{\text{Effective Formula Weight}}$$

Let us apply these equations to a real life basic detergent that was prepared from 2 moles of C_{16} alkylbenzenesulfonic acid, 12 moles of $Ca(OH)_2$, and carbon dioxide. The molecular formula for such a chemical will be $(C_{16}H_{33}C_6H_4SO_3)_2Ca \cdot 11CaCO_3$. We will rewrite the formula as $(C_{16}H_{33}C_6H_4SO_3)_2Ca_{12}(CO_3)_{11}$ in the form provided above to facilitate the above described calculations. The finished detergent in addition contains 1300 g of oil as diluent. In the formula $v=2$, $w=12$, $x=11$, and $y=0$. While the detergent may contain some unreacted $Ca(OH)_2$, its amount is ignored for two reasons: (1) it is expected to be present only in a very small amount, and (2) it is unlikely to have a significant affect on the outcome of the calculations. The *formula weight* represented by the structure $(C_{16}H_{33}C_6H_4SO_3)_2Ca$ is $(381 \times 2)+40$, or 802. This is the weight of the neutral salt, or the soap. The *effective formula weight* is the weight of all the atoms in the formula $(C_{16}H_{33}C_6H_4SO_3)_2Ca_{12}(CO_3)_{11}$, plus the weight of the diluent. This equals 3202 $\{(381 \times 2)+(40 \times 12)+11(12+48)$ plus 1300 from the diluent$\}$. The calculated values for the parameters used to describe detergents are given below.

$$\text{Metal Ratio} = \frac{2w}{v} = \frac{2 \times 12}{2} = 12 \text{ and the conversion is}$$

$$12 \times 100 \text{ or } 1200$$

$$\%\ \text{Soap} = \frac{\text{Formula Weight}[(C_{16}H_{33}SO_3)_2Ca] \times 100}{\text{Effective Formula Weight}}$$

$$= \frac{802 \times 100}{3202} = 25.05$$

$$\text{TBN (mg KOH/g)} = \frac{(2w - v) \times 56\,100}{\text{Effective Formula Weight}}$$

$$= \frac{[(2 \times 12) - 2] \times 56\,100}{3202} = 385.45$$

$$\%\ \text{Sulfated Ash} = \frac{w \times \text{Molecular Weight of } CaSO_4 \times 100}{\text{Effective Formula Weight}}$$

$$= \frac{12 \times 136 \times 100}{3202} = 50.97$$

Performance Testing

Basic detergents contain the reserve base in a colloidal form. The base, such as the carbonate, is believed to be encapsulated by the soap molecules. In this arrangement, the polar head group (sulfonate, phenate, or carboxylate) of the soap associates with the carbonate and the hydrocarbon portion of the soap associates with the oil. This is depicted in Fig. 4.90. The base neutralizes the acids that result from the oxidation of the fuel and the lubricant and the oxidation and thermal decomposition of the thermally labile additives. The soap, on the other hand, associates with products that are nonacidic and keep them solubilized in oil, as is shown in Fig. 4.91. One way to look at the situation is that the nonacidic, polar materials replace the carbonate in the detergent's micellar structure.

As mentioned during the previous discussion, detergents are used in lubricants to perform functions similar to those of the dispersants; that is, to keep oil-insoluble by-products of combustion and oil oxidation in suspension. These products, which are polar organic compounds, include alkyl nitrites and nitrates, nitroalkanes, and high molecular weight oxygenates. Such oxygenates result from the

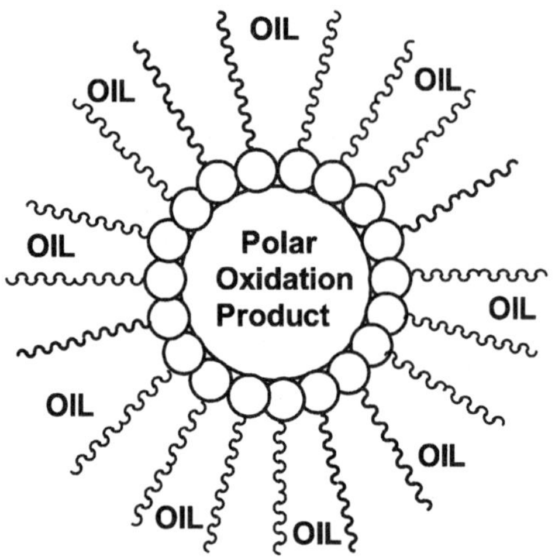

Fig. 4.91—Oil suspension of polar oxidation products.

reaction of aldehydes and ketones with each other or with the other materials, and oxidation. However, the soap component that performs the suspending function has the disadvantage of having a much lower molecular weight than that of a dispersant, which makes it less effective. This can be overcome by exploring the use of acids that have a molecular weight higher than the acids used at present, the molecular weights of which is 500 g/mol, or less. However, the low molecular weight of detergents is also an advantage since it makes them more surface active by increasing their polar character. This causes them to adsorb on metal surfaces easier to form a protective film, thereby minimizing the build-up of the undesirable materials and keeping the equipment surfaces clean. Figure 4.92 compares the performance of the dispersants and detergents in controlling the formation of the piston deposits in a two-stroke cycle gasoline engine [4]. As one can see, the calcium sulfonate at 2 % and 4 % (Part A) is more effective in keeping the piston surfaces clean than the ash-less dispersant at 5 % (Part B), as indicated by the higher ratings. It also improves the exhaust port plugging tendency relative to that of the untreated lubricant, as indicated by the lower ratings. However, the use of the heavy metal detergent (Part C) takes away the benefit provided by the dispersant, causing the worst exhaust port blocking, presumably because of the resulting ash. In reality, calcium sulfonate at 4 % provides the best balance of the piston cleanliness and the exhaust port blocking.

Detergents can also act as oxidation inhibitors, depending upon the nature of their functional group. For example, phenates, sulfurized phenates, and salicylates possess oxidation-inhibiting properties, both as salts and in the neutral form. This is primarily due to the presence of the phenolic functional group and the presence of sulfur. Oxidation inhibiting action of these functionalities as hydroperoxide decomposers was explained in the oxidation inhibitors section. The presence of sulfide and methylene substituents in the ortho position also imparts them the ability to control oxidation via the free radical scavenging mechanism. This mechanism is analogous to that of the arylamines, where the newly formed free radical cannot take part in the oxidation mechanism because of the electron delocalization. In addition, the free bridged phenols, which can be released via reaction with the acidic materials, have the ability to complex with the metal ions, such as copper and iron, thereby inhibiting their catalytic effect on oxidation as well. The mechanism of oxidation inhibition by the released bridged phenols is shown in Fig. 4.93.

Detergents are also effective corrosion inhibitors, especially the basic detergents [394,395]. This is because they not only neutralize corrosive acidic products but also form surface films that isolate the metal surfaces from corrosive agents [29,227,396]. The carbonate portion of the detergent performs the acid neutralization and the soap portion forms the protective surface films. Aristonates®, NA-SUL 729®, and NA-SUL CA-50® are examples of the commercial corrosion inhibitors belonging to this class of additives. Tests that evaluate detergent performance are described elsewhere [227,274]. Detergents derived from the fatty carboxylic acids are good friction modifiers, primarily because of the linear structure of their soaps [227].

As shown in Fig. 4.3, dispersants and detergents together made up 35 % of the total lubricant additives use in the year 2006. This is a consequence of their major use in en-

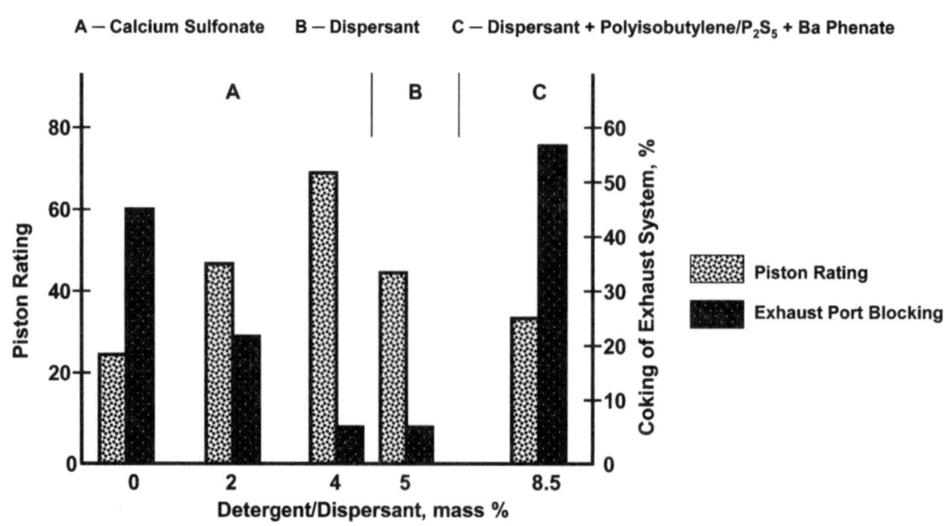

Fig. 4.92—Effect of detergent concentration on exhaust port blocking and piston cleanliness in a two-stroke cycle gasoline engine [4].

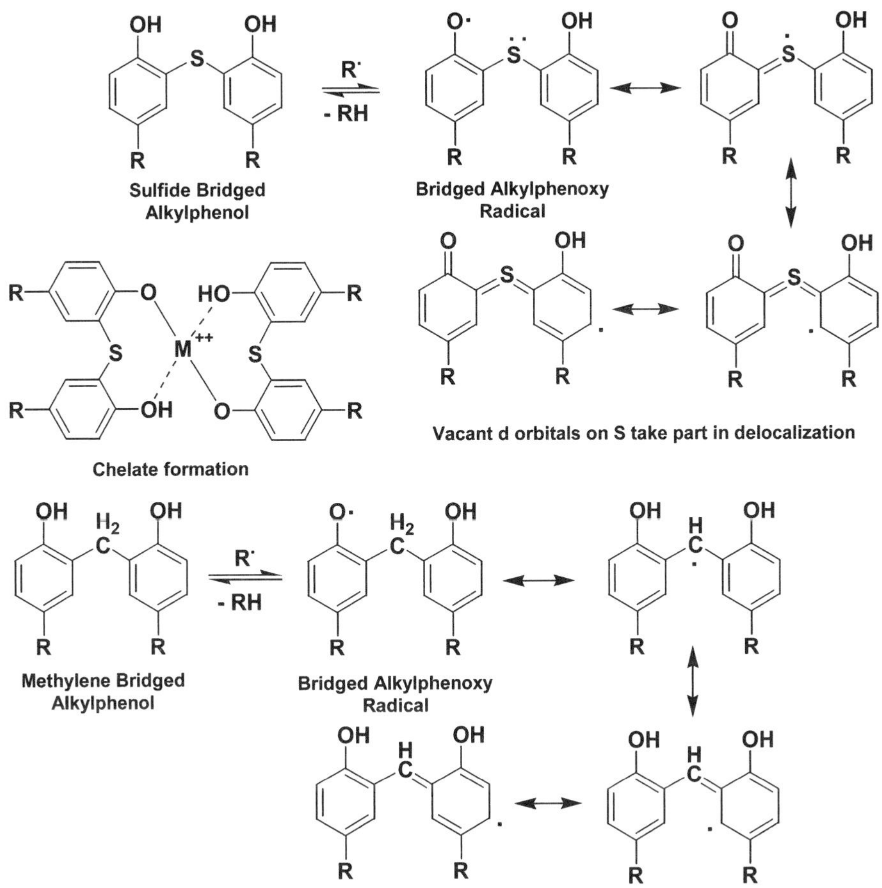

Fig. 4.93—Oxidation mechanism by sulfide- and methylene-bridged phenols..

gine oils, transmission fluids, and tractor hydraulic fluids; all of which are high-volume lubricants. Detergents have major use in engine lubricants. Gasoline and diesel engine oils account for over 75 % of the total detergent consumption. Detergent treatment levels in engine lubricant formulations are fairly high, with the marine diesel engine lubricants containing the highest detergent amounts. This is because marine engines use high-sulfur diesel or residual fuel, which leads to acidic combustion products (sulfuric acid). Therefore, the lubricants for these engines need the base reserve of the "basic" detergents for acid neutralization. Of the total detergent market, basic calcium sulfonates make up almost 65 %, followed by phenates at 31 %. At the end of the year 2008, the sulfonate use in heavy-duty diesel engine oils may see a decline. This is because the new heavy-duty diesel engines come equipped with diesel particulate filters (DPFs), some of which are sulfur intolerant. Alkenylphosphonic and alkenylthiophosphonic acid detergents also have little use in modern lubricants, primarily because of the negative effect of phosphorus on catalytic converters. As mentioned earlier, the development of the basic detergents that are not derived from the organic acids have been reported in the patent literature [381,382]. Substrates that were used to make such detergents include organic amines and polyamines, ethers, and organic sulfides (see Fig. 4.79). However, only alkali metal-derived overbased materials were reported.

A variety of proprietary and industry established tests are used to determine a detergent's effectiveness in lubricants. Because the deposit control is the joint domain of the detergents and the dispersants, the tests listed in Tables 4.6–4.8 are applicable to detergents as well. Thiosulfate overbased materials have been tested in gear oil tests, CRC L-37 and L-42. They showed good EP/anti-wear performance [387].

Film-forming Agents

Lubrication is necessary to facilitate the countermovement of two sliding surfaces. This function, which is usually performed by the base oil, can be enhanced by using high viscosity oils [397]. As shown in Fig. 4.94, a high lubricant viscosity has a positive effect on the load-carrying capacity of the gears, both at low temperatures and high temperatures. However, beyond a certain threshold temperature, the lubricant fails to form an effective film, and the friction and wear can result. The lubricant's film-forming ability under such circumstances can be made more efficient by using the film-forming agents. Such agents can interact with metal surfaces either through adsorption or the chemical reaction. Physical adsorption, or physisorption, is a weaker association of the additive with the metal than chemical adsorption, or chemisorption, which in turn is weaker than chemical reaction. During adsorption, an additive molecule generally keeps its structural integrity. In a chemical reaction, however, it has the tendency to lose it because it gets converted into new molecules. The modes of interaction of addi-

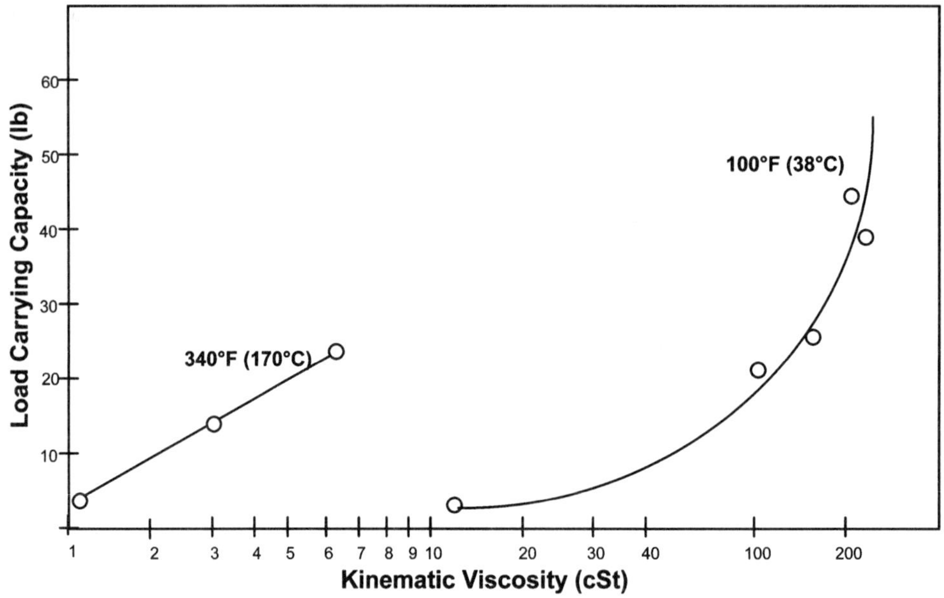

Fig. 4.94—Load-carrying capacity versus viscosity at 10,000 r/min gear speed [397].

tivemolecules with the surface are depicted in Fig. 4.95 [398,399].

Friction Modifiers

The United States is the major user and the importer of energy [400], and its usage and cost are escalating, see Fig. 4.96. Because of this and the energy's importance to the U.S. economy and the national security, the U.S. Government has established a number of regulations and measures to save energy. These include Corporate Average Fuel Economy

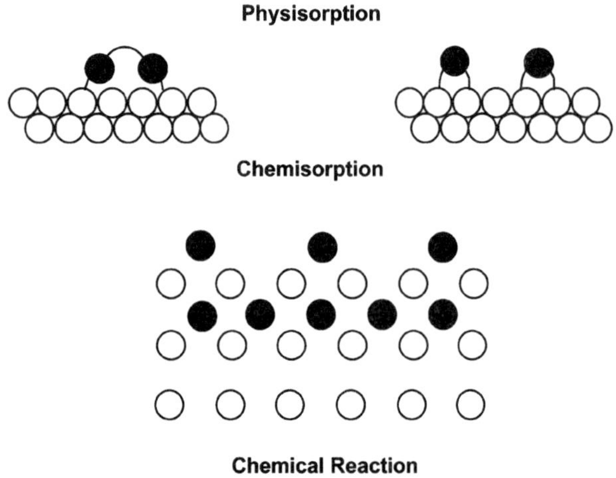

Fig. 4.95—Modes of additive—surface interactions (taken from Ref. [399] and modified).

(CAFE) Standards for passenger cars and light trucks. The light truck category includes pickup trucks, minivans, and sport utility vehicles (SUVs) [401]. The CAFE Standard for the trucks for the Model Years 2008–2011 was introduced in March 29, 2006.

Typically, only about 15 % of the energy from the fuel is converted into useful work, for example, for motion down the road or run useful accessories, such as air conditioning. The remaining 85 % of the energy is lost to engine and driveline inefficiencies and idling. Hence, there is a great deal of potential to improve fuel efficiency. The vehicle manufacturers are using a number of strategies to improve the fuel efficiency, not only to meet CAFE requirements but to surpass them [402]. These technologies include alternative fuel vehicles, hybrid electric vehicles, diesel-fueled vehicles, and the energy efficient technologies [400].

Alternative fuel vehicles are those that use E85 (85 % ethanol, 15 % gasoline), M85 (85 % methanol, 15 % gasoline), compressed natural gas (CNG), or liquefied propane gas (LPG), or any combination of these fuels. Hybrid-electric vehicles (HEVs) combine the benefits of the gasoline engines and the electric motors, and can be configured to improve fuel economy, increase power, and provide additional auxiliary power for electronic devices and power tools. Pure electric vehicles based upon fuel cell technology are not expected to become available for public use prior to the year 2010. Diesel engines are inherently more fuel efficient, by about 30–35 %, than gasoline engines, primarily because they use lean fuel-air mixtures. Their use in automobiles will save energy. Previously, the use of the diesel engines in automobiles was not common because of the issues related to their performance and efficiency, exhaust emissions, vibration, and noise. However, modern diesel engines have overcome the performance and efficiency issues through the use of fuel injection and electronic engine control technologies; and that of the exhaust emissions by the use of the low-sulfur diesel and biodiesel. Low-sulfur diesel produces less particulate matter and is the diesel particulate filter friendly. See Chap-

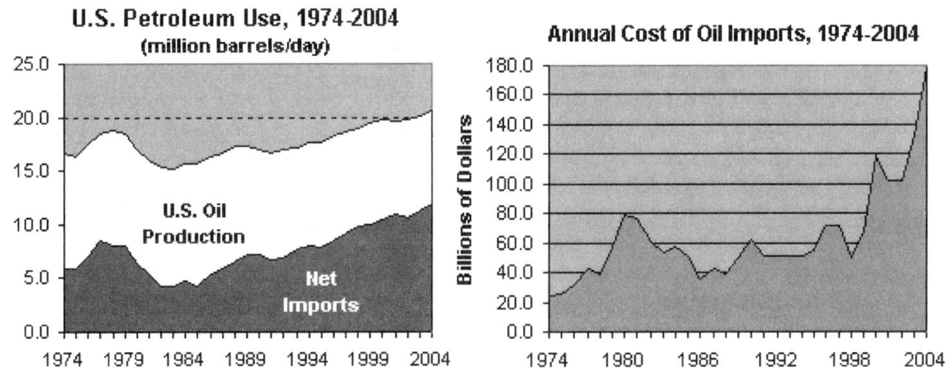

Source: Energy Information Administration (EIA) Monthly Energy Review, April 2005

Fig. 4.96—U.S. petroleum use and cost of imports [400].

ter 6 on Combustion Engine Emissions. In addition, the new engine designs come equipped with noise- and vibration-dampening systems that make them quieter and smoother. Cold-weather starting, another chronic problem in the diesel-fueled engines, has also been improved. Energy-efficient technologies for the gasoline-fueled automobiles that are presently being marketed are listed in Table 4.9 [400].

While these technologies are all well and good with respect to containing the fuel costs and the efforts to conserve resources, lubricants that can provide improved fuel economy are becoming increasingly desirable. As a matter of fact, fuel economy is one of the factors that is driving the new specifications for automotive lubricants, such as engine oils and transmission fluids. Catalyst compatibility of the lubricant is another factor.

The American Petroleum Institute (API) designates engine oil "Energy Conserving," if it shows fuel economy improvement over that of the standard reference oil, in the same engine operated under controlled conditions. While

TABLE 4.9—Energy-efficient technologies [400].

Technology	Performance Mechanism	Average Efficiency Increase
Engine Technologies		
Variable Valve Timing and Lift	Improve engine efficiency by optimizing the flow of fuel & air into the engine for various engine speeds.	5%
Cylinder Deactivation	Saves fuel by deactivating cylinders when they are not needed.	7.5%
Turbochargers and Superchargers	Increase engine power, allowing manufacturers to downsize engines without sacrificing performance or to increase performance without lowering fuel economy.	7.5%
Integrated Starter/Generator (ISG) Systems	Automatically turn the engine on/off when the vehicle is stopped to reduce fuel consumed during idling.	8%
Direct Fuel Injection (w/ turbo-charging or supercharging)	Delivers higher performance with lower fuel consumption.	11-13%
Transmission Technologies		
Continuously Variable Transmissions (CVTs)	Have an infinite number of "gears", providing seamless acceleration and improved fuel economy.	6%
Automated Manual Transmissions (AMTs)	Combine the efficiency of manual transmissions with the convenience of automatics (gears shift automatically).	7%

TABLE 4.10—Engine oil fuel economy requirements in Sequence VIB.

	Improvement over Baseline Oil, %	
SAE Viscosity Grades	16-h *Aging*	96-h *Aging*
0W-20 and 5W-20	2.3	2.0
0W-30 and 5W-30	1.8	1.5
All Other Viscosity Grades	1.1	0.8

the use of the friction modifiers helps, other factors, such as the lubricant viscosity grade, the type of viscosity modifier used, and the nature of the other additives present in the lubricant all play an important role [402]. Incidentally, new engine oil specifications ILSAC GF-3/API SL and ILSAC GF-4/API SM include Sequence VIB Fuel Economy Test (ASTM D6837), the pass/fail criteria for which are provided in Table 4.10. Fuel consumption of the candidate oil is compared at each of the five speed/load/temperature test conditions for the SAE 5W-30 baseline oil. Incidentally, the development of an updated version of the Sequence VI test, the Sequence VID, to be included in ILSAC GF-5, is already in the works [403]. ACEA 2007 Specification also includes a fuel economy test [CEC-L-54-T-96 (M111)], where a fuel economy improvement of ⩾2.5 % over the 15W-40 reference oil is needed to pass the test.

The use of the low-viscosity oils improves fuel economy since these oils minimize the energy losses due to friction that occur while cold starting and during stop-and-go type of driving operation. The friction at starting is the highest and as the equipment speed increases and the oil starts to pump, it progresses through boundary, mixed, and ultimately to hydrodynamic lubrication regimes. The amount of the start-up wear is also a function of the shut-down time; the longer the time, the higher the wear [404]. Engine oil Sequence VIB measures fuel economy after running the engine for 16 hours, which represents 1000 miles of driving; and then after running the engine for 96 hours, which represents 4500 miles of driving. The purpose of the two measurements is to determine the economy durability. Research has indicated that 0W-30 and 5W-30 oils that have lower viscosity provide better fuel economy than those that are 10W-30 and higher [405]. See Fig. 4.97 for the 16-hour fuel economy test results. Typically, the fuel economy results deteriorate at the 96-hour stage, see Fig. 4.98. This is due to the evaporation of the low-boiling components that are present in the oil, which effectively increases its viscosity. If one can limit this boiling-off, fuel economy should be maintained. NOACK volatility is one of the measures that are used to determine the oil volatility, hence the oil consumption. The problem is that the NOACK volatility limit for ILSAC GF-3 and GF-4 oils is 15 %. This implies that a 95 VI oil of greater than 15 % volatility (see Fig. 4.99) cannot be used to formulate GF-3 and GF-4 oils. However, API Group II+ oils can be used in combination with a 115 VI oil to formulate GF-3/GF-4 lubricants. This strategy is expected to provide a fuel economy benefit of around 0.5 %, which is fairly significant. Obviously, the volatility limitation to formulate 5W-30 lubricants does not apply to API Group III and Group IV oils, which have NOACK volatilities of less than 15 %. This is supported by the data in the European Transient Cycle, which indicated a fuel economy improvement of around 2 % for certain PAO-derived lubricants over the 15W-40 reference oil [406], see Fig. 4.100.

Since the fuel economy benefit derived from oil viscosity has its limits, one way to go beyond the limit is by the use of a class of additives, called the friction modifiers. Friction modifiers are materials that modify the frictional properties of a lubricant. These additives perform in the mixed-film lu-

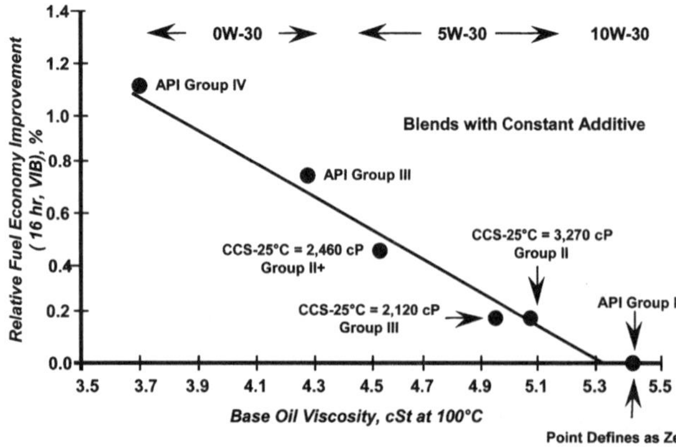

Fig. 4.97—Fuel economy improvement (FEI) versus viscosity in Sequence VIB at 16 hours [405].

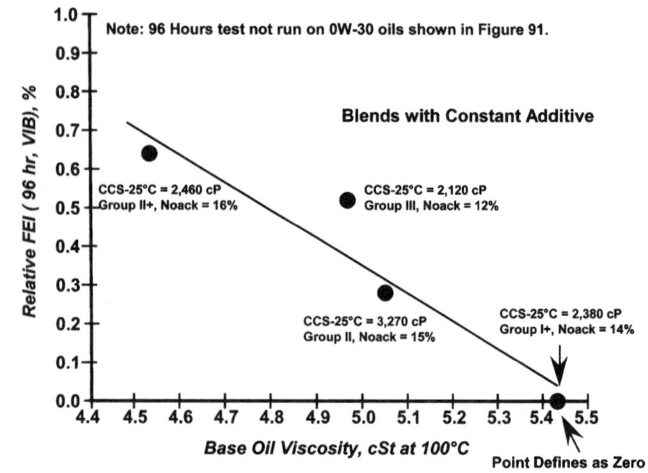

Fig. 4.98—Fuel economy improvement (FEI) versus viscosity in Sequence VIB at 96 hours [405].

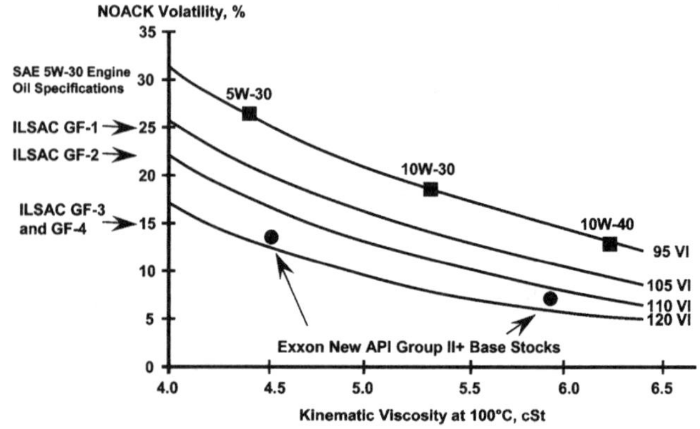

Fig. 4.99—Volatility of high VI base stocks [405].

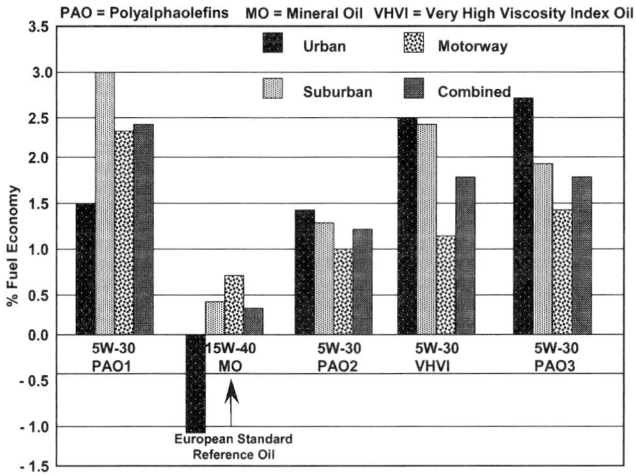

Fig. 4.100—Fuel economy benefit of PAO-derived lubricants relative to mineral oils [410a].

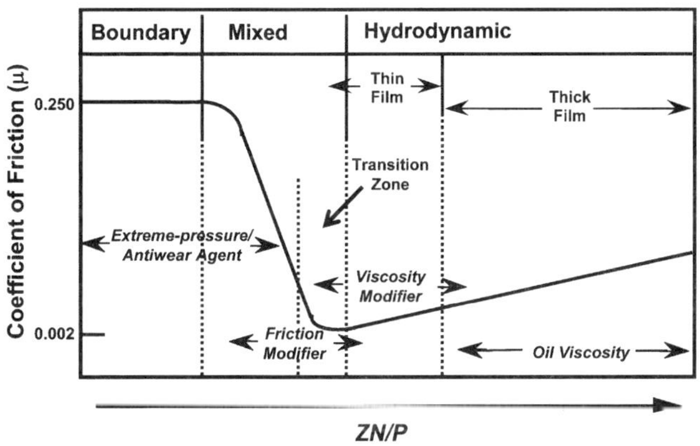

Fig. 4.101—Roll of additives in different lubrication regimes.

brication regime, as shown in Fig. 4.101. They interact with surfaces either by physical adsorption or chemical adsorption, depending upon their reactivity, see Fig. 4.95. Friction modifiers are used to decrease or increase friction, depending upon the application. In engine lubricants and gear oils, their primary functions are to reduce friction, minimize wear and noise, and improve the fuel economy by lowering the power loss. In transmission and hydraulic fluids, these additives are used to facilitate the timely engagement and disengagement of clutches and bands to assure smooth and noise-free operation. The effect of the friction modifiers on the engine power loss [407] is depicted in Fig. 4.102. At low ambient temperatures, where the lubricant viscosity is high enough to form an effective lubricating film, the friction modifiers have little effect in minimizing the power loss due to friction. However, at high temperatures, the friction modifiers become more effective because of the loss of lubricant viscosity, which may create boundary lubrication conditions. Figure 4.103 shows the friction-reducing capability of these additives in automatic transmission fluids [407]. At low sliding speeds, where greater metal to metal contact is likely, friction modifiers interact with the surfaces in the manner described above and minimize such contact and the associated rough shifting. However, at high speeds their effect is not as dramatic because such speeds promote hydrodynamic lubrication. These additives minimize brake chatter in tractors in an analogous manner.

A variety of materials are used to reduce friction. These include the following:

1. Fatty alcohols, amines, carboxylic acids and their derivatives, such as esters, amides, imides, ethoxylates, and propoxylates.
2. Linear alcohol esters of phosphoric and phosphorous acids.
3. Transition metal (molybdenum, copper, zinc) salts/complexes of linear carboxylic acids, dithiocarbamic acids, and dithiophosphoric acids.

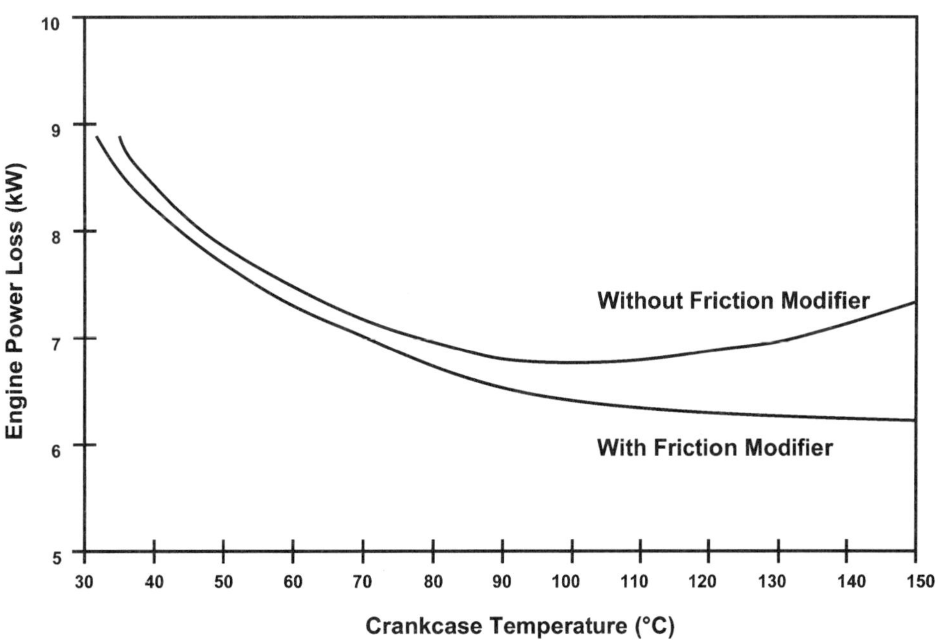

Fig. 4.102—The effect of friction-modified lubricant on engine power loss [407].

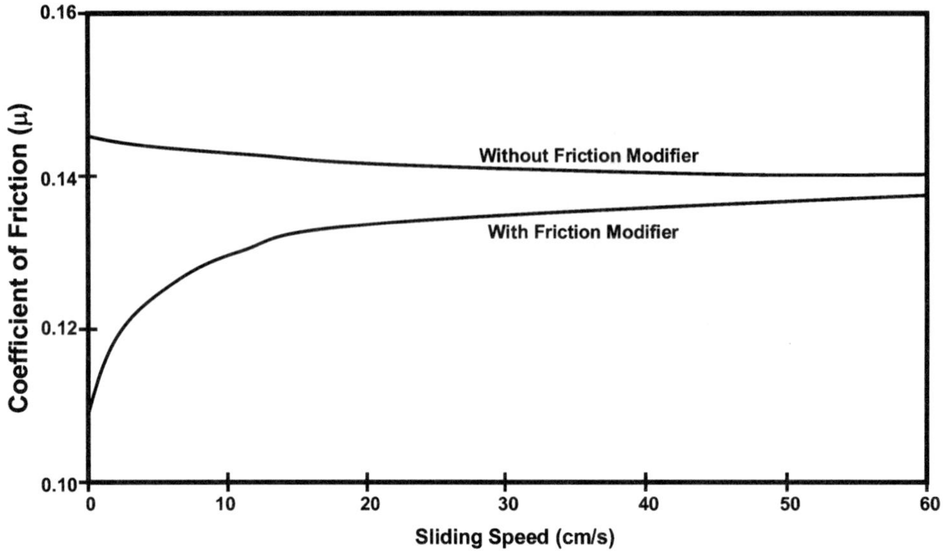

Fig. 4.103—Friction reducing ability of an automatic transmission fluid [407].

4. Sulfurized fats and hydrocarbons.
5. Organic polymers, such as PTFE.
6. Layered inorganics, such as graphite and molybdenum disulfide.

Of these, the compounds belonging to the first four groups are among those that are used most often. The others are used in specialty applications. Organic friction modifiers are long-chain molecules with a polar end group and a nonpolar linear hydrocarbon chain. The polar end groups either physically adsorb onto the metal surface or chemically react with it while the hydrocarbon chains extend into the lubricant. These chains associate with one another and the lubricant to form a strong lubricant film [408]. This is shown in Fig. 4.104. Figure 4.105 illustrates the manner of interaction of the friction modifiers with the metal surface and the lubricant, both under steady-state condition and under the influence of shear [409]. The adsorption of the additive on the metal is a function of the additive's polar/nonpolar ratio and the intensity of its interaction with the surface depends upon its reactivity. Typically, materials with a high polar/nonpolar ratio have greater affinity for the surface than those with a medium or low polar/nonpolar ratio. This is demonstrated in Fig. 4.106 by the differential rates of absorption of more polar stearic acid and less polar octadecyl alcohol on copper and the increased adsorption of the stearic acid over time [4]. As a general rule, within the same chemical class the longer the hydrocarbon portion of the friction modifier, the lower the friction. This is shown in Fig. 4.107 [4]. As the molecular

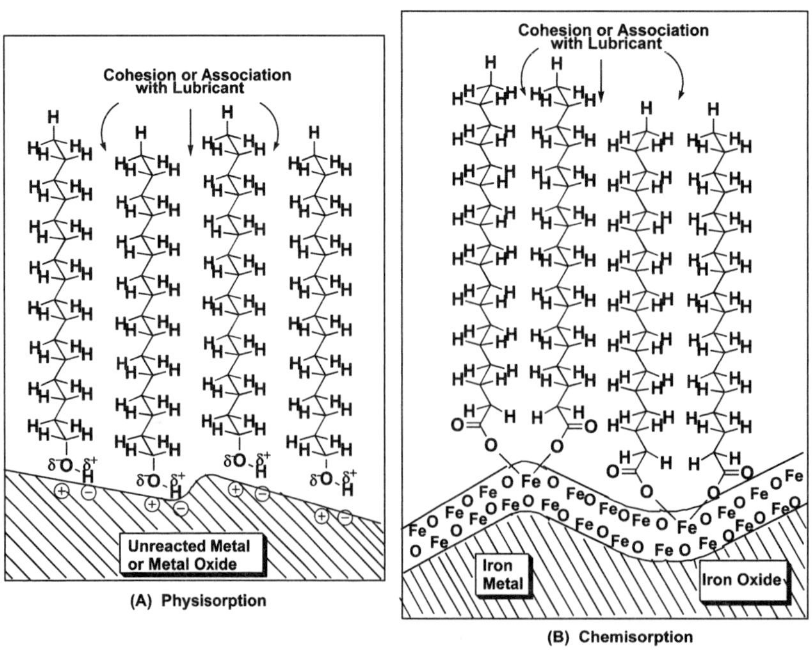

Fig. 4.104—Adsorption of polar additives on the metal surface [408 adapted].

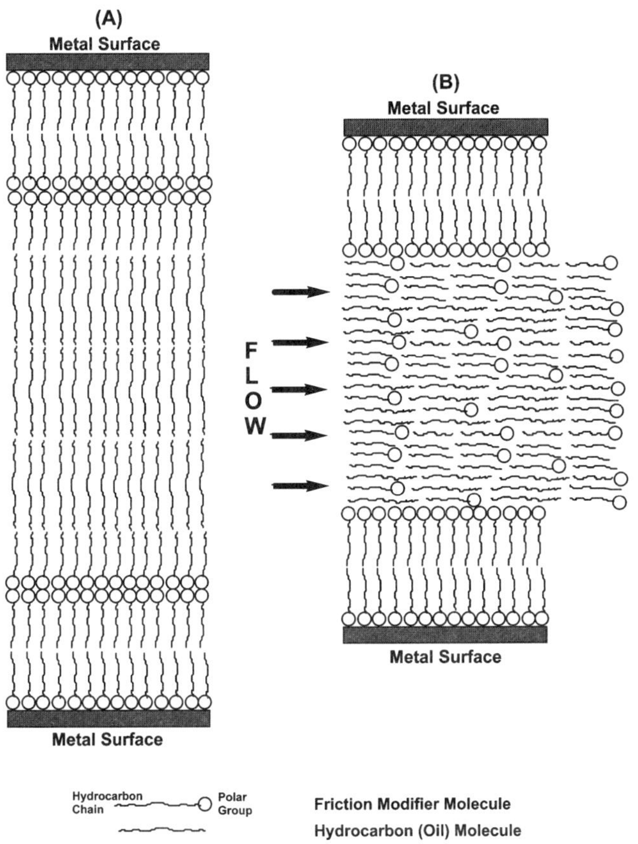

Fig. 4.105—Adsorption of friction modifiers on metal; (A) steady state, (B) under shear.

mass (molecular size) increases, the coefficient of friction decreases. Please note that the increase in molecular size decreases an additive's affinity towards the surface and the decease in coefficient of friction is a consequence of the association of the hydrocarbon portion of the molecules with each other, and oil, if present, via the van der Waals forces. The same is true of the paraffinic hydrocarbons, which also lower friction with increasing molecular size, but to a lesser degree than the carboxylic acids.

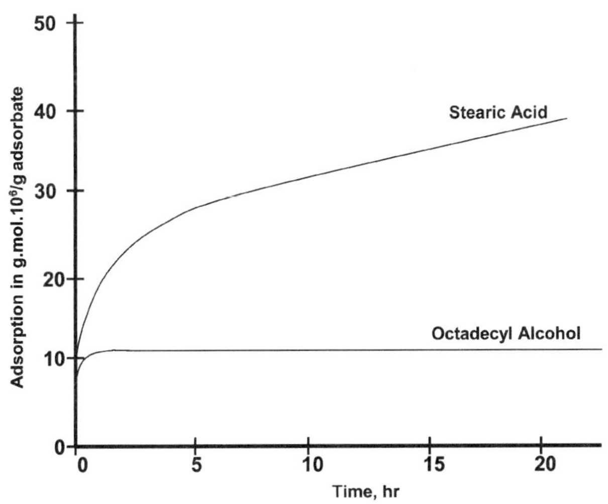

Fig. 4.106—Rate of adsorption on copper [4].

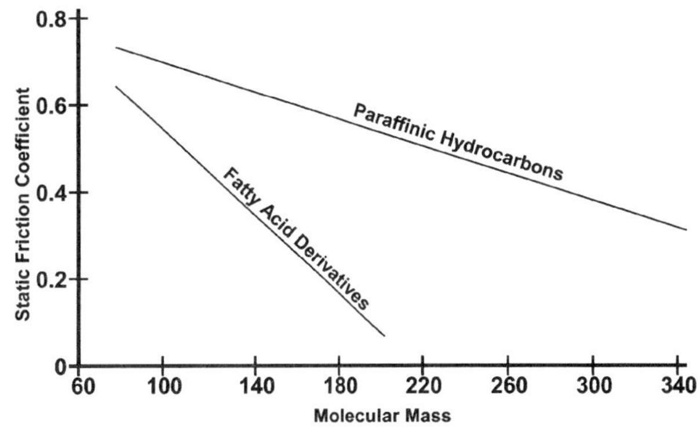

Fig. 4.107—Static coefficient of friction versus molecular mass [4].

An increase in the surface temperature leads to desorption of the organic friction modifiers. Relatively nonpolar materials, such as alcohols and oil, which adsorb on surfaces via physical adsorption, desorb more readily than those that adsorb by the chemical adsorption, such as the carboxylic acids. Figure 4.108 shows the effect of temperature on lauric acid adsorbed on the metal surface [4]. The acid stays on the surface up to 100°C, as indicated by the low coefficient of friction, above which it desorbs. Metal salt films that result from the chemisorption of the carboxylic acids with surfaces differ in their frictional properties as well as in their durability. As the data in Table 4.11 show, zinc salt has the lowest coefficient of friction but the magnesium salt has more surface affinity, as indicated by its higher film breakdown temperature [4]. This implies that the effectiveness of at least a carboxylate friction modifier is a function of the metallurgy of the machine element.

Figure 4.109 compares the frictional properties of two friction modifiers against 5W-30 baseline used for sequence

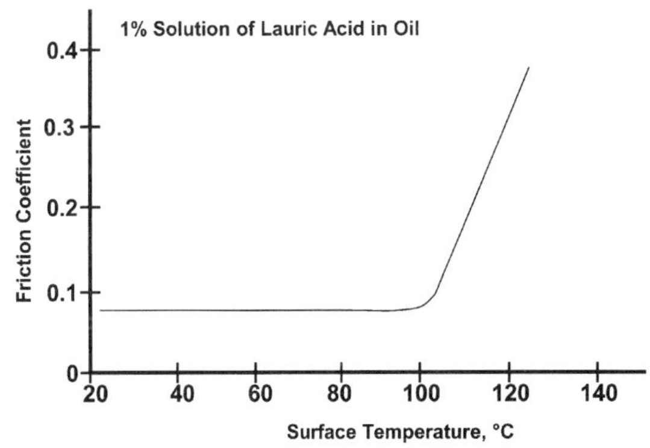

Fig. 4.108—Effect of temperature on adsorption of lauric acid on zinc surface [4].

TABLE 4.11—Friction coefficients and film breakdown temperatures [4].

Salt	Friction Coefficient at 20°C	Film Breakdown Temperature, °C
Copper Laurate[a]	0.08	100
Zinc Laurate	0.05	120
Magnesium Laurate	0.12	150

[a]Temperatures for palmitate and stearate are somewhat higher.

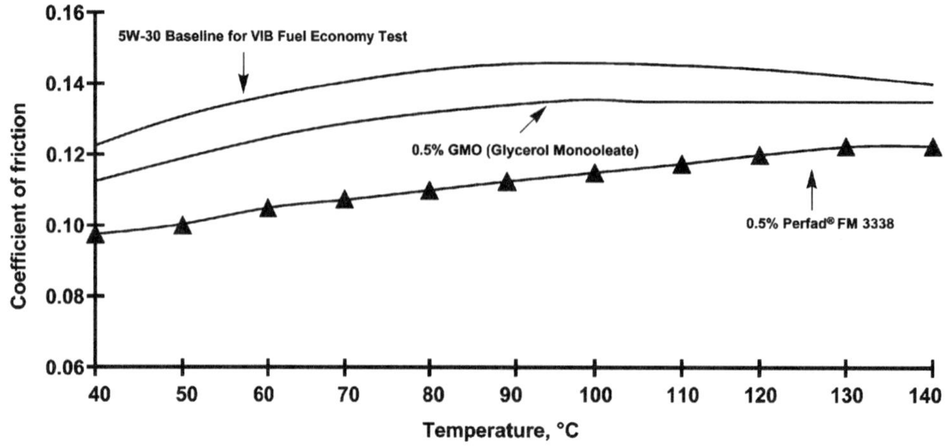

Fig. 4.109—Comparison of frictional properties of organic friction modifiers [410b].

VIB Fuel Economy Test. Glycerol monooleate (GMO) is a commonly used friction modifier and Perfad® FM 3338 is Uniquema's proprietary friction modifying additive [410]. As one can see, both GMO and Perfad® have friction coefficients lower than that of the baseline oil. Note the progressive increase in friction with an increase in temperature for the two additives, which was previously ascribed to desorption. Figure 4.110 shows the estimated fuel economy index of the two additives based on the fuel economy test [410]. Perfad® friction modifier provided a fuel economy of 2.5 %, compared to GMO at about 2.1 %, and the base oil slightly over 1.9 %. Similar results were obtained when this additive was tested in a European fuel economy test [410,411]. Perfad® has the added advantage of greater oxidation stability since unlike GMO and other oleic acid derivatives, it has no multiple bonds. Phosphoric acid and phosphorous acid esters that are used as friction modifiers include linear or fatty alcohol derivatives. Because of the phosphorus limit in engine oils, the use of the phosphorus compounds is limited primarily to nonengine lubricants, such as gear oils and hydraulic fluids. The synthesis of these materials will be described in the extreme pressure/antiwear section. For the fatty alcohol and fatty acid families, the friction-modifying properties are a function of the length and the structure of the hydrocarbon chain and the nature of the functional group. Long and linear-chain materials reduce friction more effectively than short and branched-chain materials. Also, fatty acids are better than fatty amides, which in turn are better than fatty alcohols. Saturated acids, containing a 13 to 18 carbon chain, are generally preferred. Lower molecular weight fatty acids are avoided because of their corrosivity. Fatty acid derivatives are the most commonly used friction modifiers.

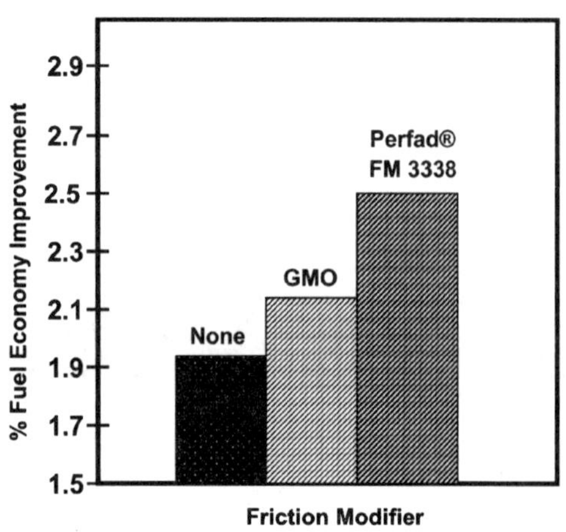

Fig. 4.110—Fuel economy index [410b].

Transition metals (molybdenum and zinc) salts/complexes of carboxylic acids, dialkyl dithiophosphoric acids, dithiocarbamic acid, and dithiophenol have been used as friction modifiers as well. Of these, molybdenum dithiocarbamate (Mo-DTC) is the most often used friction modifier in lubricants. In reality, the compounds, such as Mo-DTC, that contain sulfur perform as multipurpose additives, that is, as friction modifiers, antiwear agents, and oxidation inhibitors. Mo-DTCs are usually used in combination with zinc dialkyl dithiophosphates. This results in the formation of a mixed dithiocarbamate/dithiophosphate salt, via an exchange reaction, which has superior friction reducing properties. Sulfurized fats and hydrocarbons, depending upon the structure and the molecular weight, can also act as multifunctional additives. These will also be addressed in the extreme pressure/antiwear additives section. Organic polymers, such as PTFE, and layered inorganics, such as graphite and molybdenum disulfide—often referred to as solid lubricants, are also discussed in the latter part of this chapter.

Mechanical losses due to friction in an engine are estimated at 10 % of the total energy output. Of this, the bearings consume 31 %, valve train consumes 8 %, and pistons and rings consume 62 % [412]. This power consumption is based upon normal operating temperature and typical oil sump temperature of 190 °F (88 °C). In a cold engine, the cold oil leads to a greater power loss because of the less effective lubrication, primarily due to the higher lubricant viscosity. Figure 4.111 shows the power loss in an engine with and without a friction modifier; and at low and high temperatures [413]. While examining the figure, please note that the torque units in the right and left Y-scales are not the same. The temperature changes viscosity, which alters the nature of lubrication in different parts of the engine. At high temperatures, a decrease in viscosity increases frictional power

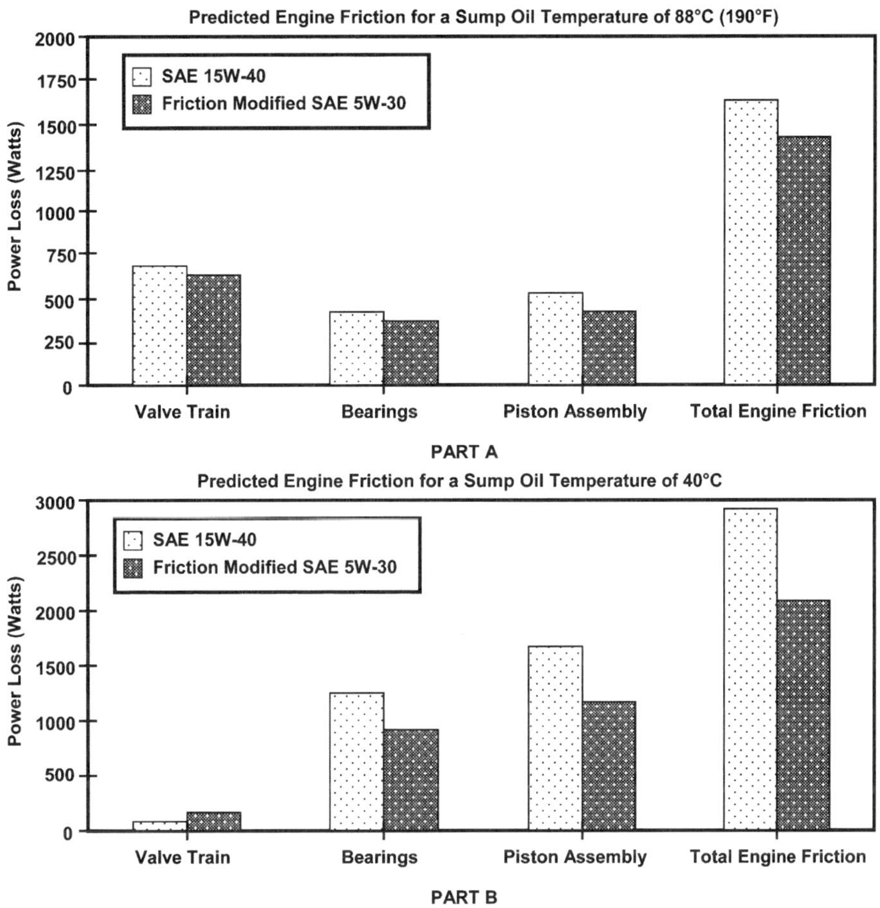

Fig. 4.111—Effect of friction modifier on engine power loss [413].

loss in the power train but lowers such loss in bearings and the piston assembly. This is because the power train operates under boundary lubrication, where higher lubricant viscosity is beneficial in an effective film-formation. Either way, the overall effect of the lower viscosity 5W-30 oil, which contains a friction modifier, is a decrease in the power loss in all parts of the engine. Also, it is important to note that at low crankcase temperatures, the lubricant viscosity overwhelms any benefits that the friction modifier may have to offer. Conversely, at high crankcase temperatures, where the low-lubricant viscosity increases boundary lubrication, the benefits of the friction modifier are more observable. This explains the discrepancy observed in Part B of the figure, where the lower viscosity oil containing the friction modifier shows a negative effect on the power loss. Figure 4.112 shows the positive effect of the friction modifier on the cranking torque [414]. As one can see, the torque required is lower when the oil contains a friction modifier than if it does not.

As described earlier, the frictional properties of the transmission fluids and the hydraulic fluids are also important. The frictional compatibility of the fluid with the functioning parts of the transmission and with wet brakes and wet clutches in tractors assures their smooth and noise-free (chatter-free) operation. Frictional properties play an equally important role in gear oils. In standard applications, friction-modified gear oils not only protect gears and axles against frictional damage but they also contribute towards fuel economy. For limited-slip axles, which contain friction clutches and cones, friction-modified gear lubricants minimize noise arising from stick-slip. LFW-1 (Alfa Laval Load, Friction, and Wear Test) and LVFA (Low-Velocity Friction Apparatus) data for gear oils showing the effect of the friction modifiers on friction [415] are presented in Table 4.12.

As can be seen from the table, friction modifiers decrease static and dynamic coefficients of friction in both single grade and multi-grade oils. In general, the friction-modified gear oils show an increase in axle-efficiency relative to normal gear oils. This is shown in Fig. 4.113 [210]. The axle efficiency is better both for urban and highway driving. Axle efficiency is believed to have a positive correlation with the fuel economy.

Under heavy loads, the EP agent replaces the friction modifier and takes over the function of preventing the wear damage. As the load decreases, the friction modifiers again come into action. Friction modifiers have a finite life, which is related to their oxidative and thermal stability. This is one of the reasons behind fuel economy durability requirement in the Sequence VIB test. These additives are commonly used in gasoline engine oils, automatic transmission fluids, tractor hydraulic fluids, power steering fluids, shock absorber fluids, and metalworking fluids. In passenger car applications, with Federal Government-mandated fuel economy, lubricant suppliers use these additives as a com-

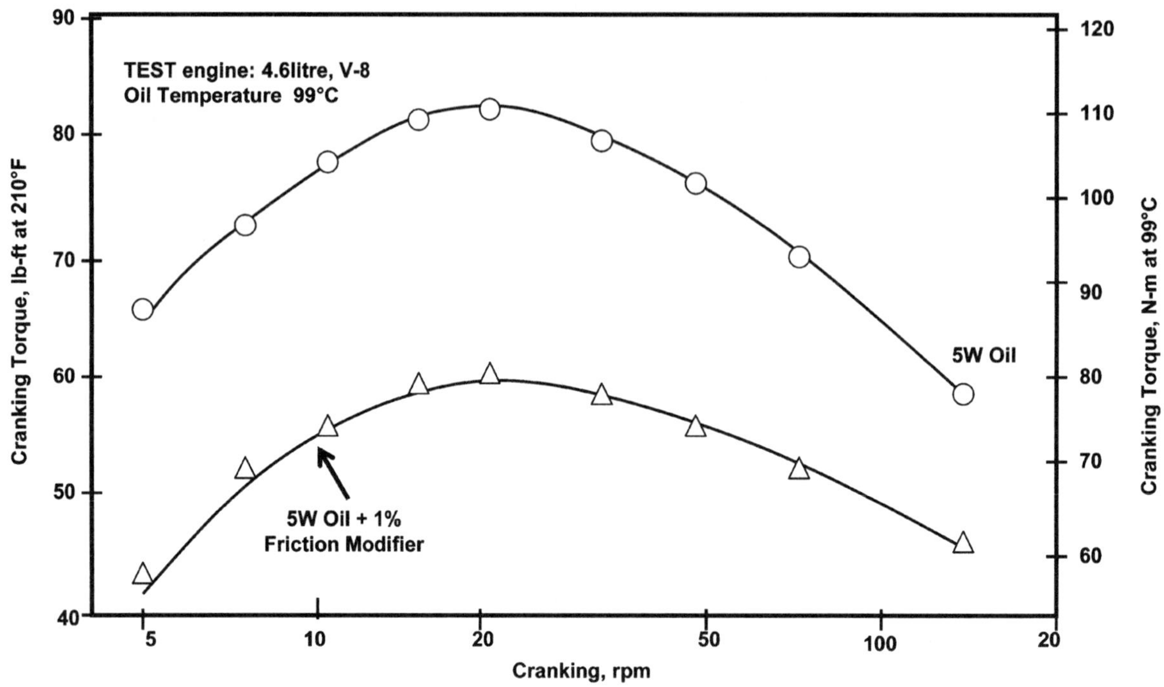

Fig. 4.112—Effect of friction modifier on cranking torque [414, adapted].

petitive marketing tool. In ATFs and limited slip axle lubricants, the friction modifiers are used to control clutch and band engagement. Anti-squawk additives, which are functionally similar to the friction modifiers, are used to reduce objectionable mechanical noise, such as squawk and chatter. Anti-squawk agents are primarily used in automatic transmission and tractor hydraulic fluids, automotive gear oils, and some industrial oils.

The effectiveness of the friction modifiers and the anti-squawk agents is judged by the lubricant's performance in LVFA and SAE #2 tests. Their load-carrying capacity is determined by the film strength tests, such as the Timken test.

Antiwear and Extreme Pressure Agents

Wear occurs in all equipment that has moving parts. Three conditions that can lead to wear are surface-to-surface contact, surface contact with foreign matter, and erosion due to corrosive materials. Wear resulting from the surface-to-surface contact is frictional or adhesive wear, from contact with foreign matter is abrasive wear, and from contact with corrosive materials is corrosive wear. Fatigue wear is an additional type of wear that is common in equipment where the surfaces are not only in contact but also experience repeated stresses for prolonged periods. Abrasive wear can be prevented by installing an efficient filtration mechanism to remove the offending debris. Corrosive wear can be controlled by using additives which neutralize the reactive species that attack the surface. The control of adhesive wear requires the use of additives called antiwear and extreme pressure (EP) agents.

Under normal conditions of speed and load, two metal surfaces are effectively separated by a lubricant film, a condition identified as hydrodynamic lubrication. An increase

TABLE 4.12—Effect of friction modifiers in gear oils [415].

LFW-1 Machine	Coefficient of Friction	
GL-5 gear oil (80W-90)	0.057	
GL-5 gear oil (80W-90) +1 % friction modifier	0.054	
GL-5 gear oil (75W)	0.055	
GL-5 gear oil (75W)+1 % friction modifier	0.052	
LVFA Machine	Coefficient of Friction	
	Static	*Dynamic*
GL-5 gear oil (75W)	0.070	0.052
GL-5 gear oil (75W)+1 % friction modifier	0.050	0.042
GL-5 gear oil (75W)+1 % oleic acid	0.052	0.043
GL-5 gear oil (75W)+1 % Armeen HT	0.062	0.050

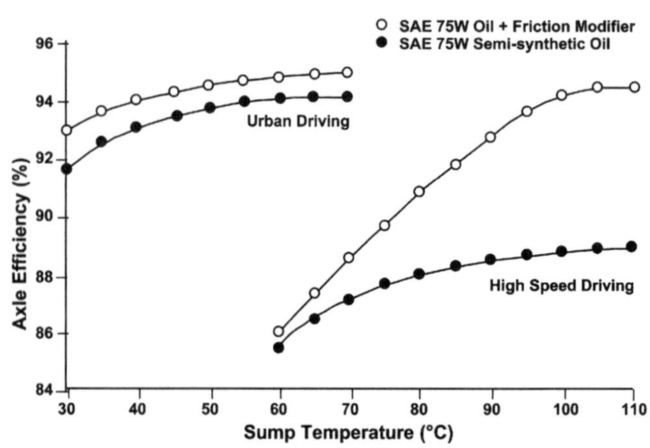

Fig. 4.113—The effect of a friction modifier on axle efficiency of gear oil [210].

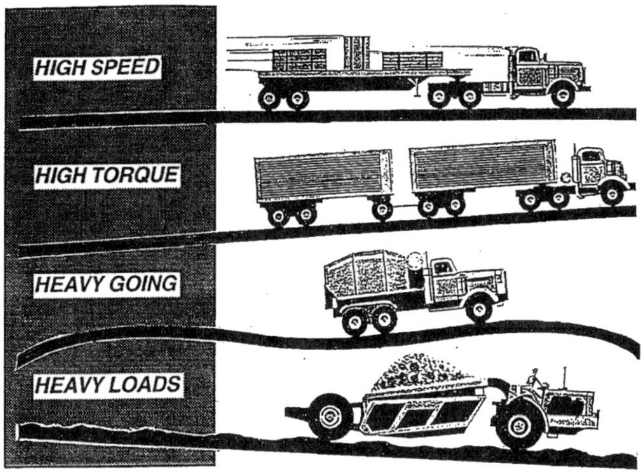

Fig. 4.114—Service conditions for automotive equipment.

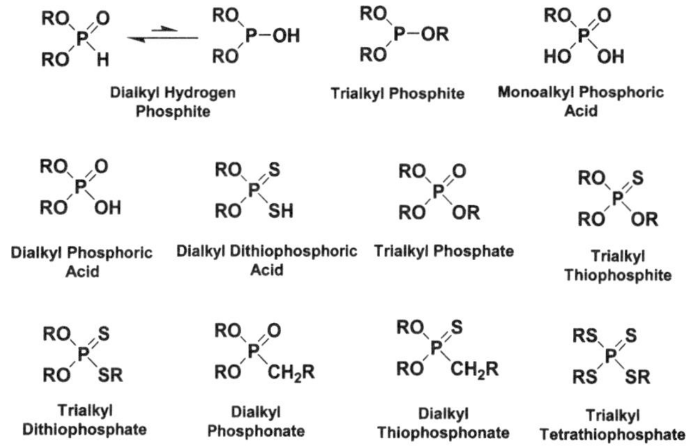

Fig. 4.115—Common phosphorus derivatives used as EP/antiwear agents.

in load or a decrease in speed promotes the metal-to-metal contact. This causes a temperature rise in the contact zone due to frictional heat, which causes the loss of lubricant viscosity and hence its film-forming ability. With a progressive increase in load or a decrease in viscosity, or both, the nature of the lubrication changes from hydrodynamic to mixed-film to boundary lubrication. In boundary lubrication, the surface asperities contact each other even though the lubricant may support most of the load. Friction depends mainly on the shearing forces necessary to cleave these adhering asperities and the wear and friction can be reduced by the use of antiwear additives and EP agents. These additives, also called the boundary lubrication additives, offer protection under *mixed-film* and *boundary lubrication* conditions [416], as shown in Fig. 4.101. Common additives that are used to provide EP/antiwear protection include metal-free and metal-containing dialkyl and diaryl dithiophosphates; alkyl phosphites, alkyl phosphates and their salts; sulfurized hydrocarbons, fats, oils, and fatty carboxylic acids; phosphorized and phospho-sulfurized fats and olefins; organic molybdenum compounds, molybdenum disulfide (solid), graphite (solid), and borate dispersions.

Antiwear and extreme pressure agents have structures that are similar to those of the friction modifiers, with the difference that their hydrocarbon chains are much shorter and are thermally labile. In other words, their polar to non-polar ratio is fairly high, which makes them more surface-active. Antiwear and EP additives both provide protection by a similar mechanism, except that EP additives typically require higher activation temperatures and load than antiwear additives. Simply stated, antiwear additives perform under mild conditions and EP additives perform under severe conditions. The severity of the conditions is determined by the "load factor" experienced by the additive and the temperature at which the additive functions. Load is a function of the equipment speed and service. Ratings of automotive service in order of increasing load are shown in Fig. 4.114. Heavy loading requires the use of extreme pressure agents and mild loading requires the use of antiwear agents. Thus, it is important to consider both the load and the temperatures that the equipment is likely to experience before selecting these additives.

Antiwear agents are commonly used in engine oils, automatic transmission fluids, power steering fluids, and tractor hydraulic fluids. Extreme pressure agents are used in other power-transmitting fluids, gear oils, shock-absorber fluids, and metalworking fluids. In engine oils, the ASTM Sequence IIIF/G and VG engine tests are used to determine the effectiveness of the antiwear agents. To determine the performance of the EP agents in gear oils, the CRC L-37 and CRC L-42 axle tests are used. The general effectiveness of these additives in the metalworking fluids is determined by Timken, 4-Ball, and Falex tests. Extreme pressure additives are usually supplemented with antiwear additives and friction modifiers to make the formulations effective at lower temperatures and under milder loading conditions as well.

Most antiwear and extreme pressure agents contain sulfur, chlorine, phosphorus, boron, or combinations thereof. The classes of compounds that inhibit adhesive wear include alkyl and aryl disulfides and polysulfides, dithiocarbamates, dimercaptothiadiazole (DMTD) derivatives [417], chlorinated hydrocarbons, and phosphorus compounds, such as alkyl phosphites, alkyl phosphates, dialkyl dithiophosphates, and alkylphosphonates. The structures of the common organophosphorus compounds, which are used either as such or are converted into compounds that are used as EP/antiwear agents, are provided in Fig. 4.115.

Both antiwear and extreme pressure additives function by thermal decomposition and by forming products that react with the metal surface to form a solid protective layer. This solid metal film fills the surface cavities and facilitates effective film formation, thereby reducing friction and preventing welding and surface wear. The metal films consist of iron halides, sulfides, or phosphates, depending upon the antiwear and EP agents used. Friction modifiers differ from the antiwear and extreme pressure agents in that they form the protective films by physical and chemical adsorption, instead of chemical reaction.

The film formation by these additives is a multi-step process. The steps involved are as follows:

1. Adsorption of the chemical onto the metal surface.
2. Formation of the chemically reactive species due to thermal decomposition or hydrolysis and their chemical reaction with the metal to form a sacrificial protective film.

3. Removal of this film by mechanical wear, exposing fresh metal surface.
4. Readsorption of the EP/antiwear agent on the freshly exposed surface.
5. Repetition of Steps 2–4.

This sequence follows the temperature profile of the contact zone. An increase in temperature causes the sequence of events to proceed in the order listed.

Antiwear Agents

Zinc salts of dithiophosphoric acids are the most widely used antiwear agents. These salts, in addition to providing antiwear protection, act as oxidation and corrosion inhibitors. They find major use in gasoline and diesel engine oils and industrial lubricants.

Zinc dialkyl dithiophosphates or zinc diaryl dithiophosphates are synthesized by the reaction of the respective dithiophosphoric acids with zinc oxide. The dithiophosphoric acid derivatives that do not produce ash on combustion are called ash-less, ash-free, or non-ash producing. These can be prepared by the reaction of dialkyl dithiophosphoric acids with amines, alkylene oxides, such as ethylene oxide and propylene oxide, or materials that contain conjugated double bonds, such as alkyl acrylates and methacrylates. Amine salts of dialkyl dithiophosphoric acids have good antiwear properties, but they are seldom used as additives because of their extreme corrosivity towards copper. Dialkyl thiophosphoryl disulfide, which can be obtained by the oxidation of the dialkyl dithiophosphoric acid, is also non ash-producing and is often used as antiwear additive and oxidation inhibitor. Synthetic scheme to prepare these materials is shown in Fig. 4.116. While in the figure we show the structure of zinc dialkyl dithiophosphate as a monomer, it has been determined by dynamic light scattering that zinc di-isobutyl dithiophosphate exists as a tetramer in the solid form [418]. The structure is provided in Fig. 4.117. During the zinc salt formation, if one uses zinc oxide in slight excess, the final salt will have a basic character, because of the excess zinc oxide being present in an associated form. Dithiophosphoric acids are the products of reaction of an alcohol or a phenol with phosphorus pentasulfide.

Thermal and hydrolytic stability of these products depend upon the nature of the organic group. Dialkyl dithiophosphates derived from primary alcohols are more thermally stable than those derived from secondary alcohols, as shown by the data in Table 4.13 [44]. Therefore, they are used extensively in formulating gasoline and automotive diesel engine oils. On the other hand, secondary alcohol derived dialkyl dithiophosphates, of somewhat lower stability, are used as EP/antiwear additives in gear oils and transmission and hydraulic fluids. Diaryl dithiophosphates, although thermally the most stable in this family, are hydrolytically the least stable and, with some exceptions, are not very effective antiwear agents. Hence, they do not get much use.

Dithiophosphoric acid derivatives decompose, generally around 200°C or so, to form thiols, olefins, polymeric alkyl thiophosphates, and hydrogen sulfide [419,420]. The antiwear performance of these derivatives depends upon their thermal stability, which in turn depends upon the nature of the alkyl group. Primary dialkyl dithiophosphates decompose via an alkyl transfer mechanism to form zinc mono-alkyl dithiophosphate and an alkyl thiophosphate ester. Through a series of steps, these materials are converted into zinc phosphate and trialkyl tetrathiophosphate, along with a variety of other products [419]. Trialkyl tetrathiophosphate appears to be the major thermal decomposition product, as shown by the ^{31}P NMR spectroscopy.

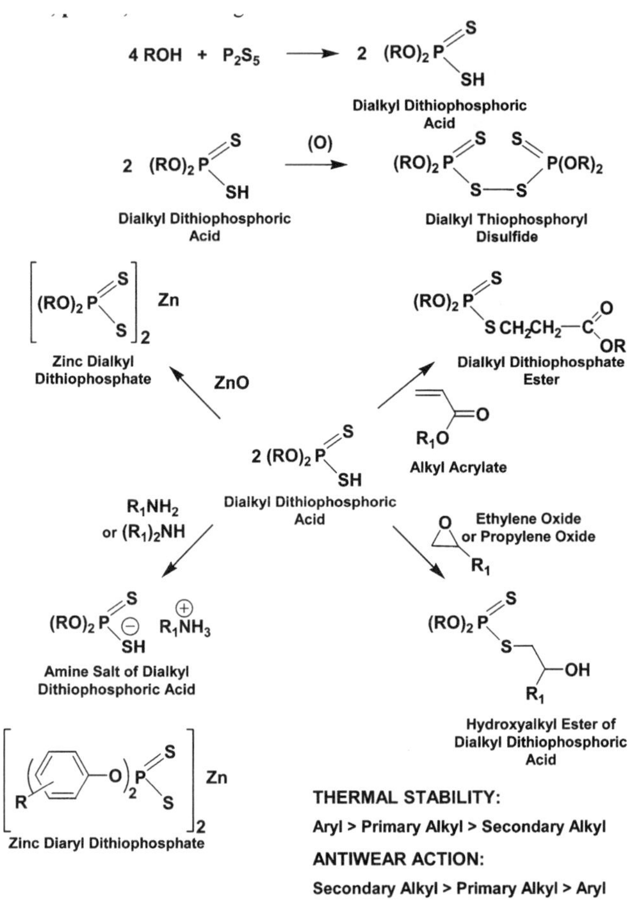

Fig. 4.116—Synthesis of dithiophosphoric acid derivatives.

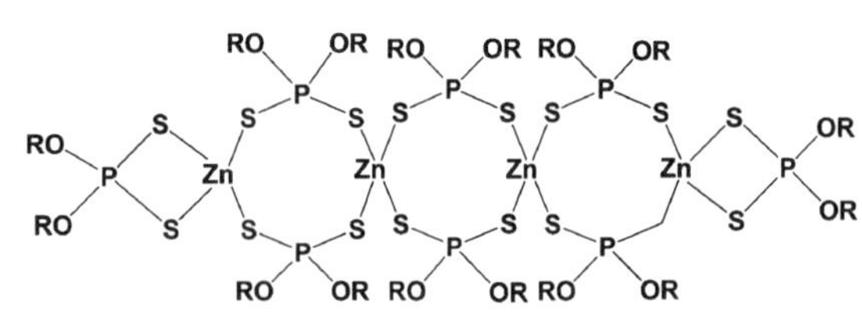

Fig. 4.117—Crystal structure of zinc dialkyl dithiophosphate [418].

TABLE 4.13—Nature of the alkyl group versus zinc dialkyl dithiophosphate stability [44]. Reprinted with permission from the Lubrizol Corporation.

Alkyl Group	Group Type	Decomposition Temperature (°C)
Isopropyl	Secondary	196
4-Methyl 2-pentyl	Secondary	197
n-Amyl	Primary	212
n-Octyl	Primary	>251

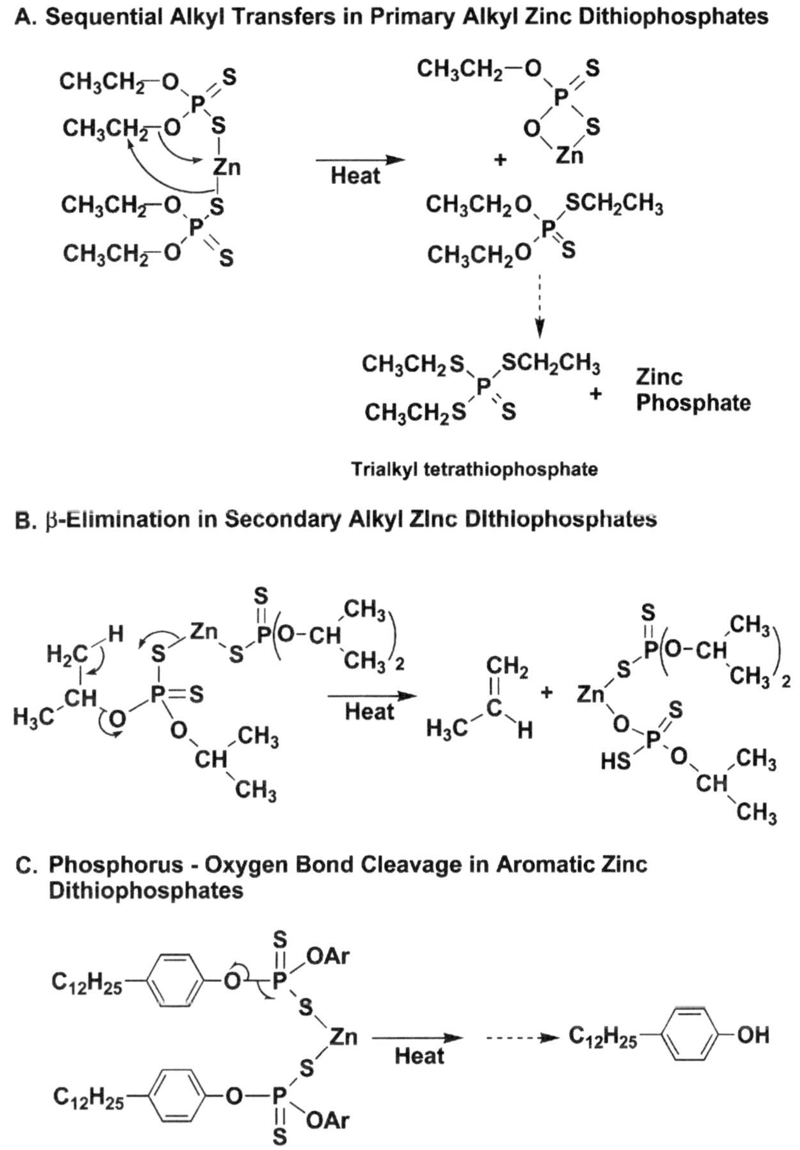

Fig. 4.118—Thermal decomposition of dithiophosphoric acid derivatives.

Secondary alkyl zinc dithiophosphates lose an olefin via β-elimination to form a product with the free dithiophosphoric acid functional group. This product can further decompose by the loss of hydrogen sulfide or another olefin to form a thioanhydride and a variety of other products. Trialkyl tetrathiophosphate is again the major product. The aromatic zinc dithiophosphates are believed to decompose by a free radical mechanism to phenol and a number of phosphorus and sulfur-containing products. These mechanisms are shown in Fig. 4.118. Besides the thermal decomposition mechanism described above, these additives can also decompose oxidatively to form products that are potent oxidation inhibitors in their own right. The details of their oxidation-inhibiting properties were discussed in the oxidation inhibitors section. It is important to note that the oxidation inhibiting action of these additives is independent of the nature of the alkyl group, but their antiwear/extreme pressure action is not. As indicated in Fig. 4.116, aliphatic zinc dialkyl dithiophosphates have better antiwear performance than the aromatic derivatives. And among aliphatics, the secondary alcohol-derived are better than those that are primary alcohol-derived.

Extreme Pressure Agents

Alkyl and aryl disulfides and polysulfides, dithiocarbamates, chlorinated and sulfochlorinated hydrocarbons, dialkyl hydrogen phosphites, and the salts of alkyl phosphoric acids are the common extreme pressure (EP) agents.

Polysulfides are synthesized from olefins either by reacting with sulfur or sulfur halides, followed by dehydrohalogenation. Sulfurization of olefins, fats and oils, or compounds with active hydrogens, with elemental sulfur, or sulfur and hydrogen sulfide, results in organic sulfides and polysulfides [421,422]. Three issues that must be addressed in relation to sulfurization using sulfur, especially of fats and oils, are color, odor and copper activity. Color is due to the presence of dithiolthione and conjugated thiophene structures, which are shown in Fig. 4.119. Odor is either due to the presence of the dissolved free sulfur; the mercaptans, which are the reac-

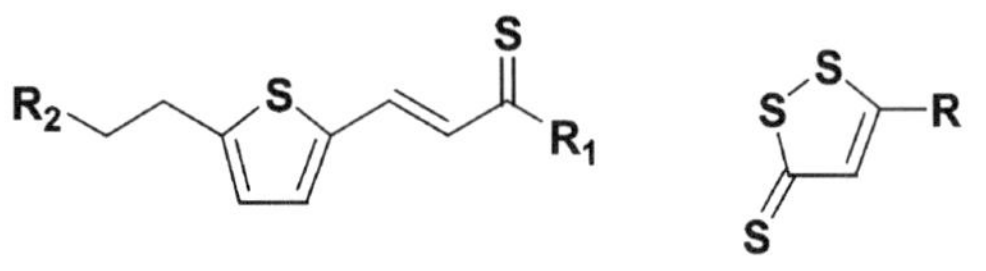

Fig. 4.119—Species responsible for dark color in sulfurized fats.

tion intermediates; or the hydrogen sulfide which may either be present in the dissolved form or may result from the disproportionation of sulfides and polysulfides. Copper activity is related to the presence of the active sulfur, which is the dissolved or easily releasable form of sulfur, for example, that present in polysulfides higher than trisulfides. Active sulfur is a concern because of its tendency to corrode yellow metals—a low temperature reaction. All three parameter depend upon the sulfurization temperature. Low temperature favors the formation of products that are of light color, low odor, and have low copper activity. The problem is the rate of reaction, which is slow, and takes a long time to achieve reasonable conversion. Hence, the challenge is either to find the optimal reaction temperature; use a sulfur transfer agent, such as metal salt of dithiocarbamic acid [423]; or post-react the product with a sulfur and hydrogen sulfide scavenger, such as an olefin or an acrylate ester [424]. Despite the active sulfur's corrosive effect on the yellow metals, its presence in applications such as metalworking fluids and greases is beneficial. Incidentally, the higher sulfur activity does show improved lubricant performance in film strength tests, such as Four-ball wear. Formulating active sulfur materials with some additives, such as zinc dialkyl dithiophosphates, can help overcome the copper corrosion problems, as is shown by the data in Table 4.14 [425]. Active sulfur in cutting oils is measured by the ASTM D1662 Test. Table 4.15 lists properties of the sulfurized products that are used for greases [425].

Dialkyldithiocarbamates are prepared either by neutralizing the dithiocarbamic acid, which results from the low-temperature reaction of a dialkylamine and carbon disulfide, with the metal bases, such as zinc oxide or antimony oxide; or by its addition to an epoxide, such as propylene oxide, or an activated olefin, such as alkyl acrylates [426].

The syntheses of these materials are depicted in Figs. 4.120–4.123. The synthesis of phosphosulfurized olefins was shown in Figs. 4.50 and 4.77. Their low molecular weight analogues have greater surface activity and hence are used as EP agents.

Alkyl and aryl phosphites are obtained by the reaction of an alcohol or a phenol with phosphorus trichloride or by a transesterification reaction [427]. Alcohols and phenols react with phosphorus pentoxide to yield a mixture of a mono-alkyl (mono-aryl) phosphoric acid and a dialkyl (diaryl) phosphoric acid [428]. These acids, when treated with bases, form salts. Alkyl phosphates and thiophosphates can also be prepared by the oxidation of phosphites. The preparation of alkyl phosphites is outlined in Fig. 4.124 and of alkyl and aryl phosphates and thiophosphates is outlined in Fig. 4.125.

Boundary lubrication requirements in equipment can be assessed by considering factors such as surface interaction, conjunction temperature, equipment load (pressure) and speed, and friction and wear requirements. Table 4.16 divides the boundary lubrication environments into four groups, based upon the combination of these criteria [6]. Temperature as a factor plays the most important role in determining the severity of the boundary lubrication environment since it represents the combined effect of all other factors, except wear, which itself is a consequence of temperature. The data in the table indicate that applications that involve greater surface interaction, higher pressures,

TABLE 4.14—Effect of zinc dialkyl dithiophosphate (ZnDTP) on copper performance of sulfurized materials [425].

Type	% Total Sulfur	% Active Sulfur	% Treatment	Copper Corr. 3 h at 100°C	% ZnDTP[a]	Copper Corr. 3 h at 100°C
Triglyceride	18	10.5	5	4c	1.5	3b
Ester	17	8.5	5	3b	1.5	Ib
Triglyceride	15	5	5	3a	1.5	Ib

[a]Thermally stabilized zinc di(2-ethylhexyl) dithiophosphate.

TABLE 4.15—Sulfurized Products for Greases [425].

Type	% Total Sulfur	% Active Sulfur	Properties
Triglyceride	8–12	0.5–3	Primarily inactive, limited EP performance
Triglyceride	13–15	4–7	Primarily active, good EP performance, hard to control copper corrosion
Hydrocarbon	45	10–15	High EP performance, distinct color, used only for encapsulated systems
Triglyceride/hydrocarbon	15	4	Primarily inactive, high EP performance
Ester	9–11	1–3	Primarily inactive, limited EP performance, excellent good low-temperature pumpability

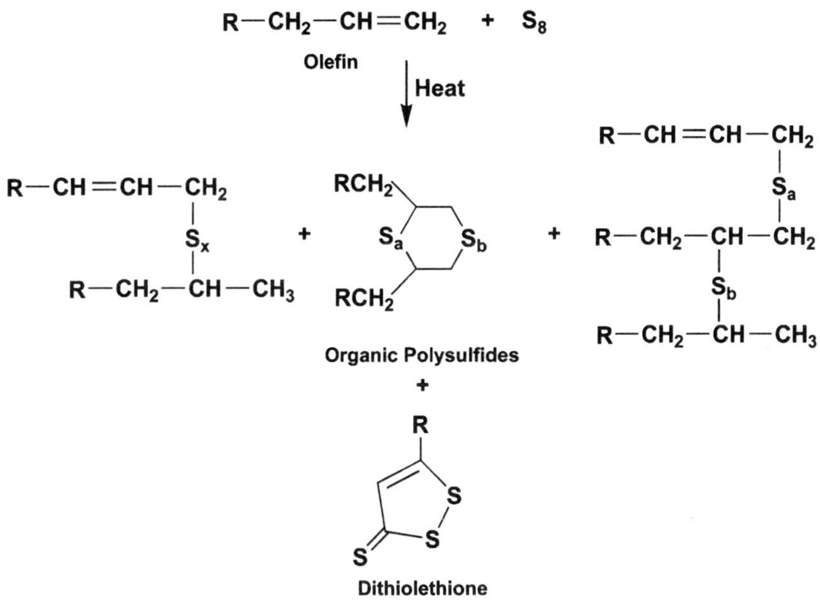

Fig. 4.120—Olefin sulfurization using elemental sulfur.

and higher temperatures will have extreme boundary lubrication situation. As shown in Fig. 4.126, there is a direct correlation between the conjunction temperature, the temperature of the two surfaces in contact, and the degree of the EP protection needed [6]. Generally, the higher the conjunction temperature, the more difficult it is to achieve the desired load-carrying capacity. Fig. 4.127 approximately correlates the antiwear numbers (AWNs) with the need for EP/antiwear performance and the ability of the hydrocarbon lubricant in achieving it. The antiwear number is inversely correlated with the wear coefficient (k), which is defined by the following relationship [6].

$$\text{Antiwear Number (AWN)} = -\log_{10} k = \log_{10}(1/k),$$

$$\text{where } k = \text{Wear Coefficient}$$

$$k = \frac{(\text{Wear Volume})(\text{Hardness})}{(\text{Normal Load})(\text{Sliding Distance})}$$

Table 4.17 provides friction coefficients, wear coefficients, and AWNs for various materials. As one can see, pure hydrocarbon materials, which are low in aromatics, have AWNs of around 7. Hence, they cannot provide a great deal of EP/antiwear performance and there is a need for friction modifiers and antiwear agents. Incidentally, if a medium has an AWN of 6 or below, extensive wear will result. Table 4.18 correlates various types of wear with wear coefficients and AWNs [429]. Comparison of the data in the two tables under-

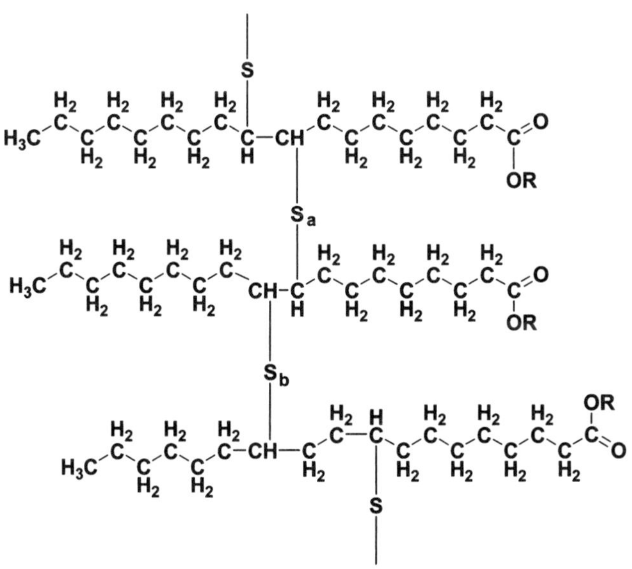

a, b = 2-5; Normally 2 or 3

Fig. 4.121—Generic structure of sulfurized fats.

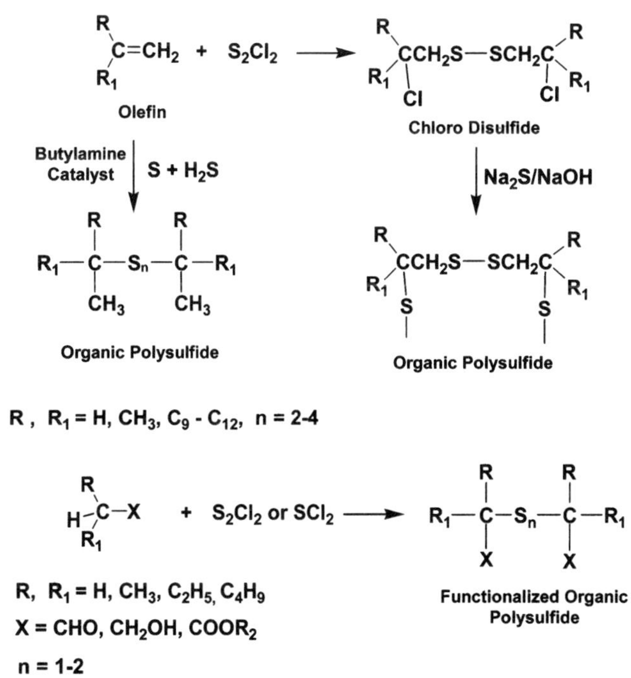

Fig. 4.122—Olefin sulfurization using sulfur chlorides, hydrogen sulfide, and elemental sulfur.

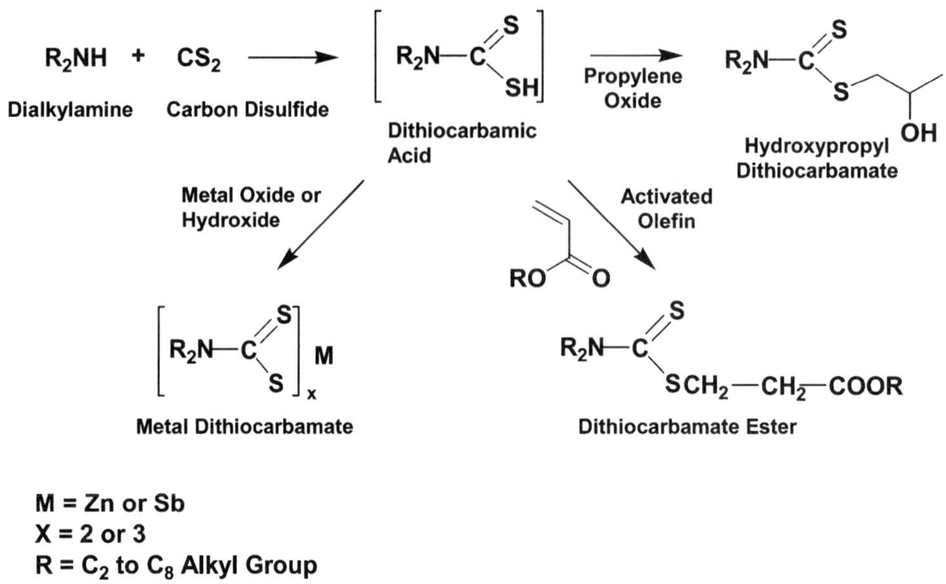

Fig. 4.123—Synthesis of dithiocarbamic acid derivatives.

scores the importance of lubrication on wear. Item 4 (C_{30} isoparaffin) is equivalent to 4 centistokes PAO fluid with respect to its molecular weight and has an AWN of 7. This implies that it can minimize wear in most parts of the mechanical equipment.

The equipment that operates at slow speeds and high loads generally requires more EP protection than the equipment that operates at high speeds and low loads. This is because the former generates higher temperatures as a consequence of the increased friction. The need for EP protection as a function of load is shown in Figs. 4.128 and 4.129 [6,430]. Figure 4.128 shows that nonlubricated equipment experiences seizure at low loads and that the equipment's load-carrying capacity progressively increases, when lubricated and film-forming additives, such as friction modifiers, anti-wear agents and EP agents are added to the lubricant [413]. Figure 4.129 shows the same in the FZG test. As the load increases, the wear increases as well and the EP agents are needed to control wear [6].

Disulfides and polysulfides decompose on metal sur-

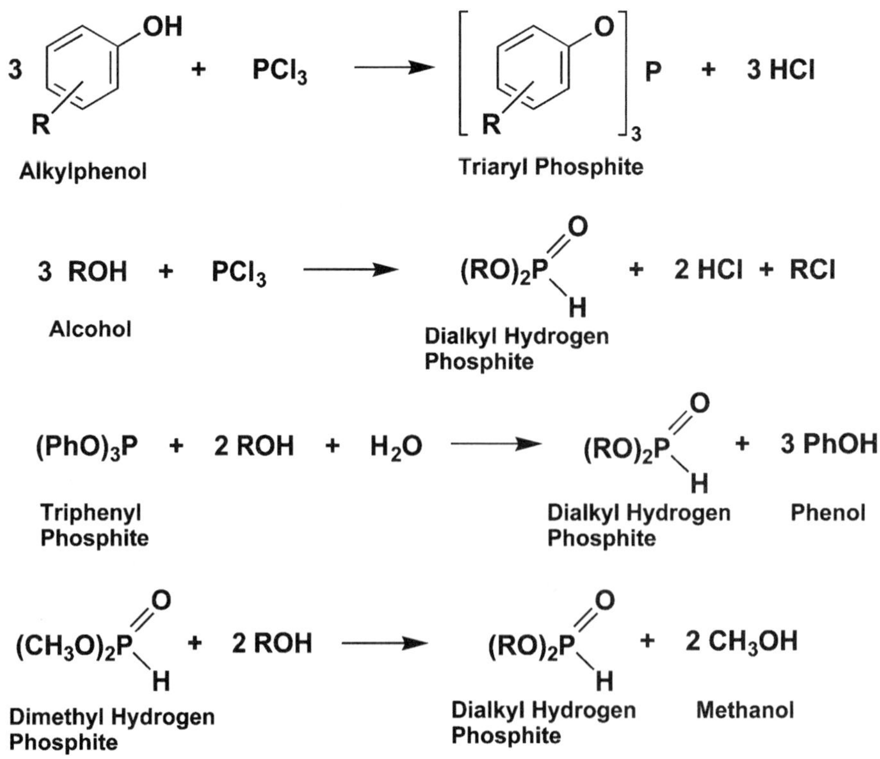

Fig. 4.124—Synthesis of alkyl and aryl phosphites.

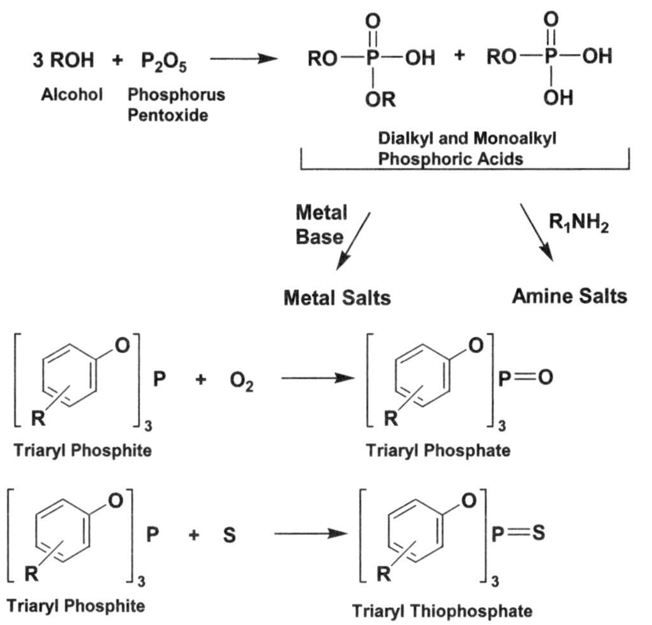

Fig. 4.125—Synthesis of alkyl and aryl phosphates and thiophosphates.

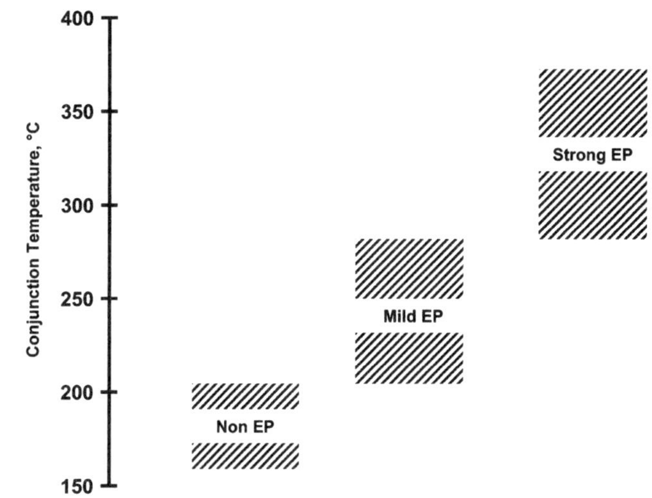

Fig. 4.126—Extreme pressure (EP) protection requirements versus the conjunction temperature [6].

TABLE 4.16—Dimensionless factors for estimating boundary lubrication requirements [6].

Factor	Definition
Interaction	$\frac{\text{Surface roughness }(\sigma)}{\text{Mean surface separation or dirt particle diameter}} = 1/\Lambda$
Thermal	
Temperature	$\frac{\text{Estimated maximum conjunction temperature (°K)}}{\text{Limiting temperature for base lubricant/material/atmosphere}}$
PV	$\frac{\text{Estimated PV for applications}}{\text{Limiting PV for base lubricant/material/atmosphere}}$
Wear	$\frac{\text{Wear Coefficient required for satisfactory wear life}}{\text{Typical wear coefficient for the base lubricant/material/atmosphere}}$
Friction	$\frac{\text{Required friction coefficient}}{\text{Measured friction coefficient}}$

	Boundary lubrication requirement			
Factor	***Probably None***	***Possible/Easy***	***Probable/Easy***	***Moderate/Difficult***
Interaction	0.2	0.5	0.7	2.0[a]
Thermal				
Temperature	0.87	0.95	1.4	1.7
PV	0.7	0.9	2.5	5.0
Wear	0.1	0.5	2.0	10.0
Friction	0.8	?	?	?

[a] Difficult only with high thermal factor

Λ = Film thickness parameter: ratio of fluid film thickness to composite roughness of wearing surfaces (or to "diameter" of particles in fluid film if they are larger than three times the composite surface roughness of the two surfaces. **P** = Apparent pressure on load support area = W/A. V = Sliding speed.

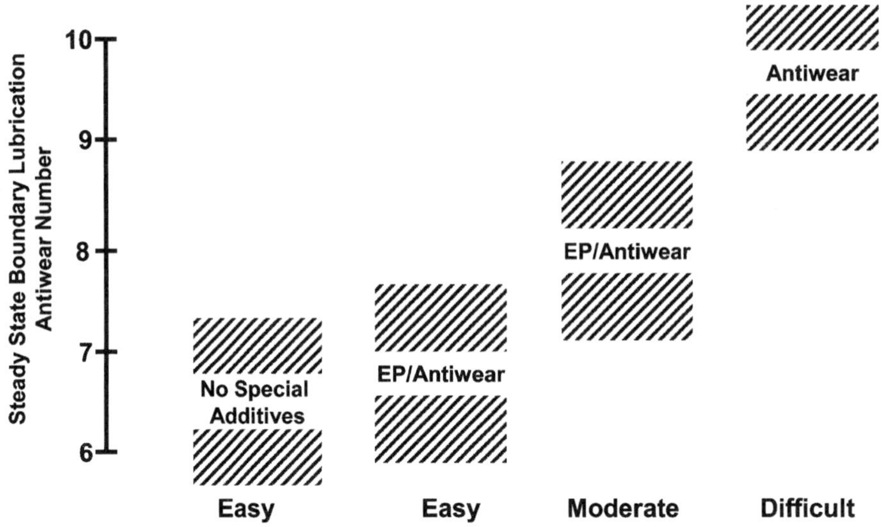

Fig. 4.127—Antiwear number versus wear control of steel-on-steel by hydrocarbon oils [6].

faces at temperatures above 200°C to form a protective sulfide layer. The thickness of this layer depends on the quantity and the lability of sulfur in the additive. Sulfurized fatty oils and sulfurized olefins are the most commonly used products in this class.

Chlorine-containing compounds provide protection under boundary lubrication conditions, via the formation of a metal chloride film. A detrimental aspect of the chlorine-based EP agents is the formation of the hydrogen chloride in the presence of moisture, which can cause severe corrosion problems. Chlorinated paraffins with 40–70 % chlorine by weight were once popular. However, environmental concerns about the negative effects of chlorine are minimizing the use of these additives, especially in Europe.

Phosphorus compounds react with the metal surface to make a metal phosphite or a metal phosphate protective film. Such films form at a much higher temperature than those formed by the sulfur EP agents. Tricresyl phosphate is the best known phosphorus EP agent. Dialkyl hydrogen phosphites [431] and phosphonic and phosphoric acid salts are the other examples of such EP agents. The effectiveness of these compounds as EP/antiwear agents is related to their affinity for the surface, that is, their polar to nonpolar ratio, and the chemical reactivity. Chemical reactivity is defined as the rate of change of radius (Δr/min) in the hot wire method. As shown in Figure 4.130, trilauryl phosphate (TLPA) and lauryl acid phosphate (LAP) show better EP/antiwear performance than trilauryl phosphite (TLPI) and dilauryl hydrogen phosphite (DLPH) [432]. Of these, TLPI has the lowest reactivity, hence shows the least wear control, and DLPH is

TABLE 4.17—Friction and wear data on various hydrocarbon materials.

Lubricant/Atmosphere	Approximate Friction Coefficient	Approximate Wear Coefficient	AWN
Dry argon	0.5	10^{-2}	2
Dry air	0.4	10^{-3}	3
Iso-paraffin (C_8)/air	0.3	10^{-5}	5
Iso-paraffin (C_{30})/air	0.12	10^{-7}	7
Aromatic oil/air	0.06	10^{-8}	8
C_{18} Fatty acid in C_{30} iso-paraffin/air	0.08	10^{-9}	9
Engine oil/air	...	$<2\times10^{-10}$	>9.7

Note: 52 100 Steel—Four-Ball Machine—230 to 690 mm/s—20 to 50 kg.

TABLE 4.18—Wear mechanism as a function of wear coefficient and AWN [429].

Wear Mechanism	Wear Coefficient Range	Antiwear Number (AWN)
Asperity Deformation and Removal	10^{-4}	4
Wear Caused by Plowing	10^{-4}	4
Delamination Wear	10^{-4}	4
Adhesive Wear	10^{-4}	4
Abrasive Wear	10^{-2} to 10^{-1}	1–2
Fretting Wear	10^{-6} to 10^{-4}	4–6
Wear by Solid Particle Impingement	...	

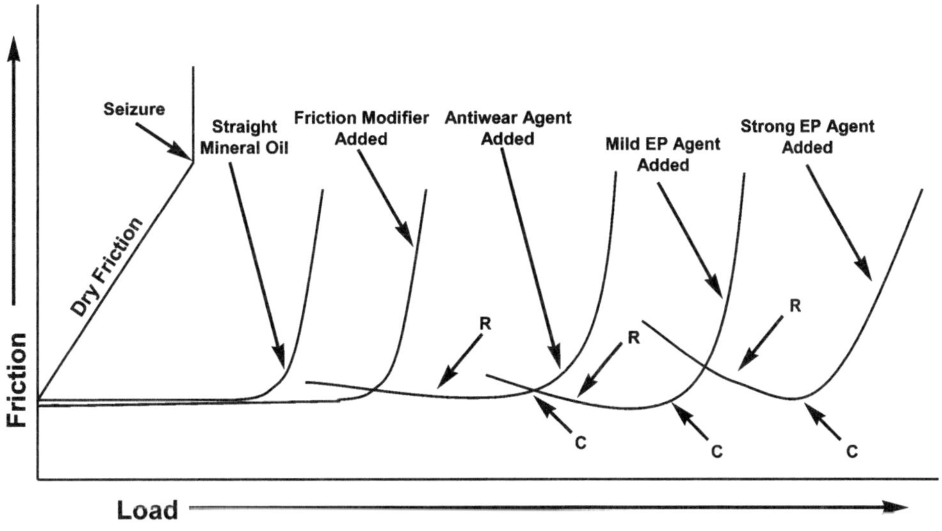

R: The point at which the EP film-forming action of the additive starts.

C: The point at which the reaction ceases and the lubricant's load-carrying ability diminishes

Fig. 4.128—Extreme pressure (EP) protection requirement versus load.

the most reactive and probably contributes towards the larger scar diameter due to corrosion. The smaller wear diameter by TLTTP, which has similar chemical reactivity, is likely to be due to sulfur. Sulfur-containing additives are known to have smaller wear scar diameter in the Timken test. The negative effect of the additive corrosivity on the EP performance is also demonstrated in Fig. 4.131 [433], where the chemically more reactive chlorine additive shows lower load-carrying capacity then the less reactive sulfur additive. And the phosphorus EP agent with medium reactivity falls in the middle. Load-carrying capacity is determined by the Mean Hertz Load (MHL) of the additive-containing lubricant relative to that of the additive-free oil. The chemical reactivity is defined the same as for the data depicted in Fig. 4.130. The data also suggest the importance of the nature and the durability of the resulting surface film, which will be described later. Incidentally, the optimum carbon chain length in organophosphorus additives differs depending upon whether the test measures friction or measures wear. Tests, which measure friction, require longer carbon chain length additives and those that measure wear require shorter carbon chain length additives [431]. The major challenge for a formulator and the tribologist is to maximize the equipment's service life by understanding the factors listed below [434].

1. Surface temperatures and pressures in the contact area.
2. Catalytic properties of the deformed metal and the newly exposed surface.
3. Chemical reaction between the metal and the additive, its kinetics, and the role of oxygen in the reaction.
4. Lubricant residence time in the contact zone.
5. Optimal surface film thickness for maximum antiwear response.
6. Properties of the surface film that control its rheology, shear strength, and adherence to the metal.

As stated earlier, the EP mechanism involves two steps: the adsorption of the EP agent onto the metal surface and the chemical reaction of the acidic materials that result from its thermal decomposition or hydrolysis due to moisture with the metal. In the case of sulfur and phosphorus com-

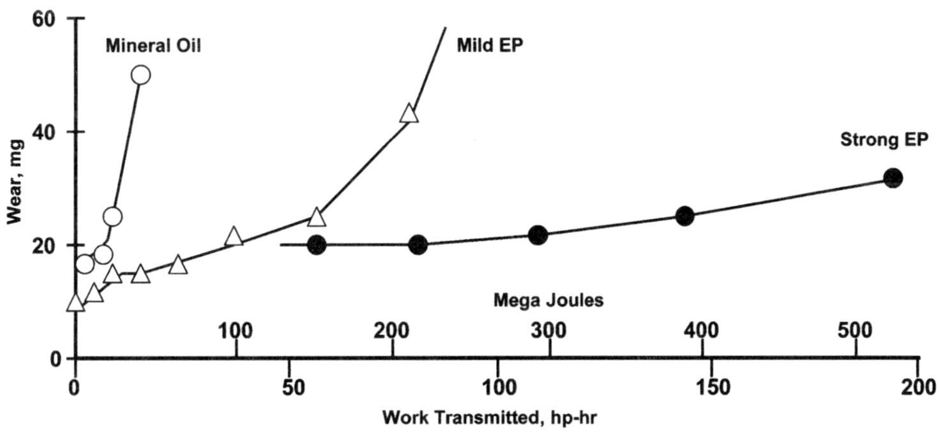

Fig. 4.129—Effect of load on wear in FGZ Gear Test A, 2175 pinion speed [430].

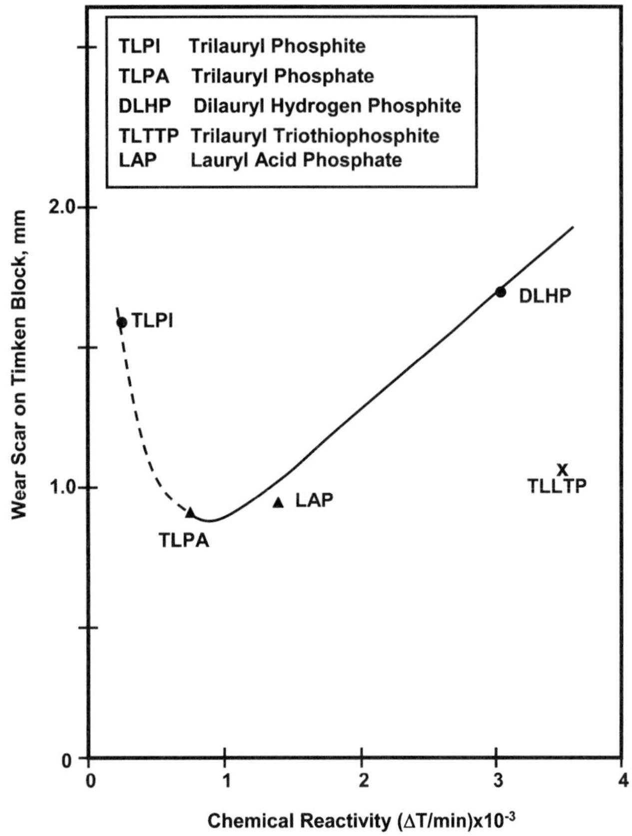

Fig. 4.130—Relationship between wear and chemical reactivity of the phosphorus compounds [432].

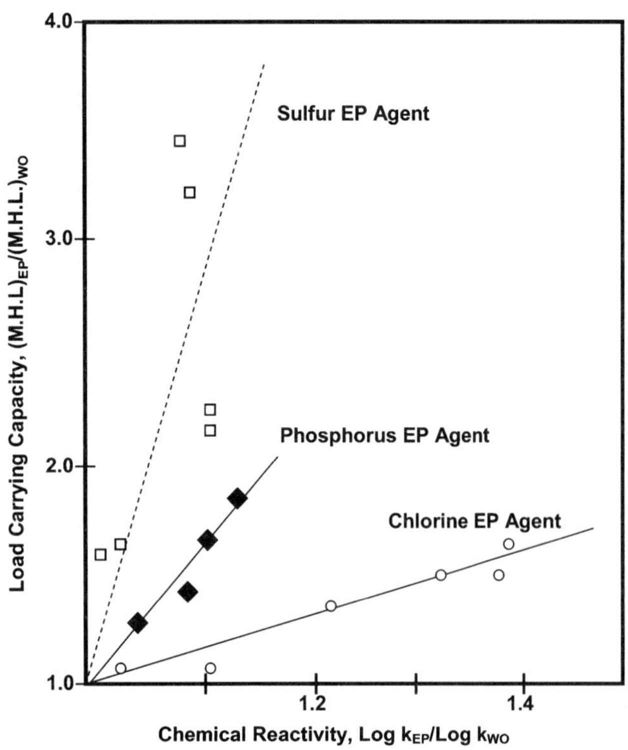

Fig. 4.131—Load-carrying capacity versus chemical reactivity [433].

pounds, the reactive materials comprise mercaptans or phosphorus compounds [435]. The probable mechanisms by which zinc dialkyl dithiophosphates, alkyl phosphites, and amine phosphates form EP films are presented in Figs. 4.132 and 4.133, respectively. Please note that the affinity of the zinc dialkyl dithiophosphate (ZnDTP) and, presumably, that of the other phosphorus and sulfur EP agents towards the metal surface, is a function of the nature of the alkyl group and the metallurgy and hardness of the surface. The degree of adsorption of the ZnDTPs with secondary alkyl groups is higher than those with primary alkyl groups and their rate of adsorption on harder metal surfaces is lower than on the softer metal surfaces [445b]. In the case of the organic halides, often not used in modern formulations because of the environmental concern pertaining to halogens, the protective film is iron halide that results from a similar mechanism [436]. The mechanism of the EP protection by disulfides and polysulfides is presented in Fig. 4.134 [437]. Organic polysulfides are converted into dialkyl disulfide, which reacts with the metal to form the metal sulfide EP film [437,438]. Molybdenum dialkyldithiocarbamates and dialkyl dithiophosphates EP/antiwear additives are believed to thermally decompose on the rubbing surfaces to form MoS_2 protective film. The frictional properties of this and other solids that are used as solid lubricants are discussed in the subsequent section.

Possible pathways by which the organic disulfides form the iron sulfide EP film are shown in Fig. 4.135 [438]. The disulfide thermally decomposes to form the alkylthio free radicals (Path *a*) or mono-sulfide and sulfur (Path *b*). The alkylthio free radical reacts either directly with iron (Path *e*), or by way of the mercaptan (Paths *k* and *j*), to form iron mercaptide, which can lose its organic portion to form iron sulfide (Path *h* and Fig. 4.134). The free sulfur from Path *b* can also react with iron to form iron sulfide (Path *i*). Alternatively, the disulfide extracts an electron from the iron surface to form a radical anion (Path *c*), which can decompose to mercaptide anion and mercaptide free radical (Path *d*). The free radical can revert to disulfide (Path *g*), form iron mercaptide directly (Path *f*), or by way of mercaptan (Paths *l* and *j*). The mercaptide anion can also react with Fe^{++} (ferrous) ions to form iron mercaptide (Path *f*).

While in Figs. 4.132 and 4.133 we dealt with ferrous surfaces and the mechanism of the FeS film formation, it is important to note that the friction-reducing ability of the metal sulfides depends upon the nature of the metals and their hardness. For example, contact between harder metal surfaces, such as those of iron, is less extensive than that between softer metal surfaces, such as copper. This is indicated by the lower coefficient friction of 0.78 for steel than that of 1.21 for copper. After reaction with sulfur-derived EP agents, the value of the coefficient of friction for the iron sulfide (FeS) coated surfaces drops to 0.39 and that for the copper sulfide (CuS) coated surfaces drops to 0.74 [Farng, L. O., "Ashless Antiwear and Extreme-Pressure Additives," *Lubricant Additives: Chemistry and Applications*, Leslie R. Rudnick Ed., Marcel Dekker, Inc., New York, Chapter 5, 2003, pp. 223–257.] These inorganic films, which are only a few molecules thick, have low shear strength and are removed during the movement of the surfaces in contact. This situation is represented in Fig. 4.136 [6]. Removal of the EP

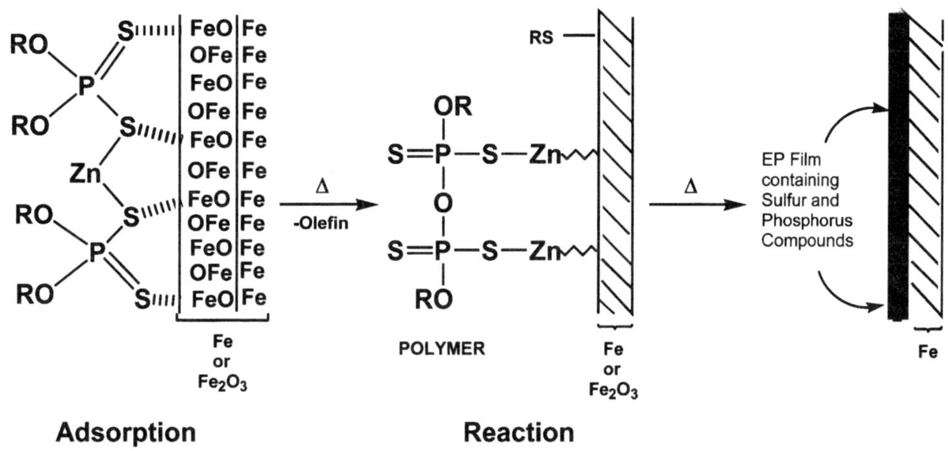

Fig. 4.132—The mechanism of boundary film formation by zinc dialkyl dithiophosphates.

film can expose fresh metal, and the film-forming process is repeated. Each time the film is removed, a molecular layer of metal is removed with it. One way of looking at the process of EP protection is the controlled wear of the rough surfaces, as shown in Fig. 4.137 [6].

In general, the formulators use different types of EP agents in combination because of the need for performance across more than one lubrication regime and to benefit from possible synergism [436,439]. Figure 4.138 shows the benefit from the combined effect of a fatty acid and an organic phosphate EP agent [440,441]. Organic phosphate is less effective in lowering the wear rate than the fatty acid, which also becomes effective somewhat earlier. A combination of the two additives provides superior wear control than either of the

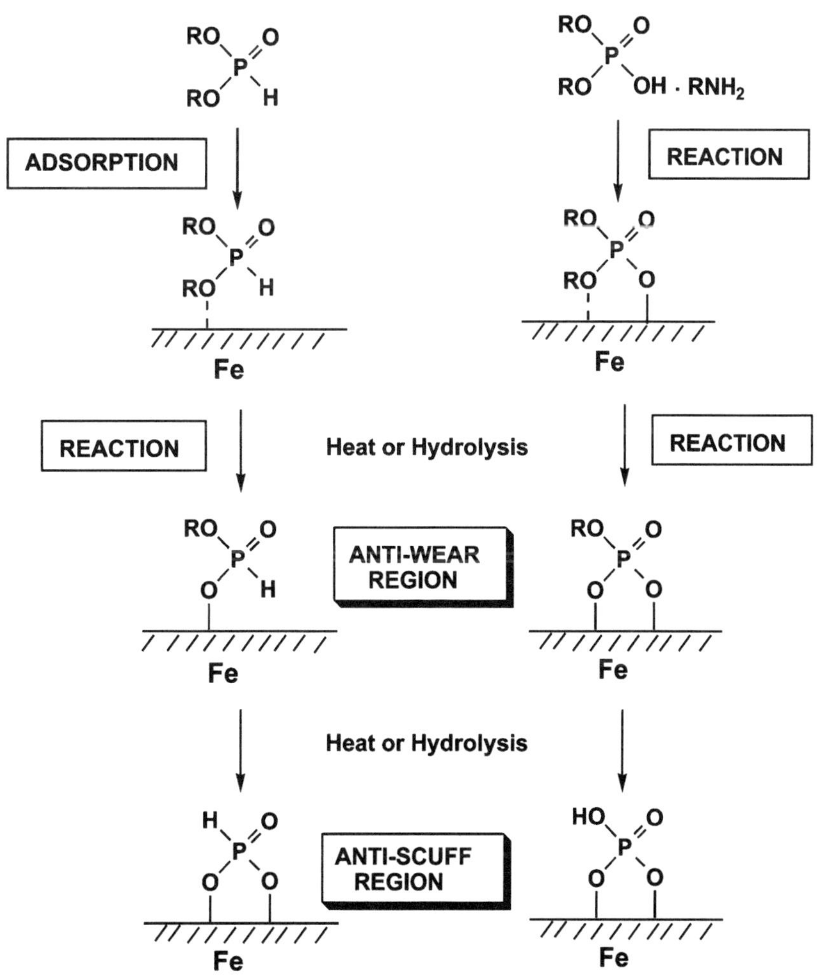

Fig. 4.133—Possible mechanism of load-carrying action of dialkyl hydrogen phosphites and amine phosphates [431, adapted].

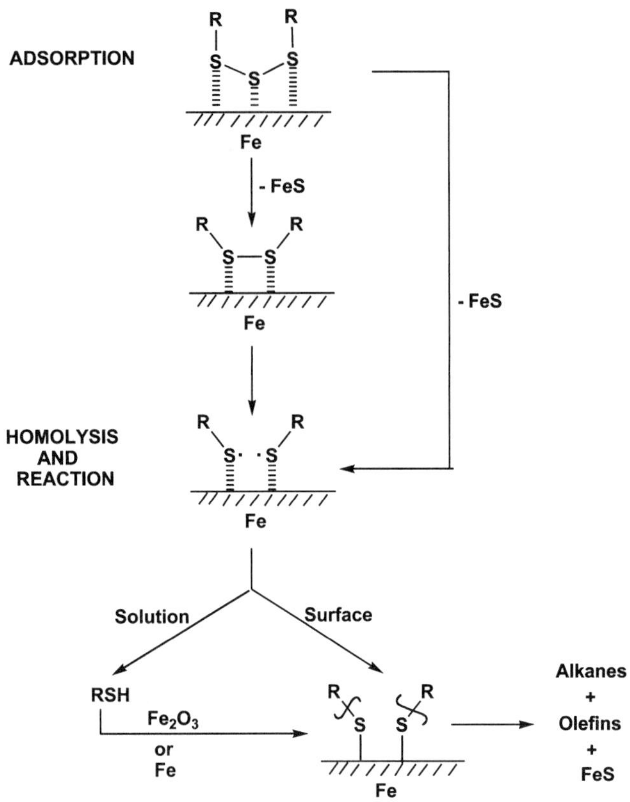

Fig. 4.134—Mechanism of iron sulfide EP film formation by organic polysulfides [437].

additives alone. The synergism between sulfur and chlorine-containing EP agents is shown in Fig. 4.139, where the average scar diameter is plotted as a function of the applied load [436]. When only the disulfide is used, weld occurs at a load of 250 kg, and the scar diameter is about 2.15 mm (Graph A). When a similar level of alkyl chloride is used, the weld load stays the same but the scar diameter improves to 1.74 mm (Graph B). Combining the two types of EP agents to deliver an amount equal to that in the previous cases increases the weld load to 350 kg and decreases the scar diameter to 1.6 mm (Graph C), thereby indicating a synergism between the two chemistries. A further increase in the amount of disulfide and chloride shows weld resistance beyond the load of 500 kg (Graph D). Similar synergism exists between phosphorus and sulfur chemistries [439]. Figure 4.140 provides the combined effect of chlorine, sulfur and phosphorus additives on the friction coefficient [440,443]. Friction modifier is the first additive to come into play, but it progressively loses its ability to control friction as the temperature approaches 400°C. At this point it has no friction control whatsoever. Around 150°C, chlorine-containing additive becomes active and keeps the friction low until around 550°C, when it abruptly loses its effectiveness. Phosphorus-containing additive comes into play around 250°C and maintains its effectiveness until around 900°C. Sulfur-containing additive has activity over 550–1000°C range. Incidentally, the melting/softening points of iron chloride, iron phosphide, and iron sulfide are 670, 1100, and 1193°C, respectively. As the temperature in the contact zone approaches these temperatures, the removal of these surface films by the rubbing asperities is facilitated. Based upon the information in the figure, combining all four types of additives is the ideal way to control friction over the broadest temperature range. The formation of the active new compounds is believed to be partly responsible for the synergism. Figure 4.141 shows the formation of trialkyl thiophosphate synergistic species from the reaction of a phosphite with an organic polysulfide [439].

Some additives in a formulation can diminish the effectiveness of EP/AW agents. These include surface-active additives, such as certain friction modifiers, oxidation inhibitors, rust inhibitors, metal deactivators, detergents, and dispers-

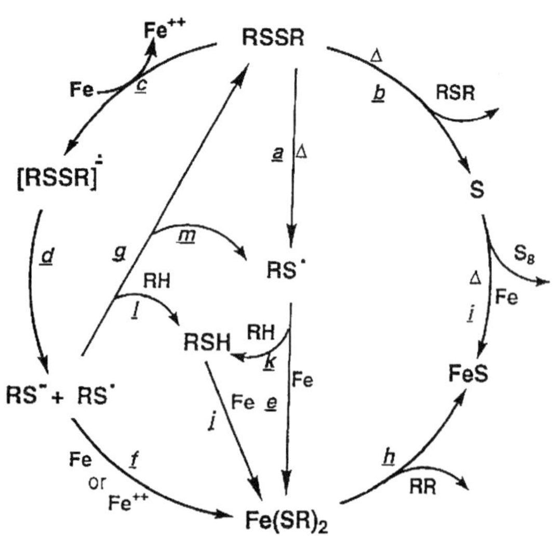

Fig. 4.135—Modes of decomposition of polysulfides to form an extreme pressure (EP) film [438].

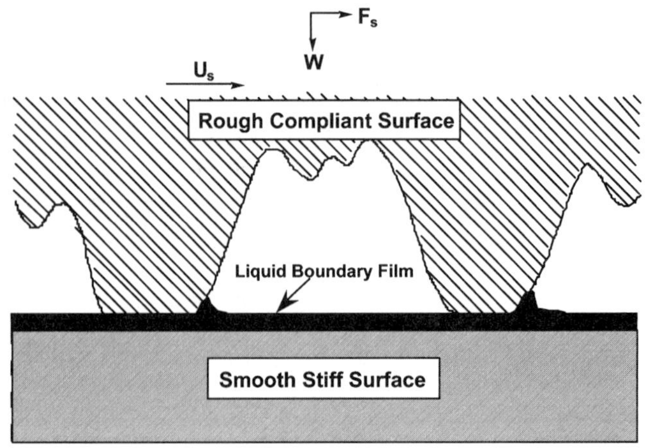

Fig. 4.136—Protective boundary film versus shear [6].

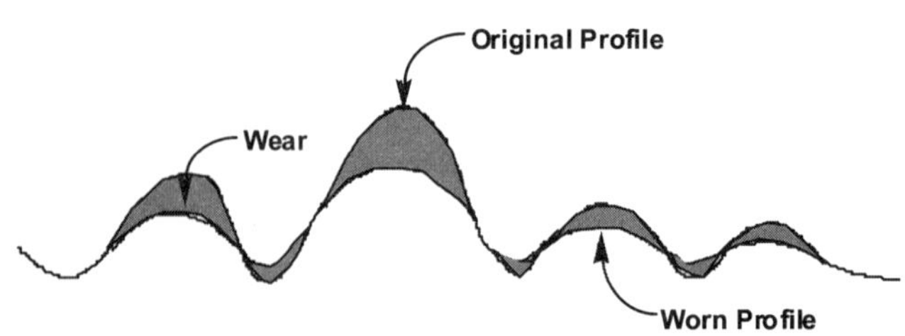

Fig. 4.137—Controlled wear of asperities to produce submicron debris [6].

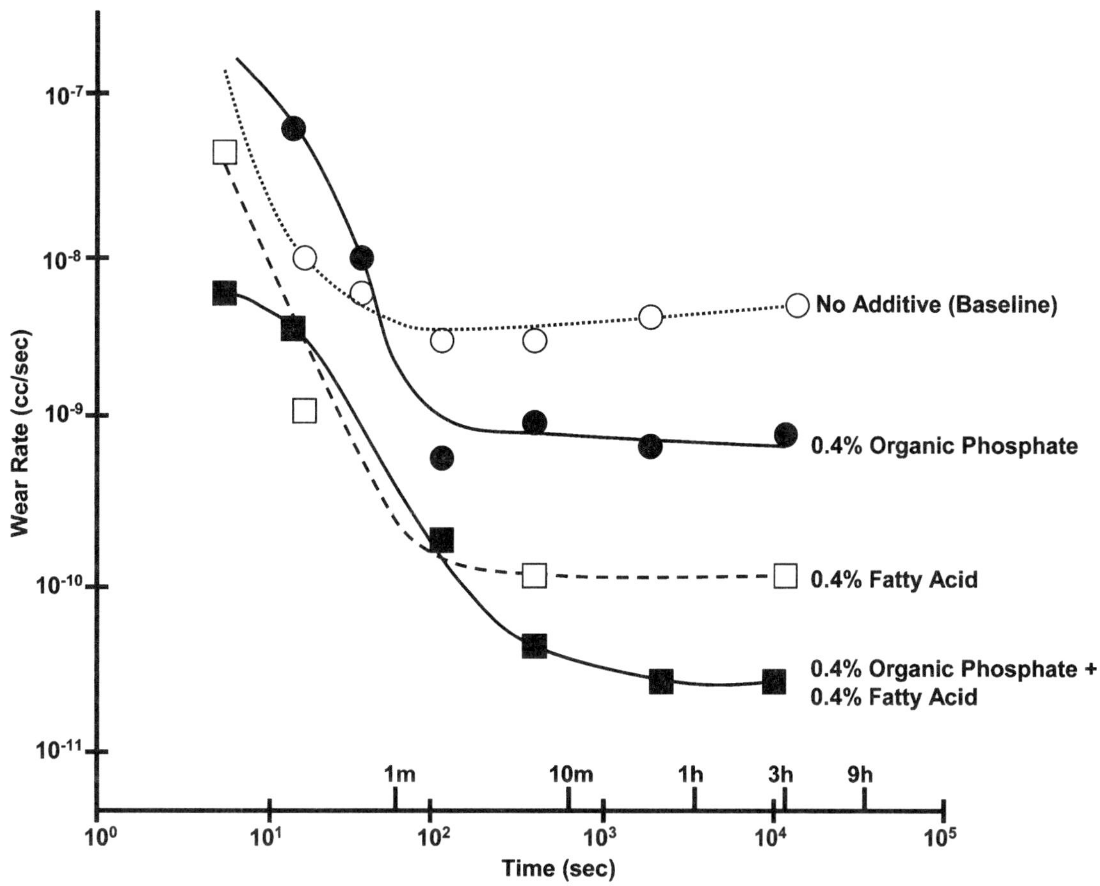

Fig. 4.138—Synergism between friction modifier and organic phosphate [441].

ants. These components interfere with the EP mechanism either by irreversibly adsorbing on the surface or by forming complexes with the EP agents, thereby rendering them inactive [444,445]. This antagonism between a dispersant, a detergent, and a zinc dialkyl dithiophosphate (ZnDTP) is depicted in Fig. 4.142. Please note that while ZnDTP in the formulation alone is very effective in controlling the viscosity of the oil, the presence of a dispersant or a sulfonate detergent has a negative effect on its performance; magnesium sulfonate causing more harm than calcium sulfonate. The same is true of some highly polar base stocks that can overwhelm the surfaces because of their shear amount. This for ester lubricants was described in the Synthetic Base Stocks chapter, Chapter 3. This type of antagonism is quite common for some lubricants, such as gear oils, where the EP agents form the core of the formulation. Many effective extreme pressure and antiwear additives are corrosive to metals. Therefore, the lubricants using them are typically formulated to optimize a balance between corrosivity and the extreme pressure and antiwear protection.

Table 4.19 compares the effectiveness of the various base fluids and various classes of film-forming agents in controlling friction and wear [440,442]. Of the compounds listed, natural oils and fats, long-chain fatty acids, amines, and alcohols, and organo-molybdenum compounds are good for friction modification. Zinc dialkyl dithiophosphates and phosphorus compounds are good as antiwear agents and organosulfur and organophosphorus compounds are good as EP agents. Some of the tribological terms that are often used in discussing friction and wear are listed in Table 4.20, along with the approximate related dimensions [6,446].

EP-additive treat level in a lubricant can range from tenth of a percent to several percent. The precise amount depends upon application, base oil quality, base oil type, other additives, equipment configuration, and the desired performance. The presence of the other additives can lead to synergism, as shown in Fig. 4.141, or to antagonism as shown in Fig. 4.142, which will reduce the effectiveness of these additives. Synergism will reduce the treat level of these additives and antagonism will increase it, if similar performance is desired. Applications that operate at high ambient temperatures, such as automotive gear systems, deplete these additives faster than applications that operate under low ambient temperatures, such as automobile engines and automatic transmissions. While discussing the synthetic base fluids, we commented on the polarity of certain base fluids. Highly polar base fluids, such as synthetic esters, compete with the polar EP/antiwear agents for the surface. Because of being present in a much larger amount, synthetic esters overwhelm the surface, thereby preventing the absorption of EP/antiwear agents on the surface. Other film-forming additives, such as rust and corrosion inhibitors, also cause a similar problem. Typically, this problem is overcome by using these additives in higher amounts. Table 4.21 provides a comparison between various classes of film-forming agents,

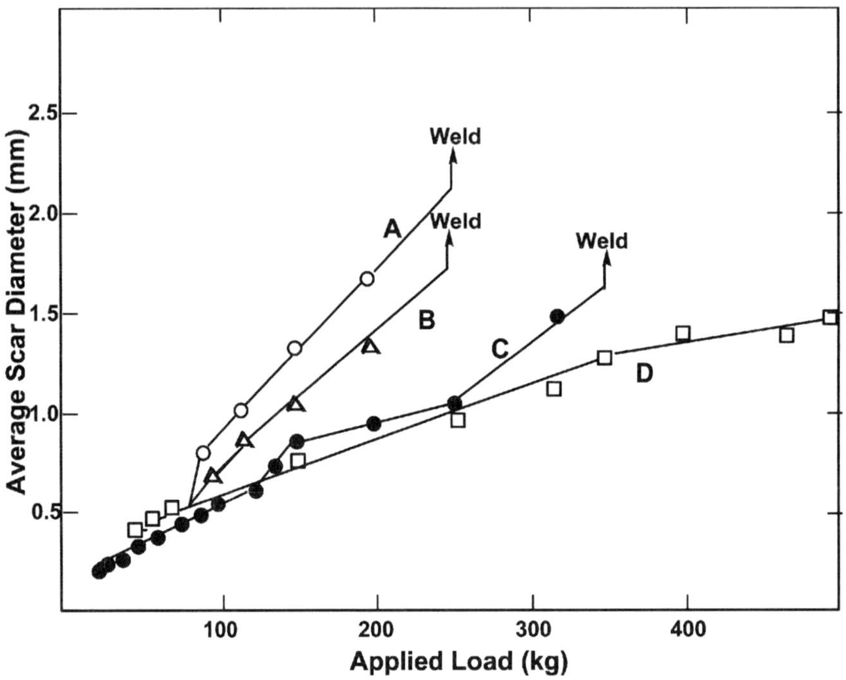

Fig. 4.139—Wear—load diagram, the ASTM procedure, showing synergism between chlorine and sulfur extreme pressure (EP) agents [436].

along with their function and the performance mechanism [440]. Table 4.22 lists typical applications of organophosphorus esters [440].

There is always a concern regarding the addition of the hetero-atom (N, S, and P) containing additives to biodegradable base oils, with respect to whether it will diminish their biodegradability characteristics. It appears from the limited data presented in Table 4.23 that some of these additives are highly biodegradable themselves; hence they are unlikely to negatively impact the biodegradability properties of the base oil [447].

Solid Lubricants

The term solid lubricant refers to finely divided solid (powder) or a thin film that protects surfaces against wear damage that are in contact and in motion. These lubricants are used in situations where fluid lubricants are either undesirable or ineffective. Examples include the following applications.

1. There is a concern for contamination of the product or the environment.
2. Maintenance is either unlikely or impossible.
3. Extreme low and high ambient temperatures are experienced.
4. Fluid lubricants are unsuitable because of the unique environment, such as vacuum/space, and that involving ionizing radiation or a high ratc of oxidation, for example, foundry operations.

A variety of materials are used as solid lubricants. For a material to function as a solid lubricant, it must possess certain specific properties. These include the ability to form surface films of low friction coefficient and low shear strength, and to control wear [448]. Solid lubricants fall under the gen-

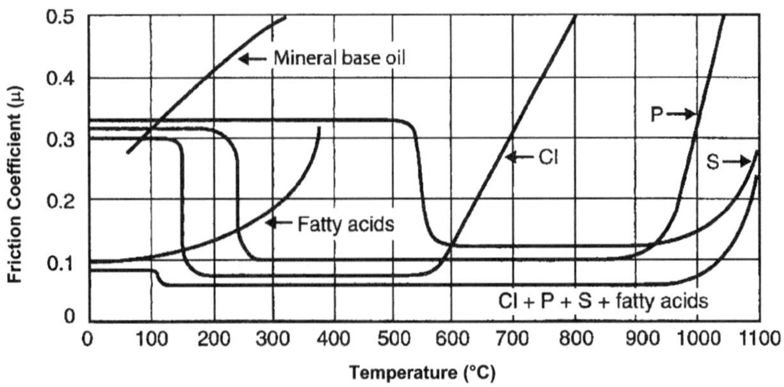

Fig. 4.140—Synergy between a friction modifier and a mixture of EP agents [443].

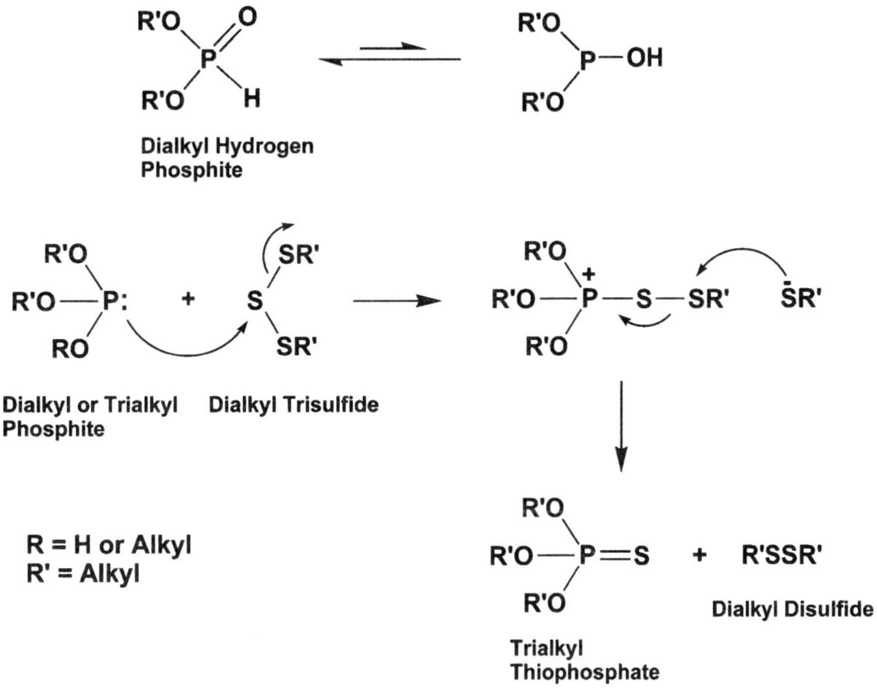

Fig. 4.141—Alkyl phosphite-polysulfide reaction leading to synergism [438].

eral classes of the lamellar solids, metal salts, reaction films, soft metal films, diffusion coatings, polymeric coatings, organic polymers, and dispersions in oils and greases.

Lamellar solids are materials that have a layered structure, where the bond strength between the atoms within the layer or the plane (intralamellar) is greater than between the atoms across the layers (interlamellar)—possibly due to van der Waals type forces [449]. This provides these materials the ability to form films of low shear stability; one of the properties desired in a good solid lubricant. Graphite and molybdenum disulfide (MoS_2) are two materials that are most commonly used as solid lubricants. The lamellar structure of the graphite is shown in Fig. 4.143. Graphite has low friction coefficient of 0.05 to 0.1, which is ascribed to its lamellar structure and easy shear. During sliding, the basal planes orient themselves almost parallel to the surface and the movement separates layers via shear, thereby lowering the friction. It is important to note that the graphite has low friction only in environments that contain water or gases, such as those present in the air. In vacuum, its friction coefficient is much higher, which is ascribed to the presence of the additional stronger electronic interaction between atoms across layers than the weak van der Waals forces alone. Graphite has high thermal stability, greater than 2000°C, but its oxidation stability is only around 500 to 600°C. While this is quite respectable for most applications, the primary limitation is the necessity of the adsorbed vapors and gases to maintain low friction. Gases desorb easier than water, which will be removed around its boiling point of 100°C. Organic vapors, present as contaminants in the surrounding environment, or the material introduced deliberately into the graphite can preserve its frictional properties at a higher temperature. Inorganic compounds, such as lead oxide (PbO), cadmium oxide (CdO), sodium sulfate (Na_2SO_4), and cadmium sulfate ($CdSO_4$) added to graphite have all been shown to be effective on nickel alloy substrates to around 550°C [448,449].

MoS_2, like graphite, also has a lamellar structure but with the difference that the interlamellar bonding is between sulfur atoms instead of being between carbon atoms. The bonding is again of the van der Waals type and hence is fairly

TABLE 4.19—Lubrication characteristics of various chemical types [440,442].

Chemical	Friction Reduction	Antiwear Performance	EP Performance
Natural oils and fats	1	4	5
Long-chain fatty acids, amines, and alcohols	1	4	5
Organo-molybdenum compounds	1	2	4
Synthetic esters	2	3	4
Organo-sulfur compounds	2	2	3
Zinc dialkyl dithiophosphates	3	1	3
Phosphorus compounds	3	1	3
Sulfur compounds	4	3	1
Chlorine compounds	5	4	1

Note: The lower the number, the better is the rating.

TABLE 4.20—Tribology terms and related dimensions [6].

Parameter	Approximate Size Range (μm)
Monomolecular layer	$0.2–2\times10^{-3}$
Sliding wear debris	0.002–0.1
Boundary film	0.002–3
EHD film	0.01–5
Asperity height	0.01–5
Rolling wear debris	0.7–10
Asperity contact	0.7–10
Hydrodynamic film	2–100
Asperity tip radius	10–1000
Concentrated contact width	30–500
Engineered counterfomal radius	$1–100\times10^{-3}$
Engineered conformal radius	$2–2500\times10^{-6}$

weak, see, Fig. 4.144. Unlike graphite, where adsorbed vapors decrease friction, in MoS_2 the adsorbed vapors usually increase friction, but the effects are relatively small. In the oxidation-free environment, thermal stability of MoS_2 is around 1100°C, but in the presence of air the stability drops to around 350 to 400°C. The air-oxidation product is MoO_3. One problem with MoS_2 is the presence of the abrasive impurities, which can contribute to abrasive wear. Despite this, it is replacing graphite in metalworking lubrication and in electrical contacts. This is because of its consistent quality, nondependence of low friction on the adsorbed vapors, low friction in vacuum, and its superior load-carrying capacity.

Boron nitride is a ceramic lubricant which is used in applications where the use of graphite and MoS_2 is less than desired. It is extremely resistant to oxidation even at temperatures of 1200°C and has high thermal conductivity.

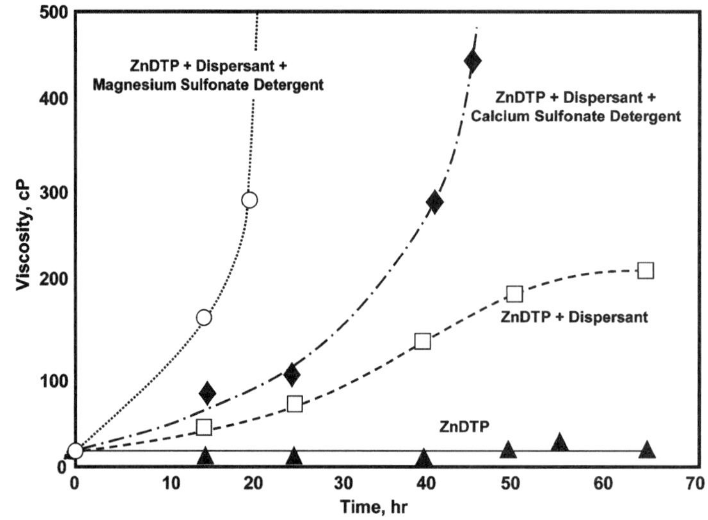

Base Fluid = 150 SN Mineral Oil (S = 0.3%), ZnDTP (0.1% P), Dispersant (4.5%), Metal Sulfonate (1%)
Test Conditions = Temperature - 165°C, Airflow Rate – 1.7 L per min, Catalyst – Ferric Acetyl Acetonate (40 ppm Fe)

Fig. 4.142—Antagonistic interactions between additives [198].

TABLE 4.21—Chemical additives used to improve lubrication performance [440, adapted].

Additive Type	Function	Typical Additives	Performance Mechanism
Friction Modifiers/Oiliness Additives	Reduce friction under near-boundary lubrication conditions	Long-chain fatty acids and esters, sulfurized fatty acids, molybdenum compounds, long-chain phosphites and phosphonates	Physical adsorption on metal surfaces via their polar end and associating with oil via their hydrocarbon chain. The result is the formation of a durable lubricant film between the surfaces
Antiwear Additives (Usually have mild EP properties)	Reduce wear at low to medium loads	Neutral organic phosphites and phosphates, zinc dialkyl dithiophosphates, metal dialkyldithiocarbamates, and low-sulfur sulfurized fats	Chemical reaction at moderate temperatures to form monomolecular layer that reduces friction at low to medium loads
Extreme Pressure Additives (Also known as load-strength additives, load-carrying additives, anti-seize additives, and anti-scuffing additives)	Protect surfaces against wear damage under high loads and expand load-carrying capacity of the equipment	High sulfur- and chlorine-containing hydrocarbons and esters; acidic organo-phosphorus compounds; and molybdenum, antimony, and lead salts of organic acids	Chemical reaction at high temperatures to form sacrificial monomolecular inorganic boundary films, such as iron chloride, phosphide, or sulfide that protect bulk metal against wear damage.

TABLE 4.22—Typical applications of organophosphorus esters [440].

Application	Triaryl Phosphates	Trialkyl Phosphates	Amine Phosphates	Acid Phosphates	Alkyl/aryl Phosphites
Automotive					
ATFs				√	√
Gear Oils			√		
Power Steering Fluids	√				
Shock Absorber Fluids	√				
Lubricants for Electric Motors	√				
Industrial					
Hydraulic Fluids	√		√		
Gear Oils			√	√	√
Turbine Oils	√	√			
Compressor Oils	√				
Gas Oils	√		√		
Universal Tractor Fluids	√				
Metalworking Fluids	√	√	√	√	√
Greases	√		√	√	
Way Oils	√				
Circulating Oils	√				
Vegetable Oils		√			
Aircraft					
Piston Engine Oils	√				
Turbine Engine Oils	√				
Greases	√		√	√	

These make its use suitable in applications that require quick dissipation of heat and high temperatures that are outside the realm of graphite and molybdenum disulfide. Boron nitride is produced by the reaction of the boric acid with urea at temperatures of 800–2000°C. The result is the formation of cubic boron nitride and hexagonal boron nitride. It is the latter form that resembles the structure of graphite and is used in lubricant applications. See Fig. 4.144 for its lamellar structure [449] and compare it with that of graphite shown in Fig. 4.143. The frictional properties of the commonly used solid lubricants in air and vacuum are provided in Table 4.24 [450]. It is important to note that in all cases, except for the molybdenum disulfide and tungsten disulfide, the friction coefficient in vacuum is higher than in air. In molybdenum disulfide and tungsten disulfide, it is lower, which makes them good lubricants for vacuum applications. Both also tolerate higher loads better than the graphite. There are many other lamellar solids, but their use in lubricant applications is limited.

Numerous inorganic salts with the low shear strength and film-forming ability have shown promise as solid lubricants. Those worth mentioning are lead oxide (PbO) and calcium fluoride (CaF_2). PbO is an effective thin film lubricant with an operating range from room temperature to about 350°C, and upwards of 500°C. Between the two temperatures, it oxidizes in air to Pb_3O_4, which has poor lubricating

TABLE 4.23—Biodegradability of film-forming agents [447].

Additive Type	Basis	CEC-L-33-T-82, %
Extreme Pressure Additive	Natural Oil and Fat	>80
(Sulfur Carriers)	Methyl Esters	>80
	Combination of Natural Oil and Hydrocarbons	>80
Antiwear Additive (Mild)	TMP Ester	>90
Corrosion Inhibitors	Succinic Acid Partial Ester	>80
	Calcium Sulfonate	>60

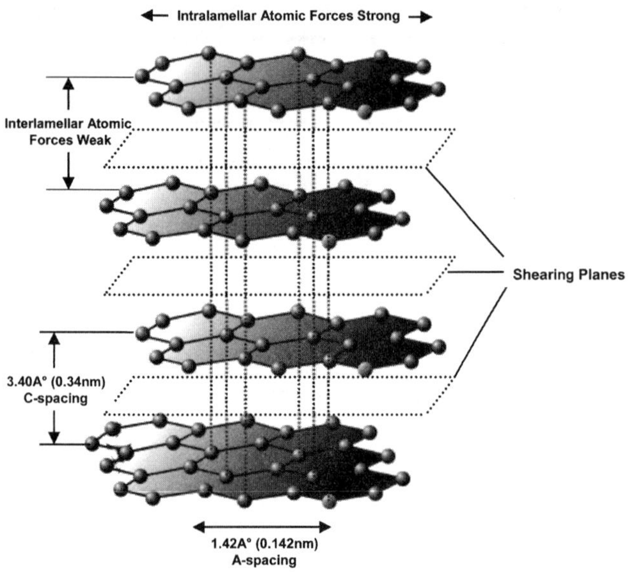

Fig. 4.143—Lamellar structure of graphite.

properties. The addition of SiO_2 results in the formation of a silicate phase containing excess PbO, which is less susceptible to oxidation and can be used in applications involving temperatures between 250 and 700°C. Unfortunately, at temperatures below 250°C, its frictional properties are not very good—the friction is too high and the film strength is too low. CaF_2 and eutectic mixtures of CaF_2/BaF_2 are also effective lubricants, over the 250 to 1000 °C temperature range. Below 250°C, their friction coefficient is around 0.3, which can be improved by the addition of silver oxide (Ag_2O).

TABLE 4.24—Friction data for commonly used solid lubricants [450].

	Coefficient of Friction (μ)	
Solid	*Air*	*Vacuum (Pressure)*
Natural Graphite (Compact)	0.19	0.44(6×10^{-9} torr)
Pyrolytic Graphite	0.18	0.50(2×10^{-9} torr)
Hot Pressed Boron Nitride	0.25	0.70(2×10^{-9} torr)
Molybdenum Disulfide	0.18	0.07(2×10^{-9} torr)
Tungsten Disulfide	0.17	0.13(3×10^{-9} torr)

The ability of oxide and other reaction films on metals to prevent intermetallic contact and reduce wear, and sometimes friction, is well known. Coefficients of friction of the oxide films are fairly high (0.4 to 0.8), but at high frictional temperatures, they result in mixed surface oxides which reduce wear by making low melting eutectic mixtures that form low-shear sacrificial films. Inclusion of the easily oxidizable elements, such as silicon (Si) and iron (Fe), into Ni alloys promotes the formation of such lubricating films [448].

Deposition of the low shear soft metals on hard substrates is another way to form effective lubricating films. This is achieved by electroplating or vacuum deposition. The metals that are commonly used for this purpose include soft metals, such as indium (In), lead (Pb), tin (Sn), silver (Ag), gold (Au), copper (Cu), zinc (Zn), thallium (Tl), barium (Ba), and bismuth (Bi). These metals have low solubility as a solid in Fe, which promotes the formation of thin low-friction films. This type of lubrication is desired in applications involving very high temperatures and or limited sliding, such as rolling element bearings. Optimum film thickness for

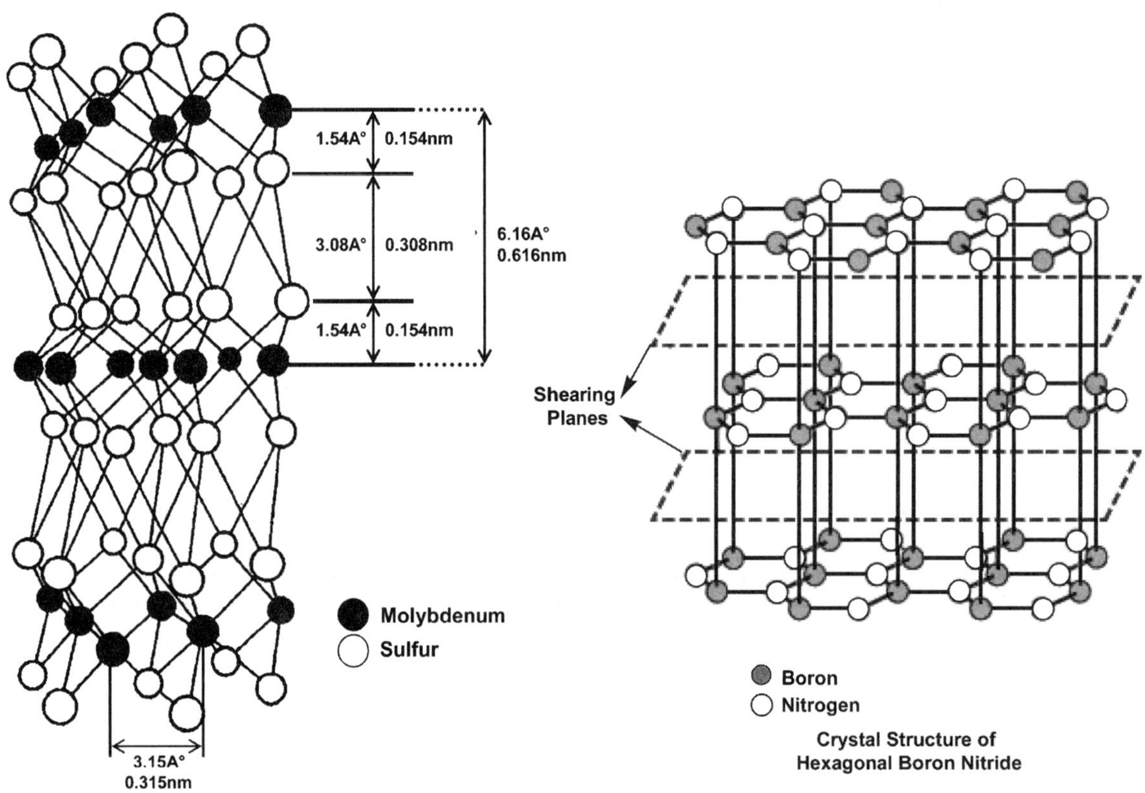

Fig. 4.144—Lamellar structure of molybdenum disulfide and boron nitride [449].

maximum wear life is 0.1 to 1 μm, which is similar to that required to give minimum coefficient of friction [448].

Diffusion coating is an alternative to deposition of the surface films for reducing friction and wear. The process involves thermal diffusion of the atoms of dissimilar elements, such as C, N, S, Cu, and Sn, into the metal surface. This is achieved by surface treatment. The result is the increased wear resistance due to increased surface hardness; production of low-shear strength surface films, which inhibit scuffing and seizure; and an increase in corrosion-resistance. Typically, these coatings are approximately 2 to 25 μm in thickness and are porous, which makes their surface ideal for depositing solids. Without the use of the supplemental solid or liquid lubricants, these coatings do not provide protection against friction and wear [448].

A variety of organic polymers are also employed to reduce friction and wear. These are used in three forms: thin films, self-lubricating materials, or binders for lamellar solids. Polytetrafluoroethylene (PTFE) is a polymer of the general formula $[C_2F_4]_n$ which is quite effective in forming thin films of low coefficient of friction, in the order of ~0.03 to 0.1. In addition, it is effective over a broad temperature range (−200 to +250°C), and is chemically inert. Its low friction coefficient is attributed to its molecular profile and orientation, which facilitate easy sliding of the coated surfaces. PTFE films are obtained by spraying the surfaces with the polymer and sintering at temperatures above 325°C. The formation of PTFE films can also be achieved by the use of a synthetic resin, which lowers the curing temperature significantly. Load-carrying capacity and durability of these films on metals are inferior to those of the MoS_2 coatings and their low thermal conductivity limits the maximum operating speed. PTFE coatings are therefore used only in applications that involve moderate conditions of sliding, for example, the food processing and plastics molding equipment. The only other polymers used as thin-film lubricants are polyimides [451]. While they can withstand higher temperatures (~300°C) than the PTFE, their frictional properties are not as good. They have a friction coefficient of 0.13 to 0.3, compared to that of 0.03 to 0.1 for PTFE. The greatest use of polymers is in self-lubricating composites as a replacement for the lubricated metals [452,453]. PTFE requires reinforcement when used in bulk, to minimize viscoelastic deformation under load. This is done by the use of thermosetting resins, such as phenolics. The presence of the lamellar solids, such as PTFE and MoS_2, in return is beneficial to the thermosetting resins since they impart to resins lower friction and wear properties. Graphite, because of being fibrous, also increases the strength of the resulting polymer composites.

The use of oil dispersions as additives in oils and greases to lower friction and to reduce wear is quite common. Graphite and MoS_2 are among the most commonly used lamellar solids. Because of their less than fluid nature, such dispersions are not suitable in applications that require hydrodynamic or elastohydrodynamic lubrication. The concentration of these additives ranges from 0.1 to 60 % by weight. The dispersions that contain high solids content are pastes, which are used only in the component assembly operations. Testing has shown that the use of MoS_2 dispersions improves the load-carrying capacity, reduces wear, and prolongs the rolling bearing life. Typical treat level in oils is 3 % by weight and in greases is 20 % by weight. The use of MoS_2 in engine oils is known to reduce wear and improve fuel economy [454]. It is important to note that the wear-reducing ability of MoS_2 and graphite can be inhibited by detergents and that their presence can affect the oxidation stability of oils and greases. In applications, such as textile machinery lubrication, which are adverse to the black color of these two additives, phosphates, oxides, and hydroxides—such as $Zn_2P_2O_7$ and $Ca(OH)_2$, and PTFE are used. PTFE-thickened fluorocarbon greases are also effective lubricants in oxidizing environments that involve broad-temperature range [455], such as rocket motors and space components.

Solid lubricants are used to control friction and wear in three types of applications: those that require thin-films, those that require self-lubricating properties, and those that require low-shear strengths films. Machines of various geometries are used to evaluate one or more of the properties of the solid lubricants. Common machines include Four-ball, hemisphere on disk, block on ring (Timken, LFW1), reciprocating pad on ring, Falex, Almen-Wieland (journal bearings), LFW-3 (thrust bearing) and LFW4 (press fit) [448]. Figure 4.145 provides their geometries. Motion is indicated by circular arrow and load is indicated by straight arrow/s. Frictional properties of some solid lubricants, as determined by the Thrust Fit Test, are provided in Table 4.25 [4]. As one can see that while all solid lubricants lower the friction of the base fluid, molybdenum disulfide and tungsten disulfide are the best performers.

Rust and Corrosion Inhibitors

Corrosion is a general term that is used to describe the destructive alteration of the metal by chemical or electrochemical action of its environment. It primarily involves a heterogeneous reaction which causes a metal to change from its nascent form (metallic state) to an oxidized form (ionic state). All metals, except noble metals, are thermodynamically unstable under atmospheric conditions and get converted into their oxidized form. Noble metals, such as gold, platinum, iridium, and palladium, on the other hand, are resistant to attack by the environment and are therefore found in nature in the free form.

There are many types of corrosion, but a lubricant formulator is primarily concerned with corrosion in the *presence* of electrolytes (electrochemical corrosion) and in the *absence* of electrolytes (chemical corrosion). Common electrolytes that lead to electrochemical corrosion include water, acids, alkalis, and salts. Chemical substances that cause chemical corrosion include acids, alkalis, and sulfur and its compounds. In alloys, corrosion can be selective or nonselective. It is selective if a particular metal is corroded in preference to another. It is nonselective if all metals in the alloy are corroded at the same rate.

Electrochemical corrosion involves the reaction of metals in the presence of electrically conducting solutions, or electrolytes, and occurs in two steps: the anodic process and the cathodic process. In the anodic process, the metal goes into the solution as ions with the extra electrons being left over. The process can be considered oxidation. The cathodic process involves the reaction of thus generated electrons with water and oxygen to form the hydroxide ions. The process can be considered reduction. In solution, the metal ions

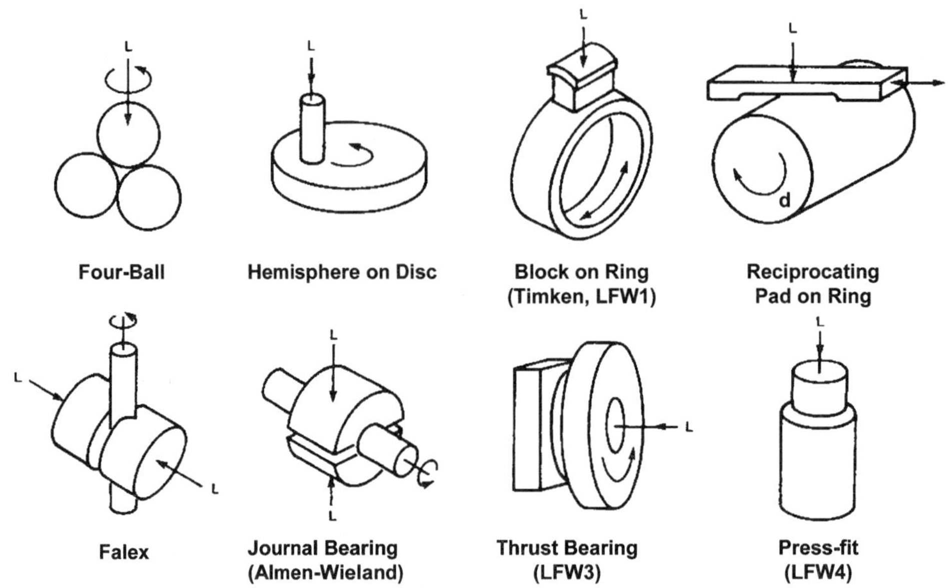

Fig. 4.145—Wear testing devices for solid lubricant [448].

combine with the hydroxide ions to form metal hydroxides, or hydrated oxides.

Chemical corrosion, on the other hand, does not need an electrolyte and can occur both in aqueous and organic media. It involves attack of the corrosive species, such as acids and sulfur, on metals. The damage occurs when the resulting salts are removed.

Factors that affect the corrosion rate include internal factors and external factors [456]. Internal factors are directly related to the metal itself, and include its composition, structure, surface condition, oxidation potential, and the presence of stresses. External factors pertain to the environment and include the nature of the oxide film, acidity, alkalinity, the presence of a polar solvent such as water, the pres-

TABLE 4.25—Friction coefficients of solid lubricants (Thrust Fit Test[a]) [4].

Solid Lubricant	Dynamic Friction	Stationary Friction	Stick-slip
None (Baseline)	0.45	-	Yes
Inorganic Lamellar (Layered Structures) Lubricants			
Molybdenum Disulfide (MoS_2), Natural	0.050	0.053	No
Molybdenum Disulfide, Synthetic	0.091	0.106	No
Tungsten Disulfide (WS_2)	0.090	0.098	No
Graphite, Colloidal in Water (22%)	0.100	-	No
Barium Hydroxide [$Ba(OH)_2$]	0.151	0163	No
Lead Chloride ($PbCl_2$)	0.191	0.214	No
Other Inorganic Compounds			
Borax	0.210	0.226	No
Lead-oxide/Zinc Oxide [PbO/ZnO]	0.250	-	Yes
Surface Layers			
Iron-manganese Phosphate	0.213	0.218	No
Iron-manganese Phosphate with Applied MoS_2	0.067	0.074	No

[a] A precision press is used t press a polished hardened steel bolt into a thick-walled steel case.

Electrode Reaction

$$M \longrightarrow M^{n+} + n\,e^{-}$$

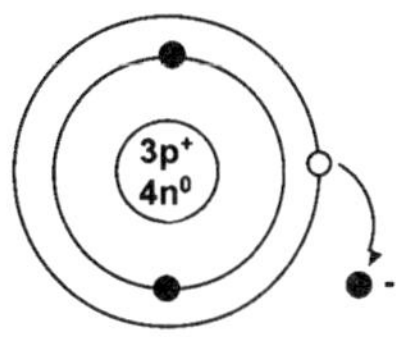

Lithium = E_0 of +3.05
Hydrogen = E_0 of 0.00

Metal	E_0 Volts	Metal	E_0 Volts
Lead (Pb)	***0.13***	Zinc (Zn)	0.76
Tin (Sn)	0.14	Chromium (Cr)	0.86/0.74
Nickel (Ni)	0.25	Manganese (Mn)	1.18/0.283
Iron (Fe)	***0.44/0.04***	Vanadium (V)	1.18
Copper (Cu)	***-0.52/-0.34***	***Aluminum (Al)***	***1.66***

Fig. 4.146—Oxidation potentials of common metals found in steel.

ence of an electrolyte (salts, acids, or bases), the presence and reactivity of aggressive species, and the temperature. While the role of these factors is considered elsewhere [456], we are especially interested in the oxidation potential of the various metals.

Oxidation potentials of the elements that make up the metallurgy of the modern mechanical equipment are provided in Fig. 4.146 [457]. The elements that are primarily used to fabricate metal parts are shown by bold italic symbols; hence their corrosion control is the primary goal. The oxidation potentials reported in the table are relative to hydrogen with a value equal to 0. Positive values indicate ease of oxidation and negative values indicate resistance to oxidation. Because of hydrogen being the reference point, we can directly compare these values to assess the relative oxidative tendency of the various metals. The higher the oxidation potential, the easier it is for the metal to oxidize. Metals in the left column of the table with the smaller values are therefore less susceptible to oxidation, hence corrosion, than those in the right column. The values with a forward slash indicate a metal's first and second oxidation potentials. For example, for iron, 0.44 represents the oxidation potential of the Fe^0 to Fe^{+2} transition and 0.04 represents the oxidation potential of the Fe^0 to Fe^{+3} transition. The same is true for copper. Of the metals of interest, aluminum is the easiest to oxidize, followed by iron and lead. Copper with the negative values is the most resistant of the metals listed in the table.

Of the external factors listed above, acidity, alkalinity, the presence of the reactive species and the temperature have the largest impact. The presence of the acids and bases can accelerate corrosion. In lubricant applications, acids result from the oxidation of the fuel sulfur and the base stock. Lubricant contains additive derived components, both organic and inorganic, which are moderately basic and therefore pose little problem by themselves. However, the reaction between the acidic and the basic species results in salts which because of being electrolytes can promote corrosion. The lubricant also contains additives that can lead to acidic decomposition products. Some of these additives are sulfur and phosphorus compounds that are added to the lubricant to improve its oxidative stability and antiwear performance. These can corrode metals either directly or by forming aggressive chemical species via decomposition. Higher temperatures that are typically encountered in the internal combustion engines can accelerate corrosion as well.

Metallurgy in automotive equipment commonly contains iron (Fe), copper (Cu), lead (Pb), chromium (Cr), manganese (Mn), antimony (Sb), aluminum (Al), vanadium (V), zinc (Zn), nickel (Ni), and tin (Sn). Protecting four of these elements against corrosion is of primary interest: iron, which is the principal metal used to forge the engine and the auxiliary equipment; copper, which is present in bearings and seals; lead, which is also present in bearings; and aluminum that is used in newer cars and trucks to make them lighter for better fuel economy.

Protection against corrosion is necessary because it can lead to a loss of metal, thereby lowering the integrity of the equipment and resulting in its malfunction. In addition, corrosion exposes fresh metal that can wear at an accelerated rate and result in metal ions that can act as oxidation promoters.

Corrosion of iron and its alloys, sometimes referred to as ferrous corrosion or rust, is primarily electrochemical in nature. It can occur both in the liquid phase and the vapor phase and needs water, electrolyte such as salt, and oxygen. In an internal combustion engine, water results from the fuel combustion, oxygen comes from the air, and electrolytes are metal salts that form by the reaction of the metals and certain additives with the combustion and oxidation-derived acids. Ferrous corrosion mainly occurs in engines that are run in short intermittent cycles (stop-and-go driving). Rust initiates when water sets up a localized electrochemical reaction between the surface iron and the iron oxide layer. Nascent iron acts as the anode, and the iron oxide layer acts as the cathode. Iron emits electrons and forms ferrous ions that are released into the medium. The electrons migrate to the cathode (oxide layer) and form the hydroxide ions by reacting with oxygen and water. Ferrous ions and hydroxide ions then combine to form ferrous hydroxide, which is subsequently oxidized to ferric hydroxide, lose water, and become rust [207]. The mechanism of the rust formation is depicted in Part A of the Fig. 4.147 [207].

Corrosion of copper and bronze (the yellow metal corrosion) is chemical in nature and occurs by the attack of the aggressive species on metals. Such species result from the oxidation and combustion of the hydrocarbon materials, such as fuel, base oil, and sulfur-containing additives. Yellow metal corrosion results when the chemically reactive materials, such as acids or sulfur, attack copper or copper oxide. The result is the formation of the ionic copper compounds that are removed, thereby causing the metal damage. This is shown in Part A of the Fig. 4.148 [207].

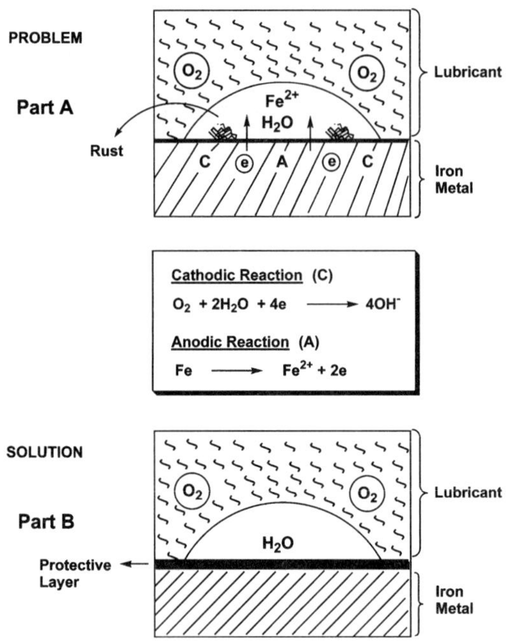

Fig. 4.147—The mechanism of rust inhibition [207].

Chemicals, called corrosion-inhibitors, are used to control corrosion. These additives fall under two general classes: acid neutralizers and film formers. Acid-neutralizing agents are additives that neutralize aggressive acidic materials and make them innocuous. Film formers either attach themselves to the metal surfaces to form impenetrable protective films [207], or they facilitate the formation of a lubricant film on the metal surface, thereby isolating it from the attack of the harmful species. Film formation can occur via physical adsorption or chemical adsorption. In the first case, the resulting film is of a somewhat transient nature. In the second case, however, it is more persistent. Film formation occurs when these additives interact with the metal surface via their polar ends and associate with the lubricant via their nonpolar ends, in a manner similar to that of the friction modifiers. Since these additives have high surface affinity, they can compete with the extreme pressure and antiwear agents and impede their function. The rust-

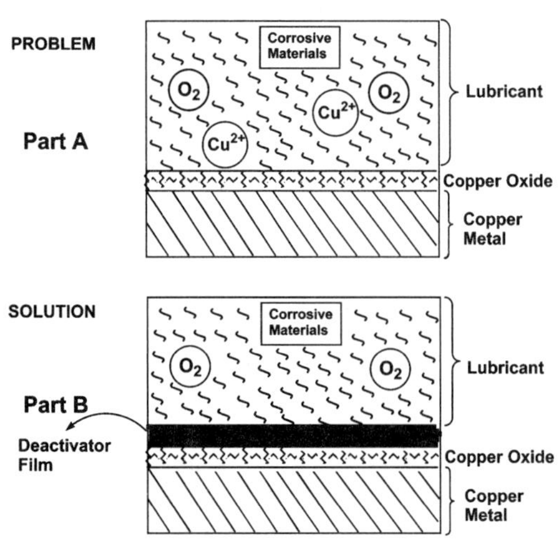

Fig. 4.148—The mechanism of copper deactivation [207].

inhibiting mechanism and the corrosion-inhibiting mechanism via the film formation are shown in Part B of Figs. 4.147 and 4.148, respectively.

Figure 4.149 depicts the manner in which an alkenylsuccinic acid makes a rust-inhibiting film. Figure 4.150 shows the probable nature of the copper corrosion-inhibiting films. Part A shows copper inhibition by a commercially available dimercaptothiadiazole (DMTD) derivative and Part B shows inhibition by tolyltriazole, another commercial product. In the case of the alkenylsuccinic acid and the DMTD derivatives, it is the lubricant associated with the hydrocarbon chain that primarily acts as a protective film and in the case of tolyltriazole, it is the adsorbed additive itself that primarily acts as a protective film.

Chemical types used to inhibit corrosion of ferrous metals include polyethoxylated alkylphenols, neutral and basic arylsulfonates and phenolsulfides, alkenylsuccinic acids and their esters, alkyl phosphites and phosphates, alkanolamines, and polymeric amines such as dispersants. Those used to inhibit yellow metal corrosion mainly include oil-soluble heterocyclic compounds, such as triazole and dimercaptothiadiazole (DMTD) derivatives.

The structures of the common corrosion inhibitors are shown in Fig. 4.151 and Figs. 4.152 and 4.153 depict the reaction schemes used to prepare some of them. It is important to note that in the hydrogen peroxide coupling of the DMTD with the mercaptan, one obtains a mixture that contains approximately 85 % of the bis-coupled product, shown in the figure, and 15 % of the mono-coupled product. The mono-coupled product is much more surface active and is used as EP/antiwear agent in some applications. Long chain organic molecules, such as alkanolamines, are examples of the physically adsorbing additives. Phosphoric acid, dithiophosphoric acid, and succinic acid derivatives are examples of the chemical film-forming additives. Basic detergents are excellent rust and corrosion inhibitors since they provide protection both by neutralizing acids and by forming the physically-adsorbed films. Figure 4.154 shows the positive effect of the lubricant base reserve on the engine rust [21]. For many applications, such as gear oils, rust and corrosion inhibitor systems are required to provide both the vapor-phase and the liquid-phase protection, that is, for surfaces above and below the lubricant level.

Lead corrosion involves preferential removal of lead, sometimes referred to as lead leaching, from the copper-lead bearings, which primarily occurs in diesel engines. While its mechanism is not well understood, lead removal may partly be due to the attack of chemically aggressive species on the bearing metal. Certain classes of additives, such as amines and amide dispersants, appear to aggravate the situation. Aluminum corrosion is very slow because the aluminum oxide film is tenacious and is not easy to remove, and acts as a barrier against the environment. No effective inhibitors are known to control lead corrosion. The only way to control lead leaching appears to be by proper balancing of the additive package.

Some additives that inhibit corrosion in some applications can cause corrosion in others. For example, zinc dialkyl dithiophosphates inhibit copper-lead bearing corrosion in the oxidative environment, but cause silver-bearing corrosion due to the presence of sulfur. When used in high

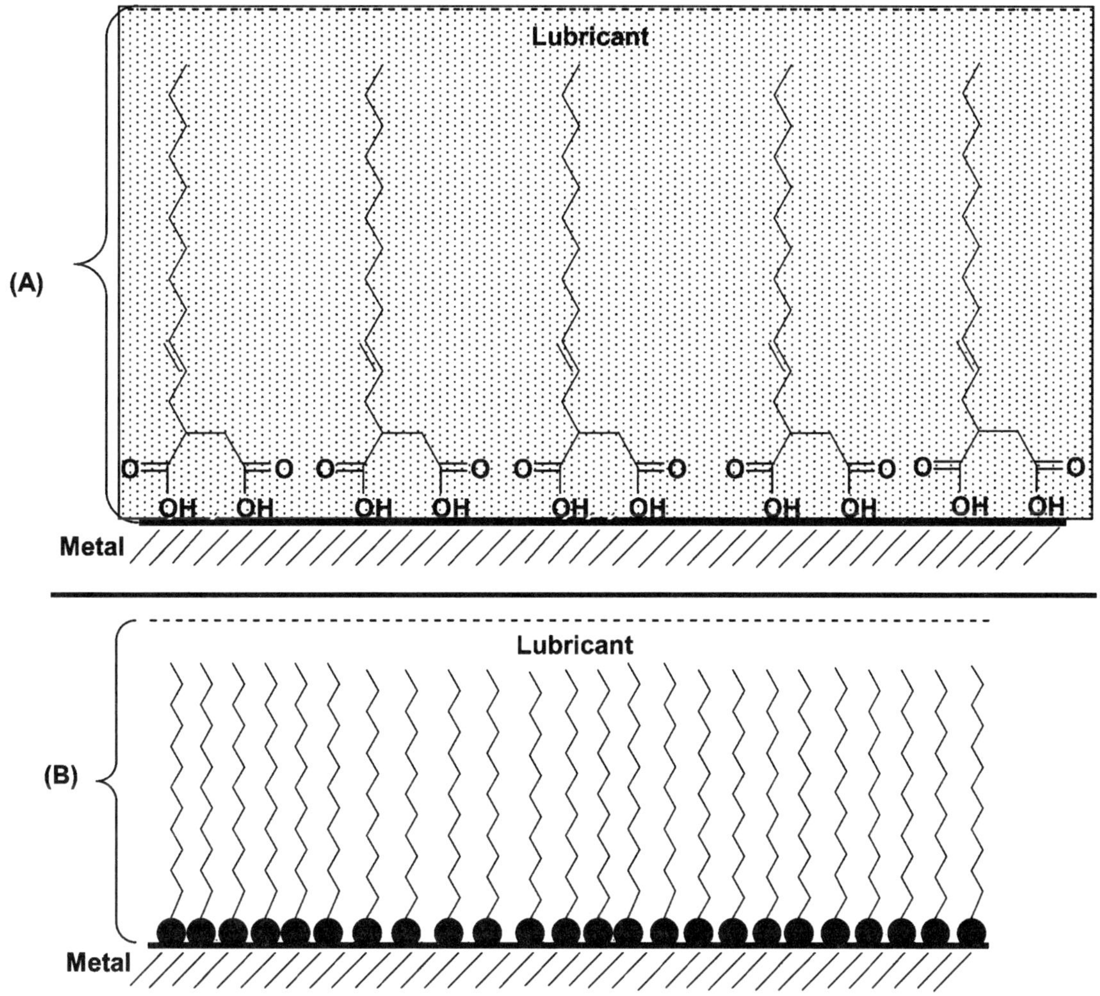

Fig. 4.149—Rust inhibitor and mode of its action.

concentrations, zinc dialkyl dithiophosphates can also lead to pitting of the some ferrous alloys.

Corrosion inhibitors have major uses in engine oils, gear oils, metalworking fluids, and greases. Thiadiazole and triazole derivatives are especially useful in protecting against nonferrous or yellow metal corrosion.

Metal corrosion is measured by a variety of tests, depending upon the application. At present for engine oils, Sequence VIII [Unleaded L-38 Test (copper/lead)], Ball Rust Test (ASTM D6557), and High Temperature Corrosion Bench Test (ASTM D6594, copper/lead/tin), also called Cummins Bench Corrosion Test, are used. For gear oils, CRC L-13, L-21, and L-33 Tests are used for rust and ASTM D130 Test is used for copper corrosion. For industrial products, ASTM D665 Test is used for rust and ASTM D130 Test is used for copper corrosion. For hydraulic and metalworking fluids, a number of ASTM specified tests are utilized [20].

Emulsifiers and Demulsifiers

Emulsifiers are chemical compounds that enable two immiscible fluids to form an intimate mixture, known as an emulsion. Oil-water emulsions are often used as lubricants in many industries and for a variety of applications. This is because such lubricants are low cost, easier to dispose of, and have fire-retardant properties. Emulsions of water and mineral oil have primary use in metalworking and hydraulic applications. To be effective, emulsions must possess a number of desirable properties. They must be stable over long periods of time, possess good lubricating properties, not attack seals and metals, and be easy to demulsify for disposal.

In the presence of water, certain lubricant formulations have an increased tendency to form emulsions. This is due to the presence of the chemical additives which act as surfactants. Demulsifiers are added to such formulations to enhance water separation and suppress foam formation.

Emulsifiers and demulsifiers are essentially surfactants (surface-active agents) that have the ability to adsorb onto surfaces to significantly alter them, as in the case of the film-forming agents, or onto interfaces to change their interfacial free energies. The term interface describes the boundary between any two immiscible phases, such as oil and water. These additives are made up of hydrophilic and hydrophobic moieties. The hydrophilic moiety is a hetero atom-derived polar functional group which is attached to a hydrophobic hydrocarbon group. Common hetero atoms are atoms of the elements nitrogen, oxygen, sulfur, and phosphorus. The hydrophobic group is usually a hydrocarbon group of sufficient chain length to provide proper solubility or dispersibility in the oil phase.

Emulsifiers and demulsifiers can be classified as non-

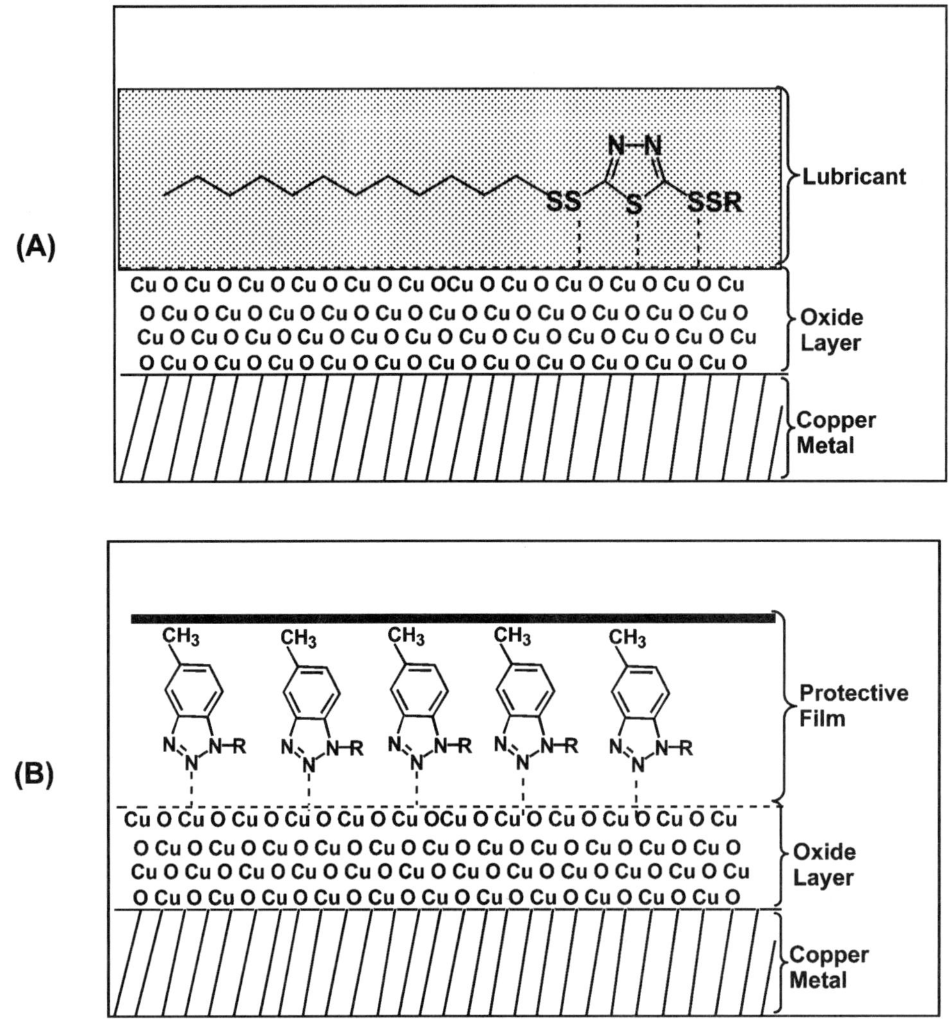

Fig. 4.150—Copper passivators and mode of their action.

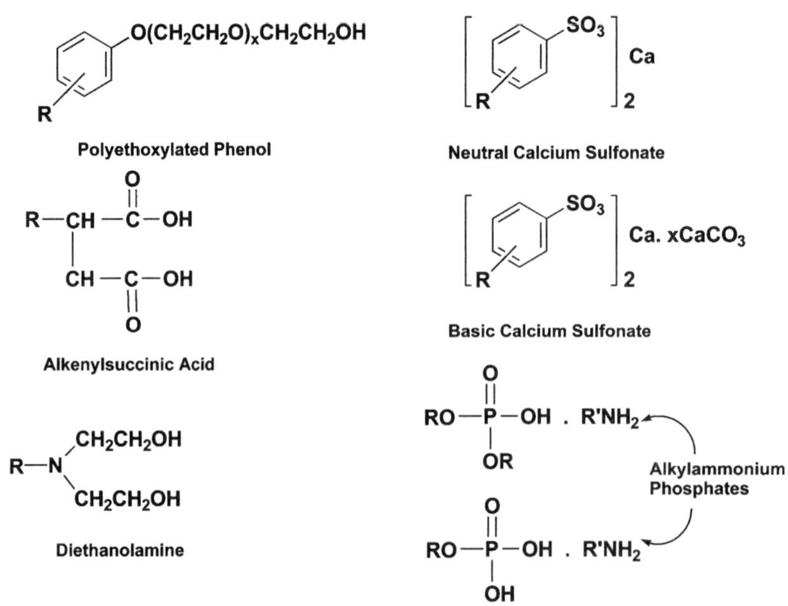

Fig. 4.151—Typical rust inhibitors.

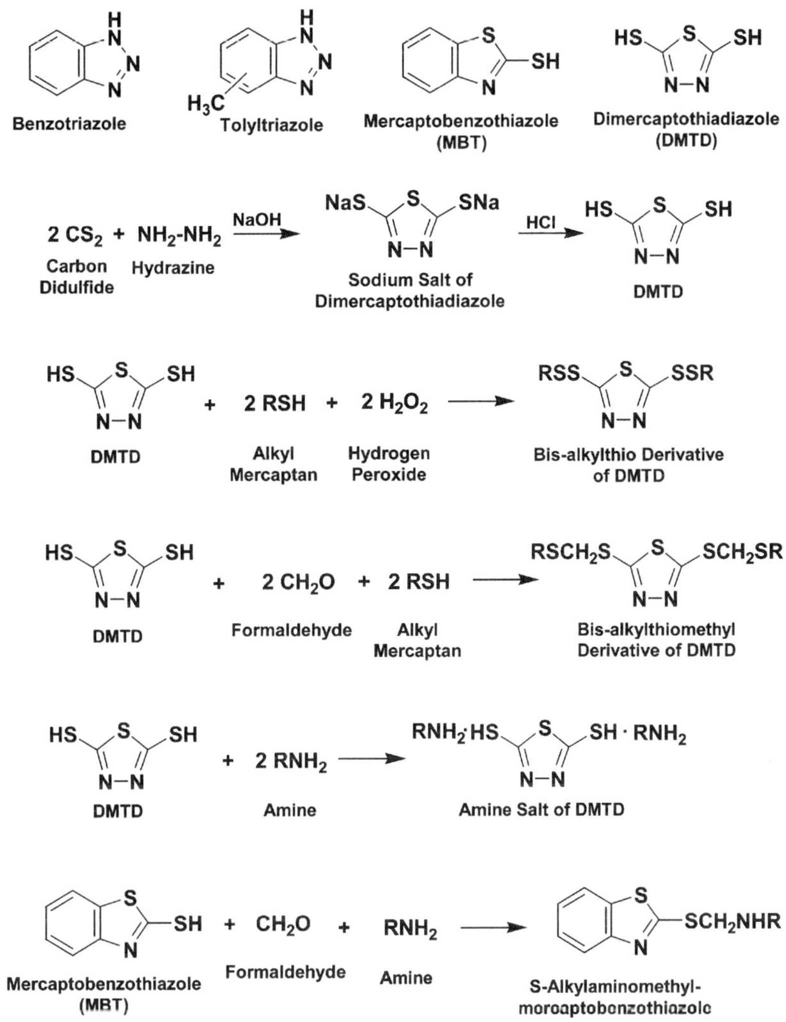

Fig. 4.152—Common copper corrosion inhibitors and their synthesis.

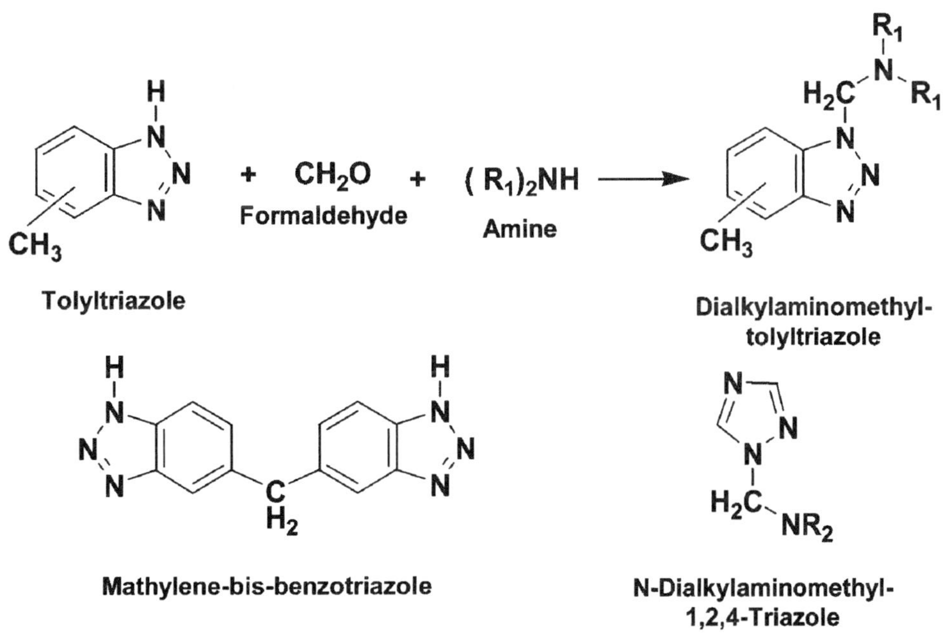

Fig. 4.153—Triazole-based copper corrosion inhibitors and their synthesis.

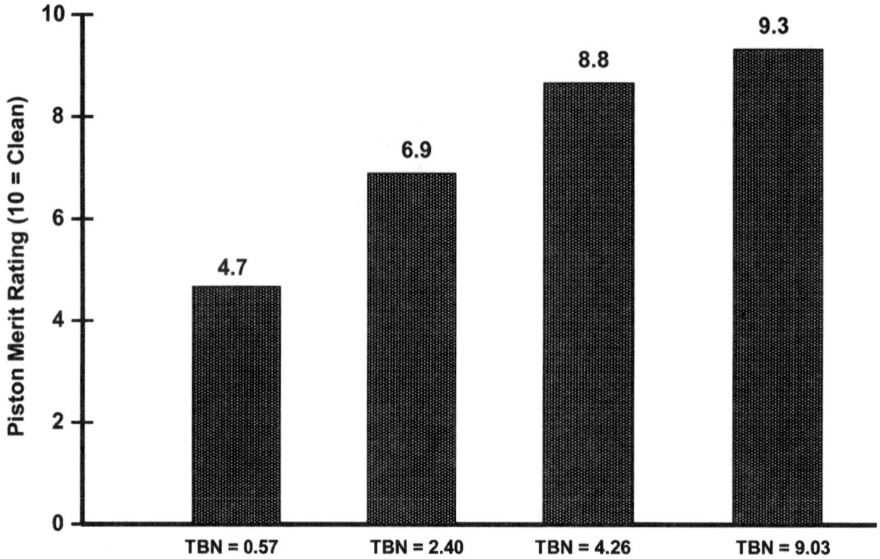

Fig. 4.154—Effect of lubricant basicity on engine rust [21].

ionic or ionic, depending upon whether the polar part is uncharged or charged. Ionic compounds can be further subdivided into cationic, if the charge is positive, and anionic, if the charge is negative [458]. Generalized structures for emulsifiers and demulsifiers [459] are given in Fig. 4.155.

It is important to note that only the charge on the functional group attached to the carbon chain is used in this classification. The charge on the counter ion, which is usually in-

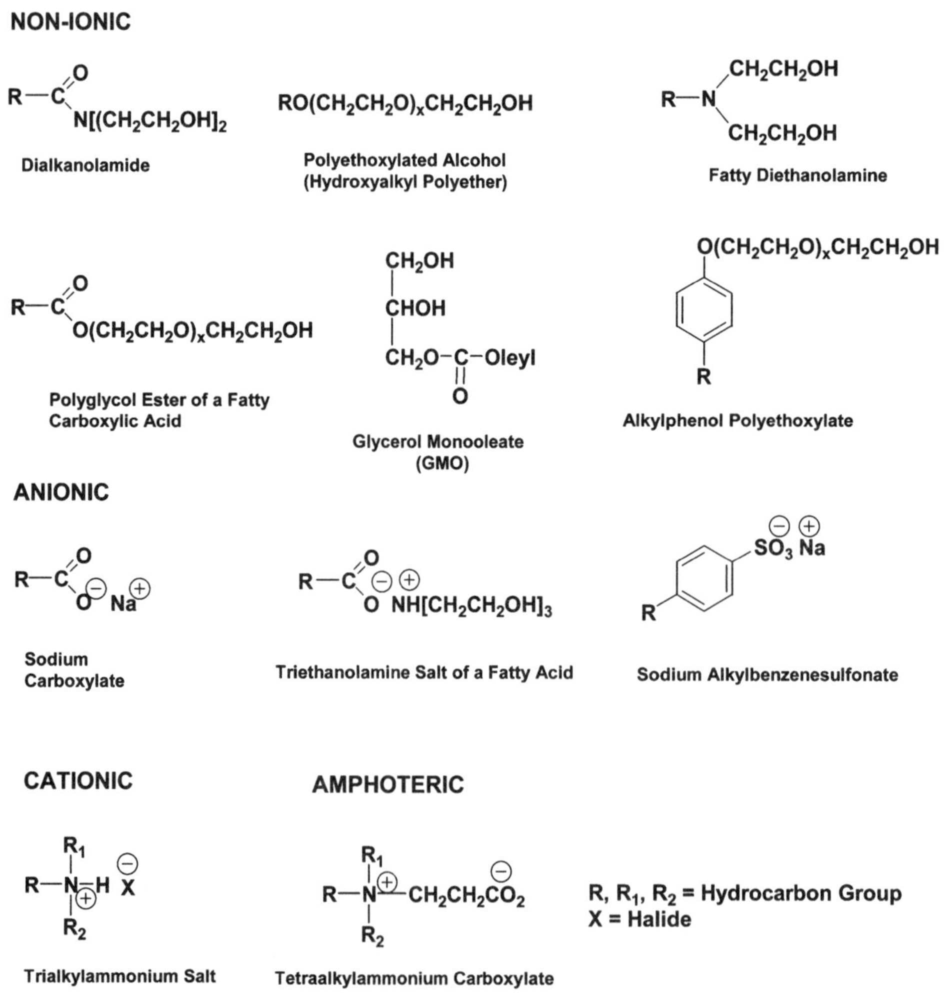

Fig. 4.155—Emulsifiers and demulsifiers.

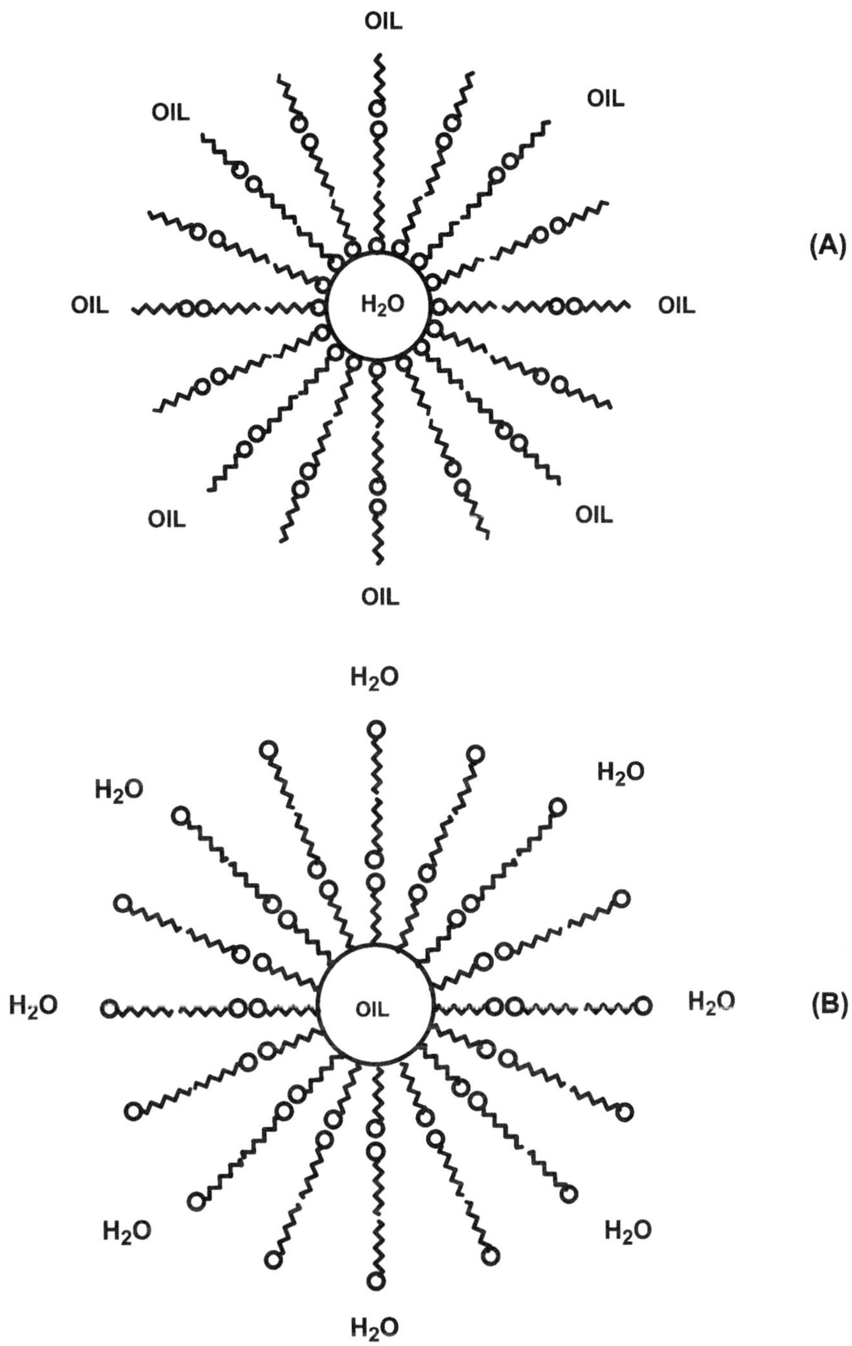

Fig. 4.156—A representation of (A) water-in-oil emulsion (b) oil-in-water Emulsion.

organic in origin, is ignored. The term amphoteric applies to a group of additives which contain both the cationic and the anionic groups of organic origin, preferably within the same molecule. They possess the structural features and the properties of both the cationic and the anionic materials grouped together.

Emulsifiers reduce the surface tension of water and, therefore, facilitate thorough mixing of the oil and water to form an emulsion. The efficiency of an emulsifier depends upon its molecular weight, which is usually less than 2000 g/mol, its HLB (hydrophile-lipophile balance) value, water *p*H and hardness, the nature of the oil, and the operating conditions, such as temperature. Emulsifiers with an HLB value of 3 to 6 are suitable for water-in-oil emulsions and those with an HLB value of 8 to 18 are suitable for oil-in-water emulsions. The manner in which these additives form emulsions is shown in Fig. 4.156. Water-in-oil emulsions form when these additives associate with water via their polar ends and with oil and other additive molecules via the nonpolar ends. This is shown in Part A of the figure. The result is water miscibility in oil, or *water-in-oil* emulsion. The mechanism of *oil-in-water* emulsion is similar, except that the additive molecules associate with oil in the reverse manner. This situation is shown in Part B of the figure. See Chapter 11 on Metalworking Fluids for further discussion on emulsifiers.

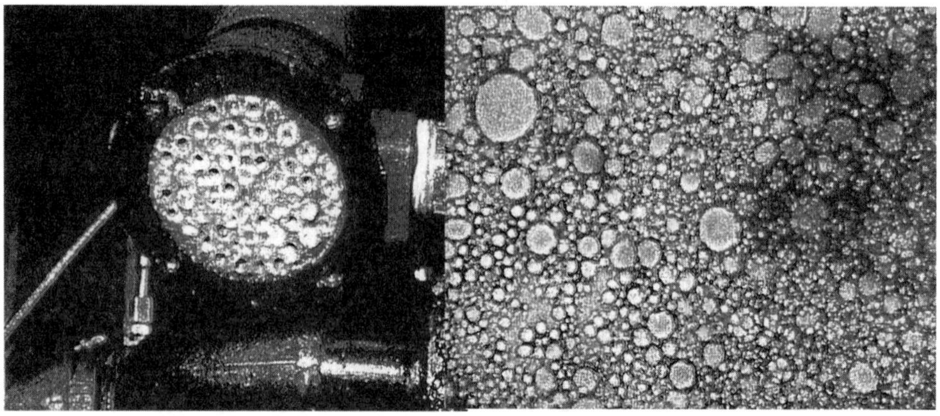

Emulsion in Breather Tube Condenser **Water-in-oil Emulsion in ASTM Sequence IID Test**

Fig. 4.157—Emulsion in ASTM Sequence IID Test [461].

Demulsifiers perform the opposite function and enhance water separation. Structurally, most demulsifiers are oligomers or polymers with a molecular weight of up to 100,000 g/mol and contain 5–50 % polyethylene oxide in a combined form. They are commonly block copolymers of propylene oxide and or ethylene oxide and initiators such as phenol-formaldehyde resins, siloxanes, polyamines, and polyols [460]. For water-in-oil emulsions, polymers containing 20–50 % ethylene oxide are suitable. These materials concentrate at the water-oil interface and create low viscosity zones, thereby promoting droplet coalescence and gravity-driven phase separation. Low molecular weight materials, such as alkali metal or alkaline earth metal salts of dialkylbenzene- and dialkylnaphthalene-sulfonic acids, are also useful in some lubricant-related applications.

As a general rule, nonionic emulsifiers are used in metalworking fluids based on naphthenic stocks, and fatty acid carboxylates are used in those based on paraffinic stocks. Poly(alkylene glycol)s, or hydroxyalkyl ethers, are sometimes avoided because their enhanced solubility in water does not allow their clean separation for disposal.

Poly(ethylene oxide) derivatives and salts of carboxylic and sulfonic acids are the most commonly used emulsifiers, primarily in metalworking fluids. Demulsifiers are used in applications where water contamination of the lubricant is a problem and quick separation of water is desired. Automatic transmission fluids, hydraulic fluids, industrial gear oils, and some engine oils, are examples of such lubricants.

Figure 4.157 shows a situation where water-in-oil emulsion essentially plugged up the breather tube in the ASTM Sequence IID, now obsolete, engine test. The problem was solved by the use of a demulsifier. The results of the field test are shown in Fig. 4.158 [461]. As can be seen, the use of a proper demulsifier led to a remarkable improvement in the emulsion performance of the lubricant. Parts that were totally covered with emulsion became emulsion-free. Compare the left half of the figure with the right half.

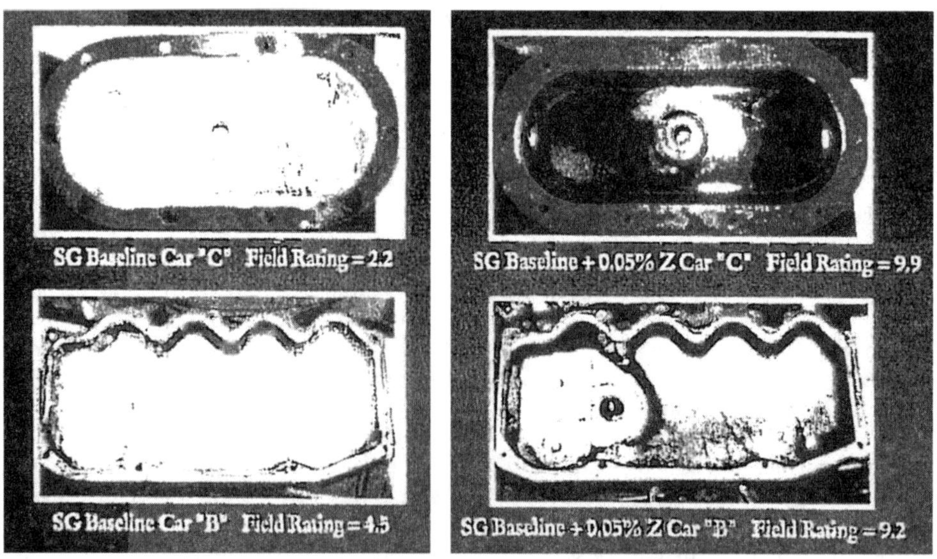

SG Quality Oil **SG Quality Oil + 0.05% Demulsifying Additive**

Fig. 4.158—Effect of demulsifier on field emulsion [461].

Polymeric Additives

Materials with polymeric structures are the key components in high-performance lubricants. They can be used as lubricant base stocks (synthetic lubricants) or to enhance a lubricant's inherent properties, such as viscosity and pour point. They can also be used as starting materials to prepare certain classes of additives, such as dispersants and detergents.

The development of the synthetic hydrocarbon polymers started with Bakelite (phenol-formaldehyde resin) in 1907, which was closely followed by the commercialization of the synthetic rubber (styrene-diene copolymer). Polystyrene became commercial around 1930 in Germany and in 1937 in the United States. Since then, a large number of synthetic polymers have become commercially available. Over time, these materials have found extensive use in a wide variety of consumer and industrial products, including lubricants. The first use of the polymeric additives (pour point depressants) to improve the low temperature properties (flow) of lubricants dates back to the mid-1930s. By the 1940s, other lubricant additives, including viscosity modifiers (VMs), were commercialized and by the 1950s their use in lubricants became a normal practice. Dispersants and foam inhibitors are two other classes of polymeric additives that are commonly used in lubricants.

Dispersant polymers, or dispersant viscosity modifiers (DVMs), are hydrocarbon polymers that have a dispersant moiety attached to them. These materials are used as viscosity modifiers with added dispersancy. The molecular weights of these materials are in the 25,000 –500,000 g/mol range. The invention of the viscosity modifiers led to the introduction of the multi-grade oils, which eliminated the need to use low viscosity oils for winter and high viscosity oils for summer. Multi-grade oils contain polymers of suitable molecular weight and physical properties in a low viscosity oil. These oils are therefore suited for all-season use. Polymers have been designed that have the viscosity-modifying ability and dispersancy or antioxidancy, or both. In this book, the multifunctional polymers are identified by the use of the letters D and DA. They precede the abbreviation for the polymer type. For example, PMA-based viscosity modifiers are referred to as DPMAs, if they are claimed to have additional dispersant action. Similarly, DAOCP refers to a polymer with both dispersant and antioxidant properties. The timeline for the development of the different types of viscosity modifiers and dispersant viscosity modifiers is shown in Fig. 4.159.

A polymer is a substance that is made up of large molecules containing repeating molecular units, derived from the monomer. Polymer molecules can be linear, branched, or network type, depending upon the arrangement of the monomer units. If the repeating units in a polymer are in a linear arrangement, the polymers are called *linear*. If the repeating units are in a branched arrangement, the polymers are called *branched*. And if these units are interconnected to form three-dimensional networks, the polymers are called *network polymers*.

The polymer chain can also contain a functional group at the end, which is denoted by the latter F in the structure. Such polymers are called *ω-functional polymers*. The functional end groups are usually introduced by terminating the homo-polymerization with a hetero atom (O, N, S, or P) containing monomer. Polymers, such as poly(alkylene oxide)s, that contain hetero atoms in the backbone are not included in this class.

Branched polymers can have a *comb-like* structure or a *star-shaped* structure. *Comb-like branched polymers* consist of a long hydrocarbon chain as the backbone and short chains as the pendent groups that appear like teeth on a comb. *Star-shaped branched polymers*, also called radial polymers, have a center from which hydrocarbon chains extend as rays. The molecular configurations of the different types of polymers [462] are presented in Fig. 4.160.

Polymers fall under two broad classes: homo-polymers and copolymers. *Homo-polymers* contain only one kind of repeating unit; that is, they are derived from a single monomer. *Copolymers* contain two or more kinds of fundamental units; that is, they are derived from two or more monomers. If different kinds of units occur randomly, the polymer is called a *random copolymer*. If they occur in an alternating fashion, the polymer is called an *alternating copolymer*. When the units occur in blocks, the polymer is called a *block copolymer*. Block polymers result from the sequential polymerization of two or more monomers. The polymerization of one monomer gives rise to one block and the polymerization of the second monomer gives rise to the other block. *Graft copolymers* are polymers where already formed polymer chains are extended by forming new blocks of new repeating units or by attaching another functional group. These polymeric arrangements for linear copolymers [462,463] are depicted in Fig. 4.161. Such arrangements are also possible for branched and network polymers.

Molecular Weight Averages and Distribution

Most polymeric materials are compositions that consist of the polymer chains of varying sizes. The bulk properties of these compositions depend upon the average molecular weight of the polymer chains. As stated earlier, the molecular weights of polymeric materials are expressed as number-average, weight-average, z-average, $(z+1)$ average, and viscosity average molecular weights [221–225,462–464]. Of these, the number-average (Mn), weight-average (Mw), and z-average (Mz) molecular weights are most often used to describe the polymer compositions. The mathematical functions representing these averages are given below. In these equations, Mi is the molecular weight and ni is the number of moles of the component molecule i.

$$\text{Number – average molecular weight:}\quad \mathrm{Mn} = \frac{\sum \mathrm{niMi}}{\sum \mathrm{ni}}$$

$$\text{Weight – average molecular weight:}\quad \mathrm{Mw} = \frac{\sum \mathrm{niMi}^2}{\sum \mathrm{niMi}}$$

$$z\text{ – Average molecular weight:}\quad \mathrm{Mz} = \frac{\sum \mathrm{niMi}^3}{\sum \mathrm{niMi}^2}$$

To calculate Mn, Mw, and Mz of an equimolar polymer composition that consists of five components of molecular weights 8000, 9000, 10,000, 11,000, and 12,000 g/mol, the following procedure can be employed. All values for the computation can be determined from ni and Mi. For a polymer composition that contains an equimolar mixture of the

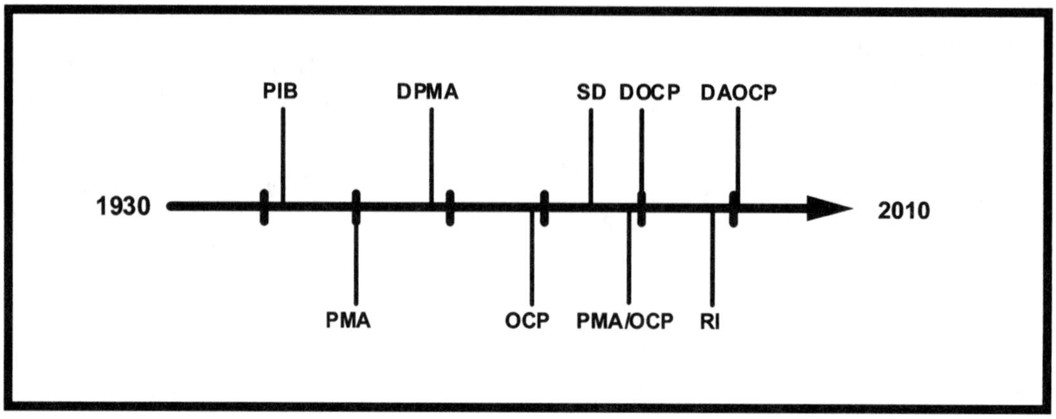

PIB = Polyisobutylene **PMA** = Polymethacrylate **DPMA** = Dispersant Polymethacrylate
SD = Styrene-diene Polymer **PMA/OCP** = Polymethacrylate-Olefin Copolymer
OCP = Olefin Copolymer **DOCP** = Dispersant Olefin Copolymer
RI = Radial Isoprene (Star Polymer) **DAOCP** = Dispersant Antioxidant Olefin Copolymer

Fig. 4.159—Polymeric additive invention timeline.

five components, these values are provided in Table 4.26. By using the values in the table, Mn, Mw, and Mz values for such a polymer composition can be calculated as follows.

$$\mathrm{Mn} = \frac{\sum \mathrm{niMi}}{\sum \mathrm{ni}} = \frac{\mathrm{n1M1}}{5} + \frac{\mathrm{n2M2}}{5} + \frac{\mathrm{n3M3}}{5} + \frac{\mathrm{n4M4}}{5} + \frac{\mathrm{n5M5}}{5}$$

$$\mathrm{Mn} = \frac{50{,}000}{5} = \frac{8000}{5} + \frac{9000}{5} + \frac{10{,}000}{5} + \frac{11{,}000}{5} + \frac{12{,}000}{5}$$
$$= 1600 + 1800 + 2000 + 2200 + 2400 = 10{,}000\ \mathrm{g/mol}$$

$$\mathrm{Mw} = \frac{\sum \mathrm{niMi}^2}{\sum \mathrm{niMi}} = \frac{\mathrm{n1M1}^2}{50{,}000} + \frac{\mathrm{n2M2}^2}{50{,}000} + \frac{\mathrm{n3M3}^2}{50{,}000} + \frac{\mathrm{n4M4}^2}{50{,}000} + \frac{\mathrm{n5M5}^2}{50{,}000}$$

$$\mathrm{Mw} = \frac{510{,}000{,}000}{50{,}000} = \frac{8000^2}{50{,}000} + \frac{9000^2}{50{,}000} + \frac{10{,}000^2}{50{,}000} + \frac{11{,}000^2}{50{,}000} + \frac{12{,}000^2}{50{,}000} = 1280 + 1620 + 2000 + 2420 + 2880$$
$$= 10{,}200\ \mathrm{g/mol}$$

$$\mathrm{Mz} = \frac{\sum \mathrm{niMi}^3}{\sum \mathrm{niMi}^2} = \frac{\mathrm{n1M1}^3}{5.10 \times 10^8} + \frac{\mathrm{n2M2}^3}{5.10 \times 10^8} + \frac{\mathrm{n3M3}^3}{5.10 \times 10^8} + \frac{\mathrm{n4M4}^3}{5.10 \times 10^8} + \frac{\mathrm{n5M5}^3}{5.10 \times 10^8} = \frac{5.30 \times 10^{12}}{510{,}000{,}000 \text{ or } 5.10 \times 10^8}$$

$$\mathrm{Mz} = \frac{8000^3}{5.10 \times 10^8} + \frac{9000^3}{5.10 \times 10^8} + \frac{10{,}000^3}{5.10 \times 10^8} + \frac{11{,}000^3}{5.10 \times 10^8} + \frac{12{,}000^3}{5.10 \times 10^8} = 1004 + 1429 + 1961 + 2610 + 3388$$
$$= 10{,}392\ \mathrm{g/mol}$$

The values of ni and Mi necessary to compute the molecular weights for a nonequimolar polymer composition are presented in Table 4.27. By using the ni and Mi values from this table, Mn, Mw, and Mz for a nonequimolar polymer composition can be computed as follows.

$$\mathrm{Mn} = \frac{\sum \mathrm{niMi}}{\sum \mathrm{ni}} = \frac{50{,}500}{5} = 10{,}000\ \mathrm{g/mol}$$

$$\mathrm{Mw} = \frac{\sum \mathrm{niMi}^2}{\sum \mathrm{niMi}} = \frac{516{,}500{,}000}{50{,}500} = 10{,}228\ \mathrm{g/mol}$$

$$\mathrm{Mz} = \frac{\sum \mathrm{niMi}^3}{\sum \mathrm{niMi}^2} = \frac{5.3445 \times 10^{12}}{516{,}500{,}000 \text{ or } 5.165 \times 10^8}$$
$$= 10{,}348\ \mathrm{g/mol}$$

The number-average molecular weight (Mn) represents the chemical stoichiometry and is useful in carrying out the chemical reactions involving polymers. The weight-average molecular weight (Mw) correlates with the mechanical properties, such as tensile strength and modulus in plastics, films, and fibers, and the viscosity improving behavior of the polymers. The *z*-average molecular weight (Mz) largely influences the polymer's viscoelastic properties, such as the melt elasticity.

While a number of methods can be employed to measure these averages, each technique has its limitations [464].

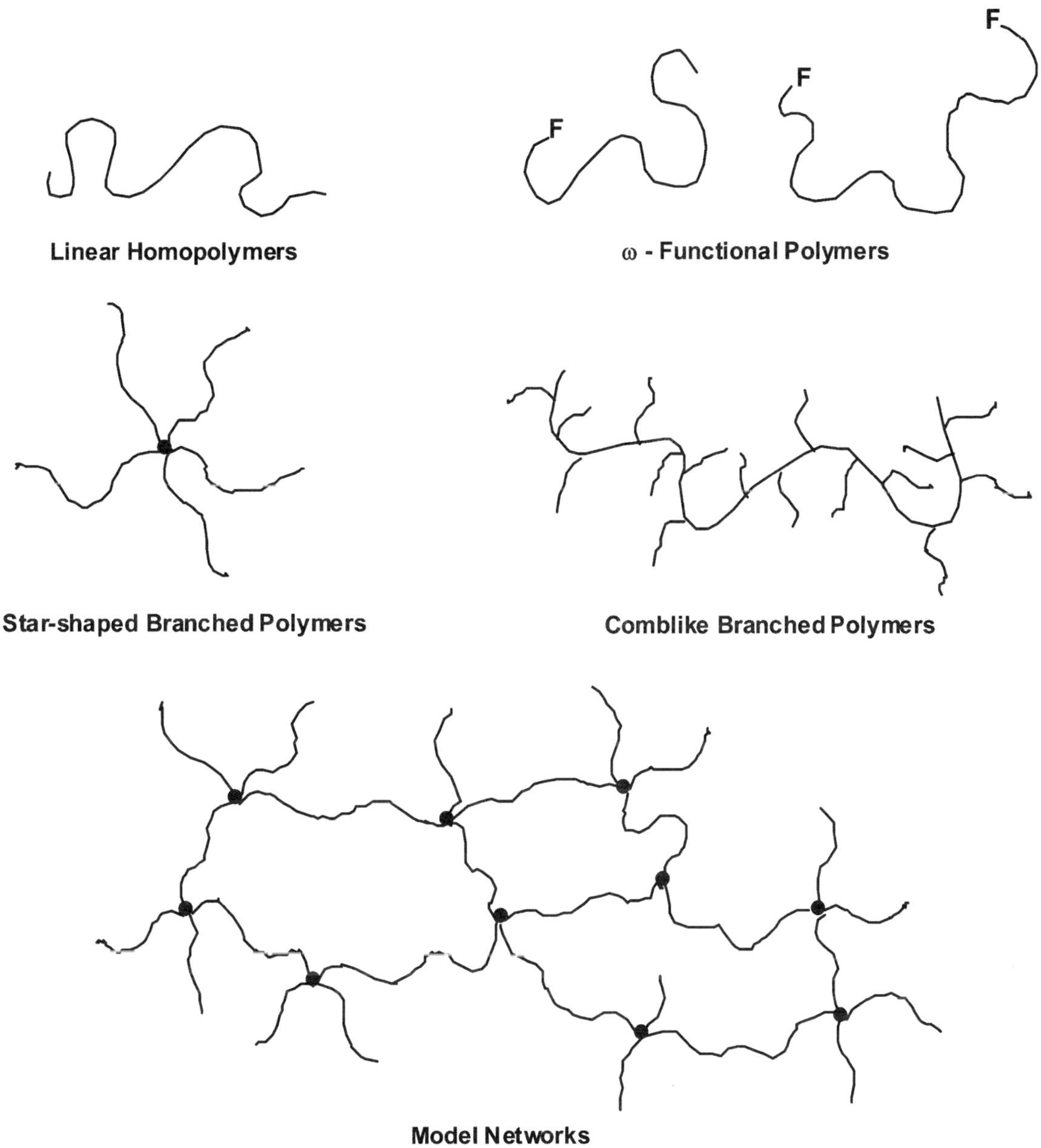

Fig. 4.160—Different polymer configurations [462].

Solution osmometry is the primary method used to determine the number-average molecular weight and light scattering is the primary method for determining the weight-average molecular weight. Gel permeation chromatography (GPC), or size-exclusion chromatography (SEC), although a secondary method, is extensively used for determining the polymer molecular weights. This technique is popular because it allows the calculation of all three molecular weight-averages from a single elution curve [223 225].

GPC makes use of a column packed with beads of porous cross-linked polystyrene gel, with an average pore size of 60 to 10^7 Angstrom. As the polymer solution passes through the column, the polymer molecules enter the pores that are large enough to accept them. Since there are a larger number of pores for smaller molecules to enter and pass through than for larger molecules, the larger molecules come out of the column (elute) faster. The result is the separation of the polymer molecules based on their molecular size. The elution is monitored by the use of a detector, a differential refractometer being the most common. This type of detector measures the difference in the refractive index of the polymer solution and the pure solvent, the difference being proportional to the concentration of the polymer in solution. Figure 4.162 shows a schematic of how different molecular weight fractions separate in a GPC column.

The GPC chromatogram, one presented in Fig. 4.163, shows a plot of the polymer concentration against the elution or retention volume. To obtain meaningful data, the retention volume scale must be calibrated to the molecular weight scale by the use of standard polymers. For polymers to be good standards, they must have narrow molecular weight distributions, and their molecular weights must be predetermined by the use of the primary methods, such as light scattering and membrane osmometry.

While average molecular weight information is important, it is the molecular weight distribution that is more use-

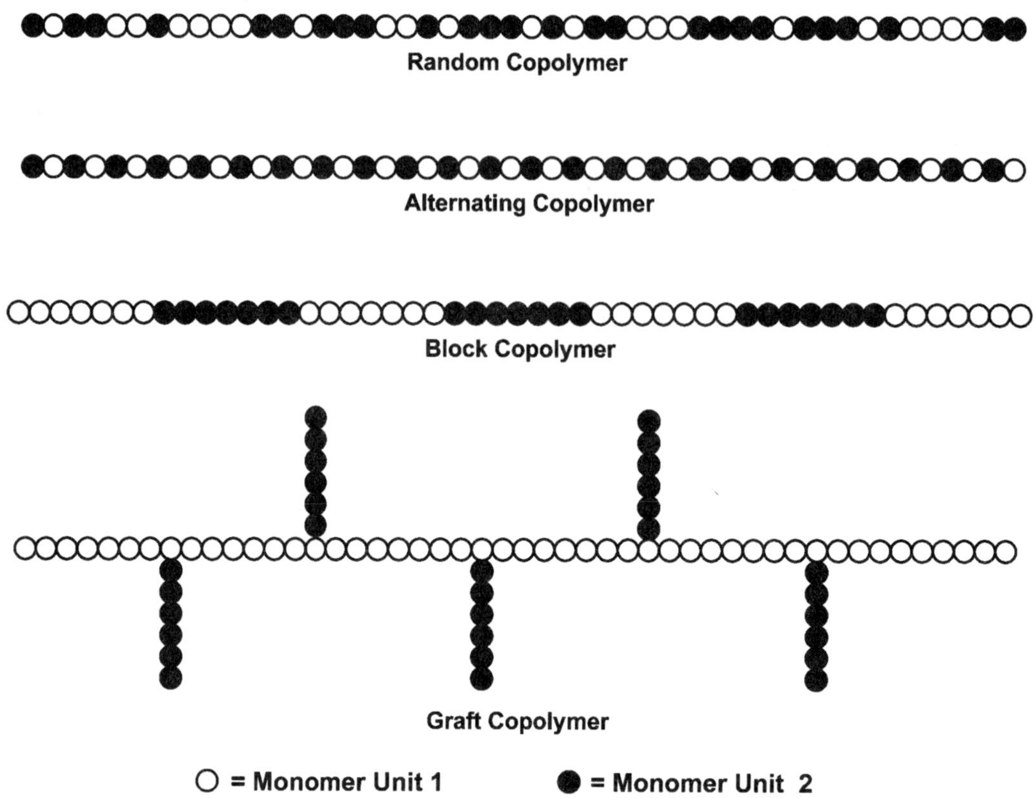

Fig. 4.161—Different types of linear polymers [463].

ful in understanding the polymer properties. A number of fractionation methods other than GPC are available for this purpose [465]. The molecular weight distribution is measured by the *polydispersity index* or *the heterogeneity index*, which is the ratio of the weight-average molecular weight to the number-average molecular weight, or Mw/Mn. For mono-disperse polymers, which contain molecules of essentially the same chain length, the value of this index is close to 1 and for poly-disperse polymers, it is greater than 1. This is because in the latter case there is a greater contribution of the higher molecular weight fractions towards the molecular weight. The relationship between different types of molecular weights is shown in Fig. 4.164 [224]. It can be seen from the figure that Mn is less than Mw, which in turn is less than Mz. The viscosity-average molecular weight (Mv), which is easy and quick to determine from the intrinsic viscosity of the polymer solution, is very close to Mw. Hence, it is sometimes used to approximate the weight-average molecular weight (Mw).

The *polydispersity index* is a function of the polymeriza-

TABLE 4.26—Parameter values to compute molecular weights of an equimolar polymer composition.

ni	Mi	niMi	niMi^2	niMi^3
1	8000	*n1M1* 1×8000=8000	1×8000^2=64,000,000	1×8000^3=5.12×10^{11}
1	9000	*n2M2* 1×9000=9000	1×9000^2=81,000,000	1×9000^3=7.29×10^{11}
1	10,000	*n3M3* 1×10,000=10,000	1×10,000^2=100,000,000	1×10,000^3=1.00×10^{12}
1	11,000	*n4M4* 1×11,000=11,000	1×11,000^2=121,000,000	1×11,000^3=1.331×10^{12}
1	12,000	*n5M5* 1×12,000=12,000	1×12,000^2=144,000,000	1×12,000^3=1.728×10^{12}
Σni=5		ΣniMi=50,000	ΣniMi^2=510,000,000	ΣniMi^3=5.300×10^{12}

TABLE 4.27—Parameter values to compute molecular weights of a nonequimolar composition.

ni	Mi	niMi	niMi^2	niMi^2
0.5	8000	*n1M1* 0.5×8000=4000	0.5×8000^2=32,000,000	0.5×8000^3=2.56×10^{11}
1.0	9000	*n2M2* 1.0×9000=9000	1.0×9000^2=81,000,000	1×9000^3=7.29×10^{11}
1.5	10,000	*n3M3* 1.5×10,000=15,000	1.5×10,000^2=150,000,000	1.5×10,000^3=1.50×10^{12}
1.5	11,000	*n4M4* 1.5×11,000=16,500	1.5×11,000^2=181,500,000	1.5×11,000^3=1.9965×10^{12}
0.5	12,000	*n5M5* 0.5×12,000=6000	0.5×12,000^2=72,000,000	0.5×12,000^3=8.64×10^{11}
Σni=5		ΣniMi=50,000	ΣniMi^2=516,500,000	ΣniMi^3=5.3445×10^{12}

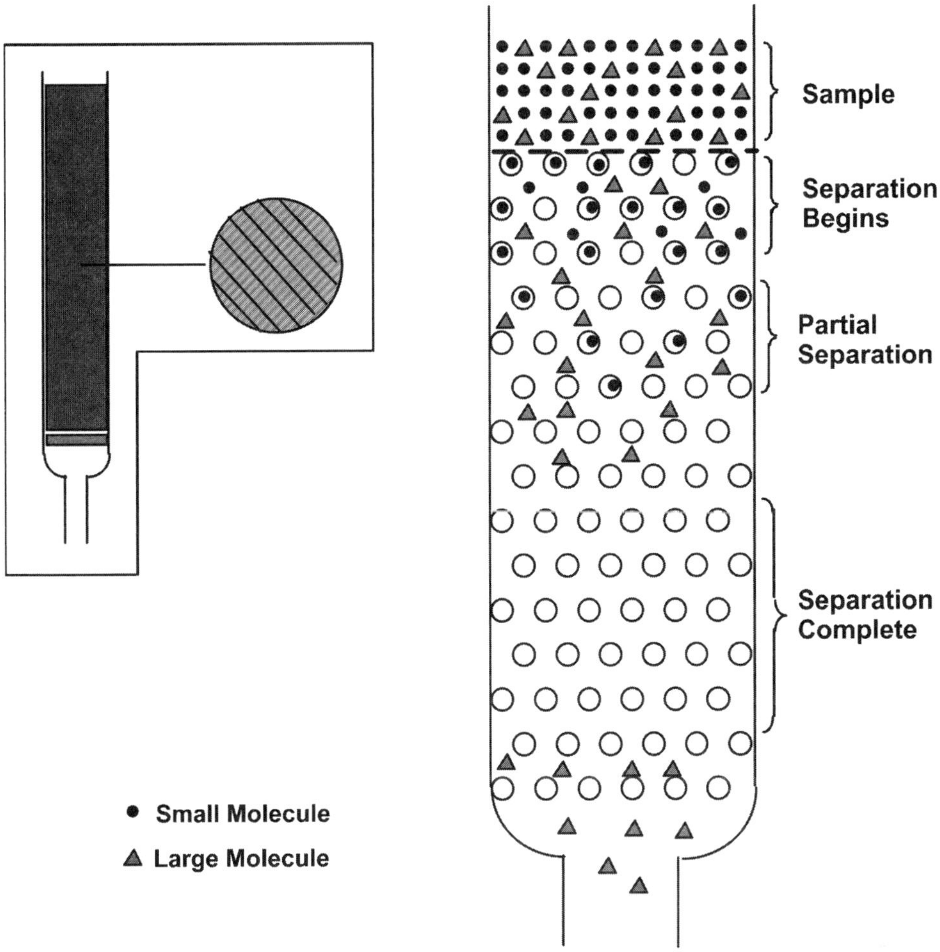

Fig. 4.162—Principle of GPC separation.

tion method. It is closer to 1 for polymers derived from anionic polymerization, 1.5–2.0 for polymers derived from step growth polymerization, between 2 and 5 for polymers derived from free radical polymerization, and above 5 for polymers derived from polymerization using the coordination catalysts.

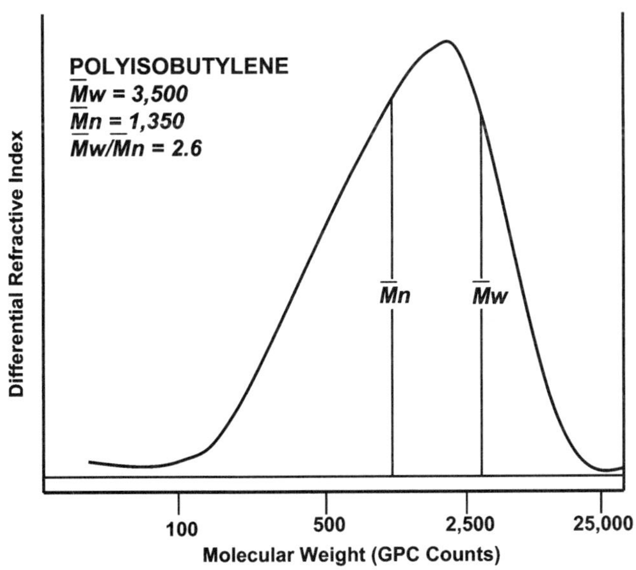

Fig. 4.163—A GPC chromatogram.

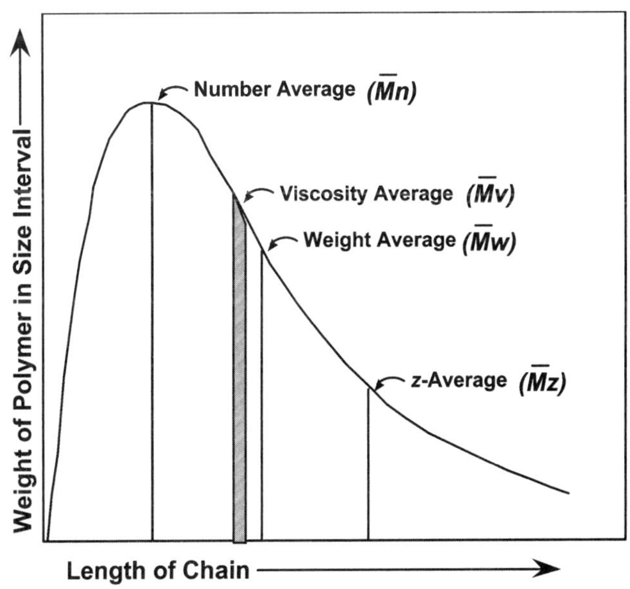

Fig. 4.164—Distribution of molecular weights in a typical polymer [224].

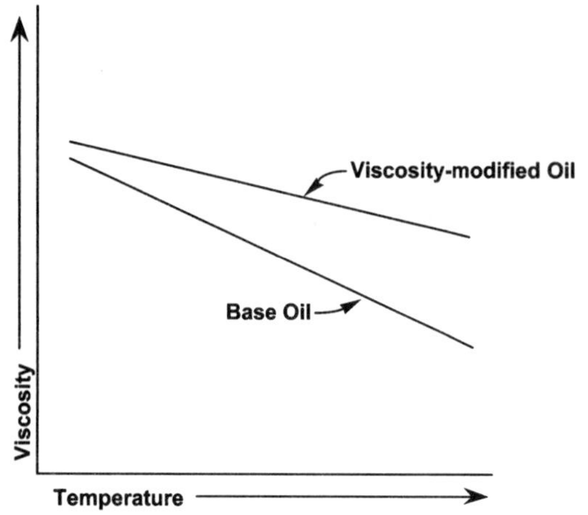

Fig. 4.165—Effect of viscosity modifier on VT relationship of an oil.

Viscosity Modifiers

The principal function of a viscosity modifier is to minimize viscosity variations with temperature. Previously, viscosity index was used as a measure of an oil's response to temperature changes. Viscosity index (VI), which is derived from the viscosity of the oil at 40°C and 100°C, is no longer meaningful. This is because most modern equipment operates at extreme temperatures of −40 to 150°C. At these temperatures, some times the viscosities do not conform to those predicted by the viscosity index [4,308].

Viscosity modifiers are polymers with average molecular weights of 10,000 to 150,000 g/mol; but those with molecular weights between 10,000 and 20,000 g/mol are used most often. These chemicals are added to the low-viscosity oils to improve their high-temperature lubricating characteristics. They minimize the oil's viscosity change with a change in temperature. This is a practical means of extending the operating range of the mineral oils to higher temperatures, without adversely affecting their low-temperature fluidity.

Viscosity modifiers cause an increase in the oil's viscosity at all temperatures, except that thickening at lower temperatures is significantly less than that at higher temperatures. At low temperatures, the polymer molecules occupy a small volume and therefore have a minimum association with the bulk oil. That is, it has small *hydrodynamic volume*. Hydrodynamic volume is the volume of the polymer and the associated oil. The effect is little viscosity increase. At high temperatures, however, the situation is reversed because the polymer chains extend or expand as a consequence of the added thermal energy. This increases the polymer's association with the bulk oil due to an increase in its surface area [50]. The result is an effective increase in viscosity. Another way to describe this phenomenon is that at higher temperatures the polymer becomes more soluble, thereby causing the viscosity to increase. The effect of a polymer on the viscosity-temperature (VT) properties of an oil is depicted in Fig. 4.165. Note that the VT line for the viscosity-modified oil has a smaller (shallower) slope than that for the base oil, thereby indicating a smaller drop in viscosity with increasing temperature. Figure 4.166 illustrates the mechanism of the oil thickening by the viscosity modifiers [50].

Variable thickening of the oil by the viscosity modifiers at low and high temperatures allows the formulation of the multi-grade oils. The multi-grade engine oils are designed to provide adequate viscosity at high temperatures for engine protection and low viscosity at low temperatures for easy startability. Figure 4.167 shows the VT characteristics of single grade and multi-grade oils [4]. It is important to note that the effect of the viscosity improver on different viscosity oils is different; the lower the initial viscosity, the higher the VI improving effect. This is shown in Fig. 4.168 [4]. Similarly, the response of the lower viscosity index oil to a viscosity modifier is larger than that of the higher viscosity index base oil, as is shown in Fig. 4.169 [4].

As mentioned while discussing viscosity, it is dangerous to assess the low-temperature viscosities of the VI-improved oils by linearly extrapolating the VI curves based on 40 and 100°C viscosities. This is because such oils can show an inflection at low temperatures, thereby leading to erroneous viscosity estimates [4]. The low-temperature viscosities of the VI-improved oils should be determined experimentally. It is also important to note that superior viscosity-temperature relationship of the multi-grade oils, as expressed by the viscosity index, is not solely due to a greater preferential swelling of the polymer at higher temperatures. The use of the low-viscosity oils with good viscosity-temperature behavior to prepare these oils makes a substantial contribution towards the overall effect [466]. In addition to affecting the VT relationship, the viscosity modifiers affect

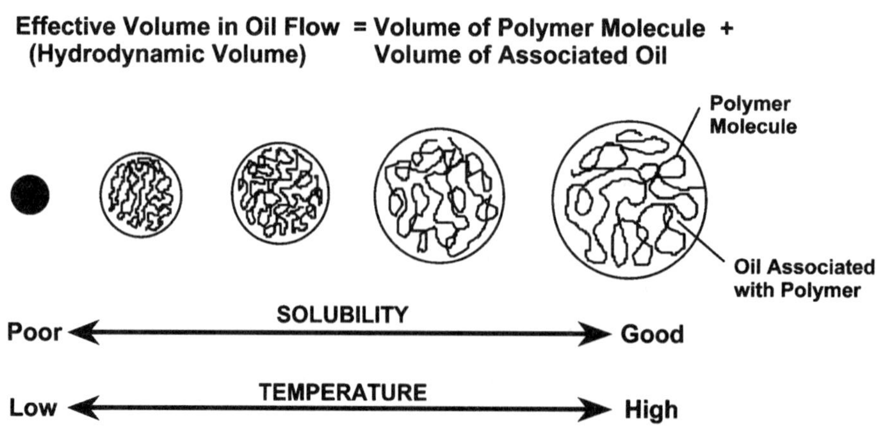

Fig. 4.166—Mechanism of oil thickening by viscosity modifiers [50].

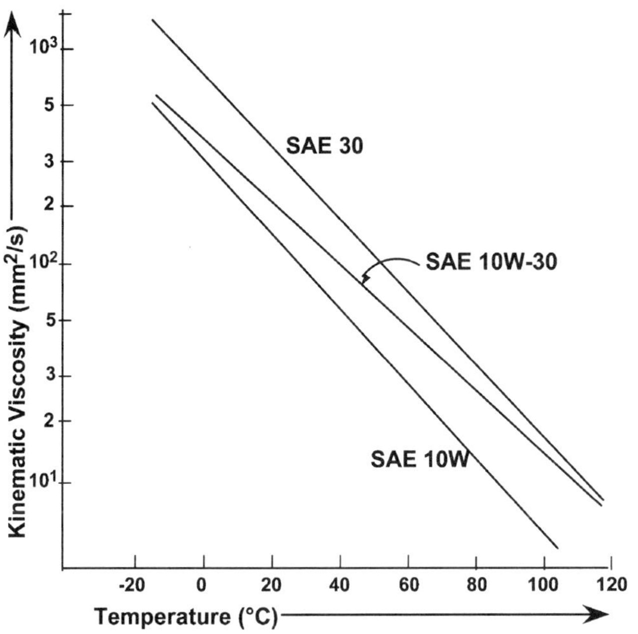

Fig. 4.167—Viscosity-temperature characteristics of single grade and multi-grade oils [4].

a lubricant's other properties. These include pour point, dispersancy, fuel economy, and indirectly the extreme pressure performance.

Commercially available viscosity modifiers belong to two general classes: olefin-based polymers and ester polymers. The olefin-based polymers include polyisobutylenes (PIBs), olefin copolymers (OCPs), and hydrogenated styrene-diene (STDs) polymers. OCP polymers from ethylene-propylene mixtures are called EPRs (ethylene propylene rubbers) and ethylene-propylene-diene mixtures are called EPDMs (ethylene propylene diene monomer rubbers). Ester polymers include polymethacrylates (PMAs) and styrene ester polymers (SEs).

Viscosity modifiers find major use in multi-grade engine oils and gear oils, transmission fluids, power steering fluids, greases, and some hydraulic fluids. Olefin copolymers are of the most popular type, followed by polymethacrylates, styrene-diene polymers, and the styrene-ester polymers. OCPs, EPRs, and EPDMs find extensive use in engine lubricants.

Olefin-based Polymers

These include polyisobutylenes, butyl rubbers, and olefin copolymers. Polyisobutylenes (PIBs) and butyl rubbers are both isobutylene-derived, except that PIBs are of somewhat lower molecular weights than butyl rubbers. Unlike PIBs that are derived from pure isobutylene, butyl rubbers are made from isobutylene containing 3–8 % diene in the monomer mixture. Both types are manufactured by the Lewis acid-catalyzed polymerization.

Olefin copolymers (OCPs) are block polymers with rubber-like properties. EPRs comprise ethylene and propylene and EPDMs contain a third nonconjugated monomer. Because of their high thickening efficiency and low cost, these polymers are widely used in passenger car motor oils and diesel engine lubricants as viscosity modifiers. They are prepared from olefin mixtures by vanadium-based Ziegler-Natta catalysis or by the use of metallocenes [467]. In EPRs and EPDMs, the ethylene to propylene ratio and their proper distribution in the backbone are critical to the polymer's low-temperature properties. An ethylene/propylene ratio of between 45/55 to 55/45 range yields a polymer that is amorphous at room temperature and cold flows. However, if the ethylene/propylene ratio increases beyond 60/40, the copolymer becomes semi-crystalline and does not cold-flow under ambient conditions. Amorphous OCPs are used as vis-

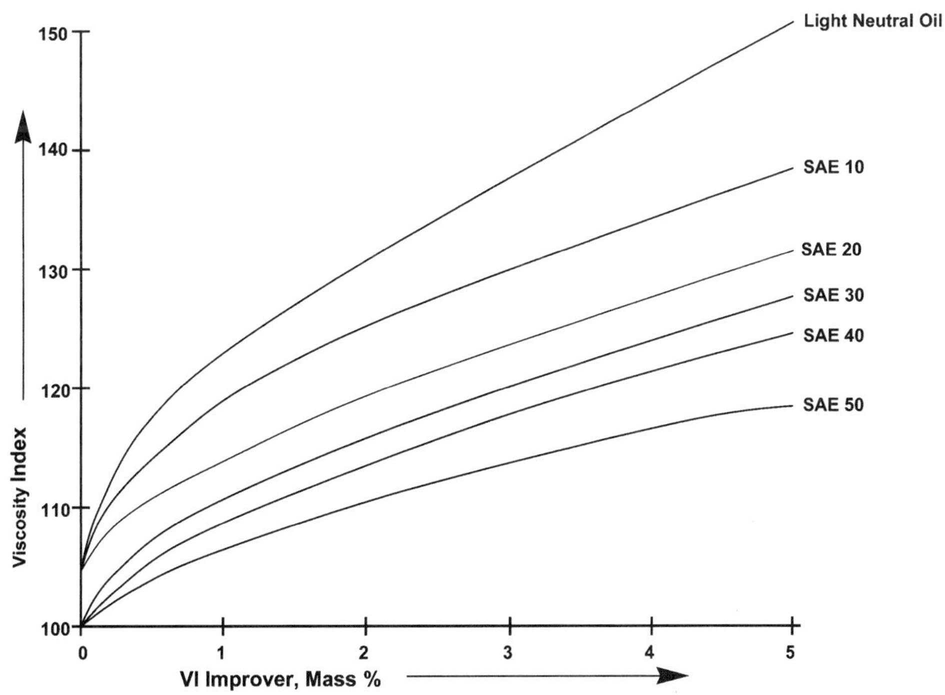

Fig. 4.168—Effect of VI improver on base oils of different viscosities [4].

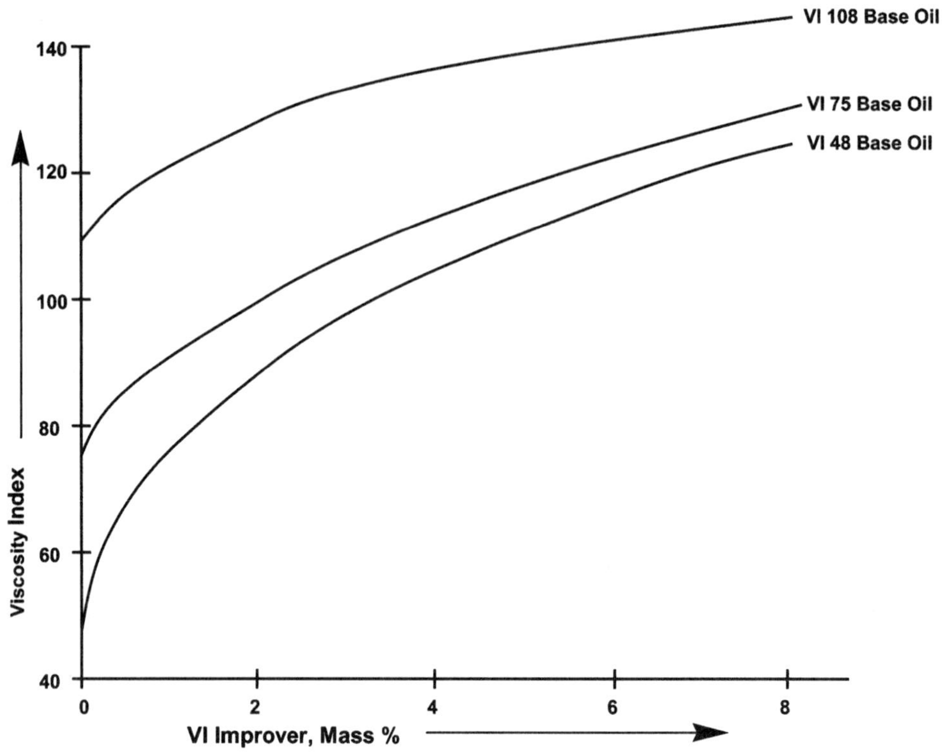

Fig. 4.169—Effect of VI improver on base oils of different viscosity indices [4c].

cosity modifiers and are available both in semi-solid form and as liquid concentrates in mineral oil. Typical concentrates contain between 10 to 15 % polymer and have 100°C viscosity of between 500 to 1500 cP (mPa·s^{-1}). Typical viscosity-concentration relationship is provided in Fig. 4.170 [467]. Generally, the higher the molecular weight of a polymer, the more prone it is to the mechanical degradation under the forces of shear. This is true for the OCPs as well.

α-Olefin (alphaolefin) copolymers, the lower molecular weight analogues of the OCPs, are prepared by polymerizing the α-olefins in the presence of a Lewis acid. See Chapter 3 on Synthetic and Natural Base Stocks. These polymers find use in lubricants, such as power steering pump and gear oils that require enhanced shear stability. The polyisobutylenes (PIBs) of the molecular weights of 2000 to 3000 g/mol are used as viscosity modifiers in gear lubricants, hydraulic fluids, and industrial oils.

OCPs are inferior to polymethacrylates with respect to

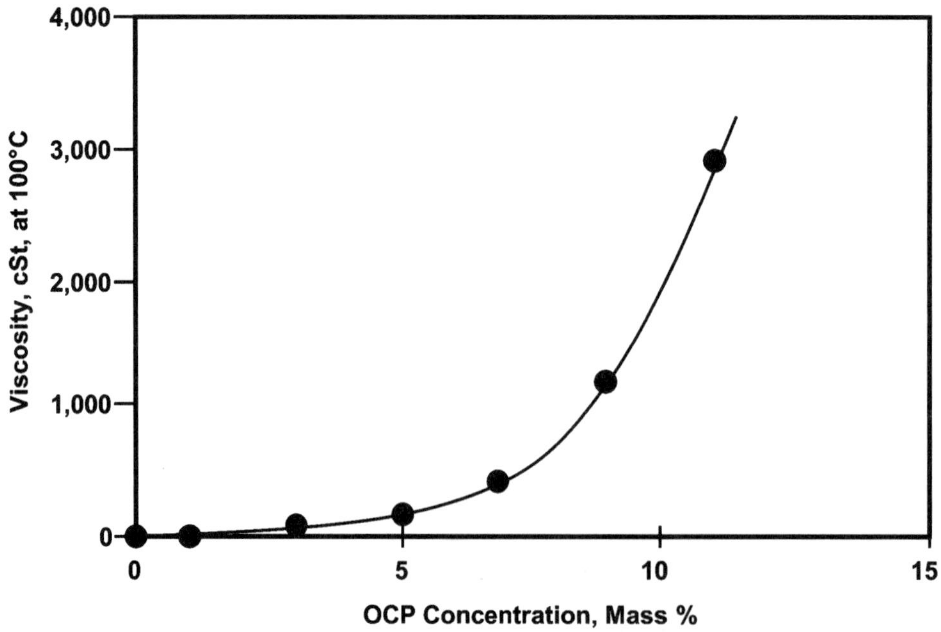

Fig. 4.170—Kinematic viscosity of 50 PSSI (Permanent Shear Stability Index) OCP in 100 N mineral oil [467].

TABLE 4.28—OCP structure versus thickening efficiency [467].[a]

Molecular Weight (M_W)	Viscosity, cSt Linear OCP	OCP with 2 % Branching Agent
230,000	13.05	12.03
180,000	11.17	10.87

[a]Thickening efficiency is defined as the kinematic viscosity at 100°C at a 1.0 % weight polymer solution in 6.05 cSt mineral oil.

TABLE 4.29—Rheological comparison of lubricants containing OCP viscosity modifiers differing in molecular weight [467].

Viscosity Modifier	Molecular Weight (M_W)	PSSI	Capillary Viscosity *cP @ 150 °C*	HTHS Viscosity *cP @ 150°C*	% TVL
OCP 1	160,000	45	5.33	3.43	36
OCP 2	80,000	30	5.33	3.77	29
OCP 3	50,000	22	5.33	3.88	27

low-temperature properties and generally require pour point depressants to improve the low-temperature performance of the lubricants containing them. Their low-temperature performance can be improved through copolymerization and or grafting with polyacrylates, which was commented upon while discussing the dispersant polymers. Besides OCPs with in-built dispersancy and pour point lowering ability, olefin copolymers that contain antioxidant and antiwear functionalities are also commercially available [467].

As mentioned earlier, the ethylene-propylene ratio in EPRs and EPDMs impact their properties. For use in engine oils, we are primarily concerned with oil solubility, thickening efficiency, viscometrics, and shear stability. When the degree of crystallinity, which is a function of their ethylene content, increases beyond 25 %, the polymer has limited solubility in oil. The degree of crystallinity of a polymer is most often determined by Differential Scanning Calorimetry, or DSC. Thickening efficiency is the amount of the polymer required to achieve a certain degree of boost in viscosity. It is largely a function of the polymer's molecular weight and its molecular weight distribution, or the polydispersity. Shear stability is the ability of a polymer to maintain its structural integrity under the forces of shear. Shear stability is inversely related to the polymer's molecular weight; the higher the molecular weight, the lower the shear stability. These and other desirable properties of the polymers will be considered in detail in the latter part of the discussion. Here, it suffices to say that the presence of the long-chain branching in a polymer has a negative effect on the thickening efficiency, hence viscosity. This is because the branching diminishes the hydrodynamic volume, which is a consequence of the less interaction of the polymer with the oil. Table 4.28 shows thickening efficiency data for amorphous (noncrystalline) OCP viscosity modifier, with and without 2 % branching agent used as co-monomer [467]. As one can see, the addition of a branching agent significantly lowers the final oil viscosity due to polymers of the both molecular weights. SAE J300 has a High-Temperature, High-Shear-Rate (HTHS) Viscosity requirement to meet the lubrication needs of the concentric journal bearings. In the test, the temporary viscosity loss is measured. A lower viscosity loss is desired. Table 4.29 shows the performance of polymers of various molecular weights in the test. The data show a direct relationship between the molecular weight and the percent temporary viscosity loss (TVL) [467]. Typically, the polymers of equal weight-average molecular weight (Mw) with narrow molecular weight distribution (lower polydispersity) undergo less temporary viscosity loss than the polymers of the broader polydispersity. HTHS viscosity can only be adjusted by increasing the viscosity of the base oil or by increasing the viscosity modifier concentration. It is difficult to adjust by other means since the engine oil must also meet the kinematic viscosity requirements. Permanent shear-related viscosity loss results when the shear forces break the longer polymer chains into shorter, lower molecular weight fragments. Permanent viscosity loss is similar to the temporary viscosity loss, but is measured by comparing the kinematic viscosity of the polymer treated fluid before and after subjecting it to shear. Permanent shear stability index, PSSI or simply SSI, is defined by ASTM D6022, as given by the equation that follows. Commercial OCP viscosity modifiers have SSI values between 23 and 55 [467].

$$\mathrm{PSSI} = \mathrm{SSI} = \frac{(\mathrm{Vo} - \mathrm{Vs}) \times 100}{(\mathrm{Vo} - \mathrm{Vb})}$$

where Vo = original oil viscosity, Vs = oil viscosity after shear, and Vb = base oil viscosity.

While the base oil and the viscosity modifier together influence low-temperature pumpability, it is the pour point depressant that primarily controls this property. SAE J300 specifies the Mini-rotary Viscometer (MRV) Test (ASTM D4684) to measure this lubricant parameter. The Scanning Brookfield Test (ASTM D5133) and the Pour Point (ASTM D97), although not part of the SAE J300, must also be considered since they are a part of the other standards, such as those of the OEMs, oil marketers, and governmental agencies, such as the U.S. Military. It is important to note that due to the advances in the base fluid technology, both mineral and synthetic, formulating them to meet all the rheological requirements of the SAE J300 is not easy. Certain types of viscosity modifiers can interact with the base oils and the pour point depressants at low temperatures to result in high MRV viscosities [467]. For example, OCPs that contain longer ethylene sequences can interact with the oil's wax crystals at low temperatures to cause higher MRV viscosity and yield stress. Such polymers may also be more sensitive to the type of pour point depressants used in the formulations. Testing of the multi-grade lubricants containing OCP viscosity modifiers in passenger car and heavy-duty truck engines has been carried out for many years and it is generally believed that the polymer-containing oils have a negative effect on engine varnish, sludge, and piston ring deposits. However, Kleiser and co-workers [468] based on taxicab fleet test data determined that an SAE 5W-30 oil that contained a higher amount of OCP viscosity modifier showed better engine deposit control than an SAE 15W-40 oil that contained a lower amount of OCP. They also observed significant improvements in sludge and varnish ratings, when a highly functionalized dispersant OCP was used. As stated before,

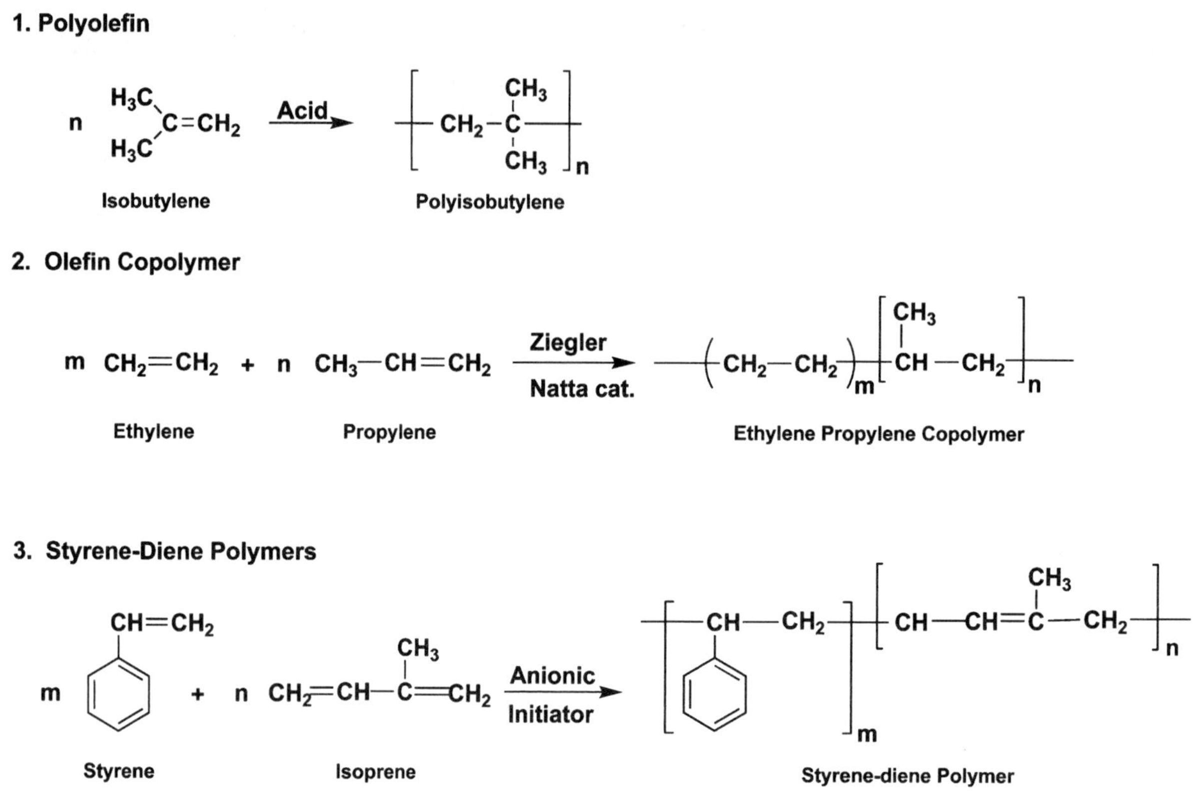

Fig. 4.171—Synthesis of olefin-based polymers.

the engine oils that use dispersant-functionalized OCPs typically need less ash-less dispersant to achieve an acceptable level of engine cleanliness than those that use nonfunctionalized OCPs.

Styrene-diene polymers (STDs) can be of the di-block or random-block type and are produced by anionic polymerization of styrene and butadiene, or isoprene. This type of polymerization produces polymers with a narrower molecular weight distribution than those obtained by the use of Ziegler and Friedel-Crafts catalysts, or the free radical initiators. That is, their Mw/Mn is closer to 1. Because of the narrow molecular weight distribution, these polymers possess the best thickening power of the types discussed so far. However, the presence of the aromatic rings and the double bonds makes these polymers more susceptible to oxidation. This problem is somewhat overcome by catalytically hydrogenating the double bonds. Figure 4.171 schematically presents the synthesis of these types of polymers.

Recently, a different type of polymer, labeled a "star" polymer (Radial Isoprene), has become commercially available. It has thickening power similar to that of the styrene-diene type, but has better shear stability. A clustered polyanion prepared from divinylbenzene and styrene forms the center of the star, and the rays are made up of the polymerized diene monomer units.

Ester Polymers

These polymers include polymethacrylates (PMAs) and styrene-ester (SE) polymers. Polymethacrylates (PMAs) are made by the polymerization of alkyl methacrylate monomers. The alkyl group either contains 1 to 7 carbons, 8 to 13 carbons, or 14 or more carbons [469]. The size of each alkyl groups affects the polymer properties differently. The smaller size alkyl group has low solubility in oil, especially at low temperatures, and it stays closely coiled. In other words, it occupies a small hydrodynamic volume; hence, it contributes little, if any, to viscosity. This is a benefit with respect to the pour point of the oil. However, as the oil temperature increases, the polymer chains open up and their association with the bulk lubricant increases, which increases the hydrodynamic volume; hence the oil viscosity. The presence of the alkyl group of 8 to 13 carbon atoms increases the solubility of the polymer in the oil, both at low and high temperatures, and contributes towards an increase in viscosity. It is important to note that the viscosity contribution at lower temperatures is a lot less than at high temperatures, primarily because of the smaller hydrodynamic volume. The inclusion of the alkyl group of 14 and more carbons imparts pour point depressing properties to the polymer since the alkyl group interacts with the wax crystals, which are of similar carbon size, during their formation. Polymethacrylates that are used solely as pour point depressants use a mixture of C_8-C_{13} and C_{14} and higher carbon monomers.

The carbon-carbon double bond in the alkyl methacrylate monomer is very reactive because of the conjugation with the ester carbonyl. This makes it amenable to addition type polymerization reaction. The presence of the methyl substituent on the double bond introduces steric hindrance, which deters attack by chemicals, including water, on the ester functional group. This imparts chemical resistance and hydrolytic stability to the polymer. The alkyl group in the methacrylate ester functional group is introduced either by direct esterification of the methacrylic acid or by transesterification. Direct esterification is commonly employed for short chain alkyl methacrylates. The process involves the re-

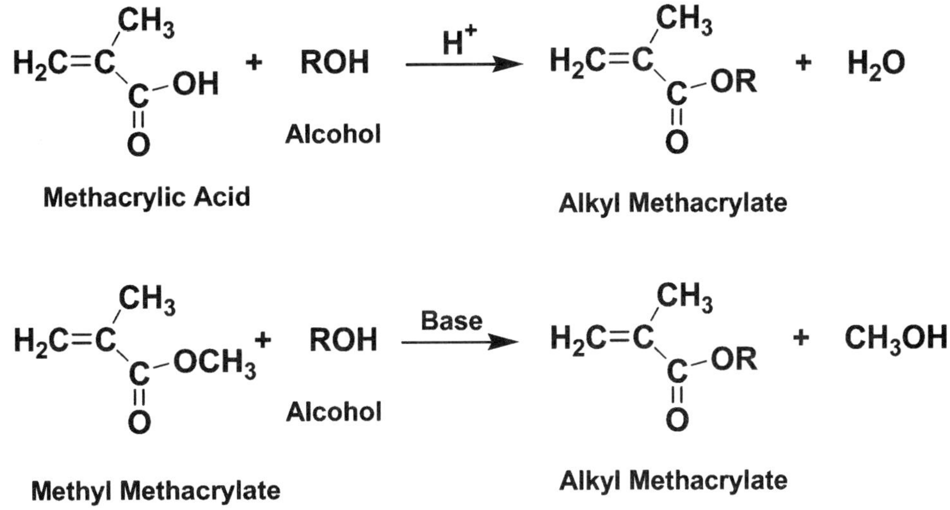

Fig. 4.172—Alkyl methacrylate synthesis.

action of an alcohol with methacrylic acid in the presence of an acid catalyst and removing the reaction water. On the other hand, transesterification is the method of choice for making long-chain alkyl methacrylates. This method involves the reaction of methyl methacrylate with an appropriate alcohol, usually in the presence of a basic catalyst. The reaction by-product, methanol, is removed to shift the equilibrium towards the product side. Both these reactions are depicted in Fig. 4.172. The methacrylic acid itself is synthesized from acetone via its cyanohydrin, followed by hydrolysis and esterification. An alternative method is via the oxidation of butylene, hydrolysis, and esterification [469].

Depending upon the polymer composition intended, either a single monomer or a mixture of monomers is treated with a free radical polymerization initiator, such as peroxide, hydroperoxide, or a per-ester. The initiator free radical attacks the terminal carbon of the double bond of the alkyl methacrylate molecule, forming a tertiary carbon free radical, which reacts with another molecule. The process continues until the process is terminated, which is done by the use of a free radical transfer agent, such as an alkyl mercaptan. The use of the chain transfer agents is necessary to control the molecular weight of the resulting polymer, so that it has proper viscometrics and shear stability. The chain transfer agent terminates polymerization by transferring hydrogen to the polymer and in the process forms a new free radical, which is not reactive enough to take part in the polymerization reaction. Polymerization temperature is typically between 60–140°C, and requires selecting an initiator which has a reasonable half life in this temperature range. Higher temperatures usually result in polymers of lower molecular weight and lower temperatures result in polymers of higher molecular weight. This is because at higher temperatures, the initiator-derived free radical formation is rapid; resulting in many free radicals and hence many polymer chains. See Fig. 4.173 for the polymerization mechanism using benzoyl peroxide as initiator. Its half life is one to ten hours, depending upon the temperature. Commercial polymethacrylates have molecular weights between 20,000 and 750,000 g/mol. Higher molecular weight polymethacrylates are almost always prepared in a solvent, which is necessary to lower their viscosity to a managing level, both during polymerization and afterwards. The solvent to be used must be nonreactive, have low volatility, avoid chain transfer reaction during polymerization, and be suitable for use in the final application. For the lubricant-related applications, mineral oil with high saturates content is quite suitable. Depending upon the application, one can choose oils of 40°C kinematic viscosity between 2.7 cSt and 24 cSt. One can also use a volatile solvent with the above-listed attributes and exchange it with the desired oil [469]. The amount of oil or the solvent used is typically between 30 to 80 %; the exact amount depends upon the molecular weight of the polymer. The purpose is to facilitate handling and pumping.

Polymethacrylate based dispersant viscosity modifiers (DVMs) are also commercially available. As stated before, DVMs provide both the viscosity improvement and the dispersancy. The dispersant moiety in PMAs is introduced both by co-polymerization [187,285] and grafting [276–294]. Copolymerization usually involves nitrogen-containing methacrylate monomers, such as dimethylaminoethyl methacrylate, as shown in Figure 4.61. Grafting involves the use of either a nitrogen-containing monomer, such as N-vinylpyrrolidinone, directly, or via the incorporation of a functionalizable moiety, such as succinic anhydride, followed by the reaction of the resulting polymer with a polyamine [238,259,288–294]. Grafting has the advantage of incorporating monomers that have different reactivity than methacrylates and hence are not copolymerizable.

Polymethacrylates are chemically inert molecules that do not easily react with most chemicals. However, the reactions that they do undergo are depolymerization, oxidation, and hydrolysis. Depolymerization is the reverse of polymerization that occurs at high temperatures. The result is the formation of the high concentration of the original monomers. The temperatures that cause depolymerization, typically around 235°C, are not normally encountered in lubricant-related applications [235,470,471]. Another thermal reaction PMAs undergo is the thermal decomposition at ~250°C, involving the alkyl group of the ester functional group. The result is the formation of an olefin and a carboxylic acid group, which can react with an adjacent carboxylate

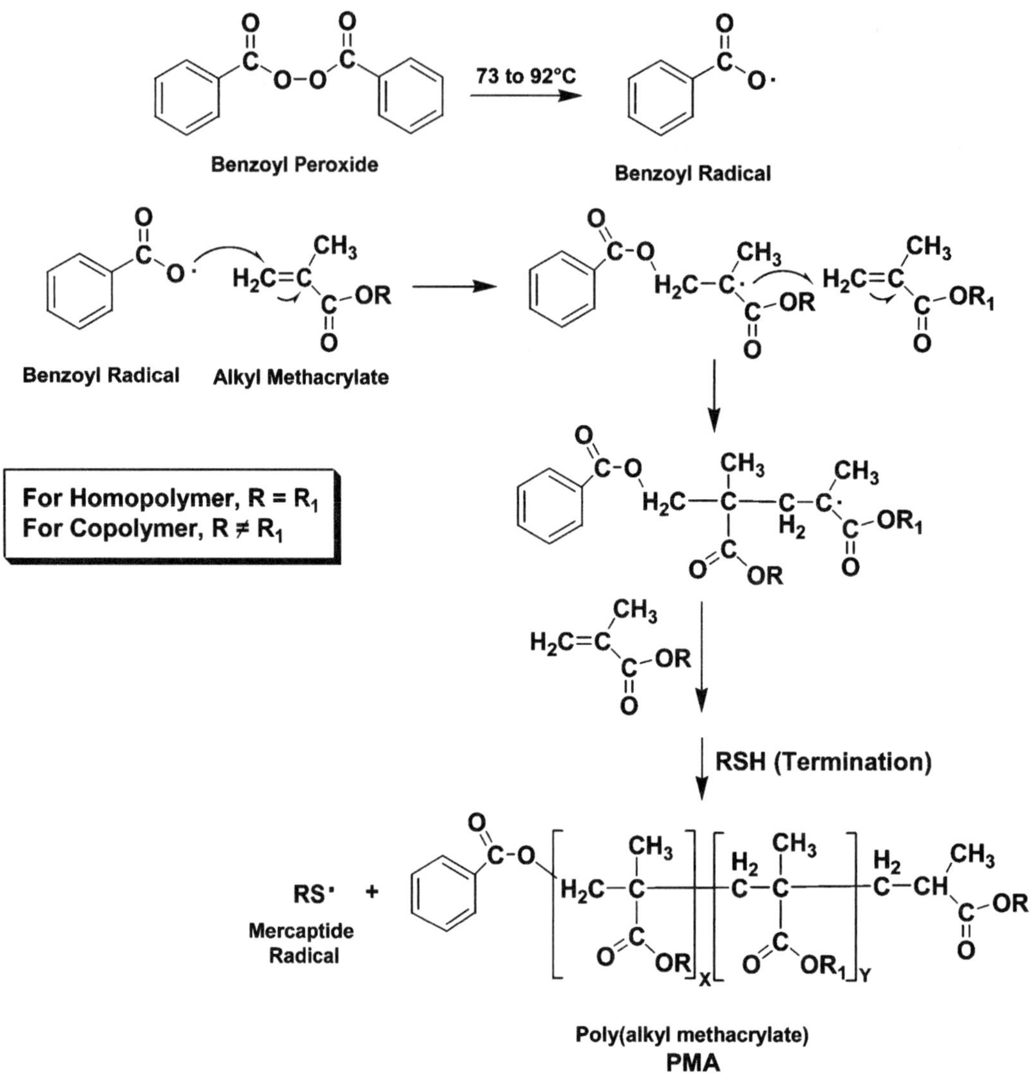

Fig. 4.173—Poly(alkyl methacrylate) synthesis.

functional group to form a cyclic anhydride, by losing a molecule of water or the alcohol. Thermal decomposition of the esters was described while discussing synthetic esters in the Chapter 3 on Synthetic and Natural Base Stocks. The ultimate result of the thermal reactions is the generation of the volatile materials and the loss of the polymer's viscosity improving and pour point depressing properties.

Like other hydrocarbon materials, polymethacrylates are also susceptible to oxidation. The result is the polymer chain scission to lower molecular weight polymeric fragments [469,470]. This lowers the thickening power of the polymer because of the drop in the molecular size. While all carbon hydrogen bonds are prone to oxidative attack, polymethacrylate structure does not contain any benzylic, allylic, and tertiary hydrogens that have the highest reactivity towards oxygen. See the oxidation inhibitor section of this chapter. Consequently, oxidative scission of the polymer is not a concern.

Polymethacrylates are quite stable to hydrolysis reactions since the hydrolyzable ester functional group is well surrounded by various groups, which hinder the approach of water towards the ester carbonyl. The presence of the medium to large hydrocarbon chains, typically present in PMAs used in lubricants, makes hydrolysis reaction even less likely because of the general hydrophobic nature of such PMAs.

Despite good overall thermal and chemical stability, PMAs are susceptible to mechanical shear, the same as OCPs and other polymers. Mechanical shearing not only decreases the size of the polymer chains but it also generates the free radicals. In lubricants, the free radicals are quickly quenched by the hydrogens of the hydrocarbon oil or by the oxidation inhibitor, if present. However, the formation of smaller polymer fragments is a concern because of their diminished effect on viscosity. As mentioned while discussing other polymers, shear stability of a polymer is related to its molecular weight, or more specifically to its linear size, and not the overall structure [472].

Since the polymethacrylates (PMAs) are produced by the free radical polymerization of alkyl methacrylates, the polymer has a relatively broad molecular weight distribution. Because of this, PMAs have low thickening efficiency and hence only moderate viscosity-improving ability. Since this ability depends upon the polymer's molecular weight, different polymers must be compared on an equal molecular

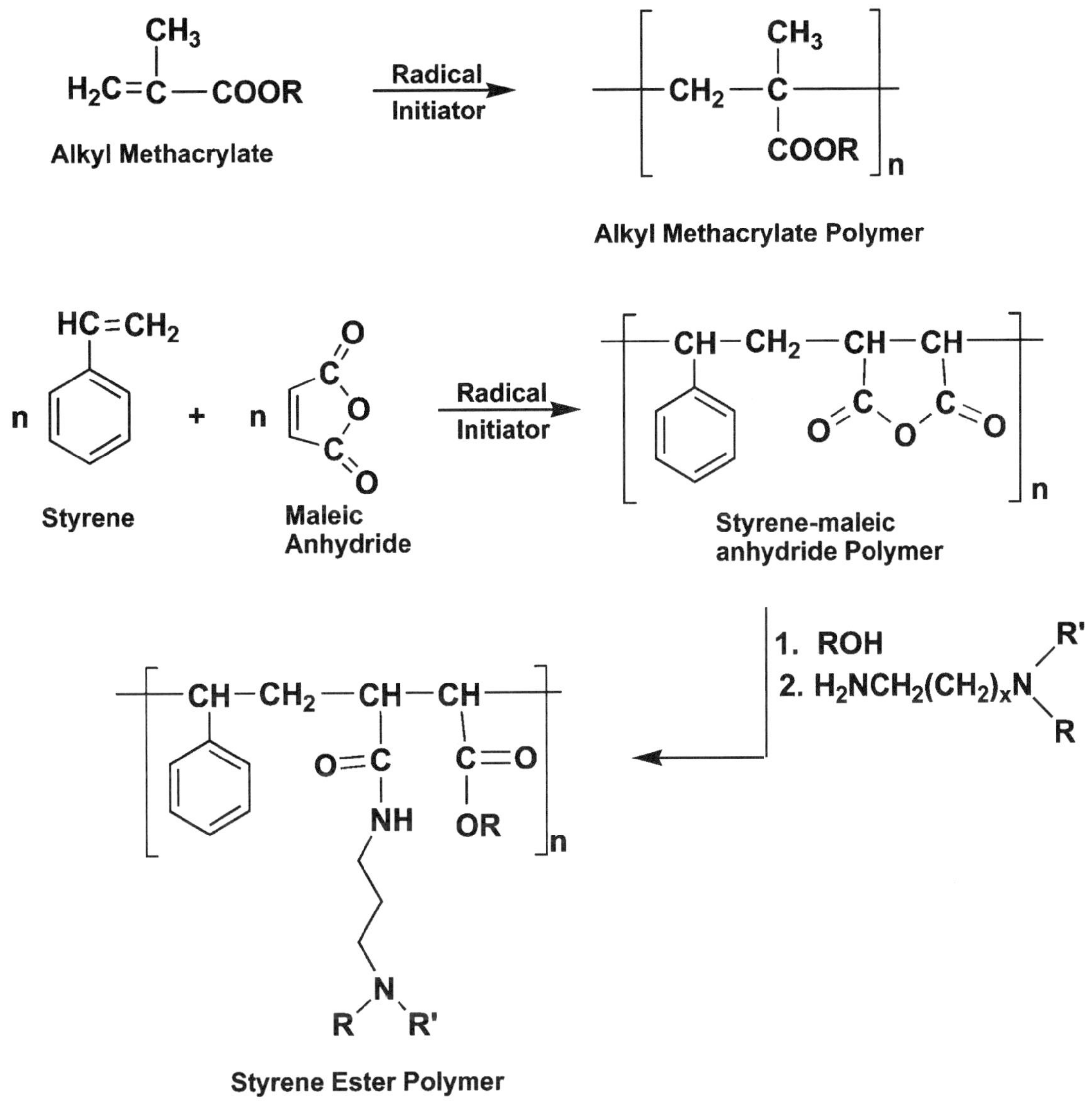

Fig. 4.174—Synthesis of alkyl methacrylate and styrene ester polymers.

weight basis. As mentioned above, there is flexibility to alter the alkyl group in the ester portion of the polymer to obtain products that have the best oil solubility and the optimum viscosity-improving properties. In addition, these polymers have good compatibility with a large number of refined and synthetic base stocks and are superior to olefin copolymers, both in oxidative and thermal stability and the low-temperature properties. Because of these attributes, PMAs find extensive use in a number of lubricants. Such lubricants include automotive engine oils, gear oils, automatic transmission fluids, hydraulic fluids, industrial oils, and greases.

Styrene-ester polymers are prepared by first copolymerizing styrene and maleic anhydride and then esterifying the intermediate alternating copolymer using a mixture of alcohols. Normally, the esterification step is taken to about 90 % or more, followed by post-neutralization using a bifunctional or polyfunctional amine. Because of the presence of the basic nitrogen, these polymers function both as a dispersant and as a viscosity modifier.

Figure 4.174 summarizes the methods of synthesis for polymethacrylate and styrene ester type viscosity modifiers. General structures of the various commercially available polymers along with their structural types are presented in Table 4.30.

Thickening Efficiency

Thickening efficiency and shear stability are two important considerations for selecting a polymer for use as a viscosity modifier. Thickening efficiency is a direct function of the polymer's molecular weight. More specifically, it is the function of the length of the polymer backbone as shown in Fig. 4.175. On an equal weight basis, a high molecular weight polymer provides higher viscosity than a low molecular weight polymer, as long as the structural features, such as branching, are similar. For a given molecular weight, the OCPs and PIBs have greater thickening power than styrene-diene polymers, which in turn have greater thickening power than poly(vinyl ether)s and polymethacrylates. In this regard, styrene ester polymers are the least effective among viscosity-modified polymers.

Shear-related Viscosity Loss

The viscosity loss in a viscosity-modified lubricant can result from mechanical, thermal, and oxidative degradation of the

TABLE 4.30—Commercial polymers.

Type	General Structure	Chemical Type	Commercial Product
Homo Polymer	—A—A—A—A—A—A—A—A—A—	**PMA** **PAO**	None PIB, POG-4, POG-6, Synthetic Fluids
Random Copolymer	—A—B—A—B—B—A—B—A—A—	**PMA** **OCP**	Acryloid 702, 716, 954, 985 1017,1019; TLA 388, 605 ECA 8899, 13112; LZ 7010, 7012; TLA 510A, 555, 656E, 6900, 7200, 7700
Alternating Copolymer	—A—B—A—B—A—B—A—B—A—	**Styrene Ester**	LZ 3702 and 3715
Di-block Copolymer	—A—A—A—A—B—B—B—B—B—	**Styrene Diene**	Shell Vis 40 and Shell Vis 50
Random Block Copolymer	—A—A—A—B—B—A—A—B—B—	**Styrene Diene**	LZ 7341 AND 7441
Graft Copolymer	—A—A—A(X)—B—B—A—A(X)—B—B— $X = \left[CH_2—C(OOR)(CH_3) \right]_n—H$	**OCP-g-PMA or Styrene-diene-g-PMA**	Viscoplex 2 Series
Star (Radial Isoprene) Polymer	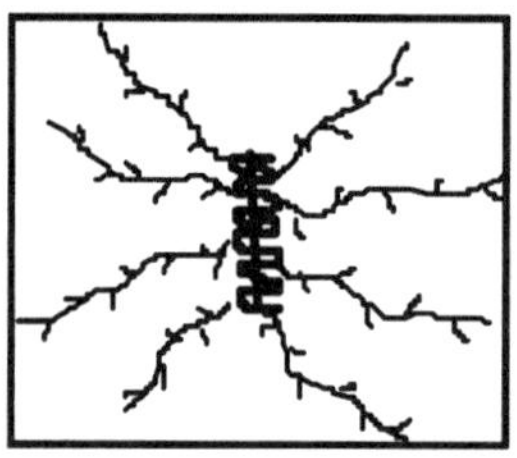	**Essentially OCP Rays**	Shell Vis 200 Series

polymer. Unlike mineral oils, which primarily exhibit Newtonian rheology, the polymer-thickened oils exhibit viscoelastic rheology. That is, their viscosity depends upon the degree of mechanical stress (shear). When oils containing the viscosity modifiers are subjected to moderate shear stress, viscosity decreases until it approaches the viscosity of the polymer-free oil. The mechanical or shear viscosity loss is generally encountered in those equipment parts that intermesh. Journal bearings, vane pumps, and gear pumps are examples of such parts. The speed of the moving surfaces also influences the shear rate and hence viscosity. The viscosity loss of a lubricant is directly proportional to the applied shear rate, as shown in Fig. 4.176 [50]. The higher the shear rate, the greater is the loss in viscosity. Piston rings experience low shear rates; hence, the lubricant in this region experiences a low viscosity loss. Conversely, rod bearings are a high shear environment; hence the lubricant undergoes a high viscosity loss. The viscosity loss in the regions of the main bearings and cylinder walls that have intermediate shear rates falls in between.

The viscosity loss can be temporary or permanent. If the viscosity bounces back to the original viscosity or close to it, when the stress is removed, it is termed as temporary viscosity loss. This type of loss is due to the reversible deformation of the polymer under the influence of the shear forces, which minimizes the association between the polymer and the lubricant. Temporary viscosity loss, shown in Fig. 4.177 [235] and Fig. 4.178, is desired in lubricants because it decreases the viscous drag at low temperatures and hence could contribute towards fuel economy.

The viscosity loss is considered permanent if after the shear forces are removed, the viscosity does not revert to its prior value [473]. Permanent viscosity loss occurs when the polymer in the viscosity-modified oil breaks down to the lower molecular weight fragments under the influence of shear, as depicted in Fig. 4.179 [235]. This type of viscosity loss is not desired because a formulated lubricant will not stay in its viscosity grade. Whether temporary or permanent, the viscosity loss depends upon the shear stability of the polymeric viscosity modifier used, which is a function of its molecular weight; more specifically it is the function of the size of the polymer backbone. This relationship, which holds true within a polymer type, is shown in Fig. 4.180.

High molecular weight polymers generally lose viscos-

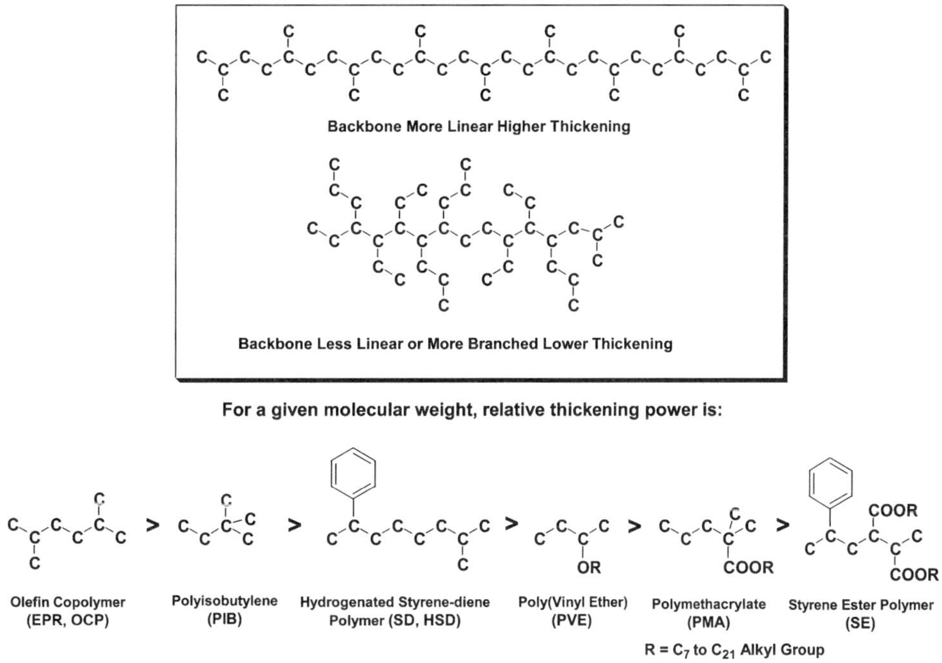

Fig. 4.175—Thickening power as a function of length and type of polymer backbone.

ity at a higher rate under shear than the low molecular weight polymers. Hence, the lubricants thickened with low molecular weight polymers are more likely to maintain their viscosity in the desired viscosity range than the lubricants thickened with high molecular weight polymers. Figures 4.181 and 4.182 depict the temporary and the permanent viscosity loss in oils containing viscosity modifiers of two different molecular weights [473]. As can be seen, in both cases, the lower molecular weight polymer experiences a lower viscosity loss.

The problem of viscosity loss due to shear can be alleviated by preshearing the high molecular weight polymer, prior to blending, or by choosing a more shear-stable polymer. Shear stability, which can be defined as the ability of a lubricant to resist viscosity loss under the influence of shear, primarily relates to the permanent viscosity loss.

Shear stability requirements for different applications parallel the severity of the lubricating environment. Engine lubrication is mostly hydrodynamic in nature; hence, the lubricants of low shear stability are adequate. Gear lubrication, on the other hand, is boundary in nature and therefore requires lubricants of high shear stability. Transmission and hydraulic fluids fall in between the two in terms of these requirements. This is shown in Fig. 4.183 [235].

The relationship between the polymer's molecular weight and its viscosity-improving effect, and the relationship between the polymer's molecular weight and its shear stability are presented in Fig. 4.184 [4]. Specific viscosity η_{sp}/c can be used as a measure of a polymer's molecular weight since the solution viscosity of the polymer-thickened lubricant is directly related to the molecular weight of the polymer. Specific viscosity is the fractional increase in vis-

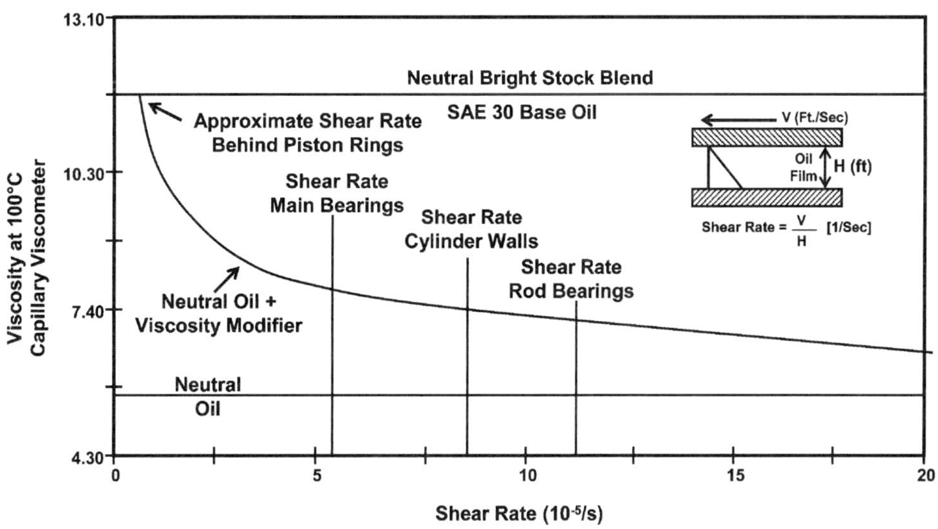

Fig. 4.176—Effect of shear rate on viscosity in different parts of an engine [50].

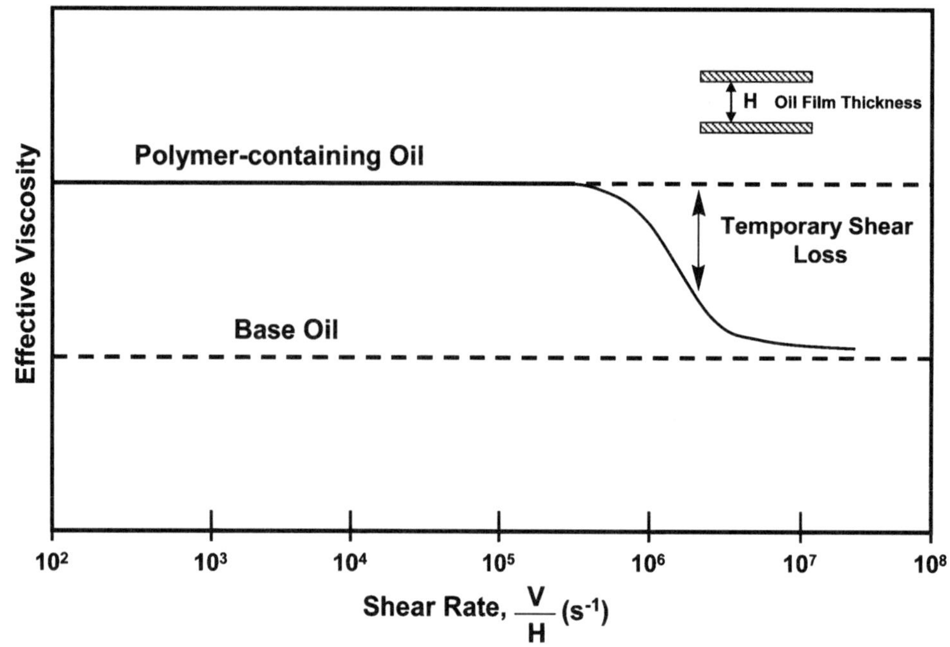

Fig. 4.177—Temporary viscosity loss due to shear [235].

cosity over that of the pure solvent caused by the addition of the polymer. As depicted in the figure, both the polymer's viscosity-improving ability and the shear sensitivity increase with an increase with its molecular weight (an increase in the specific viscosity). Higher shear sensitivity implies lower shear stability.

A number of tests are available to measure the viscosity improving properties and the shear stability of polymers in lubricants. The shear stability of a lubricant is measured by the CRC L-38 Test, FZG Shear Test, Sonic Shear Test, Orbahn Shear Test, Bosch Injector Test, and the Tapered Bearing Simulator Test. In terms of severity, the Tapered Bearing Test is the most severe, followed by the FZG Test, Orbahn Test, and the CRC L-38 Test.

Viscosity Loss Due to Polymer Degradation

In addition to mechanical breakdown (shear), polymers can also undergo thermal and oxidative degradation. Thermal degradation occurs when the polymerization process is reversed under the influence of heat. Polymers break down via chain scission to form the lower molecular-weight fragments. The consequence is a permanent loss in viscosity. This type of degradation, which is more prevalent in aromatic polymers due to their ability to form more stable allylic and benzylic free radicals, is shown in Fig. 4.185. Oxidative degradation occurs when the weak carbon hydrogen bonds in the polymer molecules react with oxygen to form hydroperoxides and peroxy free radicals. These species can disproportionate to form a number of lower molecular weight oxygenated compounds. This type of degradation not only causes a loss in viscosity but the resulting polar compounds also form varnish and coke deposits in a manner similar to that of the lubricants. The mechanism of oxidative degradation of polymers is shown in Fig. 4.186.

Table 4.31 compares the properties of interest between various classes of polymers. The data indicate OCPs to be the cheapest but in terms of the overall performance, styrene-diene types (SB and SI polymers) are the best. OCP-g-PMA is the best choice if both cost and performance are considered, which is most often the case. These polymers result from the grafting of the alkyl methacrylates onto the OCPs.

Dispersant Viscosity Modifiers (DVMs)

Dispersant viscosity modifiers (DVMs) based on polymethacrylates and OCPs are also commercially available. A new class of DVMs, made from PMA/OCP mixtures, was introduced in the 1990s. Dispersancy in these polymers is obtained by including the basic nitrogen or the surfactant type oxygen-containing monomers during the polymerization

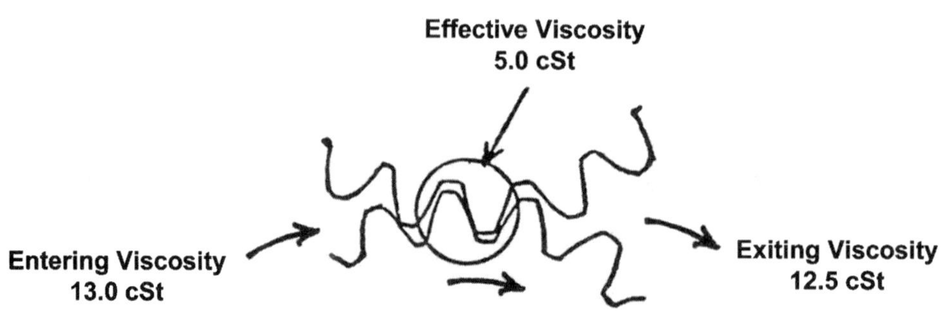

Fig. 4.178—Temporary viscosity loss in gear pumps.

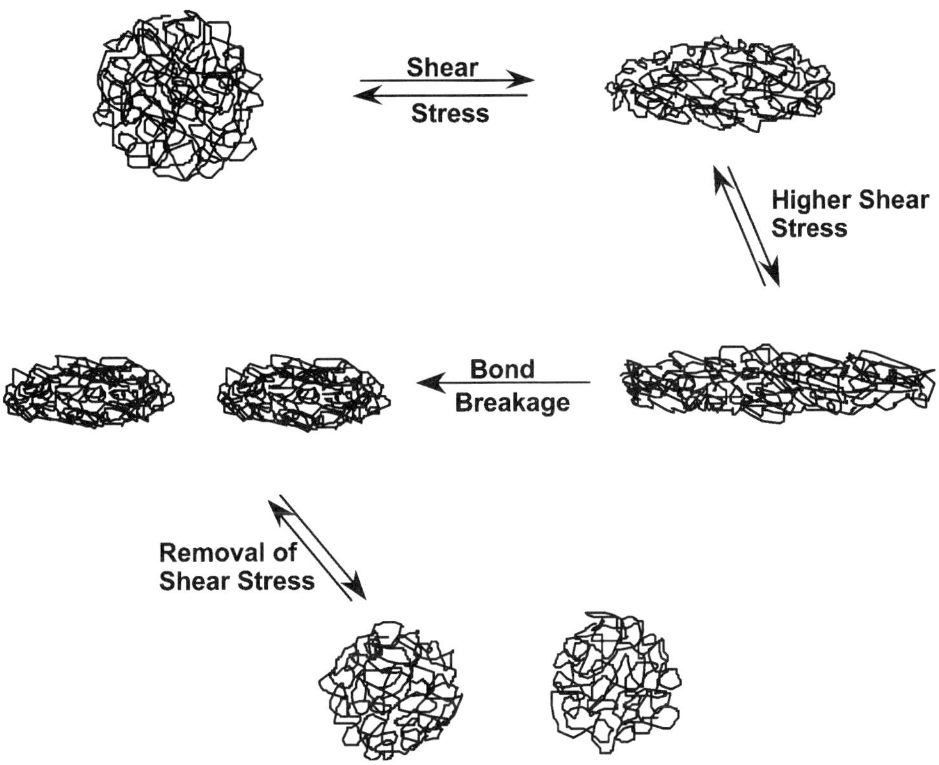

Fig. 4.179—Mechanical degradation of a polymer [235].

process. The monomers that are commonly used for this purpose are shown in Fig. 4.63. The dispersant functional group is introduced either by grafting or through copolymerization. Grafting and copolymerization with a nitrogen-containing monomer, such as 2- or 4-vinylpyridine, imparts the dispersancy directly. Grafting of maleic anhydride, on the other hand, leads to a succinic anhydride or succinic acid that must be reacted with alkylene-polyamines to form a dispersant functional group. At present, no commercial products made in this manner appear to be on the market. Antioxidant functional group can also be added in either of these two ways. Grafting is common in the case of OCP and SD type polymers but for PMAs, copolymerization is more prevalent.

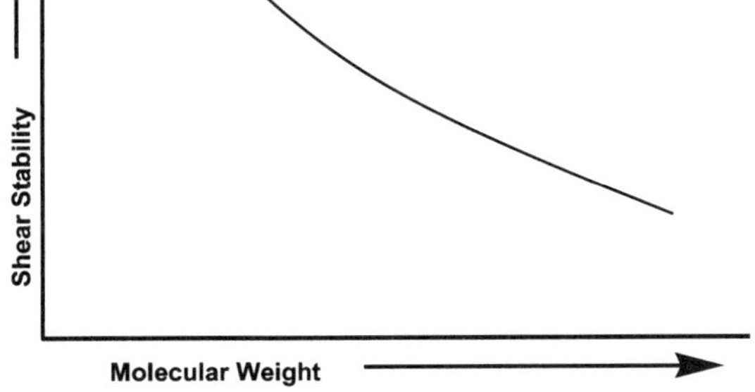

Fig. 4.180—Shear stability as a function of polymer molecular weight.

Commercially available DVMs contain two distinct functional groups: a polymeric backbone (the VM portion) and the dispersant moiety. Thus, their physical, mechanical, and chemical properties will depend upon either the inherent properties of the VM portion of the molecule, the dispersant functional group, or both. When the properties are determined by both, a possibility of synergism or antagonism exists. The properties that relate to the VM portion of the molecule are thickening efficiency, shear performance, thermal and oxidative stability, and low-temperature properties; and those pertaining to the dispersant portion are oxidative and chemical stability, and dispersancy. For a detailed discussion on the role of DVM structural features on various properties, please refer to the topic of dispersants, especially the dispersant polymers.

Considering the theorized mechanism of the dispersant action, it is reasonable to assume that the dispersancy will relate to both the basicity and the molecular weight of the DVMs. Based on the pKa values of the different functional groups, one would expect dimethylamino group to be more basic than pyridyl group, which in turn is expected to be more basic than 2-keto pyrrolidinyl group. Hence, one would anticipate dispersancy to follow the same order. Typically, formulations containing DVMs require an additional amount of dispersant since some applications, such as engine oils, need certain level of nitrogen to appropriately perform the function of suspending deposit precursors, deposits, and soot. It is generally believed that the multifunctional additives, such as DVMs, perform better than the combination of additives of the two types because both moieties are

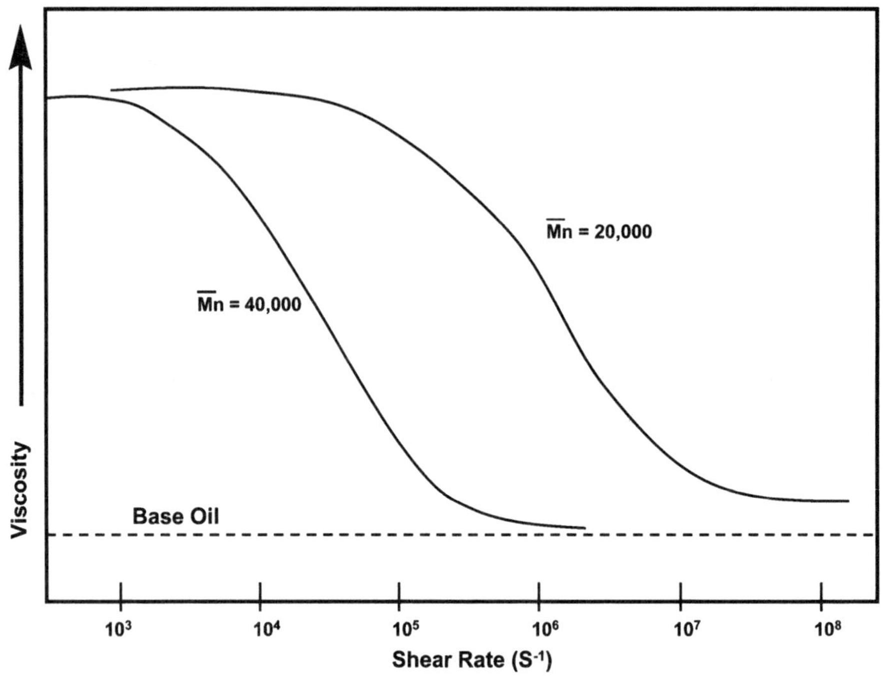

Fig. 4.181—Temporary viscosity loss due to shear [473].

concurrently present where they are needed. Data reported in Table 4.32 support this notion. The test is Sequence VG engine test that uses the SAE 5W-30 API SG quality oil [469]. The formulations contain analogous ingredients, except the viscosity modifier.

Pour Point Depressants

The pour point is the lowest temperature at which a fuel or mineral oil will pour when cooled under defined conditions. At low temperatures, the wax tends to separate as crystals with a lattice type structure. These crystals can trap a substantial amount of oil via association, thereby inhibiting the oil flow and ultimately hindering proper lubrication of the critical equipment parts.

Base oil suppliers remove most of the wax during petroleum refining. However, complete dewaxing of base oils is not practical because of the process limitations, economics,

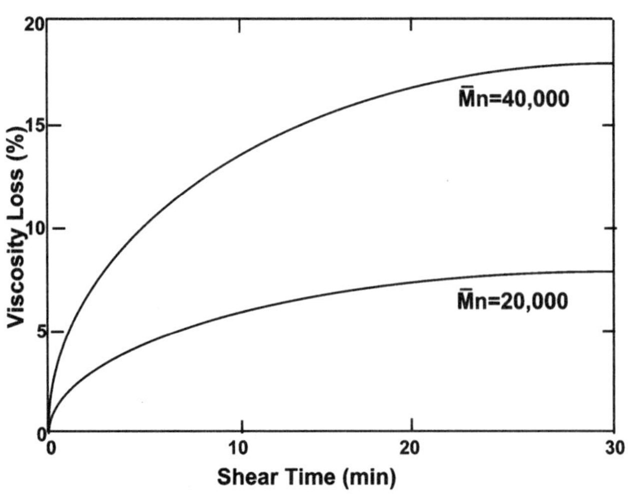

Fig. 4.182—Permanent viscosity loss due to shear [473].

and the desirable presence of wax due to its high VI character [50]. For mineral oils to function effectively at low temperatures, the additives called the pour point depressants are used. Current practice favors mild dewaxing in combination with the use of the pour point depressants. Incidentally, wax is only a problem in the API Group I oils and not in Group II or III oils.

A good pour point depressant can lower the pour point of a lubricant by as much as 40 °C. These additives are commonly used in mineral oil-based lubricants that are designed for applications with operating temperatures usually below 0 °C. Pour point depressants have virtually no effect on the temperature where the wax crystals start to precipitate (cloud point) or the amount of wax that separates. They essentially act as the wax-crystal modifiers and function by altering the crystal size. They do this either by absorption onto the surface of the newly formed crystals or by co-crystallizing with the precipitating wax. Both mechanisms inhibit lateral crystal growth and keep the bulk oil fluid. Of the commercial pour point depressants, alkylaromatics are believed to perform via the absorption mechanism and the aliphatic polymers via co-crystallization [308].

The molecular weight and the structure of the polymeric pour point depressants enable them to be effective over a wider range than their low molecular weight counterparts. The extended range of performance in the case of polymers is believed to be due to their limited solubility in the petroleum fractions. As the temperature decreases, different polymer segments become successively co-crystallizable.

A good pour point depressant must possess one or more of the following structural features.

1. Polymeric structure.
2. Waxy and nonwaxy components.
3. Comb structure—Comb structure means a short backbone with long pendent groups. This is in contrast to the

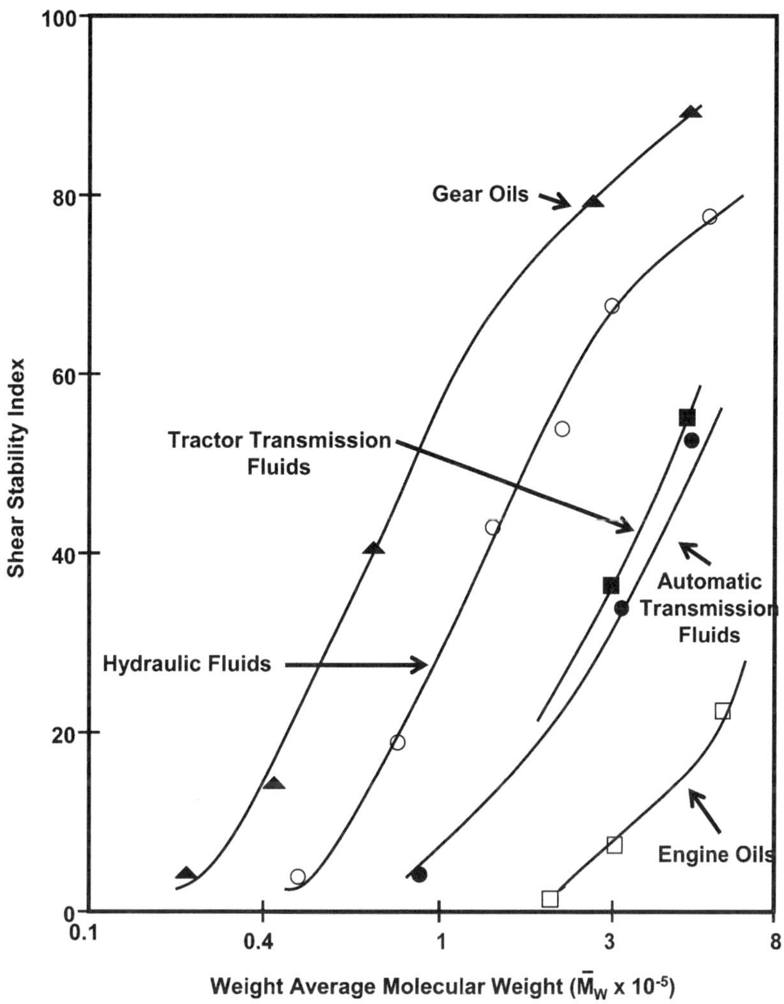

Fig. 4.183—Shear stability requirements for various lubricants [235].

structure of the viscosity modifiers where the backbone is relatively long and the pendent groups, if present, are short.

4. Broad molecular weight distribution.

Most commercial pour point depressants are organic polymers, although some nonpolymeric substances have been shown to be effective. Tetra (long-chain) alkyl silicates, phenyltristearyloxysilane, and pentaerythritol tetra-stearate are examples of the nonpolymeric type. Commercial pour point depressants include alkylated naphthalenes, poly(alkyl methacrylate)s, poly(alkyl fumarate)s, styrene esters, oligomerized alkylphenols, phthalic acid esters, ethylene-vinyl acetate copolymers, and other mixed hydrocarbon polymers.

Figure 4.187 contains the structures of the poly(alkyl methacrylate)s, alkylaromatics, and styrene-ester polymers, which are the most commonly used chemical types. The alkyl chains on these additives are largely linear and of different carbon number, from C_1 to C_{24}, or higher. Carbon distribution prevents them from crystallizing out of the oil at room temperature. At low temperatures, fatty carbon chains ($\geqslant C_{12}$) interact with the wax and interfere in the formation of the crystalline networks. For PMAs with pour point depressant properties, the molecular weight ranges between 200 and 3000 g/mol. High molecular weight polymethacrylate derivatives can act both as viscosity modifiers and pour point depressants; hence when this chemistry is used for viscosity improvement, the need for a pour point depressant is minimized.

Pour point depressants are used at treatment levels of 1 % or lower. In nearly all cases, there is an optimum concentration above and below which the pour point depressants become less effective. The essential structural difference between the pour point depressants and the viscosity improvers of the same class is that the viscosity improvers consist of long backbones with short pendent groups and the pour point depressants consist of short backbones with large pendent groups. This difference is depicted in Fig. 4.188. Figure 4.189 shows a pour point's ability to change wax crystal morphology and permit flow.

Pour point depressants are used in engine oils, automatic and power transmission fluids, automotive gear oils, tractor and industrial hydraulic fluids, and circulating oils, especially those intended for use in cold climates. The need for these additives is likely to decrease due to a general shift away from the API Groups I oils towards Group II and III oils, where the iso-dewaxing technology alters the linearity of the wax-forming components. The performance of a pour point depressant is determined in a base stock and the lubricant by one or more of the following tests.

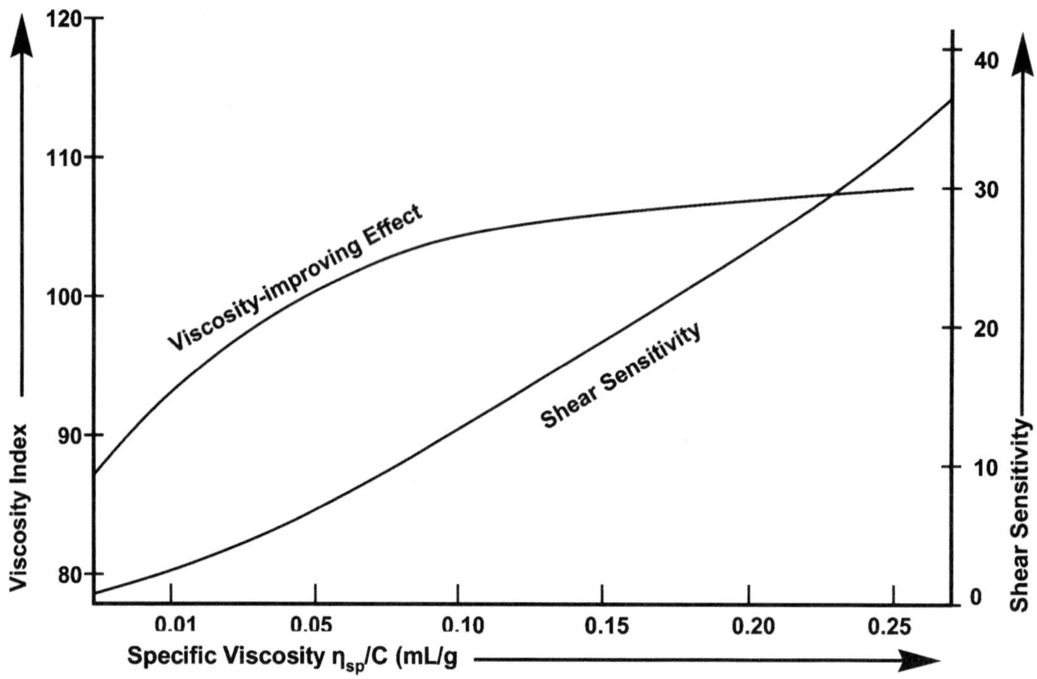

Fig. 4.184—Effect of specific viscosity on oil viscosity and shear stability [4].

1. ASTM D97—Pour point of petroleum oil
2. ASTM D3829—Borderline pumping test
3. ASTM D2602—Apparent viscosity at low temperature using cold-cranking simulator

Foam Inhibitors

Gases are soluble in mineral oils to a limited amount. The amount varies with the type of gas and the oil temperature. For example, 8 to 9 % of air by volume is soluble in mineral oil at room temperature. Dissolved gases affect a number of oil properties, such as viscosity, bulk modulus, heat transfer, oxidation, boundary lubrication, and the foaming tendency. The amount of the soluble gas in oil is measured by the ASTM D2780 Test. When the amount of a gas in oil exceeds saturation, small gas bubbles are slow to come out and the oil appears hazy for a period of time. This phenomenon is called entrainment. Foam forms when a large amount of gas, for example, air is entrained in a liquid. While foaming is desirable in certain applications, such as flotation, washing, and cleaning, it is undesirable in others, such as distillation and pumping of fluids. In lubricant-related applications, foam can act as an impediment and must be controlled. Almost every lubricant application involves some kind of agitation which encourages foam formation through air entrainment. Excessive foaming will result in ineffective lubrication due to the formation of the noncontinuous lubricating film, thereby leading to metal-to-metal contact; and, over time, will cause oxidative degradation of the lubricant. Additional problems due to foaming include over-filling of the sumps, loss of oil out of the vents, and poor performance in hydraulic systems. The foaming characteristics of an oil are measured by the ASTM D892 Test. The viscosity and the surface tension of a lubricant determine the stability of the foam. Low-viscosity oils produce foams with large bubbles, which tend to break quickly. On the other hand, high-viscosity oils generate stable foams that contain fine bubbles and are difficult to break. The presence of the surface-active materials, such as dispersants and detergents, further increases the lubricant's tendency to foam.

Foam inhibitors control the foam formation by altering the surface tension of the oil and by facilitating the separation of air bubbles from the oil phase. In general, these additives have limited solubility in oil; hence they are added as very fine dispersions. Foam inhibitors are effective at very low treatment levels, 3 to 150 parts per million.

Silicones (polysiloxanes), poly(alkyl acrylate)s, and poly(alkyl methacrylate)s are the commonly used foam inhibitors, with silicones being the more popular. ASTM D892 Test is used to assess a lubricant's foaming tendency. The structures of the three common types of additives are shown in Fig. 4.190.

Other Additives

In addition to the major classes of additives described so far in this chapter, lubricants contain a number of other additives. These include *seal-swell agents, dyes, biocides,* and *couplers*.

Seals are used in modern machinery for a variety of reasons. In lubrication systems, their functions are as follows.

1. Isolate various lubrication environments from harmful elements.
2. Help maintain hydraulic pressure.
3. Allow removal and replacement of the malfunctioning parts without the need to totally dismantle the equipment.
4. Minimize contamination and the loss of lubricant.

Seals are commonly made from polymeric materials such as fluoroelastomers, nitrile rubber, polyacrylates, and silicones. Structures of these materials are provided in Fig. 4.191. Lubricants containing certain base stocks and additive systems can cause shrinkage, brittleness, and deterioration of the seals and impair the performance of the lubricating system.

Seal-swell agents are additives which maintain the integrity of these seals. Additives belonging to this class include polyesters, some phosphorus derivatives, and proprietary chemicals. Seal-swell agents are commonly used in transmission and hydraulic fluids.

Dyes are used to color-code lubricants to ensure their use in the proper application and as a leak detection aid for the consumer. ATFs contain a red dye, and two-stroke cycle oils contain a blue or a purple dye. These dyes are oil-soluble organic compounds, mostly with an azo structure. To a limited extent, dyes are also used as a marketing tool to impart color, or the fluorescence, to lubricants that was historically perceived to indicate good performance. This is because today's refining processes remove compounds that impart this characteristic.

In general, the mineral oil-based lubricants resist microbial attack because of their high-temperature operation and the presence of the additives, many of which have biocidal action. High water-based lubricants, such as certain metalworking fluids and hydraulic fluids, are easily attacked by microbes and fungi. The control of bacterial and fungal growth is essential to minimizing product deterioration and possible health hazards. This is done by the use of the water-soluble triazine, morpholine, imidazoline, and thiazoline derivatives, which possess biocidal properties. Triazines, which owe their biocidal action to their formaldehyde-releasing ability, find extensive use in this application. The interested reader may refer to Chapter 11 on Metalworking Fluids for a more detailed discussion on biocides.

Couplers are additives that are used in water-based lubricants to help stabilize micro-emulsions. Glycol and its derivatives are commonly used for this purpose.

Metalworking fluids use a number of other additives. These include *alkalinity buffers, odor masks*, and *anti-misting agents*. *Alkalinity agents* and *odor masks* are used to control acidity and odor in water-based systems. Acidity and odor result from the decomposition of the oil and the additives due to bacterial and fungal attack. Amines and inorganic bases are used to control acidity and natural and synthetic aromatic materials are used to control odor. *Anti-misting agents* are used to suppress mist formation, primarily in oil-based fluids, which if not controlled can be harmful to workers. Polymers of various types are used for this purpose. Chapter 11 on Metalworking Fluids contains a more detailed discussion on these additives.

Multifunctional Nature of Additives

A number of additives perform more than one function. Zinc dialkyl dithiophosphates, known mainly for their antiwear action, are also potent oxidation and corrosion inhibitors. Styrene-ester polymers and functionalized polymethacrylates and can act as viscosity modifiers, dispersants, and pour point depressants. Basic sulfonates, in addition to acting as detergents, perform as rust and corrosion inhibitors. They do so by forming protective surface films and by neutralizing acids that arise from fuel combustion, lubricant oxidation, and additive degradation.

Environmental Impact of Additives

ATC (the Technical Committee of Petroleum Additive Manufacturers in Europe) in the early 1990s carried out a study that traced engine oil additives across 19 OECD (Organization for Economic Co-Operation and Development) countries with *cradle to grave* perspective. The objective was to assess the impact of additives on the consumer and the environment. The report [474] reviews the nature, development, health and safety aspects, benefits to the consumer, and the ultimate fate of the engine lubricant additives. An ac-

TABLE 4.31—Viscosity modifiers—A performance comparison.

Property	OCP	PMA	SI	RI	SB	SE	OCP-g-PMA
Thickening Efficiency (Viscosity Index of the Concentrate)	0	+	+	+	0	+	+
Viscosity Index Improvement	0	+	0	0	0	+	+
Engine Oil Pumpability	-	+	+	-	+	0	0
Cold Cranking Viscosity	0	0	+	0	+	0	0
High Temperature High Shear Rate Viscosity	0	-	-	0	+	-	+
Deposit Forming Tendency							
Diesel	+	-	+	0	+	-	0
Gasoline	0	-	+	0	+	-	0
Fuel Economy Benefits	0	0	+	-	+	+	0
Cost	+	0	+	+	-	0	0

+ = Good
0 = Average
- = Poor
Polymer

OCP=Olefin Copolymer; PMA=Polymethacrylate
SI=Styrene-isoprene Copolymer; RI=Radial Isoprene Polymer
SB=Styrene-butadiene Copolymer; SE=Styrene Ester
OCP-g-PMA=Polymethacrylate Grafted Olefin Copolymer

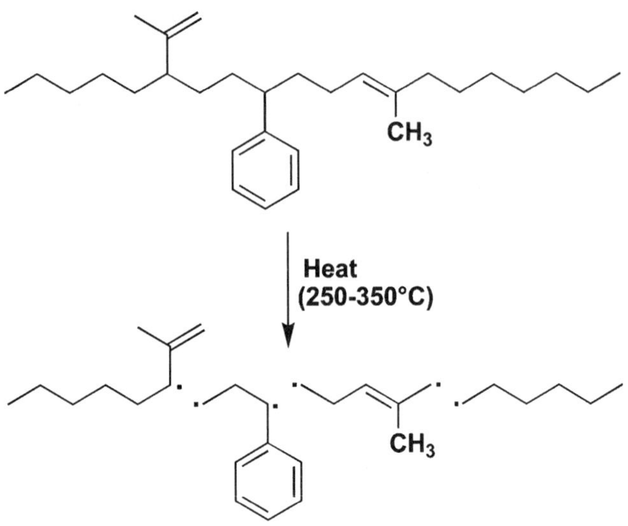

Fig. 4.185—Mechanism of thermal degradation of styrene-diene polymers.

TABLE 4.32—Performance comparison between PMA-based DVM versus PMA(5W-30 API SG formulation, Sequence VE Test) [469].

PMA VI Improver (45 SSI)	Average Sludge	Average Varnish	Average Cam Wear
Dispersant Type	9.23	6.25	1.5
Nondispersant Type	4.55	4.56	5.8
Performance Limits	>9.0	>5.0	<130

companying report [475] examined the same for fuel additives. Refer to Chapter 13 on Lubricants and the Environment for in-depth discussion of this topic.

The Introduction of a New Additive

The development of a new additive is initiated after a new product (lubricant) need is identified. The need for a new product is usually expressed by the OEMs and the end-users and relates either to the inadequate performance of the existing products in current equipment or the perceived needs of

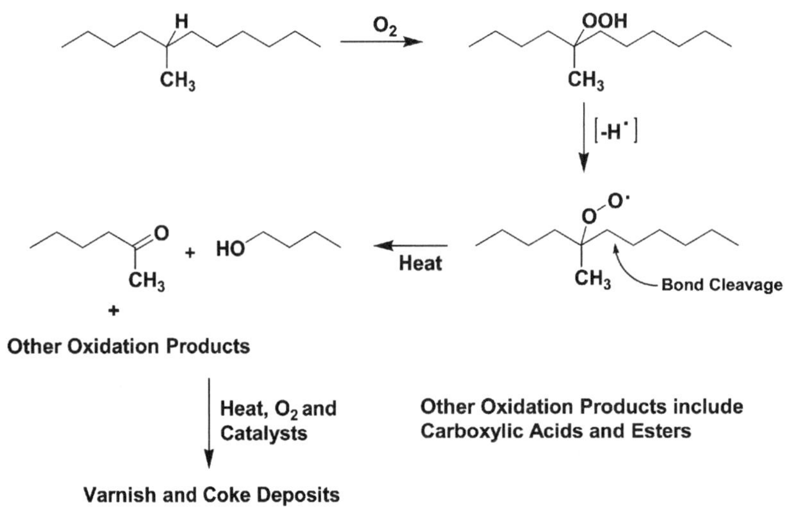

Fig. 4.186—Mechanism of oxidative degradation of polymers.

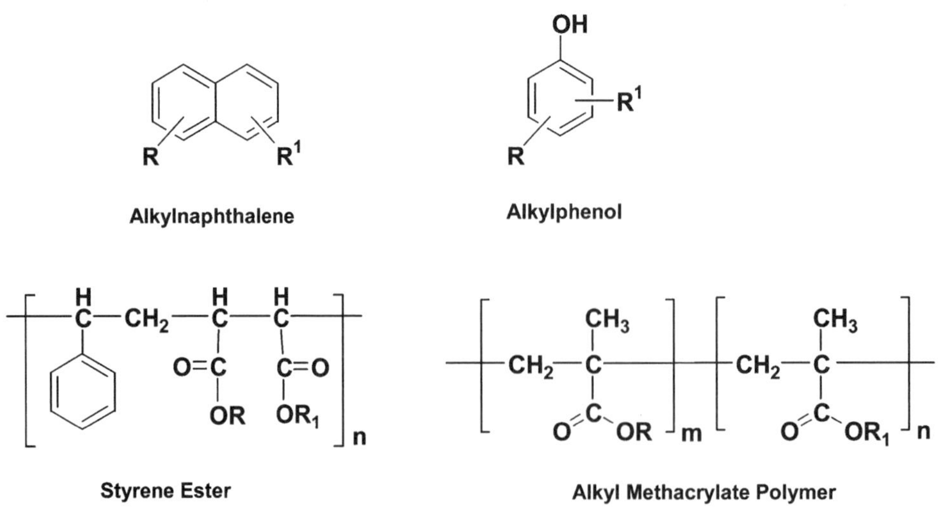

R, R_1= Waxy and Nonwaxy Alkyl Groups

Fig. 4.187—Structures of common pour point depressants.

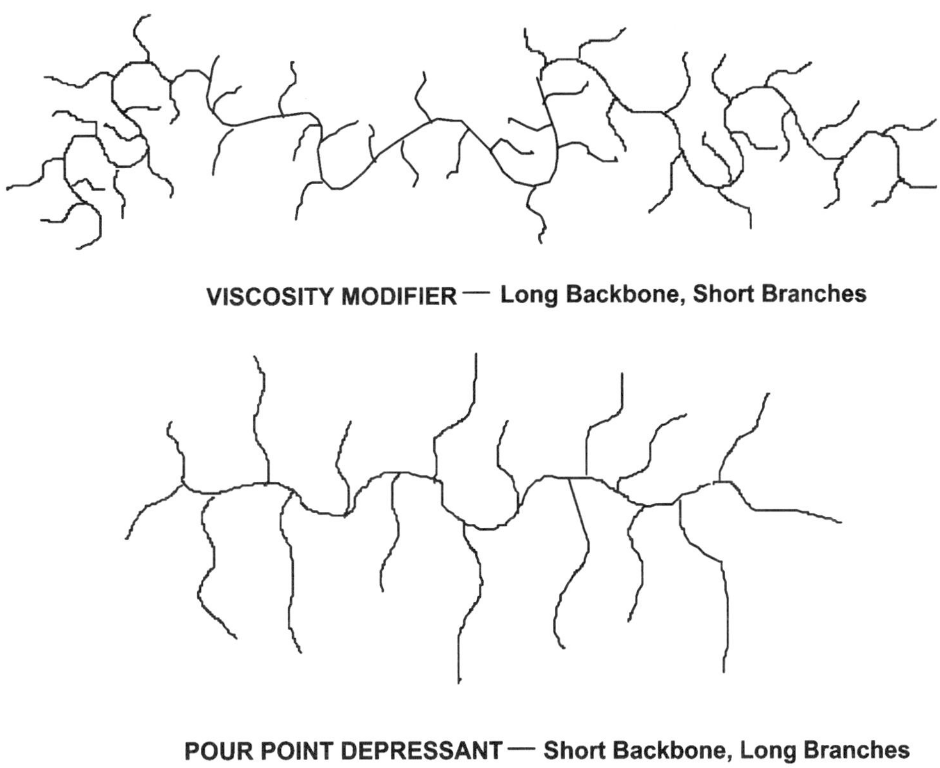

Fig. 4.188—Viscosity modifier versus pour point depressant—structural difference.

the equipment under development. To fulfill this need, various organizations, such as SAE, API, ASTM, AGMA, and OEMs, initiate the development of new performance specifications and the test methods. A flowchart leading to the development of a new category, exemplified by the development of ILSAC GF3, is depicted in Fig. 4.192 [476]. Additive companies, either alone or in collaboration with a lubricant supplier, try to satisfy the performance requirements established for the new product/category. If the additive company is unable to develop an additive system using their existing technology base, they initiate a project to develop and test a new additive.

Newly developed additives are blended with other additives in the customer's base oil and are screened in a number of proprietary bench tests. Bench tests are accelerated tests which are devised to closely simulate the conditions the lubricant is likely to experience in actual service. This kind of testing is quite common because it allows the evaluation of a large number of additives quickly and inexpensively.

Once a lubricant satisfies the performance criteria of the bench tests, full-fledged testing using actual equipment is carried out. This may be done in a laboratory or in collaboration with an end-user. For additives used in automotive products, field trials may also be necessary. The costs associ-

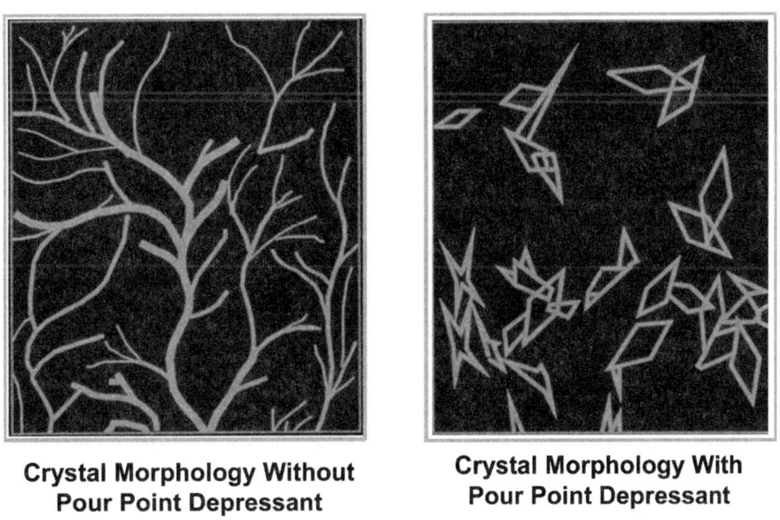

Fig. 4.189—The mechanism of the pour point depressant performance [318]. Reprinted with permission from the Lubrizol Corporation.

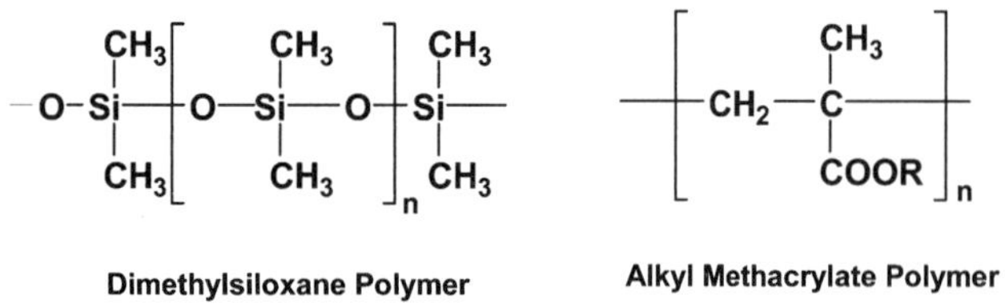

$HO(CH_2CH_2O)_xCH_2CH_2O[CH_2CH(CH_3)O]_yCH_2CH(CH_3)OH$

Ethylene Oxide - Propylene Oxide Copolymer

Fig. 4.190—Common types of foam inhibitors.

ated with the development and testing of the new additives can be phenomenal. The performance package which has successfully met all the performance requirements is ready to be marketed, either through factory-fill or service-fill lubricant blenders. Table 4.33 shows the classes of additives used to formulate engine lubricants, and Table 4.34 contains additive classes which are used to formulate nonengine lubricants. Please note that all formulations do not contain all classes of additives identified in these tables.

The Approval Process

The purpose of the approval process is to ensure that a new lubricant meets the performance criteria established by the various technical societies, OEMs, and end-users. The lubricant is therefore subjected to a variety of tests, which are described in Chapter 12 on Lubricant Testing.

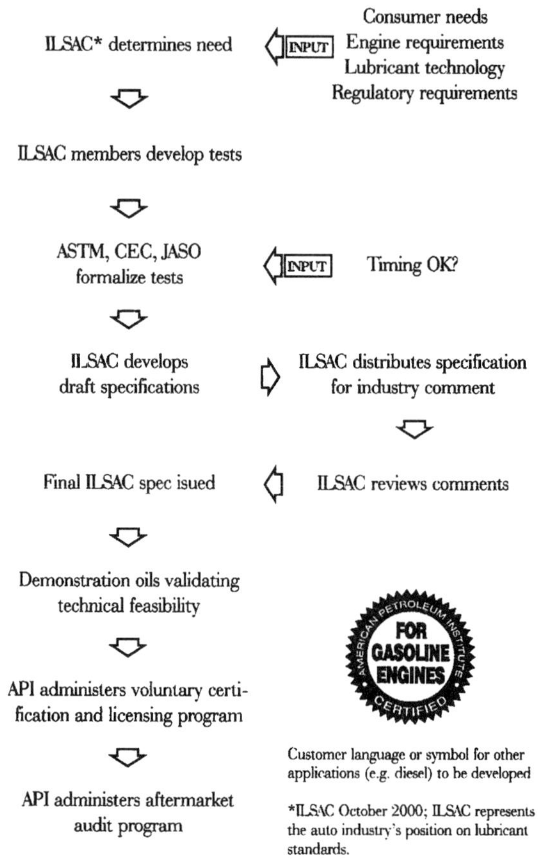

Fig. 4.192—Information flow during development of a performance category [476].

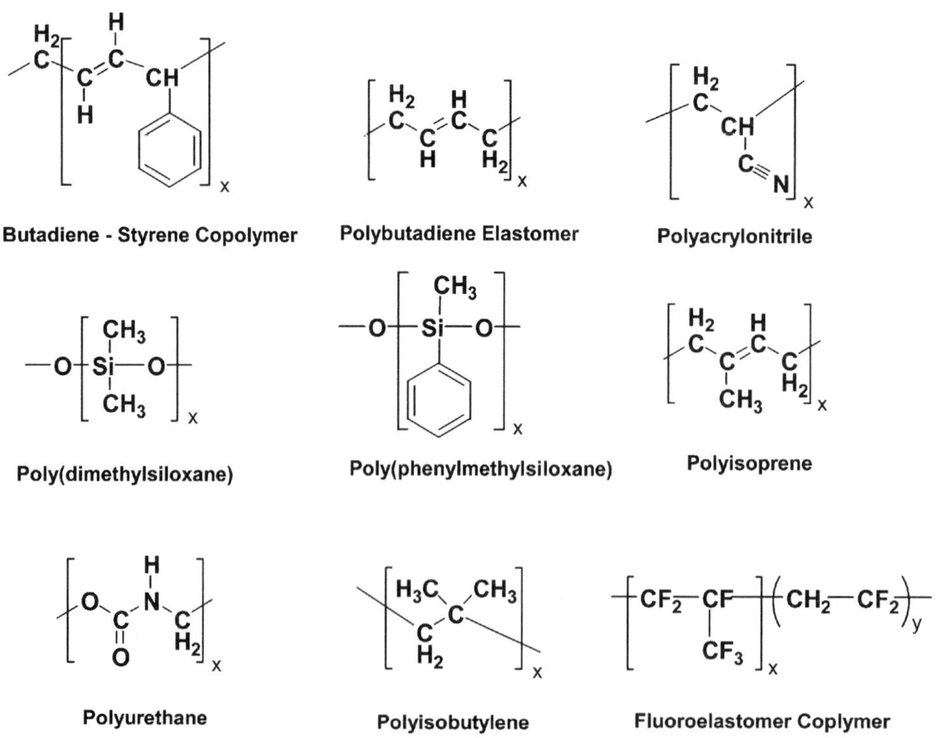

Fig. 4.191—Molecular structure of elastomers used in seals.

TABLE 4.33—Common additive types for engine lubricants.

Additive	Gasoline Engine Oils	Diesel Engine Oils	Stationary Gas Engine Oils	Aviation Engine Oils	Two-stroke Cycle Engine Oils
Dispersant	✓	✓	✓	✓	✓
Detergent	✓	✓	✓	✓	✓
Antiwear/EP Agent	✓	✓	✓	✓	✓
Oxidation Inhibitor	✓	✓	✓	✓	✓
Corrosion Inhibitor/ Metal Deactivator	✓	✓	✓	✓	✓
Friction Modifier	✓				
Pour Point Depressant	✓	✓	✓	✓	
Foam Inhibitor	✓	✓			
Viscosity Modifier	✓	✓			
Other[a]					✓

[a] Other additives include couplers, dyes, diluents, and emulsifiers.

TABLE 4.34—Common additive types used in nonengine lubricants.

Additive	Automatic Transmission Fluids	Gear Oils: Automotive Gear Oils	Gear Oils: Industrial Gear Oils	Hydraulic Fluids: Tractor Hydraulic Fluids	Hydraulic Fluids: Industrial Hydraulic Fluids	Metalworking Fluids	Greases
Dispersant	✓						
Detergent	✓			✓	✓	✓	✓
Antiwear/EP Agent	✓	✓	✓	✓	✓	✓	✓
Oxidation Inhibitor	✓	✓	✓	✓	✓	✓	✓
Corrosion Inhibitor/ Metal Deactivator	✓	✓	✓	✓	✓	✓	✓
Friction Modifier	✓	✓	✓	✓		✓	
Pour Point Depressant	✓	✓	✓	✓	✓		
Foam Inhibitor	✓	✓	✓	✓	✓	✓	
Viscosity Modifier	✓	✓	✓	✓	✓		✓
Other[a]	✓			✓	✓	✓	✓

[a] Other additives include couplers, dyes, diluents, and emulsifier

5

Combustion Engine Lubricants

DISCUSSION IN THIS CHAPTER PERTAINS TO combustion engine lubricants. The chemistry and technology of these lubricants are presented along with United States and European performance specifications and the process of establishing them. In order to facilitate understanding, various types of internal combustion engines and their operation are described. The chapter also addresses the current topics of fuel economy, emissions control, and extended service intervals. The chapter is concluded by citing examples of several engine oil formulations.

Engine lubricants, or engine oils, are designed for use in internal combustion engines. Modern engines operate on a wide variety of fuels and in environments that involve temperature extremes; hence their lubrication is quite complex. A combustion engine lubricant must possess attributes to help it perform the following functions effectively.

1. *Permit Easy Starting:* It must have low viscosity at low temperatures and be pumpable, so as to instantaneously reach the engine parts that need lubrication. This is an important attribute since most of the engine wear occurs during the start-up, primarily due to lubricant starvation.
2. *Maintain Adequate Viscosity at High Temperatures:* This is important because most oils experience a decrease in viscosity at high temperatures, such as those in and around the combustion engine. If the viscosity of the oil drops too far, the lubricant loses its ability to form the lubricating film of the appropriate thickness, which will permit metal-to-metal contact and wear will ensue.
3. *Lubricate and Prevent Wear:* This translates into the oil forming a lubricating film of appropriate thickness to prevent metal surfaces from contacting each other and experiencing wear. For most engine parts the surfaces are well separated, which makes lubrication easier. However, there are parts such as the piston rings and cam lobes, which are designed to have metal-to-metal contact and the function of the lubricant is to minimize wear by making chemical surface films.
4. *Reduce Friction:* The formation of the lubricant film of proper thickness on surfaces and its maintenance will reduce friction and the accompanied wear. This is especially true during the start-up and idle, when the lubrication is inadequate and the frictional losses occur. Therefore, controlling friction will improve the fuel economy.
5. *Protect Against Rust and Corrosion:* Water resulting from the fuel combustion, while meant to escape through the exhaust, can condense on the cylinder walls, or travel past piston rings as part of the blow-by and enter the crankcase. This typically occurs in cold weather or short distance driving because the engine and the lubricant are not hot enough for water to be removed via evaporation. Water can initiate rust and, in the presence of the acidic materials resulting from the lubricant oxidation and additive decomposition, can cause corrosion.
6. *Keep Engine Parts Clean:* Partial fuel combustion products, such as free radicals, soot, sulfur, and nitrogen oxides, enter the crankcase as the blow-by and react/interact with the lubricant to form highly polar deposit precursors and corrosive materials. These species have the tendency to separate on the hot surfaces to form deposits and to lead to corrosion. Engine lubricants are designed to prevent the formation of these species or keep them from separating on the surfaces by suspending them in the bulk lubricant, or both.
7. *Cool Engine Parts:* Cooling of the engine parts is crucial to its trouble-free operation. Parts that must be cooled include cylinder heads, cylinder walls, valves, crankshaft, main and connecting rod bearings, timing gears, pistons, and others. Certain parts of the engine can be cooled by the use of a coolant, which is typically a mixture of water and ethylene glycol. Other parts cannot be effectively cooled by the coolant, either because of their vicinity, or the part temperature is extremely high, which leads to the rapid evaporation of water. In such situations, the lubricant acts as a coolant.
8. *Seal Combustion Pressures:* Surfaces of piston rings, ring grooves, and cylinder walls do not have an ideal fit, primarily because of the machining limitations. It is important that these parts act as a good seal to prevent the loss of the high combustion and compression pressures, which are needed for the efficient engine operation. A loss into the low pressure area of the crankcase would result in a reduction of the engine power and efficiency. Engine oils therefore improve the seal by filling spaces in the above-listed parts. Typically the oil film that acts as a seal is only 0.025-mm thick; hence it is ineffective in filling spaces that are larger because of the intensive wear. Incidentally, the oil consumption in a new engine is high until the surfaces in these parts become smoother due to wear for the oil to form a better seal.
9. *Control Foam:* Foaming of the engine oil due to air entrainment occurs because of the rapidly moving engine parts which create turbulence. The result is the formation of the air bubbles, which normally rise to

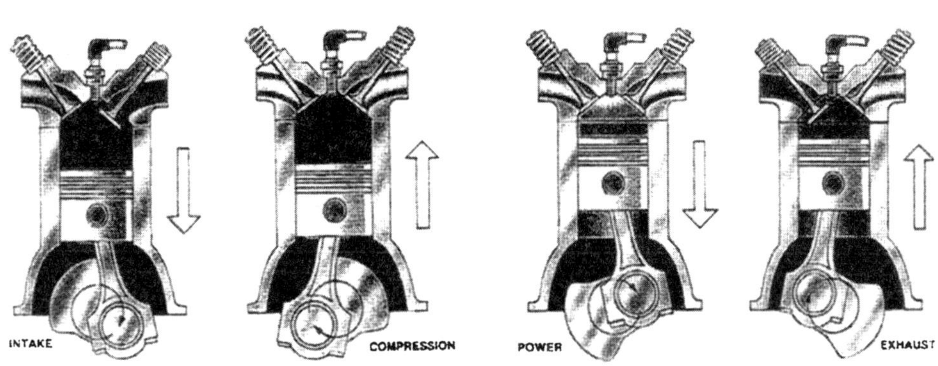

Fig. 5.1—Operation of a four-stroke cycle SI engine.

the surface of the oil and break. However, the presence of water and additives, many of which have surfactant properties, slows down this process. Foam in the engine oil is undesired because of its poor cooling ability and noncontinuous film formation, which will result in excessive engine wear.

While a good quality engine oil can perform these functions adequately, the continuing efforts of the OEMs to improve emissions quality by recycling partial combustion products from the exhaust and venting the volatiles from the fuel system and the bulk lubricant (positive crankcase ventilation) into the combustion chamber place additional demands on the lubricant. This strategy is effective in lowering the partial combustion products, such as the unburned or partially burned hydrocarbons and carbon monoxide, but at the expense of enriching the combustion mixture in NOx (nitrogen oxides), a potent oxidant. This will be discussed further in Chapter 6 dealing with Emissions in an Internal Combustion Engine.

Types Of Engines And Mode Of Their Operation

Engines are devices that help produce motion. The energy used to drive engines comes from a number of sources, but chemical sources, such as petroleum products, are the most prevalent. Engines where fossil fuels are burned to convert the fuel's latent energy into heat to do the work are called the combustion engines. They can be of external combustion type, such as steam and Stirling engines, or of the internal combustion type [477], such as gasoline and diesel-fueled engines. In an external combustion engine, the fuel is burned in a furnace that is not an integral part of the engine. The generated energy is used to heat a fluid, water or air, which does the actual work. In an internal combustion engine, the fuel is burned inside the engine itself, and the combustion products are used to drive the mechanisms that perform the work. Based upon the mode of combustion and the manner in which the energy is transformed into work, the internal combustion engines can be further classified into reciprocating-piston engines, rotary engines, and continuous-combustion gas turbine engines [23].

Reciprocating-piston Engines

These engines are used most widely and contain pistons that move up and down in cylinders; hence, the name reciprocating piston. The top of the cylinder is closed by a metal cover, called the head, which contains valves and, in gasoline-fueled engines, a spark plug. Intake valves allow the intake of air or of the air-fuel mixture and exhaust valves help remove the products of combustion. Pistons are attached to a crankshaft by the connecting rods and their motion rotates the crankshaft, which is connected to a drive mechanism by suitable gearing. Such engines can be either four-stroke cycle or two-stroke cycle, depending upon the number of strokes by the piston in one full turn, or cycle, of the crankshaft.

Reciprocating piston engines, based upon the mode of ignition, can be divided into spark-ignition (SI) and compression-ignition (CI) types. While there are a number of differences between the two types, the primary difference lies in the combustion system. The SI engines use a homogeneous air-fuel mixture, obtained by premixing the two either in a carburetor or the intake port. Combustion is initiated by a spark, and the combustion rate is primarily controlled by the shape of the combustion chamber. The CI engines, on the other hand, burn a heterogeneous mixture since the air is compressed first and then the fuel is introduced. Combustion occurs via self-ignition due to the heat generated from the compression process. Incidentally, in some CI engines, a glow plug is used to facilitate ignition. The rates of ignition and combustion in these engines are largely controlled by injection timing and the injection rate.

A graphic representation of the operation of the four-stroke cycle engine is shown in Fig. 5.1. The four strokes are intake, compression, ignition and power, and exhaust. During the intake stroke, the rotating crankshaft moves the piston down, creating a partial vacuum which facilitates the introduction of the air-fuel mixture into the cylinder through the open intake port. During the compression stroke, both intake and exhaust valves are closed and the mixture is compressed by the piston moving up. During the ignition and power stroke, ignition of the fuel-air mixture occurs and hot, expanding gases result that force the piston down and rotate the crankshaft, thereby producing power. During the exhaust stroke, the by-products of combustion are vented through the exhaust valve/s. After the exhaust stroke, the cycle starts again with the intake stroke. At the beginning of the intake stroke and the ignition and power stroke, the piston is at the top of the cylinder. At the beginning of the compression and the exhaust strokes, it is at the bottom. The number of crankshaft revolutions per minute represents the engine speed.

Unlike the four-stroke cycle engines that produce one

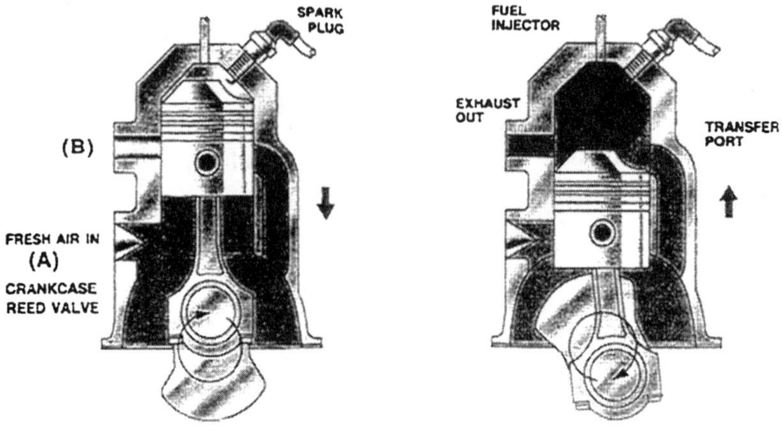

Fig. 5.2—Operation of a two-stroke cycle SI engine.

power stroke for every four strokes (two crankshaft revolutions); the two-stroke cycle engines require only two strokes (one crankshaft revolution) for each power stroke. This is achieved by combining the compression and the ignition in the first stroke and the exhaust and the intake in the second stroke. However, the two operations in each stroke are timed to achieve a higher degree of efficiency. All operations are controlled by the upper and lower edges of the pistons. The operation of a two-stroke cycle engine is shown in Fig. 5.2.

When the piston reaches the top dead center, ignition is induced either by spraying the fuel into the hot compressed air (CI) or by spark-igniting the air-fuel mixture (SI); thereby starting the power stroke. When the piston moves below the exhaust Port B, most of the exhaust is displaced. Soon after, the intake Port A is uncovered and either air or an air-fuel mixture is introduced into the cylinder. If the intake port and the exhaust port are situated across from each other, as shown in Part A of Fig. 5.3, a portion of the intake can escape

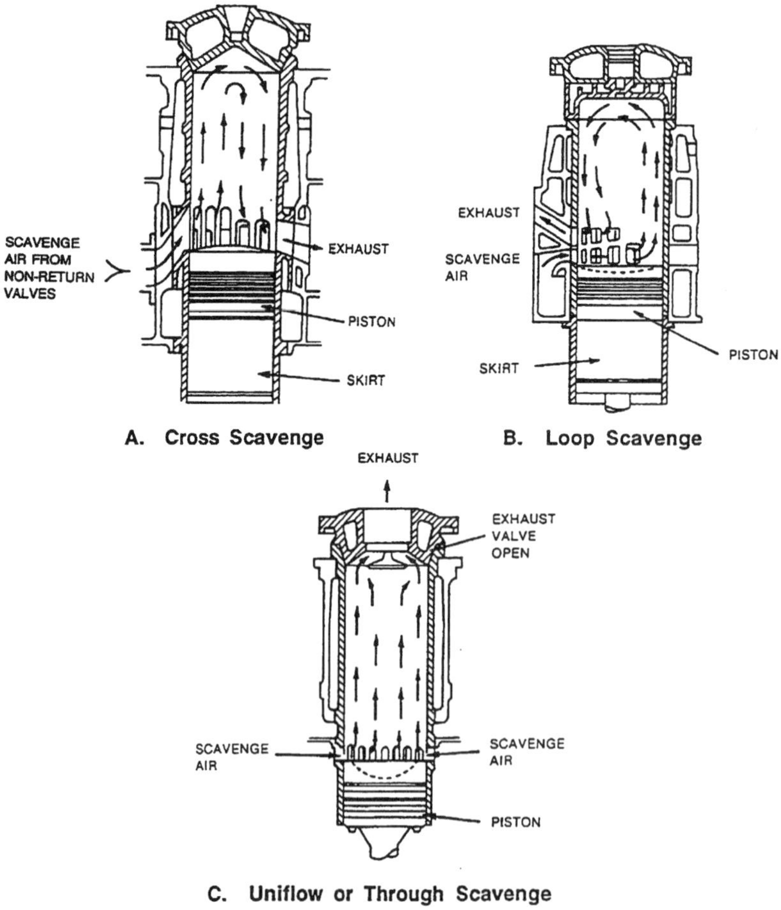

Fig. 5.3—Scavenging in two-stroke cycle engines.

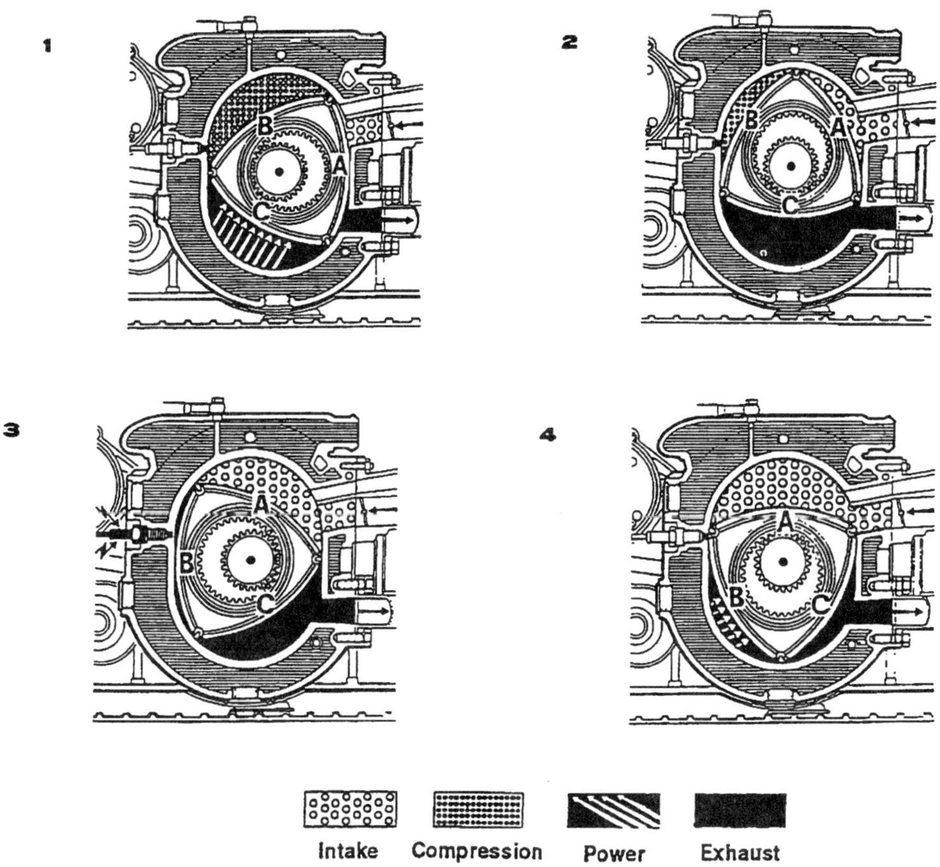

Fig. 5.4—Operation of a Wankel engine.

with the exhaust. This situation, termed cross-scavenging, can be minimized by using deflectors on the top of the pistons. Another way to minimize cross-scavenging is by loop scavenging. In this arrangement, the intake and the exhaust are placed nearby instead of across from each other (Fig. 5.3, Part B). This forces the intake to pass through a complete "loop" before reaching the exhaust port. Part C of Fig. 5.3 shows the arrangement where a combination of the exhaust valves in the head and the intake ports in the cylinder are used to achieve through or uniflow scavenging.

The rotary engine, or a Wankel engine, is also a four-stroke cycle internal combustion engine, but it differs radically from the conventional reciprocating piston engines. The cylinders and the pistons are replaced by working chambers and rotors encased in a stationary housing with ports. The working chamber and the rotor parallel the functions performed by the cylinder and the piston, the ports in the housing act as valves, and the phases take the place of strokes. The position of the rotor subdivides the working chamber into sections, each of which performs a specific function. However, unlike the piston engine where each action (intake, compression, ignition, and exhaust) takes place sequentially, in the rotary engine, several actions take place simultaneously.

The working of the Wankel engine is depicted in Fig. 5.4 [478]. In such engines, simultaneous functions are possible because of the radial location of the rotor. When the rotor is in the position shown in Part 1 of the figure, both the intake and the exhaust ports are closed by Face A, and the rotor Face B is starting to compress the air-fuel mixture. Rotor Face C is at the end of the power phase and is starting to exhaust the combustion products. As the rotor moves counterclockwise, as shown in Part 2, both the intake and the exhaust ports are open, and intake, compression, and exhaust occur simultaneously. As the rotor continues to move, these three functions are almost near completion, as shown in Part 3 of the figure. At this time, the ignition takes place and the power is generated. Further rotation leads to completion of the exhaust phase, the power phase, and the intake phase, as shown in Part 4 of the figure. At this stage, the cycle starts again. The key features of the Wankel engine are its configuration and the rotary motion. These not only facilitate designing of the advanced pollution control devices for such engines easy, but also help them achieve higher engine speeds than those ordinarily possible by the reciprocating piston engines.

The continuous-combustion gas turbine engines, the one shown in Fig. 5.5, contain a compressor, turbine, and combustion chamber [23,479]. Air is pulled into the compressor and compressed. The compressed air is split into two streams. One stream is passed into the combustion chamber where it is mixed with the fuel and the air-fuel mixture is ignited. The second stream of air is mixed with the combustion products downstream and the mixture is passed through a narrow nozzle ring to decrease pressure, hence increase velocity. The high-velocity gases are used to turn a turbine wheel, which is connected to a compressor by a shaft. Due to the continuous flow of gases, these engines achieve much

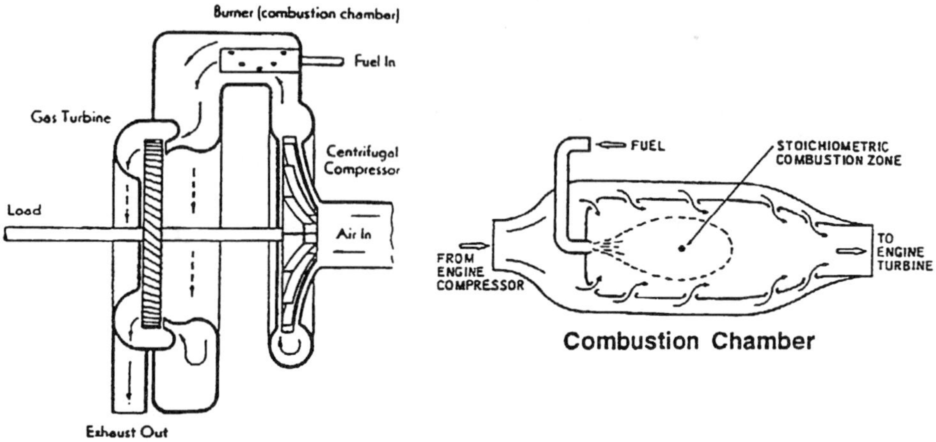

Fig. 5.5—Operation of a gas turbine engine [23,479].

higher speeds than those possible by the other types of combustion engines.

Most engines used in transport and industry today are of the internal combustion type. The prime reasons for the popularity of such engines are their high thermal efficiency and their light weight in relation to power capability. Four-stroke SI and CI engines are used in most automotive applications. Depending upon the cylinder arrangement, four-stroke cycle engines can be further classified into in-line, V (Vee), opposed-piston, and the radial types [23]. Two-stroke cycle SI engines are commonly used in lightweight applications (outboard motors, mopeds, motor scooters, motor cycles, snowmobiles, and chain saws), and two-stroke cycle CI engines find use in on-highway trucks, city buses, slow-speed marine, and some railroad diesel applications.

The manufacturers of the internal combustion engines are implementing design changes in their equipment so as to meet national and international emissions standards. Some of these are addressed in the Chapter 6 dealing with the Internal Combustion Engine Emissions.

Lubricant Specifications and Classifications

The automobile industry by far is the largest user of lubricants. Of the total lubricant use for the year 2006, estimated at 38 million metric tons, about 57 % is for the automotive applications and over two thirds of it is for automotive engine oils. Engine oil demand is slowly declining because of the OEMs' on-going efforts to improve engine designs, with the intent to reduce weight, increase fuel economy, increase power output, and meet environmental emissions guidelines. A number of other power train technologies, which are either already being marketed, such as hybrids, or are on the horizon, such as those involving the fuel cell, are likely to have additional negative impact on the demand of engine oils.

As mentioned earlier, engine oils are designed to perform a number of diverse functions in an engine. The performance of engine oils is primarily judged on their ability to reduce friction, resist oxidation, minimize deposit formation, and prevent corrosion and wear. However, there are additional requirements that relate to, for example, good viscometrics to permit easy starting at low temperatures and provide adequate lubrication at high temperatures, keeping engine parts clean, minimizing combustion chamber deposits, cooling engine parts, and sealing combustion pressures.

In the United States, performance specifications for engine lubricants are established by the collaborative efforts of a number of organizations, including the Society of Automotive Engineers (SAE), American Petroleum Institute (API), ASTM International (ASTM), American Automobile Manufacturers Association (AAMA), Engine Manufacturers Association (EMA), and Chemical Manufacturers Association (CMA) [480]. The history of development of the North American performance standard for engine oils and other lubricants is described in Ref [33]. In a number of cases, the U.S. Original Equipment Manufacturers (OEMs) and the U.S. Military have additional specifications. In Europe, ACEA (Association des Constructeurs Européens de l'Automobiles), CEC (Conseil Européens de Coordination pour les Developments des Essais de Performance des Lubrifiants et des Combustibles pour Moteurs), ATC (Technical Committee of Petroleum Additive Manufacturers), and ATIEL (Association Techniqué de l'Industries Européenne des lubrifiants) have collaborated in a manner similar to their American counterparts to come up with their own engine oil classification system, called the European Engine Lubricant Quality Management System (EELQMS). The system does include some individual OEM requirements. Prior to December 1995, Comité des Constructeurs d'Automobiles du Marché Commun (CCMC) helped ensure the availability of the lubricants that met the performance requirements of the European vehicles. In Japan and India, the Japanese Automobile Standards Organization (JASO) and the Bureau of Indian Standards perform these functions. Japanese vehicle manufacturers generally recommend engine oils for service-fill based upon the API Classification System.

The American Automobile Manufacturers Association of the United States (AAMA) and the Japan Automobile Manufacturers Association (JAMA) have collaborated to form the International Lubricant Standardization and Approval Committee (ILSAC). ILSAC has developed and approved its minimum performance standards GF-1 to GF-4 for passenger car engine oils [481], starting with GF-1 in 1992. GF-2, GF-3, and GF-4 went into effect in the year 1996, 2001, and 2004, respectively. The work has already started on the ILSAC GF-5 standard, which is slated to go into effect in the year 2009 [482]. ILSAC GF-4, the most recent standard that is currently in effect consists of three sections dealing with physical and performance properties of the lubricant, while the fourth section cites documents applicable to the standard [481]. The first section deals with viscosity and uses the SAE viscosity classification standard, SAE J300. This section also includes an ASTM Test, D5133, which measures the low temperature, low shear rate viscosity in terms of the gelation index and the temperature at which gelation occurs. Gelation index represents the maximum rate of viscosity increase. The second section deals with the performance requirements and primarily uses the API SM performance criteria. The third section contains specifications for a number of physical bench test performance parameters. These include catalyst compatibility, wear, volatility, high temperature deposits, filterability, foaming tendency, high-temperature high-shear rate viscosity, homogeneity and miscibility, and the rusting tendency.

An important feature of the ILSAC standards is that fleet testing of the lubricant to demonstrate performance is not necessary. Meeting the performance requirements in the specified engine sequence tests is considered sufficient. This is done to minimize testing time and costs. An analogous standard for diesel engine oils may be forthcoming. While the standards established by each country are important, ACEA and ILSAC standards for gasoline engine oils and the API standard for diesel engine oils are global in impact. Global specifications are also in place for heavy-duty diesel engine lubricants, DHD-1, and the proposed DHD-2.

Trends Impacting the New Performance Standards

New performance standards are developed in response to equipment changes implemented by the OEMs as a result of changing consumer needs or by the legislative action of the governmental organizations, such as the EPA. Examples of changing consumer needs are more power and low fuel use (fuel economy) and examples of the governmental actions are low emissions and high corporate average fuel economy (CAFE) legislations. New lubricant development process generally starts prior to commercialization of the new equipment because of the time and the resources involved.

Lubricants for modern gasoline engines place a major emphasis on greater oxidative stability, improved fuel economy, catalyst compatibility, and reduced emissions and those for heavy-duty engines place a major emphasis on soot handling capability, reduced emissions, and compatibility with particulate filters (DPFs) [483]. The drive for greater oxidative stability is a consequence of the desire to prolong the service life of the lubricant. At present for passenger cars in the United States, the service interval is about three months and in Europe, it is about six months to two years. Operating temperatures in modern gasoline engines are up to 130 °C at the sump (under heavy loads) and 250 °C, or higher, at the pistons [483]. At these temperatures, the rate of the lubricant oxidation is quite high. If one intends to prolong the service interval, which appears to be the case at this time, the lubricant must possess excellent oxidative stability so as to maintain its viscosity/fluidity and the lubrication properties over the intended period. While the presence of the oxidation inhbitors greatly helps, they have a finite service life. At this stage, the lubricant viscosity will see an abrupt increase. Base oil viscosity increase as a consequence of oxidation was described in Lubricant Additives, Chapter 4.

Viscosity is not the only lubricant parameter that is likely to change with oxidation; its Total Acid Number (TAN) will increase as well as the deposit-forming tendency. The former is due to the oxidation-derived acids and the latter is due to the formation of the highly oxygenated materials of low oil solubility and aldehydes and ketones that can react with each other to form polymeric materials. The mechanism of formation of the deposit precursors was explained in the section on Oxidation Inhibitors in Chapter 4. It is important to control the formation of both these species otherwise proper engine performance will be compromised due to increased corrosion and deposits. Typically oxidation-derived acids are neutralized by the use of the basic detergents. Therefore, to achieve the objective of extending the service interval, one must have the base oils, hence the lubricants, that are oxidatively more stable and additives, such as oxidation inhibitors, detergents, and dispersants that are longer lasting. The topic of drain interval is discussed further towards the end of this chapter.

The topic of fuel economy was dealt with in some detail in the Friction Modifiers Section of the Additives Chapter 4; hence we will be brief here. Many nations in the world are interested in preserving the quality of their environment, especially the air quality. Carbon dioxide (CO_2) is one of the combustion-related gases that are being scrutinized. More specifically, the drive is to decrease the vehicle-related emissions of CO_2 by lowering the use of gasoline in cars and diesel in trucks. This is precisely the objective of the Corporate Average Fuel Economy (CAFE) program in the United States. Since the OEMs have little choice but to meet the CAFÉ requirements, they are implementing engine design changes to meet them. And of course, the lubricant manufacturers are helping them achieve this objective by formulating lubricants that provide efficient fuel economy. Such lubricants are low viscosity oils (SAE 5W-30 and SAE 5W-20) with good low temperature fluidity that are supplemented with the friction-reducing additives. The realized fuel economy benefit by the use of these lubricants is 1.5 to 2.0 %.

Soot control is one of the primary reasons for design changes in the heavy-duty diesel engines. Two strategies that are presently in place to avoid exiting soot into the environment are directing it into the oil and installation of the Diesel Particulate Filters (DPFs) on the engine's exhaust system. The first strategy depends upon the dispersant additives to keep soot suspended in oil until the next service when it is removed with the used lubricant. The second strategy is to pass the soot that escapes the first strategy through a filter, prior to allowing the exit of the exhaust gases into the atmosphere. The topic of soot control is discussed in detail in Chapter 6 on Emissions Control.

Regulations pertaining to a vehicle's exhaust emissions are progressively becoming more stringent. One of the ways to meet the exhaust emissions requirements is by the use of the after-treatment devices, such as catalytic converters or particulate filters, discussed in the previous paragraph. Since they are described in detail in Chapter 6 on Combustion Engine Emissions, we will briefly comment on these here. The devices that are used to treat exhaust emissions, as in the case of gasoline engines, or capture them, as in the case of diesel engines, contain noble metal catalysts. These lose their activity by elements such as sulfur and phosphorus, which may be present in fuels and finished lubricants as additives. Because of the potential damage to the after-treatment devices, the use of such elements in fuels and lubricants is being controlled.

U.S. Standards

Physical characteristics of the engine lubricants for the North American use are defined by the SAE viscosity classifications and the performance characteristics are defined by the API service classifications. As stated before, a lubricant must have specific viscosity characteristics to ensure easy starting at low temperatures and adequate lubrication at high temperatures. The viscosity classification system helps select lubricants by taking into account the equipment's operating temperature range. The service classifications, on the other hand, assure the user that the lubricant meets the performance requirements of the equipment manufacturer. Until very recently, the SAE, ASTM, and API were the three key organizations responsible for developing such specifications. The SAE defined the need for the service category, the ASTM developed or selected the testing techniques, and the API developed the user language to define the category. However, in view of the time delays in developing new performance categories and testing constraints, a number of other organizations, such as ILSAC, CMA, AAMA and EMA, are now actively involved in the process [309]. Other countries have similar organizations that establish the performance criteria for engine oils used in their countries. As mentioned earlier, the U.S. Military and the original equipment manufacturers (OEMs) have their own performance requirements, which are over and above those of the API.

Viscosity Classification

The importance of viscosity was recognized in the earlier part of this century when in 1911 the SAE established the first engine oil classification system based on viscosity. Since then, the system has been revised many times [33]. At present, the viscosity classification systems for engine oils are described by the SAE Standards J300 and J1536 [309,485]. The SAE Standard J300 deals with the viscosity of the lubricants for four-stroke cycle engines (both CI and SI types) and two-stroke cycle CI engines. The SAE Standard J1536 specifies viscosity of oils for the two-stroke cycle SI engines only. The basic viscosity grade categories for engine oils are determined by the Crankcase Classification System devised by the SAE, which uses the test methods approved by the ASTM.

The most recent SAE Standard J300 has eleven viscosity grades, six "W" grades and five regular grades. The "W" grades range from 0W to 25W, and regular grades range from 20 to 60. Each viscosity grade must meet a number of requirements. Single grade viscosity oils with the letter "W" (winter) must meet the maximum low-temperature cranking and pumping viscosities at the prescribed temperatures and the minimum kinematic viscosity at 100 °C. Single grade oils without the letter "W" must meet the minimum and the maximum viscosities at 100 °C and the high-temperature, high-shear (HTHS) viscosity at 150 °C and 10^6 s^{-1}. The ASTM Standards D5293 and D4684 describe methods to measure the cranking and pumping viscosities, respectively. The ASTM Standard D445 is used to determine minimum and maximum viscosities at 100 °C and the ASTM test procedures D4683 and D4741 are used to determine the HTHS viscosity. The "W" or the winter grade requirements to be pumpable at low temperatures, and the non-"W" grade requirement to have sufficient viscosity at high temperatures ensure proper lubrication and the protection of the engine parts during all-season operation. The viscosity requirements relating to the different API/SAE grades are summarized in Table 5.1. Multi-grade oils should satisfy the appropriate requirements of both the "W" and the non-"W" grades, as is shown in Fig. 5.6.

The U.S. Military uses eight viscosity grades for engine oils, which are identified in three U.S. Military specifications [484], see Table 5.2. Military uses three single grades—10W, 30, and 40, and five multi-grades—0W-30, 5W-30, 10W-20, 10W-30, and 15W-40. Viscosity grade 10W requires the cranking viscosity of ⩾6600 cP at −30 °C and ⩽7000 cP at −25 °C and pumping viscosity of ⩽30,000 cP at −30 °C. Single grades 30 and 40 have minimum and maximum kinematic viscosity requirements at 100 °C and require a minimum VI of 80. Multi-grades must meet the minimum and maximum cranking viscosity requirements at −35 to −20 °C; the maximum pumping viscosity requirements at −40 to −25 °C, depending upon the grade, and the HTHS viscosity requirements. Military oils, in addition, have specifications related to pour point, stable pour point, flash point, evaporative loss, phosphorus content, and the sulfated ash [484]. Physical requirements for the military grades are provided in Table 5.3.

The SAE Standard J1536 describes the miscibility and fluidity grades for the two-stroke cycle engine oils [485]. It defines four grades, ranging from SAE-1 to SAE-4. These grades reflect a lubricant's ability to mix with the fuel at prescribed temperatures. A low-temperature viscosity requirement is associated with each of these grades. Two-stroke cycle SI engines do not usually have an oil sump, and the lubricant is mixed with the fuel in the fuel tank. The miscibility and fluidity specifications ensure that the lubricant is miscible with the fuel and that the blend meets the low-temperature viscosity requirements of the desired performance category. The SAE categories, along with the matching requirements, are presented in Table 5.4.

API and SAE Service Classifications

The API and SAE engine oil performance requirements for North America are based on the standard tests and are described by the API Engine Service Classification System, which was first introduced in 1969/1970 [33]. The API service symbol, "Donut," was established in 1983 and is presented in Fig. 5.7. This symbol, although originally designed to cover the entire lid of a round full-seal quart or a litre container, can be applied in a prominent position on a variety of other packages. The significance of the symbol is to communicate the engine oil quality and performance to the general public and to help them select oils that meet manufacturers' recommendations for use in the intended application. The upper part of the symbol displays the API service category, the center part displays the SAE viscosity grade, and the bottom part displays the energy conserving feature, if claimed.

TABLE 5.1—API/SAE engine oil viscosity classifications 2004.

SAE Viscosity Grades For Engine Oils[1,2]					
SAE Viscosity Grade	Low Temperature (°C) Cranking Viscosity[3], mPa-s, Max	Low Temperature (°C) Pumping Viscosity[4], mPa-s, Max with No Yield Stress(4)	Low-Shear-Rate Kinematic Viscosity[5](mm²/s) at 100°C, Min	Low-Shear-Rate Kinematic Viscosity[5](mm²/s) at 100°C, Max	High-Shear-Rate Viscosity[6] (mPa-s) at 150°C, Min
0W	6,200 at -35	60,000 at -40	3.8	-	-
5W	6,600 at -30	60,000 at -35	3.8	-	-
10W	7,000 at -25	60,000 at -30	4.1	-	-
15W	7,000 at -20	60,000 at -25	5.6	-	-
20W	9,500 at -15	60,000 at -20	5.6	-	-
25W	13,000 at -10	60,000 at -15	9.3	-	-
20	-	-	5.6	< 9.3	2.6
30	-	-	9.3	< 12.5	2.9
40	-	-	12.5	< 16.3	2.9 for 0W-40, 5W-40, and 10W-40 grades
40	-	-	12.5	< 16.3	3.7 for 15W-40, 20W-40, 25W-40, and 40 grades
50	-	-	16.3	< 21.9	3.7
60	-	-	21.9	< 26.1	3.7

[1] **Notes — 1cP = 1mPa-s; 1 cSt = 1mm²/S**
[2] All values are critical specifications, as defined by ASTM D3244
[3] ASTM D5293
[4] ASTM D4684: Note that the presence of any yield stress detectable by this method constitutes a failure regardless of viscosity.
[5] ASTM D445
[6] ASTM D4683, CEC L-36-A-90 (ASTM D4741) or D5481

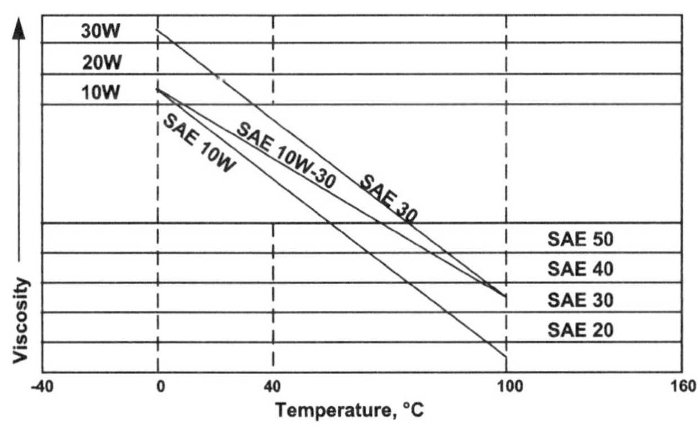

Fig. 5.6—Viscosity requirements of single grade and multi-grade oils.

Only licensees are authorized to display this symbol. The symbol is primarily used in North America, with limited use in other countries.

The API service categories for gasoline engine lubricants range from SA to SM, where S stands for "service," and are used by the service stations, garages, etc. For diesel engine lubricants, the API categories range from CA to CJ-4, where C stands for "commercial." The API SA to SH categories for gasoline engine oils and CA to CE categories for diesel engine oils are obsolete and their requirements are satisfied by API SJ and CF and higher service classes, respectively [486].

Earlier, we mentioned the ILSAC GF categories. Of the four categories, GF-1 is obsolete but GF-2, GF-3, and GF-4 are in effect. These categories use the API SJ, SL, and SM performance criteria but include a fuel economy ASTM Sequence VIB test. Field testing is only required for technologies that are radically different from those in existence. The requirements of the obsolete categories are satisfied by API

TABLE 5.2—Correspondence between military specification and SAE oil grades [484].

Specification	Specification Grade Designation	SAE Grade Classification
MIL-PRF-2104	10W	SAE 10W AND SAE 10W-20
	30	SAE 30
	40	SAE 40
	15W-40	SAE 15W-40
SAE J2362	5W-30	SAE 5W-30
	10W-30	SAE 10W-30
MIL-PRF-46167	Arctic	SAE 0W-30

TABLE 5.3—Military viscosity grades according to MIL-PRF-2104H, SAE J2362, and MIL-PRF-46167.

Parameter	10W	30	40	0W-30	5W-30	10W-20	10W-30	15W-40
Cranking Viscosity[a] (cP)								
at temperature °C, min	6,600 at -30	--	--	--	6,200 at -35	6,600 at -30	6,600 at -30	7,000 at -25
max	7,000 at -25	--	--	6,200 at -35	6,600 at -30	7,000 at -25	6,600 at -30	7,000 at -20
Pumping Viscosity[b] (cP)								
at temperature °C, max	60,000 at -30	--	--	60,000 at -40	60,000 at -35	60,000 at -30	60,000 at -30	60,000 at -25
Viscosity[c] (cSt) at 100°C								
min	4.1	9.3	12.5	9.3	9.3	5.6	9.3	12.5
max	<5.6	<12.5	<16.3	<12.5	<12.5	< 9.3	<12.5	<16.3
Viscosity Index, min	--	80	80		--	--	--	--
HTHS Viscosity[d] (cP) min	--	2.9	3.7	2.9	2.9	2.6	2.9	3.7
Flash Point[e] (°C) min	205	220	225	--	200	--	205	215
Evaporative Loss[f] (%) max	--	--	--	--	20	--	17	15
Phosphorus[g] (% mass) max	0.12	0.12	0.12	0.12	0.12	0.12	0.12	0.12
Sulfated Ash[h]	1.5	1.5	1.5	1.5	1.5	1.5	1.5	1.5

[a] ASTM D5293
[b] ASTM D4684
[c] ASTM D445
[d] ASTM D4683
[e] ASTM D92
[f] ASTM D5800
[g] ASTM D5185
[h] ASTM D874

SJ, SL, and SM, and by ILSAC GF-2, GF-3, and GF-4, which are designed for the most severe operation. The API SM and ILSAC GF-4 qualified oils are designed to have superior oxidation resistance (less oil thickening), protection against deposits, wear protection, and low-temperature performance than oils qualified under the previous performance categories. In addition, these oils have improved environmental compatibility since they extend the life of the emissions system and conserve energy. As mentioned earlier, fuel economy and emissions system compatibility are the primary factors responsible for the gasoline engine design changes and the new oils are developed to lubricate them. The "Starburst" symbol, shown in Fig. 5.8, is intended to be used for oils that meet the ILSAC performance requirements [481]. The timeline for the development of the gasoline engine oil categories is depicted in Fig. 5.9.

The licensing of both the "Donut" symbol and the "Starburst" symbol is administered by the API. The objective of these labels is to protect the public against products that do not meet the appropriate performance requirements. SAE has a similar Oil Labeling Assessment Program (OLAP) which is sponsored by the U.S. Army and the petroleum and automotive industries. SAE monitors the viscosity and the chemical properties of the publicly marketed oils. Both these organizations have installed a number of measures to discourage the illegal use of the API and the ILSAC symbols. Despite the efforts of these organizations to manage oil quality, many oils in the marketplace have lower quality than claimed by the suppliers of these lubricants.

For diesel engine oils, the API CF, CF-2, CF-4, CG-4, CH-4, CI-4, and CJ-4 categories are presently active. API CJ-4 is the newest engine oil category that was introduced to service the 2007 model heavy-duty diesel trucks fitted with diesel particulate filters (DPFs). API CF-2 is for severe-duty two-stroke cycle engines. API CF is for four-stroke cycle engines that are either turbocharged or use high sulfur fuel. Such engines are often for *off-highway* use. API CF-4 and CG-4, CH-4, CI-4, and CJ-4 categories are for four-stroke cycle engines for *on-highway* use. These engine oils are formulated for use in low-emissions engines of the model years 1991, 1994, 1998, 2004, and 2007. Emissions control is the major driving force that is causing new improvements in engine design and each major change requires the development of a lubricant that is suitable to lubricate the new engine. API CJ-4 category is designed to qualify oils for the low-emissions 2007 engines that use low-sulfur diesel fuel. These

TABLE 5.4—Two-stroke cycle engine oils miscibility/fluidity classification.

SAE Grade	Test Temperature, °C[a]	Allowable Brookfield Viscosity[a] (cP), max	Reference Oil[b]
1	0	3,500	VI-GG
2	−10	3,500	VI-FF
3	−25	7,500	VI-D
4	−40	17,000	VI-II

[a]Both miscibility and Brookfield tests must be run.
[b]Results on candidate oil must not exceed those on reference oil by more than 10 %.

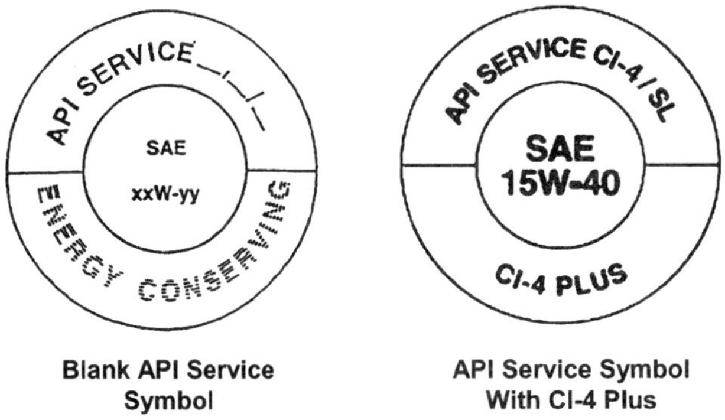

Fig. 5.7—API "Donut" service symbols [480].

lubricants must have elemental composition that will not impair the functioning of the diesel particulate filters. See Chapter 6 on Emissions Control for a detailed discussion. Other factors that are either important or gaining importance are the fuel economy and the extended service intervals. The primary objective is to control the downtime costs.

API SJ/SL/SM and CH-4/CI-4/CJ-4, the most recently introduced categories, are designed for the most severe operations [486]. The new interim categories for the next generation heavy-duty diesel engine oils are forthcoming. Again, the emphasis in the new categories is likely to be on emissions and fuel economy. The progression of the U.S. heavy-duty diesel specifications with time is provided in Fig. 5.10. In general, oils meeting the higher service requirements are suitable for use in the lower service class in the same category.

Engine test requirements for each active gasoline engine oil category are given in Table 5.5 and those for diesel engine oil categories are provided in Table 5.6. Current engine tests and the parameters the gasoline tests measure are listed in Table 5.7 and the parameters that the diesel tests measure are listed in Table 5.8. These tests are devised to measure protection against rust, varnish, deposits, sludge, wear, high-temperature oil-thickening, and ring sticking. The minimum and maximum performance limits for each parameter for various gasoline and diesel engine tests are given in Table 5.9–5.14 [487].

Figure 5.11–5.13 identify the parts of a four-stroke cycle engine that are commonly rated. Figure 5.11 shows a cross-sectional view of an in-line six-cylinder engine, and Fig. 5.12 shows a cross-sectional view of a V-8 engine. Figure 5.13 shows the various other parts of the engine that are rated. As mentioned earlier, only certain parts of the engine encounter boundary lubrication and hence experience adhesive wear damage. These are the valve train, cylinder bores, and piston rings. Figure 5.14 shows the valve train arrangements and the type of wear damage due to lubrication failure at different points [488]. Pitting and scuffing are the main types of surface damage experienced by these parts. The damage is more prevalent on the working face of the follower and on the nose and flanks of the cam.

Fig. 5.8—ILSAC Starburst symbol [480].

U.S. Military and OEM Specifications

U.S. Military equipment needs differ from those of the commercial equipment. The military operates large and diversified fleets of vehicles containing both two- and four-stroke cycle engines, air-cooled and liquid-cooled, ranging from 2 to over 1000-hp. U.S. Military specifications, designated by the prefix MIL, have performance requirements similar to those of the API service designations. However, it is important to note that the development of the military specifications does not usually follow but parallels the development of the API categories, as depicted in Fig. 5.10. Incidentally, afte1997 the MIL-L-2104 upgrades for engine oils were renamed as MIL-PRF-2104.

Most OEMs accept the performance requirements established by the API categories. However, certain OEMs have additional requirements to qualify oils for use in their equipment. Mack EO-M, Cummins CES20078, and Caterpillar's upcoming ECF-2 and ECF-3 specifications are examples of such requirements. It appears that in the future such OEM specifications are going to become more prevalent as the heavy-duty engine builders require performance above that established by the accepted industry standards, as defined by the API categories [489].

The National Marine Manufacturers Association (NMMA) has its own performance requirements for the two-stroke cycle engine oils. TC quality oils are the most specified oils for the air-cooled engines and TC-W3® specified oils are

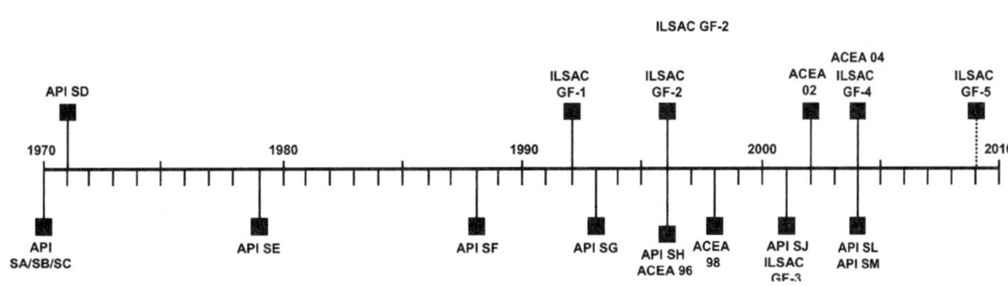

Fig. 5.9—Chronology of gasoline engine oil specification development.

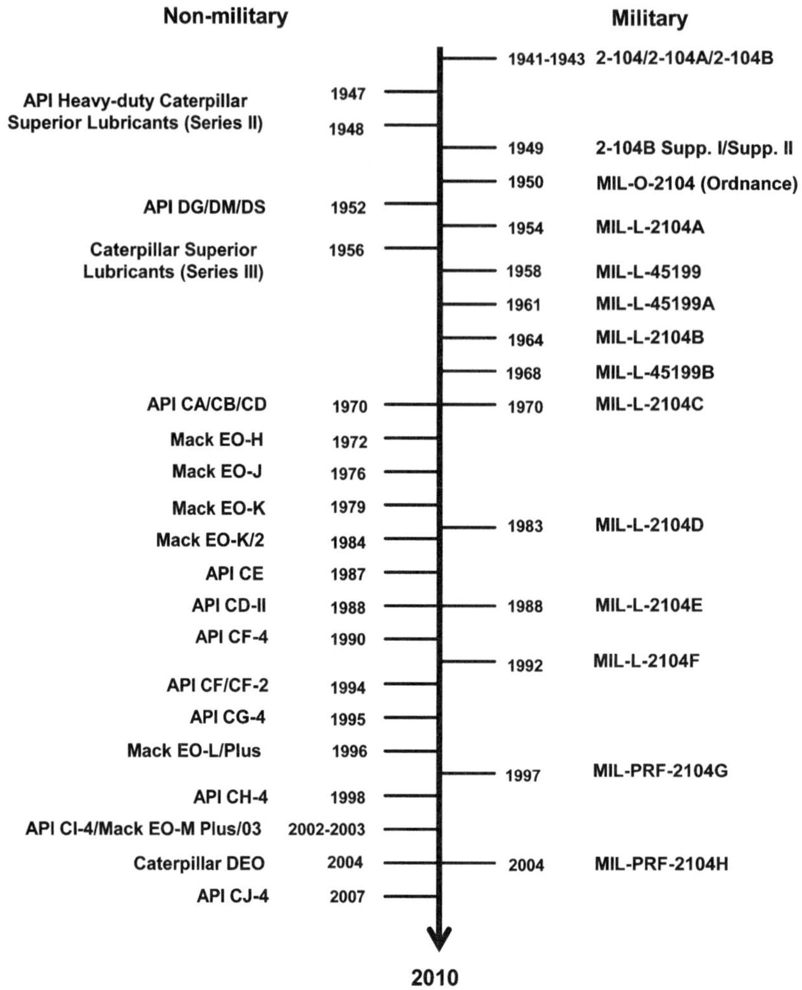

Fig. 5.10—Chronology of the U.S. diesel engine oil specification development.

the most recommended oils for use in the modern water-cooled outboard engines.

The API performance classification system, the SAE viscosity grades, and the OEM performance requirements together help design engine oils well suited for the specific end-use applications. New demands placed on the lubricants due to the changing technology and the governmental regulations require continual revisions and upgrading of the performance specifications.

European Standards

As mentioned earlier, European standards for automotive engine oils prior to December 1995 were established by the CCMC. Subsequently, ACEA assumed the responsibility of defining the lubricant quality. ACEA standards recognize that European engines differ from those in the United States, both in design and the operating conditions; hence they place different demands on the lubricant. This implies that the oils used in the European engines are unique and hence require classification system that is different from that of the API and includes European engine tests. In addition, some ACEA standards must take into account the effect of the oil on the engine emissions and emissions control systems. This is critical for engines that must meet the much tighter Euro 4 emissions standards.

All engine performance testing data to show compliance with the ACEA test sequences must be generated according to the European Engine Lubricants Quality Management System (EELQMS). This system is described in the ATIEL Code of Practice. It addresses product development testing and product performance documentation, involves the registration of all candidate and reference oil testing, and defines the compliance process. Compliance with the ATIEL Code of Practice is mandatory to claim meeting the requirements of the ACEA 2007 [490].

A number of changes have been implemented in the ACEA 2007 Engine Oil Sequences relative to those of the ACEA 2002 and 2004. These are listed below [487,490].

Light Duty Engine Sequences

1. The A (gasoline) and B (diesel) sequences were replaced by the combined A/B sequences, which contain both gasoline and diesel engine performance tests.
2. New C sequences were introduced, which again combine the gasoline and diesel engine performance, and also include the chemical limits.
3. There are four combined A/B categories (A1/B1, A3/B3, A3/B4, and A5/B5). The requirements are based upon the previous A and B category requirements.
4. There is no new category based upon A2 and B2.

TABLE 5.5—Tests associated with current U.S. gasoline engine oil classifications

TESTS	API			ILSAC			Other Categories that Include the Test
	SJ	SL	SM	GF-2	GF-3	GF-4	
CRC L-38							API CF, CF-2, CF-4, and CG-4; MIL-L-21260D, MIL-L-46167B, MIL-PRF-2104G
ASTM Sequence IID							API CG-4
ASTM Sequence IIIE (D5533)							API CG-4
ASTM Sequence IIIF (D1491)	✓	✓		✓	✓		API SJ, SL, and SM; ILSAC GF-2, GF-3, and GF-4; API CF-4, CG-4, CH-4, CI-4, CI-4 Plus, and CJ-4; MIL-PRF-2104G and 2104-H
Sequence IIIG			✓			✓	API SJ, SL, and SM; ILSAC GF-2, GF-3, and GF-4; MIL-L-21260D, MIL-L-46167B
ASTM Sequence IIIGA (D4684)			✓			✓	
ASTM Sequence IVA (D6891)	✓	✓	✓	✓	✓	✓	
ASTM Sequence VE (D5302)		✓			✓		
ASTM Sequence VG (D6593)	✓	✓	✓	✓	✓	✓	
ASTM Sequence VIB Fuel Economy (D6837)	✓	✓		✓	✓	✓	
ASTM Sequence VIII (D6709)	✓	✓	✓	✓	✓	✓	
Ball Rust Test	✓	✓	✓	✓	✓	✓	
TEOST (High Temp. Deposits)	✓			✓			
TEOST(MHT4)		✓	✓		✓	✓	

5. There are no chemical limits besides the sulfated ash in the A/B sequences.
6. Sulfur, phosphorus, and chlorine levels must be reported.
7. Sulfated ash levels are based upon the previous levels, and where the A and B levels were different, the higher limit has been used.
8. There are four new C categories (C1, C2, C3, and C4).
9. C1 defines a "low SAPS," low viscosity oil of A5/B5 performance level. SAPS is the acronym for sulfated ash, phosphorus, and sulfur.
10. C2 defines a "mid SAPS," low viscosity oil of A5/B5 performance level.
11. C3 defines a "mid SAPS," normal viscosity oil with a performance level aligned with the DaimlerChrysler p229.31 specification.
12. C4 defines normal viscosity "low SAPS" oils aligned with the Renault Long Drain specification.
13. There is no change to the engine test requirements in the A/B or C sequences, except in C4 which was introduced in 2007.
14. The XUD11BTE is obsolete.
15. The DV4 test was developed to replace the XUD11BTE.

Heavy Duty Diesel Engine Sequences

1. The E3 and E5 categories have been deleted, and the two new categories, E6 and E7 are defined.
2. The requirements for E2 and E4 remain unchanged.
3. E6 is based on the previous E4 category with the following changes.
   - Chemical limits were introduced for sulfur and phosphorus and reduced for sulfated ash.
     a. Shear stability requirements have been increased to passing after 90 cycles in the Kurt-Orbahn test.
     b. The Mack T-10/T-12 is now required.
4. E7 is based upon the previous E5 category with the following changes.
   - Shear stability requirements have been increased to passing after 90 cycles in the Kurt-Orbahn test.
     a. The Mack T-9 test is replaced by the Mack T-10/T-12 test.
     b. Cummins M11 is replaced by Cummins ISM.

ACEA 2007 European Oil Sequences for Service-fill Oils consist of the test sequences for gasoline engines, light-duty diesel engines, and for engines that are equipped with the after-treatment devices. Within each group, there are categories which reflect different performance requirements; A1/B1, A3/B3, A3/B4, and A5/B5 for gasoline and light-duty diesel engines and C1, C2, C3, and C4 for low sulfated ash, phosphorus, and sulfur oils (SAPS) for engines that are equipped with after-treatment devices.

In these designations, the letter represents the class and the number represents the category. These are to help the consumer select oils that are suitable for use in the intended

TABLE 5.6—U.S. engine oil classification system for current automotive heavy-duty diesel engine services.

	API							U. S. Military	
TESTS	**CF**	**CF-2**	**CF-4**	**CG-4**	**CH-4**	**CI-4/CI-4 Plus**	**CJ-4**	**MIL-PRF-2104G**	**MIL-PRF-2104H**
CRC L-38 (ASTM D 5119)	✓	✓		✓					
ASTM Sequence IID				✓					
High Temperature Corrosion				✓	✓	✓		✓	✓
ASTM Sequence IIIE				✓	✓			✓	
ASTM Sequence IIIG/ASTM Sequence IIIF (ASTM D 5533/D 6984)					✓	✓	✓		✓
Caterpillar 1K (ASTM RR:DO2-1273)			✓		✓	✓			✓
Caterpillar 1M-PC	✓	✓		✓				✓	
Caterpillar 1N (ASTM D 6750)				✓		✓	✓	✓	✓
Caterpillar 1P (ASTM D 6681)					✓				
Caterpillar 1R						✓			✓
Caterpillar C-13							✓		
Roller Follower Wear (ASTM D 5966)				✓	✓	✓		✓	✓
Mack T-6			✓						
Mack T-7			✓						
Mack T-8 (ASTM D 5967)				✓				✓	
Mack T-8A alternative to T-7			✓						
Mack T-8E (ASTM D 5967-96)					✓	✓			✓
Mack T-9 (ASTM D 6483)			✓		✓				
Mack T-10						✓			✓
Mack T-11 (ASTM D 7156)						✓	✓		✓
Mack T-11A							✓		
Mack T-12							✓		
Cummins NTC-400			✓						
Cummins M11 (ASTM RR:DO2-1439)					✓				
Cummins M11 EGR						✓			✓
Cummins ISM							✓		
Cummins ISB							✓		
CRC L-38	✓	✓	✓	✓				✓	
Detroit Diesel 6V92TA (ASTM D 5862)	✓	✓						✓	✓
Allison C-4/Caterpillar TO-4								✓	?
Aeration (ASTM D 6894)				✓	✓	✓	✓	✓	✓

equipment. Each category also has a two-digit number that identifies the implementation year and is for the benefit of the industry. For example, the combined designation A1/B1$_{-04}$ indicates that it was issued in the year 2004 and identifies oils for use in both gasoline engines and light-duty diesel engines. The nomenclature and the description used for various 2007 ACEA categories are summarized below.

1. *A/B—gasoline (petrol) and diesel engine oils*. *A* indicates

TABLE 5.7—U.S. gasoline engine tests.

Engine Test	Engine Type	ASTM Test	Evaluation Criteria
CRC L-38	CLR single cylinder engine	D5119	Bearing corrosion, viscosity increase
Sequence IID	1977 Oldsmobile V8 Engine	D5844 (Withdrawn)	Rust and corrosion
Ball Rust Test	Bench test	D6557	Rust
Sequence IIIE	1996/1997 GM V6 Fuel-injected engine	D5533	Viscosity increase and wear
Sequence IIIF	1996/1997 GM V-6 fuel-injected engine	D6984	Viscosity increase, varnish, deposits, and wear
Sequence IIIG	1996/1997 Series II General Motors V-6 engine.	---	Viscosity increase, deposits, and wear oil consumption
Sequence IIIGA	1996/1997 Series II General Motors V-6 engine	D7320	Aged oil low-temperature viscosity
Sequence IVA	1994 Nissan KA 24E four-cylinder in-line engine	D6891	Wear
Sequence VE	Ford 2.3-liter 4 cylinder engine	D5302	Sludge, varnish, and wear
Sequence VG	2000 Ford 4.6-liter V-8 engine	D6593	Sludge, varnish, and wear
Sequence VIA	1993 4.6-liter Ford modular V-8 engine	D6202	Fuel economy
Sequence VIB	1993 4.6-liter Ford modular V-8 engine	D6837	Fuel economy
Sequence VIII	CLR Single Cylinder engine	D6709	Bearing weight loss, viscosity change
TEOST, TEOST (MHT4)	Bench test	D6335	Thermal and oxidative stability, deposits

oil for a gasoline engine and *B* indicates oil for a light-duty diesel engine.

2. *A*1/*B*1—Oils intended for use in gasoline and diesel car and light commercial vehicles, specifically capable of using low friction, low viscosity oils with high temperature/high shear characteristics.
3. *A*3/*B*3—Oils for use in high performance gasoline and diesel cars and light commercial vehicles where extended drain intervals are specified by the vehicle manufacturer and for year-round use of the low viscosity oils and for use in severe operating conditions, or a combination thereof, as defined by the vehicle manufacturer.
4. *A*3/*B*4—Oils for use in high performance gasoline and direct injection diesel engines. Includes oils suitable for applications described under *B*3.
5. *A*5/*B*5—Oils for use at extended oil drain intervals in high performance car and light commercial gasoline and diesel engines designed for low viscosity oils.
6. *C*—Catalyst compatible oils for gasoline and diesel engines with after-treatment devices.
7. *C*1, *C*2 and *C*3—Oils for use in high performance car and light commercial gasoline and diesel engines, equipped with diesel particulate filter, three-way catalyst, and requiring low viscosity, low friction, catalyst compatible oils, or both
8. *C*4—Oils of normal viscosity suitable for use in engines that follow Renault Long Drain Specification. These oils have low sulfated ash, phosphorus, and sulfur (SAPS) limits.
9. Other classes may be added in the future if, for example, the need arises to separately issue specifications for natural gas engines.
10. *E*—Heavy-duty diesel engine oils.
11. *E*2—General purpose oil for naturally aspirated and turbocharged heavy-duty diesel engines, medium to heavy duty cycles and mostly normal oil drain intervals.
12. *E*4, *E*6, and *E*7—Stable, stay-in-grade oil for highly rated diesel engines meeting Euro 1, Euro 2, Euro 3, and Euro 4 emissions requirements and running under very severe conditions, such as significantly extended oil drain intervals. These oils are designed to provide excellent control of piston cleanliness, wear, soot handling, and lubricant stability and are suitable for engines without particulate filters, for some EGR engines, and some engines fitted with SCR (selective catalytic reduction) NOx reduction systems. *E*7 oils in addition provide effective control with respect to piston bore polishing. *E*6 oils are the oils of choice for engines fitted with particulate filters and operate on low sulfur diesel fuel (max 50 ppm). However, the recommendations may differ between engine manufacturers.

In addition to the engine oil test sequences for gasoline engine oils, light-duty diesel engine oils, and heavy-duty diesel engine oils, ACEA Standard contains physical requirements that parallel those for the U.S. engine oils. ACEA uses the SAE viscosity classification system, described in SAE J300 [309], to define the viscosity grades. It has additional requirements relating to shear stability, evaporation loss, elas-

TABLE 5.8—U.S. diesel engine tests.

Engine Test	Engine Type	Evaluation Criteria	Test included in Specifications
Caterpillar 1K	Caterpillar 1Y540 single cylinder direct injection engine (Aluminum)	Piston deposits and oil consumption	API CF-4, CH-4, and CI-4, MIL-PRF-2104H
Caterpillar 1M-PC	Caterpillar 1Y73 single cylinder indirect injection engine	Piston, rings, and liner distress	API CF, CF-2 and CG-4; MIL-PRF-2104G and 2104H
Caterpillar 1N	Caterpillar 1Y540 single-cylinder direct-injection engine (Aluminum)	Piston deposits and oil consumption	API CG-4, CI-4 and CJ-4; MIL-PRF-2104G and 2104H
Caterpillar 1P	Caterpillar 1Y3700 single-cylinder direct-injection engine	Piston, rings, and liner distress	API CH-4
Caterpillar 1R	High-speed four-stroke cycle Caterpillar 1Y3700 single cylinder engine (Steel)	Piston, rings, and liner distress; and oil consumption	API CI-4; MIL-PRF-2104H
Caterpillar C-13	Six cylinder turbocharged engine	Steel Piston Deposits, Oil Consumption	API CJ-4
Mack T-6	1991 In-line, six cylinder, Mack E7-350, turbocharged, charge air-cooled, direct injection diesel engine	Piston deposits, wear, oil consumption, and oil thickening	API CF-4
Mack T-7/8A	1991 In-line, six cylinder, Mack E7-350, turbocharged, charge air-cooled, direct injection diesel engine	Oil thickening	API CF-4
Mack T-8	Mack E7-350, in-line six cylinder 11GBA77623, direct injection engine	Soot-related viscosity increase and oil consumption	API CG-4; MIL-PRF-2104G
Mack T-9	Mack 1994, 350 bhp VMAC II in-line 6 cylinder engine	Soot thickening	API CF-4, CH-4
Mack T-10	Mack E7 E-Tech 460 engine with EGR.	Cylinder and bearing wear protection; and oil consumption	API CI-4; MIL-PRF-2104H
Mack T-11 (ASTM D7156)	Mack E-TECH V-MAC III, electronically-controlled fuel injection engine	Soot-related viscosity increase and oil consumption	API CI-4 and CJ; MIL-PRF-2104H; Mack EO-N Premium Plus and EO-N Premium Plus 03
Mack T-11A	Mack E-TECH V-MAC III, electronically-controlled fuel injection engine	Soot-related low-temperature viscosity increase	API CJ-4
Mack T-12	Mack E7 E-Tech 460 EGR engine	Piston ring and cylinder liner wear, oil consumption, and oxidation	API CJ-4
Cummins NTC-400	Cummins NTC-400 6-cylinder In-line turbocharged engine	Piston deposits, oil consumption, and wear	API CF-4
Cummins M-11 EGR	Cummins ISM 425 engine	Sliding wear, filter plugging, and sludge, crosshead weight loss	API CH-4 and CI-4; MIL-PRF-2104H
Cummins ISM	Cummins ISM straight 6, 10.8L diesel engine – same as M11 engine	Valve Train Wear, Filter, Δ P and Sludge at 6% Soot	API CJ-4
Cummins ISB	Cummins 5.9L ISB straight 4-6 cylinder turbocharged engine	Valve Train Wear	API CJ-4
GM 6.5-L RFWT (ASTM D5966 or RFWT)	General Motors 6.5 liter indirect injected engine	Roller-Follower Valve Train Wear	API CG-4, CH-4, and CI-4; MIL-PRF-2104G and 2104H
Navistar EOAT (ASTM D6894)	1994 International Truck 7.3 liter V-8, turbocharged, engine using hydraulically actuated, electronically controlled, unit injectors	Aeration	API CG-4, CH-4, CI-4, and CJ-4; MIL-PRF-2104G and 2104H
Detroit Diesel 6V-92TA	Two-stroke cycle V-6 diesel engine	Piston, liner, and ring distress	API CF-2; MIL-PRF-2104G and 2104H

TABLE 5.9—API SM/ILSAC GF-4 gasoline engine oil classifications for 2006 [487].

Requirements	Test Method	Properties	Unit	SM/EC GF-4 Limits
1. LABORATORY TESTS				
1.1 Viscosity Grades		All those that apply, typically SAE 0W-20, 0W-30, 5W-20, 5W-30, and 10W-30.		Manufacturer sets targets within SAE J300 parameter
1.2 Foam Test (Option A)	ASTM D892	Sequence I	ml initial Foam/ml after settling	10/0 max
		Sequence II		50/0 max
		Sequence III		10/0 max
	ASTM D6082	Sequence IV		100/0 max
1.3 Phosphorus	ASTM D4951	Phosphorus content	%	0.06 min 0.08 max[a]
1.4 EOFT	ASTM D6795	0.8% Water-with dry ice-%reduction in flow	% reduction	50 max
1.5 EOWTT	ASTM D6794	with 0.6% Water	% reduction	50 max
		with 1.0% Water	% reduction	50 max
		with 2.0% Water	% reduction	50 max
		with 3.0% Water	% reduction	50 max
1.6 TEOST (MHT4)		Total Deposits	mg	35 max[b]
1.7 Homogeneity and Miscibility	ASTM D6922	Oil compatibility		pass
1.8 Scanning Brookfield	ASTM D5133	Gelation Index		12 max
1.9 Volatility	ASTM D5800	Volatility (Noack)	% off at 250°C	15 max
	ASTM D6417	Volatility (GCD)	% off at 371°C	10 max
1.10 BRT (Ball Rust Test)	ASTM D6557	Rust rating	Avg. Gray Value	100 min
1.11 Sulfur	ASTM D4951 or	Sulfur content of 0W and 5W multi-grades	%	0.5 max[c]
	ASTM D2622	Sulfur content of 10W multi-grades	%	0.7 max
2. ENGINE TESTS FOR API SM				
2.1 Sequence IIIG		Viscosity increase at 100 hours	%	150 max
		Average weighted piston deposits	merits	3.5 min
		Hot stuck rings	#	none
		Average cam plus lifter wear	microns	60 max
2.2 Sequence IIIGA	ASTM D4684	Aged oil low temperature viscosity using ASTM D4684		Meet requirements of the original grade or the next higher grade[d]
2.3 Sequence IVA	ASTM D6891	Cam wear average	microns	90 max
2.4 Sequence VG	ASTM D6593	Average engine sludge	merits	7.8 min
		Rocker arm cover sludge	merits	8.0 min
		Average piston skirt varnish	merits	7.5 min
		Average engine varnish	merits	8.9 min
		Oil screen clogging	%	20 max
		Hot stuck rings	#	none
		Cold stuck rings	#	rate & report
		Oil ring clogging	%	rate & report
		Follower pin wear, cyl. #8, avg.	microns	rate & report
		Ring gap increase, cyl. #1 & #8, avg.	microns	rate & report
		Oil screen debris	% area	rate & report
2.5 Sequence VIII	ASTM D6709	Bearing weight loss	mg	26 max
		10 hr. stripped viscosity	cSt	Stay in grade
2.6 Sequence VIB	ASTM D6837	SAE 0W-20 and 5W-20 viscosity grades	% FEI1/%FEI2	2.3 / 2.0 min
		SAE 0W-30 and 5W-30 viscosity grades	% FEI1/%FEI2	1.8 / 1.5 min
		SAE 10W-30 and all other viscosity grades	% FEI1/%FEI2	1.1 / 0.8 min

Limits for SM Non-ILSAC GF-4 viscosity grades: [a]0.06 min, no max [b]45 max [c]Not Req. [d]Not Req.

Note: Limits for SM Non-ILSAC GF-4 viscosity grades:

TABLE 5.10—API SJ/ILSAC GF-2 and SL/ILSAC GF-3 gasoline engine oil classifications 2006 [487].

Requirements	Test Method	Properties	Unit	Limits	
				SJ/EC GF-2	SL/EC GF-3
1. LABORATORY TESTS					
1.1 Viscosity Grades		All those that apply, typically SAE 0W-20, 0W-30, 5W-30, AND 10W-30		Manufacturer sets targets within SAE J300 specification	
1.2 Foam Test	ASTM D892	Sequence I Sequence II Sequence III Sequence IV	ml initial Foam/ml after setting	10/0 max 50/0 max 10/0 max 200/50 max	10/0 max 50/0 max 10/0 max 100/0 max
1.3 Phosphorus	ASTM D4951 or D5185	Phosphorus Test	%	0.10 max	0.10 max[a]
1.4 EOFT	GM 9099P	0.6% Water - with dry ice-% reduction in flow	%	50 max	50 max
1.5 EOWTT	ASTM D6794	0.6% Water - without dry ice -% rate of change 1.0% Water - without dry ice -% rate of change 2.0% Water - without dry ice -% rate of change 3.0% Water - without dry ice -% rate of change	% % % %	report report report report	50 max 50 max 50 max 50 max
1.6 TEOST	ASTM D6335	Total Deposits, max	mg	60 max	NR
1.7 TEOST (MHT4)		Total Deposits, max	mg	NR	45 max
1.8 Homogeneity and Miscibility	FTM STD 791C			pass	pass
1.9 Scanning Brookfield	ASTM D5133	Gelation Index		12 max	12 max
1.10 Volatility	ASTM D5800 ASTM D6417	Volatility (Noack), % off Volatility (GCD), % off	% %	22 max 17 max	15 max 10 max
1.11 BRT (Ball Rust Test)	ASTM D6557	Rust rating	Gray value	100 min	100 min
2. ENGINE TESTS FOR API SJ AND SL					
2.1 Sequence IIIF	ASTM D1491	% Viscosity increase at 60 hours, max % Viscosity increase at 80 hours, max Average piston skirt varnish rating, min Weighted piston deposit rating, min Cam plus lifter wear avg, max Hot stuck rings Oil consumption, max Low-temperature viscosity performance	% % merits merits microns # liters Cp	325 max NR 8.5 min 3.2 min 20 max none 6.5 NR	NR 275 9.0 min 4.0 min 20 max none 5.2 max rate and report
2.2 Sequence IVA	ASTM D6891	Cam wear average, max	microns	120 max	120 max
2.3 Sequence VG	ASTM D6593	Average engine sludge, min Rocker arm cover sludge, min Average piston skirt varnish, min Average engine varnish, min Oil screen clogging, max Hot stuck compression rings Cold stuck rings Oil screen debris Oil ring clogging	merits merits merits merits % # # % %	7.8 min 8.0 min 7.5 min 8.9 min 20 max none rate and report rate and report rate and report	7.8 min 8.0 min 7.5 min 8.9 min 20 max none rate and report rate and report rate and report
2.4 Sequence VE (Only if oil Phosphorus < 0.08%)	ASTM D5302	Cam wear Average, max Maximum, max	 microns microns	 NR NR	 127 max 380 max
2.5 Sequence VIII	ASTM D6709	Bearing weight loss, max 10 hr stripped viscosity	mg cSt	26.4 max Stay in grade	26.4 max Stay in grade
2.6 Sequence VIB (FEI1 using BC before only)	ASTM D6837	For viscosity grades: 0W-20 and 5W-20 Other 0W- and 5W- multi-grades All 10W- multi-grades	 % improvement % improvement % improvement	 2.0 1.6 0.9	 NR NR NR
2.7 Sequence VIB	ASTM D6837	0W-20 and 5W-20 (FEI1/FEI2/sum), min 0W-30 and 5W-30 (FEI1/FEI2/sum), min All other viscosity grades (FEI1/FEI2/sum), min	% improvement % improvement % improvement	NR NR NR	2.0/1.7/NR 1.6/1.3/3.0 min 0.9/0.6/1.6 min

[a]0.08 min, no max

TABLE 5.11—API CJ-4 heavy-duty diesel engine oil classification 2007 [487].

Requirements	Test Method	Properties	Unit	Limits: 1 Test	2 Tests	3 Tests
1. LABORATORY TESTS FOR API CJ-4						
1.1 Viscosity Grades		SAE J300		Manufacturer specifies viscosity target within SAE J300 specification		
1.2 High Temperature Corrosion, 135°C	ASTM D6594	Copper, used oil increase, max Lead, used oil increase, max Copper strip rating, max	ppm ppm -	20 120 30	No MTAC[a,b]	
1.3 Foaming	ASTM D892	Foaming/Settling, max Sequence I Sequence II Sequence III	 mL mL mL	 10/0 20/0 10/0	No MTAC[a,b,c]	
1.4 Shear Stability	ASTM D7109	KV at 100°C after 90 passes for XW-40, min KV at 100°C after 90 passes for XW-30, min	cSt cSt	12.5 9.3	No MTAC[a,b,c]	
1.5 Noack Volatility	ASTM D5800	Evap loss at 250°C, Vis grades other than 10W-30, max Evap loss at 250°C, 10W-30, max	% %	13 15	No MTAC[a,b]	
1.6 High Temperature/ High Shear	ASTM D4683	Viscosity at 150°C, min	cP	3.5	No MTAC[a,b]	
1.7 Sooted Oil MRV	ASTM D6896	180 hour sample from Mack T-11 or T-11A viscosity at -20°C, max Yield stress	cP Pa	25,000 <35	No MTAC[a,b]	
1.8 Chemical limits (Non-critical)	ASTM D874 ASTM D4951 ASTM D4951	Sulfated Ash, max Phosphorus, max Sulfur, max	% % %	1.0 0.12 0.4	No MTAC[a,b]	

1.8 Seal Compatibility	Limits: Volume Change	Hardness	Tensile Strength	Elongation
Nitrile	+5/-3	+7/-5	+10/-TMC1006	+10/-TMC1006
Silicone	+TMC1006/-3	+5/-TMC1006	+10/-45	+20/-30
Polyacrylate	+5/-3	+8/-5	+18/-15	+10/-35
FKM	+5/-2	+7/-5	+10/-TMC1006	+10/-TMC1006
Vamac G	+TMC1006/-3	+5/-TMC1006	+10/-TMC1006	+10/-TMC1006

2. ENGINE TESTS FOR API CJ-4				1 Test	2 Tests	3 Tests
2.1 Mack T-11	ASTM D7156	Soot at 4 cSt inc, min Soot at 12 cSt inc, min Soot at 15 cSt inc., min	% % %	3.5 6.0 6.7	3.4 5.9 6.6	3.3 5.9 6.5
2.2 Mack T-12		Merits, min		1000[a]	1000[c]	1000[c]
2.3 Cummins ISB		Tappet wear, max Cam wear, max Crosshead weight loss	mg microns mg	100 55 Rate/report	108 59 Rate/report	112 61 Rate/report
2.4 Cummins ISM		Merits, min Top ring weight loss, max	 mg	1000[c] 100	1000[c] 100	1000[c] 100
2.5 Caterpillar 1N	ASTM D6750	Top land heavy carbon, max Top groove fill, max Weighted demerits, max Average oil consumption (0-252h), max Ring and liner scuffing	% % demerits g/kW-h 	3 20 286.2 0.5 None	4 23 311.7 0.5 None	5 25 323.0 0.5 None
2.6 Caterpillar C13		Merits, min Hot stuck rings		1000[c] None	1000[c] None	1000[c] None
2.7 Engine Oil Aeration Test	ASTM D6894	Oil aeration volume, max	%	8.0 (MTAC)	8.0 (MTAC)	8.0 (MTAC)
2.6 Roller Follower Wear Test	ASTM D5966	Roller follower pin wear, max	microns	7.6	8.4	9.1
2.8 Sequence IIIF[d]	ASTM D6984	Viscosity increase at EOT, max	%	275 (MTAC)	275 (MTAC)	275 (MTAC)

[a]Not an ACC Test
[b]MTAC is a statistical method for treating engine test results. Consult your sales representative for further information.
[c]Requires all individual merit ratings to be equal to or greater than zero. Consult your sales representative for further information.
[d]Passing Sequence IIIG viscosity increase at API SM is an acceptable alternative.

TABLE 5.12—API CI-4/CI-4 plus heavy-duty diesel engine oil classifications 2006 [487].

Requirements	Test Method	Properties	Unit	Limits		
				1 Test	2 Tests	3 Tests
1. LABORATORY TESTS FOR API CI-4 AND CI-4 PLUS						
1.1 Viscosity Grades		SAE J300		Manufacturer specifies viscosity target within SAE J300 specification		
1.2 High Temperature Corrosion Bench Test	ASTM D6594	Copper increase, max Lead increase, max Tin increase, max Copper strip rating, max (D130)	ppm ppm ppm	20 120 50 3	No MTAC[a,b]	
1.3 Foam Test	ASTM D892 (Option A not allowed)	Foaming/Settling, max Sequence I Sequence II Sequence III	 mL mL mL	 10/0 20/0 10/0	No MTAC[a,b]	
1.4 Shear Stability	ASTM D6278	After shear viscosity, 10W-30, min After shear viscosity, 15W-40, min	cSt cSt	9.3 12.5	No MTAC[a,b,c]	
1.5 Volatility	ASTM D5800 (Noack)	Evaporative loss at 250°C, max	%	15	No MTAC[a,b]	
1.6 High Temperature / High Shear	As allowed in SAE J300	Viscosity, min	mPa-s	3.5	No MTAC[a,b]	
1.7 Low Temperature Pumpability	ASTM D4684 (MRV TP-1)	Viscosity of 75h used oil sample from T-10 Test at -20°C, max	mPa-s	25000	No MTAC[a,b]	
	Modified D4684 (if yield stress)	Viscosity at -20°C, max Yield stress, max	mPa-s Pa	25000 35		
		Limits				
		Volume Change	**Hardness**	**Tensile Strength**	**Elongation**	
1.8 Elastomer Compatibility	Nitrile	+5/-3	+7/-5	+10/-TMC1006	+10/-TMC1006	
	Silicone	+TMC1006/-3	+5/-TMC1006	+10/-45	+20/-30	
	Polyacrylate	+5/-3	+8/-5	+18/-15	+10/-35	
	FKM	+5/-2	+7/-5	+10/-TMC1006	+10/-TMC1006	
2. ENGINE TESTS FOR API CI-4				**1 Test**	**2 Tests**	**3 Tests**
2.1 Mack T-8E	ASTM D5967 (Ext. T8-E)	Relative Viscosity at 4.8% soot, max (RV=Visc. at 4.8% soot/Visc. of new oil sheared in D6278)		1.8	1.9	2.0
2.2 Mack T-10	T10 test with EGR	Merit Rating, min		1000	1000	1000
2.3 Cummins M11-EGR	M11 high soot test with EGR	Crosshead wight loss, max Top ring weight loss, max Filter delta pressure at 250h, max Sludge rating, min	mg mg kPa merits	20.0 175 275 7.8	21.8 186 320 7.6	22.6 191 341 7.5
2.4 Caterpillar 1R	1R piston deposit test	WDR, max TGC, max TLC, max Initial Oil Consumption, max Final Oil Consumption, max Piston ring and liner scuffing Ring Sticking	demerits demerits demerits g/h g/h 	382 52 31 13.1 IOC+1.8 None None	396 57 35 13.1 IOC+1.8 None None	402 59 36 13.1 IOC+1.8 None None
2.5 Caterpillar 1K	ASTM RR: D02-1273	WDK, max TGF, max TLHC, max Oil Consumption (0-252h), max Piston ring and liner scuffing	demerits % % g/kW-h 	332 24 4 0.5 None	347 27 5 0.5 None	353 29 5 0.5 None

TABLE 5.12— *(Continued.)*

2. ENGINE TESTS FOR API CI-4				1 Test	2 Tests	3 Tests
or Caterpillar 1N	ASTM RR:D02-1321	WDN, max TGF, max TLHC, max Oil Consumption (0-252h), max Piston ring and liner scuffing Ring Sticking	demerits % % g/kW-h	286.2 20 3 0.5 None None	311.7 23 4 0.5 None None	323.0 25 5 0.5 None None
2.6 Roller Follower Wear Test	ASTM D5966	Average Pin Wear, max	μm (mils)	7.6 (0.30)	8.4 (0.33)	9.1 (0.36)
2.7 Engine Oil Aeration Test	ASTM RR:D02-1379	Aeration, max	Vol %	8.0	MTAC applies(2)	
2.8 Sequence IIIF	ASTM RR:D02-1491	Viscosity increase at 80h, max	%	275	MTAC applies(2)	
3. ENGINE TESTS FOR API CI-4 PLUS (same as API CI-4 Engine Tests above in addition to following)						
3.1 Mack T-11	T-11 Soot/ Viscosity Control	Minimum TGA Soot Content at 12.0 cSt increase at 100°C	%	6.00	5.89	5.85

[a]Not an ACC Test
[b]MTAC is a statistical method for treating engine test results. Consult your sales representative for further information.
[c]Limit of 30 pass for API CI-4, 90 pass for API CI-4 PLUS.

tomer compatibility, sulfated ash, foaming tendency, and changes in viscosity due to high shear and high temperature. Special laboratory test procedures are used to determine if a lubricant meets the desired performance requirements in these areas. These procedures are described in the SAE J2227 report [491]. ACEA specifies the use of the tests developed by the ASTM and CEC to assess the lubricant's ability to meet these requirements. The engine test requirements under ACEA 2007 are provided in Table 5.15 along with the parameters they measure and the test methods. Tables 5.16–5.18 list the actual pass/fail criteria and the test methods [490].

As stated earlier that ACEA has developed global performance standards for light-duty diesel engine oils and heavy-duty engine oils, by collaborating with the members of the Alliance of Automobile Manufacturers, Engine Manufacturers Association (EMA), and Japan Automobile Manufacturers Association, Inc. (JAMA). The objective of the specifications is to identify/develop oils with consistent performance in high-speed, four-stroke cycle light-duty and heavy-duty diesel engines used worldwide and provide the engine manufacturers the option to recommend these oils for use in their equipment. The oils for the light-duty diesel engines are designed to meet the year 2000 and newer exhaust emissions standards worldwide and those for the heavy-duty diesel engines are designed to meet 1998 and newer emissions standards. These oils are also compatible with certain older engines.

The global light-duty diesel engine oil standard consists of three specifications: DLD-1, DLD-2, and DLD-3 [492]. The heavy-duty diesel engine oil standard has only one specification, DHD-1. These specifications identify engine oils for use in applications/environment that necessitate wear control, high-temperature stability, soot handling properties, protection against oil-thickening due to oxidation and insolubles, aeration control, and minimal shear-related viscosity loss. The required engine tests for the light-duty diesel standard are listed in Table 5.19 and the specifications limits are provided in Table 5.20 [492]. The required engine tests and the specifications limits for the heavy-duty diesel standard are provided in Tables 5.21 and 5.22. Details of the standard can be accessed at JAMA's website [493].

Japanese Standards

The Japanese Automobile Standards Organization (JASO) also publishes oil standards. JASO tests use small Japanese engines, and their ratings use more stringent valve train wear standards than those used by the other countries. Japanese vehicle manufacturers use SAE Viscosity Classification System, described in SAE J300 [309], and API Classification System to recommend engine oils for service-fill applications [491]. Some of the Japanese engine tests, such as Nissan KA-24E, Nissan TD25, and Mitsubishi 4D34T4, are included in the new API, ILSAC, and ACEA specifications. While Japanese manufacturers recommend API/ILSAC and ACEA approved oils for use in their equipment, they also have their own "in-house" test procedures and performance requirements. For service-fill oils, JASO helps in coordinating these additional procedures and requirements. JASO comprises Japanese automobile and truck manufacturers, oil and additive companies, and government authorities. SAE Publication J2227 describes these procedures [491]. Current Japanese engine tests for lubricants, along with the evaluation criteria, are described in Table 5.23.

In April 2001, the Society of Automotive Engineers of Japan, Inc., in collaboration with the Japan Automobile Manufacturers Association, Inc., the Petroleum Association of Japan, and other organizations, specified the JASO DH-1 standard for engines made in Japan. The quality of JASO DH-1 diesel lubricants for four-stroke cycle diesel engines is specified under the Japanese Standard JASO M 355:2000. JASO DH-1 engine oil provides higher performance in wear and corrosion prevention, high-temperature oxidation stability, and piston deposit and soot control than the conven-

TABLE 5.13—API CH-4 heavy-duty diesel engine oil classifications 2006 [487].

Requirements	Test Method	Properties	Unit	Limits		
				1 Test	2 Tests	3 Tests
1. LABORATORY TESTS						
1.1 Viscosity Grades		SAE J300		Manufacturer specifies viscosity target within SAE J300 specification		
1.2 High Temperature Corrosion Bench Test	ASTM D6594	Copper increase, max Lead increase, max Tin increase, max Copper corrosion rating, max (D130)	ppm ppm ppm	20 120 50 3	No MTAC[a,b]	
1.3 Foam Test	ASTM D892 (Option A not allowed)	Foaming/Settling, max Sequence I Sequence II Sequence III	mL mL mL	10/0 max 20/0 max 10/0 max	No MTAC[a,b]	
1.4 Shear Stability	ASTM D6278	After shear viscosity 10W-30, min. After shear viscosity 15W-40, min.	cSt cSt	9.3 12.5	No MTAC[a,b]	
1.5 Volatility	ASTM D5800 ASTM D2887 or ASTM D5480	Noack (10W-30) Noack (15W-40) OR GCD (10W-30) GCD (15W-40)	% loss % loss % loss % loss	20 18 17 15	No MTAC[a,b]	
2. ENGINE TESTS FOR API CH-4				**1 Test**	**2 Tests**	**3 Tests**
2.1 Mack T-8E	ASTM D5967-96 Ext	Relative viscosity at 4.8% soot, max Viscosity increase at 3.8% soot, max	cSt cSt	2.1 11.5	2.2 12.5	2.3 13.0
2.2 Mack T-9	ASTM D6483	Liner wear, max Top ring weight loss, max Increase in lead level, max	microns mg ppm	25.4 120 25	26.6 136 32	27.1 144 36
2.3 Cummins M-11	ASTM RR:D02-1439	Crosshead weight loss, max Filter delta pressure, max Sludge rating, min	mg kPa merits	6.5 79 8.7	7.5 93 8.6	8.0 100 8.5
2.4 Caterpillar 1P	ASTM D6681	Weighted total demerits, max Top groove carbon, max Top land carbon, max Average oil consumption, max Final oil consumption, max Piston ring and liner scuffing	demerits % % gm/hr gm/hr	350 36 40 12.4 14.6 None	378 39 46 12.4 14.6 None	390 41 49 12.4 14.6 None
2.5 Caterpillar 1K	ASTM RR: D02-1273	Weighted total demerits, max Groove No. 1 (Top) fill, max Top land heavy carbon, max Average Oil Consumption, max Piston ring and liner scuffing	demerits % % g/bhp-hr	332 24 4 0.5 None	347 27 5 0.5 None	353 29 5 0.5 None
2.6 Roller Follower Wear Test	ASTM D5966	Pin wear, max	μm (mils)	7.6 (0.30)	8.4 (0.33)	9.1 (0.36)
2.7 Engine Oil Aeration Test	ASTM RR:D02-1379	Aeration, max	% volume	8	MTAC applies[b]	
2.8 Sequence IIIF	ASTM RR:D02-1491	Viscosity increase at 60 hours, max	%	295	MTAC applies[b]	

[a]Not an ACC Test.
[b]MTAC is a statistical method for treating engine test results. Consult your sales representative for further information.

tional API CD quality oil, which is widely used in Japan. In addition, JASO DH-1 specifies performance related to improved piston detergency, high-temperature deposits, foaming, oil consumption due to volatility, viscosity after shearing, and seal compatibility. Although JASO DH-1 was developed for new engines meeting stringent exhaust emissions regulations, it can also be applied to engines manufactured before the new exhaust emissions regulations went into effect. Test parameters and performance criteria for JASO DH-1 specification are provided in Table 5.24.

Indian Standards

Bureau of Indian Standards (BIS) has developed engine oil specifications to qualify lubricants for use in local vehicles. Physical requirements for engine oils include viscosity, pour point, flash point, evaporation loss, and foaming tendency.

TABLE 5.14—API CG-4/CF/CF2 heavy-duty diesel engine oil classifications 2006 [487].

Requirements	Test Method	Properties	Unit	Limits		
				1 Test	2 Tests	3 Tests
1. LABORATORY TESTS						
1.1 Viscosity Grades		SAE J300		Manufacturer specifies viscosity target within SAE J300 specification		
1.2 Cummins Bench Corrosion Test[d]	ASTM D5968	Copper increase, max Lead increase, max Tin increase, max Copper corrosion rating, max (D130)	ppm ppm ppm	20 60 report 3	No MTAC[a,b]	
1.3 Foam Test[d]	ASTM D892	Sequence I Sequence II Sequence III	mL mL mL	10/0 max 20/0 max 10/0 max	No MTAC[a,b]	
2. ENGINE TESTS FOR API CG-4				**1 Test**	**2 Tests**	**3 Tests**
2.1 Mack T-8	ASTM D5967	Viscosity increase at 3.8% soot, max Filter plugging, differential pressure, max Oil consumption, max	cSt kPa/(psi) g/kW-h (lb/bhp.h)	11.5 138(20) 0.304(0.0005)	12.5 138(20) 0.304(0.0005)	13.0 138(20) 0.304(0.0005)
2.2 Caterpillar 1N	ASTM D1321	Weighted total demerits, max Top groove fill, max Top land heavy carbon, max Oil consumption, max Scuffing, piston-rings-liner Number of tests allowed Stuck rings	 % volume % g/kW-h	286.2 20 3 0.5 none none	311.7 23 4 0.5 none none	323.0 25 5 0.5 none none
2.3 Roller Follower Wear Test	ASTM D5966	Wear, max	µm (mils)	11.4 (0.45)	12.4 (0.49)	12.7 (0.50)
2.4 Sequence IIIE	ASTM D5533	Hours to 375% viscosity increase, min	hours	67.5	65.1	64.0
2.5 CRC L-38	ASTM D5119	Bearing weight loss, max Used oil viscosity, cSt greater than SAE J300 lower limit for grade, min	Mg cSt	43.7 0.5	48.1 0.5	50.0 0.5
2.6 HEUI Air Entrainment Test	ASTM D1379	Aeration, max	% volume	10	-	-
3. ENGINE TESTS FOR API CF/CF-2				**1 Test**	**2 Tests**	**3 Tests**
3.1 Caterpillar 1MPC	RR:D02-1320	Weighted total demerits, max Top groove fill, max Ring side clearance loss, max Piston ring sticking Piston ring and liner scuffing Weighted total demerits, CF-2 max	% volume mm	240 70 0.013 none none 100	MTAC applies[b]	
3.2 CRC L-38	ASTM D5119	Bearing weight loss, max	mg	43.7	48.1	50.0
3.3 Detroit Diesel 6V92TA[c]	ASTM D5862	Cylinder liner scuffing area, max Average fire ring face distress Average 2 & 3 ring face distress Broken rings Cylinder liner port plugging area Average, max Single cylinder, max	% demerits demerits % %	45 0.23 0.20 none 2 5	48 0.24 0.21 none 2 5	50 0.26 0.22 none 2 5

[a]Not an ACC Test.
[b]MTAC is a statistical method for treating engine test results. Consult your sales representative for further information.
[c]This test is not required for API CF.
[d]This test is not required for CF/CF-2.

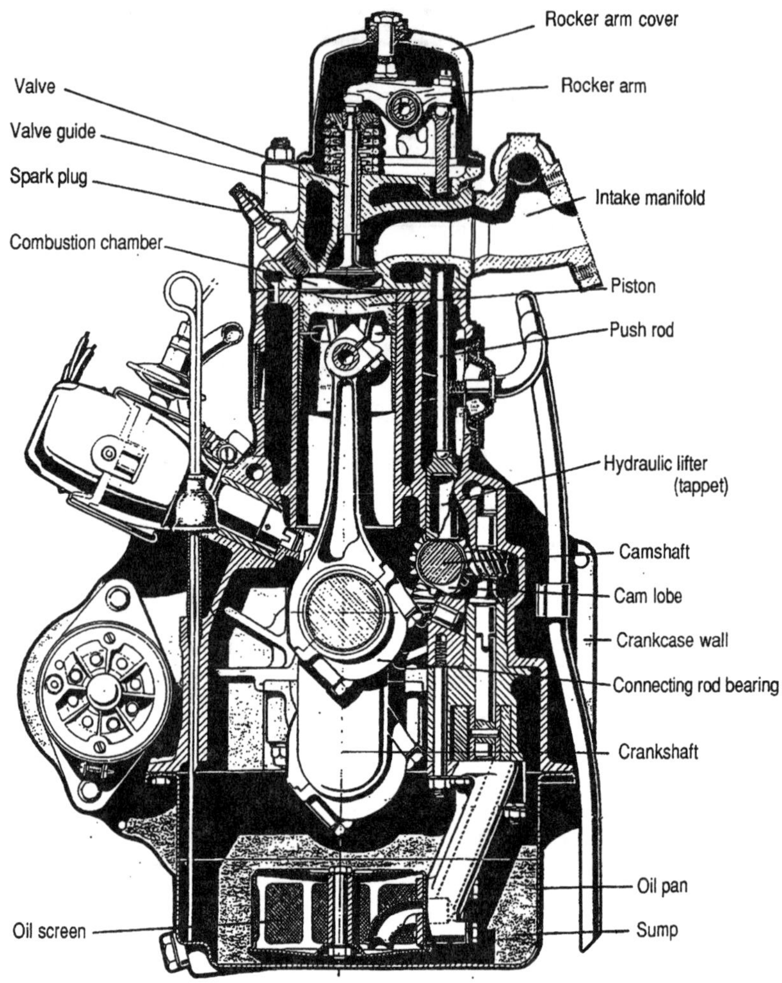

Fig. 5.11—Cross section of an in-line six-cylinder engine.

Again, the SAE viscosity classification system [309] is the basis for the Indian viscosity grades. To measure other parameters, BIS presently employs a combination of international test procedures which it plans to replace with national procedures when they are developed.

Engine Oil Classification Based on End-use

Engine lubricants can be divided into the following end-use categories:

1. Gasoline engine oils
2. Diesel engine oils
   - Automotive diesel oils
   - Stationary diesel oils
   - Railroad diesel oils
   - Marine diesel oils
3. Stationary gas engine oils
4. Aviation engine oils
5. Small engine oils

Gasoline Engine Oils

Gasoline engine (passenger car) oils for North American use are primarily defined by the API Service Classification System. As mentioned under specifications, API SM and ILSAC GF-4 qualified oils are suitable for use in the newest cars and that for Europe, ACEA, and for Japan, JASO and Japanese Industrial Standards Organization (JIS) are involved in the process.

Diesel Engine Oils

API, ACEA, and OEMs are the three key organizations which are involved in defining these oils. The oils for North American use are defined by the API Service Classifications, by the OEMs, or both. The U.S. Military also plays a role in setting specifications for these oils. In the United States, API CI-4 and CJ-4 designated oils deliver the top performance and are suitable for use in most modern engines. API CF-2 oils are designated for use in supercharged two-stroke cycle engines. ACEA categories $E7_{-04}$ and $C4_{-07}$ are designed to deliver top performance in the European engines.

The present specification-establishing process is regional and involves the development of base requirements relating to wear, deposits, corrosion, sludge, and oxidation, which are established by industry organizations; such as the API, ACEA, and JASO. To these, the OEMs add stricter limits, in-house tests, and field tests to bring up the oil performance for use in their equipment. The future trend is towards de-emphasizing the role of the industry organizations in defining lubricant quality and the OEM performance requirements playing the major role. These trends make formulating universal lubricants a challenge since invariably

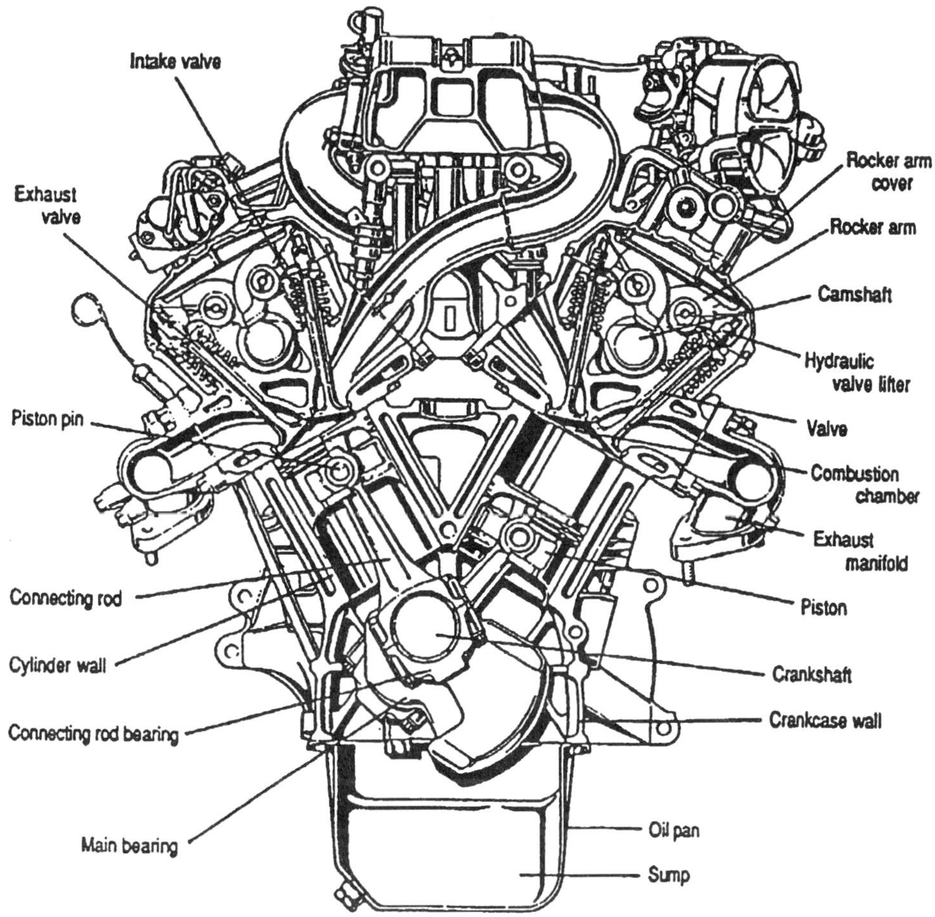

Fig. 5.12—Cross section of an eight-cylinder vee engine (V-8).

conflicts are likely to arise due to differing performance requirements across various OEMs. There is also a trend towards global specifications, which again allow a greater participation of the OEMs in the specification-establishing process [494]. Many of the OEMs are concerned about the number of worldwide heavy-duty diesel lubricant standards. However, the standards of North America and Europe are already converging and the development of the global specifications is another indication that the worldwide convergence is on the way [495].

Stationary Diesel Engine Oils

A stationary diesel engine implies any compression-ignition internal combustion engine, except combustion turbines, that converts heat energy into mechanical work and is not mobile. Such engines are commonly employed in industries, such as off-highway construction, earth moving and mining, agriculture, oil and gas exploration, chemical processing, paper-making, and for generating power to drive pumps, compressors, drilling equipment, and other auxiliary machinery. Some of these engines burn residual fuels of high sulfur content (1 to 4.0 %) and operate at high pressures, high temperatures, and have long strokes. Lubricants used in these applications are blends of highly refined base oils that have the ability to effectively perform the following functions.

1. Neutralize corrosive acids that result from the use of high sulfur fuels.
2. Minimize deposit formation on critical parts; such as cylinder ports, pistons, piston rings, ring grooves, under pistons and in general keep engine surfaces clean.
3. Lower the wear rates of the cylinders and the piston rings.
4. Be compatible with common elastomers used to make oil seals.

These oils are usually SAE 30, 40, or 50 mono-grades and are of API CF, CF2, or higher quality, depending upon if the engine is four-stroke cycle or two-stroke cycle. They typically have a high base reserve (TBN), high flash points, low pour points especially if the outdoor winter use is intended, and provide excellent wear protection, soot control, oil-thickening, and bearing corrosion protection. In April 2006, the emissions standard for stationary engines went into effect; hence the lubricants must not contribute to emissions, but instead help control them [401].

Railroad Diesel Engine Oils

Railroad diesel engines are two-stroke cycle and four-stroke cycle, naturally aspirated or turbocharged, medium-speed engines. Engines of similar type are also used in stationary and marine applications; hence some of their lubrication requirements overlap. Railroad diesel engine oils are formulated to provide year-round lubrication and typically are SAE 40 viscosity grade for single grade oils and 20W-40 grade for multi-grade oils. These oils are derived from paraffinic and naphthenic base oil blends and have a sufficient

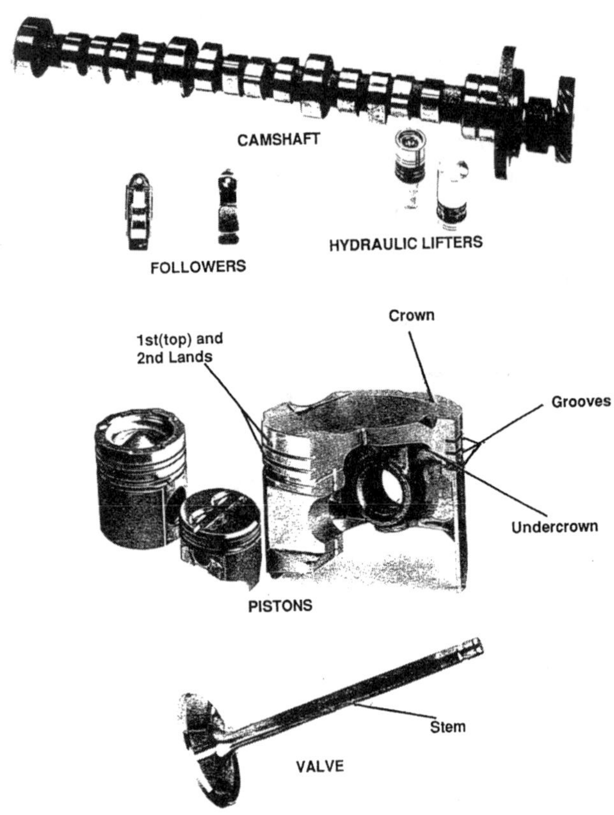

Fig. 5.13—Some engine parts that are evaluated.

base reserve (TBN of 10-20, typically 13 and 17) to neutralize the acidic combustion by-products. The need for the high TBN is also related to the use of the high sulfur fuels that produce larger amounts of sulfur acids that must be neutralized. Otherwise, high corrosion of the equipment will ensue. Seventeen TBN oils, because of the presence of the extra base, are more effective than 13 TBN oils, especially if the service intervals are longer. Extended service intervals are becoming a norm since they reduce the downtime and conserve resources. In addition to the acid-neutralizing ability, railroad

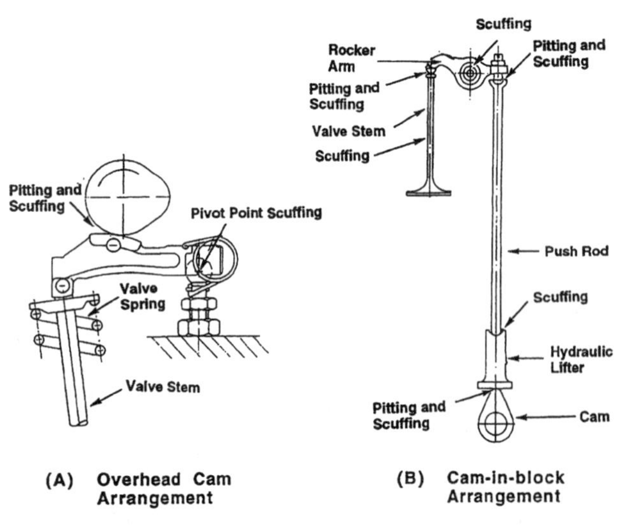

Fig. 5.14—Typical valve train arrangements showing points of possible lubrication failure [488].

engine oils must possess other properties, which include good oxidation resistance, high detergency/dispersancy for deposit control, good cold-weather startability, low volatility and flash point, antiwear control, and rust and corrosion protection. These oils have a limit on the amount of zinc they contain (10 ppm max) because of the tendency of the zinc-containing antiwear agents to attack the silver piston pin bushings, formerly used in railroad engines manufactured by General Motors. Most of these oils are formulated to meet API CF or CF2 performance, but the final approval depends on the lubricant's performance in a number of OEM specified tests. The quality levels of the locomotive lubricants are defined by the Locomotive Maintenance Officer's Association (LMOA) and by General Electric, which assure that the lubricant meets the requirements of the modern engine hardware. The newest improvements in General Electric and EMD engines include modified piston rings and cylinder liners that have significantly reduced oil consumption. This reduces the amount of the fresh oil makeup and increases the higher operating temperatures, which place an extra burden on the oil. A railroad engine oil with good base number retention and improved oxidation inhibition provides excellent service under these conditions, even with extended oil drain intervals.

The performance designations by the LMOA Fuels and Lubricants Committee for railroad engine oils are presented in Table 5.25 and Table 5.26 lists the engine builders' requirements for these lubricants [318]. Generation 5 lubricants are of the highest quality and must meet the following requirements.

- Minimum drain interval of 180 days, when used in low oil consumption engines that either average 10,000 miles/month or consume 20,000 gallons of fuel (0.3–0.5 % sulfur).
- Pass OEM required oxidation, corrosion, and friction tests.
- Meet the OEM specified engine and field test requirements.

Marine Diesel Engine Oils

Marine diesel applications predominantly employ two types of engines: two-stroke cycle, slow-speed crosshead engines and four-stroke cycle, medium-speed trunk piston engines. Hence, the ocean-going vessels require lubricants for both types of engine. Two-stroke cycle engines require two lubricants: the cylinder oil for the upper cylinder and the system oil for the crankcase. Four-stroke cycle engines require only one lubricant because they have a common sump for the crankcase and the cylinder. As mentioned earlier, four-stroke cycle diesel engines similar to marine engines are used for stationary and railroad applications. While these engines operate differently, the lubrication requirements in all cases are similar. This is because all operate at increased piston speeds, high operating temperatures, and have longer drain intervals. Operating parameters for various types of marine diesel engines are provided in Table 5.27 and their simplified pictures are shown in Fig. 5.15 [496].

Slow-speed, two-stroke cycle engines are usually of the cross-head type and use heavy fuel. These engines are equipped with diaphragms and stuffing boxes which separate power cylinders from the crankcase. This allows the use of different lubricants for the power cylinders and the crank-

TABLE 5.15—Engine tests associated with current ACEA 2007 sequences for service-fill automotive oils.

Tests	Parameter Measured	A1/B1-04	A3/B3-04	A3/B4-04	A5/B5-04	C1-04	C2-04	C3-04	C4-07	E2-96 Issue 5	E4-07	E6-04 Issue 2	E7-04 Issue 2
ASTM													
Sequence VG (D6593)	Low-temperature sludge, varnish, and wear	✓	✓	✓	✓	✓	✓	✓	✓				
Peugeot													
TU5JP-L4	Piston deposits, ring sticking, viscosity increase	✓	✓	✓	✓	✓	✓	✓	✓				
TU-3M Wear	Valve train wear	✓	✓	✓	✓	✓	✓	✓	✓				
XUD11BTE/ DV4TD	Piston deposits and viscosity increase	✓	✓	✓	✓	✓	✓	✓	✓				
Mercedes-Benz													
M111 Black Sludge	Engine sludge and cam wear	✓	✓	✓	✓	✓	✓	✓	✓				
M111 Fuel Economy	Fuel economy	✓	✓	✓	✓	✓	✓	✓	✓				
OM602A	Engine wear and cleanliness	✓	✓	✓	✓	✓	✓	✓	✓	✓	✓	✓	✓
OM364LA	Bore polishing, piston deposits, varnish, sludge, wear, and oil consumption									✓			
OM441LA	Piston deposits, bore polishing, wear, oil consumption, valve train condition, and turbo deposits										✓	✓	✓
VW													
1.6L TC Diesel	Piston deposits, varnish, and ring sticking	✓	✓			✓	✓	✓					
DI Diesel	Piston deposits, viscosity increase, and ring sticking			✓	✓	✓	✓	✓	✓				
Mack													
Mack T-8/T-8E	Soot-related oil thickening										✓	✓	✓
Mack T-10	Liner, rings, and bearings wear											✓	✓
Mack T-12 (ASTM D6987)												✓	✓
Cummins													
Cummins M11 EGR/HST	Soot-induced wear												✓
Cummins ISM													✓

case. A cross-sectional view of a cross-head marine diesel engine is shown in Fig. 5.16 [497].

The cylinder oil lubricates piston skirts and cylinders, and handles corrosive combustion products derived from the heavy fuel. It therefore provides wear protection and engine cleanliness. The crankcase oil, also called the systems oil, lubricates bearings, gears, and other engine components.

Marine cylinder lubricants are typically SAE 50 viscosity grade and have a suitable base reserve (TBN between 60 and 100; typically 70) to neutralize the acidic combustion products arising from sulfur (2–5 %) in the fuel. Cylinder oils are once-through lubricants because they are injected into the cylinder and burned with the fuel. Crankcase oils are SAE 30 viscosity grade and are formulated to handle rust, oxidation, deposits, and wear. They typically have a TBN of 5 to 10.

Medium-speed, trunk-piston engines have a single lubricating system and run on distillate or heavy fuels. They use a lubricating oil of SAE 30 or SAE 40 viscosity grade and have a TBN of 12 to 40, depending on the fuel quality. Lubricants that are suitable for use in these engines must possess superior extreme pressure/antiwear performance, high-temperature stability, oxidation resistance, detergency, and dispersancy. Again, the base reserve is necessary to neutralize the acidic combustion products due to the fuel sulfur.

For marine lubricants, water tolerance is an additional desirable property. This is to ensure that their lubricating ability will not suffer in the case of the water contamination. Table 5.28 summarizes marine diesel engine oil requirements [318]. Lubricants with API CF and CF2 quality are commonly used in this application.

At present, there are no emissions regulations on the open seas. Hence, ocean-going freighters and tankers use the lowest cost and the lowest quality fuel possible and do not worry about the resulting emissions. However, in coastal areas, environmental regulations do apply [401] and hence ferries, tug boats, and cruise ships use higher quality distillate fuels. But smaller oil sumps designed to make room for more cargo, along with emissions control devices, severely stress the lubricant.

Stationary Gas Engine Oils

Stationary engines typically operate on natural gas. Natural gas engines can be a two-stroke or four-stroke cycle, naturally aspirated or turbocharged. These engines are mainly used in stationary applications in the natural gas industry, in petroleum refining, for power generation, and in agricultural irrigation service. Primary advantages of the natural gas engines over diesel engines include lower NOx, CO, and particulate matter (PM) emissions and the lower fuel costs.

Stationary gas engines are available in various configurations and sizes. Engine characteristics include the following [498]:

1. Two- or four-stroke cycle design.

TABLE 5.16—ACEA 2007 European oil sequence for service-fill oils for gasoline and diesel engines [487].

ACEA	ACEA 2007 EUROPEAN OIL SEQUENCE FOR SERVICE-FILL OILS FOR GASOLINE and DIESEL ENGINES	Feb. 2007

This sequence defines the minimum quality level of a product for self-certification to EELQMS and for presentation to ACEA members.
Performance parameters other than those covered by the tests shown or more stringent limits may be indicated by individual member companies.

Requirement	Test Method	Properties	Unit	Limits			
				$A1/B1_{-04}$	$A3/B3_{-04}$	$A3/B4_{-04}$	$A5/B5_{-04}$
1. LABORATORY TESTS							
1.1 Viscosity grades		SAE J300 Latest active issue		No restriction except as defined by shear stability and HT/HS requirements. Manufacturers may indicate specific viscosity requirements related to ambient temperature.			
1.2 Shear stability	CEC-L-14-A-93 or ASTM D 6278	100°C Viscosity after 30 cycles	mm^2/s	xW-20 stay in grade xW-30 ≥ 8.6 xW-40 ≥ 12.0	All grades to be stay in grade	All grades to be stay in grade	All grades to be stay in grade
1.3 Viscosity at high temp. & high shear rate	CEC-L-36-A-90 (2nd Edition) (Ravenfield)	Viscosity at 150°C and 106 s^{-1} shear rate	mPa.s	max. 3.5. xW -20 2.6. min All others 2.9 min.	≥3.5	≥3.5	min 2.9 max. 3.5
1.4 Evaporative loss	CEC-L-40-A-93 (Noack)	Max. weight loss after 1 h at 250°C	%	≤ 15	≤ 13	≤ 13	≤ 13
		NOTE: the following sections apply to all sequences					
1.5 Sulphated ash	ASTM D874		% m/m	≤ 1.3 (see note 1)	≤ 1.5 (see note 1)	≤ 1.6 (see note 1)	≤ 1.6 (see note 1)
1.6 Sulphur	ASTM D5185	(see note 2)	% m/m	Report			
1.7 Phosphorus	ASTM D5185	(see note 2)	% m/m	Report			
1.8 Chlorine	ASTM D6443		ppm m/m	Report			

Requirement	Test Method	Properties	Unit	Elastomer type				
1.9 Oil / elastomer compatibility	CEC-L-39-T-96 (see note 3)	Max. variation of characteristics after immersion for 7 days in fresh oil without pre-ageing		RE1	RE2-99	RE3-04	RE4	AEM VAMAC
		Hardness DIDC	points	-1/+5	-5/+8	-22/ +1	-5/+5	
		Tensile strength	%	-40/+10	-15/+18	-30/+10	-20/+10	As per
		Elongation at rupture	%	-50/+10	-35/+10	-20/+10	-50/+10	Daimler
		Volume variation	%	-1/+5	-7/+5	-1/ +22	-5/+5	Chrysler

Requirement	Test Method	Properties	Unit	$A1/B1_{-04}$	$A3/B3_{-04}$	$A3/B4_{-04}$	$A5/B5_{-04}$
1.10 Foaming tendency	ASTM D892 without option A	Tendency - stability	ml	Sequence I (24°C) 10 - nil Sequence II (94°C) 50 - nil Sequence III (24°C) 10 - nil			
1.11 High temperature foaming tendency	ASTM D6082 High temperature foam test	Tendency - stability	ml	Sequence IV (150°C) 100 - nil			
2. ENGINE TESTS				$A1/B1_{-04}$	$A3/B3_{-04}$	$A3/B4_{-04}$	$A5/B5_{-04}$
2.1 High temperature deposits Ring sticking Oil thickening	CEC-L-88-T-02 (TU5JP-L4) 72 Hour test	Ring sticking (each part)	Merit	≥ 9.0	≥ 9.0	≥ 9.0	≥ 9.0
		Piston varnish (6 elements, average of 4 pistons)	Merit	≥ RL 216	≥ RL 216	≥ RL 216	≥ RL 216
		Absolute viscosity increase at 40°C between min and max values during test	mm^2/s	≤ RL216	≤ 0.8 x RL216	≤ 0.8 x RL216	≤ 0.8 x RL216
		Oil consumption	kg/test	Report	Report	Report	Report
2.2 Low temperature sludge	ASTM D6593-00 (Sequence VG) Under protocol & requirements for API (See Note 4)	Average engine sludge	merit	≥ 7.8	≥ 7.8	≥ 7.8	≥ 7.8
		Rocker cover sludge	merit	≥ 8.0	≥ 8.0	≥ 8.0	≥ 8.0
		Average Piston skirt varnish	merit	≥ 7.5	≥ 7.5	≥ 7.5	≥ 7.5
		Average engine varnish	merit	≥ 8.9	≥ 8.9	≥ 8.9	≥ 8.9
		Comp. ring (hot stuck)		none	none	none	none
		Oil screen clogging	%	≤ 20	≤ 20	≤ 20	≤ 20
2.3 Valve train scuffing wear	CEC-L-38-A-94 (TU3M)	Cam wear, average	µm	≤ 10	≤ 10	≤ 10	≤ 10
		Cam wear, max.	µm	≤ 15	≤ 15	≤ 15	≤ 15
		Pad merit (Ave. of 8 pads)	merit	≥ 7.5	≥ 7.5	≥ 7.5	≥ 7.5
2.4 Black sludge	CEC-L-53-T-95 (M111)	Engine sludge, average	merit	≥ RL 140	≥ RL 140	≥ RL 140	≥ RL 140

TABLE 5.16— (*Continued.*)

ACEA	ACEA 2007 EUROPEAN OIL SEQUENCE FOR SERVICE-FILL OILS FOR GASOLINE and DIESEL ENGINES						Feb. 2007
2.5 Fuel economy See Note (5)	CEC-L-54-T-96 (M111)	Fuel economy improvement vs. Reference oil RL191 (15W-40)	%	≥ 2.5	–	–	≥ 2.5
2.6 Ring sticking and piston cleanliness	CEC-L-46-T-93 (VW 1.6 TC D) see note 6	Ring sticking Piston cleanliness	Merit Merit	≥ RL 148 ≥ RL 148	≥ RL 148 ≥ RL 148		
2.7 Medium temperature dispersivity	CEC-L-093 (DV4TD) (see note 7)	Absolute viscosity increase at 100°C and 6 % soot	mm^2/s	≤ 0.60 x RL223 result	≤ 0.60 x RL223 result	≤ 0.60 x RL223 result	≤ 0.60 x RL223 result
		Piston merit (see note 8)	merit	≥ (RL223 – 2.5pts)	≥ (RL223 – 2.5pts)	≥ (RL223 –2.5pts)	≥ (RL223 – 2.5pts)
2.8 Wear, Viscosity stability & Oil consumption	OM602 A (see note 9)	Cam wear. Average (New tappet) Viscosity increase at 40°C Bore polishing Cylinder wear. Average Oil consumption	μm % % μm kg/test	≤ 50.0 ≤ 90 ≤ 7.0 ≤ 20.0 ≤ 10.0	≤ 50.0 ≤ 90 ≤ 7.0 ≤ 20.0 ≤ 10.0	≤ 50.0 ≤ 90 ≤ 7.0 ≤ 20.0 ≤ 10.0	≤ 50.0 ≤ 90 ≤ 7.0 ≤ 20.0 ≤ 10.0
2.9 DI diesel Piston cleanliness & Ring sticking	CEC-L-78-T-99 (VW DI)	Piston cleanliness	morit			≥ RL206 mlnus 3pts	≥ RL206
		Ring sticking (Rings 1 & 2) Average of all 8 rings Max. for any 1st ring Max. for any 2nd ring	 ASF ASF ASF			 ≤ 1.2 ≤ 2.5 ≤ 0.0	 ≤ 1.2 ≤ 2.5 ≤ 0.0

(1) Maximum limits, Values take into account method and production tolerances
(2) The internal standard method has to be used.
(3) Use either complete DaimlerChrysler requirements (VDA 675301, 7 days +/- 2h, 4 materials (NBR: NBR34 DIN 53538 T3 (100 °C +/- 2°C); FPM: AK6 (150 °C +/- 2°C); ACM: E7503 (150 °C +/- 2°C); AEM: D8948/200.1 (150 °C +/- 2°C)) + RE3, or complete requirements according to 1.9 above + DC requirements for AEM
(4) The limits shown are based upon those applied in U.S. market requirements. ACEA will continuously review the situation to ensure that these limits are appropriate for European vehicles and lubricants.
(5) ACEA considers the CEC-L-54-T-96 test the only valid comparator against which claims of lubricant fuel economy improvement should be made.
(6) The test according to CEC-L-78-T-99 may be run instead of CEC-L-46-T-93 for A1/B1 and A3/B3. The limits shall be as A3/B4.
(7) XUD11 BTE passing results obtained before the end of 2005 can be used instead of the DV4.
(8) Piston merit is not yet an official CEC parameter
(9) OM646LA results at an equivalent performance level can be used as soon as the test becomes available as a CEC test. In the event of OM602A and OM646 are not available, ACEA will define an alternative.

2. Less than 100 hp to 16,000 hp; 800 to 1500 hp being most common.
3. 1–20 Power cylinders.
4. Oil sump capacities of 14 to 6000 liters; 300 to 800 liters, or 80 to 200 U.S. gallons, being the most common.
5. Engine speeds of 300 r/min (slow-speed) to 2000 r/min (high-speed); mostly between 3.5 to 1200 r/min.
6. Piston bores of 22.5 in. (572 mm) in slow-speed engines to bores of 3.5 to 9.45 in. (89 to 240 mm) in high-speed engines.
7. Naturally aspirated or turbo-charged; newer engines are mostly turbo-charged.
8. Burn stoichiometric air-fuel ratio (14 parts air and 1 part fuel) or lean air-fuel ratio (greater than 14 parts air and 1 part fuel). The latter ratio will result in lower NOx emissions.
9. Engines and compressor units separate or joined end-to-end. In the latter case, they are coupled at the crankshaft by a coupling.

Stationary gas engines are of three types: spark-ignited, low-pressure gas engines; pilot-fuel ignited, low-pressure gas engines; and pilot-fuel ignited, high-pressure gas engines. While the spark-ignition engines are equipped with a spark plug, the pilot-fuel ignited engines use a small charge of the distillate fuel, approximately of 5 % of the total, to start the ignition process [318]. In high-pressure gas engines, pilot fuel is injected by the use of a fuel valve when the piston is top center. Once the ignition starts, the remaining fuel charge (usually natural gas) is introduced under high pressure.

In the low-pressure gas engines, gas and air charge is premixed either in an intake air manifold or in a prechamber, prior to entry into the cylinder. Pilot fuel is then injected to initiate combustion of the air-gas mixture. These engines use either stoichiometric or lean air-fuel mixtures. Engines that use stoichiometric mixtures are equipped with a three-way catalyst system to meet HC, CO, and NOx emissions requirements. Lean burn engines, on the other hand, do not need a catalyst system. High air (oxygen) content of the lean mixtures leads to more efficient combustion and lower combustion temperatures, which translate into lower HC, CO, and NOx emissions. Fuels used in such engines include natural gas (85 % methane), city gas (high in hydrogen), high sulfur sour gas (contains up to 8000 ppm hydrogen sulfide), sewage gas (also contains hydrogen sulfide), and landfill gas (contains corrosive organic halides). The sewage gas and the landfill gas are poor fuels since their methane content is only around 50 %. Because of the nature of the fuel and the diversity of the impurities, it is critical to select an engine oil that is not only an effective lubricant but also has the ability to protect against products resulting from the combustion of impurities.

Lubricants for natural gas engines are formulated differently than those for diesel and gasoline engines. This is because natural gas engines burn cleaner, do not cause fuel dilution because the fuel is gas, and the soot contamination of the oil; hence, they require less detergency/dispersancy,

TABLE 5.17—ACEA 2007 European oil sequence for service-fill oils for gasoline and diesel engines with after-treatment devices [490].

ACEA	ACEA 2007 EUROPEAN OIL SEQUENCE FOR SERVICE-FILL OILS FOR GASOLINE and DIESEL ENGINES WITH AFTER TREATMENT DEVICES	Feb. 2007

This sequence defines the minimum quality level of a product for self-certification to EELQMS and for presentation to ACEA members.
Performance parameters other than those covered by the tests shown or more stringent limits may be indicated by individual member companies.

Requirement	Test Method	Properties	Unit	Limits			
				$C1_{-04}$	$C2_{-04}$	$C3_{-07}$	$C4_{-07}$
1. LABORATORY TESTS							
1.1 Viscosity grades		SAE J300 Latest active issue		No restriction except as defined by shear stability and HT/HS requirements. Manufacturers may indicate specific viscosity requirements related to ambient temperature.			
1.2 Shear stability	CEC-L-14-A-93 or ASTM D 6278	100°C Viscosity after 30 cycles	mm^2/s	All grades to be stay in grade	All grades to be stay in grade	All grades to be stay in grade	All grades to be stay in grade
1.3 Viscosity at high temp. & high shear rate	CEC-L-36-A-90 (2nd Edition) (Ravenfield)	Viscosity at 150°C and 10^6 s^{-1} shear rate	mPa.s	≥ 2.9	≥2.9	≥ 3.5	≥ 3.5
1.4 Evaporative loss	CEC-L-40-A-93 (Noack)	Max. weight loss after 1 h at 250°C	%	≤ 13	≤ 13	≤ 13	≤ 11
1.5 Sulphur	ASTM D5185	(see note 1)	% m/m	≤ 0.2	≤ 0.3	≤ 0.3	≤ 0.2
1.6 Phosphorus	ASTM D5185	(see note 1)	% m/m	≤ 0.05 (2)	≥ 0.070 ≤0.090 (2)	≥ 0.070 ≤0.090 (2)	≤ 0.090 (2)
1.7 Sulphated ash	ASTM D874		% m/m	≤ 0.5 (see note 2)	≤ 0.8 (see note 2)	≤ 0.8 (see note 2)	≤ 0.5 (see note 2)
1.8 Chlorine	ASTM D6443		ppm m/m	Report	Report	Report	Report
1.9 TBN	ASTM D2896		mg KOH/g			≥ 6	≥ 6
NOTE: The following sections apply to all sequences							

Requirement	Test Method	Properties	Unit	Elastomer type				
1.10 Oil / elastomer compatibility	CEC-L-39-T-96 (see note 3)	Max. variation of characteristics after immersion for 7 days in fresh oil without pre-ageing		RE1	RE2-99	RE3-04	RE4	AEM VAMAC
		Hardness DIDC	points	-1/+5	-5/+8	-22/ +1	-5/+5	As per Daimler Chrysler
		Tensile strength	%	-40/+10	-15/+18	-30/+10	-20/+10	
		Elongation at rupture	%	-50/+10	-35/+10	-20/+10	-50/+10	
		Volume variation	%	-1/+5	-7/+5	-1/+22	-5/+5	

Requirement	Test Method	Properties	Unit	Limits			
1.11 Foaming tendency	ASTM D892 without option A	Tendency - stability	ml	Sequence I (24°C) 10 - nil Sequence II (94°C) 50 - nil Sequence III (24°C) 10 - nil			
1.12 High temperature foaming tendency	ASTM D6082 High temperature foam test	Tendency - stability	ml	Sequence IV (150°C) 100 - nil			
2. ENGINE TESTS				$C1_{-04}$	$C2_{-04}$	$C3_{-07}$	$C4_{-07}$
2.1 High temperature deposits Ring sticking Oil thickening	CEC-L-88-T-02 (TU5JP-L4) 72 Hour test	Ring sticking (each part)	Merit	≥ 9.0	≥ 9.0	≥ 9.0	≥ 9.0
		Piston varnish (6 elements, average of 4 pistons)	Merit	≥ RL 216	≥ RL 216	≥ RL 216	≥ RL 216
		Absolute viscosity increase at 40°C between min and max values during test	mm^2/s	≤ 0.8 x RL216	≤ 0.8 x RL216	≤ 0.8 x RL216	≤ 0.8 x RL216
		Oil consumption	kg/test	Report	Report	Report	Report
2.2 Low temperature sludge	ASTM D6593-00 (Sequence VG) Under protocol & requirements for API (See Note 4)	Average engine sludge	merit	≥ 7.8	≥ 7.8	≥ 7.8	≥ 7.8
		Rocker cover sludge	merit	≥ 8.0	≥ 8.0	≥ 8.0	≥ 8.0
		Average Piston skirt varnish	merit	≥ 7.5	≥ 7.5	≥ 7.5	≥ 7.5
		Average engine varnish	merit	≥ 8.9	≥ 8.9	≥ 8.9	≥ 8.9
		Comp. ring (hot stuck)	%	none	none	none	none
		Oil screen clogging		≤ 20	≤ 20	≤ 20	≤ 20
2.3 Valve train scuffing wear	CEC-L-38-A-94 (TU3M)	Cam wear, average	µm	≤ 10	≤ 10	≤ 10	≤ 10
		Cam wear, max.	µm	≤ 15	≤ 15	≤ 15	≤ 15
		Pad merit (Ave. of 8 pads)	merit	≥ 7.5	≥ 7.5	≥ 7.5	≥ 7.5
2.4 Black sludge	CEC-L-53-T-95 (M111)	Engine sludge, average	merit	≥ RL 140	≥ RL 140	≥ RL 140	≥ RL 140 + 4σ or > 9.0

TABLE 5.17— (Continued.)

ACEA	ACEA 2007 EUROPEAN OIL SEQUENCE FOR SERVICE-FILL OILS FOR GASOLINE and DIESEL ENGINES WITH AFTER TREATMENT DEVICES						Feb. 2007
2.5 Fuel economy See Note (5)	CEC-L-54-T-96 (M111)	Fuel economy improvement vs. Reference oil RL191 (15W-40)	%	≥ 2,5	≥ 2.5	≥ 1.0 for Xw-30 grade	≥ 1.0 for xW-30 grade
2.6 Medium temperature dispersivity	CEC-L-093 (DV4TD)	Absolute viscosity increase at 100°C and 6 % soot	mm^2/s	≤ 0.60 x RL223 result	≤ 0.60 x RL223 result	≤ 0.60 x RL223 result	≤ 0.60 x RL223 result
		Piston merit (see note 6)	merit	≥ (RL223 – 2.5pts)	≥ (RL223 – 2.5pts)	≥ (RL223 –2.5pts)	≥ (RL223 – 2.5pts)
2.7 DI diesel Piston cleanliness & Ring sticking	CEC-L-78-T-99 (VW DI)	Piston cleanliness	merit	≥ RL206	≥ RL206	≥ RL206	≥ RL206
		Ring sticking (Rings 1 & 2) Average of all 8 rings	ASF	≤ 1.2	≤ 1.2	≤ 1.2	≤ 1.2
		Max. for any 1st ring	ASF	≤ 2.5	≤ 2.5	≤ 2.5	≤ 2.5
		Max. for any 2nd ring	ASF	≤ 0.0	≤ 0.0	≤ 0.0	≤ 0.0
2.8 Wear, Viscosity stability & Oil consumption	OM602A (see note 7)	Cam wear average	µm	≤ 50.0	≤ 50.0	≤ 45.0	≤ 45.0
		Viscosity increase at 40°C	%	≤ 90	≤ 90	≤ 70.0	≤ 70.0
		Bore polishing	%	≤ 7.0	≤ 7.0	≤ 4,5	≤ 4,5
		Cylinder wear average	µm	≤ 20.0	≤ 20.0	≤ 15.0	≤ 15.0
		Oil consumption	kg/test	≤ 10.0	≤ 10.0	≤ 10.0	≤ 10.0

(1) The internal standard method has to be used.
(2) Maximum limits, Values take into account method and production tolerances
(3) Use either complete DaimlerChrysler requirements (VDA 675301, 7 days +/- 2h, 4 materials (NBR: NBR34 DIN 53538 T3 (100 °C +/- 2°C); FPM: AK6 (150 °C +/- 2°C); ACM: E7503 (150 °C +/- 2°C); AEM: D8948/200.1 (150 °C +/- 2°C)) + RE3, or complete requirements according to 1.10 above + DC requirements for AEM
(4) The limits shown are based upon those applied in U.S. market requirements. ACEA will continuously review the situation to ensure that these limits are appropriate for European vehicles and lubricants.
(5) ACEA considers the CEC-L-54-T-96 test the only valid comparator against which claims of lubricant fuel economy improvement should be made.
(6) Piston merit is not yet an official CEC parameter
(7) OM646LA results at an equivalent performance level can be used as soon as the test becomes available as CEC test. In the event of OM602A and OM646 are not available, ACEA will define an alternative.

which allows these lubricants to be formulated at lower ash. However, gas burns hotter than diesel; hence these engines have exhaust temperatures of 165 to 235 °C (300 to 400°F) higher than those of the diesel engines. This factor alone generates a greater amount of oxidation- and nitration related oil decomposition products and deposits. In addition, these engines operate at constant speed, which makes them retain deposits.

Natural gas engine oils are available both in mono grades and multi-grades. SAE 30 and SAE 40 are the common mono grades and 15W-40 is the common multi-grade. Multi-grade oils are useful in situations where frequent low-temperature start-ups are expected or sump heaters are either unavailable or unreliable, or both. Multi-grades not only provide easy low-temperature starting but they also have reduced oil consumption, means lower emissions, and improved fuel economy. However, because of the higher operating temperatures, the polymer in the multi-grade oil may undergo extensive thermo-oxidative degradation and may not be suitable for use in engines from some manufacturers.

Performance specifications for natural gas engine oils are generally established by the OEMs, who use various criteria to specify oils for use in their equipment. Some equipment manufacturers refer to older API "CC" or "CD" diesel engine oil designations to try to establish a minimum performance level, but the use of the diesel engine oil performance specifications to classify natural gas engine oils is not accepted by many. This is because of the different characteristics of the natural gas engines that were listed in the previous paragraph. Other OEMs have developed their own natural gas engine tests, but for the most part, the performance is measured by field testing. Dresser-Rand and Waukesha cogeneration applications have the only two published approval lists for the commercial brands of oils [499]. Sulfated ash level is another criterion that is often used by the OEMs to recommend oils for their equipment, with a preference for oils that produce ash levels of 0.5 % or lower. Oils that produce an ash level of less than 0.1 % are designated as ashless. Those that produce 0.4–0.6 % ash are designated as low ash; those that produce 0.7–1.0 % ash are designated as medium ash; and those that produce greater than 1.0 % ash are designated as high ash oils [318]. Ash primarily originates from the metal-containing additives, such as basic detergents, which are added to the lubricant to neutralize acids and suspend deposit precursors in oil. Basic detergents also protect the exhaust valves against recession by forming a sacrificial protective layer of ash. Ash results when the lubricant residue on various parts of the engine burns, leaving behind a whitish-gray deposit. While the presence of the ash-forming additives in the lubricant is beneficial, the presence of more than the optimal amount can cause excessive deposits on various parts, which can result in reduced heat transfer, preignition or detonation, or both, ring sticking or breaking, plug fouling, and valve burning.

In general, two-stroke cycle engine manufacturers prefer oils that produce no ash (ash-less) and four-stroke cycle engine manufacturers prefer low ash-producing oils. This is because the two-stroke cycle natural gas engines do not have intake or exhaust valves, but instead have oil injection ports that feed oil directly into each cylinder and a high ash-producing oil can cause exhaust port blockage. Higher ash-producing oils are necessary when severe fuels, such as those derived from landfill or sewage, are used. Older Waukesha four-stroke engines are the only engines that require high

TABLE 5.18—ACEA 2007 European oil sequence for service-fill oils for heavy-duty diesel engines [490].

ACEA	ACEA 2007 EUROPEAN OIL SEQUENCE FOR SERVICE-FILL OILS FOR HEAVY-DUTY DIESEL ENGINES	Feb. 2007

This sequence defines the minimum quality level of a product for self-certification to EELQMS and for presentation to ACEA members.
Performance parameters other than those covered by the tests shown or more stringent limits may be indicated by individual member companies.

Requirement	Test Method	Properties	Unit	Limits				
				E2-96 Issue 5	E4-07	E6-04 Issue 2	E7-04 Issue 2	
1. LABORATORY TESTS								
1.1 Viscosity		SAE J300 Latest Active Issue		No restriction except as defined by shear stability and HT/HS requirements. Manufacturers may indicate specific viscosity requirements related to ambient temperature.				
1.2 Shear stability	CEC-L-14-A-93 or ASTM D6278	Viscosity after 30 cycles measured at 100°C.	mm²/s	xW-30 ≥ 9.0 xW-40 ≥ 12.0 xW-50 ≥ 15.0 mono grades no req.	Stay in grade			
	ASTM D6278	Viscosity after 90 cycles measured at 100°C	mm²/s			Stay in grade		
1.3 Viscosity High Temperature High Shear Rate	CEC-L-36-A-90 (2nd Edition) (Ravenfield)	Viscosity at 150°C and $106\ s^{-1}$ Shear rate	mPa.s	≥ 3.5				
1.4 Evaporative Loss	CEC-L-40-A-93 (Noack)	Max. weight loss after 1 h at 250°C	%	≤ 13				
1.5 Sulphated Ash	ASTM D874		% m/m	≤ 2.0	≤ 2.0	≤ 1.0	≤ 2.0	
1.6 Phosphorus	ASTM D5185[1]		% m/m			≤ 0.08		
1.7 Sulphur	ASTM D5185[1]		% m/m			≤ 0.3		
1.8 Oil Elastomer Compatibility	CEC-L-39-T-96	Max. variation of characteristics after immersion		Elastomer Type				
See Note (2)		for 7 days in fresh oil without pre-ageing		RE1	RE2-99	RE3-04	RE4	AEM VAMAC
		Hardness DIDC	points	-1/+5	-5/+8	-25/+1	-5/+5	As per Daimler-Chrysler
		Tensile strength	%	-50/+10	-15/+18	-45/+10	-20/+10	
		Elongation rupture	%	-60/+10	-35/+10	-20/+10	-50/+10	
		Volume variation	%	-1/+5	-7/+5	-1/+30	-5/+5	
1.9 Foaming Tendency	ASTM D892 without option A	Tendency – stability	ml ml ml	Sequence I (24°C) 10 – nil Sequence II (94°C) 50 – nil Sequence III (24°C) 10 – nil				
1.10 High temperature foaming tendency	ASTM D6082	Tendency - stability	ml	Sequence IV (150°C) 200-50				
1.11 Oxidation	CEC-L-85-T-99 (PDSC)	Oxidation induction time	min				≥ 35	
1.12 Corrosion	ASTM D6594	Lead increase	ppm				≤ 100	
1.13 TBN	ASTM D2896				≥ 12			
2. ENGINE TESTS				**E2-96 Issue 5**	**E4-07**	**E6-04 Issue 2**	**E7-04 Issue 2**	
2.1 Bore polishing / Piston cleanliness See note (3)	CEC L-42-T-99 (OM364LA)	Bore polishing Piston cleanliness Average Cylinder wear Sludge Oil consumption	% merit µm merit kg/test	≤ 3.5 ≥ 40.0 ≤ 3.5 ≥ 9.4 ≤ 16.0				
2.2 Wear See note (4)	CEC L-51-A-97 (OM602A)	Cam wear Viscosity increase at 40°C Bore polishing Cylinder wear Oil consumption	µm % % µm kg/test	≤ 50.0	≤ 50.0 ≤ 90 ≤ 7.0 ≤ 20.0 ≤ 10	≤ 50.0 ≤ 90 ≤ 7.0 ≤ 20.0 ≤ 10	≤ 50.0 ≤ 90 ≤ 7.0 ≤ 20.0 ≤ 10	
2.3 Soot in oil See note (5)	ASTM D5967 (Mack T-8E) ASTM D4485 (Mack T-8)	Test duration: Relative viscosity at 1 test/2 test/3 test average Viscosity increase at 1 test/2 test/3 test average Filter plugging, Diff. pressure Oil consumption	Hours mm²/s kPa g/kWh	300 4.8% soot ≤ 2.1/2.2/2.3 3.8% soot ≤11.5/12.5/1 3.0 ≤ 138 ≤ 0.304	300 4.8% soot ≤ 2.1/2.2/2.3 3.8% soot ≤11.5/12.5/1 3.0 ≤ 138 ≤ 0.304	300 4.8% soot ≤ 2.1/2.2/2.3 3.8% soot ≤11.5/12.5/13.0 ≤ 138 ≤ 0.304		

TABLE 5.18— (*Continued.*)

ACEA	ACEA 2007 EUROPEAN OIL SEQUENCE FOR SERVICE-FILL OILS FOR HEAVY-DUTY DIESEL ENGINES					Feb. 2007
2.4 Bore polishing Piston Cleanliness Turbocharger deposits See note (6)	CEC L-52-T-97 (OM441LA)	Bore polishing Piston Cleanliness Boost pressure loss at 400 hrs Oil consumption	% merit % kg/test	≤ 2.0 ≥ 40.0 ≤ 4 ≤ 40	≤ 2.0 ≥ 40.0 ≤ 4 ≤ 40	≤ 2.0 ≥ 25.0 ≤ 4 ≤ 40
2.5. Soot induced wear see note 7	Cummins ISM	Rocker pad average weight loss at 3.9 % soot 1 test/2 test/3 test average	mg			≤ 7.5/7.8/7.9
		Oil filter diff. press at EOT 1 test/ 2 test/3 test average	kPa			≤ 55/67/74
		Engine sludge 1 test/2 test/3 test average	merit			≥ 8.1/8.0/8.0
2.6. Wear (liner-ring-bearings) See note (8)	Mack T10 ASTM D6987 (Mack T12)	Merit Avg. liner wear Average top ring weight loss End of test lead Delta lead 250-300 hrs Oil consumption (Phase II)	 μm mg ppm ppm g/hr		≥ 1000 ≤ 32 (26) ≤ 158 (117) ≤ 35 (42) ≤ 14 (18) ≤ 65 (95)	≥ 1000 ≤ 32 (26) ≤ 158 (117) ≤ 35 (42) ≤ 14 (18) ≤ 65 (95)

(1) The internal standard method has to be used.
(2) Use either the most recent complete Daimler-Chrysler requirements (VDA 675301, 7 days, 4 materials (NBR: NBR34 DIN 53538 T3 (100 °C); FPM: AK6 (150 °C); ACM: E7503 (150 °C); AEM: D8948/200.1 (150 °C)) + RE3 according to requirement 1.8 above, or complete requirements according to 1.8 above + DC requirements for AEM.
(3) Results from a CEC-L-52-T-97 (OM441LA) test as part of a DaimlerChrysler sheet 228.1 approval can be used as an alternative. Only tests according to CEC-L-52-T-97 are acceptable.
(4) OM646LA results at an equivalent performance level can be used as soon as the test becomes available as a CEC test. In the event the OM602A and the OM646 are not available, then ACEA will define an alternative.
(5) Mack T11 (ASTM D7156) results obtained as part of an API CI-4, CI-4 plus or API CJ-4 approval program, can be used in place of Mack T8E.
(6) OM501LA results at an equivalent performance level can be used as soon as the test becomes available as a CEC test.
(7) Results from Cummins a M11 HST (ASTM D6838) at API CH-4 or M11 EGR test (ASTM D6975) at API CI-4 or CI-4 Plus can be used in place of the Cummins ISM test.
(8) Mack T12 results can be used in place of Mack T10. In this case the merit scale for CI-4 Plus approvals must be applied to the Mack T12 results. Maximum allowable values for the Mack T12 test are given in parentheses.

ash oils, which is due to the high valve angles that were used. Besides ash, other parameters for these oils include their phosphorus content and resistance to oxidation and nitration. Phosphorus limit is to minimize catalyst poisoning in converters used to control emissions. More specifically, the performance of these oils is assessed in terms of the following performance criteria:

1. Improved oxidation and nitration stability and anti-coking characteristics.
2. Control of valve recession and torching (guttering).
3. Piston groove, land, and skirt cleanliness and reduction in piston ring sticking.
4. Reduced port deposits and port plugging.
5. Reduced bearing corrosion.
6. Minimized combustion chamber ash accumulation and plug fouling.
7. Catalytic converter compatibility.
8. Good TBN retention, which will allow the use of such oils in engines that are fueled by sour gas or fuel gas that contains up to 0.3 % sulfur as hydrogen sulfide and small amounts of organic halides such as chlorides.
9. Reduced wear of pistons, rings, liners, cylinder walls, valve train, and bearings.
10. Good volatility characteristics (low oil consumption).
11. Rapid circulation and pumpability at low temperatures.
12. Shear stability to stay in grade over the oil drain interval.
13. Low foaming tendency.
14. CF and CF-2 credentials for use in modern engines.

Table 5.29 summarizes the engine builders' preferences in defining lubricant quality for their equipment [318]. SAE has recognized the need for new service categories to define the performance of these oils and has requested API and ASTM to define such categories and write the user language.

TABLE 5.19—Global DLD-1, DLD-2, DLD-3 engine test summary [492].

Test Identification	Engine Speed r/min	Test Duration (h)	Operation	Power (kW)
VW IDI Intercooler	4500	10+50	Steady	55 min
XUD11 BTE	1000 and 4300	75	Cyclic	0 and 80 min
OM602A	0 to 4600	200	Cyclic	0 to 88 min
Mitsubishi 4D34T4	3200	160	Steady	120
VW TDI	Idle and 4150	2+54	Cyclic	0 to max
M111E	Idle to 3071	24	Cyclic	20 to 95
Peugeot TU5JP-L4	5500	72	Steady	62

TABLE 5.20—Global engine oil service specification DLD-1, DLD-2, DLD-3 [492].

Characteristic	Test Method	Test Name	Requirements	Limits		
				DLD-1	DLD-2	DLD-3
Engine Tests						
Ring sticking & Piston cleanliness[a]	CEC L-46-T-93	VW 1.6 TC D	Ring sticking, merit Piston cleanliness, merit.	≥ RL 148 ≥ RL 148	≥ RL 148 ≥ RL 148	
Medium temperature dispersivity	CEC L-56-T-98	Peugeot XUD11BTE	Absolute viscosity increase at 100°C and 3% soot (measurement with CEC L-83-A-97 method)	≤ 0.90 x RL197 result.	≤ 0.50 x RL197 result	≤ 0.50 x RL197 result.
			Piston merit (5 elements) (average for 4 pistons)	≥ (RL197 minus 6 pts)	≥ (RL197 minus 6 pts)	≥ RL197
Wear, Viscosity stability & Oil consumption	CEC L-51-A-98	Mercedes Benz OM602A	Cam wear. Average, μm Viscosity increase at 40°C, % Bore polishing, % Cylinder wear. Average, μm Oil consumption, kg/test	≤ 50.0 ≤ 90 ≤ 7.0 ≤ 20.0 ≤ 10.0	≤ 50.0 ≤ 90 ≤ 7.0 ≤ 20.0 ≤ 10.0	≤ 50.0 ≤ 90 ≤ 7.0 ≤ 20.0 ≤ 10.0
Wear, Valve train	JASO M354-1999	Mitsubishi 4D34T4[b] 160 Hrs	Cam Lobe Wear, μm Average	≤ 95.0	≤ 95.0	≤ 95.0
DI diesel Piston cleanliness & Ring sticking	CEC L-78-T-99	VW DI	Piston cleanliness, merit Ring sticking (Rings 1 & 2), ASF Average of all 8 rings Max. for any 1st ring Max. for any 2nd ring	----- ----- ----- -----	----- ----- ----- -----	≥ RL206 minus 3 pts ≤ 1.2 ≤ 2.5 ≤ 0.0
Fuel economy	CEC L-54-T-96	Mercedes Benz M111E	Fuel economy improvement vs. Reference oil RL191 (15W-40), %	--	≥ 2.5	--
Oxidation	CEC L-88-T-02	Peugeot TU5JP	Ring sticking (each part), merit Piston varnish (5 elements) (average of 4 pistons)	≥ 9.0 ≥ RL216	≥ 9.0 ≥ RL216	≥ 9.0 ≥ RL216
			Absolute viscosity increase at 40°C between min and max values during test	≤ 1.5 x RL216	≤ RL216	≤ 0.8 x RL216
			Oil consumption	Report	Report	Report
Bench Tests						
Viscosity grades		Kinematic Viscosity	SAE J300 Latest Active Issue	xW-50 xW-40 xW-30	xW-30 xW-20	xW-50 xW-40 xW-30
Shear stability	ASTM D3945 CEC L-14-A-93	Bosch Injector Test	Viscosity after 30 Cycles, measured at 100°C, mm²/s xW-50	≥ 15.0	–	Stay in grade
			xW-40	≥ 12.0	–	Stay in grade
			xW-30	≥ 9.0	≥ 8.6	Stay in grade
			xW-20	–	Stay in grade	–
Viscosity at high temp. & high shear rate (HTHS)	ASTM D4624 / D4683 / CEC-L-36-A-90	HP Capillary Viscometer / Tapered Bearing Simulator / Ravenfield	High Temperature / High Shear Rate Viscosity, mPa.s	3.5 Min.	xW-30 2.9 min xW-20 2.6 min	3.5 Min.
Evaporative loss	ASTM D5800 CEC L-40-A-93	NOACK	% Mass Loss, Max	15 for 10W-x or lower 13 for others	15	13
Sulfated ash	ASTM D874	Sulfated Ash	Mass %, Max.	1.8	1.3	1.6
TBN	ASTM D4739/ D2896	TBN	mg KOH/g, min	10	--	--
Corrosion	ASTM D6594	Corrosion Bench Test	Used Oil Element Content above Baseline, ppm, Max.	Copper 20, Lead 120, Tin 50	- - -	Copper 20, Lead 120, Tin 50
Corrosion	ASTM D6557	Ball Rust Test	Average gray value, min	100	-	100

Characteristic	Test Method	Test Name	Requirements	RE 1	RE 2	RE 3	RE 4
Oil / elastomer Compatibility[c]	CEC-L-39-T-96	Variation after 7 days fresh oil, No presaging					
		Hardness DIDC	Points, max.	-1/+5	-5/+8	-25/+1	-5/+5
		Tensile Strength	%. max	-40/+10	-15/+18	-45/ +10	-20/+10
		Elongation rupture	%. max	-50/10	-35/+10	-20/+10	-50/+10
		Volume variation	%. max	-1/+5	-7/+5	-1 /+30	-5/+5

TABLE 5.20— (*Continued.*)

Characteristic	Test Method	Test Name	Requirements	Limits		
				DLD-1	DLD-2	DLD-3
Foaming tendency	ASTM D892 without option A	Sequence I (24°C) Sequence II (94°C) Sequence III (24°C)	Tendency / Stability, ml. max. After 1 Min. settling	10 / nil 50 / nil 10 / nil		
Foaming - High Temperature	ASTM D6082	Sequence IV (150°C)	Tendency / Stability, ml. max. After 1 Min. settling	100 / nil		

[a] A passing result in the CEC L-78-T-99 test (VW DI) to the DLD-3 requirements may be used in place of the CEC L-46-T-93 test.
[b]Mitsubishi 4.0 liter, four-cylinder direct injection diesel engine.
[c]Use either complete Daimler-Chrysler requirements (VDA 675301, 7 days, 4 materials (NBR: NBR34 DIN 53538 T3 (100 °C); FPM: AK6 (150 °C); ACM: E7503 (150 °C); AEM: D8948/200.1 (150 °C)) + CEC RE3 according to requirements above, or complete CEC requirements according to above + DC requirements for AEM. The Elastomer Compatibility Limits are those stated in ACEA 2002 European Oil Sequences and apply to the elastomer batches available at that time. Consult the most recent ACEA Oil Sequence publication for the information on the limits with current more recent elastomer batches.
New CEC RE3 material and limits are to be developed and added to CEC requirements as soon as possible.

Aviation Engine Oils

Aviation engine oils are of two types: Those that are used in piston engines and those that are used in turbine engines. Aircraft piston engines use engine oils to lubricate the crankshaft, connecting rods, pistons, rings, bearings, and other moving parts. These oils are designed to prevent wear by appropriately lubricating the engine parts, keep deposit-forming materials and abrasives in suspension, dissipate heat, and inhibit entry of the blow-by gases into the crankcase. They must especially perform these functions under extreme operating conditions. Piston engine oils are usually SAE 30 to SAE 60 viscosity grade, are mineral oil-based, and contain nonash-producing (metal-free) additives. These oils contain oxidation inhibitors that impart oxidation stability and dispersants that impart dispersancy. Because of the presence of these additives, the piston engine oils inhibit the formation of oxidation-initiated deposit precursors and suspend them in oil, if they form; thereby preventing their separation from the oil and on hot engine parts to form varnish, sludge, and piston deposits. The suspended contaminants and abrasive particles, if present, are removed during the oil change. SAE J1899 specification, which replaced the U.S. Military Specification MIL-L-22851D, defines the quality of these oils.

Jet turbine engine oils are formulated using synthetic base stocks, usually of polyol ester and aliphatic ester types [500]. These oils are used to lubricate metal surfaces approaching temperatures of up to 675 °C and their main function is to control deposit formation on the hot surfaces. A mineral oil-based lubricant will not provide the satisfactory performance at these high temperatures due to insufficient stability of the mineral base oils. U.S. Military Specifications MIL-PRF-7808L and MIL-PRF-23699F are used to select lubricants for use in this application [501–504].

MIL-PRF-7808 specification defines lubricant quality of the low viscosity (3 cSt at 100 °C or 210°F) aircraft turbine lubricants for use in equipment employed by the U.S. Air Force. These oils are also used in commercial helicopter turbo shaft engines and in certain jet aircraft equipment/accessories that operate in extreme low-temperatures environments. Aviation engine oils for most of the newer high-performance jet engines are of higher 100 °C viscosity (5 cSt) and possess improved thermal and oxidation stability properties. The specifications for 5 cSt viscosity oils are described in U.S. MIL-PRF-23699F Specification and the U.K. Military Specification DEF STAN 91-102/2 (DERD 2499). Testing requirements of these specifications are listed in Tables 5.30–5.32 [503,504]. The required tests in these standards relate to lubricant and engine parameters that include the following:

1. Low-temperature properties, such as low-temperature viscosity and pour point, to assure low-temperature performance both on ground and at high altitudes.
2. High-temperature properties, such as flash point to assess ignitability and evaporation loss of the lubricant fractions at high temperatures and low atmospheric pressures that exist at high altitudes.
3. Vapor phase deposits resulting from the contact of the oil mist and the vapors with hot engine surfaces.
4. Load-carrying capacity, so as to prolong the life of the bearings, gears, and other highly loaded parts.
5. Cleanliness, to minimize the formation of varnish and or sludge deposits.
6. Oxidation stability, to inhibit oxidation-related lubricant degradation.
7. Compatibility with metals, elastomers, carbon seals, and other lubricants.

It is important to note that MIL-PRF-23699F Specification has two versions: the STD (Standard) version and the HTS (High Thermal Stability) version. The difference between the two primarily relates to the high-temperature performance. The STD class oils are intended for use in normal performance turbo equipment, where ferrous material (gears and bearings) corrosion induced from extended periods of nonoperation in a moist environment is not a concern. The HTS class oils are for use in hot running engine designs where evidence of oil coking or oil degradation, or both, is observed. While the two lubricant classes are interchangeable and fully compatible with each other, mixing the two will diminish the added benefits provided by the HTS oils and the mixture will revert to the STD level of performance. Lubricants meeting MIL-PRF-23699F HTS credentials must meet the additional requirements relating to thermal stability and corrosivity at 274 °C (FED-STD-791, 3411), corro-

TABLE 5.22—Global heavy-duty diesel engine oil service specification DHD-1 [493].

Characteristic	Test Method[a]	Test Name	Requirements	Limits
Engine Tests				
Aeration	ASTM RR D02-1379	International EOAT	Aeration, Vol. %, Max.	8.0
Bore Polish	CEC L-52-T-97	Mercedes Benz OM 441 LA	Bore Polish % Area, Max.	2.0
Corrosion	ASTM D6483	Mack T-9	Used Oil Lead, ppm, Max TAN Increase at EOT, max.	15[c] 2.0
Filter Plugging	ASTM RR D02-1440	Cummins M11 HST	Oil Filter Diff. Press., kPa, Max.	79 / 93 / 100
Piston Cleanliness	ASTM DXXXX	Caterpillar 1R[b]	Weighted Demerits (WDR), Max. Total Groove Carbon, % Max. Top Land Carbon, % Max. Oil Consumption g/hr, Initial Max. / Final Max.	397 / 416 / 440 40 / 42 / 44 37/ 42 / 46 13.1 / 1.5 X Initial
	CEC L-52-T-97	Mercedes Benz OM 441 LA	Weighted Merits, Min. Oil Consumption, kg/test Max.	25.0 40
Oxidation	ASTM DXXXX	Sequence IIIF, 60 Hrs[b]	Kv 40°C Viscosity Increase, % Max.	200
Turbocharger Deposits	CEC L-52-T-97	Mercedes Benz OM 441 LA	Boost Pressure Loss at 400 Hours, % Max.	4
Sludge Control	ASTM RR D02-1440	Cummins M11 HST	Eng. Sludge, CEC Merits, Min.	8.7 / 8.6 / 8.5
Soot Control	ASTM D5967	Mack T-8E	Relative Viscosity at 4.8% Soot	2.1 / 2.2 / 2.3
Wear, Valve train	JASO M354-1999 ASTM D5966-96 ASTM RR D02-1440	Mitsubishi 4D34T4 160 Hrs Roller Follower Wear Test Cummins M11 HST	Cam Lobe Wear, μm Average Pin Wear, μm maximum Rocker Pad Average Weight Loss, Normalized to 4.5% Soot, mg Max.	95.0 7.6 / 8.4 / 9.1 6.5 / 7.5 / 8.0
Wear, Ring / Liner	ASTM D6483	Mack T-9	Average Wear Normalized to 1.75% Soot Liner, μm Max. Top Ring Wt Loss, mg Max.	 25.4 / 26.6 / 27.1 120 / 136 / 144
Bench Tests				
Corrosion	ASTM D6594	Corrosion Bench Test	Used Oil Element Content above Baseline, ppm, Max.	Copper 20, Lead 120, Tin 50

Characteristic	Test Method	Test Name	Requirements	RE 1	RE 2	RE 3	RE 4
Oil / elastomer Compatibility[d]	CEC-L-39-T-96	Variation after 7 days fresh oil, No presaging		RE 1	RE 2	RE 3	RE 4
		Hardness DIDC	Points, max.	-1/+5	-5/+5	-25/+1	-5/+5
		Tensile Strength	%. Max	-50/+10	-15/+10	-45/ +10	-20/+10
		Elongation rupture	%. max	-60/10	-35/+10	-20/+10	-50/+10
		Volume variation	%. max	-1/+5	-5/+5	-1 /+30	-5/+5

Characteristic	Test Method	Test Name	Requirements	Limits
Foaming tendency	ASTM D892 without option A	Sequence I (24°C) Sequence II (94°C) Sequence III (24°C)	Tendency / Stability, After 1 Min. settling	10 / nil 50 / nil 10 / nil
Foaming - High Temperature	ASTM D6082	Sequence IV (150°C)	Tendency / Stability, ml. max. After 1 Min. settling	200 / 50
Oxidation - Hot Surface	CEC-L-85-T-99	PDSC	Oxidation Induction Time, min., Min.	35
Shear Stability	ASTM D3945 CEC L-14-A-93	Bosch Injector Test	Viscosity after 30 Cycles, measured at 100°C	Stay in Grade
Sulfated Ash	ASTM D874	Sulfated Ash	Mass %, Max.	2.0
HT/HS Viscosity	ASTM D4624/D 4683 CEC-L-36-A-90	Tapered Bearing Simulator/Ravenfield	High Temperature / High Shear Rate Viscosity, cP, Min.	3.5
Volatility	ASTM D5800 CEC L-40-A-93	NOACK	% Mass Loss, Max	15

[a] Test methods noted ASTM XXXX are under development by ASTM. A numeric designation will be provided when it becomes available
[b] The requirements for this characteristic may be met with a CH-4 level passing result in an original API CH-4 qualification.
[c] Lead maximum 25 ppm if fresh oil has TBN (ASTM D4739) greater than 10.
[d] The Elastomer Compatibility Limits are those stated in ACEA 1999 European Oil Sequences and apply to the elastomer batches available at that time. Consult the most recent ACEA Oil Sequence publication for the information on the limits with more recent elastomer batches.

sion and oxidative stability at 204°C (FED-STD-791, 5308), and sediment formation (FED-STD-791, 3010).

Most commercial aircrafts require engine oils that have a 100 °C viscosity of 7.5 cSt. These oils are formulated using synthetic base stocks and additives that inhibit wear, oxidation, and foaming, and to provide good high temperature performance and load-carrying ability. The performance of these oils is defined by the U.K. Military Specification DEF STAN 91-98/1 (DERD 2487), which is summarized in Table 5.33 [501]. There are additional military standards for avia-

TABLE 5.21—Global DHD-1 engine test summary [493].

Test Identification	Engine Speed r/min	Test Length Hours	Fuel Flow	Power Output
Mitsubishi 4D34T4	3200	160	96.0 ± 1.0 mm^3/stroke cylinder	120 kW
MB OM 441 LA	1900 / 1330 / 1140	400	160 / 181 / 191 mg/stroke	240 / 205 / 185 KW
Caterpillar 1R	1800	504	240 g/min	68 kW
Cummins M11 HST	1600 / 1800	200	117 lb/h	335 / 360 hp
Mack T-8E	1800	300	139.5 ±1% lb/h	353 hp
Mack T9	1800 / 1250	75 / 425	139.5 / 121.2 lb/h	360 / 425 hp
International 7.3L	3000	20	42 kg/h	153 kW
Roller Follower Wear Test - RFWT	1000	50	9.4 kg/h	32-36 kW

tion engine oils, some of which are available in the Air BP Lubricants brochures posted at BP's website [501–504].

Small Engine Oils

Besides gasoline and diesel engines that are used in large automotive applications, such as passenger cars and heavy trucks and buses, many applications use *small engines*. These include outboard marine engines; personal watercraft and jet skis; scooters, mopeds, and all-terrain vehicles; chain saws, lawn mowers, and other portable equipment; and motor cycles. These devices use two-stroke cycle and four-stroke cycle engines, both air-cooled and water-cooled. Specifications of lubricants for these engines are established by organizations, such as the API, JASO, ISO, TISI (Thai Industrial Institute), and NMMA (National Marine Manufacturer's Association). These engines have separate performance specifications because they have unique lubrication needs. It is important to note that the use of engine oils designed for four-stroke cycle and two-stroke cycle automotive engines in small engines is not appropriate. This is because regular engine oils contain additives, such as detergents, which are metal salts of organic acids. These on combustion with the fuel form ash which deposits on the spark plugs and forms residues in the combustion chambers. Two-stroke cycle engine oils, on the other hand, are designed to minimize spark plug fouling and residue formation. They also reduce preignition, ring sticking, scuffing, carbon formation, and the crankcase sludge. The quality of these oils is assessed on the basis of their miscibility and fluidity characteristics and performance in a number of industry tests, as defined by JASO/ISO two-cycle index and NMMA's TC-W specifications. Miscibility/fluidity classification, provided in Table 5.4, is important because the two-stroke cycle engine oils must be mixed with the fuel, prior to use. Typical gasoline to lubricant ratios are 12:1, 16:1, 24:1, 32:1, 50:1, and 100:1, depending upon the engine manufacturer's recommendation. Incidentally, the higher the amount of the oil in the mixture, the higher is its smoking tendency, see Fig. 5.17 [4]. As one can see, to attain low smoking tendency, the oil must contain smoke-suppressing additives. Without their presence a significant amount of visible smoke results, even at the fuel to oil ratio of 100:1 (1 % oil). The oil's performance is determined by an engine test that evaluates three parameters: the anti-scuff characteristics, ring sticking and engine cleanliness, and preignition.

TABLE 5.23—Current Japanese engine tests.

Engine Test	Engine Type	Evaluation Criteria
Toyota 3A	Four-cylinder Overhead Cam (OHC) engine	Valve train wear
Toyota 1G-FE	Six-cylinder double OHC engine	Piston deposits, sludge, varnish, wear, and viscosity increase
Nissan VG-20E	V-6 single OHC engine	Piston deposits, sludge, varnish, and wear
Nissan KA-24E	Four-cylinder single OHC engine	Valve train wear
Nissan TD25 (JASO 336-97)	Four-cylinder Overhead Valve Engine	Piston deposits, sludge, varnish, and wear
Mitsubishi 4D34T4 (JASO M354:1999)	Four-cylinder manual transmission	Camshaft, lifter, cylinder, piston ring, and bearing wear

Lubricant performance categories for air-cooled engines are provided in Table 5.34. Of these, TA, TB, and TC are the oldest performance categories, but API TC is still the most predominant and most demanded category since it encompasses the two-stroke cycle engine performance requirements covered by TA and TB categories. In 1994, JASO (Japan Automobile Standards Organization) proposed a performance specification for low-ash engine oils for two-stroke cycle air-cooled engines. Prior to this time, metal-free (ashless) oils were commonly recommended for these engines. One of the reasons for issuing the new specification was that

TABLE 5.24—Official JASO DH-1 specification.

Test Parameter	Test Method	Unit	Performance Criteria
Piston Detergency	JASO M336	Vol. %	60.0 max
TGF	Nissan TD-25	Merit	All Free
Piston Ring Sticking Deposits on Ring Lands		Report	Report
Valve Train Wear Protection	JASO M354	μm	95.0 max
Cam Diameter Loss (Normalized at 4.5 % wt Carbon Residue Increase)	MMC 4D34T		
Soot Dispersancy	ASTM D5967-99	cSt	0.2 max
Viscosity Increase (100–150 h) at 100 °C	Mack T-8A		
High Temperature Oxidation Stability	ASTM D5533-97	%	200 max
Viscosity Increase at 40 °C	Sequence IIIE		
Hot Surface Deposit Control	JPI 5S-55-99	Merit	7.0 min
Rating at 280 °C	Komatsu Hot Tube	Rating	
Anti-foaming	JIS K2518:1991		
Sequence I		mL/m	10/0 max
Sequence II		mL/m	50/0 max
Sequence III		mL/m	10/0 max
Volatility	JPI 5S-41-93	% mass	18.0 max
Evaporation Loss at 250°C	Noack Volatility		
Anti-Corrosion	ASTM D5968-97		
Copper	ASTM D130-94	mass ppm	20 max
Lead		mass ppm	120 max
Tin		mass ppm	50 max
Discoloration of Copper Coupon after Test at 135 °C			3 max
Shear Stability	ASTM D6278-98	cSt	Stay-In-Grade of fresh oil viscosity per SAE J300
100 °C KV of Oil after Shear			
Total Base Number	JIS K2501 6:1992 or ASTM D4739-96	mg KOH/g	10.0 min
Seal Compatibility	CEC-L-39-T-96		
RE1 (Fluoro)			
Hardness Change		Point	−1/+5
Tensile Strength, Rate of Change		%	−40/+10
Elongation, Rate of Change		%	−50/+10
Volume, Rate of Change		%	−1/+5
RE2 (Acrylic)			
Hardness Change		Point	−5/+8
Tensile Strength, Rate of Change		%	−15/+18
Elongation, Rate of Change		%	−35/+10
Volume, Rate of Change		%	−7/+5
RE1 (Silicon)			
Hardness Change		Point	−25/+1
Tensile Strength, Rate of Change		%	−45/+10
Elongation, Rate of Change		%	−20/+10
Volume, Rate of Change		%	−1/+30
RE1 (Nitrile)			
Hardness Change		Point	−5/+5
Tensile Strength, Rate of Change		%	−20/+10
Elongation, Rate of Change		%	−50/+10
Volume, Rate of Change		%	−5/+5

the ash-less oils started to experience ring-sticking, which can be corrected by the use of the low-ash oils. JASO specification has three performance categories: FA, FB, and FC; which assess the lubricant's detergency, lubricity, and exhaust port blocking and smoking tendency. In all cases, performance of the candidate oil is compared to that of a reference oil. In terms of the engine performance, these lubricant properties must accomplish the following:

1. Resist combustion chamber deposit-induced pre-ignition
2. Prevent ring sticking
3. Inhibit deposit formation
4. Reduce scuffing
5. Resist spark plug fouling

The primary goal of the JASO specification is to improve the air quality by controlling smoke in Asian cities, most of which use two-stroke cycle engine vehicles for transportation. International Standardization Organization (ISO) has also established a two-stroke cycle engine oil specification. Two of the categories in the two specifications overlap except FA in the JASO specification and GD in ISO specification. GD category represents the highest level of performance for the

TABLE 5.25—Railroad diesel engine performance designations [318]. Reprinted with permission from the Lubrizol Corporation.

Designation	Dispersant Level	Detergent Level (TBN-ASTM D2896)
Generation 1	0	4 to 5
Generation 2	Moderate	7
Generation 3	Moderate	10
Generation 4	High	13
Generation 5	High	13 and over
	OEM Evaluation Tests	
Bench	EMD Silver Corrosion and Oxidation Test	
	GE Oxidation Test	
	Bronze Friction Test	
Engine	25-h EMD 2-567	
	750-h GE 7FDL	
Field[b]	480-h Caterpillar IG2[a]	
	3 to 10 late model locomotives	
	1 year and 100,000 miles	

[a]This test is obsolete and in many cases is replaced by Caterpillar 1M-PC.
[b]Approval for field testing is granted by the OEM after the candidate oil has completed bench and engine tests.

two-stroke cycle engine oils. JASO and ISO specifications are presented in Table 5.35.

Water-cooled engines radically differ from the air-cooled two-stroke cycle engines; hence they have their own specifications. Table 5.36 compares characteristics of the two types of engines [505]. Performance specifications of water-cooled engines, which are mainly used in out-board applications, are established primarily by the National Marine Manufacturers Association (NMMA). Originally, there were three categories: TC-W, TC-WII® TC-W3®. Of these, the first two are obsolete and only TC-W3® is in effect. TC-W® qualified oils, the same as the TC oils for air-cooled engines, must prevent ring sticking, deposit-induced preignition, and piston scuffing. The difference is that the tests used are different. SAE service classifications and performance criteria for the two-stroke cycle gasoline engine oils are described in the SAE Standard J2116 [506]. Test requirements for NMMA TC-W3®, the most recent category, along with the primary performance criteria are given in Table 5.37 [318]. In the year 2004, NMMA introduced FC-W™ engine oil specifications for four-stroke cycle, water-cooled outboard gasoline engines. The primary reason for considering the use of the four-stroke cycle engines in the outboard market is to reduce the exhaust emissions. The required tests are listed in Table 5.38. The details and pass/fail criteria are available in NMMA Certification Procedure Manual [507].

TABLE 5.27—Marine diesel engine operating parameters.

Type	Speed (r/min)	Bore size (mm)
Slow Speed	65–150	260–1000
Medium Speed	230–750	300–650
Medium High Speed	600–1500	200–400
High Speed	600–2250	100–200

Figure 5.18 shows a cross-sectional view of a liquid-cooled, two-stroke cycle engine [508]. Pistons, spark plugs, and exhaust port are the parts that are usually rated for malfunction. As mentioned in the preceding discussion, the performance criteria for such engines include the lubricant's ability to provide protection against ring sticking, piston varnish, plug fouling, exhaust port blocking, and piston scuffing.

Lubricant-related Causes of Engine Malfunction

Most of the problems associated with the internal combustion engine lubrication are related to the by-products of combustion, their entry into the crankcase as the blow-by, and the subsequent lubricant decomposition. This suggests that the quality of the fuel, thermal and oxidative stability of the lubricant, and the efficiency of combustion all play an important role. The major causes of the engine malfunction due to lubricant quality are deposit formation, lubricant contamination, oil thickening, oil consumption, ring sticking, corrosion, and wear.

While discussing deposit control in the chapter on Additives (Chapter 4), we described the mechanisms which result in the formation of the engine deposits, see Figs. 4.6 to 4.12 of the Additives chapter. The deposit precursors are highly oxygenated, polar organic compounds, which have the tendency to separate out of the oil and onto the surfaces as resin, a sticky substance. Its deposition on surfaces that have high ambient temperatures, such as piston crowns, rings, and skirt, leads to dehydration or further chemistry to form varnish, lacquer, and carbon deposits. In addition to the deposit

TABLE 5.26—Railroad engine builder lubricant requirements [318]. Reprinted with permission from the Lubrizol Corporation.

Engine Builder	SAE Grade	TBN (ASTM D2896)	Sulfated Ash (% mass) max	Zinc Content max	Zinc Content min	API Classification	Road Test Requirement
General Motors EMD	40 or 20W-40	10–20	...	10 ppm	...	...	3 locomotives, 1 year
General Electric, U.S.	40 or 20W-40	13–20	...	...	...	...	3 locomotives, 100,000 miles
M.T.U.	30	...	1.5	...	0.05 %	SE/CC, SE/CD	Required
	40	...	1.5	...	0.05 %	SE/CD	Required
	15W-40	...	1.8	...	0.05 %	SE/CD	Required
G.E. Locomotive Canada (Alco 251 Engines)	40	7–13	...	...	...	CD	...
Sulzer	40	...	...	...	...	CD	Required
SEMT Pielstick	40	10 min	...	...	...	CD	Required
S.A.C.M.	40	10 min	...	...	...	CD	Required

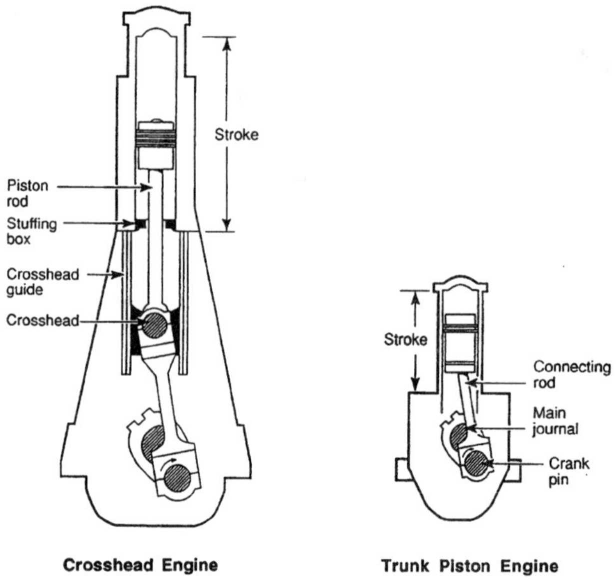

Fig. 5.15—Comparison between crosshead and trunk-piston engines [496].

precursors, lubricant also gets contaminated by the partial combustion products, for example, soot and acidic combustion products that include the sulfur and nitrogen oxides and the derived acids. Soot leads to an increase in oil viscosity, which adversely affects the oil's ability to lubricate, thereby causing increased wear. Sulfur and nitrogen oxides take part in oxidative and thermal degradation of the lubricant and in the presence of water the derived acids lead to corrosion.

The lubricant performance is assessed by analyzing various lubricant parameters for change and by inspecting various engine parts for damage. In a four-stroke cycle engine, lubricant-related parameters include viscosity increase due to oxidation or soot, sludge formation, filter plugging, fuel efficiency, and foaming. Engine parts inspected include bearings for rust and corrosion; valve train components for wear; pistons for deposits, wear, and oil consumption; and cylinder liners and piston rings for wear. These parts are identified in Figs. 5.11–5.14 and the qualifying limits for these parameters are listed in Tables 5.9–5.14, 5.16–5.18, 5.20, 5.22, and 5.24. As mentioned earlier, in two-stroke cycle and four-stroke cycle SI engines, the parameters examined include exhaust smoke, exhaust system blocking, pre-ignition, stuck rings, and piston varnish and deposits.

Rating of Engine Parts

Rating is a process that is used to describe the condition of the test components. Most lubricant and additive suppliers use one of the following methods for this purpose.

1. Coordinating Research Council (CRC) Method
2. Coordinating European Council (CEC) Method
3. Institute of Petroleum (IP) Method

While all three methods use similar basic principles, there are many differences. At present, nationally organized bodies are attempting to arrive at a uniform method. Uniformity in ratings is important in conveying meaningful information to the users. Rating is performed by qualified raters who are trained by a self-policing industry-supported group. Guidelines for rating engine parts are regularly published by CRC and the ASTM. Engine parts that are commonly rated for the most recent gasoline and diesel engine tests are listed in Table 5.39 and the test parts from some of the engines with PASS/FAIL results are shown in Figs. 5.19–5.35 [318,513,514].

Figure 5.19 shows pistons and copper/lead/tin bearings from Sequence L-38 test (ASTM D5119). Piston is evaluated for varnish deposits and the connecting rod bearing is evaluated for discoloration and weight loss. In the figure, FAIL piston shows a significant amount of varnish and fail bearing shows visible discoloration. Figure 5.20 shows pistons from Sequence IIIF and IIIG engines. Pistons are rated for varnish, deposits, and stuck rings, and the cam and lifter are rated for wear. These ratings are in addition to measuring the oxidation-related viscosity increase, oil consumption, and oil screen plugging. In the figure, PASS pistons have less varnish and carbon deposits than the FAIL pistons. Figure 5.21 shows the test camshafts from Sequence IVA engine test (ASTM D6891). This test measures the camshaft lobe wear in an overhead camshaft engine. Camshaft lobes shown in the left half of the figure have less wear, and are rated a PASS, than those shown on the right half of the figure, which are rated a FAIL. Figure 5.22 shows oil pans, baffles, and rocker arm covers from Sequence VE test (ASTM D5302). In the figure, cleaner pan, baffle, and rocker arm covers are rated a PASS while those with a lot of varnish and sludge are rated a FAIL. The test in addition rates:

1. Sludge deposits on other parts of the engine as well, which include valve deck, front seal housing, and cylinder block.
2. Varnish deposits on piston skirts, rocker arm cover, cam baffle, cylinder walls, and oil pan.
3. "Hot" stuck piston compression rings.
4. Clogging of oil pump screen, piston oil rings, and camshaft oil feed holes.

The test also measures cam lobe wear and weight loss of the follower arms. Figure 5.23 shows test pistons and baffles from Sequence VG engine test (ASTM D6593). Again, the cleaner parts have a PASS rating and those with heavier deposits have a FAIL rating. This test in addition rates sludge deposits on rocker arm covers, cam baffles, timing chain cover, oil pan and valve decks, and the varnish deposits on cam baffles. It also inspects for "hot" and "cold" stuck piston compression rings, clogging of the oil pump screen and piston oil rings, and measures the roller follower pin wear and ring gap increase. Figure 5.24 shows the connecting rod bearing from Sequence VIII engine test. Like the L-38 test, this test evaluates a lubricant's ability to protect the copper/lead/tin bearing against corrosion. However, unlike L-38, which uses leaded fuel, this test uses unleaded fuel.

Figures 5.25–5.29 show the appearance of the PASS and FAIL pistons from the various Caterpillar diesel engine tests. In all cases, the emphasis is on piston cleanliness and FAIL pistons have more carbon deposits on top lands and top grooves. Additional factors in the Caterpillar tests relate to oil consumption, change in side clearance, piston, ring, or liner distress (scuffing), and stuck rings. Figure 5.30 shows crossheads from the Cummins M-11 EGR engine test. While the PASS/FAIL decision in this case is primarily based upon wear-related weight loss, other parameters such as oil filter plugging, engine sludge, and top ring weight loss also play a

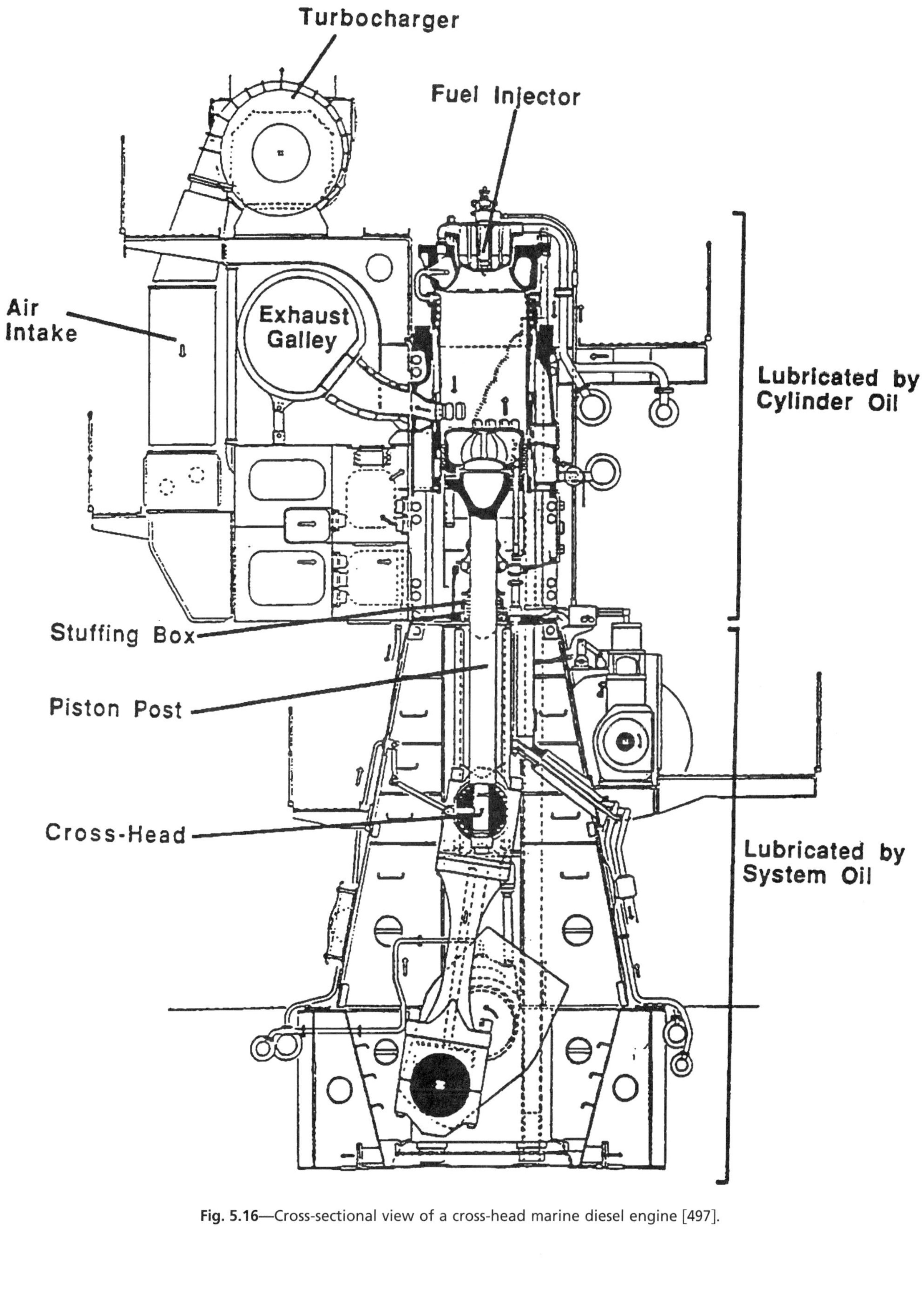

Fig. 5.16—Cross-sectional view of a cross-head marine diesel engine [497].

TABLE 5.28—Marine diesel engine oil requirements [318]. Reprinted with permission from the Lubrizol Corporation.

Slow Speed Crosshead		Medium Speed Trunk Piston
Cylinder	**Systems**	**Crankcase**
• Sulfur acid neutralization	• Lubrication of bearings, crankshaft, chains, and running gear	• Piston deposit control and ring stick prevention
• Scuffing prevention	• Detergency and thermal stability for piston under-crown cooling	• Sludge and lacquer control
• Good film strength	• Good film strength	• Stability in the presence of fuel contamination
• Detergency	• Control of drip oil and • combustion trash	• Ability to control asphaltene related deposits in engine "cold areas," i.e., crankcase and sump
• Piston, ring, and port • cleanliness	• Release of insolubles and water to purifiers	• Keep oil scraper rings clear of deposits and sludge
• Antiwear properties	• Low emulsibility and ability to separate from water and insoluble matter	• Thermal stability and oxidation control
• Systems oil compatibility	• Rust and oxidation prevention	• Rust control and alkalinity retention
• SAE 50	• SAE 30	• Combustion acid neutralization • Bearing corrosion protection • EP properties • Good water filterability and sheddability • Lubricant consumption control • SAE 30/40

role. Two new Cummins engine test procedures, Cummins ISM and Cummins ISB, are a part of the new API CJ-4 heavy-duty diesel engine oil specification. The ISM test procedure is used to evaluate a lubricant's effectiveness at reducing soot-related wear of overhead components, sludge, and oil filter plugging. The engine parts from this test are shown in Fig. 5.31. The procedure uses a Cummins ISM engine equipped with EGR and is intended to be a replacement procedure for the M-11 EGR, using newer hardware. The ISB test procedure is used to evaluate a crankcase lubricant's ability to reduce camshaft lobe and valve train wear. The test is carried out using a Cummins ISB, common rail fuel system engine equipped with EGR. Both procedures use PC-10 fuel. Parts from this test are shown in Fig. 5.32. In both these figures, the parts on the left reflect less than exemplary performance.

Four Mack tests are also included in the current heavy-duty specifications. These tests are Mack T-8, Mack T-10,

TABLE 5.29—Engine builders' preferences for stationary gas engine oils [318]. Reprinted with permission from the Lubrizol Corporation.

Engine Builder	Ash Content (% Wt)	Viscosity Grade	Other Requirements
Caterpillar	0.4–0.6	SAE 30/40 SAE 15W-40	Multi-grade oils not permitted with electric governors
Cooper-Cameron			
Ajax	<0.1	SAE 30/15W-40	No Zinc additives
Cooper-Bessemer			
2-Stroke cycle	<0.6	SAE 40	No bright stock allowed
4-Stroke cycle (<175 BMEP)	0.4	SAE 40	API CC or MIL-L-2041B
4-Stroke cycle (>175 MBEP)	0.4	SAE 40	API CD or MIL-L-2041B
Cooper-Enterprise	0.50 max	SAE 30/40	API CC
Superior	0.5 to 1.0	SAE 40	
Dressor Industries			
Clark	0.4 acceptable, 0.1 preferred	SAE 30/40	
Dresser-Rand	0.5 max	SAE 30/40	Field test required
Waukesha			
VGF	0.7 to 1.2	SAE 30/40	API CC or CD
All others	0.35 to 0.5	SAE 30/40	API CC or CD

TABLE 5.30—U.S. Military Specification MIL-PRF-7808L Grade 3 for aviation engine oils [503].

Tests	Limits		Test Method	
	Min	*Max*	*ASTM or SAE Standard*	*Fed. Test Method STD. NO. 791*
Particulate contamination, contaminant mg/L	--	5.0	Appendix A of 7808 Spec	
Filtering time, min/qt	--	30		
Total acid number, mg KOH/g	--	0.30	D664[1]	
Viscosity at 100°C, cSt	3.0	--	D445	
Viscosity at 40°C, cSt	12.0	--	D445	
Viscosity at –51°C, cSt			D2532	
at 35 minutes	--	17 000		
at 3 hours	--	17 000		
at 72 hours	--	17 000		
Viscosity stability, %, 3 hr. vs. 35-minute determination	--	6		
Flash point (C.O.C.) °C	210	--	D92	
Evaporation loss, 6½ hrs., at 205°C, wt. %	--	30	D972[1]	
Foaming characteristics (static)				3213
Foam volume, mL	--	100		
Foam collapse time, sec.	--	60		
Foaming characteristics (dynamic)				3214
Foam volume, ml/collapse time, sec.				
80°C at 1000 cc/min	--	100/60		
80°C at 1500 cc/min	--	150/60		
80°C at 2000 cc/min	--	200/60		
110°C at 1000 cc/min	--	100/60		
110°C at 1500 cc/min	--	150/60		
110°C at 2000 cc/min	--	200/60		
Deposition test, average deposition rating	--	1.5		5003
Acid no. change, mg KOH/g	--	20		
Viscosity at 40°C, % change	--	100		
Oil consumption, mL	--	100		
SOD Pb corrosion, wt. change, g/m^2	--	9.3		5321
Bronze corrosion at 232°C, wt. change, g/m^2	--	± 4.5		5305
Ag corrosion at 232°C, wt. change, g/m^2	--	± 4.5		5305
Ryder gear load, avg. % relative rating x 508 kN/m (avg. of 2 det)	420	--	D1947	
Oxidation & corrosion stability at 200°C for 96 hrs. Al, wt. change, mg/cm^2	--	± 0.2		5307
Ag, wt. change, mg/cm^2	--	± 0.2		
Bronze, wt. change, mg/cm^2	--	± 0.4		
Fe, wt. change, mg/cm^2	--	± 0.2		
M-50, wt. change, mg/cm^2	--	± 0.2		
Mg, wt. change, mg/cm^2	--	± 0.4		
Ti, wt. change, mg/cm^2	--	± 0.2		
Viscosity change, % at 40°C	- 5	+ 25		
Neut. no., change	--	4.0		
NBR-H rubber swell, % (168 hrs. at 70°C)	12	35		3604
FKM rubber swell, % (72 hrs. at 175°C)	2	25		3432
Tensile strength, % change	--	50		
Elongation, % change	--	50		
Hardness, change	--	20		
FVMQ rubber swell, % (72 hrs. at 150°C)	2	25		
Tensile strength, % change		50		
Elongation, % change		50		
Hardness, change		20		
QV I rubber swell, % (72 hrs. at 150°C)	2	30		3432
Tensile strength, % change		50		
Elongation, % change		50		
Hardness, change		20		
Bearing deposition	--	60		3450
Average deposit rating (3 determinations)	--	2.0		
Filter deposit, gm	--	1440		
Oil consumption, mL	- 5	--		
Viscosity change, %				
Neutralization number change (avg. of 3 determinations)	0.2	25		
Metal specimen weight change, mg/cm^2	--	1.0		
Compatibility, turbidity, ml sediment/200 mL of oil	--	0.005		3403[a]
Compatibility, intermixing	Pass		Sec. 3.4.1.4 of 7808 Spec.	

TABLE 5.30— (*Continued.*)

Tests	Limits		Test Method	
	Min	*Max*	*ASTM or SAE Standard*	*Fed. Test Method STD. NO. 791*
Storage stability, Pb corrosion			Sec. 4.6.6 of 7808 Spec.	
after 2 days at 110°C, g/m^2	–	40		
after 7 days at 110°C, g/m^2	–	230		
Trace metal content, ppm			Part 4.6.2 of 7808 Spec	
Al	–	2		
Fe	–	2		
Cr	–	2		
Ag	–	1		
Cu	–	1		
Sn	–	11		
Mg	–	2		
Ni	–	2		
Ti	–	1		
Si	–	2		

[a] Modified

Mack T-11, and the newest Mack T-12, which is part of the most recently introduced CJ-4 heavy-duty diesel category. Mack T-8, T-8E, and T-11 evaluate soot handling capability of an oil with regard to viscosity, oil filter plugging, and oil consumption. Mack T-10 evaluates an oil's ability to minimize cylinder liner wear and piston ring and bearing wear in engines equipped with exhaust gas recirculation (EGR). Top ring weight loss limit for Mack T-10 is 140 mg maximum, but for Mack EON-PP, it is 12 mg maximum. Mack T-12 evaluates piston ring wear, cylinder liner wear, lead bearing corrosion, oil consumption, and oxidation.

Figure 5.33 shows parts from the roller follower test. In this case, the wear on roller follower axles is the primary determinant in the PASS/FAIL decision. The wear limit for CG-4 oils is 11.4 microns (0.45 mils) and for CH-4 and CI-4 oils is 7.6 microns (0.30 mils).

Parts of the natural gas engines that are examined for varnish, deposits, or damage include pistons, piston liners, cylinder heads, valves, valve deck, connecting rod bearing, and oil pan. These for a natural gas engine oil of a suitable quality are shown in Fig. 5.34 [498,514]. Pistons from various two-stroke cycle engines are shown in Fig. 5.35 [318]. Notice the heavy deposit build-up on pistons that are unacceptable.

Formulating Engine Oils

Engine oils must possess a number of attributes to perform effectively in the tests summarized in the above-listed performance specifications. Typically an engine lubricant contains 70 % or more base oil depending upon the application, and the additives make up the balance. Base stocks manufacturers classify base stocks into four broad groups: conventional, hydroprocessed, severely hydroprocessed, and synthetics. Conventional base stocks, commonly referred to as mineral oils, are derived from petroleum refining. Hydro-processed oils are highly refined mineral oils, which are made from mineral oil stocks by hydrogen treatment under mild conditions [400 °C (750°F) and 500 psi pressure]. Base oils thus obtained are of similar quality as those traditionally made by sulfuric acid treatment, neutralizing, and clay filtration, but without generating any waste. These oils have low aromatics content and lighter color, which is due to the removal of aromatics, sulfur, and nitrogen compounds. Severely hydroprocessed oils, the same as hydrocracked oils, are made by hydrotreating under severe conditions [425–480 °C (800 to 900°F) and 1500 to 3000 psi pressure]. In addition to hydrogenating aromatics and removing sulfur and nitrogen compounds, the process involves molecular reconstruction, such as converting the linear hydrocarbons into branched hydrocarbons. Hydrocracking/hydroisomerization enables the refineries to produce oils that not only have excellent low-temperature properties, but also very high viscosity indices. Such oils are designated as VHVI or X-HVI paraffinic base stocks. These oils are comparable to synthetic base fluids in performance.

Engine oil formulations use both mineral oils and synthetic fluids. However, because of the increasing demands of a modern combustion engine on the lubricant, the conventional solvent-refined oils are not very suitable. This is because of their meager low-temperature fluidity, mediocre viscosity indices, fair volatility characteristics, and borderline thermo-oxidative stability. Hydroprocessed and hydroisomerized oils are better since they have good color, good low-temperature fluidity, uniform composition hence low volatility, and excellent thermo-oxidative stability because of being free from aromatics. These base oils are grouped under API Group II and Group III categories, which are provided in Table 5.40 along with their sulfur and saturates content and viscosity indices [515]. One of the major drawbacks of removing aromatics from the mineral oil composition to improve its properties is that the new oil lacks good solvency. That is, it has diminished ability to dissolve additives many of which are polar in character. The same issue must be addressed while selecting nonpolar synthetics such as PAOs as base fluids since their structure is analogous to that of the modified mineral oils. One way the formulators overcome the solvency problem is to add back some of the polars, such as an alkylnaphthalene, a synthetic ester, or both. Despite the fact that the aromatics removed from mineral oil via hydroprocessing are being replaced with an alkylnaphthalene, another aromatic, the new composition has superior properties since alkylnaphthalene is oxidatively more stable than the removed aromatics. If the alkylnaphthalene contains a tertiary alkyl group, which lacks easily oxidizable benzylic

TABLE 5.31—U.S. Military Specification MIL-PRF-23699F for aviation engine oils [504].

Tests	Limits Min	Limits Max	Test Method ASTM or SAE Standard	Test Method Fed. Test Method STD. NO. 791
Viscosity, cSt at - 40°C (- 40°F)	–	13,000	D2532	
% change -40°C (-40°F), 72 hrs	–	± 6	D2532	
cSt at 100°C (212°F)	4.90	5.40	D445	
cSt at 40°C (104°F)	23.0	–	D445	
Flash point, °C	246	–	D92	
Pour point, °C	–	- 54	D97	
TAN, mg KOH/g	–	1.00	SAE-ARP5088	
Evaporation loss, %, 6.5 hrs. at 204°C (400°F)	–	10	D972[a]	
Foam 1 24°C (75°F), mL/collapse time	–	25/0	D982[a]	
2 93.5°C (200°F), mL/collapse time	–	25/0		
3 24°C (75°F), mL/collapse time	–	25/0		
Rubber compatibility				3604/3633
AMS No. 3217/1, % Swell, 72 hr. at 70°C (158°F)	5	25		
AMS No. 3217/4, % Swell, 72 hr. at 204°C (400°F)	5	25		
Si Rubber, % Swell, 96 hr. & 121°C (250°F)	5	25		
Tensile strength loss, %	–	30		
Compatibility		Pass		3403
Turbidity		None		
Sediment, mg/L		20		
Storage stability, 6 wks		Pass	Part 4.4.2 of 23699 Spec	
Thermal stability and corrosivity, 274°C (525°F)				3411
% Vis change	–	5.0		
TAN change, mg KOH/g	–	6.0		
Metal wt. change, mg/cm^2	–	4.0		
Sediment 1.2 micron, mg/L	–	10		3010
Ash, mg/l.	–	1		
Corrosion and oxidative stability, 72 hrs. at 175°C (347°F)	–			5308[a]
Steel, mg/cm^2	–	± 0.2		
Ag, mg/cm^2	–	± 0.2		
Al, mg/cm^2	–	± 0.2		
Mg, mg/cm^2	–	± 0.2		
Cu, mg/cm^2	–	± 0.4		
Vis, change at 37.8°C (100°F), %	–	-5 to + 15		
TAN change, mg KOH/g.	–	2.0		
Sludge, mg/100ml	–	50		
Corrosion and oxidative stability, 72 hrs. at 204°C (400°F)				5308[a]
Steel, mg/cm^2	–	± 0.2		
Ag, mg/cm^2	–	± 0.2		
Al, mg/cm^2	–	± 0.2		
Mg, mg/cm^2	–	± 0.2		
Cu, mg/cm^2	–	± 0.4		
Vis, change at 37.8°C (100°F), %	–	-5 to + 25		
TAN change, mg KOH/g	–	± 3.0		
Sludge, mg/100ml	–	50		
Corrosion and oxidative stability, 72 hrs. at 218°C (425°F) 5308*				5308[a]
Steel, mg/cm^2	–	± 0.2		
Ag, mg/cm^2	–	± 0.2		
Al, mg/cm^2	–	± 0.2		
Ti, mg/cm^2	–	± 0.2		
Vis, change at 37.8°C (100°F), %		Report		
TAN change, mg KOH/g		Report		
Sludge, mg/100mL		50		
Trace metal content, ppm			Part 4.4.6 of 23699 Spec	
Al	–	2		
Fe	–	2		
Cr	–	2		
Ag	–	1		
Cu	–	1		
Sn	–	11		
Mg	–	2		
Ni	–	2		
Ti	–	2		
Si	–	10		
Zn	–	2		
Pb	–	2		
Mo	–	3		
Ryder gear				6508
Av. rel. rating, % of reference oil	102	–		
No. of determinations	–	–		
Shear stability			D2603	
Vis loss at 40°C (104°F), %	–	4		

[a] Modified

TABLE 5.32—U.K. Military Specification DEF STAN 91-101/2 (replaces DERD 2499) for aviation engine oils [504].

Tests	Limits		Test Method		
	Min	*Max*	*ASTM*	*IP*	*DERD*
Appearance	Clear and Bright				
Density at 15°C, kg/L	Report		D1298 or D70	160 or 190	
Flash point, °C	220	–	D93	34	
Pour point, °C	–	- 54	D97	15	
Saponification number	Report		D94[1]	136[1]	
Total acid number, mg KOH/g	Report		D664	177	
Viscosity, kinematic, cSt (mm^2/s)			D445	71	
At 100°C	4.90	5.40			
At 40°C	–	30			
At – 40°C	–	13,000			
Friction and wear				239	
Initial seizure load (60 sec.) kg	65				
Wear scar diameter at 6 kg load (60 sec.), mm		Report			
Foaming characteristics, at air flow rate of 100 mL/min.					15 Proc. A
Foam volume, mL		10			
Foam collapse time, sec.		20			
High temperature oxidative stability					9[a]
Total acid number increase, mg KOH/g		1.0			
Sediment, % mass		0.005			12 Proc. B
Trace element content, mg/kg					
Al		2	Method subject to Q.A.A.		
Cr		5			
Cu		2			
Fe		2	BP method:		
Pb		Report	USN NAVAIR 17-15-BF-62		
Mg		2			
Ni		5			
Si		Report			
Ag		2			
Ti		2			
Confined heat stability:					1
S temperature, °C		Report			
Viscosity stability at 100°C, vis. decrease, %		5.0			
Corrosivity:					
New oil, 192 hrs. at 200°C					
wt. change, mg/100 mm^2					
Aluminium alloy	- 0.1 to + 0.1				3 Proc. A
Copper	- 1.0 to + 0.1				
Titanium-copper alloy	- 0.2 to + 0.1				
Copper-nickel-silicon alloy	- 0.5 to + 0.1				
Mild steel	Report				
Lead bronze 0.1	- 0.5 to + 0.1				
High carbon chromium steel	- 0.2 to + 0.1				
Lead brass	- 1.5 to + 0.1				
Ni-Cr case hardened steel	- 0.2 to + 0.1				
High speed steel	- 0.2 to + 0.1				3 Proc. B
Nickel	- 0.1 to + 0.1				3 Proc. C
Silver	- 0.2 to + 0.1				3 Proc. D
Magnesium	Report				
Cr-Mg alloy	Report				3 Proc. A
Cr-Mg alloy, phenolic resin coated	Report				
Pretreated oil, 192 hrs. at 200°C					
wt. change, mg/100 mm^2					
Aluminum alloy	- 0.1 to + 0.1				
Copper	- 1.5 to + 0.1				
Titanium-copper alloy	- 0.1 to + 0.1				
Copper-nickel silicon alloy	- 1.5 to + 0.1				
Mild steel	Report				
Lead bronze	- 1.5 to + 0.1				
High carbon chromium steel	- 3.0 to + 0.1				
Lead brass	- 4.0 to + 0.1				
Ni-Cr case hardened steel	- 3.0 to + 0.1				
High speed steel	- 1.0 to + 0.1				
Elastomer Compatibility					4
OX-27 & OX-28					
Nitrile, 192 hrs. at 130°C					
% Swell at 1/2 hr	10	20			
% Swell at 48 hr	8.5	20.5			
Condition after band test		No Cracking			
Silicone, 192 hrs. at 175°C					
% Swell at 1/2 hr	5	15			
% Swell at 48 hr	Nil	–			
Condition after band test		No Cracking			
Reversion test: Permanent indentation mm x 10^{-2}	–	25			
OX-27					

TABLE 5.32— (*Continued.*)

Tests	Limits		Test Method		
	Min	*Max*	*ASTM*	*IP*	*DERD*
Viton, 192 hrs. at 200°C	20	35			
% Swell at 1/2 hr	20.5	32.5			
% Swell at 48 hr		No Cracking			
Condition after band test					
Viton LCS, 192 hrs. at 200°C					
% Swell at 1/2 hr		Report			
% Swell at 48 hr	12	24			
Condition after band test		No Cracking			
OX-28					
Viton, 384 hrs. at 200°C					
% Swell at 1/2 hr	20	35			
% Swell at 48 hr	28.5	40.5			
Condition after band test		No Cracking			
Viton LCS, 384 hrs. at 200°C					
% Swell at 1/2 hr	15	35			
% Swell at 48 hr	28.5	40.5			
Condition after band test		No Cracking			
Engine test				Data on file with M.O.D.	
Homogeneity:					5
High temperature, 210°C	No visible signs of separation, elation, or				Proc. A
Low temperature, - 54°C	formation of insoluble matter				Proc. B
Miscibility and compatibility				8	
Miscibility:	No visible signs of separation, elation, or				
at 210°C	formation of insoluble matter			Proc. A	
at - 54°C	–	0.05		Proc. B	
Compatibility:					
Increase in toluene insolubles. % wt.				Proc. C&D	
				Data on file with M.O.D.	
Hydrolytic stability					
Calculated time to produce increase					
in acidity at 40°C of 1.5 mg KOH/g, hr	850	–			6
Load/carrying ability					
Four ball machine assessment:				239[a]	
Mean Hertz load, Kg		Report			
Welding load, Kg		Report			
Wear load curve		Report			
Initial seizure load, Kg	65	–			
IAE Gear machine rating				166[a]	
Mean relative failure load					
at 2000 RPM, %	75	–			
at 6000 RPM, %	65	–			
90% confidence interval, %		Not greater than			
at 2000 RPM, %		0.6xmrfi at each			
at 6000 RPM, %		speed			
Oil mist coking					13
at 230°C, % of reference	Report				
at 290°C, % of reference	Report				
Volatility and high temp. oxidative stability					9
Temperature parameters at 192 hr.					
E Temperature, °C	185	–			
A Temperature, °C	190	–			
V Temperature, °C	185	–			
B Temperature, °C		Report			
Z Temperature, °C		Report			
Effective life, hours					
at 200°C					
Volatilization loss	90	–			
Acidity increase	100	–			
Viscosity increase	60	–			
Insolubles increase	225	–			
at 250°C					
Volatilization loss	2	–			
Acidity increase.	0.5	–			
Viscosity increase	1	–			
Insolubles increase	20	–			
Catalytic oxidation:					
ΔZ solidus catalytic oxidation index, °C	-15	10			14
Bearing cleanliness temperature index					
B Temperature parameter, °C	Report				9
S Temperature parameter, °C	Report				1
(2/3 B + 1/3 S) index, °C	230				
Low temperature oxidative stability	No deposition of insolubles				10
Shear stability, diesel injector rig:					
ΔViscosity (kinematic) at 100°C, %	± 2.0		Note: Sonic shear test substituted		
Appearance of oil after test	No sludge or gum				
Viscosity, kinematic, high temperature:					
at 205°C, cSt (mm2/s)	1.25	–	D445[1]	71[a]	
Viscosity stability at – 40°C, % change	–	6.0	D2532		
Auto-ignition temperature, °C	Report		D2155		

[a]Modified

TABLE 5.33—U.K. Military Specification DEF STAN 91-98/1 (DERD 2487) for aviation engine oils [501].

Tests	Limits		Test Method	
	Min	*Max*	*ASTM*	*IP*
B.S.E. 1 Bristol-Siddeley bearing rig test	Performance equal to reference oil		Part 4.1, DERD 2487	
Compatibility test	Data on file with M.O.D.		Part 4.2, DERD 2487	
Elastomer swell, 192 Hrs. at 150°C in vac.., % vol.	15.5	29.5	Appendix A of DERD 2487	
IAE-gear test				
2000 RPM, % of reference	100	--		
6000 RPM, % of reference	100	--		
Homogeneity test between – 54°C and 280°C	No separation, etc.		Part 4.6, DERD 2487	
Shear stability, % change in viscosity at 100°F	--	± 2	Appendix B of DERD 2487	
Thermal stability, % change in viscosity at 100°F				
After 6 hours	--	- 10, + 20	Appendix C of DERD 2487	
After 12 hours	--	- 10, + 20		
After 18 hours	--	- 10, + 20		
After 24 hours	--	- 10, + 20		
Viscosity at - 40°F, cSt (after 12 hrs. at – 65°F)	Report		Part 4.9, DERD 2487	
Volatility, 6½ hrs. at 200°C, % loss	Report		D972	
Mineral acidity, mg KOH/g	--	Nil		1 (Method C)
Total acid number, mg KOH/g	Report		D974	139
Total alkalinity, ml 0.01N H_2SO_4/100ml	Report		Appendix D of DERD 2487	
Oxidation corrosion test, 22 hrs at 140°C				
Cadmium plate steel, wt. change mg/cm^2	--	± 0.2	Appendix E of DERD 2487	
viscosity at 100°F, % change	--	± 5		
acidity, mg KOH/g	--	0.5		
Copper, wt. change mg/cm^2	--	± 0.2		
viscosity at 100°F, % change	--	± 5		
acidity, mg KOH/g	--	0.5		
Flash point, °F	420	--	D92	36
Specific gravity, 60/60°F	Report		D1298	160
Viscosity at – 40°F, cSt (after 35 minutes)	--	13,000	Part 4.9, DERD 2487	
Viscosity at 100°F, cSt		38	D445	71
Viscosity at 210°F, cSt	7.5	--		

hydrogens, the composition will have even better oxidation stability [516].

The major advantage of the API categories is that they minimize a marketer's retesting costs when blending licensed engine oils from base oils manufactured differently. The higher API category typically represents a higher level of performance, and so does the cost. Various properties of the API Group I to API Group IV base stocks of similar 100 °C viscosity are shown in Table 5.41 [517]. API Group V category in Table 5.40 includes base stocks that contain oxygen, phosphorus, chlorine, fluorine, etc. The only exception to the list is the alkylated aromatics, which are pure hydrocarbons. Of the API Group V base stocks, diesters, polyol esters, and alkylated aromatics are used to develop automotive and aircraft engine oils. Compared to the cost of the hydrocarbon oils shown in Table 5.41, the API Group V oils are even more expensive than the PAOs, which are grouped under API Group IV. For example, the polyesters base stocks are ten to fifteen times more expansive than conventional or hydroprocessed base oils. Semi-synthetic base stocks are a compromise where the cheaper hydrocarbon oils are mixed with PAOs and polyesters to take advantage of their superior properties at a reasonable cost. The proportion of the severely hydroprocessed PAOs or synthetic base stocks in semi-synthetic oils is usually between 10 and 25 %.

Additives that are used to formulate engine oils are listed below, along with the functions they perform.

1. Oxidation Inhibitors—Maintain oil stability over the service interval. Keep the formation of oxygenates in control, thereby minimizing an increase in the lubricant

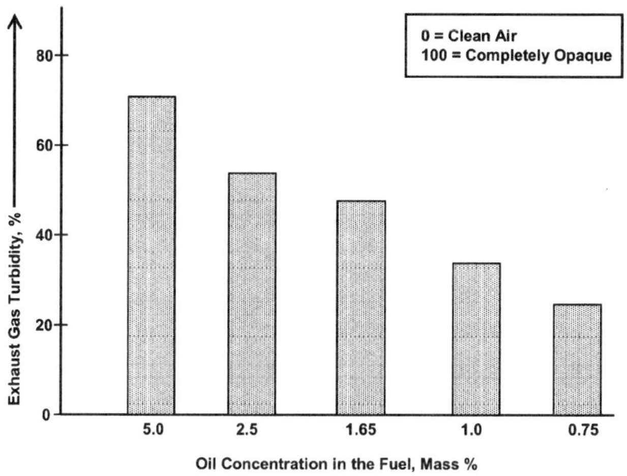

Fig. 5.17—Oil concentration versus smoking in a two-stroke cycle engine measured at full load and 110 km/h (68 mph) speed [4].

viscosity and deposit formation. These additives help the lubricant meet the performance requirements of tests that measure viscosity increase and deposit formation. Tests include Sequence IIIF/IIIG/IIIGA, Sequence VE/VG, TEOST (ASTM D6335 and MHT-4), TU5JP-L4 (CEC-L-88-T-02), and Caterpillar 1M-PC/1K/1N/1P/1R tests.

2. Detergents—Neutralize combustion-related and oxidation-related acids, thereby minimizing corrosive wear, and keep the deposit precursors solubilized in the bulk lubricant. These additives help the lubricant meet the requirements of the Sequence VE/VG and Caterpillar tests, listed in Item 1.
3. Dispersants—Keep components clean and prevent sludge and deposits formation by keeping the soot and deposit precursors suspended in the bulk lubricant. They also control soot-related lubricant viscosity increase. These additives help the lubricant meet the requirements of the Sequence VE/VG, Caterpillar 1M-PC/1K/1N/1P/1R, and Mack T-6/T-7/T-8/T-8E/T-9/T-10/T-11/T-11A, M111(CEC-L-53-T-95), VW 1.6 TC D (CEC-L-46-T-93), DV4TD (CEC-L-56-T-98), XUD11BTE (CEC-L-093), VW DI (CEC-L-78-T-99), OM364LA (CEC-L-42-T-99), and OM441LA (CEC-L-52-T-97) tests.
4. Antiwear Additives—Protect against wear due to metal-to-metal contact. These additives help the lubricant meet the requirements of the Sequence IVA, Roller Follower, Mack T-10/T-12, Detroit Diesel 6V92TA (ASTM D5862), TU3M (CEC-L-38-A-94), Cummins M-11/M-11 EGR/ISM/ISB, OM602A(CEC-L-51-A-98), and OM602A (CEC L-51-A-97) tests.
5. Corrosion Inhibitors—Protect against the effects of water condensation leading to rust and the attack of the corrosive species on metal. These additives help the lubricant meet the requirements of the CRC L-38 (ASTM D5119), Sequence IID (ASTM D5844), Ball Rust (ASTM D6557), high-temperature bench corrosion (ASTM D6594), and Sequence VIII tests.
6. Viscosity Index Improvers—Improve lubricant's ability to maintain viscosity at high temperatures. These additives help make multi-grade oils and the lubricant to meet the viscosity index requirements.
7. Pour Point Depressants—Lower the pour point of oils that contain solubilized wax, such as the API Group I oils, and keep them fluid and pumpable at low temperatures, or both. These additives help the lubricant meet the requirements of the ASTM D5133, ASTM D4684 (MRV TP-1), and modified D 4684 tests and the cranking and pumping viscosities.
8. Friction Modifiers—Minimize friction-related power losses. These additives help the lubricant meet the fuel economy requirements of the Sequence VIA/VIB and M111 (CEC-L-54-T-96) tests.
9. Demulsifiers—Minimize emulsion formation and facilitate quick water separation from the lubricant. They are used in engine oils with a high probability of water contamination, such as the marine lubricants.
10. Foam Inhibitors—Prevent oil foaming and cavitation. These additives help the lubricant meet foam requirements of the ASTM D892 and D 6082 foam tests, and the ASTM RR:D02-1379 and HEUI aeration tests.
11. Seal Swell Agents—Prevent shrinking and cracking of the seals. These additives help the lubricant meet elastomer compatibility requirements, such as in the CEC-L-39-T-96 test.

Please note that the list of tests under each additive type is not meant to be exhaustive or precise. This is because many tests measure more than one performance criterion, which may or may not be the primary one.

Additives for use in automotive lubricants, such as engine oils, are usually supplied by the additive companies as packages that meet the performance testing requirements of a particular specification. Engine oils are the largest users of lubricant additives. ILSAC GF-4, the industry's newest passenger car oils specification, requires approximately 20 % additives. These additives are packaged in two portions. The

TABLE 5.34—Performance categories for the air-cooled engines.

Designation	Normal Service Application
TA	Mopeds and other extremely small engines (typically less than 50 cc)
TB	Motor scooters and other highly loaded small engines (typically 50 cc to 200 cc)
TC	Various high performance engines (not outboards) (typically 50 cc to 500 cc)
TD	Outboard engines

TABLE 5.35—JASO and ISO two-stroke Cycle Index[a] ratings.

ISO Global Specification	...	GB	GC	GD
JASO Specification	FA	FB	FC	FD
Lubricity, min	90	95	95	95
Initial Torque, min	98	98	98	98
Detergency, min	80	85	95	125
Exhaust Smoke, min	40	45	85	85
Exhaust System Blocking, min	30	45	90	90
Piston Varnish[b], min	...	85	90	95

[a]Index=(Candidate Oil performance/Reference Oil Performance) × 100.
[b]For ISO category only.

TABLE 5.38—FC-W™ specification tests for four-stroke cycle, water-cooled gasoline engine lubricants [507].

Identification Tests	Other Tests
Kinematic Viscosity @ 40°C (ASTM D445)	Kinematic Viscosity @ 100°C (ASTM D445)
Viscosity Index (ASTM D2270)	Cold Crank Simulator Viscosity (ASTM D5293)
Specific Gravity (ASTM D1298 or D4052)	MRV-TP1 Viscosity (ASTM D4684)
Total Base Number (ASTM D2896)	Foam, Sequence I, II and III (ASTM D892)
Total Acid Number (ASTM D664)	Foam / Aeration, Sequence IV (ASTM D6082)
Elemental Analysis (ASTM D4951, D4927, D4628 or D5185)	Shear Stability (ASTM D6278)
Sulfur Content (ASTM D5453, D2622, D4294 or D6443)	High Temperature High Shear Viscosity (ASTM D4683)
Nitrogen Content (ASTM D5291 or D5762)	Corrosion (Salt fog test) (SFT) (ASTM B117)
IR Spectrum (ASTM E168)	Noack Volatility (ASTM D5800)
SAE Viscosity Grade (SAE J300)	Filter Plugging (EOFT) (ASTM D6795)
	Yamaha 115 hp General Performance Engine Test (GPET)

first portion contains polymeric additives, such as viscosity improver/s and pour-point depressant/s. The second portion, called detergent inhibitor or DI for short, contains additives that include dispersants, detergents, oxidation inhibitors, antiwear agents, corrosion inhibitors, foam inhibitors, and friction modifiers. In the case of the passenger car oils, the polymeric portion of the additive package accounts for approximately 30 % of the total additive volume, approximately 6 % of the finished oil. Polymethacrylates, styrene-isoprene polymers, and ethylene-propylene and ethylene-propylene-diene polymers are used to impart to the oil the ability to maintain appropriate viscosity across a broad temperature range. The use of the highly refined base oils, which have high viscosity indices and low pour points, greatly helps in meeting this objective. The DI portion of the additive package makes up the remaining 70 % of the total additive volume, or 14 % of the finished oil. The DI portion comprises approximately 60+ % dispersant, 20+ % detergents, 10+ % antiwear agents, and the balance is the rest of the additives. Dispersants are typically the alkenylsuccinimide type and detergents are typically alkaline-earth metal salicylate, sulfonate, and or phenate. Antiwear agents are usually of the zinc dialkyl dithiophosphates type. However, there are changes on the horizon with respect to the use of antiwear agents, corrosion inhibitors, oxidation inhibitors, and the friction modifiers. This is because automotive engine technologies are constantly being upgraded and it is critical that the new lubricants are compatible with the new changes in engine design or operation. Passenger car technologies are driven by fuel economy and maintaining catalytic converter efficiency and heavy-duty diesel technologies are being driven by emissions control via exhaust gas recirculation (EGR), fuel economy, and lowering the operating costs through extended service intervals.

The automakers' drive to protect pollution control devices is critical in meeting the increasingly stringent air emissions regulations. Automakers want to make sure that the emissions control systems on their vehicles continue to perform over the vehicle's lifetime. Therefore, they demand engine oils to have lower levels of sulfur, phosphorus, and metal compounds—elements the automakers believe deteriorate the performance of the exhaust system catalysts. While oil and additive companies do not fully agree with the automakers, they have little choice but to attempt to remove these elements from passenger car engine oils. The compounds that are especially being targeted for replacement are zinc dialkyl dithiophosphates, which have been used for many years as extremely cost-effective multipurpose additives for controlling lubricant oxidation, wear, and corrosion. These additives contain all three elements to which the automakers object. Incidentally, a statistical study on the effects of the oil phosphorus on emissions control systems appears to agree with the automakers' concern regarding the effects of phosphorus. The study concluded that phosphorus has a statistically significant effect on HC and CO emissions and the emissions degradation factors. The authors also suggested that calcium detergents may mitigate the detrimental effect of phosphorus [518]. ILSAC GF-4 and forthcoming specifications favor increasing the use of oxidation inhibitors, many of which contain sulfur. This forced GF-4 formulators to use even more expensive arylamine and hindered phenol type oxidation inhibitors.

Engine oils for heavy-duty trucks have compositions similar to those of the passenger car engine oils but they must conform to their own environmental regulations—most notably new limits on soot emissions. In the United States, it is reflected by the new heavy-duty engine oil specification, API CJ-4, which is designed to cope with higher levels of crankcase soot. This surfaced the undesirability of certain elements in heavy-duty diesel engine oils as well. Sulfur is the main element whose presence in these oils is no longer welcome because of its negative effect on diesel particulate filters and the related devices. That is why the EPA has mandated the use of ultra-low sulfur diesel, S15, in on-highway model year 2007 heavy-duty vehicles, which is going to be extended to the nonroad use by the year 2010 and to locomotive and marine engines by the year 2012. Theoretically, the lubricant sulfur should not have any effect on emissions, but it does since part of the lubricant in the internal combustion engine travels past the piston rings into the combustion chamber and burns with the lubricant. This implies that the formulators will have to formulate with additives that are either sulfur-free or have low sulfur. The major class of additives that is of concern is detergents. Basic, or overbased, metal sulfonates have been used to both neutralize acidic by-products, resulting from the high-sulfur fuel combustion and lubricant decomposition, and to suspend deposit precursors in oil, thereby controlling the deposit formation. Detergents are especially useful in formulating lubricants for rail-road diesel and marine diesel engines that burn high-sulfur residual fuels. Sulfonate technology is relatively inexpensive because alkylbenzenesulfonates are used in many consumer products as surfactants. Under the present scenario, these must be replaced with more expensive carboxylate and phenate detergents. In addition to cost, there are health and environment concerns regarding the endocrine mimicking ability (hormone disruption) of alkylphenols, especially of octylphenol and nonylphenol. Hormone disruptor is a chemical that interferes with an animal's natural hormone function. The endocrine mimicking ability of certain alkylphenols can be explained by their conformational similarity to natural hormones [519], see Fig. 5.36. Although alkylphenols are on EPA's general list of endocrine disrupting industrial chemicals [520], endocrine mimicking data on dodecylphenol, 2,6-di-t-butylphenol, and their derivatives that are commonly used in lubricant applications, are either inconclusive or negative [521].

Table 5.42 qualitatively shows the effect of various additives on engine performance. The table is useful since some additives are multifunctional and provide benefits in different areas. Other additives are performance specific and may not perform the other functions well. Information in the table shows that the dispersant is excellent in controlling deposit formation in the gasoline engines, controlling soot in diesel engines, and does not contribute to the lubricant ash. Low-ash oils are important in diesel applications since they contribute less to the formation of the carbon deposits. Basic detergents are great on gasoline engine rust and in taking care of the fuel sulfur; both because of the base reserve. This becomes evident when one compares the performance of the high base detergents with that of the low base detergents in these areas. If the deposit control in the diesel engines is an objective, which is of course true, a combination of the low base detergents and the phenate is most beneficial. This is because of the potential synergy between the two types of additives: the phenate controls the formation of deposit precursors by providing high-temperature oxidation control, and the high soap in the low base detergent keeps any that form in the bulk lubricant. Phenate, being basic, both as a neutral salt and a basic salt, is extremely effective in taking care of the fuel sulfur-derived acids. Zinc dialkyl dithiophosphate is extremely potent as an oxidation inhibitor and as a wear and corrosion control agent in gasoline engine oils. Information in this table is invaluable to a formulator to help select proper additives.

TABLE 5.36—Two-cycle engine operating conditions [505].

Parameter	Outboard	Air-cooled
Speed	Constant	Variable
Physical Use	Smooth	Active
Average Piston Temperature	Moderate	Low
Peak Piston Temperature	Moderate	High

TABLE 5.37—TC-W3® tests [318]. Reprinted with permission from the Lubrizol Corporation.

Test	Pass/Fail Criteria
Compatibility	Homogeneous after being mixed separately with each reference oil and stored 48 hours
Brookfield (Fluidity) (cP) at −25°C	<7500
Miscibility at −25°C	No more than 10 % more inversions than reference
Rust (%)	Equal to or better than reference
Filterability	Decrease in flow not greater than 20 %
OMC 40-hp (98 h)	Average piston varnish and top ring stick ratings not lower than 0.6 below same ratings of reference
OMC 70-hp (100 h)	Average piston deposits and second ring stick ratings equal to or better than same ratings of reference
Mercury 15-hp (100 h)	No stuck rings; reference run
Scuffing (%)	every 5 candidate runs 30 max
Bearing Stickiness	Needles must fall easily from
Compression Loss (psi) max	wrist pin 20
Yamaha CE50S Tightening/Lubricity	Torque drop equal to or better than reference within 90 % confidence level
Yamaha CE50S Preignition (100 h)	Major preignitions equal to or better than reference

Fuel Economy

This topic was considered in detail while discussing friction modifiers in the Additives chapter. To summarize, automakers must meet the U.S. Government's Corporate Average Fuel Economy (CAFE) requirements for passenger cars and for light trucks, which include pickup trucks, minivans, and sport utility vehicles (SUVs) [401]. The objective of the fuel economy legislation is being met by the combined efforts of the automobile manufacturers, who are implementing a number of design changes to their equipment, and the engine oil formulators, who are developing energy-conserving

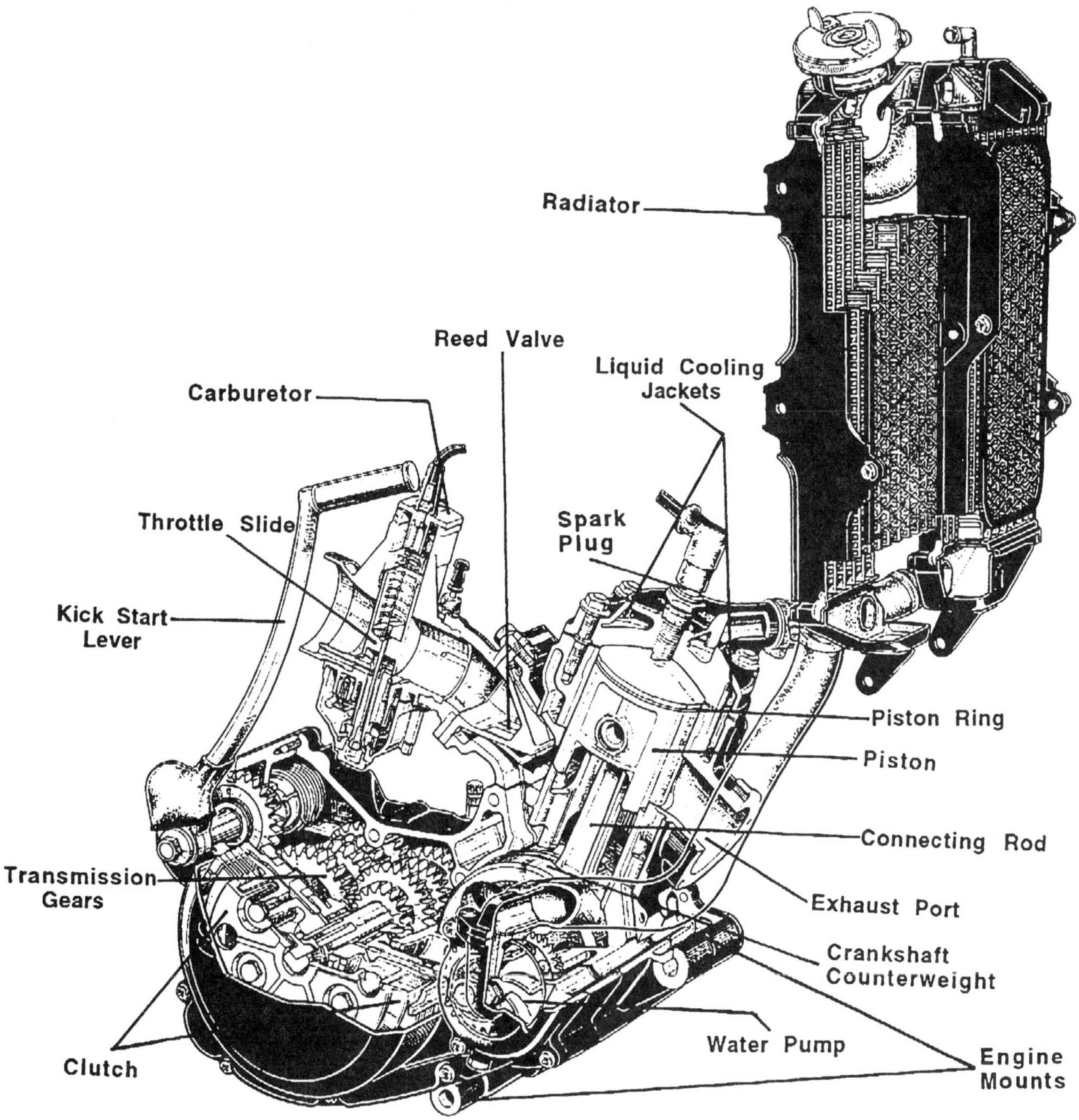

Fig. 5.18—Cross-sectional view of a liquid-cooled two-stroke cycle engine [508].

oils by the use of the low-viscosity base fluids and the friction-reducing additives. These oils must meet Sequence VIB fuel economy requirements of ILSAC GF-3/GF-4 specification. These oils must also demonstrate fuel economy durability.

Emissions Control

The topic of emissions control in engines is discussed in detail in Chapter 6, titled Emissions in an Internal Combustion Engine. To summarize, government-mandated emissions standards for both gasoline and diesel engines are met by the combined strategies of the fuel manufacturers, the OEMs, and the lubricant designers. The strategies for gasoline engines include altering fuel composition and properties, changing engine designs and operating variables, and the use of on-board diagnostics and after-treatment devices. Emissions control in diesel engines uses both prevention and correction strategies, the same as in gasoline engines. Prevention strategies include better fuel quality and delivery, increased injector and cylinder pressures, advanced combustion chamber and piston designs, and advanced timing. Correction strategies include a greater use of EGR, development of NOx reduction catalysts, and more effective particulate filters. Of course, lubricant also plays a role in lowering engine exhaust emissions by being compatible with the emissions control system and after-treatment devices, such

TABLE 5.39—Engine test part rating criteria.

Engine Test	Part	Primary Evaluation Criteria
CRC L-38	Bearing	Weight loss and discoloration
	Piston	Varnish deposits
Sequence IIIF/IIIG	Piston	Deposits, skirt varnish, and stuck rings
	Cam plus lifter	Wear
Sequence IVA	Camshaft	Lobe wear
Sequence VE	Piston	Skirt varnish, hot stuck rings, and oil ring clogging
	Rocker cover	Sludge
	Cam	Wear
	Oil screen	Clogging
	Engine	Sludge and varnish
Sequence VG	Piston	Skirt varnish, hot stuck rings, oil ring clogging, and ring gap increase
	Rocker cover	Sludge
	Cam	Wear
	Oil screen	Clogging
	Engine	Sludge and varnish
	Follower pin	Wear
Sequence VIII	Bearing	Weight loss and discoloration
Mercedes-Benz M111 Black Sludge	Pistons	Cleanliness, ring sticking, and ring wear
	Cylinder head	Sludge
	Rocker cover	Sludge
	Timing cover	Sludge
	Rocker cover	Sludge
	Oil sump	Sludge
	Cam and followers	Sludge and wear
Peugeot TU5JP-L4 High-temperature Deposits Test	Pistons	Lacquer, carbon, and ring sticking.
Caterpillar 1M-PC	Piston	Deposits, especially in the top groove, ring sticking, ring and liner scuffing
Caterpillar 1K/1N/1P/1R	Piston	Deposits, primarily in the top groove and on top land, and no stuck rings
	Liner	Scuffing
Mack T-8/T-9/T-11	Oil filter	Plugging
Mack T-10/T-12	Piston	Ring wear
	Cylinder liner	Wear
Cummins M-11 EGR	Crossheads	Weight loss
	Oil filter	Plugging
	Valve cover	Sludge
	Oil pan	Sludge
Two-stroke Cycle	Pistons	Deposits and scuffing (wear)

as catalytic converters and diesel particulate filters, and by suspending species such as soot in oil for the duration of the service interval. The compatibility of the lubricants with the emissions control devices was addressed in the previous section.

Extended Service Intervals (Extended Oil Drains)

As mentioned earlier, the OEMs and the end-users, especially diesel equipment operators, are greatly interested in extending the service intervals because it lowers their operating costs by reducing the downtime. There is an additional incentive and that is to lower the lubricant use so that millions of gallons less of used oil would require reprocessing or disposal, or both. Every organization that is involved in the design, service, or use has their own interpretation of the optimal service interval. However, oil drain intervals have been increasing in the light of the significant advances in the lubricant technology—both in the additive components and the base oils.

Service intervals for passenger cars and heavy-duty diesel engines are recommended by the OEMs. For North America, GM and Ford recommends oil service for passenger cars every 3000 to 7500 miles and Chrysler recommends service every 6000 miles. However, the typical service interval for GM vehicles is 6000 miles and for Ford vehicles it is 5000 miles. To be used in new cars, the oils must be of ILSAC GF-3 or GF-4 quality. Typical OEM service interval recommendations for use in Europe is between 9000 to 18,000 miles, with 10,000 miles being typical. For heavy-duty diesel engine oils of CH-4/CI-4 quality, Caterpillar recommends service every 200–250 hours of operation; Mack recommends service every 25,000 miles, which it downgraded from 40,000 for the pre-EGR engines; Cummins recommends service every 25,000 miles, or after 300 hours of operation; and Detroit Diesel recommends service every 22,500 miles, or after 300 hours of operation.

The lubricant suppliers tend to recommend longer intervals. This is partly to distinguish their product from that of the competition. The service companies, on the other hand, recommend a shorter interval, which is to sell their service

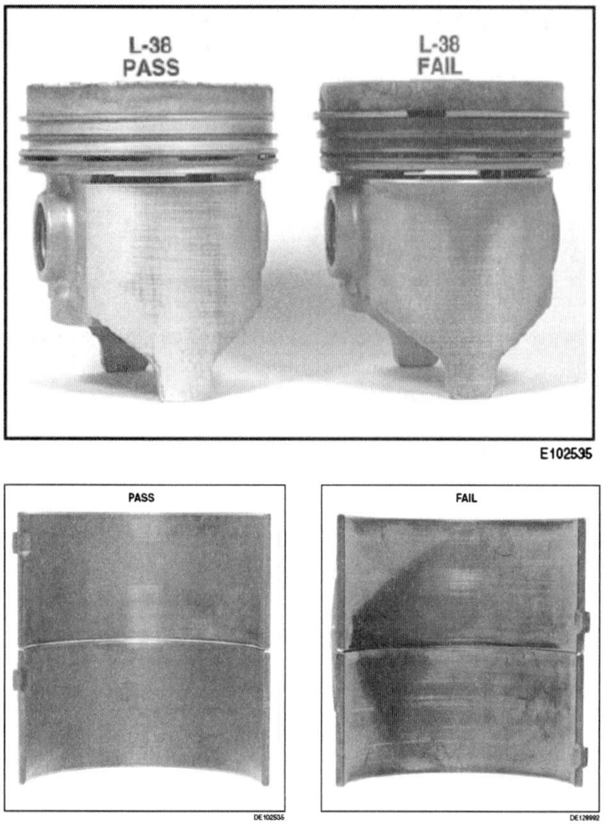

Fig. 5.19—Sequence L-38 engine test pistons and bearings [513].

more often. The OEMs tend to stay on the conservative side. Please note that the state of California and the GM consider the need for an oil service every 3000 miles a myth [522b]. Some OEMs are attempting to minimize consumer confusion by installing oil monitoring systems in their high-end automobiles. Obviously, environmental regulatory pressures will also influence whether the drain intervals increase or not, and of course in that case quick lubes, car dealerships, and the other retailers will comply.

There is on-going discussion among the OEMs and the users, especially the operators of the heavy-duty diesel vehicles to extend the service intervals. This is to minimize the yearly service cost and the downtime. However, there are a number of challenges that must be met prior to the extended service intervals becoming a reality. Almost all challenges relate to the longevity of the lubricant properties so as to provide the necessary protection to the engine during the extended service interval. If one wants to go beyond the OEM recommendations regarding the oil service for his or her vehicle, which is not advisable in the absence of oil analyses, one must try to estimate the proper drain interval. Estimation is easier for vehicles that are primarily driven on highways than those that are driven on the urban roads. This is because urban driving involves substantial stop-and-go service with long idle time, which greatly stresses the lubricant with respect to oxidation and contamination. Typically, service intervals for vehicles involving urban driving are shorter than the less-severe highway driving. This is because in urban driving, engine operating temperatures are well below the boiling point of water, which collects in the sump catalyzing lubricant degradation. However, the engine operating temperature in typical highways driving is around 220°F (110°C) and water collection does not occur, neither does the related lubricant degradation. Estimation of the oil drain intervals in off-highway engines is even more challenging. This is because for these engines, operating severity varies greatly across engines and applications. They have widely differing operating parameters and severity. Fuel consumption rate, one of the parameters that plays a dominant role in

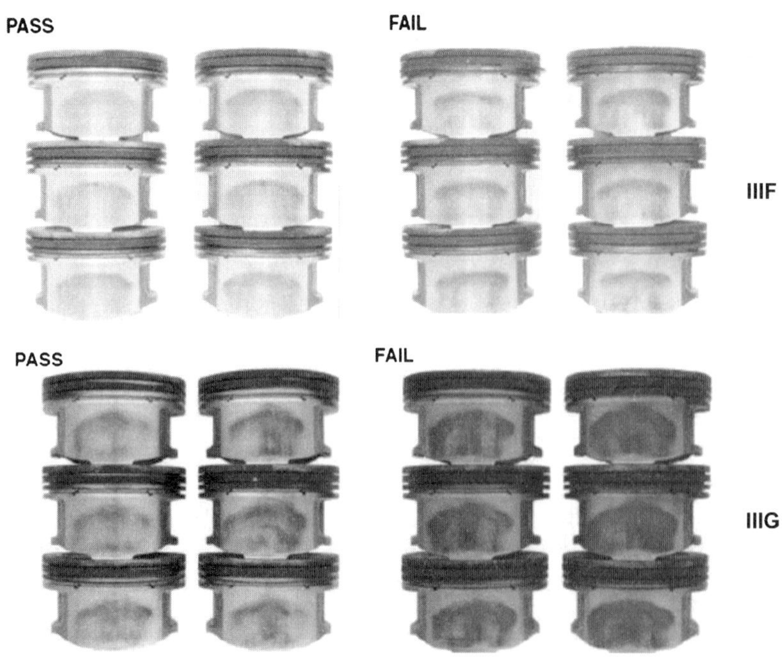

Fig. 5.20—Pistons from Sequence IIIF and IIG tests [513].

determining the oil drain intervals can vary by as much as 30 %.

Higher quality oils maintain their performance for an extended period; hence service intervals for such oils are generally longer. As exemplified in Fig. 5.37, an improvement in the oil quality is accompanied by an increase in the oil drain interval. This is because each oil upgrade approximately imparts 10 to 20 % improvement over the previous performance category, which makes the oil last longer. Factors other than oil quality that affect oil drain intervals include engine-related factors that are addressed below.

1. Increased piston deposits
2. Increased oil soot levels leading to an increase oil viscosity and wear.
3. Reduced bearing life due to corrosion/seizure, resulting from increased level of abrasives, fuel, and the coolant in the oil.
4. No allowance for a missed drain, which is likely to result in catastrophic damage to the engine.
5. Increased accumulation of the wear metals in the oil, causing increased abrasive wear and oil oxidation.
6. Elastomer seal damage due to particle accumulation at the seal lip.
7. Loss of fuel economy, especially in passenger cars.

The critical question that needs to be answered is "What is the optimal length of the extended service interval?" To answer this question, besides lubricant quality and the factors listed above, there are additional factors that need to be considered. These include the rate of the fuel use, oil consumption, and the engine sump size [522]. The OEMs use these parameters to determine the oil service interval, see Fig. 5.38 [522]. In the figure, MPQ represents oil consumption per quart. For an engine with a sump size of eleven gallons, fuel consumption rate of 5.5 miles per gallon, and oil consumption of one quart for every 1000 miles, the recommended service interval is 19,250 miles. On the other hand, for an engine of the same oil consumption and the same sump size that has a lower fuel consumption rate, that is, travels 6.5 miles per gallon, the service interval extends to 23,000 miles. The reason for this advantage is that an engine with lower fuel consumption generates less heat, which results in lower oil oxidation, hence extends the oil's useful life, and has lower piston deposits. Similarly, the service interval can be extended for engines that consume less oil. On the chart this is exemplified for the engine that has a fuel consumption rate of 5.5 miles per gallon and a high oil consumption rate of 600 miles per quart. In this case, the service interval extends to 22,000 miles. Here the advantage is due to the constant replenishment with fresh oil and additives.

The presence of high levels of soot in the oil not only leads to a viscosity increase but also causes wear of the valve train components and a loss in fuel economy. Once the oil loses its ability to suspend soot, which is likely to occur if the service interval is too long, the additional soot will be released into the environment. As a general rule of thumb, doubling the service interval will double the amount of soot in the oil [522]. Typically, an oil filter is used to remove particulates and metal debris to prevent abrasive wear, but if the soot level in the oil is too high, it will form sludge on the filter, rocker covers, and cylinder head deck. Once the filter is clogged, the oil will by-pass the filter and cause wear of all of the engine components, leading to serious damage. It is therefore imperative that the engine service interval is not extended beyond the oil's capacity to effectively suspend soot.

Bearing life is reduced by factors such as fuel dilution, coolant leaks, abrasive wear, and corrosion. Extending the service interval increases the chances of accumulation of these undesirables in oil. Fuel and coolant dilution of the oil deteriorates its ability to form an effective boundary lubrication film between the bearing and the shaft, which results in the removal of the soft metal bearing overlay and causes damage to the bearing linings. Bearing overlay performs a number of important functions, which include conformity, embedding capacity, seizure resistance, and corrosion resistance. Damage to the bearing overlay will ultimately lead to the bearing failure.

Another concern with respect to extending the drain interval is that a scheduled service cannot be missed. This is because the OEM recommended intervals are conservative, which is to promote maximum engine durability. Extending the drain intervals to the lubricant's ultimate performance limit does not allow for missing a service; otherwise catastrophic damage to the engine will occur.

Extending the oil drain interval leads to greater accumulation of the wear metals in oil, the amounts of which are directly related to the miles driven. The presence of increased metal debris in oil with a long service leads to a rapid increase in engine wear, which is shown in Fig. 5.39 [522]. The base number, on the other hand, is inversely related to the miles driven: it decreases with an increase in miles, as is shown in Fig. 5.40 [522]. If the base reserve of the oil drops too low, the oil can lose its acid-neutralizing ability, which will be evidenced by the corrosive wear of the rings, liner, and the bearings. It is imperative that the oil's base number does not drop below 3.0, as measured by ASTM D2896, or 1.5, as measured by ASTM D4739. The rate of the base number loss depends upon the fuel sulfur as well as oil consumption; high sulfur fuel and low oil consumption increases this rate and low sulfur fuel and high oil consumption decreases this rate. This is shown in Fig. 5.41 [522].

As mentioned earlier, extending the oil drain interval leads to an increase in oil contamination, which in conjunction with the loss in the oil's base reserve can result in engine power loss. This is shown in Fig. 5.42 [522]. As one can see, extending the oil drain interval from 12,000 miles to 25,000 resulted in a significant power loss, which is believed to be due to the increased ring wear that increases the blow-by [522]. Data plotted in the figure are from field testing, using the 430 hp engines.

Extending the service interval may result in the formation of increased deposits that can result in stuck compression rings, which will cause an increase in the blow-by and hence the loss of power. Stuck rings are the consequence of the depletion of the oxidation inhibitors and detergents that minimize deposit formation. If this problem is to be avoided, it is necessary to replenish these additives by periodically adding fresh oil. Crankshaft seal leakage is a potential problem that may occur if the oil drain intervals are extended. This is due to particle accumulation at the seal lip, which can accelerate wear between the seal and the crankshaft.

As depicted in Fig. 5.40, extending the service interval

TABLE 5.40—API base oil categories [515].

Base Oil	Saturates	Sulfur	Viscosity Index (VI)	Manufacturing Method
Group I	<90 %	and or >0.03 %	≥80 and <20	Solvent Refining
Group II	≥90 %	≤0.03 %	≥80 and <20	Hydroprocessing Refining
Group III	≥90 %	≤0.03 %	≥120	Hydroprocessing Refining
Group IV	Restricted to Polyalphaolefins (PAO)			Chemical Reaction
Group V	All other base oils not included in Group I, II, III, IV			Chemical Reaction

Note: Analytical Methods—ASTM D2007 for Saturates; ASTM D2270 for Viscosity Index; and ASTM D1552, D2622, D3120, D4284, D4927 for Sulfur.

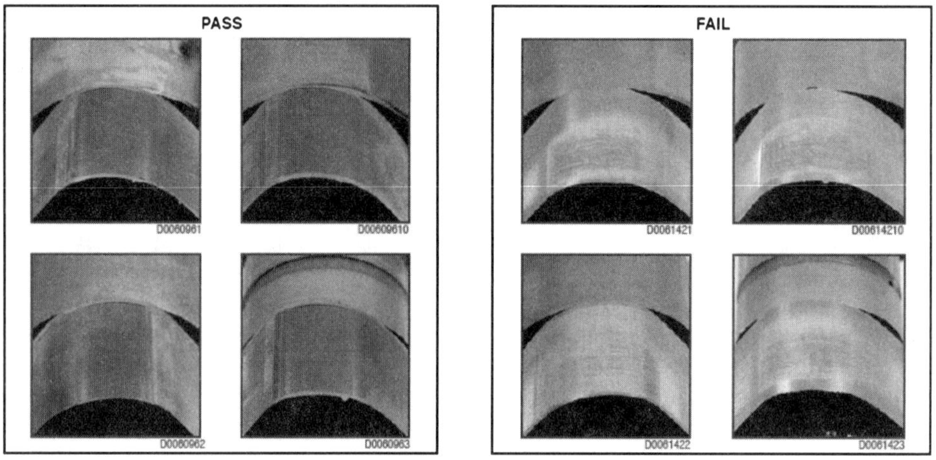

Fig. 5.21—Sequence IVA engine test parts [513].

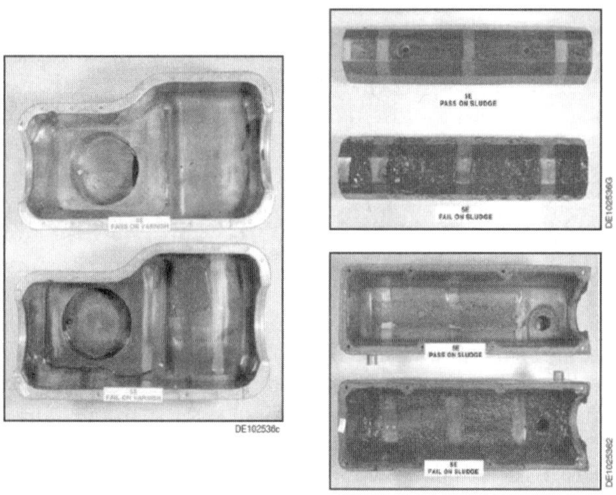

Fig. 5.22—Sequence VE engine test parts [513].

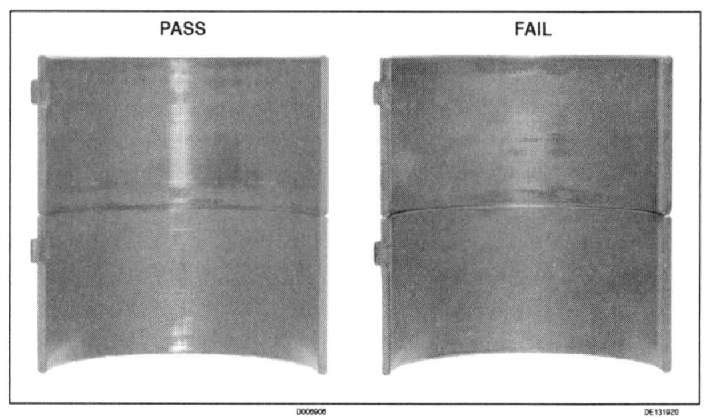

Fig. 5.24—Sequence VIII engine test parts [513].

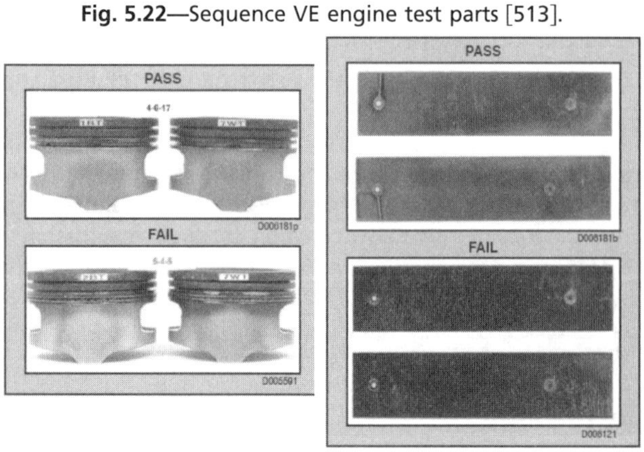

Fig. 5.23—Sequence VG engine test parts [513].

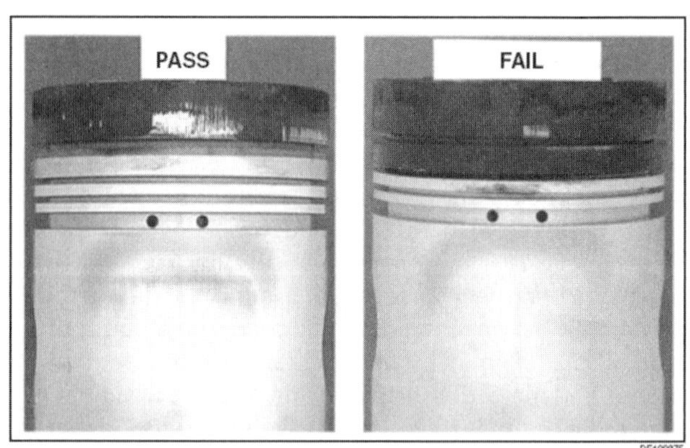

Fig. 5.25—Caterpillar 1M-PC engine test pistons [513].

TABLE 5.41—Base oil properties comparison [517].

Base Oil Category	API Group I	API Group II	API Group III	API Group IV
Viscosity at 100°C, cSt	4.1	4.1	4.1	3.9
Viscosity Index	99	104	124	124
Pour Point, °C	−18	−16	−19	−73
CCS at −25°C	1430	1440	900	360
Viscosity (Brookfield) at −40°C, cP	Solid	Solid	7000	2600
Noack Volatility at 150°C, v%	28	27	14	12
Oxidation Stability	Fair	Good	Very Good	Very Good
Deposit/Sludge Control	Good	Very Good	Excellent	Excellent
Soot Control	Fair	Very Good	Very Good	Very Good
Relative Cost	1	1	2–3	4–5

[a]With Pour Point Depressant

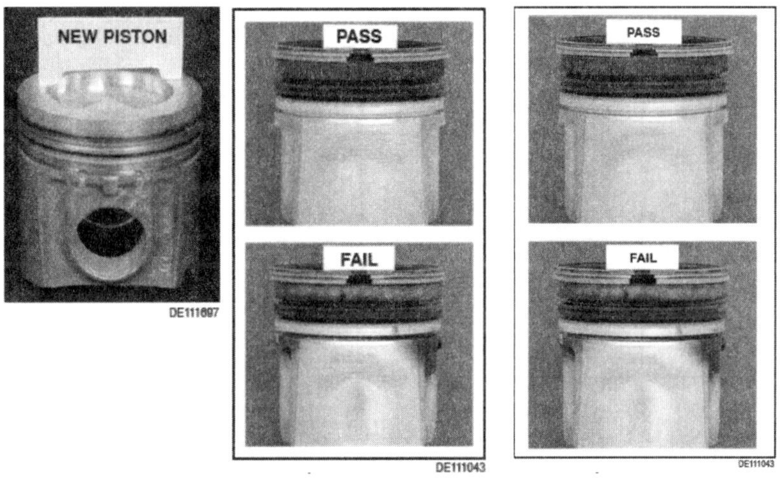

Fig. 5.26—Caterpillar engine test pistons [513].

can cause a drop in the lubricant's base number to dangerously low levels. At this time, the lubricant's acid-neutralizing ability will diminish, which will be reflected by the corrosive wear of the rings, liner, and bearings and the increased deposit formation. Replenishing the lubricant's base reserve may help correct these problems. The Cummins Centinel Advanced Oil Management System accomplishes precisely this and in an ingenious fashion. This system is an embedded system that extends oil change intervals on electronically controlled diesel engines by periodically removing a small amount of the used oil from the engine's crankcase and replacing it with the fresh oil. The used oil is sent to the engine's fuel tank, where it is blended with the fuel and burned during the normal combustion. The periodic addition of the fresh oil not only replenishes the depleted base reserve but also introduces more dispersant into the oil. The combined effect is the control of soot in oil as well as a decrease in the soot-related wear and the acid-initiated corrosion. The effects of the oil-replacing strategy are shown in Figs. 5.43 and 5.44 [523–525].

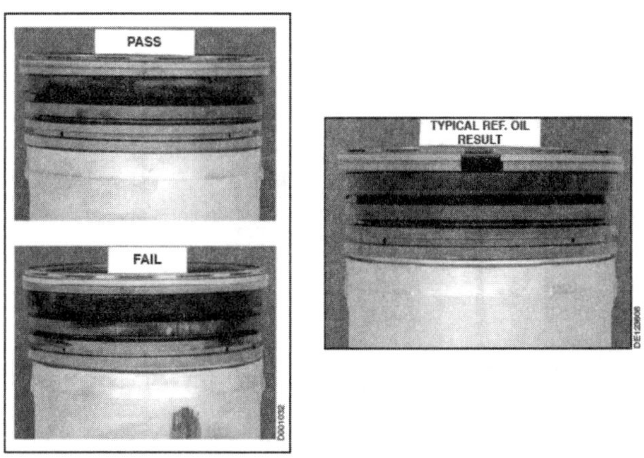

Fig. 5.27—Caterpillar 1P engine test pistons [513].

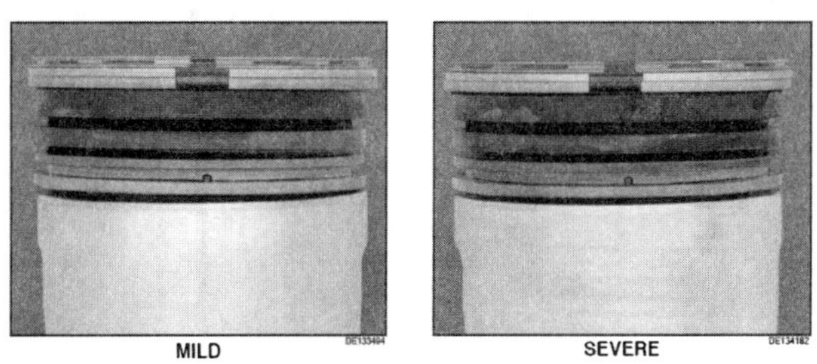

Fig. 5.28—Caterpillar 1R engine test pistons [513].

Fig. 5.29—Caterpillar C13 engine pistons [513].

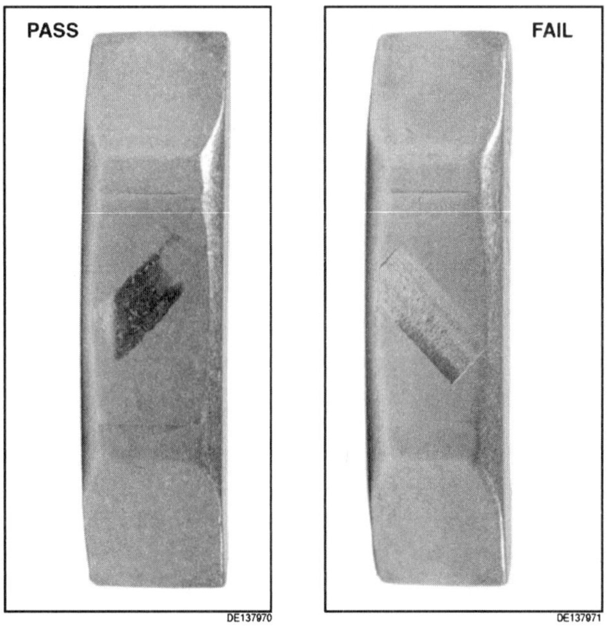

Fig. 5.30—Cummins M-11 engine test parts [513].

Another issue that will impact the drain intervals is the EPA's 2007 emissions regulations. While new engine oils are expected to be backward-compatible with earlier engines, the lubricant engineers are uncertain about where the drain intervals will wind up. They may or may not be the same for the 2007 and pre-2007 engines. Backward compatibility is good news for fleet operators since they will be able to use the new CJ-4 oils in their pre-2007 trucks without fear of the reduced protection from wear, corrosion, and piston deposits. The general feeling regarding the drain intervals is that the drain intervals for the pre-2007 engines should be different—at least initially they should be close to the original OEM recommended levels. This is because the new oils have lower chemical limits to protect the diesel particulate filters (DPFs) against damage. The CJ-4 quality oils have a sulfur content of 0.4 %; phosphorus content of 0.128 %; ash formation of 1.0 %; and the oil volatility of 13 %. The major uncertainty relates to the TBN of 8 or 9 compared to the TBN in the range of 11–13 for the CI-4 oils. This drop in the base reserve may require conservative oil drain intervals since these oils have diminished acid neutralization capability, which implies that the oil cannot last too long. Then of course the use of the ultra low-sulfur fuel may not deplete the lubricant TBN as fast since the oil needs to counteract a smaller amount of sulfur acids. While the OEMs and the manufacturers of the heavy-duty diesel engine oils are targeting a drain interval of 25,000 miles, the actual duration is hard to judge without field experience. In the absence of field experience, it is prudent to stick to the OEM recommended service intervals. Incidentally, the actual oil drain intervals for heavy-duty diesel engines in the United States range between 15,000 to 60,000 miles. Another issue related to the drain intervals is that the 2007 model engines are going to experience an increase in the operating temperatures, the effect of which on the CJ-4 oil life is uncertain. Significant increase of EGR in the 2007 engines will also increase the soot level for the oil to handle. In addition, increased EGR and DPF are going to put additional load on the engines, which will increase both the lubricant and the coolant temperatures and again their effect on the drain interval is uncertain. It is pertinent to mention that Volvo for its 2007 engines has established the oil drain intervals based upon the duty cycle: up to 30,000 miles for D11; up to 45,000 miles for D13; and, up to 50,000 miles for D16.

Synthetic oils with greater resistance to oxidative and thermal degradation are expected to outperform mineral oils with respect to the extended drain intervals. Although the use of the synthetics is growing, there are a number of issues that are a hurdle to their widespread use in engine oil formulations. These include limited additive solubility (PAOs), hydrolytic stability (synthetic esters), seal incompatibility with some seal materials, and higher cost. Polyalpha-

Fig. 5.31—Cummins ISM test parts [513].

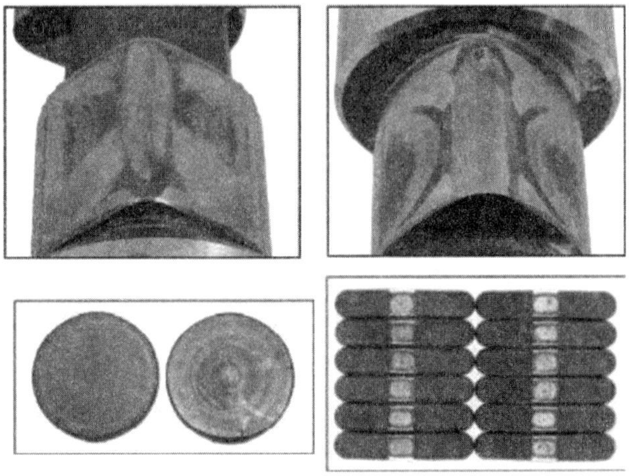

Fig. 5.32—Cummins ISB engine test parts [513].

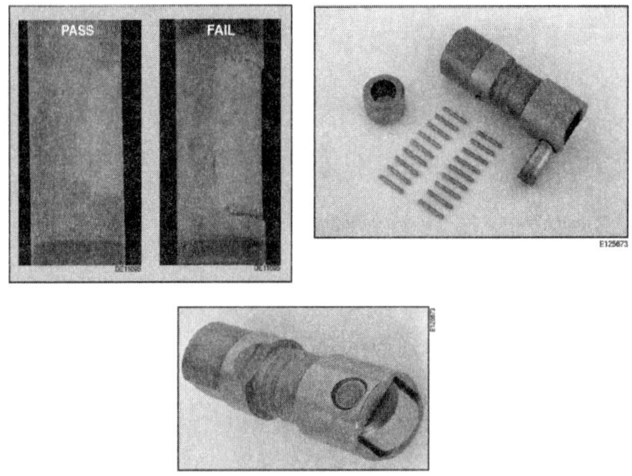

Fig. 5.33—Engine test parts from the roller-follower test [513].

olefins (PAOs) and synthetic esters have been used to formulate engine oils. These oils are proposed to have a service life of 25,000 miles. Hydrocracked oils (API Group III oils) are similar in performance to the PAOs and are only two to three times as expensive as mineral oils. Because of this, they are cheaper substitutes to PAOs in formulating engine oils. The use of these oils is likely to increases as automobile manufacturers strive for the 25,000 to 35,000 mile, or one-year, service interval. A number of lubricant suppliers are using more severely hydrocracked Group III base oils, called Very High Viscosity Index (VHVI) oils. They have a viscosity index of greater than 120 and are expected to outperform other types of mineral oils at high temperatures.

Formulation Examples

Table 5.43 lists additives that are used to formulate engine oils for various applications. Please note that all types use deposit control agents, i.e., dispersants, detergents, and oxidation inhibitors. This is expected since the lubricant oxidation and deposit formation in combustion engines is a chronic problem which must be controlled. Other additives that are present in all engine oil types are the antiwear agents and corrosion inhibitors. Friction modifiers are only added to the gasoline engine oil formulations to meet the government-mandated Corporate Average Fuel Economy (CAFE) requirements. Pour point depressants are added to all oils that are designed for use in cold climates, except to those for small two-stroke cycle engines without a sump. Viscosity modifiers are added only to automotive engine oils to make them multi-grade. These also contain foam inhibitors since they accumulate a significant amount of combustion water, which causes foam formation. Generic formulations of commonly used engine oils are provided below.

Automotive Engine Oils

These oils are formulated to possess oxidation resistance, keep engine parts such as pistons, rings, and liners clean, and protect the engine against wear, rust, and corrosion. In addition, these oils must have proper viscometrics to permit easy starting at low temperatures and to lubricate effectively at high temperatures. These oils are formulated by the use of both mineral and synthetic base stocks, primarily PAOs and synthetic esters, and contain oxidation inhibitors, dispersants, detergents, antiwear agents, rust and corrosion inhibitors, viscosity improvers, and friction modifiers.

The formulation examples are provided below.

Gasoline Engine Lubricant: 5.0 % Alkenylsuccinimide dispersant, 2.8 % basic calcium and magnesium sulfonates mixture detergent, 1.4 % primary and secondary alcohol-derived zinc dialkyl dithiophosphates mixture antiwear agent, 0.45 % phenol and arylamine mixture oxidation inhibitor, 6.5 % olefin copolymer viscosity modifier, and 10 ppm methylsilicone foam inhibitor—the balance is base oil.

Gasoline Engine Lubricant: 5.0 % Alkenylsuccinimide or Mannich dispersant, 1.6 % basic calcium and magnesium sulfonates mixture detergent, 1.8 % neutral sulfonate (soap), 1.2 % primary and secondary alcohol-derived zinc dialkyl dithiophosphates mixture antiwear agent, 1.2 % phenol and arylamine mixture oxidation inhibitor, 0.15 % corrosion inhibitor, 0.05 % sorbitan oleate/s friction modifier, 0.25 % polyethoxylated polyol demulsifier, 6.5 % olefin copolymer viscosity modifier, and 10 ppm methylsilicone foam inhibitor—the balance is base oil (formulation extracted from Ref 526).

Gasoline Engine Lubricant: 2.3 % Alkenylsuccinimide dispersant, 0.89 % basic calcium sulfonates mixture detergent, 0.88 % primary and secondary alcohol-derived zinc dialkyl dithiophosphates mixture antiwear agent, 0.90 % phenol and arylamine mixture oxidation inhibitor, 0.2 % sulfurized olefin oxidation inhibitor/antiwear agent, 0.2 % carboxylic acid-derived friction modifiers, 0.95 % olefin copolymer viscosity modifier, 0.09 % pour point depressant, and 90 ppm methylsilicone foam inhibitor—the balance is API Group II base oil. For Group I base oil, an additional amount to 0.36 to 1.2 % molybdenum dithiocarbamate oxidation inhibitor was required to obtain a pass in the Sequence IIIF test (formulation extracted from Ref 527).

Automotive Diesel Engine Oil: 5.5 % Alkenylsuccinimide dispersant, 5.5 % basic calcium and magnesium phenates and sulfonates mixture detergent, 1.0 % primary and secondary alcohol-derived zinc dialkyl dithiophosphates mixture antiwear agent, 0.6 % hindered phenol and sulfide/polysulfide mixture oxidation inhibitor, 0.3 % polymethacrylate pour point depressant, 6 % olefin copolymer or poly-

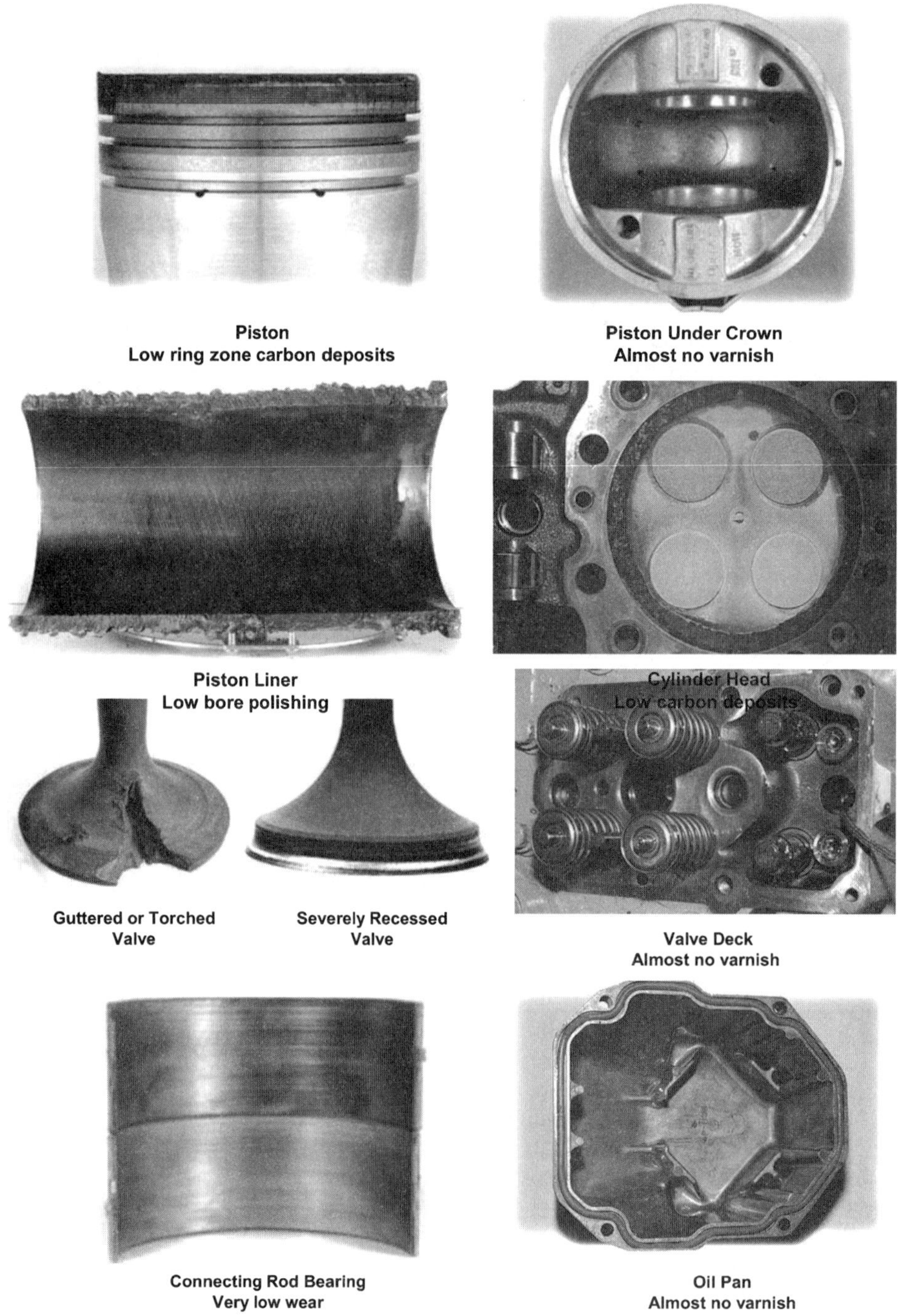

Fig. 5.34—Natural gas engine parts [498,514].

methacrylate viscosity modifier, and 30 ppm methylsilicone foam inhibitor—the balance is the base oil.

Automotive Diesel Engine Oil: 5.5 % High-molecular weight alkenylsuccinimide dispersant, 1.4 % mixture of basic calcium and magnesium sulfonate detergents, 0.85 % methylene-bridged calcium phenate, 1.3 % primary and secondary alcohol-derived zinc dialkyl dithiophosphates mixture antiwear agent, 1.6 % alkylated hindered phenol, 0.2 % polymethacrylate pour point depressant, 6.5 % olefin copolymer or polymethacrylate viscosity modifier, and 80 ppm methylsilicone foam inhibitor, 0.35 % alkenylsuccinic acid rust inhibitor, 0.15 % polyethoxylated phenol demulsifier—the balance is base oil (formulation extracted from Ref 528).

Stationary Diesel Engine Oil: Formulations analogous to those of the automotive diesel engine oil.

Railroad Engine Oils

These oils are formulated to possess oxidation resistance, protect rings and liners against scuffing, deposit control, rust and corrosion protection, and cold weather startability. They must also be zinc-free to minimize corrosion of the silver (silver-steel alloy) bushings or bearings that are used in the high-performance engine designs. These lubricants have a high TBN (13 or 17), to counter the acid resulting from the

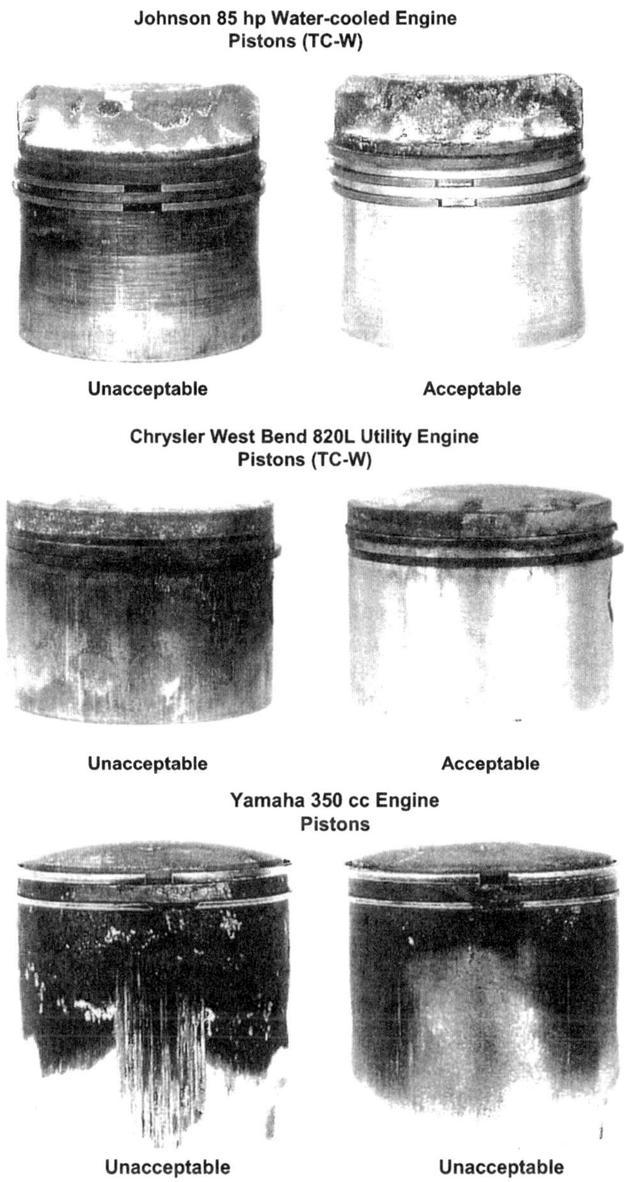

Fig. 5.35—Two-stroke cycle SI engine test pistons [318]. Reprinted with permission from the Lubrizol Corporation.

use of the high sulfur fuel, and dispersancy, which is attained by the use of detergents and dispersants. These lubricants also contain antiwear/extreme pressure agents, oxidation inhibitors, corrosion inhibitors, and foam inhibitors. While in the example formulations, chlorinated paraffins are used as EP agents, some suppliers are replacing them with the sulfur-based additives to facilitate easy disposal of the used lubricant.

Railroad Diesel Oil: 2.0 % Alkenylsuccinimide dispersant, 5.0 % basic calcium phenate detergent, 0.3 % chlorinated paraffin extreme pressure agent, 0.2 % hindered phenol oxidation inhibitor, 0.2 % triazole metal deactivator, and 10 ppm methylsilicone foam inhibitor—the balance is the base oil.

Railroad Diesel Oil: 3.3 % Alkenylsuccinimide dispersant, 7.0 % basic calcium phenate detergent, 0.05 % polymethacrylate pour point depressant, 0.05 % 2,5-dimercapto-1,3,4-thiadiazole metal deactivator, and 150 ppm methylsilicone foam inhibitor—the balance is the base oil (formulation extracted from Ref 529).

Commercial Railroad Diesel Engine Oil Concentrates

OLOA 2939 Composition: 45 % Overbased mixed calcium sulfonate/phenate detergent, 10 % polyisobutenylsuccinimide dispersant, 1.5 % polyisobutylene, 0.5 % chloroparaffin, and 43 % mineral oil.

AMOCO 6555: 28 % Overbased mixed calcium sulfonate/phenate detergent, 10 % polyisobutenylsuccinimide dispersant, 3.0 % polyisobutylene, 1.0 % chloroparaffin, 6 % branched alkylphenol, and 52 % mineral oil (formulation extracted from Ref 530).

These concentrates can be blended with appropriate base stocks to obtain finished lubricants of desired TBN and other properties for use as railroad, marine, and stationary engines. In some cases, supplemental additives, such as the antiwear agents, oxidation inhibitors, demulsifiers, and rust inhibitors may also be needed.

Marine Diesel Engine Oil

As stated earlier, marine diesel engines use three different lubricants: *cylinder oil, systems oil*, and the *trunk-piston engine oil*. The marine cylinder lubricants have a high TBN, usually between 60 and 100; typically 70, to neutralize the acidic combustion products arising from the use of the high-sulfur fuel. The systems oils have a TBN of only 2 to 10 and are formulated to handle rust, oxidation, deposits, and wear. The trunk-piston engine oils have a TBN of 12 to 40 and contain additives to provide extreme pressure/antiwear performance, high-temperature stability, oxidation resistance, detergency, and dispersancy. The additives used to formulate these oils include detergents, dispersants, oxidation inhibitors, antiwear/extreme pressure agents, rust and corrosion inhibitors, demulsifiers, and foam inhibitors. Representative formulations for these oils are provided below.

Cylinder Oil (80 TBN): 16.0–18.0 % Basic calcium sulfonate and 11.0–13.0 % basic calcium phenate mixture detergent, and 0.3–0.6 % zinc dialkyl dithiophosphate antiwear agent—the balance is the base oil.

Systems Oil (2-3 TBN): 0.8–1.0 % Basic calcium phenate detergent, 0.2–0.3 % zinc dialkyl dithiophosphate antiwear agent, and 10 ppm silicone foam inhibitor—the balance is the base oil.

Trunk-piston Engine Oil (30 TBN): 7.0–9.0 % Basic calcium sulfonate and 3.0–4.0 % basic calcium phenate mixture detergent, 0.2–0.4 % zinc dialkyl dithiophosphate antiwear agent, 0.3–0.4 % phenol oxidation inhibitor, 0.8–1.0 % triazole metal deactivator, and 10 ppm silicone foam inhibitor—the balance is the base oil.

Trunk-piston Engine Oil (30 TBN): 10 % 168 TBN basic calcium salicylate, 5.65 % 280 TBN basic calcium salicylate, 0.61 % C_8 alcohol-derived zinc dialkyl dithiophosphate antiwear agent, 10 ppm silicone foam inhibitor—the balance is the base oil (formulation extracted from Ref 531).

Stationary Gas Engine Oils

These oils are formulated to possess oxidation and nitration resistance, detergency for engine cleanliness and bearing corrosion, deposit control, soot dispersancy, and wear protection. To achieve detergency, the engine oils use a neutral or basic metal sulfonate and a phenate or a salicylate. However, in natural gas engine oils metal phenate is preferred over metal sulfonate because of its superior ability to control lubricant oxidation and deposit control. However, its acid neutralizing ability and hence rust and corrosion control is not very good. In most cases, the use of a rust inhibitor is required. Metal salicylate, in view of having phenol functional group, has excellent oxidation control. It also has a basic carboxylate functional group, which greatly helps in controlling rust and corrosion. Phenol and carboxylate functionalities are next to each other; that is, they have ortho orientation, which provides salicylate with great complexing ability that assists in good deposit control [532]. Examples of the stationary gas engine oil formulations are provided below.

Ashless Stationary Gas Engine Oils: 3.5 % Alkenylsuccinimide dispersant, 0.5–1.0 % tricresyl phosphate antiwear agent, 0.3–0.5 % arylamine oxidation inhibitor (NOx-inhibiting agent), 0.5–0.8 % triazole metal deactivator, and 5 ppm silicone foam inhibitor. The balance is the base oil.

Ash-producing Stationary Gas Engine Oil: 3.0–3.5 % Alkenylsuccinimide dispersant, 4.0–5.0 % calcium sulfonate and phenate mixture detergent, 0.2 % tricresyl phosphate antiwear agent, 0.4–0.5 % arylamine oxidation inhibitor, 0.5–0.6 % triazole metal deactivator, and 5 ppm silicone foam inhibitor. The balance is the base oil.

Aviation Engine Oils

These lubricants fall under two general categories: lubricants for *jet turbine engines* and lubricants for *aircraft piston engines*. Jet turbine oils possess excellent thermal and oxidative stability, load-carrying capacity, deposit control, and anti-corrosion protection. They are based upon synthetic ester base stocks and contain oxidation inhibitors, antiwear agents, and corrosion control additives. Piston engine lubricants are designed to prevent wear of the engine parts, keep abrasives and deposit-forming materials in suspension, dissipate heat, and seal off blow-by of the combustion chamber gases into the crankcase. These oils are based upon highly refined mineral oils, synthetics, or semi-synthetics and contain dispersants, antioxidants, antiwear agents, and rust and corrosion inhibitors. Esters are superior to mineral oil-containing formulations with respect to low temperature viscometrics, viscosity index, volatility, and the Ryder gear test (EP) performance [214]. For turbo-propeller use, viscosity modified diesters are used. However, for turbofan and turbojet applications, polyol esters without the viscosity modifiers are preferred. Turbofan lubricants employ high viscosity esters and turbojet lubricants employ low to medium viscosity esters. Representative formulations for the aviation engine oils are provided below.

Jet Turbine Oil: 1.0–3.0 % Tricresyl phosphate and a proprietary additive mixture antiwear agent, 1.0–3.0 % hindered phenol oxidation inhibitor, and 0.5–1.0 % quinizarin metal deactivator. The balance is the synthetic ester base stock.

Ashless Piston-engine Oil: 2.0–3.0 % Polymethacrylate-based dispersant, 0.5–1.0 % tricresyl phosphate antiwear agent, and 0.3–0.5 % hindered phenol oxidation inhibitor. The balance is the base oil.

Small Engine Oils

Small engines are both the two-stroke cycle and the four-stroke cycle types. Two-stroke cycle engine oils are formulated to reduce smoke, prevent ring sticking, and improve cleanliness and lubricity. These oils are mineral oil-based and contain additives, such as dispersants, oxidation inhibitors, antiwear agents, and corrosion inhibitors. Four-stroke cycle engine oils are formulated to improve cleanliness, wear, corrosion, oxidation stability, and the clutch friction performance. These lubricants are formulated using detergents, dispersants, rust inhibitors, and oxidation inhibitors. The functions performed by these additives are provided in Table 5.44 [533]. Representative formulation examples for these oils are provided below.

Ashless Two-stroke Cycle Engine Oil: 3.0–5.0 % Alkenylsuccinimide dispersant, 0.1–0.2 % sulfurized fat antiwear agent, 0.2–1.0 % hindered phenol oxidation inhibitor, 0.5–1.0 % fatty acid and alkenylsuccinic acid-based corrosion inhibitor, 0.1–0.2 % long-chain alcohol coupler or emulsifier, 2.0–2.5 % Stoddard solvent, and 0.01–0.05 % blue or purple dye. Stoddard solvent is added to the formulation to enhance fuel compatibility. Sometimes the formulation contains high molecular weight amines that act cleanliness aids. The balance is the base oil.

Ash-producing Two-stroke Cycle Engine Oil: 1.0–1.5 % Neutral calcium sulfonate or fatty acid-based corrosion inhibitor, 0.1–0.2 % long-chain alcohol coupler or emulsifier, 1.5–2.0 % Stoddard solvent, and 0.01–0.05 % blue or purple dye. The balance is the base oil.

Low-ash Two-stroke Cycle Engine Oil: 4.75 % Alkenylsuccinimide dispersant, 3.0–4.5 % alkylphenol—alkaline earth metal alkylsalicylate mixture detergent, 0.25 % sulfur-free organo-molybdenum derivative friction modifier/detergent, 3 % polyisobutylene lubricity agent, 0.5 % fatty acid imidazoline, 0.2 % long-chain alcohol coupler or emulsifier, 2.0–2.5 % kerosine, and 0.01–0.05 % blue or purple dye. The balance is the base oil. Kerosine is added to the formulation to enhance fuel compatibility (formulation extracted from Ref 534).

6

Emissions in an Internal Combustion Engine

UNDESIRABLE EMISSIONS IN INTERNAL COMBUS-tion engines are of major concern beause of their negative impact on air quality, human health, and global warming. Therefore, there is a concerted effort by most governments to control them. Undesirable emissions include unburned hydrocarbons (HC), carbon monoxide (CO), nitrogen oxides (NOx), and particulate matter (PM). In this chapter, we present the U.S. and European emissions standards, both for gasoline and diesel operated engines, and strategies to control the undesirable emissions. The role of engine design, vehicle operating variables, fuel quality, and emission control devices in minimizing the above-listed pollutants are also detailed in this chapter.

"Emissions" is a collective term that is used to describe the undesired gases and particles which are released into the air or emitted by various sources. Its amount and the type change with a change in the industrial activity, technology, and a number of other factors, such as air pollution regulations and emissions controls [535].

The U.S. Environmental Protection Agency (EPA) is primarily concerned with emissions that are or can be harmful to the public at large. EPA considers carbon monoxide (CO), lead (Pb), nitrogen dioxide (NO_2), ozone (O_3), particulate matter (PM), and sulfur dioxide (SO_2) as the pollutants of primary concern, called the *Criteria Pollutants*. These pollutants originate from the following four types of sources.

1. Point sources, which include facilities such as factories and electric power plants.
2. Mobile sources, which include cars and trucks but also lawn mowers, airplanes, and anything else that moves and releases pollutants into the air.
3. Biogenic sources, which include trees and vegetation, gas seeps, and microbial activity.
4. Area sources, which consist of smaller stationary sources such as dry cleaners and degreasing operations.

The primary drivers that are used to maintain or improve the country's air quality include the Clean Air Act (CAA), Office of Air and Radiation (OAR) Rules, and Air Toxics Rules. We will primarily focus on pollutants resulting from the mobile sources since they are most pertinent to our discussion [536].

Air pollution from the mobile sources occurs through two processes: combustion and fuel evaporation. The pollutants that are considered a significant threat to human health and the environment are carbon monoxide (CO), hydrocarbons (HC), nitrogen oxides (NOx), and particulate matter (PM). In addition to these, there are other air pollutants, such as air toxics and greenhouse gases, which are also of concern because of their indirect threat to the environment. In the United States, mobile sources are the largest contributor to the air toxics. Air toxics consist of pollutants that are either known or the suspected carcinogens, or are a serious hazard to the public health or the environment. These include materials, such as benzene and formaldehyde, and greenhouse gases, such as carbon dioxide (CO_2), that trap heat in the Earth's atmosphere, thereby leading to the global climate change.

Carbon monoxide is a colorless, odorless, poisonous gas that is a major air pollutant in most urban areas. It results from the incomplete combustion of the hydrocarbon fuel. According to the EPA studies, up to 95 % of the carbon monoxide is due to vehicle emissions. *Hydrocarbons* are another serious air pollutant. It is not only a component of the smog, but it also contributes to the formation of the ground-level ozone by reacting with nitrogen oxides in the presence of sunlight. Hydrocarbon emissions result from the incomplete combustion of the fuel as well its evaporation from the tank and other fuel supply components. *Nitrogen oxides* result from the reaction of nitrogen and oxygen in the air at high combustion temperatures, such as those present in an internal combustion engine. More than half of all nitrogen oxide emissions come from the mobile sources, both on-road and non-road vehicles. *Particulate matter* is a term used to describe solid or liquid particles found in the air. Some particles are quite large, around ~10 micron(μ) or more in size, or are dark enough to be seen as soot or smoke, or both. While the particles of this size can lead to irritation of the nose and throat and coughing, it is the particles of the smaller size, which are generally invisible to the naked eye, that have serious health effects. Mobile sources, both on-road and non-road, of particulate emissions consist mainly of particles that have a size of 2.5 μ or less. This is especially true of the diesel-powered vehicles, which are responsible for more than 50 % of the mobile source particulate emissions. The particulate matter of the diameter 2.5 μ or less (PM2.5) can travel deep into the lungs and cause chronic lung problems, such as asthma, difficulty in breathing, bronchitis, and sometimes even cancer. Fine particulates can also cause haze, thereby reducing visibility.

Mobile source air toxics (MSATs), also known as hazardous air pollutants, are also of concern since these chemicals are known or suspected to cause cancer or other serious health effects, such as reproductive problems or birth defects. These chemicals enter the atmosphere either by direct evaporation of the fuel, or its passage through the engine via combustion. These components may also be present in the fuel as additives, result during combustion,

TABLE 6.1—Mobile source contribution to 1999 NATA risk drivers [539].

1999 NATA Risk Drivers	Percent Contribution from All Mobile Sources	Percent Contribution from On-road Mobile Sources
Benzene	68 %	49 %
1,3-Butadiene	58 %	41 %
Formaldehyde	47 %	27 %
Acrolein	25 %	14 %
Polycyclic organic matter[a]	6 %	3 %
Naphthalene	27 %	21 %
Diesel PM and Diesel exhaust organic gases	100 %	38 %

[a]This POM inventory includes the 15 POM compounds: benzo[b]fluoranthene, benz[a]anthracene, indeno(1,2,3-c,d)pyrene, benzo[k]fluoranthene, chrysene, benzo[a]pyrene, dibenz(a,h) anthracene, anthracene, pyrene, benzo(g,h,i)perylene, fluoranthene, acenaphthylene, phenanthrene, fluorene, and acenaphthene.

or the atmospheric oxidation. Examples of such chemicals include formaldehyde and acetaldehyde. Some air toxics, such as metals, can result from engine wear or from impurities in the oil or the fuel. A detailed list of the compounds emitted from the mobile sources and their cancer and non-cancer risk to the regional and national population is listed in the Integrated Risk Information System (IRIS) database [537] and the EPA February 28, 2006 proposed rule document, titled "Control of Hazardous Air Pollutants from Mobile Sources" [538]. Among the national- and regional-scale cancer and non-cancer risk drivers identified in the 1999 NATA, seven compounds have significant contributions from the mobile sources. These include benzene, 1,3-butadiene, formaldehyde, acrolein (acrylaldehyde), polycyclic organic matter (POM), naphthalene, diesel particulate matter (PM), and diesel exhaust organic gases. These air toxics can cause a variety of cancer and non-cancer health effects. Non-cancer health effects include neurological, cardiovascular, liver, kidney, and respiratory effects as well as effects on the immune and reproductive systems. Table 6.1 provides mobile source contribution to the 1999 NATA risk drivers [539].

These projected risks were estimated using the same tools and methods as the 1999 NATA (National-Scale Air Toxics Assessment). NATA assesses human health impacts from the chronic inhalation exposures to the outdoor sources of the air toxics. It assesses lifetime risks assuming continuous exposure to levels of the air toxics estimated at a particular point in time. The most recent NATA was carried out for the year 1999 [539]. Figure 6.1 summarizes the changes in average population inhalation cancer risk for the MSATs listed in Table 6.1 [539]. Despite significant reductions in risk from these pollutants, average inhalation cancer risks are expected to remain well above 1 in 100,000. According to the EPA, the emissions reductions from the February 28, 2006 proposed standards for the motor vehicles and their fuels, combined with the standards already in place, represent the maximum achievable reductions of emissions from motor vehicles by the application of technology that is or will be available. These standards propose controls on gasoline-fueled passenger vehicles, and portable gasoline containers (gas cans). Among MSATs, benzene is a known human carcinogen, and mobile sources are responsible for the majority of the benzene emissions. Hence, the EPA is proposing to limit the benzene content of the gasoline to an annual average of 0.62 % by volume, beginning in the year 2011. The EPA is also proposing to limit the exhaust emissions of hydrocarbons from the passenger vehicles, when they are operated at cold temperatures. This standard will be phased in during the 2010 to 2015 period. Further proposals involve the evaporative emissions standards for passenger vehicles, which are equivalent to those in California, and hydrocarbon emissions standard for gas cans beginning in 2009. This standard will reduce evaporation and spillage of the gasoline from these containers. These controls will significantly reduce emissions of benzene and other MSATs, such as 1,3-butadiene, formaldehyde, acetaldehyde, acrolein, and naphthalene. In addition, these standards will significantly reduce emissions of the particulate matter from passenger vehicles, thereby offering substantial benefits to the public health. In the absence of these and other existing antipollution measures, the pollution is expected to increase significantly in the future, as shown in Fig. 6.2 [539]. The figure shows contribution of the source categories to the air toxics emissions from 1990 to 2020, not including the diesel particulate matter. The dashed line represents the projected emissions without the controls enacted by the Clean Air Act. If the diesel PM emissions were added to the mobile source total, mobile sources would account for 48 % of the total 5,398,000 tons in 1999. Figure 6.3 summarizes the trend in the diesel PM between 1999 and 2020, by source category [540]. Diesel PM emissions will be reduced from 368,000 tons in 1999 to 114,000 tons in 2020, a decrease of 70 %. As the controls on highway diesel engines and non-road diesel engines phase in, the diesel-powered locomotives and commercial marine vessels increase from 11 % of the inventory in 1999 to 27 % in 2020.

Exhaust Emissions Of Concern

As mentioned earlier, emissions in an internal combustion engine are of two types: evaporative and combustion-related. Fuel system and crankcase are the sources of the evaporative emissions. These emissions are hydrocarbon in nature and can enter the environment directly. Combustion-related emissions, on the other hand, arise from the oxidation of the hydrocarbon material that is the source of the

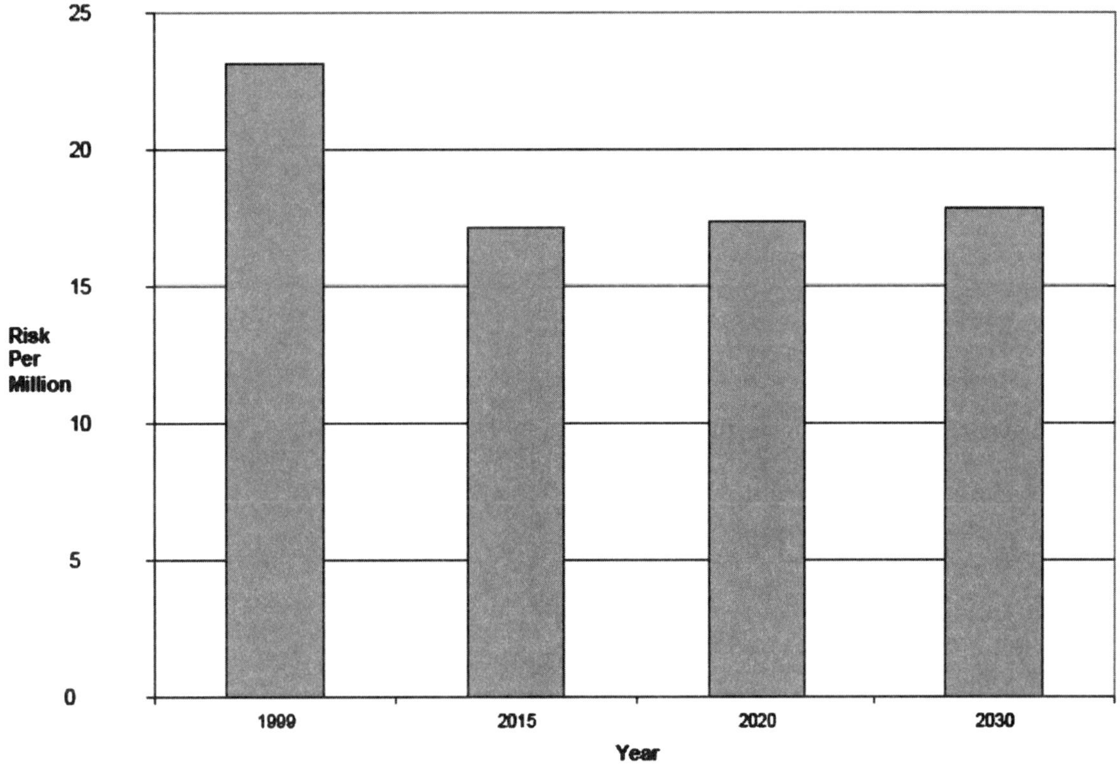

Fig. 6.1—Trends in nationwide average population cancer risk from inhalation exposure to outdoor sources of mobile source air toxics, 1999 to 2030 [539].

chemical energy, which through combustion gets converted into thermal energy, and ultimately into mechanical work. Complete combustion results in the formation of the fully oxidized products, such as carbon dioxide (CO_2) and water (H_2O). Incomplete combustion, however, leads to the formation of the partially oxidized products, such as carbon or particulate matter (PM), carbon monoxide (CO), aldehydes (C_nH_mCHO), ketones [$(C_nH_m)_2CO$], and carboxylic acids

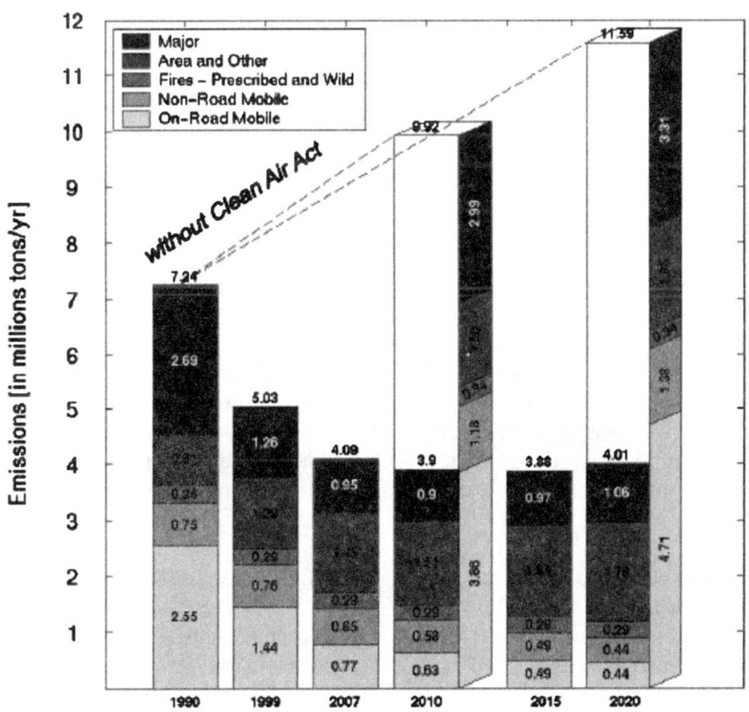

Fig. 6.2—Contribution of source categories to air toxic emissions, 1990 to 2020 (not including diesel particulate matter) [539].

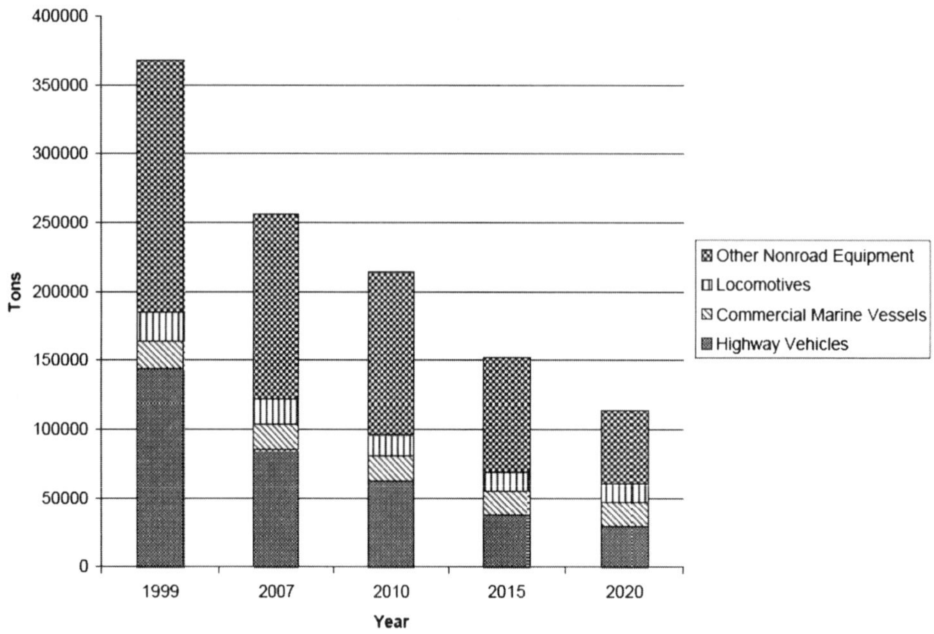

Fig. 6.3—Contribution of mobile source categories to diesel particulate matter emissions 1999 to 2020 [540].

(C_nH_mCOOH). Fuel components that escape combustion become part of the exhaust as the unburned hydrocarbons (HC). The exhaust, in addition, contains small amounts of hydrogen, ethylene, acetylene, and polycyclics that result from thermal cracking of the hydrocarbons in the fuel. Sulfur oxides (SO_2 and SO_3) and NOx, also present in the exhaust, are produced from the reaction of oxygen with the fuel sulfur and from the reaction of oxygen and nitrogen, both of which are present in the air. NOx is primarily a mixture of nitric oxide (NO) and nitrogen dioxide (NO_2). NOx formation requires temperatures of about 1370 °C. Combustion-related emissions of concern comprise unburned hydrocarbons (HC), carbon monoxide (CO), nitrogen oxides (NOx), aldehydes, and particulate matter (PM).

The amount of pollutants in the exhaust gas depends upon the type and the quality of the fuel, the composition of the air-fuel mixture, combustion quality, the type of ignition (spark versus compression), engine operating conditions, and the presence or absence of the emissions-control devices. Emissions of concern in gasoline engines primarily consist of HC, CO, NOx, and CO_2, and for diesel engines, they, in addition, include aldehydes, PM, and odor constituents.

Besides water and carbon dioxide, which are the essential products of combustion, all others are considered undesirable. Public concern towards emissions is growing at a rapid rate and their control is being mandated by various government organizations and a number of legislations in the United States and other countries exist to control their release into the atmosphere. Even the amount of CO_2 is being scrutinized because of its contribution to the greenhouse effect. However, its control is difficult because of the widespread use of the fossil fuels.

There are a number of ways to reduce undesirables in the exhaust. These include fuel reformulation, changing engine operating conditions, and modifying engine designs. Since the future emissions standards are progressively becoming more stringent, combining of all these strategies will be necessary to meet them. The OEMs' efforts in this regard pertain to engine designs that minimize the formation of the undesirable emissions and developing effective emissions control systems, which are primarily based upon after-treatment devices that convert undesirables into innocuous products.

Before considering the emissions control strategies, individual emissions, their properties, and their sources must be carefully examined. After that we will discuss general performance of the internal combustion engines, which will be followed by the emissions standards, both for gasoline and diesel engines, and the strategies to control emissions. And in this regard, the impact of the fuel properties, engine operating conditions, and the engine design changes will be considered.

Unburned Hydrocarbons (HC)

HC is a collective term that is used to describe all hydrocarbon material present in the exhaust gas. In the presence of sunlight and NOx, these compounds can form oxidants, such as organic peroxides, ozone, and peroxy-acetyl nitrates, some of which may cause irritation of the mucous membrane. Polycyclic hydrocarbons, if present, are also of concern, due to their carcinogenic activity.

Besides evaporation, these emissions in a four-stroke cycle engine arise from two sources, incomplete combustion of the fuel and wall quenching. In a two-stroke cycle engine, the additional source is exhaust scavenging of the fresh incoming mixture. Incomplete combustion is a consequence of the ineffective flame propagation that can result when the fuel-air mixture is too rich, too lean, or is highly diluted by the exhaust gas. This problem is more prevalent during the warm-up and deceleration when the fuel-air mixture is of improper composition for efficient combustion. This is sometimes indicated by misfiring. Under these conditions, the amount of HC is very high. As mentioned earlier, wall

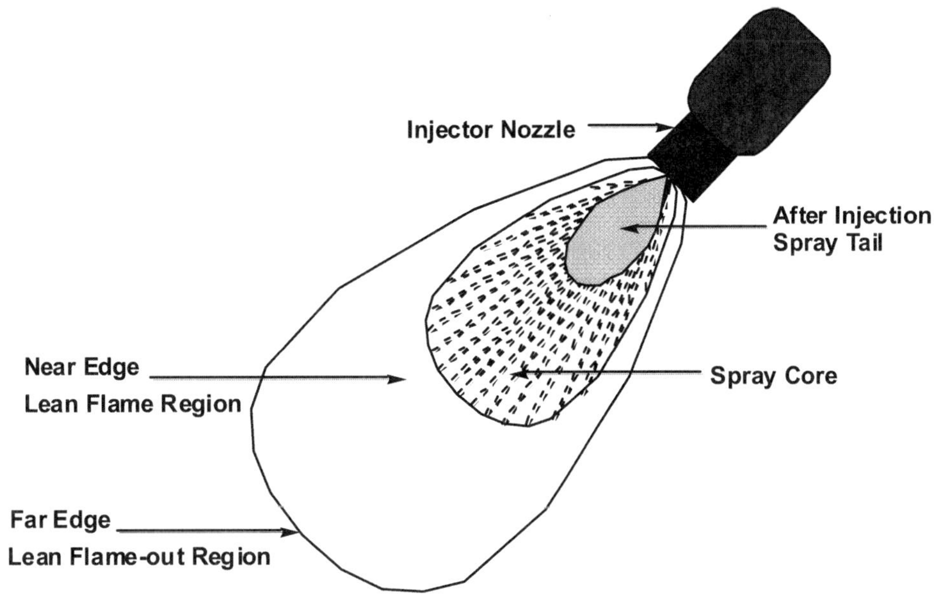

Fig. 6.4—Fuel spray pattern from a diesel injector [23].

quenching is the process where free radicals, essential to combustion, become inactive (quench) after colliding with the wall, and the wall temperature is too low to support combustion. Scavenging-related hydrocarbon emissions occur because of the tendency of the unburned fuel-air mixture to escape directly into the exhaust. The amount of HC emissions in the cross-scavenged two-stroke cycle engines is the highest.

Carbon Monoxide (CO)

This pollutant results when the temperature and the oxygen content of the fuel-air mixture are not high enough to affect complete combustion. The low-temperature situation typically exists between the lean flame region and the lean flame-out region, usually at the onset of combustion and at low loads. Figure 6.4 identifies the various flame regions. The situation changes as the combustion process proceeds. The CO formation under high loads is not a problem since they require richer fuel-air mixtures, which translates into high reaction temperatures that facilitate its oxidation to CO_2. However, too rich a combustion mixture can increase CO by making the combustion less efficient which is due to the inadequate oxygen content. The problem can be corrected by using close to stoichiometric combustion mixtures and by ensuring homogeneous mixing of the fuel and air.

Carbon monoxide (CO), a colorless, odorless gas, is highly toxic. Inhalation of air contaminated with 0.3 volume percent carbon monoxide can cause death within 30 minutes. The toxicity of the CO is related to its ability to permanently block hemoglobin's oxygen-complexing site, thereby leading to asphyxiation. Its concentration in the exhaust gas from the spark-ignition engines is the highest at idle speed and in colder months.

Nitrogen Oxides (NOx)

NOx is a collective term used for the oxides of nitrogen that are many. These are nitrous oxide (N_2O), nitric oxide (NO), nitrogen dioxide (NO_2), dinitrogen trioxide (N_2O_3), and unstable nitrogen pentoxide (N_2O_5). With respect to engine emissions, we are primarily concerned with the formation of the nitric oxide (NO) and nitrogen dioxide (NO_2). Nitric oxide is a colorless, odorless gas that, at least in principle, can result from the direct reaction of nitrogen and oxygen that constitute air. NO_2, a reddish-brown gas with a penetrating odor, which can cause irritation of the mucous membranes, results when the NO oxidizes further. The mechanism of NOx formation is shown by the chemical equations given below.

$$N_2 + O_2 \rightleftharpoons 2NO$$

$$2NO + O_2 \rightleftharpoons 2NO_2$$

The rate of NO formation by this mechanism is slow [542]; hence the equilibrium quantities produced at peak temperatures and pressures in the cylinders cannot be reached. Zeldovich proposed an alternative mechanism based upon atomic oxygen as the initiating species [543]. Lavoie et al. [544] extended it by including the reaction with hydroxyl free radicals (·OH) derived from the oxidation of the fuel. See the oxidation inhibitors section in Chapter 4 on Additives. The combined mechanism is shown by the chemical equations below.

$$O_2 \rightleftharpoons 2O$$

$$N_2 + O \rightleftharpoons NO + N$$

$$N_2 + O \rightleftharpoons N_2O$$

$$N + O_2 \rightleftharpoons NO + O$$

$$N + \cdot OH \rightleftharpoons NO + H\cdot$$

$$H\cdot + N_2O \rightleftharpoons N_2 + \cdot OH$$

$$O + N_2O \rightleftharpoons N_2 + O_2$$

$$O + N_2O \rightleftharpoons 2NO$$

Empirical data [545] suggest that the rate of NO formation is higher for the fuel-rich mixtures than for the stoichiometric or the lean mixtures. This is not surprising since rich mixtures result in higher temperatures and pressures, both of which facilitate the reaction of the nitrogen and oxygen to NOx. NOx is also affected by the air-fuel (A/F) ratio and by the engine design factors that influence the temperature, such as compression ratio, spark timing, and the cooling system performance. Combustion chamber deposits are generally believed to have a modest insulating effect, thus raising the combustion temperatures and increasing the NOx emissions slightly. While the original air quality standard was implemented to control the release of nitrogen dioxide (NO_2) into the atmosphere, the concern for NO is equally valid since it oxidizes in the air to form NO_2. NOx emissions primarily result from the motor vehicles and the industrial and commercial industries that burn fossil fuels. NO is nontoxic, but contributes to ozone formation. However, NO_2, its oxidation product, is quite toxic and in the worst case can cause lung damage.

Ground Level Ozone

Ozone is an indirect contaminant, that is, it is not part of the exhaust emissions but results as a consequence of these emissions. Ozone is a colorless gas that is the chief component of the urban smog. It is formed by the interaction of the reactive organic gases, also called volatile organic carbon (VOC), with the oxides of nitrogen (NOx) in the presence of sunlight. Most hydrocarbon emissions (HC) are VOCs. That is why there are on-going efforts to reduce them. Not all hydrocarbons contribute equally to ozone formation. Olefins and aromatics are more likely to take part in ozone formation than the paraffinic hydrocarbons. Ozone's health effects include damage to the lung tissue, reduced lung function, and the sensitization of lungs to other irritants.

Sulfur Dioxide

Sulfur dioxide (SO_2) results from the combustion of fuels that contain sulfur. On-road and off-road engine fuels are estimated to contribute less than 3 % towards the total SO_2 emissions. SO_2 is a moderate lung irritant and it along with NOx are the major contributors towards acid rain, most of which occurs in and around the highly industrialized cities.

Aldehydes

Aldehydes occur in emissions of both the CI (compression ignition) and the SI (spark ignition) engines. The mechanism by which they result parallels that of the CO formation and the combustion temperature and oxygen content of the combustion mixture affect their formation in an analogous manner. These pollutants are produced in the combustion chamber during the preflame stage when the temperature is low and only partial oxidation of the hydrocarbon fuel takes place. Aldehydes also form in air-injected exhaust systems. Two common aldehydes that are found in emissions are formaldehyde and acrylaldehyde, or acrolein. See structures below. Data show formaldehyde formation to precede knock in the gasoline engines [546], suggesting thereby that partial oxidation may be the cause of auto-ignition.

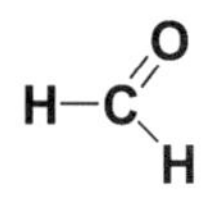

Formaldehyde

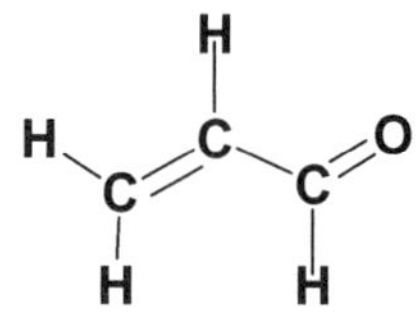

Acrolein
Acryldehyde

Particulate Matter (PM)

Particulates (particulate matter or PM) are defined as substances that are present in the exhaust gases in a solid (ash, carbon) or a semi-solid form. Particulate matter is composed of different size particles. Some are as big as 10 microns (PM_{10}) in size; the others are 2.5 microns ($PM_{2.5}$) or smaller. According to the EPA estimate, less than 2% of the total PM_{10} is attributable to on-road and off-road engines. Vehicle exhaust-related PM_{10} includes both the primary carbon particles, mainly from the diesel fuel, and the secondary sulfate and nitrate aerosols, formed by the reactions of SO_2 and NOx in the atmosphere. Both the EPA and California have established the PM_{10} standards. $PM_{2.5}$, on the other hand, poses a much serious health threat since it is small enough to travel far into the lungs and may contain other associated materials that may also be harmful.

Particulates are not a serious problem in gasoline engines but they are in diesel engines. Diesel particulates are of two types: liquid particulates and soot. Liquid particulates, which appear as white smoke, result during cold starting and idling, and at low loads. These usually consist of the fuel and the low boiling fractions of the lubricating oil that escape combustion. The white smoke tends to cease as the load increases. Soot, or black smoke, which results from the incomplete combustion, is much more persistent and is the cause of great concern.

Soot is primarily made up of fine carbon particles in an agglomerated form and results from the stripping of the hydrogen atoms from the hydrocarbon molecules of the fuel or the lubricant. Soot at low loads forms during the initial stage of combustion and both in the lean flame zone and the spray core, see Fig. 6.4. It goes away as the combustion proceeds. However, soot formation in the spray core at high loads is persistent and is primarily a consequence of the pyrolysis reactions that occur due to oxygen deficiency. Fuel deposition on the cylinder walls can also result in soot and for the same reason.

Odor

Kerosine type odor of the diesel fuel, while not of primary concern, arises from the aromatic organic compounds, such as alkylbenzenes, naphthalenes, indenes and their partially hydrogenated derivatives that escape combustion. Odor intensity increases at high loads where the combustion is inefficient. Strategies that improve combustion also reduce odor.

The governments of many world nations, including that of the United States are aware of the impact of these pollutants on the health and welfare of their populations and have installed measures to control their release into the environment. These measures are continuously being revised to stay abreast of this problem. In the United States, the Environmental Protection Agency (EPA) is assigned to monitor, control, and or improve the nation's ambient air quality. Present U.S. ambient air quality standards are provided in Table 6.2 [540, 541].

General Engine Performance Considerations

Traditionally, three key parameters were used to evaluate engine performance: power, fuel consumption (thermal efficiency), and driveability [23,547]. In recent years, another measure, namely emissions control, has been added to the list. Engines that produce more power, consume less fuel, and generate little or no undesirable emissions are highly desired. All of the listed parameters relate to combustion, hence improving combustion efficiency is one of the ways to improve the engine performance.

Fuel requires a certain amount of oxygen, or air, to burn completely. The air-fuel ratio, denoted by AF, indicates the relative proportion of air and the fuel in the combustion mixture. The term fuel-air ratio, FA, which is also used, is the reciprocal of the air-fuel ratio. The exact amount of air necessary to convert the fuel into completely oxidized products is called theoretical or stoichiometric amount [23,479]. This amount equals 14.7 parts by weight of air for one part by weight of the hydrocarbon fuel, that is, an AF of 14.7:1, which is also called a stoichiometric air-fuel mixture. Air-fuel mixtures with ratios less than this are labeled rich and those greater than this are labeled lean. This is shown in Fig. 6.5 [479]. Rich and lean pertain to the amount of fuel present in such mixtures. While nitrogen does not take part in the combustion process itself, it does oxidize to form nitrogen oxides, which constitute undesirable emissions. Equivalence ratio λ, a ratio of AF *stoichiometric* and AF *actual*, is another term used to describe the composition of the combustion mixtures [23]. The value of λ equals one for the stoichiometric case. In this chapter, all three ratios will be used, depending upon the concepts being explained.

While it is possible to operate an internal combustion engine by the use of the air-fuel mixtures of varying composition, the maximum combustion temperature, hence the maximum mechanical efficiency, is obtained when the air-to-fuel ratio is 14.7 to 1, which is indicated by the *ideal air air-fuel mixture* in Fig. 6.5 [479]. However, the combustion products from such a mixture contain substantial amounts of unburned hydrocarbons (HC), carbon monoxide (CO), and nitrogen oxides (NOx). Figure 6.6 shows the effect of the air-fuel mixtures of varying composition on emissions [479]. As we can see, burning of the leaner mixtures (air-to-fuel ratios of >14.7), in general, tends to lower the HC and CO emissions and increases the NOx emissions. However, when the mixtures become ultra lean, the situation is reversed. Lean mixtures are also preferable because of the fuel economy. Nonetheless, the power generation often requires burning rich mixtures, which produce higher HC and CO emissions.

Combustion efficiency depends upon the composition of the combustion mixture as well as the engine type. Since torque and mean effective pressure (MEP) are directly related, the combustion mixtures that result in maximum MEP result in maximum torque at a given speed. Maximum MEP is reached at each engine speed when all of the air in the cylinder is effectively consumed. This can be achieved by the use of the richer than stoichiometric mixtures. The use of the higher amounts of fuel to boost the engine performance is impractical for two reasons. First, the amount of fuel in a mixture can be increased until a maximum is reached; beyond which there is not enough air in the combustion mixture to ensure efficient combustion [23]. As a consequence, there is both a loss of the generated power and an increase in HC emissions. Second, at low air-fuel ratios, the cost of the fuel becomes an issue. Fuel consumption is expressed in terms of specific fuel consumption (SFC). A minimum SFC is reached when all fuel in the cylinder is consumed most effectively. To increase fuel economy, leaner mixtures are preferable. However, such mixtures generate less power because they tend to burn slowly. A minimum is reached beyond which the flame is difficult to maintain [23].

The effect of the air-fuel composition on MEP and SFC in an SI engine (compression ratio = 10) at a specific speed is shown in Fig. 6.7 [23]. The data show that as we progressively move from the stoichiometric mixture towards rich mixtures, MEP increases (top curve) until it reaches its maximum at AF of 13:1. Conversely, moving towards lean mixtures decreases the SFC (bottom curve) until a minimum is reached at AF of ~16:1, after which it increases. Because MEP and SFC are inversely related, achieving both simultaneously is not an easy task. Since the low MEP generation is ascribed to lean mixtures burning slowly, increasing their combustion efficiency can help correct this problem. The easiest way to obtain high efficiency is to use a high octane fuel in a stoichiometric air-fuel ratio, employ optimized spark timing, and operate at temperatures that will ensure the complete combustion. In general, the higher the gasoline octane, the higher is the MEP. This can be seen in the figure by relating the MEP values to the fuel octane numbers. This strategy works as indicated by the lower amount of HC and CO in emissions. However, it increases the amount of NOx formation, as shown in Fig. 6.6, which is a consequence of the higher temperatures that result from the complete combustion. As mentioned earlier, HC in the SI engines primarily results from wall quenching, incomplete combustion, and inefficient scavenging [547]. Wall quenching, a situation where a flame does not propagate well in the vicinity of a wall, occurs for two reasons: (1) the lower wall temperatures promote fuel deposition, mainly the high boiling fractions, hence poor combustion; and (2) free radicals that are responsible for combustion terminate at the wall, thereby slowing down the combustion chain reaction. Wall quenching is the primary source of the unburned hydrocarbons in the four-stroke cycle engines. These factors will be revisited in the latter section of the discussion.

Combustion efficiency, hence emissions quality, also depends upon engine design. In general, the emissions from the piston engines have higher amounts of pollutants than those from the gas turbine engines. Data for NOx levels in emissions from the various types of engines are graphically presented in Fig. 6.8 [479]. As can be seen, gasoline and diesel engines emit the highest amount of NOx and gas turbine engines emit the lowest amount. The amount of NOx in emissions from the piston engines fluctuates because of the design differences, such as revolutions per minute, fuel injection method, etc.

There are two main reasons for the presence of a lower

TABLE 6.2—Ambient air quality standards [540].

Pollutant	Averaging Time	California Standards [1]		Federal Standards [2]		
		Concentration [3]	Method [4]	Primary [3,5]	Secondary [3,6]	Method [7]
Ozone (O_3)	1 Hour	0.09 ppm (180 μg/m³)	Ultraviolet Photometry	—	Same as Primary Standard	Ultraviolet Photometry
	8 Hour	0.070 ppm (137 μg/m³)*		0.08 ppm (157 μg/m³)[8]		
Respirable Particulate Matter (PM10)	24 Hour	50 μg/m³	Gravimetric or Beta Attenuation	150 μg/m³	Same as Primary Standard	Inertial Separation and Gravimetric Analysis
	Annual Arithmetic Mean	20 μg/m³		50 μg/m³		
Fine Particulate Matter (PM2.5)	24 Hour	No Separate State Standard		65 μg/m³	Same as Primary Standard	Inertial Separation and Gravimetric Analysis
	Annual Arithmetic Mean	12 μg/m³	Gravimetric or Beta Attenuation	15 μg/m³		
Carbon Monoxide (CO)	8 Hour	9.0 ppm (10mg/m³)	Non-Dispersive Infrared Photometry (NDIR)	9 ppm (10 mg/m³)	None	Non-Dispersive Infrared Photometry (NDIR)
	1 Hour	20 ppm (23 mg/m³)		35 ppm (40 mg/m³)		
	8 Hour (Lake Tahoe)	6 ppm (7 mg/m³)		—	—	—
Nitrogen Dioxide (NO_2)	Annual Arithmetic Mean	—	Gas Phase Chemiluminescence	0.053 ppm (100 μg/m³)	Same as Primary Standard	Gas Phase Chemiluminescence
	1 Hour	0.25 ppm (470 μg/m³)		—		
Sulfur Dioxide (SO_2)	Annual Arithmetic Mean	—	Ultraviolet Fluorescence	0.030 ppm (80 μg/m³)	—	Spectrophotometry (Pararosaniline Method)
	24 Hour	0.04 ppm (105 μg/m³)		0.14 ppm (365 μg/m³)	—	
	3 Hour	—		—	0.5 ppm (1300 μg/m³)	
	1 Hour	0.25 ppm (655 μg/m³)		—	—	—
Lead[9]	30 Day Average	1.5 μg/m³	Atomic Absorption	—	—	—
	Calendar Quarter	—		1.5 μg/m³	Same as Primary Standard	High Volume Sampler and Atomic Absorption
Visibility Reducing Particles	8 Hour	Extinction coefficient of 0.23 per kilometer — visibility of ten miles or more (0.07 — 30 miles or more for Lake Tahoe) due to particles when relative humidity is less than 70 percent. Method: Beta Attenuation and Transmittance through Filter Tape.		No Federal Standards		
Sulfates	24 Hour	25 μg/m³	Ion Chromatography			
Hydrogen Sulfide	1 Hour	0.03 ppm (42 μg/m³)	Ultraviolet Fluorescence			
Vinyl Chloride[9]	24 Hour	0.01 ppm (26 μg/m³)	Gas Chromatography			

*This concentration was approved by the Air Resources Board on April 28, 2005 and is expected to become effective in early 2006.

See footnotes provided below

1. California standards for ozone, carbon monoxide (except Lake Tahoe), sulfur dioxide (1 and 24 hour), nitrogen dioxide, suspended particulate matter—PM_{10}, $PM_{2.5}$, and visibility reducing particles, are values that are not to be exceeded. All others are not to be equaled or exceeded. California ambient air quality standards are listed in the Table of Standards in Section 70200 of Title 17 of the California Code of Regulations.
2. National standards (other than ozone, particulate matter, and those based on annual averages or annual arithmetic mean) are not to be exceeded more than once a year. The ozone standard is attained when the fourth highest eight hour concentration in a year, averaged over three years, is equal to or less than the standard. For PM_{10}, the 24 hour standard is attained when the expected number of days per calendar year with a 24-hour average concentration greater than150 μg/m³ is equal to or less than one. For $PM_{2.5}$, the 24 hour standard is attained when 98 percent of the daily concentrations, averaged over three years, are equal to or less than the standard.
Contact U.S. EPA for further clarification and current federal policies.
3. Concentration expressed first in units in which it was promulgated. Equivalent units given in parentheses are based upon a reference temperature of 25°C and a reference pressure of 760 torr. Most measurements of air quality are to be corrected to a reference temperature of 25°C and a reference pressure of 760 torr; ppm in this table refers to ppm by volume, or micromoles of pollutant per mole of gas.
4. Any equivalent procedure which can be shown to the satisfaction of the ARB to give equivalent results at or near the level of the air quality standard may be used.
5. National Primary Standards: The levels of air quality necessary, with an adequate margin of safety to protect the public health.
6. National Secondary Standards: The levels of air quality necessary to protect the public welfare from any known or anticipated adverse effects of a pollutant.
7. Reference method as described by the EPA. An "equivalent method" of measurement may be used but must have a "consistent relationship to the reference method" and must be approved by the EPA.
8. New federal 8-hour ozone and fine particulate matter standards were promulgated by U.S. EPA on July 18, 1997. Contact U.S. EPA for further clarification and current federal policies.
9. The ARB has identified lead and vinyl chloride as 'toxic air contaminants' with no threshold level of exposure for adverse health effects determined. These actions allow for the implementation of control measures at levels below the ambient concentrations specified for these pollutants.

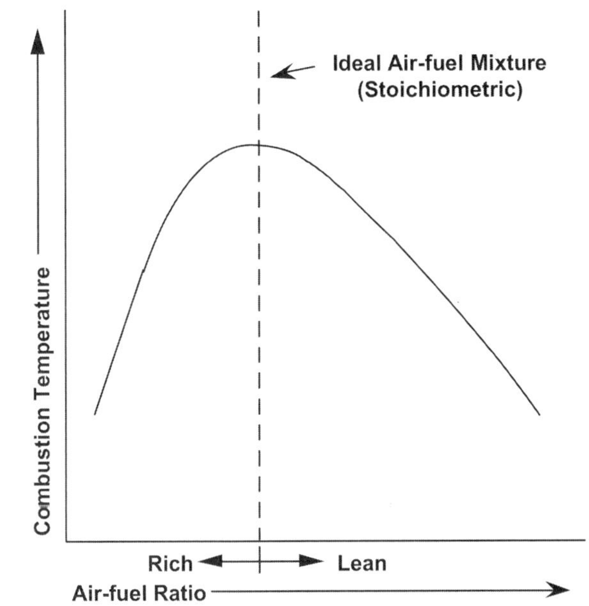

Fig. 6.5—Relationship between air-fuel mixture composition and combustion temperature [479].

amount of NOx pollutants in the exhaust of the gas turbine engines. First, the combustion temperatures are low because gas turbines typically use air-fuel ratios of 50:1, and higher. Second, these engines operate continuously, which allows ample time for efficient fuel combustion, see Fig. 6.9 [479]. Piston engine design is such that it requires high temperatures to run smoothly and efficiently, which are only achievable by operating at or close to ideal, or stoichiometric, air-fuel ratios. The SI engines generate peak combustion temperatures of about 2600 °C, or 4500 °F, by burning rich mixtures (air-fuel ratio of 14.5:1). Although the CI engines use leaner mixtures (air-fuel ratio of 25:1), they also generate high combustion temperatures, which is primarily

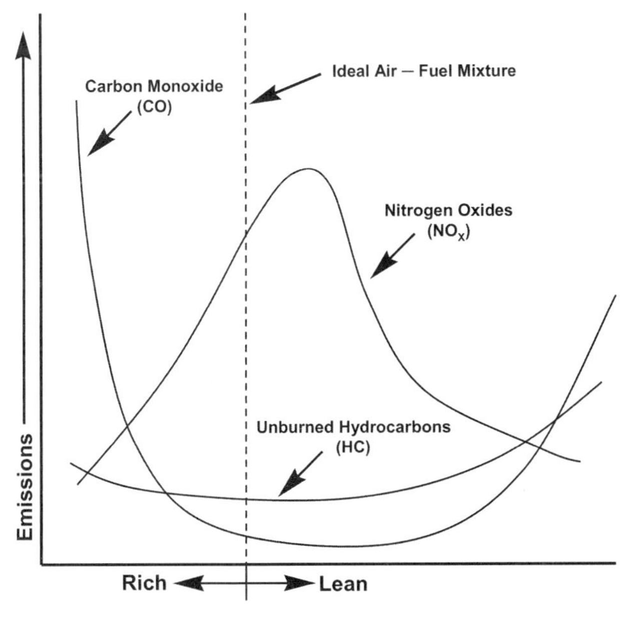

Fig. 6.6—Effect of air-fuel ratio on emissions [479].

because of the higher combustion pressures. Hence, both SI and CI engines have high levels of NOx in their emissions. NOx emissions as a function of the air-fuel mixture from different types of engines are shown in Fig. 6.10 [479]. In addition, because of the design, it is easier to develop combustion systems that generate lower amounts of pollutants for the turbine engines than for the piston engines. Due to the difficulties in modifying the piston engine design to improve the emissions quality, a number of alternative techniques are used to alter the combustion process itself [547]. These techniques are discussed in the section on engine design and operating variables.

As mentioned earlier, the CI engines suffer from the same problems as the SI engines. Diesel emissions contain HC, CO, NOx, aldehydes, particulate matter (PM), and odor constituents. The presence of both the HC and the formaldehyde is related to the incomplete oxidation of the fuel at high loads, which require burning rich combustion mixtures to generate power. The smoke particulates are also an outcome of the inefficient combustion. Their formation is hard to control, even by using lean combustion mixtures (as low as AF of 30:1). To minimize smoking and fouling of the engine, the CI engines are usually operated below the point of maximum power. Thus, the points of maximum power (a function of MEP) and economy are shifted toward the higher AF ratios [23], as depicted in Fig. 6.11. For a 16:1 compression ratio engine, an AF ratio of ~22:1 or 23:1 appears to be optimal with respect to smoke and fuel consumption, but obviously it generates less than the maximum power.

Emissions Standards

U.S. Emissions Standards

Most developed and developing countries have emissions standards in place. The concern for air quality is clearly the impetus. The United States is one of the first countries to legally establish emissions standards and California being one of its first states. The Clean Air Act of 1963 initiated the U.S. Federal Government's regulation for air pollution. It was amended in 1967, 1970, 1977, and, most recently, in the 1990. The stated purpose of the act is *...to protect and enhance the quality of the nation's air resources...* As the purpose suggests, the act addresses a wide range of air pollution issues, not just vehicle emissions. The act forbids the states from setting separate vehicle emissions standards, so that auto manufacturers did not have to produce cars with different emissions control systems to meet the different state standards. However, the restriction was waived for California because of its more severe smog levels and its long history of working to control vehicle emissions. California is allowed to establish its own regulations for controlling the vehicle emissions. However, they are subject to federal approval. Under certain circumstances, other states are allowed to require the sale of the new vehicles that meet the more stringent California standards. California's laws covering the vehicle emissions are administered by the California Air Resources Board (California ARB), which was established by the legislature in 1969. The sections that follow provide standards, federal and California, for the gasoline engines first and provides the standards for the diesel engines afterwards.

Please note that the diesel standards use weight per unit

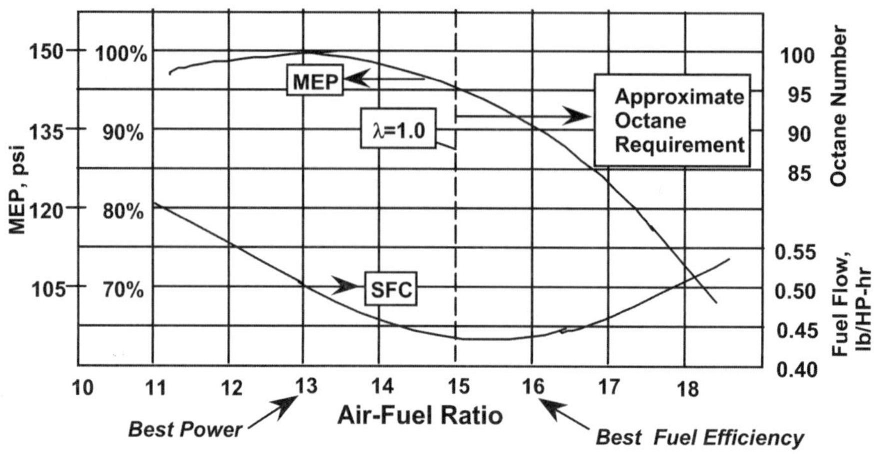

Fig. 6.7—The effect of air-fuel ratio on MEP and SFC in an SI engine [23].

of work (grams/brake horsepower-hour—g/bhp-h), which allows the use of a single standard for engines of all sizes. A larger engine generates a higher volume of exhaust and a higher absolute amount of emissions than a smaller engine, but it also can do more work. The emissions of the gasoline-powered vehicles are also expressed as weight per unit of work, but the units are grams per mile (g/mile). This measure of emissions is not suitable for use for diesel emissions stan-

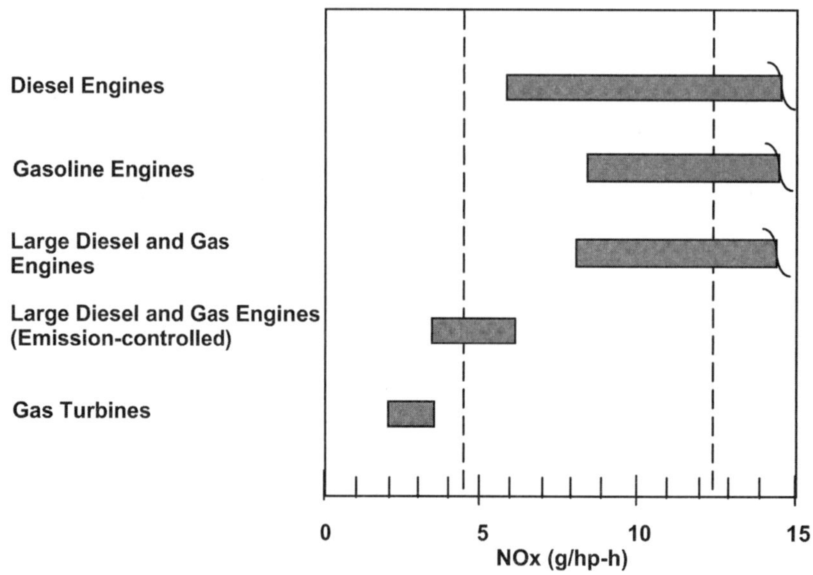

Fig. 6.8—NO_X emissions in different engine types [479].

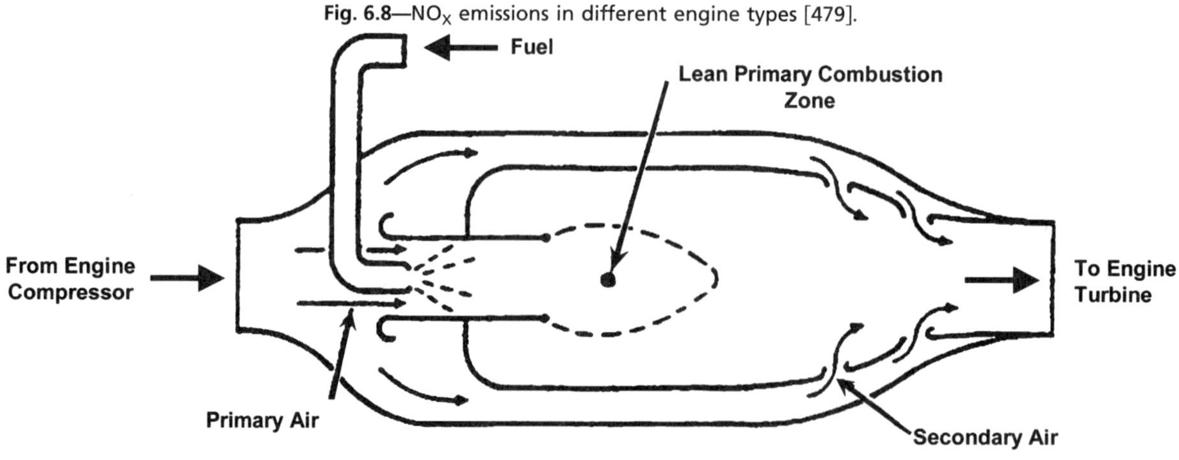

Fig. 6.9—The operation of a gas turbine engine [479].

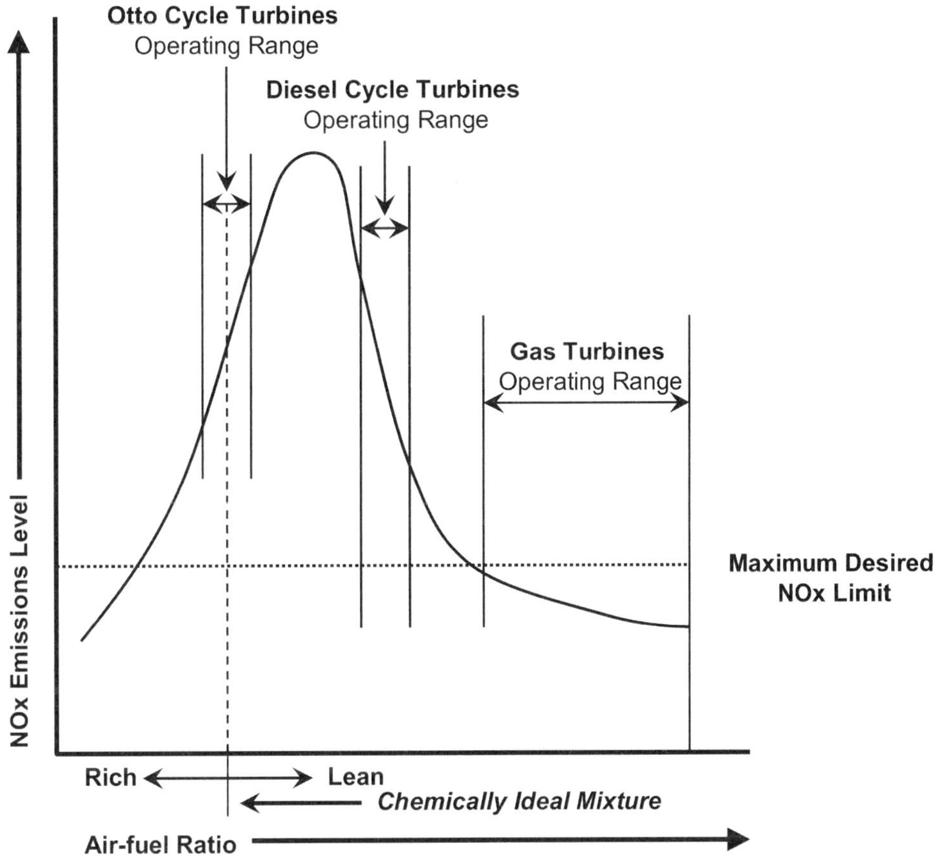

Fig. 6.10—NO_X emissions as a function of air-fuel composition [479].

dards because the diesel emissions are tested in an engine that is stationary and is not in a vehicle. Also, there is more variation in the sizes and loads of the diesel vehicles than is in the gasoline vehicles.

Gasoline Engine Standards

Acting on the provisions of the Clean Air Act, the EPA set standards, or limits, for exhaust emissions from the gasoline-powered cars, starting with the 1968 model year. These first limits applied to CO and hydrocarbons (HC) emissions, the limits for NOx were added starting with the 1973 model year. The emissions limit data for these pollutants up to the year 2003 are presented in Figs. 6.12–6.14 [548]. Examination of the plotted data underscores the effectiveness of the EPA emissions standards in lowering HC, NOx, CO,

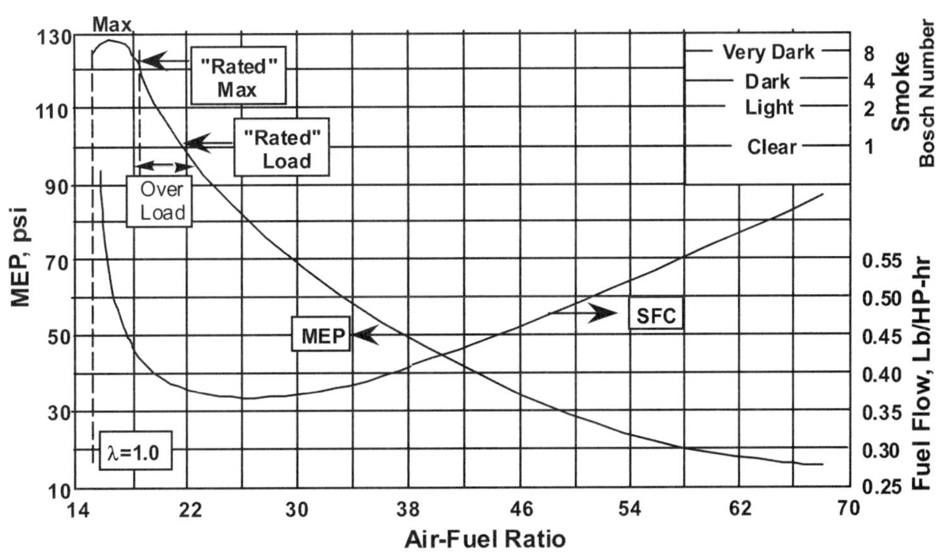

Fig. 6.11—The effect of air-fuel ratio on MEP and SFC in a CI engine [23].

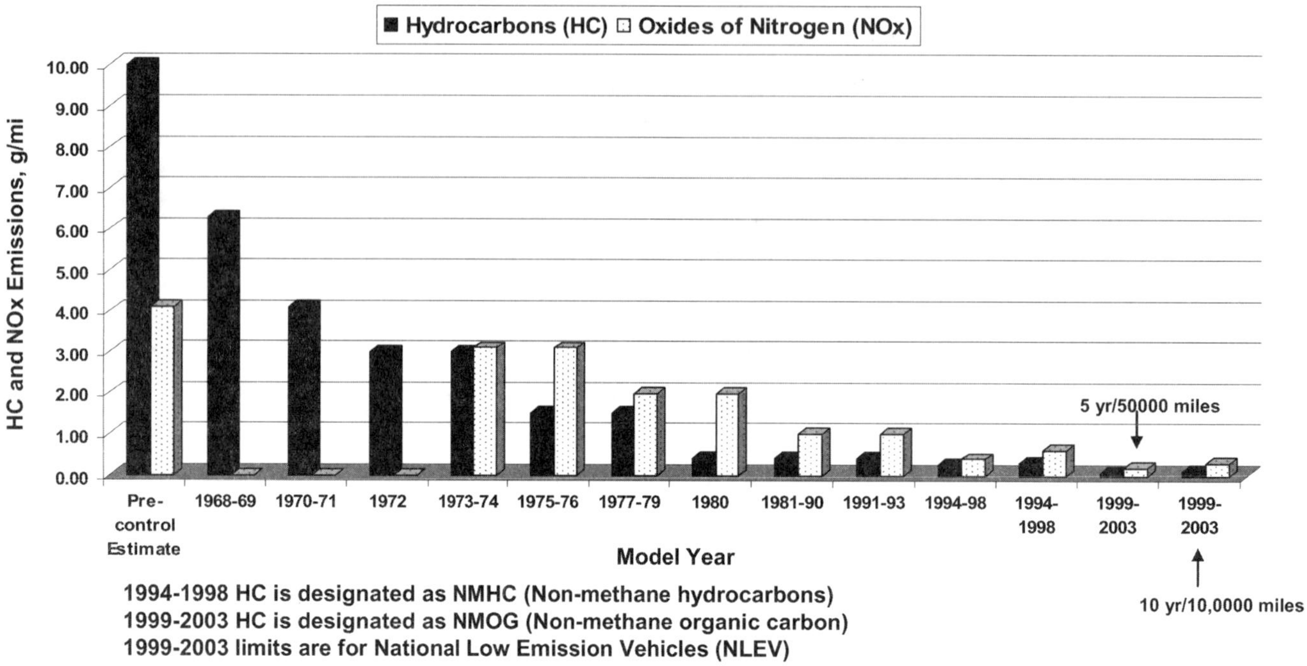

Fig. 6.12—Evolution of U.S. Federal passenger car hydrocarbon (HC) and nitrogen oxides (NO_X) exhaust emission standards.

and PM exhaust emissions from the gasoline-fueled vehicles.

Federal Standards

For light-duty vehicles, there are two sets of standards: Tier 1 and Tier 2. Tier 1 standards were fully implemented in the year 1997 and Tier 2 standards were phased in beginning in the year 2004. Tier 1 standards apply to new light-duty vehicles (LDV), such as passenger cars, light-duty trucks, sport utility vehicles (SUV), minivans, and pick-up trucks. The LDV category includes all vehicles of less than 8500 lb gross vehicle weight rating, or GVWR. These standards apply to

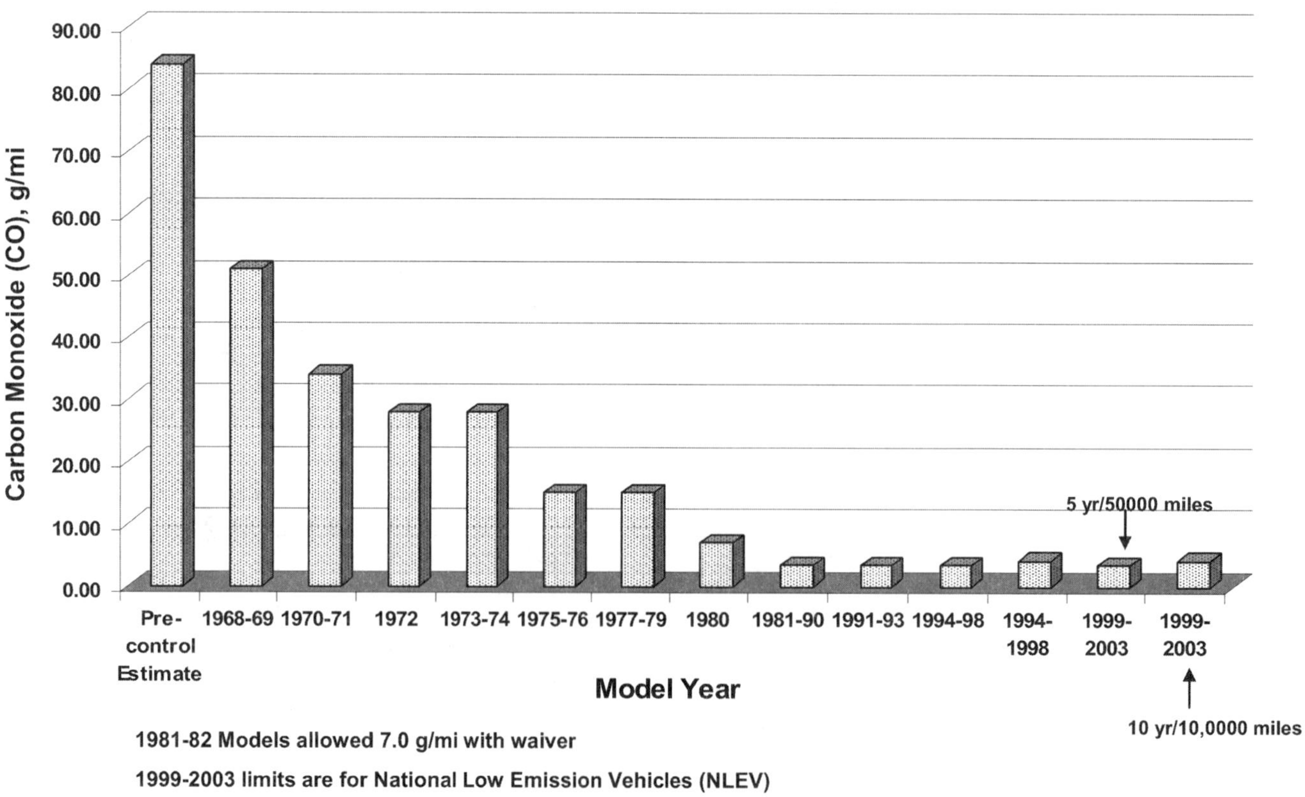

Fig. 6.13—Evolution of U.S. Federal passenger car carbon monoxide (CO) exhaust emission standards.

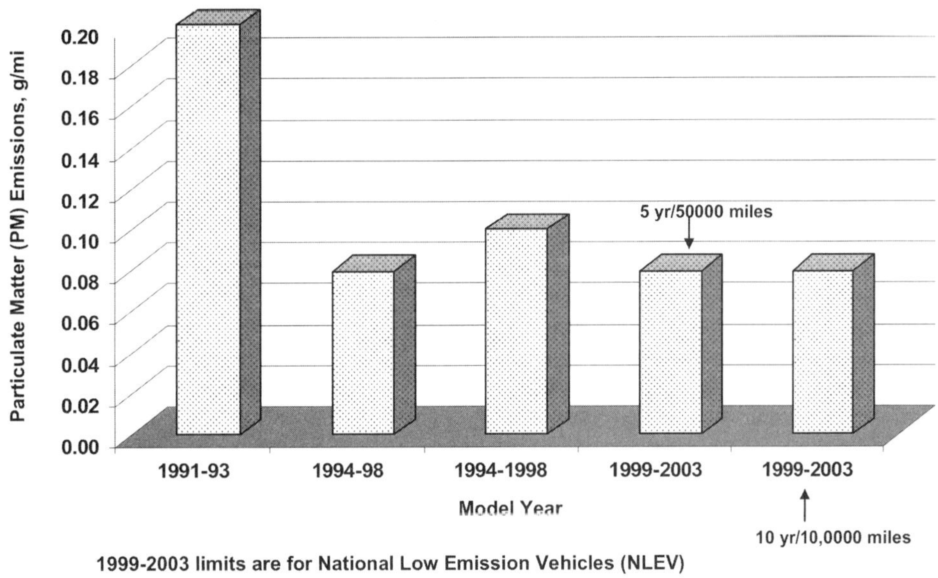

Fig. 6.14—Evolution of U.S. Federal passenger car particulate matter (PM) exhaust emission standards.

the full useful life of the vehicle, which is considered to be 100,000 miles. The standards also provide limits that must be met at 50,000 miles. Car and light truck emissions are measured by using the Federal Test Procedure (FTP) 75 and the results are expressed in g/mile. In addition, a Supplemental Federal Test Procedure (SFTP) has also been phased in between the years 2000 and 2004. The SFTP includes test cycles that measure emissions during aggressive highway driving and urban driving while the vehicle's air conditioning system is being used. Tier 1 standards are provided in Table 6.3 [401].

National Low Emissions Vehicle (NLEV) Program is a voluntary program, under which the northeastern states and the auto manufacturers agreed to follow the more stringent emissions standards between the year 1997, when Tier 1 standards were implemented, and the year 2004, when Tier 2 standards went into effect. Northeastern states in the model year 1999 and the states across the nation in the model year 2001 followed such a standard. The standard is similar to that used by California for its low emissions vehicle program; see Table 6.5. The National LEV program extends only to lighter vehicles and does not include Heavy Light-duty Truck [HLDT, gross vehicle weight (GVW) >6000 lb] vehicles.

Tier 2 standards are provided in Table 6.4 [401]. In addition to the lower limits than those of the Tier 1, the limits apply equally to all vehicle weight categories, i.e., cars, minivans, light-duty trucks, and SUVs, irrespective of if they operate on gasoline, diesel, or the alternate fuel. This means that the large engines, such as those used in light trucks or SUVs, will need more advanced emissions control technologies than that used in cars. The standard also applies to "Medium-Duty Passenger Vehicles" (MDPV) that are used for personal transportation, with GVWR between 8500 and 10,000 lb.

The Tier 2 regulation introduces new requirements for the fuel quality. Newer fuels must be compatible with the advanced emissions after-treatment devices, for example, catalysts that are needed to meet these and upcoming regulations. Lower sulfur level in gasoline is one such requirement. Beginning in the year 2004, the average gasoline sulfur level

TABLE 6.3—EPA Tier 1 emission standards for passenger cars and light-duty trucks, FTP 75, g/mile [401].

	50,000 miles/5 years						100,000 miles/10 years[a]					
Category	THC	NMHC	CO	NOx (diesel)	NOx (gasoline)	PM	THC	NMHC	CO	NOx diesel	NOx gasoline	PM
Passenger cars	0.41	0.25	3.4	1.0	0.4	0.08	—	0.31	4.2	1.25	0.6	0.10
LLDT, LVW <3750 lb	...	0.25	3.4	1.0	0.4	0.08	0.80	0.31	4.2	1.25	0.6	0.10
LLDT, LVW >3750 lb	...	0.32	4.4	—	0.7	0.08	0.80	0.40	5.5	0.97	0.97	0.10
HLDT, ALVW <5750 lb	0.32	—	4.4	—	0.7	—	0.80	0.46	6.4	0.98	0.98	0.10
HLDT, ALVW >5750 lb	0.39	—	5.0	—	1.1	—	0.80	0.56	7.3	1.53	1.53	0.12

[a]Useful life 120,000 miles/11 years for all HLDT standards and for THC standards for LDT.
Abbreviations:
LVW—loaded vehicle weight (curb weight +300 lb)
ALVW—adjusted LVW [the numerical average of the curb weight and the gross vehicle weight rating (GVWR)]
LLDT—light light-duty truck (below 6000 lb GVWR)
HLDT—heavy light-duty truck (above 6000 lb GVWR)

TABLE 6.4—EPA Tier 2 emission standards, FTP 75, g/mile [401].

	50,000 miles					120,000 miles				
Bin #	NMOG	CO	NOx	PM	HCHO	NMOG	CO	NOx*	PM	HCHO
Temporary Bins										
MDPVc						0.280	7.3	0.9	0.12	0.032
10[a]	0.125 (0.160)	3.4 (4.4)	0.4	—	0.015 (0.018)	0.156 (0.230)	4.2 (6.4)	0.6	0.08	0.018 (0.027)
9[a]	0.075 (0.140)	3.4	0.2	—	0.015	0.090 (0.180)	4.2	0.3	0.06	0.018
Permanent Bins										
8b	0.100 (0.125)	3.4	0.14	—	0.015	0.125 (0.156)	4.2	0.20	0.02	0.018
7	0.075	3.4	0.11	—	0.015	0.090	4.2	0.15	0.02	0.018
6	0.075	3.4	0.08	—	0.015	0.090	4.2	0.10	0.01	0.018
5	0.075	3.4	0.05	—	0.015	0.090	4.2	0.07	0.01	0.018
4	—	—	—	—	—	0.070	2.1	0.04	0.01	0.011
3	—	—	—	—	—	0.055	2.1	0.03	0.01	0.011
2	—	—	—	—	—	0.010	2.1	0.02	0.01	0.004
1	—	—	—	—	—	0.000	0.0	0.00	0.00	0.000

* Average manufacturer fleet NOx standard is 0.07 g/mile.
[a]Bin deleted at end of 2006 model year (2008 for HLDTs).
[b]The higher temporary NMOG, CO and HCHO values apply only to HLDTs and expire after 2008.
[c]An additional temporary bin restricted to MDPVs, expires after model year 2008.
[d]Optional temporary NMOG standard of 0.195 g/mile (50,000) and 0.280 g/mile (120,000) applies for qualifying LDT4s and MDPVs only.
[e]Optional temporary NMOG standard of 0.100 g/mile (50,000) and 0.130 g/mile (120,000) applies for qualifying LDT2s only.
[f]50,000 mile standard optional for diesels certified to Bin 10.

was lowered to 120–300 ppm, which has dropped further to 30–80 ppm.

California Standards

Tier 1/LEV California emissions standards were in effect through the year 2003. In the year 2004, LEV II regulations went into effect. The current California emissions standards are defined through the following emissions categories:

Tier 1

1. Transitional Low Emissions Vehicles (TLEV)
2. Low Emissions Vehicles (LEV)
3. Ultra Low Emissions Vehicles (ULEV)
4. Super Ultra Low Emissions Vehicles (SULEV)
5. Zero Emissions Vehicles (ZEV)

Table 6.5 summarizes the California ARB standards for the new light-duty vehicles and Table 6.6 summarizes the standards for the medium-duty vehicles [401]. After the year 2003, Tier 1 and TLEV standards were eliminated. The same standards apply to gaseous pollutants for the diesel- and gasoline-fueled vehicles. PM standards apply to diesel vehicles only. Emissions are measured using the FTP 75 test and the results are expressed in g/mile. The additional SFTP

TABLE 6.5—California emission standards for light-duty vehicles, FTP 75, g/mile [401].

	50,000 miles/5 years					100,000 miles/10 years				
Category	NMOG[a]	CO	NOx	PM	HCHO	NMOG[a]	CO	NOx	PM	HCHO
Passenger Cars										
Tier 1	0.25	3.4	0.4	0.08	—	0.31	4.2	0.6	—	—
TLEV	0.125	3.4	0.4	—	0.015	0.156	4.2	0.6	0.08	0.018
LEV	0.075	3.4	0.2	—	0.015	0.090	4.2	0.3	0.08	0.018
ULEV	0.040	1.7	0.2	—	0.008	0.055	2.1	0.3	0.04	0.011
LDT1, LVW <3750 lb										
Tier 1	0.25	3.4	0.4	0.08	—	0.31	4.2	0.6	—	—
TLEV	0.125	3.4	0.4	—	0.015	0.156	4.2	0.6	0.08	0.018
LEV	0.075	3.4	0.2	—	0.015	0.090	4.2	0.3	0.08	0.018
ULEV	0.040	1.7	0.2	—	0.008	0.055	2.1	0.3	0.04	0.011
LDT2, LVW >3750 lb										
Tier 1	0.32	4.4	0.7	0.08	—	0.40	5.5	0.97	—	—
TLEV	0.160	4.4	0.7	—	0.018	0.200	5.5	0.9	0.10	0.023
LEV	0.100	4.4	0.4	—	0.018	0.130	5.5	0.5	0.10	0.023
ULEV	0.050	2.2	0.4	—	0.009	0.070	2.8	0.5	0.05	0.013

[a]NMHC for all Tier 1 standards.
Abbreviations:
LVW—loaded vehicle weight (curb weight +300 lb)
LDT—light-duty truck
NMOG—nonmethane organic gases
HCHO—formaldehyde

TABLE 6.6—California emission standards for medium-duty vehicles, FTP 75, g/mile [401].

	50,000 miles/5 years					120,000 miles/11 years				
Category	NMOG[a]	CO	NOx	PM	HCHO	NMOG[a]	CO	NOx	PM	HCHO
MDV1, 0–3750 lb										
Tier 1	0.25	3.4	0.4	—	—	0.36	5.0	0.55	0.08	—
LEV	0.125	3.4	0.4	—	0.015	0.180	5.0	0.6	0.08	0.022
ULEV	0.075	1.7	0.2	—	0.008	0.107	2.5	0.3	0.04	0.012
MDV2, 3751–5750 lb										
Tier 1	0.32	4.4	0.7	—	—	0.46	6.4	0.98	0.10	—
LEV	0.160	4.4	0.4	—	0.018	0.230	6.4	0.6	0.10	0.027
ULEV	0.100	4.4	0.4	—	0.009	0.143	6.4	0.6	0.05	0.013
SULEV	0.050	2.2	0.2	—	0.004	0.072	3.2	0.3	0.05	0.006
MDV3, 5751–8500 lb										
Tier 1	0.39	5.0	1.1	—	—	0.56	7.3	1.53	0.12	—
LEV	0.195	5.0	0.6	—	0.022	0.280	7.3	0.9	0.12	0.032
ULEV	0.117	5.0	0.6	—	0.011	0.167	7.3	0.9	0.06	0.016
SULEV	0.059	2.5	0.3		0.006	0.084	3.7	0.45	0.06	0.008
MDV4, 8501 10,000 lb										
Tier 1	0.46	5.5	1.3	—	0.028	0.66	8.1	1.81	0.12	—
LEV	0.230	5.5	0.7	—	0.028	0.330	8.1	1.0	0.12	0.040
ULEV	0.138	5.5	0.7	—	0.014	0.197	8.1	1.0	0.06	0.021
SULEV	0.069	2.8	0.35	—	0.007	0.100	4.1	0.5	0.06	0.010
MDV5, 10,000–14,000 lb										
Tier 1	0.60	7.0	2.0	—	—	0.86	10.3	2.77	0.12	—
LEV	0.300	7.0	1.0	—	0.036	0.430	10.3	1.5	0.12	0.052
ULEV	0.180	7.0	1.0	—	0.018	0.257	10.3	1.5	0.06	0.026
SULEV	0.090	3.5	0.5	—	0.009	0.130	5.2	0.7	0.06	0.013

[a]NMHC for all Tier 1 standards.
Abbreviations:
MDV—medium-duty vehicle (the maximum GVWR from 8500 to 14,000 lb). The MDV category is divided into five classes.
MDV1—MDV5, based on vehicle test weight. The definition of "test weight" in California is identical to the Federal ALVW.
NMOG—nonmethane organic gases
HCHO—formaldehyde

procedure was implemented in California between the years 2001 and 2005.

Low Emissions Vehicle II (LEV II) standards, provided in Table 6.7 [401], apply to the model years 2004 to 2010. In this regulation, light-duty truck and medium-duty vehicle categories of below 8500 lb gross weight were reclassified by in the year 2007. At present, most pick-up trucks and sport utility vehicles are required to meet the passenger car emissions standards. Medium-duty vehicles above 8500 lb gross weight (old MDV4 and MDV5) will still certify to the medium-duty vehicle standard; see the second half of Table 6.7. Under this standard, the same limits apply to both gasoline and diesel vehicles, including the PM standard.

California—Greenhouse Gas Emissions for Cars

As mentioned earlier, there is a concern towards the increase of the greenhouse gases, such as carbon dioxide, in the atmo-

TABLE 6.7—California LEV II emission standards [401].

Passenger Cars and LDVs <8500 lb, **g/mi**

	50,000 miles/5 years					120,000 miles/11 years				
Category	NMOG	CO	NOx	PM	HCHO	NMOG	CO	NOx	PM	HCHO
LEV	0.075	3.4	0.05	—	0.015	0.090	4.2	0.07	0.01	0.018
ULEV	0.040	1.7	0.05	—	0.008	0.055	2.1	0.07	0.01	0.011
SULEV	—	—	—	—	—	0.010	1.0	0.02	0.01	0.004

Medium Duty Vehicles, Durability 120,000 miles, g/mi

Weight (GVWR), lbs.	Category	NMOG	CO	NOx	PM	HCHO
8500–10,000	LEV	0.195	6.4	0.2	0.12	0.032
	ULEV	0.143	6.4	0.2	0.06	0.016
	SULEV	0.100	3.2	0.1	0.06	0.008
10,001–14,000	LEV	0.230	7.3	0.4	0.12	0.040
	ULEV	0.167	7.3	0.4	0.06	0.021
	SULEV	0.117	3.7	0.2	0.06	0.010

TABLE 6.8—California fleet average GHG emission standards [401].

		GHG Standard, g CO_2/mile (g CO_2/km)		CAFE Equivalent, mpg (*l*/100 km)	
Time Frame	**Year**	**PC/LDT1**	**LDT2**	**PC/LDT1**	**LDT2**
Near Term	2009	323 (201)	439 (274)	27.6 (8.52)	20.3 (11.59)
	2010	301 (188)	420 (262)	29.6 (7.95)	21.2 (11.10)
	2011	267 (166)	390 (243)	33.3 (7.06)	22.8 (10.32)
	2012	233 (145)	361 (225)	38.2 (6.16)	24.7 (9.52)
Medium Term	2013	227 (142)	355 (221)	39.2 (6.00)	25.1 (9.37)
	2014	222 (138)	350 (218)	40.1 (5.87)	25.4 (9.26)
	2015	213 (133)	341 (213)	41.8 (5.63)	26.1 (9.01)
	2016	205 (128)	332 (207)	43.4 (5.42)	26.8 (8.78)

sphere because of their global warming effect. In order to offset this, the California Air Resources Board (CARB) has implemented the greenhouse gases (GHG) emissions standard that became effective on January 1, 2006. The standard is to be phased in over the 2009 to 2016 period, as shown in Table 6.8 [401]. The implementation of these standards is expected to result in an average decrease in greenhouse gases of about 22 % in the year 2012 and about 30 % in the year 2016, relative to the level of GHG emissions in the year 2004.

The GHG standards are part of the California low emissions vehicle (LEV) legislation. There are two fleet average GHG requirements. One is for the passenger car/light-duty truck 1 (PC/LDT1) category, which includes all passenger cars and light-duty trucks below 3750 lb equivalent test weight (ETW). The other is for the light-duty truck 2 (LDT2) category, including light trucks between 3751 lb ETW and 8500 lb gross vehicle weight (GVW). Medium-duty passenger vehicles (MDPVs) from 8500 to 10,000 lb GVW are included in the LDT2 category.

Diesel Emissions Standards [401]

Emissions standards for the diesel engines relate to the control of CO, HC, NOx, and the particulates. EPA emissions standards call for progressively lower NOx and particulate emissions. This is evident from Fig. 6.15. In this figure, we did not include the CO emissions since they were constant at 15.5 g/bhp-h over the span of the last 20 or so years. Figure 6.16 is visually more apt in reflecting the chronology and tightening of the emissions standards over this time span [549].

The following emissions standards apply to the new diesel (compression-ignition) engines used in the heavy-duty highway vehicles. Heavy-duty vehicles are the vehicles of GVWR (gross vehicle weight rating) of above 8500 lb in the federal jurisdiction and above 14, 000 lb in California (model year 1995 and later). Diesel engines used in the heavy-duty vehicles are further divided into service classes by GVWR, which are described below.

- Light heavy-duty diesel engines: 8500 < LHDDE < 19,500 (14,000 < LHDDE < 19,500 in California, 1995+)
- Medium heavy-duty diesel engines: 19,500 ≤ MHDDE ≤ 33,000
- Heavy heavy-duty diesel engines (including urban bus): HHDDE > 33,000

Under the Federal light-duty Tier 2 regulation, phased in beginning in 2004, the vehicles of GVWR up to 10,000 lb used for personal transportation have been reclassified as "medium-duty passenger vehicles" (MDPV—primarily larger SUVs and passenger vans) and are subject to the light-duty vehicle legislation. For the model years 1988 to 2003,

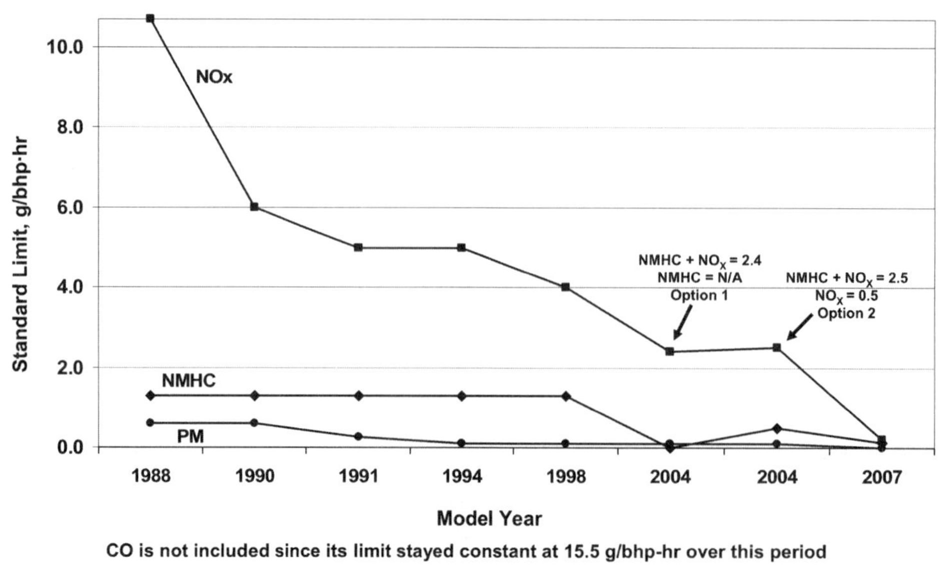

Fig. 6.15—Progression of the heavy-duty diesel standard.

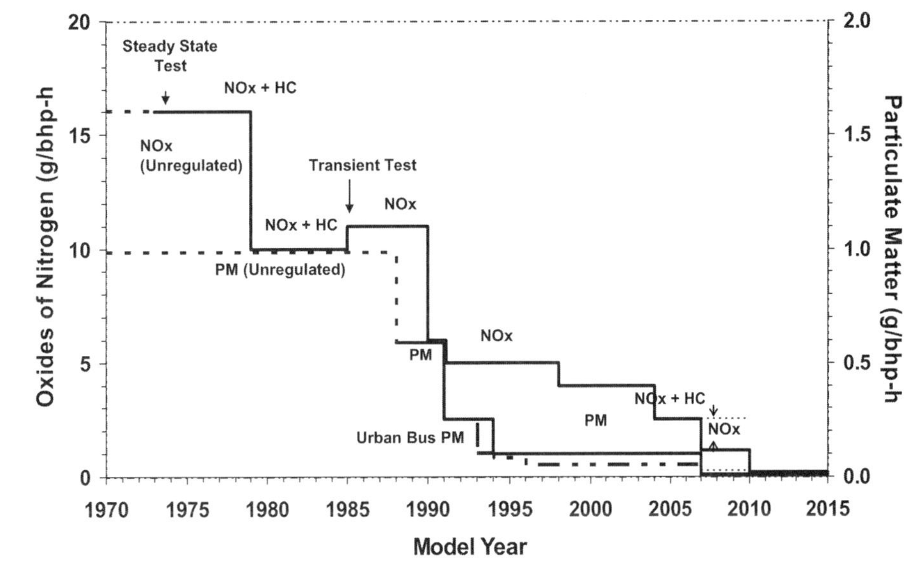

Fig. 6.16—Chronology of EPA heavy-duty engine emission standards [549].

U.S. Federal (EPA) and California (ARB) emissions standards for the heavy-duty diesel truck and bus engines are summarized in Table 6.9 [401]. The certification fuel sulfur content for the year 1994 and the following years have been established at 500 ppm by weight. Compliance with the emissions standards must be demonstrated over the useful life of the engine, which is defined as follows.

1. LHDDE—10 years/110,000 miles (whichever occurs first)
2. MHDDE—10 years/185,000 miles
3. HHDDE—10 years/435,000 miles/22,000 hours
4. Emissions warranty—5 years/100,000 miles

Voluntary Clean Fuel Fleet (CFF) emissions standard is the federal standard that applies to the 1998 to 2003 model year engines, both CI and SI, over 8500 lb GVWR, see Table 6.10 [401]. Vehicles must meet this standard as well as other applicable conventional standards for other pollutants. The emissions standards for the model year 2004 and beyond are provided in Table 6.11 [401]. For the year 2004, manufacturers have the flexibility to certify their engines by using either of the two options. The Federal 2004 standards for highway trucks are harmonized with California standards. However, California certifications for the model years 2005 to 2007 require the Supplemental Emissions Test and NTE (Not-to-exceed) limits of 1.25 times the FTP standards. California also has more stringent standards for the 2004 to 2006 model year engines for the public urban bus fleets. For the model year 2007 and later heavy-duty highway diesel engines, the regulation consists of two components: (1) emissions standards and (2) diesel fuel regulation.

The standard for the 2004 model year engines combines the non-methane hydrocarbons (NMHC) and NOx together, instead of considering them separately. This makes sense since both these pollutants are of concern because of their role in the generation of ozone. The concern about carbon dioxide emissions and their relationship to global warming is leading to a serious consideration of using diesel engines in light-duty trucks, vans, sport utility vehicles, and U.S. passenger cars. The inherent fuel efficiency of the diesel engine relative to that of the gasoline engine results in substantially

TABLE 6.9—EPA and California emission standards for heavy-duty diesel engines, g/bhp·h [401].

	EPA Emission Standards					California Emission Standards			
Year	HC	CO	NO_x	PM	NMHC	THC	CO	NO_x	PM
Heavy-Duty Diesel Truck Engines									
1987						1.3	15.5	6.0	0.60
1988	1.3	15.5	10.7	0.60					
1990	1.3	15.5	6.0	0.60					
1991	1.3	15.5	5.0	0.25	1.2	1.3	15.5	5.0	0.25
1994	1.3	15.5	5.0	0.10	1.2	1.3	15.5	5.0	0.10
1998	1.3	15.5	4.0	0.10					
Urban Bus Engines									
1991	1.3	15.5	5.0	0.25	1.2	1.3	15.5	5.0	0.10
1993	1.3	15.5	5.0	0.10			15.5	5.0	0.10
1994	1.3	15.5	5.0	0.07	1.2	1.3	15.5	5.0	0.10
1996	1.3	15.5	5.0	0.05[a]	1.2	1.3	15.5	4.0	0.05
1998	1.3	15.5	4.0	0.05[a]					

[a]In-use PM Standard 0.07.

TABLE 6.10—Clean fuel fleet program for heavy-duty SI and CI engines of the model year 1998-2003 [401].

Category[a]	CO (g/bhp·h)	NMHC+NOx (g/bhp·h)	PM (g/bhp·h)	HCHO (g/bhp·h)
LEV (Federal Fuel)	—	3.8	—	—
LEV (California Fuel)	—	3.5	—	—
ILEV	14.4	2.5		0.050
ULEV	7.2	2.5	0.05	0.025
ZLEV	0	0	0	0

[a]LEV—low emission vehicle; ILEV—inherently low emission vehicle; ULEV—ultra low emission vehicle; ZEV—zero emission vehicle.

lower emissions of CO_2 per driven mile. However, there may be emissions control challenges for such vehicles to meet the EPA standards. While the Federal 2004 standards for highway trucks are harmonized with the California standards, California certification for the model years 2005–2007 additionally requires the Supplemental Emissions Test (SET) and not to exceed (NTE) limits of 1.25 times the FTP (Federal test procedure) standards. California standards for the model years 2004 to 2006 engines for the public urban bus fleets are more stringent than the EPA standards.

The EPA is also limiting the diesel fuel sulfur content for on-highway diesel fuel to 15 ppm by weight, down from the previous limit of 500 ppm. Ultra low sulfur diesel (ULSD) fuel has been introduced as a "technology enabler" to pave the way for advanced, sulfur-intolerant exhaust emissions control technologies, such as catalytic diesel particulate filters and NOx catalysts. Such technologies were necessary to meet the 2007 emissions standards. Since the designation ULSD may refer to different maximum sulfur content in other parts of the world, it is advisable to use the designations S15, S500, and S5000 to describe diesel fuels that meet 15 ppm, 500 ppm, and 5000 ppm maximum sulfur content, respectively; as defined in the ASTM Standard D975. In North America (U.S. and Canada), ULSD and S15 are often used interchangeably.

Ultra Low Sulfur Diesel—ULSD (S15) regulation is based upon U.S. Diesel Emissions Control–Sulfur Effects (DECSE) Project. The project was a joint effort of the U.S. Department of Energy, two national laboratories, and the manufacturers of the heavy-duty compression ignition engines and of the emissions control systems [550]. Researchers conducted tests to determine the effects of the various levels of the fuel sulfur on emissions exhaust control systems to lower NOx and particulate matter (PM). The objective of the study was to determine if the EPA's 2007 emissions standards for heavy-duty diesel engines pertaining to lower NOx and PM emissions were achievable. The study employed compression ignition, direct injection (CIDI) diesel-cycle vehicles and four emissions control systems, which were as follows:

1. Diesel Oxidation Catalysts (DOCs)
2. Lean-NOx Catalysts (LNCs)
3. NOx Adsorber Catalysts (NACs)
4. Diesel Particulate Filters (DPFs)—two types were used: continuously regenerating DPF (CR-DPF) and a catalyzed DPF (CDPF).

The key findings of the study are summarized below.

1. *Diesel oxidation catalyst* (*DOC*) did not control PM emissions well enough to meet the EPA's 2007 standards. The DOC may be effective when used in combination with selective catalytic reduction (SCR) technology, either as a pre-catalyst for converting NO to NO_2 or as a post-catalyst to control ammonia slip. Ammonia slip is the unwanted ammonia, which is a by-product of the NOx reduction processes, such as SCR. Ammonia slip must be controlled to minimize its downstream impacts, such as corrosion, fouling, and undesirable emissions.
2. *Lean-NOx catalyst* had limited reduction efficiency, about ~20 %; hence it did not meet the EPA's 2007 emissions standards. However, it did meet the 2004 emissions regulations for light- and heavy-duty diesel engines. A DOC may be used to clean up HC slip when this approach is used. HC slip is the amount of HC (hydrocarbon) that is in excess of the amount needed to reduce NOx, which is exhausted into the air.
3. *NOx adsorber catalyst* may prove promising for meeting

TABLE 6.11—Emissions standards for the model year 2004 and later [401].

Pollutant	EPA Emission Standards for MY 2004		EPA and CARB Emission Standards for MY 2007 and Later[a]
	Option 1	Option 2	
$NMHC+NO_X$ (g/bhp-h)	2.4	2.5	—
NMHC (g/bhp-h)	n/a	0.5	0.14
PM (g/bhp-h)	—	—	0.01
NO_X (g/bhp-h)	-	-	0.20

[a]The PM emission standard took full effect in the 2007 heavy-duty engine model year. The NOx and NMHC standards will be phased in for diesel engines between 2007 and 2010. The phase-in would be on a percent-of-sales basis: 50 % from 2007 to 2009 and 100 % in 2010 (gasoline engines are subject to these standards based on a phase-in requiring 50 % compliance in 2008 and 100 % compliance in 2009).

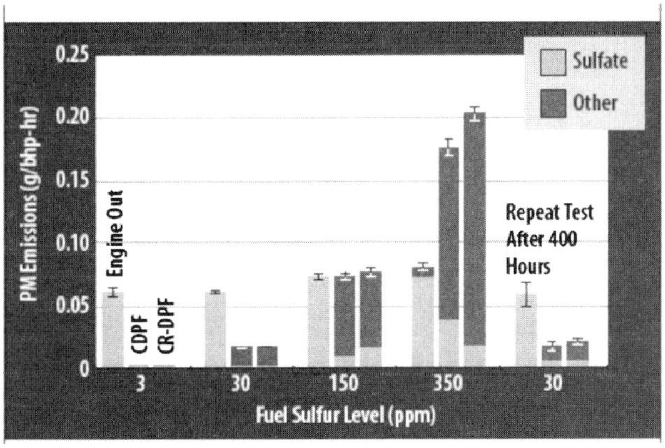

Fig. 6.17—Engine-out and post-DPF emissions of total PM and components versus fuel sulfur level for the OICA cycle (with 95% confidence intervals on estimated PM) [550].

future NOx standards. However, more study is needed to investigate the frequency of desulfurization and thermal degradation associated with the high-temperature desulfurization. More studies are also needed to address the long-term operation of the NOx adsorber catalyst, the durability of the engine and the catalyst, and other exhaust constituents, such as smoke levels during regeneration and its reduction and control.

4. *Diesel particulate filter*, when used in combination with the low-sulfur fuel, is capable of meeting future PM standards. Research is needed to demonstrate DPFs' effectiveness in combination with SCR and a NOx adsorber. Additional research on measurements for PM mass, size, and composition, as well as for air toxics must also be conducted.

The effect of the fuel sulfur on PM emissions is provided in Fig. 6.17 [550]. As one can see, at a fuel sulfur level of less than 30 ppm, particulate emissions primarily consist of sulfates. This was confirmed by a repeat test after 400 hours of operation.

The results and the test experiences from the DECSE project are being used by its successor, the Advanced Petroleum Based Fuels–Diesel Emissions Control (APBF-DEC) Project. The goal of the project is to identify the optimal combinations of fuels, lubricants, diesel engines, and emissions control systems which will meet the projected EPA emissions standards for the years 2008 to 2010. APBF-DEC will also identify the properties of the fuels and vehicle systems that could lead to even lower emissions beyond the year 2010 [550]. The project selected two emissions control technology systems for further study.

1. Selective Catalytic Reduction (SCR)/Diesel Particulate Filter (DPF)—The SCR is an emissions reduction device that, combined with a DPF and advanced fuel formulations, may reduce regulated (especially NOx), unregulated, and toxic emissions. Two types of SCR-based catalysts are being evaluated in combination with DPFs and possibly DOCs.
2. NOx Adsorber Catalyst/Diesel Particulate Filter—The NOx adsorber may significantly reduce NOx, HC, and CO emissions from the diesel engine exhaust. Combined with a DPF, the NOx adsorber can also effectively oxidize the PM and other unregulated emissions from the diesel exhaust. Two systems are being evaluated on light-, medium-, and heavy-duty engines; and light- and medium-duty vehicles.

The EPA has mandated the use of the S15 fuel beginning June 2006 and of high-efficiency catalyst exhaust emissions devices, or comparably effective technology, in the diesel-operated model year 2007 vehicles. The EPA considers that these actions will help vehicles meet the future nitrous oxide (NOx) and particulate matter (PM) emissions standards as well. It is important to note that the lowering of the diesel sulfur will impact its lubricity, cetane number, and the energy content. Lubricity is the diesel fuel's ability to protect various parts of the engine's fuel injection system against wear. The process required to reduce sulfur is likely to remove the naturally-occurring lubricity agents in the diesel fuel; which, if not replaced, can increase wear of the engine's fuel delivery system. The sulfur-removal process reduces the aromatics content as well, which negatively impacts fuel economy, only slightly, but boosts the cetane number. The American Society for Testing and Materials (ASTM International) has established a lubricity specification that is defined in its standard ASTM D975, which has been in effect since January 1, 2005.

While our discussion so far has been limited to the gasoline engines and on-highway heavy-duty diesel engines, it is important to note that the emissions standards for non-road, marine, and recreational engines also exist. Table 6.12 provides Tier 2 marine emissions standards [401]. There exists a voluntary "Blue Sky Series" program, which permits the manufacturers to certify their engines to these more strin-

TABLE 6.12—Tier 2[a] marine emission standards [401].

Category	Displacement (D) dm^3 per cylinder	CO g/kWh	NOx+THC g/kWh	PM g/kWh	Date
1	Power ≥37 kW; D<0.9	5.0	7.5	0.40	2005
	0.9≤D<1.2	5.0	7.2	0.30	2004
	1.2≤D<2.5	5.0	7.2	0.20	2004
	2.5≤D<5.0	5.0	7.2	0.20	2007[b]
2	5.0≤D<15	5.0	7.8	0.27	2007[b]
	Power <3300 kW; 15≤D<20	5.0	8.7	0.50	2007[b]
	Power ≥3300 kW; 15≤D<20	5.0	9.8	0.50	2007[b]
	20≤D<25	5.0	9.8	0.50	2007[b]
	25≤D<30	5.0	11.0	0.50	2007[b]

[a]Tier 1 standards are equivalent to the MARPOL Annex VI NOx limits.
[b]Tier 1 certification requirement starts in 2004.

TABLE 6.13—"Blue Sky Series" voluntary emission standards [401].

Displacement (D) dm³ per cylinder	NOx+THC g/kWh	PM g/kWh
Power ⩾37 kW & D<0.9	4.0	0.24
0.9⩽D<1.2	4.0	0.18
1.2⩽D<2.5	4.0	0.12
2.5⩽D<5.0	5.0	0.12
5.0⩽D<15	5.0	0.16
15⩽D<20 and Power <3300 kW	5.2	0.30
15⩽D<20 and Power ⩾3300 kW	5.9	0.30
20⩽D<25	5.9	0.30
25⩽D<30	6.6	0.30

gent emissions standards. The qualifying emissions limits for this program are listed in Table 6.13 [401]. The Blue Sky program begins upon the publication of the rule and extends up to the year 2010.

Some recreational vessel standards became effective at the beginning of the year 2006. Others will become effective in the years that will be specified later. The phase-in schedule, listed in Table 6.14, depends on the size of the engine [401]. Recreational engines are also subject to not-to-exceed limits. There are no smoke requirements for the recreational marine diesel engines. Similar to the commercial vessels, voluntary "Blue Sky Series" limits exist for the recreational vessels also, which are based on a 45 % emissions reduction beyond the mandatory standards.

Useful life and warranty periods for marine engines are listed in Table 6.15 [401]. The periods are specified in operating hours and in years, whichever occurs first. The relatively short useful life period for Category 3 engines is based on the time the engines operate before being rebuilt for the first time. For non-road diesel standards, please check out Ref [401].

Proposed New Emissions Standards

In February 2006, the EPA proposed new standards to control hazardous air pollutants, especially benzene, from mobile sources to further improve the nation's air quality [538].

TABLE 6.14—Recreational marine diesel engines standards [401].

Displacement (D) dm³ per cylinder	CO g/kWh	NOx+HC g/kWh	PM g/kWh	Date
0.5⩽D<0.9	5.0	7.5	0.40	2007
0.9⩽D<1.2	5.0	7.2	0.30	2006
1.2⩽D<2.5	5.0	7.2	0.20	2006
D⩾2.5	5.0	7.2	0.20	2009

TABLE 6.15—Useful life and emission warranty periods [401].

	Useful Life		Warranty Period	
Category	hours	years	hours	years
Category 3	10,000	3	10,000	3
Category 2	20,000	10	10,000	5
Category 1	10,000	10	5000	5
Recreational	1000	10	500	3

TABLE 6.16—Proposed 20° F FTP exhaust emission standards [538].

Vehicle GVWR and Category	NMHC Sales-Weighted Fleet Average Standard (grams/mile)
⩽6000 lb: Light-duty vehicles (LDV) and Light light-duty trucks (LLDT)	0.3
>6000 lb: Heavy light-duty trucks (HLDT) up to 8500 lbs and Medium-duty passenger vehicles (MDPV) up to 10,000 lb	0.5

Essentially, the EPA is proposing controls on gasoline, passenger vehicles, and portable gasoline containers (gas cans) to significantly reduce emissions of benzene and other hazardous air pollutants, called mobile source air toxics (MSATs). The proposal consists of the three sets of standards, which are as follows.

1. Light-Duty Vehicle Emissions Standards propose control of both the exhaust and the evaporative emissions from passenger vehicles. These will significantly reduce nonmethane hydrocarbon (NMHC) emissions from passenger vehicles at cold temperatures. These hydrocarbons include many mobile source air toxics, including benzene and VOC.
2. Gasoline Fuel Standards to limit the benzene content of all gasoline, both reformulated and conventional.
3. Portable Gasoline Container (Gas Can) Controls to reduce the hydrocarbon emissions from evaporation, permeation, and spillage.

Light-Duty Vehicle Cold Temperature Emissions Standards

These standards apply to Tier 2 gasoline-fueled vehicles and will achieve proportional NMHC control from the 75 °F Tier 2 standards to the 20 °F test point. The purpose of these standards, which are provided in Table 6.16 [538], is to achieve the maximum hydrocarbon emissions reductions, by utilizing the current emissions control hardware. These standards are expected to reduce PM resulting from the cold-temperature operation as well. The implementation of these standards is proposed to begin in the 2010 model year for LDVs/LLDTs and 2012 model year for HLDTs/MDPVs.

Evaporative Emissions Standards

These standards apply to all light-duty vehicles, light-trucks, and medium-duty passenger vehicles. The proposed standards are equivalent to California's LEV II standards and are shown in Table 6.17 [538]. Theses standards are expected to lead to a 20 to 50 % reduction in evaporative emissions, depending upon the vehicle weight class and the type of test, relative to the Tier 2 standards that will be in effect in the years immediately preceding the implementation of the presently proposed standards. The implementation of these standards is proposed in the model year 2009 for LDVs/LLDTs and the model year 2010 for HLDTs/MDPVs.

Gasoline Benzene Control Program

The standard is proposed to be implemented on January 1, 2011. The refiners are expected to meet an average gasoline benzene content of 0.62 % by volume on the gasolines, both reformulated and conventional, that they produce. The

TABLE 6.17—Proposed evaporative emission standards (g-HC/test) [538].

Vehicle Class	3-Day Diurnal[a] Plus Hot Soak[b]	Supplemental 2-Day Diurnal[a] Plus Hot Soak[b]
LDVs	0.50	0.65
LLDTs	0.65	0.85
HLDTs	0.90	1.15
MDPVs	1.00	1.25

[a]Diurnal emissions (or diurnal breathing losses) mean evaporative emissions as a result of daily temperature cycles or fluctuations for successive days of parking in hot weather.
[b]Hot soak emissions (or hot soak losses) are the evaporative emissions from a parked vehicle immediately after turning off the hot engine. For the evaporative emissions test procedure, diurnal and hot soak emissions are measured in an enclosure commonly called the SHED (Sealed Housing for Evaporative Determination). Larger vehicles may have greater nonfuel evaporative emissions, probably due to an increased amount of interior trim, vehicle body surface area, and larger tires.

implementation will result in the largest feasible overall reductions in benzene.

Proposed Emissions Control Program for Gas Cans

The purpose of this standard is to minimize the entry of the volatile organic compounds (VOCs) into the atmosphere, since they are ozone precursors, and the VOC-based toxics, such as benzene and toluene. Gas cans are to meet a maximum of 0.3 grams per gallon per day (g/gal/day) of HC as evaporative and permeation losses, as measured over a diurnal test cycle. The cans would be tested as a system with their spouts attached. The test is to be carried out by placing the cans in an environmental chamber which simulates summertime ambient temperature conditions and cycling the cans through the 24-hour temperature profile (72–96 °F). The test procedures would ensure that gas cans meet the emissions standard over a range of in-use conditions, such as different temperatures, different fuels, and taking into consideration factors affecting durability. As an aspect of considering the proposed standard's technological feasibility, EPA is proposing to require manufacturers to meet the standard beginning January 1, 2009.

Emissions Standards—European Union

European emissions regulations for new heavy-duty diesel engines are commonly referred to as Euro I to Euro V. Arabic numerals are also used (Euro 1 to Euro 5). In the discussion that follows, the Roman numerals are used in heavy-duty engine standards, and the Arabic numerals are used for light-duty vehicle standards.

Emissions Regulations for New Light Duty Vehicles

These regulations cover cars and light commercial vehicles, as specified in the Directive 70/220/EEC. The chronology of the standards that were implemented is provided below.

1. Euro 1 standards (also known as EC 93).
2. Euro 2 standards (EC 96).
3. Euro 3/4 standards (2000/2005). These standards included minimum diesel cetane number of 51 (year 2000), maximum diesel sulfur content of 350 ppm in the year 2000 and 50 ppm in the year 2005, and maximum petrol (gasoline) sulfur content of 150 ppm in the year 2000 and 50 ppm in the year 2005. "Sulfur-free" diesel and gasoline fuels (≤10 ppm S) must be available from the year 2005, and become mandatory from the year 2009.

Emissions standards for passenger cars and light commercial vehicles, vehicle categories M1 and N1, respectively, are summarized in Tables 6.18 and 6.19 [401]. Please note that Euro 2 imposes limits for diesel engines that differ from those for the gasoline engines. Diesel engines have more stringent CO standards but are allowed higher NOx. Gasoline vehicles are exempt from PM standards through Euro 4. Euro 5 proposes PM standards for lean-burning gasoline cars. Values listed in the tables are the new approval emissions limits, unless stated otherwise. Vehicles must stay in compliance with emissions standards for a vehicle's useful life, i.e., emissions durability, which for regulation purposes is defined as follows.

1. Euro 3 stage—80,000 km or five years, whichever occurs first. Instead of an actual run, the manufacturers may use the following deterioration factors: 1.2 for CO, HC, and NOx for gasoline or 1.1 for CO, NOx, HC+NOx, and 1.2 for PM for diesel.

TABLE 6.18—EU emission standards for passenger cars (Category M1[a]), g/km [401].

Tier	Date	CO	HC	HC+NOx	NOx	PM
Diesel						
Euro 1[a]	1992.07	2.72 (3.16)	—	0.97 (1.13)	—	0.14 (0.18)
Euro 2, IDI	1996.01	1.0	—	0.7	—	0.08
Euro 2, DI	1996.01[d]	1.0	—	0.9	—	0.10
Euro 3	2000.01	0.64	—	0.56	0.50	0.05
Euro 4	2005.01	0.50	—	0.30	0.25	0.025
Euro 5[c]	mid-2008	0.50	—	0.25	0.20	0.005
Petrol (Gasoline)						
Euro 1[b]	1992.07	2.72 (3.16)	—	0.97 (1.13)	—	—
Euro 2	1996.01	2.2	—	0.5	—	—
Euro 3	2000.01	2.30	0.20	—	0.15	—
Euro 4	2005.01	1.0	0.10	—	0.08	—
Euro 5[c]	mid-2008	1.0	0.075	—	0.06	0.005[e]

[a]Before Euro 5, passenger vehicles >2500 kg were type approved as Category N1 vehicles.
[b]Values in brackets are conformity of production (COP) limits.
[c]Proposed.
[d]Until 1999.09.30 (after that date DI engines must meet the IDI limits).
[e]Applicable only to vehicles using lean burn DI engines.

TABLE 6.19—EU emission standards for light commercial vehicles (Category N1), g/km [401].

Class[a]	Tier	Date	CO	HC	HC + NOx	NOx	PM
Diesel							
N1, Class I <1305 kg	Euro 1	1994.10	2.72	-	0.97	-	0.14
	Euro 2, IDI	1998.01	1.0	-	0.70	-	0.08
	Euro 2, DI	1998.01[c]	1.0	-	0.90	-	0.10
	Euro 3	2000.01	0.64	-	0.56	0.50	0.05
	Euro 4	2005.01	0.50	-	0.30	0.25	0.025
	Euro 5[b]	mid-2008	0.50	-	0.25	0.20	0.005
N1, Class II 1305-1760 kg	Euro 1	1994.10	5.17	-	1.40	-	0.19
	Euro 2, IDI	1998.01	1.25	-	1.0	-	0.12
	Euro 2, DI	1998.01[c]	1.25	-	1.30	-	0.14
	Euro 3	2001.01	0.80	-	0.72	0.65	0.07
	Euro 4	2006.01	0.63	-	0.39	0.33	0.04
	Euro 5[b]	mid-2008	0.63	-	0.32	0.26	0.005
N1, Class III >1760 kg	Euro 1	1994.10	6.90	-	1.70	-	0.25
	Euro 2, IDI	1998.01	1.5	-	1.20	-	0.17
	Euro 2, DI	1998.01[c]	1.5	-	1.60	-	0.20
Class[a]	**Tier**	**Date**	**CO**	**HC**	**HC + NOx**	**NOx**	**PM**
	Euro 3	2001.01	0.95	-	0.86	0.78	0.10
	Euro 4	2006.01	0.74	-	0.46	0.39	0.06
	Euro 5[b]	mid-2008	0.74	-	0.38	0.31	0.005
Petrol (Gasoline)							
N1, Class I <1305 kg	Euro 1	1994.10	2.72	-	0.97	-	-
	Euro 2	1998.01	2.2	-	0.50	-	-
	Euro 3	2000.01	2.3	0.20	-	0.15	-
	Euro 4	2005.01	1.0	0.1	-	0.08	-
	Euro 5[b]	mid-2008	1.0	0.075	-	0.060	0.005[d]
N1, Class II 1305-1760 kg	Euro 1	1994.10	5.17	-	1.40	-	-
	Euro 2	1998.01	4.0	-	0.65	-	-
	Euro 3	2001.01	4.17	0.25	-	0.18	-
	Euro 4	2006.01	1.81	0.13	-	0.10	-
	Euro 5[b]	mid-2008	1.81	0.10	-	0.075	0.005[d]
N1, Class III >1760 kg	Euro 1	1994.10	6.90	-	1.70	-	-
	Euro 2	1998.01	5.0	-	0.80	-	-
	Euro 3	2001.01	5.22	0.29	-	0.21	-
	Euro 4	2006.01	2.27	0.16	-	0.11	-
	Euro 5[b]	mid-2008	2.27	0.12	-	0.082	0.005[d]

[a]For Euro 1/2 the reference mass classes were Class I < 1250 kg, Class II 1250-1700 kg, Class III > 1700 kg.
[b]Proposed
[c]until 1999.09.30 (after that date DI engines must meet the IDI limits)
[d]applicable only to vehicles using lean burn DI engines

2. Euro 4 stage—100,000 km or five years, whichever occurs first.
3. Euro 5 (draft proposal)—160,000 km or five years, whichever occurs first.

The year 2000/2005 regulations had provisions to include emissions requirements for on-board diagnostics (OBD) equipped vehicles and for a low temperature emissions test (−7 °C) for gasoline vehicles, effective year 2002. The limits for cars are 15 g/km for CO and 1.8 g/km for HC, measured over the urban part of the test only.

TABLE 6.20—European OBD threshold limits, g/km [401].

Category	Class	Tie[a]	Date	CO	HC	NOx	PM
Diesel							
M_1		EU 3	2003	3.20	0.40	1.20	0.18
		EU 4	2005	3.20	0.40	1.20	0.18
N_1	I	EU 3	2005	3.20	0.40	1.20	0.18
		EU 4	2005	3.20	0.40	1.20	0.18
	II	EU 3	2006	4.00	0.50	1.60	0.23
		EU 4	2006	4.00	0.50	1.60	0.23
	III	EU 3	2006	4.80	0.60	1.90	0.28
		EU 4	2006	4.80	0.60	1.90	0.28
Petrol (*Gasoline*)							
M_1		EU 3	2000	3.20	0.40	0.60	...
		EU 4	2005	1.90	0.30	0.53	...
N_1	I	EU 3	2000	3.20	0.40	0.60	...
		EU 4	2005	1.90	0.30	0.53	...
	II	EU 3	2001	5.80	0.50	0.70	...
		EU 4	2005	3.44	0.38	0.62	...
	III	EU 3	2001	7.30	0.60	0.80	...
		EU 4	2005	4.35	0.47	0.70	...

[a]Euro 4 threshold limits are proposed values, still under discussion.
Note: Passenger cars category M_1 > 2500 kg or with more than six seats meet OBD requirements for Category N_1.

Starting from the Euro 3, vehicles are required to be equipped with an on-board diagnostic system for emissions control. The proposed emissions limits for such vehicles are provided in Table 6.20 [401]. The thresholds are based on the NEDC (cold start ECE + EUDC) test. NEDC is an abbreviation for New European Driving Cycle and ECDE is an abbreviation for Extra Urban Driving Cycle. To distinguish from the U.S. OBD, the European limits are also referred to as the EOBD (European OBD).

European Heavy-duty Diesel Truck and Bus Engine Emissions Standards

The emissions standards provided in this section are from consolidated European Directive 05/55/EC and apply to all motor vehicles with a "technically permissible maximum laden mass" over 3500 kg (~7700 lb), equipped with compression ignition engines, or positive ignition natural gas or LPG engines. The chronology and highlights of the Euro emissions standards for this type of equipment are as follows.

1. Euro I standards were introduced in the year 1992, followed by the introduction of the Euro II standards in 1996. These applied to both truck engines and urban buses; the urban bus standards, however, were voluntary.
2. Euro III standards were introduced in the year 2000, followed by the introduction of Euro IV standards in the year 2005. Euro V is slated to go into effect for the model year 2008 vehicles. The standards have the voluntary compliance of stricter emissions limits for extra low emissions vehicles, known as "enhanced environmentally friendly vehicles," or EEVs.

TABLE 6.22—Emission standards for diesel and gas engines, ETC Test, g/kWh [401].

Tier	Date	Test	CO	NMHC	CH4[a]	NOx	PM[b]
Euro III	1999.10, EEVs only	ETC	3.0	0.40	0.65	2.0	0.02
	2000.10	ETC	5.45	0.78	1.6	5.0	0.16 0.21[c]
Euro IV	2005.10		4.0	0.55	1.1	3.5	0.03
Euro V	2008.10		4.0	0.55	1.1	2.0	0.03

[a]For natural gas engines only.
[b]Not applicable for gas fueled engines at the year 2000 and 2005 stages.
[c]For engines of less than 0.75 dm^3 swept volume per cylinder and a rated power speed of more than 3000 min^{-1}.

The European Commission has submitted a proposal for Euro VI emissions standards in the year 2007, which in addition to having more stringent emissions limits may include new standards for pollutants that are as yet unregulated. The standard is expected to become effective for heavy duty engines for the model year 2013. The unregulated pollutants this standard plans to limit may result from the use of the alternative fuels or additive-based emissions control systems, or both. The commission may also assess the need to instill additional limits on particle levels and size. Table 6.21 contains a summary of the emissions standards and their implementation dates [401]. Dates in the tables refer to the new approvals; the dates for all type approvals are in most cases one year later (EU type approvals are valid longer than one year).

Engine test change in Euro III Standard was made in the year 2000—the old steady-state engine test cycle ECE R-49 was replaced by two cycles: the European Stationary Cycle (ESC) and the European Transient Cycle (ETC). Smoke opacity is measured on the European Load Response (ELR) test. For approval of new diesel vehicles with engines accord-

TABLE 6.21—EU emission standards for HD diesel engines, g/kWh (smoke in m^{-1}) [401].

Tier	Date	Test	CO	HC	NOx	PM	Smoke
Euro I	1992, <85 kW	ECE R-49	4.5	1.1	8.0	0.612	
	1992, >85 kW		4.5	1.1	8.0	0.36	
Euro II	1996.10		4.0	1.1	7.0	0.25	
	1998.10		4.0	1.1	7.0	0.15	
Euro III	*1999.10, EEVs only*	*ESC & ELR*	*1.5*	*0.25*	*2.0*	*0.02*	*0.15*
	2000.10	ESC & ELR	2.1	0.66	5.0	0.10 0.13[a]	0.8
Euro IV	2005.10		1.5	0.46	3.5	0.02	0.5
Euro V	2008.10		1.5	0.46	2.0	0.02	0.5

[a]For engines of less than 0.75 dm^3 swept volume per cylinder and a rated power speed of more than 3000 min^{-1}.

TABLE 6.23—Emission durability periods [401].

Period[a]	Vehicle Category[b]
100,000 km or 5 years	N1 and M2
200,000 km or 6 years	N2
	N3 ≤ 16 ton
	M3 Class I, Class II, Class A, and Class B ≤ 7.5 ton
500,000 km or 7 years	N3 > 16 ton
	M3 Class III, and Class B > 7.5 ton

[a]km or year period, whichever comes first.
[b]Mass designations (in tons) are "maximum technically permissible mass."

ing to the Euro III standard, the manufacturers have the option of selecting either of these tests. However, for approval according to Euro IV (2005) and later and for EEVs emissions, limit values must be determined on both the ETC and the ESC/ELR tests. Emissions standards for diesel engines that are tested on the ETC test cycle, as well as for heavy-duty gas engines, are summarized in Table 6.22 [401]. Effective October 2005 for new approvals and October 2006 for all approvals, the manufacturers are required to demonstrate that the engines will comply with the emissions limits for the useful life of the equipment, which is defined in Table 6.23. [401].

Greenhouse Gas Emissions

Like the United States, Europe is also concerned about the greenhouse gas (GHG) emissions from the transportation sector. Standards for CO_2 emissions are provided in Table 6.24 [401]. The European Commission has signed voluntary agreements with the automotive industry to reduce the emissions of carbon dioxide (CO_2). The automobile manufacturers include ACEA, JAMA (Japanese Automobile Manufacturers Association), and KAMA (Korean Automobile Manufacturers Association). These automaker groups, who supply 90 % the cars sold in Europe, are facing new challenges to reach the 140 g/km CO_2 target by the year 2008/09.

As mentioned in the earlier part of this chapter, emissions control is a priority in most developed and developing countries. Here we have summarized the emissions control for the United States and the European Union. The reader interested in such standards for other nations should check Ref [401], which provides an excellent and comprehensive discussion on the United States and the worldwide emissions standards.

Gasoline Engine Emissions Control

Gasoline Properties versus Emissions

Gasoline properties that impact emissions are its volatility, antiknock quality, and the ability to inhibit the intake system deposit formation. *Volatility* depends upon the molecular weight and the degree of branching of the gasoline's hydrocarbon components. Since the demand for gasoline is greater than the amount naturally present in the crude (straight run gasoline), refiners produce the additional amount by chemical processes, such as cracking and polymerization. Cracking is the process that breaks down the larger molecules into the smaller molecules and polymerization is the process that combines the smaller molecules into the larger molecules. Cracking could be thermal or catalytic. Catalytic cracking is better because it yields gasoline with better antiknock properties. These processes were discussed in the Chapter 2, the chapter on Mineral Base Oils.

Volatility of gasoline is measured by a number of techniques. These include Reid vapor pressure (RVP), the ASTM distillation curve (distillation profile), and equilibrium air distillation (EAD). RVP is used to calculate the vapor-liquid ratio (V/L) that indicates a modern car's tendency to vapor lock. The ASTM distillation curve, based on the percent of gasoline evaporated as a function of temperature, is a measure of gasoline composition based on the boiling point, see Fig. 6.18. Different parts of the curve relate to different performance parameters [551]. These are appropriately marked in the figure. The front end of the curve relates to the ease of starting, the middle part relates to warm-up, and the tail end relates to engine performance after the warm-up. Mid values for percent evaporated that impact these properties are 10, 50, and 90, respectively. Typical percent ranges, along with approximate temperatures, are shown in Table 6.25. A large departure from these values in each range can lead to problems. For example, too high a volatility at the front end could lead to poor hot starting, vapor lock, and high evaporation losses. Too low a volatility, on the other hand, will result in starting difficulty, especially at low temperatures. Similarly, in the mid range, there is a compromise between icing, for cars equipped with carburetors, if the volatility is high; and poor warm-up, rough acceleration, and poor short trip economy, if the volatility is low. A similar compromise between poor long trip economy and combustion chamber deposit-forming tendency exists at the tail end of the boiling range. Since volatility is a function of the weather and the altitude, it needs to be adjusted for gasoline to vaporize and burn properly. Otherwise, an increase in undesirable emissions will result.

Fuel parameters that affect cold weather driveability differ amongst gasolines from different sources. Hence, an overall *Driveability Index* (DI) was developed [552,553]. While there are many versions, two that are most commonly used are given in equations that follow, where T10, T50, and T90 represent 10, 50, and 90 % evaporated gasoline temperatures.

$$DI = 0.5\ T10 + T50 + 0.5\ T90$$

$$DI = T10 + 3\ T50 + T90$$

The use of the driveability index is confined to the United States. The Europeans use either T50 or E100 point (percent evaporated at 100°C) to assess cold weather driveability [553]. Unlike the ASTM distillation method where the evaporation occurs in the presence of the fuel vapor only, in the Equivalent Air Distillation (EAD) process, the fuel is va-

TABLE 6.24—Carbon dioxide emissions and interim targets (2003), g/km [401].

	CO_2 in 2003			Reduction since		
	Total	Gasoline	Diesel	1995	2002	Interim Target
ACEA	163	171	154	11.9 %	1.2 %	165–170 (2003)
JAMA	172	170	177	12.2 %	1.1 %	165–175 (2003)
KAMA	179	171	201	9.1 %	2.2 %	165–170 (2004)
EU-15	164	171	157	11.8 %	1.2 %	...

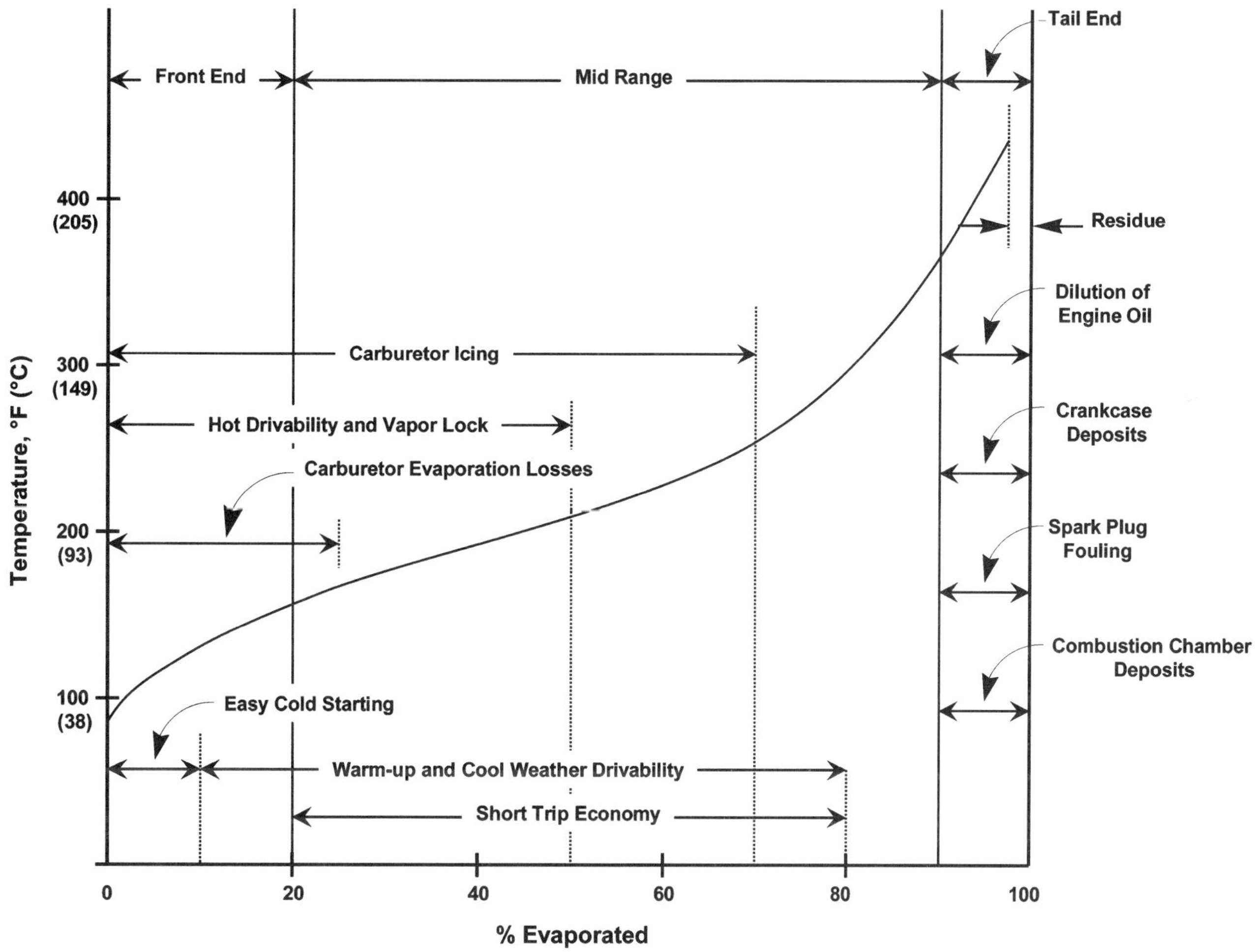

Fig. 6.18—Gasoline properties versus boiling point [551].

porized with the assistance of air flow. This is an attempt to duplicate the evaporation in an engine where the fuel is vaporized under the influence of the intake air. In this method, air and liquid fuel are supplied at a steady rate to an open system that is maintained at a specific temperature. Percent fuel evaporated is calculated by taking into account the total fuel introduced and the amount of fuel that passes through the apparatus. The results are shown as temperatures plotted against percent vaporized fuel. While in principle these charts can be used to calculate the air-vapor ratios, the results do not parallel those found in the actual engine because of the geometrical differences and the non-equilibrium conditions.

TABLE 6.25—Boiling ranges versus performance.

Boiling Range (°C)	Percent Evaporated	Performance Parameter
~30–70	0–20	Ease of starting
~71–150	21–80	Driving in cold weather, especially during warm-up, and fuel economy
~151–215	81–100	Driveability after warm-up, fouling and deposit-forming tendency

Fuel volatility can affect engine performance in many ways. Ignition requires the air-fuel ratio between 6:1 (rich mixture) and 20:1 (lean mixture), with 14.5:1 being optimal. During cold starting, the temperature is too low for proper fuel evaporation which necessitates the use of a rich mixture. This, in a carbureted engine, is achieved by the use of a choke to reduce the air supply. One would expect the higher fuel volatility to favor the cold starts, which is indeed the case when the ambient temperature is <20 °F (−6.7 °C). At higher temperatures, however, the volatility has little or no effect. The effect of volatility during the warm-up and the acceleration, on the other hand, is quite large. A correlation exists between the warm-up performance and the temperature at which 50 % evaporation occurs. Ambient temperature also affects this performance. In general, the lower the ambient temperature, the higher is the difficulty to achieve smooth acceleration. This may be partly related to the rate of the choke opening. A slow opening choke is likely to provide smooth running at low temperatures but can lead to higher HC emissions. A fast opening choke may lead to rough running at low ambient temperatures but will provide better fuel economy and improved HC emissions.

High fuel volatility may lead to icing problems in cars equipped with carburetors. Icing occurs in and below the venturi and is a consequence of evaporation of the low boil-

TABLE 6.26—The effect of gasoline modification on emissions.

Gasoline Modification	Regulated Emissions			Unregulated Emissions		Green House Emissions
	CO	HC	NO_X	Benzene	Aldehyde	CO_2
Benzene Reduction	=	=	=	+	-	-
Aromatics Reduction	+	+	=/-	+	-	?
Olefins Reduction	=	-	+	=	=	-
Sulfur Reduction	+	+	+	+	=/+	-
Vapor Pressure Reduction	+	+	=	=	=	-
Lower Volatility	+	+	=	+	+	-
Addition of Ethers/Oxygenates	+	+	+/-	+	-	?
Cleanliness Additives	+	+	+	+	=	+

Notes: + Positive benefit - Negative benefit = No benefit ? Benefit unknown

ing components of the fuel. Icing could result in inefficient running and stalling of the engine, primarily at idle engine speeds. Vapor lock is also related to high volatility. It occurs when at high temperatures the pressure build-up due to fuel vapors equals that of the pressure in the fuel system. This impairs the fuel flow and the engine loses its ability to run. The ASTM 10 % distillation point and RVP correlate with a fuel's tendency to vapor lock. Other problems associated with fuel volatility include evaporation losses, primarily from the fuel tank and the carburetor, and the engine lubricant dilution. Fuel loss from the tank relates to the ASTM distillation temperature around 71 °C (160 °F) and from the carburetor relates to the RVP, measured at ~38 °C (100 °F). Fuel dilution of the lubricant relates to the high boiling fraction or the back end, the components in the 90 % distillation temperature range, of the ASTM distillation curve.

The newer gasoline vehicles come equipped with injectors instead of carburetors. Fuel injection systems belong to three general groups. If the injector is installed in the throttle body on top of the air intake manifold, it is called a throttle body injection (TBI) system. A port fuel injector system (PFI) uses an injector at the intake port of each cylinder. It sprays fuel on the intake valves. A central port fuel injection system uses a central distribution valve connected to spring-loaded poppet nozzles that spray fuel on the intake valves of each cylinder. Fuel injectors are more precise in their fuel delivery than the carburetors; hence they provide better emissions control.

Antiknock quality is the primary determinant of the ignition quality in a gasoline engine. Knock is an audible noise that results from spontaneous ignition (detonation) of the end gases at *high loads* and *slow speeds*. This type of ignition is undesired because it not only results in the loss of power and the fuel economy but it can also lead to mechanical damage. Engine knock relates to a fuel's octane rating. Fuel with a high octane rating has a lower tendency to knock than the one with a low octane rating. The factors that influence an engine's tendency to knock include intake pressure and temperature, compression ratio, spark advance, and the coolant temperature. The higher the value of these parameters, the greater is the engine's tendency to knock. Near stoichiometric air-fuel mixtures also result in knock. While one can use reformulated gasoline to control knock, it is usually controlled by the use of the antiknock additives, or the octane boosters.

Engine knock is also a function of an engine's compression ratio, or CR. CR is defined as the ratio of the volume of gases in the cylinder at the beginning of the compression stroke and the volume of gases at the end of the compression stroke. Mathematically, it can be represented by the equation below, where V_D is the displaced volume, or the swept volume, that is, the volume swept by the piston in one stroke and equals the volume of the cylinder; and V_C is the volume of the compressed gases when the ignition takes places, that is, the volume of the combustion chamber.

$$CR = \frac{V_D + V_C}{V_C}$$

The usual compression ratio ranges for the gasoline engines are from 8:1 to 11:1, and for the diesel engines are from 15:1 to 25:1.

As mentioned above, the octane rating of the fuel determines the engine's ability to run knock-free. Higher compression ratios are desired because of the higher thermal efficiency, but the fuel octane rating is the limiting factor. If the octane rating of the fuel is not appropriate, high compression engines will tend to knock. For example, for a 6.5 CR engine, a fuel with an octane rating of 60 is fine; but for a 10 CR engine, a fuel with a minimum octane rating of 95 is necessary. After the onset of knock, thermal efficiency drops rapidly. However, an increase in the compression ratio results in higher cycle temperatures, which increase the formation of NOx.

Deposits can form in the intake system as well as in the combustion chamber. These are undesired because they not

TABLE 6.27—Chronology of United States gasoline regulations [548].

Year	Federal Action	California Action
1960		**Bromine Number** Control/reduce the formation of ozone
1971		**Summer Vapor Pressure** Reduce evaporative hydrocarbon emissions from vehicles and fuel distribution system, and reduce ozone.
1974	**Unleaded Gasoline** Reduce health risks from airborne lead oxides and enable use of catalytic converters on vehicles to reduce emissions	
1976-1980		**Sulfur Content** Reduce sulfur oxide emissions and particulates
1989	**Summer Vapor Pressure** Reduce evaporative hydrocarbon emissions from vehicles and fuel distribution system, and reduce ozone.	
1992	**Summer Vapor Pressure** Further Reduce evaporative hydrocarbon emissions and reduce ozone **Winter Oxygen** Reduce carbon monoxide emissions from vehicles in non-attainment areas	**Phase 1 Reformulated (Summer Vapor Pressure)** Further reduce evaporative hydrocarbon emissions and ozone – more restrictive than federal requirements **Winter Oxygen** Reduce carbon monoxide emissions from vehicles **Leaded Fuels Banned** Protect public health **Deposit Control Additives** Reduce carbon monoxide emissions
1995	**Phase I Reformulated (Summer Vapor Pressure)** Reduce ozone and toxics in non-attainment areas **Deposit Control Additives** Reduce vehicles emissions caused by fuel system deposits	
1996	**Leaded Fuels Banned** Protect public health	**Phase 2 Reformulated** Achieve maximum "cost-effective" reductions in vehicle emissions, including toxics
2000	**Phase II Reformulated** Further reduce ozone and toxics in non-attainment areas	
2004	**Tier 2 Sulfur Regulation** Ensure effectiveness of advanced emission control technologies	**Phase 3 Reformulated** Bans MTBE and provides more refinery blending flexibility

only have an adverse effect on driveability, power, and fuel consumption, but they also lead to an increase in undesirable emissions due to inefficient combustion. Therefore, it is important that the injectors, intake valves, and combustion chamber are deposit-free. Both the fuel composition and the additives, such as octane boosters, can affect deposit formation. A variety of deposit control additives are commercially available. These will be discussed in the section dealing with the fuel additives.

Formulated Gasoline

Attempts have been made to optimize gasoline performance with respect to the air quality. These include controlling its chemical composition and using additives. A number of gasoline components negatively impact emissions. These include oxygenates, aromatics, olefins, sulfur, heavy metals, and the high boiling fractions. The decision to remove lead from the gasoline emphasized the need for alternative octane boosters. Alcohols and ethers are among the most commonly used octane boosters. While these additives control CO, like other oxygenates in gasoline, they result in increased deposit formation that adversely affects the other emissions [554]. Hence, their amount in gasoline must also be controlled. Table 6.26 shows the effect of changes in the gasoline composition on emissions. The data suggest that a gasoline which has low volatility, contains low levels of sulfur, and is treated with ethers and cleanliness additives results in the lowest amount of undesirable emissions. Other parameters are somewhat less effective in controlling emissions.

Deposits as a problem is universally recognized by most OEMs, who recommend the use of fuels that contain deposit control additives. Such additives improve emissions, increase power, and lead to better fuel economy [555]. They do so by minimizing deposit formation and by cleaning the already formed deposits. Because of the described benefits, many OEMs have their own standards for intake system cleanliness.

Reformulated Gasolines (RFGs)

Acting on the above relationships, the EPA and the California Air Resources Board have established a number of regulations that control the gasoline properties to reduce emissions from the gasoline-fueled vehicles. Table 6.27 shows the chronology of the EPA-mandated changes to the gasoline composition [548]. Reformulated gasoline (RFG) is a gasoline whose composition is optimized to meet the specified emissions requirements. RFG is either Simple Model or Complex Model. Incidentally, the terms simple model and complex model to describe gasoline are used only in the United States.

Simple Model gasoline, introduced in 1995, was designed to decrease the volatile organic carbon (VOC) in gasoline by 15 % and no NOx change relative to the 1990 baseline fuel. Simple model gasoline contains minimum 2 weight percent oxygen and maximum 1 volume percent benzene and has an RVP of 7.2 for Class B and 8.1 for Class C gasoline.

TABLE 6.28—Federal Phase I and Phase II reformulated gasoline programs [551].

Phase	Effective Date	Reduction in Emissions, %[a] VOC	Toxics	NO_x
Phase I				
Simple Model	1995	Vapor pressure limits	≥16.5	No Increase
Complex Model	1998[b]	$17.1^2 \geq 36.6$[c]	≥16.5	≥1.5
Phase II				
Complex Model	2000	$\geq 27.4^2 \geq 29.0^3$	≥21.5	≥6.8

[a]Base line is 1990 refinery gasoline.
[b]Northern states.
[c]Southern states.

Complex Model gasoline, Phase I introduced in 1998 and Phase II introduced in the year 2000, has the objective of further reducing VOC, toxics (such as CO and aromatics), and NOx. Both simple and complex model gasolines contain deposit control additives. These gasolines were developed through regression modeling of the effects of the physical and chemical properties of the gasoline on VOC, toxics, and NOx. Table 6.28 provides the details on the reformulated gasolines [551].

In January 1, 2004, California changed from Phase II to Phase III reformulated gasoline. Phase III RFG prohibits the intentional blending of the MTBE into California gasoline and disallows its presence and that of the other oxygenates, except ethanol, above a specified small amount. This was done to stop the MTBE, which imparts an odor and a taste to water, from entering the fresh water resources of the state. Ethanol is the only oxygenate that the refiners can use to satisfy the federal oxygen mandate for most gasoline sold in California. In addition, Phase III RFG sets lower limits for sulfur from 30 ppm to 20 ppm and for benzene to 0.8 % by volume.

Reformulated gasoline has many advantages over simple gasoline. In addition to the lower probability of the potential problems, reformulated gasoline provides better driveability at all operating temperatures. This is achieved by controlling the intake valve and the combustion chamber deposits, which also result in improved emissions performance.

Starting in the year 2005, the European Union, through the Committee for Standardization Gasoline Specification CEN EN 228, has reduced sulfur to 50 ppm, and 10 ppm on a "balanced geographical basis, determined by each country" and has limited aromatics to a maximum content of 35.0 % by volume. All gasoline will be at 10 ppm maximum sulfur by the year 2009. Beginning in 2005, the Japanese Industrial Standard gasoline specification JIS K 2202 is also limiting the sulfur content to 50 ppm maximum and summertime vapor pressure to 65 kPa (9.4 psi) maximum.

Effect of Engine Design and Operating Variables on Emissions

Engine design and operating variables that impact emissions of a gasoline engine include the following [547]:

- Air-fuel Ratio
- Speed
- Exhaust Back Pressure
- Intake Manifold Pressure
- Surface Temperature
- Combustion Chamber Design
- Displacement per Cylinder
- Air Intake Temperature Control
- Load or Power Level
- Ignition Timing
- Valve Overlap
- Combustion Chamber Deposit Buildup
- Surface to Volume Ratio
- Stroke-to-bore Ratio
- Compression Ratio
- Exhaust Gas Recirculation

Air-fuel Ratio

As mentioned earlier, the air-fuel ratio of 14.7:1 is called the stoichiometric ratio. While calculating this ratio, only oxygen is considered as a reactant and the nitrogen in the air is considered inert. But from emissions standpoint, nitrogen is not inert because at high temperatures it does combine with oxygen to form the undesired NOx. Hydrocarbons in gasoline and diesel fuel may be represented by the general formula $(CH_2)_n$, a carbon to hydrogen weight ratio of 6:1. Please note that the use of the formula $(CH_2)_n$ for the fuels is an approximation because it ignores the additional hydrogens in the saturated hydrocarbons. The true representation for such hydrocarbons will be C_nH_{2n+2}, or $(CH_2)_n+2H$. Based on this, *iso*-octane will have carbon to hydrogen weight ratio of 5.33:1 and not 6:1 and the air-fuel ratio for complete combustion is 15:1 and not 14.7:1. Calculation of the air-fuel ratio using the carbon-hydrogen ratio is described in Ref [547]. This ratio for hydrocarbons in commercial gasolines ranges from 6:1 to 6.8:1, which translates into 14.7 to 14.4 lb air being required for the combustion of each pound of fuel, or the AF of 14.7:1 to 14.4:1. As discussed before, SFC and MEP, hence power, both depend upon air-fuel ratio. For fuel economy, the engine is run at lower fuel-air ratios (fuel lean; AF > 14.7) and for greater power, i.e., maximum cycle temperatures; it is run at higher fuel-air ratio (fuel rich; AF < 14.7). If the fuel-air mixture is too rich, there is a chance of incomplete combustion and a greater possibility of increasing the HC and CO emissions. Lean mixtures can overcome this problem to a degree, but they can lead to poor driveability. Beyond an AF of 16:1, the emissions increase because there is not enough fuel to sustain flame propagation at a reasonable rate [557].

In order to analyze the effect of the air-fuel mixture on NOx, the equivalence ratio λ, or the air factor, defined earlier, is a more appropriate measure. Stoichiometric air-fuel mixture has $\lambda = 1$, lean mixtures have $\lambda > 1$ (means more air), and rich mixtures have $\lambda < 1$ (means less air). The amount of NOx formation parallels peak combustion temperatures that not only depend upon the composition of the combustion mixture but also on the load. Air-fuel mixtures with $\lambda = 0.95 - 1.15$ (AF ratios of 14–17) lead to higher NOx concentrations than those outside this range. Extremely rich and lean mixtures both lead to low peak temperatures, hence higher HC and CO emissions. This is because of the low oxygen content in the first case and the slow rate of combustion in the second case. Also, NOx formation increases with load because of the higher peak temperatures.

Load or Power Level

In most engines, load has no effect on HC and CO emissions, if the speed and the mixture composition are unchanged and the spark timing is adjusted to obtain the best torque. However, some engines show an increase in HC with load, which primarily relates to a shorter residence time for HC in the exhaust for the oxidation chemistry to complete.

Engine Speed

HC emissions decrease with increasing engine speed. This is attributed to improved combustion because of the increased turbulence. Higher speeds also cause higher turbulence in the exhaust system, thereby leading to a more effective oxidation of HC to water and carbon dioxide. Carbon monoxide (CO), on the other hand, is not affected by speed.

The effect of engine speed on NOx depends upon the AF ratio of the combustion mixture. Increasing engine speeds lead to an increase in NOx, if rich mixtures are burned, and to a decrease in NOx, if lean mixtures are burned. This is because higher speeds increase the rate of combustion due to increased turbulence and by reducing heat losses per cycle, both of which increase compression hence combustion temperatures and pressures. However, the difference in the rate of NOx formation between rich and lean mixtures is due to the different rates of combustion. Rich mixtures, which burn faster, result in higher temperatures and pressures than the lean mixtures that burn slower. The amount of NOx in the former case is therefore higher.

Ignition Timing

In an internal combustion engine, work is performed by the expanding combustion gases. Combustion of the fuel-air mixture at maximum compression, i.e., when the piston is at top dead center, help attain maximum work. The problem is that combustion is not instantaneous and in an SI engine takes 1–2 milliseconds to complete. This is overcome by initiating combustion before as well as after the piston is at top dead center. Initiating ignition before piston is top dead center is called *advanced* ignition timing and after top dead center is called *delayed* ignition timing. If ignition is initiated too early, the pressure buildup precedes completion of combustion, which will lead to an increase in temperature and the fuel-air mixture's tendency to auto-ignite (knock). Also, the effect of high temperature on emissions will be the formation of a greater amount of NOx at all speeds and loads. The delayed ignition timing, on the other hand, will lower the release of maximum pressure that in turn will lower the engine efficiency. It is also likely to form the higher amounts of the partial combustion products, such as HC and CO. However, this strategy transfers part of the energy into the exhaust system where it facilitates reactions that lower HC and NOx. The overall effect is the reduction of emissions, but at the expense of fuel economy. Ordinarily, the effect of the delayed timing on CO alone is not as striking and little, if any, improvement is observed. Too much retard can lead to higher CO emissions by not allowing sufficient time for CO to oxidize to CO_2.

Exhaust Back Pressure

It is the pressure exerted by the exhaust gases on gases exiting the combustion chamber. The amount of the back pressure determines the amount of the residual gas in the combustion chamber. If the back pressure is low and does not lead to substantial dilution of the combustion mixture, combustion enhancement occurs, thereby lowering the HC emissions. This is because the tail end exhaust gases, primarily responsible for the back pressure, are rich in HC and burn with the fresh charge. If, however, the back pressure is high and charge dilution occurs, inefficient combustion will result, which will lead to higher HC emissions. The charge dilution, on the other hand, will decrease the amount of NOx.

Valve Overlap

Opening and closing of the intake and exhaust valves in an internal combustion engine is timed to obtain the maximum power and on a continuous basis. Conceptually, the intake valve should open when the piston is top center and close when the piston is bottom center. The exhaust valve, on the other hand, should open when the piston is bottom center and close when it is top center. For high-speed engines, the closing of the exhaust valve and opening of the intake valve is overlapped to obtain a higher output. Overlapping involves two valves to be open simultaneously. This strategy takes the advantage of fluid dynamics to scavenge the residual gases more effectively, which allows a greater amount of fresh charge to enter the cylinder, therefore resulting in increased output.

The degree of overlap varies among engines and its effect on emissions is analogous to that of the exhaust back pressure; that is, it primarily affects HC emissions. It lowers them at very low overlap (2°), beyond which they increase. It has no effect on CO emissions, unless the combustion mixture becomes richer at which time it leads to increased emissions.

Intake Manifold Pressure

The function of the intake manifold is to uniformly distribute the air-fuel mixture to all cylinders. This is a difficult task, especially when the fuel in the combustion mixture is not in a fully vaporized state. Intake manifold pressure that facilitates proper distribution to cylinders varies with the load and is a function of the throttle opening. While intake manifold pressure varies over a wide range, the values between 9 and 20 mm of mercury are the best for low HC and CO emissions. Both tend to be higher outside this range.

The lower the manifold pressure, the lower is the amount of NOx formation. This is because the lower pressures decrease the temperature and increase the amount of the residual gases, thereby leading to a greater ignition delay and reducing the rate of combustion. Both decrease the amount of the NOx formation.

Combustion Chamber Deposit (CCD) Buildup

CCDs are irregular in shape and porous in texture. Their shape increases a combustion chamber's surface area, thereby leading to more quenching, and their texture makes them trap high boiling fuel fractions that are released during the next cycle. Both these factors contribute towards an increase in HC emissions. Again, CCDs have little or no effect on CO emissions.

Surface Temperature

Combustion chamber surface temperature determines the amount of HC in emissions; the higher the temperature, the lower the HC emissions. However, the engine design changes to achieve this are likely to affect an engine's octane requirement, volumetric efficiency, and the need for effective lubrication.

Surface-to-volume Ratio (s/v)

HC emissions primarily result from quenching; hence the combustion chamber designs with lower surface area should improve emissions. Surface-to-volume ratio, or s/v, is one of the measures that can be used to assess the effectiveness of the combustion chamber design on emissions. Engines with low s/v ratios are better on HC emissions than those with high s/v ratios. Carbon monoxide (CO) emissions, on the other hand, are insensitive to these ratios.

Combustion Chamber Design

In general, the lower the surface area of a combustion chamber, the lower are the HC emissions. For the same clearance volume, the double hemisphere design was found to be superior to others.

Stroke-to-bore Ratio (s/b)

This parameter is related to surface-to-volume ratio (s/v ratio). Engines with long strokes and small bores have low s/v ratio, hence they result in lower HC emissions. Modern engines, however, do not favor this design because such engines tend to be large, costly, low in fuel efficiency, and have low peak power.

Displacement per Cylinder

Everything else being constant (compression ratio and s/b ratio), an increase in the displacement of a cylinder will lower its s/v ratio and hence HC emissions. The reason for this is described earlier while discussing surface-to-volume ratio.

Compression Ratio

This parameter was defined while discussing gasoline properties. It is the ratio of the volume of the cylinder at the beginning of the compression stroke to the volume of the cylinder at the end of the compression stroke. The higher the compression ratio, the higher is the air temperature in the cylinder at the end of the compression stroke. In general, higher compression ratios lead to higher thermal efficiency; hence improved fuel economy. While in diesel engines higher compression ratios are required for fuel auto-ignition, in gasoline engines, they lead to auto-ignition related engine knock, a ping sound. The side effect of too high a compression ratio in a diesel engine is the increased HC emissions and the increased formation of NOx. Compression ratio affects emissions by influencing the s/v ratio. A decrease in the compression ratio leads to a large decrease in the s/v ratio, hence a decrease in HC emissions, but at the expense of lowering thermal efficiency and engine power. While the lower thermal efficiency is undesired, it is beneficial in lowering the HC emissions by increasing the exhaust temperature that facilitates oxidation of HC to water and carbon dioxide in the exhaust system. One strategy that is being pursued to boost the output of an engine, while maintaining the decrease in HC, is to vary the compression ratio according to the operating conditions. Combining this strategy with super-charging or turbo-charging has the advantage of even a greater increase in output.

The role of the spark retard and the combustion rate in reducing the HC emissions can partly be explained in terms of the s/v ratio. The spark retard implies ignition, hence combustion, when the piston is well down the cylinder, at which time the s/v ratio is much lower than when the piston is top center. Similarly, the air-fuel ratio change and any design change that will slow the combustion process will have an analogous effect. The downside of the spark retardation is a decrease in power.

HC emissions comprise hydrocarbons that survive oxidation, which occurs during combustion and in the exhaust system. Henein and Patterson [547] summarize the effect of various engine parameters on HC during their passage from the combustion chamber to the exhaust.

Controlled Auto-ignition Combustion

This technology, at present at the conceptual stage, is analogous to homogeneous charge compression ignition (HCCI) technology in a diesel engine. The perceived benefits of this technology are fuel economy and exhaust emissions control. The technology involves compressing an air-fuel mixture to a high enough temperature to initiate auto-ignition. This type of auto-ignition differs from that produced from knocking combustion in that it is controlled and occurs at a slower rate, except at high loads where its control is difficult.

Effect of Coolant Temperature

The coolant's influence on the amount of HC, CO, and NOx arises from its ability to affect combustion temperature by way of the combustion mixture. High coolant temperatures lead to lower HC and CO emissions but higher NOx. Low coolant temperatures, on the other hand, will cool the cylinder walls and the combustion mixture more effectively than the high coolant temperatures; hence they will result in lower NOx.

Effect of Humidity

Its effect on the amount of NOx relates to its effect on the flame temperature. High humidity lowers this temperature and hence the amount of NOx formation.

Air Intake Temperature Control

Air intake temperature affects emissions via its effect on composition and the homogeneity of the combustion (air-fuel) mixture [558]. Low ambient temperatures, especially at low intake air speeds, lead to nonhomogeneous air fuel mixtures, with poor burning ability and tendency to misfire. The likely result will be the formation of the incomplete combustion products, such as HC, CO, and the particulates. Too high an ambient temperature, on the other hand, will lead to an expansion of the fuel-air mixture, thereby limiting its amount in the cylinder and hence a drop in output. Air intake temperature control help maintain fuel-air temperature in the optimum range by regulating the temperature of the intake air.

Effect of Exhaust Gas Recirculation (EGR)

As described earlier, controlling emissions by modifying the air-fuel mixture composition is not an effective strategy because its effect across all pollutants is not uniform. While altering the combustion mixture composition from rich (AF <14.7) to stoichiometric (AF$=14.7$) to lean (AF>14.7 to ~16) results in a progressive drop in the amount of HC and CO, it leads to a progressive increase in the amount of NOx formation, until very lean mixtures (AF of ~18 to 20) are reached. That is when a substantial decrease in the amount of NOx takes place. The problem is that at such air-fuel ratios, the engine performance is less than desired. Since NOx occurs at very high temperatures, lowering the combustion temperature should help minimize its formation. Exhaust gas recirculation (EGR) and introducing water vapor or an inert gas, such as CO_2, into the combustion mixture help achieve this. The drop in temperature is due to the dilution of the combustion mixture which reduces the flame speed and the ensuing temperatures. The principle of EGR is fairly simple. It involves introducing part of the exhaust gas into the intake manifold (recirculation), where it gets mixed with air and the fuel. When the mixture burns, the combustion temperature is not as high as in the absence of the EGR. Because the NOx formation is an exponential function of the combustion temperature, a small decrease in this temperature will greatly reduce its amount. A 16 % drop in peak temperature is estimated to drop the amount of NOx by as much as 85 % [558]. EGR also decreases the amount of HC and CO emissions, if present in the exhaust, via re-entry into the combustion chamber and facilitating their oxidation to H_2O and CO_2.

While the NOx-lowering potential of the EGR can not be refuted, it has a negative effect on power and fuel economy. In general, the higher the amount of the EGR, the higher is the specific fuel consumption and the lower is the brake power. In addition, after 15 % EGR, NOx reduction tends to level off. However, this is not a problem because at 15 % EGR the amount of NOx is lowered by 88 % while the power loss and the fuel consumption increase is only 16 % and 14 %, respectively. The negative effect of the EGR on power and the fuel consumption can be offset by ignition advance and manipulation of the air-fuel ratio. However, these parameters have an adverse impact on NOx and after all the adjustments the realized benefit is only 60 % reduction, which is quite respectable [558].

Effect of Turbo-Charging

Turbo-charging has no significant effect on HC and CO emissions. It is the process in which a compressor, driven by an exhaust gas turbine, forces compressed air into the intake of an engine. This increases the flow of the fuel and hence results in a gain in power output. Turbo-charging, therefore, makes it possible to obtain a higher output from smaller and lighter compact engines. In addition, because turbo-charging is on demand, engines fitted with turbochargers usually have better overall fuel economy. Turbo-charging results in higher pressures and temperatures that can lead to self-ignition or knock in the SI engines and a corresponding increase in NOx. Inter-cooling is an option where the intake air is cooled in an effort to lower the combustion temperatures. However, this option is rarely used. Instead, knock is usually controlled by using higher octane fuels, rich mixtures, or by retarding ignition timing, or both. Unfortunately, all of these strategies partially offset the mechanical efficiency gained through turbo-charging.

On-Board Diagnostics (OBD)

Most modern vehicles come equipped with on-board diagnostic systems (OBDs). These are designed to constantly monitor the performance of a variety of components and systems in a vehicle, including those dealing with the emissions control. The emissions control performance relates primarily to the catalyst system efficiency, improved ignition quality, effective fuel delivery to minimize the undesirable emissions, fuel's evaporative control, effective exhaust gas recirculation, and proper functioning of the emissions control systems. The engine components that are monitored include the catalytic converter, oxygen sensor, spark plugs, EGR system, coolant system, and evaporative emissions control system. The OBD system contains an electronic control module (ECM), which has the ability to correct the above-listed problems, if they surface, by electronic optimization of the engine's operating parameters. Such parameters comprise fuel and spark requirements, metering of the fuel, consistency of the intake manifold pressure and temperature with the vehicle's operation, ignition control (electronic spark timing) to minimize the engine knock, EGR adjustment, control of the idle speed, and optimization of the in-cylinder charge, such as heating and cooling of the intake air and adjustment of the air-fuel ratio.

Emerging New Engine Technologies

The impetus behind the development of these technologies is to improve fuel economy or lower emissions, or both. These technologies come under two general classes: hydrocarbon-fueled technologies and petroleum-free technologies. Direct fuel injection (DFI) vehicles and hybrid gasoline-electric vehicles belong to the former class and the fuel cell technology and the electric vehicles belong to the latter class. Vehicles equipped with DFI technology and hybrid cars are already being marketed.

Direct fuel injection technology was already discussed. Hybrid vehicles are either already available or are being developed for introduction into the marketplace. A hybrid power train combines a gasoline (or diesel) internal combustion engine with an electric motor, a generator, and a storage battery. There are two types of hybrid drive systems, series and parallel. In the series configuration, a gasoline engine drives a generator that provides the electrical energy for the storage batteries and the electric motor, but only the electric motor drives the vehicle. In the parallel configuration, the gasoline engine drives a generator and the vehicle under low-load conditions, and the electric motor is used to provide the additional power for driving the vehicle under higher-load conditions. Both types of systems recover the energy that is normally lost during decelerations and braking and use it to charge the storage batteries. The hybrids being offered today and those under development are typically parallel configuration hybrids.

Gasoline hybrids are likely to be replaced with the fuel cell electric vehicles in the future, when the fuel cell technology becomes commercially viable. This is because this technology is expected to have extremely high efficiency in generating electrical energy from the fuel and generate the lowest amount of emissions. Fuel cell technology generates power in the form of electricity and heat by converting hydrogen, obtained from the fuel by the use of on-board reformers, and oxygen into water. It essentially converts chemical energy of the fuel into electricity. There are two fuel options: hydrocarbon fuel and methanol. At present, the hydrocarbon fuel is preferred because of the existence of the infrastructure for its delivery, production, cost, etc. In a fuel is cell, hydrogen is the reactant on the anode side, oxygen is the reactant on the cathode side, and water is their final reaction product. Typically, the reactants flow in and the reaction product(s) flow

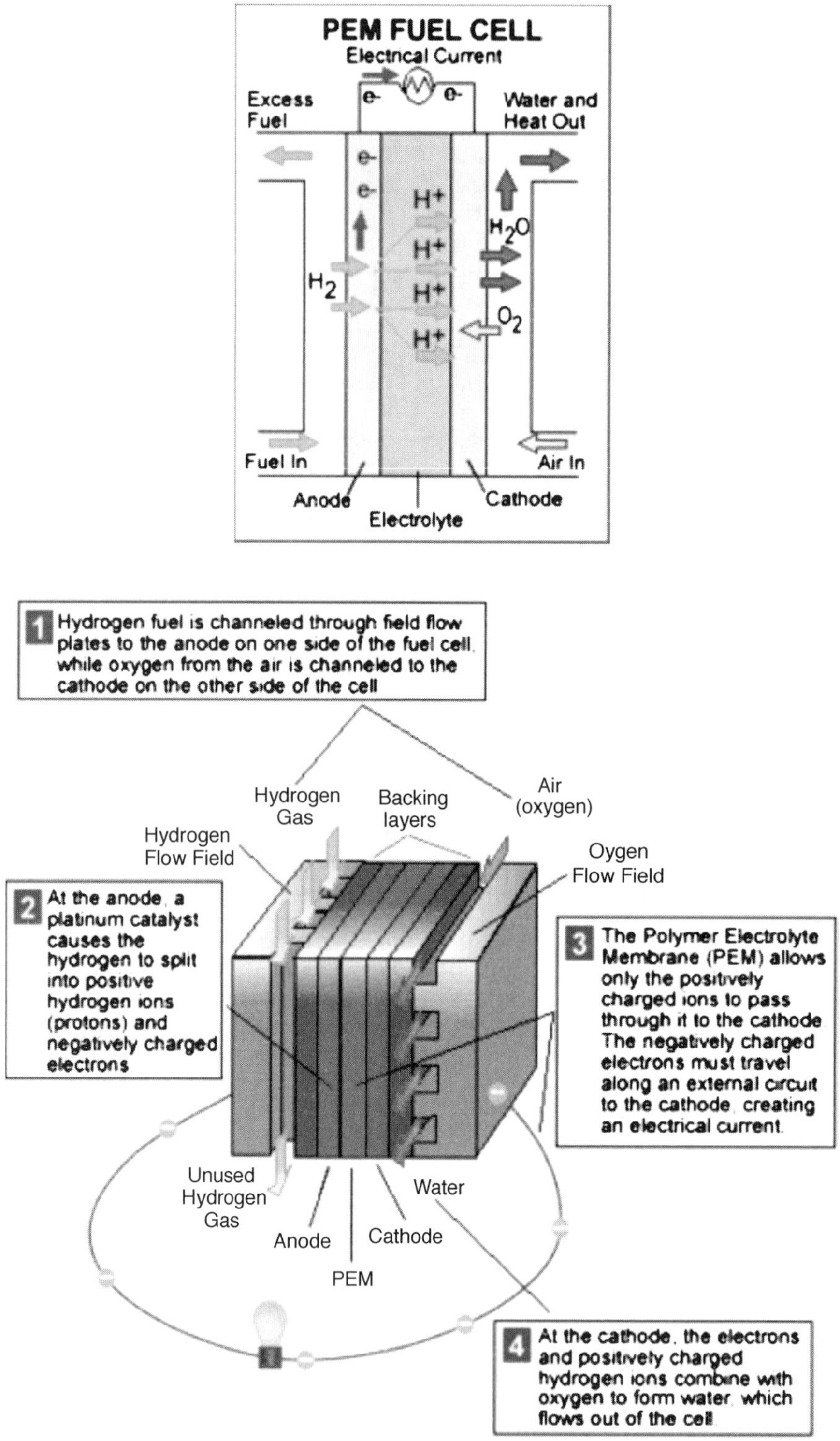

Fig. 6.19—Schematics of a fuel cell [559].

out. In principle, virtually continuous long-term operation is feasible as long as these flows are maintained.

Chemically, the fuel cell is a hydrogen cell, which like any other cell consists of an anode and a cathode that are separated by a polymer electrolyte membrane, see Fig. 6.19 [559]. A stream of hydrogen is delivered to the anode side of the membrane-electrode assembly (MEA). At the anode side, the hydrogen is catalytically converted into hydrogen ions (protons) and electrons. This oxidation half-cell reaction is represented by the equation below.

$$H_2 \rightarrow 2H^+ + 2e^- \quad E^o = 0\ V$$

The newly formed protons permeate through the polymer electrolyte membrane to the cathode side. However, the elec-

TABLE 6.29—The effect of design and operating variables on gasoline engine emissions.

Variable Increased	HC	CO	NO_x	Aldehydes
Air-fuel Ratio (14.7≥AF≤16)	+	+	−	−
Load	=	=	−	
Speed	+	=		
Spark Retard	+	=		+
Exhaust Back Pressure	+	=		
Valve Overlap	+	=		
Intake Manifold Pressure (9 mm≥IMF≤20 mm)	+	+		
Combustion Chamber Deposits	−	=		
Surface to Volume (s/v) Ratio	−	=		
Combustion Chamber Surface Area	−	+		
Stroke to Bore (s/b) Ratio	+	=		
Displacement Per Cylinder	+	=		
Compression Ratio	−	=		
Air Injection	+	+	−	−
Fuel Injection	+	+	−	
Combustion Chamber Surface Area	−	=		
Coolant Temperature	+	+	−	
Air Intake Temperature Control	+	+		
Turbo-charging	=	=	−	
Exhaust Gas Recirculation	+	+	+	

Note: + Positive benefit; − Negative benefit; = No benefit.

trons travel along an external load circuit to the cathode side of the MEA, thus creating the electric current output of the fuel cell. In the meantime, a stream of oxygen is delivered to the cathode side of the MEA. At the cathode side, the oxygen molecules react with the protons permeating through the polymer electrolyte membrane and the electrons arriving through the external circuit to form the water molecules. This reduction half-cell reaction is represented by the following equation.

$$4H^+ + 4e^- + O_2 \rightarrow 2H_2O \quad E^o = 1.229 \text{ V}$$

Besides a fuel cell stack that converts hydrogen into electrical energy, a fuel cell powertrain consists of an electrical motor/generator drive unit, storage battery, and hydrogen fuel storage—with high-pressure liquid, or hydride options. Alternatively, an on-board fuel reformer fed by the liquid hydrocarbon fuels might be used to produce the hydrogen. Because of the many hurdles that need to be overcome, the timing of use of the fuel cell technology in commercial vehicles is hard to forecast.

Table 6.29 summarizes the effect of various design and operating variables on emissions.

Diesel Engine Emissions Control

Control of the exhaust emissions from a diesel engine is a challenge. Two major components of the diesel engine exhaust are NOx and particulates (PM), both of which are a serious health hazard. Of the two, NOx is invisible and PM can be visible (smoke) or invisible, depending upon its particle size. Contrary to the exhaust of the gasoline engines, which contains large amounts of unburned or partially burned hydrocarbons (HC) and carbon monoxide (CO), their amount in the diesel exhaust is minimal. The formation of NOx and PM is related to the mechanism of the diesel combustion. In a diesel engine, the fuel at high pressure is injected into the cylinder that contains compressed air. Because of the design, the fuel-air ratio in different parts of the cylinder is different and also when the injected fuel hits the compressed air. There are pockets of fuel-rich zones and air-rich zones, which makes the combustion nonuniform. Efficient combustion occurs only in zones which have the fuel-air ratio within a certain range. The zones where this ratio is too low or too high, the combustion either does not occur or is inefficient and the result is higher CO, HC, and PM emissions. It is important to note that NOx (NO and NO_2), HC, CO, and PM result under different conditions and via different mechanisms.

NOx forms in the stoichiometric and slightly lean fuel-air mixture regions, where the ambient temperature and the concentration of oxygen are high. HC results in regions where the temperatures are low or oxygen is insufficient to initiate and propagate combustion. The upper cylinder walls and regions around the top of the piston edge are examples of the low temperature regions. As mentioned earlier, in diesel engines HC and CO emissions are not a major problem since they tend to operate on lean fuel-air ratio. Particulate matter (PM) results from fuel droplets that do not vaporize, hence do not burn properly. PM results mainly from oxidation of the fuel during preignition stage and not combustion. Oxidation essentially strips fuel hydrocarbons, partially or fully, of hydrogen leaving behind soot, which is the primary component of the diesel exhaust. The formation of soot commonly occurs when the last portion of the fuel enters the cylinder where the combustion has already ensued, or the engine is being operated at high load and or high speed. In the former situation, the fuel gets oxidized prior to hitting the combustion zone; and in the latter situation, the fuel is being pumped into the cylinder at a faster rate than the combustion rate. Additional contribution to soot comes from the lubricant that travels past the piston rings, either by design or inadvertently, and burns partially.

A number of strategies have been used to control emissions from the diesel engines. Those that deal with manipulating the operational parameters are described below.

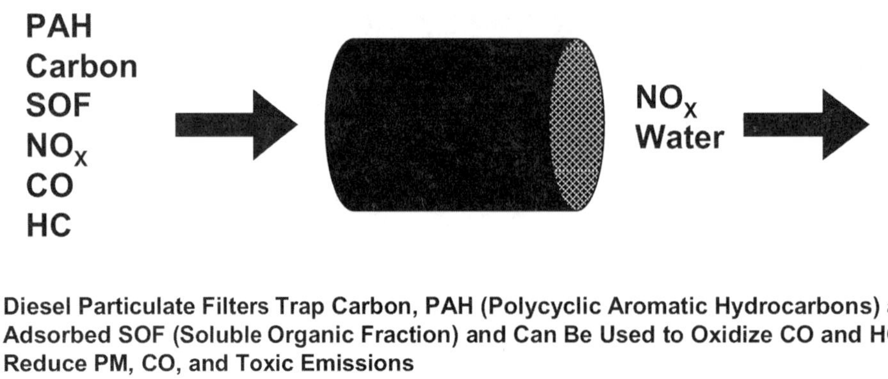

Fig. 6.20—Diesel particulate filter operation [561].

- Increasing injection pressures—The result is better atomization of the fuel, hence improved combustion.
- Directional injection—Control of the position and the angle of the injector in the cylinder head and better nozzle design to deliver fuel in the optimal combustion zone.
- Fuel charge control—Controlling the rate of fuel entry during injection to increase the combustion efficiency. In some cases, a small amount of fuel is charged first (pilot injection), which initiates the combustion. This is followed by the main injection.
- Generating turbulence—Cylinder head, air intake valve, and piston head are designed to improve fuel-air mixing.
- Charge cooling (inter-cooling)—Most diesel engines employ supercharging or turbo-charging to increase power. The strategy is to introduce more air (oxygen) into the engine, which improves the combustion efficiency, hence power. These engines have a mechanically-driven (supercharging) or an exhaust-driven (turbo-charging) pump that compresses or pressurizes the intake air. Since compression increases the intake air temperature, the NOx formation tends to increase as well. This problem can be overcome by cooling the intake air by passing it through a heat exchanger, prior to allowing it to enter the cylinder. A key advantage of the turbochargers is that they increase the engine power with only a slight increase in weight.
- Lower oil consumption—All engines have the tendency to burn the lubricant that travels past piston rings into the combustion chamber. This contributes to exhaust emissions. New diesel engines use piston designs with low crevice volume. Such pistons minimize lubricant travel into the combustion chamber, thereby controlling the lubricant-related contribution to emissions.
- Exhaust gas recirculation (EGR)—As mentioned earlier, NOx formation increases at high temperatures. The temperatures can be decreased by diluting the reaction mixture with an inert gas, which effectively decreases the temperature of the combustion mixture. Exhaust gas is an ideal option since it is essentially inert and the engine does not need an additional source of such a gas. The EGR process directs a portion of the exhaust gas into the air intake manifold. It is important to note that while this strategy reduces the NOx formation, a small loss of power does result.

Because of the stringency of the modern emissions control regulations, strategies discussed so far are not sufficient. In order to meet the present and future government-mandated requirements, additional steps are necessary. These involve installation of devices that either remove exhaust pollutants or treat them to make them innocuous. Those worth mentioning include particulate filters and catalytic converters. Diesel particulate filters (DPFs) are devices installed on the diesel engine vehicles that collect particulate matter from the exhaust without obstructing the flow of the exhaust gases or damaging the vehicle, or both. Figure 6.20 shows a conceptual diagram of the functioning of a DPF [561]. If desired, a number of technologies can be installed in conjunction with a DPF to reduce the amount of harmful exhaust gases expelled into the atmosphere.

Several types of diesel particulate filters, also known as trap oxidizers, are either commercially available or are under development. Wall-flow ceramic cordierite, woven fiber cartridges, and temperature resistant paper (disposable) are the three types of filters that are currently in commercial use. These filters have collection efficiencies ranging from 50 to 90 %. In addition, noble-metal catalyzed, fuel additive catalyzed, and externally regenerated filter systems are also available, with filter efficiencies well over 90 %. The PM removal process involves the passage of the exhaust gases through a porous cell, which removes the particulate matter. After some period, such filters reach their limit and need to be replaced, if disposable, or regenerated. The regeneration is usually carried out by burning, i.e., the oxidation of the PM, which is primarily carbon. Some designs have built-in burners or electrical resistance heaters that raise the trap's internal temperature to burn off the particulates as soon as they reach the device. Figure 6.21 shows the schematic of a regular filter and a continuously regenerating filter [550]. In addition to removing more than 90 % of the PM, DPFs also destroy the soluble organic fraction (SOF) of the PM, the carbon particles, CO, HC, and the toxic emissions, such as aldehydes, from the diesel engines. EPA describes changes in engine performance that can occur when retrofitting this program [560]. These include the following:

1. Possible back pressure increase above the manufacturer specifications, which can lead to a decrease in power.
2. A slight increase in fuel consumption.
3. Noise attenuation and muffler warranty—in almost all cases, retrofitting with a DPF will require replacement

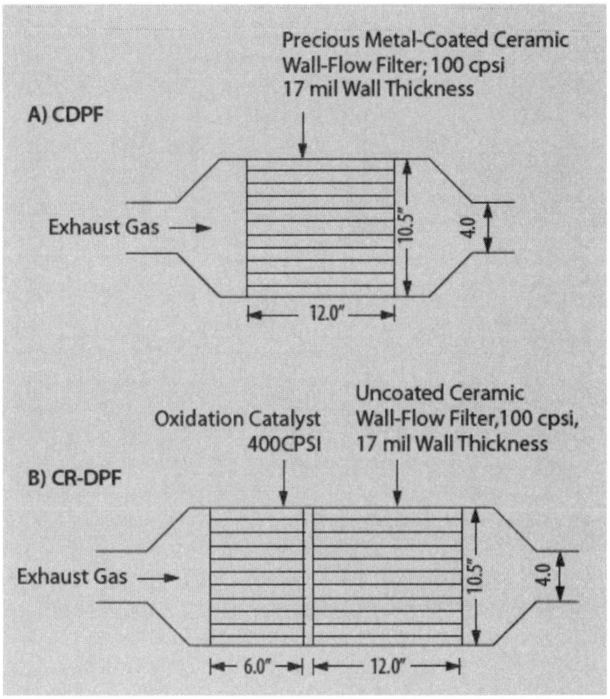

Fig. 6.21—Diagram of catalyzed diesel particulate filter (CDPF) and continuously regenerating diesel particulate filter (CR-DPF) [550].

of the existing muffler. In this case, the noise attenuation for the replacement muffler will be the same as with the stock muffler.

Catalytic converters remove pollutants by altering their structure, primarily through oxidation. Oxidation catalysts that are used to treat diesel exhaust are similar to those used for the gasoline engines. They oxidize unburned hydrocarbons (HC) and carbon monoxide (CO) to carbon dioxide and water. However, diesel catalytic converters suffer from a number of problems. These include deactivation due to fuel sulfur; particulate build up on the catalyst, thereby impairing the flow of gases; and lower catalyst efficiency on account of the diesel engine exhaust being cooler than the gasoline engine exhaust. Since catalytic converters primarily address HC and CO emissions, the quantity of which in diesel engine exhaust is low, catalytic after treatment is not generally needed. As mentioned earlier, the removal of NOx from the diesel engines is the real challenge. While lowering NOx levels by the use of the EGR is possible, its complete removal from the exhaust is not highly likely. The reason is that it will involve reduction conditions, which in diesel engine exhaust do not exist. As a matter of fact, the environment in diesel exhaust is oxidizing because of the presence of an excess amount of air (oxygen) in the air-fuel mixture. However, this problem can be overcome by the use of selective catalytic reduction (SCR) technology. In this approach, a nitrogen compound with N-H bonds, such as ammonia or urea, is injected into the exhaust gas in an amount proportional to the NOx present. Nitrogen oxides (NOx) get reduced to nitrogen and water, and ammonia and urea gets oxidized to nitrogen, water, and carbon dioxide, as is shown in the reaction scheme in Fig. 6.22. Mercedes-Benz is one company that is

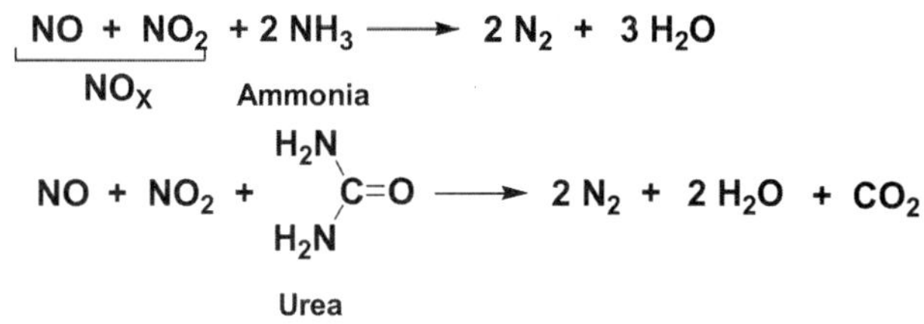

Fig. 6.22—Selective catalytic reduction of NO_X.

marketing this technology under the trade name of Blue Tec® and installs it on the vehicles it manufactures. The BlueTec® system uses aqueous urea solution that is dispensed from a renewable canister. The system is depicted in Fig. 6.23 [562]. Integrated systems combining catalysts, filter, SCR, and oxidation catalyst technologies are also proposed. Oxidation catalysts can remove ammonia (NH_3) slip, if required [561].

Diesel Combustion

Hydrocarbons (HC) and CO in the exhaust of a diesel engine are lower than those present in the exhaust of a gasoline engine, except at low operating temperatures and high loads. This is because diesel engines burn lean combustion mixtures, i.e., they operate on excess air ($\lambda > 1$). The main concern is NOx, smoke, and aldehydes. Even the NOx levels for the CI engines are lower than those for the SI engines. Major problem in the diesel engines is smoke, especially gray or black smoke that contains soot particles, or PM.

Emissions in a diesel engine largely relate to the use of a heterogeneous combustion mixture and the diesel fuel's higher boiling range (150 to 380°C). As a result, diesel engines are always run on combustion mixtures with excess air ($\lambda > 1$). Otherwise, combustion will be incomplete and large amounts of HC, CO, and particulates, such as soot, will result. Spraying of the diesel fuel into the swirling air introduces heterogeneity and results in zones of varying combustibility. Henein and Patterson suggest a combustion efficiency model to explain the formation of undesirable emissions in diesel engines [547]. They classify combustion zones of high, medium, and low flammability. Moving away from the injector nozzle, there is a spray zone with a spray core and a spray tail (and after injection), and near and far edges of the nonspray zone. Figure 6.4 shows these zones in the fuel spray from a diesel injector [23].

The spray core has fuel droplets of good size and there is enough oxygen to have complete combustion, which is true under part load. This results in high combustion temperatures that facilitate NOx formation. However, if the temperature and the oxygen content of the fuel-air mixture are not high enough to cause complete oxidation, CO and aldehydes will result. This situation exists between the lean flame region and the lean flame-out region at the onset of combustion and at low loads. Incomplete oxidation is more likely under high loads when the core becomes more fuel rich and the combustion becomes relatively inefficient. This leads to the increased formation of CO, aldehydes, and carbon. The spray tail is the last part of the fuel that is injected and because of the low pressure it has poor penetration and mixing with air. The result is the formation of large droplets with inability to burn effectively. Therefore, they act as a source of

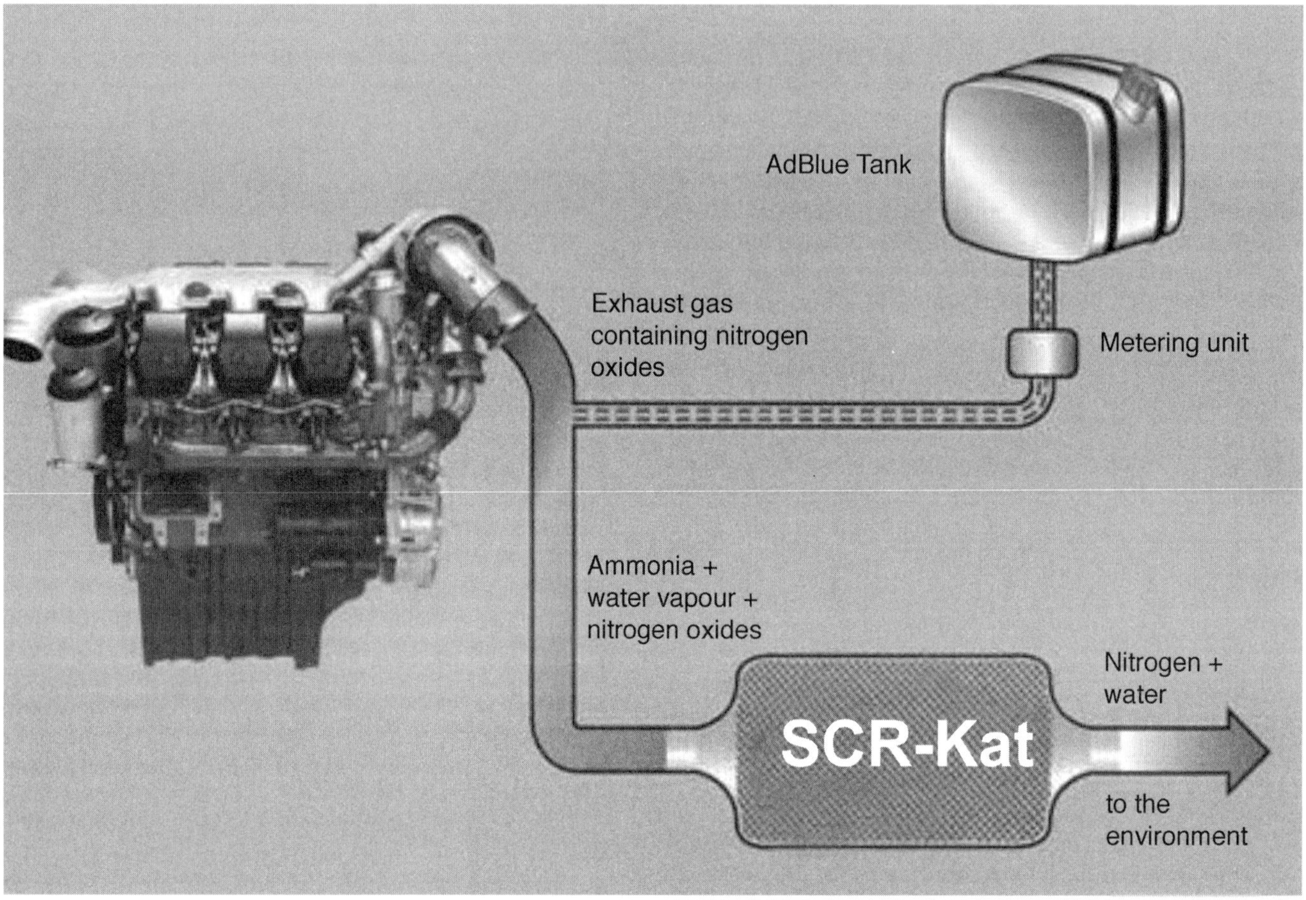

Fig. 6.23—Mechanism of selective catalytic reduction of NO_X [562].

HC, CO, carbon, and oxygenates. After-injection occurs under medium and high load conditions when a small amount of fuel bleeds after the main injection and has similar consequences. The situation changes as the combustion process proceeds. Carbon monoxide (CO) formation under high loads is not a problem since such loads require richer fuel-air mixtures. This translates into high reaction temperatures that facilitate oxidation of CO to CO_2. However, too rich a combustion mixture can increase CO by making the combustion less efficient because of the inadequate oxygen content.

The near edge of the nonspray zone, called the lean flame region, has a reasonable amount of fuel vapor to undergo combustion. However, there are only pockets of fuel-air mixture that are appropriate for combustion. Once the ignition occurs, the flame spreads throughout the region, generating NOx. The far edge of the nonspray zone, called the lean flame-out region, is fuel lean and therefore can not support efficient combustion and is primarily responsible for the HC emissions. Wall quenching also leads to poor combustion and is another source of HC, partial oxidation products, and carbon.

In addition to the above discussed spray-related factors, fuel deposited on the cylinder walls is another factor that plays a role in determining the quality of the exhaust emissions. Fuel deposition occurs because of the shorter spray path, as in the case of small diesel engines, or because of poor penetration arising from the low fuel injector pressures. If the deposited fuel can evaporate, because of the high velocity of the surrounding gas and high wall temperatures, and there is ample oxygen for the vapors to burn, there is little problem. However, if the volatility of the deposited hydrocarbon components is low and the wall temperature and the velocity of the surrounding gas are not suitable for efficient combustion, HC, partial oxidation products, and carbon particles result.

Diesel Fuel Properties

Fuel-related parameters that impact diesel vehicle emissions include the fuel's cetane number/Index, density, volatility, distillation range, viscosity, the amount of aromatics and sulfur, and the additive usage. An inadequate cetane number leads to less efficient combustion, hence higher HC and CO emissions.

Cetane Number

Cetane number measures the ignition quality of a diesel fuel. The number is based upon a scale of zero to 100; where α-methylnaphthalene has a cetane number of 0 and cetane (hexadecane) has a cetane number of 100. Low cetane numbers lead to misfiring, engine deposits, and rough running. Increasing the cetane number improves fuel combustion and hence reduces NOx and PM emissions. The cetane number effects are not linear, unlike the fuel sulfur effects, and

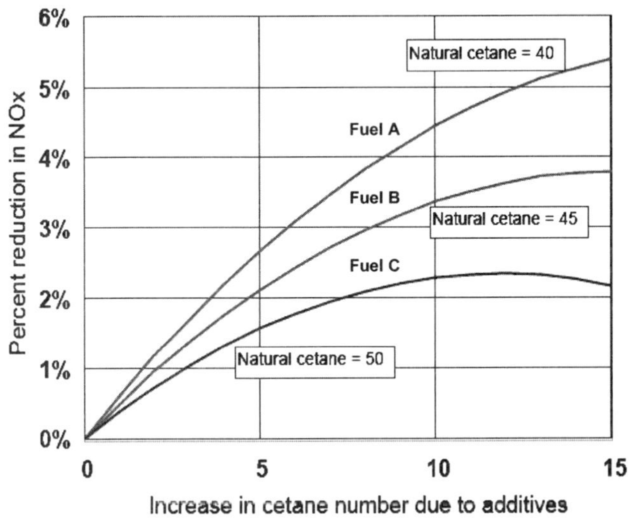

Fig. 6.24—Predicted effect of cetane difference on NOx for heavy-duty highway engines [563].

cetane number improvers do most good if the starting fuel has a low cetane number, see Fig. 6.24 [563]. In the figure, fuel A, with low natural cetane number, has the smallest response to cetane-improving additives and fuel C, with a high natural cetane number, has the lowest response. The EPA has established a minimum cetane index of 40 and a maximum aromatics content of 35 %. The California Clean Air Act of 1993 requires the use of a reformulated diesel fuel that contains no more than 0.05 % sulfur and 10 % aromatics, or of an "emissions equivalent" fuel. A fuel is considered emissions equivalent if its emissions levels, when tested in a DDC Series 60 engine, are the same as those of the reference fuel containing 0.05 % sulfur and 10 % aromatics content.

Density and Aromatics

Fuel density is related to its energy content that comes into the engine. Studies indicate that the reduction in the fuel density decreases NOx emissions but only in old engines and not in the modern engines that use electronic injection and computer control. Reducing the amount of aromatics, especially those that are polynuclear, in diesel fuel reduces NOx and PM_{10}, but only in some engines.

Volatility

A typical diesel fuel volatility curve is depicted in Fig. 6.25 [551]. Various boiling fractions affect different performance parameters, except the cetane quality, which is a function of the whole boiling range. Front-end volatility affects the flash point; mid-range volatility affects the ease of starting and white smoking tendency; and back-end volatility affects the yield/price of the fuel and its cold flow properties. T95 is the temperature at which 95 % of a particular diesel fuel distills in a standardized distillation test (ASTM D86). Reducing T95 decreases NOx emissions slightly, but increases the hydrocarbon and CO emissions, and without affecting PM_{10}.

Diesel Sulfur

The sulfur content of the diesel fuel affects PM emissions because some of the sulfur in the fuel is converted into sulfate particles in the exhaust. There is a direct correlation between the fuel sulfur and the PM emissions; hence diesel sulfur

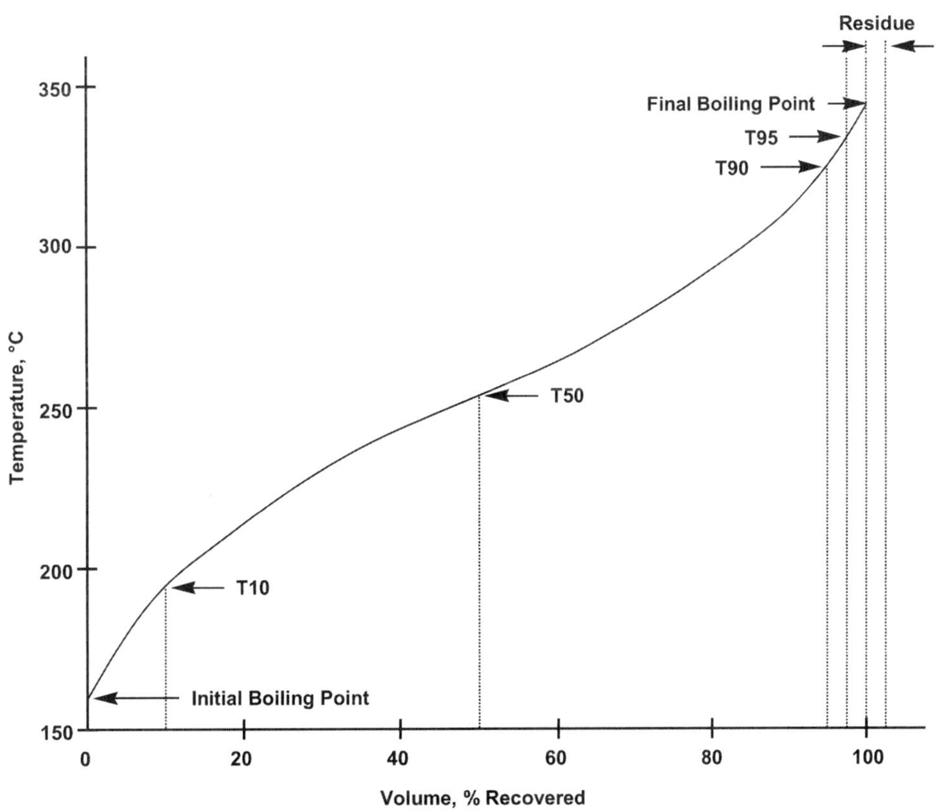

Fig. 6.25—Diesel fuel volatility curve [551].

TABLE 6.30—Diesel fuel properties versus emissions.

Diesel Modification	CO	NO_x	HC	Particulates
Reduced Sulfur	=	=	=	+
Reduced Aromatics	=	=	=	=
Reduced Polyaromatics	=	?	=	?
Reduced Density	=	=	=	+
Increased Cetane Number	+	+	+	+
Reduced T90/95	?	=	?	?

Note: + Positive benefit; = No benefit; ? Effect presently unknown.

content for on-road use has been progressively decreasing since 1982, when the diesel sulfur was around 0.85 %. The 2004 diesel sulfur limit, both in Europe and North America was 0.05 % (500 ppm). As stated above, the EPA has lowered it to 15 ppm (SD15) in June 2006 and SD15 fuel has become available in the market place. Typically, the national average sulfur content of the low sulfur diesel fuel being used today is between 300 to 350 ppm. At a sulfur level of 300 ppm, the sulfate particles make up about 10 % of the total PM emissions for an engine emitting 0.1 g PM/bhp-h.

However, there are a number of issues pertaining to the use of the ultra-low sulfur diesel fuel. These include the following:

1. Concern for a drop in lubricity—This is because the removal of sulfur is likely to involve desulfurization via hydroprocessing, which is apt to remove or alter the amount of chemical functionalities, such as those that are linear. Such functionalities impart lubricity to the diesel fuel.
2. Soft seal compatibility.
3. Loss of oxidative stability/increased corrosion of the fuel system.
4. Low-temperature operability—Inability to blend high sulfur kerosine that is used in diesel during winter, a current practice. There is a need to develop new low-temperature operability additives.
5. May pose conductivity problems—The electrostatic charges develop when diesel moves through the diesel handling equipment which can lead to fire, explosion, or both. Hydrotreating to reduce the sulfur content also decreases the fuel's polar components, the removal of which lowers the diesel's conductivity. This can result in a potentially dangerous situation. The conductivity of a high-sulfur diesel fuel is around 100 picosiemens per metre (pS/m), while that of the SD15 may be as low as 0 pS/m. Addition of the antistatic additives can correct this problem.

Table 6.30 summarizes the effect of these properties on emissions. The effect appears to be minimal except that from cetane number and the sulfur level. A cetane number increase lowers all emissions and the reduced sulfur lowers only particulates.

Two major projects were recently initiated to demonstrate the effect of the fuel properties on emissions from heavy-duty diesel engines. The first project (VE-10), initiated by the Coordinating Research Council utilized two engines—1994 Navistar DTA-466 with a catalyst and 1994 DDC Series 60 without a catalyst. Each engine was calibrated for 5g and 4g NOx emissions per bhp-h and the effect of the cetane number, aromatics content, and oxygen content (amount of oxygenates) on HC, CO, NOx, and PM determined. Fuels of 0, 2, and 4 % oxygen content, cetane numbers of between 45 and 60, and aromatics in the 10–30 % range were utilized. The results are provided in Table 6.31. The data suggest a positive impact of the cetane number increase and the aromatics content decrease on emissions. Oxygen content appears to have mixed results, with a negative effect on HC and NOx. A follow up study determined the effect of the cetane number on emissions of a prototype 1998 DDC Series 60 engine. It also supported the previous findings that a high cetane fuel is highly effective in lowering the undesirable emissions.

The second project was carried out under the European Programme on Emissions, Fuels, and Engine Technologies (EPEFE) and with a similar objective. That is, to identify the relationship between the diesel fuel properties, engine technologies, and emissions (regulated and unregulated). The findings of this study are revealed in Table 6.32. Cetane number increase, density reduction, and aromatics reduction appear to have the best overall effect on emissions.

Reformulated Diesel Fuel

California air resources board (CARB) has established a maximum sulfur content of 0.05 % mass for both off-road and on-road vehicular diesel fuel. However, in other states, the sulfur content of the off-road fuel, regular No. 2-D diesel fuel, is 0.5 % by weight, which is the ASTM D975 limit for the

TABLE 6.31—The effect of diesel fuel parameters on emissions.

	Cetane Increase		Aromatic Reduction		Oxygen Increase	
Emission Type	Navistar DTA-466	DDC Series 60	Navistar DTA-466	DDC Series 60	Navistar DTA-466	DDC Series 60
NOX Calibration 5g/bhp-h						
HC	+	=	+	=	–	=
CO	+	+	=	=	=	+
NO_X	+	+	+	+	–	–
PM	=	=	+	=	+	+
NO_X Calibration 4g/bhp-h						
HC	+	=	=	=	=	=
CO	+	+	=	=	=	+
NO_X	+	+	=	+	=	=
PM	+	=	+	=	+	+

Note: +Positive benefit; –Negative benefit; =No benefit.

TABLE 6.32—Diesel fuel properties versus emissions.

Parameter	Light-duty Engines				Heavy-duty Engines			
Emission Type	HC	CO	NO_X	PM	HC	CO	NO_X	PM
Cetane No. Increase	+	+	=	–	+	+	=	=
Density Reduction	+	+	–	+	–	+	+	+
Polyaromatic Reduction	–	–	+	+	+	=	+	+
T95 Reduction	=	=	–	+	–	–	+	=

Note: +Positive benefit; –Negative benefit; =No benefit.

high sulfur diesel fuel. CARB has also limited the aromatics content for on-road diesel to 10 % by volume, maximum. Alternative formulations with higher aromatics contents are allowed, if they achieve the same or lower emissions in a standardized engine test than the reference fuel with 10 % aromatics. In these formulations, five properties are examined, which are the sulfur content, nitrogen content, aromatics content, polycyclic aromatics content, and the cetane number. If a candidate formulation passes the emissions test and receives the CARB approval, commercial formulations based on these credentials must not exceed the sulfur, nitrogen, aromatics, and polycyclic aromatics contents of the approved formulation.

Biodiesel

Recently, biodiesel has been getting a lot of publicity as a possible alternative to the petroleum-derived diesel. The main incentive for this interest is that the biodiesel can be obtained from sources that can be replenished and the environmental compatibility. In the United States, soybean oil is the largest source of biodiesel, although oils from other biological sources may also be used. Biodiesel is a mixture of fatty methyl carboxylates, which can be obtained from catalyzed trans-esterification reaction of triglycerides (oils and fats) with methanol. The chemistry involved is depicted in Fig. 6.26. This methyl ester mixture has properties that are comparable to those of the conventional diesel fuel hence it is considered a good substitute. See Table 6.33 for comparison of the typical properties [564].

While diesel engines can run on 100 % biodiesel, Biodiesel B100, most of the testing in the United States was carried out on 20:80 blend with low sulfur diesel, Biodiesel B20. Limited testing has shown this fuel to produce lower HC, CO, and PM emissions than the petroleum-derived diesel. Pure biodiesel has good lubricity properties and is devoid of sulfur or aromatics. However, its higher pour point limits its use in cold climates. Bio-diesel is highly biodegradable, and hence is prone to attack by microorganisms, which can become an issue during storage. Since the structure of its components contains unsaturation, its oxidation stability is lower than that of the petroleum diesel as well. Its major disadvantage is its higher cost; hence its widespread use in the United States in the very near future is not likely. However, its use is gaining popularity in Europe because of its potential environmental benefits.

TABLE 6.33—Low-sulfur diesel versus Biodiesel (B100)—Properties comparison [564].

Property	Biodiesel (typical)	Low-sulfur Diesel (typical)	ASTM D975 Specification
Flash Point, °C	100	60	52 min
Viscosity, 40 °C, cSt	4.7	3.2	1.9–4.1
Sulfur, % mass	<0.01	0.03	0.05 max
Cetane Number	48–52	45	40 min
Heating Value net, Btu/gal	128,000	130,000	...
Relative Density, 15 °C	0.88	0.83-0.86	...

PuriNOx™ Technology

Recently, another pollution-reducing diesel fuel technology (PuriNOx™) was commercialized by Lubrizol. The product is a well-blended, safe-to-use, "water-in-diesel" fuel emulsion, which is made from normal diesel fuel, purified water, and proprietary PuriNOx™ additive chemistry. The PuriNOx™ fuel reduces NOx emissions up to 30 % and particulate matter (PM) emissions up to 65 %, compared to the conventional No. 2 diesel fuel. The beneficial effect of water on diesel combustion has been known for many years, except its method of delivery into the combustion chamber was a challenge until recently.

Diesel Engine Design and Operating Variables

Design and operating variables that impact diesel emissions include the following:

- Fuel-air Ratio/Air-fuel Ratio
- Cetane Number

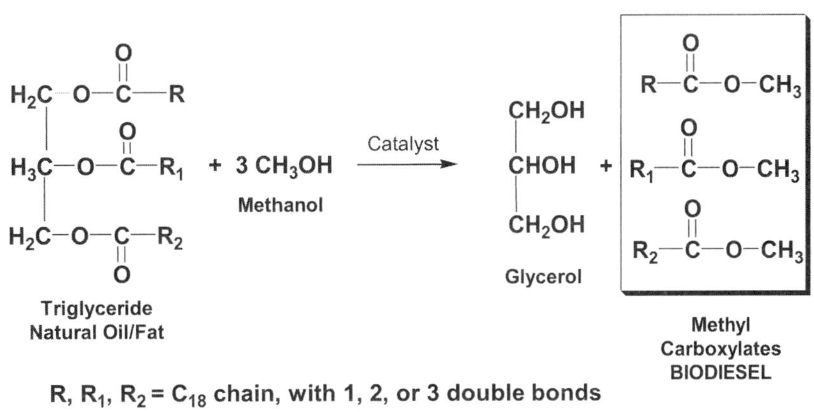

Fig. 6.26—Trans-esterification of triglycerides.

- Turbo-charging
- Intake Air Temperature
- Injection Timing/Rate
- After Injection
- Swirl
- Load and Speed

Fuel-air Ratio

Hydrocarbons in the diesel exhaust comprise both low and high molecular weight materials. High molecular weight components are fuel or lubricant-derived hydrocarbons that either escape combustion in the lean flame-out region, or result from the recombination of the lower molecular weight fragments in the spray core at higher temperatures. The flame-out region is also responsible for the formation of aldehydes, other oxygenated materials, and lower molecular weight hydrocarbons; as a consequence of partial combustion and fragmentation. The effect of the fuel-air ratio on HC emissions is a function of the load. At low loads and idling speeds, an increase in the air-fuel ratio leads to an increase in HC, presumably because fuel concentration in the core is small and most of the fuel resides in the lean flame-out region. However, under moderate loads, the situation reverses and the HC and oxygenates concentration decreases. Now more fuel is in the spray core, which in the presence of oxygen and higher temperatures burns more efficiently. If the fuel-air mixture needs to be enriched further because of the higher loads, the combustion of the fuel in the spray core becomes deficient in oxygen which results in increased HC emissions.

The NOx primarily results from the combination of nitrogen and oxygen at high temperatures. Because of the high-temperature requirement, NOx formation occurs in the spray core, especially when rich fuel-air mixtures are burned. However, it can also form in the lean flame region, where the pockets of rich fuel-air mixture exist, but at a slower rate. All in all, the final concentration of NOx in this region is higher than in the spray core where there is a greater possibility for NOx to undergo chemical reduction by the fuel hydrocarbons. An increase in the fuel-air ratio increases NOx because of the higher combustion temperatures. NO_X in the CI engines primarily results from the nonuniform combustion of the fuel-rich zones of the combustion mixture that generate high temperatures. NOx in the diesel engines is usually lower than in the SI engines, due to averaging across all combustion zones.

Soot results from the high boiling fraction of the fuel when its hydrocarbon molecules lose hydrogen through cracking and become charged and agglomerate. Soot is not a problem if it oxidizes during combustion to carbon dioxide. However, if the rate of its formation is greater than the rate of its oxidation, it can be released into the environment as black smoke. This situation arises when higher output is desired and the fuel is injected faster.

Cetane Number

Low cetane fuels do not burn readily and as a result the fuel accumulates before ignition and there is more fuel present at the time of ignition. This can lead to inefficient combustion; hence higher HC emissions. In addition, low cetane fuels require higher temperatures to ignite which are associated with NOx formation. Some data suggest that high cetane fuels, designed for use in high-speed engines, lead to more black smoke [565]. This observation is justified based on the lower stability of such fuels. While one would expect diesel fuel containing more volatile components to produce more soot, conclusive data pertaining to this are lacking [547].

Turbo-charging

Turbo-charging leads to an increase in average gas temperature, thereby increasing the efficiency of combustion and further oxidation in the exhaust system and the turbocharger. Additional benefits achieved include better spray formation that leads to more complete combustion. This lowers the amount of HC and partial oxidation products but at the expense of increasing the amount of NOx.

Injection Timing

An advance in injection timing leads to higher HC emissions, presumably because it increases the ignition delay period and the fuel vapor and the droplets have more time to be carried away by the swirling air to reside in the lean flame-out region. Advancing the start of injection, on the other hand, lowers particulate formation, which is due to higher temperatures being generated during combustion and the longer fuel residence time. Both factors facilitate burning of the soot as it forms. However, the drawbacks of this technique to eliminate soot are excessive combustion noise, an increase in thermal and mechanical stresses, and a higher amount of NOx formation. Higher rates of injection eliminate soot by increasing combustion temperatures. Since the formation of the soot is related to the oxygen deficiency, an improper spray pattern can aggravate the situation. Therefore, injector designs that result in more atomization are less likely to lead to soot formation because they increase the homogeneity of the air-fuel mixture; hence increasing the effectiveness of the combustion.

Swirl

To a degree, an increase in swirl lowers the HC emissions. It does so by improving fuel-air mixing and hence oxidation. However, too much swirl can increase HC emissions by increasing the amount of fuel in the lean flame-out region [547].

Intake Air Temperature

Higher intake air temperatures result in higher combustion temperatures, which in turn lead to higher NOx emissions. In turbocharged engines, the amount of NOx can be very high. This problem is offset by carrying out inter-cooling of the compressed intake air. The effect of intake air temperature on PM depends upon the volatility of the fuel. For high-volatility fuels, high air temperature lowers the spray penetration, which increases the heterogeneity of the combustion mixture near the injector nozzle and produces fuel-rich zones. This and the fact that the high temperatures support decomposition reactions lead to increased smoke. For low volatility fuels, however, high air temperatures do not adversely affect dispersal of the fuel but instead accelerate the oxidation reactions, thereby minimizing smoke.

After-injection

After-injection occurs when a needle valve leaks fuel after the main injection has taken place. The lower the amount of the after-injection, the lower is the amount of smoke.

Other factors that impact HC emissions include injection system design, timing, and the rate of injection. These will be discussed in the section dealing with new engine design strategies to lower emissions.

Noise in a diesel engine, or diesel knock, although not directly related to emissions is also of concern because it indirectly contributes to the undesirable emissions. Diesel knock is the combustion noise that results from high pressures that are generated during the rapid initial combustion. During the delay period, i.e., the period between the start of injection and the start of combustion, the fuel evaporates and mixes with the air to form a flammable mixture that has the ability to ignite at many sites simultaneously [566]. When this occurs, the result is the diesel knock. The problem can be overcome by controlling the formation of such a mixture during the delay period. This can be done either by retarding the injection timing or reducing the delay period itself. Retarding timing will allow the fuel to be injected closer to the end of the compression stroke but will lead to higher fuel consumption and higher smoke. However, this strategy, in conjunction with increasing the injection rate (higher combustion temperatures), can help overcome this problem. The reduction of the delay period, a more common approach, employs higher temperatures, pressures, and the use of spontaneously ignitable (high cetane) fuels.

Higher temperatures can be achieved by the use of turbo-charging or low heat rejection. While these strategies help overcome the diesel knock problem, they lead to higher NOx emissions. As mentioned earlier, NOx results from the high-temperature reaction of nitrogen and oxygen. Its amount also relates to the mode of injection. Indirect injection engines, on account of having higher compression ratios, generate higher combustion temperatures than direct injection engines, and therefore produce more NOx. Indirect injection engines have a CR of ~23:1 versus CR of 15:1 or 16:1 for the direct injection engines. Exhaust gas recirculation (EGR) in an indirect injection engine can help lower these emissions.

Emissions Control via After-treatment

Emissions control strategies fall under two general classes: preventive and corrective. Manipulating the fuel properties by changing its composition, altering engine design, optimizing operating parameters, and using additives are preventive in nature. Corrective strategies include crankcase ventilation, exhaust gas recirculation, and exhaust after-treatment.

Crankcase Ventilation

This strategy involves directing the volatile hydrocarbons (HC) from the crankcase into the combustion chamber to burn them. Closed ventilation systems, used for this purpose, recirculate the blow-by gases into the combustion chamber (Fig. 6.27). These systems use air intake to direct hydrocarbons into the combustion chamber. PCV is the common acronym used for such systems, often referred to as positive crankcase ventilation systems.

Exhaust-gas Recirculation (EGR)

EGR involves redirecting the exhaust gases into the combustion chamber by mixing with the intake air, thereby allowing complete oxidation of HC and CO to carbon dioxide and water. The key benefit of this strategy is that it lowers HC and

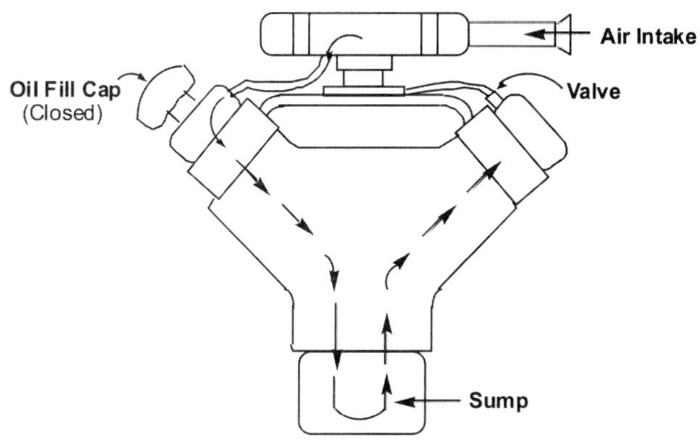

Fig. 6.27—Closed manifold PCV system.

CO without increasing the amount of NOx. This is because the EGR lowers the amount of oxygen in the fresh charge via a dilution effect and also increases its specific heat, both of which lower the combustion temperatures. The result is a lower amount of NOx production. The amount of recirculated exhaust should be limited so as not to lower the oxygen content of the fuel-air mixture to a point where the combustion becomes inefficient. Otherwise, an increase in soot, CO, and HC will occur.

Exhaust After-treatment

This strategy involves completing the oxidation reactions in the exhaust system instead of the combustion chamber, as is the case in PCV and EGR. After-treatment employs thermal afterburning and catalyst-assisted removal of pollutants. *Thermal afterburning* was the method of choice before the installation of catalytic converters in automobiles became popular. Basically, the process involved holding the exhaust components (that escaped combustion) at high temperatures and in the presence of excess air to complete their oxidation to CO_2 and H_2O. Today, the method is not used because it has no effect on NOx. However, it has the potential in lowering the HC and CO emissions during the warm-up phase when the catalytic converter is below its optimum operating temperature. It could, therefore, be used in conjunction with the catalytic converters to help comply with more stringent emissions requirements of the future [567].

Catalytic After-treatment

Catalytic after-treatment involves converting HC and CO to the fully oxidized species and NOx to nitrogen by the use of a catalyst. Most modern automobiles are equipped with catalytic converters, which is the most important pollution control device on a gasoline-fueled vehicle. The catalyst, normally a mixture of noble metals, such as platinum, palladium, and rhodium, is supported on a surface to provide a large surface area. Since lead and phosphorus deactivate these catalysts, the new automobiles use unleaded gasoline and low-phosphorus lubricants. The search for more stable and poison-resistant catalysts continues. Minimum temperature for the converter to be effective is ~250 °C, with 400–800 °C to be the ideal operating range. The converters are installed somewhat away from the engine's immediate vicinity to guard against very high exhaust temperatures (>1000 °C) that are beyond the operating limit of the converters. The use of the electrically heated catalysts is being explored to overcome the waiting period for the converter to become operational. Close coupled catalytic converters are used in most modern ultra-low emissions vehicles (ULEV). Close coupled means closer to the engine, thus faster heating. Electrically heated catalysts are not in commercial use yet.

Catalytic converters are of three types, oxidation catalytic converters, reduction catalytic converters, and dual-bed catalytic converters. *Oxidation catalytic converters* oxidize CO to CO_2 and HC (VOC) to CO_2 and water. Lean mixtures have enough air to complete the oxidation but rich mixtures require excess air through a secondary air injection. In air-injected exhaust systems, the air is injected into each exhaust port, downstream from the exhaust valve/s, by the use of an air pump [558]. High exhaust gas temperatures catalyze oxidation of HC and CO to innocuous water and carbon dioxide. The effectiveness of this strategy depends upon a number of factors including the composition of the air-fuel mixture, the amount of the air injected, and the injection pressure. While the amount of both HC and CO decreases with air injection, the effect is more pronounced in the case of rich fuel-air mixtures, with the magnitude of HC reduction being larger than that of the CO. An increase in the amount of air injected and the injection pressure also reduce the amount of these pollutants. However, the HC decrease with injected air levels off at about 20 % of the engine air flow. *Reduction catalytic converters*, used to lower NOx, operate with little or no air. The chemistry in this case involves the use of the unburned hydrocarbons in the engine exhaust as the chemical reducing agent, which translates into a higher fuel consumption ($\lambda = 0.9$). Because the oxidation type converters lower HC and CO and the reduction type converters lower NOx, in a *dual-bed converter* the two types are combined in a series.

Three-way Catalytic Converter Exhaust Treatment

The purpose of the treatment is to achieve oxidation and reduction within one converter, i.e., oxidize CO to CO_2, HC to H_2O and CO_2, and reduce NOx to N_2. Ordinarily, these reactions are sluggish. However, their rates can be increased by the use of the noble metals catalysts, such as platinum and rhodium. Platinum is a good oxidation catalyst and rhodium is a good reduction catalyst. Their reactivity is the highest when the fuel-air mixture is stoichiometric or rich. Since the two metals in combination remove all three pollutants, a catalyst system containing both is called a three-way catalyst system. Because the reactions occur in a gas phase, an adsorbent surface, commonly provided by the inert alumina (Al_2O_3), is necessary. Catalyst metals are imbedded in the surface as very fine particles to provide a large surface area. A minimum exhaust temperature of 250–300 °C is required for the reactions to occur. Typical exhaust temperatures are 300–400 °C at idle and up to 900 °C at full load. The normal operating range of the catalytic converters is 400–800 °C. Below this range, they are not very effective and above this range, they lose their activity due to sintering.

For converters to be effective over an extended period, the engine must operate at very close to stoichiometric (AF = 14.7; $\lambda = 1$) or at slightly richer (AF = 14.48–14.62) air-fuel ratios. This is because at high loads when the exhaust contains high amounts of NOx, unburned hydrocarbons (HC) facilitate its reductive reaction to nitrogen and water. The process involves three steps: adsorption of pollutants on the catalyst surface at low temperatures, oxidation and reduction of pollutants to innocuous products, and desorption of these products at high temperatures.

The exhaust gas contains oxygen in varying amounts. Its amount can be correlated with the quantity of air in the air-fuel mixture. A *three-way* or *selective catalytic converter* with a lambda closed-loop control system is the most advanced of the converters. The system helps in precisely controlling the composition of the fuel-air mixture to near stoichiometric. A lambda sensor is an electronic device that measures the amount of oxygen in the exhaust. It generates an electronic signal based on the oxygen level in the exhaust and sends it to the engine's electronic unit. The unit makes adjustments to the air-fuel mixture composition if λ is different from 1. The lambda closed-loop control system interconnects the lambda sensor, the electronic control unit, the engine, and the fuel metering system [558]. It signals adjustments to the fuel metering system. Such sensors can also be used to determine the catalyst activity. This is accomplished by installing one sensor before and one after the catalyst system. These sensors have a threshold operating temperature of >280 °C and a maximum temperature of ~850 °C, beyond which their electrodes get damaged. Hence, they, like catalytic converters, are installed close enough to the engine to benefit from the combustion heat to attain activation temperatures quickly but far enough to minimize damage due to sintering in case of extensive heat. This problem can be overcome by installing the unit farther away from the engine but using a heater for the sensor to achieve its operating temperature. The converter catalysts get deactivated by lead and phosphorus that block the active sites, a situation referred to as poisoning.

Emissions Control via Engine Design Changes

Minimizing harmful emissions by manipulating fuel composition and engine operating parameters obviously has its limits, so has the use of deposit control/cleanliness additives to improve combustion efficiency via a uniform flow. The OEMs realize the challenges of meeting the future emissions standards by the use of these strategies and are actively involved in designing new engines that have better combustion efficiency, improved fuel economy, and lower exhaust emissions. They plan to achieve these goals through engine modifications, electronics/fuel management, and exhaust treatment [558]. New technologies will be discussed for the gasoline engines first, followed by those for the diesel engines. It is important to note that the emissions control challenges are resulting in a loss of distinction between traditional gasoline and diesel engine technologies and they are merging together. The key requirement of the technologies is that it must work. In other words, every known technique is being considered to improve emissions from the internal combustion engines.

Gasoline Engines

For gasoline engines, the prevention technologies include more efficient delivery systems, on-board diagnostics, and new combustion chamber designs. Correction technologies include a greater use of exhaust gas circulation (EGR) and exhaust gas after-treatment, both of which were briefly touched on during the preceding discussion.

Efficient Fuel Delivery

Gasoline engines are designed to burn premixed air-fuel mixtures that are prepared either through carburetion or by fuel injection. Modern automobiles favor fuel injection because it results in more efficient combustion, hence maximum output. The fuel is usually injected into the intake manifold rather than directly into the cylinder. This is to avoid combustion deposit buildup and high-temperature damage to injectors. However, the direct-injection engines are better for fuel economy. This is because direct injection uses stratified charge and also results in better homogeneity due to turbulence, hence more complete combustion. OEMs are pursuing this technology primarily because of the fuel economy benefit. The use of the high pressure swirl injectors and the invention of effective deposit control additives do help preclude deposit formation on the injector tips. Deposit infested injectors increase both the fuel consumption and the amount of pollutants (HC, CO, and NOx). The amount of these emissions varies among engines and depends upon the air-fuel ratio of the combustion mixture; the engine's operating variables, such as ignition timing, load, and speed; and the presence of the deposit control additives. Since rich fuel-air mixtures result in CO and HC due to partial combustion and in NOx on complete combustion, the use of lean combustion mixtures should minimize their formation. However, lean mixtures result in lower power, therefore other strategies, such as the use of catalytic converters and thermal reactors as the after-treatment devices are often preferred. Strategies to burn lean mixtures without losing power are described below.

The use of additional valves results in better turbulence and fuel-air mixture control, which translates into more complete combustion. This, in conjunction with the use of the high pressure swirl injectors, electronic swirl control valves, and intelligent variable valve timing, can improve the combustion efficiency and lower undesirable emissions. Variable valve timing controls the amount of the air intake. It can also minimize the need for the valve overlap to achieve smooth engine performance. Extensive valve overlap has a tendency to increase HC and CO emissions due to the escape of the raw fuel into the exhaust (scavenging). Most modern vehicles are equipped with an electronic control module (ECM), which facilitates variable valve timing that is adjusted to the engine's operating conditions.

On-board Diagnostics

As mentioned earlier, combustion mixture composition and ignition timing are the two key parameters that largely contribute towards emissions and fuel consumption. Typically, combustion mixtures are defined in terms of the fuel-air ratio. However, it is the amount of air that is critical to engine performance and emissions. The use of the excess air factor, λ, is therefore more appropriate. We know that rich mixtures ($\lambda < 1$) lead to higher HC, CO, and NOx because of the high combustion temperatures, and lean mixtures ($\lambda > 1$) lead to lower emissions and good fuel economy. We also know that the amount of air required for proper combustion depends upon the engine operation. For example, at idle, λ of 0.9–1.05 is necessary to minimize misfire, which will lead to an increase in HC emissions. At low loads and high speeds, λ of 1.1 is optimal for fuel economy, and at heavy loads, $\lambda < 0.9$ is needed to maintain smooth engine operation. Hence, the delivery of the combustion mixture with the right amount of air is imperative to minimizing emissions while maintaining proper engine operation and fuel economy. In modern cars, on-board diagnostics help achieve this. They measure a number of parameters, such as engine speed, manifold pressure, and ignition timing, and use them to formulate air-fuel mixtures that will lead to optimum output and low emissions. The diagnostics also include knock sensors and oxygen sensors. A knock sensor detects the onset of the engine vibration resulting from knock and adjusts the ignition timing by taking into account the fuel density and the octane rating. An oxygen sensor (Lambda sensor) measures the exhaust oxygen level to determine the air-fuel ratio and adjusts it to within 1 % of the stoichiometric ratio. This leads to maximum power, fuel economy, and low emissions.

New Combustion Chamber Designs

Engine torque, power output, and fuel consumption are a function of an engine's compression ratio, the shape of the combustion chamber and piston crown, the number and the size of the intake and exhaust valves, and the position of the spark plug. Both thermal efficiency, hence fuel economy, and power output can be increased by raising the compression ratio and minimizing the combustion time. However, raising the compression ratio not only results in higher NOx but is also limited by the onset of knock. The latter can be corrected by the use of the high octane fuels. An alternative way to increase fuel economy and output is by the use of high turbulence, lean burn, and compact combustion chamber engines.

Good combustion chamber designs are those that synchronize intake and exhaust with the engine's operation, allow thorough mixing of the fuel and air, and burn the charge rapidly and smoothly. Design features that meet these requirements include low surface-to-volume ratio (s/v ratio), adequate swirl to allow proper mixing of the fuel and air, squish zones to create turbulence, little or no preignition, the highest possible compression ratio without promoting detonation, short flame travel distance, a centrally located spark plug, and adequate number of valves to make efficient charging of the combustion mixture and the removal of the burned gases [558]. Most of these features increase the combustion rate, which translates into lower HC emissions.

Lean or Fast Burn Combustion Chamber Technology

Most pollutants result from incomplete combustion. If the combustion is complete, the formation of these pollutants is minimized. Lean burn technology will allow complete or near complete combustion, thereby reducing the pollutants and increasing the fuel economy. The SI engines require AF ratio of 12.5–13.5:1 for maximum power, which leads to higher HC and CO emissions, and AF ratio of 15.5–16.5:1 for fuel economy, which will result in high NOx levels although HC and CO emissions decrease. Therefore, there is a trade-off between decreasing NOx and decreasing HC and CO. Combustion chamber design research has produced engines that can operate consistently with lean combustion mixtures, that is, air-fuel ratios of 18:1 and 19:1. These produce lower NOx but increase HC only slightly. The main problem is that engines equipped with such combustion chambers often experience hesitation and misfiring. This is because of the difficulty in consistently producing homogeneous mixtures of similar strength, swirl, and turbulence, and generating accurate and equal spark energy to produce similar cylinder temperatures across all cylinders at varying speeds, loads, and throttle openings. Thus, for multi-cylinder lean burn engines, mixture preparation, distribution, and ignition energy release between cylinders must be near perfect. With further improvements in promoting induction swirl and cylinder squish, and with a greater emphasis in matching mixture strength with the injection timing, it will be possible to reliably burn mixtures that are leaner than even an AF ratio of 18:1 and 19:1.

Advances in combustion chamber designs are based on the understanding that rich mixtures burn quickly, which is needed for power, and lean mixtures burn slowly. New lean burn designs produce a high degree of swirl and turbulence, both of which initiate and propagate fast combustion, making a lean burning chamber to be a fast burning chamber. Turbulence allows the leaner mixtures to burn completely and at a faster rate, which minimizes both knock and the loss of power. Lean burn is achieved through structured flow, that is, by stratifying charge and direct injection technology. Charge stratification involves introducing a small amount of fuel into the cylinder to initiate combustion that spreads through the main charge of the lean combustion mixture. Pressure development, a prerequisite to efficient combustion, not only depends upon the shape of the combustion chamber but also the position of the spark plug and the ignition timing. Certain combustion chamber designs allow faster burning of the charge and hence combustion can be initiated later to develop the same or higher peak pressures. A properly situated spark plug can facilitate combustion in an analogous manner.

Fast burn combustion chambers usually require fuels of moderate octane ratings because the time for auto-ignition is relatively short. New combustion chamber designs take all these factors into account. They minimize the distance traveled by the flame via structured flow, charge is concentrated at the center, by locating spark plugs and exhaust valves in close proximity, and by having the end gas in the cooler part of the combustion chamber [566]. Also, the size of the squish area, the area between the cylinder head and piston at the end of the compression stroke, in such designs is just sufficient to produce enough turbulence to ensure a reasonable rate of combustion but not large enough to produce knock arising from too rapid a combustion. Most of these technologies decrease HC and CO emissions by improving combustion efficiency but lead to higher NOx formation. Therefore, these technologies require the use of a higher degree of EGR and of the NOx storage catalysts. Such catalysts store the NOx produced at high loads so as to reduce them at low loads.

Valve Size and Arrangement

Air intake and removal of the exhaust can be greatly improved either by the use of large valves or by the use of multiple intake and exhaust valves. These strategies enhance volumetric efficiency, hence improve combustion. Multiple valves, in addition, increase a combustion chamber's thermal efficiency by decreasing the exposed space around their rims. Data show that at high speeds, a greater number of valves result in more power and more torque. At low to moderate speeds, however, there is little advantage [558].

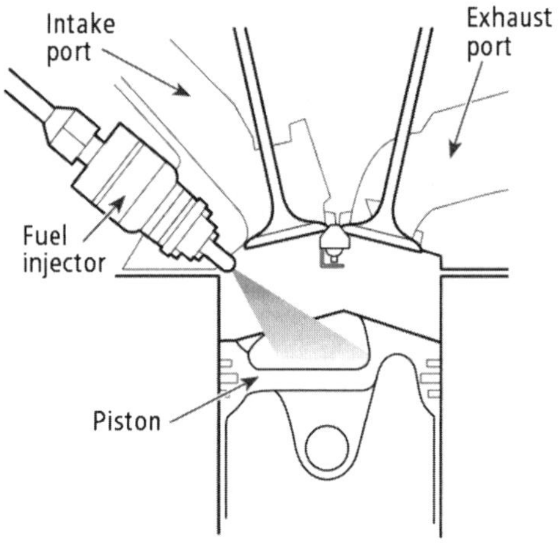

Fig. 6.28—Illustration of direct injection in a gasoline engine [556,568].

Direct Injection

In this technology, the fuel injector is directly mounted into the cylinder head, making it possible for the fuel spray to be directly injected into the combustion chamber, as shown in Fig. 6.28 [556, 568]. This type of fuel delivery requires a higher pressure than port fuel injection. This is to overcome the compressed air cylinder counter-pressure. The advantage of the direct injection is that the fuel delivery at high pressures helps in atomizing fuel and mixing it with air more thoroughly. Direct injection also permits more valve overlap. Both these factors result in complete and rapid combustion, hence better fuel economy, decreased HC and CO emissions, and a lower need for the exhaust gas treatment. The direct injection combustion uses a combination of homogeneous charge and a stratified charge. The former is used for heavy acceleration operation and the latter is used for cruising conditions. Homogeneous charge delivers combustion mixtures that are near stoichiometric. Stratified charge consists of pockets of the fuel-rich combustible mixture within a large volume of fuel-lean mixture [556]. The fuel-rich mixture is concentrated around the spark plug. The stratified charge strategy has a number of unique advantages, which include the following:

1. Resistance to preignition—even with extremely lean (20:1) mixtures.
2. Improved throttle response—a direct injection, stratified charge engine output is determined by the fuel flow and not the airflow as is the case in conventional spark-ignition engines.
3. Improved fuel economy—a stratified charge improves an engine's ability to safely tolerate lean mixtures, which translates into a gain in fuel economy, of as much as 15–20 %.
4. Combustion cleanliness—stratified charge burns in a more progressive fashion and the end gases are ignitable. This can result in up to 10 % decrease in vehicle emissions. However, if the engine uses a higher compression ratio, the amount of NOx will also be higher. NOx can be removed by the use of a reduction type or three-way catalytic converter.

Variable Valve Timing

At least in theory, the intake and exhaust valves must open and close in concert with intake and exhaust strokes. The purpose is to facilitate charge entry and exhaust exit. However, for high-speed engines, opening and closing of the intake and exhaust valves is overlapped. The period during which both types of valves are open concurrently is called the overlap period.

Valve overlap also has its drawbacks, the most pronounced of which is its effect on emissions. Delay in closing the exhaust valve allows fresh charge to escape with the exiting gases thereby increasing HC and CO emissions. Early opening of the intake valve increases the possibility of the exhaust gas entering the intake manifold and leading to charge dilution. This will lead to less effective combustion, hence higher HC and CO emissions. The higher the degree of overlap, the greater is the amount of undesirable emissions. The degree of effective overlap depends upon an engine's operation. Maximum benefits are obtained when the engine is operating at high speeds and at high loads but not during start-up and slow speed operations [558]. The degree of the overlap also has a bearing on NOx emissions [569]. Variable valve timing makes it possible to adjust opening and closing of the intake valves based upon speed and throttle conditions.

In diesel engines, valve overlap is not as beneficial because they generally run slower than gasoline engines. In addition, the possibility of the fuel loss directly into the exhaust does not exist because the intake in diesel engines is only air. However, high-speed diesel engines could take advantage of the more effective scavenging through valve overlap without the risk of high HC emissions.

Honda power train technology has incorporated many of the above described design parameters and have come up with an ultra-low emissions vehicle (ULEV). It was tested against standard technology using CARB Phase II fuel and 100,000 miles aged catalyst. It produced NMOG of 0.035, CO of 0.527, and NOx of 0.054 gm per mile, much lower than those of the standard vehicle that produced 0.055 NMOG, 2.1 CO, and NOx of 0.3 gm per mile.

Diesel Engines

Emissions control in diesel engines also uses prevention and correction strategies. Prevention strategies include better fuel delivery, increased injector and cylinder pressures, advanced combustion chamber and piston designs, and advanced timing. Correction strategies include a greater use of EGR, development of NOx reduction catalysts, and more effective particulate filters.

Low Emissions Piston Designs

These include articulated pistons and pistons with low crevice volume. Articulated pistons have the thrust pad portion, or more commonly the skirt, carried independently by the piston pin. For high-speed engines, the crown is of iron or a ferrous alloy and the skirt is of aluminum. This allows the clearance to be small which makes them low lubricant consumption engines. Pistons with low crevice volume have the same characteristics because of the less cut back crowns and smaller first lands, which was discussed in the oil consumption section. Such engines result in lower HC and particulate emissions.

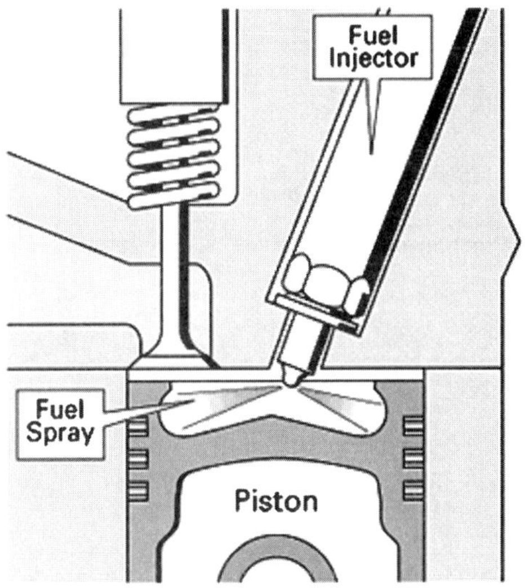

Fig. 6.29—Direct injection (DI) process [570,571].

Fuel Injection

As described earlier, in a diesel engine the liquid fuel is injected into the hot, compressed air late in the compression stroke, shortly before the piston reaches TDC. In order to burn, the fuel must vaporize and thoroughly mix with the air. Otherwise, complete combustion will not occur and the undesirable HC and PM emissions will result. Since proper fuel-air mixing in a diesel engine is difficult because of its design, it can be improved by injecting less than stoichiometric amount of fuel. Typically, diesel engines use an optimum fuel-air ratio to limit the formation of the particulates emissions. The drawback of this strategy is that it lowers the engine's output. Fortunately, there are design features that can facilitate rapid fuel-air mixing. These include the following:

1. High pressure (up to 30,000 psi, 200 MPa) injection.
2. Injection system design that will produce a fine fuel spray.
3. Optimized position and angle of the nozzle in the cylinder head.
4. Sculpting the piston tops and intake ports to generate a swirling motion of the gases in the cylinder.

Direct-injection and Indirect-injection

Diesel engines employ, two fuel injection processes: direct-injection (DI) and indirect-injection (IDI). These are depicted in Figs. 6.29 and 6.30, respectively [570, 571]. The DI process results in fuel injection directly into the cylinder, above the piston. The IDI process, on the other hand, delivers fuel into a small prechamber, which is connected to the cylinder via a narrow passage. During compression, air is forced through this passage, generating a strong swirling motion in the prechamber, where the fuel is injected and the ignition occurs. The rapidly swirling air in the prechamber and the jet-like expansion of the combustion gases enter into the cylinder, which enhances mixing and combustion of the fuel and air [570].

Each method has its advantages and disadvantages. IDI engines suffer from the disadvantages of harder starting and lower efficiency, but they run at high engine speeds. Hence, they are commonly used in smaller automotive and light-duty truck applications. The DI diesel engines, on the other

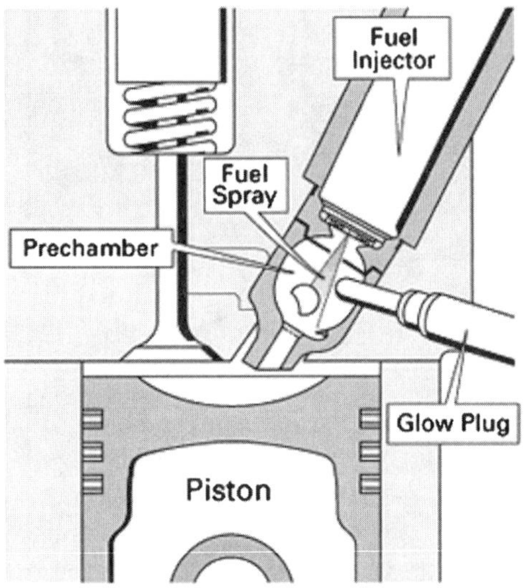

Fig. 6.30—Indirect injection (IDI) process [570,571].

hand, have a fundamental fuel economy advantage of up to 20 % over the equivalent indirect injection (IDI) engine. The continuing drive towards lower emissions will ultimately lead to the use of high-pressure fuel injection equipment (FIE) on DI engines [570].

Fuel Delivery

Combustion mixture composition, injection timing and sequence, and the degree of the fuel's atomization are all important from the emissions perspective. In diesel engines, the start of the fuel injection is the start of combustion. A delay in injection timing reduces NOx by lowering the peak temperatures. However, too much delay can increase the amount of HC emissions. Since, a small deviation in injection timing has a large impact on emissions, timing precision is important. Electronic control systems make this job easier. Injector nozzle sensors based on movement of the injector needle during injection are even more precise and hence can improve the combustion quality. Fuel leakage, dribble or after-injection, which directly contributes to HC emissions, is usually avoided by keeping the volume of the fuel in the injector nozzle low. On-board diagnostics also help reduce in-use emissions resulting from malfunction by early detection and signaling repair.

Since the fuel in a highly atomized state mixes with air more thoroughly, the resultant mixture burns more rapidly. The result is the lowering of HC and particulate emissions. The degree of atomization is a function of the injection pressures and the injector tip design. High injector pressures improve combustion efficiency, hence decrease black smoke substantially. However, it lowers NOx only slightly. Deposits on the injector tip will impair the degree of atomization; hence an increase in emissions will result. Deposit control additives, discussed later in the chapter, are used to correct this problem, when and if it arises. Figure 6.31 shows the effectiveness of these additives in increasing the fuel flow through the injector.

Another way to minimize soot formation is to use extra lean combustion mixtures. An excess air factor is therefore maintained in the combustion chamber. Advanced injection timing also helps in lowering HC and particulate emissions. Selective use of ceramics is also being explored. These materials control emissions by lowering heat losses and resulting in higher combustion temperatures. At present, activity in this area is largely confined to engines for military use because of cost. High cylinder pressures can increase combustion efficiency which translates into higher power, better fuel economy, and lower HC and particulate emissions, but with an increase in NOx.

Fuel Injection Strategies

These include pilot injection and the rate-shaping injection. These approaches allow burning the lean air-fuel mixtures and result both in better fuel economy and lower HC, CO, and particulate emissions. Pilot injection is similar to the stratified charge technology used in the gasoline engines. In such engines, a small amount of fuel is injected into the cylinder to produce a layer of rich mixture so as to initiate combustion that spreads through the main charge of the lean combustion mixture [572]. Because diesel engines do not use a premixed combustion mixture, the same effect is achieved by the dual injection [558]. In this strategy, a small amount of fuel is sprayed into the compressed air prior to the piston's

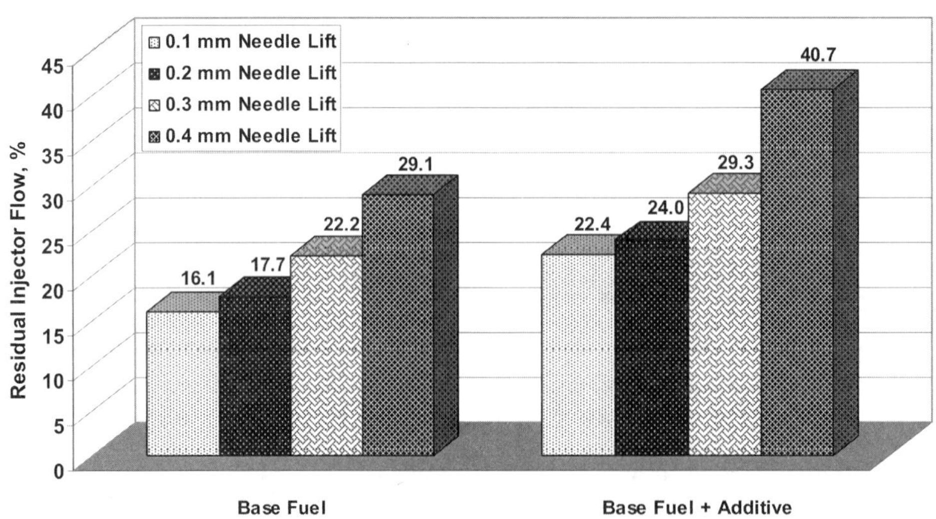

Fig. 6.31—DB OM616 injector nozzle coking test for diesel fuel Test engine—4-cylinder, 2.4 litre; test duration—3 h.

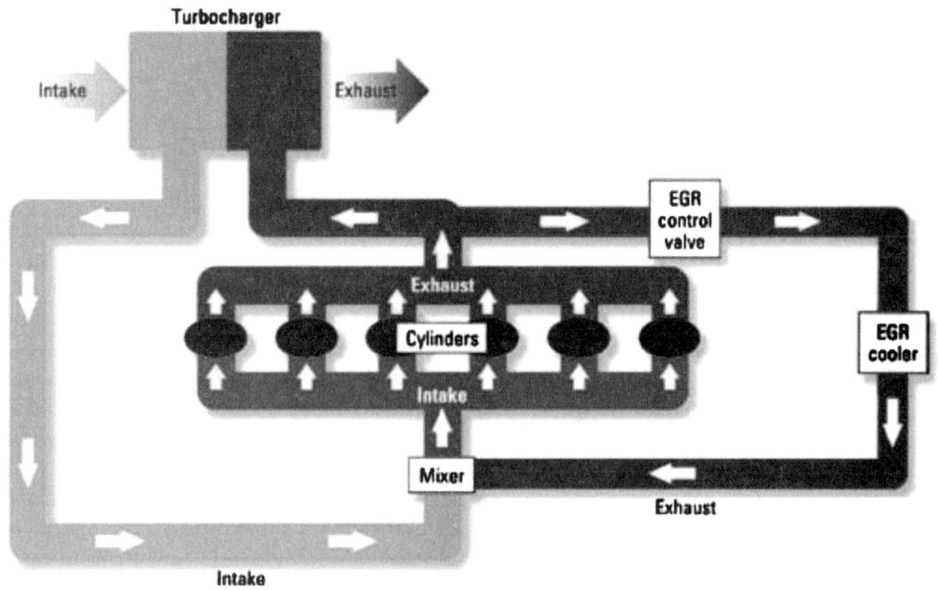

Fig. 6.32—EGR circulation system in a turbocharged engine [573].

TDC position, that is, between −10 and 0°, to initiate combustion, which is followed by the main fuel charge at between 0 and 10°. The flow in the rate-shaping fuel injector is controlled by adjusting the needle valve lift. This type of flow has the advantage of being continuous and therefore the engine experiences less hesitation and noise.

Exhaust Gas Recirculation (EGR)

EGR recirculates a portion of the engine exhaust gases back into the engine. Intermixing of the incoming air with the recirculated exhaust gases dilutes the combustion mixture in the engine with the inert gas. This lowers the peak combustion temperatures and the amount of excess oxygen, thereby reducing the generation of NOx. However, this is achieved at the expense of an increase in particulates.

Figure 6.32 shows the flow of the intake air and the exhaust gases through a turbocharged engine, by the use of arrows [573]. A portion of the exhaust from the engine is directed back into the engine by the use of an EGR valve, which regulates and times the gas flow (right loop). The purpose of the EGR cooler, also called the intercooler, is to lower the temperature of the exhaust gases prior to their mixing with the intake air (left loop) and their entry into the engine. Despite the benefits, EGR has its limitations. These include the following:

1. Increased soot accumulation in the lubricant, as reflected by the viscosity increase.
2. Soot-derived wear.
3. Increased engine temperatures, which causes higher lubricant oxidation and possible loss of some lubricant components through volatilization.
4. Increased corrosion because of condensation and the higher amount of acid formation.

Increase in acid formation with the increased EGR is shown in Fig. 6.33 and increased wear and lubricant viscosity data are provided in Table 6.34 [574]. It appears that approximately 100 h of EGR is the upper time limit, after

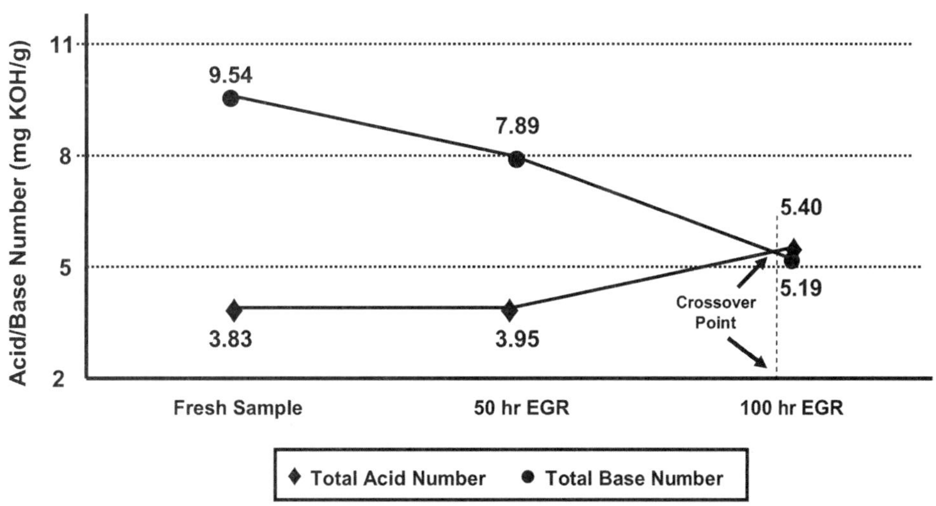

Fig. 6.33—EGR duration versus acidity [574].

TABLE 6.34—Viscosity and wear versus EGR duration [574].

EGR, hours	0	50	100[a]
Soot Level/Particle Size	0/None	Medium/Large	High/Small
40 °C Lubricant Viscosity, cSt	117.6	118.0	126.2
Wear	Baseline	High	Medium

[a]This time equals approximately half the drain interval without EGR.

which the equipment corrosion will ensue. Soot-derived wear around 50-h is higher than at 100-h, presumably because of the larger size of the soot particles. The next question that needs to be answered is what percent of EGR is optimal? This is a valid question because there is a NOx—HC trade-off. Too much EGR increases particulate matter (PM), due to inefficient combustion (dilution), and an increase in NOx, as shown in Fig. 6.34 [575]. It appears from the figure that ~40–45 % EGR is optimum based upon the test parameters. At this level, both HC and the NOx are at their lowest levels. Interestingly, at this EGR concentration, particulate accumulation rate is also at its lowest, as shown in Fig. 6.35 [575].

Exhaust After-treatment

Heat sensitivity of the after-treatment catalysts was commented upon while discussing gasoline emissions. Consideration of the thermal degradation aspects of the catalyst system is especially important in diesel engines because on average they generate more heat. In order to guard against the extensive degradation of the catalyst, there is a move towards the use of palladium, instead of platinum, and a trimetallic system, which comprises palladium, rhodium, and platinum. Alternative catalyst technologies include electrically heated catalysts; by-pass catalyst systems; fuel burner/exhaust gas ignition, which was commented while discussing gasoline emissions control technologies; and lean NOx catalyst systems that are active at low NOx levels. Metal hydride-cold start heaters using hydrogen gas are also under consideration. These help catalysts attain suitable operating temperatures within seconds.

Diesel Particulate Filters

Particulates from combustion refer to any substance, other than water, that can be removed by the use of a filtering device. The amount of particulates from a gasoline engine is too small to be of serious concern, but from diesel engines it is quite large and therefore is a matter of great concern, primarily because of their adverse health effects. Diesel exhaust mainly consists of soot particles with adsorbed hydrocarbon material/s. Hence, they can be removed by passing the exhaust through a filtering device, which has the ability to burn them by hot exhaust gases and regenerate itself. The problem is that normal ignition temperature of the soot is around 500 to 600 °C and the exhaust temperature is between 200 to 500 °C. One strategy that has been successfully employed to overcome this difficulty involves coating the device with an oxidation catalyst that will lower its activation temperature. A good catalyst can reduce this temperature by almost 200 °C. Filter and burn strategy is most profound since it not only burns the soot but also lowers the amounts of HC and CO, by oxidizing them to water and carbon dioxide.

We briefly touched on the topic of the particulate filters in the emissions section while discussing reasons behind the low sulfur diesel specification. Here we talk about particulate filters in more detail. That is because in order to meet emissions standards for heavy-duty diesel vehicles for the model year 2007 and beyond, diesel particulate filters (DPFs) are the principal strategy. The new API CJ-4 performance category is specifically designed to protect these after-

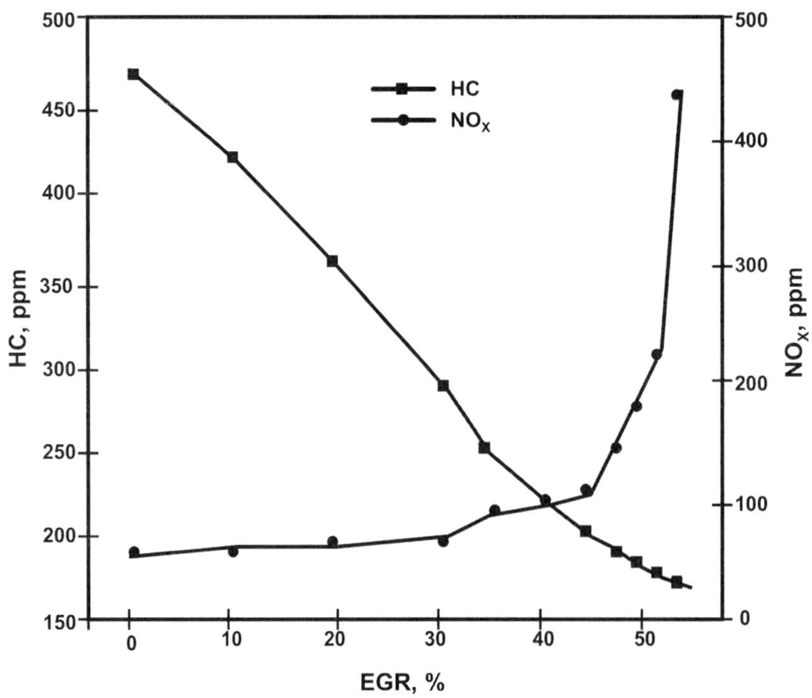

Fig. 6.34—HC and NOX versus % EGR [575].

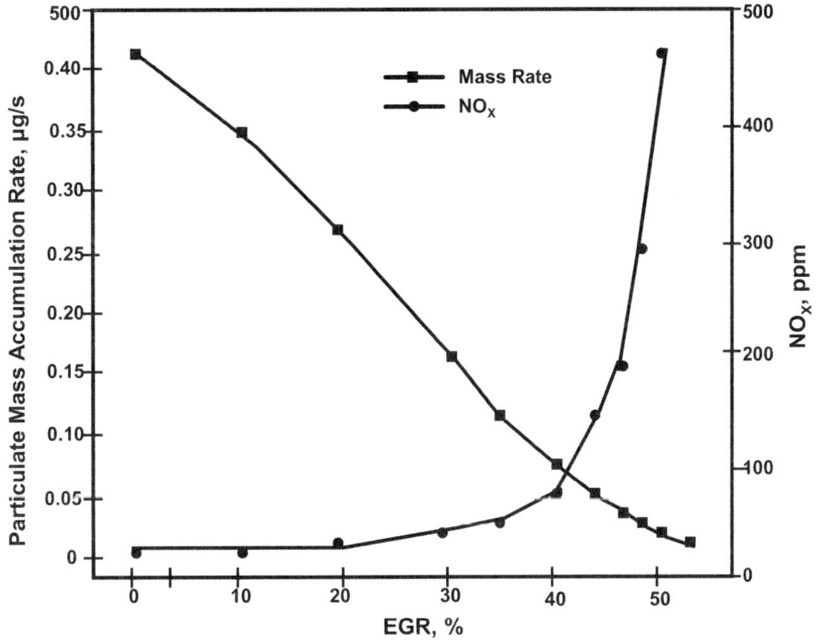

Fig. 6.35—Particulate matter and NOX versus % EGR [575].

treatment devices, whose primary role is to lower particulate and nitrogen oxide emissions from the model year 2007 on-road engines.

As mentioned in the earlier part of the discussion, DPFs remove particulates from the diesel exhaust, prior to its discharge into the atmosphere. The collected particulates can either be burned externally (disposable DPFs) or internally in the device itself (catalyzed DPFs or CDFs). DPFs of the latter type are either precatalyzed, or can use fuel additives as catalysts. A precatalyzed DPF uses noble metal catalysts, such as platinum, palladium, and rhodium, to reduce the temperature necessary for the particulate burn off.

Fuel additive-catalyzed filters lower the temperature required for soot burning, but the ash from the additive/s remains in the filter after the burn off, which creates the problem of the ash build-up. This problem is overcome in the continuously regenerating DPF (CR-DPF), which contains a diesel oxidation catalyst (DOC) upstream of the diesel particulate filter. This catalyst oxidizes the oxides of nitrogen other than NO_2 to NO_2 before the exhaust enters the DPF. The NO_2 lowers the temperature required to oxidize or burn off the particulates, thereby allowing the CDPF to continuously clear itself of the trapped particulates. DOCs are platinum or palladium compounds that catalyze the oxidation of the organic particulates from the unburned fuel and the oil, hydrocarbons, and carbon monoxide to carbon dioxide (CO_2) and water. Besides DOCs, there are two other catalyst technologies that are worth mentioning. These are lean-NOx catalysts (LNCs), which are capable of converting NOx into nitrogen (N_2) in the presence of oxygen, and NOx adsorber catalysts (NACs), which first store (adsorb) NOx and then reduce it under fuel-rich conditions.

In an ideal situation, soot and other organic matter collected in the DPF is burned as stated. However, engine oils contain a number of metal-containing additives that do not burn completely, leaving behind ash. Ash forms when certain phosphorus and sulfur compounds present in the engine oil, react with metals. This can deposit in the DPF and closing its pores, which will lead to plugging. Extensive filter blockage can lead to back pressure and improper functioning of the engine, which will be reflected by increased fuel consumption and a loss in power. Ash deposits on the surfaces of the catalysts can deactivate them since they can no longer facilitate oxidation chemistry. Phosphorus-derived ash is particularly detrimental to noble metals as they block their active sites, a process referred to as poisoning. Sulfur in the form of sulfur trioxide (SO_3), on the other hand, blocks the NOx storage sites or the reaction sites of the NACs and LNCs, thereby reducing the efficiency of these catalysts to convert NOx into nitrogen. In addition, since all catalysts involve oxidation, sulfur will be converted into metal sulfates by way of its oxidation to sulfur trioxide. As a result, excessive amount of sulfur will show up in the form of the increased particulate emissions. Sulfur is also detrimental to the selective catalyst reduction (SCR) technology because the catalyst can be fouled by sulfates, especially ammonium sulfate, which forms at low exhaust temperatures. Because of these reasons, the EPA mandated the use of ultra-low sulfur diesel (SD15) in 2007 model year heavy-duty trucks, which are equipped with the emissions reducing technologies discussed in this section. Please note that these technologies can affect fuel consumption: SCR improves fuel economy but DPF, NAC, and LNC result in a fuel penalty.

Turbo-charging—Inter-cooling

Power output in an engine is proportional to the amount of air it takes in. In general, the greater the amount of air, the greater is the power output. Two ways to introduce a greater amount of air is through turbo-charging or supercharging. Either technique increases the density of air or the air-fuel mixture, via compression. This allows the introduction of a greater mass of charge into the cylinder under pressure. The consequence is a more efficient combustion; hence lower HC

TABLE 6.35—Classes of fuel additives.

Performance Additives	Distribution Additives	Fuel Quality Additives
• Deposit Control Additives/ Cleanliness Agents • Injector Cleanliness Additives (Diesel) • Fluidizers • Anti-icing Agents • Octane Improvers (Gasoline) • Lubricity Agents (Diesel) • Cetane Improvers (Diesel) • Combustion Modifiers/Smoke Suppressants (Diesel) • Low-temperature Operability Additives (Diesel)	• Flow Improvers/ Wax Modifiers (Diesel) • Corrosion Inhibitors • Foam Inhibitors (Diesel) • Demulsifiers • Biocides • Antistatic Agents • Drag Reducers • Dyes and Markers	• Oxidation Inhibitors/ Stabilizers • Metal Deactivators

and CO emissions. HC and CO emissions from the turbocharged engines and low heat rejection engines are lower than those from the naturally aspirated engines. However, the resulting high temperatures increase the amount of NOx formation. This can be controlled by installing an intercooler, see Fig. 6.32, which reduces the temperature in the interim and hence lowers the NOx emissions.

Electronic Engine Controls

The key requirements of a modern diesel engine include high power output, fuel economy, durability, and low levels of emissions. These objectives can be met primarily by improving the combustion efficiency of the engine. This involves the control of the amount, timing, and preciseness of the fuel delivery during each engine cycle. Today's vehicles are equipped with the electronic engine controls that make this task easier. These systems use a variety of sensors to monitor the functioning of the various components and have the ability to manipulate various operating parameters to correct any malfunction that may occur. Common operating parameters that are monitored by these devices include the engine speed, load demand, engine coolant and exhaust temperatures, and the air temperature and pressure. A microprocessor interprets the sensor outputs and generates signals to operate the electronic fuel injectors. In this way, both the total quantity of the fuel and the rate at which it is injected, i.e., the injection profile, are optimized for the instantaneous engine conditions.

Fuel Additives

Fuel properties during manufacture are carefully monitored so as to ascertain their proper performance during use. In the earlier part of this chapter, we presented data showing the effects of various fuel parameters on the deposit formation, see Tables 6.26 and 6.30. A number of specifications dealing with these properties exist [576]. A variety of chemicals, listed in Table 6.35, are added to fuels to correct fuel deficiencies and enhance performance [551]. These chemicals can be broadly classified into performance additives, distribution additives, and those that maintain fuel quality. The treatment levels of these additives are usually very low, in parts per billion to parts per million. The use of fuel additives is growing fast, primarily because of a concern for emissions and the OEM endorsement who are interested in the long-term durability of their equipment.

Deposit Formation

Harmful deposit formation in an engine is unavoidable unless the fuel contains effective deposit control additives. The function of these additives is to keep engines clean, both by removing the existing deposits and preventing the formation of the new ones. As a matter of fact, good engine deposits control is the distinguishing feature of a good commercial gasoline. The potential deposit sites and the effects of deposits are shown in Fig. 6.36 [556]. The function of the deposit control additives is not only to keep combustion-related parts clean but also to keep the fuel delivery system (injectors, pumps, filters, tanks, and lines) clean.

The function of the fuel injectors is to deliver a precisely metered amount, in the form of a fine spray, into the combustion chamber. For maximum combustion efficiency, the spray pattern must be uniform. Because of fuel passage through the injector is narrow, any deposits on the injector tip will make the spray pattern nonuniform. This in turn will

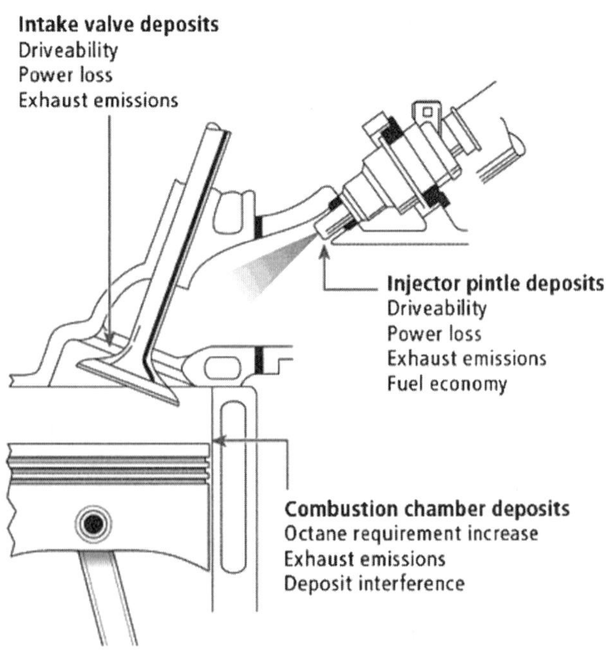

Fig. 6.36—Location and performance effects of engine deposits [556].

Fig. 6.37—Port fuel injector director plate deposits [556].

decrease the combustion efficiency, hence decrease power, degrade driveability, lower fuel economy, and increase undesirable exhaust emissions. Deposits can cause similar problems in the older carbureted engines because they block narrow channels and orifices in the carburetor that meter the fuel. Figure 6.37 shows a close up of the deposits in the metering holes of a gasoline injector tip [556].

The formation of deposits on the intake valves, ports, and in the combustion chamber has similar effects. Deposits on these parts also form because of the high ambient temperatures. Deposits not only increase the undesirable exhaust emissions by impairing the proper functioning of the parts [577], but they also cause part damage, e.g., valve sticking and burned valves. The magnitude of the emissions increase can be substantial. This not only decreases the combustion efficiency but also decreases the efficiency of the catalytic converter. Figure 6.38 shows different levels of the intake valve deposits common in today's engines [556] and Fig. 6.39 depicts the effect of the intake valve deposits (IVDs) on emissions. As the figure shows, the cleaner valves produce lower emissions than the valves with deposits. In Fig. 6.38 LAC detergency rating corresponds to detergency resulting from the lowest additive concentration.

The development of the cleanliness/deposit-control agents is the most active area of fuel's research. As mentioned earlier, deposits can form on many surfaces that see high temperatures, but the fuel additive supplier is primarily concerned with the intake system deposits (primarily IVDs)

Fig. 6.38—Typical intake valve deposits [556].

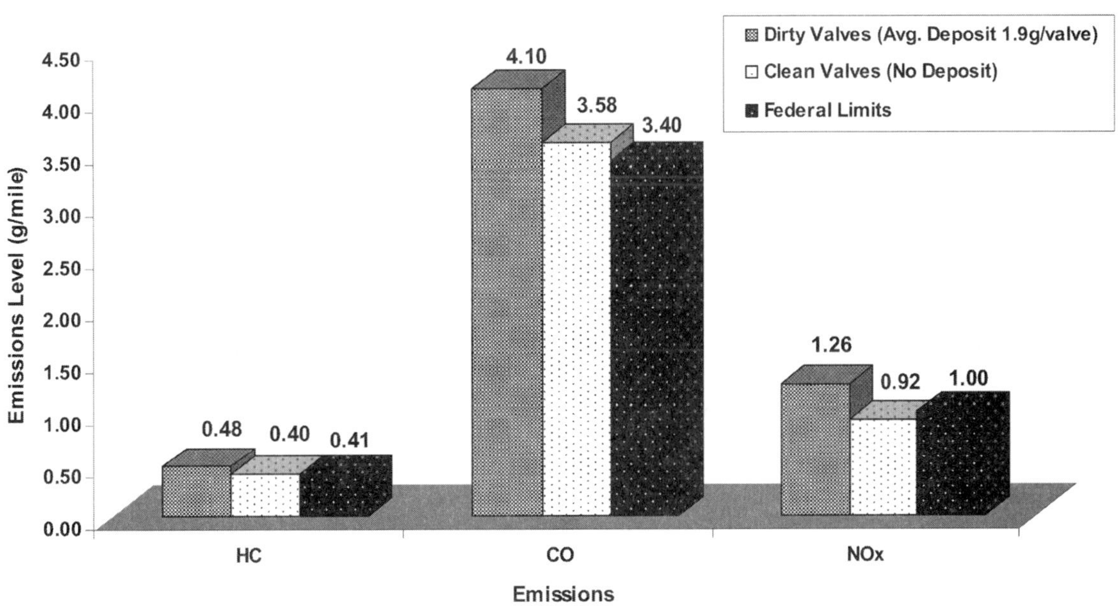

Fig. 6.39—Effect of additives on exhaust emissions in a BMW 1.8 litre 4-cylinder engine.

Fig. 6.40—Typical combustion chamber deposits [556].

and combustion chamber deposits (CCDs). Both the IVDs and the CCDs result from the higher boiling fuel fractions and the lubricant via a thermo-oxidative process [578–582]. The high boiling hydrocarbons form a surface layer either on the intake valves or on the combustion chamber wall and oxidize to form the deposit precursors. Port fuel injector deposits result after the engine has been turned off. The residual stationary gasoline in the injector tip is exposed to a higher temperature for an extended period than the gasoline that flowed through the injector during the engine operation. The result is thermo-oxidative degradation of the gasoline that results in deposits.

Obviously, a combustion chamber in a new vehicle is clean. However, with use the engine will develop combustion chamber deposits (CCDs). Beyond a certain limit, the CCDs start affecting engine performance, which is reflected by an increase in the octane number requirement (ONR); that is, a higher octane number gasoline is required to prevent engine knock. If CCDs are extensive, it is difficult to correct knocking by the use of the commercial gasoline. In vehicles that are equipped with a knock sensor, a substantial loss of power will be observed. Figure 6.40 shows a CCD-infested combustion chamber. The CCDs increase ONR, either by increasing the combustion temperature or by increasing the compression ratio. The temperature increase is a consequence of the CCDs acting as an insulator and not allowing the heat loss via the surrounding metal surface. The compression ratio increase is related to the CCDs' bulk volume in the combustion chamber. The compression ratio is the ratio of the volume of the cylinder at the beginning of the compression stroke, when the piston is at BDC, to the volume of the cylinder at the end of the compression stroke, when the piston is at TDC. The reduction in volume at TDC due to the presence of CCDs therefore leads to an increase in the compression ratio. Generally, the compression ratio change has a much smaller effect on octane requirement increase (ORI) than the temperature increase.

Precursors to the combustion chamber deposits have their origin in the fuel, some fuel additives, and the engine oil. Combustion chamber deposit formation is believed to involve five stages [581], which are listed below.

1. Initiation of Zone 2 Deposits.
2. Growth of Zone 2 Deposits.

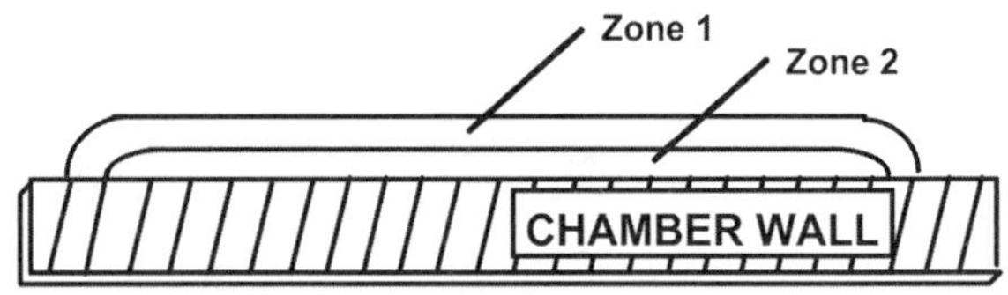

Fig. 6.41—Mechanism of combustion chamber deposits formation [581].

3. Initiation of Zone 1 Deposits.
4. Growth of Zone 1 Deposits.
5. Equilibrium between Zone 1 and Zone 2 Deposits.

Deposit formation initiates in Zone 2 (Fig. 6.41), the area close to the chamber wall, where the high boiling components of the fuel, lubricant, and or the fuel additives condense. Subsequent vaporization, oxidation, and the interaction between molecules result in "lacquer-type" deposits that are difficult to remove. Further condensation of the high boiling fractions on these deposits and the subsequent degradation in the flame environment result in the Zone 1 deposits. A change in morphology gives rise to carbonaceous and "soot-like" deposits that grow by interactions with the other deposit molecules. Ultimately, the rate of deposit growth equals the rate of removal due to combustion, producing a state of equilibrium. At this stage, a fixed ratio of Zone 2 to Zone 1 deposits exists.

While deposits have both advantages and disadvantages, the disadvantages outweigh the advantages. The key advantage is their ability to insulate the combustion chamber and minimize the heat loss, which translates into higher thermal efficiency and hence improved fuel economy [580]. As far as disadvantages, deposits lead to uncontrolled ignitions; power loss; poor driveability; undesirable noise (knocking or rap due to CCD interference); higher NOx, HC, and CO emissions; poor fuel economy; and the octane requirement increase (ORI). All these are interrelated. At present, effective technology to control CCDs is not commercially available, although some additives that are used to control IVDs have a positive effect on CCDs. A good dcposit control additive is desired to have the following attributes.

- It must be free of phosphorus, sulfur, chlorine, and metals. Phosphorus is known to poison catalysts in the catalytic converters; sulfur forms acidic products that can lead to chemical corrosion of the metal parts; chlorine is associated with the formation of the toxic dioxins; and metals on combustion form ash that can lead to an increase in the particulate emissions.
- Its molecular weight must be appropriate for dispersancy (~800–2000 g/mol).
- It should possess basic nitrogen because the base number (TBN) is associated with the deposit cleanup.
- It must have enough thermal stability to survive intake valve and combustion chamber deposit temperatures to either coat or adsorb but combust to form components that do not form dirt, resinous material, or deposit precursors.

Since certain fuel and engine oil components form more deposits than others, the effective means of removing these deposits is to treat the gasoline with aftermarket deposit control additives. A typical treat level of these additives is 10 to 20 times higher than that in the service station gasolines.

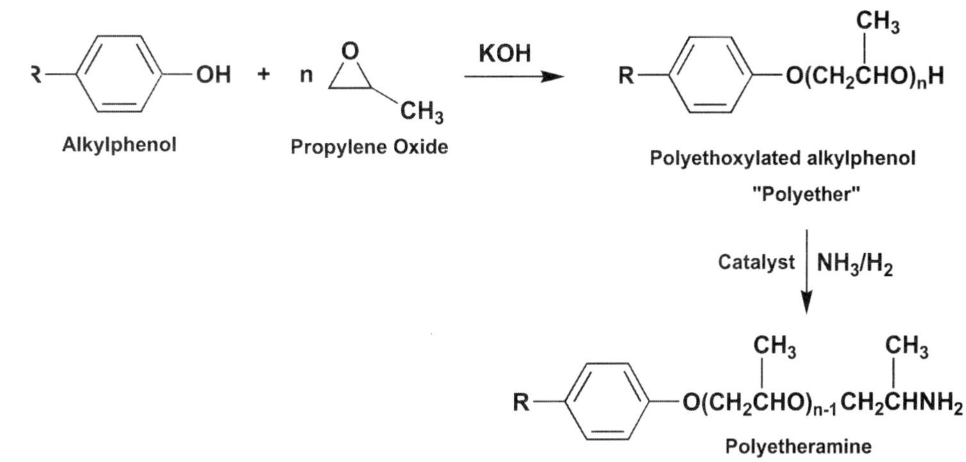

Fig. 6.42—Polyetheramine synthesis.

One tankful of gasoline treated with these additives can decrease ORI by 30 to 40 %. Subsequent periodic treatments with deposit control additives are necessary to maintain the deposit-free combustion chambers, otherwise the deposits will reappear and ONR will revert to its previous equilibrium level.

Two CCD-related problems that occur in some engines are chamber deposit interference (CCDI) and combustion chamber deposit flaking (CCDF). The former results from the physical contact of the piston top deposits and the cylinder head deposits. The contact produces a loud, metallic banging sound. This problem normally occurs in engines that are designed to have minimal clearances to reduce emissions. Combustion chamber deposit flaking can lead to low compression pressures, if the flakes end up between the valve face and the valve seat, thereby resulting in a poor seal. Difficulties in starting and rough running, when the engine is cold, indicate a CCDF problem.

Deposit Control Additives/Cleanliness Agents

Commercially available deposit control additives belong to two general classes: detergents and dispersants. Detergents include basic materials, such as low molecular weight amines, alkanolamines, amido-amines, and imidazolines. These additives differ from detergents used in lubricants in that they do not contain any metals. Dispersants are also basic nitrogen-containing additives, but they differ from detergents in being derived from higher molecular weight polymers. Examples are polyisobutylene derived amines and polyetheramines. Some refiners perceive the latter type to be superior because of the presence of oxygen which they believe leads to lower particulates, NOx, and CCDs. In addition, some of these amines do not need fluidizer oil to overcome the valve stick problems which is common with polyisobutylene-amines.

Commercially available polyetheramines consist of a hydrocarbon portion derived from an alcohol or alkylphenol which is reacted with an epoxide to yield a polyether alcohol. The resulting alcohol is then reacted with ammonia, in the presence of hydrogen, to introduce the nitrogen-containing moiety. The reaction sequence is shown in Fig. 6.42.

General structures of the commercial deposit control agents are given in Figs. 6.43 and 6.44. Polyisobutylene derived amines are prepared by the reaction of polyisobutylene or polyisobutenyl-succinic anhydride with a polyamine. Mannich types are the reaction products of polyisobutylphenol, formaldehyde, and a polyamine. The methods of their synthesis were described under dispersants in Chapter 4, the Lubricant Additive chapter.

Figure 6.45 shows the effect of the cleanliness additive on intake valve deposits. The greater the level of the additive, the lower is the amount of deposits. An effective deposit control additive must pass a number of performance standards, including those that are listed below. The first four were developed by four automobile manufacturers (BMW, General

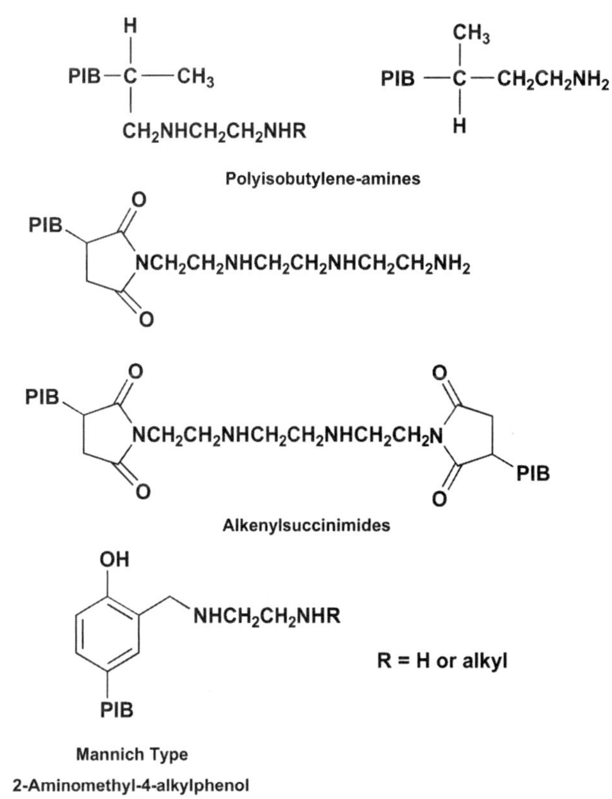

Fig. 6.43—Commercial polyisobutylene-derived amines.

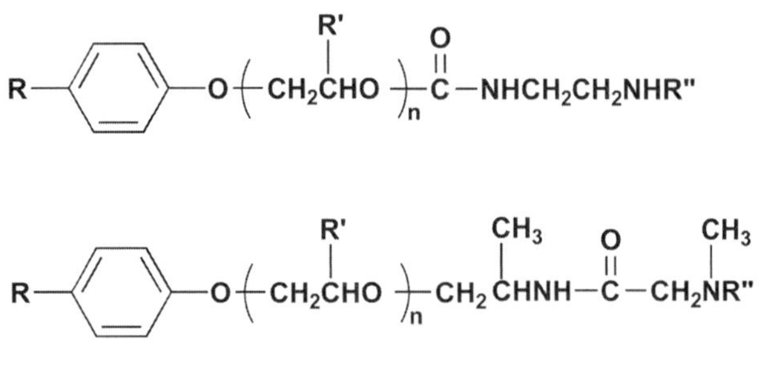

Fig. 6.44—Commercial polyetheramines.

Motors, Honda, and Toyota) to assess the ability of Top Tier unleaded detergent gasoline to minimize deposits on fuel injectors, intake valves, and combustion chambers.

1. Intake valve deposits (IVDs) keep-clean performance by using ASTM D6201, Standard Test Method for Dynamometer Evaluation of Unleaded Spark-ignition Engine Fuel for Intake Valve Deposit Formation.
2. Combustion chamber deposits (CCDs) measurement by collecting and weighing along with IVD using ASTM D6201. This ASTM standard does not contain a procedure for collecting and measuring CCDs. The scrape and weigh procedure was developed by CARB and results for the individual cylinders and an average are reported.
3. Fuel Injector Fouling Initial Performance Standard—Fuel injector fouling is measured by using the Top Tier fuel injector fouling vehicle test, available from GM.
4. Intake Valve Sticking Initial Performance Standard—Intake valve sticking tendency is determined using either the 1.9 L Volkswagen engine (Waterboxer, according to CEC F-16-T-96) or the 5.0 L 1990-95 General Motors V-8 engine (SWRI IVS test).
5. No fuel octane requirement increase (ORI) or an increase in exhaust emissions.
6. No increase in CCDs resulting from the base fuel alone.
7. Must have acceptable corrosion and demulsibility performance with or without a rust inhibitor and a demulsifier.

Since the deposit control additives are used at concentrations that are 20 to 50 times higher than that of the other gasoline additives, they can alter certain gasoline properties or harm fuel system materials, or both. Gasoline containing the deposit control additives is therefore tested for the absence of the negative attributes (no harm) as well as for the positive attribute of controlling deposits. Attributes of an ideal deposit control additive include the following:

1. Deposit control
2. Water tolerance
3. No spark plug fouling
4. No contribution to engine sludge
5. No sticking of the intake and exhaust valves
6. Innocuousness to elastomers and metals that the fuel will be in contact
7. Compatibility with other gasoline additives

Fuel or engine oil, or both can cause deposit formation on injector nozzles, because of the high-temperature mediated oxidation and decomposition. While the degree of the deposit formation depends upon the engine design, fuel composition, lubricant composition, and the operating conditions, deposits can adversely affect the fuel spray pattern. This may result in poor fuel-air mixing and hence can result in the loss of fuel economy and an increase in undesirable emissions. Figure 6.46 shows spray patterns from a clean injector and a deposits-infested injector [570].

Ashless (metal-free) high molecular weight polymeric detergents that are used to clean the injectors as well as keep them clean have a polar functional group and a nonpolar functional group. The polar functional group complexes with the polar decomposition products on the injector tip and the nonpolar functional group associates with the fuel, thereby bringing the deposit forming species into the fuel. By a similar mechanism, they keep potential deposit-forming species in the fuel and prevent them from depositing on the injectors. The fuel treatment level with these additives is in the range of 50 ppm to 300 ppm. Figure 6.47 shows the effectiveness of these additives in keeping the injectors clean [570].

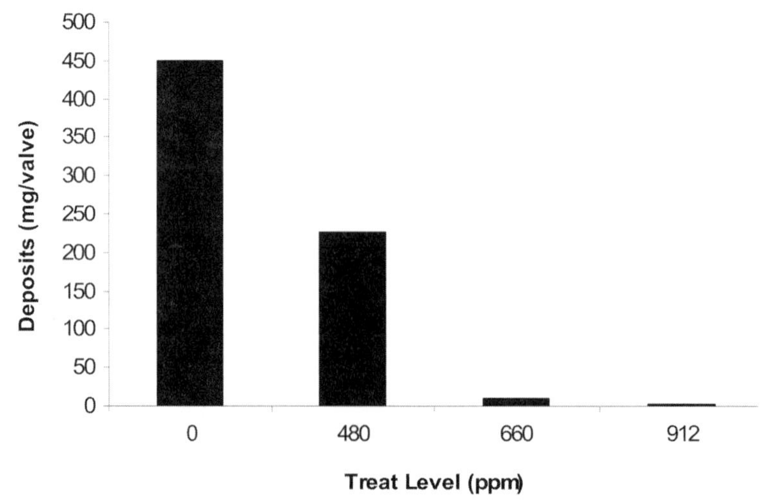

Fig. 6.45—The effect of additives on the intake valve deposits.

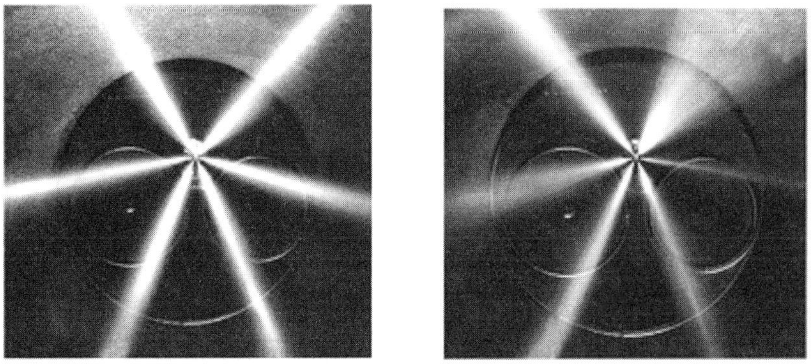

Fig. 6.46—Fuel spray pattern—clean versus the one with deposits [570].

Fluidizers

These additives are used in gasoline to improve its compatibility with other components in the formulation. These are usually solvents or diluents, such as alcohols, aromatics (xylene or toluene), refinery streams (containing aromatic and aliphatic hydrocarbons), kerosine, light mineral oils, and high flash commercial solvents. Aromatics are more effective than aliphatics, i.e., xylene and toluene are more effective than stoddard solvent and kerosine. Hydroxyalkyl ether type fluidizers are necessary to compatibilize some polyisobutylene-derived deposit control agents with the hydrocarbon fuel. The fluidizers are prepared by the reaction of an epoxide with an alcohol or a diol. Fluidizers must not have flash points lower than that of the base fuel which can hurt its performance. It may also require diesel fuel, considered nonflammable for shipping purposes, to be shipped as a flammable liquid.

Anti-icing Agents

Icing of a carburetor or a throttle body commonly occurs during the cool humid weather and is due to a drop in temperature by the evaporating gasoline. Additives that minimize this problem include alcohols, glycols, and alkylamine salts of carboxylic acids. Alcohols and glycols perform by lowering the freezing point of water and amine carboxylates perform by forming a monomolecular surface layer that prevents the ice build up. These additives are no longer used in bulk gasoline blending because newer cars use fuel injectors instead of carburetors. However, these are available in smaller containers for the do-it-yourself market. Diesel engines do not suffer from gasoline engine type icing problems because they do not have the carburetors either, but have injectors. Diesel fuel contains more water, which forms crystals at low temperatures. These plug fuel lines and filters, thereby impeding the fuel flow. The low molecular weight alcohols or glycols are used to prevent this problem.

Octane Improvers

Octane improvers are additives that minimize gasoline-related knock. Common types of additives used for this purpose include alcohols and ethers. The EPA's decision to remove lead from the gasoline prompted the use of the oxygen-containing compounds (oxygenates) as octane boosters. They are called oxygenates because they increase the oxygen

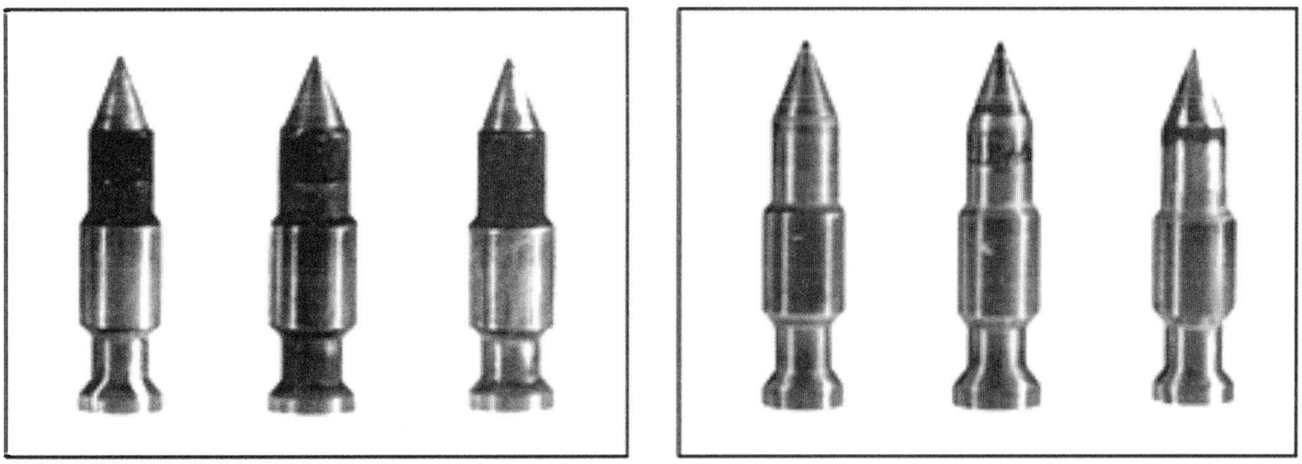

Fig. 6.47—The effectiveness of a deposit control additive [570].

TABLE 6.36—EN 228 oxygenate requirements.

Oxygenate	Volume %, Max
Methanol	3
Ethanol	5
Isopropyl Alcohol	10
Isobutyl Alcohol	1
tert-Butyl Alcohol	7
Ether (5 or more C atoms)	15
Other Oxygenates	10

level of the motor gasoline. Commercial oxygenates include methanol (MeOH), ethanol (EtOH), iso-propanol (IPA), tert-butanol (TBA), iso-butanol (IBA), Oxinol (TBA-MeOH mixture), methyl t-butyl ether (MTBE), t-amyl methyl ether (TAME), and ethyl t-butyl ether (ETBE). Of these, MTBE is the most popular among refiners, primarily because of its consistent quality (purity) and it having no adverse effect on Reid Vapor Pressure (RVP). MTBE is manufactured by the reaction of methanol and isobutylene. The use of ethanol as an octane improver is also common because of its cost effectiveness. Oxygen facilitates gasoline combustion, thereby reducing harmful tailpipe emissions from the motor vehicles. Oxygenates are used in reformulated gasoline (RFG), which is blended to burn cleaner and reduce smog-forming and toxic pollutants in the air. The Clean Air Act requires RFG to be used in the cities with the worst smog pollution to reduce harmful emissions of ozone. The act also specified RFG to contain 2% oxygen by weight. Maximum levels of oxygenates allowed by the European Union's specification EN 228 are provided in Table 6.36. The Japanese Industrial Standard (JIS) K 2202 allows MTBE up to 7 volume % maximum. Canada limits methanol to 0.3 % by volume, other oxygenates to 2.7 mass % oxygen, and ethanol specifically to 3.7 % by weight oxygen which nominally equates to 10 % by volume. Brazil allows up to 24 volume % ethanol, but the vehicles are calibrated to use this high level.

MTBE (methyl tertiary butyl ether) and ethanol are the two most commonly used substances that add oxygen to gasoline. As stated earlier, most refiners have chosen to use MTBE over other oxygenates, primarily due to its blending characteristics and economic reasons. However, MTBE has high water solubility (4.3 %); hence it finds its way into the nation's drinking water supply. Water contaminated with low levels of MTBE is undrinkable because of the odor and taste. To make matters worse, research has shown that when the research animals inhaled high concentrations of MTBE, some developed cancer or experienced other noncancerous health effects. To date, independent expert review groups, who have assessed MTBE inhalation health risks, e.g., Interagency Assessment of Oxygenated Fuels, have not questioned the use of MTBE in gasoline. Similarly, the EPA based on its 1997 assessment, concluded that there was insufficient information to set health advisory limits in drinking water. The document also indicates that there is little likelihood that MTBE in drinking water will cause adverse health effects at concentrations between 20 and 40 ppb or below [583].

There are limits on the amount of oxygenates in gasoline because of their deposit-forming tendencies and the propensity to alter RVP. The United States and the E.C. limits by volume are listed in Table 6.37. In the United States, the limits relate to an oxygen content of 2.7 % by weight for ethers and 3.5 % by weight for ethanol. The realized octane values for additized gasolines are provided in Fig. 6.48.

TABLE 6.37—Oxygenate limits in gasoline.

Oxygenate	U.S. Limit (% Volume)	E.C. Limit (% Volume)	Proposed E.C. Limit (% Volume)
Ethanol	10.0	5.0	10.0
MTBE	15.0	10.0	...
TAME	16.7	—	—
ETBE	17.2	—	—
Methanol	—	3.0	—
IPA	—	5.0	—
TBA	—	7.0	—
IBA	—	7.0	—
Oxinol	—	5.0	—
Ethers containing ⩾5 carbons	—	7.0	15.0
Other Organic Oxygenates	—	2.5 %(w) Oxygen	3.5 %(w) Oxygen

The examination of the effect of the various oxygenates on the vapor pressure of the gasoline, shown in Fig. 6.49, indicates only ethanol to have an adverse effect. Ethers appear to lower it, with MTBE having the lowest effect, which is another reason for it being preferred. Besides those shown in the exhibit, other oxygenates used as octane boosters include Arcanol® (*tert*-Butyl alcohol or TBA), Oxinol™, Dupont's proprietary formulation—*Octamix*, and various other aliphatic alcohols and ethers.

Lubricity Agents

Lubricity of diesel fuel is deemed important. Proper lubricity assures minimum wear (corrosive or scuffing) damage to the injection pump. Diesel fuel used in very cold climates is usually lower boiling than that used in other areas. This is to minimize the cold flow problems during winter. Since such fuel has inferior film-forming ability due to low viscosity, it can lead to increased pump wear. The problem can be overcome by the use of long chain surface-active compounds, such as fatty acid derivatives, that improve a diesel's film-forming ability. Typical treat levels for diesel are in the range of 10 ppm to 250 ppm. Many U.S. organizations, such as California Air Resources Board (CARB), Engine Manufacturers Association (EMA), and American Petroleum Institute (API) have recommendations pertaining to this fuel property. Diesel lubricity is determined by the Ball-on-Cylinder Lubricity Evaluator (BOCLE). ASTM D5001 procedure is used to measure the wear scar diameter on the ball and correlate it to the injection pump performance. The correlation is fairly linear. Figure 6.50 compares the effectiveness of two lubricity additives in terms of wear control. Additive B meets the target performance at the 200 ppm treat level while Additive A requires 500 ppm. The High Frequency Reciprocating Rig (HFRR) Test is another test used to determine diesel fuel's lubricity. This test is accepted by ISO, has an ASTM number (ASTM D5001), and is extensively used in Europe and Asia.

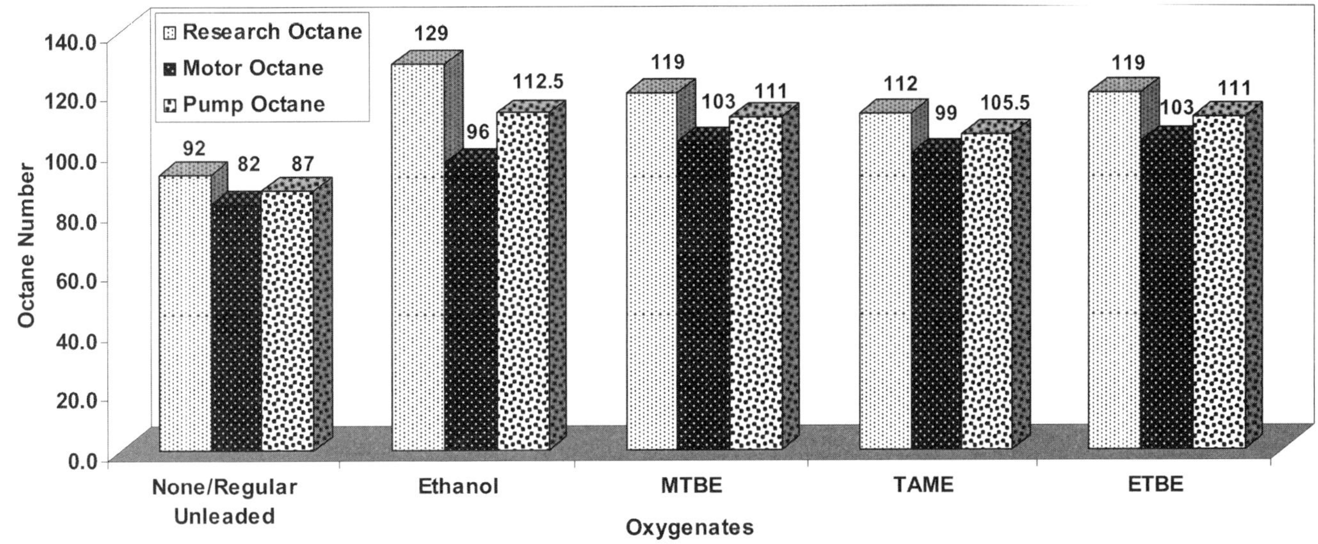

Fig. 6.48—Octane boosting tendency of various oxygenates.

Cetane Improvers (Diesel Ignition Improvers)

These additives enhance ignitability of the diesel, a low volatility fuel, reduce combustion noise, and smoke. The magnitude of the benefits depends upon the engine design and the operating mode. A variety of thermally unstable compounds are used for this purpose including alkyl nitrites and nitrates, nitro and nitroso compounds, organic peroxides, and a number of other oxidizing agents. Of these, 2-ethylhexyl nitrate, sometimes called the octyl nitrate, is the most often used cetane improver because of it being the most effective. See Fig. 6.51 for chemical structures. These materials thermally decompose at high temperatures in the combustion chamber to form species that facilitate fuel combustion. The increase in cetane number from 2-ethylhexyl nitrate depends upon its concentration and the quality of the diesel fuel. If the diesel fuel being treated has high initial cetane number, the benefit from the additive is high, as is shown in Fig. 6.52 [584]. The boost in cetane number gets progressively smaller as the amount of 2-ethylehxyl nitrate is increased. Typically, ethylhexyl nitrate concentration in diesel fuel ranges 0.05 % by weight to 0.4 % by weight. This usually leads to a cetane number increase of 3 to 8 units. Di-tertiary butyl peroxide has been a recent addition to the family of the commercial cetane number improvers.

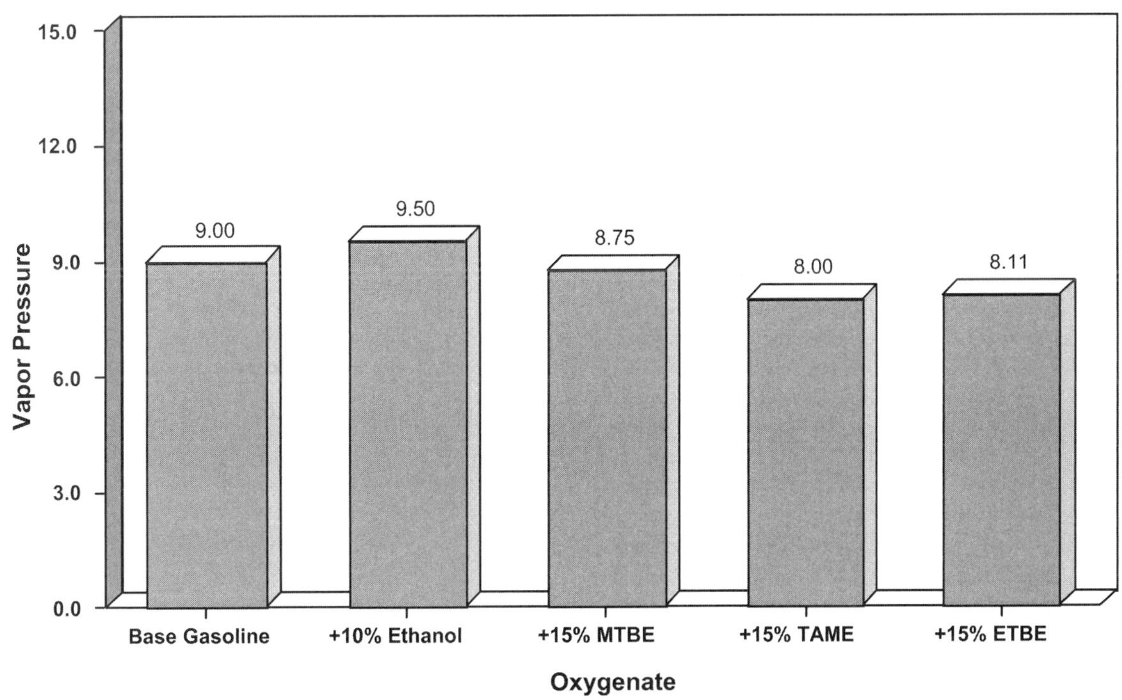

Fig. 6.49—Oxygenates versus vapor pressure.

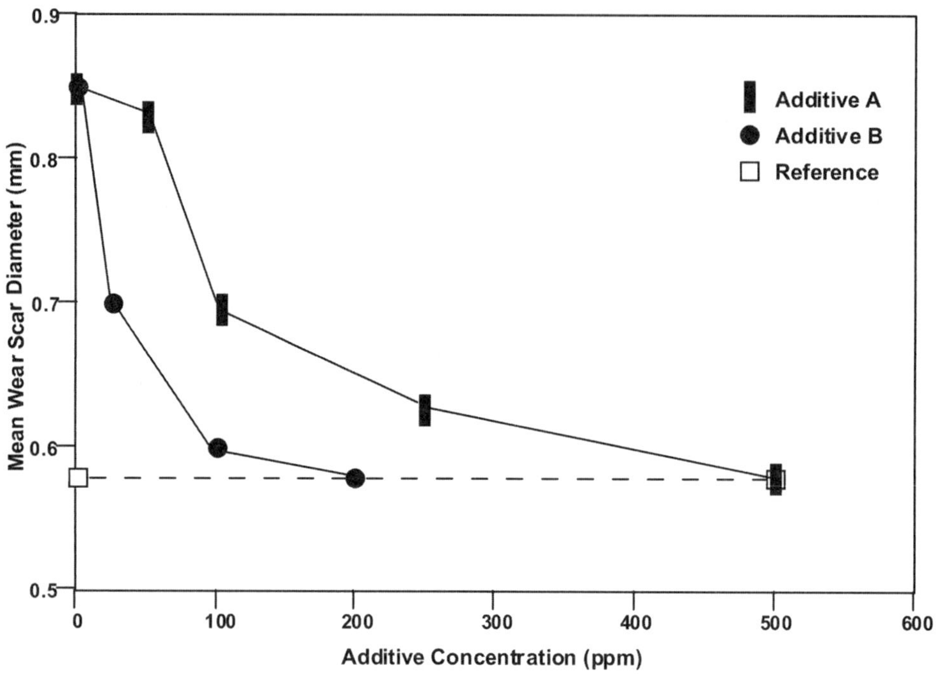

Fig. 6.50—Modified BOCLE wear versus additive concentration in a Class II diesel fuel.

Combustion Modifiers/Smoke Suppressants

These additives improve combustion efficiency of the diesel fuel, thereby minimizing undesirable smoke emissions. These are primarily calcium, cerium, platinum, manganese, iron, and copper containing organic compounds. While previously the use of these additives was discontinued, there is a possibility of a comeback due to their ability to lower the auto-ignition temperature of soot. It is important to note that while these additives are used in other parts of the world, they have not yet been approved by the EPA for use in the United States, which may be due to their ash-forming tendency. See the discussion on DPFs. Just the same, such additives are often used in vehicles equipped with diesel particulate filters to further improve emissions control.

Low-temperature Operability Additives

These additives are used to lower the diesel fuel's pour point (*gel point*), cloud point, and improve its cold flow properties. These additives are analogous to pour-point depressants that are used to improve the low-temperature properties of the mineral oils. They are polymeric additives that interact with wax crystals that form in the diesel fuel at low temperatures. Different classes of additives affect various fuels differently; hence the best additive and the treat rate for a particular fuel must be determined experimentally. The benefits that can be expected from this class of additives are provided in Table 6.38 [570].

Flow Improvers/Wax Modifiers

At low temperatures the diesel fuel separates wax which can impede its flow as well as lead to filter plugging by wax crystals. These additives modify crystal growth so that only fine crystals result and include ethylene-vinyl acetate polymers and polyolefins. These usually work in conjunction with the low-temperature operability additives.

TABLE 6.38—Typical performance of low-temperature operability additives [570].

Test	Typical Treat Rate, ppm	Typical Benefit Relative to Additive-free Fuel	
		°C	°F
Cloud Point	200-2000	3-4	5-7
LTFT	50-2000	8-12	15-25
CFPP	100-2000	15-20	25-35
Pour Point	100-300	30-40	50-70

LTFT = Low Temperature Flow Test CFPP= Cold Filter Plugging Point

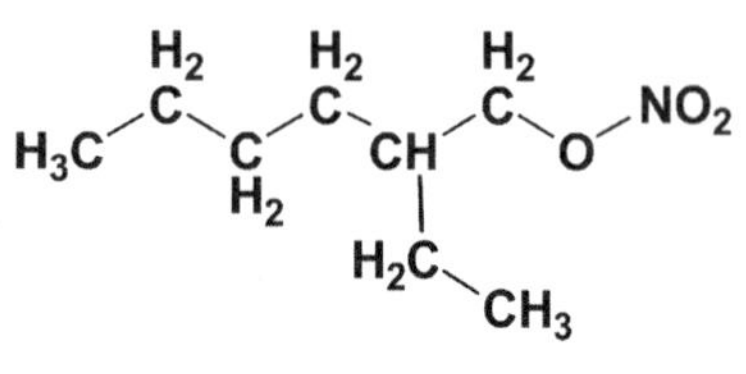

2-Ethylhexyl Nitrate [EHN]

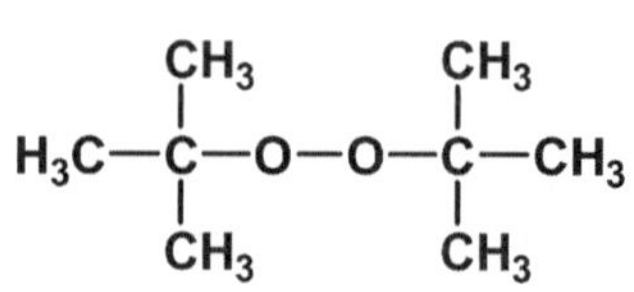

Di-tert-butyl Peroxide [DTBP]

Fig. 6.51—Chemical structures of commonly used cetane improvers.

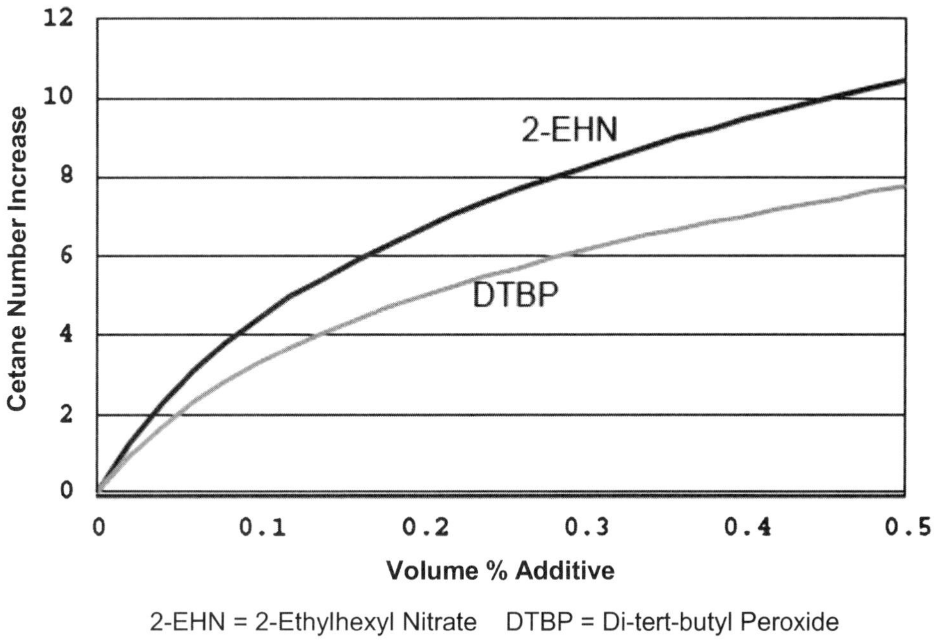

Fig. 6.52—Additive impact on cetane number [584].

Foam Inhibitors

Some diesel fuels foam during pumping into the vehicle tanks, which can hinder complete filling of the tank or may even result in a spill. This can be precluded by the use of the polysiloxane type of foam inhibitors. They are added to the fuel at concentrations of 10 ppm or lower.

Corrosion Inhibitors

These additives minimize corrosion of tanks, fuel lines, injectors, and pumps that primarily results from rusting because of the contamination of the fuel with water. The presence of the dissimilar metals in the vehicle's fuel system is partly responsible for the rust, a form of electrochemical corrosion. Rust starts at sites of the manufacturing faults, such as pinholes in the fuel tank. The presence of the oxygenates in gasoline leads to chemical corrosion of the tank's lead lining. If not controlled, corrosion can lead to blockage of the filters as well as deterioration of the automobile's fuel system, thereby creating a fire hazard.

Most corrosion inhibitors are metal-free surface active materials that function by making surface films that keep corrosive materials away. See the corrosion inhibitor section of Chapter 4 on Lubricant Additives. The presence of the oxygenates in gasoline warrants a higher additive treatment level. This is because oxygenates, also being surface active, compete with the corrosion inhibitors for the surface. Carboxylic acids and organic derivatives of the phosphoric acids and sulfonic acids are among those commonly used. These additives are also used in refineries, tanks, and pipelines to protect equipment, which is typically constructed of uncoated steel. Rust in pipelines can cause reduced flow rates and suspended particles that can result in blocked filters. Once the fuel is in the vehicle, corrosion control is not as critical since the metal parts in the fuel systems of today's vehicles are made of corrosion-resistant alloys or of steel coated with corrosion-resistant coatings. More and more plastic and elastomeric parts are replacing metals in the fuel systems. Furthermore, the service station systems and operations are designed to prevent free water from entering a vehicle's fuel tank. The fuel treat level with these additives is 5 to 15 ppm.

The mechanism by which these additives perform involves their adsorption on the metal surface via their polar end and associating with the fuel via their non-polar end (Fig. 6.53, Part B). This results in the formation of a film that acts as a barrier against surface attack by the corrosive species. Figure 6.53, Part A depicts the mechanism of rust inhibition by alkenylsuccinic acids.

Demulsifiers

Typically, fuel and water do not stay mixed. However, emulsion can result from the contamination of the fuel with water and passage through high-shear portion of a centrifugal pump. The presence of the polar materials, such as additives that are present in the fuel, further exacerbates this problem. If not controlled, the emulsion will not only lead to the extensive corrosion of the various metal parts but also will cause filter plugging. Demulsifiers facilitate water separation from the emulsions, if they form, and therefore prevent filter plugging and also make it easy to siphon off the separated water. Demulsifiers are added to the fuel in very small amounts, at a concentration ~5 to 30 ppm. They are usually nonionic compounds with a long chain hydrophobic group and at least one polar or the hydrophilic group. Polyethoxylated alcohols (polyglycols), phenols, and amines are the common types used.

Biocides

While high temperatures during refinery processing to make fuels effectively sterilize them, they can quickly get contaminated with microorganisms that are present in air or water. These microorganisms include bacteria and fungi, such as

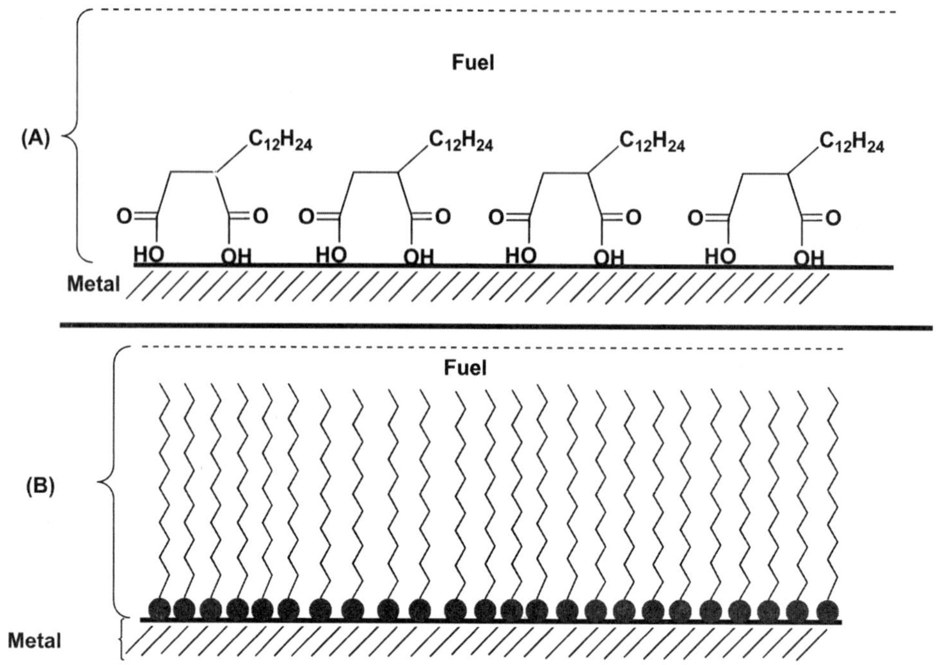

Fig. 6.53—Mechanism of rust inhibition.

yeasts and molds. Bacterial growth in the water-organic mixtures is common and leads to the formation of slime and undesirable odor. Most microorganisms grow at the fuel-water interface, when one exists. Higher ambient temperatures also favor growth. Microorganism growth must be controlled since it results in the acidic by-products, which will accelerate metal corrosion, and slime that will plug filters. Static equipment is more amenable to microbial growth than that in constant use. The best approach to limit microbial growth is prevention, which can be easily achieved by eliminating water from the fuel storage tanks. Biocides must be used when microorganism growth becomes evident. These are chemicals that discourage microbial attack and stop their growth. The best biocides are those that are soluble both in water and fuel, so that they can counteract microbes in both phases. Typical use concentration of these additives is in the 200 ppm to 600 ppm range. Organoborates, nickel-amine complexes, glycol ethers, and quaternary salts of salicylic acid are among the most commonly used biocides.

Antistatic Agents

An electrostatic charge develops during the fast pumping of the fuel, an operation commonly used for loading and unloading large quantities. The charge poses an explosion hazard and must be controlled. Antistatic agents are chromium based materials that are added to the fuel at few ppm levels to control static.

Drag Reducers

These are high-molecular-weight polymeric additives that improve the flow characteristics of the low-viscosity petroleum products. Their use lowers the energy required to pump fuels through the pipelines. They reduce the frictional drag between the flowing fuel and the walls of the pipe and turbulence during pumping. The result is an increase in the fuel flow rate of 20% to 40%. Typical use concentration of these additives is below 15 ppm.

Dyes and Markers

Dyes are primarily colored aromatic compounds with azo or quinonoid structures that are used in fuels to visually differentiate batches, grades, or applications of the gasoline products. For example, gasoline for general aviation, which is manufactured to different and more exacting requirements, is dyed blue to distinguish it from motor gasoline for safety

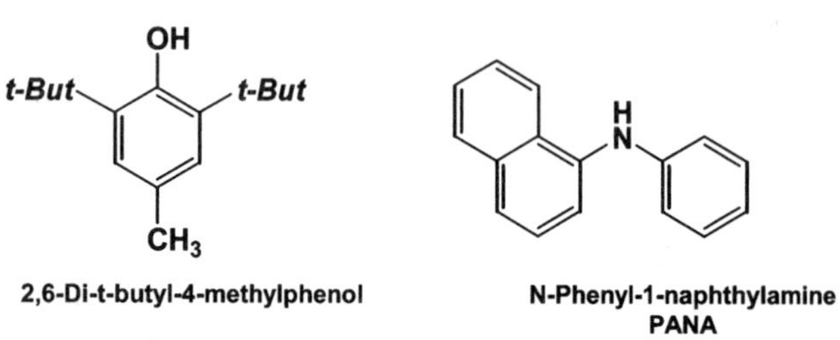

Fig. 6.54—Commonly used fuel oxidation inhibitors.

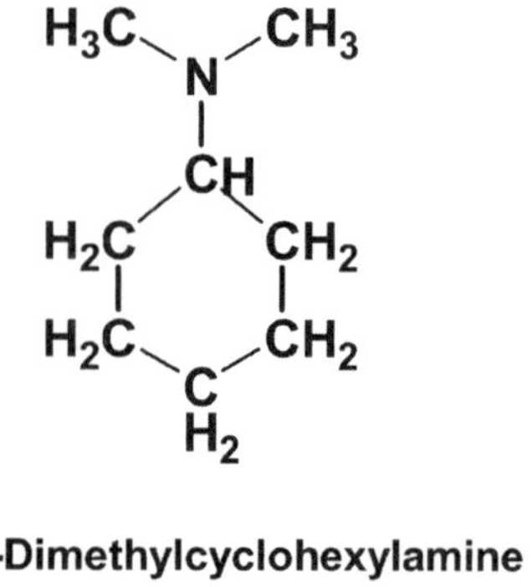

Fig. 6.55—Amine stabilizer for diesel fuel.

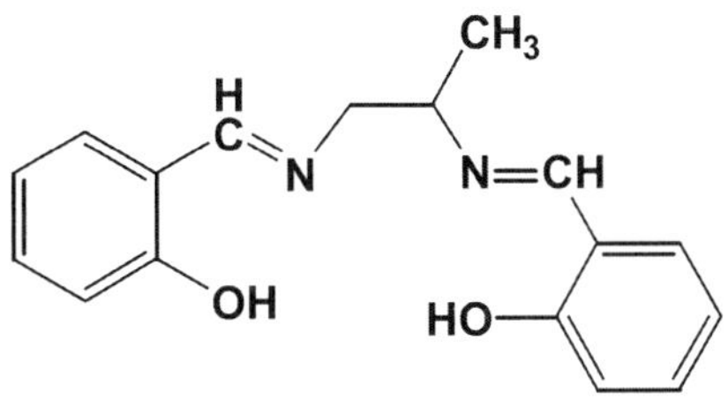

N, N-Disalicylidene-1,2-propanediamine (DPD)

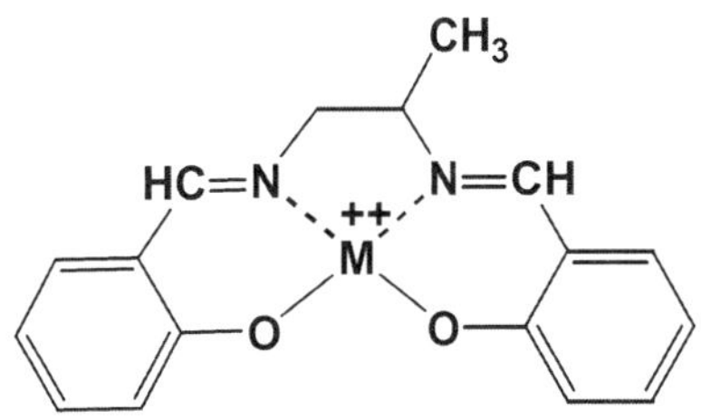

DPD-Metal Ion Complex

Fig. 6.56—Structure of a commonly used metal deactivator.

reasons, and they are used to indicate sulfur level in off-highway diesel fuel. Markers, on the other hand, do not color the fuel and their presence can only be detected by chemical or spectroscopic means. A refiner may use a marker in gasoline to identify its movement through the distribution system. Furfural and diphenylamine are among those used often.

Oxidation Inhibitors/Stabilizers

These additives minimize color degradation, control high-temperature deposit formation, improve long-term fuel storage stability, and lower fuel filter plugging tendency. Most of these problems relate to the attack of the atmospheric oxygen on fuel to form peroxides and gums. Peroxides degrade the antiknock quality, cause fuel pump wear, and attack plastic and elastomeric fuel system parts. Soluble gums can lead to engine deposits and insoluble gums can plug fuel filters or the fuel injection system. Inhibiting oxidation is important for fuels designed for use in modern fuel-injected vehicles, as their fuel recirculation design may subject fuel to higher temperatures and increased oxygen exposure. Fuels of high olefinic and aromatic content are more susceptible to oxidative attack than those of the high paraffinic content. The classes of materials that are commonly used as inhibitors include alkyl/aryl substituted amines and phenylene-diamines, alkylphenols, and alkylaminophenols. Structures of the two common inhibitors are given in Fig. 6.54. Typical use concentration for these additives is between 10 ppm to 80 ppm.

Fuels contain certain components that get converted into high molecular weight species with little or no solubility in the fuel. These include nitrogen- or sulfur-containing compounds, or both, organic acids, and reactive olefins. While the formation of the polymeric species primarily involves oxidation, the acid catalysis can also result in their formation. One way to minimize their formation is to control the fuel acidity. This is achieved by the use of strongly basic stabilizers, such as amines. One such stabilizer is N, N-Dimethylcyclohexylamine, whose structure is provided in Fig. 6.55. The typical treat level of the stabilizers ranges from 50 ppm to 150 ppm. They react with weakly acidic compounds to form products that are fuel soluble.

Metal Deactivators

Trace amounts of metals and metal ions can catalyze oxidative degradation of the hydrocarbon materials, such as the fuels, which can lead to problems of all sorts. These additives chelate with the metal ions, such as those of copper and zinc, and make them ineffective as oxidation catalysts. N, N′-Disalicylidene-1,2-propanediamine is an example of such an additive. They are typically used in the concentration range of 1 ppm to 15 ppm. Figure 6.56 shows this deactivator both in the free form and in the complexed form.

7

Hydraulic and Transmission Fluids

IN THIS CHAPTER WE PRESENT TECHNOLOGY relating to hydraulic and transmission fluids. Under the topic of hydraulic fluids, we describe fluids both for tractor-type equipment and industrial hydraulics. We list the performance requirements of various equipment manufacturers and teach the formulating of an appropriate fluid. Under the topic of transmission fluids, we describe fluid requirements for automatic transmissions, power transmissions, and continuously variable transmissions. We conclude the chapter by citing formulation examples for each type of fluid.

Hydraulic Fluids

Hydraulics is a branch of science that deals with mechanical properties of fluids, usually liquids, and the hydraulic machinery. The latter comprises machines and tools which use the fluid power to perform work. Many industries employ these devices. Common examples of such machines include bulldozers, excavators, industrial cranes, harvesters, tractors, and aircrafts; examples of the tools that employ hydraulics include backhoes and automotive brakes and transmissions; and examples of industries that use these devices include agriculture, transport, construction, excavation, forestry, iron and steel, paper, food processing, and mining.

The principle of hydraulic power is simple. It is based upon the knowledge that a force applied on a small area can result in a much larger force by a larger area due to hydrostatic pressure, and a large amount of energy can be carried by a small flow of a highly pressurized fluid. In hydraulic machines, an almost incompressible fluid (the hydraulic fluid) at a high pressure is transmitted through small tubes and flexible hoses to various parts that perform the actual work.

While the primary function of the hydraulic fluid is to transfer power, it must also perform the functions that are typical of all lubricants, which are to reduce friction, reduce wear, and remove heat. The pump is the central component in most hydraulic systems, whose function is to collect the fluid from a reservoir, pressurize it, and transfer it to various system parts that perform the work. Most of the components of the hydraulic system operate at high speeds, high pressures, and high temperatures. Hence these parts require lubrication and effective cooling to extend the equipment's useful life.

Incompressibility, or low compressibility, is one of the key requirements for an efficient power transfer medium, because this permits maximum transfer of the generated power. Bulk modulus of a fluid, which is the reverse of compressibility, is the change in fluid volume as a function of the applied pressure. The bulk modulus can be estimated from a lubricant's viscosity and density [154,585,586]. A high bulk modulus implies minimum compressibility.

Compressibility is the rate of decrease in the fluid volume with an increase in pressure. Below 1000 bars, the volume decrease is approximately 0.5 % per 100 bars at 20 °C and it is approximately 0.7 % per 100 bars at 80 °C. At higher pressures, the volume decrease per 100 bars pressure becomes progressively smaller [4]. The decrease in volume due to compression is because of an increase in density. A related concern in high-pressure hydraulic systems is an increase in air entrainment, which is undesired since it can lead to a variety of problems, including the possibility of cavitation damage. Solubility coefficients for different gases at temperatures of 0–100 °C are provided in Table 7.1 [4]. These coefficients are valid up to approximately 300 bars, above which the solubility of gases decreases. Dissolved air is no problem if it stays dissolved. However, when the pressure is eased off, the entrained air comes out as bubbles in pipes and cavities or forms surface foam in valves, pumps, and storage tanks. When the pressure is reapplied, cavitation can occur. For petroleum oils, compressibility is estimated to be 0.5 % for each 1000 psi pressure increase up to 4000 psi [585,586].

Heat in hydraulic systems is a normal occurrence and arises from the following sources [587]:

1. Friction between the moving parts of the pump and or the hydraulic motor.
2. Fluid's interaction with surfaces of valves, pipes, and other devices.
3. Heat release due to compression of the fluid.

Thermal conductivity and specific heat of a fluid indicate its ability to absorb heat. ASTM D2717 and ASTM D2766 Test Methods are used to measure these parameters. In view of the fact that friction of one kind or another is the primary source of heat, a lubricant with the ability to reduce friction, hence wear, will be extremely beneficial as a hydraulic fluid. For discussion pertaining to various lubricant-related parameters, such as the viscosity and film-forming ability, that enhance lubrication, refer to the Lubrication Fundamentals and Additives chapters, Chapters 1 and 4, respectively.

TABLE 7.1—Solubility of different gases in medium viscosity mineral oil [4].

	Bunsen Coefficient α^a		
Gas	0 °C	50 °C	100 °C
Air	0.092	0.091	0.091
Oxygen	0.150	0.137	0.130
Nitrogen	0.081	0.088	0.090
Hydrogen	0.047	0.053	0.067
Carbon Dioxide	1.0	...	...

[a]With low-viscosity fluids, α values are higher

The same as the other lubricants, hydraulic fluids are blends of base stocks/fluids and additives. The term hydraulic is a composed word with its origin in the Greek language: *hydor* meaning water and *aulos* meaning pipe. So, the term hydraulic fluid represents water (or a water-based fluid) that travels through a pipe from one point to another point. Despite the origin of the name hydraulic, modern hydraulic fluids are derived from many other incompressible liquids, which are the basis for their composition based classification. Common base stocks used to formulate hydraulic fluids include the following:

1. Solvent-refined paraffinic base oils (API Group I).
2. Severely hydrotreated naphthenic base oils.
3. Mildly hydrotreated paraffinic base oils (API Group II).
4. Severely hydrotreated paraffinic base oils (API Group III).
5. Synthetic base stocks, such as polyalphaolefins (PAOs, API Group IV).
6. Other synthetic base stocks (API Group V).
   - Synthetic esters
   - Poly(alkylene glycol)s
   - Organophosphate esters
   - Silicones
7. Natural base stocks, such as canola oil and castor oil.

Mineral base stocks (Items 1–4) are used most often to formulate hydraulic fluids. Synthetic fluids and natural oils are used only for specialized applications. For a detailed discussion on their manufacture or properties, please refer to Chapters 2 and 3.

Like most commercial lubricants, hydraulic fluids need additives that enhance their performance. Table 7.2 lists the commonly used lubricant additives along with their functions, mechanism of performance, and the test methods used to determine their effectiveness. For further details on additives, please refer to Chapter 4.

The typical additive treatment rate for hydraulic fluids ranges between 0.5 and 10 % and in the case of multi-grade hydraulic fluids, the amount of the viscosity modifier is additional. Hydraulic fluids for use in tractors contain additives in the amount of 6–9 % and for those to be used in industrial applications, the additives treatment is between 0.5 and 2.0 %. In transmission fluids, the additive treatment rate is in the vicinity of 6–12 % and that of the viscosity modifier is 3–14 %. The additives used in these applications may be the same or different. Additives that are typically used in industrial hydraulics include the following:

1. Oxidation Inhibitors
2. Anti-wear Agents
3. Extreme-pressure (EP) Agents
4. Rust and Corrosion Inhibitors
5. Foam Inhibitors
6. Demulsifiers
7. Metal Passivators

In some formulations, other types of additives may also be present. For example, the emulsion-type hydraulic fluids contain emulsifiers and biocides and the multi-grades contain viscosity modifiers. In addition to many of the additive listed above, tractor hydraulic fluids contain a detergent, friction modifier, pour point depressant, and viscosity modifier.

Hydraulic fluids are classified in two ways: based upon application and based upon composition. Application-based classification includes tractor hydraulic fluids, industrial hydraulic fluids, transmission fluids, brake fluids, shock absorber fluids, and power steering fluids.

Tractor Hydraulic Fluids

Tractor hydraulic fluids (THFs) are multipurpose lubricants that are used to lubricate tractor transmissions, final drives, hydraulic systems, wet brakes, and wet clutches [588–590]. To perform these functions properly, THFs must combine hydraulic and transmission properties with extreme-pressure properties. Their function as a transmission fluid and as a lubricant for wet brakes and wet clutches requires them to possess proper frictional characteristics. A schematic representation of a tractor's lubrication requirements is provided in Fig. 7.1. Tractors use a wide variety of fluids to lubricate various parts. The list is provided in Table 7.3 [588,591].

Tractor hydraulic fluids differ widely in performance requirements because the OEMs cannot agree on common specifications for a universal tractor hydraulic fluid. The specifications for these fluids, in general, deal with extreme pressure (EP) and antiwear properties, and with matching the frictional requirements of the equipment. The quality of these fluids is assessed on the basis of their ability to meet individual OEM specifications, as well as API GL-4 for EP and Allison C-4 for friction, oxidation, and wear performance requirements. Physical characteristics of these fluids are described in Table 7.4 and their performance specifications are described in Tables 7.5 and 7.6, respectively [318].

There are eight major OEM specifications, and most tractor fluids are formulated to meet them. These specifications are:

- JI Case MS1207
- John Deere J20C/D, and J27
- New Holland FNHA-2-C-201.00 and M2C159B/C
- AGCO Massey-Ferguson M1135, M1139, and M1141

In addition to a hydraulic fluid, farm tractors and related equipment need an engine oil and a transmission fluid. In order to reduce the number of lubricants handled by the farmer, the concepts of Universal Tractor Transmission Oil (UTTO) and Super Tractor Oil Universal (STOU) were developed. UTTOs have the ability to lubricate transmissions, wet brakes, and hydraulic systems. STOUs have the additional ability to be used as engine oils and meet major performance criteria of the leading equipment manufacturers. Tables 7.7 and 7.8 provide a list of the performance requirements of these lubricants [318].

Industrial Hydraulic Fluids

These fluids are used in industries, such as automotive, manufacturing, material handling, construction, chemical, mining, textile, food, rubber, and agriculture.

As mentioned earlier, the pump is the central component in an industrial hydraulic system. Pumps create both the flow and the pressure necessary for the hydraulic system to work. Common pump designs used in hydraulic systems include piston pumps, vane pumps, and gear pumps. Each type has unique lubrication needs because of design. Piston pumps primarily operate under hydrodynamic lubrication; hence fluids for them do not need the antiwear additives for

TABLE 7.2—Additives and their functions.

Additives	Function	Chemical Types	Performance Mechanism	Test Method (s)
Stabilizers/Deposit Control Additives				
Oxidation Inhibitors	Minimize oxidative breakdown of the lubricant	Hindered phenols, aromatic amines, sulfurized alkylphenols, phenothiazine, metal and ash-free dialkyldithiocarbamates, metal dialkyl dithiophosphates, and organic complexing agents, containing nitrogen, oxygen, and sulfur	Decompose peroxides and remove oxidation-promoting free radicals and reduce catalytic effect of metals and their ions by complexing with them.	ASTM D1743 ASTM D2070 ASTM D2272 ASTM D4310 ASTM D6186 IP 280 DIN 51 352
Detergents	Neutralize acids and keep surfaces free of deposits	Neutral and basic sodium, calcium, and magnesium sulfonates, metal phenates, and metal phosphonates	Chemical reaction with sludge and varnish precursors to neutralize them and physical interaction to keep the products suspended in oil	ASTM D4310
Dispersants	Keep insoluble contaminants dispersed in the lubricant	Alkenylsuccinimides, alkenylsuccinic acid esters, and Mannich reaction products	Suspending deposit-forming species in oil through polar interaction, thereby preventing them from agglomerating and coming out of oil	ASTM D4310
Surface Protecting Agents				
Friction Modifiers	Modify friction to match it with the frictional requirements of the equipment	Fatty acids and amides, lard oil and high molecular weight organic phosphorus and phosphoric acid esters	Adsorb on the metal surface to help maintain an oil film on surfaces to control their countermovement	Some of the antiwear tests listed below
Antiwear and Extreme Pressure (EP) Agents	Reduce friction and wear and prevent scoring and seizure	Dialkyl dithiophosphoric acid esters and their metal and amine salts, metal dialkyldithiocarbamates, organic phosphates and acid phosphates, organic sulfur and chlorine compounds, sulfurized fats, sulfides and disulfides, derivatives of 2-5-dimercapto-3,4-thiadiazoles, and molybdenum carboxylate	Chemical reaction with metal surfaces to form low friction, low shear strength sacrificial film, thereby minimizing metal-to-metal contact	ASTM D4172 ASTM D2882 ASTM D5182
Performance-enhancing Additives				
Corrosion and Rust Inhibitors	Prevent corrosion and rusting of the metal parts that are in contact with the lubricant	Alkenylsuccinic acid derivatives, ethoxylated phenols, fatty amines, salts of fatty acids and amines, aromatic metal and amine sulfonates, imidazoline derivatives, zinc dialkyl dithiophosphates, metal phenates, fatty acids and fatty amines, benzotriazoles and tolyltriazoles (tolutriazoles), mercaptobenzimidazoles, 2-mercaptobenzothiazoles, and their derivatives	Preferential adsorption on metal surfaces to form a protective films and neutralize corrosive acids	ASTM D665 ASTM D130
Pour Point Depressants	Enable the lubricant to flow at low temperatures	Alkylated naphthalene and phenolic polymers, polymethacrylates, alkyl maleate-fumerate ester copolymers	Modify wax crystals to prevent the formation of the crystalline networks	ASTM D97 ASTM D5949
Viscosity Modifiers	Reduce the rate of viscosity change with temperature	Polymers and copolymers of olefins, poly(alkyl methacrylate)s, olefin copolymers, and styrene-diene polymers	Increased physical association with oil at high temperatures, thereby minimizing temperature-related viscosity drop	ASTM 5621
Foam Inhibitors	Prevent formation of a persistent lubricant foam	Silicone polymers (polydimethylsiloxanes), organic copolymers	Reduce surface tension to speed the collapse of foam	ASTM D892
Seal Swell Agents	Swell elastomer seals	Organic phosphates and aromatic hydrocarbons	In some cases, diffuse into the seal material to prevent shrinkage and in other cases, prevent the removal of plasticizer from the elastomer	ASTM D6546 ASTM D412 ASTM D471 ASTM D2240

TABLE 7.2— (Continued.)

Additives	Function	Chemical Types	Performance Mechanism	Test Method (s)
Demulsifiers	Promote water separation	Polyethoxylated phenols Polyethoxylated polyols Polyethoxylated polyamines Metal salts of alkylated arylsulfonic acids	Alter surface tension of the oil-water interface and help break emulsions	ASTM D1401 ASTM D2711

good performance. Typically, oils containing oxidation and rust inhibitors, R&O oils, are satisfactory and so are the Denison HF-0 quality fluids that contain thermally stable zinc dialkyl dithiophosphate (ZDTP) antiwear agents. Thermal stability of the ZDTP is important since these pumps contain bronze piston shoes and the hydraulic systems contain copper and brass. The acidic decomposition products from thermally unstable ZDTP can lead to serious corrosion of these parts. Nonzinc antiwear agents, based on sulfur-phosphorus chemistry, are also acceptable for use in these pumps. Vane pumps, on the other hand, operate under boundary lubrication and there is extensive continuous metal-to-metal contact. Therefore, hydraulic fluids for these pumps contain antiwear agents to control wear. For low-pressure applications, those involving less than 500 psi, R&O fluids work quite well but for high pressure applications, antiwear agents are necessary for satisfactory performance. Gear pumps operate under either full film (hydrodynamic) or mixed film lubrication and typically under mild to medium loads. There is little or no metal-to-metal contact between the drive and the idler gears. Unlike vane pumps that require fluid cleanliness, axial piston and radial piston pumps are not contamination-sensitive. Hence, they can be used in dirty environments, such as steel mills and mining operations. R&O oils provide the necessary performance in these pumps [591,592].

Primary functions that industrial hydraulic fluids perform include the following:

1. Power transmission
2. Pressure transmission
3. Heat removal
4. Corrosion protection
5. Elastomer compatibility
6. Wear reduction

For a hydraulic fluid to perform these functions effectively, they must possess proper viscometrics (viscosity, viscosity index, and low temperature fluidity), excellent EP properties, low compressibility, high heat capacity and thermal conductivity, low flammability, good oxidation resistance (aging stability), low foaming tendency, good demulsibility (the ability to shed water quickly), high biodegradability, and noncorrosivity. A variety of chemicals are added to the base fluids of various kinds to enhance or impart these properties. The performance requirements of the finished fluids are described in major industry manufacturers' specifications; such as Denison HF-0, HF-1, HF-2;

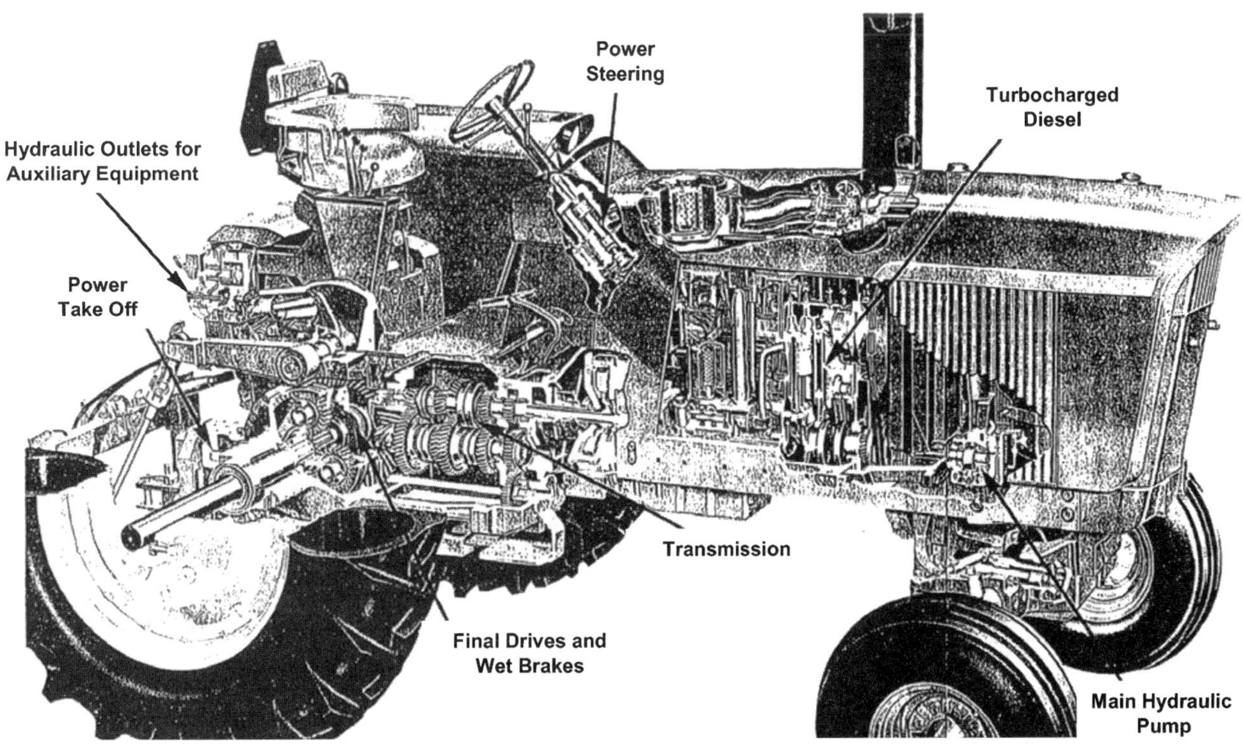

Fig. 7.1—Primary lubrication requirements of a farm tractor.

TABLE 7.3—Lubricants used in various parts of a farm tractor [591].

Symbol	Lubricant Type	Symbol	Lubricant Type
EO	Engine Oils—API Service Classes SH, SM, CF, CF4	HYDO	Hydraulic Oils, Industrial (Inhibited), Industrial (Inhibited, Antiwear)
ATF	Automatic Transmission Fluids DEXRON®-III, MERCON®, MERCON® -V	FRF	Fire Resistant Hydraulic Fluids, Oil/Water Emulsions, Water/Glycol, Phosphate Ester Type
CTF	Commercial Transmission Fluids Allison C-4, Caterpillar TO-4	CG	Cup Greases, Without Solid Lubricant, With Solid Lubricant
TC	Torque Converter Fluids	MPG	Multipurpose Type Greases, Without Solid Lubricant and with With Solid Lubricant
TF	Tractor Hydraulic Transmission Fluids	WBG	Wheel Bearing Greases
RGL	Regular Type Gear Lubricants GL-1, Straight Mineral AGMA (Inhibited)	HTG	High Temperature Greases
WGL	Worm Gear Lubricants GL-2 AGMA (Compounded)	EPG	Extreme Pressure Greases
MPL	Multipurpose Gear Lubricants GL-5, GL-4, GL-3 AGMA (Extreme Pressure)	PTL	Pneumatic Tool Lubricants
OGL	Open Gear Lubricants	TSC	Two-stroke Cycle Oils (Air-cooled Engines)
TRL	Track Roller Lubricants	TO	Transformer Oils
BF	Brake Fluids J1703, J1702	HTO	Heat Transfer Oils
		SPL	Special Lubricants (Greases, Oils), Synthesized Hydrocarbons, Silicones, Government Specifications, Proprietary Specifications

Eaton-Vickers M-2950-S, 1-286-S; Cincinnati Lamb Landis P-68, P-69, P-70; General Motors (1S2) 1H-03-1, 1H-04-1, 1H-06-1; DIN 51524, Part 2; and U.S. Steel 127. While in the United States, the original equipment manufacturers (OEMs) primarily establish the product requirements; in Europe, the major country standards organizations set the requirements [592]. In other nations, even governmental agencies get involved in setting such standards. The standards for the United States and Europe are provided in Tables 7.9 and 7.10 [592]. The desired properties that are listed in the table are evaluated by using standardized and nonstandardized tests. These are briefly described below.

Oxidation/thermal Stability

The ability of a hydraulic fluid to withstand high temperatures is extremely critical. This is because hydraulic fluid is designed to act as a heat transfer medium. The resulting high temperatures can cause hydraulic fluids to rapidly react with oxygen to form polar products that can cause thickening of the fluid, formation of acidic by-products, and deposit formation. Fluids that are designed for use in high temperature environments contain oxidation inhibitors, such as zinc dialkyl dithiophosphate and hindered phenol, which retard the fluid break down. The tests that are commonly used to assess fluid stability include the following:

1. ASTM D943, Standard Test Method for Oxidation Characteristics of Inhibited Mineral Oils (also known as the Turbine Oil Oxidation Stability Test-TOST).
2. ASTM D4310, Test Method for Determination of the Sludging Tendencies of Inhibited Mineral Oils.
3. ASTM D2070, Standard Test Method for Thermal Stability of Hydraulic Oils.
4. ASTM D2272, Rotary Pressure Vessel Oxidation Test–RPVOT (Formerly Rotary Bomb Test, RBOT).
5. IP 280 and DIN 51 352, Cigre and Pneurop Oxidation Tests.

Besides the use of the oxidation inhibitors, which act to prevent the formation of sludge and deposits, one can also use detergents/dispersants that act as corrective additives. They suspend the deposit-forming species in the bulk fluid in a soluble form and do not allow them to separate on surfaces to form deposits. Detergents are neutral and basic alkaline earth metal (calcium and magnesium) salts of alkylphenols and alkylsalicylic and sulfonic acids. Dispersants are metal-free derivatives of alkylphenol-derived amines and alkenylsuccinimides. See the Additives chapter, Chapter 4, for further details.

Wear Control

Friction reduction and wear prevention are the basic functions of all lubricants. Both these objectives can be attained by the use of a fluid with suitable viscosity and the use of the film-forming additives, such as friction modifiers and EP/antiwear agents. Fluid viscosity is of extreme importance in hydraulic applications; hence a match with the equipment's operating requirements is critical. A fluid of too low a viscosity will cause low volumetric efficiency, fluid overheating, and increased pump wear; and a fluid of too high a viscosity will cause poor mechanical efficiency, difficulty in starting, and wear due to insufficient fluid flow [593]. Since viscosity is a function of temperature, the temperature operating window for a particular viscosity grade must be considered. Incidentally, hydraulic fluid viscosity is expressed in terms of International Standardization Organization (ISO) viscosity

TABLE 7.4—Physical characteristics of farm tractor hydraulic/transmission fluids [318]. Reprinted with permission from the Lubrizol Corporation.

Specification	Viscosity at 100 °C (cSt)	Viscosity Index	Brookfield Viscosity (Cp), max				Pour Point (°C) max	Flash Point (°C) min
			At −18 °C	At −29 °C	At −35 °C	At −40 °C		
AGCO Deutz-Allis								
PF821 XL	9.0 min	135 (ref)	4000	19,000 (ref)	...	...	−32	193
Massey-Ferguson								
M1110[a]	5.75 min	...	2400	...	...	...	−30	160
M1127A[a]	11.1 max	120 min	6000	...	...	...	−30	...
M1127B[a]	9.6 max	120 min	4000	...	...	...	−30	...
M1129A[a]	9.65 max	120 min	4000	...	...	...	−30	...
M1135	10.0–11.5	...	9000	...	...	...	−25	...
M1139	10.2–11.9	...	8000	...	...	...	−30	...
M1141	9.6 min	130 min	4000	...	...	70,000	−37	200
White Farm								
Q1826	8.5 9.6	...	4000	...	...	...	−37	193
Allison C-4				*Conform to SAE J300 Grade*				
J. I. Case								
MS 1204[a]	8.75 min	130 min	2700 (CCS)	...	...	...	−34	193
MS1205[a]	11.1 min	140 min	5600	...	...	...	−32	193
MS 1206[a]	8.8 min	140 min	4000 (−20 °C)	...	...	...	−34	190
MS 1207	6.2 min	95–115	3500 (−20 °C)	15,000 (−30 °C)	...	...	−37	195
MS 1210	6.65 min	120 min	1800 (CCS)	...	...	...	−46	182
Hy-Tran[a]	6.2 min	95–115	2600 (CCS)	14,000 (CCS)	...	...	−37	195
John Deere								
J14C[a]	5.4 min	...	3500	20,000	...	...	−35	180
J20C	9.1 min	...	5500 (−20 °C)	...	70,000	...	−36[b]	200
J20D	7.0 min	...	1500 (−20 °C)	...	...	20,000	−45[b]	150
J21A[a]	5.0 min	...	1400	...	...	...	−40	180
J27				*Conform to SAE J300 Grade*				
New Holland								
M2C41B	7.0 min	...	1400	...	...	...	−35	177
M2C48B	SAE 10W-30	135	...	...	...	...	−30	190
M2C48C	ISO 32	...	...	...	...	...	...	...
M2C86B	10.5–11.6	105 min	9230	...	...	...	−27	219
M2C134D	9.0 min	...	4000	...	...	...	−37	190
M2C159-B1/C1	SAE 10W-30	...	...	...	...	...	...	190
M2C159-B2/C2	SAE 15W-30	...	...	...	...	...	...	190
M2C159-B3/C3	SAE 20W-40	...	...	...	...	...	...	190
Kubota UDT	8.8 min	140	4800 (−20 °C)	...	70,000 (−34 °C)	...	−37	200

[a]Obsolete specification.
[b]Stable pour point.

grades and the older Seconds Saybolt Universal (SSU) viscosity grades. ISO viscosity grades are based upon midpoint viscosity of the fluid at 40 °C. Common ISO viscosity grades (ISO VG) along with corresponding SSU grades are provided in Table 7.11. Figure 7.2 shows the effective viscosity range of ISO VG fluids, as defined by ASTM D2422, Classification of Industrial Fluid Lubricants by Viscosity System [594]. For example, ISO VG 32 hydraulic oil will provide satisfactory performance over −8 to 64 °C temperature range. Viscosity of hydraulic fluids is measured according to ASTM D445, Standard Test Method for Kinematic Viscosity of Transparent and Opaque Liquids.

In high-pressure hydraulic applications, the nature of lubrication changes from being hydrodynamic to mixed-film and boundary. This allows metal-to-metal contact and the wear can ensue. Additives used to control wear in hydraulic systems include zinc dialkyl dithiophosphates (ZDTP), tricresyl phosphate (TCP), sulfur compounds, amine phosphates, dithiocarbamates, and chlorinated phosphorus/sulfur and molybdenum compounds. Details regarding the functioning of these additives are provided in Chapter 4, the Additives chapter. The tests commonly used to assess the wear control properties of a hydraulic fluid include the following:

1. ASTM D2782, Standard Test Method for Measurement of Extreme-pressure Properties of Lubricant Fluids (Timken Method).
2. ASTM D2783, Standard Test Method for Measurement of Extreme-Pressure Properties of Lubricating Fluids (Four-Ball Method).
3. ASTM D4172, Standard Test Method for Wear Preventive Characteristics of Lubricating Fluids (Four-Ball Method).
4. ASTM D5182, Standard Test Method for Evaluating the Scuffing (Scoring) Load Capacity of Oils (the FZG Test).
5. DIN 51 354, Part 2—Four-square Gear Test Rig (FZG Test).
6. ASTM D2882, Standard Test Method for Indicating the Wear Characteristics of Petroleum and Non-petroleum

TABLE 7.5—J. I. Case and John Deere tractor hydraulic specifications [318]. Reprinted with permission from the Lubrizol Corporation.

Test	Typical Results	J I Case	John Deere	
		MS1207	J20C/D	J27
Four Ball Wear, Scar diam (mm)				
1 h, 65°C, 1500 rpm, 40 kg	0.35	—	—	0.45 max
Load Carrying Capacity (ASTM D 2733)				
Load Wear Index (kg)	57	—	—	—
Weld Load (kg)	251	—	—	—
Wet Brake Chatter/Squawk (various)	Pass	Pass	Pass	Pass
PTO Clutch (various)	Pass	Pass	Pass	Pass
High Torque Axle (various)	Pass	Pass	Pass	Pass
Plessey Pump	Pass	—	—	—
Allison C-4				
Oxidation	Pass	—	Pass	Pass
Vickers Pump	Pass	—	Pass	Pass
Graphite Friction	Pass	—	Pass	Pass
Paper Friction	Pass	—	Pass	Pass
Rust Protection	Pass	Pass	Pass	Pass
Falex Pin Corrosion	Pass	—	—	—
Copper Strip Corrosion				
3 h at 150°C	1a	—	—	—
Foaming (tendency/stability)				
Sequence I (ml)	0/0	50/10 max	25/0 max	25/0
Sequence II (ml)	30/0	50/10 max	50/0 max	25/0
Sequence III (ml)	0/0	50/10 max	25/0 max	25/0
Oxidation (100 h at 149°C)			J20C J20D	
Evaporation Loss (%)	0.6	Pass	5 max 10 max	5 max
Viscosity Increase at 99°C (%)	2.5	Pass	10 max 20 max	10 max
Separation/Sludge	None	Pass	None None	Trace/none
Seal (70 h at 100°C)				
Volume Change (%)	+0.9	0 to +5	Allison	0 to +5
Durometer Change (%)	-2	—	C-4	0 to -5
Precipitation	None	—	Test	None
Water Tolerance				
Water Volume (%)	—	10	0.4	0.4
Sediment Volume (% max)	—	—	0.1	0.1
Additive Loss (% mass)	—	—	15 max	—
Water Separation	—	—	—	—

Hydraulic Fluids in a Constant Volume Vane Pump (Vickers 104C).

7. ASTM D2271, Standard Test Method for Preliminary Examination of Hydraulic Fluids (Wear Test).
8. Vickers V-104C and 35VQ25 Vane Pump Tests for screening hydraulic fluid wear performance for higher pressure and mobile applications.
9. Denison T6HZOC Piston Vane Pump Test.
10. Komatsu HPV35+35 Twin-pistons Pump Test using cycled pressure for biodegradable vegetable oil-based hydraulic fluids.
11. The Sundstrand Series 90 Piston Pump (water stability) Test to determine the effect of water contamination on mineral oil hydraulic performance and yellow metal corrosion.
12. Three-stage Piston Pump Test based on the Brueninghaus A4VSO piston pump, proposed by Rexroth [592].

Water Content and Hydrolytic Stability

In many hydraulic systems, the fluid either contains water or is susceptible to contamination with water. In water-free fluids, contamination with water can lead to a drop in lubricity, increased corrosion, additive degradation, filter plugging, and cavitation. Hence, OEMs attempt to limit the amount of water that enters their hydraulic systems and the fluid formulators try to design hydraulic fluids that resist chemical degradation and or hydrolysis in the presence of water. Hydrolytic stability is the key factor in the wet filterability behavior of the hydraulic fluids [595]. There are no specific additives that can improve hydrolytic stability of a fluid; it is the selection of the stable components that results in a hydrolytically stable hydraulic fluid. Methods used to assess water content and hydrolytic stability of the hydraulic fluids include the following:

1. ASTM D95, Standard Test Method for Water in Petro-

TABLE 7.6—New Holland and AGCO Massey-Ferguson tractor hydraulic specifications [318]. Reprinted with permission from the Lubrizol Corporation.

	New Holland		AGCO Massey-Ferguson		
Test	**FNHA-2-C-201.00[a]**	**M2C159B/C**	**M1135**	**M1139**	**M1141**
Four Ball Wear, Scar diam. (mm)					(1800 rpm)
1 h, 65 °C, 1500 rpm, 40kg	0.40 max	0.40 max	—	0.40 max	0.40 max
Load-carrying Capacity (ASTM D2733)					
Load Wear Index (kg)	—	—	55 min	55 min	35 min
Weld Load (kg)	—	—	—	—	200 min
Wet Brake Chatter/Squawk (various)	Pass	Pass	Pass	Pass	Pass
PTO Clutch (various)	Pass	Pass	Pass	Pass	—
High Torque Axle (various)	Pass	Pass	Pass	Pass	—
Plessey Pump	—	—	—	—	—
Allison C-4					
Oxidation	—	Pass	—	—	—
Vickers Pump	—	Pass	—	—	—
Graphite Friction	Pass	Pass	—	—	—
Paper Friction	—	Pass	—	—	—
Rust Protection	Pass	—	Pass	Pass	Pass
Falex Pin Corrosion	—	Pass	—	—	—
Copper Strip Corrosion					
3 h at 150 °C	2b max	1b max	1a max	1a max	1b max
Foaming (tendency/stability)					
Sequence I (mL)	20/0 max	20/0	100/0 max[b]	100/0 max[b]	50/0[b]
Sequence II (mL)	50/0 max	50/10	100/0 max	100/0 max	50/0
Sequence III (mL)	20/0max	20/10	100/0 max	100/0 max	50/0
Oxidation (100 h at 149 °C)					
Evaporation Loss (%)	—	—	—	—	—
Viscosity Increase at 99 °C (%)	10 max	10 max	10 max	10 max	15 max
Separation/Sludge	None	None	None	—	None
Seal (70 h at 100 °C)					
Volume Change (%)	—	—	—	—	—
Durometer Change (%)	—	—	—	—	—
Precipitation	—	—	—	—	—
Water Tolerance					
Water Volume (%)	0.5	0.5	—	0.5	—
Sediment Volume (% max)	0.1	0.1	—	0.1	—
Additive Loss (% mass)	—	—	—	—	—
Water Separation	Trace	Trace	—	—	—

[a]Same as Ford M2C134D.
[b]1.0 % volume water added to Massey-Ferguson foam tests.

leum Products and Bituminous Materials by Distillation.

2. ASTM D96, Standard Test Method for Water and Sediment in Crude Oil by Centrifuge Method.
3. ASTM D1744, Standard Test Method for Determination of Water in Liquid Petroleum Products by Karl Fischer Reagent.
4. ASTM D2619, Standard Test Method for Hydrolytic Stability of Hydraulic Fluids (Beverage Bottle Method).

Demulsibility

Demulsibility is the term used to describe a fluid's ability to separate from water. As discussed in the previous section, water contamination can lead to deterioration of the fluid as well as it can affect equipment durability. It is therefore desirable for a hydraulic fluid to shed water as quickly as possible, so that it settles to the bottom of the reservoir and it can be drained off, if desired. For fluids with poor demulsibility characteristics, the separation is either very slow or unlikely. To improve the properties of such fluids, chemicals called demulsifiers, are added. These additives alter the surface tension of the oil-water interface, which facilitates separation, and include alkylphenol ethers and low to medium molecular weight synthetic metal sulfonates. Demulsibility in a hydraulic fluid is measured by ASTM D1401, Standard Test Method for Water Separability of Petroleum Oils and Synthetic Fluids and ASTM D2711, Demulsibility Characteristics of Lubricating Oils (Wheeling Steel Demulsibility).

Aeration and Foam

The presence of air in a hydraulic fluid is quite common. Its amount depends upon the fluid's viscosity and polarity and the ambient temperature and pressure. Low viscosity fluids contain a higher amount of air than those with higher viscosity, but they also lose it quicker, if the other factors do not change. Fluids with higher polarity, such as synthetic esters and water-based fluids, are likely to have a higher amount of dissolved air than the less polar mineral oils. Higher temperatures and lower pressures decrease the amount and the lower temperatures and higher pressures increase it. In addition, the fluid circulation through the hydraulic system may result in additional introduction of air into the hydrau-

TABLE 7.7—Universal tractor transmission oil *(UTTO)* [318]. Reprinted with permission from the Lubrizol Corporation.

Viscosity	Summer	SAE 10W-30 — 9.0 to 11.6 cSt at 100°C 50,000 to 70,000 cP max at -35°C Brookfield
	Winter	SAE 5W-20 — 6.2 to 7.0 cSt at 100°C 20,000 cP max at -40°C Brookfield

Application	Specification	Test
Transmission	API GL-4	L-20 High-torque Low-speed
Wet Brakes	Massey- Ferguson M1141/M1135	Independent Power Take-off, 4-Square Axle, Pump
	New Holland FNHA-2-C-201.00 (M2C134)	Jenkins Tractor Transmission Stall and High Energy
	John Deere J20C/D	Brake Chatter/Capacity, Powershift Transmission, High-Torque Axle
	Caterpillar TO-2[a] Allison C-4	Powershift Transmission Friction, Wear, Oxidation, Friction
Hydraulic Equipment	Vickers Vane Pump Denison HF-0 Vickers Sheet 1-286-S	Wear (ASTM D2882) Vane and Piston Pump Vickers 35VQ25

[a]This test or category is obsolete; parts, fuel, or reference oils are no longer generally available, or the test is no longer monitored by the test developer or ASTM.

lic fluid, especially if the reservoir size, which is the case in the new equipment, or the design does not allow sufficient residence time for air separation to occur. Aeration is a problem because it will result in foam formation, which under pressure can cause cavitation damage [596]. Cavitation is the process where the bubbles resulting from the air entrainment of the fluid implode under compression. In addition to cavitation, other reasons that make aeration and foam formation in a hydraulic fluid undesirable include vibration, noise, sluggish component response (due to spongy behavior of the fluid), reduced thermal conductivity and dielectric properties, and fluid degradation. While cavitation commonly occurs at the pump, it can occur anywhere within the hydraulic system. Cavitation damage leads to metal erosion, thereby causing physical damage to hydraulic system components and contaminating the fluid. In extreme cases, it can even cause catastrophic failure of the hydraulic pumps, control valves, and motors. Sometimes an air-ignitable mixture is present within the bubble, which will ignite due to a rise in temperature that accompanies compression. Although the

TABLE 7.8—Super tractor oil universal *(STOU)* [318]. Reprinted with permission from the Lubrizol Corporation.

Viscosity: SAE 10W-30
9.3 to 12.5 cSt at 100°C
3,500 cP max at -20°C (Cold Cranking Simulator)

Application	Specification	Test
Engine	API CD[a]/CE[a]/SF[a]	Caterpillar 1G2[a], Sequence IID, Sequence IIID[a], Sequence VD[a], L-38, Mack T-6 and T-7, NTC-400[a]
	CCMC D-4	M-B OM364A, CCMC quality must be demonstrated with API or MIL data as background information
Transmission/Wet Brakes	API GL-4, Massey- Ferguson M1139, New Holland M2C159B/C	L-20 High-torque Low-speed, Independent Power Take-off, 4-Square Axle, Pump
	John Deere J27	Brake Chatter/Capacity, Powershift Transmission, High-Torque Axle
	Caterpillar TO-2* Allison C-4	Powershift Transmission Friction, Oxidation, Wear, Friction
Hydraulic Equipment	Vickers Vane Pump Denison HF-0 Vickers Sheet I-286-S	Wear (ASTM D2882) Vane and Piston Pump Vickers 35VQ25

[a]This test or category is obsolete; parts, fuel, or reference oils are no longer generally available, or the test is no longer monitored by the test developer or ASTM.

TABLE 7.9—Original equipment manufacturers (OEM) viscosity guidelines for hydraulic fluids [592]. Reprinted with permission from the Lubrizol Corporation.

Manufacture	Equipment	Operating Viscosity (cSt) Min	Max	Max Start-Up Viscosity Under Load (cSt)	Optimum Viscosity (cSt)
Bosch					
Form No. S/106 US	FA;RA;K	15	216	864	26–45
	Q; Q-6; SV-10, 15,20,25; VPV 16, 25, 32	21	216	864	32–54
	SV-40, 80, 100; VPV 45, 63	32	216	864	43–64
	Radial Piston (SECO)	10	65	162	21–54
	Axial & RKP Piston	14	450	647	32–65
Commercial Intertech	Roller & Sleeve Bearing Gear Pumps	10		1600	20
Sauer Danfoss	All	10	...	1618	21–39
Denison	Piston Pumps	13	...	...	24–31
Bulletin 440	Vane Pumps	10	107	860[a]	30
Dynex/Rivett	PF4200 series	1.5	372	372	20–70
Axial Piston Pumps	PF2006/8, PF/PV4000, PF/PV6000 Series	2.3	413	413	20–70
	PF1000, PF2000, PF3000 Series	3.5	342	342	20–70
Eaton	Heavy-duty Piston Pumps & Motors, Medium-duty Piston Pumps & Motors, Charged Systems, Light-duty Pumps	6	...	2158	10–39
	Medium-duty Piston Pumps & Motors, Uncharged Systems	6	...	432	10–39
	Gear Pumps, Motors & Cylinders	6	...	2158	10–43
Eaton-Vickers	Mobile Piston Pumps	10	200	860	16–40
	Industrial Piston Pumps	13	54	220	16–40
	Mobile Vane Pumps	9	54	860	16–40
	Industrial Vane Pumps	13	54	860	16–40
Eaton-Char-Lynn	J, R & S Series Motors & Disk Valve Motors	13	...	2158	20–43
	A & H Series Motors	20	...	2158	20–43
Haldex Barnes	W Series Gear Pumps	11	...	750	21
Kawasaki	Staffa Radial Piston Motors	25	150	2000 (no led)	50
P-969-0026					
P-969-0190	K3V/G Axial Piston Pumps	10	200	1000	
Linde	All	10	80	1000	15–30
Rexroth	V3, V4, V5, V7 Pumps	25	...	800	25–160
	V2 Pumps	16	160	800	25–160
	R4 Radial Piston Pumps	10	200	...	25–160
	G2, G3, G4 Pumps & Motors, G8, G9, G10 Pumps	10	300	1000	25–160
Parker Hannifin	Gerotor Motors	8	...	...	12–60
	PGH. D/H/M Series Gear Pumps	...	...	1000	17–180
	Hydraulic Steering	8	...	...	12–60
	PFVH/PFVI Vane Pumps	...	...	1000	17–180
	Series T1	10	...	1000	10–400
	VCR2 Series	13	...	1000	...
	LSHT Motors	10	...	...	...
	Variable-Volume Piston Pumps	...	...	1000	17–180
	PVP & PVAC	...	...	1000	17–180
	Axial Fixed Piston Pumps	...	...	850	12–100
	Variable-Volume Vane—PW	...	...	440	16–110
Poclain Hydraulics	H & S Series Motors	9		1500	20–100
Sauer-Sundstrand USA	All	6.4	...	1600	13
Sauer-Sundstrand GmbH	Series 10 & 20 RMF Hydrostatic Motor	7	...	1000	12–60
	Series 15 Open Circuit	12	...	860	12–60
	Series 40. 42. 51 & 90 CW S-8 Hydrostatic Motor	7	...	1600	12–60
	Series 45	9	...	1000	12–60
	Series 60LPM Hydrostatic Motor	9	...	1600	12–60
	Gear Pumps & Motors	10	...	1000	12–60

[a]Low speed and pressure.

TABLE 7.10—Major hydraulic fluid specifications [592]. Reprinted with permission from the Lubrizol Corporation.

Performance and Physical Tests	Denison (ISO 46 unless otherwise noted)	Eaton-Vickers		Cincinnati Lamb Landis			GM (LS-2) LH-03, 04, 06	DIN 51 524 Part 1	DIN 51 524 Part 2	DIN 51 524 Part 3
Pump Tests	HF-0	M-2950-5	1-286-5	P-68	P-70	P-69				
Denison T6H20C Piston/Vane										
Phase 1: 1700 rpm, 230° F [110°C]										
Phase 2: 1700 rpm, 176° F [80°C] + 1 % water Both phases are cyclic in pressure < 50-4061 psi) < 4-280 bar and run for 300 hours each.										
Pump Wear [vanes & pins] 15 mg max										
Denison Assessment	Pass									
Eaton-Vickers 35VQ/25 Vane										
(3000 psi, 2400 rpm, 203°F [95°C])										
Total Ring & Vane Wt. Loss, mg		90	90				60			
IP 281 Vane Pump Test (V104C)										
(2050 psi, 1440 rpm, 250 hours)										
Ring, mg									120 max	120 max
Vane, mg									30 max	30 max
Oxidation Tests										
Turbine Oil Oxidation Test (D943)										
Hrs. to 2.0 NNA							1500 min			
1000 Hour Sludge Test (D4310)										
NNA, mg KOH/g	1 max							2.0 max	2.0 max	≤ 2.0
Insoluble Sludge, mg	100 max									
Total Copper, mg	200 max									
Physical Tests										
Demulsibility										
Turbine Oil Demulsibility (D1401) (130° F [54.4°C)) ISO VG 32/46										
ml: Oil-Water-Cuff	40-37-3	(b)	(b)				40-40-0	40-37-3	40-37-3	40-37-3
Separation Time (minutes)	(30)						(30)	(40)	(40)	(40)
Rust										
Turbine Oil Rust Test (D665A/B)										
A - Distilled Water	Pass	(b)	(b)	Pass	Pass	Pass	Pass	Pass	Pass	
B - Synthetic Sea Water	Pass	(b)	(b)				Pass			
Hydrolytic Stability										
Hydrolytic Stability (D2619)										
Water NNA, mg KOH/g	4.0 max						4.0 max			
Copper Wt. Loss, mg/cm^2	0.2 max						0.2 max			
Thermal Stability										
Cincinnati Lamb Landis Thermal Stability (168 hrs, 275°F [135°C])										
Viscosity Change, max				5%	5%	5%	5%			
NNA Change, max				±50%	±50%	±50%	±50%			
Sludge, mg/100 ml, max	100				25	25	25	25		
Copper Wt. Loss, mg, max	10				10	10	10	10		
Copper Rod Appearance, max	Report				5	5	5	5		
Steel Wt. Loss, max					1.0	1.0	1.0			
Steel Rod Appearance					1.5 max	1.5 max	1.5 max	No Discolor		
Filterability										
Denison TP-02100-A										
A - No Water (see)	600 max						600 max			
B - 2% Water (see)	2xA max						2xA max			
ISO 13357										
Part 2 without water										
Stage I F1 min. %								80		
Stage II F2 min. %								60		
Part 1 with water										
Stage I F1 min. %								80		
Stage II F2 min. %								60		
Foam										
Foam (D892) (Sequence I, II & III)										
Tendency/Stability	NR/0	NR/0	NR/0				50/0			
DIN 51 566										
Sequence I								≤ 150/0	≤ 150/0	≤ 150/0
Sequence II								≤ 75/0	≤ 75/0	≤ 75/0
Sequence III								≤ 150/0	≤ 150/0	≤ 150/0
FZG	Report									
DIN 51 354, Part 2	ISO 32									
Fail Load Stage, min	only						10		10	10
Copper Strip (D130)										
100°C/3 hrs							1b max	2 max	2 max	2 max

TABLE 7.10— (Continued.)

Performance and Physical Tests	Denison (ISO 46 unless otherwise noted)	Eaton-Vickers	Cincinnati Lamb Landis	GM (LS-2) LH-03, 04, 06	DIN 51 524 Part 1	DIN 51 524 Part 2	DIN 51 524 Part 3
Air Separation							
(D3427/DIN 51 381)				(c)	(c)	(c)	(c)
Minutes ISO VG 46/68	7/10 max						
Seal Test							
(DIN 53 521/DIN 53 538, Part 1:							
Specifies Seal Material)							
Volume Change (%)							
ISOVG 32/46	(d)			0 to 12	0 to 12	0 to 12	0 to 12
ISOVG 68				0 to 10	0 to 10	0 to 10	0 to 10
Shore A Hardness Change (%)							
ISOVG 32/46				0 to -7	0 to -7	0 to -7	0 to -7
ISOVG68				0 to -6	0 to -6	0 to -6	0 to -6
Cleanliness							
ISO 4406				16/13		21/19	
Viscosity (D445)							
cSt at 40°C	(a)	(a)	32 46 68	(a)		(a)	
cSt at 100°C	Report	Report					
Viscosity Index (D2270)							
Minimum	90	90	90 90 90	95			140 (e)
Aniline Point (D611)	100°C (d)						
Shear Test (Multi-grades only) (f)							
KRL 20 hrs							
Max Shear Loss	15%						
Gravity API	840-900						
% Zinc	Report						
Pour Point (D97) (g)	Report						
Flash Point (D92) (h)	Report						
Acid Number (D664)	Report						

(a) ISO 32: 28.8 – 35.2; ISO 46: 41.4 – 50.6; ISO 68: 61.2 – 74.8
(b) Evidence of satisfactory performance required, tests not specified
(c) Up to ISO 32: 5; ISO 46 & 68: 10; ISO 100: 14
(d) If aniline point is less than 100°C, seal test data must be reported (e) ≥ 120 or greater at ISO VG 100
(f) KRL required for fluids containing a viscosity index improver/high VI oils (g) Pour point varies with viscosity grade
(h) Flash point varies with viscosity grade

process is of brief duration, lasting only nanoseconds, the local temperatures can be as high as 2012 °F (1100 °C), or higher, which will lead to thermal and oxidative degradation of the fluid, resulting in varnish and sludge formation. This will be discussed later. Foam-inhibiting additives, such as polydimethylsiloxanes and poly(alkyl acrylate)s are used to break the foam bubbles. The tests used to evaluate these fluid parameters include the following:

1. ASTM D892, Standard Test Method for Foaming Characteristics of Lubricating Oils.
2. ASTM D3519, Standard Test Method for Foam in Aqueous Media (Blender Test).
3. IP 313, DIN 51381 or ASTM D3427, Standard Test Method for Air Release Properties of Petroleum Oils.

Filterability

Proper fluid selection and fluid maintenance are important in achieving reliability and durability of the hydraulic equipment. Fluid maintenance includes assuring fluid cleanliness and filtration. Fluid should easily flow through the filter, with minimum pressure drop and little or no depletion of additives. ISO 4406-99, Fluid Cleanliness Code, and SAE AS 4059 are the most widely used methods for determining particle counts in turbine oils and hydraulic lubricants. ISO 4406 uses a three-range number system. The first number corresponds to particles larger than 4 μm; the second number corresponds to particles larger than 6 μm; and the third number corresponds to particles larger than 14 μm. An ISO code of 18/16/13 for a hydraulic fluid indicates that 1 mL of the fluid contains 1300 to 2500 particles larger than 4 μm, 320 to 640 particles larger than 6 μm and 40 to 80 particles larger than 14 μm [597]. Typical cleanliness rating to meet Denison HF-0 and Cincinnati Lamb requirements is 17/15/11 [598], to meet GM LS-2 requirement is 19/16/13 [599], and to meet Rexroth requirement is 20/18/15 [600]. Table 7.12 lists the ISO 4406:1999 cleanliness codes [597]. SAE AS 4059 is an Aerospace Standard (AS) that defines cleanliness classes for particulate contamination of hydraulic fluids and includes methods of reporting the related data. The contamination classes selected are based on the widely accepted NAS 1638 cleanliness classes. The SAE Standard has been revised five times, with the E version being introduced in May of 2005. SAE AS 4059 offers significant advantages over NAS 1638, which include the following [601]:

1. It uses ISO 11171 for calibration of the automatic particle counters (APCs), which provides increased precision and improved repeatability and reproducibility.
2. It presents data in terms of cumulative counts (>Xμm; >Y μm) rather than range mode (X-Y μm). This is the manner in which particle counters generate the data.
3. It introduces a new cleanest class, 000.
4. It extends the size range to smaller sizes [>4 μm(c)] for increased sensitivity,

The SAE AS 4059E Standard is presented in Tables 7.13A and 7.13B. Part A of the table provides cleanliness classes for cumulative particle counts and Part B of the table

TABLE 7.11—ISO viscosity grades.

ISO Viscosity Grade[a]	Kinematic Viscosity at 40 °C (cSt)	Saybolt Viscosity at 100 °F (SSU)
2	1.98–2.42	32
3	2.88–3.52	36
5	4.14–5.06	40
7	6.12–7.48	50
10	9–11	60
15	13.5–16.5	75
22	19.8–24.2	105
32	28.8–35.2	150
46	41.4–50.6	215
68	61.2–74.8	315
100	90–110	465
150	135–165	700
220	198–242	1000
320	288–352	1500
460	414–506	2150

[a]Viscosity grade numbers of the ISO are the same as those of ASTM (The American Society of Testing and Materials) and BSI (British Standards Institution), with the difference that the viscosities for the ISO grades are measured at 40 °C, while those of the ASTM and BSI are measured at 100 °F (37.8 °C). Lubricants of a given ASTM or BSI grade are slightly more viscous than lubricants of the corresponding ISO grade.

provides cleanliness classes for differential particle counts. Please note that contrary to Part A that uses individual particle sizes to assign a class, in Part B particle size ranges are used for this purpose. Part B is analogous to the NAS 1638 Cleanliness specification.

Data below are presented for a hypothetical fluid that employed Method 1 to determine the sizes and the number of particles. Cumulative class rating for such a fluid according to 4059E Standard will be 2A/3B/2C/3D/3E/5F. Please note that ISO Code is based on the number of particles per milliliter (mL) but SAE 4059 is based on number of particles per 100 mL.

- 2563 particles in >1 μm size = 2A
- 162 particles in >15 μm size = 2C

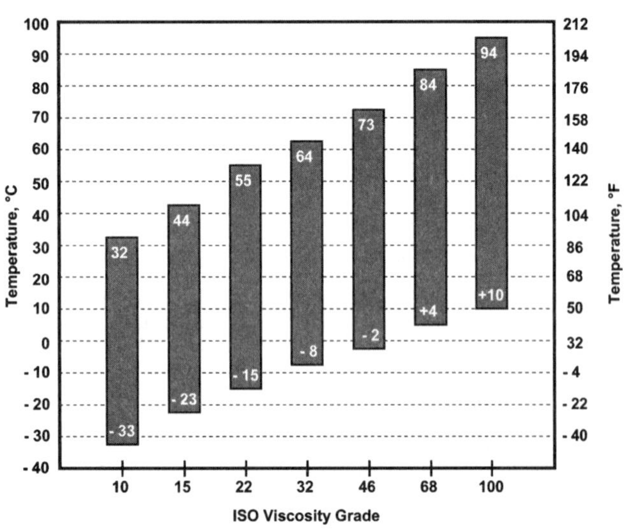

Fig. 7.2—Operating temperature ranges of ISO viscosity grades [594].

TABLE 7.12—ISO 4406, SAE J2472 fluid cleanliness codes number of particles per mL [597].

ISO Code	More Than	Up to and Including	ISO Code	More Than	Up to and Including
1	0.01	0.02	15	160	320
2	0.02	0.04	16	320	640
3	0.04	0.08	17	640	1300
4	0.08	0.16	18	1300	2500
5	0.16	0.32	19	2500	5000
6	0.32	0.64	20	5000	10,000
7	0.64	1.3	21	10,000	20,000
8	1.3	2.5	22	20,000	40,000
9	2.5	5.0	23	40,000	80,000
10	5.0	10.0	24	80,000	160,000
11	10.0	20.0	25	160,000	320,000
12	20.0	40.0	26	320,000	640,000
13	40.0	80.0	27	640,000	1,300,000
14	80.0	160.0	28	1,300,000	2,500,000

- 62 particles in >25 μm size = 3D
- 1241 particles in >5 μm size = 3B
- 12 particles in >50 μm size = 3E
- 6 particles in >100 μm size = 5F

For hydraulic fluids that are likely to encounter filtration problems because of the environment or the chemistry, high molecular weight alkenylsuccinimide and Mannich dispersant additives are used. These additives keep particulates suspended in the fluid, thereby preventing their deposition on the filter. Filterability of the hydraulic fluid is assessed by tests that consist of filtering a specified quantity of fluid, with or without water, through a standard medium while monitoring changes in flow rate. The use of water is based upon the fact that in many hydraulic fluids, its presence usually makes filterability worse, which is a consequence of the hydrolysis and the interaction of some of the additives. Denison TP 02100 Test, Pall Filterability Test, and AFNOR Filterability Test are used to assess filterability of the hydraulic fluids.

Corrosion Protection

Corrosion-causing species, such as acids, originate from the oxidation of the fluid or oxidation and decomposition of the additives. These materials, if present, can attack metal components and cause corrosion. Ferrous metal corrosion, primarily rust, is often due to water contamination, while corrosion of copper and its alloys (yellow metals) is due to sulfur compounds or acidic species that are either present in the fluid or result from the high temperature degradation of the fluid and its components. For a discussion on rust inhibitors and the mechanism of rusting, please refer to Chapter 4 on Additives. Ferrous metal corrosion control in hydraulic systems is achieved by the use of acid-neutralizing and film-forming agents, such as neutral and basic arylsulfonates, alkenylsuccinic acid derivatives, and fatty amines. Yellow metal corrosion, on the other hand, is controlled by the use of soluble sulfur and nitrogen-containing heterocyclic compounds, such as tolyltriazole and dimercaptothiadiazole (DMTD). The ASTM methods used for evaluating corrosion inhibition properties of hydraulic fluids include the following:

1. ASTM D665, Rust-Preventing Characteristics of Inhibited Mineral Oil in the Presence of Water.

TABLE 7.13A—SAE AS 4059E Standard—Cleanliness classes for cumulative counts (Particles per 100 mL) [601].

Method 1 [ISO 4402]	>1 μm	>5 μm	>15 μm	>25 μm	>50 μm	>100 μm
Method 2 [ISO 11171]	>4 μm(c)	>6 μm(c)	>14 μm(c)	>21 μm(c)	>38 μm(c)	>70 μm(c)
Size Codes	**A**	**B**	**C**	**D**	**E**	**F**
Class 000	195	76	14	3	1	0
Class 00	390	152	27	5	1	0
Class 0	780	304	54	10	2	0
Class 1	1560	609	109	20	4	1
Class 2	3120	1217	217	39	7	1
Class 3	6250	2432	432	76	13	2
Class 4	12,500	4864	864	152	26	4
Class 5	25,000	9731	1731	306	53	8
Class 6	50,000	19,462	3462	612	106	16
Class 7	100,000	38,924	6924	1224	212	32
Class 8	200,000	77,849	13,849	2449	424	64
Class 9	400,000	155,698	27,698	4898	848	128
Class 10	800,000	311,396	55,396	9796	1696	256
Class 11	1,600,000	622,792	110,792	19,592	3392	512
Class 12	3,200,000	1,245,584	221,584	39,184	6784	1024

Note: **Method 1**: Size range, Optical microscope, based on longest dimension as measured per ARP598 or APC Calibrated per ISO 4402:1991.
Method 2: Size range, APC Calibrated per ISO 11171 or Electron Microscope, based on projected area equivalent diameter.

2. ASTM D5534, Test Method for Vapor-Phase Rust-Preventing Characteristics of Hydraulic Fluids.
3. ASTM D3603 is the Horizontal Disk Method for Rust-Preventing Characteristics of Steam Turbine Oils in the Presence of Water.
4. ASTM D6547, Test Method for Corrosiveness of a Lubricating Fluid to a Bi-Metallic Couple.
5. ASTM D2070, Standard Test Method for Thermal Stability of Hydraulic Oils (Cincinnati Thermal Stability Test).
6. ASTM D130, Standard Test Method for Detection of Copper Corrosion from Petroleum Products by the Copper Strip Tarnish Test. Copper strip rating criteria are provided in Table 7.14 [594].

Seal Compatibility

Hydraulic systems contain a variety of seals whose function is to maintain hydraulic pressure, isolate sections to facilitate removal and replacement of the malfunctioning parts, and isolate the fluid from harmful elements. Hydraulic fluids at high temperatures and high pressures can interact with rubber and elastomer seals causing them to swell, shrink, harden, or crack, depending upon whether the fluid travels into or the plasticizer is removed out of the elastomer mate-

TABLE 7.13B—SAE AS 4059E Standard—For differential particle counts [a] (particles per 100 mL)

Method 1 [ISO 4402]	5–15 μm	15–25 μm	25–50 μm	50–100 μm	>100 μm
Method 2 [ISO 11171]	6–14 μm(c)	14–21 μm(c)	21–38 μm(c)	38–70 μm(c)	>70 μm(c)
Class 00	125	22	4	1	0
Class 0	250	44	8	2	0
Class 1	500	89	16	3	1
Class 2	1000	178	32	6	1
Class 3	2000	356	63	11	2
Class 4	4000	712	126	22	4
Class 5	8000	1425	253	45	8
Class 6	16,000	2850	506	90	16
Class 7	32,000	5700	1012	180	32
Class 8	64,000	11,400	2025	360	64
Class 9	128,000	22,800	4050	720	128
Class 10	256,000	45,600	8100	1440	256
Class 11	512,000	91,200	16,200	2880	512
Class 12	1,024,000	182,400	32,400	5760	1024

Note: **Method 1**: Size range, Optical microscope, based on longest dimension as measured per ARP598 or APC Calibrated per ISO 4402:1991.
Method 2: Size range, APC Calibrated per ISO 11171 or Electron Microscope, based on projected area equivalent diameter.
[a]Classes and contamination limits identical to NAS 1638.

TABLE 7.14—Copper strip classifications [594].

Designation	Description
Slight tarnish	Light orange, almost the same as freshly polished strip
Slight tarnish	Dark orange
Moderate tarnish	Claret red
Moderate tarnish	Lavender
Moderate tarnish	Multicolored with lavender blue or silver overlaid on claret red
Moderate tarnish	Silvery
Moderate tarnish	Brassy or gold
Dark tarnish	Magenta overcast on brassy strip
Dark tarnish	Multicolored with red and green showing (peacock), but no gray
Corrosion	Transparent black, dark gray, or brown with a trace of peacock
Corrosion	Graphite or lusterless black
Corrosion	Glossy or jet black

rial. A reduction in seal volume or seal damage can result in the loss of fluid, contamination, loss of pressure, and environmental damage and fire hazard, depending on the severity of the spill. In general, seal swelling is preferred over seal shrinkage because the fluid loss, if it occurs, will be minimal and so will be the adverse consequences. Chemicals, called seal swell agents, slightly swell or soften elastomer seals to make them pliable to counter the effects of temperature and mechanical stress. They are usually needed when the hydraulic fluid is low in aromatics, that is, it is based upon severely refined API Group III oils or PAOs. Seal swell agents are normally aromatic compounds and phosphate esters. Seal compatibility of the hydraulic fluids is tested by using tests that include ISO 7619, ISO 6072, DIN 53 538, and the ASTM test methods listed below:

1. D 6546, Standard Test Method for and Suggested Limits for Determining Compatibility of Elastomer Seals for Industrial Hydraulic Fluid Applications.
2. D412, Standard Test Methods for Vulcanized Rubber and Thermo plastic Elastomers Tension.
3. D471, Standard Test Method for Rubber Property-Effect of Liquids.
4. D2240, Standard Test Method for Rubber Property-Durometer Hardness.

Another test that is indirectly useful is the ASTM D611 test, which measures the aniline point of the fluid. Aniline point determines the aromatic content of the fluid. Fluids with lower values contain a higher aromatic content and therefore are more compatible with seals than those with the higher values, such as paraffinics.

Allowed change limits in elastomer seal properties used in hydraulic systems are provided in Table 7.15 [594]. Ideally, the elastomer should show a 3 to 5 % increase in volume and a slight softening of about 1 to 4 Shore A Durometer points. Tensile and elongation values that reflect an increase in elongation and a slight decrease in tensile strength are desirable. Since seal geometry and mechanical conditions of the application can also affect seals, it is prudent to test seal materials under conditions that closely simulate the actual application [602].

Coolant Separability

Hydraulic systems that are used in machine tool operations can get contaminated by the aqueous cutting fluids (coolants). These fluids contain components that have poor oxidation resistance, high deposit forming tendency, or high corrosivity, or a combination thereof. The cross-contamination of the metalworking fluid by the hydraulic fluid is also a concern since it can lower the effectiveness of the metalworking fluid. While cross-contamination of the hydraulic fluid may or may not hurt its performance, its ability to readily separate coolant is very desirable. Unfortunately, there are no additives that can facilitate separation because of the widely varying coolant chemistries. Since water is the common factor in these coolants, one approach is to use a potent demulsifier and mix it with the coolant of interest to test separability. There is no standard industry test to evaluate this property, but some large industrial manufacturers and lubricant suppliers use in-house procedures. All involve mixing the hydraulic fluid with a known percentage of the coolant and holding it at a specified temperature to evaluate the rate of separation.

Viscosity and Shear Stability

Proper viscosity of all fluids, including that of the hydraulic fluids, is extremely important for lubrication and minimizing wear. Hydraulic fluids that are used in mobile hydraulic equipment, such as excavators, farm tractors, cranes, and timber harvesters, often operate under extreme high and low temperature conditions. In these applications, information on viscosity outside the normally measured viscosity range of 40 °C and 100 °C is required. Usually, regular fluids do not have the attributes necessary to suitably perform at extreme temperatures; hence they are formulated with viscosity modifiers and pour point depressants to extend their performance range. Hydraulic fluids for extreme applications contain hydrogenated styrene-diene polymers, olefin copolymers, and polymethacrylates that help them maintain suitable viscosity at high temperatures. Common pour point depressants include alkylated naphthalenes and poly-

TABLE 7.15—Recommended property change limits for elastomer seals used in industrial hydraulics [594].

Time in Hours	% Volume Swell, max.	% Volume Shrinkage, max.	Hardness Change, Shore A Points	% Tensile Strength Change, max.
24	15	−3	±7	−20
70	15	−3	±7	−20
100	15	−3	±8	−20
250	15	−4	±8	−20
500	20	−4	±10	−25
1000	20	−5	±10	−30

TABLE 7.16—Low temperature viscosity grades for hydraulic fluid classifications [594].

ASTM Viscosity Grade	Minimum Viscosity	Maximum Viscosity
L 5	...	−50
L 7	−49	−42
L10	−41	−33
L 15	−32	−23
L 22	−22	−15
L 32	−14	−8
L 46	−7	−2
L 68	−1	4

methacrylates, which improve the low-temperature fluidity of the hydraulic fluids. The use of polymers to improve the viscometrics of the hydraulic fluids raises a concern for their shear stability. This lubricant parameter relates to mechanical flattening of the polymer, by devices such as gear pumps and vane pumps that involve closely contacting surfaces. Flattening of the polymer molecules decreases the interaction of the polymer with the fluid, thereby causing a drop in viscosity. This is a temporary situation because when the shear forces ease off, the fluid regains its viscosity (temporary viscosity loss). However, in some instances, the shear forces are extremely high, as in the case of high-pressure hydraulic systems where they approach the shear rates of $10^7\ s^{-1}$ [603], the polymer chains can get caught between surfaces and shear to smaller size polymer fragments. These smaller fragments do not boost the fluid viscosity to the same extent as the original longer chain polymer and since the process is not reversible, a permanent drop in viscosity is observed. Since the multi-grade fluids are the ones that contain viscosity modifiers, they are subjected to shear stability tests. For a detailed discussion, refer to the Chapter 4 on Additives. Some of the shear stability and low-temperature fluidity tests used for multi-grade hydraulic fluids are the same as those used for other multi-grade lubricants. These include the following:

1. ASTM D445, Absolute and Kinematic Viscosity.
2. ASTM D2270, Standard Practice for Calculating Viscosity Index from Kinematic Viscosity at 40 and 100 °C.
3. ASTM D6278, Test Method for Shear Stability of Polymer Containing Fluids Using a European Diesel Injector.
4. ASTM D5621, Standard Test Method for Sonic Shear Stability of Hydraulic Fluid.
5. ASTM D97, Standard Test Method for Pour Point of Petroleum Products.
6. ASTM D6351, Test Method for Determination of Low Temperature Fluidity and Appearance of Hydraulic Fluids.
7. ASTM D2983, Brookfield Viscosity.
8. ASTM D5133, Scanning Brookfield (SBV).
9. ASTM D4684, Mini Rotary Viscometer (MRV).
10. ASTM D2882, Vane Pump Test.
11. ASTM D6080, Practice for Defining the Viscosity Characteristics of Hydraulic Fluids.

Performance specifications that include low temperature pumpability requirements, such as ASTM D6080, specify a temperature range for the different viscosity grades. These are provided in Table 7.16 [594]. Please note that the temperature range for a given L-grade is approximately equivalent to that of the ISO viscosity grade of the same numerical designation and having a viscosity index of 100. For instance, the temperature range for the L32 oil is approximately the same as an ISO VG 32 grade with a Viscosity Index of 100. Compare these values with those provided in Fig. 7.2.

A major challenge for all lubricants, including the hydraulic fluids, is to develop test methods that simulate the equipment's operating environments. Bench test methods that are often used to screen a lot of materials quickly and inexpensively do not or cannot take into consideration all the critical operating variables of the actual application. As a result, the information obtained from these tests is less than perfect. Hence, testing using full scale test stands that simulate field conditions better is often necessary.

Types of Hydraulic Fluids

As stated earlier, hydraulic fluids are classified in two ways: based upon composition and based upon application. Hydraulic fluids belong to three general classes based upon the base fluid used: Petroleum or mineral oil base fluids, synthetic and natural fluids, and water-based fluids. These classes are shown in Fig. 7.3, while identifying fluids based

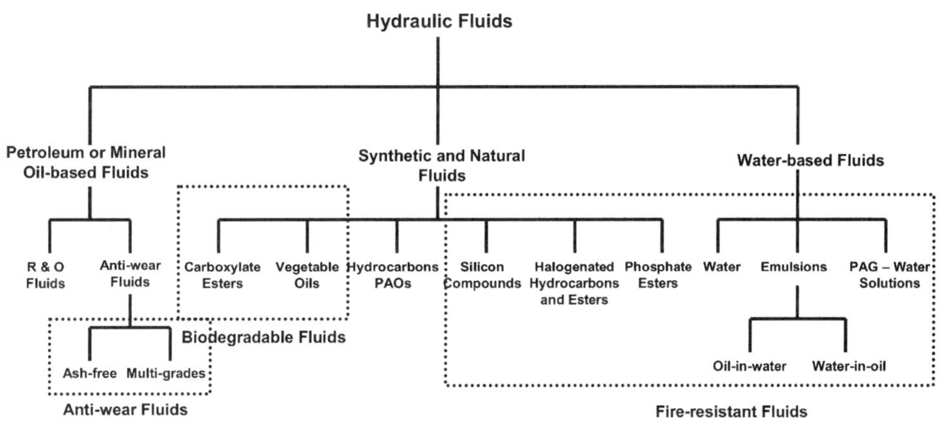

Fig. 7.3—Hydraulic fluid classification.

TABLE 7.17—Water solubilities for organophosphate ester hydraulic fluid components [604].

	Chemical Name	Water Solubility (mg/L)
1.	Triphenyl phosphate	1.9
2.	Tricresyl phosphate, mixed isomers	0.36
3.	Trixylenyl phosphate, mixed isomers	0.89
4.	Isopropylphenyl diphenyl phosphate, mixed isomers	2.2
5.	Nonylphenyl diphenyl phosphate, mixed isomers	0.77
6.	Cumylphenyl diphenyl phosphate, mixed isomers	0.063
7.	t-Butylphenyl diphenyl phosphate, mixed isomers	3.2
8.	2-Ethylhexyl diphenyl phosphate	1.9
9.	Tributyl phosphate	280.0
10.	Dibutyl phenyl phosphate	280.0

TABLE 7.18—Thermal conductivity and specific heat data for materials [4.154].

Material	Thermal Conductivity W/m·K at 373 °K (Btu/h/ft^2/F/Ft at 212 °F)	Specific Heat J/kg·K 293 K (Btu/lb °F·68 °F)
Steel	46.7 (27)	460 (0.11)
Oil	0.14 (0.08)	1966 (0.47)
Water	0.67 (0.39)	4184 (1.0)

upon their unique properties, i.e., antiwear fluids, fire-resistant fluids, and biodegradable fluids.

As shown in Fig. 7.3, hydraulic fluids belong to the general classes of petroleum-based, synthetic-based, and water-based fluids. Water is the oldest hydraulic fluid known to man and despite the fact that its use in modern applications has diminished; it is still in use in some systems. This is because of its inherent advantages, which include extremely low cost, relative abundance, high-viscosity index (maintains its viscosity with an increase in temperature), high compressibility (least affected by pressure), high cooling ability, and compatibility with all sealing materials, except leather and the porous cellular type. The primary limitations resulting in its declining use are:

1. Its narrow operating range of 3 °C to about 50 °C.
2. Its freezing point of 0 °C and boiling point of 100 °C.
3. Its rate of evaporation above 60 °C to be too high to contain it in the hydraulic system without difficulties.
4. Its inability to provide protection against rust and corrosion.
5. Its low lubricity and susceptibility to bacterial growth, after being contaminated with organic material.

Despite the disadvantages, a renewed interest in the use of water in hydraulic fluids/systems has been developing. This is being facilitated by new formulating concepts and new pump designs.

Petroleum-based fluids are also low cost, respond well to additives, have a broad operating range of about −40 to 100 °C, and provide excellent lubricity, except that they lack adequate fire resistance. Synthetic fluids have a higher cost, but are preferred in many applications because of the much broader operating range than that of the petroleum base fluids. Synthetics that are used to formulate hydraulic fluids include PAOs, diesters, silicones, and phosphate esters. Some of these base stocks exhibit a high degree of fire resistance, which is useful, but many are incompatible with packings, paints, and finishes that are used in hydraulic systems. Incompatibility with certain metals may also be a problem, especially with ester-type fluids that in the presence of water will hydrolyze to generate corrosive acids. Phosphate ester fluids and phosphate ester-mineral oil blends are often used as nonaqueous fire-resistant fluids. They have high flash and fire point, but above the recommended use temperature, they thermally decompose to form harmful compounds. Phosphate ester fluids are susceptible to hydrolysis, which is further facilitated by their affinity for water. The rate of water absorption and the rate of hydrolysis are a function of the organic moiety. When the alky group is hindered, both rates are slow, which is highly desirable. One way to assess water affinity of a material is to consider its water solubility. Materials with low water solubility will be the best for use in hydraulic fluids. Table 7.17 provides water solubility data for various phosphate esters [604]. Data suggest that while items 1–8 are better than items 9 and 10, items 2, 3, 5, and 6 are the best. The lubricity characteristics of these fluids are comparable to those of the petroleum-based hydraulic fluids and hence they can be used in most commercial hydraulic pumps. These materials are very aggressive to packings, seals, paints, and pipe joint compounds. However, butyl, silicone, Viton, and PTFE (Teflon®) seals are okay to use with these fluids. These fluids are usually compatible with all metals that are normally used in hydraulic systems and can be filtered through most types of filters. However, they must be handled with care and prolonged exposure and repeated skin contact must be avoided. In general, phosphate esters are considered to have excellent fire resistance and their fire resistance is an inherent chemical property and not a property provided by an outside agent such as water. Sometimes mineral oil blends with esters, such as phosphates esters, are used as hydraulic fluids. This is done to lower the cost and improve the hydrolytic stability of the phosphate esters, and hence decrease their corrosivity.

Another class of fire-resistant fluids that are used in hydraulic applications comprises emulsions. Soluble oils are excellent coolants and extremely fire resistant since they contain 80 to 95 % water, but their use is extremely limited because of the low lubricity and poor additive solubility. The cooling ability of these fluids is due to the presence of water which has high thermal conductivity and specific heat relative to mineral oil. Comparative data in Table 7.18 [154] indicate the advantage of water over all other fluids with respect to heat dissipation. High values of both parameters are better with respect to cooling ability. Water-in-oil emulsions based upon petroleum oil are generally preferred. While these fluids can have any amount of oil in the composition, typically these emulsions contain 55 to 65 % oil and the balance water and additives. Because of the higher oil content, these fluids have good lubricity, and since they contain functional additives, they have good rust and corrosion and antiwear properties as well. Their lubricity is suitable under many conditions and for many pumps. Their fire resistance is due to their high-water content. In addition, these fluids are compatible with most conventional packing and sealing

TABLE 7.19—ISO 6743/4 hydraulic fluid classification (ISO 6743/4).

Hydraulic systems	H
Automatic transmissions	HA
Fire resistant fluids	HF
Oil in water emulsions (soluble oils)	HFAE
Chemical solutions in water	HFAS
Water in oil emulsions (invert emulsions)	HFB
Water polymer solutions (water glycol)	HFC
Synthetic fluids containing no water	HFD
Phosphate ester	HFDR
Chlorinated hydrocarbons	HFDS
Mixtures of HFDR and HFDS fluids	HFDT
Other compositions	HFDU
Hydraulic-slide way	HG
Refined mineral oil	HH
R&O	HL
Antiwear	HM
Couplers & converters	HN
High VI R&O	HR
High VI AW	HV
Synthetic w/o fire resistance	HS

materials, with the exception of the butyl rubber. Leather and porous material may also swell due to water absorption. Water-glycol fluids are solutions of water, glycols, and/or poly(alkylene glycol), or a combination thereof, which contain rust and corrosion inhibitors, lubricity agents, and antiwear additives to improve their performance. These fluids are quite stable, have long service life, and are suitable for use in most high-pressure hydraulic pumps. They are compatible with most types of packing and seals, but not porous materials. They are not very compatible with many paints and pipe compounds, but when properly inhibited, they are compatible with all metals except zinc, cadmium, and nonanodized aluminum. It is important to note that these fluids must not be filtered through diatomaceous (Fuller's earth) filters.

ISO Classification

ISO (ISO 6743/4, 1999) has its own classification system for hydraulic fluids. Hydraulic fluids come under the lubricant class *Lubricants, Industrial Oils and Related Products*, class symbol L, and under hydraulic, the family symbol H. This classification is provided in Table 7.19. For complete ISO 6743 Classification, see Table A.1 in the Appendix. ASTM used this classification as a basis for creating ASTM D6158, Standard Specification for Mineral Hydraulic Oils, which defines the physical properties and the performance requirements of mineral hydraulic fluids. The description of these fluids is provided below [594].

HA fluids are mineral oil- or synthetic-based fluids that are used in automatic transmissions. For a detailed discussion, please consult the section below on automatic transmission fluids.

HFA fluids contain greater than 80 % water. The ISO 6743-4 classification divides HFA fluids into HFAE (oil-in-water emulsions) and HFAS (chemical solutions or blends of selected additives in water) categories. These products are sold as concentrates and diluted with water, prior to use. Because of the high water concentration, the maximum use temperature for the HFA fluids is limited around 50 °C [605].

HFB fluids are water-in-oil emulsions consisting of petroleum oil, emulsifiers, selected additives, and water. These fluids are invert emulsions. The continuous oil phase provides lubricity and rust protection and the encapsulated water phase provides the fire resistance. Water content in these fluids is around 43 to 45 % and when it drops below 38 %, which may be due to evaporation, their fire resistance diminishes and the emulsion destabilizes. ASTM methods, such as D3709, Standard Test Method for Stability of Water-in-Oil Emulsions under Low to Ambient Temperature Cycling Conditions, and D3707, Standard Test Method for Storage Stability of Water-in-Oil Emulsions by the Oven Test Method, are used to determine emulsion stability.

HFC fluids are solutions of water, glycols, additives, and thickening agents. These water-glycol fluids are formulated using diethylene glycol or propylene glycol and a poly(alkylene glycol) based thickening agent [606]. Combining glycols with the polyglycol thickening agent improves lubricating properties of the fluid and reduces the possibility of the cavitation-related erosion. Because of the high water content, the maximum operating temperature of these fluids is around 50 °C, which is no better than the HFA type fluids [607]. These fluids are corrosive to metals, such as zinc, cadmium, magnesium, and nonanodized aluminum, resulting in the formation of gum-like residue. This is due to the presence of the highly alkaline amine based corrosion inhibitors.

HFD fluids are water-free fire resistant fluids. The 1982 version of the ISO 6743-4 standard included HFDS and HFDT categories that were for fluids based upon polychlorinated biphenyls (PCBs) or other chlorinated aromatic compounds. However, the 1999 revision got rid of these classes because these fluids are no longer in the market, due to the adverse health and environmental effects of the chlorinated compounds. The fluids belonging to the remaining classes HFDR (based upon phosphate esters, especially triaryl phosphates) and HFDU [polyol esters and poly(alkylene glycol)s] are used widely in many commercial and military hydraulic applications. As mentioned earlier, phosphate esters burn but do not propagate combustion. Nonflame propagating properties of the hydraulic fluids are demonstrated by ASTM D5306, Standard Test Method for Linear Flame Propagation Rate of Lubricating Oils and Hydraulic Fluids. Aryl phosphate esters also possess anti-wear properties because of which they are used as antiwear additives in mineral oil-based hydraulic fluids [608]. With these materials, hydrolytic stability is an issue which was commented upon earlier. Phosphate esters are compatible with all common metals, except aluminum, but are not compatible with water-based fire resistant fluids and paints and coatings. To formulate HFDU type fluids, trimethylolpropane oleate, neopentyl glycol oleate, and pentaerythritol esters are the most commonly used polyol esters. Triglycerides derived from soybean, sunflower, and rapeseed, which are naturally occurring polyol esters, are also used. Polyol esters derive their fire resistance from high flash, fire, and auto-ignition points and because of the low hydrocarbon content (molecules contains oxygen) relative to the mineral oil. Polyol esters have excellent lubricating properties but can hydrolyze in the presence of water [609]. Their susceptibility to oxidation is also high because of the unsaturation in the fatty acid portion of the ester. However, they enjoy the advantages of higher biodegradability, compatibility with most common metals, except lead, zinc, and cadmium, and with common filter construction materials.

Relative to the mineral oils, triaryl phosphates have superior VT properties, lower vapor pressure, lower compressibility, and higher shear strength. However, they may be harmful to some seal materials, such as acrylonitrile-butadiene rubbers and butyl and ethylene/propylene elastomers, depending upon the temperature. Overall, the fluoro-hydrocarbon elastomers have proven to be the most resistant materials. A compatibility test with the fluid is therefore recommended with each polymer type.

HG type fluids are for slide way use. These are formulated using a variety of base stocks to have good frictional and antiwear performance and metal cutting fluid compatibility. This is because slide ways are often used in metal cutting operations.

HH type fluids are mineral oils without additives. They are suitable for use in air-over-oil hydraulic systems, such as those used in car lifts at automotive service centers, manual hydraulic pumps, jacks, and other low-pressure hydraulic systems. While these fluids are effective in transmitting power, they lack the high temperature performance and have limited lubricity.

HL type fluids are also mineral oil-based, but they contain rust and oxidation inhibitors to extend the life of the hydraulic fluid and protect equipment against water-promoted rust damage. These fluids, also known as R & O oils, are recommended for use in machine tool applications, where system pressures are ≥2000 psi, and for some piston pump applications, such as Denison piston pumps [610], where the zinc dialkyl dithiophosphate containing oils may not be suitable, due to being aggressive to yellow metal (brass and bronze) and silver alloyed components. Typically, rust inhibitors used in these fluids are alkenylsuccinic acid derivatives [226], which are not always compatible with the basic metal sulfonate or phenate rust inhibitors and zinc dialkyl dithiophosphates that are used in many antiwear hydraulic fluids. The metals from these additives react with the alkenylsuccinic acid to form oil insoluble metal succinates, which can cause hydraulic valve sticking and filter plugging problems [611]. HLP fluids are HL type fluids with improved antiwear properties and HLPD fluids are HLP fluids that contain a detergent additive. Please note that HLP and HLPD are not part of the ISO 6743/4 (1999) Hydraulic Fluid Standard.

HM type fluids contain antiwear additives in addition to the rust and oxidation inhibitors used in HL fluids. HL fluids are the most widely used mineral oil-based hydraulic fluids, since they possess the antiwear performance necessary in most high-pressure hydraulic applications and have the thermal stability necessary for their use in piston pumps. Zinc dialkyl dithiophosphate is the most widely used antiwear additive for hydraulic applications. Because of the recent interest in removing the heavy metals from lubricants, due to their impact on the environment, zinc-free (ash-less) technology has been developed to replace zinc dialkyl dithiophosphates. The new technology is based upon the sulfur/phosphorus chemistry.

HN type fluids are used for hydraulic couplers and converters. They are usually light mineral oil-based with film-forming additives, similar to those used in automatic transmission fluids. As a matter of fact, many of the coupler and converter mechanisms use ATFs.

HR type fluids are HL fluids that contain viscosity modifiers, which improve their viscosity, giving them broad-temperature performance capability. Also see description under HV oils below.

HV type fluids contain the same chemistry as HM fluids, except that they contain the additional viscosity index (VI) improver. Viscosity improvers make these fluids multi-grades. These fluids are produced by adding a polymer to a low viscosity fluid. The polymer increases the viscosity of the oil more at high temperatures than at low temperatures. The result is easy flow at low temperatures and the ability to counter heat-related viscosity loss at high temperatures. This improves the fluid's performance at both low and high temperatures [611]. See Chapter 4 on Additives for a more detailed discussion. A wide variety of polymers are used to achieve this.

HS type fluids are petroleum oil-based fluids that have good water stability, low aeration and foaming tendency, suitable low and high temperature viscometrics, good oxidation resistance, good materials compatibility, and excellent protection against rust. Many HLP and HVLP fluids (HV type fluids with improved rust and oxidation inhibition and anti-wear properties), engine oils, and automatic transmission fluids fulfill these requirements.

Tractor Fluids, ATF, and Engine Oils form a miscellaneous group, which besides ATFs (HA) are not directly classified under ISO classification for hydraulic fluids, but they are often used as hydraulic oils. Groups of oils used as tractor hydraulic fluids were discussed in the earlier part of this chapter, ATFs are discussed below, and engine oils were discussed in Chapter 5; hence discussing them here will be redundant.

Four classes of fluids that are of special interest are R&O oils, fire resistant fluids, antiwear fluids, and biodegradable fluids.

R&O Oils

For simple applications, the fluids essentially have no performance requirements and refined mineral oils are often used. R&O oils, on the other hand, have many requirements. Table 7.20 lists the requirements of HL and HLP type hydraulic fluids specified in DIN 51 524, Part 1 [4]. As stated earlier, HL fluids are straight R&O oils and HLP fluids are oils that contain additional antiwear additives. These oils are mineral oil-based that are used in high-pressure hydraulic systems which operate at high temperatures and require extended service life. The service life of these fluids is measured by the standard industry tests, such as ASTM D943 and Cigre and Pneurop oxidation tests (IP 280 and DIN 51 352). Newer hydraulic pumps produce higher temperatures than the older pumps and hence they require additives that will provide oxidation and corrosion protection at elevated temperatures. These fluids contain oxidation inhibitors, rust and corrosion inhibitors, and possibly demulsifiers and foam inhibitors. HL hydraulic fluids are used in systems that demand greater thermal and oxidative stability and protection against corrosion from the fluid, as well as longer service life. HLP fluids are used in applications that in addition demand the antiwear protection. Good demulsifying properties are

TABLE 7.20—Requirements for hydraulic oils according to DIN 51 524, Part 1 [4].

	Requirements						Test
HL Hydraulic Oils	**HL 10**	**HL 22**	**HL 32**	**HL 46**	**HL 68**	**HL 100**	
ISO Viscosity Grade according to DIN 51 519	ISO VG 10	ISO VG 22	ISO VG 32	ISO VG 46	ISO VG 68	ISO VG 100	
1. Viscosity at -20°C max., mm^2/s (cSt) at 0°C max., mm^2/s (cSt) at 40°C, mm^2/s (cSt) at 100°C min., mm^2/s (cSt)	600 90 10 2.4	- 230 22 4.1	- 120 32 5.0	- 180 46 6.1	- 1400 68 7.8	- 2560 100 9.9	DIN 51 550 in relation to DIN 51 561 DIN 51 562 Part 1 or DIN 51 569
2. Pour point (°C), equal to or lower than	-30	-21	-18	-15	-12	-12	DIN 51 597 DIN ISO 3016
3. Flash point (°C) , higher than	125	165	175	185	195	205	DIN 51 376
4. Insolubles (g/100 g)	Below quantitative detection limit						DIN 51 592
5. Water content (g/100 g)	Below quantitative detection limit						DIN ISO 3733
6. Rust-preventing characteristics towards steel, max.	Degree of corrosion 0 - A						DIN 51 585
7. Copper corrosion, max.	Degree of corrosion 2 -100 A 3						DIN 51 759
8. Air release properties at 50°C, min. max.		5			10		DIN 51 381
9. Foaming characteristics ml, max.(Procedure B) S1 S2 S3	Limits in Preparation						DIN 51 566
10. Demulsification ability[2] at 54°C, min. max.	30		40			60	DIN 51 599
11. Aging characteristics[a] Increase in neutralization number after 1000h mg KOH/g			2.0[c]				DIN 51 587
HLP Hydraulic Oils	**HLP 10**	**HLP 22**	**HLP 32**	**HLP 46**	**HLP 68**	**HLP 100**	
Requirements 1 – 10 the same as for HL oils, additional requirements 12 – 14							
12. Aging characteristics[a]	To be stated by supplier						Standard in preparation
13. Mechanical Testing in FZG gear testing rig Seizure load stage min. Weight loss as a function of work. mg/kWh max.	Note[c]			10 0.27			DIN 51 354, Part 2
14. Mechanical testing in vane pump (mg abrasion)[b] Ball race, max. Vance, max	Note[c]			120 30			DIN 51 389, Part2
Sea compatibility of HL and HPL oils determined with SRE-NR1 rubber							DIN 53 538 Part 1

[a]Due to the long test duration these parameters are measured only when testing the oil categories.
[b]Limiting values are not valid for hydraulic oils with detergent characteristics.
[c]Test can be carried out with ISO VG 10 and 22; suitable additive treatments must be used in these viscosity grades, as in ISO VG 32 to 100.

required for both classes since in some applications the fluid is exposed to water due to water leak, condensation, or contamination; for example, by the water-based metalworking fluids. The hydraulic fluid is expected to separate water quickly. HLPD hydraulic fluids are modified HLP fluids that contain detergents in applications that require reserve base, for example, for mobile hydraulic systems in road building machinery. Instead of these oils, one can use engine oils that in most cases provide satisfactory performance. Table 7.21 provides viscometrics and temperature limits of engine oils for use as hydraulic fluids [4]. It is important to note that engine oils are not designed for use in hydraulic systems; hence they may not meet all the performance requirements.

TABLE 7.21—Typical temperature limits for the use of motor oils as hydraulic oils [4].

	Approximate Temperature (°C) for a Max. Start-up Viscosity of		Approximate Temperature (°C) for a Working Viscosity of			Approximate Temperature (°C) for Limiting Viscosity
SAE Grade	2000 mm^2/s	1000 mm^2/s	500 mm^2/s	25 mm^2/s	15 mm^2/s	10 mm^2/s
IOW	−15	−8	0	55	65	80
20	−10	0	8	62	78	95
30	0	8	15	75	90	107
15W–40	−7	2	12	82	96	115
10W–30	−10	0	8	68	87	103

Fire-resistant Fluids

Fire-resistant fluids are produced either by the use of the synthetic base stocks or are water-based. While mineral oil-based fluids have most of the properties necessary in a hydraulic liquid, they are flammable under normal conditions and can cause explosion under the influence of high pressures and high temperatures. Nonflammable synthetic liquids have been developed for use in hydraulic systems where the fire hazards exist. Fire-resistant hydraulic fluids are formulated using materials with lower BTU content than the mineral oils, such as polyol esters, phosphate esters, and water-glycol solutions. Hence, when they burn, less heat is generated. Water-based fluids are largely nonflammable, but synthetic ester-based fluids, such as phosphate esters type, will burn, if sufficient heat and flame are applied; but they do not support combustion. These fluids include soluble oils, invert emulsions, water-glycol mixtures, and those based upon phosphate esters and polyol esters. The major drawbacks of these fluids include their deleterious effect on paints and adhesives, insulations used in electrical cables, and many gasket and seal materials. Industrial uses of these fluids include hydraulic environments that are in close proximity to flame or high temperatures, such as foundries, steel mill operations, furnace hydraulics, aircraft systems, and turbines for generating power. Fire hazard occurs, if a flammable hydraulic fluid leaks on hot surfaces due to damage to the hydraulic systems during operation, or if overheating of the pressurized hydraulic system causes leakage and self-ignition of the fluid. The use of the fire-resistant hydraulic fluids minimizes this hazard. The fire resistance of the fluid is assessed not by a single test but a battery of tests which attempt to simulate different hazard situations. One of the tests even deals with slow leaks on hot pipelines and is also capable of distinguishing between different kinds of hydraulic fluids. These tests are listed below under the section on fire-resistant hydraulic fluids. The ISO classification for these fluids, shown in Table 7.19, is summarized below:

1. HFAE—Oil-in-water emulsions, typically containing $>80\%$ water.
2. HFAS—Chemical solutions in water, typically containing $>80\%$ water.
3. HFB—Water-in-oil emulsions containing approximately 45 % water.
4. HFC—Water-polymer solutions containing approximately 45 % of water.
5. HFDR—Synthetic fluids containing no water and consisting of phosphate esters.
6. HFDU—Synthetic fluids containing no water and of other compositions.

Water-Miscible (Aqueous) Fire-Resistant Hydraulic Fluids

Aqueous fire-resistant hydraulic fluids fall under three categories: HFA, HFB, and HFC. These categories include emulsions and solution type fluids that contain large amounts of water. The majority of the water-based fluids are either the oil-in-water-type, HFAE, or the polymer-thickened aqueous solutions, HFC. Their high water content is responsible for their low flammability. Additional advantages include their lower cost and greater compatibility with the environment, compared to the nonaqueous fluids. These attributes are important if there is leakage or a spill. Because of their high water content, their properties are influenced by the properties of water. Table 7.22 lists the typical properties of the aqueous fire-resistant hydraulic fluids [4]. As shown in the table, the operating range for emulsions is 2 to 60 °C. Below 2 °C, they freeze and above 60 °C the water evaporation rate is too high for their use. The same is true for the HFC fluids, although they have much lower freezing points that permit their use in most low-temperature applications. The use of the aqueous fluids above 60 to 65 °C can result in cavitation-related damage. Type HFA emulsions have very low viscosities, $\sim 1.5\ mm^2/s$ at 20 °C and $\sim 0.5\ mm^2/s$ at 80 °C, and hence they have high leakage rates. This is corrected by making clearances in the equipment narrower than those intended for use with more viscous fluids. This in turn will create a greater need for filtration to remove the dirt particles which can cause jamming and wear of the mobile parts.

All aqueous hydraulic fluids have VT characteristics inferior to those of the mineral oils and so is the compressibility. These fluids reflect poor wear control relative to mineral oil-based fluids, because of the lack of appropriate lubricity. HFC fluids are somewhat better in controlling wear since their viscosity relative to HFA fluids is higher. Aqueous fluids, especially of the HFA type, suffer from corrosion problems and are highly susceptible to bacterial attack. Soluble oils, another class of aqueous fire-resistant fluids, also suffer from the same problems. These problems are corrected by the use of the rust and corrosion inhibitors and biocides and fungicides.

The DIN 24 320 specification demands good emulsion stability, good corrosion protection, low metal attack, and a pH of 6.5 to 8.5. If the emulsion is designed to be biodegradable, the emulsifier(s) and the base fluid used must have a high degree of biodegradability. HFB type fluids, which are water-in-oil emulsions, are used less often. Their need for the rust-inhibiting additives and bactericides is much less. Polymer-thickened HFC fluids are mixtures of glycol, water-soluble glycol ethers, and polyglycols with oxidation inhibitors and rust control agents. Their water content is 35 %, or more, which imparts to them the desired fire resistance. Emulsions require continuous quality control and purification (filtration) to maintain stability.

Nonaqueous Fire-resistant Hydraulic Fluids

These fluids are used in extremely high temperature and high pressure environments and hence they must possess good thermal stability, excellent boundary lubrication properties, low volatility, fire resistance, and moderate hydrolytic stability. These fluids include polysiloxanes (silicones), polychlorinated biphenyls (PCBs) containing 35–50 % chlorine, and phosphoric acid esters. PCBs only see a limited use, and only in military applications, because of the chlorine's role in carcinogenic dioxin formation. Some high boiling and high flash oleic acid esters and polyglycols are also used in these applications, but with limited effectiveness. For large volume use, silicones are not considered because of their high cost and poor lubrication characteristics. These fluids use specialized tests to assess their fire resistance. These tests are listed below, along with some of the specifications that include these tests [4].

TABLE 7.22—Properties of aqueous fire-resistant hydraulic fluids [4].

Fluid Category	HFAE	HFB	HFC
Type	Oil-in-Water Emulsion	Water-in-Oil Emulsion	Aqueous Polymer Solution
Self-Ignition Temperature	Depends on test procedure, but significantly higher than of mineral oils or similar substances, > 500°C on hot surface		
Environmental Protection/ Biodegradability	Good with suitable raw materials	Like mineral oil	Moderate to poor
Lubricating Properties	Moderate	Medium to good	Good
Application Temperature	2 to 55°C	2 to 60°C	-30 to 65°C
Corrosion Protection	satisfactory	Good	Good
Special Seal Requirements	None	None	None
Seal Compatibility	Good	Good	Good
Attack on Metal	Can be negligible		
Heat Transfer	Very good	Good	Good
*p*H	7.5-9	7	7.5-11
Water Content	Minimum 80%	Around 45%	Minimum 35 %
Properties to be monitored	pH Concentration Water Hardness Microorganisms	Water Content Viscosity Emulsion Character	Water Content Viscosity *p*H

Fire Resistance Fluid Tests

1. Determination of the flammability of the fluid, atomized under pressure.
2. Determination of flame propagation in a mixture of coal dust and fluid.
3. Determination of flammability of atomized fluid.
   a. Wick test
   b. B13C flash test
4. Test with hot metal surface (Hot Manifold Test).
5. Test with molten light alloy.
6. Self-ignition temperature.
7. Ignition temperature.

Fire Resistant Fluid Specifications

1. Luxemburg Report (Community of Six), Part III 3.1
2. Luxemburg Report (Community of Six), Part III 3.2
3. Luxemburg Report (Great Britain), Part VI 6.15 NCB 570/1270
4. Luxemburg Report (Great Britain), Part VI 6.16 NCB 570/1270
5. DIN 51 515-2
6. SAE AMS 350
7. BBC Flash Test (BBC ZLC 2-4-3)
8. ASTM D2155, DIN 51 794 Requirements
   a. No flash appears, or flame must not reach screen.
   b. Flame must not exceed arithmetic mean value of 10 cm (propagation).
   c. Atomized jet must not continue to burn for more than 30 s after removal of igniter flame.
   d. Mean duration of burning must not exceed 60 s.
   e. "Self-extinguishing" (duration of burning <5 s), no flash or short flash only at 704 °C.

The oxidation stability of the phosphoric acid tri-esters is satisfactory and is comparable to that of the mineral oils, but their hydrolytic stability is of concern. Aryl esters and short-chain alcohol esters hydrolyze more rapidly than the long-chain esters. Sterically hindered aryl and alkyl groups diminish the rate of hydrolysis. Since the hydrolysis and the oxidation of these materials generate acidic products with the ability to further catalyze hydrolysis, continuous filtration of the fluid through bleaching clay is normally employed to remove these deleterious materials. Table 7.23 contains some of the relevant properties of the polychlorinated biphenyls and a phosphoric acid ester [4] and Table 7.24 compares the various characteristics of all classes of fire-resistant fluids that are commonly used [613]. Phosphoric acid ester type fire-resistant fluids are not compatible with the mineral oil-based fluids, and mixing will cause a drop in their resistance and an increase in their oxidation susceptibility. They also harm some elastomers, such as Buna-N or nitrile, PVC coatings, and paints.

The U.S. Military, especially the U.S. Navy, is one of the principle users of the fire-resistant hydraulic fluids. The standards for various fluids used by the U.S. Navy and their properties are provided in Table 7.25. We previously discussed the hydrolytic stability issues of the phosphate esters and concluded that the phosphates that are hydrolytically stable will be the base fluids of choice. This is borne out by examining the U.S. Military Standards, provided in Table 7.26, where specifications MIL-H-19457B and MIL-H-19457C that employ phosphate esters use the hydrolytically stable aryl esters [604].

Antiwear Hydraulic Fluids

These fluids contain additives that impart them antiwear properties and, as stated earlier, these are the most commonly used hydraulic fluids. Metal dialkyl dithiophosphates, which are often used as oxidation inhibitors, for example in R&O oils, also act as antiwear agents to control wear that results from metal-to-metal contact in hydraulic pumps and the related equipment. Zinc dialkyl dithiophosphates are shown to be good in cases of steel-to-steel contact, but not in the case of copper and bronze. This is because most zinc dialkyl dithiophosphates suffer from inadequate thermal stability, leading to the formation of products that are aggressive to yellow metals.

TABLE 7.23—Properties of nonaqueous fire-resistant hydraulic fluids [4].

Properties	Polychlorinated Biphenyls	Triaryl Ester of Phosphoric Acid
Density at 15 °C, g/mL	1.43	1.135
Flash Point (COC), °C	212	256
Fire Point (COC), °C	343	352
Pour Point, °C	−1	−24
Viscosity at 40 °C, mm^2/s (cSt)	28	38
Viscosity at 100 °C, mm^2/s (cSt)	2.85	5.0
Viscosity Index	Less than −400	19
Neutralization Number, mg KOH/g	1.1	0.1
Self-ignition Temperature, °C	650 (ASTM D286)	600 (ASTM D286) 555 (ASTM D2155)
Spray Ignition Test with Acetylene-oxygen Flame (5. Lux. Rep. Part III)	Pass	Fail
Flame Propagation in Coal Dust/Fluid (5. Lux. Rep. Part III)	Pass	Pass
BBC Flash Test (BBC ZLC 2-4-3)	Pass	Pass
Wick Test (5. Lux. Rep. Part IV, NCB 570/1270)	Pass	Pass
Hydrolytic Stability	Very good (No ester groups present)	Moderate to good
Oxidation Stability	Very good	Good
Lubrication Properties	Poor	Good
Toxicological Properties	Strongly Toxic	Depend upon structure

TABLE 7.24—Hydraulic fluid comparison[a] [613].

Property	Antiwear Hydraulic Fluids	Phosphate Ester	Oil/Synthetic Blend	Polyol Ester	Water-in-Oil Emulsion (Invert)	Oil-in-Water Emulsion (Soluble Oil)	Water-Glycol
ISO Designation	**HM**	**HFDR**	**HFDU**	**HFDU**	**HFB**	**HFAE**	**HFC**
Specific gravity	0.85–0.89	1.15	1.0	0.91–0.96	0.96	1.0	1.1
Water Content, %	0.05	0.05	0.05	0.1	43	80–95	43
Viscosity, cSt at 40 °C	32–68	22–100	64.5	46–68	80–100	1.8	43
Viscosity index	Good	Poor	Fair/good	Fair/good	Good	NA[b]	Excellent
Low-temperature properties	Good	Fair	Fair/good	Fair/good	Fair	Poor	Excellent
Vapor Pressure	Low	Low	Low	Low/medium	Medium	High	High
Operating Temperature Range							
°C	−5/65	−5/65	−5/65	−50/66	5/50	5/50	−20/50
°F	20/150	20/150	20/150	−45/150	40/120	40/120	0/120
Spontaneous Ignition							
°C	355	>590	470	399	443[b]	NA[b]	NA[b]
°F	675	>1100	875	750	830[b]	NA[b]	NA[b]
Fire Resistance	Poor	Excellent	Fair	Poor/Fair	Good	Excellent	Excellent
Heat of Combustion[a]	29.1 kJ/g	19.0 kJ/g	...	21.1 kJ/g	16.3 kJ/g	NA	5.3 kJ/g
Lubricating Quality	Excellent	Excellent	Good	Excellent	Fair/good	Limited	Good
Heat Transfer Properties	Fair	Fair	Fair	Fair	Fair/good	Excellent	Good
Corrosion Protection	Excellent	Excellent	Excellent	Excellent	Good	Fair	Good
Seal Compatibility	Good	Fair	Poor	Fair	Good	Good	Excellent
Metals Attacked	None	None	None	None	None	None	Zn, Cd
Vane Pump Rating[c]	100 %	67 %	100 %	100 %	33 %	Not Available	67 %
Relative Cost[d]	1.0	6.5	5.0	6.0	1.5	0.10	3.5

[a]All comparisons are broad generalizations. Performance and quality of specific fluids will vary depending on make or manufacturer of fluids. Viscosity may vary according to application. Viscosity shown is most common industrial viscosity employed. Consult manufacturer's recommendation before selecting viscosity.

[b]NA=Not applicable/available.

[c]Roberts and Brooks Flammability Data, NFPA T2.13.8-l997, a calculated estimate was used for HFDU.

[d]Cost assumes fluid is diluted with water.

TABLE 7.25—Hydraulic fluids used by the U.S. Navy.

Fluid Type	Fluid Use	Military Specification
Mineral Oil	Hydraulic systems on aircraft	MIL-H-5606
	Gun mechanisms, aiming devices, gyrocompass, and general lubrication at moderate temperatures	BS 4475/1975 grade CSB-68 (OM-65)
	Mooring winches, steering gear, and submarine central hydraulic systems	MIL-L-17672
	Power transmission fluid and naval ordnance equipment	MIL-F-17111
	Hydraulically operated rockets	DEF STAN 91/30 (OM-33)
Synthetic Hydrocarbon	Aircraft and weapons and other systems with synthetic seals	MIL-H-83282
	Aiming mechanisms for guns	MIL-H-46170
Phosphate Ester	Catapult systems and deck elevators on aircraft carriers	MIL-H-19457
Silicone Oil	Brake fluid for combat vehicles	MIL-B-46176
Chlorophenyl Silicone Oil	Lubrication of gyrocompass and rocket systems over wide temperature range	MIL-S-81087
Water-Glycol	Catapult systems on-board system carriers	MIL-H-22072
Glycol Ether and Polyglycol	Brake fluid for administrative vehicles and certain submarine systems	SAE J1703

Zinc dialkyl dithiophosphates (ZnDTP) have been in use since the 1950s. First, they were used as antiwear/antioxidant component of the mono-grade engine oils that were used at that time as hydraulic fluids. Later, the industry discovered that some components of the engine oils, such as dispersants and detergents, had a detrimental effect on pump performance because they prevented the oil from separating water readily (demulsification). Good demulsibility properties were necessary for hydraulic fluids used in industrial applications since many of them are exposed to water, which contributed to the formation of sediment, deposits, and other undesirables. This led to the development of the fluids for use specifically in hydraulics. Zinc dialkyl dithiophosphates were used in these fluids as well. Unlike vane pumps, piston pumps do not need the antiwear protection of the zinc dialkyl dithiophosphates. As a matter of fact, their presence in fluids caused extensive wear of the copper components used in pumps, piston shoes, and cylinder liners. The development of the stabilized ZnDTP chemistry changed all that and almost all hydraulic fluids used today contain ZnDTP and the stabilized ZnDTP additives.

The primary incentive for using the dialkyl dithiophosphate chemistry in hydraulic fluids is its performance, both as an antiwear agent and as an oxidation inhibitor, which ex-

TABLE 7.26—Military standards for hydraulic fluids [604].

Property	MIL-H-5606[a]	MIL-H-5606F[b]	MIL-H-19457B[c]	MIL-H-19457C[c]	MIL-H-83282[b,d]	MIL-H-83306[e]
Fluid Class	Petroleum Base[f]	Petroleum Base[f]	Phosphate Ester[f]	Phosphate Ester[f]	Hydrogenated Polyalphaolefin[f]	Phosphate Ester[f]
Product Description	No data	No data	No data	Butylated triphenyl phosphate mixture	No data	No data
Synonyms	MLO 82-585	No data	No data	MIL-H-19457D	MIL-H-83282C	No data
Major Components	Naphthenic type petroleum oil	Petroleum products with additives	Trixylyl phosphate	Triphenyl phosphate; t-butylphenyl diphenyl phosphate; di-t-butylphenyl phenyl phosphate, contains little or no tri(t-butylphenyl) phosphate	Trimers of polyalphaolefin[e]	Tributyl phosphate and dibutyl phenyl base stock
Operational Temperature Range	No data	-54 to 135°C	No data	No data	-40 to 205°C	No data
NATO Code	No data	H515	No data	No data	H-537	No data
Other	No data	No data	Obsolete since 1981	No data	No data	No data

[a]Kinkead et al. 1985
[b]Department of Defense 1993
[c]FMC 1995
[d]MIL-H-83282 LT is a candidate low-temperature hydraulic fluid
[e]Mattie et al. 1993
[f]Numerous chemicals may be added to the hydraulic fluid base stocks to improve fluid characteristics. Tricresyl or triphenyl phosphate may be added as anti-wear additives (Department of Defense 1993). The amount of ortho isomer of tricresyl phosphate may not exceed 1% of total Tricresyl phosphate Mattie et al. 1993)

TABLE 7.27—Effect of ester group on zinc dialkyl dithiophosphate performance [614].

Property	Aryl	Primary Alkyl	Secondary Alkyl
Thermal Stability	Best	Medium	Worst
Anti-wear Protection	Medium	Medium	Best
Hydrolytic Stability	Worst	Medium	Best
Relative Cost	High	Low	Low

tends the pump and equipment life and that of the oil. Zinc dithiophosphates belong to three general classes, each differing in properties. Table 7.27 compares these classes with respect to thermal stability, antiwear performance, hydraulic stability, and relative cost [614].

Both aryl and alkyl derivatives are employed commercially. As mentioned earlier, zinc dialkyl dithiophosphates are aggressive to yellow metals, but fortunately by selecting the right combination of the alkyl and aryl groups and the presence of ligands bonded to zinc, such as excess zinc oxide or a zinc carboxylate, modify their thermal stability favorably.

Hydraulic fluids containing these additives continue to perform well; hence they are still extensively used in many demanding applications. At present, there is a move towards the use of the zinc-free technology because of a concern for zinc being a heavy metal and its presence in the groundwater being an issue. Since there is a no mandate and zinc-based chemistry has many advantages, including a reasonable cost, the progress to replace zinc is slow. Incidentally, zinc-free fluids have been developed and work quite well, but they're expensive and not yet fully tested [614]. The new metal-free (ash-less) sulfur/phosphorus technology has also been developed which allows the hydraulic fluid to behave well with respect to both the thermal stability and the antiwear properties. In addition, the delivered performance of this technology is superior to that of the dithiophosphate derivatives. Table 7.28 compares the relative effectiveness of the two technologies. Table 7.29 provides international specifications for the antiwear hydraulic fluids [594].

DIN 51 524 classifies HLPD fluids as antiwear hydraulic fluids that contain detergents and dispersants. These fluids, which are approved by most major equipment manufacturers, have the ability to emulsify water and disperse and suspend contaminants, such as varnish and sludge. This keeps surfaces of many of the equipment components free of deposits. The contaminant-dispersing properties are an advantage in mobile hydraulic systems, which have little opportunity for the removal of the contaminants by settling and precipitation, due to the relatively small volume. However, for industrial systems, these contaminants must be filtered out. The major concern regarding the use of these fluids is their water emulsifying ability, which, if present, accelerates the aging of the fluid, reduces its lubricity and filterability, reduces seal life, and leads to corrosion and cavitation. These problems can be overcome by keeping the water level below the saturation point of the fluid at the operating temperature.

A mentioned earlier, the presence of the antiwear additives facilitates protection under boundary conditions and that zinc dialkyl dithiophosphates are the most common antiwear agents used in hydraulic fluids. The presence of these additives can reduce filterability due to the interaction of their thermal decomposition products with other additives, for example, metal detergents. This problem can be alleviated by the use of stable zinc dialkyl dithiophosphates, which contain zinc oxide or zinc carboxylate as the excess base.

Biodegradable Fluids

Many suppliers of lubricants and fluids promote their brand as being environmentally friendly. Other terms that are used in the same context are environmentally acceptable, environmentally responsive, and environmentally compatible. Whatever these terms mean, biodegradability is the fundamental consideration when making this claim. The challenge is defining the extent of biodegradability to call a fluid biodegradable since most fluids have some degree of biodegradability. To add to the confusion, there are a dozen differ-

TABLE 7.28—Typical oxidation and thermal stability performance of various antiwear hydraulic fluids [198]

Testing	Metal-Containing Formulation		Metal-free Formulation	
Total Additive Treatment	0.9%-1.2%	0.9%-1.2%	0.4-0.5%	0.9-1.2%
Antiwear Additive	Normal ZnDTP (0.1% Phosphorus)	Stabilized ZnDTP (0.035% Phosphorus)	Ashless Antiwear Agent	Ashless Antiwear Agent
Thermo-oxidative Stability Test - TOST (ASTM D943)[a]				
Time (hr) to TAN 2.0 mg KOH/g Oil	1500-1800	2500-3400	1800-2800	2800-3800
Sludge (mg) (ASTM D4310)	130-430	45-90	140-320	35-90
Thermal Stability Test – Cincinnati Milacron Test (ASTM D2070)[a]				
Sludge (mg/100 ml)	150-263	5-46	6-40	6-30
Cu appearance (CM Color)	10-12	1-3 (10)[c]	1	1-3
Steel appearance (CM Color)	8	1-4	2	1-4

[a]ASTM D943 Test Conditions – 95°C, H_2O, Fe and Cu catalysts, Air = 3 l/hr.
[b]ASTM D2070 Test Conditions – Procedure A, 168 hr at 135°C.
[c]One fluid gave a rating of 10 (black) in four different base stocks.

TABLE 7.29—International specifications for antiwear hydraulic fluids [594].

Properties	ASTM Method	ISO 11158 HM (Anti-wear)				DIN 51524 Part 2 Requirements				ASTM D 6158 HM (Anti-wear)			
ISO Viscosity Grade		*22*	*32*	*46*	*68*	*22*	*32*	*46*	*68*	*22*	*32*	*46*	*68*
Kinematic Viscosity, cSt	D445												
40°C max.		24.2	35.2	50.6	74.8	24.2	35.2	50.6	74.8	24.2	35.2	50.6	74.8
40°C min.		19.8	28.8	41.4	61.2	19.8	28.8	41.4	61.2	19.8	28.8	41.4	61.2
Flash Point. "C min.	D92	140	160	180	180	165	175	185	195	165	175	185	195
Flash Point. "C min.	D93	128	148	168	168								
Brookfield Vis < 750 cp, max.	D2983									-15	-8	-2	4
Pour Point, °C, max.	D97	-18	-15	-12	-12	-21	-18	-15	-12	-21	-18	-15	-12
Appearance	Visual	< Report >								< C & B >			
Water Content, wt%	D1744	< 0.05 >				< Report >							
Density @ 15°C	D1298	< Report >				< Report >				< Report >			
TAN, mg KOH/g, max.	D664 or D974	< Report >				< Report >				< Report >			
Rust Test A Rust Test B (Salt Water)	D665 A D665 B	< Pass >				< Pass >				< Pass > < Pass >			
Cu Rating (3 hr, 100"C), max.	D130	2	2	2	2	2	2	2	2	2	2	2	2
TOST Oxidation, hours	D943									1000	1000	1000	1000
AN after 1000 h, max.		2	2	2	2	2	2	2	2				
Air Release @ 50°C, minutes, max.	D3427	5	5	10	13	5	5	10	10	5	5	10	13
Foam tendency/stability	D892												
Seq I max.		< 15/00 >				< 15/00 >				< 15/00 >			
Seq II max.		< 75/0 >				< 75/0 >				< 75/0 >			
Seq III max.		< 150/0 >				< 150/0 >				< 150/0 >			
Demulsibility @ 54"C Minutes to 37 mL water	D1401	30	30	30	30	40	40	40	60	30	30	30	30
FZG Fail Stage	D5182	10	10	10	10	-	10	10	10				
Change in Hardness	D471	168 hrs @ 100°C				168 hrs @ 100°C				288 hrs @ 100°C			
NBR1	ISO 7619	< Report >				0 to -8	0 to -7	0 to -7	0 to -6	0 to -8	0 to -7	0 to -7	0 to -6
Change in Volume (%) NBR1		< Report >				0 to 15	0 to 12	0 to 12	0 to 10	0 to 15	0 to 12	0 to 12	0 to 10
Viscosity Index, min.	D2270	< Report >								< 90>			
CM Thermal Stability	D2070A												
Sludge. mg/100 ml max										Report	25	25	25
Copper rod appearance, rating (max.)										Report	5	5	5
Oxidation (1000 h)	D4310												
AN. mg KOH/g, max.		2	2	2	2								
Total sludge, mg max.										200	200	200	200
Total Metals in oil/water/sludge		< Report >								< Report >			
Vickers 104C mg. wear, max.	D2882	< Report >				< 150 in extended test				50	50	50	50

ent tests that are used worldwide to determine biodegradability and they all provide different estimates. There is a move by the United States and the European standard developing organizations to distinguish between environmentally friendly and biodegradability. Biodegradability is easier to define than environmentally friendly since physical tests to determine biodegradability exist. One interpretation of environmentally friendly is that the technology is nontoxic, which means to life, and hence is biodegradable. The question is which organisms should be used in the fundamental

TABLE 7.30—Classification of base fluids based on composition [115, 616].

Base Fluids	Composition	Description
Bio-Based Natural Esters (HETG[a])	Triglycerides	Rapeseed and sunflower seed oils
Bio-Based Synthetic Esters (HEES[a])	Either acid or alcohol, or both bio-based (animal or vegetable)	At least one component should be bio-based; e.g., oleic acid derivatives of TMP, NPG or PE
Mineral Oil-Based Synthetic Esters (HEES[a])	Both acid and alcohol from petrochemicals	For example, derivatives of adipic acid and 2-ethylhexanol
PAOs (Polyalphaolefins) (HEPR[a])	Based on 1-decene (purely petrochemical)	For example, so-called fully synthetic oils resulting from polymerization and hydrogenation
Mineral Oils (HV[a])	Widely varying fractions of mineral oils	Pure petrochemicals
Polyglycols (HEPG[a])	Esters derived from polyethylene glycols	Not mixable with other oils

[a]Terminology of ISO 15380; VDMA 24 568

tests to clearly arrive at the conclusion. Another interpretation of the term is that the technology has no adverse effects on the environment. This involves assessing the impact on the environment, isolating various components to determine the cause and effect relationship, and then eliminating the harmful or potentially harmful component/s from the formulation. The term environmentally acceptable implies that the formulation does not contain chemicals that are known to be harmful to life and the environment and may be more appropriate to use. Either way, it is important to question the manufacturer about the label used and what tests were carried out to arrive at the conclusion.

Environmentally acceptable hydraulic fluids are used in hydraulic applications where there is risk of fluid leaks and spills entering the environment, especially waterways, and affecting aquatic and terrestrial life. Examples of such applications include forestry, construction, locks and dams, heavy-duty lawn equipment, amusement parks/entertainment industry, off-shore drilling, and maritime. These fluids must be nontoxic to aquatic life and possess aerobic biodegradability. Organizations such as the Organization for Economic Co-operation and Development (OECD), the Coordinating European Council (CEC), and the U.S. Environmental Protection Agency (EPA) have developed standard test methods to determine the toxicity and the biodegradability of substances. ASTM and ISO have also developed Guide for Assessing Biodegradability of Hydraulic Fluids (ASTM D6006), Classification of Hydraulic Fluids for Environmental Impact (ASTM D6046), and ISO environmental hydraulic fluid classifications based upon the above organizations' methods. ISO Environmental Hydraulic Fluid Classifications are shown in Table 7.30 [115,616] and are described in the subsequent paragraphs. ISO, in collaboration with regional environmental organizations, also awards Eco Labels [168,169], such as German Blue Angel, Nordic Swan, and Japanese EcoMark.

The biodegradability estimates of various base fluids are provided in Table 7.31. Three of the tests that are used to determine biodegradability are CEC-L-33-T-82; EPA 560/6-82-003, "Shake Flask" Test; and OECD 301B, "Modified Sturm" Test. CEC-L-33-T-82 Test is 21-day test, which is used for oil-based substances. The test involves adding natural salts and activated sewage sludge sample (bacteria) to the test sample and water. The sample is considered biodegradable if 70 % of the oil is consumed in 21 days. EPA 560/6-82-003 Test is a 28-day test that involves shaking a mixture of the sample, water, sewage sample, and oil. The sample is considered biodegradable if in 28 days 60 % of the sample is converted into carbon dioxide. OECD 301B Test, also a 28-day test, is used for water-insoluble oil samples. The sample is considered biodegradable if over the period of 28 days 60 % of the sample is converted into carbon dioxide. Of the fluids listed in the table, vegetable oils and synthetic fluids have a high degree of biodegradability, but others are not as good. Incidentally, PAGs [poly(alkylene glycol)s], that are derived from ethylene oxide have higher degree of biodegradability than those derived from propylene oxide. Biodegradability of the PAGs derived from ethylene oxide/propylene oxide mixtures varies, depending upon their ethylene oxide content. This suggests that if one is seeking a hydraulic fluid with high biodegradability, one must use vegetable oil or a carboxylate ester as the base fluid.

While selecting a biodegradable lubricant one must also consider other performance criteria, such as oxidation stability, antiwear protection, hydrolytic stability, viscosity-temperature properties and the cost. Biodegradable lubricants use biodegradable base stocks, such as vegetable oils and synthetics, and biodegradable additives, where possible. While the base fluid makes up the major proportion of the

TABLE 7.31—Biodegradability of various base fluids according to CEC-L-22-T-82 biodegradability test.

Fluid	% Biodegradability
Mineral Oil	15–35
White OIl	25–45
Polyalphaolefin (PAO)	5–30
Polyisobutylene (PIB)	0–25
Polyethers	0–25
Phthalate and Trimellitate Esters	5–80
Polyol and Diesters	55–100
Natural and Vegetable Oil	70–100

fluid, the nonbiodegradable additives may significantly harm its biodegradability.

HETG type fluids are based on naturally occurring vegetable oils or triglyceride esters. Without the addition of a thickener, vegetable oils are limited to a narrow viscosity range between that of ISO 32 and 46. While HETG fluids biodegrade rapidly, have excellent natural lubricity, and have a natural VI in excess of 200, they are unsuitable for use at high and low temperature extremes. This is because they tend to gel at low temperatures and oxidize quickly at high temperatures. The practical temperature limits for using HETG fluids is -25 °F to 165 °F (-32 °C to 74 °C) [594].

HEES type fluids are based on unsaturated to fully saturated synthetic esters. Common ester chemistries utilized in hydraulic fluids consist of TMP oleates, neopentyl glycol and pentaerythritol ester-based HEES fluids adipate esters, and complex esters. The synthetic esters-based HEES fluids have better performance than the HETG type hydraulic fluids, with respect to broader operating temperature ranges, greater number of ISO viscosity grades, and better oxidation stability while still maintaining biodegradability [594].

HEPG type fluids are poly(ethylene glycol)s (PEGs), which possess good oxidation stability and low temperature flow characteristics. At molecular weights of up to 600–800 g/mol, HEPG type fluids are ecotoxicologically harmless and readily biodegradable (>90 % in 21 days) [615,616]. Some disadvantages of this class of fluids include miscibility with water, incompatibility with mineral oils, and aggressiveness toward some types of elastomer seal materials.

HEPR type fluids are polyalphaolefins (PAO), or synthesized hydrocarbon (SHC) based fluids, which have significantly better viscometric properties over a broader range of temperatures than mineral base fluids with the same standard viscosity classification. Some low viscosity PAOs have shown acceptable primary biodegradability, though not as rapid as vegetable or synthetic ester base fluids. Additional advantages claimed for synthetic lubricants over comparable petroleum-based fluids include improved thermal and oxidative stability, superior volatility characteristics, and preferred frictional properties [594].

The major worldwide performance specifications for biodegradable fluids are listed below:

1. ISO/CD15380
2. German VDMA 24568, 24569, 24570
3. Swedish Standard 155434
4. German RAL-UZ 79 Blue Angel
5. Nordic White Swan 0002/3.1
6. Austrian Onorm C 2027
7. Netherlands Vamil
8. Caterpillar BF-1
9. MIL-PRF-32073
10. ASTM D6046
11. Canadian ECP-05-94
12. Komatsu BO
13. Rexroth RE 90221
14. Vickers Guidelines

Vegetable oils that are commonly used for formulating hydraulic fluids, or other biodegradable lubricants, include corn oil, soybean oil, rapeseed (canola) oil, sunflower seed oil, peanut oil, olive oil, and others. In their natural form, these oils have performance deficiencies, such as poor thermal, hydrolytic, and oxidation stability, and high pour point. They do have the advantages of good lubrication ability, good surfactancy, and high viscosity indices. Some of the new natural oil sources use genetically engineered strains to provide oils with superior thermo-oxidative properties. Mixing vegetable oils with synthetic diesters help maintain their biodegradability as well as improve their low-temperature viscometrics and oxidative stability.

Formulating a Hydraulic Fluid

For a hydraulic fluid to function properly, it must transfer fluid power throughout the hydraulic system, act as heat transfer medium to remove heat from the system, lubricate all moving parts, provide internal seals, remain stable in most environments, and be compatible with all components of the system. While hydraulic fluids have changed little over the years, there are changes on the horizon, because the operating environment of these fluids is changing due to a drive towards more compact systems. This translates into higher pressures, smaller lines, smaller reservoirs, and higher temperatures due to less cooling of the fluid. While the system requirements depend upon design and the application involved, the most desirable hydraulic fluid properties are listed below:

1. Good flow properties over a broad temperature range (proper low and high temperature viscometrics, high viscosity index, low pour point, shear stability, etc.).
2. Good film-forming and friction-reducing ability.
3. Ability to inhibit corrosion.
4. Good mechanical, thermal, and chemical stability.
5. High bulk modulus, or low compressibility.
6. Low foaming tendency and air entrainment properties, i.e., rapid deaeration and demulsification.
7. Low specific gravity and low vapor pressure.
8. Low coefficient of thermal expansion.
9. High heat-transfer rate (high heat capacity and thermal conductivity).
10. Adequate fire resistance.
11. Material compatibility with system materials, such as paint, metals, plastics, and elastomers.
12. Biodegradability, environmental compatibility, and nontoxicity, both when new and after use and upon decomposition.
13. Low cost and easy availability.

A hydraulic fluid with suitable properties will help fulfill the equipment manufacturers' requirements, which are summarized in Tables 7.9 and 7.10. Many of the properties in a hydraulic fluid, for example 1, 2, 5, 7, 8, and 9 in the above list, can be achieved by the use of the appropriate base fluid/s. Achieving other properties in most cases requires the use of the performance-enhancing chemicals, or additives. A list of additives that are commonly used along with their function and the mechanism by which they perform is presented in Table 7.2. Different types of fluids contain different types of additives.

Modern hydraulic fluids last longer, provide greater protection, and perform better than their predecessors. As mentioned earlier, the need for durable fluids of superior quality arose because of the modern hydraulic equipment becoming smaller and more efficient with respect to higher speeds,

TABLE 7.32—Typical temperature limits for ISO VG grades for hydraulic oils [4].

ISO Viscosity Grade[c]	Viscosity Index	Approximate Temperature (°C) for a Maximum Start-up Viscosity of[b]			Approximate Temperature (°C) for a Working Viscosity of		Approximate Temperature (°C) for Limiting Viscosity[b]
		2000 mm^2/s	1000 mm^2/s	500 mm^2/s	25 mm^2/s	15 mm^2/s	10 mm^2/s
10	50	−39.0 ±2.0	−33.5 ±2.0	−27.0 ±2.0	16.5 ±3.0	28.5 ±3.0	40.0 ±3.0
	100	−44.0 ±2.5	−38.0 ±2.5	−31.5 ±2.5	14.5 ±3.0	27.5 ±3.0	40.0 ±3.0
	150	−50.0 ±2.5	−44.0 ±2.5	−37.0 ±2.5	12.5 ±3.0	26.5 ±3.0	40.0 ±3.5
15	50	−30.5 ±2.0	−26.5 ±2.0	−16.0 ±2.0	27.0 ±2.5	40.0 ±2.5	52.0 ±3.0
	100	−34.5 ±2.0	−28.5 ±2.0	−21.5 ±2.0	26.5 ±3.0	40.0 ±3.0	52.5 ±3.0
	150	−41.0 ±2.0	−34.5 ±2.0	−27.0 ±2.5	25.0 ±3.0	40.0 ±3.0	53.0 ±3.5
22	50	−23.0 ±2.0	−17.0 ±2.0	−10.0 ±2.0	36.5 ±2.5	52.0 ±3.0	62.0 ±3.0
	100	−28.5 ±2.0	−20.0 ±2.0	−13.0 ±2.0	36.0 ±3.0	53.0 ±3.5	64.0 ±3.5
	150	−35.0 ±2.0	−26.0 ±2.0	−18.0 ±2.0	35.5 ±3.5	54.5 ±4.0	66.0 ±4.5
32	50	−15.0 ±2.0	−9.0 ±2.0	−2.0 ±2.0	45.5 ±2.0	59.0 ±2.0	71.0 ±2.5
	100	−19.5 ±2.0	−13.0 ±2.0	−7.0 ±2.0	46.0 ±2.5	60.5 ±2.5	74.0 ±2.5
	150	−25.5 ±1.5	−18.5 ±1.5	−10.0 ±2.0	47.0 ±3.0	63.0 ±3.0	78.0 ±3.5
46	50	−9.0 ±1.5	−2.5 ±1.5	+5.0 ±1.5	53.0 ±2.0	66.5 ±2.0	79.5 ±2.0
	100	−13.5 ±1.5	−6.5 ±1.5	−1.5 ±2.0	54.5 ±2.5	69.5 ±2.5	83.5 ±2.5
	150	−20.0 ±1.5	−12.0 ±1.5	−3.5 ±2.0	56.5 ±2.5	74.0 ±3.0	90.0 ±3.5
68	50	−2.5 ±1.5	4.0 ±1.5	11.0 ±1.5	61.0 ±2.0	75.0 ±2.0	88.0 ±2.0
	100	−7.5 ±1.5	0 ±1.5	8.0 ±1.5	64.0 ±2.5	79.5 ±2.5	94.5 ±2.5
	150	−14.0 ±1.5	−6.0 ±1.5	3.5 ±1.5	67.5 ±3.0	86.0 ±3.0	103.0 ±3.5
100	50	3.0 ±1.5	9.5 ±1.3	17.5 ±1.5	67.0 ±2.0	83.0 ±2.0	96.5 ±2.0
	100	−2.0 ±1.5	6.0 ±1.5	14.5 ±1.5	70.5 ±2.5	89.5 ±2.5	105.0 ±3.0
	150	−8.0 ±1.5	0.5 ±1.5	10.0 ±2.0	76.0 ±3.0	98.5 ±3.0	107.0 ±4.0

[a]For oils with Newtonian characteristics at low temperatures and sufficiently low cloud and pour points; temperature tolerances refer to upper and lower ISO viscosity range.
[b]Wear-free operation can generally be expected above 10 mm^2/s
[c]Includes ISO viscosity tolerance of ±10 %.

higher operating pressures, and smaller sump reservoirs. The increase in operating temperatures causes fluids to degrade rapidly. This requires the use of longer lasting antiwear additives, oxidation inhibitors, rust and corrosion inhibitors, foam inhibitors, demulsibility additives, and other chemical components. Conventional additives lack the ability to retain performance over the duration of the intended service. This is because the established hydraulic fluid specifications are not enough to ensure adequate protection of the equipment during use, since they define the lubricant's *minimum* performance quality. That is why the industry generally requires tests that are more consistent with the modern equipment's performance needs. Some fluids suppliers use modified industry tests and proprietary tests to assess the quality of their fluids. These tests deal with the fluid's thermo-oxidative properties, foaming, and air entrainment tendency [592].

One way is to design fluids whose viscosity matches the viscosity needs of the hydraulic components. This is critical because the viscosity changes with the operating temperature and the fluid must provide proper lubrication over the operating temperature range of the equipment. Therefore, it must meet both the minimum and the maximum temperature needs of the hydraulic system components. Otherwise, wear will ensue. The other way is to test the durability of the fluid by extending the duration of the current standard pump tests and or run the tests at higher temperatures [614].

The viscosity requirements of a hydraulic fluid depend upon a number of factors. For conventional systems, these include the working temperatures and the type of pump being used. The working temperature of a stationary hydraulic system for low or moderate pressures is usually 40 to 50 °C above the ambient temperature. At this temperature, the hydraulic fluid must have a working viscosity of 13–16 mm^2/s (13–16 cSt), which must not drop below 10 mm^2/s, so as to avoid wear. However, in high-pressure systems (above 400 bars), the working temperature is higher by about 10 °C than that in the low pressure-systems; hence the working viscosity of about 25 mm^2/s is recommended. Since the required viscosity range is also a function of the pump used; the gear pumps usually require a working viscosity of at least 20 mm^2/s. It may even be necessary to install a cooling unit to minimize the heat-related viscosity loss and maintain the viscosity in the optimal range. Consideration of maximum viscosity is also important since the start-up temperature after prolonged inactivity is quite low and the fluid will have much higher viscosity than at higher operating temperatures. Maximum intake (start-up) viscosity guidelines are as follows:

1. Gear pumps—2000 mm^2/s (2000 cSt)
2. Piston pumps—1000 mm^2/s (1000 cSt)
3. Vane pumps—500–700 mm^2/s (500–700 cSt)

The start-up and the working viscosities of the hydraulic fluids belonging to different ISO viscosity grades and having different viscosity-temperature (VI) characteristics are shown in Table 7.32 [4]. Commercial mineral oil fluids have VIs of about 100. Hence, the working viscosity between 10 and 25 cSt in the 20 to 80 °C temperature range and maximum start-up viscosity of between 500 to 2000 cSt are of interest. Considering the data presented in Table 7.32, ISO VG 15 has the limiting working viscosity of 10 cSt around 54 °C and working viscosity of 25 cSt 28 °C; ISO VG 22 has these between 38 and 66 °C, ISO VG 32 has these between 48 and 75 °C, and ISO VG 46 has these between 56 and 85 °C. Of

TABLE 7.33—Recommended viscosity requirements for industrial hydraulic oil selection [613].

Unit Type	Viscosity		Minimum Starting Temperature		Maximum Running Temperature		Oil Type
	SSU at 100°F	cSt at 40°C	°C	°F	°C	°F	
In-line piston pumps and motors	150	32	6	43	62	144	Antiwear hydraulic oils or automotive engine oils
	215	46	12	54	71	160	
	315	68	19	66	80	176	
Angle piston, vane, and gear motors	150	32	-12	10	62	144	Antiwear hydraulic oils or automotive engine oils
	215	46	-6	21	71	160	
	315	68	0	32	80	176	
High- torque low-speed vane motors	150	32	14	57	62	144	Antiwear hydraulic oils or automotive engine oils
	215	46	22	72	71	160	
	315	68	29	84	80	176	
Hydraulic adjustable speed drives	315	68	Note 1	Note 1	Note 1	Note 1	Antiwear hydraulic oils or automotive engine oils

Note 1: Consult Manufacturers Recommendations.

course, none of these grades exceed the maximum viscosity requirements, except at very low temperatures. Hence, for most applications, ISO VG 32 and ISO VG 46 work well. However, for very low-temperature operations, the lower viscosity grades are better, and for very high temperature applications, higher viscosity grades are better. Since the higher viscosity indices expand the operating temperature range, the use of the very high VI oils and or the viscosity modified oils can be beneficial in some applications.

It is important to recognize that under high pressures, the oils decrease their volume, hence increase their viscosity; the viscosity increase being greater for naphthenic oils than for paraffinic oils. However, the pressure must be greater than 200 bars, below which its effect on viscosity is minimal. This implies that in high pressure applications, naphthenic oils will offset the temperature-related viscosity loss by gaining viscosity because of the high pressures. In addition, the naphthenic oils have good low temperature properties (low pour points), which permit their use in applications where the paraffinic oils that have higher pour point cannot be used.

Tables 7.33 and 7.34 provide minimum permissible needs and minimum optimal needs of the various hydraulic components [613,614]. The selection of correct fluid viscosity requires the knowledge of the equipment's lowest operating temperature and the highest operating temperature. If the operation is in winter-like conditions, the use of a multigrade oil is appropriate. This will permit flow at freezing temperatures and ensure maintenance of suitable viscosity at high temperatures, to provide proper lubrication. Optimal viscosity range for the all-temperature operation is between 25 and 36 cSt. While the use of the multi-grades may reduce the power consumption, the presence of the viscosity improvers, used to make the multi-grades, can hinder air separation from the fluid. In addition, the viscosity-improving polymers can lose their efficiency over time due to the high shear rates and turbulent flow conditions that are often present in hydraulic systems. A guideline for selecting a high VI or multi-grade fluid is to increase the hydraulic component manufacturers' minimum permissible viscosity values by 30 %, which is to compensate for the viscosity improver's shear-related viscosity loss. If the hydraulic system has a narrow operating temperature range, choosing the optimum fluid viscosity based upon mono-grade oil will work.

TABLE 7.34—Typical minimum viscosity values for hydraulic components [614].

Component	Minimum Permissible Viscosity (cSt)	Minimum Optimum Viscosity (cSt)
Vane	25	25
External Gear	10	25
Internal Gear	20	25
Radial Piston	18	30
Axial Piston	10	16

Formulation Examples

Tractor Hydraulic Fluid: 4.0–6.0 % Basic metal sulfonate detergent, 0.5–1.5 % zinc dialkyl dithiophosphate antiwear agent, 0.5–1.5 % hindered phenol oxidation inhibitor, 0.3–0.5 % benzotriazole metal deactivator, 0.3–0.5 % long-chain organic acid derivative friction modifier, 0.1–0.5 % polymethacrylate pour point depressant, 200–250 ppm methyl silicone foam inhibitor, and 0–4.0 % olefin copolymer viscosity modifier. The balance is the mineral base oil.

Mineral Oil Hydraulic Fluid (Ash-producing): 0.3–0.8 % Primary alkyl and aryl zinc dialkyl dithiophosphate antiwear agent, 0.25 % hindered phenol oxidation inhibitor, 0.2–0.5 % calcium sulfonate corrosion inhibitor, 0.02–0.03 % benzotriazole metal deactivator, and 50–100 ppm methyl silicone or 500–1000 ppm polyacrylate foam inhibitor. The balance is mineral cabe oil.

Mineral Oil Hydraulic Fluid (Ash-producing): 0.3–0.8 % Stabilized zinc dialkyl dithiophosphate antiwear agent, 0.01–0.12 % basic metal sulfonate detergent, 0.01–0.1 % neutral calcium sulfonate demulsifier, 0.1 to 0.25 % hindered phenol oxidation inhibitor, 0.2–0.5 % calcium sulfonate corrosion inhibitor, 0.02 to 0.08 % alkenylsuccinimide dispersant, 0.02–0.08 % benzotriazole metal deactivator, and 50–100 ppm methyl silicone or 500–2000 ppm polyacrylate foam inhibitor. The balance is mineral base oil.

Mineral Oil Hydraulic Fluid (Ash-free): 0.5–0.8 % Sulfur-phosphorus antiwear agents, 0.01–0.1 % polyethoxylated phenol demulsifier, 0.1 to 0.25 % hindered phenol oxidation inhibitor, 0.1–0.2 % alkylated diphenylamine oxidation inhibitor, 0.2–0.5 % fatty amine corrosion inhibitor, 0.02 to 0.08 % alkenylsuccinimide dispersant, 0.02–0.08 % ben-

TABLE 7.35—Hydraulic fluid monitoring requirements [613].

Antiwear Petroleum Oil	Water-Glycol	Phosphate Ester	Oil/synthetic Blend	Water-in-Oil Emulsion (Invert)	Oil-in-water Emulsion (Soluble Oil)
Viscosity	Viscosity	Viscosity	Viscosity	Viscosity	*p*H
Neutralization number	Water content	Neutralization number	Neutralization number	Water content	Emulsion stability
Moisture	Fungi	Moisture	Moisture	Emulsion stability	Oil content
		Specific gravity	Specific gravity	Fungi	Fungi

zotriazole metal deactivator, and 50–100 ppm methyl silicone or 500–2000 ppm polyacrylate foam inhibitor.

Mineral Oil Hydraulic Fluid (Ash-free R&O plus Antiwear Performance): 0.3–0.5 % Trialkyl or triaryl thiophosphate (phosphorothionate), 0.2–0.4 % trialkyl mixed ester of dithiophosphoric acid, 0.02–0.05 % ethoxylated fatty amine salt of dialkyl dithiophosphoric acid, 0.1 % Arylamine antioxidant 0.5 % phenolic antioxidant, 0.1 % benzotriazole metal deactivator, 0.05 % DMTD type of metal deactivator, and 0.05–0.1 % imidazoline or alkenylsuccinate type rust inhibitor. The balance is water (formulation extracted from Ref [617]).

Oil-in-Water Type Water-based Hydraulic Fluid: 1.17 % Polyisobutenyl succinic ester amide dispersing agent, 0.9 % phosphate-modified diethanolamine stearic/oleic acid amide, 0.38 % zinc dialkyl dithiophosphate or metal dialkyldithiocarbamate EP agent, 0.5 % ether glycol emulsifier, 4.8–12.5 % polyglycol thickener, 0.5 % diethylethanolamine rust inhibitor, 0.06 % benzotriazole metal deactivator and silicone foam inhibitor (formulation extracted from Ref [618]).

Soluble-oil Hydraulic Fluid: 0.15 % Sulfurized fat antiwear agent, 0.25 % alkanolamines corrosion inhibitor, 0.1 % triazole metal deactivator, 0.02 % methyl silicone foam inhibitor, 0.05 % triazine biocide, and 0.25 % petroleum sulfonate emulsifier. The balance is water.

Condition-Monitoring of Hydraulic Fluids

All hydraulic fluids must be condition-monitored to ascertain that there is no deterioration in properties. Otherwise, the fluid will lose its effectiveness and the equipment damage or failure will occur. Table 7.35 lists the fluid parameters for various types of hydraulic fluids that must be monitored [613]. Condition monitoring of the phosphate ester-derived fluids is especially important since they are extremely expensive and the loss of their structural integrity will result in a loss of many of their desirable properties. Hydrolysis and thermal degradation of the phosphate esters, two of the common problems with these fluids, result in acidic products that catalyze further fluid degradation. With proper maintenance, they can remain in service for 15 years, or more.

Condition monitoring of the phosphate ester fluids is carried out by the use of tests that are similar to those recommended for the mineral oil fluid condition monitoring. The following parameters are typically examined [619]:

1. Color
2. Appearance
3. Viscosity
4. Acidity/acid number/neutralization number
5. Chlorine content
6. Water content
7. Particle count
8. Mineral oil content
9. Electrical resistivity
10. Elemental spectroscopy
11. Foaming
12. Air release properties

If a parameter is trending out of specification, it is prudent to take corrective action. Stabilizing the acid number (Total Acid Number, TAN), with its hydrolysis catalyzing effect, is the most critical requirement in the ester fluids. Most turbine manufacturers allow a maximum TAN limit of 0.20 mg KOH/g in the phosphate ester fluid. Obviously, the hydrolysis of the ester fluids can also be controlled by controlling the amount of water, which is of course a challenge. Controlling the water content is not only important to minimize the fluid degradation but also to control rusting and corrosion of the metal components of the hydraulic system. Hence, it is important to maintain the moisture levels as low as is reasonably achievable. Some fluids are more tolerant of water than others. Moisture control devices for industrial hydraulic systems are also available.

Particles, if present in a fluid, not only clog small orifices (which impairs flow) but they can also cause abrasive wear. Most turbine OEMs recommend fluid ISO 4406 cleanliness levels of 18/16/13 or NAS 1638 cleanliness level of 7. Particles in the hydraulic systems originate from the usual sources of machining debris, common in the newly machined equipment, dirt and dust from the environment, and the wear metals. In antiwear hydraulic fluids, the particles also arise from the hydrolysis-related acid corrosion and the interaction of the hydrolysis acids with the detergent additives, if present. In the present context, an increase in the particle count is indicative of the fluid break down. However, it is important to develop effective sampling, collection, and analysis procedures that provide meaningful data with as little variability as possible. Ways to remove particles include filtration and magnetic devices. Obviously, these devices will need periodic cleaning or filter replacement to revive them.

Transmission Fluids

Transmission fluids are mineral oil or synthetic-based hydraulic fluids with special properties. The key functions of these fluids are lubrication and wear control, cooling, and to act as a hydraulic medium to transmit power. These fluids are of three types: Automatic transmission fluids, manual transmission fluids, and power transmission fluids. There is no official API classification system for these fluids. Performance requirements for transmission fluids are primarily established by the OEMs [620]. The most important features of these fluids are their frictional consistency (durability)

TABLE 7.36—Automatic transmission fluid specifications.

Fluid Type	Remarks
Ford Specifications	
Type F	Introduced by Ford in 1967 for their automatics. Also used by Toyota.
Type CJ	Special Ford fluid for C6 transmissions. Similar to DEXRON®-II. Must not be used in automatics that require Type F.
Type H	Another limited Ford spec that differs from both DEXRON® and Type F. Can be replaced with MERCON®.
MERCON®	Ford fluid introduced in 1987, very similar to DEXRON®-II. Okay for all earlier Fords except those that require Type F. As of July 1, 2007 Ford has discontinued production and licensing of MERCON® fluid and the transmissions using this fluid will be serviced by MERCON®-V.
MERCON®-V	Ford's newest type, introduced in 1997 for Ranger, Explorer V6 and Aerostar, and 1998 & up Windstar, Taurus/Sable and Continental. Must not be used in 1997 or earlier Fords.
GM Specifications	
DEXRON®	General Motors original ATF for automatics.
DEXRON®-II	Improved GM formula with better viscosity control and additional oxidation inhibitors. Can be used in place of DEXRON®.
DEXRON®-IIE	GM fluid for electronic transmissions.
DEXRON®-III	Replaces DEXRON®-IIE and adds improved oxidation and corrosion control in GM electronic automatics.
DEXRON®-III (H)	Improved version of DEXRON®-III released in 2003.
DEXRON®-III/SATURN	A special fluid spec for Saturns.
DEXRON®-VI	For 2006 GM Hydra-Matic 6L80 6-speed rear-wheel drive transmissions, can also be used in earlier transmissions that require DEXRON®-III and III (H).
Chrysler Specifications	
Chrysler 7176	For Chrysler FWD transaxles.
Chrysler 7176D (ATF+2)	Adds improved cold temperature flow and oxidation resistance. Introduced 1997.
Chrysler 7176E (ATF+3)	Adds improved shear stability and uses a higher quality base oil. Required for four-speed automatics (do NOT use DEXRON® or MERCON® as a substitute).
Chrysler ATF+4 (ATE)	Introduced in 1998, ATF+4 is synthetic and replaces the previous ATF+3 fluid. It can be used in earlier Chrysler transmissions (except 1999 and earlier minivans with 41TE/AE transmission). ATF+3 should continue to be used for 1999 and earlier minivans because of the potential for torque converter shudder during break in.
Foreign OEM Specifications	
BMW LT71141 or LA2634	Special formula for BMW transmissions.
Genuine Honda ZL ATF	Special ATF for Honda automatics (except CVT applications).
Mitsubishi Diamond SP II & SP III	Special formula ATFs for Mitsubishi transmissions.
Nissan J Matic	Special formula for Nissan transmissions.
Toyota Type T, T III & T IV	Special formula ATFs for Toyota and Lexus transmissions.

and frictional compatibility with the transmission's components. In automatic transmissions, such components include clutches and bands, and in manual transmissions and manual transaxles, they include cone or plate type synchronizers. Unlike automatic transmissions, which use transmission fluids recommended only by the OEMs, manual transmissions use a wide variety of fluids. Such fluids include automatic transmission fluids, engine oils (5W-30), some gear lubricants, and a variety of specialty fluids. In addition to frictional properties, certain OEMs require transmission fluids for their equipment to have improved shear stability, low-temperature fluidity, and other specific characteristics.

Passenger car automatic transmissions were introduced in the United States in 1939. Although special lubricant needs were realized as early as 1937, in the absence of such lubricants, widely available engine oils and mineral oils were used. This practice continued until 1949 when GM introduced its first specification for these fluids [181–183]. Recently, another type of transmission is being introduced in the market place. These transmissions are called Continuously Variable Transmissions, or CVTs. Despite the advantages of the CVTs, the Hydra-Matic transmissions still occupy the largest share in the U.S. cars and SUVs.

Automatic Transmission Fluids

Generally, the performance of these fluids is defined by service-fill specifications of the passenger car and commercial vehicle transmission manufacturers. As stated earlier, the frictional compatibility of the automatic transmission fluids (ATFs) with the transmission's clutch and band system is their most important feature. Over the years, the ATFs have evolved along with the evolution of the automatic transmissions. For the present transmission fluid specifications and their suggested use, see Table 7.36 [621].

DEXRON® and MERCON® are the two major types of automatic transmission fluids that are presently in use in the United States DEXRON® fluids meet General Motors' performance specifications and are primarily designed for use in its transmissions. MERCON® fluids meet Ford's performance requirements and are used in its transmissions. DEXRON-III®, DEXRON®-VI, and MERCON-V® fluids are the most current specifications for the transmission fluids.

In response to a growing need for better fuel economy and improved emissions, GM in the 1990s designed a new transmission [622]. This transmission required a lubricant that had superior performance with respect to low-temperature properties, oxidative stability, antiwear characteristics, shift-feel smoothness, and frictional durability. DEXRON®-III/III (H) specification was introduced to qualify fluids to service these transmissions. In 2006, GM issued DEXRON®-VI specification for a new automatic transmission fluid for its 2006 and later model year cars and trucks equipped with Hydra-Matic transmissions. This specification is an advance over the other major existing ATF specifications with respect to performance, requiring running the fleet tests, and demonstrating new chemistries to be compatible with those that were previously approved. This fluid upgrade over DEXRON®-III (H) is to accommodate the needs of Hydra-Matic 6L80, GM's first six-speed automatic transmission. The viscosity profile of this fluid is such that it would perform more consistently in extreme conditions and would degrade less over time. These fluids have better oxidative and shear stability, foam performance, and protection against pitting. GM considers DEXRON®-VI to deliver more than twice the durability and stability in friction tests than do the other existing fluids [623].

MERCON® specification of 1987 was introduced by Ford Motor Company to meet the lubrication requirements of the 4L60 transmissions. MERCON® fluids were required to have better thermal stability and improved performance in the plate clutch and cycling tests than the Ford's previously specified Type F and Type H fluids [622]. This specification underwent several revisions, the most recent of which occurred in 1994. Ford released MERCON® V upgrade for use in 1996 and later model year vehicles. The fluids meeting this specification must have better shear stability, antiwear performance, low-temperature fluidity, oxidation resistance, friction retention, and anti-shudder durability than that of the MERCON® qualified fluids. Previously, MERCON® V fluids were not recommended for use in older cars because of the backward compatibility issues. The use of the MERCON® quality fluids was recommended for cars produced between 1981 and 1993, but for cars produced prior to 1981 Ford Type F fluids were to be used. It appears that Ford has solved the backward compatibility issue since as of July 1, 2007, production and licensing of the MERCON® fluids have stopped and MERCON® V will service transmissions that previously required the use of the MERCON® fluids [624]. Recently, Ford introduced a new transmission fluid, MERCON® LV, intended for fill-for-life. MERCON® V and MERCON® LV are not interchangeable [624].

In Table 7.37 we present physical properties and performance requirements for DEXRON®-III and MERCON®-V fluids. GM's 2005 upgraded specification for factory-fill/service-fill DEXRON®-VI automatic transmission fluids is superior to DEXRON®-III and DEXRON®-III(H) regarding low temperature viscosity/fluidity, oxidation resistance, wear performance, and frictional durability. The new specification also contains volatility, cold cranking viscosity, and shear stability requirements that were not part of the previous specifications. Data in Table 7.38 provides a comparison between the three DEXRON® specifications [625]. In addition to the properties listed in these tables, some OEMs, such as Mercedes-Benz, Ford International, and certain Japanese auto makers, require ATFs with even higher shear stability and better low-temperature fluidity.

As is evident from the data in the two tables, suitable frictional properties, oxidation resistance, and antiwear characteristics are the key performance parameters for ATFs. These fluids must not only possess the proper frictional characteristics but must also maintain them for a certain number of cycles (frictional durability). The frictional suitability is determined by the use of the SAE #2 Friction Test. Figure 7.4 shows a schematic diagram of the SAE #2 test rig.

The test consists of repeated starting and stopping of a clutch pack through 20,000 or more cycles and determining the degree of change in the fluid's frictional properties. The clutch pack consists of multiple composite and steel disks or plates that alternate in a sandwich-type arrangement. Typical modern composites include Kevlar®, graphite, and paper type materials. The steel disks are splined externally and are held stationary in the clutch head. Composite disks are splined internally and can be rotated by an electric motor. The clutch pack is spun at 3600 r/min and the hydraulic pressure is applied. This pressure is generated by the use of a piston. As soon as the clutch pack engages, the motor is turned off and the transferred power is recorded by the use of a computer. The engagement occurs when the rotating composite disks are pushed against the stationary steel disks. This process is repeated the desired number of times (cycles) over a 100 to 200-hour test period. The fluid temperature is as prescribed, usually 135 or 140 °C, and the air is blown into the clutch head at a rate of 50 mL/minute. The data generated are plotted as an XY graph, where X-axis represents the lock-up time in seconds and the Y-axis represents the torque in Newton-meters. Examples of three such graphs are shown in Fig. 7.5.

The parameters of interest are midpoint dynamic torque, end torque, maximum torque, delta torque, and the stop time.

1. Midpoint torque is the torque value centered half way between the start of the clutch engagement and the lockup.
2. End torque is calculated using the north-east corner torque.
3. Maximum torque is the torque value between 500 r/min and the clutch lockup.
4. Delta torque is maximum torque minus the midpoint dynamic torque.

TABLE 7.37—Physical and performance requirements for automatic transmission fluids.

Property	DEXRON®-III		MERCON®-V	
Color	6.0-8.0 (red)		6.5-7.5 (red)	
Miscibility	No separation or color change at end of test, using reference fluid		No separation or color change at end of test	
Kinematic Viscosity (cSt)	Report at 40°C and 100°C		6.0 min at 100°C	
Viscosity Stability, 100°C	No Requirement		Kinematic viscosity 6.8 cSt (40 passes)	
FISST (ASTM D5273)			Undegraded: rate and report	
Savant TBS VLT (ASTM D4683)			Degraded: rate and report	
Shear rate 200/s, 150°C			Undegraded: rate and report	
Shear rate 100,000/s, 150°C			Degraded: 2.6 cP min (40 passes)	
Shear rate 100,000/s, 100°C			Degraded: 5.4 cP min (40 passes)	
Brookfield Viscosity (cP)	Report at -10°C; 1,500 max at -20°C; 5,000 max at -30°C; 20,000 max at -40°C		1,500 max at -20°C; 5,000 to 13,000 at -40°C	
Evaporation Loss (%)	Not Required		5% max after 2 h at 150°C; End of test	
NOACK ,1 h at 150°C			Brookfield at -40°C EOT = 2,000 cP max change	
Flash Point (°C) min	170		180	
Fire Point (°C) min	185		Not Required	
Copper Strip	1b max		1b max	
Rust and Corrosion	No rust or corrosion on any test surface		No visible rust	
Foam Test	1. No foam at 95°C 2. 5 mm max at 135°C 3. 15 s break time at 135°C		ASTM D892 Seq. I – 50/0 ml Seq. II – 50/0 ml Seq. III – 50/0 ml Seq. IV – 100/0 ml	
Elastomer Compatibility	Vol. Ch (%)	Hardness (%)	Vol. Ch (%)	Hardness (%)
Polyacrylate (A)	+5 to +12	-8 to +1	+3 to +8	-7 to +7
Polyacrylate (C)	+2 to +7	-4 to +4	—	—
Silicone	+23 to +45	-30 to -13	+33 to +43	-5 to +5
Viton	+0.5 to +5	-5 to +6	+1 to +6	-8 to +8
Vamac	+13 to +27	-17 to -7	+6 to +16	-5 to +5
Buna-N	+1 to +6	-3 to +6	+1 to +6	-7 to +7
Hypalon	—	—	+17 to +37	-10 to -34
Hydrogenated Nitrile	—	—	0 to +10	-5 to +5
Vacuum Diaphragm Compatibility	Not required		Zero fails	
Elemental Analysis	Report Al, B, Ba, Ca, Cl, Cu, Fe, Mg, N, Na, P, Pb, S, Si, Zn		Report Al, B, Ba, Ca, Cl, Cu, Fe, Mg, N, Na, P, Pb, S, Si, Zn	
Infrared Spectrum	ASTM E168 — Report		Report	
Wear Tests	*Vickers Pump Wear [(ASTM D2882), 80°C, 6.9 mPa]*: 15 mg max wt loss		*Vickers Pump Wear (ASTM D2882):* 10 mg max wt loss at 80°C and at 150°C *FZG Wear (ASTM D5182)* 11 Load stage pass @ 150°C/1450 rpm *4-Ball Wear* 0.61 mm max wear scar 2 run avg at 100°C and 150°C, 600 rpm *Timken* 0.6 mm 2 run avg burnish width at 150°C: no scoring *Falex EP* No seizure load @ 100°C and 150°C: 750 lbs (2 run avg)	
Plate Clutch Friction Test (SAE #2)	Between 10 and 100 h of operation — No unusual plate wear or flaking; midpoint		Satisfactory operation for 20,000 cycles — midpoint dynamic coefficient 0.140-0.160, 25-	
SD-1777 Plates	dynamic torque between 150 and 180 N-m; delta torque (max-mid.) <30 N-m; maximum torque >150 N-m; stop time between 0.50 and 0.60 s; report end torque		20,000 cycles (127-157 N-m); static breakaway coefficient 0.120-0.150, 100-20,000 cycles (97-147 N-m); low-speed dynamic coefficient 0.135-0.160, 25-20,000 cycles; stop time 0.70-0.90 s, 25-20,000 cycles; low-speed dynamic ratio 0.9-1.0, 25-20,000 cycles; breakaway/midpoint ratio - report, 200-15,000 cycles (B/M formerly S_2/D); individual plate wear limit 0.1524 mm max; average plate wear limit 0.0762 mm max; used fluid analysis — report pentane insolubles, ΔTAN; ΔIR, viscosity at 100°C and -40°C	

5. Stop time is the elapsed time between the start of the clutch engagement and the complete clutch lockup.
   Each OEM has its own specifications for these parameters. Graph (A) represents the ideal torque trace. Graph (B) is the torque trace of a lubricant that does not possess the proper frictional characteristics. The engagement time is

TABLE 7.37— *(Continued.)*

Plate Clutch Friction Test (SAE #2) BW-4400 Plates	No requirement	Satisfactory operation for 20,000 cycles — dynamic coefficient 0.105-0.140, 100-20,000 cycles; static breakaway coefficient 0.105-0.135, 100-20,000 cycles; low-speed dynamic coefficient 0.120-0.155, 25-20,000 cycles; stop time 0.50-1.05 s, 25-20,000 cycles; low-speed dynamic ratio 1.10-1.35, 25-20,000 cycles (E/M formerly S_1/D); breakaway/midpoint ratio - report, 200-20,000 cycles; individual plate wear limit 0.1524 mm max, average plate wear limit 0.0762 mm max; used fluid analysis — report pentane insolubles, ΔTAN; ΔIR, viscosity at 100°C and -40°C
Band Clutch Test	Between 10 and 100 h of operation — No unusual wear or flaking on drum and band; midpoint dynamic torque between 180 and 225 N-m; delta torque <80 N-m; end torque >170 N-m; stop time between 0.35 and 0.55 s; report maximum torque	Not required
Anti-shudder Friction Test	Not required	Torque ratio R1 (2 rpm/20 rpm) >1.0; torque ratio R2 (40 rpm/120 rpm) >1.0; negative slope 135 h duration; individual plate wear limit 0.1524 mm max, average plate wear limit 0.0762 mm max
Oxidation Test	4L60 Transmission — Satisfactory operation for 300 h; transmission part cleanliness and physical condition equal to or better than that obtained with reference fluid; TAN increase <3.25; carbonyl absorbance increase <0.45; min 4% oxygen content of transmission effluent gas; used fluid viscosity at -20°C <2,000 cP; used fluid viscosity at 100°C >5.5 cSt; no cooler braze alloy corrosion	ABOT — Satisfactory operation for 300 h; pentane insolubles 0.35% max at 300 h; TAN change 3.5 max at 300 h; Differential IR 40 max at 300 h; viscosity increase 25% max at 300 h; copper strip 3b max at 50 and 300 h; report EOT Brookfield at -40°C; report gravimetric sample loss
Cycling Test	4L60 Transmission Tune Port Injected (TPI) Engine — Operating temperature 275°F; satisfactory operation for 20,000 cycles; transmission part cleanliness and physical condition equal to or better than that obtained with reference fluid; 1-2 shift time between 0.30 and 0.75 s; 2-3 shift time between 0.30 and 0.75 s; report 3-4 shift time; TAN increase <2.0; carbonyl absorbance increase <0.30; used fluid viscosity at -20°C <2,000 cP; used fluid viscosity at 100°C >5.0 cSt	4L60 Transmission (TPI) — Satisfactory operation for 20,000 cycles; transmission part cleanliness and physical condition equal to or better than that obtained with reference fluid; transmission must shift properly; no excess wear; no clutch plate deterioration; clean test parts; used fluid viscosity at -20°C <1,500 cP; used fluid viscosity at 100°C >6.0 cSt
Vehicle Performance	1999 Chevrolet Tahoe with 4L60 transmission — Shift performance essentially equal to that obtained with reference fluid	Ford Taurus — Shift performance essentially equal to that obtained with reference fluid

short and the shifting is jerky, as indicated by the abrupt increase in torque (rooster tail) prior to the clutch engagement. Graph (C) represents a friction-modified lubricant. While the midpoint torque and the lockup time are close to what is desired {compare with graph (A)}, the clutch engagement is sluggish indicating too much slippage.

The frictional characteristics of the lubricant, determined by the use of SAE #2 test, simulate smooth engagement and disengagement of the clutch and proper functioning of the band and the input drum in the actual transmission. The ultimate test of a transmission fluid's frictional properties is the "Shift-feel Test," where a qualified operator judges the lubricant's suitability under the actual service conditions.

Figure 7.6 shows a cross-sectional view of a Hydra-Matic transmission and Figs. 7.7 and 7.8 show the parts that are commonly rated. The antiwear properties of the transmission fluid are assessed by examining the sun gear and the vane pump parts.

Turbo-Hydramatic Oxidation Test (THOT) is the test used to determine a transmission fluid's oxidative stability. If adequate oxidative resistance is lacking, the oxidation of the fluid over its extended use will result in products that can cause corrosion as well as lead to the formation of sludge and varnish. The sludge and varnish are likely to adhere to the forward clutch housing, clutch piston, control valve body, and oil screen components and impair their function. Figure 7.9 compares the performance of a fluid with good oxidative stability against one with poor oxidative stability [626].

The use of the ATFs is not limited to automatic transmissions. A significant amount of the total (~50 %) is used in nonautomotive applications, thereby making these fluids highly versatile lubricants. Nonautomotive applications in-

TABLE 7.38—DEXRON® automatic transmission fluids comparison.

Property	DEXRON®–III	DEXRON®–III (H)	DEXRON®–VI
Flash Point, °C/°F, min	170/338	170/338	180/356
Fire Point, min	185/365	195/383	195/383
Viscosity, cSt , max		Report	
40°C	Report	Report	32
100°C	Report		6.4
150°C			Report
Viscosity Index, min	-	-	145
Pour Point, °C/°F	-	-51/-60	-54/-65
Brookfield Viscosity (cP)			
-10	Report	Report	Report
-20°C	1,500	1,500	1,500
-30°C	5,000	5,000	5,000
-40°C	20,000	20,000	15,000
High Temperature, High Shear (HTHS) Viscosity [150°C, 10^6 s^{-1}] cP, min	-	-	2.00
Tapered Roller Bearing Shear	-	-	♦ KV100, 5.5 cSt (min) at end of test ♦ 10% Decrease (max) at end of test ♦ (BOV + EOTV) / 2 > 5.0 cSt. at 100°C
Cold Cranking Simulator (CCS) Temperature = -30°C Shear Rate = 10^5 to 10^4 s^{-1} Kinematic Viscosity, max	-	-	3,300 cP
Copper Strip, max	1b	1b	1b
Noack Volatility, 200°C, 1h	Not Required	Not Required	10% max
Oxidation Procedure	300 h	450 h	450 h
Cycling Procedure	20,000 cycles	32,000 Cycles	42,000 Cycles
Plate Clutch Procedure	100 h; 23,000 cycles	150 h; 27,000 cycles	200 h; 36,000 cycles
Band Clutch Procedure	100 h; 24,000 cycles	100 h; 24,000 cycles	150 h; 36,000 cycles
Vane Pump Wear Procedure (100 h, 80°C), Wt Loss, mg max	15 mg	10mg	10mg
Vehicle Performance Procedure (Shift Feel)	Uses 1999 Chevrolet Tahoe	Uses 2005 Chevrolet Tahoe	Uses 2005 Chevrolet Tahoe Within 10% of the reference fluid
ECCC (electronically controlled capacity clutch) Vehicle Performance Test	-	1997 Buick Riviera	1997 Buick Riviera
Sprag Clutch Overrunning Wear Procedure, (1 h at 120°C and 16,800 rpm)	-		
Inner Race Weight Loss, mg max		60	50
Aeration Test Air Entrainment at 40°C, 90°C, and 120°C	-	Results for information purposes only	New and used fluid (from Cycling Test) must be equal to or better than Reference Fluid.
Low-Speed Carbon Fiber Friction Procedure (Anti-shudder Test)	-	Results for information purposes only	New and used fluid (from Cycling Test) must be equal to or better than Reference Fluid.

clude use in power shift transmissions in off-highway construction, agriculture, mining equipment, automotive, industrial, mobile, and marine hydraulic systems, power steering fluids, and rotary screw compressors.

Power Transmission Fluids

Power transmission fluids (PTFs) are used in heavy-duty automatic transmissions and the torque converters in off-highway equipment. Such equipment is commonly used in agriculture and construction industries. Viscosity and frictional properties of these fluids are critical to their performance. Just like the ATFs, the SAE and OEM performance specifications are used to describe these fluids [626].

Power transmission fluids (PTFs) have a viscosity range between the SAE 5W and the SAE 50 and are classified on the basis of their performance in Allison C-4 and Caterpillar TO-4 friction tests. Fluids meeting the Allison C-4 requirements are designed for equipment that has both the torque conversion and the automatic transmission features. Many engine oils meeting API's most recent service requirements are formulated to also meet Allison C-4 and Caterpillar TO-4 specifications, so that they can be used both in the engine and the transmission (See MIL-PRF-2104H specification in Table 5.6). Allison C-4 and Caterpillar TO-4 specifications were introduced to address the demands placed on the PTFs due to changes in the transmission design. Caterpillar requires the use of the straight grade fluids for its transmissions, wet disk brakes, and final drives. Fluids meeting this specification provide superior frictional properties and wear protection. Allison C-4 specifications are presented in Table 7.39 and Caterpillar TO-4 requirements are presented in Table 7.40.

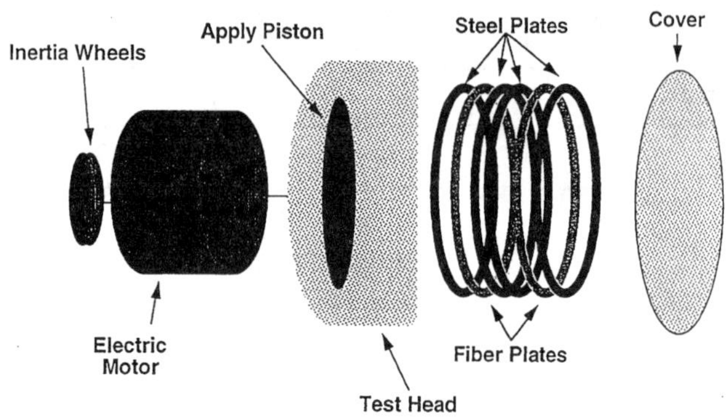

Fig. 7.4—Components of SAE #2 test rig.

Some European OEMs use automatic transmissions supplied by GM, Allison, and others; hence they can be serviced by the fluids recommended by these manufacturers. The other OEMs have their own performance specifications. Such OEMs include Voith, Zahnradfabrik Friedrichshafen (ZF), and DaimlerChrysler. Voith has two specifications: G 607, which is for mineral and semisynthetic lubricants, and G 1363, which is for full synthetic lubricants. These specifications are provided in Table 7.41 [318].

Both ZF and Voith require additive system approvals, which they grant based upon physical data, some in-house tests, and may even carry out field trials. ZF specification for passenger car automatic transmission fluids is designated as ZF-TE-ML-11 and that for the automatic transmissions used in commercial vehicle is designated as ZF-TE-ML-14. ZF-TE-ML-11A fluids are DEXRON®-IID quality and ZF-TE-ML-11B fluids are DEXRON®-III quality. Both fluid types are for use in vehicles not fitted with a continuously slipping torque converter. ZF-TE-ML-14 fluids are for commercial vehicle transmissions. Specifications for these fluids are listed in Table 7.42 [318].

DaimlerChrysler no longer manufactures passenger car transmissions without continuously slipping torque converters and commercial automatic transmissions; hence it recommends GM quality fluids for these applications. The automaker has issued a listing (236.X) of the approved ATFs with specific additives packages from an additives manufacturer that qualify DaimlerChrysler's in-house requirements and have demonstrated satisfactory performance in field trials. The listing is provided in Table 7.43 [318]. Incidentally, DaimlerChrysler does manufacture passenger car transmissions that are fitted with continuously slipping torque converters. Lubricants for these transmissions are dedicated

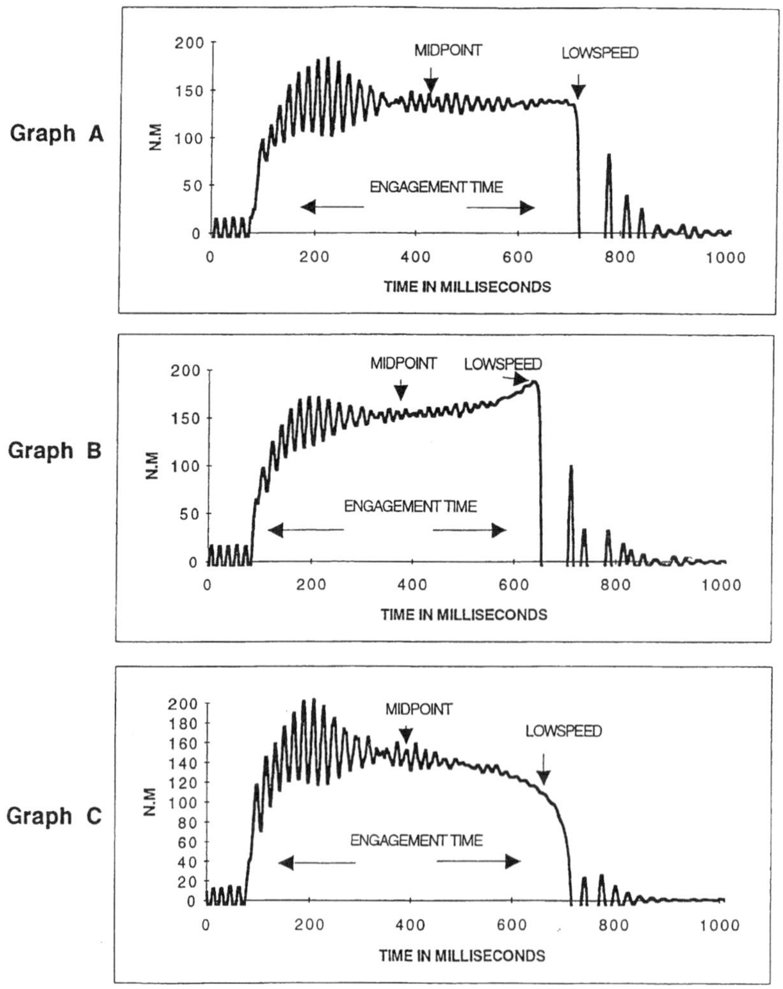

Fig. 7.5—SAE #2 torque traces for automatic transmission fluids.

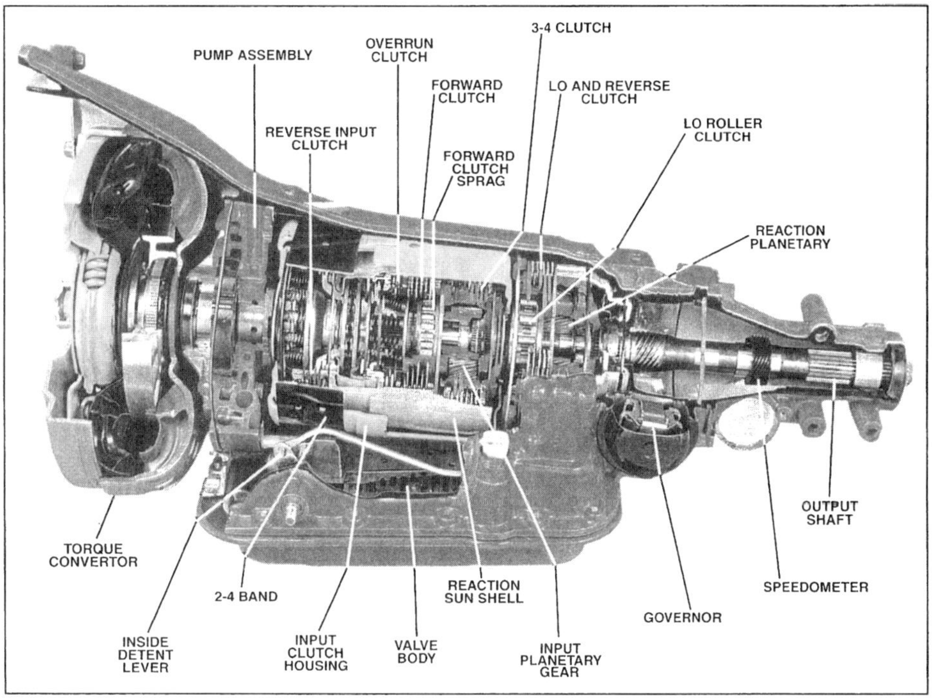

Fig. 7.6—Cross-sectional view of a Hydra-Matic transmission.

filled-for-life fluids. To be approved for the listing, the fluid must meet GM's Type A Suffix A, DEXRON®-IID, and DEXRON®-III requirements, as well as well as the test requirements of DaimlerChrysler.

Transmission Fluid Composition and Testing

As mentioned earlier, automatic transmission fluids perform many functions, which include lubrication and wear control, cooling, and to act as a hydraulic medium to transmit power. Wear and frictional heat in the transmission occurs due to metal-to-metal contact of the friction plates, gears, and bearings. Frictional properties of an ATF are not only to reduce friction, which is to minimize wear, but also to increase it so as to promote engagement of clutches during operation. Most OEMs use proprietary frictional materials; hence each ATF type is usually designed to meet the performance requirements of the transmissions from a specific manufacturer. In addition, ATFs must be compatible with all transmission components, operate at both low and high temperature extremes, and maintain consistent performance for extended periods. Most ATFs can withstand normal operating temperatures of around 200 °F (~94 °C) and

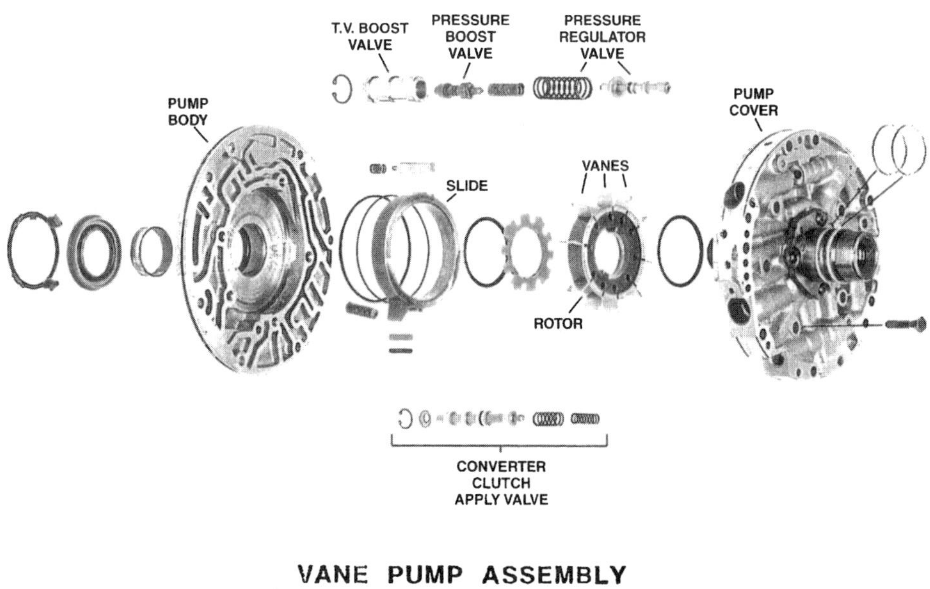

Fig. 7.7—Commonly rated automatic transmission parts.

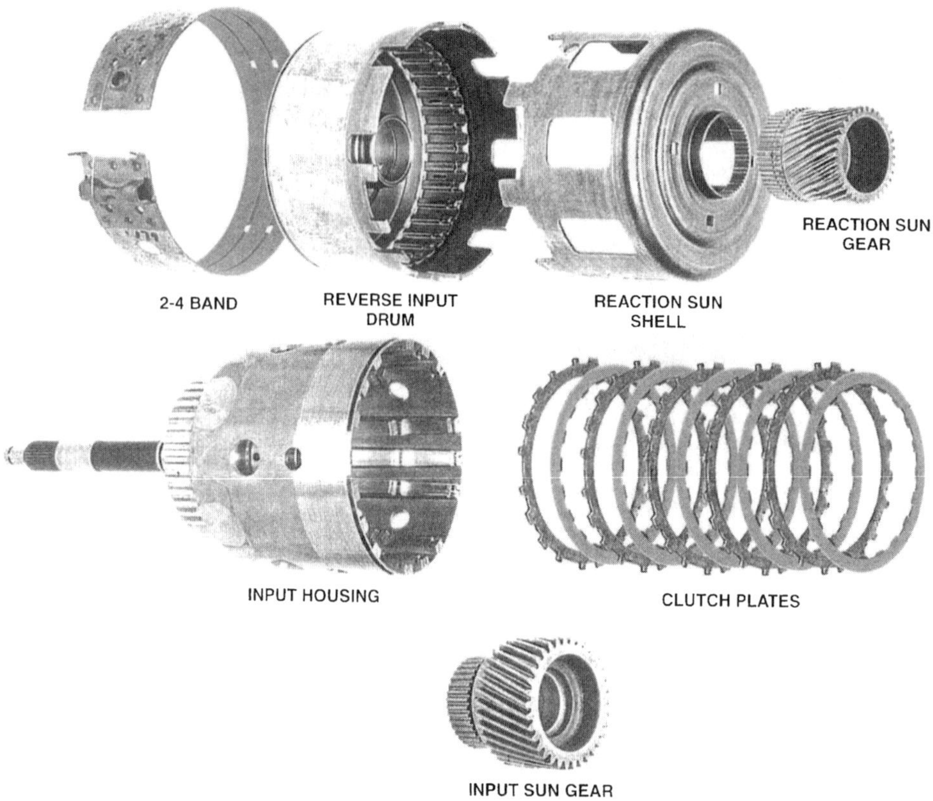

Fig. 7.8—Commonly rated automatic transmission parts (continued).

for tens of thousands of miles. But if the temperature of the fluid rises above 220 °F (105 °C), the fluid starts to break down quickly.

To meet the OEM established performance requirements, the transmission fluids are formulated by the use of mineral and synthetic base stocks of suitable viscosity and a number of additives. Synthetic oils have superior thermo-oxidative properties than mineral oil-based lubricants, but both types need additives to attain and maintain the desired frictional properties, minimize wear, impart oxidation resistance, and provide corrosion inhibition to the transmission components. Synthetic transmission fluids can be designed by the use of the PAOs or the PAO/ester blends. Both standardized and the OEM tests are used to test various lubricant parameters. The advantage of using a synthetic fluid, such as PAO, in the ATF over that of the mineral oil is demonstrated by data presented in Fig. 7.10 [87]. The superior oxidative stability is reflected by almost no viscosity increase for PAO under the test conditions. On the other hand, mineral oil after 16 hours experiences a sharp viscosity increase as a consequence of oxidation. THOT (Turbo-Hydramatic Oxidation Test) and ABOT (Aluminum Beaker Oxidation Test) data suggest that synthetic ATFs last twice as long as the mineral oil-based ATFs [87, 627]. Suitability of the various synthetic-base stocks in formulating ATFs with respect to the commonly sought ATF properties are listed in Table 7.44 [628]. It is important to note that besides the listed properties, there are other factors which must also be considered while selecting a suitable synthetic base stock for use in the ATFs. Those worth additional consideration are listed in Table 7.45 [628]. The objective is to realize the most effective formulation at a minimal cost.

Standard ASTM tests used to analyze and test transmission fluids are listed in the SAE International Surface Vehicle Information Report, SAE J311. Some of these are listed in Table 7.46. Additives used in transmission fluids are briefly described below:

1. *Viscosity improvers* to reduce the rate of change of viscosity with temperature. It is important to maintain the viscosity in-grade at high temperatures to provide proper lubrication. Styrene ester and poly(alkyl methacrylate) type polymers are commonly used.
2. *Pour point depressants* to improve the low temperature fluidity, especially for use in cold climate countries of North America and Northern Europe. Although the low viscosity base stocks work pretty well in this regard, to attain the required Brookfield viscosity at −40 °C, mineral oil (API Group I oil) derived ATFs are treated with pour point depressants, such as alkylated naphthalene, alkylated phenols, and poly(alkyl methacrylate)s that contain fatty alkyl groups.
3. *Oxidation inhibitors* to slow down the lubricant degradation due to oxidation. Hindered phenol and its derivatives and arylamines help minimize oxidative breakdown of the transmission fluid.
4. *Dispersants* to keep deposit precursors suspended in the bulk lubricant, to prevent their separation on hot surfaces to form sludge and varnish. Polyisobutenylsuccinimides and polyisobutenylsuccinate esters are used to accomplish this.

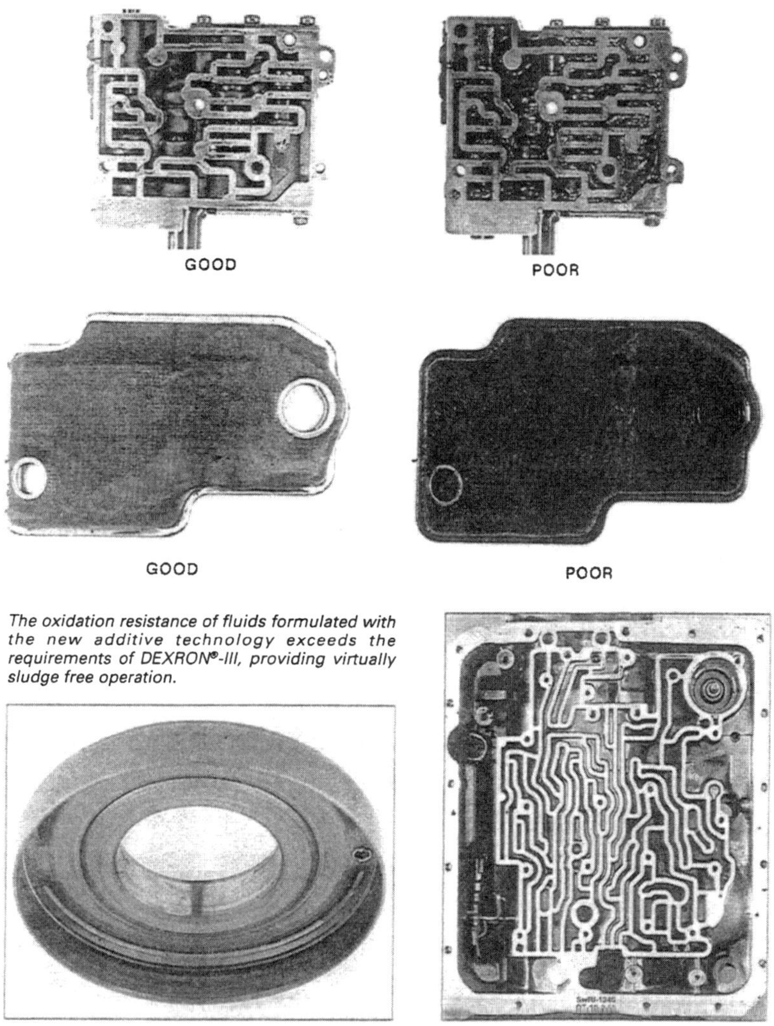

Fig. 7.9—Valve body and oil screen from transmission oxidation test [626].

5. *Antiwear agents* to protect planetary gears, bushings, and thrust washers against wear. It is important for the wear rates to remain low so that excessive loss of material does not cause the shift characteristics to deteriorate. Common antiwear agents used include metal and amine salts of dialkyl dithiophosphoric acids, alkyl phosphoric acids, and dialkyldithiocarbamic acids.
6. *Friction modifiers* to match the lubricant's frictional characteristics with those of the clutch plates and the band. A poor match is reflected by the generation of shudder, or vibration. Long-chain fatty materials; such as oleic acid esters, amides, and imidazolines are used to meet the frictional requirements of the ATFs. However, too much friction reduction will cause the transmission components not to engage.
7. *Corrosion inhibitors* to protect the metal parts against rust and corrosion due to acidic and oxygenated products from the lubricant oxidation, thermal degradation, or the sulfur-active additives. Fatty amines, alkenylsuccinic acid derivatives, and metal salts of arylsulfonic acids are used to protect against rust and DMTD and triazole derivatives are used to protect against the yellow metal corrosion.
8. *Seal swell agents* to minimize deformation of the elastomer seals. These materials, usually alkylaromatics, protect elastomers seals against damage due to lubricant attack.
9. *Foam inhibitors* to facilitate the collapse of foam, when it forms. These additives are usually polysiloxanes (silicones) and alkyl acrylate polymers.
10. *Red dye* for identification purposes.

Continuously Variable Transmissions (CVTs)

Use of Continuously Variable Transmissions (CVTs) in automobiles is a recent development, although the first CVT equipped car was introduced in 1958 by DAF of The Netherlands. Subaru offered a CVT transmission in the Justy in 1989, Saturn in the VUE in 2002, and Ford in the European version of the Fiesta in 2004. Honda also uses CVT for certain Civic models, so does Nissan in Japan.

A continuously variable transmission (CVT) is a type of automatic transmission that can change the "gear ratio" to any arbitrary setting within the limits. CVTs achieve this by the use a flexible metal belt and pulleys to constantly shift the gear ratios. The metal belts are composed of several (typically 9 or 12) thin bands of steel that hold together high-

TABLE 7.39—Allison C-4 automatic transmission fluid specifications.

Test	Requirements
Kinematic Viscosity (cSt) at 40°C and 100°C	Fluids must meet SAE J300 viscosity grades, except that heavy-duty diesel oils must meet MIL-L-2104E and ATFs must meet DEXRON®-III and MERCON® requirements
Brookfield Viscosity (cP) at -18°C	Report temperature at which fluid viscosity equals 3,500 cP
Apparent Viscosity	Report
Metals Content	Within application limits
Non-metals Content	Within application limits
Acid and Base Number	Report
Infrared Spectrum	Report.
Flash point, °C, min	160
Fire point, °C, min	175
Foaming tendency	No foam @ 95°C, 10 mm max foam @ 135°C with 23 s max to break
Rust protection	No rust or corrosion on test panels
Copper Corrosion	No blackening with flaking
Rust/copper Protection	No visible rust on test pins
Elastomer Compatibility	
Nitrile (Buna-N)[a]	
Volume change (%)	+1 to +5[b]
Hardness change (pts)	-5 to +5
Dip Cycle (Polyacrylate)[a]	
Volume change (%)	0 to +10[b]
Hardness change (pts)	0 to +5
Tip Cycle (Silicone)[a]	
Volume change (%)	0 to +5[b]
Hardness change (pts)	0 to -5
Fluoroelastomer (Viton)[a]	
Volume change (%)	0 to +4[b]
Hardness change (pts)	-4 to +4
Ethylene Acrylic (Vamac)[a]	
Volume change (%)	+12 to +28[b]
Hardness change (pts)	-6 to -18
Vickers Wear[a]	Total weight loss of cam ring and vanes not to exceed 15 mg
Graphite Friction[a]	Defined by minimum performance reference fluid
Paper Friction[a]	Defined by minimum performance reference fluid
Oxidation Stability (300 h at 163°C, 90 cc/min air)	TAN increase 4.0 max, carbonyl absorbance difference 0.75 max, Viton seal shall not frail, copper bushings shall not cause mechanical failure, parts free of sludge and varnish

[a] Required for MIL-L-2104F approval.
[b] Actual limits adjusted to account for batch-to-batch elastomer variation.

strength, bow tie-shaped pieces of metal. These belts do not slip and are highly durable, reliable, and efficient, enabling the CVTs to handle a substantial amount of torque from the engine. Incidentally, the metal belt-driven CVTs are quieter than the rubber-belt-driven older CVTs. Unlike typical automotive transmissions that are constrained to a small number of gear ratios, such as the 4 to 6 forward ratios, CVTs do not suffer from this constraint. This is because they do not generally have gears and the term "gear ratio" refers to the ratio of the engine shaft speed to the driveshaft speed.

CVTs have a much smoother operation than the conventional automatic transmissions, which gives a perception of low power to many drivers who expect a jerk when they begin to move the vehicle. The jerk indicates the changing of gears and in CVTs is simulated by the use of the CVT control software, especially at slow speeds. This is primarily to please the drivers. [629].

CVT transmissions have been refined over the years and are much improved from their originals. Compared to the hydraulic automatic transmissions, which at present are used in >90 % of the cars and trucks sold in the United States, CVTs can smoothly compensate for changing vehicle speeds, allowing the engine speed to remain at its level of peak efficiency. This and their smaller size and lighter weight improves both the fuel economy and the engine exhaust emissions. The fuel efficiency advantage is between 10 and 20 % over the commonly used four-speed Hydra-Matic automatic transmissions. Most CVTs are simple to build and repair. For friction-driven CVTs, their torque handling capability is limited by the strength of the belt or the chain and by

TABLE 7.40—Caterpillar TO-4 transmission and drive train fluid requirements. Reprinted with permission from the Lubrizol Corporation.

Test	10W	30	50	Comments
Viscosity at 100°C (cSt)	4.1 min	9.3-12.5	16.3-21.9	
Max Temp (°C) for Brookfield Viscosity of 150,000 cP	-35	-25	-15	
Max Temp (°C) for Low-temperature Pumpability of 30,000 cP max	-25	-15	-5	
Min High-temperature/High-shear Viscosity (cP) at 150°C	2.4	2.9	4.5	
Flash Point (°C)		X[a]		160 min
Fire Point (°C)		X[a]		175 min
Rust Control, BT-9		X[a]		Pass 2 of 3 rods at 175 h
Fluid Compatibility		X[a]		No sedimentation/precipitation
Homogeneity		X[a]		No sedimentation/precipitation
Copper Corrosion (2 h at 100°C)		X[a]		1A=Pass, 1B - 4C=Fail
Foaming				
Sequence I		X[a]		25/0
Sequence II		X[a]		50/0
Sequence III		X[a]		25/0
Sequence I with 0.1% water		X[a]		25/0
Sequence II with 0.1% water		X[a]		50/0
Sequence III with 0.1% water		X[a]		25/0
Fluoro-elastomer Seal Test		X[a]		≤ Reference plus 10%
Allison C-4 Seals				Limits based on batch
Nitrile (Buna-N)		X[a]		
Volume Change (%)				
Hardness Change (pts)				
Dip Cycle (Polyacrylate)		X[a]		
Volume Change (%)				
Hardness Change (pts)				
Tip Cycle (Silicone)		X[a]		
Volume Change (%)				
Hardness Change (pts)				
Fluoroelastomer		X[a]		
Volume Change (%)				
Hardness Change (pts)				
FZG Gear Wear	X[a]	X[a]		Average of three separate runs less than 100 mg
Vickers Pump	X[a]	X[a]		15 mg max vane weight loss; 75 mg max ring weight loss
C-4 THOT		X[a]		
TAN Increase				7.0 max
Carbonyl Absorbance				0.9 max
Viton Seal				Pass
Sludge				None
VC70 Friction	X[a]	X[a]	X[a]	Compared to reference

[a]Must be run.

TABLE 7.41—Voith ATF specifications [318]. Reprinted with permission from the Lubrizol Corporation.

Specification	Base Fluid	Performance Level	Transmission Type	Drain Interval (km)
G 607	Mineral	DEXRON®--IID/MERCON®	Midmat, DIWA	30,000
	Mineral	DEXRON®--IID/MERCON®	DIWA	60,000
	Semi-Synthetic	DEXRON®--IID/MERCON®	DIWA	60,000
	Hydrocracked			
G 1363	Synthetic	DEXRON®--IIE, III/MERCON®	DIWA	120,000

TABLE 7.42—ZF approved commercial ATF specifications for service fill [318]. Reprinted with permission from the Lubrizol Corporation.

Specification	Base Fluid	Performance Level	Shear Stability (cSt), min	Drain Interval (km)
TE-ML-14A	Mineral	DEXRON®-IID/DEXRON®-III	5.3	30,000
TE-ML-14B	Semi-synthetic	DEXRON®-III	5.5	60,000
TE-ML-14C	Synthetic	DEXRON®-IIE/III	5.7	120,000
TE-ML-14D		For recommendation purposes only		

their ability to withstand frictional wear between the torque source and the transmission medium. Hence, their use was confined to the low-powered cars. However, more advanced IVTs (Infinitely Variable Transmissions) that use advanced lubricants can handle torque in production vehicles, such as buses, heavy trucks, and earth moving equipment.

Because of the size, weight, and the cost advantages, and because they are easier to build and install than the traditional stick-shift and automatic transmissions, CVTs are being considered as transmissions of choice for automotive use by many OEMs. However, the technology is still evolving. A cross-sectional view of a CVT showing its essential parts is presented in Fig. 7.11 [629].

Lubricants for the Continuously Variable Transmissions (CVTs)

All CVTs are considered traction drives. Traction is the friction between a drive member and the surface it moves upon, where the friction is used to provide motion. Traction is exemplified by a wheel moving on the road surface. In CVTs, motion is created by the use of the pulleys and the belt or a chain, or by the use of rotating rollers, as in the case of Nissan's extroid transmission. Additional transmission parts include a vane pump, bearings, a differential gear, an oil pan, and an enclosure for these parts.

Traction occurs at the contact between the roller and the shaft, a cone and a ring, between a ball and a disk, or between a toroidal element and a roller. Because of the materials involved and the various types of contacts the working elements experience, the use of a suitable lubricant is critical to minimizing wear. CVTs are subjected to all lubrication regimes: hydrodynamic, elasto-hydrodynamic, mixed-film, and boundary. Critical areas of contact include pulley-sheave contact, nutating cone/control point, and traction drive/roller. A lubricant designed for use in CVTs must possess the following properties:

1. Ability to control rolling and sliding friction.
2. Suitable viscosity to provide hydrodynamic and elasto-hydrodynamic lubrication.
3. Thermo-oxidative resistance.
4. Foaming resistance.
5. Corrosion protection.
6. Low-temperature fluidity.
7. Extreme-pressure properties to protect against wear.
8. Protection against fatigue wear, if possible.
9. Elastomer compatibility.

Since CVTs involve a number of rolling elements, lubricants that provide elasto-hydrodynamic (EHD) lubrication are greatly desired. The ability of a lubricant to form an EHD film of suitable thickness can be determined by measuring its traction coefficient (friction coefficient). Rolling elements involve high contact pressures, which will flatten the surfaces in the contact zone, thereby leading to adhesion and seizure. The presence of a lubricating film in the contact zone prevents this from happening. The film under these extreme pressures becomes solid, elastically deforming metal, and minimizing metal-to-metal contact. However, if the friction coefficient is too low, slip will occur and as a result power transfer will be inefficient. The capability of transferring power is described as the traction coefficient. There is a proportional relationship between the traction force and the traction coefficient. As a result, to increase the traction force, a larger traction coefficient is required. CVTs require fluids that have the highest possible traction coefficient. Such fluids are called the traction fluids.

Traction coefficient of a fluid is determined by the use of a twin-disc machine that uses high-speed rolling contact under high pressure of at least 300,000 psi [630]. Figure 7.11 shows a plot of the traction coefficients of six fluids [630]. High traction coefficient is desired since it correlates with the fluid's ability to form an effective EHD film. Cycloaliphatic hydrocarbons, such as naphthenics, are the best and PAOs, almost linear, are the worst with respect to the EHD film-forming ability. This is not too surprising since we already know that naphthenics and aromatics rich mineral oils have the best and the paraffinics have the worst pressure-viscosity relationship. The paraffinic mineral oil has the next best traction coefficient, which is due to its minor but significant naphthenic and aromatic content. In the Synthetic and Natural Fluids Chapter, Chapter 3, we discussed a group of hydrocarbon synthetics called multiply alkylated cyclopentanes and bridged cyclohexanes. These base stocks, because of their high cyclic content, are likely to have even higher traction coefficients [631]. The finished lubricant based upon these cycloaliphatic hydrocarbons did not experience a negative effect on their traction coefficients due to the presence of the additives. In addition, these lubricants showed excellent performance as bearings lubricants [630]. These base fluids are commercially available under the trade names of Penzane® and Santotrac®. Other base stocks that are extensively covered in the patent literature to formulate CVT lubricants are polysiloxanes [631–635]. Incidentally, the traction coefficient of a DEXRON®-III fluid is only 0.03, which implies that it is not suitable for use as a lubricant in CVTs [630].

The data presented in Fig. 7.12 have an important bearing in formulating lubricants for CVTs. This is because while cycloaliphatic hydrocarbons have superior traction properties and low pour points, they have low viscosity indices and inferior oxidation resistance. This implies a higher treat-

TABLE 7.43—DaimlerChrysler listing of approved ATFs [318]. Reprinted with permission from the Lubrizol Corporation.

Application	236.2	236.3	236.6	236.7	236.8	236.1	236.81	236.9
Passenger Cars								
Transfer Case (4 Matic)			✓					
Mechanical Steering Lo75 Z		✓						
Power Steering		✓						
Manual Transmission	✓		✓					
Automatic MB Transmission			✓	✓	✓	✓		✓
Cross Country Vehicles								
Mechanical Steering		✓						
Power Steering		✓	✓	✓				
Manual MB Transmission	✓							
Automatic MB Transmission			✓	✓	✓	✓		✓
Commercial Vehicles								
Mechanical Steering	✓	✓						
Allison Transmission	✓		✓	✓	✓	✓		
Voith, ZF Automatic Transmission			✓	✓	✓	✓	✓	
W4 B035			✓					
Power Steering	✓	✓		✓				
Hydraulic Fan Drive For Auxiliary Radiator	✓	✓	✓					
MB Automatic Transmission				✓		✓		
UNIMOG, MB-Trac				✓		✓		
Differential Lock	✓	✓						
Power Steering	✓							

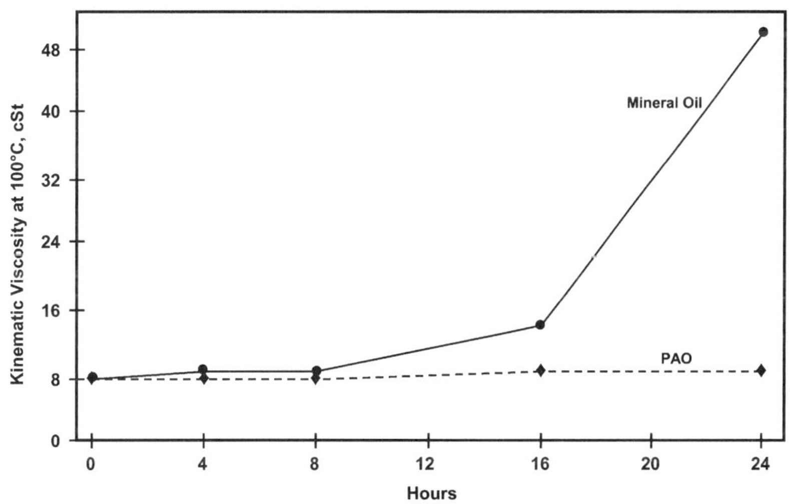

Fig. 7.10—ATF hot oil oxidation test 200 °C—PAO versus mineral oil [87].

TABLE 7.44—Synthetic base fluids—Properties comparison [628].

Property	Diester	Polyol Ester	Polyalphaolefin PAO	Poly(alkylene glycol) (PAG)	Phosphate Ester	Poly(phenyl ether)
High Temperature Thermal Stability	5	6	6	4	5	10
Oxidation Stability	7	8	6	4	5	6
Low Temperature Fluidity	10	10	8	6	0	0
Volatility	7	7	7	6	6	8
Viscosity Index	8	6	6	10	2	0
Hydrolytic Stability	6	6	10	6	4	10
Fluid Range	7	8	7	6	4	2
Overall Rating[a]	50	52	50	42	26	30

[a]The ratings are arbitrary. For reference, overall rating for mineral oil is 44. Performance Scale—10=Excellent, 0=Poor.

TABLE 7.45—Finished lubricants—Properties comparison [628].

Property	Mineral Oil	Super-refined Mineral Oil	Polyalphaolefin PAO	Ester	Silicone	Poly(alkylene glycol) (PAG)	Phosphate Ester	Poly(phenyl ether)	Fluorinated Hydrocarbons
Antiwear/Lubricity	2	6	6	8	0	6	10	2	6
Viscosity-Temperature Relationship (VI)	0	6	8	8	10	8	2	0	0
Volatility	6	6	8	6	6	2	2	6	4
Additive Solvency/Response	10	8	6	10	0	6	4	1	0
Compatibility									
Hydrocarbon Lubricants	4	10	10	8	0	2	6	0	0
Paint, plastics, elastomers	2	6	5	1	6	2	0	2	6
Relative Cost	Low	Medium	Medium/High	High	High	Medium	Medium	Very High	Very High

Note: Rating Scale—10=Excellent, 0=Poor.

ment of viscosity modifier and oxidation inhibitors. An alternative method is to use traction increasing additives and depend less on the traction properties of the base fluid itself. Such additives are used in ATFs to facilitate engagement of the clutches and include aromatic metal sulfonates, i.e., the detergents.

Other desirable properties in the CVT lubricants are the same as for ATFs and include good high and low-temperature viscosity, good oxidation stability and thermal resistance, improved low temperature operation, specified friction control, improved load-carrying ability, rust and corrosion resistance, and low foaming tendency. These properties are imparted by the use of the additives listed above under ATF formulation. Transmission fluid properties are evaluated by the use of the standardized tests, which are listed in Table 7.46, and the OEM required tests.

TABLE 7.46—ASTM test methods used for transmission fluids.

ASTM Test Method	Description
D974	Acidity
D664	Acid Number—Total
D4628, D4927, D4951, D5185	Additive Elements
D287	API Gravity, Specific Gravity
D2896, D4739	Base Number
D1500	Color
D130	Copper Corrosion
D1298, D4052	Density, Relative Density, or API Gravity
D93	Flash Point
D92	Fire Point
D665	Rust Prevention
D1552	Sulfur Content
D445	Viscosity Kinematic
D2983	Brookfield Viscosity, Low Temperature
D97	Pour Point
D2270	Viscosity Index
D892	Foam Test
D2619	Hydrolytic Stability Test
D5273, D4683	Viscosity Stability
D5800	NOACK Volatility
D2882	Vickers Pump Wear
D5579	Thermal Stability (Cyclic Durability Test)
D5704	Thermal and Oxidative Stability

Formulation Examples

DEXRON®-III Transmission Fluid: 2.0–6.0 % Alkenylsuccinimide dispersant, 0.5–1.5 % zinc dialkyl dithiophosphate antiwear agent, 0.4–1.0 % arylamine or substituted phenol oxidation inhibitor, 0.2–0.4 % basic sulfonate and triazole mixture corrosion inhibitor/metal deactivator, 0.3–0.8 % sulfurized olefin friction modifier, 0.1–0.5 % alkylnaphthalene pour-point depressant, 3.0–6.0 % styrene ester or polymethacrylate viscosity modifier, 200–500 ppm methyl silicone foam inhibitor, 0–3.0 % adipate esters seal-swell agent, and a red dye. The balance is mineral oil.

MERCON® Transmission Fluid: 2.0–6.0 % Alkenylsuccinimide dispersant, 0.2–0.5 % alkyl phosphite/phosphate antiwear agent, 0.2–0.5 % arylamine or substituted phenol oxidation inhibitor, 0.1–0.2 % basic sulfonate and triazole mixture corrosion inhibitor/metal deactivator, 0.1–0.4 % sulfurized olefin or carboxylic ester friction modifier, 0.1–0.5 % polymethacrylate or alkylnaphthalene pour-point depressant, 3.0–6.0 % styrene ester or polymethacrylate viscosity modifier, 200–250 ppm methyl silicone foam inhibitor, 0–3.0 % adipate esters or a proprietary additive seal-swell agent, and a red dye. The balance mineral is oil.

Continuously Variable Transmission (CVT) Fluid: Borated succinimide dispersant 3.5 %, 0.7 % overbased calcium sulfonate, 0.4 % calcium salicylate, 0.3 % alkyl phosphite+phosphoric acid, 0.6 % antioxidant(s), 1.1 % sulfur-containing agents, 0.2 % borated epoxides, 6.8 % polymethacrylate viscosity modifier, <0.1 % antifoam agent(s), and <0.1 % red dye. The balance is 100N mineral oil (formulation extracted from Ref [637]).

Automatic Transmission Fluid or a Continuously Variable Transmission Fluid: 1.0 % Oleyl polyglyceryl ether and 1.0 % calcium phenate friction modifiers, 0.5 % phenyl-1-naphthylamine oxidation inhibitor, 10 % polymethacrylate viscosity modifier, 1.5 % zinc dioctyl dithiophosphate antiwear/EP agent, and 1.0 % foam inhibitor. The balance is mineral or synthetic base fluid (formulation extracted from Reference [638]).

Miscellaneous Hydraulic Fluids

Brake Fluids for Road Vehicles

Brake fluids typically behave like a hydraulic fluid, since the fluid power is used to apply force to the brakes. Two types of braking systems that are typically used in vehicles are drum brakes and disk brakes. In a drum brake, the drum rotates

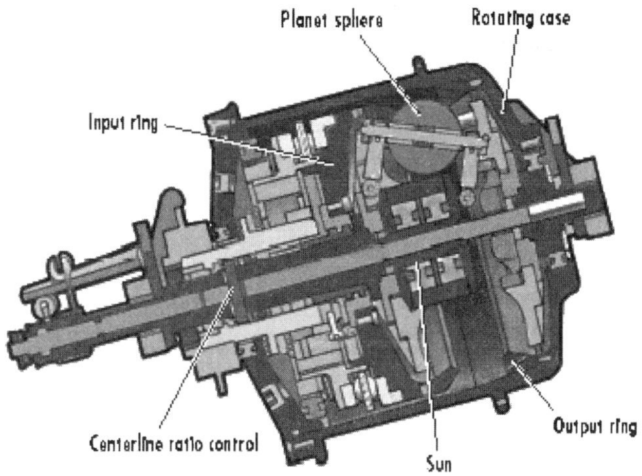

Fig. 7.11—Cross-sectional view of a continuously variable transmission [629].

with the wheel. Curved brake shoes that are inside the drum are separated by a slight air gap from the drum. When a driver steps on the brake pedal, the brake fluid, acting like a hydraulic fluid, is sent to the braking system to press the brake shoes onto the rotating drum, thereby stopping the drum and the wheel. A disk brake has a metal disk and flat shoes or pads that are pressed against the disk to provide the force necessary to stop the vehicle, by a hydraulic mechanism similar to that of the drum brakes. Thus, the braking mechanism is controlled by a hydraulic system that includes a pump, a fluid, and lines (tubes) that transmit the hydraulic force to the brake components.

Brake fluids used in vehicles is a sub-type of the hydraulic fluids, with a high boiling point and a low freezing point. Brake fluids must fulfill many other requirements to assure safe functioning under extreme climatic and operating conditions. Modern brake fluids are typically formulated by blending poly(glycol ether)s and polyglycols with rust and oxidation inhibitors, and may be a seal-swell agent. In some cases, low-viscosity silicone oils, silicone oil-phosphate ester blends, and mineral oils that contain a shear-stable viscosity modifier are also used. These fluids demonstrate extreme broad temperature performance. The additive treatment level in brake fluids is typically 1–2 %. A brake fluid must satisfy the following requirements:

1. Low viscosity and high fluidity at low temperatures.
2. Low bubble-forming tendency at elevated temperatures.
3. High thermal and oxidation stability.
4. Low volatility, reserve alkalinity, and corrosion resistance.
5. Compatibility with metals and elastomers, that is, must not react with the components of the brake system.
6. Compatibility with water and easy mixing at all temperatures, without affecting the functioning of the brake system.

A number of standard ASTM test methods are used to test these brake fluid parameters.

Brake fluids based on polyglycols are covered by a number of specifications, which include SAE J1702F (ARCTIC), SAE J1703 JAN 80, DOT 3, DOT 4, DOT 5, and ISO 4925. These specifications primarily differ in requirements pertaining to the low-temperature viscosity and the minimum boiling point of the brake fluid when new and after treatment with moisture, see Table 7.47 [4]. The boiling point is meant to provide information on the bubble forming tendency in the brake system at high loads. Since the procedure is not too precise, the plans are to replace it with the vapor pressure measurement. The established maximum low-temperature viscosity limit guarantees trouble-free functioning of the brakes in the wintery conditions.

Since polyglycol fluids are hygroscopic, they absorb water. An increase in the water content causes an increase in the low-temperature viscosity of the fluid, as shown in Fig. 7.13,

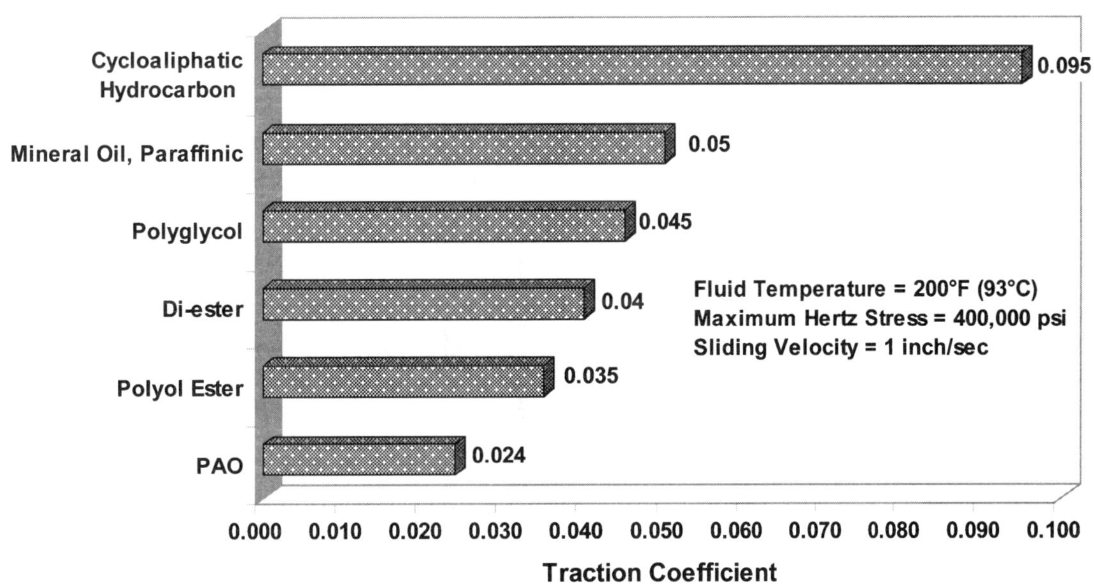

Fig. 7.12—Traction coefficients of various base fluids [630].

TABLE 7.47—Viscosities and boiling points of brake fluids based on polyglycols, required by different specifications.

Property	SAE J1702f (Arctic)	SAE J1703 Jan. 80	DOT[a] 3	DOT[a] 4	DOT[a] 5	ISO 4925
Kinematic Viscosity at −55 °C (mm^2/s), max.	1500	...	—	—	900	—
Kinematic Viscosity at −40 °C (mm^2/s), max.	—	1800	1500	1300	—	1500
Kinematic Viscosity at 50 °C (mm^2/s), min.	—	—	4.2	4.2	—	—
Kinematic Viscosity at 100 °C (mm^2/s), min.	3.5	1.5	1.5	1.5	1.5	1.5
Boiling Point according to ASTM D1120 (fresh brake fluid), °C, min.	150	205	205	230	260	205
Boiling Point according to ISO 4925 ("moistened" brake fluid), °C, min.	—	—	140	155	180	140
Ignition point, °C, min.	—	—	82	100	—	
pH	—	—	7.0 to 11.5	7.0 to 11.5	—	—

[a]DOT=U.S. Department of Transportation.

a drop in the boiling point; see the top curve in Fig. 7.14, and an increase in corrosive attack on metals [4]. Hence, occasional replacement of the fluid is warranted. The operating temperature in the brake systems depends upon many factors, such as vehicle design, weight, operation, and the type of brakes. Temperatures can reach up to 150 °C during normal operation and up to 180 °C during mountain driving. They can even reach higher under high-speed driving. The temperatures reached in the disk brakes are lower than those reached in the drum brakes. This is because the disk brakes are largely open and have the opportunity to be cooled by air. Under the influence of heat, the fluid that contains water will lead to vapor lock due to bubble formation, the result being spongy braking. As shown by the bottom curve in Fig. 7.13, for a fluid with high water content, the bubble formation occurs in the normal operating temperature range, but for a fluid of low water content it occurs at temperatures that are well outside this range. Silicone fluids are well suited for use in brake systems because they are chemically inert, have excellent viscosity-temperature properties, and good oxidation stability and elastomer compatibility. Silicone fluids are marked with a different dye so as not to mix them with the blue polyglycol brake fluids since they can not be used together.

Shock Absorber Fluids

These fluids are used in hydraulic systems that are utilized in vehicles, machinery, and equipment to dampen the vibration. Such devices include shock absorbers for wheels and dampeners for steering systems. These fluids must possess good VT characteristics, proper frictional properties, thermal and oxidative stability, rust control, and good antiwear properties and elastomer compatibility. These fluids are exposed to temperatures between 60 to 150 °C, depending upon the load. The heat from these fluids is removed by the air stream of the moving vehicle. Since some of the devices operate at low ambient temperatures, they must also have good low-temperature properties. Some ATFs meet these criteria and can be used in this application. The shock absorber producer and the OEMs jointly establish performance criteria for these fluids.

Typical properties of the shock absorber fluids of various viscosities are provided in Table 7.48 [4]. These fluids employ naphthenic oils because they have good low-

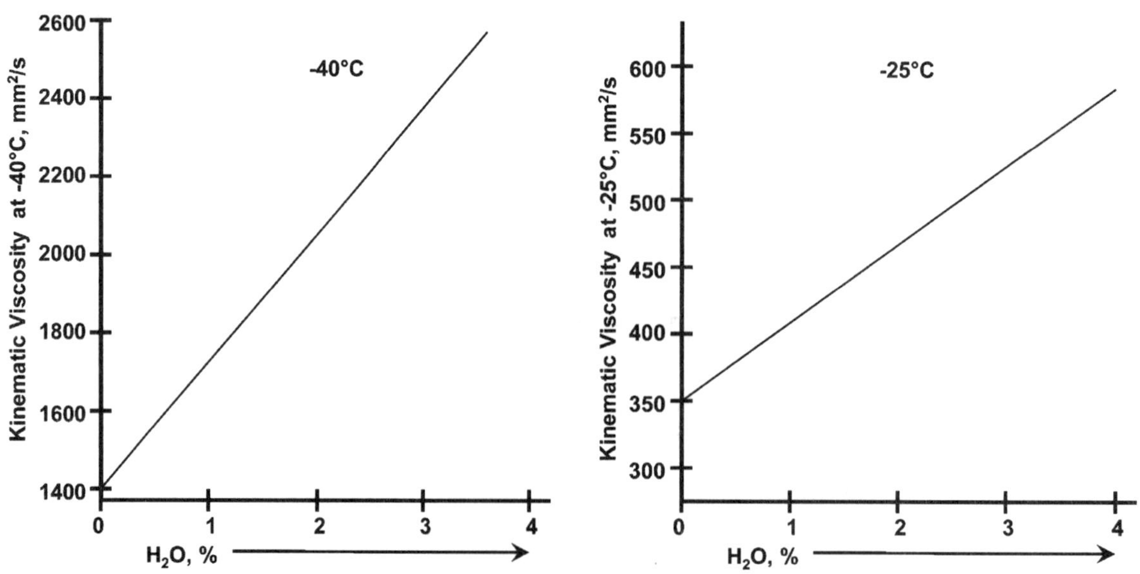

Fig. 7.13—Effect of water content on the low-temperature viscosity of polyglycol brake fluids (DOT 3) [4].

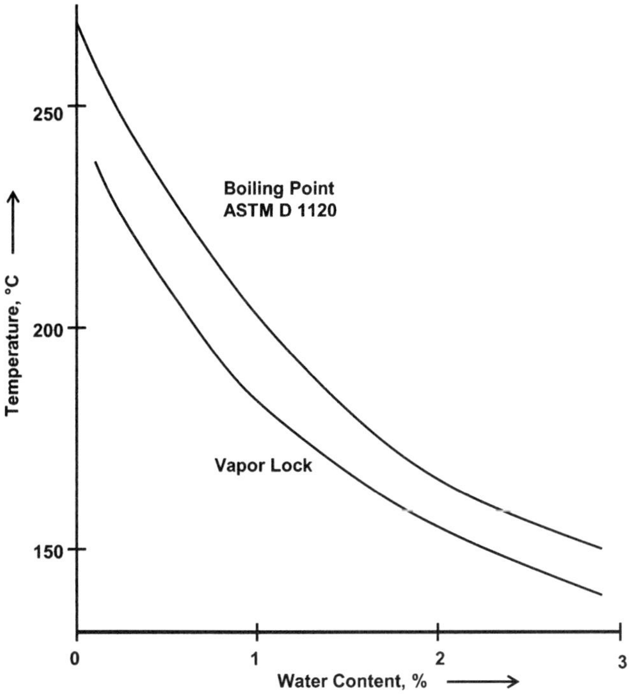

Fig. 7.14—Effect of water content and boiling point of polyglycol brake fluids (DOT 3) [4].

temperature properties. High-temperature properties, such as the viscosity index, are attained by the use of the shear-stable viscosity improvers. However, the amount of the viscosity improving polymers used is limited by the required low temperature viscosity, which may increase due to the presence of these polymers. Alternatively, synthetic fluids with good low and high-temperature viscometrics can be used, if their higher cost is not a concern. High operating

TABLE 7.48—Typical requirements shock absorber oils [4].

Properties	Type A	Type B	Type C
Viscosity at −40 °C, mm^2/s (cSt)	2500	5000	250
Viscosity at 40 °C, mm^2/s (cSt)	10	15	10
Viscosity at 100°C, mm^2/s (cSt)	2.5	3.5	3.8
Viscosity Index, minimum	90	130	300
Pour Point, °C	−50	−50	−50
Flash Point, °C	145	100	100
Density at 15 °C, g/mL	0.87	0.87	0.86
Rust Prevention	+	+	+
Antiwear	+	+	+
Shear Stability	+	+	+
Evaporation Loss	+		+
Foaming Tendency	+	+	+

temperatures require protection against oxidation, which is accomplished by the use of oxidation inhibitors. Many of the fluids also contain foam inhibitors and corrosion inhibitors, which are necessary to counter the effects of the water contamination that can occur when seals are damaged. Other additives that are used include phosphorus-containing antiwear additives and friction-reducing agents, which prevent excessive noise in the case of high lateral loads. The overflow noise (hissing), which is heard in some cases, is eliminated by mechanical measures which influence sound generation and sound velocity.

Power Steering Fluids

These fluids are used to protect the power steering parts against wear and rust damage. Performance requirements for these fluids are established by the OEMs. The quality of these fluids is assessed by their frictional properties, seal compatibility, oxidation resistance, and rust control. Some ATFs meet these criteria and can be used in this application.

8

Gear Lubricants

IN THIS CHAPTER WE DISCUSS GEAR LUBRICANT technology. We deal with the topics of gear types, gear metallurgy, and types of gear damage. An understanding of all these subjects is important to formulating a suitable gear lubricant. Performance standards and testing of the gear lubricants, both automotive and industrial, are also described. The chapter concludes with a number of formulation examples.

The primary functions of a gear oil are to reduce friction between the gear surfaces in contact, dissipate heat, and provide extreme pressure protection to gears and axles against fatigue, scoring, and wear damage, under boundary lubrication conditions [12]. In automotive applications, the gear lubricant must also minimize shock and the noise resulting from the motion of the meshed gears in the rear axles, thereby providing a smooth and quiet operation during acceleration.

Gear Types

Gears are mechanical devices that help transfer rotating motion and power from one part of a machine to another. In mechanical equipment, gears are used for increasing and decreasing torque, changing speed, or changing direction of movement. When two gears mesh or run together, the larger gear is called the ring gear and the smaller gear is called the pinion gear. Because of the differences in design, each gear type places different demands on the lubricant. Different kind of gears and gear arrangements that are used in automobiles and industry are shown in Fig. 8.1 [12].

Spur gears, shown in the top part of the figure, are the simplest. They are of three types: external, internal, and rack and pinion. All have straight teeth that are cut parallel to the axis on a cylindrical surface. Because only one set of teeth is in contact at any one time, the load-carrying capacity of these gears is limited. They are most commonly used in industrial machines that operate at moderate speeds and loads. Applications include construction equipment, machine tools, indexing equipment, multi-spindle drives, roller feeds, and conveyors. They are not used in automobiles because they produce sound when the teeth of the two gears move against each other, which also increases stress on the gear teeth.

Helical gears are a modification of the spur gears in that the teeth are cut at an angle to the axis of rotation, i.e., they are twisted in a helix. This allows more than two teeth to be in contact at a time, thereby increasing their load-carrying capacity. Because of this and the fact that they run smoothly and quietly, even at high speeds, and are durable, helical gears are extensively used in high speed transmissions. Potential industrial uses of these types of gears are in textile machinery, blowers, feeders, rubber and plastics processing equipment, sugar manufacture, rolling mills, food industry, elevators, conveyors, compressors, and others. However, since the load in these gears is transmitted through a helix, the generated thrust along the axis of the gear must be handled by the use of the appropriate thrust bearings. A double helical gear, commonly known as herring bone gear, has two opposed helices and the end thrust is therefore balanced. With this arrangement, one of the gears must be free to move axially to align the two helices. Double helical gears have greater load-carrying capacity because at any one time a greater number of teeth are in contact. Spur and helical gears can take several forms. They can be external or internal, depending if the teeth are inside or outside the gear diameter. Most planetary gear sets use a combination of internal and external gears, as is shown in the figure of the internal spur gears. Spur and helical gears generally operate on parallel shafts, as shown in the figure of the helical gear.

Bevel gears are cut to produce teeth as truncated cones. The shafts in bevel gears are not parallel but intersect. While bevel gears are made for any included angle between the shafts, the most common shaft angle is 90 degrees. The teeth on bevel gears can be straight, helical, or spiral. In straight or simple bevel gears, the teeth are straight and radiate from the points of the cones. The gears in these gear sets are of equal size. A *helical bevel* gear is a toothed gear in an angular design. Spiral bevel gears have spiral angles, which results in certain performance improvements. The contact between the teeth starts at one end of the gear and then spreads across the whole tooth. In all types of bevel gears, the shaft must be perpendicular to each other and be in the same plane. Since spiral bevel gears have curved teeth that are at an angle to the radial lines, the area of contact increases, which also increases their load-carrying capability. In addition, these gears are quieter in operation at all speeds. Bevel gears are used in locomotives, marine equipment, automobiles, printing presses, power plants, and steel plants. Spiral bevel gears can handle high speeds, high loads, and an extremely large number of load cycles and are used in rotorcraft drive systems to redirect the shaft from the horizontal gas turbine engine to the vertical rotor. They are also used in power windows and power seats. They are used where speed and strength are desirable, along with the change in angle of the power flow.

Worm gears are used where the shafts are at right angles and have maximum offset. The long gear is the worm, or the pinion gear, and the round gear is the wheel gear. The wheel gear has a large number of teeth, typically 359 or 360, which mesh with a single tooth, spirally wrapped around the worm gear face. For one full rotation of the worm, the gear will ro-

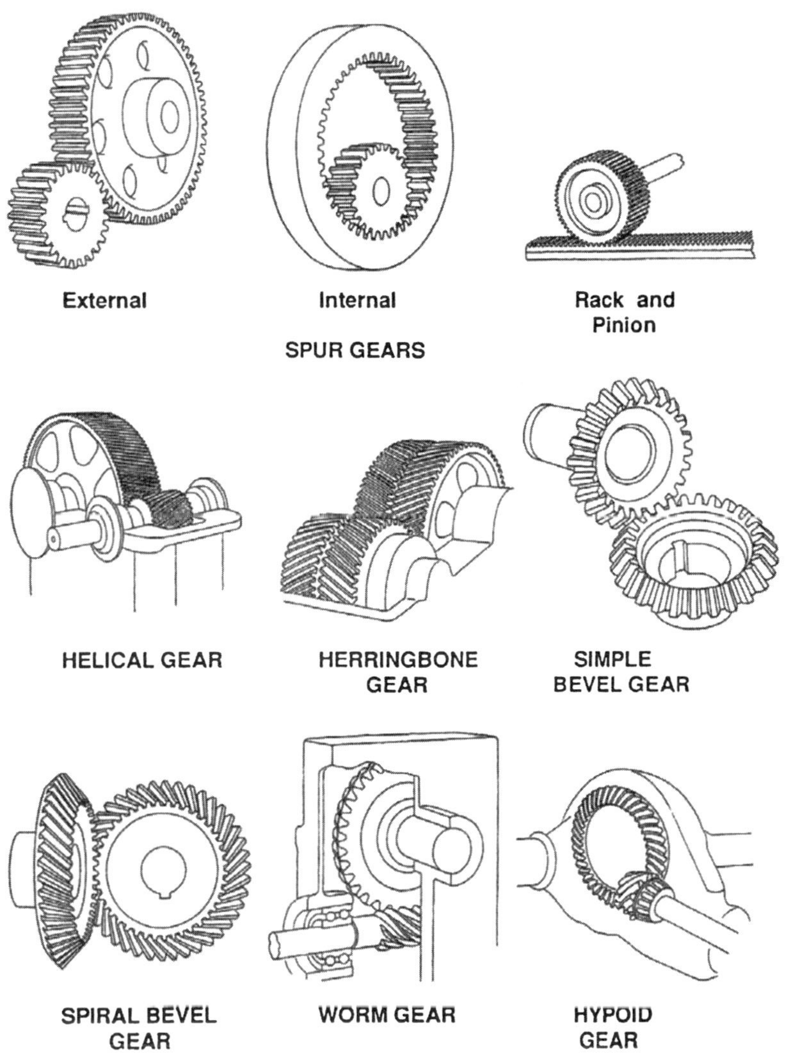

Fig. 8.1—Gear arrangements [12].

tate 1/359 (or 1/360) of a revolution and thus drive the right ascension axle the same amount. The worm gear is always used as the input gear. The worm gear and the worm shaft are supported by antifriction roller bearings. The torque is applied to the input end of the worm shaft by a driven sprocket or an electric motor and the resulting motion drives the ring gear. Because the concave faces of the worm gear tooth fit the curvature of the worm gear, it provides a line contact, instead of a point contact. These gears can provide a high angular velocity between nonintersecting shafts at right angles and are capable of transmitting high tooth loads. Hence, they provide the ultimate power ratio. Their main disadvantage is that they experience high friction due to the maximum sliding and extensive contact. Their use in high-torque applications results in extensive wear of the gear teeth and erosion of the restraining surface. Proper lubrication of these gears is therefore a challenge. Worm gears are widely used in packaging machinery, material handling, machine tools, indexing, and food processing. They are also used in Torsen® differentials that are used in some high-performance cars and trucks.

Hypoid gears resemble spiral bevel gears and have the characteristics of both the spiral bevel gears and the worm gears. However, their pinion, or the axle, is offset so that its axis does not intersect the gear axis. This design provides these gears with very high load-carrying capacity and quieter operation. The axle offset permits the lowering of the drive shaft, which promotes their use in automotive applications. Tooth-contact conditions of the hypoid gears are considerably more severe than those of the spur, bevel, and helical gears in that the offset of the axes introduces additional sliding between teeth. This, along with the high loads attainable, creates an extreme lubrication environment. Because of the increased areas of contact and sliding, these gears are difficult to lubricate. Worm gears, described earlier, can be considered an extreme type of hypoid gear since they have the maximum offset possible. Because of this, they have even greater contact and sliding and are therefore more difficult to lubricate than the hypoid gears.

Gear Metallurgy

Modern gears are made from a variety of materials ranging from steel to plastics. For gears used for power transmis-

TABLE 8.1—Surface roughness versus working technique [639].

Machining Technique	Working Operation	Average Surface Roughness		
		Microns	Millimeters	Micro-inches
Normal Machining	Sawing	254 to 25.4	0.254 to 0.0254	10,000 to 1,000
Precision Machining	Turning	5.08	0.00508	200
	Coarse Grinding	1.27	0.00127	50
Ultra-precision Machining	Fine Grinding	0.127	0.000127	5
	Mechanical Polishing	0.508	0.0000508	2
	Electrolytic Polishing	0.0254	0.0000254	1

sion, steel is often the best material, since it is strong and relatively inexpensive. Any deficiencies, if they emerge, are corrected by the use an effective lubricant. However, for some types of gears, such as worm gears, where extensive metal-to-metal contact results in extensive wear, the main goal is to use material that is largely wear resistant since a lubricant may not provide the necessary protection.

If steel is being used to manufacture gears, it is important to consider its carbon content since it determines the steel's hardness. The best steel has the carbon content of 0.8 %. It provides maximum hardness after heating and quenching. Small quantities of steel alloying elements, such as chromium, manganese, nickel, and molybdenum, are added to modify the steel's response to heat treatment. Since extremely hard metals are not easy to machine, many times it is desirable to form the part, in this case the gear, from ductile metal and then heat treating or chemical treating it to achieve the desired degree of hardness. However, it is important to note that unhardened steels may work harden, depending on their chemical composition. Chemical composition also influences the wear properties. For example, stainless steel and nickel-alloyed steel gears show poor scoring properties, but the presence of chromium and molybdenum appears to improve them. Nitriding also appears to help. Nitriding is a surface-hardening heat treatment that involves introducing nitrogen into the surface of the steel at a temperature of 500–550 °C or 930–1020 °F. The use of the hardened-steel gears is predominant since hardening does improve wear resistance. For a maximum benefit, it is important to harden both the gear and the pinion to achieve a similar degree of hardness, although sometimes hardening the pinion only provides the best wear results. Cast iron, which is also used to produce gears, has a carbon content of 2.5 to 4.0 %. Some of the carbon is in the free form and the rest is in the combined form. Free carbon, which is in graphite form, reduces the metal's strength, ductility, and elasticity, but imparts cast iron the abilities to self-lubricate and dampen vibration. Sintered-iron gears are formed from iron powder by compacting in a die, then heating in a furnace to fuse the particles together. Copper powder is often mixed in to improve the bonding of the iron particles. This technique is usually used to make small spur gears by the use of a die. The advantages are the ability to make gears with good wear resistance and to mix in a solid lubricant to make them self-lubricating.

Gears are also made from nonferrous metals. Copper is too soft a material to be used to make gears, but its alloys, such as bronze, which is 90 % copper and 10 % tin, is the most commonly used nonferrous alloy. The desirability of bronze as gear material is due to its excellent corrosion resistance and the ability to withstand high sliding loads. Hence, it is quite useful in making wheel gear in worm gear set that predominantly involves sliding motion. Other nonferrous alloys are also employed in forming gears through die-casting. Such gears are used in small mechanisms where only moderate power-transmitting capacity and life are desired.

Nonmetallic gears are used in systems where low noise is desired. Noise is a consequence of the inaccuracies in tooth profile and spacing. Low-modulus materials, such as polyethylene, polystyrene, Nylon® and Teflon® are used to make these gears. In many cases, it is possible to obtain load capacity that is comparable to that of the metallic gears. The load capacity is even higher for nonmetallic gears that mesh with metallic gears and most of these require little or no lubrication.

Surface finish and the surface coating, while not directly related to the topic being discussed here, do influence the wear properties of the finished gears. A number of methods are available for the finishing step and include grinding, shaving, hobbing, broaching, and casting. With respect to finish, grinding is the most accurate and casting is the least accurate. Selection of a finishing method depends upon the objective. Too high a surface roughness will lead to scoring and wear damage, at least initially, but after the break-in, the surface finish becomes the same, irrespective of the initial finish. In the worm gear sets, where different materials are used for the ring gear and the worm gear, the use of harder and finely finished worm gear prolongs the life of the gear set. Another technique that is used to improve wear and scoring resistance of the gears is surface coating. Easily removable sacrificial coatings, whether of soft metal or hard chemical, produced by physical or chemical means, have been attempted. There is some evidence in favor of their short-term effectiveness, but their long-term effectiveness is doubtful. This is not too surprising since such surface films quickly wear off during service. Table 8.1 shows surface roughness as a function of the machining technique.

Lubrication requirements for different types of gears are related to their load-carrying capacity, which in turn depends upon the area of contact. The higher the area of contact, the higher is the load-carrying capacity, and hence the more demanding the lubrication requirements. Spur gears, with the lowest load-carrying capacity, are at the one extreme and the hypoid gears, with the highest load-carrying capacity, are at the other extreme, with respect to the lubrication requirements. The lubrication requirements for other types of gears fall in between.

Gear Failure

Gear Failure Modes

Metal gears fail for a number of reasons. Some are related to gear design and manufacture and others are due to service-related factors. Common gear failure modes are fatigue (bending and Hertzian), pitting and micro-pitting, scuffing, and wear [640].

Fatigue-related failure results in the development of cracks, which under extreme situations may result in the separation of the tooth from the body of the gear. Pitting, sometimes referred to as macropitting to distinguish it from micropitting, is surface damage that results from the cyclic contact stresses being transmitted through the lubricant film which is in or near elasto-hydrodynamic lubrication regime. Pitting is one of the most common causes of gear failure as well as of other machine components, such as antifriction bearings and cams, which experience rolling/sliding contact under heavy loads. Pitting starts with small surface or subsurface cracks that grow under repeated contact loading. When such cracks reach the tooth surface and intersect, a small amount of material is lost from the surface, leaving behind a pit. If pitting damage is significant, vibrations and noise will be evident. This type of failure occurs in both through-hardened and surface-hardened gears. Micropitting, a related type of failure, is the formation of the much smaller craters on the tooth surface than those observed from pitting. In both cases, the removed material can cause abrasive wear. However, the abrasive wear more often accompanies micropitting than macropitting. As a consequence, engineers incorrectly identify micropitting damage as abrasive wear. Micropitting is tooth surface or subsurface damage resulting from rolling/sliding contact fatigue, where fatigue arises from repeated normal and tangential loads in a boundary or mixed-film lubrication regime.

Scuffing, sometimes called scoring, is a severe type of adhesive wear, which damages the tooth surfaces that are in relative motion. Scuffing welds unprotected surfaces that are in metal-to-metal contact together. Due to motion, metal particles detach and transfer from one or both meshing teeth. During successive rotations, these particles can scratch teeth flanks in the sliding direction. This type of damage generally occurs in areas of high contact pressure and sliding velocity that are far from the pitch surface. This is because the conditions in these areas do not allow effective lubricant film formation to prevent the direct metal-to-metal contact. Protective film may either be a thick oil film (greater than the surface roughness) as a consequence of high lubricant viscosity or an adsorbed or chemically formed film established by the lubricant additives. Scuffing damage is related to lubricant and lubricating conditions and not to material strength. Scuffing commonly occurs when the gears are new and hence tooth surfaces are not well run-in. The risk of scuffing also goes up as the lubricant degrades over time or becomes contaminated with metal particles or water [640].

Wear is a continuous, abrasive process where the material is removed from the mating gear teeth. This occurs either due to abrasive particles in the oil or rubbing of the surfaces with the scuffing-derived transferred material. Continual wear of tooth roots weakens the gear until it breaks. Wear typically occurs under boundary and mixed lubrication conditions, where the lubricant film thickness is not suitable to separate the tooth surfaces. The presence of the antiwear additives will help prevent this type of damage, by forming a durable chemical protective film.

Gear Diagnostics

Assessing in-service performance of gears has its challenges since it is not always practical to stop the equipment to examine the condition of gears. However, this difficulty must not preclude the importance of determining the in-service gear condition. This is because many times corrective measures can be implemented if the problem is lubricant-related or service-related. The objective of the exercise is to prevent or delay gear failure. This is because gear failure will cause an unexpected shut down of the machine which is both inconvenient and expensive. A gear system fails when it ceases to efficiently perform the function for which it was designed. This can be the result of a single catastrophic event or an accumulation of the initially undetected and rather innocuous events that may have occurred in the gear system. As stated previously, common modes of gear failure are macropitting, micropitting, scuffing, tooth wear, and breakage.

Gear diagnostics is a set of new techniques, the objective of which is to identify factors and events, which may appear innocuous or are hard to detect, or both. Once they are identified, attempts are made to correct the damage or slow down its rate to prevent or delay catastrophic failure. An engineer has many options to choose a diagnostic technique, such as thermal imaging, oil analysis, wear metal analysis, vibration analysis, etc. Of these, vibration and noise measurement and analysis, alternatively called mechanical signature analysis, is one of the most common techniques used [641,642]. Mechanical signature analysis is a process of monitoring the vibration signatures of the operating equipment or its components to determine their condition. The technique may also be used to diagnose the cause of a problem. Signature deviation of the various parts of the equipment is compared to a baseline signature in the observed domain, to identify a potential problem. Many statistical techniques help facilitate the collected data analysis and interpretation. These are listed in Refs [643,644]. It is important to note that vibration harmonics-based diagnostic-methods for fault detection in gear-type couplings with defects and slow-speed gears do not always work satisfactorily [645].

In addition to vibrational analysis, other techniques used to assess in-service gear performance include monitoring temperature [646] and identifying wear metals in oil and on oil filters [647]. A greater than expected temperature rise and the presence of wear metals are also indicative of either a potential or a real problem. Depending upon the results from these techniques, one can either decide to continue running the equipment, or stop it to analyze surface damage to gears and identify its cause. The latter objective is achieved by examining the gear alignment and gear tooth contact patterns and surface condition. Combining these techniques with other surface analysis techniques, such as

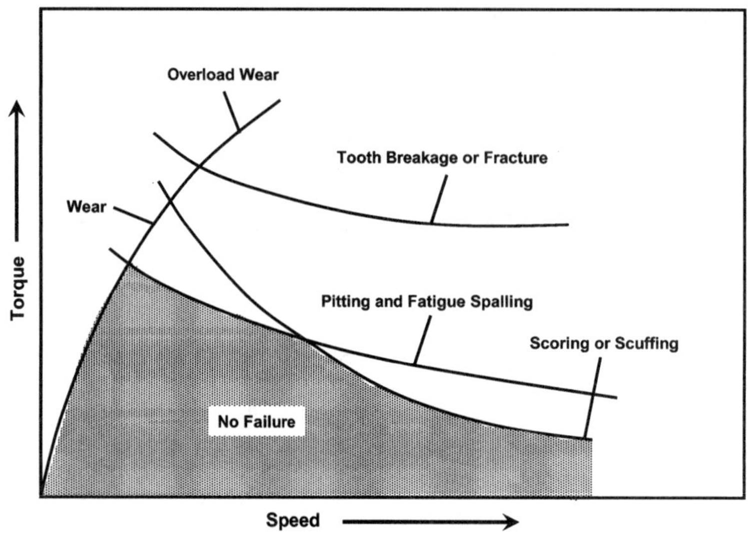

Fig. 8.2—Gear performance limits [647].

Scanning Electron Microscopy (SEM) [648], will help clarify the nature and root causes of the gear damage. Besides common gear failure modes discussed above, another mode of gear failures which is worth mentioning is the gear overload.

Gear geometry data assist in estimating tooth bending stress, contact stress, lubricant film thickness, and gear tooth contact temperature, based on the transmitted loads. Typical values can be calculated according to ANSI/AGMA 2001 and AGMA 925 standards. These can be compared with AGMA's allowable values to help determine risks of bending fatigue, micropitting, macropitting, wear, and scuffing. Light microscope is also useful for confirming the failure mode, identifying the origin of a fatigue crack or disclosing a material flaw such as a nonmetallic inclusion [649]. If necessary, SEM may also be used. SEM uses energy dispersive X-ray which helps in identifying corrosion, contamination, and inclusions [648]. Other nondestructive tests that can be used to detect material or manufacturing defects include the following:

1. Measuring surface hardness and roughness
2. Looking for magnetic particles
3. Determining gear tooth accuracy

Destructive tests include:

1. SEM microscopy to study fracture surfaces
2. Determination of nonmetallic inclusions
3. Micro-hardness survey
4. Micro-structural determination using acid etches
5. Determination of grain size

Data from these evaluations ought to provide reasonable insight into the probable mode of the gear failure.

Causes of Gear Failure

Durability, reliability, efficiency, and in-service performance of a mechanical drive system, such as gears, depend on a number of factors. These include factors such as design, metallurgy, and mechanical integrity when they are new; operating variables such as speed and load; and the lubricant quality [639,649,650].

Mechanical Factors

These factors relate to metallurgy, forging, cutting, grinding, and finishing of the gears. Many advances have been made in gear design, manufacturing capabilities, and materials technology. These pertain especially to very large gear sets and to technology related to the improved steel hardness [640,651,652]. The result is an increase in durability of 100 %, or higher, and an increase in strength of over 50 % [653]. Similarly, greater tooth accuracy of the pinions and gears is significantly improved. In this regard, the positive influence of the American National Standards Institute/American Gear Manufacturers Association (ANSI/AGMA) rating standards, such as ANSI/AGMA 1103-H07 and ANSI/AGMA 1003-H-07 cannot be underestimated [654].

Operation-related Factors

Figure 8.2 shows the effects of load (torque) and speed on the types of gear failures (*wear, scoring, pitting, fatigue spalling*, and *tooth breakage or fracture*) identified above. The exhibit identifies the load-speed combinations leading to these types of gear damage [647,655].

Wear and *scoring* are two forms of adhesive wear. Wear reduces the tooth thickness and can change the contour of the teeth. Wear initiates at slow equipment speed and low torque and progressively increases with an increase in load, as is shown to the left of the overload wear curve in the slow speed section of the figure. At slow speeds, the formation of a continuous lubricant film is difficult because only a small amount of the lubricant is available in the contact zone. However, higher speeds facilitate the formation of partial, mixed, or full lubricant film, i.e., there is partial metal-to-metal contact or the film is thick enough to prevent metal-to-metal contact; which will minimize wear. Viscosity and temperature, with its effect on viscosity, are two additional variables that determine the integrity of the lubricant film and hence should be taken into account. High equipment speed, high oil viscosity, and low operating temperatures, which facilitate hydrodynamic film formation, increase the load that a gear set can handle without failure. In other words, these factors *increase* both the slope of the wear line

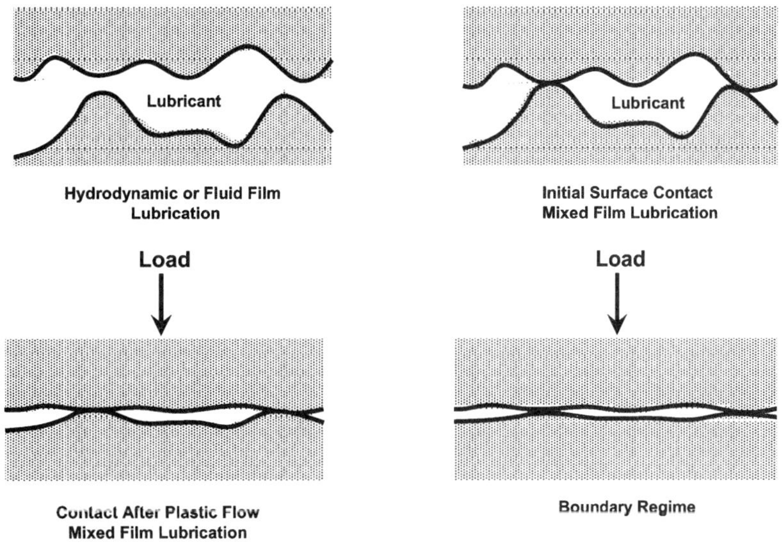

Fig. 8.3—Lubrication regimes and effect of load on asperities.

and the dimensions of the no-wear region. The use of the extreme-pressure additives produces the same effect. This type of wear is apt to occur in high-ratio gear sets, such as worm gears, that operate at slow speeds. An increase in speed will shift the wear regime to the right of the *scoring* or the *scuffing* line. *Scoring*, or scuffing is the wear damage to gear surfaces at high equipment speeds, shown in the high speed region of the figure, and is a consequence of the oil-film failure due to high temperatures that arise from sliding of the tooth surfaces across the oil film. This causes tooth surfaces of the two mating gears to rub against each other and weld. It is the shearing of the welds at high equipment speeds that leads to scoring. Since scoring roughens the tooth surfaces, it further increases the wear rate. The regions of the gear teeth affected are between the pitch line and both the gear root and tip. Scoring usually causes damage to many teeth simultaneously, which results in a large volume of wear debris. Because of the thermal origin of scuffing, oxidation of the metal surfaces to metal oxides also takes place. The degree of oxidation depends upon the lubricant and the severity of scuffing. Since high speeds generate greater heat, the load required to initiate the *scoring* failure is lower than that to initiate *wear* failure. *Scoring* can be minimized by improving the gear design and by using high viscosity lubricants and chemical additives, such as friction modifiers and extreme-pressure agents.

Above the *pitting* and *fatigue spalling* line, wear primarily depends upon the strength of the gear material and not because the lubricant film is inadequate. In reality, the wear damage occurs due to the load being transmitted through the oil film, and if the load is high, fatigue particles result from the gear pitch line. However, if the load is excessive, tooth breakage may occur. The quality of the lubricant has little effect, if any, since this event primarily depends upon the gear material and load [647]. *Pitting*, *spalling*, and *tooth breakage* are gear fatigue failures that occur at fairly high loads and across all speeds. However, the load for *breakage* failure is higher than that required for *pitting*. Nonetheless, as the speed increases, the susceptibility of the gears to experience these types of failure decreases. This is shown in the figure by the gentle drop in the slope of the *pitting* and *breakage* lines with increasing speed. While improved gear design and material hardness are used to minimize this kind of wear damage, chemical additives also help to some degree.

Lubrication-related Factors

Gear lubricants are designed to perform in all three types of lubrication environments, that is, boundary, mixed film, and full film. Boundary lubrication occurs when the gear sets start or stop. When the gears are operating at slow speeds, they are in mixed lubrication regime and when they are operating at high speeds, they are in full film lubrication regime. Of course, the introduction of pressure or load into this equation alters the nature of the lubrication. For example, high loads on gears operating in full film regime will change the lubrication regime to mixed film and the higher loads for those operating in the mixed film regime will alter the regime to boundary, see Fig. 8.3. For a detailed discussion on lubrication regimes, refer to the Chapter 4 on Additives.

The fluid film under boundary conditions is negligible and the metal-to-metal contact is extensive. If the gears are operated under these conditions for extended periods, without adequate protection, rapid and severe wear damage will occur. Protection in this type of environment is provided by the extreme-pressure/antiwear additives. These are thermally labile organo-sulfur and organo-phosphorus compounds that form protective chemical films on metal surfaces. These films protect the metal surfaces against the scoring damage. In mixed film lubrication environment, there are zones of metal-to-metal contact and zones where metal surfaces are well separated by a lubricant film of adequate thickness. Friction resulting from the contacting asperities not only induces wear directly but also indirectly. This is because the frictional heat leads to a drop in lubricant viscosity, which impairs the lubricant's ability to form an effective lubricating film. In the mixed-film regime, wear will occur at a slower rate than in the boundary regime. The additives that are effective in controlling the wear damage are lu-

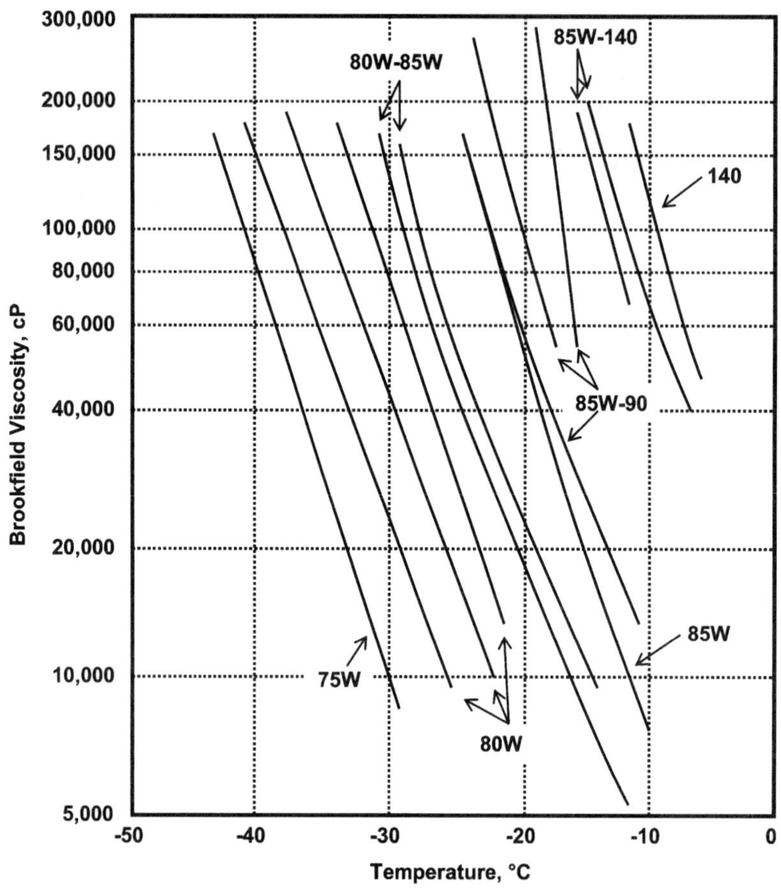

Fig. 8.4—Brookfield viscosity versus temperature of typical automotive gear lubricants [656].

bricity agents, which are natural or synthetic fatty materials, and antiwear agents, which are derivatives of dialkyl dithiophosphoric acids and dialkyldithiocarbamic acids. Fatty materials increase the durability of the lubricating film via physical association or a weak chemical reaction. Antiwear agents form protective chemical films in a manner analogous to that of the extreme-pressure agents, i.e., through chemical reaction, with the difference that they do so at a lower temperature than the EP agents.

Full-film, or hydrodynamic lubrication, is the optimal type of lubrication. In this case, the fluid film is thicker than the asperities of the surfaces. The result is that the film is thick enough to keep the surfaces well apart, with no metal-to-metal contact whatsoever. However, in gear systems, this type of lubrication is not very common because of the gear design and the intended function of efficiently transmitting power. Attempting to keep surfaces apart to achieve hydrodynamic lubrication will lower their power transferring efficiency. Effective transfer of power requires the gear teeth of one gear to mesh with those of the other gear in the gear set. This produces boundary and mixed-film lubrication environment when the gears are static or when they are moving slowly. However, when the gears are moving at high speeds, the lubricant is pulled into the contact zone and under pressure, especially in the case of heavy loaded gears, it gains viscosity to become semisolid (grease like) in behavior. The semisolid lubricant is quite effective in minimizing extensive metal-to-metal contact; hence in minimizing wear. This type of hydrodynamic lubrication is called elasto-hydrodynamic (EHD) lubrication. The elasto-hydrodynamic (EHD) oil film is at least two to three times as thick as the composite surface roughness. The lubricant viscosity and the nature of the base fluid used to formulate the gear lubricant affect the lubricant's ability to provide this type of lubrication. Base fluids of high pressure-viscosity relationship, such as alkylaromatics or naphthenics, are the most effective. Incidentally, lubricant viscosity also affects other lubrication regimes. While the high-viscosity lubricants will invariably form more durable lubricating films, they will have lower fluidity at low temperatures. This will make it more difficult to deliver the lubricant to gears, which will experience increased wear due to lubricant starvation. This is especially the case in equipment that is operated in cold wintry environment, such as that of North America and Northern Europe. That is why, due care must be taken to select automotive lubricants with proper low temperature and high temperature viscometrics.

Low-temperature viscosity of the gear oils is determined by the use of Brookfield viscosity measurement (ASTM D2983). Maximum temperature to attain a viscosity of 150,000 cP is reported. Figure 8.4 shows Brookfield viscosities of viscosity grades that are often used in automotive gear lubricants [656]. As the figure shows, the lower winter (W) grades reach the required viscosity at a lower temperature than the higher winter grades. Straight grade, such as the SAE 140, as shown in the figure, attains the required viscosity of 150,000 cP at −12 °C. It is also important to note that

TABLE 8.2—SAE J306 and MIL-PRF-2105E (SAE J2360) viscosity classifications for automotive gear lubricants.

SAE J 306 Viscosity Classification

Properties	70W	75W	80W	85W	80	85	90	140	250
Viscosity at 100°C, cSt[a]									
Min[b]	4.1	4.1	7.0	11.0	7.0	11.0	13.5	24.0	41.0
Max.	None	None	None	None	< 11.0	< 13.5	< 24.0	< 41.0	None
Viscosity of 150,000 cP[c,d]									
Max Temp, °C	-55[e]	-40	-26	-12	None	None	None	None	None

MIL-PRF-2105E Military Specification (SAE J2360)

Properties	75W	80W90	85W140
Viscosity at 100°C, cSt[a]			
Min[b]	4.1	13.5	24.0
Max	...	<24.0	<41.0
Viscosity of 150,000 cP			
Max Temp, °C[c,d]	-40	-26	-12
Channel Point, min, °C	-45	-35	-20
Flash Point, min °C	150	165	180

[a]Using ASTM D445.
[b]Limit must also be met after testing in CEC L-45-T-93, Method C (20 hours).
[c]Using ASTM D2983.
[d]Additional low-temperature viscosity requirements may be appropriate for fluids intended for use in light-duty synchronized manual transmissions.
[e]The precision of ASTM D2983 has not been established for determinations made at temperatures below -40°C. This fact should be taken into consideration in any producer consumer relationship.

when a viscosity modifier is added to a low viscosity winter grade oil, there is an increase in temperature at which the required low-temperature viscosity is attained. In general, the broader the viscosity grade, the larger is the discrepancy. This is obvious when one compares the viscosity-temperature curves for 85W-90 and 85W-140 multi-grades. The 85W-140 multi-grade attains the required viscosity at −15 °C to −16 °C and 85W-90 attains it at −18 °C to −24 °C. Compare these with the temperature of −24 °C for the straight winter grade 85W. Please note that the Brookfield viscosity (ASTM D2983) requirement is for multi-grades and not for single grades, see Table 8.2. In the case of the mineral oil-derived gear lubricants, it is often necessary to add pour point-depressants to the multi-grades as well. Most commonly used automotive gear oils are of 75W-90, 80W-90, and 85W-140 viscosity grades.

While abrasion, corrosion, fatigue, and adhesion all can cause gear damage, fatigue and adhesive wear play the prominent roles. Figure 8.5 depicts the kinds of surface damage that can occur in gears as a result of inappropriate lubrication [657].

Scratching is the alteration of the tooth surface in the form of scratches of random length in the direction of sliding. It either results from the hard particles that are lodged between the moving surfaces or by rubbing of the asperities of one surface against the other surface. *Corrosive wear* can cause pitting or polishing of the metal surface and results from the attack of the metal surface by water or the lubricant. *Rippling* is an alteration of the tooth surface to give an appearance of a more or less regular pattern resembling ripples on water, or fish scales. *Ridging* is the alteration of the gear tooth to give a series of parallel raised and polished ridges running diagonally in the direction of the sliding motion. These ridges run either partially or completely across the tooth surface of the hypoid gears. *Pitting* is the damage reflected by the appearance of the small irregular cavities in the tooth surface. These result from the removal of the small areas of the surface metal. *Spalling* can be the last stage of pitting or it can occur without previous pitting, as in case-hardened gears. In the latter situation, spalling failure is subsurface in origin. *Pitting* and *spalling* are a consequence of surface fatigue and occur after an extended operation. *Scoring* is the displacement of the metal from one tooth surface to another, resulting in the development of a matte (dull) surface.

Rolling, *peening*, *rippling*, and *ridging* arise from plastic flow. Plastic flow results from overloading of the gear beyond its yield stress. *Rolling* and *peening* damage, shown in the figure, usually occurs in gears that are made of soft materials. *Rippling*, also an indication of the excessive loads, is caused by shearing stresses at the metal surface and can be remedied by the use of a lubricant with a lower coefficient of friction. *Ridging* occurs when the high spots plow the contacting surface.

Protection against wear depends upon the quality of the lubricating film. This film can be physical or chemical in nature. Physical films, somewhat temporary in nature, result from the wetting ability of the lubricant, with or without a friction modifier. Chemical films, somewhat more permanent, result from the reaction of the metal surfaces with chemicals that are present in the lubricant. Viscosity of a gear oil is directly related to its film-forming ability, hence load-carrying capacity; the higher the load-carrying capacity of the fluid, the greater the protection against wear. Consequently, high viscosity oils minimize wear. For example, SAE 140 viscosity grade lubricants allow less wear than SAE 90 viscosity grade oils [658]. Figure 4.96 shows the effect of temperature on the load-carrying capacity of a lubricant. As the

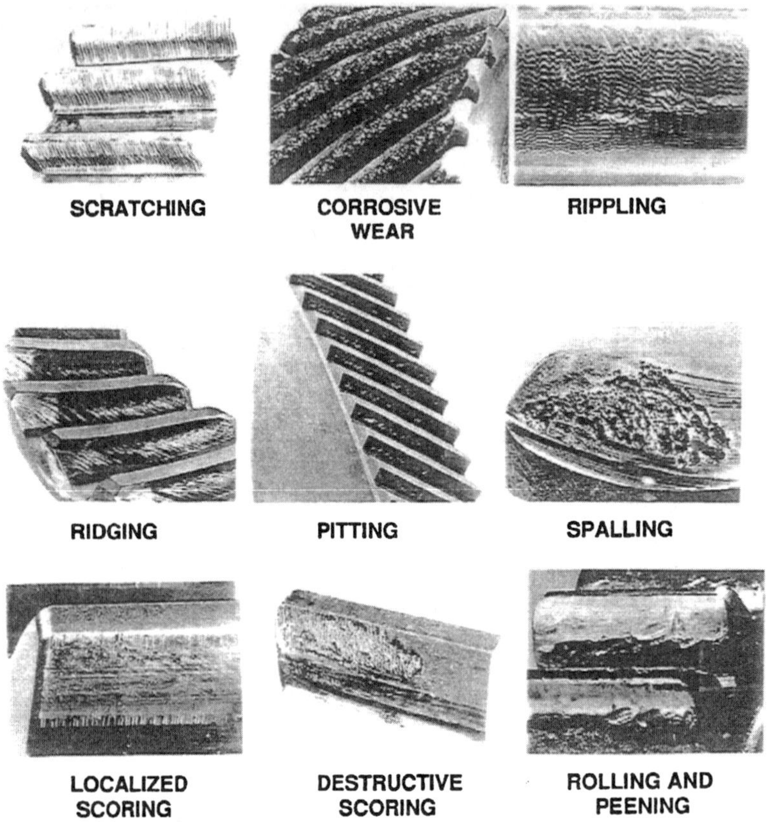

Fig. 8.5—Types of gear damage [657,656].

viscosity of the lubricant drops as a consequence of the higher temperature, its load-carrying capacity also drops and correlates with the lubricant's film-forming ability [397].

As mentioned earlier, fatigue and adhesive wear are the two primary causes of gear failure. Adhesive wear can be considered to take place in two stages: adhesion and welding. A representation of the two processes for ferrous (iron) surfaces is given in Fig. 8.6. Iron forms body-centered crystals which are shown as two crystal lattices, each representing a metal surface. The bonds between the metal atoms are based on coordination and are shown in the exhibit in a manner that facilitates explanation. In reality, though, bulk metals consist of positively charged ions (nuclei) immersed in a sea of electrons [399], as depicted in Fig. 8.7. That is why, metals have good thermal and electrical conductivity. During adhesion, some of the *intrafacial* bonds are replaced by *interfacial* bonds. The *intrafacial* bonds are coordination bonds between atoms of the same surface, and *interfacial* bonds are bonds between the atoms of one surface and the other. This situation is presented in the middle part of Fig. 8.6, where the dotted bonds show adhesion. It is important to note that the bond formation in metals is different from

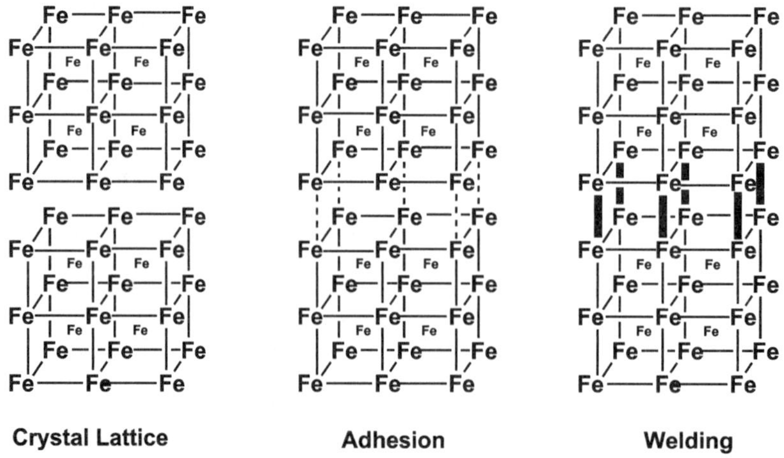

Fig. 8.6—Metal-to-metal interaction—A simplistic view.

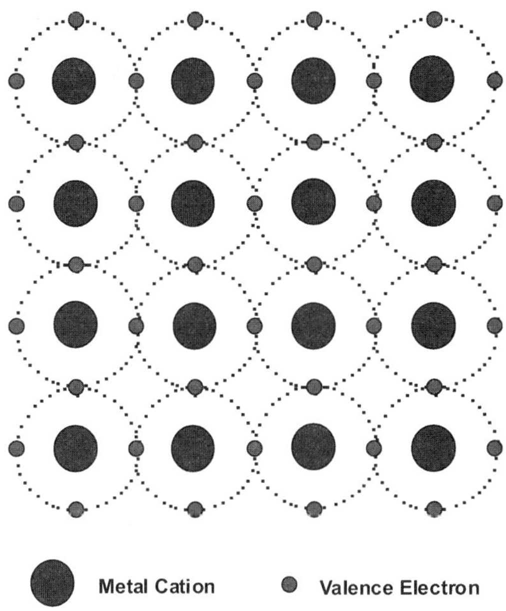

Fig. 8.7—Nature of the metal-metal bond.

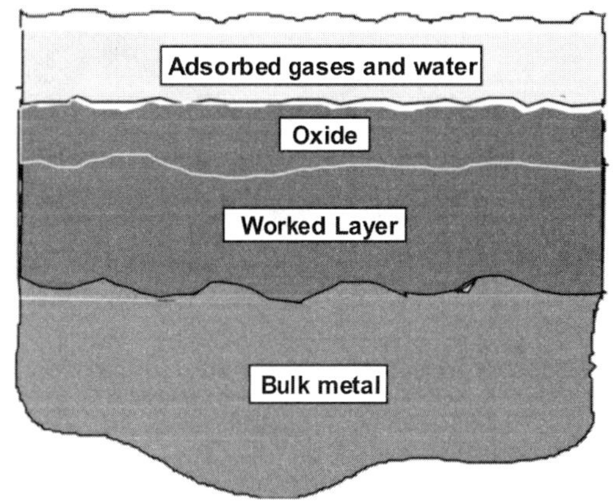

Fig. 8.8—A schematic representation of a metal surface [399].

that in organic compounds. Organic compounds have well-defined bonds resulting from sharing of the electrons by the atoms involved. The bonds in metal atoms, on the other hand, are not defined at all and refer to the bonding nuclei holding on to their own or their neighbors' valence electrons. In transition metals, such as iron, each nucleus can hold on to many electrons because of the incompletely filled *d* or *p* orbitals in their atoms. The force of adhesion, the force necessary to separate the two metal surfaces, is a function of the atomic structure of the metals, their inter-atomic bond energies, their tendency to form solid solutions, and their hardness [8].

Except under very specific circumstances, the metal surfaces, in addition to containing an oxide layer, are contaminated with water and adsorbed gases [399], as shown in Fig. 8.8. Consequently, the surfaces do not strongly adhere to each other. However, the situation changes when they either lose their adsorbed films because of heat, or fresh metal is exposed as a consequence of the wear damage. The force of adhesion is the strongest when the metal surfaces are clean. Nonetheless, as the surfaces get contaminated, the force of adhesion weakens. This is shown in Fig. 8.9 [659]. In the figure, the data for the atomically clean surfaces are off the scale and are not shown.

For gears, the load has an additional impact on fusion or welding. All surfaces, irrespective of the quality of the finish, have surface roughness or asperities. The size of these asperities depends upon the finishing method, see Table 8.1. As the load is applied, the local pressure on the asperities increases. When this pressure exceeds the metal's yield stress, the asperities are flattened via plastic deformation to support the load [7], see Fig. 8.3. The result is an increased surface-to-surface contact, and hence a greater degree of adhesion. This is depicted in Fig. 8.10 [7]. Fusion occurs at the areas of increased contact to form "welded junctions," which are points where the two surfaces are fastened more or less irreversibly. That is, they cannot be separated from each other without the transfer of metal from one surface to another. This situation is shown in the right most part in Fig. 8.6, where the welded junctions are indicated by the heavy bonds. Figure 8.11 illustrates the difference between adhesion, also called cold welding, and metallurgical welding [660]. Sliding or movement of the surfaces causes shearing of these junctions leading to fracture, the generation of the wear particles, and the interfacial metal transfer [Fig. 1.27].

Gear Lubrication

The objective of lubrication in gears is to prevent tooth wear and premature failure of the gear system. A proper lubricant help achieve this by promoting sliding between teeth to reduce the coefficient of friction (μ) and limiting the temperature rise caused by rolling and sliding friction.

Lubrication Methods

Three general methods used to lubricate gears are grease lubrication, splash lubrication (oil bath method), and forced oil circulation lubrication [661,662]. The application

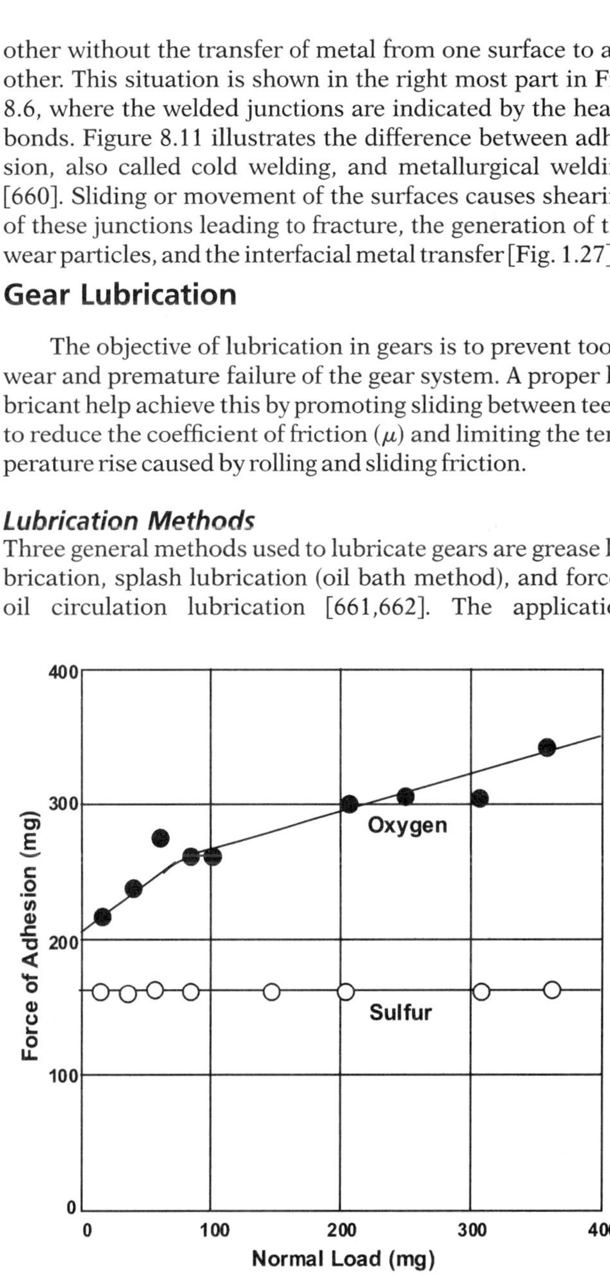

Fig. 8.9—Adhesion of iron surfaces [659].

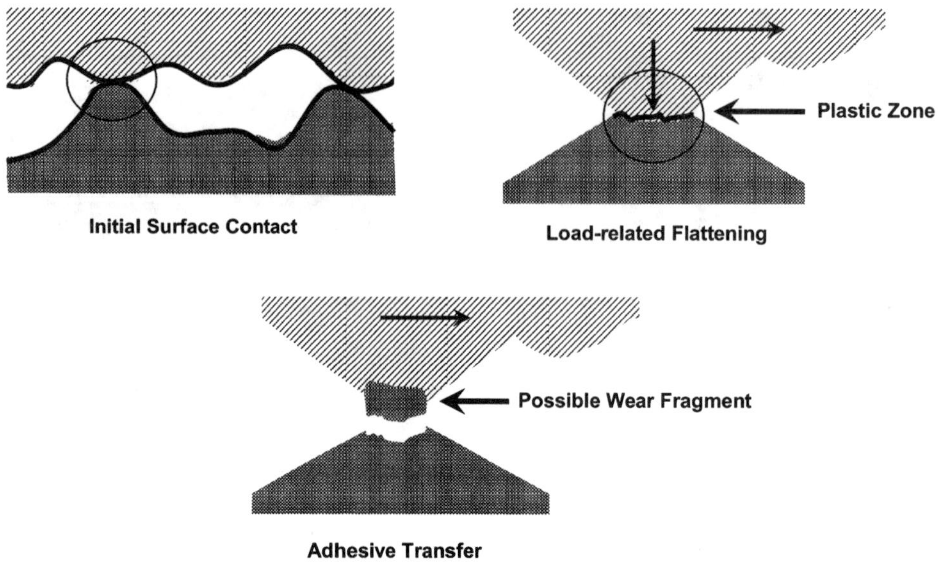

Fig. 8.10—Mechanism leading to adhesive wear [7].

method depends upon the tangential speed in m/s (metres/second) and the rotating speed in r/min (revolutions per minute). At slow operating speeds, grease lubrication is a good choice. For medium and high speed operation, splash lubrication and forced circulation lubrication are more appropriate. Of course, there are exceptions. Sometimes, for maintenance reasons, a grease lubricant is used even for high speed operations. Table 8.3 lists suitable lubrication methods for various gear speeds [662].

Grease lubrication is suitable for any gear system that is open or enclosed, as long as it runs at slow speed. However, there are three considerations before selecting grease to lubricate gears. These are listed below [662].

1. Grease must have a suitable viscosity, especially when used to lubricate the enclosed gear systems which require good lubricant fluidity.

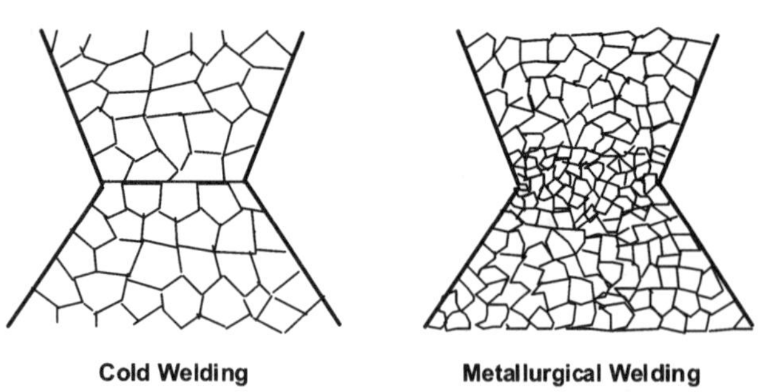

Fig. 8.11—Cold welding versus metallurgical welding [660].

TABLE 8.3—Lubrication method versus gear speed [662].

Lubrication Method	Spur and Bevel Gears Tangential Speed (m/s)	Worm Gears Sliding Speed (m/s)
Grease	0.1 to 6	0.1 to 4
Splash	4 to 15	3 to 10
Forced Circulation	13 to >25	8 to >25

2. Lubricating grease is not suitable for use under high load and continuous operation. This is because the cooling effect of grease is not as good as that of the lubricating oil and the gear sets may run at higher temperatures, which can lead to extensive wear damage.
3. The amount of grease must be sufficient to ensure proper lubrication. Too little will allow wear damage to occur and too much may impair proper functioning of the gears due to viscous drag and the accompanying power loss.

Splash lubrication is commonly used in an enclosed system. The rotating gears splash the lubricant onto the gear system and the bearings. For effective lubrication, a minimum tangential speed of 3 m/s is required. Splash lubrication suffers from several problems, two of which are oil level and temperature limitations. Too high an oil level will result in excessive agitation losses and too low an oil level will not deliver an effective amount of lubricant to lubricate and cool the gear system. It is therefore important to monitor the oil level during the operation to make sure that the oil level is suitable. This is because the oil level will drop when the gears are in motion. This problem may be corrected by raising the static level of lubricant or by installing an oil pan [662]. The temperature of a gear system may rise because of the friction increase in gears and bearings and due to lubricant agitation. Rising temperature may cause lowering of the lubricant viscosity, acceleration of lubricant degradation due to oxidation, and deformation of the housing, gears and shafts. High-performance gear lubricants typically endure temperatures up to 80 to 90 °C. If the lubricant's temperature is expected to exceed this limit, cooling fins must be added to the gear box, or a cooling fan incorporated into the system [662].

Forced circulation lubrication supplies lubricant to the contact portion of the gear teeth by means of an oil pump. The lubricant is delivered into the contact zone in the form of a drop, spray, or an oil mist using compressed air. The oil mist method is especially suitable for high-speed gearing.

TABLE 8.4—Desired gear lubricant properties [662].

No.	Property	Description
1	Suitable Viscosity	Lubricant must maintain proper viscosity to form a stable oil film at the specified temperature and speed of operation. Refer to ANSI/AGMA 9005-E02 Standard and Reference [654]
2	Anti-wear	Lubricant must have the property to prevent scoring failure of the tooth surface while under high-pressure or load.
3	Thermal and Oxidation Stability	Lubricant must not oxidize easily and must perform in moist and high-temperature environment for a long duration.
4	Water Tolerance	Moisture tends to condense due to temperature change when the gears are stopped. The lubricant must be resistant to moisture and have the ability to quickly shed water.
5	Foam Inhibition	Lubricant must resist foam formation resulting from agitation and when and if it forms, to break it quickly. Foam inhibition in a lubricant is important because a lubricant with foam will not form a proper lubricating film.
6	Corrosion Resistance	Lubricant must be neutral so as not to cause corrosion of iron and copper and must have the ability to prevent corrosion resulting from corrosive species.

Forced circulation system is considered the best way to lubricate gears.

Selection of a Gear Lubricant

For a gear lubricant to be effective, it must not only have suitable viscosity but also a variety of other properties, some of which are listed in Table 8.4 [662].

Improper lubricant use will contributes towards gear failure. Lubricant-related failures stem from contamination, oil film collapse, additive depletion and use of a lubricant that is unsuitable for the application. The most common failure is due to particle contamination of the lubricant. Dust and wear particles are highly abrasive and can penetrate through the oil film, causing "plowing" wear or ridging on metal surfaces. Water contamination can cause rust on working surfaces of metallic gears and eventually destroy their structural integrity.

Oil film collapse will result in pitting which is due to tooth surface and subsurface fatigue cracks resulting from metal-to-metal contact of asperities or defects. High-speed gears with smooth surfaces and good film thickness may experience pitting due to subsurface cracks. Pits are formed when these cracks break through the tooth surface and cause material separation. When several pits join, a larger pit, or a spall, is formed. Another possible cause of pitting is hydrogen embrittlement of the metal due to water contamination of the lubricant. Pitting can also be caused by foreign particles being present in the lubricant.

Proper oil viscosity will ensure quiet operation by keeping the surfaces well apart and dampening the noise. If the oil viscosity is too high, drag-related power losses and a temperature rise will occur. The latter will lead to increased degradation of the lubricant. In cold climates gear lubricants must flow at low temperatures. Typically, gear oils must have a minimum pour point of 5 °C (9 °F) lower than the lowest expected temperature. The pour point for mineral gear oil is typically –7 °C (20 °F) [663]. When lower pour points are required, synthetic gear oils with pour points of –40°C (40 °F) are usually used. Gear speed is also a consideration while selecting lubricant viscosity. High velocities are generally associated with light loads and very short contact times. For these applications, the low-viscosity oils are usually adequate. In contrast, slow speeds are associated with high loads and long contact times. These conditions require higher-viscosity oils. EP additives may also be required, if the loads are very high [663].

Ambient and operating temperatures also influence the selection of a gear lubricant. Normal gear oil operating temperature ranges from 50 to 55 °C (90 to 100 °F) above ambient temperatures. Oils operating at high temperatures require good viscosity and high resistance to oxidation and foaming. High operating temperatures are also indicative of oils that are too viscous for the application, which was mentioned earlier, excess oil in the housing, or an overloaded condition. Oils for gears operating at low ambient temperatures must be able to flow easily and provide adequate viscosity. Therefore, these gear oils must possess high viscosity indices and low pour points.

Gear Lubricant Classification

Traditionally, gear lubricants are classified into two broad groups: automotive gear oils and industrial gear oils. This classification distinguishes between the types of gear systems they lubricate and the system's operating environments. Automotive gear oils primarily lubricate spiral-bevel gears in on-highway and off-highway truck axles, and hypoid gears in vehicles that include trucks and passenger cars. Conversely, industrial gear oils are used to lubricate gear systems in nonautomotive applications. These are predominantly used in industries engaged in manufacturing of one kind or another. We will discuss the automotive gear oils first, which will be followed by the discussion on the industrial gear oils.

Automotive Gear Oils

Automotive gear oils are classified in a manner similar to that of the engine oils, i.e., through the SAE viscosity grades [664], API service designations [665, 666], U.S. Military specifications, and the OEM performance requirements. However, it is important to note that the SAE viscosity grades for gear oils are different from the SAE viscosity grades for engine oils and are independent of each other, see Fig. 1.15. A numerical comparison may therefore be mis-

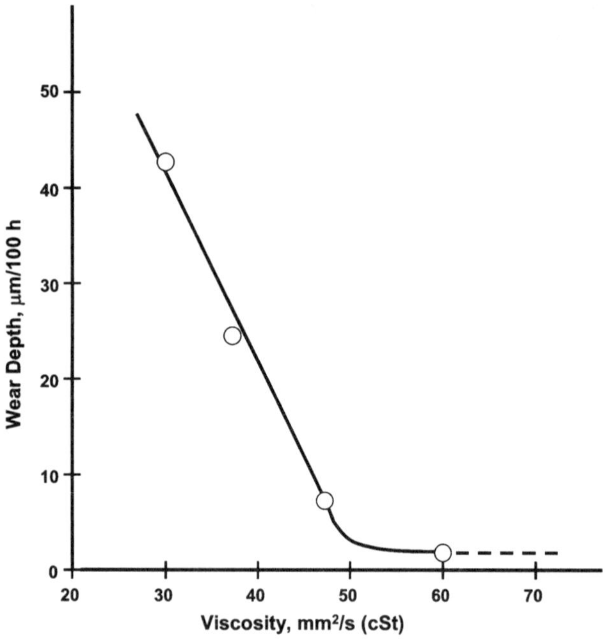

Fig. 8.12—Viscosity versus wear [210].

leading. For example, the viscosity range for an SAE 85W grade gear oil is not higher than that for the SAE 40 grade engine oil, as the higher numerical rating may suggest. Similarly, the viscosity of the SAE 90 grade gear oil is not higher than that of the SAE 50 grade engine oil, but is almost equivalent.

Table 8.2 provides physical requirements for the gear oil viscosity grades according to the SAE Standard J306 and U.S. Military specification MIL-PRF-2105E (SAE J2360). Viscosity plays a crucial role in controlling the load-carrying capacity and wear in gear systems; hence its importance in gear lubrication cannot be overemphasized. The load-carrying capacity is the maximum load that a sliding or rolling system can support without failure. Figure 4.94 shows the load-carrying capacity as a function of the lubricant viscosity at two temperatures [397] and Fig. 8.12 depicts the relationship between viscosity and wear [210]. The data presented in these figures show that the load-carrying capacity increases with an increase in viscosity, and the wear decreases. Field data substantiate greater wear resistance when high viscosity gear oils, such as SAE 140, are used than when low viscosity gear oils, such as SAE 90, are used [658]. However, from the fuel conservation point of view, the lower viscosity grades are better than the higher viscosity grades [668,669]. This observation is based upon the axle efficiency data. Axle efficiency represents the ability of an axle to convert input power into output power. Axle efficiency is temperature-dependent and tends to increase with increasing temperature until it reaches an optimum, after which it drops. This is shown in Fig. 8.13. The figure also shows that at high temperatures the low-viscosity oils lose their effectiveness much more rapidly than the high-viscosity oils. This suggests that the fuel economy advantage due to viscosity characteristics is likely to be different at different operating temperatures.

While a wide variety of multi-grade gear oils are possible, the U.S. Military specifications require 80W-90 and 85W-140 only. Multi-grade oils, in general, are superior to single grade oils due to their effective film-forming ability over a broad temperature range, thus providing an enhanced wear protection. Figure 8.14 shows the relative effectiveness of the various viscosity grades with respect to temperature.

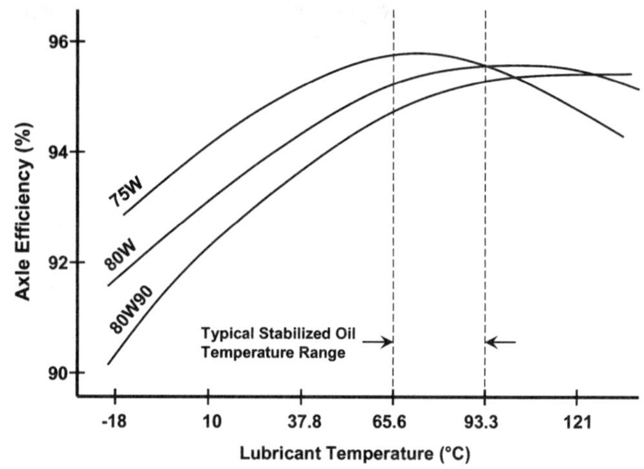

Fig. 8.13—Axle efficiency versus lubricant temperature.

As mentioned earlier, the primary function of the gear oils is to protect gears and axles against fatigue, scoring, and wear under boundary lubrication conditions. To achieve this function effectively, gear oils must have the ability to form physical and chemical films of the appropriate film thickness and strength. The quality of such films is a function of the operating temperatures, lubricant viscosity, and the presence of chemical additives, such as friction modifiers and extreme-pressure agents.

The performance classifications for automotive gear oils range from GL-1 to GL-6, specifying oils in increasing order of load-carrying capacity. The abbreviation "GL" stands for gear lubricant. The GL-6 classification, which was previously used to describe anti-scoring performance over and above that provided by the GL-5 lubricants, is technically obsolete. The GL-4 and GL-5 categories correspond to U.S. Military specifications MIL-L-2105 and MIL-L-2105D, respectively, and define oils for service-fill only. The specification MIL-PRF-2105E, issued in 1995, combines the GL-5 requirements of the military specification MIL-L-2105D and

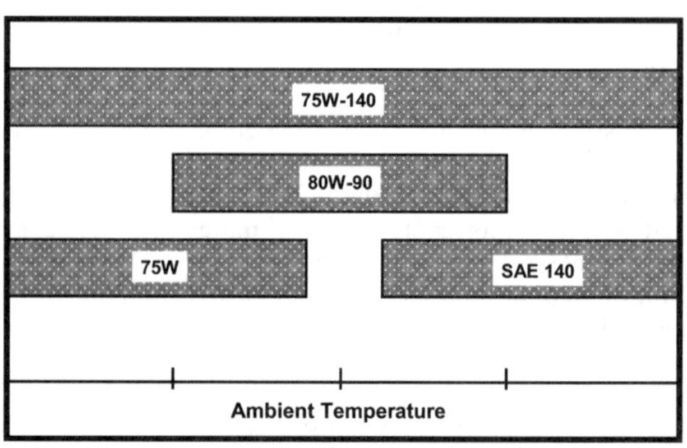

Fig. 8.14—Useful gear oil viscosity grades.

TABLE 8.5—API service designations for gear oils.[a]

API Classification	Service Severity	Lubricant Type	Applications
GL-1	Mild	Straight petroleum oils or mineral oils[b].	Manual transmissions that operate under mild condition, such as truck and tractor manual transmissions
GL-2 Presently not in use	Moderate	Products suited for this type of service contain anti-wear or film-strength improvers specifically designed to protect worm gears.	Automotive worm-gear axles operating under conditions of load, temperature, and sliding velocities where lubricants satisfactory for API GL-1 service will not suffice.
GL-3 Presently not in use	Moderate to Severe	Oils containing mild EP additives to impart load-carrying capabilities exceeding those by API GL-1 lubricants but below those of API GL-4 lubricants. These lubricants are not intended for axles with hypoid gears. Some transmission and axle manufacturers specify API Category CC or CD motor oils for this service.	Manual transmissions operating under moderate to severe conditions and spiral-bevel axles operating under mild to moderate conditions of speed and load.
GL-4	Mild to Severe	Oil quality equivalent to now obsolete MIL-L-2105; performance requirements are usually satisfied with 50% GL-5 additive level	Axles with spiral bevel gears operating under moderate to severe conditions of speed and load or axles with hypoid[c] gears operating under moderate speeds and loads. These oils may be used in selected manual transmission and transaxle applications where MT-1 lubricants are unsuitable.
GL-5	Severe	Virtually equivalent to MIL-L-2105D; primary field service recommendation of most passenger car and truck builders worldwide. However, the API designation does not require military approval. Performance test details are provided in ASTM Publication STP-512A.	Gears, particularly hypoid[c] gears, in axles operating under various combinations of high-speed/shock load and low-speed/high-torque conditions.
GL-6 Presently not in use	Most Severe	A shift to more modest pinion offsets and the lack of original API GL-6 test equipment and procedures have reduced commercial use of API GL-6 gear lubricants.	Gears designed with a very high pinion offset. Such designs typically require protection from gear scoring in excess of that provided by API GL-5 gear oils.
MT-1	Moderate to Severe	The lubricant must provide protection against thermal degradation, component wear, and oil-seal deterioration, which is not provided by lubricants meeting the requirements of API GL-1, GL-4, or GL-5. API MT-1 lubricants should not be mixed with engine oils in the same transmission unit. Transmission manufacturers' specific lubricant quality recommendations must be followed.	Non-synchronized manual transmissions used in heavy-duty service, such as those used in buses and heavy-duty trucks. Not suitable for use in synchronized transmissions and transaxles used in passenger car and heavy-duty applications.

[a]For full description see API Publication 1560

[b]However, they may be supplemented with additives; such as oxidation inhibitors, rust inhibitors, foam inhibitors, and pour point depressants, to improve their performance.

[c]Frictional requirements for axles equipped with limited-slip differentials are normally defined by the axle manufacturer.

thermal oxidation stability, antiwear, and seal compatibility requirements of the API specification MT-1. The new specification defines lubricants for nonsynchronized manual transmissions used in buses and heavy-duty trucks. Factory-fill oils are defined by the major car and truck manufacturers. Such oils have performance characteristics that are critical to the satisfactory operation of a particular drive train and may include break-in, bearing preload, and limited

TABLE 8.6—Bench and axle test requirements for gear oil categories [670].

API GL-5	MIL-L-2105D	MIL-PRF-2105E/ SAE J2360	API MT-1	Description	Characteristics Measured
ASTM L-33	ASTM L-33	ASTM L-33	—	Gear test using differential assembly	Resistance to corrosion in the presence of moisture
ASTM D6121 (L-37)	ASTM D6121 (L-37)	ASTM D6121 (L-37)	—	Gear test using complete axle assembly	Resistance to gear distress under low-speed, high-torque conditions
ASTM L-42	ASTM L-42	ASTM L-42	—	Gear test using complete axle assembly	Resistance to gear distress (scoring) under high-speed, shock-load conditions
ASTM L-60	ASTM L-60	—	—	Bench test using spur gears	Thermal oxidation stability
—	—	ASTM D5704 (L-60-1)	ASTM D704 (L-60-1)	Bench test using spur gears	Thermal and oxidation stability, and deposits
ASTM D892	ASTM D892	ASTM D892	ASTM D892	Bench test	Foaming tendency
	ASTM D130	ASTM D130	ASTM D130	Bench test	Stability in the presence of copper and copper alloys
—	—	ASTM D5662	ASTM D5662	Bench test	Seal compatibility
—	—		ASTM D5182	Gear test	Spur gear wear
—	—	ASTM D5579	ASTM D5579	Gear test	Transmission cyclic durability (waived for approved lubricants)
—	SS&C FED-791 and FTM 3430		—	Compatibility with existing gear lubricants and storage stability of the finished gear oil	Compatibility with existing oils and long term solubility
—	Controlled Field Test		—	—	Field Performance
—	LRI Committee Review		—	—	—

TABLE 8.7—Automotive gear oil performance criteria for API GL-4, GL-5, and MT-1 categories.

Performance	API GL-4 MIL-L-2105	API GL-5 MIL-L-2105D/E	API MT-1
Scoring resistance under high-speed shock load conditions	CRC L-19[a] test or FTM 6504T: equal to or better than RGO-105	L-42 test: gear/pinion coast side scoring equal to or better than RGO-110	No requirement
Resistance to gear distress under high-torque, low-speed conditions	CRC L-20 test: no tooth disturbance, such as rippling, ridging, pitting, or severe wear	ASTM D6121 (L-37 test): no tooth disturbance, such as rippling, ridging, pitting, or severe wear	No requirement
Cyclic durability	No requirement	No requirement for API GL-5/MIL-L-2105D, better than reference oil[b] for MIL-PRF-2105E	Better than reference oil[b]
Corrosion resistance in the presence of water	1. CRC L-13 or FTM 5313.1 2. CRC L-21 No evidence of rusting	L-33 test: no evidence of rusting after 7 days exposure on any working surface; maximum 0.5 in[b] rust on cover plate (1% of surface area)	No requirement
Thermal and oxidation stability/component cleanliness	No requirement	L-60-1 test: 100% max viscosity increase; 3% max pentane insolubles; 2% max toluene insolubles *For MIL-PRF-2105E* 100% max viscosity increase; 3% max pentane insolubles; 2% max toluene insolubles; 7.5 min carbon/varnish rating on large gear; 9.4 min sludge rating on all gears	L-60-1 test: 100% max viscosity increase; 3% max pentane insolubles; 2% max toluene insolubles; 7.5 min carbon/varnish rating on large gear; 9.4 min sludge rating on all gears
Antiwear	No requirement	No requirement	FZG (A/8.3/90): minimum 10 stage pass
High-temperature lubricant stability	No requirement	No requirement	Mack Transmission Test T-2180: Equal to or better than reference
Oil seal compatibility	No requirement	No requirement for API GL-5 and MIL-L-2105D *For MIL-PRF-2105E* (ASTM D5662) *Polyacrylate* -60 to 0% elongation change, -35 to +5 pts hardness change, -5 to +30% volume change; *Fluoro-elastomer* -75 to 0% elongation change, -5 to +10 pts hardness change, -5 to +15% volume change	ASTM D5662 *Polyacrylate* -60 to 0% elongation change, -20 to +5 pts hardness change, -5 to +30% volume change; *Fluoro-elastomer* -75 to 0% elongation change, -5 to +10 pts hardness change, -5 to +15% volume change
Antifoaming characteristics	CRC L-12 test: readings taken immediately after 5 min aeration; Sequence 1 — 23.9°C, 650 ml; Sequence 2 — 93.3°C, 650 ml	ASTM D892: readings taken immediately after 5 min aeration; Sequence I — 20 ml; Sequence II — 50 ml max; Sequence III — 20 ml max	ASTM D892: readings taken immediately after 5 min aeration; Sequence I — 20 ml max; Sequence II — 50 ml max; Sequence III — 20 ml max
Copper corrosion	ASTM D130: 3B max after 1 h at 121.1°C	ASTM D130: 3 max after 3 h at 121.1°C, 2a max for MIL-PRF-2105E	ASTM D130: 2A max after 3 h at 121.1°C
Channeling characteristics	No requirement	FTM 3456.1 Modified: SAE 75 = -45°C max; SAE 80W-90 = -35°C max; SAE 85W-140 = -40°C max	No requirement
Compatibility with existing gear lubricants	SS&C FED-791	SS&C FED-791	SS&C FED-791
Solubility — measure separated material after centrifuging oil stored for 30 days at room temperature (29.4±9.5°C)	FTM 3430: 0.25% wt max of original non-petroleum material in sample	FTM 3430: 0.25% wt max of original non-petroleum material in sample	FTM 3430: 0.25% wt max .of original non-petroleum material in sample
Compatibility — same as solubility, except mixed 50/50 with each of six reference oils	FTM 3440: 0.50% wt max of original non-petroleum material in sample	FTM 3440: 0.50% wt max of original non-petroleum material in sample	FTM 3440:0.50% wt max of original non-petroleum material in sample

[a]Equipment no longer available; impossible to conduct test per original procedure.
[b]Waived for approved lubricants.

slip durability. The majority of passenger cars now use a transaxle drive train arrangement, reducing the need for the rear axle lubricants. These vehicles are filled for life at the factory. Conventional API GL-4 lubricants are being replaced by more specialized manual transmission fluids. These fluids have excellent thermal stability and carefully tailored frictional characteristics to provide smooth synchronization and good shift quality. A performance area not addressed by the industry specifications is limited slip. Because of the hardware differences among the various

TABLE 8.8—Proposed light-duty manual transmission fluid standard [649].

Test	Test Method	Conditions/Modifications	Limits
Synchromesh SSP 180	CEC L-66-X-96	Under Development	No clash at 100 000 cycles
Wear	ASTM D5182	90°C, 3600 rpm	10 load stage pass
Thermal Oxidation	GFC Procedure A	160°C, 192 hours	50% vis inc. max 2.5 mg KOH/g
Viscosity after shear loss	CEC-L-45-T-93	20 hrs, 60°C, 5000 N	25% max, 5.8 cSt min at 100°C
Pour Point	ASTM D97	-	Less than - 39°C
Brookfield Viscosity	ASTM D2983	-40°C	50 000 cP max
Cold Crank Simulator	ASTM D5293	- 30°C	5000 cP max
Foam	ASTM D892 ASTM D6082	Seq I, II, III Seq IV	20,50,20/0 ml 100/0 ml
Rust Prevention	ASTM D665 (Method A)		Pass
Copper Corrosion	ASTM D130	3 hrs, 121°C	1b max
Seal Compatibility	ASTM D5662		Pass

limited-slip differentials, no standard industry-wide test is available to evaluate a lubricant's ability to prevent chatter in this application. Lubricant requirements, therefore, are based on performance in an individual manufacturer's test rig or vehicle. The European OEMs use API GL-5 and MIL-L-2105D to define minimum performance requirements for oils used in their equipment. They have the additional requirements pertaining to surface fatigue, component cleanliness, synchromesh durability, and viscometrics, depending upon their specific needs. Japanese OEMs recommend API GL-5 lubricants for vehicles fitted with hypoid and spiral bevel axles and API GL-3 and GL-4 lubricants for cars and trucks equipped with manual transmissions. As stated above, most modern cars no longer need rear axle lubricants.

Gear oils are tested according to methods established by the American Coordinating Research Council (CRC) and in specified axles. Table 8.5 presents the API service designations and Table 8.6 lists the affiliated bench and axle tests [670]. Table 8.7 describes the performance criteria for GL-4, GL-5, MIL-PRF-2105G, and MT-1 service categories. As mentioned earlier, the API GL-4 quality lubricants are used

TABLE 8.9—Current/Future OEM Axle Specifications[a] [649].

OEM	Specification	SAE Viscosity Grades	Performance Requirements	Recommended Drain Period (miles)
Mack	GO-J GO-J Plus	80W-90, 85W-140	Mil-PRF-2105E Seal immersion test Mack spalling test[b] 500 000 mile field test 100 hour L-60-1 Extended Mack cycling test[a]	250,000 500,000
Eaton/Dana	PS-163/ SHAES-256	75W, 80W-90, 85W-140	Mil-PRF-2105E Copper & bronze corrosion tests Dynamic seal tests Spalling test 500,000 mile field test Wet L-37 Test	500,000
ArvinMeritor	076-D & A	80W-90, 85W-140	MiL-PRF-2105E	100,000 miles without pump and filter to 500,000 miles with pump, filter and appropriate additives
Mercedes Benz	235	90, 85W-90, 85W-I40	API GL-5 In-house axle tests	60,000 (long haul) for all without flexible service system (FSS); 220,000 for FSS
Mercedes Benz	235.6	80W-90 85W-90, 90	API GL-5 In-house axle tests Dynamic seal tests 50 ppm Cl max.	60,000 (long haul), for all without flexible service system (FSS) 220,000 for FSS
Volvo	97310 97312	75W, 80W, 90, 85W-140, 80W-I40 75W-90	API GL-5 API GL-5/Pitting test/field test	75,000 (long haul) 110,000 current (long haul) , 250 000 target
MAN	342N 342ML 342SL	90 80W-90 75W-90, 80W-90	API GL-5 API GL-5, FZG PITS field test API GL-5, FZG PITS field test	50,000 100,000 (long haul) 200-250,000
Iveco		90, 85WI-40	API GL-5	75,000
ZF	TEML05	90	API GL-5 In-house axle and FZG tests	50,000

[a]This oil is used for axles and transmissions.
[b]Test currently not available

for spiral bevel and hypoid gears in axles, operating under moderate speeds and loads. These lubricants may also be used in selected manual transmissions and transaxles. While this service category is commercially used, some of the original equipment for qualifying against this service designation is no longer available. ASTM is therefore investigating the possibility of redefining this service designation using the modern test equipment. This service category is called Passenger Manual Transmission (PM-I). A summary of the proposed tests for this category are given in Table 8.8 [649]. The OEM performance specifications for automotive gear oils are listed in Table 8.9 [649].

The L-37 and L-42 axle tests assess extreme-pressure performance; the L-33 test determines rust inhibiting characteristics; and the L-60 test measures thermal and oxidative stability of the GL-5 lubricants. The L-60-1 test in addition evaluates the deposit-forming tendency of the gear lubricant. Besides L-60 and L-60-1 tests, another test that is used to determine a gear lubricant's suitability for high-temperature applications is the thermal durability test. Additional tests deal with foaming tendency (ASTM D892), yellow metal corrosion (ASTM D130), seal compatibility (ASTM D471 modified and ASTM D5662), spur gear wear (ASTM D5182), and cycling durability (ASTM D5579) of these lubricants. Also, for use in limited-slip differentials, the gear oil's frictional properties are important.

Of the two extreme pressure tests, the L-42 test evaluates a gear lubricant's performance under high-speed, shock load conditions, and the L-37 test evaluates this performance under slow-speed high-torque conditions. A high-temperature gear lubricant is designed to protect transmissions and final drives that experience operating temperatures of up to 300 °F, or 149 °C. Major truck and car manufacturers around the world use high power-density engines and aerodynamic body designs to improve drag coefficients and fuel economy. This generates high temperatures that place extreme thermal stresses on the drive train components and the lubricant.

Conventional differentials, one shown in the top half of Fig. 8.15, tend to *stall* under certain conditions. Stalling is a situation where one wheel loses traction and the vehicle does not move. This is because the driving torque is divided equally between the two rear wheels. During stalling, the wheel with less traction takes away all the power and spins. This drawback is corrected in limited-slip differentials which are essentially the same as conventional differentials, shown in the bottom half Fig. 8.15, but contain clutch plates or friction cones [671]. These are shown in Fig. 8.16. These devices store energy during engagement, which is suddenly released during disengagement. This allows the transfer of more torque to the wheel with traction. The result is that both wheels spin and the vehicle moves. The common problem with limited-slip differentials is the noise or chatter resulting from stick-slip (engagement-disengagement) that occurs between the clutch elements at slow speeds. Gear oils with proper frictional characteristics minimize this noise.

Gleason Works has developed a unique locking differential under the name of Gleason Torsen® (acronym for torque sensing) differential [672,673]. This design, shown in Fig. 8.17, does not employ clutch plates or friction cones; hence it does not need lubricants with special frictional properties.

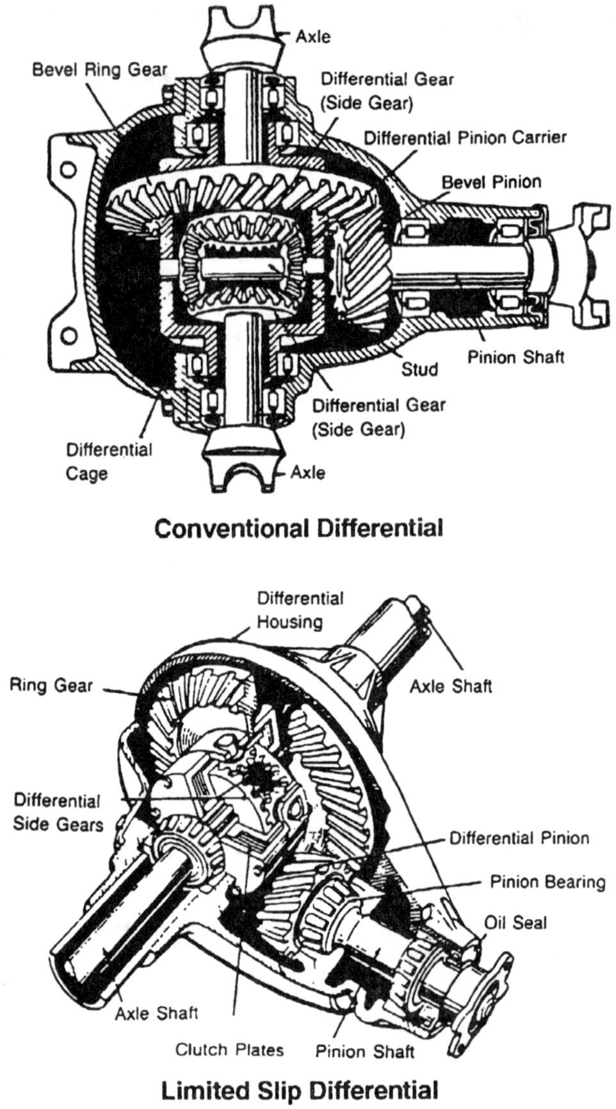

Fig. 8.15—Conventional and limited slip differentials [671].

This differential maintains power to both sides of the axle while turning, without allowing one wheel to spin.

The performance requirements of the U.S. Military specifications MIL-L-2105 and MIL-L-2105C match those of the API GL-4 and GL-5 categories for single grade gear oils and the performance requirements of the military specification MIL-L-2105D almost matches those of the API GL-5 category for 80W-90 and 85W-140 multi-grade gear oils. MIL-PRF-2105E, the newest U.S. Military specification, includes additional tests to evaluate a gear lubricant's deposit-forming tendency, seal compatibility, and transmission's cyclic durability.

A new performance specification, PG-2, for automotive gear oils is presently under consideration [674]. It is a temporary designation for axle lubricants to be used in heavy-duty truck and bus final drives that employ spiral bevel and hypoid gears. This category is designed to qualify lubricants with higher thermal durability, seal compatibility, and surface fatigue performance than the existing GL-5 lubricants. The performance requirements and the recommended tests

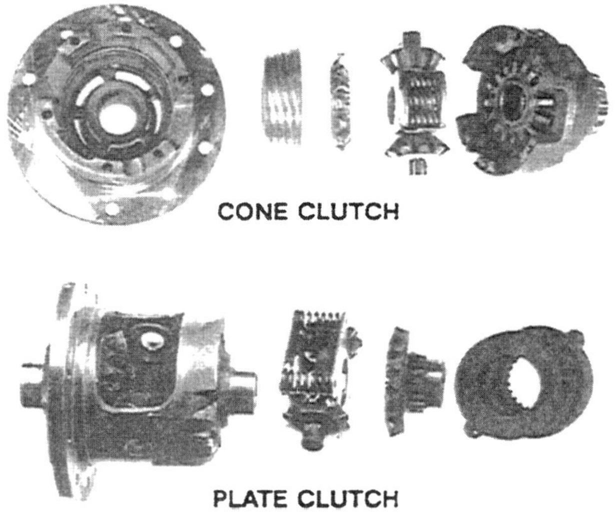

Fig. 8.16—Limited slip differential parts.

for this category are listed in Table 8.10. High-temperature performance and seal compatibility are the key performance features of this and the new military specification, MIL-PRF-2105E. As stated earlier, high temperatures are a direct consequence of the improved aerodynamic design of the modern vehicles. Such designs limit the amount of airflow in the vicinity of the transmissions and transaxles, leading to less effective cooling. Thermal durability of the lubricant *can be* assessed by measuring its wear performance at a higher temperature. A modified L-37 test, called the high-temperature L-37 test (ASTM D6121), is used for this purpose. This test is run at 325 °C for 16 hours in contrast to the regular L-37 test that is run at 275 °C for 24 hours.

Industrial Gear Oils

The diversity of the operating conditions and the performance requirements for the industrial gears necessitate the use of a variety of lubricants. These include both solid and liquid lubricants. Liquid lubricants are obtained by adding the appropriate additives to mineral (petroleum-derived) oils, synthetic fluids, or the oils of biological origin, such as vegetable and animal oils. However, for most gear applications, mineral oil-based lubricants easily meet the performance requirements and at the lowest cost. For other applications, synthetic or biological base fluids may be necessary. Industrial gear lubricants are deigned for gear systems that operate under moderate loading. They are used for both open and enclosed gears. Industrial gear oils are used to lubricate equipment that is used in industries, such as steel mills, construction and mining, kilns, and furnaces. Service requirements of these lubricants are established by organizations such as Association of Iron and Steel Engineers [AISE (U.S. Steel 224)], American Gear Manufacturers Association (AGMA), General Motors, David Brown, DIN 51 517/3m, Cincinnati Milacron, the Society of Tribologists and Lubrication Engineers (STLE), and a variety of other organizations, such as Cincinnati Milacron, and Alcoa [318]. The key functions of these lubricants are to reduce friction and wear.

Industrial gear lubricants can be classified based upon function as well as composition. Function-based classes include solid lubricants, lubricants for open gears, and lubricants for enclosed gears. These classes are briefly described below.

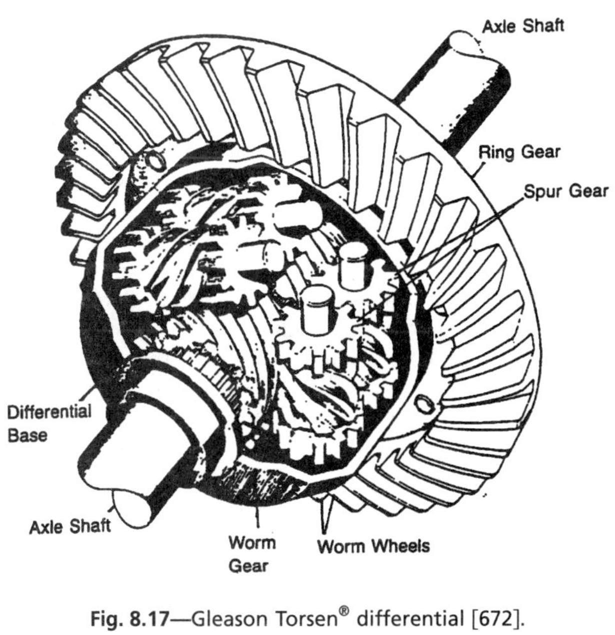

Fig. 8.17—Gleason Torsen® differential [672].

Function-based Classes

Solid lubricants are used in applications either where the frictional heat in the metal-to-metal contact zone is so small that the heat is easily dissipated via conduction through metal, or in space applications where the liquid lubricants will vaporize because of the reduced pressure. Solid lubricants that are commonly used in these situations include graphite, molybdenum disulfide, lead oxide (white lead), polytetrafluoroethylene (PTFE), nylon, and soft metal coatings, such as lead, zinc, and under some conditions, gold and silver. These lubricants form physical films on contacting surfaces and minimize scoring under high speeds. These materials were discussed in some detail under the extreme-pressure section in the Chapter 4 on Additives.

Open gear lubricants, as the name indicates, are open to the environment. These are often used for applications involving large slow-moving gears or gears which operate for short periods of time; and it is impractical or unnecessary to build an enclosure around them. In these cases, the lubricant must form a durable film and should resist being thrown off the metal or being wiped off. These lubricants are very high viscosity residual oils or greases. Straight petroleum or residual mineral oils, which are heavy fractions left after the removal of the hydrocarbons that are suitable for use as fuel and mineral base oils. Typical viscosity of residual oils used in the open gear applications is in the range of 6 to 25 cSt at 100 °C. Some of the high viscosity oils need to be heated or thinned with a low flash solvent, prior to application. In some cases, the application needs to be via painting or brushing. Solvent-thinned oils may be applied through spraying. However, the use of solvent-thinning is losing its appeal due to the implementation of the more stringent air quality standards. While high viscosity of these lubricants provides adequate overall protection to gears, fatty materials and extreme-pressure (EP) additives are sometimes added to improve their performance in high-

TABLE 8.10—PG-2 specifications for gear oils [674]. Reprinted with permission from the Lubrizol Corporation.

REQUIRED PERFORMANCE	RECOMMENDED TEST
Thermal Stability/Component Cleanliness	L-60 Deposit (Intevep rating procedure)
Oil Seal Compatibility	ASTM D471
Copper Compatibility	ASTM D130 (3 h at 121°C)
Gear Surface Fatigue	Mack Spalling
API GL-5 Performance	Current API GL-5 Tests
Compatibility With Existing Gear Lubricants	SS & C FED-STD-791 Methods D3430 and D3440

severity applications. Greases are used when the ambient temperatures are high and leakage is a problem. Low viscosity greases are also used in some high-temperature enclosed gear applications and for the same reasons.

Enclosed gear lubricants are used for enclosed gear systems which involve long and continuous periods of operation with severe stirring and agitation of the lubricant. These gears operate over a wide range of loads, speeds, and temperatures; hence the lubricants chosen must have proper viscosity and the EP characteristics. Enclosed systems have the advantages of low lubricant loss and the ability to use a wider variety of lubrication methods. These include splash, spray, and oil mists. For large operations, such in steel mills, the use of a circulating oil system is more economical. Fluid lubricants of medium viscosity can fulfill the needs of the most gear applications. For high-speed applications, heat removal and flow properties of the lubricant are important. Fluid gear lubricants are made from base oil/s of suitable viscosity and one or more additives to improve service life and suitability under various operating conditions. Because of the relationship of the viscosity to the load-carrying capacity as well as pumpability, gear lubricant must have good low-temperature fluidity as well as a high viscosity index, to provide protection under high-load high-temperature conditions. At low temperatures, the oils may be too viscous to flow, leading to lubricant starvation in certain parts, leading to extensive wear. This is typically a problem when the conventionally refined mineral oil base stocks are used to formulate lubricants. This is because such base stocks contain straight long-chain molecules, which wax out at low temperatures to form crystalline networks. The wax networks absorb most of the oil, thereby impairing its fluidity. This problem can be overcome by the use of the pour point depressants, as stated in the Additives chapter. The use of the multi-grades, which are made by adding a polymeric viscosity modifier to a low viscosity oil in automotive gear oils is quite common, but not in industrial gear oils. Enclosed gear lubricants contain other additives to improve oxidation, EP properties, foaming, corrosion, demulsibility, and so on. Typical viscosity grades for these lubricants range from ISO 46 to ISO 1500, but for selecting a lubricant for use, the lowest viscosity permitted by the gear manufacturer should be used since it will minimize the energy consumption. Typical viscosity grade recommendations are as follows:

1. ISO 68, 100 and 150 for gears and bearings in small, high-speed gear reducers and larger units where the temperatures are less than 10 °C.
2. ISO 220 and 320 for gears and bearings in medium-sized units, where speeds are low and the temperatures are normal (30–40 °C).
3. ISO 460 and 680 for gears and bearings operating in large units, having heavy loads and moderate speeds. These can also be used in smaller units where temperatures exceed 40 °C.

Composition-based Classes

R & O oils are mineral oils that contain rust and oxidation inhibitors to control lubricant's oxidative break down under high operating temperatures and protect the metal surfaces against corrosion, resulting from the attack of the metal by aggressive chemical species. Such species are either present in the environment or the lubricant, or result from the thermal and oxidative decomposition of the lubricant. These oils in addition contain foam inhibitors and antiwear additives. R&O oils are commonly used to lubricate high-speed single helical, herringbone reduction gear sets in applications that experience light to moderate loads. The effective temperature range of these lubricants is −5 °F to 250 °F (−15 °C to 121 °C).

Extreme pressure oils contain additives that provide protection to gears under extreme pressure, or the boundary lubrication conditions. These additives are usually blends of highly surface-active sulfur and phosphorus compounds that reduce friction and control wear damage to the gears. These lubricants may also contain inorganic solids, such as graphite, molybdenum disulfide, and potassium tri-borate, as a colloidal suspension. Ideal particle size of the suspended solids must be no more than 0.5 microns. Otherwise, they will have the tendency to separate from the bulk lubricant. These materials make physical protective films rather than chemical protective films and further improve the gear lubricant's load-carrying capabilities. EP oils are used to lubricate spur, straight bevel, spiral-bevel, and helical gears, where the loads are too high and sliding conditions moderate to high for R&O oils to provide adequate protection. Some EP oils are supplemented with dispersants and detergents, which make them suitable for use in hypoid axles that are exposed to heavy loads or shock loading. Shock loading occurs when heavy acceleration is followed by quick deceleration. It promotes scoring damage to the gear surfaces. While typical operating range for EP oils is between −20 and 120°C (−4 to 250°F), some oils can perform adequately even over −54 to 170°C (−65 to 338 °F) range. This is especially true of the oils that are based on synthetic base fluids.

Compounded oils contain fat, tallow, or synthetic fatty materials as lubricity agents. These materials are added to

TABLE 8.11—Types of gear lubricants used with various gear applications [649].

	Gear Types				
Lubricant	**Spur**	**Helical**	**Worm**	**Bevel**	**Hypoid**
R & O oils (non-EP)	Normal loads	Normal loads	Light loads, slow speeds only	Normal loads	Not recommended
EP oils	Heavy or shock loading	Heavy or shock loading	Satisfactory for most applications	Heavy or shock loading	Required for most applications
Compounded oils (ca. 5 % tallow)	Not normally used	Not normally used	Preferred by most gear manufacturers	Not normally used	For light loading only
Heavy-bodied open gear oils	Slow-speed open gearing	Slow-speed open gearing	Slow-speeds only EP additive desirable	Slow-speeds, open gearing	Slow speeds only EP additive required
Greases	Slow-speed open gearing	Slow-speed open gearing	Slow-speeds only EP additive desirable	Slow-speeds open gearing	Not recommended

the heavy cylinder oil base stocks to reduce friction. Compounded oils are suitable for use in worm gear drives, where high sliding speeds between the gear teeth requires lubricity additives to reduce friction and improve torque efficiency. Typical effective temperature for these oils is between 5 and 82°C (40 to 180°F). Because of the extensive contact between the worm gear and the wheel gear, the lubricant gets constantly removed; hence the need for relubrication is constant. Most worm gear drives normally require an ISO 460 or 680 compounded oil, and in some cases ISO 1000 oil. The viscosity grade required depends upon the worm gear drive's speed and operating temperature. Generally, the lower the worm gear's speed, the higher is the required viscosity grade.

Open gear compounds were already commented upon earlier. Essentially, these are materials that are used to lubricate large gear sets that operate at slow speeds. These lubricants are supplemented with tackifiers to make the lubricant adhere to gears and resist easy removal. Typical operating range for these lubricants is 5 to 120°C (40 to 250 °F). Because of their low fluidity, these lubricants cannot act as coolants.

Greases were also commented upon earlier. The primary advantage of a grease as a lubricant is that it stays in place. The use of lubricating greases is restricted to gears that operate at slow speeds and under light loads. Greases are ideal lubricants for small gear boxes used in household appliances. Operating temperatures range for greases is

TABLE 8.12—Viscosity ranges for AGMA gear lubricants [649].

Rust and Oxidation Inhibited Gear Oils	Viscosity Range[a] mm^2/s (cSt) at 40 °C	Equivalent ISO Grade[a]	Extreme Pressure Gear Lubricants[b]	Synthetic Gear Oils[c]
AGMA Lubricant No.			*AGMA Lubricant No.*	*AGMA Lubricant No.*
0	28.8-35.2	32	...	0S
1	41.4-50.6	46	...	1S
2	61.2-74.8	68	2 EP	2S
3	90-110	100	3 EP	3S
4	135-165	150	4 EP	4S
5	198-242	220	5 EP	5S
6	288-352	320	6EP	6S
7,7 Comp[d]	414-506	460	7 EP	7S
8,8 Comp[d]	612-748	680	8 EP	8S
8A Comp[d]	900-11 00	1000	8 AEP	...
9	1350-1650	1500	9 EP	9S
10	2880-3520	...	10 EP	10 S
11	4140-5060	...	11 EP	11 S
12	6120-7480	...	12 EP	12 S
13	190-220 cSt at 100 °C (212 °F)	...	13 EP	13 S
Residual compounds[f]	**Viscosity ranges**[e]	...	...	...
	cSt at 100 °C (212 °F)	...	...	...
14R	428.5-857.0	...	...	...
15R	857.0-1714.0	...	...	...

[a]Per ISO 3448, Industrial Liquid Lubricants—ISO Viscosity Classification, also ASTM D2422 and British Standards Institution B.S. 4231.
[b]Extreme pressure lubricants should be used only when recommended by the gear manufacturer.
[c]Synthetic gear oils 9S -13S are available but not yet in wide use.
[d]Oils marked Comp are compounded with 3 % to 10 % fatty or synthetic fatty oils.
[e]Viscosities of AGMA Lubricant Number 13 and above are specified at 100 °C (212 °F) as measurement of viscosities of these heavy lubricants at 40 °C (100 °F) would not be practical.
[f]Residual compounds—diluent type, commonly knows as solvent cutbacks, are heavy oils containing a volatile, nonflammable diluent for ease of applications. The diluent evaporates leaving a thick film of lubricant on the gear teeth. Viscosities listed are for the base compound with diluent. CAUTION: These lubricants may require special handling and storage procedures. Diluent can be toxic or irritating to the skin. Do not use these lubricants without proper ventilation. Consult lubricant supplier's instructions.

TABLE 8.13—Minimum physical and performance specifications for inhibited and compounded gear lubricants [649].

Property	Test Procedure	Acceptance Criteria
Viscosity	ISO 3104 ASTM D445	Must be as specified in Table 36 (ISO Grades Table)
Viscosity Index	ISO 2909 ASTM D2270	90 minimum
Oxidation Stability	ISO 4263 ASTM D943	Hours to reach a neutralization number of 2.0[a] *AGMA Grade* — *Hours (minimum)* 0,1,2 — 1500 3,4 — 750 5,6 — 500
Rust Protection	ISO 7120 ASTM D655B	No rust after 24 hours with synthetic sea water
Corrosion Protection	ISO 2160 ASTM D130	1b Copper strip after 3 hours at 121°C (250°F)
Foam Suppression	ASTM D892	Must be within these limits: Max volume of foam (ml) after: *Temperature* — *5 min. blow* — *10 min. rest* Sequence I — 24°C (75°F) — 75 — 10 Sequence II — 93.5°C (200°F) — 75 — 10 Sequence III — 24°C (75°F) — 75 — 30
Demulsibility	ASTM D2711	Must be within these limits. Max percent water in the oil after 5 hour test = 0.5% Max cuff after centrifuging = 2.0 ml Min total free water collected during entire test = 30.0 ml (start with 90 ml of water)
Cleanliness	None	Must be free of visible suspended or settled contamination.

[a]The criteria for acceptance indicated for oxidation stability and demulsibility is not applicable to compounded gear oils.

−30 to 120 °C, depending upon the base oil and the soap used.

Table 8.11 summarizes the various types of gear lubricants and their application [649]. As one can see from the table that for applications that involve heavy loads and or high speeds, the EP oils are the best and for light loads, compounded oils deliver reasonable performance. For most automotive applications that use hypoid gears, EP oils are the most suitable gear lubricants. Gear lubricant viscometrics and additives technology used will be addressed later.

Industrial Gear Oils Specifications and Tests—As mentioned earlier, the performance of the industrial gear oils is established by the OEMs and industrial organizations. Like other lubricants, industrial gear oils have viscosity grades as well as performance classifications. AGMA viscosity grades along with the ISO viscosity grades for these lubricants are provided in Table 8.12 [649]. These grades apply to R&O oils, EP gear lubricants, and synthetic industrial gear lubricants. In the table, for the R&O oils, the designation either has no suffix attached or it has the suffix Comp. No suffix indicates that the lubricants contain R&O additives only and suffix Comp indicates that the lubricants in addition contain fatty or synthetic fatty oils. The suffix EP indicates that the lubricants are supplemented with the extreme pressure additives and the suffix S indicates these lubricants to be synthetic in nature. Table 8.13 lists the minimum physical and performance specifications for inhibited and compounded gear lubricants and Table 8.14 provides the same for the EP lubricants [649]. The two most common performance specifications for industrial gear oils are AISE 224 and AGMA 9005. These along with the other major industrial gear oil specifications and tests are provided in Table 8.15 [318]. The tests included in these specifications are described in the books on the ASTM Standards, which the ASTM publishes annually, and elsewhere.

Gear Oil Formulation

The primary functions of a gear lubricant are to reduce friction, remove heat, control wear, suspend contaminants, and protect against rust and corrosion. In addition, they must not foam and readily separate water. In order to meet these performance objectives, the gear oils are formulated by adding antiwear and extreme pressure additives, oxidation inhibitors, rust inhibitors, and foam inhibitors to a mineral oil, synthetic fluid, or a vegetable oil.

Base Fluids

As mentioned earlier, gear oils for most applications are formulated by the use of the mineral oils. This is because of their reasonable cost and adequate performance. However, for some automotive applications, gear oils with good low temperature fluidity and high temperature thermo-oxidative stability are required. Such oils are formulated by the use of the synthetic base fluids. For automotive gear oils, the normal 100 °C viscosity range is between 4.1 cSt (SAE 70W) and 41 cSt (SAE 250) and for industrial gear oils, it is between 4.5 cSt to well over 200 cSt, or a 40 °C viscosity of between 28 cSt and over 7500 cSt, see Tables 8.2 and 8.12, and Fig. 1.15. Because of the need for high viscosity lubricants, gear oils use medium to heavy base stocks. Physical properties of the common mineral oil and synthetic base stocks are provided in Table 3.17 [318]. For gear oils, blends of heavy neutral and synthetic base stocks are most useful since aro-

TABLE 8.14—Minimum physical and performance specifications for EP gear lubricants [649].

Property	Test Procedure	Acceptance Criteria: EP Gear Lubricants	Acceptance Criteria: Synthetic Gear Lubricants
Viscosity	ISO 3104 ASTM D445	Must be as specified in Table 9 (ISO Grades Table)	Must be as specified in Table 9 (ISO Grades Table)
Viscosity Index	ISO 2909 ASTM D2270	90 minimum (applies to viscosity grades 2 EP thru 8 EP only)	120 minimum[a]
Maximum Pour Point	ASTM D97	No Requirement	-32°C (22°F)
Oxidation Stability	ASTM D2893	Increase in kinematic viscosity of an oil sample at 121°C (250°F) should not exceed *6%*	Increase in kinematic viscosity of an oil sample at 121°C (250°F) should not exceed *6%*
Rust Protection	ISO 7120 ASTM D655B	No rust after 24 hours with synthetic sea water	No rust after 24 hours with synthetic sea water)
Corrosion Protection	ISO 2160 ASTM D130	1b Copper strip after 3 hours at 100°C (212°F)	1b Copper strip after 3 hours at 121°C (250°F)
Foam Suppression	ASTM D892	Must be within these limits: Max volume of foam (ml) after: *Temperature* / *5 min. blow* / *10 min. rest* Sequence I 24°C (75°F) 75 10 Sequence II 93.5°C (200°F) 75 10 Sequence III 24°C (75°F) 75 10	Must be within these limits: Max volume of foam (ml) after: *Temperature* / *5 min. blow* / *10 min. rest* Sequence I 24°C (75°F) 75 10 Sequence II 93.5°C (200°F) 75 10 Sequence III 24°C (75°F) 75 10
Demulsibility	ASTM D2711MOD	Must be within these limits: *2 EP-7 EP* / *8 EP-13 EP* Max percent water 5-hour test: 20% / 20% Max cuff after centrifuging: 1.0 ml / 4.0 ml Min total free water collected during entire test (start with 90 ml of water): 80 ml / 50 ml	Must be within these limits[b]. Max percent water in the oil after 5 hour test = 1.0% Max cuff after centrifuging = 2.0 ml Min. total free water collected during entire test = 60.0 ml (start with 90 ml of water)
Cleanliness	None	Must be free of visible suspended or settled contamination.	Must be free of visible suspended or settled contamination.
Load carrying property	ASTM D2782 (Timken Test) and DIN 51 354 (FZG Test)	An oil must meet both; a 60 pound Timken OK load, and fail stage greater than 12 on the FZG machine with A/8.3/90°C parameters for acceptance.	Timken test not required. An oil must meet 11 stage fail (10 stage pass) on FZG machine with A/8.3/90°C parameters for acceptance.
Filterability	None	Must be filterable to 25μ (microns) (wet or dry) without loss of EP additive (β25 = 200 filter rating)	Must be filterable to 25μ (microns) (wet or dry) without loss of EP additive (β25 = 200 filter rating)

[a]Esters having a lower viscosity index and meeting all other requirements of this specification may be used in specific applications where proper viscosity at operating temperature has been verified.

[b]Polyglycols which will not pass the demulsibility test. but meet all other requirements of this specification, may be used in specific applications where there is no danger of water contamination.

matic and naphthenic components in such oils have viscosity-pressure relationship that is much superior to that of the paraffinic oils. Note the lower pour points of synthetics and their high viscosity indices, both of which are of great benefit for developing gear oils with the broad-temperature performance.

High viscosity in gear oils, both for automotive and industrial applications, is needed not only to improve their film-forming ability under heavy loads, which will protect the gears against failure, but also to muffle the noise from the meshing gear teeth. The final gear lubricant viscosity is achieved either by the use of a mixture of base fluids of different viscosities and or by the use of a viscosity modifier. The former strategy is used to develop narrow viscosity range multi-grades, such as 75W-90, and the latter strategy is often used to develop broad-viscosity range multi-grades, such as 85W-140. Such multi-grades are primarily for automotive use. Because of the high shear forces experienced by the lubricant in the contact zone, high molecular weight polymeric viscosity modifiers are often less effective than low molecular weight viscosity modifiers. This is because of the propensity of the high molecular weight polymers to degrade rapidly under shear, a parameter that is measured by a polymer's shear stability index (SSI); low values are desired in gear type applications. The SSIs of a number of alkyl methacrylate polymers of varying molecular weights used in different applications are depicted in Fig. 8.18 [469]. Note that within each application, the SSI values increase with the increasing molecular weight (MW) of the polymer, indicating a decrease in its shear stability. Across applications, the SSI-MW lines are steeper for gear oils and hydraulic fluids, which are used for high shear applications, than for automatic transmission fluids (ATFs) and engine oils. Shear-related viscosity loss in polymer-thickened lubricants is determined by the use of many tests. These include KRL Shear Stability Test (CEC L-45-A-99), ASTM D6278, DIN 51 382, and ASTM D3945. To minimize the shear-related viscosity loss and to keep the lubricant within grade during its service life, gear lubricants often use low molecular weight viscosity modifiers. However, because of the low thickening efficiency of the low molecular weight polymers, the lubricant treatment level with these additives is quite high. It is between 25 to 35 % for automotive gear oils and 2 to 12 % for industrial gear oils, depending upon the viscosity grade. The high levels of polymer are important so that even after shearing during service, the reduced viscosity is within the desired range to form the lubricant film of suitable thickness to prevent premature equipment failure. As mentioned earlier, the multi-grade lubricants meet both the low-temperature viscosity and the high-temperature viscosity requirements of the gear systems, which not only help prevent wear but also improve the fuel economy, shiftability, and service life. Com-

TABLE 8.15—Major industrial gear oil specifications and tests [318]. Reprinted with permission from the Lubrizol Corporation.

Performance and Physical Tests	AISE 224 (U.S. Steel)	AGMA 9005-E02 (5 EP)	DIN 51 517 Part 3 January 2004	David Brown 51.53.101 (5E)	Cincinnati Lamb Landis P-74 (C-220)	General Motors LS-2
Timken (ASTM D2782)						
OK Load, lbs, min.	60			Report	45	60
4 Ball EP (ASTM D2783)						
Weld, kg, min.	250			Report		250
LWI, min.	45					45
4 Ball Wear (ASTM D2266)						
54°C/1800 rpm/20 Kg/1 Hour						
Scar Diameter, mm, max.	0.35					0.35
FZG (A18.3/90)						
Fail Load Stage, min.	11	12	12	12		12
Demulsibility (ASTM D1401, IP 19)			30 minutes			Report
Copper Strip (ASTM D130)						
3 Hours/100°C, max	1b	1b	1	1		1b
Rust (ASTM D665, IP 135)						
A - Distilled Water	Pass		Pass	Pass	Pass	
B - Synthetic Sea Water	Pass	Pass		Report		Pass
Oxidation (ASTM D2893B)						
(121°C/312 Hours/10 L Air/ Hour)		Up to ISO 200				
% Viscosity Increase, max	6	= 6[a]	6	Report		6
Demulsibility (ASTM D2711)						
Modified for 90 ml Water						
% Water in Oil, max	2	2		2		1.0
Total Free Water, ml, min	80	80		90		60
Emulsion, ml, max	1	1[1]		1		2.0
Pour Point (ASTM D97), max	+15°F		< ISO 150 = -12°C ISO 150-460 = -9°C > ISO 460 = -3°C	-6°C	+10°F	-10°C
Foam Tendency/Stability (ASTM D892, IP 146)						
Sequence I		50/0	100/10	75/10		50/0
Sequence II		50/0	100/10	75/10		50/0
Sequence III		50/0	100/10	75/10		50/0
Air Release (DIN 51 381, IP 313) (ASTM D3427)						
At 90°C, minutes, max.				9[a]		
Thermal Stability Test (CM Procedure B)						
% Viscosity Increase (ASTM D2161), max.					5	5
Sludge					None	2.5
Condition of Steel Rod					1.5 max	No Discoloration
Condition of Copper Rod					5 max	5 max
Copper Weight Loss, mg					(CM Color)	10 max
Seal Test						
7 Days at 100°C using SRE-NBR 28			-5/+10			-10/+10
% Volume Change			-10/+10			-7/+10
% Shore A Hardness Change			30			
% Tensile Strength			30			
% Rupture Elongation						

[a]Test targets vary with viscosity.

mon types of polymers used in gear lubricants include styrene esters and alkyl methacrylate polymers.

While additives can improve many of the properties of the mineral oils, there is a limit to this improvement. Synthetic gear oils are used to extend this limit. Synthetic lubricants have an advantage over those based upon mineral oils in applications where the operating temperatures are either very low or very high; the loads are extremely high; and the volatility and flammability are a concern. Synthetic gear oils surpass mineral oil-based lubricants with respect to the following properties.

1. Heat resistance (thermal and oxidative stability) and hence a lower tendency to form residues and deposits.
2. Film strength (high viscosity indices).
3. Low temperature fluidity (pour point and pumpability).
4. Volatility (flash point and evaporation loss).
5. Lubricity (at temperatures greater than 185 °C, or 365 °F).

Synthetics, such as esters and PAGs, also prevent the start-up gear wear, because of their greater surface affinity which causes them to leave a residual lubricant film at shut down. They also reach the critical surfaces quickly due to their easy low-temperature pumpability. However, they do suffer from the disadvantages of the higher price; hydrolytic instability to form acidic products that can cause metal corrosion, incompatibility with paints, elastomers, and certain metals; and limited miscibility with the mineral oils. The operating range of the synthetics is approximately −73 to 260 °C.

The most common synthetic base fluids used to formulate gear lubricants include synthetic hydrocarbons, both PAOs and alkylaromatics, poly(alkylene glycol)s or PAGs, and synthetic diesters and polyol esters. PAO-derived gear lubricants are used in food processing and pharmaceutical manufacturing machines. While alkylaromatic base stocks are highly prone to oxidation, they still see some use in gear oils because of their excellent viscosity-pressure relationship. However, they need a large amount of oxidation inhibitors to improve their oxidation resistance.

PAG-based gear lubricants have extremely low friction coefficients, which make them suitable for use in hypoid and worm gears that involve high sliding action. Once they are fortified with suitable additives, they are very effective in providing antiwear protection to steel/bronze worm gears and extreme pressure protection to hypoid gears. Because of their high oxygen content PAGs are highly surface active; hence they form durable films on gear surfaces. Because of this, PAGs provide some extreme pressure protection without the use of additives. Unfortunately, PAGs are extremely aggressive towards elastomers and may even dissolve some paints. The only seals that resist these lubricants are fluoroelastomer seals made of polytetrafuoroethylene (PTFE). In addition, PAG-derived gear lubricants have poor compatibility with the mineral oils; hence special care must be taken to prevent mixing. While in the bench tests PAGs are innocuous to ferrous and nonferrous metals, they must not be used in worm gears made of aluminum-bronze alloy, because they can cause increased wear due to a chemical reaction in the contact zone.

Synthetic esters have high thermal stability and good low-temperature fluidity. For use in industrial applications, especially in Europe, rapidly biodegradable ester-based lubricants are gaining importance. Hydrolytic stability in some esters can be a problem, which will result in corrosive and hydrolysis-catalyzing acids. Esters also have high surface affinity, but not as high as PAGs; hence they are somewhat better suited for use in formulating gear oils. The use of PAG, ester-based, and other synthetics to design gear lubricants can boost the overall efficiency of an operation by 10 to 30 %. Hence, their higher cost may be offset by the lower energy consumption and the extended drain intervals.

Synthetic gear oils are primarily used in spur, straight bevel, spiral bevel, helical, herringbone and hypoid worm enclosed gear drives. Synthetic gear lubricants may contain rust and oxidation inhibitors or EP/antiwear agents. These fluids are used in enclosed gear applications where very low or very high ambient and operating temperatures are encountered.

Additives

In addition to the viscosity modifiers discussed above in the base fluids section, gear oils contain the following additive types.

1. Friction Modifiers
2. Extreme-pressure (EP) Antiwear Agents
3. Oxidation Inhibitors
4. Amines/Dispersants
5. Detergents
6. Rust and Corrosion Inhibitors
7. Foam Inhibitors
8. Demulsifiers
9. Pour Point Depressants
10. Seal-swell Agents

The interested reader may like to refer to Chapter 4 to learn about the chemical types used under each of these classes and the mechanism of their performance.

Combining the right additives in a single gear lubricant is not easy. To achieve this, the formulator must understand the effect of the individual additives on the various performance parameters of a gear lubricant, learn to reconcile high speed, high-torque performance, and be cognizant of the synergism and antagonism that may exist between additives. In addition, the formulator must understand the base oil effects on performance. For example, highly polar synthetic base stocks, such as PAGs and synthetic esters, may compete with the additives for the surface. And because of being in a larger amount they may overwhelm the surfaces, thereby preventing the access and the absorption of the EP/antiwear additives to perform their function. Similarly, other additives present in the formulation, such as certain friction modifiers, oxidation inhibitors, rust inhibitors, metal deactivators, detergents, and dispersants, may hinder the performance of the EP/antiwear agents [444,445]. See Fig. 4.142. These types of antagonisms are quite common in gear oils because the surface-active EP agents form the core of the formulation. The best strategy is to run full-scale passenger car and truck axle tests, both in the laboratory and on the road, to reveal any deficiencies in gear lubricant formulations. Obviously, in view of the cost associated with such a venture, it is important that only the best formulations are subjected to the full-fledged testing. Many chemical and bench tests are available that can be used to screen out the least promising candidates.

Lubricant Selection

Both industrial and automotive gear oils are formulated to meet the operating requirements of particular gear sets. Automotive gear oils, designed for use in spiral bevel and hypoid gear sets, contain almost all additive classes described in the earlier part of this section. Table 8.16 shows the effectiveness of the various additives in meeting the automotive gear performance tests. Large industrial spur gears place a lower demand on the load-carrying capacity of the gear oil than the hypoid gears. Also, they operate at slow speeds and at low ambient temperatures of 60 to 80 °C. Because of these factors, additive-free or EP gear oils are usually specified for industrial gears. The oxidation stability and pour point can be improved by the addition of additives at a low treatment level. For enclosed industrial gear sets, high-speed gear oils are required, which are lubricants that contain rust and oxidation inhibitors and antiwear agents.

Worm gears use a blend of mineral base oils with synthetic fatty oils to provide lubricity for sliding motion under heavy pressures. These lubricants contain from 4 to 6 % fatty

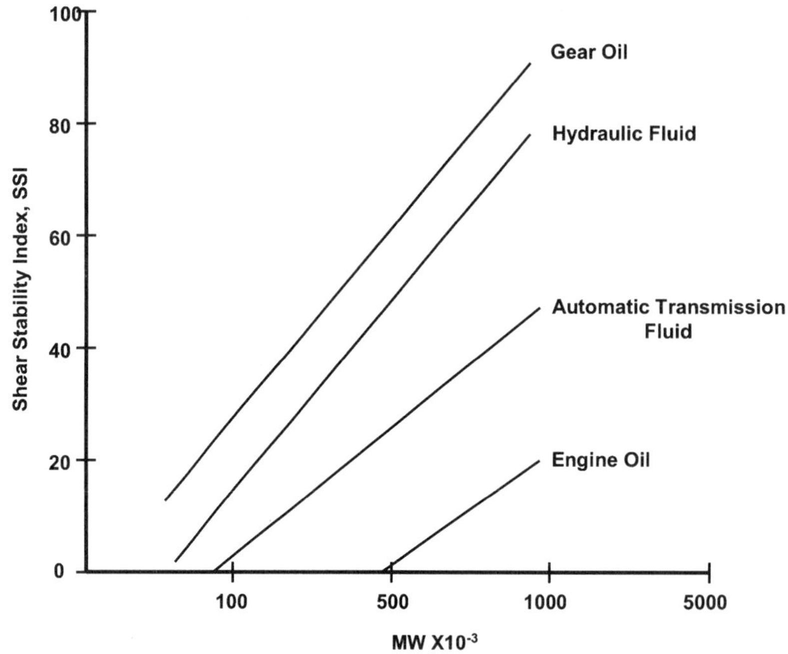

Fig. 8.18—Poly(alkyl methacrylate) viscosity modifier—molecular weight versus shear stability index [469].

materials and may be supplemented with rust and oxidation inhibitors. EP oils of low additive treatment level also work well. Such oils are useful for gears operating under high pressure/high temperature conditions, where the EP agents form the chemical protective films. However, the additives must be free of active sulfur, which is to minimize corrosion of the bronze ring gear. In this application, the PAO and PAG derived synthetic gear lubricants are also used. PAO-based lubricants have good low and high temperature properties, are elastomer compatible, and are miscible with the mineral oils. They are usually supplemented with a small amount of synthetic esters or the antiwear additives to improve their boundary lubrication ability. PAG-based lubricants are superior to those based on mineral oils because they are not only better coolants, but they also possess good low and high temperature properties and control friction and wear more effectively. PAGs also have high viscosity indices, approaching 280, which makes it possible to use a lower initial viscosity grade. This minimizes the energy usage at low temperatures, because of easy flow, and at the same time providing adequate film-forming ability at high temperatures. As stated earlier, most PAGs possess inherent EP/antiwear properties and hence do not need any EP additives. However, open gear sets require high viscosity lubricants that contain extreme-pressure and antiwear additives. These oils belong to AGMA viscosity grades 7 and 8. Besides cost, they have the disadvantages of incompatibility with the other fluids and aggressiveness to paints, elastomer seals, and polycarbonate which is used as a sight glass.

Running-in gear oils are another class of lubricants that are used to improve the surface finish of the new gears chemically and increase their load-carrying capacity. Oils containing sulfur-phosphorus additives and chlorinated paraffins are suitable for use as running-in gear oils. Sulfur-

TABLE 8.16—Additives versus automotive gear test performance.

Test	Performance Criteria	Chemical Type	Specific Additives
ASTM L-37	High torque, low speed – measures load-carrying capacity	Phosphorus	Amine salts of alkyl phosphoric and dithiophosphoric acids
ASTM L-42	High speed, shock loading – measures anti-scoring performance	Sulfur	Di-isobutylene polysulfides made from isobutylene/S and isobutylene/S_2Cl_2/NaOH reaction
ASTM L-33	7-Day moisture corrosion test	Amines	Long-chain amines and nitrogen dispersants
ASTM L-60/L-60-1 (ASTM D5704)	High temperature Stability	Boron	Borated and boron-free succinimides dispersants
ASTM D892	Foaming tendency	Silicones and Polyacrylates	Polydimethylsiloxanes
Cu Corrosion (ASTM D130)	Cu strip	DMTD Derivatives	DMTD/alkyl mercaptans coupled products, tolyltriazole and its condensates with formaldehyde and dialkylamine
Limited Slip		Fatty Amines	Oleylamine and di-ethoxylated fatty amine
Cross-shaft Score	Buick cross shaft test	Sulfur	High sulfur di-isobutyl polysulfides

TABLE 8.17—Typical data for gear oils.

Properties	Manual Gear Oil	Hypoid Gear Oil	Hypoid GearOil	Industrial Gear Oil	Industrial Gear Oil
Density at 15 °C g/mL	0.900	0.909	0.913	0.892	0.903
Viscosity at 40 °C, mm^2/s	87	200	335	140	440
Viscosity at 100 °C, mm^2/s	10.0	17.4	25.1	13	28
Viscosity Index	94	93	97	96	97
Pour point, °C	−30	−24	−18	−24	−12
Phosphorus content, mass%	0.07	0.11	0.11	0.05	0.05
Sulfated ash, mass%	0.1	0.1	0.1	0.03	0.04
FZG A/16.6/140 (Damage load stage)	>12	>12	>12	>12	>12

phosphorus additives, which have both running-in and normal-service characteristics, are used for automotive hypoid axles. Chlorinated paraffins containing 40 % chlorine are often used for the industrial gears. These oils need to be replaced with normal oils at the end of the running-in period. The major problem associated with these oils is that they are very corrosive, which can be controlled by the use of inhibitors. Chlorinated paraffins function by reacting with the metal to form ferrous chloride at points where sufficient temperatures are generated due to frictional heat to cause a chemical reaction. The resulting iron (II) chloride, with a melting point of 677 °C, has lower shear strength than the metal itself and hence gets removed easily. This help smooth surfaces to some degree.

Most industrial gear oils can be defined as extreme pressure (EP) gear oils and contain the same sulfur-phosphorus EP additives that are used to formulate the automotive gear oils, but at a lower treat level. A typical automotive gear lubricant treat level is 6 % EP additive, while the industrial EP gear oils usually contain about 2 % additive. Industrial gear oils contain some components that are not present in the automotive gear oils, for example, a demulsifier. However, compounded industrial gear oils do not contain sulfur/phosphorus EP additives but contain a less active and less corrosive additive, such as a fatty acid derivative. These oils are often used to lubricate bronze-on-steel worm gears where darkening or corrosion of the yellow metal is a concern. Typical properties of the gear oils for various gear sets are provided in Table 8.17.

While gear oils perform many important functions, the primary criteria to assess their effectiveness pertain to their ability to form low friction physical or chemical films to protect gears against wear and seizure damage. Figure 8.19 demonstrates the performance of properly formulated gear lubricants in the FZG Test [4]. Oil a is a plain gear oil that contains no EP additives. It fails at Stage 8 due to scoring. Oil b, which contains an EP additive, even at a low treatment level goes to Stage 11, prior to failing due to pitting. However, oils c, d, and e, which have the proper viscometrics and a higher amount of EP additive treatment achieve performance up to 12th Stage.

Gear Lubricant Tests

Gear lubricants are subjected to a wide variety of tests to evaluate their properties. Some are physical and analytical tests that determine physical and chemical properties of these lubricants. Others are performance tests that evaluate a lubricant's ability to perform effectively in the intended equipment. These tests comprise both bench tests, lab tests, and field performance tests. Tests used for gear lubricants are listed below.

Physical Tests

1. API Gravity (ASTM D287)
2. Kinematic Viscosity (ASTM D445)
3. Viscosity Index (ASTM D2270)
4. Shear Stability CEC-L45-T-93
5. Low Temperature Brookfield Viscosity (ASTM D2983)
6. Pour Point (ASTM D97)
7. Cold Crank Simulator (ASTM D5293)
8. Channel Point FTM 3456
9. Flash and Fire Points by Cleveland Open Cup (ASTM D92)
10. Precipitation Number (ASTM D91)

Analytical Tests

1. Elemental Content ASTM D4951
2. Elemental Analysis of Lubricants by WDXRF (ASTM D6443)
3. Chlorine Level (ASTM D808)
4. Phosphorus Level (ASTM D1091)
5. Sulfur Level (ASTM D1552)
6. Nitrogen Level—Kjeldahl Method (ASTM D3228)
7. Acid Number (ASTM D664)
8. Base Number (ASTM D4739)
9. Water Content (ASTM D1744)
10. Insolubles in Used Oils (ASTM D893)

Performance Tests

EP/Antiwear Tests

1. Four-Ball EP (ASTM D2783)
2. Four-Ball Wear Test (ASTM D4172)
3. Falex Continuous Load (ASTM D3233)
4. FZG (ASTM D5182, DIN 51 354)
5. Timken EP Test (ASTM D2782)
6. Cyclic Durability Test (ASTM D5579) [Automotive Gear Oils]
7. L-37 Low Speed High Torque Hypoid Test (ASTM D6121) [Automotive Gear Oils]
8. Gear Scoring Test L-42 (Automotive Gear Oils)
9. Synchronizer SSP 180 Test (Automotive Manual Transmission Test)
10. FZG Pits C i80 TS(Automotive Gear Oils)

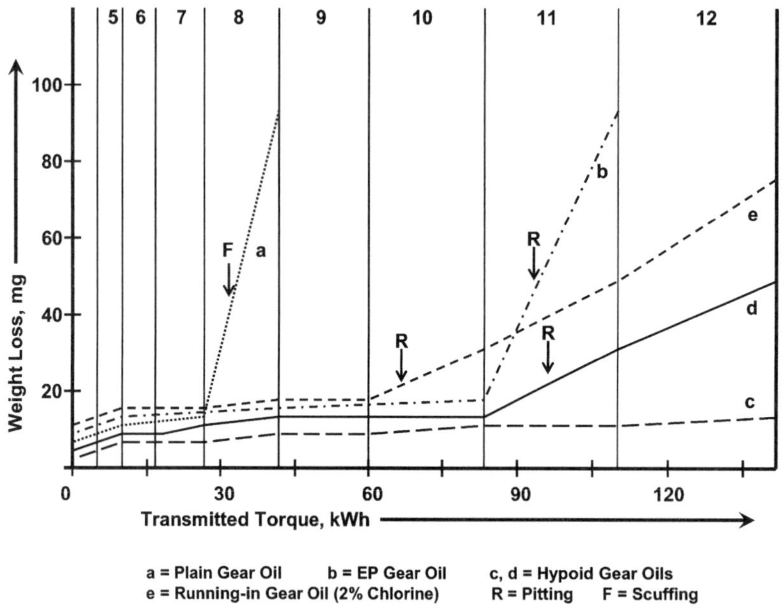

Fig. 8.19—Effect of gear oil quality on FZG wear test (A/8.3/90) performance [4].

Demulsibility/Water Separability Tests

1. Water Separability Characteristics of Petroleum and Synthetic Fluids (ASTM D1401)
2. Demulsibility Characteristics of Lubricating Oils (ASTM D2711)

Rust and Corrosion Tests

1. Copper Strip Corrosion (tarnish) Test (ASTM D130)
2. Turbine Oil Rust Test (ASTM D665)
3. Gear Corrosion Test L-33 (Automotive Gear Oils)

Foaming/Air Release Tests

1. Foaming Characteristics of Lubricating Oils Test (ASTM D892)
2. High Temperature Foam Inhibition (ASTM D6082)
3. Storage Solubility of Gear Lubricants (FTM 3440)
4. Air Release (ASTM D3427 or DIN 51 381)

Oxidation Tests

1. Oxidation of EP Oils ASTM D2893 (S-200)
2. Thermal and Oxidation Stability (L-60-1) (ASTM D5704) [Automotive Gear Oils]
3. GFC Oxidation

Material Compatibility Tests

1. Dynamic Seals Test
2. Seal Compatibility (ASTM D5662)
3. Compatibility of Gear Lubricants [FTM 3430 (Automotive Gear Oils)]

Formulation Examples

Automotive Gear Oil: 4.0–6.0 % Sulfurized isobutylene and alkyl phosphite/phosphate EP agents, 0.25–0.5 % phenolic oxidation inhibitor, 0.3–0.4 % triazole metal deactivator, 0.6–1.0 % fatty amine friction modifier (anti-squawk agent), 0.1–0.5 % poly(alkyl methacrylate) type pour point depressant, and 0.05–0.1 % poly(alkyl methacrylate) foam inhibitor. The rest is mineral oil of viscosity range suitable for use as a gear lubricant.

Automotive Gear Oil (low phosphorus): 4.0 to 8.0 % Sulfurized isobutylene or sulfurized polyisobutylene (primarily tri-isobutylene) EP agent, 0.6 % secondary alcohol-derived zinc dialkyl dithiophosphate antiwear agent, 3 % ethylene carbonate treated bis-succinimide dispersant, 1 % low TBN calcium sulfonate detergent, 2.4 % high TBN calcium phenate detergent, 0.5 % diphenylamine oxidation inhibitor, 0.04 to 1.0 wt. % in terms of boron content of an alkali-metal borate, 0.1 to 0.8 % copper corrosion inhibitor; 0.01 to 0.1 % foam inhibitor; 0.01 to 0.1 % rust inhibitor, and a viscosity index improver [poly(alkyl methacrylate)]. The rest is API Group II mineral oil of viscosity range suitable for use as a gear lubricant (formulation extracted from Ref [676]).

Automotive Gear Oil: 4.4 % Di-alkyl polysulfide and 0.9 % alkyl phosphates/phosphites EP/Antiwear agents, 1.14 % alkylamine rust inhibitor, 0.76 % alkenylsuccinimide and 0.38 % borated alkenylsuccinimide dispersants, 0.1% bis(dialkylthio)dimercaptothiadiazole corrosion inhibitor, 0.1 % poly(butyl acryl ate) foam inhibitor, 1 % pour point depressant. The rest is mineral oil of viscosity range suitable for use as a gear lubricant (formulation extracted from Ref [677]).

Automotive Gear Oil for Manual Transmissions: 0.35 % of an amine phosphate extreme pressure agent, 0.2 % of nonylated diphenylamine oxidation inhibitor, 0.7 to 3.0 % by weight overbased magnesium sulfonate, 0.15 % dimercaptothiadiazole derivative metal deactivator, 2.0 % borated alkenylsuccinimide dispersant, 0.2 % of oleylamine friction modifier, 4.67 % of an olefin copolymer viscosity modifier 150 ppm of poly(alkyl acrylate) and 60 ppm of silicone foam inhibitors. The balance is hydrogenated polyalphaolefin (PAO 6) (formulation extracted from Ref [678]).

Automotive Gear Oil: 2.3 to 2.5 % Sulfurized isobutylene, 0.24 % dibutyl hydrogen phosphite, and 0.65 % 2-ethylhexyl acid phosphate EP agents; 1.08 % ash-less dialkyl dithiophosphoric acid ester antiwear agent/oxidation inhibitor; 0.30 to 0.35 % t-alkylamine rust inhibitor; 0.165 % 2,5-dimercapto-1,3,4-thiadiazole metal deactivator; 0.04 %

poly(alkyl acrylate) foam inhibitor; and 0.04 % caprylic acid. The balance is mineral oil of viscosity range suitable for use as a gear lubricant.

Industrial Gear Oil:: 0.9 to 1.0 % Sulfurized isobutylene, 0.10 % dibutyl hydrogen phosphite, and 0.025 % 2-ethylhexyl acid phosphate EP agents; 0.0422 % ash-less dialkyl dithiophosphoric acid ester antiwear agent/oxidation inhibitor; 0.12 % t-alkylamine rust inhibitor; 0.0645 % 2,5-dimercapto-1,3,4-thiadiazole metal deactivator; 0.015 % poly(alkyl acrylate) foam inhibitor; and 0.016 % caprylic acid. The balance is mineral oil of viscosity range suitable for use as a gear lubricant (formulation extracted from Ref [679]).

Industrial Gear Oil: 1.0–2.0 % Sulfurized olefin and alkyl phosphite/ phosphate EP agents, 0.08–0.2 % phenolic oxidation inhibitor, 0.1–0.13 % triazole metal deactivator, 0.2–0.3 % fatty amine friction modifier (anti-squawk agent), 0.03–0.2 % poly(alkyl methacrylate) type pour point depressant, and 0.02–0.03 % poly(alkyl methacrylate) foam inhibitor. The rest is mineral oil of viscosity range suitable for use as a gear lubricant.

9

Miscellaneous Industrial Lubricants

DISCUSSION IN THIS CHAPTER DEALS WITH lubricants that are used in various industrial applications. The lubricants described include turbine lubricants, compressor and refrigeration oils, food grade lubricants, and a variety of others. For each type of lubricant, criteria for selection, performance requirements, and testing are considered. The chapter is concluded by presenting representative formulations for some of these lubricant types.

The lubricant market is divided into two major segments: Automotive and industrial. Automotive lubricants include engine oils, transmission fluids, automotive gear oils, tractor hydraulic fluids, and automotive greases. Industrial lubricants are used to lubricate industrial equipment which help produce a variety of materials and products for the consumer use. Such lubricants include metalworking fluids, general machine oils, slide way lubricants, industrial gear oils, hydraulic fluids, turbine lubricants, rust prevention oils, compressor oils, refrigeration oils, industrial greases, and miscellaneous others. Distinction between the two lubricant segments is not clear cut since some applications can be included in either of the two groups. For example, aviation turbines, marine engines, and railroad engines are considered by some as industrial applications and not automotive. Worldwide share of industrial lubricants is around 45 % of the total lubricant market. Physical properties of these lubricants are usually specified by ISO viscosity grades, and their performance requirements are established by the U.S. Military, OEMs, and end-users.

Types Of Industrial Oils

Industrial lubricants are classified either based on application or function. Application-based classes include hydraulic fluids, gear oils, turbine oils, metalworking fluids, and greases. Of these, hydraulic fluids and gear oils were described in Chapters 7 and 8, respectively. Others are being discussed in this chapter and the subsequent chapter. Function-based classes are many, but those that are of primary interest to us include rust and oxidation (R&O) oils, EP oils, circulating oils, compounded oils, and fire-resistant fluids.

As stated earlier, most modern lubricants are based upon either mineral oil or a synthetic fluid, whose properties are enhanced by the use of additives. Synthetic fluids are used where the performance of mineral oils have reached their upper limit. Table 9.1 shows a comparison of the various base stocks with respect to some of the desirable properties [447]. For further details regarding the chemistry and performance of the synthetic fluids, please refer to Chapter 3 on Synthetic Fluids. Figure 9.1 shows the operating temperature advantage of the synthetics over mineral oil [447]. Temperature advantage is important, especially for formulating lubricants for use in extreme temperature applications. Please note that the finished lubricant properties may improve or deteriorate somewhat, depending upon the type, the quality, and the quantity of the additives used.

Synthetic lubricants provide the industry a means to increase productivity, efficiency, and energy conservation. The

TABLE 9.1—Base oil characteristics (adapted from Ref [447]).

Property	Mineral Oil	Diester	Perfluorinated Aliphatic Ether	Phosphate Ester	Polyalpha-olefin	Polyglycol	Polyol Ester	Poly(phenyl ether)	Silicone
Viscosity range	4	4	1	1	3	3	4	1	5
Viscosity index	2	3	3	1	4	4	3	1	5
Pour point	3	4	4	3	4	3	4	1	5
Oxidation stability	2	3	5	3	4	2	4	5	5
Hydrolytic stability	5	2	3	2	5	3	2	5	3
Thermal stability	2	3	5	3	4	3	4	5	5
Compatibility with mineral oil	...	3	1	3	5	1	3	3	1
Solvency of additives	5	4	1	3	4	2	4	1	1
Compatibility with varnish and paint	5	2	5	1	5	2	3	3	3
Volatility	2	3	5	3	3	3	4	5	5
Corrosion protection	5	2	...	1	5	3	2	...	3
Boundary lubrication	3	4	3	4	3	5	4	3	1
Fire resistance	1	2	5	5	1	2	2	4	2
Elastomer swell, BUNA	1	2	1	1	1	1	2	1	1
Cost	Low	Moderate	Very high	Moderate	Moderate	Moderate	Moderate	Very high	High

Note: 1—None/Low; 2—Moderate; 3—Good; 4—Very Good; 5—Excellent.

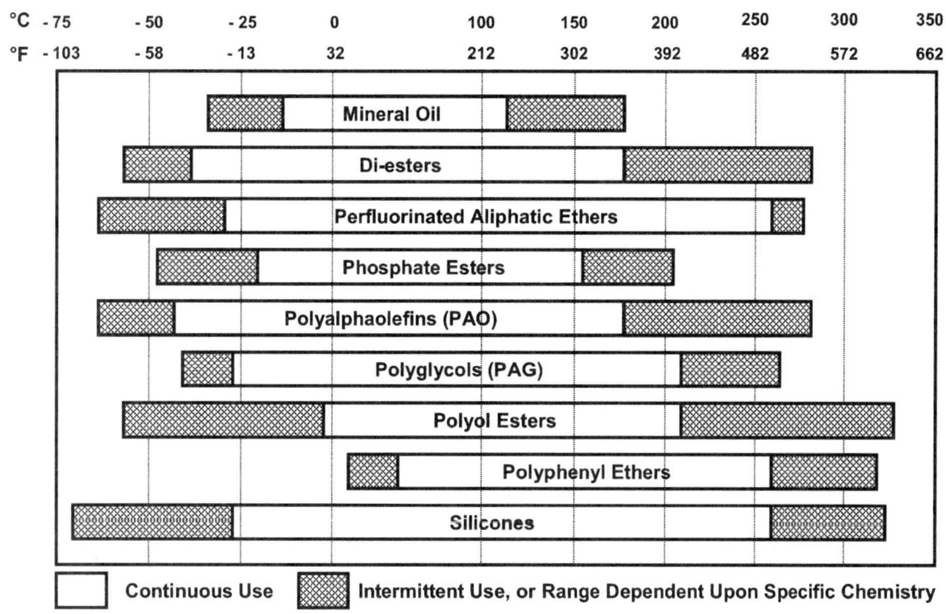

Fig. 9.1—Effective temperature range of various base fluids [447].

selection of a proper synthetic lubricant for a specific application is based upon the tribological needs of the equipment; operating conditions, such as temperature, speed, load, sealed-for-life; and the operating environment.

Turbine Lubricants

Turbine Types

Turbines are devices that convert kinetic energy of a fluid into mechanical energy. Turbines employ a variety of fluids to produce power. These include steam, gas, water, and wind. Steam turbine is a mechanical device that extracts thermal energy from pressurized steam, and converts it into mechanical work. Gas turbine, also called a combustion turbine, is a rotary engine that extracts energy from the flow of combustion gases resulting from the combustion of the air-fuel mixture. These gases are directed through a nozzle over the turbine's blades, spinning the turbine that does the work. These types of turbines are used to power aircraft, trains, ships, generators, and even tanks. Turbines for nonaircraft use are usually of rugged construction and involve on-site maintenance. Such turbines are used in power generation, pipeline, and petrochemical applications. In the petrochemical industry, they are used to drive pumps and compressors.

Water turbine, or hydro turbine, is a rotary engine that takes energy from the moving water. Water turbines are one of the oldest types and have been used to produce power for industrial use. Today, they are used to generate hydroelectric power. A wind turbine converts the kinetic energy of the wind into mechanical energy. If the mechanical energy is used directly by the machinery, such as a pump or grinding stones, the wind turbine is usually called a windmill. If the mechanical energy is then converted into electricity, the turbine is called a wind generator. Water and wind turbines are most environmentally compatible since they use clean and renewable energy sources. However, their operation is a function of the weather.

Turbines are usually coupled with alternators for electricity generation or to a pump, compressor, or fan which performs the work. For steam, gas, and wind turbines, there may be a need to install a reduction gear between the turbine and the driven equipment, if the speed of the turbine is too high. Aero-engine gas turbines are not directly coupled to other equipment, but are stand alone units, which use the energy of the expanding gas to provide the forward motion. In some cases, the main turbine is used to drive a second turbine, called a power turbine, which is attached to an alternator or compressor. In each class, turbines are subdivided based upon size, output, and application. Some of the information on turbines is available from Refs [214,680].

Lubrication Requirements

The primary function of a turbine oil is to lubricate the turbine and the generator bearings. However, it must also perform the additional functions and possess the attributes listed below.

- Act as a hydraulic medium for hydraulic cylinders, accumulators, servo valves, high-pressure pumps, and torque converters.
- Lubricate the reduction gear, if used, and the coupling between the turbine and the alternator.
- Lubricate bearings, gearing, and gas seals and must also act as a heat transfer fluid (coolant) and remove heat from these devices and the other hot surfaces.
- Act as a shaft seal for hydrogen-cooled generators.
- Minimize friction and wear.
- Control rust and corrosion of the metal parts.
- Be compatible with the materials that are used in constructing turbines.
- Be fire resistant, if used in applications that pose a fire hazard, and also be easy to recondition.
- Easily gets rid of contaminants, such as air and water, and is filterable.
- Possess all the properties necessary for trouble-free storage and use, such as stability, viscosity, viscosity index,

specific gravity, specific heat, thermal conductivity, pour point, and ignition temperature.

- Has a reasonable cost and long service life.

Almost all of these functions are the same across most turbine lubricants, irrespective of the turbine size or type. Lubricant systems, except for some military aviation applications, are closed systems consisting of a reservoir; a pump train with main, auxiliary, and emergency pumps; filters; a cooler; and may be a centrifuge or a purifier to remove the excess water and dirt [214]. Most turbines employ combined lubricating and hydraulic oil systems, which normally use mineral lubricating oil, with the lubricating oil supply pressure of 1–2 bars and the hydraulic oil pressure of 3–4 bars. Some turbines use separate hydraulic systems where the pressure is as high as 160 bars.

In large steam turbines, the temperatures reach 600 °C (1112 °F) and the presence of the mineral oil in the hydraulic system poses a fire hazard. To eliminate this, such turbines are designed with separate hydraulic system that uses a fire-resistant fluid. The use of two different lubricants helps contain the overall cost. We described the various classes of hydraulic fluids in Chapter 7. Because of the extreme operating temperatures of turbines, fire-resistant fluids based upon triaryl phosphates are the most appropriate. Incidentally, a phosphate ester fluid that can be used both as a hydraulic fluid and a main bearing lubricant for steam turbines of up to 1000 MW output is commercially available.

Aero-derivative gas turbines use synthetic ester-based fluids because mineral oils do not possess the low-temperature properties and the high-temperature stability required for use in modern turbine engines, see Fig. 9.1. Synthetic esters are used both for aviation and industrial applications where they function as a hydraulic fluid as well as a lubricant. Mineral oils still find use in older and smaller aero gas turbines used in military aviation and in industrial applications, where thermo-oxidative stresses are low. In some aero-derivative units, both mineral oil lubricants and synthetic lubricants are used; a turbine engine is lubricated by the ester lubricant and the driven equipment is lubricated by the mineral oil, again to contain the overall cost.

Selection of a suitable lubricant requires consideration of a number of factors, which include system design, duty cycle (continuous versus intermittent), lubricant stability, system maintenance, and the rate of the lubricant top-up. The first three factors determine the thermal and oxidative stresses to which the lubricant is exposed and the last two deal with maintaining the quality and the effectiveness of the lubricant. The operating temperatures of bearings in large steam turbines reach 320 °C (608 °F) and for the gas turbines, they reach 115–149 °C (239–302 °F) for plain bearings and 300 °C (572 °F) for roller bearings. For aero-derivative gas turbines, the lubricant comes in direct contact with the metal surfaces that have temperatures ranging from 204 to 316 °C (400 to 600 °F). The sump lube oil temperatures may range from 71 to 121 °C (160 to 250 °F). The duty cycle is important because the intermittent operation is less demanding on the lubricant than a continuous operation over extended periods, if all other factors are assumed to be the same. Since both the high temperatures and the continuous operation promote faster lubricant oxidation and thermal breakdown, the lubricant's thermo-oxidative stability will have an influence on its useful life.

Fluid maintenance and system maintenance can have an indirect effect on fluid quality. For example, minimizing air and water entry into the lubricant can considerably prolong its service life. Aeration of the lubricant can directly increase its oxidation rate by increasing its oxygen content, which is one of the factors that promote hydrocarbon lubricant oxidation; the others being temperature and metals and metal salts. Aeration and foaming cause additional problems, which include poor lubrication, loss of fluidity, and cavitation, if the aerated fluid is used as a hydraulic fluid. Top-up rate is the rate at which the fresh fluid is added to the fluid in the equipment. Typical top-up rates are 3–10 % per year for steam turbines, up to 33 % for industrial gas turbines that use mineral oils, and 0.25 L/h for aero engines. Top-up rates influence lubricant quality since they help replenish used additives to the lubricant in service. High top-up rates reflect a loss of lubricant or its components due to evaporation or misting. In industrial operations, the oil consumption can be lowered by the use of efficient oil demisters.

Previously, the OEMs recommended the lubricant service by establishing acidity, water, viscosity increase, and contamination limits. However, modern filtration devices greatly reduce the level of contaminants such as water and the particulates. This, combined with the practice of replenishing the depleted additives, has helped in prolonging the lubricant service life greatly. The depleted additives can be identified by the use of the modern analytical and spectroscopic techniques. Useful lubricant life for steam turbines, both for mineral oil and synthetic hydrocarbon types, is between 2 and 25 years. For industrial gas turbines, it is between 1.5 and 6 years and for new generation combined cycle gas turbines, it is between 4 and 5 years. For aviation gas turbines, the synthetic ester-based oil is continually topped up and the whole charge is almost never replaced. In industrial turbines, the lubricant is changed annually, or after 8000 hours of service. The service life of the fire-resistant hydraulic fluids and lubricants is between 5 and 20 plus years, depending upon the type of in situ conditioning or purification [214].

Lubricant Performance Specifications

Turbine oils are classified by ISO viscosity grades and by ISO Standard 6743-5, Class L (Lubricants) and Family T (Turbines). ISO viscosity grades need not be addressed here since they were already discussed in Chapters 7 and 8 on Hydraulic Fluids and Gear Oils, respectively. ISO Standard 6743-5 classifies turbine lubricants based on turbine type and the type of service they provide. The standard is duplicated in Table 9.2 [440]. As one can see, the standard broadly classifies turbine lubricants into five groups, which are lubricants for steam turbines, lubricants for gas turbines, control system fluids, aircraft turbine lubricants, and hydraulic oils. The first two classes are subdivided to identify lubricants for applications with specialized lubrication needs. Turbine oils for steam and gas turbines, which are used in marine and stationary applications, are designed to lubricate parts, such as sealed electric motors, sintered bearings, plain and anti-friction bearings, fans, compressors, and gears. Circulating oils are turbine lubricants that are used in turbines equipped

TABLE 9.2—ISO classification for turbine lubricants (ISO Standard 6743-5) [214].

Application	Composition and Properties	Symbol ISO-L-	Typical Uses
Steam Turbines, Direct coupled or geared to the load			
Normal Service	Highly refined petroleum oil with rust protection and oxidation stability	TSA	Power generation and industrial drives and their associated control systems. Marine drives where improved load-carrying ability is not required for the gearing
Special Properties	Synthetic fluids with no specific fire-resistant properties	TSC	Power generation and industrial drives and their control systems where special properties of the fluid are advantageous, for example oxidation stability, low temperature properties.
Fire-resistant	Phosphate ester lubricant	TSD	Power generation and industrial drives and their control systems where special properties of the fluid are advantageous, for example oxidation stability, low temperature properties.
High Load-carrying Ability	Highly refined petroleum oil with rust protection, oxidation stability, and enhanced load-carrying ability	TSE	Power generation and industrial drives; marine geared drives and their associated control systems where the gearing requires improved load-carrying ability.
Gas Turbines, Direct coupled or geared to the load			
Normal Service	Highly refined petroleum oil with rust protection and oxidation stability	TGA	Power generation and industrial drives and their associated control systems. Marine drives where improved load-carrying ability is not required for the gearing
High-temperature Service	Highly refined petroleum oil with rust protection and oxidation stability	TGB	Power generation and industrial drives and their associated control systems where high temperature resistance is required due to hot spot temperatures
Special Properties	Synthetic fluids with no specific fire-resistance properties[a,b]	TGC	Power generation and industrial drives and their control systems where special properties are advantageous, for example oxidation stability, low temperature properties.
Fire-resistant	Phosphate ester lubricant	TGD	Power generation and industrial drives and their associated control systems with need for fire-resistance
High Load-carrying Ability	Highly refined petroleum oil with rust protection, oxidation stability, and enhanced load-carrying ability	TGE	Power generation and industrial drives; marine geared drives and their associated control systems where the gearing requires improved load-carrying ability
Miscellaneous			
Control System – Fire Resistant	Phosphate ester control fluid	TCD	Steam, gas, hydraulic turbine control mechanisms where the fluid supply is separate from the lubricant and fire-resistance is needed
Aircraft[c]	...	HA	...
Hydraulic[c]	...	TH	...

[a]These products may not be compatible with petroleum-based products.
[b]This category includes synthetic hydrocarbons as well as other chemical types.
[c]Classifications for these categories have not been established.

with circulating oil systems. Such systems are installed to facilitate quick removal of the large quantities of heat or where heavy contamination of oil is a concern. As a rule, all lubrication and hydraulic functions are satisfied by the circulating oil. The exception is water turbines, where the lubrication and heat control systems are separate because the same viscosity grade lubricant cannot satisfy the needs of both. *Control system fluids* are highly fire-resistant phosphate ester fluids, which have the ability to endure extreme hot temperatures and high oil pressures in large-scale turbine hydraulic control systems. *Aircraft lubricants* are for lubricating turbines used in the jet engines. *Hydraulic fluids* are used to lubricate the mechanisms that are used to operate the turbine's peripheral equipment. For details pertaining to the function and the properties of these fluids, refer to the Chapter 7 on Hydraulic Fluids.

Turbine oils, in a manner analogous to that of the hydraulic fluids, are available in ISO viscosity grades and are classified as R&O oils, non-EP oils, EP oils, and the fire-resistant fluids. R&O oils are formulated to provide rust and oxidation protection and EP oils are formulated to provide EP protection, depending upon the intended end use. EP oils contain R&O packages enhanced with antiwear additives. Some applications, such as the load equipment, operate at high speeds and high loads and hence need both EP and antiwear protection. Addition of 2 % tricresyl phosphate or zinc dialkyl dithiophosphate to a mineral oil lubricant helps in controlling gear wear in such applications. However, these additives can be corrosive and increase sludge formation, hence they must be used only when a real need exists. Fire-resistant fluids are used in operations that pose a potential fire hazard, such as in turbines used for pumping oil and gas. Originally, chlorinated biphenyls were used in these turbines but for the environmental reasons they are being replaced by triaryl phosphates.

U.S. Military and OEM specifications are used to determine the quality of the turbine oils. These specifications pertain to the fluid's oxidation resistance, thermal stability, rust and corrosion inhibition, demulsibility, yellow metal protection, foam inhibition, air release characteristics, seal compatibility, low-temperature fluidity, and volatility. The EP/antiwear performance is usually demonstrated by the use of the 4-Ball EP, Falex, and Ryder gear tests.

The United States and European OEM performance specifications for R&O, EP, and non-EP turbine oils are provided in Tables 9.3 and 9.4, respectively [318]. R&O turbine oils inhibit rust and oxidation and are used in high-speed and heavy-duty steam and gas turbines. These oils are also used in light-duty or non-antiwear industrial hydraulic and gear systems. Several specifications covering R&O and turbine oils have been issued by the equipment builders, users, military, and the technical societies in the United States and Europe. An exhaustive list of national and international specifications is provided in Ref. [214]. Table 9.5 compares the specific technical requirements for steam/industrial gas turbine oils, air-derivative turbine oils, and fire-resistant hydraulic fluids. Please note that the air-derivative turbine oil requirements are based on the U.S. MIL-PRF-23699F specification.

Desirable Turbine Oil Properties

The desired properties in a turbine fluid include proper low and high temperature viscometrics, volatility, acidity, oxidation resistance, water and chlorine content, fire resistance, thermal stability, lubrication and wear prevention, rust and corrosion inhibition, water separation and demulsibility, foam inhibition, air release characteristics, and seal and materials compatibility [214]. While some of these properties can be attained by the use of a proper base stock, others need the help of additives. Additives that are commonly used in turbine oils include the following:

1. Oxidation inhibitors
2. Rust and corrosion inhibitors
3. Detergents/dispersants
4. EP/antiwear agents
5. Foam inhibitors
6. Demulsifiers

Viscometrics

Each lubricant is designed to possess viscosity characteristics that are consistent with the equipment needs under all operating conditions, so do these fluids. Kinematic viscosities of turbine lubricants at 40°C and 100 °C are measured by using ASTM D445 test method and are converted into viscosity index according to the ASTM D2270 procedure. Another ASTM method, D2532, measures viscosity changes on prolonged exposure to low temperatures. Figure 9.2 shows such a plot for VT behavior, a surrogate of VI, of a number of fluids below 40°C [440]. Please note that the slopes of the viscosity-temperature lines in this temperature range are in agreement with the VIs of the different fluids. ASTM D2422 and ISO 3448 standards for industrial lubricants classify lubricants according to their viscosity at 40°C. For industrial steam and gas turbine oils, ISO VG 32 and 46 and for aero turbines, and ISO VG 10 and 22 are the most commonly used viscosity grades.

Low-temperature Properties

It is critical that a lubricant flows and pumps at low temperatures, otherwise operating problems will set in. Unlike mineral base oils (API Group I oils) that have poor low temperature performance because of the presence of the waxy components, synthetic fluids do not usually suffer from this disadvantage. However, synthetic esters, when wet, become turbid at low temperatures and also gain viscosity on storage. Therefore, aviation gas turbine oils, which employ ester-derived lubricants, are tested for extended storage stability. MIL-PRF-23699F, for example, requires storage stability at −40°C to +60°C for a period of three years. ASTM D97 is used to determine the pour point.

Volatility

In some military aviation applications, a 3 cSt (low viscosity oil) is used in a high temperature environment. Hence, there is a concern for volatility, which is determined by the use of ASTM D972, Standard Test Method for Evaporation Loss of Lubricating Greases and Oils.

Acidity

Acidity in a lubricant, whether from manufacture or the presence of the acidic additives and fluid degradation products, is not desirable. This is because it is likely to lead to problems during storage and use, for example, extensive foaming (ASTM D892) and poor air release (ASTM D3427). It may even cause corrosion (ASTM D665) and promote oxidation (ASTM D2272). For phosphate ester lubricants, it will increase the rate of hydrolysis, which will destroy the integrity of the lubricant. Acidity in lubricants is monitored by the use of ASTM D974 and ASTM D664 standards. For aviation lubricants, MIL-PRF-23699F includes a potentiometric method. ASTM D943 uses an acidity limit which indicates the rate of the lubricant breakdown due to oxidation. Two other ASTM standards (D3339 and D5770) are also used to determine acidity, but during the test rather than at the end of the test.

Oxidation Resistance

Heat reduces turbine oil life due to increased oxidation. For every 10 °C (18 °F) rise in temperature above 60°C (140°F), the oxidation rate of hydrocarbon materials doubles. A conventional mineral oil will start to rapidly oxidize at temperatures above 180 °F (82 °C). Most tin-Babbitt journal bearings begin to fail at 121 °C (250 °F), which is well above the temperature limit of the conventional turbine oils. High-quality oxidation inhibitors can delay oxidation, but excess heat and water must be minimized to gain long turbine oil life.

As mentioned in the Additives chapter, the function of the *oxidation inhibitors* is to slow down the oxidative degradation of the lubricant; hence control viscosity increase, the formation of the harmful deposit precursors, and the corrosion-causing acids. They achieve this by facilitating the hydroperoxide decomposition to innocuous materials and quenching the reactivity of the peroxides. Sulfur-containing compounds, such as sulfides and polysulfides and dialkyl dithiophosphoric acid and dithiocarbamic derivatives, belong to the first group and hindered phenols and alkylated diphenylamines belong to the second group. The combined use of the additives of the two types allows a formulator to benefit from synergy, if it exists. However, the use of additives that contain both types of functionalities within their structure is often more advantageous than the use of two mono-functional additives.

While zinc dialkyl dithiophosphates are excellent oxidation inhibitors and antiwear agents, their use in turbine oils is not preferred because of their tendency to hydrolyze in the presence of water to form zinc oxide/hydroxide. Because of this, some turbine makers have imposed a limit on the amount of zinc in lubricants for use in their equipment. For metal-free (ash-less) turbine lubricants, hindered phenols and arylamines are used, either solely or in combination.

TABLE 9.3—Typical U.S. and European R&O specifications [318]. Reprinted with permission from the Lubrizol Corporation.

Performance Test	U.S. Steel 126	Cincinnati Lamb Landis P-38/55/54	Denison HF-1	General Electric GEK 32568F	AFNOR NFE 48-603 HL	DIN 51 524 Part 1
ASTM D2882 Vane Pump						
Ring & Vane Wt. Loss (mg), max	250					
4-Ball Wear (ASTM D2266)						
Scar Diameter (mm), max	0.8					
Denison P-46 Piston Pump			Pass			
Rotary Vessel Pressure Oxidation (ASTM D2272)						
Time to 25-psi Loss (min)	120 min		450 min	500 – 85% of the original		
Turbine Oil Oxidation (ASTM D943)						
Time to 2.0 NNA (h), min	2000			3000		1000
1000-hour Sludge (ASTM D4310)						
Sludge (mg), max			100			
NNA Increase, max			0.2			2.0
Rust (ASTM D665)						
A-Distilled Water	Pass	Pass	Pass		Pass	Pass
B-Synthetic Sea Water	Pass	—	Pass	Pass	—	—
Turbine Oil Demulsibility (ASTM D1401)						
Oil-Water-Cuff (ml)	40-37-3				40-37-3	40-37-3
Separation Time (min)	20				30	20
Cincinnati Lamb Thermal Stability						
Copper, max		5				
Steel, max		1.5				
Sludge (mg/100 ml), max		25				
Viscosity Change (%), max		±5				
Acid Number Change, max		±0.15				
Copper Wt. Loss (mg), max		10				
Neutralization Number, max		0.15		0.2	Report	Report
Copper Strip (ASTM D130)						
3 h @ 100°C				1 max	1 max	2 max
Foam (tendency/stability) (ASTM D892)						
Sequence I (ml)				50/0	100/10	150/0
Sequence II (ml)				50/0	100/10	75/0
Sequence III (ml)				50/0	100/10	150/0
Air Release (DIN 51 381, ASTM D3427)						
Time (min)				5	< 5-14[a]	5-14[a]
Seal Testing (DIN 53 538, DIN 53 521)						
% Volume Change						0-10
% Shore A Hardness Change						0-7

[a]Depends on viscosity grade.

One of the most preferred phenol is 2,6-dibutyl-4-methylphenol (butylated hydroxytoluene, or BHT), because of its lower cost and high efficiency. In some turbine formulations, it is being replaced by higher molecular weight analogues that have lower volatility.

Turbine oil performance data by the use of various oxidation inhibitors are provided in Table 9.6 [214]. Data in *Part A* pertain to two phenolic inhibitors in a solvent neutral oil (API Group I oil) and a hydrotreated oil (API Group II oil). The data demonstrate superior performance of the bi-functional inhibitor 1 to that of the mono-functional inhibitor 2, and a better response of both in solvent neutral oil than in the hydrotreated oil. The somewhat inferior performance in the latter case is due to the absence of the additional sulfur which is present in the solvent neutral mineral oil. Data in *Part B* pertain to alkylated diphenylamine in API Group I oil. Fluid 1, which contains only the alkylated diphenylamine, has a reasonably good oxidation performance. Fluid 2, which contains only a diester sulfide, has very poor performance, but Fluid 3, which contains a synergistic combination of the two inhibitors, has a much superior performance. The structures of the phenolic and aminic inhibitors used are provided in Fig. 9.3. While in general arylamines are more effective than hindered phenols across a wide range of conditions, they are costlier than phenols and impart color to the lubricant. It is important to note that the different base stocks respond differently to these additives. API Group I oils, which have some aromatic and sulfur content, usually respond better to hindered phenols and API Group II and Group III oils that are devoid of aromatics and sulfur respond better to arylamines. Alkylated aromatic amines are the inhibitors of choice in aviation gas turbine oils, which is primarily due to their superior effectiveness at high temperatures.

The effectiveness of the oxidation inhibitors is assessed by monitoring the lubricant's viscosity increase and acidity,

TABLE 9.4—Typical U.S. and European turbine oil requirements [318]. Reprinted with permission from the Lubrizol Corporation.

Performance Test	NON-EP OILS			EP OILS
	British Standard BS 489	DIN 51 515	General Electric GEK 32568A	Alstom HTGD 90117
Viscosity Index, min	90			90
Flash Point (°C), min	185	185	215	200
Pour Point (°C), min	-6	-6	-12	-9
Neutralization Number, max	0.45	Report		0.30
Air Release (DIN 51 381, ASTM D3427)				
Time (min)	5-10[a]	5[a]		4
Foam (tendency/stability) (ASTM D892)				
Sequence I (ml)	400/0[b]	450/10	10/0	300/0
Sequence II (ml)	50/0	100/0	20/0	50/0
Sequence III (ml)	400/0	450/0	10/0	300/0
Demulsibility				
Steam (s), max	300	300		300
Turbine Oil Demulsibility (ASTM D1401)				40-37-3
Separation Time (min), max	—	—		30
Oxidation Stability				
ASTM D943 Time to 2.0 TAN (h)	—	2000	2000[3]	2000
ASTM D4310 Sludge (mg), max	—	—	—	100
IP 280 TOP, max	0.70	—	—	1.8
IP 280 Sludge (%), max	0.30	—	—	0.4
ASTM D2272 Time to 25 psi Loss (min)	—	—	450	300
Copper Strip (ASTM D130, 3 h, 100°C)	1 max	2 max	1 max	2 max
Rust (ASTM D665)				
A-Distilled Water	—	Pass	Pass	—
B-Synthetic Sea Water	Pass[c]	—	—	Pass
FZG (A/8.3/90)				
Pass Load Stage, min				7

[a]Depends on viscosity grade.
[b]ISO VG 68/100 Requirements: *Sequence I* 450/40, *Sequence II* 100/10, *Sequence III* 450/40.
[c]Modified procedure.

and by the formation of sludge. Two major tests that are used to measure oxidative stability of the inhibited turbine oils are ASTM D2272 and ASTM D943. ASTM D943, also known as TOST (turbine oil stability test), is a very long test, hence some specifications, such as MIL-L-17331, has placed a time limit of 1000 hours but have added limits on the metal content and sludge [214]. ASTM D4310, which also uses the D943 apparatus, measures the amount of sludge formed after 1000 hours. Another oxidation test, the Universal Oxidation Test (ASTM D5846), measures the acidity and sludge when the acid number reaches 0.5 mg/g or the level of sludge becomes unacceptable. The sludge rating is sometimes improved by the use of *detergents* and *dispersants*, which keep surfaces clean by suspending the contaminants in the lubricant.

The stability testing of the aero turbine oils typically involves higher temperatures than for steam turbine oils and uses variations of FED-STD-791, Method 5308. ASTM D4636 was later issued that combined the 5307 and 5308 methods offering three alternative procedures. In the 5308 procedure, the fluid viscosity and catalyst coupons are evaluated at the end of the test, but in the 5307 procedure, the fluid samples are monitored for viscosity and acidity increase during the test. Variations of 5308 differ from each other in the OEM specified test temperatures and the test duration [214].

Water Content

Water in a lubricant is undesirable, both in dissolved and dispersed form. Solubility of water in mineral oils is very low (about 30–100 mg/kg); hence its presence can be easily detected in mineral turbine lubricants as turbidity. However, water solubility in polar ester lubricants at ambient temperatures is much higher (2000–2500 mg/kg); hence analytical methods must be used to measure its amount. The common method used for this purpose is Karl-Fischer reagent (ASTM D1744). While a small amount of water in turbine oils is not a problem, large amounts will cause a drop in viscosity, hydrolysis of the additives, and even of the ester base fluids.

Density

The knowledge of density, or specific gravity, of a lubricant is important to a user since it indicates the energy required to pump it. Phosphate esters have densities ~30 % higher than those of the mineral oils, 1.17 versus 0.9 for an ISO VG 32 fluid, and hence they require more power to circulate. Fluids of higher density also hold more insolubles in suspension, which necessitates the use of an efficient filtration system. ASTM D1298 and D5002 are used for determining density. Although values are usually determined at 15 and 20 °C, they can be extrapolated to other temperatures. API gravity

TABLE 9.5—Technical requirements for turbine oils and fire-resistant hydraulic fluids.

Property	Steam/Industrial Gas Turbine Oil	Aero-Derivative Gas Turbine Oil	Fire-Resistant Hydraulic Fluid	Test Method/s
Performance Standard	***ASTM D4304 for Type 1 Oils***	***MIL-PRF-23699F***	***IEC 61221 (ISO-L-TCD) Triaryl Phosphate Ester***	
ISO Viscosity Grade	32	22	46	ASTM D2422, ISO 3448
Appearance (20°C)	Clear and bright	...	...	
Kinematic Viscosity (cSt)				ASTM D445, ISO 3104
at 40°C	28.8-35.2	> 23	41.4-50.6	
100°C	...	4.9-5.4	...	
-40°C (cP), max	...	13000	...	
Viscosity Change after 72 h at -40°C (%), max	...	6	...	
Pour point (°C), max	-6	-54	-18	ASTM D97, ISO 3016
Flash Point (°C), max	180	246	...	ASTM D92
Density at 15°C, max	...	...	1.2	ISO 3675
Total Acid No (mg KOH/g), max	Report	1.0	0.1	ASTM D974, ISO 6619, SAE ARP 5088
Foaming Tendency/Stability				ASTM D892, ISO 6247
Sequence I max	50/0	25/0	150/0	
Sequence II max	...	25/0	25/0	
Sequence III max	...	25/0	150/0	
Air Release at 50°C, max	5	...	6	ASTM D3427, ISO 9120
Chlorine Content (mg/kg), max	...	...	50	ISO 5597
Water Content (%), max	...	...	0.1	ISO 760
Volume Resistivity at 20°C (MΩm) 60247, min	...	...	40	IEC
Emulsion Characteristics at 54°C, minutes to 3 ml emulsion, max	30	...	15	ASTM D1401, ISO 6614
Evaporation Loss (%) , max	...	10	...	ASTM D972
Rust Preventing Properties, Procedure A	Pass	...	...	ASTM D665
Rubber Compatibility (% swell)	...	25-May		ASTM D6546, FED-STD-791-3604/3433, ISO 6072
Corrosiveness to Copper Rating, max	1	...	...	ASTM D130
Fluid Compatibility-sediment (mg/L)	...	20	...	FED-STD-791-3403
Trace Metal Content (ppm), max	...	Various (1-11)	...	Atomic Emission Spectroscopy
Shear Stability-Viscosity Loss at 100°F (%), max	...	4	...	ASTM D2603
Bearing Deposition-demerit rating, max	...	80	...	FED-STD-791-3410
Thermal Stability & Corrosivity at 274°C	...	...	...	FED-STD-791-3411
-Viscosity Change (%), max		5.0		
-Acid Number Change (mg KOH/g), max		6.0		
-Metal Weight Change (mg/cm^2), max		4.0		
Oxidation Stability, hrs to acid number 2.0, min	2000	...	...	ASTM D943
Oxidation Stability	...	...	...	FED-STD-791-5308
Viscosity change (%), max	...	-5 to + 25		
Total Acid Number Change, max	...	3.0		
Sludge Content (mg/100 mL), max	...	25/50		
Wt. change (mg/cm^2)-Fe, max	...	±0.2		
Wt. change (mg/cm^2)-Cu, max	...	±0.4		
Wt. change (mg/cm^2)-Ag, max	...	±0.2		
Wt. change (mg/cm^2)-Mg, max	...	±0.2		
Wt. change (mg/cm^2)-Ti, max	...	±0.2		
Wt. change (mg/cm^2)-Al, max	...	±0.2		
Oxidation Stability, max	...	...		ISO15595
Acid number increase (mg KOH/g)			1.5	
Wt. loss-Fe (mg)			1.0	
Wl. loss-Cu (mg)			2.0	

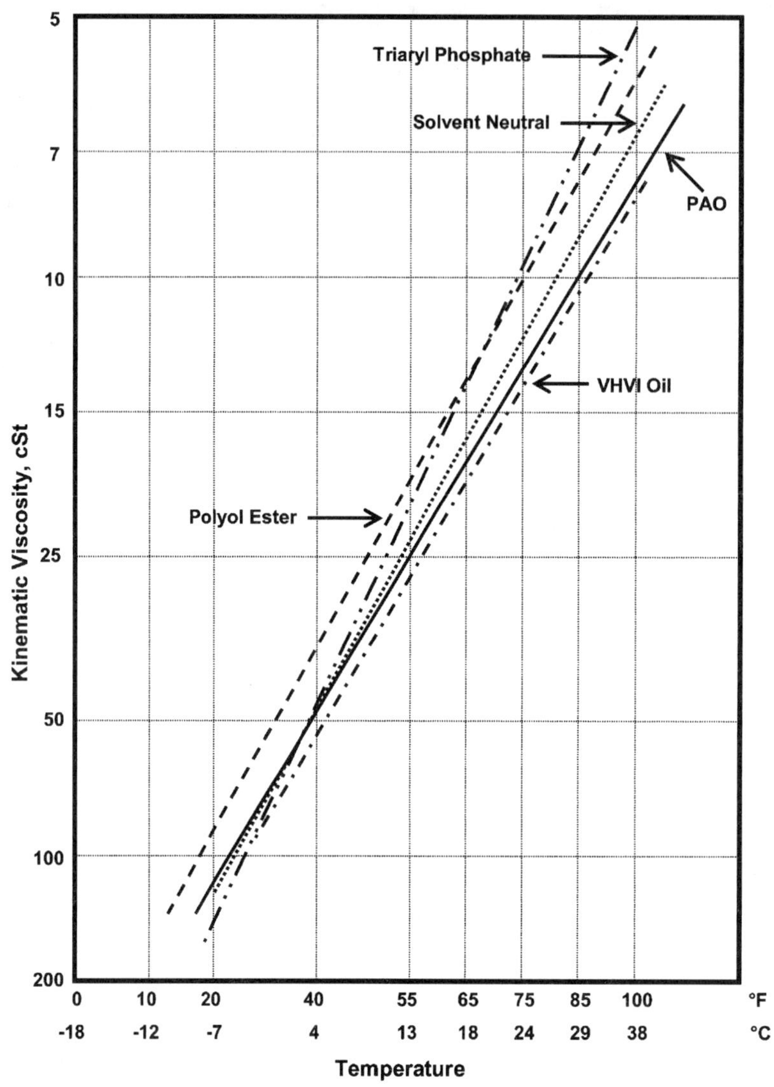

Fig. 9.2—Viscosity-temperature relationship of different turbine oils and fluids [214].

is a special function of the relative density and is obtained by the hydrometer measurement carried out at 60 °F (15.6 °C). Its significance was discussed in Chapter 4, the Mineral base Oils chapter.

Chlorine Content

One of the requirements for the phosphate ester control system fluids is the measurement of volume resistivity, which relates to the tendency of the fluid to produce servo-valve erosion. This is linked to the chemical structure of the fluid as well as the presence of the contaminants, such as water, acid, chloride ions, dirt, and metal soaps. The imposed limit on chlorine in the new fluids is 50 ppm, but accurate measurement of such low levels of chlorine is not easy. ASTM D808, Standard Test Method for Chlorine in New and Used Petroleum Products (Bomb Method), works well on samples with chlorine levels of 0.1–50 %. However, commercial phosphate esters are of high purity and contain extremely low chlorine levels [214].

Fire Resistance

In turbines, extremely hot surfaces create a potential fire hazard; hence a number of tests are used to assess the flammability of the turbine lubricants and include the following:

1. Spray ignition tests, such as those described in ISO 15029 Standard and Factory Mutual Test Standard 6930.
2. Hot surface tests, such as ASTM D286 or D2155 (now obsolete) or E 659, Standard Test Method for Autoignition Temperature of Liquid Chemicals. Another hot surface ignition method, The Hot Manifold Test, based on FED-STD-791 Method 6053, is currently under development as ISO Standard 20823.
3. Wick Tests, such as ASTM D5306, Standard Test Method for the Linear Flame Propagation Rate of Lubricating Oils and Hydraulic Fluids, or ISO 14935.

TABLE 9.6—Oxidation inhibitor performance in turbine oils (adapted from [214]).

Part A				
Test Fluids	**Test Fluid 1**	**Test Fluid 2**	**Test Fluid 3**	**Test Fluid 4**
Base Oil Purification Technique	*Solvent Neutral*[a]	*Solvent Neutral*[a]	*Hydrotreated*	*Hydrotreated*
ISO Viscosity Grade	*32*	*32*	*32*	*32*
Test Formulation				
Oxidation Inhibitor	*0.25% Inhibitor* **1**	*0.25% Inhibitor* **2**	*0.25% Inhibitor* **1**	*0.25% Inhibitor* **2**
Corrosion Inhibitor	0.05%	0.05%	0.05%	0.05%
Base Oil	99.70%	99.70%	99.70%	99.70%
Oxidation Stability TOST (ASTM D943) -- Test Conditions - 93°C, Water, Fe and Cu Catalysts, Air at 3L/Hour Sludge Formation (ASTM D4310)				
Time to Total Acid Number 2.0 mg KOH/g oil, hrs	4300	2400	2200	1100
Sludge (mg)	6	39	28	47

Part B			
Test Formulation – ISO VG 32 Base Stock (6.5% aromatics, 0.54% Sulfur) + Oxidation Inhibitor Package + Rust Inhibitor			
Test Fluid Composition	***Fluid 1***	***Fluid 2***	***Fluid 3***
Alkylated Diphenylamine, *Inhibitor 3*	0.25%	--	0.20%
Sulfur compound – $S(CH_2CH_2COOR)_2$	--	0.25%	0.05%
Rust Inhibitor	0.05%	0.05%	0.05%
Base Stock	99.70%	99.70%	99.70%
Oxidation Stability TOST (ASTM D943) -- Test Conditions - 93°C, Water, Fe and Cu Catalysts, Air at 3L/Hour Sludge Formation (ASTM D4310)			
Time to Total Acid Number 2.0 mg KOH/g oil, hrs	2,000	200	3,300
Sludge (mg) (ASTM D4310)	172	> 5,000	89

[a]Contains 6.5% aromatics and 0.54% sulfur

Thermal Stability

Concern for thermal stability is due to high temperature operating environments of the turbines and is one of the many parameters of concern. The others are oxidative stability and flammability. Decomposition, especially of the ester-based lubricants, will not only destroy their structure thereby impairing their function, but also will result in volatile products which can pose a fire hazard. Tests used to evaluate this lubricant parameter include FED-STD-791 method 3411, which is primarily used in aviation turbine oil specifications such as MIL-PRF-23699F and ASTM D2070, Standard Test Method for Thermal Stability of Hydraulic Oils.

Lubrication and Wear Prevention

The lubrication performance of a fluid is its ability to form an effective lubricating film on surfaces, thereby reducing metal-to-metal contact and wear. Typically, EP/antiwear additives are added to the turbine oils to provide protection against wear under mixed-film and boundary lubrication conditions, which are encountered in pumps and gears at medium to heavy loads. These materials comprise sulfur and phosphorus additives that form chemical protective films of low shear strength. Examples of these additives include organic polysulfides; alkyl and aryl phosphites, phosphates

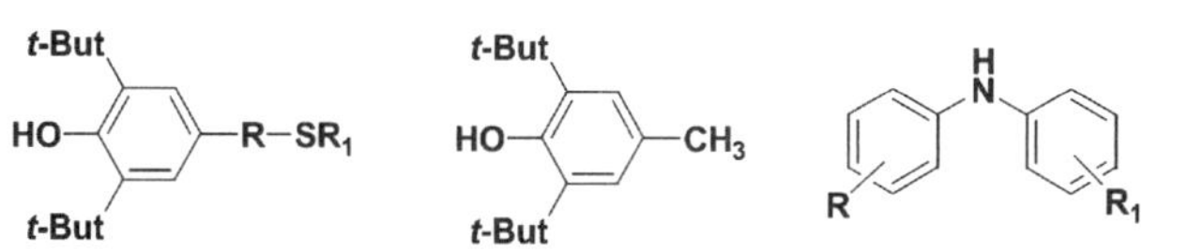

Inhibitor 1
MW = 492

Inhibitor 2
BHT

Inhibitor 3
Alkylated Diphenylamine

Fig. 9.3—Commonly used oxidation inhibitors.

and thiophosphates; and dialkyl dithiophosphoric and dialkyldithiocarbamic acid derivatives. See the Additives chapter for further details. Load-carrying requirements pertain only to gas turbine oils and where the oil is used to consecutively lubricate a steam and a gas turbine. Aviation lubricants require gear testing using Ryder gear test (FED-STD-791-6508) and the FZG test (ASTM D5182).

Rust and Corrosion Inhibition

Rust and corrosion problems in turbines are associated with water contamination of the steam turbine oils and corrosion can result from the attack of the acids generated by the thermo-oxidative degradation of the lubricant. *Inhibitors* are added to turbine oils to control rust of ferrous metals and corrosion of yellow metals, such as copper and bronze. Typical rust inhibitors are dodecenylsuccinic acid esters and acid amides and calcium and barium salts of alkylbenzene- and alkylnaphthalene-sulfonic acids. Since many turbines operate in salty environments, rust protection against saltwater is often needed. Polyol ester and aryl phosphate lubricants, because of their high surface affinity, provide some rust protection, without the aid of additives. However, rust inhibitors can be used to further improve their rust performance, if necessary. It is important to note that the rust inhibitors, especially those that are acidic, can be removed by adsorbents that are used to lower the acidic degradation products from the phosphate ester fluids. ASTM D665 Test Procedure is used to assess the rust-inhibiting ability of the turbine fluids. Additives used to inhibit corrosion of yellow metals due to acids and sulfur and its compounds, either present in the lubricants or resulting from the degradation of lubricants, are called metal deactivators. Yellow metal corrosion, the same as rusting, is the oxidation of the copper surfaces to form copper salts. Yellow metal deactivators are usually deriva-

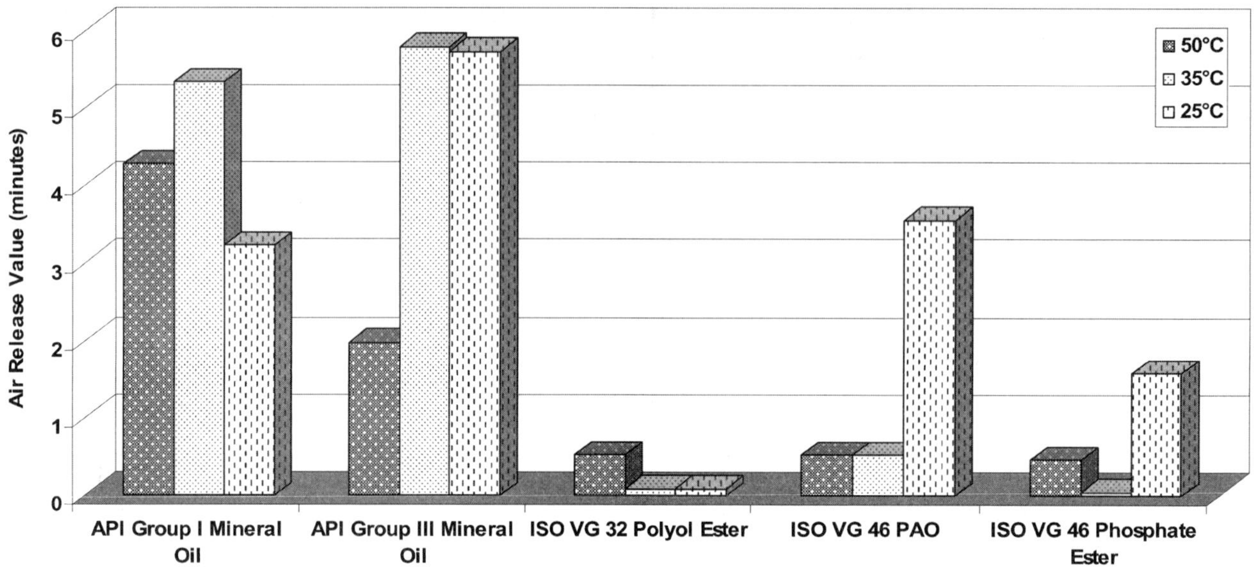

Fig. 9.4—Turbine oil air release rates as a function of temperature (adapted from Ref [214]).

tives of tolyltriazole or dimercaptothiadiazole (DMTD). ASTM D130 copper strip test is used to assess a lubricant's copper protecting ability. Turbine oil specifications require Cu strip rating of 1b or better.

The function of the *foam inhibitors* in turbine oils is to inhibit foam formation and facilitate quick collapse of the foam, if it forms. This is important since foam can lead to improper lubrication and increased lubricant oxidation. Foam inhibitors are added to the lubricant in a very small amount, i.e., at ppm level. At a higher level, these additives are harmful since they adversely affect air release properties of the turbine oil. Besides silicones (polysiloxane)s, poly(glycol ether)s and poly(vinyl ether)s are used for controlling foam. However, they are not as effective as silicones and are needed in a larger amount. ASTM D892 (ISO 6247) is the common method to test a lubricant's foam-forming tendency.

Turbine oils, such as those used in steam turbines and hydro turbines, get contaminated with water. Water contamination can lead to many problems in addition to reducing the oxidation life of the lubricant and promoting rust formation. These problems were stated while discussing water contamination. It is therefore imperative that water be removed as quickly as possible. *Demulsifiers* are additive that facilitate water separation and removal. Common chemical types include alkaline earth metal salts of aromatic sulfonic acids and block copolymers of propylene oxide or ethylene oxide with a polyhydric alcohol, such as glycerol, as an initiator. ASTM D1401 Standard is used to determine the demulsibility characteristics of the turbine oils.

Foaming and Air Release

All turbine oils contain air, either in dissolved form or as dispersed bubbles. Dissolved air is not a concern, except in the reduced pressure environment where it can instantaneously come out, causing cavitation. Dispersed or entrained air, on the other hand, is a problem since it can result in a loss of compressibility, increased oxidation, and even poor lubrication. The difference between foam and dispersed air is that the foam occurs on the surface of the liquid and consists of air bubbles surrounded by a thin film of oil and the dispersed air is present within the bulk fluid and a thick film of oil separates the bubbles. A number of factors affect these properties and include circulation rates, fluid viscosity and fluid surface tension, polar impurities, and the fluid temperature. Tests used to measure a lubricant's foaming tendency include ASTM D892 and the dynamic foam test, FED-STD 791-3214 test method. Air-release properties are assessed by the use of the ASTM D3427 (ISO 9120, IP 313, and DIN 51 381) test method. While this method is designed specifically for mineral-based oils, it can be used for some synthetic fluids as well. Figure 9.4 shows the air release properties of the various turbine fluids and at different temperatures [214]. Incidentally, 50 °C is the normal recommended temperature for the test. While the temperature effect on air release is not possible to interpret, the data in the figure do indicate that the air release from hydrocarbon oils is slow but from synthetic fluids it is fast.

Water Separability and Demulsibility

Steam turbine oils often get contaminated by water, primarily via steam penetration through seals, coolant leaks, and condensed water dripping into the hydraulic actuators [214]. Water in turbine oils leads to rusting, emulsion formation, additive hydrolysis, and bacterial growth. Hence it is important to remove water from the lubricant as quickly as possible. Emulsion formation, which mainly occurs due to the presence of the polar additives, such as detergents, must be minimized or emulsions broken to facilitate water separation. ASTM D1401 is the common test used to evaluate emulsion forming tendency of the lubricant and additives called demulsifiers are used to quickly break emulsions.

Compatibility with System Materials

It is imperative that the selected fluid is compatible with the system's constructional materials. These include seal and gasket materials and paints used to cover various system parts. Incompatibility or poor compatibility may cause

TABLE 9.7—Elastomer compatibilities of different turbine oils and fluids [214].

Seal/Hose Material	Mineral Oil and PAOs	Phosphate Ester	Polyol Ester
Neoprene	Yes	No	No
Nitrile rubber	Yes	No	Yes[a]
Chlorosulfonated polyethylene	Yes	Yes[a]	Yes
Epichlorohydrin polymer	Yes	No	No
Styrene-butadiene polymer	No	No	No
Ethylene-propylene polymer	No	Yes[a]	No
Butyl rubber	No	Yes	No
Fluorocarbon polymer	Yes	Yes[b]	Yes
Perfluoroether polymer	Yes	Yes	Yes
Polytetrafuoroethylene (PTFE)	Yes	Yes	Yes

[a]Variations in the composition of these elastomers in particular can cause differences in fluid compatibility and the seal manufacturers should be consulted before use.
[b]Fluorocarbon rubbers are compatible with aryl phosphates, but not with alkyl phosphates.

swelling, softening, or cracking of the elastomer seals, leading to a fluid leak. Fluid can attack paint causing flaking, which can lead to filter blockage or expose metal surfaces to harsh chemical environments. Highly refined mineral oils (API Group III oils) and synthetic hydrocarbons, such as PAOs, lead to seal shrinkage. Hence, they are supplemented with carboxylate esters or aromatic compounds to improve their elastomer compatibility. Phosphate esters are usually aggressive towards paints and some metals. The extreme temperature operating environment, for example, that experienced by aviation gas turbine lubricants, requires highly resistant fluorocarbon and perfluoro-elastomer seals. For nonaerospace applications, turbine oils are tested by using the ISO 6072 procedure. For testing aero-gas turbine oils, FED-STD-791 methods 3604 and 3433 or procedures listed by the equipment manufacturers are used. No standard methods in turbine oil specifications are known for evaluating paint compatibility and one must follow the recommendations of the paint manufacturer [214]. Elastomer compatibilities of the various tubrine oils and fluids are provided in Table 9.7.

Turbine Oil Formulation

Turbine oils are based upon both the solvent-refined paraffinic mineral oils and the hydrocracked oils. For industrial gas turbines and the combined cycle units that have high operating temperatures, the use of the hydrocracked oils increases the operating life of the lubricant up to 50,000 hours. In steam turbines, mineral hydraulic oil is completely replaced by triaryl phosphate type fire-resistant fluids to reduce the fire hazard. Table 9.8 provides typical operating conditions that a turbine oil encounters [214]. Let us consider the individual lubricant parameters listed in the table. Higher lubricant system capacity implies availability of a higher amount of lubricant to cool the bearings and the higher circulation rate implies a higher rate of cooling. Efficiency of cooling depends upon the combined effect of both these parameters and is reflected by the remaining lubricant parameters in the table, the bulk lubricant temperature and the bearing oil return temperature. The differential between the values of these parameters indicates the efficiency of the lubricant to cool the bearings; the greater the differential, the higher the effectiveness of cooling. Based on this, we conclude that with respect to the thermal stress on the lubricant, the turbine types follow the order shown in Fig. 9.5. Of these, the industrial turbines, marine turbines, and aviation turbines have the highest bulk lubricant temperature and the bearing oil return temperatures; hence the lubricants in these turbine applications are expected to experience the greatest degree of thermo-oxidative degradation. For these applications, either heavily inhibited mineral oils are needed or the use of the thermally stable synthetic lubricants will be beneficial.

TABLE 9.8—Typical operating conditions for turbine oils [214].

Parameter	Turbine Type					
Power Source	*Steam*	*Industrial Gas*		*Aero-derivative Gas*		
Size	Medium-Large	Small	Medium-Large	Aviation	Marine	Industrial
Power (Capacity)	250-1500 MW	1-14 MW	30-300 MW	11,000-110,000 lb Thrust	4 MW	31 MW
Lubrication System Capacity (m^3)	19-115	1-3	15-34	0.006-0.02	0.11	0.9
Circulation Rate (changes/hour)	5.5-10	17-21	6-10	60-240	60	5
Bulk Fluid Temperature (°C)	60-65	50-70[a]	45-80[a]	60-70	70	150[a]
Bearing Oil Return Temperature (°C)	70	60-115	55-85	120-130	150	150-180

[a]This temperature in the tank may be similar to bearing return oil temperature if the cooler is placed on the pressure side of the pump. In this case the bearing oil supply temperature will be significantly lower than that in the tank.

TABLE 9.9—Hydrocarbon base oils—Physical properties comparison [214].

Parameter	Test Method	PAO	160 Hydro-treated	240 Neutral	200 Solvent Neutral	Very High Viscosity Index (VHVI) Oils	
API Group	-	IV	II	I	I	III	III
Viscosity (cSt)	ASTM D665						
100°C		5.98	5.77	6.98	6.31	5.14	5.9
40°C		30.9	33.1	47.4	40.8	24.1	-
Viscosity Index		143	116	103	102	149	127
-40°C (cP)		7830	Solid	Solid	Solid	Solid	NA
Pour Point (°C)	ASTM D97	-64	-15	-12	-6	-15	-12
Flash Point (°C)	ASTM D92	235	220	235	212	230	235
Evaporation Loss (NOACK), %	DIN 51581	6.1	16.6	10.3	18.8	8.8	6.0

Base Stocks

Turbine oils are formulated by the use of the conventional solvent-refined base stocks, hydrotreated and hydrocracked base stocks, and synthetic base stocks, such as PAOs, synthetic esters, and triaryl phosphates.

Mineral Base Oils

Physical properties of the hydrocarbon base oils that are used in turbine lubricants are presented in Table 9.9 [214]. Consideration of the data clearly identifies hydrocracked and severely hydrocracked API Group II and Group III oils to be most suitable to formulate turbine oils. In addition to being thermo-oxidatively stable, these oils have the advantages of the low deposit-forming tendency, improved demulsibility, reduced volatility, low toxicity, and high biodegradability. Their major disadvantage is their high cost.

Synthetic Base Fluids

Three types of fluids that find use in turbine oils are polyalphaolefins (PAOs), synthetic esters, and triaryl phosphates. PAOs have properties that are very similar to those of the API Group II and Group III oils and even the disadvantages. The major disadvantage is their lower solvency, which can be corrected by blending a small amount of synthetic ester in the formulation. Another disadvantage of the PAOs is that they do not penetrate rubber seals and hence need seal swell agents. See Table 9.9 for comparative physical properties of the various hydrocarbon base stocks. Of these, PAO-based fluids are used in gas turbines, primarily because of their superior thermo-oxidative stability.

Synthetic esters are either used by themselves, or sometimes in combination with the PAGs, to lubricate aero-derivative gas turbines in aviation applications. Synthetic ester-based turbine oils for aviation use conform to four viscosity specifications of 3 cSt, 4 cSt, 5 cSt, and 7.5 cSt at 100 °C. Each fluid is designed for its own specific use. For example, the high viscosity oils are used to lubricate turbo-propellers where a thicker lubricating film is needed to reduce wear at heavy loads and the low viscosity fluids are used in military turbojet engines that need to start quickly at low temperatures. Synthetic ester-PAG blends are primarily used for turbo-propeller aircrafts.

Table 9.10 compares some of the properties of the mineral oils with those of the ester base stocks [214]. Esters are superior to mineral oils of similar viscosity with respect to viscosity index, load-carrying capacity, low-temperature viscosity, and volatility, as is reflected from the data in the table. The last two properties are especially useful in the use of the ester base stocks in aviation applications, which involve extreme low temperature and low pressure ambient conditions. Compare data for low viscosity oils in Columns 2 and 3 of the table. Columns 4 and 5 in the table evaluate the properties of high viscosity ester and mineral oil fluids. They demonstrate similar Ryder gear test performance, volatility, and flash point characteristics. However, in the case of the mineral oil, good low-temperature performance is lacking, making it less suitable than the ester, which has the desired low-temperature performance. While the esters possess excellent thermal stability, they are susceptible to hydrolysis in the presence of acids, which is a concern in turbine applications. However, long-chain polyol esters have excellent water tolerance due to the presence of the sterically crowded ester functional group and the lower affinity for water. In addition, they are considered to have low flammability because of their high flash and fire points. Hence, the ester reaction product of trimethylolpropane and oleic acid is used in steam turbine lubricants. This is in contrast to short-chain products used in aviation gas turbine oils.

Polyol esters are supplemented with polymeric thickening agents, which increase their droplet size in the presence of blowing air or gases, thereby suppressing mist formation during use. This also reduces ignitibility, which is due to a decrease in surface area of the large droplets and hence a decreased exposure to oxygen. The thickening effect of the polymer is not long lasting and is lost due to shear encountered in gears and pumps during use, and so does the fire resistance. Incidentally, fire resistance due to polymer is limited only to spray flammability and not under other

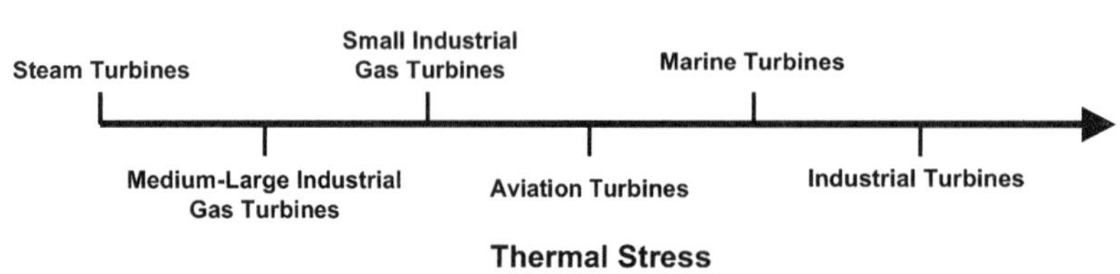

Fig. 9.5—Turbine type versus thermal stress on lubricant.

TABLE 9.10—Properties comparison—Synthetic esters versus mineral oils [214].

Parameter	Mineral Oil	Ester	Mineral Oil	Ester
Kinematic Viscosity (cSt) at 210°F (99°C)	2.5	3.3	20.1	7.5
-40°F (-40°C)	3,000	1,700	Solid	12,000
-65°F (-54°C)	25,000	11,000	Solid	Solid
Viscosity Index	75	154	95	150
Pour Point, °F(°C)	Less than -75 (-59)	-80 (-62)	0 (-18)	-65 (-54)
Vapor Pressure (psi) at				
250°F (121°C)	0.15	-	-	-
350°F (177°C)	1.3	0.01	-	-
400°F (204°C)	3.5	0.016	0.01	<0.01
550°F (288°C)	-	0.52	0.065	0.28
Ryder gear machine failure load lb/in tooth face width	600	2,000	~3,000	3,200
Flash Point (COC), °F(°C)	295 (146)	430 (221)	485 (252)	465 (241)

conditions, such as contact with hot surfaces. Esters are not compatible with neoprene seals. Table 9.11 lists specifications and the tests required for the fire-resistant fluids [214].

Triaryl phosphates are produced by the reaction of alkylphenols with phosphorus oxychloride, or phosphoryl chloride. Alkylphenols are either petroleum in origin or synthetic in origin. Petroleum-derived alkylphenols are obtained from the crude oil through distillation and are mixtures of cresols and xylenols, commonly known as cresylic acids. The phosphates derived from them are also mixtures that contain tricresyl phosphate and trixylyl phosphate in varying ratios, which depend upon the composition of the cresylic acids starting materials. Synthetic alkylphenols are made by alkylating phenol with propylene, butylene, or its oligomers. See Fig. 9.6 for the synthetic sequence. Please note that although in the figure only a single structure for each alkylphenol is provided, many other isomers exist. All aryl phosphates are not alike and they differ in properties, which are listed in Table 9.12 [214]. As one can see, trixylyl derivative, which contains two methyl groups, outperforms others in most properties. Properties of phosphate esters are compared with those of the mineral oil of similar viscosity in Table 9.13 [214]. The data indicate phosphate esters to be better than mineral oils in all properties, except water content and rust prevention. Phosphate esters also possess excellent antiwear performance, as indicated by the hydraulic pump test (ASTM D2882). Phosphate esters have a number of deficiencies, which include their low viscosity indices, which greatly diminish their lubricating ability at high temperatures; their high density, which requires more power for circulation; sensitivity to moisture; and aggressiveness towards materials, such as paints, coatings, and elastomers.

We commented earlier that polyol esters are promoted as low flammability lubricants. Table 9.14 compares the fire resistant properties of two esters with each other and with the mineral oil [214]. The data indicate phosphate esters to be the best, polyol esters to be the next best, and the mineral oil to be the worst. The superiority of the phosphate esters over polyol esters as fire-resistant fluid is demonstrated by its excellent performance in the hot manifold ignition test, wick ignition test, and the compression ignition test. Since polyol esters meets all the requirements listed in the table, except that of the wick test, for most applications, the polyol esters may do an adequate job and at a lower cost. Since all phosphate esters are fire resistant, there are additional properties that help select the right phosphate ester for the right application. For example, in steam turbine applications, where contact with water is most likely, hydrolytically more stable tri-xylyl phosphate ester may be more suitable, and in gas turbines where superior oxidative stability is important, the use of tri-*t*-butyl phosphate may be warranted, see Table 9.12 [214]. It is important to note that like most organic materials phosphate esters are not completely nonflammable. They will burn if enough energy is supplied, resulting in the for-

TABLE 9.11—Fire-resistant properties of different base fluids [214].

Property	Test Method	Mineral Oil	Polyol Ester	Phosphate Ester
Viscosity Grade		ISO VG 46	ISO VG 68	ISO VG 46
Flash point – COC (°C)	ASTM D92	220	258	270
Fire Point (°C)	ASTM D92	245	316	350
Auto-ignition Temperature (°C)	ASTM D2155	340	425	580
Hot Manifold Ignition (°C)	AMS 3150C	350	430	> 800
Wick Ignition	ISO 14935	Fail	Fail	Pass
Compression Ignition	MIL-PERF-19457D			
Ignition Ratio		10	--	> 42
Spray Ignition				
Persistence of Burning	F.M. Std 6930	Fail	Pass	Pass
Persistence of Burning	F.M. Std 6930	Fail	Pass	Pass
Spray Flammability Parameter	ISO 15029-1	Group 3 (Worst)	Group 1-2	Group 1
Ignitability Parameter	ISO 15029-2	Class H (Worst)	Class H	Class D/E

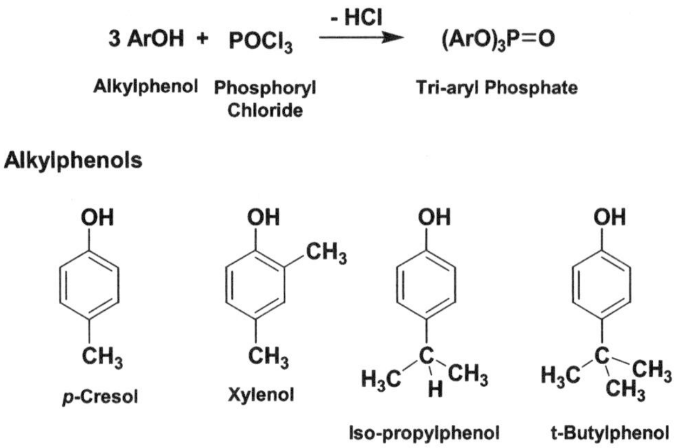

Fig. 9.6—Synthesis of triaryl phosphate.

mation of toxic products. Technology dealing with the synthetic base fluids, including that of esters, is described in Chapter 3 on Synthetic Base Stocks.

Turbine Oil Additives and Testing

As mentioned earlier, the conventional mineral oils have limitations, some of which have been corrected by the new isomerization techniques, such as hydrotreating and hydrocracking. Primary improvements due to these refining techniques pertain to low temperature fluidity, viscosity index, and oxidation stability. At least at present it appears that further improvement in the properties of mineral base stocks is not possible. Synthetic base fluids have also been in use for many years and their strengths have been well exploited. Hence, if one is seeking to improve the lubricant properties beyond what the synthetic base stocks have to offer, the primary avenue appears to be through formulating, i.e., by the use of suitable additives and in the right amount.

Compressor and Refrigeration Oils

Compressors are mechanical devices that are used to transport a gas, or air, from one place to another. They do so by increasing the pressure of the gas by reducing its volume. The compression process is accompanied by an increase in temperature. Some of this heat is removed by the air, water, or oil cooling. Compressors are closely related to pumps: Both increase the pressure on a fluid and both transport fluid through a pipe. However, they differ from each other in that the pumps only raise the pressure of a liquid to allow it to be transported elsewhere, but the compressors also reduce the volume of the gas.

Compressors are of many types and include reciprocating compressors, rotary screw compressors, centrifugal compressors, axial flow compressors, diagonal or mixed flow compressors, and scroll compressors. Compressors are used in a variety of applications, some of which are listed below [681].

1. For pressurizing aircraft to provide a breathable atmosphere of higher than ambient pressure.
2. To power pneumatic tools in industrial, manufacturing, and building processes.
3. To move heat from one place to another, as in refrigeration and air conditioning equipment.
4. In pipeline transporting of the domestic gas from the production site to the consumer.
5. In turbo-charging and super-charging to increase the performance of the internal combustion engines by concentrating oxygen.
6. To provide compressed air for filling pneumatic tires.

In industrial operations, compressors are used to compress a variety of gases, including natural gas, ethylene, air, and ammonia. Natural gas is widely used to heat homes, generate electricity, and as a basic material in the manufac-

TABLE 9.12—Triaryl phosphate properties comparison [214].

Phosphate type	Fire resistance	Oxidation Stability	Hydrolytic Stability	Air release
Tri-xylyl phosphate	Good	Good	Good-Excellent	Excellent
Tri-t-butylphenyl Phosphate	Good	Excellent	Poor	Moderate
Tri-isopropyl Phosphate	Moderate	Moderate	Moderate	Moderate

TABLE 9.13—Properties comparison—mineral oil versus phosphate ester [214].

Property	Test Method	Mineral Oil	Phosphate Ester
Kinematic Viscosity (cSt)	ASTM D445		
100°C		6.7	5.0
40°C		46	43
0°C (cP)		450	1700
Pour Point (°C)	ASTM D97	-9	-20
Acid Number (mg KOH/g)	ASTM D664	0.15	0.05
Specific Gravity (20/20°C)	ASTM D1298	0.87	1.135
Air Release (min)	ASTM D3427	3	1
Foaming Tendency/stability at 24°C (ml)	ASTM D892	250/0	30/0
Water Content (%)	ASTM D1744	0.001	0.06
Rust Prevention (Distilled Water)	ASTM D665	Pass	Fail
Thermal Conductivity (W/mK)	--	0.134	0.134
Specific Heat at 20°C (J/gK)	--	1.76	1.69

TABLE 9.14—Fire-resistant properties of different base fluids [214].

Property	Test Method	Mineral Oil	Polyol Ester	Phosphate Ester
Viscosity Grade		ISO VG 46	ISO VG 68	ISO VG 46
Flash point – COC (°C)	ASTM D92	220	258	270
Fire Point (°C)	ASTM D92	245	316	350
Auto-ignition Temperature (°C)	ASTM D2155	340	425	580
Hot Manifold Ignition (°C)	AMS 3150C	350	430	> 800
Wick Ignition	ISO 14935	Fail	Fail	Pass
Compression Ignition	MIL-PERF-19457D			
Ignition Ratio		10	--	> 42
Spray Ignition				
Persistence of Burning	F.M. Std 6930	Fail	Pass	Pass
Persistence of Burning	F.M. Std 6930	Fail	Pass	Pass
Spray Flammability Parameter	ISO 15029-1	Group 3 (Worst)	Group 1-2	Group 1
Ignitability Parameter	ISO 15029-2	Class H (Worst)	Class H	Class D/E

ture of the many types of chemicals. The compressors that move gas are located in ships and drilling fields, in chemical and process plants, and in the maze of pipes that make up the distribution network to bring gas to the market in a pure, useable form. For transportation and storage, natural gas must be compressed to save space.

Compressors can be classified into two basic categories: Reciprocating and rotary. Reciprocating compressors are used for compressing air, natural gases, and other process gases when the gas flow rates are low and the desired pressures are high. These compressors compress gas to increase pressure by physically reducing the volume of the gas contained in a cylinder by the use of a piston. The piston may be driven by a crankshaft. The compressors can be stationary or portable, can be single or multi-staged, and can be driven by electric motors or the internal combustion engines. Rotary compressors use two meshed rotating positive displacement helical screws to force the gas into a smaller space. These are usually for continuous, commercial, and industrial applications, both stationary and portable. They are commonly used by the roadside repair crews for powering their air tools. This type of compressors is also used in many automobile engine super-chargers because of the ease in matching the induction capacity of the piston engine. Rotary compressors are further classified into positive displacement and the dynamic types. A positive displacement compressor utilizes gas volume reduction to increase the gas pressure. Examples of this type of compressors include rotary screw, lobe, and the vane types [682].

There are many options to power a compressor. These include the use of gas turbines (jet engines), steam turbines, or water turbines for large compressors; electric motors for static compressors; and alternating current, diesel engines or gas engines for portable compressors and for compressors used for super-charging and turbo-charging. Super-charging uses the crankshaft power and turbo-charging uses the exhaust gas energy.

As mentioned earlier, the compression process causes an increase in temperature. This is due to the release of the kinetic energy of the air or the gas molecules as their motion is confined to a small space. Most of this heat must be removed by means other than the lubricant since the extreme temperature will cause rapid degradation of lubricant, both thermally and due to oxidation. Hence, cylinders in the reciprocating type compressors are equipped with cooling jackets and water or water-glycol refrigerant is circulated through them. Despite this, the lubricant for these types of compressors must perform the function of a coolant as well. The reciprocating type compressors contain two sets of parts: The cylinder parts and the running parts [682]. The cylinder parts include pistons, piston rings, cylinder liners, cylinder packing, and valves. The running parts include cross-head guides, the main bearing, and the wristpin, crankpin, and cross-head pin bearings. Although in some cases, it is possible to use the cylinder lubricant to cool both types of parts, in other cases different lubricants are used for each type. This is to avoid exposure of the cylinder lubricant to the compressed gas at high temperatures.

The screw type compressors are either dry or wet (oil-flooded). Dry screw type compressors have rotors that run inside a stator, without a lubricant (or a coolant). For the oil-flooded type, the lubricant is injected into the gas that is trapped inside of the stator. The ambient temperature that the lubricant is exposed to is 80 to 115 °C (180 to 240°F). The lubricant and the gas mixture from the compressor discharge line goes into a gas/lubricant separator where the compressed gas is separated from the lubricant. After separation, the lubricant is cooled and filtered, then pumped back into the compressor housing and bearings. The function of the lubricant is to cool, seal, and lubricate. Unlike the reciprocating type compressors, which are once through, rotary compressors, such as the screw compressor, continuously recirculate the lubricant-gas mixture to facilitate gas cooling and separation [683]. Lubrication of the rotary vane compressors is also once through and the lubricant injected into the compressor exits with the compressed gas and is not recirculated.

Table 9.15 shows the operating temperatures of the various types of compressors and the associated lubrication-related problems [682]. It is obvious that the compressor lubricants must possess excellent thermal and oxidative stability.

Compressor Lubricants

Compressor lubricant must be selected based upon the equipment's lubrication and cooling needs which are defined by the compressor design, pressures involved, operating temperatures, and the gas being compressed. Piston com-

TABLE 9.15—Comparison of the operating temperature ranges for different compressor types [682].

Compressor Type	Operating Temperature Range	Lubrication-related Problems
Rotary Screw	80 to 115°C (180 to 240°F)	Deposits block filter, separator elements Varnish on bearings
Vane	80 to 150°C (180 to 300°F)	Deposits block filters Vane wear increases filter deposits and varnishing
Reciprocating	Single Stage: Up to 270°C (500°F)	Varnish and carbon deposits on exhaust and inlet valves
	Multi-stage: 160 to 210°C (325 to 425°F	Piston ring wear increase leakage and deposits

pressors generate the highest gas pressures; hence have the highest ambient temperatures. Their needs are usually met by the use of the R&O oils. Vane type compressors have extensive metal-to-metal contact; hence they need good EP protection. If the lubricant selected for a particular compressor is not appropriate, the following problems will surface.

1. Increase in oil viscosity and total acidity
2. Copper corrosion
3. Sludge deposits
4. Substantial oil entrainment in discharge gas (air) due to decreased efficiency of the demister element
5. Oil strainer plugging
6. Bearing failure

Desirable properties in a compressor lubricant are almost analogous to those of the other industrial lubricants, such as hydraulic fluids and turbine oils, discussed so far and include the following [685–687].

1. Oxidation resistance
2. Thermal stability
3. Suitable low and high-temperature viscosity
4. Low pour point and good low-temperature pumpability
5. High viscosity index
6. Low volatility/high flash point
7. Superior antiwear performance
8. Good demulsibility
9. Rust and corrosion inhibition
10. Good hydrolytic stability
11. Materials compatibility
12. No sludge formation
13. Low foaming tendency
14. Good cooling ability
15. Nontoxicity and environmental compatibility

Compressor lubricants are usually formulated by the use of low viscosity base stocks so as to facilitate lubrication through narrow valve and piston clearances and tight fitting vanes. These oils generally possess good thermal stability and oxidation and corrosion resistance. Compressors use a variety of lubricants, including engine oils, hydraulic fluids, and automatic transmission fluids (ATFs). In most cases, the OEMs recommend custom formulated fluids for their systems and because of this the compressor lubricants lack the standard performance tests. Most specifications define only physical characteristics, such as viscosity, flash and fire points, and pour point; but none defines parameters such as oxidation life, corrosion protection, or the pump performance.

Mobile compressors are generally lubricated with the same lubricant as that used in the power unit which is driving the compressor. Hence, a compressor driven by a gasoline engine will be lubricated with an engine oil and a compressor driven by a natural gas engine will be lubricated by a natural gas engine oil. Industrial compressors, on the other hand, use R&O oils and antiwear hydraulic fluids but the ideal choice is that recommended by the compressor manufacturer. Centrifugal compressors, often called dynamic or kinetic compressors, are driven by a high-speed impeller or bladed rotor turning inside a close fitting shroud. These are used in applications, such as in a steel mill, where high air volumes are required. Gear drive type compressors use antiwear oils and direct drive compressors use R&O oils. Reciprocating compressors, which operate under fixed conditions, due to sliding friction of the piston have even higher operating temperatures. This can cause a build up of the carbon-like deposits on the upper cylinder heads, which can cause fire and explosion. R&O or anti-wear oils are normally recommended for use in these compressors. These lubricants are often formulated from naphthenic base oils because they produce softer deposits on the cylinder heads, which lowers the maintenance and operating costs.

Rotary sliding vane compressors involve boundary lubrication conditions in addition to the high temperatures; hence they require lubricants that possess oxidation stability as well as antiwear performance. As mentioned earlier, screw compressors are either of the "dry" type or of the "flooded or wet" type. In dry compressors, screws do not contact and hence they do not need a lubricant and the compressed gas is delivered oil-free. However, these compressors do need gear oils. In a "wet" screw compressor, the incoming air is mixed with the lubricating oil, which helps cool the compressed air and provides a seal between the rotating screws. The oil is separated from the compressed air, cooled, and recycled. Synthetic lubricants that have greater thermo-oxidative stability are often used in these compressors.

Compressor lubricants are formulated by the use of the mineral oils, both solvent-neutral and hydrotreated. Naphthenic oils are often preferred because of their superior pressure-viscosity relationship and good low temperature properties. These base stocks are supplemented with oxidation inhibitors to improve their oxidation performance and the sludge forming tendency, foam inhibitors, rust and corrosion inhibitors, demulsifiers, and EP/antiwear additives [688,689]. ISO lubricant classification for oil-lubricated air compressors is provided in Table 9.16 [682].

As mentioned before, some applications require perfor-

TABLE 9.16—ISO 6743, Part 3A—*Family D* compressor lubricant classification for oil-lubricated air compressors [682].

Application	ISO Symbol L-	Service Severity	Service Characteristics	Remarks
Positive displacement air compressors with oil-lubricated compression chambers				
Reciprocating (Crosshead and Trunk Pistons or Rotary Drip Feed –Vane)	DAA	Light-duty	Intermittent Operation	Sufficient rime to allow cooling between periods of operation Compressor stop and start Variable discharge capacity
			Continuous Operation	a. Discharge pressure > 10^3 kPa (10 bar), discharge temperature > 160°C, stage pressure ratio > 3:1 or b. Discharge pressure > I0^3 kPa (I0 bar), discharge temperature > 140°C, stage pressure ratio > 3:1
	DAB	Medium-duty	Intermittent Operation	Sufficient time to allow cooling between periods of operation a. Discharge pressure > 10^3 kPa (I0 bar), discharge temperature> 160°C or c. Discharge pressure > 10^3 kPa (10 bar), discharge temperature > 140°or stage pressure ratio > 3:1 6
	DAC	Heavy-duty	Intermittent or Continuous Operation	As for "medium," when conditions a, b, or c above are fulfilled and where severe coke formation in a discharge system might be anticipated as a result of previous experience with a medium-duty oil.
Rotary Oil-flooded (Vane and Screw Compressors)	DAG	Light-duty	---	Air and air/oil discharge temperature < 90°C, discharge pressure < 800 kPa[1] (< 8 bar)
	DAH	Medium-duty	---	Air and air/oil discharge temperature < 100°C, discharge pressure 800-1500 kPa (8-15 bar) or Air and air/oil discharge temperature 100-110°C, discharge pressure < 800 kPa[1] (< 8 bar)
	DAJ	Heavy-duty	---	Air and air/oil discharge temperature > 100°C, discharge pressure < 800 kPa[1] (< 8 bar) or Air and air/oil discharge temperature of 100°C, discharge pressure 800-1500 kPa[1] (8-15 bar) or Discharge pressure > 1500 kPa[1] (> 15 bar)
Positive Displacement Air Compressors with Oil-free Combustion Chambers				
Liquid Ring Compressors, Water-flooded Vane and Screw Compressors, Reciprocating oil-free Compressors, Rotary oil-free Compressors	---	---		Lubricants suitable for gears, bearings and transmissions
Dynamic Compressors				
Radial and Axial Turbo Compressors	---	---		Lubricants suitable for bearings and gears

[a] Under favorable conditions, light-duty oil may be used at discharge pressures higher than 800 kPa.(8 bar)

mance that is outside the limits of the mineral oils, necessitating the use of the synthetic base stocks which are superior to mineral base oils in many respects. Common synthetic base stocks that are used to formulate compressor oils include poly(alkylene glycol)s (PAGs), carboxylate esters, phosphate esters, polyalphaolefins (PAOs), silicones, fluorocarbons, and alkylbenzenes [690]. While choosing a lubricant, it is important to consider its reactivity towards the gas being compressed as well. Some of the gases, such as methyl chloride, sulfur dioxide, hydrochloric acid, and ammonia, are quite reactive towards some of the base stocks, especially those that are not of hydrocarbon types [694].

PAG-based compressor lubricants are often used in hydrocarbon gas compression applications because of their resistance to dilution by hydrocarbons [695,696]. This is because of the poor solubility of the hydrocarbon gases in these base stocks, which also prevents washing away of the lubricant from the lubricated surfaces that will result in metal-to-metal contact, referred to as dry running [690]. However, this can lead to a build up of the condensed gases due to high operating pressures in the tank as a separate layer. While PAGs have high VIs, they can depolymerize at temperatures over 200 °C [690]; hence their use in applications with an ambient temperature above this must be avoided. PAG-based lubricants work well in compressors used for natural gas, nitrogen, carbon dioxide, hydrogen, helium, and ethylene [697]. PAOs combine the strengths of the mineral oils, such as low gas reactivity and materials compatibility, with good viscosity indices, excellent thermal and oxidative stability, and hydrolytic stability. Also, their cost is the lowest among the synthetic base fluids. Hence, they are used extensively in formulating rotary screw compressor lubricants [695]. For compressors used in the food industry, the food-grade compressor lubricants must be used, which are based on white mineral oil [698] or food-grade PAOs [700]. The topic of food grade lubricants is discussed in latter part of the chapter.

Ester base stocks used in compressor oil formulations include diesters, polyol esters, alkyl phthalates, and alkyl trimellitates [699]. Esters have excellent oxidative resistance, high-temperature stability, and solvency; hence they are often used as cylinder lubricants for the reciprocating compressors. In addition, the use of the esters minimizes deposit formation on hot pistons and discharge valves [695]. Silicones are used as compressor fluids for compressing hydrogen chloride and chlorine [694], which is due to their inertness towards these gases and most other chemicals. However, they cannot tolerate strong oxidizing agents, such as pure oxygen. Some silicones possess corrosion protection properties and are sometimes used as additives in the PAO-derived compressor fluids [694]. Silicones possess excellent viscosity-temperature properties and thermal, oxidation, and hydrolytic stability [696]. Fluorinated base stocks used in compressor lubricants include chlorofluorocarbons, perfluoroalkylpolyethers (PFPEs), and fluorosilicones. They are extremely expensive and their use is limited to applications where the complete chemical inertness is a prerequisite [696]. Phosphate esters are primarily used for their fire-resistant properties, particularly in the mining industry. In addition to fire resistance, they exhibit high flash points, excellent resistance to aging, and minimal coke formation. However, they suffer from poor materials compatibility and hydrolytic stability, which results in the formation of the corrosive products. Alkylated aromatics (alkylbenzenes) have limited use in compressor lubricants because of their high volatility, low viscosity indices, and borderline oxidation resistance. Despite the drawbacks, they are used to formulate lubricants for reciprocating compressors, especially those used for ammonia [701]. This is because the alkylated aromatics do not form carbonaceous sludge.

Gas solubility in compressor lubricants is a concern because of the pressures involved; in general, the higher the pressure, the higher the solubility. Gas solubility is a function of the polarity of the gas and the base stock. Low polarity materials, such as natural gas and other hydrocarbons, are highly soluble in nonpolar mineral oil and PAO base stocks, but not as much in polar synthetics, such as PAGs. In reciprocating and rotary screw compressors, the gas being compressed and the lubricant come into direct contact with each other; hence the gas solubility becomes an issue. This is because if the gas solubility is high, lubricant dilution will occur; which will be reflected by a drop in the lubricant viscosity. This can result in lubrication failure. Solubility of a gas increases with increasing pressure and decreases with increasing temperature. Figure 9.7 shows the hydrocarbon gas solubility in ISO 220 PAG [poly(propylene oxide)] with increasing pressure at a constant temperature [682]. As one can see that at each temperature the hydrocarbon solubility increases with an increase in pressure. The reason for choosing PAG over mineral oil to demonstrate this is because the hydrocarbon solubility in PAG is low, which makes it easier to observe the change, if it occurs. Figure 9.8 shows a drop in viscosity with an increase in temperature, at each concentration of the solubilized gas [682]. Earlier, we mentioned that the nonpolar gases, such as natural gas which has methane content of ~90 %, are very soluble in mineral oils and PAOs—the materials of low polarity. This is demonstrated in Fig. 9.9 and Fig. 9.10 [682]. The solubility of methane in PAG, a more polar material, is almost half that in the mineral oil and PAO. Nitrogen, hydrogen, ethylene, propane, and carbon dioxide reflect a similar behavior [682]. The reverse of gas solubility in oil is the oil solubility in the gas, that is, the lubricant carry-over by the gas. Data exist that suggest that the gas absorbs a significant amount of oil [685].

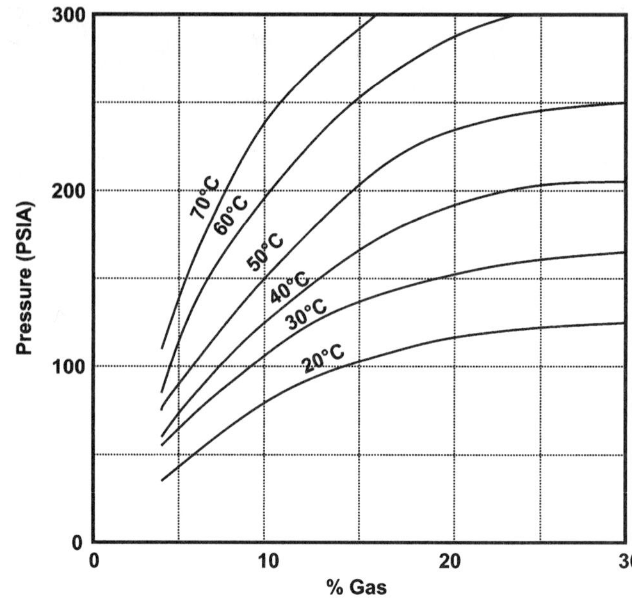

Fig. 9.7—Hydrocarbon gas solubility in ISO 220 poly(propylene oxide) [682].

Properties that are of importance in the compressor lubricants are listed below along with the appropriate tests.

1. Kinematic Viscosity (ASTM D445)
2. Specific Gravity (ASTM D287 and D1298)
3. Pour Point (ASTM D97)
4. Water Content (ASTM D95, D1744, and D4007)
5. Demulsibility (ASTM D1401 or D2711)
6. Hydrolytic Stability (ASTM D2619)
7. Foaming Tendency (ASTM D892)
8. Air Release (ASTM D3427)
9. Corrosion Tests
   a. Corrosion (ASTM D4310)
   b. Copper Corrosion (ASTM D130)
   c. Iron Corrosion (ASTM D665)
10. Flash and Fire Points (ASTM D92)
11. Auto-ignition (ASTM E659)
12. Evaporation Tests
    a. Evaporation Loss (ASTM D972)
    b. NOACK Evaporation Loss (ASTM D5800 and D6375)
13. Carbon Residue Tests
    a. Conradson Carbon Residue (ASTM D189)
    b. Ramsbottom Carbon Residue (ASTM D524)
    c. "Micro Method" for Carbon Residue (ASTM D4530)
14. Precipitation Number (ASTM D91)
15. Acid Number (ASTM D664 and D974)
16. Infra-Red Spectroscopy
17. Fluid Oxidation Tests
    a. Turbine Oil Stability Test-TOST (ASTM D943)
    b. Rotating Pressure Vessel Oxidation Test-RPVOT (ASTM D2272)
    c. Indiana Stirring Oxidation Test-ISOT (Test Method JIS K 2514)
    d. Wolf Strip Oxidation Test [685]
18. Liquid Heptane Washing Test for Oil Film Wash-Off Resistance [685]
19. Tests for Antiwear Properties

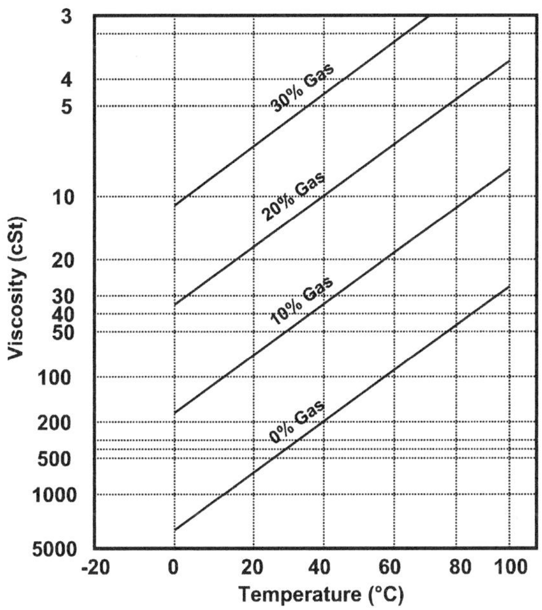

Fig. 9.8—Effect of hydrocarbon gas dilution on viscosity of ISO 220 poly(propylene oxide) [682].

   a. 4-Ball EP and Wear Test (Test Methods ASTM D2783 and D4172 and IP 239)
   b. FZG Visual Method (ASTM D5182)
20. Compressor Tests
   a. Broomwade 2050H Compressor Rig Test [685]
   b. Reavell VHP 15 Compressor Rig Test [685]
   c. Coalescer Blocking Tendency (CBT) Test [702]
21. Gas Solubility into the Lubricant
   a. Gas Solubility in Petroleum Oils at Atmospheric Pressure (ASTM D2779)
   b. Estimation of Gas Solubility in Petroleum and Other Organic Liquids (ASTM D3827)
   c. Experimental Determination of Gas Solubility in Liquids (ASTM D2780)
   d. Viscosity of Gas/Liquid Mixtures Under Pressure
   e. Absorption of the Lubricant into the Gas Phase [685]
22. Heat Transfer Efficiency
   a. Thermal Conductivity (ASTM D2717)
   b. Specific Heat (ASTM D2766)
23. Materials Compatibility
   a. Elastomer Seals Compatibility (ASTM D471)
   b. Paint Compatibility Test [685]
24. Filter Bowl Compatibility Test [685]

Refrigeration Lubricants

Refrigeration is the process of removing heat from an enclosed space or from a substance, by lowering its temperature and maintaining the lower temperature. The removed heat is rejected elsewhere. The most widely used current applications of refrigeration are the air-conditioning of private homes and public buildings and the refrigeration of foodstuffs in homes, restaurants, and large storage warehouses. Refrigeration is also used to liquefy gases, such as oxygen, nitrogen, propane and methane, and in oil refineries, chemical manufacturing, and petrochemical plants. In petrochemical plants, it is used to maintain certain processes at their required low temperatures, such as alkylation of butenes and butane to produce a high octane gasoline component.

Typical air conditioning (A/C) and refrigeration systems operate under elasto-hydrodynamic and boundary lubrication regimes. This requires the lubricant to form physical or chemical films to reduce friction and wear. Other lubricant functions include the transfer of heat from one region to another; to act as a sealant, especially in pressurized systems; protect surfaces from corrosion; and to suspend system debris. In many cases, refrigeration lubricants are required to be compatible with the refrigerant with respect to solubility, miscibility, and chemical interaction [703]. The solubility of gases in various lubricants and its effect on viscosity was already addressed while discussing compressor lubricants. The compatibility issue is important because many of the lubricants that were previously used with chlorofluorocarbons, such as CCl_3F, CCl_2F_2, and $CHClF_2$, are no longer usable with new hydrofluorocarbon (HFCs) refrigerants. This is because of the incompatibility of the HFCs with the mineral oils and the different lubricating characteristics of the nonmineral lubricants. Hence new lubricants need additives to improve their lubricity and antiwear properties. These additives are especially useful in instances where the lubricant

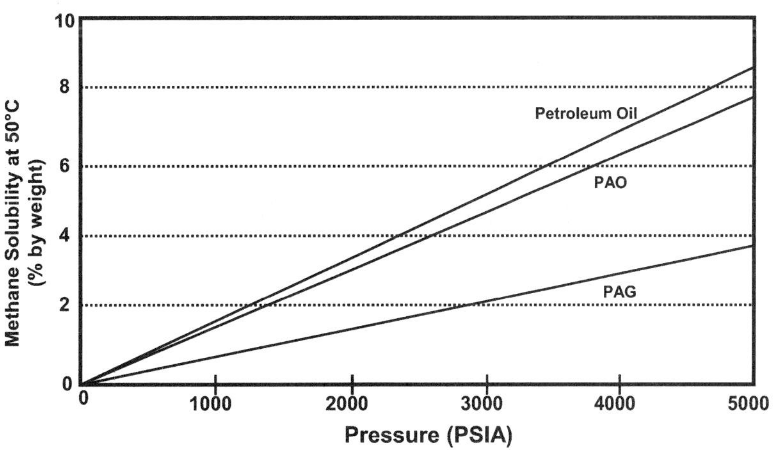

Fig. 9.9—Methane gas solubility in various base fluids [682].

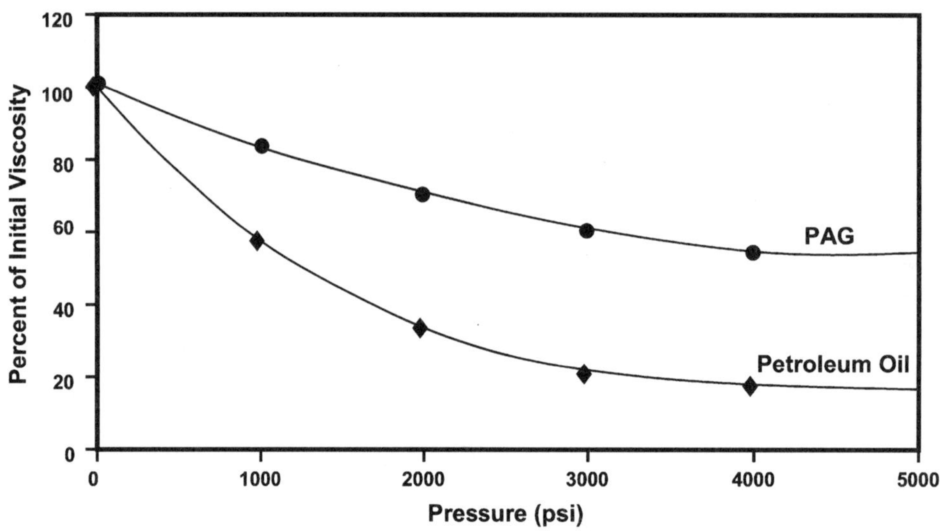

Fig. 9.10—Viscosity decrease due to methane at 50 °C—PAG versus petroleum oil [682].

viscosity drops due to gas dilution and during shut down and start-up operations. Since the new lubricants, such as polyol esters and PAGs are highly polar, they have affinity for water. This can cause the formation of insolubles via hydrolysis of some of the base stocks and additives and the cross reaction of the resulting products. The newly formed products can cause plugging of the capillary tubes and the other expansion devices through deposition, resulting in a loss of performance [704]. Other properties that are desirable in refrigeration lubricants include foaming tendency and the effect of foam on lubrication, on inlet ports of the compressors, and the lubricant carry-over during start-up [705]. Polysiloxanes that modify the surface tension of the bubbles can correct this problem.

The initial decision of the industry to use poly(alkylene glycol)s (PAGs) and polyol esters (POEs) as lubricants with the new HFC refrigerants was based on the assumption that mutual solubility and miscibility of the POEs and HFCs due to similar polarity will facilitate lubricant circulation and compressor operation. However, this has been proven incorrect by recent lubricant circulation studies [706,707].

Automotive air-conditioning lubricants face the same issues as the home and industrial refrigeration lubricants. These are to find a lubricant that is compatible with the refrigerant and has suitable lubricity to provide long life to the pump. Because of the compactness of the air conditioning system in an automobile and its high-temperature operating environments, these lubricants require even higher thermal stability and better seal performance and other materials compatibility. Unfortunately at present no standard tests specific to this application exist.

Refrigeration Lubricant Selection and Composition

The quality of a refrigeration lubricant, like other lubricants, is defined by its ability to meet the equipment's operational requirements, as stipulated by the OEMs. These requirements are translated into lubricant specific parameters and the related tests, which will help the lubricant perform its function as desired. Major factors that govern lubricant selection include the refrigerant (whether CFCs, HFCs, CO_2, or ammonia), compressor's low- and high-temperature viscosity requirements, and lubricant-refrigerant compatibility.

Refrigeration lubricants are formulated by the use of both the mineral oils and synthetic base stocks. Of mineral oils, paraffinic base stocks are preferred because of their high viscosity indices. However, because of a concern for wax crystallization on the expansion valve, capillary line plugging, and oil trapping in the evaporators, only paraffinic oils of low wax content are selected. Naphthenic oils, which are almost wax-free and hence have more favorable pour points, are also used.

TABLE 9.17—Typical properties of refrigeration fluids [704].

Property	Mineral Oil		Alkylbenzene	Polyalkylene Glycol	Polyol Ester	Polyvinyl Ether
	Naphthenic	*Paraffinic*				
Density @ 15°C (g/cc)	0.92	0.86	0.87	0.99	0.97	0.92
Viscosity @ 40°C (cSt)	33	34	32	30	30	30
Viscosity Index	12	95	27	210	90	69
Pressure Viscosity Coefficient @ 60°C (GPa^{-1})	25	30	25	14	15-18	20
Average Molecular Weight	300	378	320	750	500-600	500-600
Pour Point (°C)	-43	-18	-46	-46	-48	-45
Flash Point (°C)	170	200	178	200	234	212
Lower Critical Solution Temperature with R-134a or R-22a (°C)	0[a]	52[a]	-70[a]	--	≤ 34	≤ 50

[a]Solutions with mineral oil or alkylbenzene.

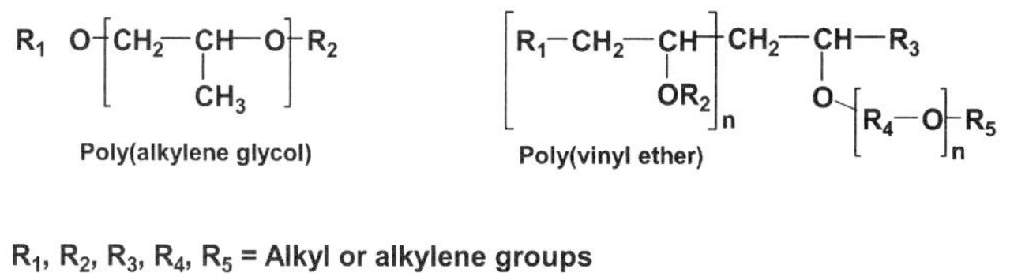

Fig. 9.11—Poly(alkylene glycol) and poly(vinyl ether) structures.

Synthetics used in these lubricants include alkylbenzenes, olefin oligomers (PAOs), dibasic acid esters, polyol esters, and poly(alkylene glycol)s [708]. Typical properties of some of the common refrigerant lubricants are provided in Table 9.17 [704]. Alkylbenzenes have excellent low temperature fluidity and low pour points and are reasonably stable against oxidation, high temperature decomposition, and hydrolysis, and are completely compatible with the mineral oils [709]. They are commonly used with R-22 and R-502 refrigerants. PAOs have very high viscosity indices, good thermal and oxidation stability, and low pour points. They are used in severe service refrigeration systems that employ screw compressors that use R22 as the refrigerant, and in the ammonia systems. They, like mineral oils, are not miscible with the HFC type refrigerants, hence they have limited use. Polyol esters have good high temperature stability, good low temperature fluidity, low pour point, and good HFC compatibility. However, they have lower pressure-viscosity coefficients than the mineral oils and alkylbenzenes, see the data in Table 9.17. Pressure-viscosity coefficient is a measure of an increase in lubricant viscosity with pressure at a constant temperature. A higher value of this lubricant parameter is desirable. The major disadvantage of polyol esters is their lower hydrolytic stability, but compared to that of the diesters it is much better. Despite this, these base fluids have the highest use in HFC containing A/C and refrigeration systems. PAGs, on account of their high polarity, are fully miscible with most HFC refrigerants. They also possess good lubricity, high viscosity indices, low temperature fluidity, and low pour points. However, they are not miscible with the mineral oils and are chemically less stable than the other synthetic base fluids. Their highly hygroscopic nature, which greatly reduces the electrical resistivity of these lubricants, precludes their use in hermetically sealed or closed systems where the motor windings are directly exposed to the refrigerant. This problem can be solved by careful water control in the system. In this application, PAGs are being replaced by the polyol esters that are much less hygroscopic. PAGs are effectively used in systems with external motor drives for the compressor and in the automobile A/C market that has adopted R-134a as the replacement refrigerant for R-12.

Poly(vinyl ether)s or PVEs, a new class of ether type synthetic base stocks, are structurally similar to PAGs, with the difference that the ether linkages are part of the pendent group and not of the back bone, as in the case of PAGs [710]. Figure 9.11 compares the structures of the two [705]. In the PVE structure, the groups labeled m and n are adjusted to control lubricity and the refrigerant solubility characteristics. The lubricity characteristics are discussed in ASTM F2161 and a test procedure is available in ASTM D2670. Like PAGs, PVEs do not hydrolyze to form organic acids and are hygroscopic. Their major strengths are their viscosity—pressure coefficients, which are comparable to those of the mineral oils, and their high miscibility with most HFC refrigerants.

Additives used to formulate these lubricants are the same or similar to those used for compressor lubricants. These include pour point depressants, viscosity index improvers, foam inhibitors, detergents and dispersants, oxidation inhibitors, rust and corrosion inhibitors, and antiwear agents. Since most of these additives were considered in the earlier part of this chapter, here we will consider only those additives that have advantages or issues specific to their use in refrigeration lubricants. These are briefly discussed below.

It is important to note that when using the synthetic base stocks one must consider potential chemical reaction between the base stock and the additives. Base stocks which are likely to react with the additives include synthetic esters and poly(alkylene glycol)s. Similar considerations should be extended to the reactivity of the refrigerants, such as ammonia which is still used in some large commercial plants, towards the base stocks. It is also important to note that many additive formulations that were developed or previously used for chlorofluorocarbon-mineral oil systems are not effective in the HFC-synthetic lubricant systems. Because of this, some OEMs avoid the use of the additives altogether, relying purely on the system design changes for reliability.

Oxidation Inhibitors

Hydrocarbon materials or largely hydrocarbon materials such as those used as lubricant base stocks oxidize in the presence of oxygen, the oxidation rate being faster at high temperatures. Hence, lubricants for high temperature applications, such as heat pumps, are treated with oxidation inhibitors. This is irrespective of whether they are mineral oil-based or synthetic-based. Alkylphenols, such as butylated hydroxytoluene (BHT), are used for polyol ester lubricants and both arylamines and alkylphenols are used in mineral oil formulations. Typical treatment level of these additives for refrigeration lubricants is 0.1 to 0.5 %, depending upon the system's operating conditions [705]. Tests used to assess the effectiveness of these additives are the same as used for the turbine oils, ASTM D943 and D5846.

EP/Antiwear Additives

These additives are needed in these lubricants primarily to counter the effects of the lubricant's dilution by refrigerant gas that diminishes the lubricant's ability to form effective surface films, thereby causing wear to occur. This problem is more pronounced in HFC systems than in chlorofluorocarbon (CFC) systems where the chlorine present in the CFC refrigerants can react with the bearing metal under pressure and frictional heat to provide the metal chloride protective films. Hence, the lubricants for HFC systems require EP/antiwear additives to protect parts against wear. Antiwear agents provide protection under normal (low to medium load) operating conditions and the extreme-pressure (EP) agents provide protection during system break-in and extreme operating conditions of load and temperature (>200 °C). Dialkyl dithiophosphate derivatives and dialkyldithiocarbamates are the commonly used antiwear agents and the dialkyl disulfides and polysulfides and organophosphorus compounds are among those that provide the

EP performance. However, the products of reaction of some of these additives with metal surfaces produce materials of low lubricant solubility, which can impair the function of the capillaries and the expansion devices due to deposition. Reaction of tricresyl phosphate with polyol esters in the presence of water also produces materials that after reaction with metals lead to similar problems. This necessities the need to install filter driers and the control of moisture to overcome this problem. Moisture content of the lubricant is determined by the use of the ASTM Standards E203 and D6304. Typical treatment level of EP/antiwear additives is between 1.0 and 3.0 %. The performance of these additives is evaluated by the use of the ASTM D3233 test.

Acid Scavengers

These compounds are basic materials that are used to neutralize acids that may form in polyol esters and poly(alkylene glycol)s due to the absorption of water or the decomposition of the EP/antiwear additives. Alkanolamines, long chain amides and imines, carbonates, and epoxides are among the most often used acid scavengers. Typical use concentration of these additives is 0.1 to 0.5 %. The major drawback pertaining to the use of these additives is their reactivity towards other additives, such as the antiwear agents, which decreases their effectiveness. ASTM Standards D664, D2896, and D4739 are used to determine the effectiveness of these additives.

Foam Inhibitors

In refrigeration systems, foaming results from mechanical mixing of the lubricant and the refrigerant and by the sudden release of the low boiling refrigerant from the lubricant at reduced pressures. While such foam in CFC-mineral oil systems is persistent and does not collapse easily, in HFC-synthetic lubricant systems it collapses readily. Foam control agents that are used in these lubricants are polydimethylsiloxanes (silicones) and polyacrylates. Their typical treatment level is 100 to 1000 ppm. ASTM D892 Standard is used to determine the foaming tendency of the lubricants.

Miscellaneous Industrial Applications

Other uses of industrial lubricants are in applications listed below along with their performance requirements.

Food-grade Lubricants

These lubricants are used in machinery that is used in food and beverage production and processing and the manufacture of food packaging. Machines employed in these operations have many moving parts that require lubricants to operate reliably and efficiently. Food and beverage contamination can occur from drips off the chains, hydraulic hose failure, oil leaks from seals and gearboxes, or a release of compressed air containing an oil mist. The use of nonfood-grade industrial oils and greases is therefore inappropriate in these settings. If a plant uses a nonfood-grade lubricant, the U.S. Food and Drug Administration (USDA) allows zero amount of lubricant to come into contact with the food. If the lubricant accidentally comes into contact with food, the batch must be discarded. Conversely, if the plant uses food-grade lubricants, which are nontoxic, odorless, colorless, and tasteless, the USDA allows lubricant contamination of up to 10 parts per million [711]. The USDA groups food-grade lubricants into three general classes [712], which are:

1. H1 lubricants: Those that are used in food-processing environments where there is a possibility of incidental food contact.
2. H2 lubricants: Nonfood-grade lubricants that are used on equipment and machine parts in locations where there is no possibility of food contact.
3. H3 lubricants: Food-grade lubricants, typically edible oils, which are used to prevent rust on hooks, trolleys, and similar equipment.

Base fluids and additives allowed for H1 lubricants are identified in Federal Government issued CFR 178.3570 Standard [713]. National Science Foundation (NSF) oversees the registration of the new products and maintains records of Class H1 approved lubricants. For an inventory of the Effective Food Contact Substance Notifications see Ref [714]. For further discussion of this topic, please see the food processing section of Chapter 10 on Lubricating Greases.

Transformer Oils

A transformer is an electrical device that transfers energy from one circuit to another by a magnetic coupling with no moving parts. A transformer comprises two or more coupled windings, or a single tapped winding and, in most cases, a magnetic core to concentrate the magnetic flux. An alternating current in one winding creates a time-varying magnetic flux in the core, which induces a voltage in the other windings. Transformers are used to convert between high and low voltages and to provide electrical isolation between circuits.

Large transformers generate a substantial amount of heat and need to be cooled. Power transformers, rated up to a few kilowatts, are cooled by natural air convection but for cooling transformers of higher power, a number of alternative methods are employed. These include fan cooling, cooling by nitrogen or sulfur hexafluoride gas, or by the use of a transformer oil [715]. This oil is a highly-refined mineral oil that is stable at high temperatures. Large transformers, intended for use indoors must use a nonflammable liquid. Previously, polychlorinated biphenyls (PCBs) were used for this purpose because of their fire resistance and thermo-oxidative stability. However, due to their tendency to accumulate in the environment and the negative impact, PCBs are no longer used in the new equipment. They are replaced by highly stable nontoxic silicone-based oils or fluorinated hydrocarbons. Other less flammable fluids, such as canola oil, may also be used.

The function of the oil is to cool the transformer and act as an electrical insulation between the internal electrically active parts. Hence, it must be stable at high temperatures so that a small short or arc will not cause a breakdown or fire. A number of methods are used to keep the oil temperature down and include convection cooling, cooling fans, oil pumps, and even water via the heat exchangers. Oil-filled transformers undergo prolonged drying processes to make sure that the oil is free of water. This helps in preventing the electrical breakdown under load.

Specifications for the insulating oils for general application, based primarily on naphthenic base stocks are provided in Table 9.18 [716]. These specifications may also apply to paraffinic oils, with the exception of the aniline point requirement and performance at low temperatures. Uninhibited oils do not contain any additives. Inhibited oils are

TABLE 9.18—Transformer oil purchase specification—test limits [716].

Type of Test	ASTM Method	Un-inhibited	Inhibited	
			Type I	Type II
Aniline Point, °C	D611	63-84 max	63-84 max	63-84 max
Carbon Type Composition % Aromatics, Naphthenics, Paraffinics	D2140	No limits	No limits	No limits
Color[a]	D1500	0.5 max	0.5 max	0.5 max
Corrosive Sulfur	D1275	non-corrosive	non-corrosive	non-corrosive
Dielectric Breakdown, kV	D877	30 min	30 min	30 min
Dielectric Breakdown, kV	D1816 (0.04" gap)	20 min	20 min	20 min
Water Content, ppm	D1533 (as received)	30 max	30 max	30 max
Flash Point, °C	D92	145 min	145 min	145 min
Furan Compounds[b] (optional test)	D5837	25 μ g/L max	25 μ g/L max	25 μ g/L max
Impulse Breakdown Voltage, kV @ 25°C[c]	D3300	145 min	145 min	145 min
Interfacial Tension, dynes/cm, 25°C	D971 (unfiltered)	40 min	40 min	40 min
Neutralization Number, mg KOH/g	D974 (modified)[d]	0.015 max	0.015 max	0.015 max
Pour Point, °C	D97	-40 max	-40 max	-40 max
Power Factor, 100°C, %	D924	0.30 max	0.30 max	0.30 max
Power Factor, 25°C, %	D924	0.05 max	0.05 max	0.05 max
Specific Gravity, 60/60	D1298	0.910 max	0.910 max	0.910 max
Viscosity: Kinematic cSt, max	D445			
100°C		3.0	3.0	3.0
40°C		11.0	11.0	11.0
0°C		76.0	76.0	76.0
Oxidation Inhibitor Content, % by wt.	D2668 or D4768[e]	0.00 max	0.08 max	0.3 max
Sludge-Free Life (SFL), measured at 8-hr periods, hrs, +8 hrs.	(Doble Procedure)[f]	40 min	64 min	80 min
Power-Factor Valued Oxidation (PFVO) (optional test)	(Doble Procedure)[f]	See attached graph for limit curve.	See attached graph for limit curve.	See attached graph for limit curve.
Oxidation Stability (acid sludge)	D2440[f]			
72 hours: % sludge by wt.		0.15 max	0.15 max	0.1 max
Total acid no., mg KOH/g		0.5 max	0.5 max	0.3 max
164 hours: % sludge by wt.	-	0.3 max	0.3 max	0.2 max
Total acid no., mg KOH/g		0.6 max	0.6 max	0.4 max
Oxidation Stability (Rotating Bomb), minutes	D2112[f]	-	195 min	220 min
Polychlorinated Biphenyls	D4059	ND[g]	ND[g]	ND[g]
OPTIONAL TEST Gassing Tendency Under Electrical Stress, micro L/min @ 80°C, hydrogen	D2300	Negative[h]	Negative[h]	Negative[h]

[a]Oil must be clear and bright.
[b]The test is for five furan derivatives, 5-hydroxymethyl-2-furfural, furfuryl alcohol, 2-furfural, acetyl furan, 5-methyl-2-furfural. The limit of 25 μ g/L maximum applies to each compound.
[c]Needle negative to sphere grounded, 1-in (25.4-mm) gap.
[d]Neutralization number is measured by using 1/100 normal potassium hydroxide standard alcoholic solution.
[e]Both 2, 6-di-tertiarybutyl-paracresol and 2, 6-ditertiary butyl-phenol have been found to be suitable oxidation inhibitors for use in oils meeting this specification.
[f]Sludge-Free Life and Power Factor Valued Oxidation (PFVO) tests are performed utilizing Doble Methods. The Sludge-Free Life of an oil, sampled at 8-hour intervals, is the number of hours which have elapsed between the start of the test and the time of taking the last sample which showed a sludge-free precipitation test.
[g]ND = none detected.
[h]The characteristic is positive if gas is evolved under the conditions of the test, and negative if gas is absorbed.

supplemented with either 2,6-di-tertiary-butylphenol, 2,6-di-tertiary-butyl-*p*-cresol (BHT), or any other specified and acceptable oxidation inhibitor. They do not contain any other additives. The tests performed on these oils include the following [716].

1. Aniline Point (ASTM D611)—measures the oil's total aromatic content, which relates to the solvency of the oil for materials in contact with the oil. The lower the aniline point, the greater is the solvency effect.
2. Carbon Composition (ASTM D2140)—the test deter-

mines the aromatic, naphthenic, and paraffinic content of the oil. A change in composition can affect oil's properties, such as oxidation stability, low-temperature fluidity, and the viscosity-temperature relationship (VI).

3. Color (ASTM D1500)—reflects oil quality in the new oil and the deterioration level of the oil in service.
4. Corrosive Sulfur (ASTM D1275)—measures the amounts of the dissolved elemental and thermally unstable sulfur, which causes corrosion of the nonferrous transformer metals, such as copper and silver.
5. Dielectric Breakdown (ASTM D877 and D1816)—this problem occurs due to the electrical flashover in an oil. These tests measure the ability of the oil to withstand electrical stress at power frequencies without failure. A low value for the breakdown voltage indicates the presence of contaminants, such as water, dirt, and other conducting impurities in the oil.
6. Water Content (ASTM D1533)—measures the water content in the lubricant. Low water content is necessary for acceptable electrical strength and low dielectric losses in the insulation systems.
7. Flash Point (ASTM D92)—high flash point minimizes the formation of the combustible air-oil mixture except at very high temperatures.
8. Furan Compounds (ASTM D5837)—these compounds are generated as by-products of degradation of the cellulosic materials, such as insulating paper, pressboard, and wood. The presence of these compounds is indicative of the insulation degradation.
9. Impulse Breakdown Voltage (ASTM D3300)—indicates the oil's ability to handle electrical flashover under impulse conditions, such as those caused by the nearby lightning strikes and high-voltage switching surges.
10. Interfacial Tension (ASTM D971)—this parameter measures the ease of rupture of the oil film at the oil-water interface. This is related to the foam formation; the lower the surface tension, which occurs due to the presence of soaps, paints, varnishes, and oxidation products; the easier the foam formation. An increase in surface tension during service indicates an increase in the formation of the oxidation products that can attack the insulation and interfere in the cooling of the transformer windings.
11. Neutralization Number (ASTM D974)—this oil parameter measures the amount of acidic or alkaline materials. During service, the acidity and hence acid neutralization number of the oil increases. This indicates the presence of materials that can attack metals used in transformers
12. Power Factor (ASTM D924)—the power factor of the insulating oil indicates the dielectric loss in an oil. A high power factor suggests the presence of the contaminants or deterioration products.
13. Specific Gravity (ASTM D1298)—specific gravity of the mineral oil influences its heat transfer rates. Oils of vastly different specific gravities may not mix well.
14. Oxidation Inhibitor Content (ASTM D2668 and D4760)—reflects the amount of oxidation inhibitor left in an inhibited oil and can affect service life of the inhibited insulating oils.
15. Power Factor Valued Oxidation (PFVO)—this test, developed by the Doble Engineering Company, measures the power factor of an oil while it is being aged at 95 °C in the presence of copper and air. It indicates the dielectric-loss characteristics of the insulating oil on aging.
16. Oxidation Stability (acid/sludge) (ASTM D2440)—this test measures the oxidation resistance of an oil by determining the amount of acid/sludge formed under prescribed conditions.
17. Oxidation Stability (ASTM D2112)—this test evaluates the oxidation stability of the new inhibited insulating oils. Good oxidation stability is a principal requirement for long service life of the transformer oils.
18. Gassing under Electrical Stress (ASTM D2300)—gassing tendency is the rate of gas evolved or absorbed by an insulating oil when subjected to electrical stress of sufficient intensity to cause ionization. This characteristic is positive if gas is evolved and negative if gas is absorbed.
19. Polychlorinated Biphenyls (ASTM D4059)—today's regulations prohibit the presence or the use of the polychlorinated biphenyls (PCBs). PCB contamination levels must be monitored to ensure their absence.
20. Viscosity (ASTM D445)—viscosity of an oil affects its flow, its cooling ability, and the speed of moving parts in tap changers and circuit breakers. High viscosity oils are less desirable, especially in cold climates.

Details of these tests are available from ASTM Standards Books, published by ASTM International.

Lubricants for Air Tools (Pneumatics)

Air tools are devices that are used to accomplish diverse mechanical processes, such as drilling, hammering, grinding, sanding, and spraying. These tools use compressed air as the source of power, which is generated by the use of air compressors. Common tools include drills, ratchets, impact wrenches, hammers, grinders, sanders, and sprayers. Air tools are portable, versatile, convenient, and relatively light weight, and provide the power and speed to accomplish the intended operations quickly and efficiently.

Air tools employ cylinders and valves that require lubrication. The cylinders are used to convert the air pressure into mechanical energy, which is used to perform linear movement, such as lifting or moving tools and work pieces. The valves control the starts, stops, direction, and pressures to ensure that the pressurized air follows the intended path. Lubricants for air tools must possess many of the following properties.

1. Suitable viscosity, consistent with the operating temperatures of approximately −30°C to +50°C.
2. Protection at extreme operating temperatures.
3. Lubricate the valves, pistons, and the other tool components to protect them against friction and wear.
4. Varnish, carbon, sludge, and deposit control (high detergency).
5. Elastomer seal and O-ring compatibility.
6. Rust and corrosion protection of the tool's internal parts.
7. Demulsibility.
8. Reduce icing.
9. Noise reduction.
10. Environmental compatibility.

11. Cost-effectiveness.

Air cylinders are lubricated by the use of the ISO viscosity grade 22 and 32 R&O oils that provide good corrosion protection [592]. Air motors, of both vane and piston type, are lubricated either by the oil mist or the built-in splash mechanism. Most air motors work via gear reduction boxes, to reduce high speeds to more suitable slow speeds. These are lubricated by the use of lubricants that are similar to those used to lubricate compressors, although in some cases gear oils may also be used. For the reciprocating type tools, lubricants containing extreme pressure agents are the most beneficial. These lubricants possess good oxidation stability and are able to handle water better, either by dissolving it or by shedding it. They normally have low odor and are non-toxic, and belong to ISO viscosity grades 32 to 100. ISO 32 and ISO 46 viscosity grades are for small and medium size drills, grinders, vibrators, wrenches, screw drivers, air nailers and staplers, air cylinders, motors, valves and chucks, and ISO 68 and higher viscosity grades are for large drills, grinders, vibrators, hand-held tools, and jackhammers.

Lubricants for Chain Drives

Chain drive is a mechanical device that is used to transmit mechanical power from one location to another. While these devices are used in many machines, their use to convey power to the wheels of a bicycle or a motorcycle is the most familiar example. Other uses of these devices are to transmit power, move materials, and operate equipment.

A chain drive system typically comprises a chain (the drive chain) and sprockets where the chain is used to transmit motion between the rotating shafts via sprockets mounted on the shafts. The chain is made up of many metallic links with rollers, bushings, and interconnecting links. Chains belong to three broad classes: stamped steel or malleable iron, roller or block type, and silent or inverted tooth. Of these, the roller type chains are the most common. Roller chains are generally manufactured from high specification steels and are therefore capable of transmitting high torques within compact space envelopes. Sometimes the power output is obtained by simply rotating the chain, which can be used to lift or drag objects. In other situations, a second gear is placed and the power is recovered by attaching shafts or hubs to this gear. Though drive chains are often simple oval loops, they can also go around corners by placing more than two gears along the chain; gears that do not put power into the system or transmit it out are generally known as idler-wheels. By varying the diameter of the input and output gears with respect to each other, the gear ratio can be altered. Industrial chain drives are generally designed to operate in enclosed cases with installed lubrication systems.

Fatigue and wear of the sprocket teeth are the normal modes of failure in chains, which causes the chains to jump the teeth. This can be minimized by reducing friction, cooling, and lowering impact resistance at higher chain speeds. The OEMs generally provide recommendations for the lubrication requirements for their chain drives. If suitable lubrication is not provided the capacity of the chain drive is greatly reduced. Four basic modes of lubrication that are used for chain drives are:

1. Manual/Drip lubrication: In manual lubrication, the lubricant is generously applied to the chain drive about every 8 operating hours and in drip lubrication, the oil is continuously dripped on the chain center line.
2. Bath/Disc lubrication: In bath lubrication, the lower strand of the chain runs through a sump containing the lubricant. The oil level is above the lowest pitch line of the chain when it is operating normally, excessive immersion may result in turbulence of the oil bath. Disc lubrication is based on a disc attached to one sprocket which is immersed in an oil bath. As the disc rotates it picks up the oil and deposits it onto the chain. A trough is often used to direct the oil to the optimum point on the chain. A peripheral disc speed of between 3 and 40 m/s is normal.
3. Oil stream lubrication: In this type of lubrication, a continuous stream of filtered oil is circulated by a pump to be spread evenly across the width of the slack side of the chain.
4. Oil-mist lubrication: This type of lubrication is used for high speed chain drive and is based on the chain case being filled with an oil mist.

New developments in chain drive technology include self-lubricated chains and plastic chains. These either do not require lubrication or only intermitted lubrication. Please note that plastic chains drives have reduced operating capabilities compared to steel chains. Non-lubricated chains are essential for industries that require controlled environments, such as paper, packaging, electronics, and white and brown goods manufacture.

Chain drive lubricants are selected on this basis of the type of housing, speed, load, clearances, degree of bending at sprocket, and environmental conditions. The lubricant must hae low viscocity to reach the internal surfaces, but the viscosity must be high enough to maintain an oil film under bearing pressures. In addition, the lubricant must be non-corrosive, maintain lubricating properties over a broad-temperature range and under moisture, and must not foam. However, a chain drive lubricant's primary function is to provide rust and EP/antiwear protection to the device [592]. The chains when used in dusty environments, such as in cement, paper, or flour mills, will develop deposits resulting from the accumulated dust and the lubricant. These will decrease the chain speed, increase power consumption, and destroy the efficiency of the chain drive. For these environments, dry-film or solid lubricants are used. They are applied to the chain by spraying a solution of the lubricant in a low flash solvent. The solvent evaporates, thereby leaving behind a thin layer of the solid lubricant. To minimize dripping during application and use, these lubricants contain a tackifier. Oven conveyor chains are lubricated with synthetics containing solid lubricants to impart oxidative stability (minimize sludge formation and gumming), and provide antiwear, EP, and rust protection.

Chain Saw Lubricants

Portable chain saws are widely used for felling, delimbing, topping, and bucking (cutting logs to a desired length) operations. These saws usually have a two-stroke cycle air-cooled engine which drives a toothed chain around an elongated guide bar through a safety clutch, either directly from the engine crankshaft or through a gear box. Besides lubricating the engine, the lubricant is also needed to lubricate the chain, bar, and the sprocket. These lubricants contain a tackifier to minimize the amount of lubricant being thrown off the chain during the operation. These lubricants also contain EP/antiwear agents, solid film-forming agents, or solid lubricants to minimize wear. Because of the extensive use of these devices in natural environments, high biodegradability of the lubricant is a desired trait. Hence, these lubricants use highly biodegradable vegetable oils and synthetic esters. The typical viscosity range for these lubricants is from ISO VG 46 to ISO VG 100.

Lubricants for Concrete Molds

Concrete is the most commonly used building material, which is used both for construction and to add decorative qualities to buildings and on structures. Whether it is precast or poured on-site, concrete is shaped with shuttering or molds. After the concrete has set, it must be possible to remove the shuttering or mold, without damaging the set concrete. The surface of the concrete must have a smooth finish as well as accurately duplicate the shape of the mold. Mold-release agents are therefore indispensable for casting concrete or cement.

In concrete molds, the mold release agents or form oils, are used to minimize the tendency of the concrete to stick to surfaces in contact. These oils are applied to formers and shutters, needed to support the curing concrete. These lubricants are of two types: oil-solvent systems and emulsifiable oils. Oil-solvent systems use naphthenic oils that contain fatty acids, which react with calcium in the concrete to form soaps that provide the needed separation properties and rust protection to the metal shutters. Emulsifiable oils are water-in-oil emulsions (invert emulsions).

Continuing interest in environmentally compatible lubricants, especially in Europe, has led to the development and marketing of the bio-based mold-release fluids. Most of them are soy-based, although any natural oil can be used to formulate them [717]. In order to maintain a high degree of biodegradability, many are formulated with highly biodegradable corrosion inhibitors and viscosity-reducing additives.

Glass Molding Lubricants

Commercially produced glass belongs to five broad classes, which are soda-lime glass, lead glass, fused silica glass, borosilicate glass, and 96% silica glass. Of these, soda-lime glass accounts for 77 percent of the total glass production and is manufactured from sand, limestone, soda ash, and cullet (broken glass). Glass production involves four steps: (1) preparation of raw material, (2) melting in a furnace, (3) forming, and (4) finishing. The final products made from soda-lime glass are flat glass, container glass, and pressed and blown glass. Crushed sand, limestone, and soda ash raw materials are mixed with cullet, to ensure homogeneous melting, and the mixture is conveyed to a batch storage bin where it is held until dropped into the feeder to the melting furnace. The furnace most commonly used is a continuous regenerative furnace capable of producing between 45 and 272 megagrams (50 and 300 tons) of glass per day. As the material enters the melting furnace through the feeder, it floats on the top of the molten glass already present in the furnace and as is melts, it passes to the front of the melter and eventually flows through a throat leading to the refiner. In the refiner, the molten glass is heat-conditioned for delivery to the forming process. After refining, the molten glass leaves the furnace to be shaped by pressing, blowing, pressing and blowing, drawing, rolling, or floating to produce the desired products. Pressing and blowing are performed mechanically, using blank molds and glass cut into sections (gobs) by a set of shears. In the drawing process, molten glass is drawn upward in a sheet through rollers, with thickness of the sheet determined by the speed of the draw and the configuration of the draw bar. The rolling process is similar to the drawing process except that the glass is drawn horizontally on plain or patterned rollers and, for plate glass, requires grinding and polishing. In the float process, the molten glass from the refiner moves to the molten tin bath over which the glass is drawn and formed into a finely finished surface requiring no grinding or polishing. The end product undergoes finishing (decorating or coating) and annealing (removing unwanted stress areas in the glass) as required, and it is then inspected and prepared for shipment to market. Any damaged or undesirable glass is transferred back to the batch plant to be used as cullet [Ref. www.epa.gov/ttn/chief/ap42/ch11/final/c11s15.pdf]

Glass manufacture employs a lubricant at many stages and since it is a high-temperature operation, the mineral oil-based lubricants with low thermo-oxidative stability are not appropriate. This is because such lubricants on burning will generate carbon and ash in scoops and troughs, thereby adversely affecting loading, clarity, and the quality of the glass. Hence, synthetic oils, natural oils, or solid lubricants, such as graphite, are commonly used. Solid lubricant is used as a suspension in a carrier fluid which burns cleanly leaving behind a film of the solid lubricant. For example, in order to lubricate shears that are used to cut the stream of the molten glass into gobs (temperature ~900°C), a lard oil-based lubricant is used which provides both lubrication and cooling. The next stage requiring lubrication is the blank mold, where the gob obtained from the first stage is transformed into a hollow container. The lubricants for this stage, called the swabbing agents, contain wax thickeners and suspending agents, which although wax-like melt when stirred or rubbed into the swab. The waxes have the advantage of burning without residue or color. Besides swabbing agents, water-based or cured thermoset polysiloxane coatings containing graphite are another way to coat the molds. These are applied by spraying and form hard durable coating on curing. Low concentration graphite and sulfer suspensions in oil are also used as the oven chain lubricants and lubricants for conveyers, rollers, hinge pins, and bearings [592]. When applied to a hot metal surface the oil vehicle boils off cleanly to leave behind a silvery-black solid lubricant film. All lubricants in glass production must be environmentally friendly as well as have excellent water separating ability.

Paper Mill/Paper Machine Oils

Paper mill lubricants include greases, gear oils, and hydraulic fluids. The dryer sections of the mill experience extreme temperatures of the superheated steam. In this section, there are a large number of roller bearings that require lubrication. Lubricants for these bearings must possess excellent thermo-oxidative stability, rust performance, detergency, demulsibility, and long service life. Premium quality antiwear hydraulic oils and EP gear oils provide the needed performance in this application. There is a growing preference for oils that have the ability to simultaneously provide antiwear and the extreme-pressure protection.

Lubricants for Rock Crushers

Rock crushers are devices that help reduce the size of rocks that contain minerals. A rock crusher requires several lubricants to lubricate hydraulics, gears, bearings, etc. [592]. Engine oils are usually of 5W-40 viscosity grade. For gears and bearings, the EP gear oils having USS 224, AGMA 250.04, and DIN 51 254, Part 2 performance are used. OEMs recommend oils both for gears and hydraulics to have good oxidation stability, EP/antiwear performance, rust and corrosion inhibition, demulsibility, and low foaming tendency. Gear oils are of ISO viscosity grades 150 to 680 (SAE 80W to 140) and hydraulic fluids are of ISO viscosity grades 46 to 68 with ATF type performance.

Rock Drill Lubricants

These lubricants are used for percussion air tools, which include rock drills, jackhammers, stoppers, drifters, wagon drills, pavement breakers, pneumatic pile drivers, ballast tampers, and chipper hammers. The moving parts in rock drills are lubricated by the oil mist which is carried through the drill by the passing air. These lubricants must possess good thermal and oxidative stability, provide good EP/antiwear performance, inhibit rust, corrosion, and foam, and possess emulsifying properties to absorb the excess water that may be present in the air that is passing through the system [592]. In addition, these lubricants must be compatible with steel, copper, and alloy parts, and have low odor and be non-toxic. Non-toxicity is important while drilling in confined spaces because of the fine oil mist that is emitted along with the exhaust. Viscosity grades for these lubricants are between ISO VG 46 and ISO VG 320. While selecting a viscosity grade, it is important to take into consideration the ambient temperature since the air exiting the tool cools the lubricant. These lubricants are usually formulated using high VI paraffinic base oils, tackiness agents, and select additives to provide rust and corrosion inhibition, resistance to foaming, excellent oxidation stability, and superior load-carrying ability.

Slide Way Lubricants

A slide way is a track upon which a machine tool slides back and forth during certain manufacturing processes. Slide ways are also found in lumber mills, on the saws and carriages used to produce lumber. Slide way lubricants help provide smooth tool movement and eliminate sticking and chatter. Heavy loads squeeze the lubricant film out, thereby producing boundary lubrication conditions [592]. These oils must perform at extreme temperatures, high loads, in the presence of moisture, and in poor ambient air quality environment. Hence, they must possess EP activity and rust and corrosion-inhibiting properties. Slide way lubricants therefore contain EP/antiwear agents and rust and corrosion inhibitors. For use in vertical slide ways, the lubricants are supplemented with tackifiers that minimize run-off. Viscosity range for these lubricants is between ISO VG 46 and ISO VG 100. For more information on these lubricants, please refer to Chapter 11 on Metalworking and Machining Fluids.

Lubricants for Textile Mills

Lubricants used in the fiber and textile industries are either used for the lubrication of the textile machinery or are used as process oils. Process oils are used for processing the natural fibers, the production and processing of the synthetic fibers, and for dressing or finishing of the intermediate or final products obtained from these fibers. Since these oils are close to or in contact with the fibers, yarns, and fabrics, there is a concern for contamination of these materials by the oil in the form of oily residue and non-removable stains, which may hinder their dying.

Textile mill lubricants are used to lubricate various parts of the textile machinery and include R&O type oils of ISO viscosity grade 22. They are formulated with light-colored base oils or highly-refined technical white oils so as not to mar the fiber and/or the fabric. Some operations need friction reduction and wear control; hence these oils are formulated with fatty oils to offer wear protection under high speed, start-up conditions. Other oils, such as the coning lubricants, are formulated with emulsifiers and antistatic agents to reduce snagging and pulling of the yarn when it comes off the cone [592]. For fast-running textile machinery medium to high-viscosity lubricants of ISO viscosity grade 68 and higher are used. This is to minimize their spattering on the fiber and the fabric by spindles rotating at speeds of up to 15,000 rpm. Sometimes tackifiers are added to improve adhesion properties of these oils. Some formulations contain aging inhibitors that insure prolonged service of up to 5000 h and facilitate the removal of the oils by washing, even when they have been in prolonged use. Premium products also contain additives that facilitate the removal of the oil stains from the fabrics.

For hard-to-process fibers, as those of jute, oil-in-water emulsions are used to improve working. Such emulsions are based upon low-viscosity, lightly-colored spindle oils of low odor that contain oleic acid to facilitate emulsion formation with alkaline water, which is responsible for softening the fibers. In the manufacture of some synthetic fibers the thread that leaves the nozzle is coated with a fine film of a colorless petroleum fraction of 260 to 330°C boiling point to permit further processing. While this oil is designed to evaporate during the subsequent steps, sometimes an oily residue remains which may lead to yellow discoloration of the products. This yellow color may be hard to remove if the oil has inferior ant-aging characteristics. Staple fibers are treated after spinning with white-oil based formulations (winding oils), in order to give the thread or yarn favorable processing characteristics. An important role of the mineral oil component is to lubricate the thread so that the fiber experiences a uniform pulling force. This is achieved by the use of the tackifiers. Another group of lubricants that is used is called "tearing oils," which are light-colored mineral oils that contain friction-reducing additives. Their function is to prevent tearing of the used yarns and threads during re-spinning to obtain better quality yarns with long fibers.

Vacuum Pump Fluids

Vacuum pumps reduce pressure of an environment below atmospheric and in effect are compressors working in a reverse manner. Vacuum pumps are of three functional types: mechanical, diffusion, and ejector. Mechanical type includes reciprocating, rotary, and dynamic units which are identical

TABLE 9.19—Typical properties of some wire rope lubricants [720].

Property	Test Method	Asphalt	Petrolatum	Grease
Viscosity at 100°C, cSt	ASTM D445	84.5	24	-
Dropping Point, °C (°F)	ASTM D566 (IP 132, ISO 2176) and ASTM D2265	74 (166)	72 (162)	85 (185)
Penetration (worked 60 strokes)	ASTM D217, IP 50	72	86	209
Adhesion, kPa (psi)	Bethlehem Steel	77.9 (11.3)	66.8 (9.7)	-
Water Spray-off, %	Bethlehem Steel	0.03	1-5	-
Rust Test Rating	Bethlehem Steel	1	1	1
Four-Ball EP Test	ASTM D2596			
Weld Point		315	-	160
Load Wear Index		46.7	-	31.1
Four-Ball Wear Test	ASTM D2266			
Scar at 20 kg (mm)		0.54	0.69	0.30
Scar at 40 kg (mm)		0.94	0.72	0.61
Throw – off Temperature, °C (°F)	Bethlehem Steel	-	71.1 (160)	-

to compressors of the corresponding type. In general, the lubrication requirements of mechanical vacuum pumps are similar to those of their compressor counterparts. These are lubricated with mineral oil, turbine oils, or other high quality R&O circulating oils. Mineral oil-based lubricants cool, lubricate, and provide a seal in piston and rotary vane air vacuum pumps. The lubricant must have low vapor pressures for maximum vacuum efficiency. ISO viscosity grades of these lubricants range from ISO VG 32 to ISO VG 100. For pumps that handle moisture, 3 to 7 % fatty oil may be added to counteract the effect of water condensing on the cylinder walls. High vacuum rotary oil-sealed pumps may be lubricated with ISO viscosity grade 68 straight mineral oils. Chemical stability and low-vapor pressure of the diesters and the phosphate esters make them good candidates for applications in cases where petroleum-based oils are not suitable [592,719]. Diffusion pumps and ejector pumps have no moving parts to be lubricated. Diffusion pumps utilize a high velocity vapor stream to entrain and remove the molecules of the gas being evacuated from the system. Oil or another suitable substance is evaporated to produce the vapor stream. Ejector pumps use high-pressure steam or another gas, which is discharged through a nozzle that directs a high velocity jet across a suction chamber into a venturi-shaped diffuser case. Molecules of the gas being removed are picked up and entrained in the jet stream and discharged with it.

Vacuum pumps, especially of the diffusion type, are used in a number of research and instrumentation related applications. Examples include electron microscopes and vacuum evaporators and the associated instrumentation. Pump fluids are classified according to the vacuum pump type, for example, mechanical versus diffusion; or by the fluid type. Fluids used in these pumps are based on hydrocarbons, poly(phenyl ether)s, perfluorinated polyethers, silicone, or diesters [718,719].

Lubricants for Wire Ropes or Wire Cables

Wire ropes are used in many machines and structures to perform a number of functions, such as pulling, dragging, and hoisting. Applications include drag lines, cranes, elevators, shovels, drilling rigs, suspension bridges, and cable-stayed towers. There are many kinds of wire ropes, each type designed for a specific application. Most ropes are made of continuous wire strands wound around a central core. The core may be made of steel, fiber, and even plastic. However, most wire ropes are made from high carbon steel for strength, versatility, resilience, availability, and cost. Wire ropes may be uncoated or galvanized.

Lubricating wire ropes is not easy, irrespective of their construction and material. The ropes with fiber cores are somewhat easier to lubricate than those made exclusively from steel. Wire rope lubricants are used to reduce friction, as the individual wires move against one another and other surfaces, and rust and corrosion protection. These lubricants are of two types: those that penetrate and those that coat. Penetrating lubricants contain a petroleum solvent that carries the lubricant into the core of the wire rope and then evaporates, leaving behind a heavy lubricating film to protect and lubricate each strand. Coating lubricants penetrate slightly and seal the outside of the cable against moisture and harmful elements. This reduces rust, wear, and fretting corrosion of the external surfaces. Since most wire ropes fail from the inside, it is important to make sure that the center core receives sufficient lubricant. A combination approach in which a penetrating lubricant is used to saturate the core, followed with a coating lubricant to seal and protect the outer surface, is recommended. Wire rope lubricants can be petrolatum, asphalt, grease, petroleum oil, or the vegetable oil-based [720]. It is important to note that when the asphaltic materials are used to enhance adhesion, lubricant hardening and chipping off can occur at cold temperatures, for example, those encountered in mining operations.

Wire rope lubricants are formulated with rust inhibitors and EP/antiwear agents. Typical viscosity grades for wire rope lubricants are between ISO VG 32 and ISO VG 100. Tests used to assess suitable performance of a wire rope lubricant include four-ball EP (ASTM D2783), salt spray (ASTM B117), and humidity cabinet (ASTM D1748) [721]. Table 9.19 compares typical properties of three materials which are often used as wire rope lubricants [720].

Machine Tool Lubricants

Machine tools are a set of basic tools that the industry uses to create many of its products. These include milling machines, lathes, grinders, planers, broaches, drills, and a wide variety of other multipurpose and specialized tools. The quality of a machine tool is determined by its precision which is greatly facilitated by a quality lubricant. A lubricant helps achieve this by reducing friction, heat, and wear. Depending on the complexity of the machine, it may require one lubricant or many lubricants. For example, if the machine contains a power transmission, it will require a transmission fluid. If it contains bearings, it will require a bearing lubricant or lubricating grease. In some cases, no lubricant is necessary since the machine tool component has a wear-resistant surface coating of chromium or phosphate.

When a lubricant is necessary, one must consider a number of factors before selecting a suitable lubricant. These include ambient and operating temperatures, humidity, and particles in the air. This is because these factors can profoundly affect precision and surface finish. Depending on the machine element, the lubrication method may be manual, partial immersion, splash, centralized forced feed, or oil mist type. These and other lubrication methods were described in Chapter 1 on Lubrication Fundamentals. The viscosity of these oils is defined by ISO viscosity grades.

Gear oils used in machine tool applications are based on mineral base stocks that contain extreme pressure additives to provide high-load carrying capability. The additive may be animal tallow or preferably of sulfur-phosphorus type. These lubricants are required to lubricate highly loaded gears or gears, such as worm gears, that have extensive sliding contact. For lightly loaded high-speed gears, hydraulic fluids or spindle lubricants suffice. Typical viscosity ranges for gear lubricants are ISO 68, 150, 320, and 460 (SAE 80W to 140). Spindle oils are highly refined mineral oils with good oxidation resistance. These oils typically are of low viscosity (2 to 22 cSt at 40°C) since they are used to lubricate parts, such as antifriction and hydrostatic bearings, electromagnetic clutches, and lightly loaded gears, that have high operating velocities. Other common lubricants that are used in machine tools are slide way lubricant, hydraulic fluids, and greases. As mentioned in the earlier part of the chapter, the primary function of slide way lubricants is to prevent stick-slip; hence they are formulated with anti-squawk and tackiness additives to lubricate plain bearing slide ways. These lubricants usually are of ISO viscosity grades 32, 68, and 220. Hydraulic oils used in this application are either R&O (rust and oxidation inhibited) oils or antiwear hydraulic fluids. While the basic function of the hydraulic oils is to transmit power, they also lubricate pumps, valves, and other system components. R&O oils have adequate performance in low-pressure systems but for high-pressure systems, especially those employing vane and rotor type pumps, fluids with good oxidation and antiwear performance may be required. These oils contain either metal-free sulfur-phosphorus compounds or stabilized zinc dialkyl dithiophosphates as antiwear additives. Normal zinc dialkyl dithiophosphates are too unstable to be useful in controlling oxidation and wear since they easily decompose in the presence of moisture, copper and bronze, and elevated operating temperatures and lead to sludge formation due to oxidative degradation of the lubricant. Hydraulic fluids typically are of ISO viscosity grades 32, 46, 68, and 150. Lubricating greases used in this industry are multipurpose and are used to lubricate plain and roller element bearings and miscellaneous other parts. Lithium soap greases are often preferred, primarily because of their excellent resistance to softening upon working and good resistance to removal by water. Extreme pressure greases containing EP additives are also used in lubricating heavily-loaded ball and roller element bearings. Greases containing MoS_2 are also used in extreme pressure applications, especially in machine parts that operate at very low speeds. Normally greases used in machine tools do not contain fillers such as clay, mica, or asbestos. Lubricating greases of NLGI Grades 1 and 2 take care of the lubrication needs of most applications.

Lubricants Used in Mining Industry

Lubricating mining equipment poses a challenge, primarily because of the distinctive operating environment. Because of the space constraint, mining equipment dimensions are small, which tends to limit the size of the lubricant sumps or reservoirs, the amount of air space around them, and the size of the bearings and bearing surfaces. These factors increase the lubricant and bearing temperatures, gear pressure, and loadings, and makes servicing of these parts difficult because of the space limitations. In addition, the presence of dust in the air further exacerbates the situation.

Mining environment determines the extent of the lubrication challenges, If the ore is located near the surface of the earth, the mining is called "surface" or "strip" mining and "open pit" techniques are used. The lubrication needs of machinery in this case are different from that used in underground mining where the ore is located in the earth's crust at some depth. Primary factors that alter the nature of lubrication in the two cases are the amount of dust and the temperature differential. Dust is the particulate matter, airborne or otherwise, that may contaminate the lubricant and impede its flow through small orifices. If the dust is acidic, it may cause an additional concern since it may hydrolyze the lubricant components, thereby leading to corrosion of the mining equipment. Temperature effects are related to the lubricant's viscosity change. Too high a temperature will lower the lubricant viscosity, which will impair its ability to form a suitable lubricating film to protect surfaces against friction and wear. Too low a temperature will increase lubricant viscosity, causing "drag" or even failure to flow to the critical parts. High temperatures will also increase the oxidation rate of the lubricant which may result in an increase in its viscosity and also produce deposit-forming species. An additional factor that may need to be considered when mining in closed environment is potential fire hazard. This implies that the lubricants used in this type of environment must have a high flash point.

Mining operation uses both automotive and industrial equipment and includes engines operated on diesel, gasoline, LPG, and natural gas fuels; gear systems; hydraulics; and various other devices that need to be lubricated. Engines are lubricated with good quality four–stroke cycle and two-stroke cycle engine oils of suitable viscosity. Please note that the oil and the oil filter change interval is likely to be narrower for engines used in mining than for those used in on-road and other off-road applications. This is because of the dusty and the enclosed environment. A typical way of determining a lubricant's in-sevice life is to analyze it for wear metals; the amount of water, fuel, coolant and silica that is part of the dust; oil thickening; and base reserve.

Automotive gear lubricants are primarily used in automotive equipment utilized in open pit mining and are of GL-4 or GL-5 quality. While for enclosed industrial gears the

R&O oils usually suffice they may be substituted with an EP product if an unusual rise above ambient temperature is observed. This is to offset any drop in lubricant viscosity due to the higher temperature which will adversely affect a lubricant's film-forming ability. Open industrial gear lubrication may involve continuous or intermittent lubrication. Continuous lubrication methods include splash or idler immersion, gravity feed, drip, and recirculating spray. Intermittent lubricating includes manual application and maually-activated or automatically-timed mechanical sprays.

Hydraulic fluids used in mining include mineral hydraulic fluids, R&O oils, antiwear-R&O oils, and fire-resistant hydraulic fluids. For details on these fluids, please refer to Chapter 7 on Hydraulic and Transmission Fluids. Underground mining operations utilize most types of fire-resistant fluids, which is due to the enclosed nature of the operation and sometimes due to government mandate. Lubricating greases used in mining equipment are both specialty and general-purpose types. However, the growing trend is to use only general-purpose greases, which is to manage inventory storage limitiations and costs. In addition to the lubricants discussed so far there are additioanl lubricants that are used in the mining industry [Okon, L. W., "Mining Industry," CRC Handbook, Vol. I, pp. 405–430]. These are listed below. At least three of these were discussed in the earlier part of this chapter.

1. Dragline cam lubes.
2. Rope oils.
3. Rock drill oils.
4. Compressor oils.

Industrial Uses of Synthetic Fluids

While discussing turbine oils, compressor lubricants, and refrigeration lubricants, we commented on the use of the synthetics and their unique attributes. There are many other applications that can benefit from the use of synthetics. In addition, there are many industries which use processes that involve temperatures between 150 °C and 1000 °C to manufacture and finish their products. These temperatures are either at the borderline or outside the effective range of the mineral oil-based lubricants; hence synthetic fluids must be used by necessity. In this section, we will discuss the uses of the synthetics in these applications and industries.

Synthetic lubricants do not have a direct advantage over mineral oils in most applications, but they do possess certain unique properties that give them a performance edge. These include higher viscosity-pressure coefficients and higher surface affinity, for example in synthetic esters, which help in the formation of thicker lubricant films in ball bearings and gears, operating under elasto-hydrodynamic and mixed-film lubrication conditions. In addition, synthetics have higher oxidation resistance, which makes their use suitable in high-speed environments where significant agitation and aeration occurs. Some synthetics handle heavy loads better than other synthetic and mineral oil-derived lubricants. For example, the use of the synthetic hydrocarbon fluids and PAGs in gear drives operating at high speeds results in lower frictional losses; hence higher efficiency, lower energy consumption, and longer life [447].

In some applications, it is impractical or impossible to relubricate and the initial lubricant becomes a sealed-for-life lubricant. Such lubricants must have increased oxidation stability, lower volatility, and higher viscosity index; all these properties exist in synthetic lubricants. Hence, they are ideal for sealed-for-life applications. Examples include sintered metal plain bearings that are used in appliances, hand tools, and automotives, and high-speed spindle bearings that are used in many applications. Lubricants based upon PAGs, esters, and perfluorinated aliphatic ether (PFAE) provide excellent performance in sintered bearings and the lubricants based on esters and PAOs perform well in high-speed spindle bearings.

Some synthetic lubricants have the ability to survive specialized environments, such as vacuum, pure oxygen, acids and alkalis, and solvents. For example, lubricants based upon PFAE and PTFE (polytetrafluoroethylene) are most suitable for aero-space and vacuum applications. This is because of their extremely low vapor pressure ($<10^{-9}$ mbar at room temperature), high thermal stability without deposit formation (up to a temperature of 350 °C, or 650 °F), resistance to electron and ion bombardment, and ignition resistance. Lubricants based upon polyalphaolefins, silicones, and polyglycols offer varying degree of resistance towards the aggressive species listed above and are therefore suitable for use in environments that contain them. PAGs are less prone to hydrocarbon dilution; hence they maintain their viscosity in high hydrocarbon environments. Therefore, they are lubricants of choice in propane production and refrigeration. PFAE, in view of being inert, can be used in almost all aggressive environments [447].

Concern for chemical accumulation in the environment has led to an interest in developing environmentally degradable lubricants. One of the tests that are used to determine the environmental compatibility of a substance is a 21-day biodegradability test, CEC-L 33-T 82. A lubricant is considered highly biodegradable if 90 % or more of the sample degrades during the test. Typically, mineral oils and PAGs degrade 20 to 40 %, diesters and polyol esters degrade >90 %, and poly(ethylene glycol)s degrade >90 %. For greases, all currently used thickeners are suitable.

Examples of some of the industrial operations that can benefit from the superior properties of the synthetic lubricants are described below [447].

Textiles

Operations in this industry that have specialized lubrication needs include fabric transport through tenter oven to dry and stretch, fabric finishing by the use of a calender machine that has two or more heavy rolls, and synthetic fiber production—extrusion of the melted polymer or the polymer solution through the spinneret to form continuous fiber filaments. Fabric transport uses ball bearings, roller chains, and sliding chains that operate at temperatures of 80 to 250 °C. For ball bearing rollers, PFAE/PFPE-derived grease, thickened with polytetrafluoroethylene (PTFE) particles, is most suitable since it performs well and requires only an annual service. For lubricating roller chains and sliding chains, both diester and polyol ester lubricants are employed. Diesters are used if the operating temperature is between 120 and 200 °C, and polyol esters are used if the temperature is above 200 °C. These fluids have the advantages of low evaporation, minimal residue build-up, and lower lubricant consumption [447].

Fabric finishing involves passing the fabric through a calender machine that has two or more heavy rollers, which are used to produce special effects, such as luster and embossing. The rollers are heated to around 200 °C by the use of a polyglycol circulating fluid, which acts as a lubricant as well as survives the high temperatures for a reasonable period of time. The bearings on the rollers are lubricated with ester-based grease, if their temperature is 150 °C or less; silicon oil grease, if it is between 150 and 200 °C; and PFAE grease, if it is above 200 °C. In synthetic fiber production, the melted synthetic polymer is converted into fiber by extruding through a metal disk with small holes, called a spinnerette. This device, which has a face temperature of 250 °C, is coated with a thin silicone lubricant film, or the separating agent, to prevent build-up of the polymer residue and clogging of the holes [447].

Textiles also involve high-speed operations. For example, separator rollers used in fiber manufacturing can reach speeds of up to 30,000 r/min and the false twist tubes in textile machines operate at speeds as high as 80,000 r/min. In these operations, the use of the ester- and PAO-derived greases eliminates the problems associated with the use of the oil mist method of lubrication.

Wood Products Industry

Wood processing has many operations where the use of the synthetic lubricants is beneficial. Board production from wood particles and fibers mixed with resins is a prime example. The process involves supporting the particles-resin mixture between two steel belts and passing it through a machine that uses high temperatures (~240 °C) and high pressures to convert the mixture into a wooden board. Parts that need lubrication are rollers that are between the hot plates and the steel belts, and the chains that drive the rollers and the feed guide chains. All these parts are lubricated by an ester-based fluid, which being thermally stable minimizes the residue build-up due to lubricant oxidation. The chain sprocket bearings are lubricated with a PFAE grease to prolong their service life.

Pulp and Paper Industry

Many of the processes in this industry are similar to those of the textile industry; hence the lubricants employed are similar. Calenders and rotating unions, which operate at high temperatures, are lubricated by polyglycol based circulating oil and for rotating union bearings, ester grease, silicone grease, or PFAE/PTFE grease are used, depending on the temperature.

Plastic Film Industry

Two of the many plastic film production processes that involve high temperatures are extrusion and passing the film through a die and the film stretching and width control. Extrusion produces a sheet or a tube that is forced through a heated die that is at a temperature of 240 °C. The sheet film is then subjected to a film stretching and width control machine that also operates at high temperatures. The die bearings are lubricated with a PFAE-derived grease, the chains on the film stretching machine are lubricated by an ester type fluid, and the width control spindles are lubricated with silicone grease [447].

Automotive Manufacturing Industry

Curing of the paints and coatings after the finished component assembly is an important part of the automobile manufacturing process. The trolley wheel bearings of the conveyor, which are exposed to temperatures of up to 250 °C, are lubricated with a PFAE-derived grease. The oil-lubricated overhead conveyor systems employ an ester-based lubricant.

Brick and Ceramics Industry

Two major operations in the production of bricks and ceramics are kiln curing and exhaust removal. Premanufactured bricks are moved into the kiln on a large, heavy-duty cart running on steel tracks. Since the wheel bearings in these carts approach a temperature of 1000 °C, the fluid portion of the lubricant will either evaporate or burn. Polyglycol (PAG) type fluid, supplemented with solid lubricant to form a grease-like paste, is used to lubricate these bearings. PAG fluids decompose/burn cleanly and leave little or no carbon residue to hinder the rotation of the wheel bearings. This is because PAG depolymerizes around a temperature of 200 °C to form low flash products, leaving behind the dry lubricant on the bearings that help remove the cart from the kiln. Exhaust fans that are used to remove the hot air and fumes from the process contain sealed-for-life motor bearings, which are lubricated by a PFAE fluid thickened with PTFE particles [447].

Food Processing Industry

Baking, painting, decoration of the beverage cans, and heat-curing, prior to filling, and freezer storage are some of the operations of the food industry that benefit from the use of the synthetic lubricants. Baking involves continually conveying the food product through the oven. The chains that drive the conveyor are equipped with sealed bearings to withstand the oven temperatures of 200 °C and above. These bearings are usually packed with a PFAE-derived grease. Painted beverage cans are heat-cured in an oven that typically operates at a temperature of around 200 °C. The high speed chains used to bring the cans in the curing oven are lubricated by an ester type oil to minimize residue formation, resulting from the lubricant oxidation. Bearings used in blast freezers for food storage help maintain a temperature of −40 °C. These bearings are lubricated with silicone or ester greases, using thickener systems that provide low apparent dynamic viscosities, to facilitate smooth motion and low torque operation.

Metal Industries

These industries employ a variety of operations where the use of the synthetic fluids is beneficial. Such operations include part (piece) forming/die casting, mining/steel production, and metalworking. Synthetic hydrocarbon and silicone oils are used as die lubricants and separating agents during the pressure die casting operation. Water-miscible synthetic waxes are also used as the separating agents in die casting. Synthetic hydrocarbon oils mixed with metallic solids pastes are used as ladle dressing. When the ladle is exposed to the molten metal and the synthetic hydrocarbon oil oxidizes and carbonizes to become part of the solid bonding matrix, it acts as a protective insulation. Phosphate esters are used as fire-resistant hydraulic fluids in steel mills, foundries, and underground mines. Synthetic fluids are often used as water-soluble cutting fluids, grinding coolants, and rolling fluids. The use of the synthetic fluids in place of petroleum-based products results in a drop in the lubricant consumption and parts cleaning costs and increases tool life. For detailed discussion on metalworking fluids, refer to Chapter 11 on the Metalworking Fluids.

Machine Tool Industries

Machine tool manufacturers continuously strive to increase speed for improving the efficiency of their cutting operation. Bearings used in the cutting devices are typically lubricated with an oil mist that creates the worker hazard. Many manufacturers are switching to an ester or polyalphaolefin-based grease as a lubricant, which help them achieve similar speeds and eliminates the worker hazard. These greases can also be used to lubricate high speed ball screws on the machine tools.

High-speed gears in the gear head of the multi-spindle drives, lubricated with mineral oil, suffer from the sealing difficulties. The result is leakage into the high-speed spindle bearings, which reduces their service life. The use of the ester—or PAO-based grease eliminates this problem.

Formulation Examples

Turbine Oil/R&O Oil: 0.2 % Ashless dithiocarbamate EP/Antiwear agent, 0.05 % tolyltriazole derivative corrosion inhibitor, 0.25 % alkylated diphenylamine oxidation inhibitor. Balance is API Group II oil (formulation extracted from Ref [722]).

Industrial R&O Oil: 0.375 % Alkylated diphenylamine and 0.1 % hydroxyethyl n-dodecyl sulfide oxidation inhibitors, 0.05 % ethylenediamine salt of dinonylnaphthalene-sulfonic acid and 0.002 % tolyltriazole rust and corrosion inhibitors. The balance is a 90:10 mixture of 220N and 600N API Group II base oils. The addition of foam inhibitor; demulsifier, pour point depressant, and a viscosity modifier is optional (formulation extracted from Ref [723]).

Industrial R&O Oil: 0.75 % Alkylated diphenylamine and 0.18 % hydroxypropyl t-dodecyl sulfide oxidation inhibitors, 0.05 % alkylamine salts of alkyl phosphoric acids and 0.002 % tolyltriazole rust and corrosion inhibitors, ethylene-propylene oxide copolymer demulsifier, and 0.02 % (2-Ethylhexyl/Ethyl) acrylate copolymer foam inhibitor. The balance is a 90:10 mixture of 220N and 600N API Group II base oils. The addition of a pour point depressant and a viscosity modifier is optional (formulation extracted from Ref [723]).

Turbine Lubricant (EP): 2.0 % Synthetic triaryl phosphate EP/antiwear agent, 0.3 % 2,6-di-t-butylphenol oxidation inhibitor, 0.075 % caprylic acid vapor phase rust inhibitor, 0.017 % benzotriazole corrosion inhibitor, 50 ppm poly(alkyl acrylate) foam inhibitor, and 0.015 % of alkylmaleic acid and dodecyl di-hydrogen phosphate mixture. The balance is mineral oil (formulation extracted from Ref [724]).

Compressor Lubricant: 0.5 % Phenolic and alkylated diphenylamine oxidation inhibitors, 0.1 % alkenylsuccinic acid half ester rust inhibitor, 0.1 % triazole derivative corrosion inhibitor. The balance is a base fluid blend comprising PAO and polyol ester of ISO VG 68 (formulation extracted from Ref [725]).

Refrigeration Lubricant: 0.2 % Ethoxylated phosphate ester and 0.2 % mercaptobenzothiazole EP/antiwearagents, 0.3 % propoxylated amine such as Jeffamine® rust inhibitor, 0.1 % tolyltriazole corrosion inhibitor. The balance is PAG fluid of 100 to 1200 SSU (20.5 to 259 cSt) at 100°F (38°C). For automotive air conditioning a viscosity of 500 SSU (108 cSt) at 100°F is preferred (formulation extracted from Ref [726]).

Pneumatic Tool Lubricant: 0.79 % Sodium petroleum sulfonate detergent, 0.2 % sulfur, 0.1 % mercaptobenzothiazole copper corrosion inhibitor, 0.5 % polyisobutylene antimisting agent, and 10 ppm polysiloxane foam inhibitor. The balance is naphthenic base oil (formulation extracted from Ref [726B]).

10

Lubricating Greases

IN THIS CHAPTER WE REVIEW THE TECHNOLOGY of lubricating greases. Because of their distinct physical form relative to liquid lubricants, greases are used in specialized applications, such as roller bearings and slow speed gear systems, where a liquid lubricant either cannot be used or has inadequate performance. Composition, chemistry, properties, manufacture, and methods to produce lubricating greases are discussed. The chapter also includes discussion pertaining to selection criteria, testing requirements, and handling and disposal of greases. We close the chapter with examples of grease formulations.

The use of animal fat in combination with lime, a crude form of the modern lubricating grease dates back to 1400 B.C. [33]. In fact, the word "grease" is derived from the Latin word "crassus" for fat. Lubricating grease is one of the oldest lubricants used by man. The use of the animal fat for lubrication continued until the late 19th/early 20th century when mineral oil-derived greases were developed. This occurred after the discovery of petroleum in 1859. Incidentally, in this chapter, grease and lubricating grease and oil and fluid are interchangeable terms and imply the same.

Mineral oil thickened with calcium carboxylate (lime soap) was the first grease that was marketed in volume. This was followed by aluminum stearate grease, sodium soap grease, calcium complex soap grease, and lithium and barium soap greases. A great break through occurred in 1942 when it was discovered that lithium 12-hydroxystearate-derived greases possess superior properties. The evolution of the lubricating greases is depicted in Fig. 10.1. One way to look at the timeline is that a particular lubricating grease had some inherent deficiencies; hence another grease was developed to overcome them.

The development of the current generation of lubricating greases was first reported in the early 1940s, after the invention of lithium12-hydroxystearate-derived grease, which was almost 65 years ago. During this duration, many new developments have occurred that have shaped the lubricating grease industry to its present state. The lubricating grease has been defined in many ways [727–731], but the two that we find of importance are stated below.

1. The ASTM definition, provided in ASTM Standard D288, defines grease in terms of its physical appearance and composition. This definition characterizes the lubricating grease as *"a solid-to-semifluid product of dispersion of a thickening agent in a liquid lubricant. Other ingredients imparting special properties may be included."* Incidentally, because of its inadequacies, this definition of the lubricating grease no longer appears in the book of ASTM standards.
2. Sinitsyn's definition is based upon rheology rather than the appearance and composition. According to this definition, the lubricating grease is *"a lubricant which under certain loads and within its range of temperature application, exhibits the properties of a solid body, undergoes plastic strain and starts to flow like a fluid should the load reach the critical point and regains solid body properties after the removal of the stress."* [731]

The ASTM definition acknowledges the lubricating grease to be thickened oil, comprising at least a two-component system, containing a thickener and a liquid lubricant. Sinitsyn's definition establishes the dual nature of the grease as being both a solid and a liquid, depending on the physical conditions of temperature, stress, etc. This definition also recognizes an additional property of the grease that is not present in a conventional liquid lubricant, a yield value. Yield value is the minimum shear stress that produces flow. Shear stress is the force per unit area that causes shearing of a structure. However, shear stress differs from stress in that it acts parallel to the surface (cross section), while normal stress acts perpendicular (normal) to the cross section. Since each of the two definitions describes different but important properties of the lubricating grease, a combined definition is probably more appropriate. The combined definition reads "Lubricating grease is a solid-to-semisolid lubricant that results from dispersing a thickening agent in a liquid lubricant. It loses its structure under load or stress, thereby releasing the liquid lubricant, and regains its solid-to-semisolid appearance after the stress is removed."

Obviously, in this definition, we are envisioning the lubricating grease to be a two phase system consisting of:

1. A liquid phase that may or may not contain performance-enhancing chemicals, or additives.
2. A solid phase that either has a network structure, such as a metal salt or soap, that has intimate association with the oil or fluid, or is simply dispersed in oil, as in the case of the nonsoap thickeners, such as polyurea, modified clay, and graphite.

Soap thickeners possess interlocked fibrous structures that are responsible for the above-mentioned affiliation between the soap and the lubricant. The mechanism by which this occurs will be addressed under the section on soaps. However, the nonsoap thickeners do not usually possess fibrous structures and hence do not interact with the oil by the same mechanism and to the same degree as soaps. It is important to note that such a two-phase system must perform the functions of a lubricant. This primarily occurs when under operating conditions the thickener re-

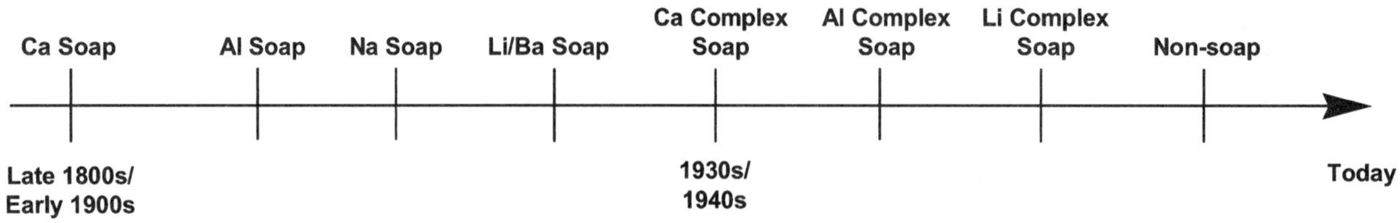

Fig. 10.1—Timeline for lubricating grease development.

leases the lubricant to fulfill the function of lubrication.

There are many solid-oil mixtures that look like greases, for example, pastes. However, they differ from lubricating greases in two major aspects, their solids content is very high, in the vicinity of 70–95 %, and the affinity between the solid and the oil is not essential. The latter factor makes such systems unstable and makes them poor lubricants in applications where greases do an excellent job.

For a lubricating grease to meet the specifications of the above definitions, the grease must comprise three components: A base fluid, a thickening agent, and an additive package; and in most cases it must have a network structure. While a wide variety of agents can disperse in oil to yield grease-like dispersions, only those that form dispersions with lubricating properties are useful. In addition, grease contains additives that impart other desirable properties, such as EP, water resistance, etc. The lubrication function is carried out by the small amount of oil that is released during the equipment's operation. Because of their semisolid nature, greases are used when fluid lubricants are inefficient, the need for lubrication is infrequent, and the lubricant is required to maintain its original position in a mechanism, or combinations thereof.

Lubricating grease is a unique product and its creation, manufacture, and use requires knowledge in chemistry, physics, tribology, rheology, formulation, manufacturing, chemical engineering, and health and environmental sciences. For example, the lubricating grease manufacture is based on chemistry, but its function as a mechanical barrier between two moving surfaces comes under the domain of physics; its structural design so that it possesses suitable viscosity to adhere to and stay between the surfaces falls under rheology (the science of flow); and its impact when it enters the environment is a concern of health and environmental sciences.

Since lubricant grease is used in a large number of machine components, it is important to consider it as an integral part of the equipment design rather than an after thought necessity.

Lubricating Grease Versus Liquid Lubricant

As stated above, lubricating greases are thickened lubricants with less mobility than their liquid counterparts. Because of their physical characteristics, the lubricating greases offer a number of advantages over liquid lubricants, when used in certain specialty applications. Some of the advantages of lubricating greases are listed below.

1. Lubricating greases resist leakage through dripping, splattering, loss due to high pressure, and centrifugal removal. Dripping is a gravity-induced loss, splattering is common to the vertically positioned equipment, and centrifugal removal usually occurs in equipment that operates at high speeds.
2. Lubricating greases have the ability to act as a seal against dirt, water, and other contaminants. Examples include industrial bearings that are commonly exposed to water and exposed gears that encounter a significant amount of airborne dust.
3. Lubricating greases have thicker film-forming ability and better adhesion properties, both of which are useful in lubricating equipment/parts that operate at high temperatures and under extreme conditions of shock-loading, reversible operation, slow speeds, or very high speeds. In this regard, the use of the lubricating grease assures better friction and wear protection and rust and corrosion protection than the liquid lubricants. Examples of parts where lubricating grease is a suitable lubricant include journal bearings that encounter heavy shock loading and mining equipment that is usually operated on slow speeds and heavy loads. Shock loading is an instantaneous and severe increase in stress. It tends to rupture the thin lubricant film, leading to rapid wear. Greases form durable films. Shock loading is common in the rolling mill operation when a thick steel slab hits the rollers which have been preset to a lesser thickness.
4. Lubricating greases reduce noise and vibration due to worn parts. If their use in equipment designed to lubricate with a liquid lubricant is possible, they extend the life of the worn parts by coating them with a thick lubricating film.
5. Lubricating greases are suitable to lubricate equipment, such as ball bearings, roller bearings, and sealed for life bearings, where the operation is intermittent and the relubrication need is infrequent or not necessary.
6. Lubricating greases are useful in applications where it is necessary for a lubricant to maintain its properties over long periods in spite of unfavorable temperature, pressure, and environmental influences; for example, to lubricate aerospace equipment.
7. Lubricating greases are largely impervious to water and can easily handle small amounts of water.
8. Because lubricating greases are designed to have good compatibility with seal materials, in some cases they can be used as inexpensive seals. This is partly due to their viscoelastic properties because of which the seals are not constantly subjected to a fresh amount of lubricant.
9. Lubricating greases facilitate the use of the solid additives or those that have limited solubility in traditional mineral base oils or synthetic fluids.
10. Lubricating greases, especially those derived from the

natural acids and fats, can be considered environmentally compatible products. Many are biodegradable.

Of course, lubricating greases also possess the major disadvantage of lacking little or no ability to provide cooling and cleaning. In additions, its removal and replenishment after it is past its usefulness is not a trivial task.

History Of Lubricating Grease Development

The use of a lubricant to minimize friction dates back in time. As mentioned earlier, the use of fat-based grease dates back to about 1400 BC. The evidence regarding its use is based on the analyses of the residues from the wooden axle hubs of chariots and carriages of the ancient Egyptians of that time. The use of animal fat and vegetable oil-derived greases continued until the mid-1800s at which time the composition of the lubricating greases changed. The primary driving force behind this development was the discovery of mineral oil in 1859 [727].

Lime or calcium soap (calcium carboxylate) greases date back to the 1880s [728]. They were among the first to be produced and marketed in volume. These greases were made by the cold mixing of lime (calcium hydroxide) with rosin oil (an acid) solution in mineral oil and dispersing the resulting product in water and oil. Such lubricating grease while adequate to lubricate the slow-moving machinery of the late 19th and early 20th centuries has little use in the machines of today.

Modern greases owe their invention to the industrial age when diverse equipment needing superior lubrication was developed. The invention of the modern machines, such as automobiles and railroad engines, underscored the need for lubricants, including greases, which could survive higher operating speeds and temperatures. Developing the calcium soap greases made by the reaction of animal fats with calcium bases at high temperatures was the next evolutionary step. The greases thus formed had a smooth texture and better water tolerance than those made by the cold reaction. However, these greases still required the presence of 1–2 % water to impart structural stability. As a result, their use was limited to applications that will not experience a temperature of 100°C or more. Otherwise, the fate of these greases will be no different than that of their counterparts made by cold mixing of the ingredients. The next generation of greases were invented to overcome the use temperature limitation.

Aluminum stearate soaps led to the formation of lubricating grease that was clear and smooth, with good water tolerance and fair rust resistance. However, its heat resistance is not that much better than lime soap greases. Limiting temperature in this case was around 175°F (79°C), above which it lost its clarity and smooth texture and became rubbery. The rubbery texture on cooling became brittle and lost most of the oil. Poor heat resistance of these greases is not related to the water loss since these greases are anhydrous. They are still produced for use in a limited number of applications where their strengths are beneficial.

Anhydrous sodium soap greases were also developed in an attempt to replace lime soap greases, to overcome the structural instability due to water loss. These greases, despite being anhydrous, did not prove very useful since their water tolerance is low, owing to the substantial water solubility of the sodium soaps. Some of these in the presence of water emulsify at room temperature while others need higher temperatures for emulsification, although in both cases the lubricating grease loses its soap and hence its texture. Despite this disadvantage, sodium soap greases have a temperature tolerance of about 100°F (56°C) above that of the calcium soap greases, as long as the use conditions stay anhydrous. They have the additional advantages of better mechanical handling, called working, and better stability than both the calcium and the aluminum soap greases. Sodium soap greases have been used to lubricate automotive wheel bearings, electric motors, and a variety of industrial machinery. However, they are now being replaced by nonwater washable products with equal or better heat resistance.

In the 1930s and 1940s, search for new thickeners that resulted in better lubricating greases, especially those for use in multi-purpose greases, ensued. Calcium complex soaps were developed first which was closely followed by the development of lithium soaps and barium soaps. Today, lithium soap greases constitute almost 38 % of the total world production and use. The reasons for the popularity of these products are their superior thermal stability and water tolerance. Previously, animal fats and fatty acids were used as the soap precursors. In some cases, hydrogenated fatty acids were subsequently added in an effort to improve the properties of the resulting grease. A similar approach was extended to lithium soaps. During experimentation, it was discovered that the use of hydrogenated castor oil, a triglyceride of 12-hydroxystearic acid, yielded a soap that led to a grease with superior overall properties than the greases made from the other fatty acids or esters.

Barium soap greases, although similar to lithium soap greases in terms of water tolerance and heat resistance, suffer from poor low-temperature performance. This is partly a consequence of the high molecular weight of the soap molecule. As a result, these greases are not as popular as the lithium-based products. In addition, their use is diminishing because of the concern for the toxicity of barium and its impact on the environment, due to it being a heavy metal. Most barium soap greases in use are complex soap greases rather than the simple soap greases. Many other grease thickeners were developed during the 1930–1950 period, but most of the derived greases were short-lived, with little or no commercial success.

Calcium complex soap grease, the first of the complex soap greases discovered, has been in use since the 1950s but its share is progressively decreasing. The reason for their initial success was their good thermal stability and the ability to carry heavy loads in bearings. Unfortunately their water sensitivity is one of the reasons for their demise. In the presence of water, they change their consistency to softer or harder, depending upon the conditions. Aluminum complex soap grease was the next complex soap grease to become commercial. While these greases have a small share of the overall grease market, their use appears to be growing. The primary reasons for their increasing demand is their superior water tolerance, making them the lubricating grease of choice in applications that see a substantial amount of water, such as in rolling mills. Lithium complex soap greases, which at 30 % enjoy the largest share of today's complex soap grease market, were first commercialized in 1962. The reasons for

TABLE 10.1—NLGI classification system based on consistency.

NLGI Grade	Penetration Range @ 25°C (ASTM Worked)	Visual Characteristics	Application
000	445–475	Very soft—just enough thickener	Open gear lubrication
00	400–430	to keep oil from running	
0	335–385	Soft	Open gear lubrication and centralized lubrication systems requiring low-temperature pumpability
1	310–340	Soft	Needle and multiple row roller bearings; flexible chain coupling
2	265–295	Creamy texture	Plain and antifriction bearings operating under moderate load and at medium speeds; most flexible couplings
3	220–250	Semi-solid	Antifriction bearings; automotive wheel bearings; prelubed ball bearings—double sealed and double shielded type
4	175–205	Stiff	Water pumps and other high-speed, lightly-loaded applications
5	130–160	Stiff	High-speed applications
6	85–115	Hard solid	Pillow-block lubrication

their overwhelming success are their good water resistance and thermal stability.

Nonsoap thickeners, like lithium complex soap greases, are also a somewhat recent development. This class of thickeners comprises organic thickeners and inorganic thickeners. Organic thickeners include polyureas and substituted ureas; salts of terephthalic acids, phosphoric acids, thiophosphoric acids, and phosphonic acids; and polyethylenes and halogenated polyethylenes, such as Teflon® polycarbohydrates; and pigments. None of these, except polyureas, have been used in greases that are commercially available in large volume. Originally, arylureas were used to make lubricating greases but the new greases use alkyl aryl polyureas instead. Polyurea greases have good oxidation stability which in part may be due their low hydrocarbon content. These greases are ideally suited for lubricating ball bearings that are exposed to high temperatures, such as those used in electric motors. Polyurea complex greases made from mixed polyurea-calcium acetate thickeners are also available. These have extreme pressure properties equal to those of the calcium complex greases.

A variety of inorganic materials have also been used as thickeners for lubricating greases. They include modified clays, molybdenum disulfide (MoS_2), and graphite. Natural clays are hydrophilic and as such are not useful as thickeners for lubricating grease. However, their structure can be modified to make them hydrophobic, or oleophilic, by exchanging the alkali metal or alkaline earth cation with a quaternary ammonium cation. This is achieved by reacting or coating the clay with suitable quaternary ammonium salt. The resulting modified clays are easy to disperse in oil and the resulting greases are resistant to heat and somewhat resistant to water.

Lubricating Grease Classification

There are a number of ways in which the lubricating greases are classified. Some of these are listed below.

- Application—industrial, automotive
- Application Temperature—low-temperature grease, normal temperature grease, and high-temperature grease.
- Application Range—multi-purpose grease, normal grease, and specialty grease.
- Soap Type—soap-free or nonsoap grease and soap-containing grease. While the use amount of the nonsoap greases is smaller than that of the soap-containing greases, the nonsoap greases have a larger variety than the soap-containing greases. Soap-containing greases are classified based on the metal ion of the soap or the thickener; for example, lithium, sodium, calcium, barium, aluminum soap greases. The soap in greases can be simple soap, complex soap, or mixed soap. Simple soap greases are made from a single carboxylic acid. Complex soap greases are made from a mixture of carboxylic acids. Usually, one of the acids in such a mixture is a low molecular weight acid. Mixed soap greases contain more than one metal ion. Lubricating grease classification according to the metal ion is quite useful since in many cases metal ion imparts certain important properties, such as high dropping point, to the lubricating grease. For example, the dropping point of the lime (calcium) soap greases is around 100°C, but that of lithium soap greases is around 180°C.
- Base Oil—mineral oil-based grease or synthetic fluid-based grease. A number of important properties of lubricating greases depend on the oil component [4], which is discussed later in the discussion.
- Load Level—normal load grease or high load or extreme pressure (EP) grease.
- Consistency—consistency is the resistance of the lubricating grease to deform under load, or in other words it measures the degree of hardness of the grease. Consistency is usually measured by the ASTM Cone Penetration Test, ASTM D217. The higher the penetration, the softer is the grease. Various classes of lubricating greases based on consistency are provided in the National Lubricating Grease Institute's (NLGI) classification system. Under this system, the lubricating greases are classified into nine classes, which are provided in Table 10.1, along with potential applications. This classification is old but is still widely used.

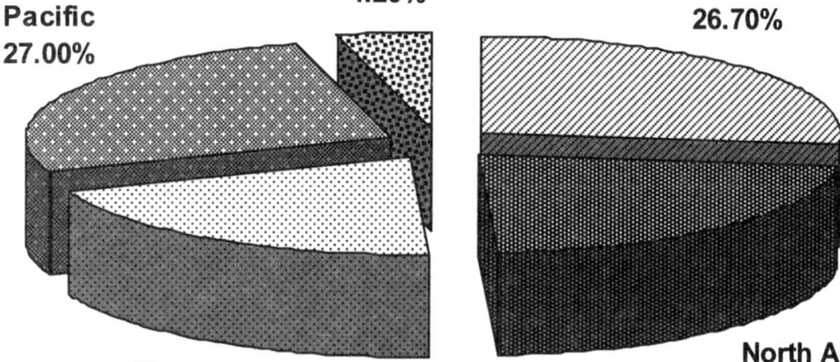

Fig. 10.2—2007 worldwide grease production and use [732].

Lubricating Grease Market

The actual size of the world market for lubricating greases is hard to estimate since there is ongoing development of new products to meet the constantly changing performance specifications and environmental compatibility requirements. The worldwide production and usage of the lubricating grease for the year 2007 is estimated at 2.31 billion pounds, or almost 1050 thousand metric tons [732]. Figure 10.2 provides use of the lubricating grease by region and Table 10.2 shows its use by thickener type. Lubricating greases represent about 3 % of the total lubricant market; of which 7–15 % comprises synthetic greases, depending on their definition. With respect to the soap greases, lithium soap greases amount to 68 % of the worldwide lubricating grease market, followed by the aluminum soap grease at 9 % and the calcium soap greases at 6 %. The use of the nonsoap greases is about 15 % and is equally divided between organoclay greases and polyurea greases. Approximately 51 % of the lubricating greases used in the United States and 78 % of the lubricating greases used in Europe are based on mineral oils. Regarding the main consumers of the lubricating grease, about 50 to 60 % of the greases are used for industrial applications and the rest is needed for automotive applications. Trucks and buses, automobiles, and agriculture and construction are largely responsible for the automotive grease use. Figure 10.3 shows the break down of the automotive grease market. The use share of the greases for various industries is shown in Fig. 10.4. There are indications that compared to the United States the percentage of the industrial grease consumption in Europe is slightly higher.

It is important to note that various regions of the world use different quality products because their uses differ. For example, North America, Europe, and Japan use higher quality products than the rest of the world because of their primary use in more demanding applications. Africa uses products that are specifically suited to lubricate mining equipment and Asia primarily uses polyurea greases.

While in North America and Europe, the use of the low cost/low performance greases also called commodity greases predominates; their value in terms of dollars is less than that of the high performance greases. Commodity greases are based on lithium, calcium, and sodium soaps and are normally used as general purpose products. Because of their high volume use and production, their cost is low. Besides commodity greases, some multi-purpose greases, such as some extreme-pressure (EP) and lithium complex greases, are low cost as well and for the same reasons. Conversely, specialty greases which as the name indicates are not

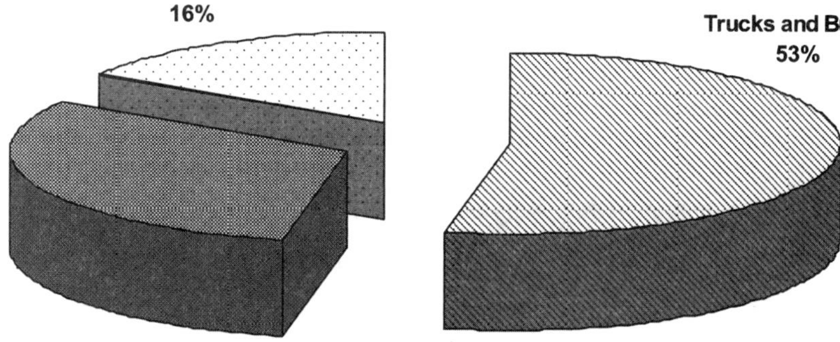

Fig. 10.3—Worldwide automotive grease market.

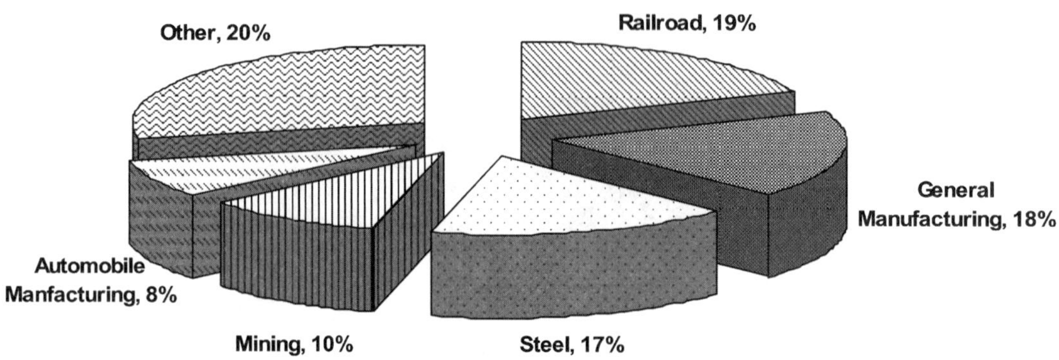

Fig. 10.4—Worldwide industrial lubricating grease market.

general purpose are designed to meet superior performance. Their higher cost is due to the presence of high quality performance additives that are necessary to meet the higher performance requirements of some applications. As will be described later in the formulation section, the additives can be liquid or solid and are used in small amounts to impart the desired properties to the lubricating grease. Because of their somewhat higher cost of production and greater use in high performance products, the additives account for the higher value of the specialty greases.

Specialty greases, sometimes referred to as high-specification greases, because of their ability to meet the higher standards or specifications, are primarily based on complex soaps and highly refined mineral oils or synthetic base fluids. Specialty greases are typically used in applications whose performance requirements cannot be met by the commodity greases. As mentioned earlier, North American and European use of lubricating greases is largely for highly demanding applications and therefore it is not surprising that the use of the commodity greases in these regions is progressively decreasing and that of the specialty greases is increasing.

Grease Composition

Lubricating greases are composed of three components: These are thickener, base fluid, and additives, which are used to enhance grease properties and performance. While the amount of each differs from grease to grease, as a general guideline the lubricating grease contains 70–90 % base fluid, which can be mineral, synthetic, or natural in origin; about 5–25 % thickener; and the additives make up the difference. It is the base fluid and the additives that perform the lubrication function and impart various desirable performance properties to the lubricating grease. Soap's or thickener's function is to provide structure to the grease. It contributes little, if any, towards the lubrication function of the grease. Common thickeners used to make lubricating greases are listed below.

1. Soap Thickeners
   - Simple soaps
   - Mixed soaps
   - Complex soaps
2. Nonsoap Thickeners
   - Organic thickeners
     a. Polyureas
     b. Polytetrafluoroethylene (PTFE)
     c. Fluoroethylenepropylene Polymer
   - Inorganic Thickeners
     a. Bentonite
     b. Silica gel

Thickener is the component that is primarily responsible for the gel-like structure of the lubricating grease. Thickeners can be classified into two general classes: Metal soap or metal carboxylate thickeners and the nonsoap thickeners. Nonsoap thickeners can be further subdivided into organic thickeners and inorganic thickeners. Modern lubricating greases predominantly use metallic soaps, of which lithium soaps top the list in terms of the market share. According to a conservative estimate, soap-thickened lubricating greases account for well over 60% of the greases manufactured and used.

Soaps are the metal salts of the naturally occurring carboxylic acids. The term soap does not apply to the metal salts of the man-made, or synthetic, carboxylic acids. The metals that have been found suitable to make soaps for use as thickeners in greases are lithium, sodium, calcium, barium, and aluminum. The carboxylic acids or fats and oils that are commonly used to make these soaps are those that are natural in origin. For example, stearic acid is available from the animal fat, 12-hydroxystearic acid is available from castor beans, and oleic acid is available from soybean, cottonseed,

TABLE 10.2—Regional grease use by thickener type.

	Aluminum Soap	Calcium Soap	Lithium Soap	Polyurea	Other Thickeners
N. America	9 %	7 %	70 %	6 %	8 %
Europe	5 %	15 %	70 %	3 %	7 %
China	2 %	9 %	79 %	3 %	7 %
Japan	2 %	11 %	60 %	21 %	6 %
Indian sub-continent	<1 %	6 %	85 %	0 %	9 %

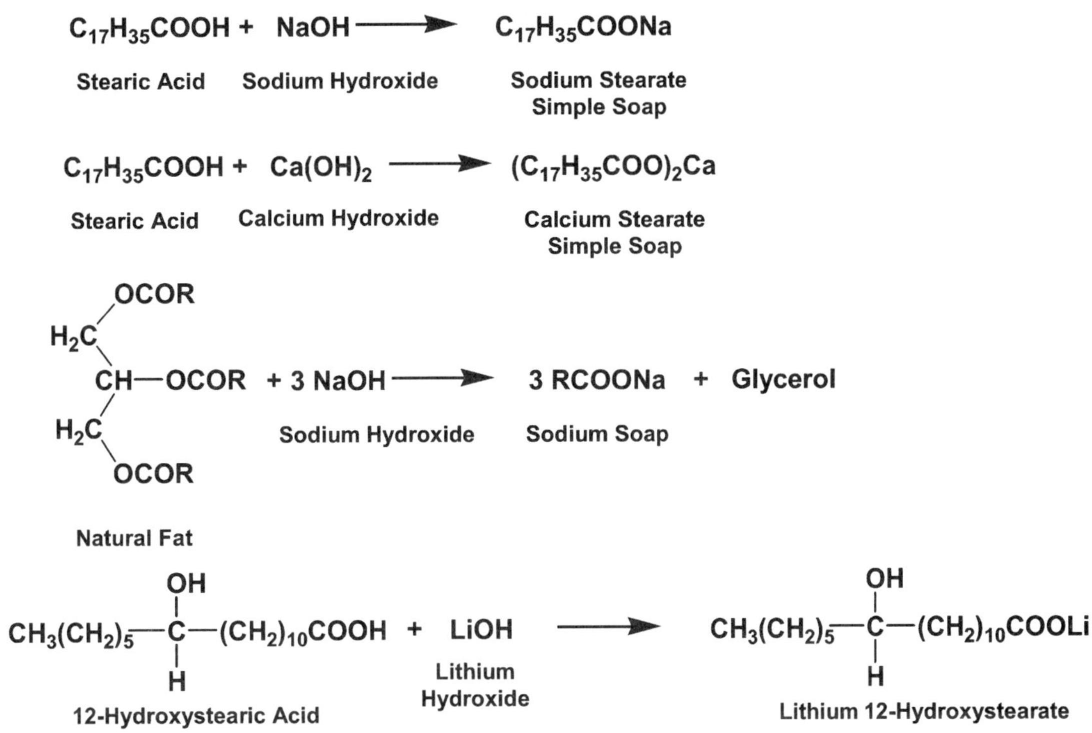

Fig. 10.5—General soap formation reaction.

and many other vegetable oils. These natural fatty acids normally contain an even number of carbons in a straight chain arrangement and preferably contain zero to one double bond.

Fats and oils that are directly used to make these soaps include animal fats such as beef tallow and bovine fat, and vegetable oils and fats such as those available from natural seeds and fruits, for example coconut oil, palm oil, soybean oil, and cottonseed oil. The use of soaps derived from fish oils is rare. Frequently, hydrogenation is used to remove the unsaturation present in many natural fats and acids derived from vegetable oils since it improves the oxidation stability of the derived lubricating grease. Hydrogenated castor oil used to make lithium 12-hydroxystearate is one such example. The metal to form soaps is in the form of metal oxides or hydroxides, the hydroxide form being more prevalent. When the reaction involves a carboxylic acid and a metal base, the reaction is called a neutralization reaction. However, when the reaction involves vegetable oils or animal fats, which are also called triglycerides; the reaction is called a saponification reaction. Each reaction is presented in Fig. 10.5 by a chemical equation.

Metal ion, or cation, in the soap determines many of the critical properties of the lubricating grease. It determines the thickening ability, water resistance, and dropping point. The carboxylate portion of the grease influences other properties. For example, the length and branching of the carboxylate portion affects its oil solubility and hence consistency and the surface properties of the grease. In order to achieve optimum consistency there is an optimal carboxylate chain length. If the chain length is too long or too short, the thickening efficiency of the soap will be affected. This is because the oil solubility of the soaps of very long chain length is very high and that of soaps of very short chain lengths is very low. The ideal soaps are those that have borderline solubility in oil so as to maintain its network structure, which is required in lubricating grease. Optimal chain length for the carboxylate group in soaps is C_{18}. The presence of the extensive branching in the carboxylate portion increases oil solubility and lowers the melting point of the soap. The result is a drop in thickening efficiency and the dropping point of the derived grease, which is undesired. Unsaturation in the soap molecule also increases oil solubility and has a similar negative effect on thickening and the dropping point of the grease. Unsaturation in soap is in addition lowers the oxidation stability of grease, thereby limiting its use in most high-temperature applications. Polar groups, such as hydroxyls, on the other hand, decrease the oil solubility and increase the melting point of the soaps. The result is an increase in thickening efficiency and the dropping point of the derived greases. Of course, the position of these groups on the side chain is also critical to the grease performance. 12-Hydroxystearic acid, which possesses all the advantages listed above, is the most widely used carboxylic acid for lubricating greases. Lithium and calcium soap greases are among the most desirable of the soap containing greases. The use of aluminum, barium, and sodium soap greases is progressively decreasing.

The thickener structure can range from being linear, as in the case of metal carboxylates (soaps), to more complex circular structures, as in the case of the nonsoap thickeners. Approximate dimensions of the thickener particles along with the appearance they impart to lubricating greases are provided in Table 10.3 [4].

TABLE 10.3—Microstructure of grease thickeners [4].

Thickener	Mean Dimension Diameter × Length 10^{-12} m	Grease Appearance	
		Microscopic	*Macroscopic*
Sodium Soaps	1 × 100	Long fibers	Long fibers, strings
Lithium Soaps	0.2 × 25	Long fibers; spirals, mostly double	Medium fibers
Lithium Soaps	0.2 × 2.0	Short fibers, rod clusters	Short fibers, smooth
Sodium Soaps	0.15 × 1.5	Short fibers, short threads	Short fibers, smooth
Calcium Soaps	0.1 × 1	Fine threads, short rings	Short fibers, smooth
Aluminum Soaps	0.1	Spheres	Short fibers, smooth
Organophilic Bentonites	0.1 × 0.5	Platelets, card house structure	Short fibers, smooth

Soap-thickened Lubricating Greases

Soaps used in the grease manufacture are classified as simple soaps, complex soaps, and mixed soaps. Detailed description of each type of soaps and the derived lubricating greases is provided in the following sections.

Simple Soap Greases

Simple soaps are prepared by the reaction of a single fatty acid and a single metal base. Metal bases of interest belong to the groups of alkali metals; such as lithium and sodium, alkaline earth metals; such as calcium and barium; and aluminum that belongs to Group III in the periodic table of elements. The bases can be in the form of oxides, hydroxides, or alcoholates. Metal hydroxides are usually preferred. Alcoholate is only used in making aluminum soaps because of its higher reactivity. The soap formation reaction usually requires heat, sometimes pressure, and mixing. Water is present in the reaction either as added water, as in the case of the saponification reaction, or as a by-product of neutralization. Some water may remain in the grease. In some cases, for example in the case of calcium grease, its presence is beneficial since it help stabilize the grease structure.

Simple soap greases make up the largest proportion of the lubricating grease production. These greases contain the three components in the following ratios: 4–20 % by weight soap, 75–96 % by weight base oil, and 0–5 % by weight additives. The amount of soap used to make the grease also affects its consistency: the higher the amount of soap, the higher the consistency of the lubricating grease.

Aluminum Soap Greases

Aluminum soaps have a high thickening effect because of their greater network forming tendency, owing to the presence of three carboxylate chains for each aluminum atom and their borderline solubility in oil. Carboxylic acids used in making the aluminum soap greases include ethylhexanoic acid, dimerized polyunsaturated fatty acids, stearic acid, and other fatty acid mixtures. Sometimes the hydrogenated

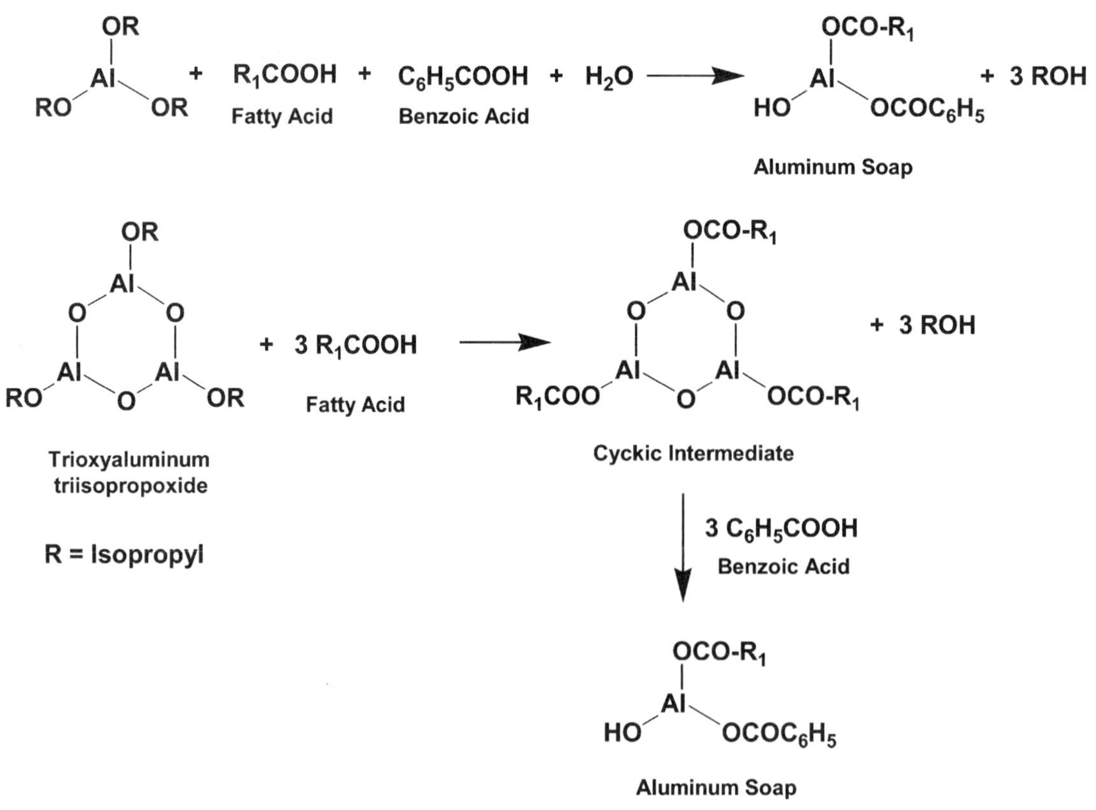

Fig. 10.6—Aluminum soap formation [4].

fish oil is also used to make greases. Unlike alkali and alkaline earth metal oxides or hydroxides, which are highly reactive towards weak acids, such as carboxylic acids, aluminum oxide and hydroxide are not. Hence, they are not good bases for making aluminum soaps. However, aluminum alcoholate, with a higher reactivity towards carboxylic acids, is a good starting material. Another way to obtain water-tolerant aluminum grease is by the reaction of the water-soluble soaps with aluminum salts. See Fig. 10.6 for the reaction scheme [4]. Water tolerance in lubricating grease is a highly desirable property. Aluminum soap greases are usually made by dissolving the *preformed* soap in oil and not by making the soap in the presence of oil, as is the case in other greases. Aluminum stearate is the most common conventional aluminum soap that is used to make the aluminum soap greases. The manufacture of the aluminum soap lubricating grease does not present any problems; the soap is simply added to the stirring oil and the mixture held at 240 to 350°F, or 116 to 177°C, until complete homogeneity is achieved. Cooling of the mixture to room temperature yields the grease. The cooling can be carried out without mixing, which will result in a firm grease. However, if the cooling rate is slow and is accompanied by mixing, a softer grease is obtained. Aluminum greases are smooth clear gels, slightly stringy in texture, especially if made with high-viscosity oils. Their stringiness is often enhanced with additives.

The use of the paraffinic base oils in making aluminium soap greases is preferred since they have a stronger thickening effect than naphthenic oils [4]. Aluminum soaps possess good adhesion and water-resisting properties. However, they suffer from the disadvantages of low shear stability, low dropping point (120°C), pronounced thixotropic characteristics, and a gel-forming tendency. Also, the aluminum soap greases have low maximum worked temperature of 60°C to 100°C [4]. In addition, when they are exposed to temperatures above 170°F (77°C), they lose their smooth texture and become rubbery. This makes the grease lose its adhesion and hence its lubricating ability. Aluminum soap greases also suffer from serious aerating tendency on vigorous mixing. Because of these limitations, these greases are being replaced by other greases, for example, lithium soap greases.

Barium Soap Greases

For many years, barium soap greases were the most widely used greases in the United States, primarily because of their very unique properties. They are prepared by the reaction of barium hydroxide with various fatty acids or natural fats in mineral oil. Sometimes oxidation inhibitors, corrosion inhibitors, and EP agents are added to improve their oxidation stability and corrosion protecting and load-carrying properties. They have buttery to fibrous texture and are reddish-yellow or green in color.

Barium soap greases have the advantages of good water resistance, good shear stability, and a high dropping point 350°F (177°C). Hence, they were among the first multipurpose greases used. These greases are often used to lubricate nearly all types of bearings and electrical cables for power transmission lines. Barium soap greases suffer from a number of serious disadvantages. These include the following:

1. Difficult large-scale manufacture.
2. Need to use a higher amount of soap to obtain a given consistency, presumably a consequence of the high atomic weight of barium.
3. Poor low-temperature properties.
4. High raw-material costs.
5. General concern towards toxicological properties of barium and its impact on the environment.

Because of these drawbacks, barium soap greases are not suitable for use at low temperatures and in very high speed applications; hence they are being replaced by the other types of greases [4]. The same is true about strontium soap greases, which are used in some specialty applications.

Calcium (Lime) Soap Greases

These greases are among those used most often. They are relatively inexpensive to produce. They are used as axle grease, water pump grease, and general-purpose grease for machinery.

These greases are usually produced in situ and not by making the soap separately and dissolving it in oil. For making calcium soap greases, the usual fatty component is tallow. The use of the fatty acids derived from beef tallow or another similar source leads to the formation of unstable grease. However, this problem can be overcome by using a blend of fat and fatty acid. The alkali needed for making calcium soap grease is calcium hydroxide, also called hydrated lime. This is why, calcium soap greases are often called lime soap greases. For making the calcium soap, the amount of lime is used in excess of the stoichiometric amount needed. This assures completion of the soap formation reaction. If fat alone is used as the starting material, the addition of a certain amount of water is necessary to facilitate saponification. If a mixture of an acid and fat is used, the water resulting from the neutralization reaction fulfills this function.

To manufacture calcium soap grease, hydrated lime and water are added to a stirring mixture of fat, fatty acid, if used, and oil. The reaction temperature is raised to a temperature that the reactor permits. A higher temperature is preferred since it causes the saponification reaction to proceed faster. A temperature of around 300°F or 149°C is typical, if a pressure reactor is used. In case of an open kettle, the temperature must be kept below the boiling point of water to minimize its loss. The reaction is checked for completion by checking the alkalinity (strong base number) of the mixture. If the reaction is complete or close to completion, the alkalinity should be low. Once the reaction is complete, the cooling can be started. If initially less oil was used than needed for the finished grease, adding additional oil will facilitate cooling. At first, the oil may need to be added in smaller portions until the grease is fluid enough to accept larger amounts. Once the batch cools down to 230°F (110°C), water is added to break the sticky gel-like texture of the grease. This is called the hydration step; hence the name hydrated calcium soap grease. The amount of water that leads to the formation of a stable grease structure is about 10 wt %. If the water level is too low, a grainy product with strong oil separating tendency will be obtained. On the other hand, if the water level is too high, an opaque grease of low soap content results. The 10 wt % water is just a guideline; the optimum amount of water depends upon the presence of the so-called structure modifiers. These include glycerol, free fatty

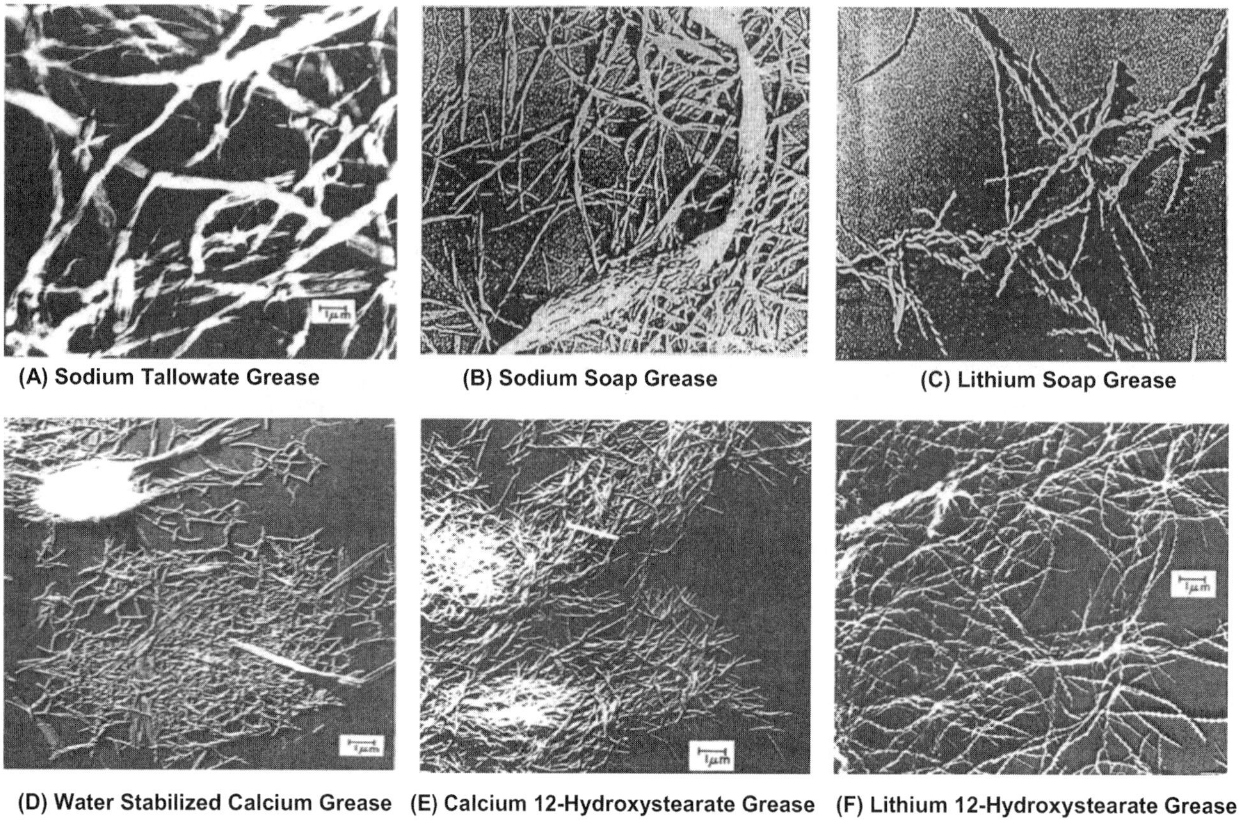

Fig. 10.7—SEM micrographs of various simple soap greases [727,733].

acid, and glycol. While the first two of the three will invariably be present during the soap formation, glycol will need to be added. It is believed that water associates with the calcium soap to enhance its thickening ability. Complete removal of the hydration water leads to a complete grease structure break down, resulting in the formation of an oil layer and a soap layer. In the manufacture of the calcium soap greases, naphthenic and aromatic mineral oils are the oils of choice. Calcium-soap greases are yellow or reddish in color, and have a smooth buttery texture.

As commented earlier, the consistency of the lubricating grease is affected by the type and the quantity of the soap, the base oil, and the type of additives. If one wishes to develop an NLGI Grade 2 consistency grease, 11 to 16 wt % calcium soap is required. Calcium soap greases possess a smooth structure; good low-temperature properties, they do not phase transform and remain pumpable; good water resistance; and good adhesion properties. In addition, they do not form emulsions with water; hence they resist washout from the bearings. Their smooth texture is due to small, closely packed, fibrous network of the soap, see Fig. 10.7, Part D [727,733]. These greases suffer from a number of disadvantages which include low maximum working temperature of 175°F (~80°C), low dropping point of 195 to 210°F (90 to 100°C), and insufficient stability at high speeds when used in antifriction bearings. The low dropping point is due to the breakdown of the calcium soap-water thickening system. Common uses of calcium soap greases include lubrication of the machinery and water pumps that either do not contain antifriction or plain bearings or they are not being operated at high speeds. Hence, there is little or no concern for the greater oxidation stability of these greases.

The overall performance of the calcium soap greases is improved if they are made by the reaction of (85:15) weight mixture of 12-hydroxystearic acid and stearic acid. The greases thus made can be used at temperatures of up to 120°C and at higher bearing speeds. The dropping point of these greases also increases to almost 150°C, or 300°F. The higher dropping point of these greases and their enhanced oxidation stability are ascribed to the presence of only trace amounts of water, in the range of 0.1–1 wt % [4]. For the same reason, these greases are called anhydrous calcium soap greases and can be used as multi-purpose greases within their temperature limitations.

Because of their excellent water resistance, calcium soap greases find use in water pumps, wire ropes, food plants, wet industrial and sewage plant machinery, equipment exposed to weather, harvesting equipment for damp crops, marine hardware, and chassis lubrication. In general, the calcium soap greases are suitable to lubricate machine components that operate at temperatures of no more than 125 to 150°F, or 52 to 66°C. For water pumps, the use of the NLGI Grade 4 grease is preferred. However, to lubricate light-duty equipment, such as tractor track rollers, farm equipment, mine cars, and textile machinery, a softer, more easily pumpable grease is the lubricant of choice. For use in food plants, calcium 12-hydroxystearate soap grease containing zinc oxide is commonly used. However, the base oil used in these greases must be acceptable for incidental food contact [713]. The lower thermal stability and shear stability of the calcium soap greases preclude their use in plain and roller bearings.

Lithium Soap Greases

The use of lithium soap greases has increased manifold since their discovery in the early 1940s. According to a recent NLGI estimate, these greases account for almost 70 % of the total grease use. Lithium soaps are made by the reaction of lithium hydroxide with fatty acids, or natural fats, in mineral oils or synthetic fluids. Since lithium hydroxide is a weaker base than sodium hydroxide and calcium hydroxide, lithium soap formation requires temperatures in the range of 160 to 220°C. These temperatures are substantially higher than the temperatures required to make the sodium and calcium soaps.

The fatty acids commonly used in lithium soap production are either stearic acid, 12-hydroxystearic acid, or their triglycerides. Lithium soap greases are made by dispersing the soap in oil, either mineral oil or synthetic fluid such as an ester, ether, and silicone oil. Lithium 12-hydroxystearate soap can be dispersed at temperatures around 200°F (93°C) but other lithium soaps require much higher temperatures, generally in the range of 400°F (204°C), or more. After the soap dispersion, the resulting greases are cooled. This can be carried out without stirring, for example in shallow pans, by circulating through coolers, or by stirring during cooling. Slow cooling results in long fibers. This improves mechanical stability of the grease, but degrades its ability to retain oil. Fast cooling, on the other hand, does the opposite. That is, it forms short fibers, because of which the grease has good oil retention but poor mechanical stability. Lithium soap grease is buttery in texture with a brownish-red color. One way to develop lubricating grease with good mechanical stability as well as good oil retention is to mix the soaps of different fiber lengths. The variation of the fiber structure in greases is also a function of the starting acid or the fat.

Lithium stearate greases exhibit a smooth, stringy texture. Their advantages include the following:

- Exceptional shear stability, which makes them suitable for use in high-speed plain and rolling-element bearings.
- High dropping points; almost as high as 350°F, or 177°C, which are significantly higher than those for the sodium, calcium, and aluminum simple soap greases; and good thermal stability.
- Maximum service temperature approaches 300°F, or 150°C.
- Good low-temperature properties, when made by using oils of low pour point. Can be used at temperatures as low as -60°F, or -51°C.
- Good tolerance towards water; which makes them suitable for use in mill bearings where water washout is a concern.
- Rust and corrosion protection is at least equal to that of the sodium greases.
- Response to the performance additives, such as rust inhibitors, oxidation inhibitors, and EP agents is superior to that of the other metal soap greases. This factor is beneficial in custom designing products for use in diverse applications.
- Excellent sealing properties.
- Readily pumpable.

Replacing the common fatty acid or natural fats by 12-hydroxystearic acid or its triglyceride leads to a product whose dropping point is even higher, ~190°C, or 374°F. High dropping point is very desirable since it greatly improves the working stability of the grease. This means that the grease does not soften when worked. This advantage is among many others that are responsible for the widespread use of the lithium 12-hydroxystearate greases. Lithium greases have good overall performance and are extremely cost effective.

The grease properties that most interest an end user are listed below:

- Decomposition temperature
- Resistance to aggressive elements
- Radiation stability
- Corrosion protection properties
- Oxidation stability
- Oil separating tendency
- Structural stability
- Dropping point
- Load-carrying capacity
- Antiwear properties
- Viscosity-temperature characteristics (VI)
- Suitability of use in different bearing types that operate over a wide speed range
- Adhesion properties
- Noise characteristics in bearings
- Sealing ability
- Service life

As mentioned earlier, both the base oil and the thickener determine the properties of the lubricating greases. Of the properties listed, the properties that are solely dependent upon the thickener include the dropping point, water resistance, and sealing ability and those that solely depend upon the base oil include the low-temperature behavior and the compatibility with seal materials [4]. Lithium soap greases possess a great number of these very desirable characteristics; hence they are the lubricating greases of choice in most applications. The properties that are especially worth mentioning are dropping points of above 180°C, good water resistance, and excellent structural and shearing stabilities, even in applications that involve high-speed bearings. Their other properties, such as oxidation stability, corrosion protection, and EP and antiwear performance are mediocre, but these can be enhanced by the use of the appropriate additives.

If one wished to manufacture an NLGI grease of Consistency 2, one needs to use approximately 6 wt % of the lithium soap in a naphthenic oil or approximately 9 wt % in a paraffinic oil. The base oil to formulate multi-purpose grease typically has a 40°C viscosity of 60 to 129 cSt (mm^2/s). It is interesting to note that to obtain equivalent thickening the amount of lithium soap necessary is substantially less for napthemic base oils than for paraffinic base oils. Besides the oil viscosity, the thickening effect of the soap also depends upon the composition of the base oil; the higher the naphthenic or the aromatic content, the greater the soap-related thickening. Figure 10.8 adequately demonstrates the base oil effect [4]. In this figure, Oil A is 95VI solvent refined paraffinic oil, Oil B is 50 VI solvent raffinate—a highly aromatic oil, and Oil C is 50 VI acid-treated, aromatics-free, primarily naphthenic oil. In all three cases

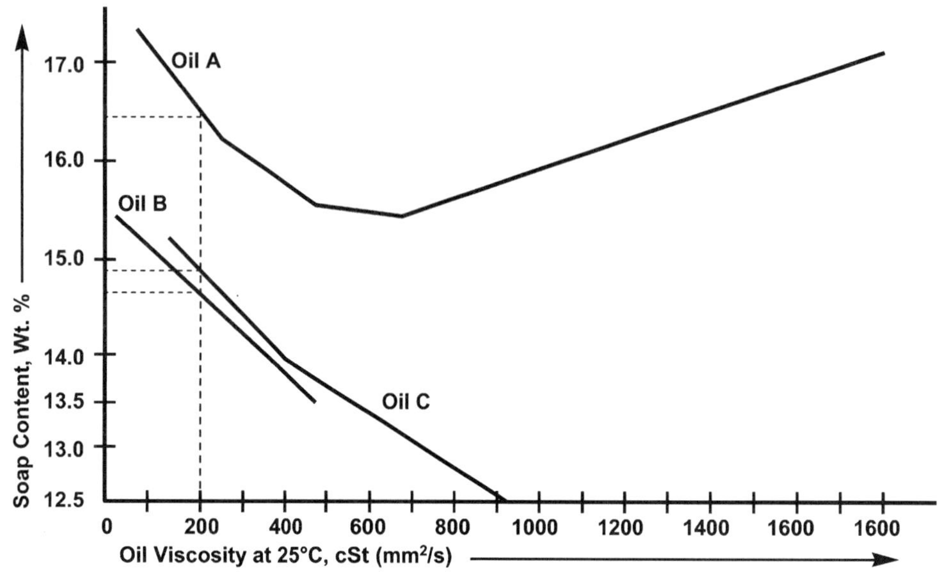

Fig. 10.8—Soap content versus viscosity for different base oil types in an NLGI Grade No. 3 grease [worked penetration 220 (tenth of a millimetre)] [4].

the objective is to come up with a grease of NLGI Consistency Grade 3 with a worked penetration rating of 220. Examination of the graph suggests the following.

- In the case of the primarily paraffinic Oil A, as the base oil viscosity (shown on x-axis) increases, the amount of soap necessary to make a Consistency Grade 3 grease decreases. It is at least true until the oil viscosity of 640 cSt is reached. After that the curve turns upwards and the amount of soap necessary increases.
- In the case of the highly aromatic Oil B and the largely naphthenic Oil C, such a transition does not occur; and the amount of soap necessary to make Consistency Grade 3 grease decreases with an increase in the base oil viscosity.

In order to meaningfully assess the base oil effect, the oil viscosity of 200 cSt (mm^2/s) at 25°C was chosen. This is indicated in Fig. 10.8 by vertical dotted line. This value is fairly close to the real life viscosity of 60 to 129 cSt at 40°C for oils used to formulate lithium soap greases. This is so because as the temperature increases from 25 to 40°C, the viscosity of the oil is expected to fall from 200 cSt to the 60 to 129 cSt viscosity range. As indicated by the horizontal dotted lines, the amount of soap necessary to make the Consistency Grade 3 grease from the aromatic Oil B is ~14.7 %, which is the lowest; for the naphthenic Oil C, it is ~14.9 %, the next highest; and for the paraffinic Oil A, it is ~16.5 % is the highest.

Lithium soap greases were the first multi-purpose greases invented. They provide both the good water resistance, which is similar to that of the calcium soap greases, and the high-temperature properties that are superior to even the sodium soap greases. Lithium soap greases are suitable for use in both automotive and industrial equipment. They have provided satisfactory performance in journal and antifriction bearings, and lubricated-for-life rolling contact bearings. However, their use in some devices, such as those involving sliding and reciprocating action, is not recommended. This is because of their smooth texture due to which lithium soap greases lack sufficient adhesion for use in such applications. Overall, lithium soap greases do not suffer from many serious disadvantages. However, their selection for use in applications that involve extremely high temperatures, speeds, loads, and pressures needs careful consideration.

Lithium-12-hydroxystearate greases are used in many other applications as multi-purpose greases. They contain no unsaturation in the carboxylate portion of the soap; hence they have good oxidation stability. They also possess good shear stability, which is ascribed to the hydrogen bonding ability of the hydroxyl group. NLGI Consistency Grades 1, 2, and 3 are based on mineral oils of 40°C viscosity of 60–120 cSt (mm^2/s). The EP greases for high load-carrying applications use oils of somewhat higher viscosity, presumably to boost the film-forming ability (adhesion) of the grease. Oils of 40°C viscosities of 100–350 mm^2/s are often employed for this purpose. Lithium 12-hydroxystearate greases for specialty uses are created by the use of the synthetic base fluids. For use in aviation and space travel, diesters are the base fluids of choice; for use in gears poly(alkylene glycol)s are used as the base fluids.

Lithium simple soap greases suffer from a number of disadvantages, including high raw material costs, high production costs because of the higher production temperatures, and the dropping point of around 180°C, which is much lower than 240°C for most modern complex soap greases. This leads to the rapid softening of the grease with an increase in temperature.

Sodium Soap Greases

The use of the sodium soap greases is miniscule compared to that of the lithium and calcium soap greases. However, these greases are of interest for specialty applications, such as lubrication of gears and high-speed bearings. Sodium soap greases are manufactured in situ by the reaction of the fatty acids or natural fats with an excess of sodium hydroxide in oil at a temperature of 300–500°F, or 150–260°C. One problem that typically occurs during the making of the sodium soap is the foam formation; hence heating the ingredients slowly is advisable. Sometimes polysiloxane (silicone) foam inhibitors are used to contain the foam. Sodium soap greases are made in a variety of structures. The structure obtained depends upon a number of factors. These include the type of fat or the fatty acid, their ratio in the starting materials, the presence of the residual moisture, presence of structure modifiers, effectiveness of mixing and milling, cooling rate, and so on. When the reaction temperature is high, for example above 400°F (204°C), the soap will be completely in solution and the resulting greases will be completely anhydrous. The presence of the residual moisture generally leads to a fibrous grease. Most sodium soap greases are of spongy texture. When the grease is made from animal or vegetable fat, which are triglycerides, the glycerol by-product acts as a stabilizer for the fibrous structure of the grease. However, it impairs the grease's oxidation stability. Since the fibrous structure in a grease is desirable because of its superior shear stability, the presence of some water and glycerol in the lubricating greases is beneficial.

The base oil selected depends on the end use. As usual, viscosity is an important consideration. With a high-viscosity index, medium to light viscosity base oil the sodium based greases can equal or surpass the low-temperature performance of the calcium and lithium soap greases. However, the high viscosity oils give rise to variations in other grease characteristics. Naphthenic oils are oils of choice but sometimes the grease can gel. Both short and long fiber sodium greases are commercially available.

Sodium soap greases have relatively high dropping points of ~165°C or 175°C (330–350°F), which is the highest among the simple soap greases. Hence, they can be used in antifriction bearings that experience temperatures of up to 120°C. Short fiber greases, which have good mechanical stability, can be used to lubricate antifriction bearings that operate at speeds of up to 500 metres/min. Sodium soap greases do possess a number of desirable properties, which include good rust and corrosion inhibition, high temperature stability, and fair shear stability. Because of these properties, sodium soap greases are used to lubricate high-speed spindle bearings, sometimes gears, some rolling element bearings, and electric motor bearings, which experience moderately high temperatures. Lighter consistency sodium soap greases are used in textile plants so that if the grease leaks on to the cloth, it is easily removed during the normal washing process. For many years, sodium soap greases were used in automotive wheel bearings, but they are being replaced by better quality multi-purpose greases with superior water tolerance. Sodium soap greases made with lighter oils are also used for ball and roller bearing lubrication, as are the combinations (mixed base) of the calcium and sodium greases. Because of their working stability and intermediate melting point, sodium soap greases are used for lubricating wheel bearings, other than those used with disk brakes, and for general-purpose industrial applications. Typical examples include rough, heavy bearings operating at slow speeds, as well as skids, track curves, and heavy-duty conveyors.

Sodium soap greases have the additional advantages of good gear lubricating ability and superior corrosion protection properties. They are cheaper to make since the raw material costs are low and the process to manufacture them is fairly straightforward. Despite these advantages, sodium soap greases suffer from a number of serious disadvantages as well. These include water sensitivity, gel-forming tendency, and incompatibility with most other grease. Water sensitivity relates to high solubility or emulsifiability of the sodium soaps in water. However, they do tolerate the presence of a small amount of water, without an adverse effect on consistency. Because of the soap's high water solubility, these greases are not normally used in industrial applications and steel rolling mill applications, where bearings come in contact with copious amounts of water. Sodium soap greases are also susceptible to phase transformations and hardening. Their compatibility with other soap greases, especially with calcium soap greases, is poor and results in a mixture with fluid-like consistency. The resulting mixture is fluid-like in consistency. Since sodium soap greases possess inherent EP properties, they are formulated solely by the use of oxidation inhibitors to lubricate ball and roller bearings. However, the EP additives, such as sulfurized fatty oils, organic sulfides and polysulfides, and organo-chlorine derivatives can also be used to enhance the load-carrying properties of these greases. Tackiness additives and rust inhibitors are usually not needed.

Water resistance of these greases can be improved by adding a small amount of the calcium soap. However, when used in completely anhydrous environments, sodium soap greases perform well even at a temperature of 156°C, which is substantially higher than the temperature that most simple soap and some complex soap greases can withstand. Sodium 12-hydroxystearate greases possess greater thermal and shear stabilities than the regular sodium soap greases.

Mixed Soap or Mixed Base Greases

Mixed soap greases are those that contain two or more metal-derived soaps. In other words, they contain more than one metal ion. Typical examples of such greases are those that contain sodium and calcium, lithium and calcium, or sodium/lithium/calcium mixed soaps. The final properties of these greases depend upon the actual proportion of the each type of soap present in the grease. In most cases, the presence of an additional metal ion improves specific property or properties of a particular soap grease. The examples below adequately demonstrate this fact.

- The presence of the calcium soap in the sodium soap greases improves their water resistance. In addition, it decrease the overall cost of the sodium soap greases.
- The presence of the sodium soap in the calcium soap greases improves their thermal stability, thereby making their use possible in applications that experience high temperatures. In such applications, the calcium soap greases do not perform well.
- Calcium-sodium mixed soap greases can be used at higher temperatures than the calcium soap greases alone.
- Lithium-calcium mixed soap greases have better water tolerance than the pure lithium soap greases and have a lower overall cost. In addition, the dropping point of ~150°C of the mixed soap grease is significantly higher than 85–105°C for the calcium soap greases.

Many of the cited properties can be optimized by producing the desired metal soaps in situ and in proper

TABLE 10.4—Thickener-based classification for greases.

Thickener Type/Soap Base	Texture	Dropping Point (°C)	Condition After Heating to 200°C and Cooling	Max Temp. (°C) (Prolonged Use)	Effect of Water	Resistance to Softening Upon Working at Room Temperature
Calcium Soap (Lime)–water Stabilized	Buttery	85–105	Soap and Oil Separate	70–80	Highly Resistant, Water Repelling	Fair to Good
Calcium 12–Hydroxystearate	Buttery	140–150	Soap and Oil Separate	110–120	Highly Resistant, Water Repelling	Excellent
Calcium Complex	Buttery or Fibrous	260–300	Hardens, Approaches Original Consistency on Working	120–150	Highly Resistant	Excellent
Sodium Soap (Soda)	Buttery or Fibrous	175–300	Hardens, Approaches Original Consistency on Working	120–150	Susceptible (Emulsifies)	Fair
Sodium Mixed Base/soap	Buttery	175–200	Hardens, Approaches Original Consistency on Working	120–150	Susceptible (Emulsifies)	Fair
Sodium Synthetic	Buttery	260–300	Little Change if Worked	150–175	Highly Resistant, Water Repelling	Excellent
Aluminum Soap	Buttery	90–110	Becomes Brittle Upon Cooling	70–80	Good Resistance	Fair to Poor
Aluminum Complex	Buttery	240–270	Slight Hardening	110–135	Resistant	Good to Excellent
Lithium Mixed Base/soap	Buttery	170–200	Soap and Oil May Separate	120–140	Resistant	Fair to Good, Depends onSoap Content
Lithium 12-Hydroxystearate	Buttery	175–200	Soap and Oil May Separate	120–140	Resistant	Excellent
Lithium Complex	Buttery	260–300	Little Change if Worked	150–175	Resistant	Excellent
Synthetic Organic (Polyurea)	Buttery	240–260	Little Change if Worked	150–175	Highly Resistant, Water Repelling	Good to Excellent
Polyurea Complex	Buttery	240–260	Little Change if Worked	150–175	Highly Resistant	Good to Excellent
Inorganic Thickeners	Buttery	260+	Little Change if Worked	120–140	Resistant	Fair to Excellent
Barium	Buttery or Fibrous	200–260	Little Change if Worked	120–140	Highly Water Resistant	Good

amounts. This is achieved by the reaction of the carboxylic acid with one metal base first and then reacting the residual acid with the other metal base. Usually the primary metal base is charged in larger amount than the secondary metal base and the two together equal the stoichiometric amount necessary to completely neutralize the acid. The mixed soap thus obtained can then be mixed with the base oil and the additives to make the mixed soap grease. Alternatively, one can make a mixed soap grease by mixing the finished simple soaps in the desired amount at room temperature. The latter approach does not always work well and if it does, the obtained grease has less stable properties.

As mentioned earlier, the sodium soap greases are incompatible with many other soap greases. However, mixed soap greases involving sodium soaps do not suffer from this problem. Sodium-calcium soap mixed base greases are products with good structure and are used to lubricate wheel bearings and ball and roller bearings. In this mixed soap grease, the calcium soap is claimed to shorten the fibers. Sodium-aluminum soap greases, which can also be made without difficulty, are used in many industrial applications. The aluminum soap provides the smooth texture to the grease and the presence of the sodium soap boosts its dropping point to 375 to 425°F (191 to 218°C). Incidentally, aluminum simple soap grease has a dropping point of around 90–110°C. Similarly, the presence of the aluminum soap overcomes the poor water tolerance of the sodium soap greases, at least at low temperatures.

Complex Soap Greases

The major driving force behind the invention of the complex soap greases was to fulfill the lubrication needs of the modern equipment that operates at higher temperatures. In such applications, the performance of most conventional greases was inadequate. See the dropping point data and the maximum use temperatures provided in Table 10.4. The complex soap greases exemplify the best in the soap-based grease technology and because of their high dropping points they are used in applications such as those involving high temperatures where the simple soap greases do not suffice. Of all the complex soap greases that can be made, those based upon aluminum, barium, calcium, and lithium are commercially available.

A complex soap is prepared by the reaction of a mixture of two or more carboxylic acids with a single metal hydroxide base. The complexing acids usually have a lower molecular weight than the fatty acid components and are short chain acids, such as formic acid and acetic acid. The higher molecular weight fatty acid is usually stearic acid or 12-hydroxyxtearic acid. See Fig. 10.9 for the reaction schemes. It is important to note that the each high molecular weight soap requires a different complexing agent to yield grease with superior thermal properties. According to the ASTM [4], the soap crystals or soap fibers in a “complex soap” are formed by the co-crystallization of two or more compounds. If one reacts a monovalent metal base, such as lithium or sodium hydroxide, with a mixture of stearic acid and acetic

$$C_{17}H_{35}COOH + LiOH \longrightarrow C_{17}H_{35}COOLi$$

Stearic Acid, Lithium Hydroxide → Lithium Stearate

$$CH_3COOH + LiOH \longrightarrow CH_3COOLi$$

Acetic Acid → Lithium Acetate

(Components of Lithium Complex Soap)

$$C_{17}H_{35}COOH + Ca(OH)_2 \longrightarrow (C_{17}H_{35}COO)_2Ca$$

Stearic Acid, Calcium Hydroxide → Calcium Stearate

$$CH_3COOH + Ca(OH)_2 \longrightarrow (CH_3COO)_2Ca$$

Acetic Acid → Calcium Acetate

$$C_{17}H_{35}COOH + CH_3COOH + Ca(OH)_2 \longrightarrow C_{17}H_{35}COO\text{-}Ca\text{-}OCOCH_3$$

Stearic Acid, Acetic Acid → Calcium Mixed Acid Soap

(Components of Calcium Complex Soap)

Fig. 10.9—Components of the complex soaps.

acid, the complex soap will have only two components as part of the crystal structure. These are lithium or sodium stearate and lithium or sodium acetate. However, if the metal is divalent or polyvalent, such as calcium or aluminum, the crystal structure will have more than two components. If only two fatty acids are used, the resulting soap will be a mixture of two simple soaps; each derived from the individual acid, and one complex soap which will be the cross-salt of the two acids. For example, a calcium complex grease will have three components: calcium acetate, a simple soap; calcium stearate, again a simple soap; and calcium acetate-stearate, a cross-acid soap. Of course, the amount of each component in this soap composition will depend upon the ratio of the two acids present in the starting material. The combination of soap components imparts unique properties to the derived grease. Incidentally, physical mixing of the two soaps have little or no effect on the properties of the grease because it is the co-crystallization of the two soaps made in situ that produces the desired effect. Figure 10.10 shows the variations of lithium grease structures and a comparison of the lithium simple soap structure with that of the lithium complex grease structure. Some complex soap greases can also contain inorganic salts, such as metal carbonates. Dropping point is the normal measure used to assess the heat resistance of a grease. The presence of another acid or an inorganic salt during the complex soap formation increases the dropping point of the grease by about 80–130°F, or 30–50°C. As a consequence, the complex soap greases are more suitable for use in high temperature applications than the corresponding simple soap greases. Complex soap lubricating greases are made both by the use of the mineral base oils and the synthetic base fluids.

Aluminum Complex Soap Greases

Aluminum complex soap greases have dropping points of 250°C or higher, excellent mechanical stability, good water tolerance, and low oil-separation (bleeding) tendency. Unfortunately, they lack the corrosion protecting ability and have a higher production cost relative to that of the lithium soap based multi-purpose greases.

There are two ways to obtain the complex aluminum soaps. They can be obtained by the reaction of a mixture of a water-soluble salt of a low molecular aromatic acid and a water-soluble salt of a high molecular weight aliphatic carboxylic acid with an aluminum salt, such as the aluminum chloride. When the reaction is complete, the aluminum complex soap which has low water solubility separates. This is collected and dispersed in oil to form the lubricating grease [4]. However, a better method is to dissolve the mixed carboxylic acids in oil and to react them with aluminum isopropoxide, or its trimer. When the aluminum isopropoxide is used for making the soap, the reaction is completed by the addition of water. This generates a free hydroxyl (OH) group

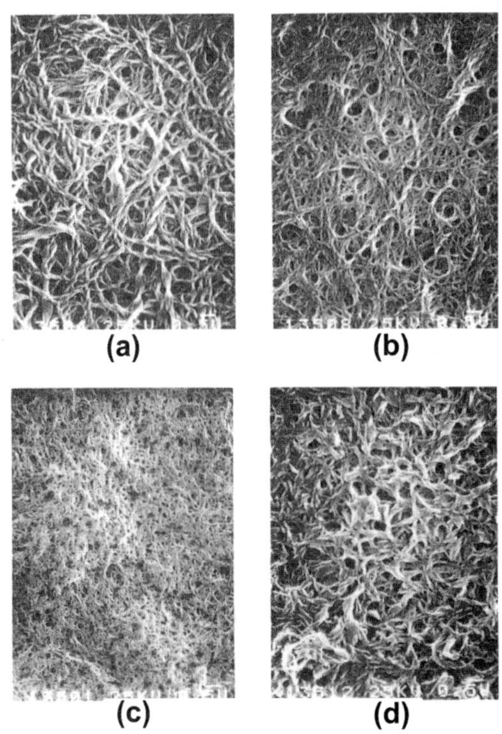

Fig. 10.10—SEM micrographs of various lithium 12-hydroxystearate soap structures in mineral oil. (a) Lithium soap normal, coarse structure, (b) lithium soap fine structure, (c) lithium soap very fine structure, and (d) lithium 12-hydroxystearate complex soap structure [730b].

in the soap structure. However, when the aluminum isopropoxide trimer is used, there is no need to add the water. In order to make the aluminum complex soap grease, a mixture of stearic acid, benzoic acid, and part of the oil in the desired proportion are heated until the mixture becomes homogeneous. The trimer is then added and the mixture heated to 380 to 400°F, or 193 to 204°C. When the reaction is complete, the reaction is cooled by turning the heat off and adding the rest of the oil in portions. The grease is finished by the addition of the additives and milling to make it smooth.

The thickening efficiency of the complex aluminum soap decreases as the aniline point of the mineral oil increases [4]. Since high aniline point equates with low aromatic character, or the high paraffinic character, of the base oil, this behavior of the aluminum soap is not atypical and resembles that of the lithium soap. See Fig. 10.8 which shows paraffinic oils to lead to the least thickening of the grease. However, this problem can be alleviated by increasing the fatty acid/benzoic acid ratio. Since the fatty acid is the high molecular weight acid, it will effectively increase the yield hence the amount of the soap in oil. The consequence will be increased thickening. However, the drawback of this strategy is that although the grease is thicker, it is due to the higher soap content and not because of the improved structural stability. Such a grease is likely to have a low dropping point. Another way to improve the structural stability and increase the yield is by using fatty acids that contain 20 or greater carbon atoms [4].

As stated earlier, the aluminum complex soap greases have high dropping points and excellent water tolerance and because of these, they are used in steel mills and other wet environment applications. These greases are not always compatible with other greases.

Barium Complex Soap Greases

In the past, barium complex soap greases were extensively used in the United States as general purpose greases. However, because of the concern for the toxicological properties of barium, their use has been declining. At present, their use is limited to applications that can benefit from their superior EP properties, good water resistance, and high dropping points. These greases are made in a manner similar to that of the calcium complex soap greases. The short chain acid used to make the complex soap is the acetic acid. The energy requirement for making these greases is higher because the complex soap attains firm and stringy consistency during manufacture.

Sodium Complex Soap Greases

Despite the fact that sodium complex soap greases suffer from the disadvantage of water solubility, they are still in use in a limited number of specialty applications. In this regard, the sodium complex soap greases with soap contents of ~25 wt % or higher, are of special interest. Their high dropping points (~240°C), low oil separation tendency, and good adhesion properties make them suitable for use in the lubrication of the wheel bearings. In this application, these greases are superior to the conventional lithium soap greases. Sodium complex greases containing 35 wt % soap and a worked penetration of 210 (tenth of a millimeter) are ideal for lubricating the outer ring runners, for example in braiding spindles, which can experience a speed of 25,000 per min, or higher [4].

Sodium complex soaps are made by the reaction of a mixture of C_2-C_6 acid, such as acrylic acid, with a fatty acid. As a general practice, the fatty acids in a base oil are reacted with the sodium hydroxide first and then with the acrylic acid. The reaction temperature can be as high as 260°C [4].

Calcium Complex Soap Greases

In terms of use, calcium complex soap greases are the second highest after the lithium soap greases. They have the dropping points of 500–570°F (260–300°C), good shear stabilities, good water tolerance, and a low oil separating tendency. Properly formulated calcium complex soap greases can be used in antifriction bearings that experience temperatures of up to 160°C. The performance of these greases starts to drop off beyond this temperature, due to the decomposition of the soap. These greases possess excellent extreme-pressure properties and provide good friction and wear performance, but they have the tendency to harden when worked.

These greases are manufactured by the reaction of a mixture of stearic acid and usually acetic acid in mineral oil with excess calcium hydroxide. A small excess of the base is beneficial in increasing the rate of reaction, but the presence of a large amount of the excess base has a negative effect on the consistency of the grease; it may lead to the formation of softer grease. Initially, the reaction temperature is kept below 80°C so as to minimize the loss of the highly volatile acetic acid. Alternatively, a closed pressure vessel can be used to minimize its loss and exposure to the manufacturing personnel. After the acetic acid has reacted, the temperature is increased to remove the reaction water. It is important that all water be removed from the finished grease. If it is not, grease can change its consistency on storage. The maximum reaction temperature is in the 160 to 250°C range. The molar ratio of the two acids can be altered to optimize the desired properties. Lowering the amount of the acetic acid in the mixture increases the structural stability of the grease but increasing it raises the hardening tendency. The strategy is analogous to that used for the aluminum complex soap greases. As a positive note, the higher amount of acetic acid does improve EP properties of the grease, as indicated by Timken OK load. In these greases, the calcium acetate is the primary component that makes the soap a complex soap and hence is responsible for improving the properties of the grease. Other components, such as the excess calcium hydroxide and calcium carbonate, present in the product, may also contribute towards the overall properties. Sometimes other medium molecular weight acids, such as caprylic acid (octanoic acid) or caproic acid (hexanoic acid), are also added to further impact grease properties. Calcium soap greases are usually made with a thickener content of 20–30 % because of the ease and the belief that the high soap content is necessary to obtain reasonable thickening efficiency. However, the grease with a lower thickener content but the same consistency can be made by controlling the processing variables. Calcium complex soap greases with the higher thickener content have poor low-temperature performance and have the tendency to harden on long-term storage and in high pressure lubrication systems. Also, such greases can sometimes lead to the bearing failures.

Since this type of grease has a dropping point of 260°C or above, it is used to lubricate the rolling-element bearings that operate at temperatures of 160–200 °C. If the operating temperature is higher than this, the grease deteriorates, releasing ketones. Ketones are the decarbonylation products of the calcium carboxylates at high temperatures. While the calcium complex grease is still in use in some applications, its demand is declining. These greases are not always compatible with other greases.

Lithium Complex Soap Greases

Lithium complex soap greases enjoy the largest share of the complex grease market and are second only to lithium simple soap greases with respect to overall use. Previously, lithium complex soap greases were made from the in situ reaction of the stearic acid, fish oil, and acetic acid with lithium hydroxide [4]. These greases suffered from the reversibility of consistency and were not suitable for many applications. Today, some lithium complex greases are made from mixtures of the hydroxy fatty acids and aliphatic fatty acids. Others are made by the use of the lithium borate and the lithium soaps of the hydroxy fatty acids [4], with and without lithium salicylate. These greases have properties superior to those made by the use of the previous methodology.

The performance of the lithium complex greases is similar to that of the lithium simple soap greases, except that their dropping points are about 50 °C higher. The higher dropping point in this case can not be due to the presence of a cross-bridged soap molecule containing both acids since lithium is a monovalent metal, and such a salt is not possible. Lithium can have only a single carboxylate attached to it. This situation is analogous to that of the sodium complex grease. The possibility of co-crystallization of the long chain soap and the short chain soap to boost the dropping point of the grease is a reasonable possibility. Other strategy that is effective in enhancing the dropping point of a grease is to add a dibasic acid, such as azelaic or sebacic acid, or their esters to the starting acid mixture. Azelaic acid is nonanedioic acid and sebacic acid is decanedioic acid. Their structures are given below:

$$\underset{\text{Azelaic Acid}}{HOOC(CH_2)_7COOH} \quad \underset{\text{Sebacic Acid}}{HOOC(CH_2)_8COOH}$$

In this case again we are depending upon the co-crystallization mechanism and not the cross-bridged salt formation to improve the dropping point. Lithium complex soap greases have good all temperature performance when used to lubricate the tapered roller bearings.

The manufacture of the lithium complex soap greases from 12-hydroxystearic acid involves its reaction with concentrated lithium hydroxide solution by heating the mixture in oil. The reaction temperature is held at 149 °C until all the water is removed. The reaction mixture is then cooled to 100–110 °C and azelaic acid is added. This is followed by the addition of an additional amount of the saturated aqueous lithium hydroxide solution. The reaction temperature is raised first to 149 °C and then to 199 °C to completely remove the water. At this stage, the product is cooled and the remaining oil and additives are added to complete the formation of the grease. It is the presence of the azelaic acid, or its monolithium salt that is responsible for the complex structure of this grease [4].

Lithium complex soap grease that involves the use of lithium borate is manufactured as follows. Aqueous solutions of boric acid and lithium hydroxide are added to a 50% solution of the 12-hydroxystearic acid in oil at 82–88 °C. The reaction temperature is brought to 194 °C with stirring and the remaining oil is added to the grease. The grease is then cooled and homogenized to obtain the finished product. Methyl salicylate is added to the reaction mixture to improve the dropping point. However the increase is only moderate.

Lithium complex soap greases possess good lubricating properties, high dropping points, stable consistency, and low oil separation tendency. Because of these characteristics, these greases have a longer service life than the lithium simple soap greases in the antifriction bearings that operate at high temperatures [4]. The maximum service temperature for these greases is around 175 °C, which is significantly higher than 140 °C for the lithium simple soap greases. Synthetic ester-based lithium complex soap greases when formulated with proper additives are effective across all temperatures, low, medium, and high.

Thickening Mechanism

For a soap to be an effective thickener it must have the correct particle size and the shape and it must have the ability to associate with the oil. As stated earlier, the materials that are good thickeners have only a marginal solubility in oil. Too much or too little solubility will not allow the formation of a good grease. Association between the soap particles and the oil requires week attractive forces, called the van der Waals interactions. The van der Waals interactions involve only surfaces. In general, the larger the surface area of the soap molecule, more extensive is the association with the oil, which results in greater thickening. Similarly, the larger the number of the soap molecules, that is, higher the soap concentration, the greater is the thickening. Again in this case the overall surface area is large. Linear soap molecules, because of their geometry, associate with the oil to a greater degree than the branched chain soaps; hence they cause greater thickening. Many other factors affect this association between a thickener and the oil. Twisted fibers or interlocked three-dimensional molecular structures (networks), in addition, trap oil much more effectively. Further details on this are provided in the section on lubricating grease structure.

The formation of the lubricating grease with optimal performance requires the formation of the thickener in situ. Making soap separately and then suspending it in oil does not always result in a grease with stable structure and its properties usually deteriorate over time, or during use. The presence of the oil during grease formation is beneficial because it facilitates the formation of fibers, or particles, of proper length and geometry for maximum association with the base fluid. Besides the presence of the oil, many other factors play a role in defining the quality of lubricating grease. These include the processing temperature, cooling rate, intensity of mixing, and milling.

Nonsoap Greases

Soap thickened greases make up the largest share of the lubricating grease market. According to a more recent estimate, the worldwide share of the soap-based greases is 83 %. The balance is made up by the nonsoap greases. The use of a number of nonsoap thickeners in greases has been attempted, but only a few have gained significant use. Organoclay and polyurea derived greases, with a combined market share of about 17 %, top the list. Compared to the ordinary metal soap greases some of these thickeners result in greases that may be superior with respect to the following properties:

- Higher dropping points
- Better cold flow behavior
- Broader temperature service range
- Better consistency stability
- Better water stability
- Better corrosion protection behavior

Non-soap thickener derived greases suffer from a number of disadvantages as well and include the following:

- Lack of resistance to steam
- No compatibility with other thickeners
- Poorer thickening efficiency
- Poorer consistency recovery
- Higher noise generating behavior
- More problematic production processes

Two classes of thickeners that are used to make the non-soap greases are organic thickeners and inorganic thickeners. Organic thickeners can be subdivided into (1) pre-made, largely oil-insoluble materials, and (2) those that are produced in situ by the reaction of the two or more ingredients. As a matter of fact, there is only one thickener worth mentioning that falls under the latter group and it is polyurea.

Organic Thickeners

1. Carboxylate soap-like salts
2. Noncarboxylate-based organic thickeners
3. Pigments
4. Medium to high molecular weight polymers
5. Polyureas

Carboxylate soap-like salts include alkali metal and alkaline-earth metal salts of terephthalic acid, sebacic acid N-laurylamide, *n*-octadecyl terephthalate, and N-lauroyl-6-aminocaproate. Noncarboxylate-based organic thickeners include calcium salt of N-stearoyl-sulfanilic acid, lithium propyl phosphate, lithium stearamidomethane phosphonate, and some anilides. Some metal alkyl thiophosphates serve as thickeners as well as EP additives [4].

Pigments, such as alizarin, anthraquinone, indigo, azo compounds, indanthrene and phthalocyanine dyes, and Ultramarine Blue [4] are primarily used for silicone greases. Copper phthalocyanine, which has an intense blue color, is available as very fine particles and is used to thicken specialty silicone fluids. Some of these products are used by the U.S. military in aircraft applications [734].

A number of medium to high molecular weight polymers has also been tried as thickeners for greases. These include linear, branched, and partially branched polyethylenes; halogenated polyethylene, for example polytetrafluoroethylene (PTFE); isotactic polypropylenes; polyisobutenes; poly(4-methyl-1-pentene)s; alkyl acrylate-acrylamide copolymers; condensation products of alkylphenol, fatty acid, and formaldehyde; polycarbohydrates; alkyl hydroxyethylcellulose; and polyureas [4]. A few of these are used in commercial products; others are just an academic curiosity. Most of these materials are chemically and thermally stable; hence they form greases that have use across a broad temperature range. Also, since these thickeners are metal free, they do not deteriorate oxidation properties of the base oil. The greases based on these thickeners have good low-temperature properties and have dropping points of over 500°F, or 260°C. They are primarily used in high-temperature ball bearing and aerospace applications.

Polyurea Greases

Urea itself is not oil soluble, but is water soluble. However, substituted ureas, where the hydrogens on the nitrogen have been replaced with oleophilic groups, such as a long-chain or cyclic hydrocarbon groups, are oil compatible/oil soluble. Nevertheless, these materials are not good thickeners because of their low molecular weight. However, when two or more of these moieties are linked together, there is significant improvement in thickening ability. The term polyurea refers to *diureas* and *tetraureas*. Polyureas are prepared by the reaction of proper amounts of amine or diamine with an organic diisocyanate in oil.

$$\underset{\text{Urea}}{H_2N-CO-NH_2} \quad \underset{\text{Substituted Urea}}{RHN-CO-NHR}$$

R = Hydrocarbon Substituent

Polyureas are the most predominantly used organic, non-soap greases thickeners. Polyurea is a low molecular weight organic polymer that is generated in situ in oil from the raw materials, an organic isocyanate and an amine. Reaction schemes for polyurea synthesis are provided in Fig. 10.11, along with the possible polyurea structures. Polyurea possesses borderline solubility in oil and a gel-forming tendency, both of which are necessary attributes for a good lubricating grease thickener. During the production of a polyurea greases due care must be practiced because the isocyanates and the amines are hazardous and toxic. It is important to point out that while the reactants are hazardous, the product polyurea is not. The reaction does not require any heat and proceeds at a good rate at room temperature. There are no by-products of the reaction to be concerned about. The use of polyfunctional isocyanates provides products of higher molecular weights. If a diisocyanate is used, two moles of a monofunctional amine, or a monoamine, are required to form a product called diurea, which is a urea dimer. This is a popular thickener for greases destined for the Japanese market. However, in the United States, tetraurea is the polyurea that is used to make the polyurea greases. Tetraurea is made from the reaction of a bifunctional isocyanate with a bifunctional amine, charged in the stoichiometric amounts. The term polyurea is normally applied to tetraurea.

The properties of diurea and polyurea greases depend upon the structure of the isocyanate and the amine starting materials. They also depend upon the average molecular weight and the molecular weight distribution, or polydispersity. This is because both these factors affect thickening and hence determines the amount of thickener needed to develop the lubricating grease of a certain consistency. If the polyurea has a greater weight-average molecular weight (Mw) and has low polydispersity (closer to 1), it will not only have good thickening ability but will also have good mechanical properties. Polyurea greases normally have a smooth "buttery" texture, but can be made to possess a fibrous texture for heavy duty applications. The amount of the polyurea needed to make lubricating greases of NLGI Consistency Grade 2 is about 7–12 %. If the diurea is used as a thickener, its amount will be higher because of its lower mo-

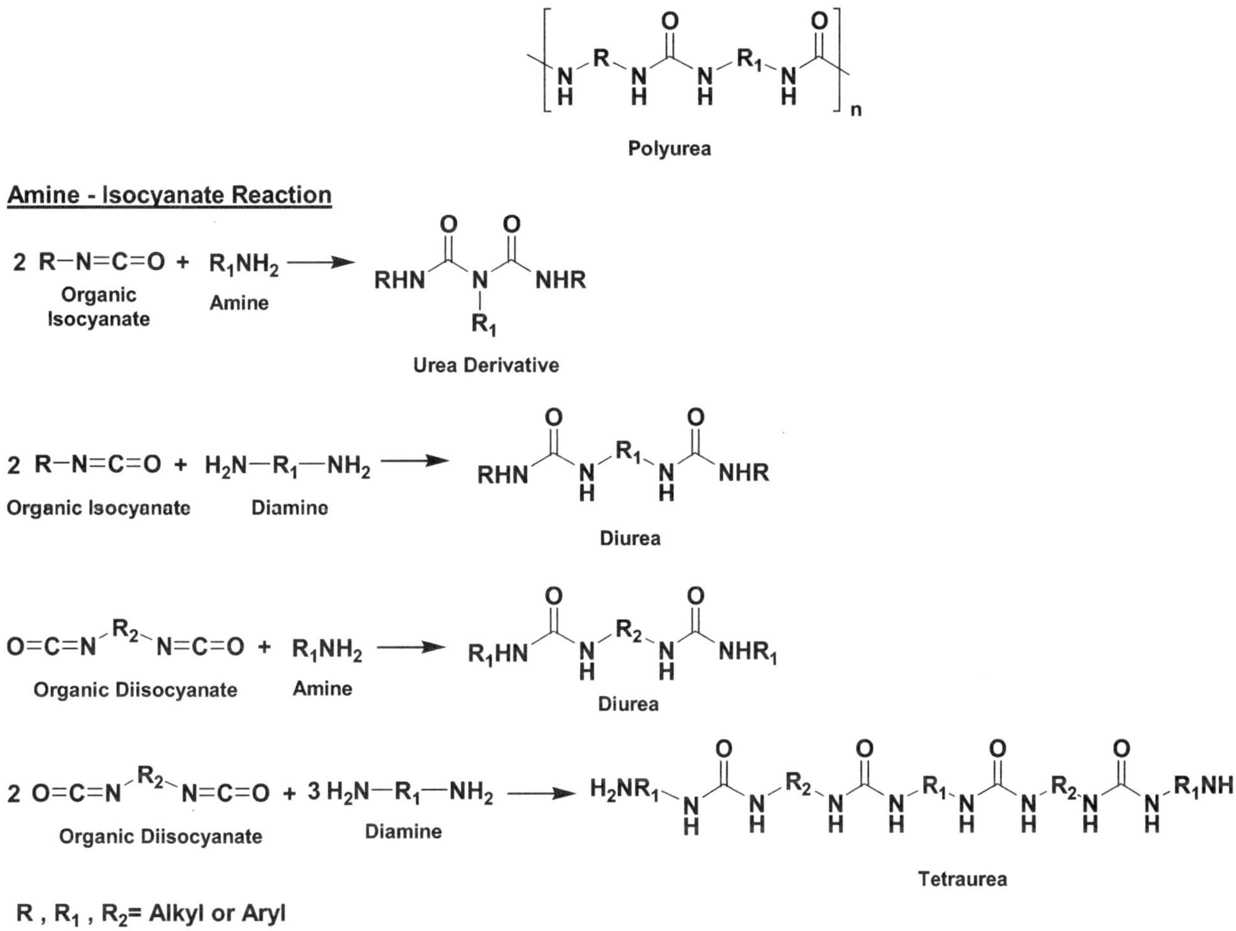

Fig. 10.11—Polyurea synthesis.

lecular weight [735,736]. Polyurea greases possess many desirable properties, which include the following:

- High dropping point, ~250°C (480°F)
- Good thermal stability
- Excellent water resistance
- Excellent compatibility
- Excellent pumpability
- Low oil separation
- Long bearing life
- Excellent oxidation stability

Excellent oxidation stability of the polyurea greases can be ascribed to their somewhat lower hydrocarbon content and being devoid of metals which are known to promote oxidation. Because of their superior oxidation resistance and the high dropping points, polyurea greases are extensively used in sealed-for-life ball bearings and as factory-fill in electric motor bearings, alternators, water pump bearings, and constant velocity universal joints. Sealed-for-life bearings are filled during assembly, permanently sealed, and operated without relubrication for the life of the equipment. Polyurea greases are also used in automotive, industrial, and aviation applications as multi-purpose greases. These greases can also be used to lubricate food machinery since they can easily be formulated for USDA Class H1 approval. Polyurea greases also have a number of deficiencies. These include poor shear stability, poor storage stability, and incompatibility with other greases. However, some of these deficiencies can be overcome by means of proper formulation.

Polyurea greases are more expensive than the soap-based greases because of the higher cost of the raw materials and complex processing. These greases are not suitable for use in centralized lubrication systems because of their poor pumpability characteristics.

Polyurea Complex Greases

Polyurea complex greases are made by incorporating calcium acetate or calcium phosphate complexing agents into the polyurea greases. Such greases possess all the attributes of the polyurea greases and in addition have excellent water resistance, good low temperature characteristics, and superb EP/antiwear and load-carrying properties. Polyurea complex greases are used in a number of industrial and automotive applications, especially those that can benefit from their unique advantages.

Inorganic Thickeners

Clays, such as bentonite and hectorite, silica-gel, and graphite are examples of this type of grease thickeners. Most inorganic thickeners appear as fine, almost amorphous powders, but on a microscopic scale they consist of either tiny spheres or platelets. Nevertheless, it is their small particle size that provides them large surface area to associate with a large amount of oil, which is the reason for the thickening effect.

Since these materials do not have a fibrous structure, like soaps, the mechanism by which they thicken is different from that of the soaps. These thickeners consist of electrochemically charged colloidal particles (platelets), which in oil form a matrix that holds the oil. They essentially form gels in oil with grease-like appearance and properties. Out of the above listed inorganic thickeners, silica and modified-clays have found extensive use in greases. Unfortunately, both in their natural state have affinity towards water, making them unsuitable for use in oil. This hindrance is overcome through modifying them by coating with an appropriate organic reagent, hence the name organo-clay or modified clay. The coatings resist removal until the temperature reaches 300 °F (149°C), or above.

In its natural form, silica is a hard, abrasive material that has no affinity for oil, but by the use of the specialized processes, such as precipitation or deposition as fine particles from the gas phase, it is converted into a light, fluffy, noncrystalline powder called fumed silica. In this form, it has extremely large surface area which is suitable for its use as a lubricating grease thickener. However, the problem is that it still lacks affinity towards oil, which is absolutely necessary for it to form a lubricating grease. This problem is solved by its reaction with surface altering chemicals, such as diisocyanates or epoxides [4]. Fumed silica thickeners are used in various base fluids to create high-melting greases. These greases are inherently water sensitive, but when they are made in silicone fluids, their water-resistance greatly improves. They are costly to produce and hence are used only in certain specialty applications.

Bentonite clays (montmorillonite, hectorite), are made water resistant by exchanging sodium or potassium ions of the natural clays with quaternary ammonium ions. This is achieved by the reaction of the clay with quaternary ammonium chloride or bromide [4].

Specialty Thickeners

Besides the thickeners discussed so far, there are many other important thickeners that are used commercially, though in relatively small amounts. Some fine grades of carbon black have good thickening power. Greases made from this thickener are good for high temperature applications. However, they suffer from low water resistance. Again, their water-resistance problems can be alleviated by making them in silicone and other synthetic fluids. In some applications, carbon black is combined with soaps to form mixed greases. Carbon black greases are black in color, which the users do not always appreciate. These greases find use in specialty applications, such as those requiring good oxidation stability, heat resistance, and lubricity. These can also conduct electricity and can be used in bearings used in xerography machines.

Making Organo-clay Lubricating Grease

Unlike soap greases that require heating the reactants, usually in the presence of oil, the organo-clay greases are made by cold dispersing the clay in oil, but under vigorous (high-shear) mechanical stirring to ensure a uniform dispersion. In most cases, the chemicals called dispersing aids are needed. Dispersing aids, also called propelling agents and dispersants, are polar compounds that aid clay dispersion by absorbing on to the surface of the clay particles to make them swell and increase their affinity towards oil. Commonly used dispersing aids include acetone, ethanol, methanol, and propylene carbonate. The presence of a small amount of water facilitates swelling of the clay particles, hence the formation of the dispersion. Some of the water comes from the organo-clay itself, some from the dispersing aid, and the rest must be added. Since the presence of too much water is not beneficial, it is advisable that the proper amount of water to be added is determined by experimentation. Dispersing aids are all volatile, combustible liquids and pose a fire hazard; hence proper precautions must be taken. After the dispersing step, the batch is milled at around 250°F (121°C). This causes the dispersing aids and water to evaporate. In the final step, the batch is cooled to around 180°F (82°C), 0.1 % water, additives, and balance of the oil are added. Organo-clay greases are smooth textured greases with excellent overall properties.

The clays used to make these greases either do not melt or have extremely high melting points. Hence, greases thickened by clays have essentially no melting point, and their service temperatures solely depend upon the oxidation stability of the base oil and the additives. Since the oxidation resistance of the base oil is the sole limitation, the dropping points for these greases are estimated to be greater than 500°F (260°C) but the maximum use temperature is estimated as 350°F (177°C). These are similar to those reported for other high-temperature greases of the complex soap and nonsoap types. In the absence of the inhibitors, the mineral oil-based clay greases start to decompose around 250°F (121°C). However, with the use of the proper oxidation inhibitors these greases can be effectively used in most high-temperature applications, such as the aerospace lubrication.

Organo-clay greases are versatile lubricants and are used both in industrial and automotive applications. Industrial uses include lubrication of the rolling contact bearings operating at moderate speeds and temperatures, and automotive uses include both as a general purpose grease as well as to lubricate the wheel bearings. These greases can even be used in some high temperature applications, if the equipment relubrication is an option. While the low-temperature properties of these greases are acceptable, many organo-clay greases are formulated for high-temperature applications. High viscosity oils are normally used to make such products. Of course, such formulations will have poor low-temperature properties. Work stability of these greases is rated as fair to good. Since these greases do not have a fibrous structure, they have better pumpability than the soap based lubricating greases but lack the ability to retain oil, especially under high pressures. To make NLGI Consistency Grade 2 grease, one needs to use 5–8 % of the organo-clay thickener.

Consistency of these greases depends on the type of the shearing forces applied. If the shear force is two-dimensional and laminar, the clay platelets slide, interrupting the attractive forces between the clay particles and the oil, leading to a loss in penetration. However, if the grease is in a turbulent three-dimensional environment, such as that of a rotary mill, the loss of interaction due to movement is re-established when the shearing stops. In other words, the loss of penetration, that is, the apparent viscosity loss, is temporary. This is because unlike soaps, clay platelets do not shear.

Organo-clay greases suffer from a number of deficiencies. These are as follows:

- They have poor work stability relative to the soap greases.
- They can not be formulated with some performance additives, such as metal sulfonates, lead naphthenates, and some organophosphorus compounds. This is because these additives disrupt the gel structure of the grease.
- Although these greases are water resistant, they are susceptible to degradation from other contaminants, such as brine.
- While their low-temperature performance in low to medium viscosity oils is acceptable, their performance in high-viscosity base oils is far from satisfactory. These oils are usually used to formulate greases for high-temperature applications.
- They lack compatibility with some of the other types of greases.

As mentioned earlier, silica is made oil compatible by reacting it with diisocyanates or epoxides. Silica-thickened greases produced in aromatic oils are used in roller element bearings employed in the nuclear power plants. This is because of their tolerance to radiation.

Greases containing organo-clay thickeners in both the mineral oils and the synthetic fluids are useful in high temperature applications where the soap-based greases may decompose. Greases prepared by the use of the carbon black or the colloidal silica serve for the lubrication of hot, dust-exposed open gears. Some organic thickeners are used in silicone oils for the manufacture of the high-temperature greases. However, they can cause fouling of the machine parts.

Grease Chemistry

As stated earlier, lubricating grease is a lubricant that is immobilized by the use of a thickener and like any other lubricant, it is made by blending an assortment of additives in a base fluid. The thickener, the immobilizing component in lubricating grease, is usually a metal salt of a fatty carboxylic acid or metal salts of a mixture of carboxylic acids. These salts are obtained from the animal and vegetable fats either by a direct reaction with a metal base or via hydrolysis of the fat to yield the fatty acid which is then reacted with the metal base. Acids in uncombined form are not plentiful in nature but fats are; hence sometimes mixtures of fats and acids are used to make lubricating greases. Commonly used acids in the manufacture of grease include acetic acid (CH_3COOH), stearic acid, ($C_{17}H_{35}COOH$), and 12-hydroxystearic acid.

Base fluid is an integral part of the lubricating grease, both with respect to structure as well as performance. It is either petroleum-derived (mineral oil), synthetic in origin, or is obtained from plants and animals. Since the base fluid in the lubricating grease makes up about 70–90 % by weight of its composition, proper selection of the base oil is important to the grease's overall properties.

Base Fluids

The base fluid properties that influence the properties of the finished lubricating grease include the following:

- Viscosity and viscosity index
- Oxidation and thermal stability
- Chemical stability
- Additive solubility
- Volatility and flash point
- Behavior towards elastomers
- Environmental compatibility

Most of these properties were discussed in detail in Chapters 2 and 3 while discussing base fluids. Hence, here we will only comment on properties that impact the properties of the grease.

Mineral Oils

Base fluid viscosity has the most profound effect on the viscosity of the finished grease. Mineral oils used to manufacture lubricating greases range from being 90N (neutral) to 600N; which covers almost the whole range of the petroleum-derived base stocks. Some of the greases even use bright stocks, which are extremely viscous oils with high aromatics content and excellent solvency. There are many applications where the conventional oil-based greases do not work well. For such applications, high performance greases made from base oils of higher oxidative and thermal stability are employed. Such oils include highly refined, chemically modified mineral oils, that is, API Group II and Group III oils and synthetic base fluids.

Mineral oil's overall chemical composition, that is its paraffinic, naphthenic, and aromatic content affects seal compatibility, thermal and oxidation stability, and tendency of the grease to separate oil. The effect of greases on seal materials (swelling, tensile strength) is largely a function of the oil's aromatic content. In general, the higher the aromatic content of the oil, the greater is the grease-related damage to seals. This is because some aromatic compounds migrate into the seal material and cause swelling and the loss of its tensile strength. Other compounds tend to remove the plasticizer from the seal and cause seal shrinkage and cracking. Oxidation stability and the decomposition temperature of the base oil impact the maximum working temperature and the service life of the grease, especially with respect to its use in antifriction bearings. Oil separation increases as the aromatic character of the oil decreases. This is because the soap has better solubility in the aromatic component of the oil and hence maintains its structural integrity.

Paraffinic mineral oils have excellent oxidative stability, high VI, and good behavior towards elastomers, so do the greases made from these oils. Because of these characteristics, the greases have the ability to perform over a broad (wide) temperature range. However, the highly paraffinic oils suffer from lower ability to dissolve polar additives (poor solvency), meager low-temperature properties, and low thickening response to most soap type thickeners, as is shown in Fig. 10.8. Low thickening due to soap is related to it being polar and these oils being nonpolar and therefore lacking good association with each other. These factors make formulation of the high performance greases from these oils somewhat of a challenge. The low-temperature properties in greases derived from these oils are not of great concern because many thickeners function as pour point depressants as well. Colorless base stocks, called the white oils, are used in food grade, USDA H1 greases that are formulated for incidental food contact.

TABLE 10.5—Properties of petroleum oils versus grease properties [729].

Properties	Paraffinic	Naphthenic	Aromatic
Molecular form	Saturated	Saturated	Unsaturated
Oxidation stability	Best	Moderate	Poor
Thermal stability	Best	Moderate	Poor
Additive solubility	Poorest	Moderate	Best
Higher grease yields (use of less soap)	Poorest	Moderate	Best
Viscosity index	Highest	Low	Lowest
Low temperature	Fair	Good	Best
Flash points	Highest	Lower	Lowest
Boiling range	Highest	Lower	Lowest
Evaporation	Low	Moderate	Highest
Carbonization due to oxidation	Normally hard[a]	Soft	Soft
Toxicity	Minimal	Minimal	High
Elastomer seal swell[b]	Minimal	More swell	Most swell

[a]Can be corrected by better compounding.
[b]Oils do affect elastomers or rubber seals. Choice of seals with various greases must be considered.

As stated previously, for mineral oil-based greases naphthenic oils are the oils of choice, although aromatic base oils are also quite suitable and are often used. While naphthenic oils have lower viscosity indices than paraffinic oils, their use is still preferred in greases because of their low-temperature fluidity and better thickening by soaps. Aromatic oils, or bright stocks, although have good solvency and boost the thickening power of the soaps well, are oxidatively less stable; hence they cannot be effectively used to formulate high-temperature greases. When using naphthenic or aromatic base stocks, the use of the oxidation inhibitors is essential. Elastomer-seal compatibility of the lubricating grease is also important and is negatively impacted by naphthenic and aromatic oils; more by the aromatic oils than the naphthenic oils.

Oil blends of a viscosity between ISO VG 100 and VG 220 are among those that are used most often, but for greases designed for specialty uses, the oils of lower or higher viscosity grades may also be employed. Greases for low-temperature and or high-speed use utilize lower viscosity base oils and greases for slow speed, high loads, and shock loading benefit from the use of the oils of higher viscosity. Lubricating greases made from low-viscosity oils possess good low-temperature properties and good transportability. They also have low working and equilibrium temperatures and hence find use in plain bearings and joints and high speed antifriction bearings. Lubricating greases from base oils of high viscosities are employed in slow-running bearings and gears that operate under high loads and experience high ambient temperatures. An increase in the oil viscosity minimizes its evaporative loss, enhances adhesion and the corrosion preventing properties, controls noise, and improves water tolerance.

Mineral base oil properties and their influence on the behavior of the lubricating greases are summarized in Table 10.5 [729].

Synthetic Fluids

Synthetic fluid-derived greases are used when the petroleum-derived products do not perform the intended functions. Some situations where mineral oil-derived greases are deficient include extreme-temperature applications and applications where there is a need for a longer lubricant life or a cleaner operation, i.e., the formation of the less carbon, sludge, and varnish. Synthetic fluid-based greases are useful in high temperatures applications (95 to 315°C, or 200 to 600°F) and low temperature applications (−40 to −75°C, or −40 to −100°F). Unfortunately, the cost of the synthetics is significantly higher than that of the mineral oils, which may act a deterrent towards their use in greases other than high-performance products. Relative cost comparison of the various synthetics with mineral oils is provided in Table 10.6 [4]. See Table 10.7 to examine the strength and weaknesses of the each type of base stock [737].

Of the base oils listed in Table 10.6, diesters, poly(alkylene glycol)s or poly(glycol ether)s, silicones, phosphoric acid esters, perfluoroalkyl and chlorofluoroalkyl ethers, and hydrocarbons are of the greatest importance with respect to use in grease. The greases developed from these synthetic oils have certain advantages over mineral oil-based greases, which will be pointed out during different parts of the subsequent discussion. Most synthetics are available in different viscosity grades and have excellent viscosity-temperature and low-temperature properties. However, some fluids, such as poly(phenyl ether)s and silicones, either have high pour points or have a strong creeping tendency on the metal surfaces and are therefore not useful in typical grease applications. Greases made from some synthetic base stocks are negligible in use volume, which is partly due to their higher

TABLE 10.6—Relative cost of synthetic fluids [4].

Oil/Fluid	Relative Cost
Solvent Raffinates	1
Hydrocracked oils	2...3
Polyolefins	3...10
Dicarboxylic acid esters	4...10
Phosphoric acid esters	5...10
Poly(glycol ether)s [Poly(alkylene glycol)s]	6...10
Silicic acid esters	20...30
Silicone oils	30...100
Poly(phenyl ether)s	200...500
Polychlorofluorohydrocarbons	400...600

Base Fluid	Density, g/cm³	Viscosity at 40°C (104°F), mm²/s	Viscosity Index	Flash Point °C	Flash Point °F	Pour Point °C	Pour Point °F	Volatility[1]	Oxidation Resistance	Lubricity[a]	Thermal Stability	Applications
Mineral Oil								5	5	5	5	Standard lubricant components, either as oil or grease base
Paraffinic	0.881	95	100	210	410	-7	19	–	–	–	–	
Naphthenic	0.894	70	65	180	356	-18	0	–	–	–	–	
Mixed Base	0.884	80	99	218	424	-12	10	–	–	–	–	
VI Improved	0.831	33	242	127	261	-40	-40	–	–	–	–	
Polyalphaolefin	0.853	32	135	227	441	-54	-65	4	4	5	4	
Esters								3-5	3-7	3-6	4-7	Jet turbine lubricants, hydraulic fluids, heat transfer products; used as bases for low volatility, low viscosity grease
Dibasic	0.945	14	152	232	450	<-60	<-76	3	5	5	5	
Polyol	0.971	60	132	275	527	-54	-65	–	–	–	–	
Tricresyl Phosphate	1.160	37	-65	235	455	-23	-9	3	3	3	7	Phosphate esters fire resistant
Polyglycol Ether	0.984	36	150	210	410	-46	-51	3-5	7	4	5	
Silicate	0.909	6.5	150	188	370	<-60	<-76	3	7	6	4	
Silicones								1	1	7	1-5	High-temperatures/low-volatility applications, either as oil or grease base; lightly loaded bearings; excellent thermal stability
Dimethyl	0.968	100	400	>300	>572	<-60	<-76	–	–	–	–	
Phenyl Methyl	0.990	75	350	260	500	<-60	<-76	–	–	–	–	
Chlorophenyl	1.050	55	160	288	550	<-60	<-76	–	–	–	–	
Perfluoroethyl	1.230	44	158	>300	>572	<-60	<-76	–	–	–	–	
Phenyl Ether								1	1-3	5-7	1	
Low Viscosity	1.180	75	-20	263	505	10	50	–	–	–	–	
High Viscosity	1.210	355	-74	343	650	5	41	–	–	–	–	
Perfluoroalkyl Ether	1.910	320	138	–	–	-32	-26	3-7	1	5-7	1-3	Extreme-temperature fluid; used in very low volatility applications

[a]The value 5 characterizes highly refined mineral oil. Values of < 5 reflect superior performance, whereas values of > 5 reflect inferior performance to mineral oil with respect to lubrication properties.

TABLE 10.8—U.S. Military specification requiring synthetic lubricating greases [738].

Specification	Thickener	Temperature Range, °F (°C)	Synthetic Fluid Type	Application
MIL-G-4343	Lithium soap	−60 to 250 (−52 to 121)	Dicarboxylic acid esters/ silicones	Pneumatic systems
MIL-G-6032	Soap or nonsoap	20 (93)	Animal/vegetable ester and or silicone	Systems where the resistance to various fluids is required
MIL-G-21164	Lithium soap/clay	−100 to 250 (−73 to 121)	Dicarboxylic acid esters	MoS_2-containing lubricating grease for antifriction bearings
MIL-G-23827	Lithium soap/clay	−100 to 250 (−73 to 121)	Dicarboxylic acid esters	Ball, roller, and needle bearings used in instruments, cameras, electronic gear, and general use in aircrafts
MIL-G-25013	Nonsoap	−100 to 450 (−73 to 232)	Silicones	Roller and ball bearings in aircraft
MIL-G-27617	Inorganic	−30 to 400 (−34 to 204)	Perfluoroalkylpolyethers	Ball, roller, and needle bearings used in aircrafts
MIL-G-81322	Clay	−65 to 350 (−54 to 177)	Synthetic hydrocarbons or esters	Aircraft grease
MIL-G-81827	Clay	−65 to 350 (−54 to 177)	Synthetic hydrocarbons	MoS_2-containing lubricating grease for antifriction bearings operating at a broad temperature range
MIL-G-81937	Lithium soap	−65 to 250 (−54 to 121)	Dicarboxylic acid esters	Miniature and instrument bearings requiring broad temperature performance and ultra-clean lubrication
MIL-G-83261	PTFE	−100 to 450 (−73 to 232)	Fluorinated polysiloxane	Aircraft actuators, gearboxes, oscillators, and other applications involving heavy loads and extreme temperature ranges
MIL-G-83363	FEP/clay	−60 to 300 (−51 to 149)	Aliphatic polyol esters and fluorinated polysiloxane	Helicopter, tail rotor and transmissions, and gearboxes
DOD-G-24508	Clay	−65 to 300 (−54 to 149)	Synthetic hydrocarbons	Multi-purpose ball and roller bearings

cost and limited use only in highly specialized products. Such fluids include fluorinated hydrocarbons and perfluoropolyethers, which are extremely resistant to oxidation, and poly(phenyl ether)s, which have high thermal stability and radiation resistance.

Synthetic fluids have well defined structures and hence have precise physical and chemical properties. The lubricant greases made from synthetic fluids have the following attributes:

1. Good broad temperature behavior
2. Good to excellent chemical resistance
3. Superior corrosion protection
4. Low volatility at high temperatures
5. Excellent oxidation stability
6. Excellent extreme-pressure/antiwear properties
7. Good compatibility with other greases
8. Good electrical properties
9. Ability to lubricate nonmetallic surfaces
10. Environmental compatibility; no or low toxicity, reduced fire hazard
11. Biodegradability

Synthetic fluid-based greases find extensive use in high-performance aircraft, missiles, and space vehicles. When the thickener and the fluid are both synthetic, the grease is used almost exclusively in high-performance equipment. For some missile applications, a service life of minutes, or less, may be adequate. Military specifications for a variety of synthetic greases and their temperature ranges are shown in Table 10.8 [738].

With the exception of the poly(alkylene glycol)s, all other synthetic fluids used in greases have viscosities in the range of the lighter HVI neutral mineral oils. However, viscosity indices and flash points of the synthetics are higher and the pour points are considerably lower than those of the comparable mineral base oils. In addition to the higher cost, esters have the additional disadvantage of having greater seal-swelling tendencies. It is therefore important to pay attention to the type of the seal material used in the equipment, prior to formulating an ester-based grease. Ester-based greases are used in applications that experience broad temperature ranges, such as for aircraft lubrication. The greases made from poly(alkylene glycol)s have reasonable thermal stability, are nonaggressive to most elastomer seals, and on decomposition leave only the thickener-derived residue.

Silicones, or polysiloxanes, possess excellent fluidity at low temperatures, low volatility, good oxidation resistance, good water resistance, good elastomer seal compatibility, and excellent thermal stability. These properties make these fluids useful base stocks for making greases that can be used for high-temperature and broad-temperature applications. However, silicone-derived greases are not suitable for applications involving high loads since they do not provide protection against wear. To make things worse, their response to the load-carrying additives is also weak. Another disadvantage of these fluids is that if they get on metal surfaces they prevent the surface coatings, such as paint, to adhere. However, they do not harm an already painted surface. In terms of cost, silicone fluids are quite expensive.

Synthetic hydrocarbons that have utility in greases in-

TABLE 10.9—Typical additives used in lubricating greases (adapted from Ref [742]).

Additive Type	Purpose	Typical Compounds	Typical Treat Level (%)
Anti-wear/EP Agents	Reduce friction and wear and prevent scoring and seizure	Zinc dithiophosphates, metal (zinc, molybdenum, and bismuth) dithiocarbamates, organic phosphates and acid phosphates, organic sulfur and chlorine compounds, sulfurized fats, sulfides and disulfides, MoS_2, PTFE, and graphite	0.1-15.0
Rust Inhibitors	Rust prevention of the metal parts in contact with the lubricant	Metal phenolates; basic metal sulfonates; fatty acids, fatty amines, and their salts	0.01-6.0
Copper Corrosion Inhibitors	Corrosion prevention of the copper and bronze	Mercaptobenzothiazole, dimercaptothiadiazole and its derivatives	0.1-1.0
Film-strength Improvers/ friction Modifiers	Alter the coefficient of friction	Organic fatty acids and amides, lard oil, high molecular weight organic phosphorus and phosphoric acid esters	0.1-5.0
Adhesion Promoters/tackifiers	Improve adhesion	Polyisobutylenes, olefin oligomers, and resins	0.5-2.0
Viscosity Modifiers	Increase base oil viscosity	Polymers and copolymers of methacrylates, butadiene, olefins, or alkylated styrenes	0.1-1.0
Structure Modifiers	Stabilize soap-oil systems	Glycols, glycerol, higher carbon alcohols, monoglycerides, fatty acids, alkalis, water, and alkali metal sulfo-naphthenates	0.1-1.0
Sludge Control Agents	Control sludge formation	Alkenylsuccinimides and alkenylsuccinates, polysiloxanes	0.001-1.0
Water Repellants	Improve water sensitivity/minimize water washout	Oleic and other fatty acids	0.1-2.0
Oxidation Inhibitors	Retard oxidative decomposition	Zinc dithiophosphates and dithiocarbamates, hindered phenols, aromatic amines, sulfurized phenols, dialkyl sulfides and selenides	0.1-5.0
Metal Deactivator	Reduce catalytic effect of metals on oxidation rate	Organic complexes containing nitrogen or sulfur, amines, sulfides and phosphites	0.1-1.0
Odor Masks	Minimize offensive odor	Perfumes, pine oil	0.05-0.5
Color Stabilizers	Preserve color	Substituted hydroquinones, furfural azine	0.01-0.10
Dyes	Impart color	Oil-soluble aniline dyes	0.01-0.03

clude polyalphaolefins (PAOs) and alkylated aromatics. These materials do not contain any hydrolyzable functional groups, hence they are hydrolytically stable. In addition, they are not harmful to the surface coatings, again because of the absence of the highly polar functional groups. Relative to the mineral oils, these fluids have low volatility, better heat resistance, greater oxidation resistance, and good low and high temperature viscometrics. Because of these properties, these fluids are ideally suited to make greases for use in broad-temperature applications, such as airplanes and ships, and at a reasonable cost. The PAO-derived greases are also used in some industrial applications. Greases made from alkylated aromatics possess good hydrolytic stability, thermal stability, harmlessness to surface coatings, and low volatility. However, they suffer from poor oxidation stability, primarily due to their aromatic structure.

Incidentally, there is a move towards the greater use of the vegetable oil-based greases, especially in Europe, due to their high environmental compatibility.

Additives

A variety of additives are used in lubricating greases to enhance their properties. The properties of interest that need to be improved are structural, rheological, and chemical. Additives that are used to achieve this include oxidation inhibitors, metal deactivators, extreme pressure (EP)/antiwear agents, film-strength improvers (friction modifiers), rust inhibitors, copper deactivators, viscosity modifiers, adhesion promoters, sludge control agents, tackifiers, water repellants, odor masks, and specialty solids.

Table 10.9 lists the properties of the various additives that are used in lubricating greases along with their typical treat level and their chemistry [742]. The additives in a lubricating grease may interact with one another or with the thickener and the interaction may be synergistic or antagonistic. The knowledge of the possible interactions will help design better lubricating grease. The assessment of the additives needed in a grease requires a number of considerations, some of which are listed below:

- Performance requirements consistent with the intended application/s.
- Compatibility, synergistic or antagonistic reactions with the soaps, the additives, and the other greases.
- Environmental considerations; such as the product use, odor, disposal, and biodegradability.
- Color.
- The overall cost.

While the lubricating greases are formulated to deliver specific performance, the overall cost of their components, that is the base oil, thickener, and additives, cannot be ignored. For a detailed discussion on the role of additives in lubricants, please refer to Chapter 4.

Solid additives, also called fillers or physical additives are occasionally used in lubricating greases. They are more suitable for use in greases than in liquid lubricants since they do not need be soluble but be dispersible. The purpose of using these solids is to impart friction reduction and extreme pressure properties to grease. Materials that are commonly used include the following [738]:

1. Graphite

TABLE 10.10—Lubricating grease characteristics needed for industrial applications [744]. Reprinted with permission from the Lubrizol Corporation.

Type Thickener	Grease Characteristics	Applications
Simple Soap-Thickened Greases		
Aluminum	Smooth, gel-like appearance Low dropping point Excellent water resistance Softening/hardening tendencies Greatly dependent on shear rate	Lubricate low-speed bearings Wet applications Usage decreasing
Sodium	Rough, fibrous appearance Moderately high dropping point Poor water resistance Good adhesive (cohesive) properties	Older industrial equipment where re-lubrication is frequent Rolling-element bearings
Calcium	Smooth, buttery appearance Low dropping points Good water resistance	Bearings in wet applications Railroad rail lubricant
Lithium	Smooth, buttery to slightly stringy appearance High dropping point Resistant to softening and leakage Moderate water resistance	Automotive chassis and wheel bearings General industrial grease Thread lubricants for the oil drilling industry
Complex Soap-Thickened Greases		
Aluminum Complex	Smooth, slight gel-like appearance Dropping points above 500°F Good water resistance Resistant to softening Shorter life at high temperature	Steel mill roll neck, rolling, and plain bearings
Calcium Complex	Smooth, buttery appearance Dropping points above 500°F Good water resistance Inherent EP/load-carrying capability	High-temperature industrial and automotive bearing applications
Lithium Complex	Smooth, buttery appearance Dropping points above 500°F Resistant to softening and leakage Moderate water resistance	Automotive wheel bearings High-temperature industrial service including various rolling-element applications
Non-soap Greases		
Polyurea	Smooth, slight opaque appearance Dropping points above 450°F Good water resistance Oxidation resistant Less resistant to softening and leakage	Industrial rolling-element bearings Automotive constant velocity joints
Organo-Clay	Smooth and buttery appearance Dropping point above 500°F Resistant to leakage Good water resistance	High-temperature bearings with frequent re-lubrication Steel mill roll neck bearings

2. Molybdenum disulfide (MoS_2)
3. Zinc oxide, magnesium oxide, and cerium fluoride
4. Copper and nickel powders
5. Carbon black, talc, and mica
6. Inorganic boron compounds, such as boric acid, borax, boron nitride, and metal borates
7. Bismuth, zinc, and lead naphthenates; and zinc stearate
8. Inorganic sulfur-phosphorus compounds, phosphate-thiosulfate blends [739]
9. Calcium acetate, carbonate, and phosphate
10. Phosphate glasses
11. Fluorinated polymers.

Lubricating greases derived from these materials help protect heavily loaded bearings from galling and seizing. These materials do no provide the EP protection by reacting with the metal at high temperatures like the conventional oregano-sulfur and organo-phosphorus compounds. Instead they do so by physically depositing the fine solids, suspended in the grease, on the metal surface, which acts as a low-shear film that provides the boundary lubrication. Particle size of the fillers is important since in some bearings, such as rolling element bearings, the clearance may be as small as 0.0001 inch. If the particle size is too large, it can lead to abrasive wear. Solid additives are also well suited in space applications where vacuum can cause the loss of liquid additives. This is because these have low volatility and excellent lubricity. In these applications, the greases containing fluorinated polymers and perfluoropolyalkylethers have been used successfully.

Graphite-based greases are effective in minimizing metal-to-metal contact and wear in sliding surface bearings but are not too useful in rolling element bearings. For rolling element bearings, molybdenum disulfide is the filler of choice. Grease containing 3 % or more of this filler provides an effective protective film. Molybdenum disulfide and graphite are sometimes used for back-up lubrication in high-

TABLE 10.11—Critical grease properties and related performance characteristics.

Property	ASTM Test Method	Description
Shear Stability	ASTM D217	Multi-stroke penetration
	ASTM D1831	Roll stability
	ASTM D4290	Wheel bearing leakage
Oxidation Resistance	ASTM D942	Bomb oxidation
	ASTM D3527	Wheel bearing life
	ASTM D3336	High-temperature performance
Water Resistance	ASTM D1264	Water washout
	ASTM D4049	Water spray-off
Bleed Resistance	FTM 321.3	Oil separation (static)
	ASTM D1742	Pressure oil separation
Extreme Pressure/Antiwear	ASTM D2596	Four-ball EP
	ASTM D2509	Timken method
	ASTM D2266	Four-ball wear
	ASTM D4170	Fretting protection
Corrosion	ASTM D1743	Rust test
	IP 220	EMCOR
	ASTM D4048	Copper corrosion
Pumpability	ASTM D4693	Low-temperature torque
	US Steel LT37	Mobility
Elastomer Compatibility	ASTM D4289	Seal Compatibility
Identification and Quality Control	ASTM D566 or D2265	Dropping point

temperature applications over 315°C (600°F). Zinc and magnesium oxides are usually used in greases for food processing industries, primarily because they are lighter in color and have the ability to neutralize acids. Metal flakes and powdered soft metals, such as lead, tin, zinc, and aluminum, are used in pipe threading and anti-seize compounds. Talc has use in greases for die and drawing and roll neck bearing applications. The best way to incorporate many of the fillers in a grease is during the milling step. Carbon black, although commonly used as a filler, also acts as a thickener for greases. It is important to note that the fillers can accumulate in bearings and cause abrasive wear damage. Accumulation of the filler can result due to evaporative loss of the oil because of heat or bleeding. Therefore, it is imperative that when using solids-based grease proper regreasing schedule is followed to prevent this from happening.

EP properties of the phosphate glasses, inexpensive white powders, in lubricating greases were compared with those of the molybdenum disulfide, graphite, molybdenum dithiocarbamate, polytetrafluoroethylene (PTFE), and boron nitride. Under severe conditions, the phosphate glass greases demonstrated superior load-carrying capacity compared to others [740,741].

Grease types, along with the thickener used and their performance capabilities, are provided in Table 10.10 [744]. Table 10.11 summarizes the critical grease properties and the ASTM test procedures which are used to determine them.

Grease Manufacture

Production of the modern lubricating greases on a large scale is quite complex. The complexity increases when the production process involves saponification, which is the reaction of a natural fat with a base. Most thickener systems are produced by this process. The properties of the final grease depend upon a number of factors. These include the quality and type of fats and carboxylic acids starting materials, the metal cation/s, the base oil, the soap concentration, and the pH during the reaction (excess alkali or acid); and most importantly the manufacturing process. So much so that many of the grease manufacturers consider the grease production an art and the details of the grease manufacture are kept confidential. Lubricating greases can be manufactured either by a two-step process or by a one-step process. The two-step process involves producing the soap first, by the reaction of the acid and or the fat with a metal base and isolating it. The soap is then dissolved in the hot oil, and the mixture cooled under prescribed conditions to obtain the grease. In the one-step process, the soap is created directly in the oil. The fatty acid or the triglyceride is mixed with a portion of the base oil and the mixture reacted with the aqueous base. After the salt formation or the saponification reaction is complete, the reaction mixture is held at an elevated temperature until either a crystalline fluid is formed, as in the case of lithium soap greases, or the mixture becomes homogeneous, as in some sodium complex soap greases. It is important to note that in some greases the maximum production temperature is well below the temperature where the crystalline fluids are formed. This is the case in the sodium soap greases and choosing different maximum production temperatures can lead to the greases with different properties, even though the soap is the same.

Most modern greases are made by batch production. The simplest way to make the lubricating grease is to employ a heated open reactor, commonly called a kettle. Alternatively, a contactor, a pressurized reactor, can be used. A kettle reactor is well suited for greases that are produced in small volume and is often used by small grease manufacturers. Such reactors typically have a capacity of 1500 to 2000 gallons (5.7 to 7.6 m^3). However, to manufacture the low volume products, smaller kettles can be used. The contactor, on the other hand, is necessary in the manufacture of the complex soap greases, since the contactor help keep volatile low molecular weight acids in the reaction. The contactor is useful for producing greases in medium quantities. Batch process is preferred because the product quality is easier to monitor. However, there are a limited number of continuous grease manufacturing plants in operation. Continuous production is at least four times cheaper than the batch process.

Batch Production

A number of factors determine the number of steps involved in the batch production of the lubricating grease, the thickener being the most important. In order to demonstrate steps in a batch production, we use the example of a lithium simple soap grease. The steps involved are as follows:

- The carboxylic acid or the natural fat raw materials are dispersed in oil.
- The lithium hydroxide base dissolved or suspended in water is added.
- The mixture is heated to facilitate salt or soap formation, either through neutralization or through saponification.
- Removal of water by further increasing the temperature and further addition of the oil.
- Cooling of the reaction mixture to help crystallize the completely or partially dissolved soap.
- Mixing in of the additives.
- Homogenizing the resulting grease.
- Adjusting the grease to specific worked penetration.
- Checking the quality parameters.

If a fatty acid is used in the manufacture of the lubricating grease, the salt or soap formation is fairly fast. However, if a natural fat is used as the starting material, one must consider factors that influence the saponification reaction. Saponification is a process by which triglycerides (natural fats) are reacted with an alkali metal (lithium or sodium) hydroxide to produce glycerol and a fatty acid salt, or the soap. The rate of the saponification reaction depends upon the reaction temperature, the concentration of the base, the intensity of mixing, the kind of fat used, and the catalyst, if used. While the higher temperatures increase the rate of the saponification reaction, a temperature of over 100°C requires the use of a pressurized reactor. This is to keep the water in the aqueous base in the reactor until the reaction is complete. Since the intimate contact between the reactants facilitates reaction, good mixing by the use of an efficient stirrer is extremely effective. The batch process usually involves the use of at least one reactor and a number of tanks. The reactor is used to mix the ingredients and convert them into soap or the soap concentrate. This reactor must have high temperature capability and should be able to handle pressures of greater than one atmosphere. The soap concentrate from the reactor is passed through different tanks where the soap is diluted, treated with additives, and the final grease adjusted for consistency. Since all these steps require mixing, these tanks are equipped with stirrers and some tanks also have heating capabilities. Multiple-reactor and multiple-tank capability adds flexibility and minimizes the possibility of cross contamination from other types of greases that were manufactured in the system previously.

Continuous Production

Technical challenges to produce lubricating greases on an industrial scale have been overcome during the past few years. The steps involved are the same as listed under the batch production but in continuous production everything is automatic. Again citing the example of making a lithium simple soap grease, measured quantities of the prefabricated soap or of the fatty raw materials, alkali, and mineral oil are introduced into the reactor by means of adjustable reciprocating pumps. The desired temperature and pressure are dialed in to initiate the soap formation. Good heat transfer and short residence times can be achieved by preheating the raw materials. The reactor residence time is in the order of five minutes, or less. The grease at this time is pumped into the next reactor, where dehydration under vacuum is carried out to remove the volatiles and water. Next, the dehydrated soap is moved to the finishing section, where the additional base fluid and the additives are combined and dispersed.

Finishing

Deaeration, milling, and filtration are the final steps necessary to obtain the finished grease. Deaeration is used only on some products. Greases made in the kettle often contain entrained air because of the vigorous mixing. Air must be removed to improve clarity and the brightness of the lubricating grease.

Milling is important since its influence on the size and shape, hence the surface area, of the fibers is the largest. The primary purpose of the milling step is to break down the solid particles, or the fibers, and to uniformly disperse the resultant small particles in the liquid (homogenization). During manufacture after the soap formation, the grease is almost always obtained as a heterogeneous mixture, irrespective of the rate of cooling. To obtain grease of good quality, it needs to be homogenized, which is accomplished by the use of the pressure valves, tooth-colloid mills, and high pressure homogenizers. In all cases, the process involves high shear mixing, which converts the fibers of different sizes into fibers of similar sizes, thereby resulting in grease with a uniform creamy texture. However, too much milling can be harmful in that it will sever the soap fibers, or decrease the particle size to an extent that will have less association with the oil and hence weakening the lubricating grease structure. Greases that are milled are more durable, that is, they maintain their consistency longer during use.

Filtration removes the unwanted particulate matter from the grease. Most of the greases are filtered prior to packaging and use; however, the firm greases cannot easily be filtered. NLGI recommends installation of the 40 mesh filters on the centralized grease dispensing systems, suggesting that the filtration through 40-mesh or finer screen is appropriate.

Incorporation of Additives

Most chemically active additives used in lubricating greases are sensitive to heat and will decompose at high temperatures and hence become ineffective. They may also lose their structure or the activity in the presence of the strong bases, calcium, sodium, or lithium hydroxide, which are used to make soaps. Therefore, the normal practice is to incorporate the additives after the soap formation and when the partly finished grease is being cooled. At this time, the batch temperature is around 185°F (85°C), or below. There are some exceptions to this guideline. This is the case for additives that are not heat sensitive and need to be heated to facilitate blending in the grease. Therefore, it is important to know which additives can be added at different stages of the grease manufacture, so that the finished grease has all the additives in tact.

Desirable Grease Properties

For grease to be effective in the intended application, it must possess certain specific properties, which include the following:

- Consistency and Consistency Stability
- Penetration
- Rheological Properties
- Thermal Stability (Heat Resistance)
- Oxidation Stability
- Storage Stability
- Sensitivity to Water
- Corrosion Protection
- Load-carrying Capacity
- Wear Control

Consistency

Consistency is the degree of hardness of the lubricating grease and is a measure of the resistance of the lubricating grease against deformation under load. Consistency of the lubricating grease is a surrogate measure of its rheological properties. It is an important measure since the greases are classified and adjusted according to consistency. Greases are available in a variety of consistencies, which are defined by the National Lubricating Grease Institute's Consistency Grades. The grades consist of numbers 000, 00, 0, and 1 to 6 which are based on *worked* penetration of the grease at 25 °C (77 °F). Worked penetration is measured according to the procedures described in the ASTM Standard D217 and DIN Standard 51 804, Part 1. It is important to note that if the temperature falls, say to 32 °F (0 °C), the grease will be firmer by one or two NLGI consistency numbers. Conversely, if the temperature rises to 110 °F (43 °C), the grease will be at least one consistency number softer. NLGI classifies greases into eight classes, but DIN standard classifies them into seven. These classes along with their recommended uses are provided in Table 10.1.

Consistency of the lubricating grease depends upon a number of factors. These include the amount and the nature of the thickener; its particle size; the structure of the fatty acid salt; i.e., the number of carbon atoms, degree of unsaturation, branching, or the presence of another polar group, if soap is the thickener; and the presence of the additional rheology modifiers. A greater amount of thickener will lead to a higher consistency grade. However, the acid structure has a varying effect. The soaps of higher than C_{18} chain length, the C_{18} chain length appears to be ideal, will lead to a thinner grease; so will the soaps with a higher degree of unsaturation or the branching. Lower molecular weight acid derived soaps, on the other hand, are less effective thickeners since they associate with the lubricant to a lesser degree. Incidentally, thickening results from the van der Waals type of interaction/association of the hydrocarbon portion of the soap with the hydrocarbon lubricant. One way to express the situation is that to be an effective thickener, a soap must have reasonable association with the oil; too much or too little will lead to a thinner grease, either because of the high solubility of the soap in the oil or the limited association of the soap with the oil. The presence of a polar group, such as hydroxyl that is present in 12-hydroxystearic acid soaps, leads to improved thickening, which is a consequence of the hydrogen bonding in the soap molecules. This effectively increases the soap's molecular weight; hence its hydrodynamic volume. Certain polar chemical agents, referred to as structure modifiers, facilitate dispersibility of the soap in the oil, thereby improving the grease consistency. A decrease in the particle size of the soap also improves consistency because the smaller particles, if the soap content of the grease is the same, have a greater surface area; hence a greater association with the oil resulting in increased thickening. Many grease manufacturers take advantage of this property of the soap by creating small particles by controlling the temperature and the stirring rates during the production of grease.

Consistency of lubricating greases ranges from almost fluid, or semi-fluid, to very firm, or hard. Too soft a grease will leak out of the equipment and too hard a grease will cause trouble while using. Consistency also influences pumpability; the softer greases generally being easier to pump. When the equipment is to be greased by the use of a dispensing system, the consistency used may be a compromise between that required for lubrication and that required for dispensing.

Consistency is commonly measured by the ASTM Cone Penetration Test (D217). Penetration is the depth, in tenths of millimetres, to which a standard cone, weighing 150 g, sinks into the grease under prescribed conditions. Thus, the higher penetration numbers indicate softer greases, since the cone sinks in the sample deeper. Penetration can be unworked or worked. Unworked penetration is carried out on the grease sample at 77 °F (25 °C) without subjecting the grease to any mechanical treatment. Worked penetration is measuring the effect of manipulation on the lubricating grease since *working* can have a significant effect on its consistency; hence the service behavior. A significant difference between unworked and worked penetration numbers indicates poor shear stability of the lubricating grease. The details of both of these tests are provided in the ASTM Standard D217. Incidentally, the penetration value is commonly expressed without units, which are in tenths of a millimetre. Firm greases have low-penetration values and soft greases have high-penetration values. Generally, the unworked consistency measurement is less reliable than the worked consistency measurement. This is because the first parameter is measured by simply transferring the lubricating grease from a container to the measurement device, with minimum disturbance. Unfortunately, the disturbance cannot be totally eliminated and some invariably occurs. The reason for *inconsistency* in the unworked consistency numbers is the inability to reproduce the same amount of disturbance in each measurement because it cannot be controlled or repeated to the same extent. On the other hand, worked consistency data are easier to reproduce since each sample is worked 60 double strokes, in the standard grease worker, prior to measurement. This helps in standardizing the disturbance in each sample.

A full-scale penetration test requires a large sample,

about 500 lb. For smaller samples, an alternative ASTM Method, D1403 (IP 310) is used. However, the data obtained are less precise than those obtained from the ASTM D217 procedure. Because of this problem, these data are mathematically converted to reflect as if the data were obtained by a full-scale measurement.

There are three other modifications that are used to measure consistency that are worth mentioning. These are prolonged worked penetration, undisturbed penetration, and block penetration. In prolonged worked penetration, the grease sample is first worked for a long period in the grease worker, for example, by subjecting it to 10,000, 50,000, or 100,000 double strokes, cooling it to the penetration test temperature of 25°C (77°F), and then working it 60 double strokes, prior to the penetration measurement. Many consider this test to truly assess the shear stability of a lubricating grease. Block penetration is used to estimate the consistency of the grease that is too firm to transfer to a worker cup. The penetration is determined on three faces of a freshly-cut, 50-mm (2-in.) cube of grease. Undisturbed penetration is measured on the grease sample in a container as received, without any disturbance. This measurement used to be a requirement in ASTM D217, but because of the difficulty in obtaining analogous results on another sample of the same grease, this measurement is no longer a requirement. This measurement was used to control consistency during the grease manufacture and to evaluate the grease's tendency to change on prolonged storage.

As stated above, based upon worked penetration NLGI has classified lubricating greases in nine grades, from 000, 00, and 0 to 6. In general, the greases of Consistency Grades 1–3 are used in automotive and industrial applications, and in the largest amount. Others are used in small volume and only in specialty applications.

Consistency Stability

Greases can change consistency during service or storage. Service-related change is primarily due to the mechanical shear, which can alter the size and the shape of the thickener particles. This will affect their association with the oil, thereby changing the grease's apparent viscosity, or consistency. A change in consistency during service will lead to inadequate lubrication and may lead to equipment damage. It is therefore imperative that a grease that maintains its consistency during use be chosen and the shear forces have a minimum effect on its thickener consistency. Alternative terms used for consistency stability are shear stability, work stability, and mechanical stability. Consistency change in the lubricating grease depends upon the thickener used and the application. The grease may harden or soften during use, but the change should not be too drastic to impair the lubricating ability of the grease. Aging and wide temperature fluctuations can also change the consistency of a lubricating grease. This tendency can be determined by carrying out controlled experiments.

Consistency stability is assessed by the prolonged worked penetration test (ASTM D217) or by the roll stability test (ASTM D1831). These tests are low shear tests and the results are only meaningful for low shear applications. If shear rate in a particular application is higher, more specialized tests may be required. In the roll stability test, a cylindrical roller weighing 11 pounds (5 kg) is inserted in a chamber along with 50 g sample of the test grease. The chamber revolves at 165 r/min for two hours at room temperature. The roller mashes, kneads, and works the grease. Worked penetration after rolling is compared with that before rolling. The difference is indicative of the mechanical stability of the grease. There is a recent trend to run this test at higher temperatures, sometimes above 100°C, instead of at room temperature.

Penetration

As stated under consistency, penetration is the depth in tenths of a millimetre that a standard cone penetrates under the prescribed conditions defined in the ASTM Cone Penetration Test (ASTM D217 and D1403). Examination of the NLGI Consistency Numbers, provided in Table 10.1, suggest each NLGI Consistency Grade to differ from the other by 30 units. The grease of Consistency Grade 2 is the most commonly used grease. Other grades are chosen if an application requires softer or harder grease. Softer greases are good with respect to pumpability and if the equipment operates at lower temperatures. Conversely, harder greases are better in lubricating high-speed bearings and applications where leakage is likely to occur.

Rheological Properties

Rheology is the study of the deformation and the flow of matter. In the case of the lubricating grease, we are interested in observing its behavior under stress. Factors influencing the rheology of grease include the shear stress, shear rate, temperature, and time. Shear stress is the per unit load tending to cause relative movement between the adjacent layers of grease, and shear rate is the rate at which this movement occurs. Fluid viscosity is defined as the ratio of the shear rate to the shear stress. Fluids are of two general types: Those that exhibit viscosity that is directly proportional to shear stress (τ) over shear rate (γ), or τ/γ, and those that do not exhibit a direct relationship to shear stress over shear rate. The former types of fluids are called Newtonian and the latter types of fluids are called non-Newtonian. The non-Newtonian fluids are further divided into dilatant, pseudoplastic, and Bingham plastic, depending upon their response to the shear stress and the shear rate. Figures 10.12 and 10.13 shows the behavior of different types of fluids under the influence of shear rate and shear stress. Greases, like many plastic solids, are non-Newtonian. The non-Newtonian behavior of greases is due to the presence of thickeners that are used to develop the grease's gel structure. Because the greases are not true fluids, they do not exhibit true viscosity, but apparent viscosity. If a fluid's apparent viscosity *increases* with the increasing shear rate, it is a dilatant fluid; and if its apparent viscosity *decreases* with the increasing shear rate, it is a pseudoplastic fluid. Plastic fluids are similar to pseudoplastic fluids, except that they have a yield point, beyond which flow takes place. Lubricating grease is a plastic solid, and like other non-Newtonian fluids, its apparent viscosity changes with varying shear rate. Thus its viscosity must be specified at a specific shear rate, as indicated in Fig. 10.13.

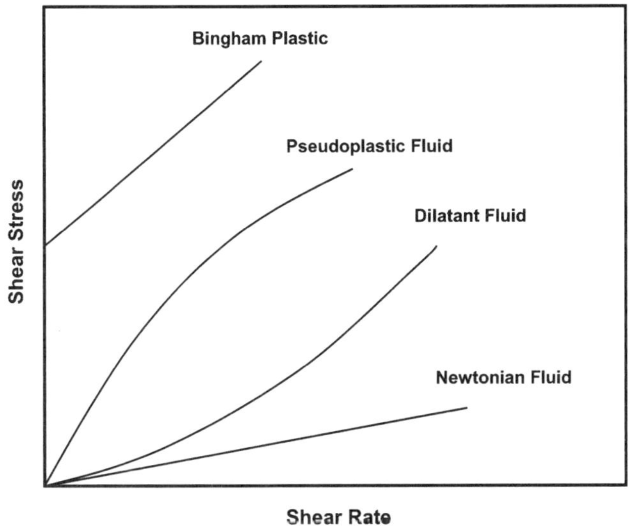

Fig. 10.12—Flow characteristics of Newtonian and non-Newtonian fluids.

Lubricating Grease Structure

Grease is neither a true liquid nor a true solid, but is a semi-solid with very specialized properties. Grease owes its physical appearance to a matrix structure that results from the thickener's association with oil. The proper term used to describe the lubricating grease is that it is a pseudoplastic fluid. This means that it loses its structure, hence viscosity, under shear and becomes a liquid. Depending upon the amount of shear, the grease matrix structure can be completely lost and the viscosity drops down to that of oil, as shown in Fig. 10.14. Though this structural loss is temporary and when the shear forces are removed the viscosity reverts back.

The lubricating grease can also be considered a viscoelastic solid, i.e., a solid that is both viscous and elastic. A *simple* definition of viscosity is a body's resistance to flow, or the "thickness." Elasticity is a property of an object to deform in response to stress (force per unit area) resulting from the application of an external force and when the force is removed, the object regains its original shape. Both these concepts appear to apply to lubricating grease.

There are a number of methods used to determine the rheological properties of lubricating greases. These include Apparent Viscosity (ASTM D1092), Measurement of Flow Properties at High Temperatures (ASTM D3232), Determination of Flow Pressure (DIN 51 805) and, of course, Cone Penetration (ASTM D217). In recent years, new methods to measure rheology have been developed, some of which may be better than those presently in use.

The flow characteristics of lubricating greases are a function of their structure, which largely depends upon the form and the shape of the thickener particles. Soap-based lubricating greases contain soaps as fibrous structures and nonsoap greases contain the thickener as nonfibrous structures of various shapes and sizes. These structures profoundly impact the grease properties. Optimum structure requires the thickener, which exists in the grease as a separate entity, to have a network structure. It should neither be completely soluble in the base fluid nor completely insoluble, but rather co-exist as a part of the heterogeneous system. The invention of the scanning electron microscope (SEM) in the late 1930s made it possible to obtain high resolution images of various materials, including those of the thickeners in lubricating greases. The SEM micrographs of the various lubricating greases are provided in Figs. 10.7, 10.10, and 10.15 [727]. Sodium soap is usually fibrous and stringy and contains long, thick fibers in the form of an interconnected network, as revealed by Micrographs A and B in Fig. 10.7 for sodium tallowate and other sodium simple soap lubricating greases. Water-stabilized calcium soap greases, on the other hand, contain soap fibers that are short with little or no tangling, as is shown by Micrograph D in the figure. Similarly, micrograph E shows the anhydrous calcium 12-hydroxystearate fibers to be somewhat longer and more tangled, but small enough to form a smooth grease. Lithium 12-hydroxystearate greases contain longer, visibly twisted, and well-tangled fibers, but much smaller than those of the sodium soap greases (Micrograph F). The other electron micrographs shown illustrate other types of thickener particles—platelets, clumps, small fibers, and small spheres. The shape and size difference in the thickener particles are responsible for the differences in the rheological properties of the various lubricating greases.

The particle size data provided in Table 10.3 are based upon the electron microscopic determination. While the individual thickener systems differ in their particle shapes, it is the association of the oleophilic portion of the soap with the base fluid or the oil that is primarily responsible for the consistency of the lubricating grease. As stated before, this association can be explained in terms of van der Waals forces. van der Waals forces, also called London dispersion forces, are attractive forces between molecules. These forces are the weakest of the intermolecular forces and result when the electrons in two adjacent atoms belonging to different molecules develop temporary dipoles. A detailed explanation on this concept is provided under the friction modifiers in the additives chapter, Chapter 4. In the case of the lubricating grease these forces involve (1) the multiple soap molecules, (2) the soap molecules and the base oil, and (3) the soap molecules and the additives, when present. In the case of the multiple soap molecules, the attractive forces have two components: attraction between the nonpolar groups of two or more soap molecules (van der Waals forces) and electrostatic attraction involving the permanent dipoles of the metal carboxylate functional groups. The result is the formation of an aggregate, or a micelle, which is a much larger structure; hence has a greater surface area than the individual soap molecule. The presence of a polar group, such as hydroxyl, further promotes association due to polarity. Incidentally, the electrostatic attraction involving the polar carboxylate functionalities is stronger than the van der Waals attraction that exists between the nonpolar hydrocarbon portions of the soap molecules. The interaction between the soap molecule and the base oil is also of the van der Waals type, because the base oil is largely nonpolar, except in a few cases, and it is the hydrocarbon group in the soap that is primarily involved in the interaction with the base oil. Most mineral base oils are a mixture of paraffinic, naphthenic, and aromatic hydrocarbons; hence they are not highly polar in character. However, there is some difference in their relative polarity. This is reflected by their 68°F (20°C) dielectric

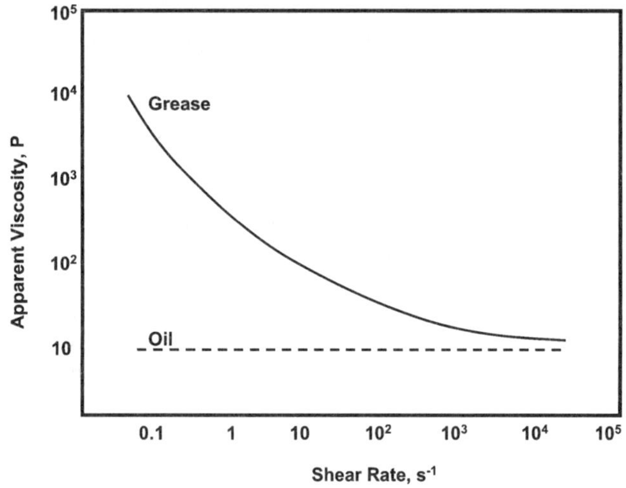

Fig. 10.13—Apparent viscosity versus shear rate for a typical soap-thickened grease.

constants. The dielectric constant of hexane is 1.9, of cyclohexane is 2.0, and of benzene is 2.3 [743]. The dielectric constant is one of the parameters that measure polarity; the higher the dielectric constant, the higher the substance's polarity. The dielectric constants of the hydrocarbons hexane, cyclohexane, and benzene are meant to serve as surrogates for the polarity of the paraffinic, naphthenic, and aromatic base oils or their components. Paraffinics, having the lowest polarity, will associate with the soap primarily through its nonpolar hydrocarbon portion. Alkylaromatics, being the most polar, will associate with the soap's hydrocarbon and the polar carboxylate functional groups via their hydrocarbon portion as well as the aromatic ring. The naphthenics, with intermediate polarity, will fall in between the two with respect to association with the soap. To summarize, all things being equal, the same soap will lead to a thicker grease in aromatic base oils and a thinner grease in aliphatic base oils. This is consistent with the data presented in Fig. 10.8, which show a greater thickening by the soap in a highly aromatic oil and a lower thickening by the soap in a highly aliphatic oil.

Since the magnitude of the van der Waals forces also depends upon the molecular size and the molecular shape, the larger soap molecules and those that have linear hydrocarbon chains, as is the case for most fatty soaps, will promote greater association with the oil and hence will show greater thickening. Of course, in this regard the larger micelles resulting from the polar interaction of the multiple soap molecules have the largest contribution towards thickening of the lubricating grease. Polyurea, being highly polar and with substantially less hydrocarbon content, acts as a thickener primarily by self association and the borderline solubility of the resulting composite structure in the oil. Clays being charged may also cause oil thickening via a similar mechanism.

In the previous discussion, we suggested that the physical properties of the lubricating grease are affected by all three components of the grease formulation, viz., thickener, base oil, and additives. Service-related properties that are of primary interest are consistency, shear stability, viscosity, oil

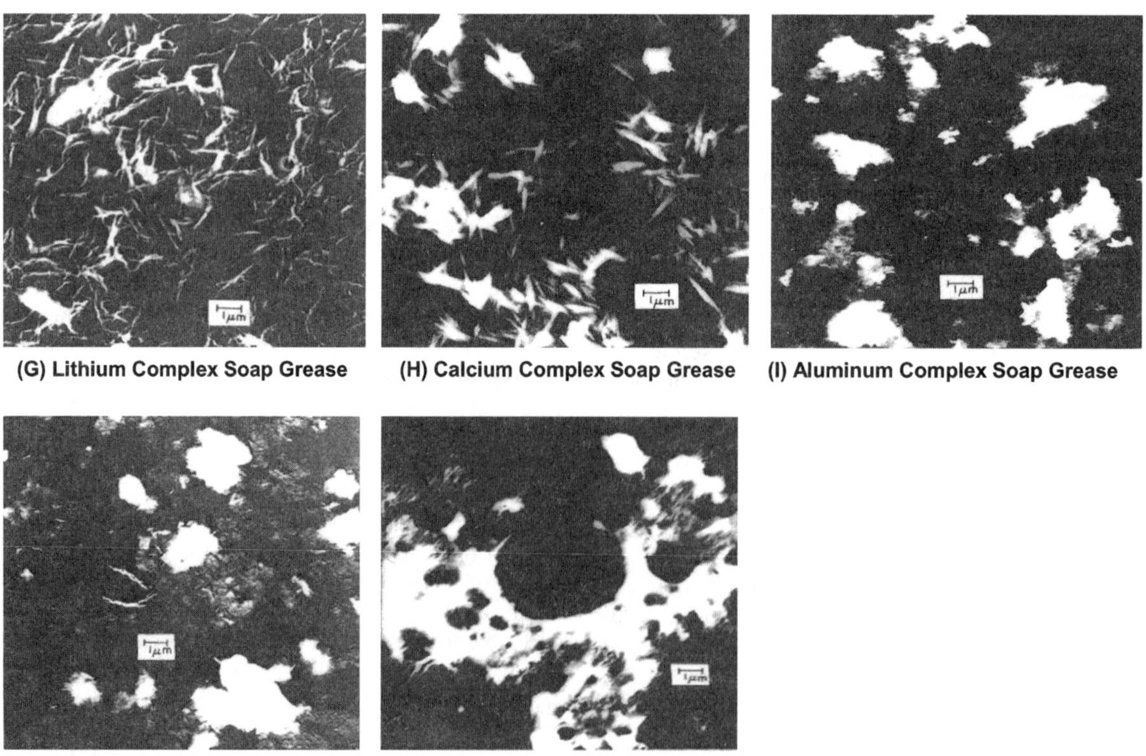

(G) Lithium Complex Soap Grease (H) Calcium Complex Soap Grease (I) Aluminum Complex Soap Grease

(J) Organo-clay Grease (K) Polyurea Grease

Fig. 10.14—SEM micrographs of various complex soap and non-soap greases [727].

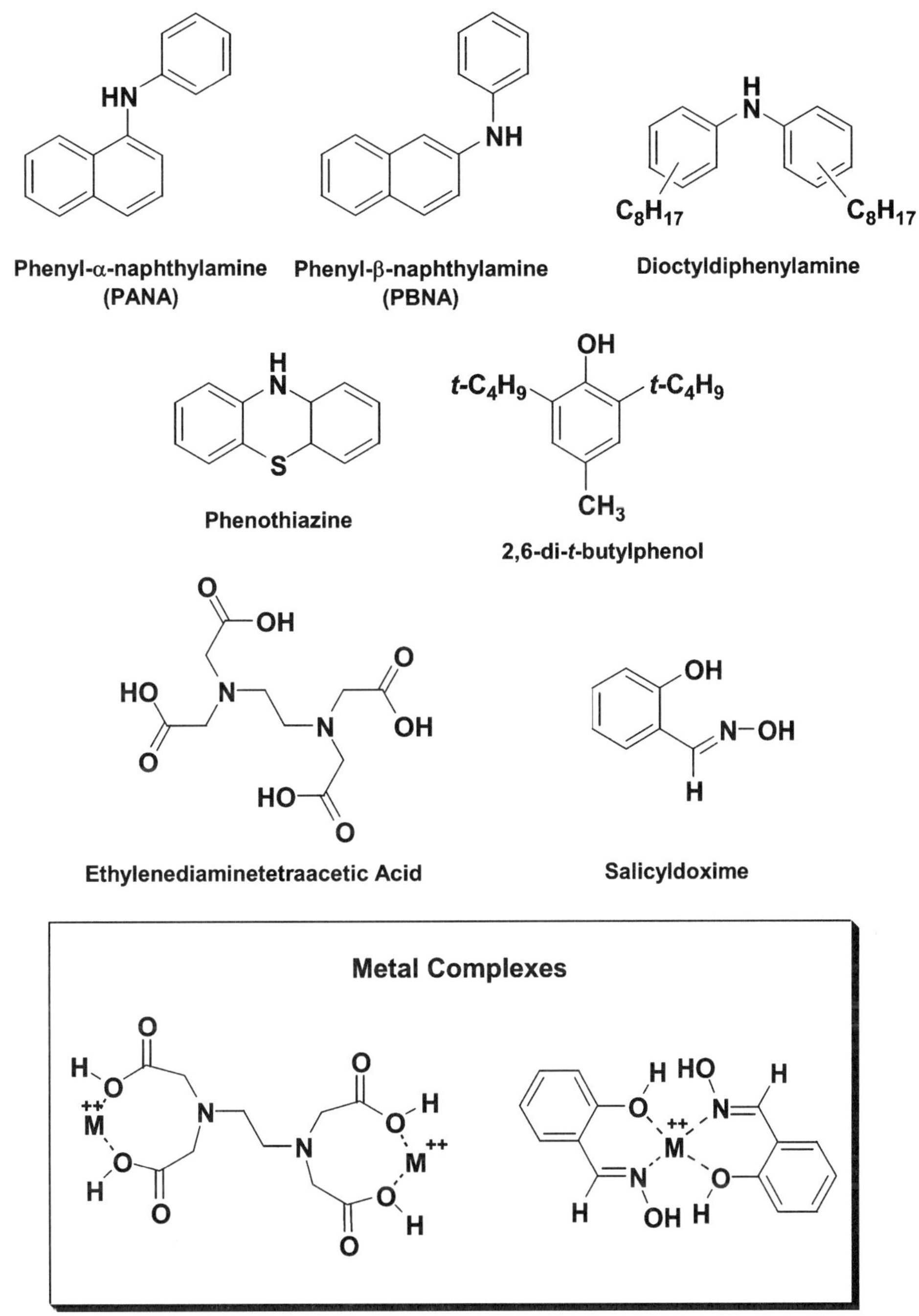

Fig. 10.15—Structures of oxidation inhibitors used in greases.

separation tendency, and its useful temperature range. In the case of the specialty greases there are additional performance-related properties that are of interest. The lower NLGI grade greases, such as 000 and 00, contain only a small amount of soap and their performance is primarily governed by the quality of the oil. The greases with the NLGI grades of 1–3, contain a substantially higher amount of soap, which contributes towards the friction-reduction and extreme-pressure/antiwear performance of these greases. The amount of soap in hard greases, such as NLGI Consistency Grades 5 and 6, is close to 50 %, or more, and in this case, it is primarily the soap that performs the lubrication function. In the case of the greases of NLGI Grades of 3 and lower, the thickener is responsible for the slow release of the oil; which actually performs the lubrication function. The oil release occurs when the thickener network collapses, which primarily occurs due to shear.

Flow Properties

Viscosity and Apparent Viscosity

Although some grades of lubricating grease are firm they must flow so that they can lubricate. As mentioned earlier, the greases flow when the applied stress is beyond their yield point. And since the greases are non-Newtonian fluids, their flow is not directly proportional to the applied stress. This is in contrast to the Newtonian fluids where flow is directly proportional to the applied stress, which is the true definition of viscosity. To distinguish the grease's behavior from that of the Newtonian fluids, the term apparent viscosity is used. Apparent viscosity of the grease varies with the shear rate but that of a Newtonian fluid, such as an additive-free oil, does not. Of course, both types of viscosities vary with temperature. Apparent viscosity is measured by the ASTM D1092 test, as mentioned in the test methods section. The information on apparent viscosity is useful in determining pumpability and dispensability and other related parameters of the lubricating grease. NLGI provides charts that help in converting the apparent viscosity data to pipe flow data. Another test by U.S. Steel, called the mobility test, is used to predict pumpability of the grease at low temperatures. The method employs the same equipment as ASTM D1092, but the piston is activated by nitrogen instead of by hydraulic actuation.

Low-Temperature Torque

For the bearing lubrication, grease flows during the operation of the bearing. Because of the grease structure when the bearing moves through grease it encounters increased resistance, or torque, relative to what it will experience if a free flowing fluid was being used as a lubricant. This needs additional energy, which is not a problem in most applications since they have ample power to spare. However, in some devices this can pose a problem since they lack the extra energy. In some applications, such as aircraft control devices, the low-temperature torque can restrict the use of the equipment. ASTM D1478 is used to measure the low-temperature torque of grease using a ball bearing at −55°F (−54°C). Other test temperatures may also be used. The method suffers from poor precision, presumably because of the low temperature. Just the same, the test is useful in designing devices for aerospace applications.

Thermal Stability (Heat Resistance)

High heat affects a number of important properties of lubricating grease, namely, consistency, bleeding (oil separation), oil evaporation, and grease oxidation. High temperatures make greases softer, as well as increase their tendency to separate oil, which can evaporate. Also, at high temperatures, the rate of oxidation increases, which adversely affects all three components of the grease: the soap thickener, the base fluid, and the additives. Neither of the changes is desired since it can seriously diminish grease's ability to perform satisfactorily.

Too drastic a consistency loss in grease implies that it has lost its structural integrity, hence ability to act as a thickened lubricant. The temperature at which grease loses its gel structure more or less completely is called the dropping point. Two ASTM methods are used to measure the dropping point of grease: ASTM D566 and ASTM D2265. The details of these tests are described in the pertinent ASTM Standards. It is important to note that the dropping point is beyond a grease's use temperature.

Some users of the lubricating grease consider the melting point of the thickener to be the threshold limit for the usefulness of a grease. Beyond this temperature, the thickener loses its ability to impart network structure to the grease, thereby making it ineffective in the grease-related applications. Of course, the thickener must be first removed from other components of the grease to determine the melting point, which is virtually impossible. However, most others feel that the grease properties depend upon it being a system and the soap properties alone will provide little, if any, clue to the usefulness of the grease. Hence, they give credence to the properties of the greases as a whole.

As stated earlier, grease structure is a result of association of oil and the soap. In some cases, the association between the two is strong, but in other cases it is weak. All greases release oil; some to a lesser degree and others to a greater degree. The release of the loosely held oil is the basis of the grease's performance as a lubricant. If no oil is released, the grease fails to perform most of its functions. The loss of the oil can even occur at room temperature and under storage conditions, which is often called bleeding. This is of concern because it will alter the consistency of the grease, minimizing its usefulness. The ASTM D1742 test is used to determine the grease's bleeding tendency. As the temperature increases, the oil separation increases as well. Federal Test Method Document 791, Method 321, is also used to measure oil separation tendency of a grease, but at higher temperatures.

Evaporation of the oil component of the grease is a frequent occurrence in most applications. This is because most of them either have high operating temperatures or have high frictional heat. However, the rate of evaporation in most cases is fairly slow. If there is a major reason for concern, synthetic base stocks with high boiling points are an alternative. Typically the greases used in low-temperature or broad-temperature applications, such as those used by the military, are based on synthetics. This is because these oils have excellent low and high temperature properties.

Two tests that are commonly used to measure evaporation are ASTM D972 and ASTM D2595. ASTM D972 uses a temperature between 210°F (99°C) to 300°F (149°C) and D2595 uses a temperature range between 200°F (93°C) to 600°F (316°C).

Storage Stability

A number of undesirable things happen to the lubricating greases on storage. These include deterioration of performance, oil separation (bleeding), and age hardening—a consequence of the thixotropic nature of the grease. A rare storage-related change that can occur in arid climates is the loss of water, for example, from hydrated calcium soap greases. These greases need water to maintain the stable structure and its loss due to low humidity causes a loss in their consistency. In some cases, the addition of water helps in reverting consistency. The most serious consequence of the long-term storage is oxidative breakdown of the grease.

Oxidation Stability

Lubricating greases are largely comprised of oil; hence they have the same tendency towards oxidation as the liquid lubricants. In lubricant greases, the oxidation results in a change in one or more of the grease properties. These include drying and cracking of the grease, increase in penetration, lowering of the dropping point, increased acidity, and the deposits formation on various parts of the equipment, especially those that experience high temperatures. Hence, it is important that its oxidation be controlled. Oxidation of

TABLE 10.12—Testing conditions comparison between the Four-Ball Test Rig, the FALEX Tester, and the ALMEN-WIELAND Rig [4].

Parameter	Units	Four-Ball Rig	Falex Tester	ALMEN-WIELAND Rig
Speed	rpm	1450	330	200
Sliding speed	Cm/s	55.7	10.4	6.6
Hardness of stationary parts	HV	750-850	730-850	150
Hardness of the moving parts	HV	750-850	240-300	170
Average specified surface load[a]	kN/cm^2	400	100	20

[a]The surface pressures were calculated from mean HERTZ load

grease occurs both on storage and during service and a variety of oxidation inhibitors are used to alleviate it. These additives include peroxide decomposers, radical scavengers, and metal deactivators. These were discussed in some detail in Chapter 4 on Lubricant Additives. The structures of some of the inhibitors used in lubricating greases are provided in Fig. 10.15. The figure also shows the possible mechanism by which metal deactivators remove oxidation-promoting metal ions from the oxidation mechanism.

Inhibiting the grease oxidation is a challenge. Soap-thickened greases are difficult to inhibit because of the presence of the oxidation-catalyzing soaps. Clay-thickened greases are hard to inhibit because the bentonite thickeners have an affinity towards the aromatic amines, which get absorbed on clay's platelet surfaces. This prevents the additive to be present in the oil phase where it is meant to perform its function as an oxidation inhibitor. Inhibitor response of the grease also depends upon the type of the oil used to make the grease. Low viscosity index naphthenic oils and bright stocks, because of the types of structures they contain, are more prone to oxidation than the largely paraffinic oils. Therefore, greases made from the latter group of oils are oxidatively more stable and respond better to the presence of the oxidation inhibitors.

Greases differ in their composition, i.e., with respect to thickener and the base oil, application, and the level of performance. Hence it is important to choose the oxidation inhibitor/s that meet use and performance objectives. The other considerations that should dictate the inhibitor selection are protection during storage and whether it is normal purpose, multi-purpose, or specialty grease. Multi-purpose and specialty greases are designed for high-temperature use; hence it is important that the additives survive the designated service life. Incidentally, one should also consider whether the grease was made from natural fat (triglycerides) via saponification. Such greases contain a significant amount of glycerol which oxidizes faster that the oil; hence the inhibitor/s will be depleted much faster.

In order to measure damage due to oxidation during storage, ASTM D942 (Pressure Vessel Oxidation Test) is used. Although used extensively, the test appears to have its limitations; the major one being that it is not applicable to all products. The following tests can also provide some information on this grease parameter:

1. ASTM D1741, functional life of the ball bearing greases, which is run at 125°C (257°F).
2. ASTM D3336 is run at a temperature of up to 700°F (371°C).
3. Another bearing test is ASTM D3337, which deals with the greases suitable for very small bearings
4. For automotive service or for use in tapered roller bearings, two service-related tests are available. The older test, ASTM D1263, is run at 220°F (104°C) and a newer test, ASTM D3527, is run at 150°C (302°F).

Sensitivity to Water

Water contamination of the grease can alter many of its properties. These include a change in consistency, becoming softer or firmer, change in texture, loss of adhesion, emulsion formation, which will not only diminish the grease's lubrication capability, but may also lead to washout or loss of rust and corrosion protection, or both. There are two tests that are used to determine the effects of water on lubricating grease: The water washout test and the water spray-off test. A number of nonstandard tests are also used to analyze the effect of water on lubricating grease.

ASTM D1264, Test for Water Washout Characteristics of Lubricating Greases, evaluates the resistance of a lubricating grease in a bearing to be washed out by water. The test temperature is either 38°C (100°F) or 79°C (175°F). The test suffers from a number of deficiencies, which include poor test precision and the results depend upon the grease texture and consistency.

ASTM D4049, Test Method for Resistance of Lubricating Grease to Water Spray, is used to assess the ability of grease to adhere to a metal panel, when subjected to direct, intense water spray. The test measures the amount of grease removed. Grease removal occurs by two mechanisms: Solubility and impingement. Water soluble greases are not generally evaluated in this test. Greases that do not have good adhesion, such as those that are soft or are made by the use of the low viscosity oils, are easily removed. This test has good correlation with the operations involving direct water impingement, such as steel mill roll neck bearing service and certain automotive body hardware applications. The test uses water at a temperature of 38°C (100°F).

Corrosion Protection

Corrosion involves deterioration of the metal surfaces by chemical or electrochemical attack. Yellow metal corrosion is chemical in nature and ferrous corrosion (rusting) is electrochemical in nature. Yellow metal corrosion occurs in copper and its alloys, such as brass and bronze, due to sulfur containing additives and is normally measured by using ASTM D130 test method. The problem with this method is that it was devised for liquids and hence is not easy to extend

TABLE 10.13—NLGI service classification for automotive use.

Application	NLGI Classification	Service
Chassis	LA	Mild duty, frequent relubrication
	LB	Infrequent relubrication, high loads, water exposure
Wheel Bearings	GA	Mild duty
	GB	Moderate duty, typical of most vehicles
	GC	Severe duty, high temperatures, frequent stop and go service

to greases because of their semi-solid consistency. A new ASTM Method, D4048, has been introduced to measure copper corrosion in grease. Yellow metal inhibitors are useful in controlling this type of surface damage. Ferrous (iron) corrosion, commonly called rusting, occurs when iron or steel comes in contact with water, oxygen, and trace amounts salt, acids, or alkalis. These compounds catalyze rusting. Rusting tendency of the greases is evaluated by the ASTM D1743 Test. Rusting can be controlled, and even prevented, by adding rust inhibitors to the lubricant.

Load-carrying Capacity

The foremost function of a lubricant is to form a low/no friction barrier between surfaces to keep them apart. This will minimize metal-to-metal contact, hence wear. Wear occurs when (1) high spots, called asperities, of one surface rub against those of another surface, (2) metal breaks, due to fatigue, or (3) cross-surface weld spots resulting from the extreme temperatures in the contact zone shear. The resulting wear debris may cause additional wear through abrasion. This is because unlike liquid lubricants where there is a chance for wear debris to be removed through filtration or an oil change, in greases that have minimal fluidity there is little chance for this to occur. Therefore, it is critical that the formation of the wear debris in grease-related applications is minimized, so as to prevent abrasion of the metal surfaces.

Lubricating greases differ in their load-carrying capability. In this regard, greases formulated from high viscosity oils are better than those formulated from low-viscosity oils. Of course, this is only valid when the loads are low to moderate. When the loads are high, the lubricant viscosity becomes irrelevant since there is little lubricant in the contact zone to form a lubricating film. As a consequence, extensive surface contact, welding, and wear will occur. In these situations, the use of sulfur and phosphorus-based antiwear/EP agents is warranted.

A number of tests are used to evaluate the load-carrying ability of a grease. Three that are most widely used are: Four-Ball Wear (ASTM D2266), Four-Ball EP (ASTM D2596), and Timken Lubricant and Wear Test (ASTM D2509). The first test measures wear at relatively light loads and the test results are reported as average scar diameter; low wear is indicated by a smaller scar. The results only pertain to steel-on-steel and not to other metal combinations. Since the test

TABLE 10.14—Guide to requirements for grease categories (ASTM D4950).

Test	Description	LA	LB	GA	GB	GC
D217	Penetration	√	√	√	√	√
D566[a]	Dropping Point	√	√	√	√	√
D1264	Water Washout				√	√
D1742	Oil Separation		√		√	√
D1743	Rust Protection		√		√	√
D2266	4 Ball Wear	√	√		√	√
D2596	4 Ball EP		√			√
D3527	High Temperature Life				√	√
D4170	Fretting Wear		√			
D4289	Elastomer Compatibility	√	√		√	√
D4290	Leakage				√	√
D4693	Low Temperature Torque		√	√	√	√

[a]D2665 may be substituted.

TABLE 10.15—ASTM4950 automotive grease specifications [745].

Properties	ASTM Test Method	LA	LB	GA	GB	GC	LB/GC
Worked Penetration	D217	220-340	220-340	220-340	220-340	220-340	220-340
Dropping Point (°C)	D556 or D2265	80	150	80	175	220	220
Four Ball Wear (mm) max	D2266	0.9	0.6	—	0.9	0.9	0.6
Four Ball EP	D2596						
LWI (kg) min		—	30	—	—	30	30
Weld Point (kg) min		—	200	—	—	200	200
Elastomer Compatibility	D4289	0 to 30	0 to 30	—	—	—	0 to 30
CR, Volume Change (%)		0 to -10	0 to -10	—	—	—	0 to -10
CR, Hardness Change		—	—	—	-5 to 30	-5 to 30	-5 to 30
NBR-1, Volume Change (%)		—	—	—	2 to -15	2 to -15	2 to -15
NBR-1, Hardness Change							
Oil Separation (%) max	D1742	—	10	—	10	6	6
Rust, max	D1743	—	Pass	—	Pass	Pass	Pass
Fretting Protection (mg loss)	D4170	—	10	—	—	—	10
Low Temperature Torque at 40°C (N-m)	D4693	—	15.5	15.5	15.5	15.5	15.5
Wear Resistance at 80°C (%) max	D1264	—	—	—	15	15	15
High Temperature Life (h) min	D3527	—	—	—	40	80	80
Leakage Tendency (g) max	D4290	—	—	—	24	10	10

TABLE 10.16—Formulation and application requirements.

Service	Requirements
High temperature	1. High-temperature thickener, high dropping point.
	2. Oil of high viscosity--petroleum or synthetic.
	3. Oil of high flash point--low evaporation.
	4. Firm consistency (preferred).
	5. Oxidation resistance.
Low temperature	1. Low percentage of thickener.
	2. Soft consistency.
	3. Oil of low viscosity.
	4. Oil of high viscosity index (preferred).
	5. Oil of low pour point (at very low temperatures).
	6. Rust resistance (water may condense).
	7. Good low-temperature torque.
	8. Good pumpability.
Wide temperature range	1. High-temperature thickener.
	2. Low percentage of thickener.
	3. Good low-temperature torque.
	4. Good pumpability.
	5. Low evaporation.
	6. Oil of low viscosity and high VI, preferably synthetic.
	7. Oil of low pour point.
	8. Oxidation resistance.
	9. Rust resistance.
	10. EP-antiwear often needed.
Exposure to water	1. Low washout.
	2. Low spray-off.
	3. Firm consistency.
	4. Water-resistant thickener High-viscosity oil.
	5. Viscosity booster (helpful, not required)-polymer, wax, asphalt, etc.
	6. Rust resistance.
Long pumping distances	1. Low apparent viscosity—good pumpability.
	2. Good slumpability or feedability.
	3. Low percentage of thickener.
	4. Soft consistency.
	5. Oil of low viscosity (preferred).
Heavy loads	1. Low wear test values.
	2. High EP test values.
	3. EP-antiwear additives.
	4. Solid additives, if indicated.
	5. Oil of high viscosity (preferred).
Contaminating environment	1. Frequent re-lubrication--good pumpability.
	2. Low-cost formulation.
	3. Firm consistency (for better sealing).
Multi-service	1. High-temperature thickener.
	2. Oil of medium viscosity.
	3. EP-antiwear usual.
	4. Oxidation resistance.
	5. Rust resistance.
	6. Acceptable pumpability.
	7. Water resistance.

conditions involve only light loads, seizure or welding does not occur. Correlation with the real service needs to be established. The other two tests measure severe wear and the welding tendency. Four-Ball EP results are reported as seizure load, weld load, and load wear index (LWI). It is important to be prudent in interpreting the results from this test since again they apply only to steel-on-steel and at one speed and one load. The extent of correlation of this test with the field performance is also uncertain. The Timken test can be used to measure wear at low loads as well as at high loads, to assess a grease's load-carrying capacity. The test measures two parameters, OK Load and Score Value, although usually OK Load is the only parameter reported. OK Load is the load at which scoring is not observed and Score Value is a load at which scoring occurs. The Timken test, like the other two, has poor precision. The results from this test do not coincide with those obtained in the Four-Ball tests since the speeds, loads, test duration, and geometry of contact are different. This is shown in Table 10.12.

Many bearings are used in applications that create oscillatory motion, which is small amplitude vibration, in rolling elements while contacting a bearing race. This leads to fretting wear, which is characterized by the removal of the fine particles from the surfaces in contact. A number of tests are designed to evaluate the grease's performance in this type of environment. The tests include the following:

TABLE 10.17—Properties of mineral oil-based grease with various thickeners.

Thickener	Dropping Point °C	Dropping Point °F	Maximum Temperature Continuous °C	Maximum Temperature Continuous °F	Maximum Temperature Peak °C	Maximum Temperature Peak °F	Low-temperature Limit °C	Low-temperature Limit °F	Water Resistance[a]	Load-carrying Capability[a]	Corrosion Protection
Soap Type											
Aluminum Complex	>230	>446	150	302	177	350	−30	−22	G-E	G	G-E
Barium Complex	>200	>392	150	302	177	350	−30	−22	E	G-E	E
Calcium	90	194	60	140	77	170	−20	−4	E	G	G-E
Calcium Complex	>250	>482	140	284	177	350	−30	−22	G-E	G-E	E
Lithium	180	356	121	250	150	302	−35	−31	F-G	F-G	G-E
Lithium Complex	>250	>482	150	302	177	350	−35	−31	G-E	G	E
Sodium	180	356	100	212	121	250	−20	−4	P-F	F-G	P-F
Sodium Complex	240	464	121	250	150	302	−30	−22	G	G	E
Nonsoap Type											
Polyurea	>225	>437	150	302	177	350	−20	−4	E	F	E
Inorganic (Clay)	>250	>482	150	302	177	350	−20	−4	E	F	G-E

[a]P=Poor; F=Fair; G=Good; E=Excellent.

1. ASTM D3704—Wear Preventive Properties of Lubricating Greases Using the (Falex) Block on Ring Test Machine in Oscillation Motion.
2. ASTM D4170—Fretting Wear Protection by Lubricating Greases.

Fretting failures commonly occur in wheel bearings of automobiles shipped across long distances before operation and in equipment stored for emergency service.

Grease Classifications

Greases are described by the National Lubricating Grease Institute (NLGI) Consistency Grades and NLGI Service Classification System, first implemented in 1991 [318,727]. As stated earlier, NLGI classifies the lubricating greases based upon consistency, which is described in the ASTM Standard D217. NLGI grades range from 000 to 6 and are based upon the Cone Penetration Test. The lower numbers are for softer greases and the higher numbers are for firmer greases; the Consistency Grade 2 greases, with the medium hardness, provide the line of demarcation. Each consistency number has a range of 30 points, and between numbers is a space of 15 points. The spacing was established with the consideration that the penetration repeatability is within five units. NLGI Grade 2 grease is the most commonly used grease and for use in central lubricating systems and at lower temperatures, greases of the lower consistency are more suitable. Heavier grade greases have limited use, except to feed grease to journal bearings in some paper mills and to lubricate high-speed bearings that operate at speeds of above 5000 r/min. Softer grades greases are used for good pumpability or low-temperature service. Firmer grades are used for certain high-speed bearings and to preclude leakage concerns. Table 10.11 shows ASTM laboratory testing procedures used for lubricating greases.

Table 10.1 summarizes greases based on NLGI consistency grades and the matching specifications for industrial use and Table 10.10 lists the lubricating greases based on the thickener type and suitable characteristics for various applications [744]. Table 10.13 provides the NLGI service classification for automotive use and Table 10.14 describes the chassis and wheel bearings greases according to ASTM D4950, published in 1989. LA and LB are classes for chassis greases and GA, GB, and GC are classes for wheel bearing greases. LA is for mild duty, frequent relubrication service and LB is for infrequent relubrication, high loads, and water exposure type service. GA is for mild duty, GB is for moderate duty, and GC is for severe duty service. Prior to this classification, the SAE recommended practice, published in SAE information report J310, was used for this purpose. The report, first introduced in 1951, had several revisions [745]. Table 10.15 summarizes the performance requirements of the various classes of the chassis and bearing greases.

Characteristics Of Modern Greases

Modern greases are based upon flexibility in formulating, preferably for use in many applications. In order to achieve this goal, the lubricating grease must be developed from thickeners, oils, and additives that have the ability to deliver optimal performance. Table 10.16 lists many of the requirements of the lubricating greases in relation to service [727]. We already mentioned that many of the grease properties are determined by the combination of the thickener, the base oil, and the additives; and the others are determined by the additives alone. Hence, the optimal grease formulation is possible only if each component possesses the appropriate characteristics to contribute towards the overall properties of the grease. Thickener influences the dropping point, low-temperature limit, maximum use temperature, water resistance, load-carrying capacity, and corrosion protection. Table 10.17 provides the effect of the common thickener types on these properties. Also see Grease Application Guide provided in NLGI Lubricating Grease Guide [727]. Base oil properties that influence the properties of grease include its viscosity, viscosity index (VI), pour point, and pumpability. For formulating the low-temperature greases, we need to use low viscosity oils because they possess better low-temperature performance than the high viscosity oils. On the other hand, too low an oil viscosity will not form lubricating film of proper thickness, resulting in extensive equipment wear.

For high-temperature greases, the viscosity index of the oil becomes important. A high VI oil is more suitable in such applications than a low VI oil since at high temperatures the former maintains its viscosity better; which translates into

TABLE 10.18—Inter-grease compatibility [746].

Grease	Aluminum Complex	Barium	Barium Complex	Calcium	Calcium 12-Hydroxystearate	Calcium Complex	Silica Gel	Lithium	Lithium 12-hydroxystearate	Lithium Complex	Sodium	Polyurea	Bentonite
Aluminum Complex		–	+	–	+	–	–	–	–	+	–	–	–
Barium	–		?	–	+	–	?	–	–	–	?	–	–
Barium Complex	+	?		+	+	+	+	+	+	+	+	?	?
Calcium	–	–	+		+	0	+	+	0	+	-	?	+
Calcium 12-Hydroxystearate	+	+	+	+		0	+	+	+	+	-	-	+
Calcium Complex	–	–	+	0	0		+	0	-	+	-	+	-
Silica Gel	–	?	+	+	+	+		+	+	+	+	+	?
Lithium	–	–	+	+	+	0	+		+	+	-	-	-
Lithium 12-hydroxystearate	–	–	+	0	+	-	+	+		+	-	-	-
Lithium Complex	+	–	+	+	+	+	+	+	+		-	-	-
Sodium	–	?	+	-	-	-	+	-	-	-		?	?
Polyurea	–	–	?	?	-	+	+	-	-	-	?		-
Bentonite	–	–	?	+	+	-	?	-	-	-	?	-	

Note: +Compatible; 0 Limited Compatibility; –Not Compatible.

better durability of the lubricating film. Similarly, for a heavily loaded bearing, where because of the high pressures the lubricating film is too thin to be effective, we must use oils of as high a viscosity as possible, as long as they do not interfere in pumping or dispensing of the grease. Paraffinic oils have better viscosity indices than naphthenic and aromatic oils. The VI of paraffinics is around 100 and of naphthenics is from 0 to 50; however a VI of between 25 and 50 is more reasonable for use in greases. Highly refined mineral oils, or synthetics, have viscosity indices of well over 100. Hence, with respect to the high temperature or the broad temperature applications, paraffinic and highly refined oils are better. Synthetics can also be used, if their high price does not pose a constraint. Most mineral oil-based greases, if they have a suitable dropping point, can be used in equipment that operates up to a temperature of 250°F (121°C). A few can also be used in applications with an ambient temperature of up to 350°F (177°C). Around or beyond this temperature, the synthetic fluids may be necessary. This is especially the case if relubrication of the machine part is sporadic.

While considering the use in low-temperature applications, pour point of an oil is an important criterion to consider. Paraffinic base oils are poorer in this property than naphthenics and aromatics; hence they are least suitable of the three for use in low-temperature applications. However, this deficiency may be corrected by the use of the pour point depressants.

For applications that require good load-carrying capacity, such as universal joints and needle bearings, naphthenics are better since they have a superior viscosity-pressure relationship. This helps in forming a more durable film at higher contact pressures.

Many times the greases need to be pumped into a reservoir, distribution system, an applicator, or directly into a machine part. If the lubricating grease is made from the high viscosity oils, it will have poor pumpability, or slumpability. This property depends upon the grease structure as well. If the grease is fibrous or stringy, it will be easier to feed into the application device. On the other hand, if the grease is buttery, the converse is true.

The rheological properties of the lubricating greases can be considered to relate to their viscosity, the same as in the case of the lubricating oils. However, it is the apparent viscosity and not the viscosity which is used as a surrogate measure of the rheological properties. The apparent viscosity depends upon the temperature, shear rate, shearing time, and the history of the sample [4]. The apparent viscosity decreases rapidly with increasing temperature, shear rate, and shearing time. This is because these parameters destroy the grease's network structure. The result is the reversal of the grease's viscosity loss to that of the base oil [4]. The mechanism by which the grease loses its structure is simply that these parameters overcome the weak van der Waals forces that are responsible for the formation of the soap-oil complex. Some lubricating greases, such as sodium soap greases, are more susceptible to shear-related viscosity loss than other greases, for example, the lithium soap greases. Sometimes the greases undergo unexpected consistency changes, such as hardening, softening, and the like, with temperature. These are related to alteration of the grease structure due to phase transformation or chemical reaction between the soap and the chemically reactive additives, or both.

Inter-Grease Compatibility

All greases are not compatible with each other. The incompatibility of the two greases is indicated by a deterioration of certain desired properties, such as structural integrity, consistency (worked penetration), oil separation, and the dropping point [4]. The incompatibility may occur either because of the soap or the additives used to formulate a grease. Compatibility or incompatibility is hard to predict

TABLE 10.19—Grease characteristics guide [727].

Properties	Sodium	Calcium (Convent.)	Calcium (Anhydrous)	Lithium	Aluminum Complex	Calcium Complex	Lithium Complex	Polyurea	Organo-Clay
Dropping point (°F)	325–350	205–220	275–290	350-400	500+	500+	500+	470	500+
Dropping point (°C)	163–177	96–104	135–143	177–204	260+	260+	260+	243	260+
Maximum usable temperature (°F)	250	200	230	275	350	350	350	350	350
Maximum usable temperature (°C)	121	93	110	135	177	177	177	177	177
Water resistance[a]	P-F	G-E	E	G	G-E	F-E	G-E	G-E	F-E
Work stability	F	F-G	G-E	G-E	G-E	F-G	G-E	P-G	F-G
Oxidation stability	P-G	P-F	F-E	F-E	F-E	P-G	F-E	G-E	G
Rust protection	G-E	P-E	P-E	P-E	G-E	F-E	F-E	F-E	G
Pumpability (in centralized systems)	P-F	G-E	F-E	F-E	F-G	P-F	G-E	G-E	G
Oil separation	F-G	P-G	G	G-E	G-E	G-E	G-E	G-E	G-E
Appearance	Smooth to fibrous	Smooth, buttery	Smooth, buttery	Smooth, buttery	Smooth, buttery	Smooth, buttery	Smooth, buttery	Smooth, buttery	Smooth, buttery
Other properties	Adhesive, cohesive	EP available	EP available	EP available, reversible	EP available, reversible	EP, antiwear inherent	EP available	EP available	
Production volume and trend	Declining	Declining	No change	The leader	Increasing	Declining	Increasing	Increasing	Declining

[a]P=poor; F=fair; G=good; E=excellent.

without carrying out the actual experimentation. However, based upon the previous experience, the following is true:

1. Greases containing the same type of soap are usually compatible.
2. Lithium soap greases are incompatible with sodium soap greases and the calcium soap greases are incompatible with calcium complex soap and sodium complex soap greases.
3. The addition of small quantities of calcium (lime) soap grease to the complex soap grease does not cause any changes in properties, but the addition of a small quantity of a calcium complex soap grease to a lime soap grease causes a change in the pH of the mixture which leads to significant structural change.
4. When soap greases are added to bentonite (nonsoap) greases, the exchange reaction takes place between the soap cations, the cations of the additives, and the quaternary ammonium ions of the clay. This results in the deterioration of the structure of the bentonite grease, resulting in its softening.

Table 10.18 provides intercompatibility of the mineral oil-based greases containing different thickeners [733,746].

Applications Involving Lubricating Greases

Greases are used to lubricate a number of machine parts in many industries. Common use applications include the following:

- Bearings (rolling element and plain)
- Gears (open and enclosed)
- Universal joints
- Chassis
- Electric motors
- Couplings
- Centralized lubricators
- Wipe ropes
- Chains
- Pipe threads as sealants
- Tool joints as thread compounds

Grease Selection

For lubricating grease to be effective, its properties must be matched with the lubrication needs of the equipment. See Table 10.19 for the grease selection guide [727]. Grease is the lubricant of choice when the equipment runs intermittently, is stored for long periods of time, experiences extreme operating conditions such as high temperatures, pressures, shock loads, slow speeds, or is severely worn. Lubricating greases are used in these types of equipment to perform the following functions:

1. Prevent wear
2. Reduce relubrication frequency
3. Act as a sealant
4. Provide rust and corrosion protection
5. Inhibit oxidation
6. Suspend or act as a reservoir for the solid additives
7. Protect elastomer seals
8. Reduce noise and vibration
9. Minimize leakage, dripping, and spattering

Each application has specialized lubrication needs which one must determine before selecting an appropriate lubricating grease. One must also assess the grease properties that will be necessary for a good match. These include consistency, penetration, dropping point, bleeding tendency, mechanical stability, and various other characteristics, such as oxidation resistance, volatility, and viscometrics of the oil component. The other needed properties can be imparted to grease by the use of the additives.

Bearing Lubrication

Industrial machinery has a number of components that are amenable to grease lubrication. These are bearings, couplings, open gears, and a variety of other moving parts. Bearings are extensively used in many industries to reduce fric-

TABLE 10.20—Important antifriction bearing greases and their properties [4].

Thickener	Base Oil	Working Temperature, °C[a]	Behavior Towards Water	Remarks
1. Sodium Soap	Mineral Oil	−20 to 100	Unstable	Emulsifies with water, can become liquid
2. Lithium Soap[b]	Mineral Oil	−20 to 130	Stable up to 90°C	Emulsifies with small quantities of water, softens with large quantities, multipurpose grease
3. Lithium Complex Soap	Mineral Oil	−30 to 150	Stable	Multi-purpose grease, highly temperature resistant
4. Calcium Soap	Mineral Oil	−20 to 50	Very stable	Good sealing against water, penetrated water is not picked up
5. Aluminum Soap	Mineral Oil	−20 to 70	Stable	Good sealing against water
6. Sodium Complex Soap	Mineral Oil	−20 to 130	Stable up to ~40°C	Suited for high temperatures and loads
7. Calcium Complex Soap[b]	Mineral Oil	−20 to 130	Very stable	Multi-purpose greases suited for high temperatures and loads
8. Barium Complex Soap[b]	Mineral Oil	−20 to 150	Stable	Suited for high temperatures, loads, and speeds (depending upon base oils viscosity), steam resistant
9. Polyurea[b]	Mineral Oil	−20 to 150	Stable	Suited for high temperatures, loads, and speeds
10. Aluminum Complex Soap[b]	Mineral Oil	−20 to 150	Stable	Suited for high temperatures, loads, and speeds (depending upon base oils viscosity)
11. Organ-clay (Bentonite)	Mineral Oil or Synthetic Ester Fluid	−20 to 150	Stable	Gel grease, suited for high temperatures at low speeds
12. Lithium Soap[b]	Synthetic Ester Fluid	−60 to 130	Stable	Suited for low temperatures and high speeds
13. Lithium Complex Soap	Synthetic Ester Fluid	−50 to 220	Stable	Multi-grade greases for wide temperature range
14. Barium Complex Soap[b]	Synthetic Ester Fluid	−60 to 130	Stable	Suited for high speeds and low temperatures, steam resistant
15. Lithium Soap[b]	Silicone Fluid	−40 to 170	Very stable	Suited for high and low temperatures at low loads and medium speeds

[a]Depends on type of bearing and lubrication period. Cold properties of greases 1–10 can be improved by appropriate selection of mineral base oils (e.g., to −30°C, in special cases to −55°C.
[b]May contain EP additives.

tion, for converting the linear motion into the rotary motion. Such industries include steel mills, mining, construction, and transportation. Because of their design and use configurations, the use a liquid lubricant in these machine elements is not suitable, but the use of the lubricating grease is.

Bearings are of two basic types: plain bearings and rolling-element bearings. Plain bearings are based on sliding motion between a stationary and a moving element; rolling-element bearings have either balls or rollers which separate motion between the stationary part and the moving part. In either case, a film of lubricant separating the moving surfaces is essential for a long service life. As a general guideline, the grease dropping point should be about 27°C (80°F) above the bearing temperature, so that the grease does not liquefy. Also, the base oil used to make such a grease must be selected by taking into consideration the bearing's operating temperatures. If the temperatures are too high, the base oil evaporation rate may be too high for the grease to be suitable for long term use.

Plain Bearings

A plain bearing is the most basic type of bearing and contains no moving parts. The bearing is held in a stationary machine element and consists of a sleeve or a bushing and a moving part. The bushing is made of a material, or an alloy, that is softer than that of the part that slides or moves against it. Examples of metals that are suitable as bearing material include bronze and Babbitt metal, which is an alloy of tin, lead, copper, and antimony. The advantage is that in this arrangement the bearing assumes most of the wear. This is an important economic advantage because the bearings are more conveniently replaced or adjusted than the relatively inaccessible moving components. Plain bearings can be described by their configuration, by their motion, or by the type of loading they accommodate. Three major categories of plain bearings are: Journal, guide, and thrust bearings. Details to their design and functions are available elsewhere and will not be discussed here [744].

Plain bearings can be divided into three main designs: hydrodynamic plain bearings, hydrostatic plain bearings, and maintenance-free plain bearings. The lubrication needs of each type are different. For example, for the hydrodynamic type, if the conditions are not suitable to create a fluid film that will separate the journal and the bearing surface, the bearing is in the boundary or mixed-film lubrication regime. In these regimes, the surfaces are not fully separated and have some metal-to-metal contact. This situation commonly occurs when the shaft is at rest or moving at slow speeds, which typically occurs at start-up and at high loads. This means that the lubricating grease needs to provide the wear protection. This can be achieved by improving the lubricant's film-forming ability, either by adjusting the grease viscosity or by using surface-active or EP additives. Obviously, as the speed increases, the boundary lubrication progressively changes first to mixed-film and then to hydrodynamic lubrication, and the need for the film-forming additives is not as critical. In the case of the hydrostatic plain bearings, the need for the lubricating grease does not exist since these are lubricated with oil, which is continuously supplied under pressure to the bearing. In the case of the maintenance-free plain bearings, again the need for lubrication does not exist since the surface of the bushing has a coated layer of a polymer, such as polytetrafluoroethylene (PTFE); a low friction material.

Antifriction Bearings

Most antifriction bearings are lubricated with greases of NLGI Consistency Grades 1, 2, and 3. See Table 10.1 for the recommended applications for the lubricating greases of various consistencies. Table 10.20 lists various greases that are used for lubricating antifriction bearings, along with their properties [4]. The type of grease is chosen by considering a number service and operation-related factors. These include the service life, speed, working temperature, position of the bearings, starting torque, and bearing sealing. In antifriction bearings, the greases are exposed to mechanical, thermal, and chemical stresses which can alter the grease's structure over time. One thing that commonly occurs is the loss of oil through the process of bleeding. Bleeding occurs when the liquid lubricant separates from the thickener due to the grease's exposure to high temperatures, a consequence of the inability of the grease to effectively dissipate heat. When the loss of oil from the grease due to bleeding is 50 %, the grease is beyond its useful life [4]. This class of bearings includes rolling-element bearings and ball bearings, both of which operate on the concept of the rolling friction.

Rolling-Element Bearings

These bearings consist of rollers, a retainer, races, a shaft, a bearing housing, and seals. The contact surface of these bearings has a fine surface finish, which increases their efficiency. Also, these bearings are made of hard steel alloys because their small rolling elements must carry a wide range of loads, which makes the stresses on the contact surfaces very high. These bearings may contain one or more rows of rollers, which are either cylindrical, spherical, or tapered (conical). When the rollers are long and of small diameter, bearings are called needle bearings. Such bearings have a higher load-carrying capacity than the ball bearings of the similar size. They overcome frictional resistance by a rolling contact and are suited to large, heavy assemblies. Roller bearings are more efficient in reducing friction than sliding bearings, have the ability to operate at high speeds, and are easier to lubricate.

Cylindrical roller bearings, the most basic type of rolling element bearings, are designed to carry heavy radial loads, and can operate at high speeds. Tapered roller bearings, because of the shape of their rollers and the race geometry, have the ability to carry both heavy radial and thrust loads, making them especially useful for use in the automotive applications. Spherical roller bearings use convex or barrel-shaped rollers, which accommodate high radial and shock loads but they are limited on speed. Needle bearings, because of containing the cylindrical rollers with a high length-to-diameter ratio, have the highest load-carrying capacity among the rolling-element bearings. These bearings are designed to accommodate the oscillating motion and have high load-carrying capacity, but again they are speed limited. Needle bearings that contain fewer rollers handle lower load capacity, but have the advantage of operating at higher speeds. These bearings encounter high rolling and sliding friction, hence the grease used to lubricate them must offer protection against the wear damage. Additional requirements that must be met by the lubricating grease are preservation of the surface finish against corrosion and pitting and to act as a sealant.

Ball Bearings

Ball bearings are machine components which comprise an outer ring, an inner ring, balls, retainers, shields, and snap rings. The balls are made of hardened steel, ground to a true sphere, and polished to a fine finish. These bearings are normally used in light precision machinery where high speeds are maintained. Friction encountered is minimal because of the rolling action of the hard steel balls as well as the small contact area. Radial ball bearings are functionally similar to plain journal bearings and the thrust ball bearings are functionally similar to plain thrust bearings. Bearing housings differ in design, depending on the application, and serve to support the bearing and contain the lubricant. Suitable seals are usually installed to exclude water, dust, dirt, or other external contaminants from the bearing components and to prevent leakage of the lubricant from the housing.

Gear Lubrication

Greases used to lubricate gears are fluid greases that have NLGI Consistency Grades of 1 or lower. These greases are usually made from the sodium soaps with long fibers, a high viscosity oil, and the EP additives. All three components contribute towards the load-carrying properties of the grease, which are necessary to avert wear damage and welding that commonly occurs in gears as a result of the extensive metal-to-metal contact. EP greases, if they can be used, offer a number of advantages over gear oils. These include good sealing, low losses due to leakage, and good adhesion, which is clearly beneficial during the start-up where the boundary conditions prevail and the presence of an effective lubricant film between surfaces is critical. It is important to note that soft greases are only suitable for lubricating slow running, poorly sealed gears, such as those used in cranes, geared engines, and nonstationary drives in the mining industry. For equipment that uses toothed gears, such as drills and small tools, soft greases are not adequate and the greases of NLGI Consistency Grade of at least 2 are needed. Again, because of the superior load-carrying capacity, the sodium soap greases containing long fibers are the lubricants of choice. They are sometimes applied to the gear teeth via a spray. Because of their less fluid nature they have better adhesion to the tooth surfaces. Incidentally, the selection of an appropriate gear grease requires the same considerations as the selection of a good gear oil.

Automotive Aftermarket

For automotive applications, lubricating greases of NLGI Consistency Grade 2 are used most often. This is because these greases stay soft in cold weather. Common automobile parts that are lubricated by grease include chassis, power train, and wheel bearings. These greases have the following attributes:

- High temperature stability.

- Shear stability—resist breakdown due to mechanical action.
- Water resistance—do not easily wash out.
- Oxidation resistance.
- Rust and wear protection, especially against fretting wear.
- Easy to handle.
- Universal use—ability to lubricate all mechanical parts in a vehicle. For lubricating wheel bearings in cars with disk brakes, greases with thermal stability of 350°F (177°C) or higher, are required.

Other properties shown in Table 10.15 must be matched with the needs of the application, to select the appropriate grease. For use in the heavily loaded trucks, high load-carrying capability is also needed. Original Equipment Manufacturers' (OEMs) recommendations can also be helpful in selecting proper grease. For example, for wheel bearings, one builder uses lithium grease containing polyethylene and molybdenum disulfide, while another prefers lithium complex soap grease instead.

Primary Metals—Steel Mills

Major steel producers use at least 15 or more different greases. They specify their own compositions and performance for lubricating greases that suit their needs. In steel mills, the EP multi-purpose greases based upon lithium and aluminum complex soaps are used in the highest volume. These are used to lubricate machinery with typical operating temperatures of less than 135°C (275°F). Depending upon the plant size, the greases are pumped over a long distance and applied by the use of complex dispensing systems. Obviously, this requires good pumpability. Since copious amount of water is employed around the bearings for cooling or the scale removal, the lubricating greases must be water resistant. In addition, the greases used in steel mills must possess high load-carrying capacity and be inexpensive.

The steel industry also uses a variety of bearings that are lubricated with grease. These include plain journal bearings, especially in rolling mill operations, rolling-element bearings, and table and roll neck bearings. Rolling-element bearings are more effective than the plain journal bearings in transmitting power and withstanding loads, hence they are used in the newer mills. They are also used in cranes, unloaders, conveyor belts, and furnaces. Some steel mills have separate lubrication systems for bearings operating at lower temperatures. In these systems calcium soap greases, which are cheaper, are sometimes used to minimize cost.

Food Processing

Lubricating greases are also used in machinery employed in food-related industries, such as bakeries, breweries, dairies, fruit and vegetable packaging, soft drink canning, and metal can manufacturing. The United States Department of Agriculture (USDA) regulates "nonfood compounds," such as lubricants, that come in contact with the food. One needs to obtain permission from the Food Ingredient Assessment Division (FIAD) for use in such plants. Lubricants authorized for the incidental food contact are identified in Publication 1419 as Category Code H1. Category Code H2 is for lubricants used to lubricate parts that have no food contact. Food grade lubricants are regular commercial products that contain no poisonous ingredients. For approval as Category H1 lubricants, they must be developed from the ingredients that are deemed safe. These are identified in another USDA publication, AH562, under "Lubricants." This publication identifies a number of thickeners that are safe to use in food grade greases. The thickeners include aluminum stearate, aluminum complex (aluminum stearoyl, benzoyl hydroxide), organo-clay (dialkyldimethylammonium aluminum silicate), and polyurea. Additives that can be used to formulate lubricating greases are also identified. These must be used in the least amount needed to fulfill the performance objective sought. The base fluid must be pure as is defined in the Code of Federal Regulation (CFR) under Title 21, Part 178, Section 178.3620(a) or (b). The oil is specified by color and ultraviolet absorbance. Commercial products meeting H1 requirements are often marketed as "food industry" greases or "food machinery" greases. Please note that while the USDA regulates food grade lubricants, the National Science Foundation (NSF) oversees the approval process and maintains records. Liquid food grade lubricants were discussed in Chapter 9 on Miscellaneous Industrial Lubricants.

Textiles

The textile industry primarily uses multi-purpose greases. However, because of the unique needs of this industry, the same as others, the customized products are preferred. Due to a concern for staining by the grease, the light-colored greases that are easier to clean and leave no stains are favored. Since humidity in the textile plants is kept high to minimize the formation of the static charges, the greases with rust-preventive properties are highly desired.

Additional properties for greases used in the textile industry include load-carrying capacity, oxidation resistance, and good adhesion. The last requirement is specifically for the semi-fluid, NLGI Consistency Grade 000 grease, which is used as a loom lubricant. This lubricant has the additional requirements of having low resistance to flow and develop low torque; the latter requirement is to save on energy cost. Greases used for dryers require products of high dropping point because of the high temperatures in their immediate vicinity.

Grease Storage

Greases during storage can suffer from the problems of oxidation, bleeding, contamination, change in appearance and texture, and the loss of consistency. The degree of change of the grease properties depends upon the duration of storage, temperature, and the nature of the grease. Lubricating grease is a thixotropic fluid and like all other fluids of this type shows a time dependent change in viscosity, or the consistency. They lose their consistency (soften) under the influence of shear and when the shear forces are removed they see a reversal *towards* the original consistency. If the duration of the grease being subjected to the shear forces is short, the reversal to the original consistency is easy. However, the longer the grease is under the influence of shear, the harder the reversal becomes when the shear forces are removed. This is commonly referred to as age-hardening. However, all greases do not harden to the same degree once the shear forces are removed. The degree of age-hardening after work softening varies from grease to grease. Some greases show a higher degree of thixotropy than others, meaning that they become firmer under long-term storage. Another long-term storage effect is a greater tendency of the grease to lose oil, or bleeding, which gets further aggravated by high temperatures. Contaminants, such as moisture, can deteriorate the lubricating grease. Sodium soap greases that have low water tolerance will become soft in the presence of moisture. Conversely, clay and calcium complex soap greases become firmer. Water contamination can also make some lubricating greases to look hazy. While the oxidation of greases under storage is not a common problem, but when it occurs, darkening of the grease surface may result.

Handling and Disposal of Used/ Waste-Greases

Proper handling and disposal of the used or waste greases and other lubricants are covered by the United States Federal laws, through the guidelines issued by the Environmental Protection Agency, and by the state laws. Currently, the final composition of the used or waste greases determines whether it is a hazardous material or not. The four characteristics of ignitability, corrosivity, reactivity, and toxicity are used by the EPA as criteria for determining the hazard status of a waste material. The definition of these terms is available elsewhere.

Environmentally Compatible Greases

These are greases that combine the high-performance properties with the environmental safety and compliance. These greases are used both for civilian industrial and automotive applications as well in military applications. The main objective of these greases is to minimize their negative impact on the environment. This implies no harm or damage to water, soil, or life. One raw material that helps in this regard is vegetable oil. It can be used both to make soap as well as the base oil component in grease. Vegetable oils, because of being natural products, are inherently biodegradable. Two of the problems with these oils and the derived thickeners are that they have polyunsaturation and have high pour points. These make the greases made from them to have poor oxidation stability and poor low temperature properties. One can improve the properties of the greases by replacing the vegetable base oil with largely paraffinic mineral base stocks and some synthetic fluids; many of which have good to acceptable biodegradability. A vegetable oil of high oleic content, that is, 75–85 % mono-unsaturation, is commercially available. Greases derived from the use of this base oil have better properties than those made by the use of the regular vegetable oils. Biodegradable greases from clay, polyurea, and aluminum complex soaps have also been developed, but they all suffer from performance and cost limitations.

If one wants to design a lubricating grease which is environmentally compatible, one must also be concerned with the toxicity and biodegradability of the third component of the grease, the additive package. In some instances, toxicity of the additives is related to the presence of metals and the sulfur and phosphorus. Replacement of such additives with metal-free (ash-less) additives will reduce pollution. Nevertheless, identifying the biodegradable additives for use in these products is a challenge. Environmentally compatible grease technology is used in the following industries:

1. Waste water purification plants
2. Food processing machinery
3. Automobiles (chassis and wheel bearing lubricants)
4. Railroads (curve and flange greases)
5. Mining (open-gear greases)
6. Steel (rolling mill lubricants)
7. Agriculture (cotton picker-spindle lubricants)
8. Construction (dipper-stick lubricants)
9. Marine and nature preserves (construction and excavating equipment lubricants)
10. Forestry (cam lubricants)

At present, the biodegradability of the lubricant greases is tested by the use of the OECD 301B Modified Sturum, ASTM D5864, and CEC L-33-T-82 tests, which are designed to test lubricants in general. The CEC L-33-T-82 test is now listed as CEC L-33-A-94 test. In the United States, the ASTM committee has adopted the OECD 301 Modified Sturum procedure within the ASTM D5864 test. The highest level of biodegradability is attained when the amount of carbon dioxide (CO_2) evolved is 60 %, or greater, of the theoretical amount [747]. The pass criterion in the OECD 301 Modified Sturum Test is 70 % removal of dissolved organic carbon (DOC) and 60 % of theoretical oxygen demand (THOD). Also consult Chapter 13 on Lubricants and the Environment.

While biodegradability of is the primary consideration in environmental compatibility, by far it is not the only one. There are other actions which can be taken to minimize the environmental impact. These include the following:

- Lowering toxicity of the finished products by the use of the environmentally innocuous chemicals.
- Curb the use of the heavy metals, such as lead, bismuth, and antimony.
- Minimize or eliminate the use of chlorine to improve the EP performance. Chlorine leads to the formation of the carcinogenic dioxins.
- Sealed-for-life applications to reduce the amount of grease used.
- Greater use of synthetic greases, such as those based upon polymers. They will have a lower tendency to wash off or leak.
- Use leakage-resistant greases for surface mining applications.

Grease Testing Requirements

As mentioned earlier, the properties of the lubricating grease must be matched with the needs of the equipment to be lubricated. The producer or the marketer of the lubricating grease must find a way to let the user know the suitability of the lubricating grease for a particular application. This is done by meeting the performance or use specifications. There are many national and international standards that define the grease quality. The standardized test specifications provide a means to determine and verify the lubricating grease performance. Typical automotive, industrial, and military grease specifications are available elsewhere [4,744]. These references also includes common international (European and Japanese) grease specifications and standards, such as DIN 51 502 and DIN 51 825. Besides these, there are additional specifications for use in automotive application as well as in military equipment.

TABLE 10.21—Index to grease tests.

Characteristic	Test Specification	IP[a]	Common Designation
Bleed Resistance	FTM 321.3		Oil Separation (Static)
	ASTM D1742		Pressures Oil separation
Corrosion	ASTM D1743	220	Rust Test
	ASTM D5969		Rust Test
	EMCOR (D6138)		Steel Corrosion
	ASTM D4048		Copper Corrosion
Extreme Pressure/ Antiwear	ASTM D2596, D2783		Four-Ball
	ASTM D2509	326	Timken Method
	ASTM D2266	239	Four-Ball Wear
	ASTM D3233		Falex Test
	Optimal SRV		Oscillation
Dropping Point	ASTM D2265, D566		...
Oxidation Resistance	ASTM D942	142	Dropping Point
	ASTM D3527		Wheel Bearing Life
	ASTM D3336		High-Temp. Performance
	DIN-51806		Roller Bearings
	SKF RDF		High Temps. & Speeds
Apparent Viscosity	ASTM D1092		At 16 Shear Rates
Pumpability	ASTM D4693, D1478		Low-Temperature Torque
	U.S. Steel LT37		Mobility
Shear (mechanical) Stability	ASTM D217, D1403	50	Multi-stroke Penetration
	ASTM D1831		Roll Stability
	ASTM D4290		...
Leakage	ASTM D1263		Wheel Bearing Leakage
Evaporation	ASTM D972, D2595		Evaporation Loss
Water Resistance	ASTM D1264		Water Washout
	ASTM D4049	215	Water Spray-Off
Constant Velocity Joints	CVJ		CVJ Unit Performance

[a]Institute of Petroleum (UK) designation.

Major organizations that are responsible for developing specifications or standards for the lubricating greases include NLGI, ASTM, IP (The Institute of Petroleum), AFNOR (Association Française de Normalisation), and TGL (state standards of the GDR). Each specification is accompanied by a number of tests that must be qualified to assure grease's performance in that application. The tests that accompany these standards are numerous, some of which are standardized but the others are not and are based on consumer feedback. Nonstandardized tests are carried out to evaluate the grease performance in a specific application or in specific operating environments. While some of the tests are carried out under laboratory conditions, others must be carried out in the real use environment, that is, in the field. Test methods are designed to properly assess the appropriate property or properties. The two major organizations that are instrumental in designing and standardizing tests for the lubricating greases are ASTM and IP. These tests measure a number of performance parameters of grease, including flow properties, heat resistance, oxidation stability, antiwear/EP performance, and corrosion control. Most of the tests for greases are the standardized ASTM tests, the details of which are described in the books of ASTM Standards [27]. Lists of the common grease tests used in the United States are provided

TABLE 10.22—Simulated operational tests for automotive greases.

Evaluation	ASTM Method	Comments
Ball Joints	D3428	Measures a chassis grease's ability to prevent wear, stick-slip, and noise in an automotive ball joint
Wheel Bearing Leakage	D1741	Measures the leakage tendency of wheel bearing greases at a given temperature
Wheel Bearing Life	D3527	Measures the life performance characteristics of automotive wheel bearing greases
High-temperature Ball Bearing Life	D3336	Measures the life performance characteristics of greases in bearings operating at high temperatures
Small Bearing Life	D3337	Measures the life performance characteristics of greases in small, high-speed bearings
Rubber Swell	Fed. 791a 3603.4	Measure the rubber swelling tendencies of automotive chassis greases

TABLE 10.23—Guide to principal European grease standards [748].

No.	Characteristic / Property	ISO	DIN	IP	NF-T 60	ASTM D	FTM 791 b	GOST
1	Determination of Dropping Point	2176	51 801	132	102	566	1421	6793
2	Determination of cone Penetration 1/1 – ½ - ¼ Cone	2137	51 804/1-2	50	132	217	311	5346
3	Determination of Flow Pressure		51 805					
4	Corrosion Preventing Properties		51 802	220	135	1743	40001.2	5757
5	Corrosive Effects on Copper	2160	51 811	112		1261	5309.4	
6	Determination of ash of Greases (incl. Sulfate)		51 803	5	M 07-037	128		1461 6474
7	Neutralization Number		51 809/1-2	37/137/ 139	133 112	128 974		
8	Content of Solid Foreign Matters		51 813	134		(1404)	3005.3	6370
9	Effect of Water							
	9.1 Water Washout Test		51 807/2	215		1264		
	9.2 Static Test		51 807/1					
10	Oxidation Stability		51 808	142		942	3453	5734
11	Oil Separation		51 817	121		1742	321/2	7142 1631
12	Content of Base Oil and Soap		51 814			128		
13	Content of Li/Na/Ca by Atomic Absorption Spectroscopy		51 815		M 07-38			
14	Determination of Li/Na by Flame Photometer			199		3340		
15	Determination of Solids (Graphite or Molybdenum Disulfide)		51 831				3720/22	
16	Determination of Particle Size of Solid Lubricants		51 832					9270
17	Density			59				
18	Evaporation Loss			183		972		9566
19	Content of Water	3733		74	113	95		1044
20	Pumpability Properties							2077
	20.1 Shell-DeLimon Rheometer		51 816/1					
	20.2 Decompression Characteristics		51 816/2					
	20.3 Apparent Viscosity				139	1092		
21	Roll Stability					1831		
22	Extreme Pressure Properties							
	22.1 Shell Four-Ball Test		51 350-4/5	239		2266	6503.2	
	22.2 Timken Test			326		2509		
23	Mechanical Dynamic Testing							
	23.1 SKF R2F Method		51 806				331.2/333.1	
	23.2 Rolling Bearing Performance Test				168			
	23.3 Churning Test				266			
	23.4 FAG FE 8 (Extreme Pressure Greases)		51 821/2					
	23.5 FAG FE 9		51 821/1					
	23.6 Wheel Bearing Leakage Test					1263		
	23.7 FAG KSM Test							
24	Low Temperature Torque				186	1478		
25	Elastomer Compatibility							
	25.1 Change of Hardness (Shore A)	R 868	53 505					
	25.2 Change of Volume	1817	53 521				3603.3	
26	Tests on Base Oil							
	26.1 Viscosity	3104	51 562	71		445		
	26.2 Pour Point	3016	51 597	15		97		
	26.3 Flash Point	2592	51 376	36		92		
	26.4 Aniline Point	2977	51 775/ 51 787	2		611		
27	Infrared Analysis		51 820 E					

in Tables 10.21 and 10.22 and the list for the tests used in Europe is given in Table 10.23 [748]. Please note that the lists are representative and not exhaustive. The properties tested are briefly described below [748].

Flow Properties

As mentioned earlier, the consistency of grease is one of the most important properties of the lubricating grease that determines its performance in various applications. The limitations of the NLGI consistency scale is that it reflects performance only at 25°C (77°F) but most applications involve temperatures other than 25°C. The effect of temperature on consistency can occur due to phase changes in the thickener or the degree of its association with the oil. Similarly, mechanical stresses in most real use applications differ from those simulated by the worked penetration. These differences must be recognized, so that one can predict the performance of the grease at different temperatures and at different operating conditions. There are methods that determine the flow measurements under different conditions than those used in the ASTM D217 test. These methods are listed below. The details of these test methods are available in the appropriate ASTM Standards:

1. ASTM D3232, Test Method for Measurement of Flow Properties of Lubricating Greases at High Temperatures.
2. ASTM D1092, Test Method for Apparent Viscosity of Lubricating Greases. Since apparent viscosity varies with the temperature and the shear rate, the specific temperature and shear rate must be reported along with the measured viscosity.
3. ASTM D1478, Test Method for Low-Temperature Torque for Ball Bearing Greases, measures the starting and running torque of lubricating greases packed in small ball bearings at temperatures as low as -54°C (-65°F).
4. ASTM D4693, Test Method for Low-temperature Torque of Grease Lubricated Wheel Bearings. This method is better suited to test low-temperature flow properties than the ASTM D1478 Method.

Heat Resistance

Heat affects the lubricating grease in a number of ways. It increases its rate of oxidation, causes it to lose oil due to volatilization, causes melting of the thickener, and lowers the thickener's association with oil. The tests that are used to evaluate these events are provided below:

1. Dropping Point—Two procedures are used to determine the dropping point.
   ASTM D566 (IP 132), Test Method for Dropping Point of Lubricating Grease and ASTM D2265, Test Method for Dropping Point of Lubricating Grease Over Broad Temperature Range. The results from the two methods concur up to about 260°C (500°F).
2. Evaporation Loss—Two ASTM test methods are used to measure this grease parameter. ASTM D972, Evaporation Loss of Greases and Oils, determines mass evaporative losses at a temperature in the range of l00–150°C (210–300°F). ASTM D2595, Test Method for Evaporation Loss of Lubricating Greases over Broad Temperature Range, determines the evaporative loss over a temperature range of 93–316°C (200–600°F).
3. Oil Separation—Oil separation, or bleeding, in storage is a common occurrence in greases, but they differ from grease to grease. Too much oil separation will harden the grease, which will impair its ability to lubricate effectively. On the other hand, greases that do not separate some oil during operation can be noisy in service. Bleeding in grease is a function of the gel structure, the nature and the viscosity of the lubricating fluid, and the applied pressure and temperature.
   Oil Separation (Static Test)—ASTM D1742, Test Method for Oil Separation from Lubricating Greases is used to determine the bleeding tendency of the grease to simulate oil loss during storage.
   Oil Separation (Centrifuge Test)—ASTM D4425 describes a procedure for determining the tendency of the lubricating grease to separate oil when subjected to high centrifugal forces. This test is used to assess grease performance in shaft couplings, universal joints, and rolling element thrust bearings, which are subjected to large or prolonged centrifugal forces.
4. Leakage from Wheel Bearings—Two tests are used to determine leakage of grease from wheel bearings at high temperatures.
   ASTM D1263, Test Method for Leakage Tendencies of Automotive Wheel Bearing Greases, an old test, utilizes a modified automotive front hub assembly (1940s vintage design and bearings). ASTM Test D1263, the new test, differentiates grease products with differing leakage characteristics.
5. Accelerated Leakage from Wheel Bearings—ASTM D4290, Test Method for Determining Leakage Tendencies of Automotive Wheel Bearing Grease under Accelerated Conditions, is a modern test that uses a model front wheel-hub-spindle assembly employing current production, tapered roller bearings.

Oxidation Stability

Oil, additives, and the thickener, if it is organic, all are susceptible to oxidation. The rate of oxidation depends upon a number of factors; those worth mentioning include the oil characteristics, whether it is paraffinic, naphthenic, or aromatic; the soap, whether it contains unsaturation; high temperature; and the presence of metals. A number of tests are used to assess the oxidation stability of greases. Some of these are listed below:

1. Bomb Oxidation Test—The Standard Test Method for Oxidation Stability of Lubricating Greases by the Oxygen Pressure Vessel Method, ASTM D942 (IP142), determines the resistance of the lubricating greases to oxidation when stored statically in an oxygen atmosphere in a sealed system at an elevated temperature. This test suffers from many deficiencies; the lack of correlation with field performance is at the top of the list. At present, there are no standard, dynamic oxidation tests.
2. PDSC Oxidation Test—A recently developed grease test uses Pressure Differential Scanning Calorimetry (PDSC) to evaluate oxidation stability of grease.
3. Greases in Ball Bearings at Elevated Temperatures—ASTM D3336, Test Method for Performance Characteristics of Lubricating Greases in Ball Bearings at El-

evated Temperatures, is used to evaluate the performance of the lubricating greases in ball bearings, operating under light loads at high speeds and elevated temperatures for extended periods.

4. Greases in Small Bearings—ASTM D3337, Test Method for Evaluation of Greases in Small Bearings, such as those used in computer and aircraft industries. This test involves temperatures in the 250°C to 315°C (500°F to 600°F) range.
5. Wheel Bearing Grease Life—ASTM D3527, Standard Test Method for Life Performance of Automotive Wheel Bearing Greases, evaluates grease life in tapered roller, wheel bearings. The normal test temperature is 160°C.

Extreme Pressure and Wear

A lubricant's main function is to lubricate. This a lubricant achieves by making a lubricant film between surfaces, thereby minimizing the metal-to-metal contact. Unfortunately, perfect lubricating film formation, referred to as a hydrodynamic film, is not always possible, either through design or because of the speeds and loads. The result is metal-to-metal contact, which must be controlled to avoid wear damage to the surfaces. In many cases, this problem is overcome by the use of an oil with good viscometrics and or by the use of friction and wear control additives, such as friction modifiers, antiwear agents, and extreme-pressure (EP) agents. The tests listed below are used to evaluate the antiwear and load-carrying properties of lubricating greases:

1. Extreme Pressure Timken Method—ASTM D2509, Test Method for Measurement of Extreme Pressure Properties of Lubricating Grease (Timken Method), can be used to determine the load carrying capacity of grease at high loads.
2. Extreme Pressure Four-Ball Test—ASTM D2596, Test Method for Measurement of Extreme Pressure Properties of Lubricating Grease (Four-Ball Method), is another test used to determine the load-carrying properties of the lubricating greases.
3. Wear Preventive Characteristics of Grease—ASTM D2266, Test Method for Wear Preventive Characteristics of Lubricating Grease (Four-Ball Method), is used to determine the wear-preventive characteristics of greases in sliding steel-on-steel applications.
4. Fretting Wear—ASTM D4170, Test Method for Fretting Wear Protection by Lubricating Grease, evaluates grease performance in a proprietary test machine (Fafnir Friction Oxidation Tester). Fretting wear is degradation of the two contacting surfaces due to repeated sliding or by vibratory or oscillatory motion of small amplitude. Fretting wear damage involves removal of small particles from the contacting surfaces.
5. Oscillating Motion—ASTM D3704, Test Method for Wear Preventive Properties of Lubricating Greases, is another wear test involving oscillatory motion.
6. Oscillating Wear (SRV) Tests: ASTM D5706 and ASTM D5707—Both procedures are based upon ball-on-disk configuration.
7. ASTM D5706—Standard Test Method for Measuring Wear Properties of Lubricating Greases using High Frequency Linear-Oscillating Test Machine (EP), measures wear-protection qualities and coefficient of friction in a grease.
8. ASTM D5707—Standard Test Method for Determining Wear Properties of Lubricating Greases using High Frequency Linear-Oscillating Test Machine (Wear), measures load-carrying properties of lubricating grease.

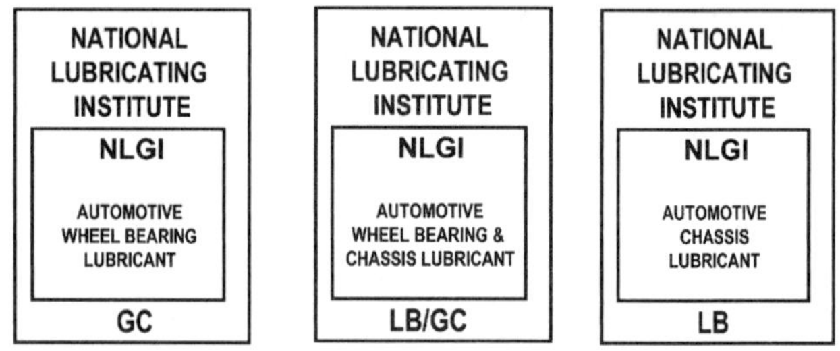

Fig. 10.16—NLGI automotive grease symbols [727].

Corrosion

Corrosion is the oxidation of metal, either by air/water/electrolytes or by chemically reactive acidic materials and nonmetals, such as sulfur. Corrosion to nonferrous metal surfaces, such as copper and bronze, is referred to as yellow metal corrosion; and corrosion to ferrous metal surfaces is called rusting. Many equipment parts contain copper or copper alloys, such as bronze, which are susceptible to corrosion. Greases are formulated to be noncorrosive to metals as well as protect them against corrosion by the harmful degradation products resulting from grease during service. The tests listed below are used to assess the corrosion protecting ability of the lubricating grease:

1. Copper Corrosion—ASTM D4048, Test Method for Detection of Copper Corrosion from Lubricating Grease by the Copper Strip Tarnish Test. The method is analogous to ASTM D130 used to evaluate oils.
2. Rust Protection—ASTM D1743, Test Method for Corrosion Preventive Properties of Lubricating Greases. This test uses distilled water and is useful in assessing rust and corrosion properties of the grease under static or storage conditions.
3. Accelerated Corrosion Tests—Two test methods are used to determine corrosion protection properties of greases under severe conditions. ASTM D5969, a version of D1743 using synthetic sea water and two procedures of the IP 220/DIN 51 802 dynamic rust test. This test, commonly known as the EMCOR test, is adopted as ASTM D6138 and uses distilled water, synthetic sea water, or sodium chloride solution [749].

Seal Compatibility

Good seal and elastomer compatibility in lubricating grease are important so that seal damage does not occur. The test listed below is used to determine this grease parameter:

1. ASTM D4289—Test Method for Compatibility of Lubricating Grease with Elastomers. In this test elastomer specimens cut from standard ASTM sheets (D3182) are immersed in test grease for 70 h at either 100 or 150°C and the changes in volume and Durometer A hardness (D 2240) are measured. The results are used to judge service characteristic of lubricating greases. ASTM D4950

TABLE 10.24—Properties of synthetic greases [733].

Soap	Base Fluid	Dropping Point (°C)	Operating Temperature Range (°C)		Water Resistance	Corrosion Protection	Comments
Lithium	Ester	>190	−40	130	++	+	Low Temperature Grease
	Polyalphaolefin	>190	−60	140	++	+	Low Temperature Grease
	Silicon Oil	>190	−40	170	+++	−	Low Temperature Grease
Lithium Complex	Ester	>260	−40	160	+++	+	Low Temperature Grease
Barium Complex	Ester	>260	−40	130	++	+++	Low Temperature Grease
	Polyalphaolefin	>260	−60	150	++	+++	Low Tepmerature Grease
Sodium Complex	Silicon Oil	>220	−40	200	+	+	Wide Temperature Grease

Note: In these tables, information on aluminum complex synthetic grease is not available.

requires testing using reference elastomers CR (polychloroprene) and NBR-L (acrylonitrile-butadiene).

Stability

1. Dropping Point—A standard measure of the resistance of grease to flow as temperature is increased is the Dropping Point (ASTM D2265) [16].
2. High Temperature Stability—High Temperature Bleed (ASTM D1742), Trident Probe (ASTM D3232), Cone Bleed (FTM 791B), Evaporation (ASTM D972 and D2595), Rolling Stability (ASTM D1831), Oxidative Stability (ASTM D942), and a High Speed Bearing Test (ASTM D3336).

Water Tolerance Tests

1. Water washout is measured using ASTM D1264, Water Washout Characteristics of Lubricating Greases.
2. Water spray-off is measured using ASTM D4049, Resistance of Lubricating Greases to Water Spray.
3. The Wet Roll Stability of Grease is evaluated using a modified ASTM D1831 test.

Bench Performance Tests

A number of bench performance tests are used to evaluate greases prior to field evaluation. These include the following:

1. High-Temperature Wheel Bearing Test (ASTM D3527), an SKF R2F Test that simulates paper mill applications, the FE-8 Test, CEM Electric Motor Test, and a GE Electric Motor Test.
2. Low temperature torque is evaluated using ASTM D1478 and ASTM D4693 at a temperature of −54°C.

Grease Compatibility

As mentioned earlier, all greases are not compatible with each other. When the greases made from different thickeners are mixed, sometimes the mixture may reflect a deterioration of the physical properties, or a drop in service performance. The term incompatibility is commonly used to describe this phenomenon. Common consequences of the incompatibility include a drop in heat resistance, a change in consistency, usually softening, and a loss in shear stability. Incompatibility is not always due to thickeners, but can also arise from all three components of the grease, the thickener, fluid, or additives. Incompatibility is not predictable and must be determined either through experimentation or by monitoring performance during service.

1. ASTM Standard D6185, Practice for Evaluating Compatibility of Binary Mixtures of Lubricating Greases, can be used to judge grease incompatibility.
2. There are several nonstandard ways to determine grease incompatibility. One involves mixing two or more greases in different ratios and subjecting the mixture/s to various grease tests to determine deterioration in performance. The tests that can be used for this purpose include measuring the dropping point (ASTM D566 or D2265), shear stability (either by prolonged working or by the roll test, ASTM D1831), oil separation (ASTM D1742), and storage stability (change in consistency after prolonged storage, such as one to six months). Other tests, such as ASTM D3527 (life performance), ASTM D4290 (leakage), ASTM D4049 (water spray-off), or

TABLE 10.25—NLGI consistency grades used in DIN 51 502 coding of greases [733].

NLGI Grade	Penetration Range at 25°C (ASTM Worked)
000	445–475
00	400–430
0	335–385
1	310–340
2	265–295
3	220–250
4	175–205
5	130–160
6	85–115

TABLE 10.26—High operating temperature and behavior towards water ratings used in DIN 51 502 coding of synthetic greases [733].

Letter	Highest Operating Temperature	Behavior Towards Water (DIN 51 807)
C	+60°C	0-40 or 1-40
D		2-40 or 3-40
E	+80°C	0-40 or 1-40
F		2-40 or 3-40
G	+100°C	0-90 or 1-90
H		2-90 or 3-90
K	+120°C	0-90 or 1-90
M		2-90 or 3-90
N	+140°C	According to Agreement
P	+160°C	
R	+180°C	
S	+200°C	
T	+220°C	
U	Over 220°C	

TABLE 10.27—High operating temperature ratings used in DIN 51 502 coding of synthetic greases [733].

Number	Lowest Operating Temperature
−10	−10°C
−20	−20°C
−30	−30°C
−40	−40°C
−50	−50°C
−60	−60°C

ASTM D6184 (Oil Separation by Conical Sieve Method), can also be considered, if deemed necessary.

NLGI Certification Mark

In 1989, the NLGI developed a licensing protocol and certification marks for the highest performance categories of chassis greases and wheel bearing greases meeting the ASTM D4950 Standard. The certification marks used for the NLGI service classifications include only the following:

1. GC Automotive Wheel Bearing Grease
2. GC-LB Automotive Wheel Bearing and Chassis Grease
3. LB Automotive Chassis Grease

The NLGI symbols are shown in Fig. 10.16 [727]. In the future, if higher ASTM standards are established, the Marks will likewise reflect such changes. Industry response to the licensing system has grown over the last decade and containers of the grease bearing the certification symbols are commonly available in the aftermarket, and their availability is expected to expand. Beginning with the 1992 model year, most U.S. auto makers began recommending the use of NLGI Service Greases GC, LB, and GC/LB for scheduled maintenance of chassis and wheel bearings of passenger cars and light-duty trucks.

Grease Analysis

A number of analytical methods are used to determine the chemical composition of the lubricating grease. This knowledge is important since it helps in identifying changes, intentional or unintentional, that occur during manufacture, formulation, or use of lubricating grease. The major components of the grease, viz., soap, base fluid, and additives, can be separated and their structures determined by a number of techniques. The identity of the soap can be determined by treating the grease with hydrochloric acid or potassium hydrogen sulfate and separating the products thus obtained into hexane-soluble and hexane-insoluble fractions. The components of these fractions are then identified either by the use of the classical methods or by chromatographic and spectroscopic techniques. In most cases, ASTM D128 can be used to determine the identity and the quantity of soap and the thin layer, gas, or liquid column chromatography can be used to separate the organic components and identifying them by the use of the infrared (IR), ultraviolet (UV), and nuclear magnetic resonance (NMR) spectroscopy, or mass spectrometry. Dialysis (DIN 51 814) is an alternative technique that can be used to separate the grease components, which can then be characterized by the use of the techniques mentioned above. Although ASTM D128 is the general analytical method for greases, there are other test methods for specific constituents and include the following:

1. ASTM D95-Test Method for Water in Petroleum Products and Bituminous Materials by Distillation
2. ASTM D129-Test Method for Sulfur in Petroleum Products (General Bomb Method)
3. ASTM D808-Test Method for Chlorine in New and Used Petroleum Products (Bomb Method)
4. ASTM D1317-Test Method for Chlorine in New and Used Lubricants (Sodium Alcoholate Method)
5. ASTM D3340-Test Method for Lithium and Sodium in Lubricating Greases by Flame Photometer.

Specialty Greases

Synthetic Lubricating Greases and Their Coding

Synthetic greases are made by the use of synthetic fluids in contrast to conventional greases that are made by the use of mineral oils. Synthetic greases are used in many industries and many applications. The industries include automotive, industrial equipment manufacture, steel mills, cooling and refrigeration, railroad, other transportation, and military equipment. The aircraft industry and the U.S. military are the two major users of the synthetic greases. This is because their equipment operates over a broad temperature range of

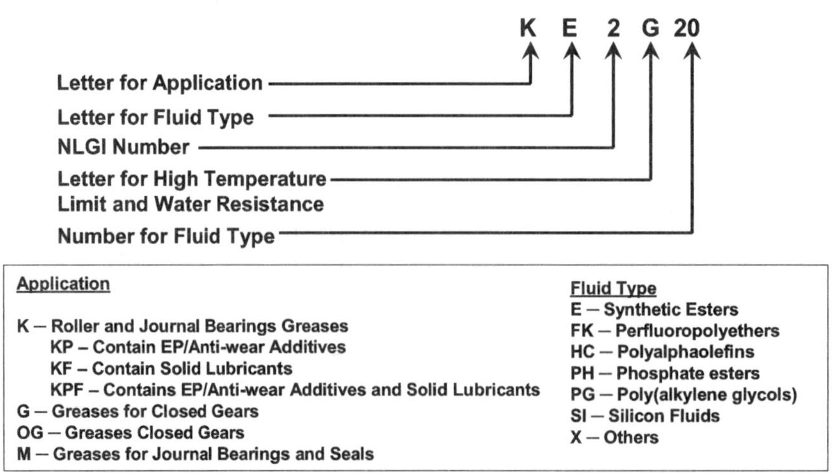

Fig. 10.17—Grease coding according to DIN 51 502 [733].

TABLE 10.28—Properties of soap-based synthetic greases [733].

Soap	Base Fluid	Dropping Point (°C)	Operating Temperature Range (°C)		Water Resistance	Corrosion Protection	Inherent EP/Antiwear properties	Suited for Roller Bearings	Suited for Journal Bearings	Comments
Lithium	Ester	>190	–40	130	++	+	+	+++	+	Low Temperature Grease
	Polyaphaolefin	>190	–60	140	++	+	+	+++	+	Low Temperature Grease
	Silicon Oil	>190	–40	170	+++	–	–	+++	+	Broad Temperature Grease
Lithium Complex	Ester	>260	–40	160	+++	+	+	+++	+	Low Temperature Grease
Barium Complex	Ester	>260	–40	130	++	+++	+++	+++	++	Low Temperature Grease
	Polyalphaolefin	>260	–60	150	++	+++	+++	+++	++	Low Temperature Grease
Sodium Complex	Silicon Oil	>220	–40	200	+	+	–	++	+	Broad Temperature Greas

–54 to 315°C, or –65 to 600°F; which is beyond the capabilities of the petroleum-based greases. Besides extreme temperatures, there are a number of other factors that determine the need for synthetic grease. These include the following:

1. Extreme operating temperatures, both low and high.
2. Low-pressure applications, for example those involving vacuum.
3. Stability against the surrounding mediums, such as those containing reactive compounds and or corrosive vapors.
4. Special requirements regarding low friction and low noise.
5. Physiologically harmless (food, beverages, pharmaceuticals, cosmetics).
6. Rapidly biodegradable.
7. For life lubrication.

Applications include fan motor bearings, clutch release bearings, paper machines, drying installations, water pumps, power generators, power tools, conveyors for drying, painting and burning-in machines, conveyors in textile machines, high-speed spindle bearings in machine tools, aerospace, and automobile applications. Also, please refer to section on industrial uses of synthetic fluids in Chapter 9 on Miscellaneous Industrial Lubricants.

Comparison of the properties of the synthetic greases with those of the mineral oil-derived greases reveals the synthetic greases to be superior, see Table 10.24 [733]. The superior properties include dropping point, range of operating temperatures, water resistance, corrosion protection, EP/antiwear performance, and suitability for use in roller and journal bearings. In these tables, the mineral oil grease rating is 0. Of course, the synthetic greases suffer from a number of deficiencies as well, which were already listed.

Synthetic greases are coded according to DIN 51 502, especially for the European use. Coding uses both letters and numbers, see Fig. 10.17 [733]. The letters are used to indicate the recommended application, fluid type, high-temperature limit, and water resistance. The numbers are used to represent the NLGI Consistency Grade and the low temperature limit. Tables 10.25–10.27 provide the designations along with their significance [733]. Table 10.25 provides the numbers for the NLGI Consistency Grade. Table 10.26 provides the high temperature limit and the water resistance behavior and Table 10.27 provides the low temperature limit. To cite an example, a synthetic lubricating grease that is recommended for use in journal bearings and contains EP/antiwear additives, is made from a synthetic ester base fluid, is of the NLGI Consistency Grade 2, has a high temperature limit of 100°C (212°F), the water tolerance behavior of 0–90 or 1–90 (DIN 51 807), and the lowest operating temperature of –20°C will be coded as KE2G-20. In addition, the grease

TABLE 10.29—Properties of nonsoap synthetic greases [733].

Soap	Base Fluid	Dropping Point (°C)	Operating Temperature Range (°C)		Water Resistance	Corrosion Protection	Inherent EP/Anti-wear Properties	Suited for Roller Bearings	Suited for Journal Bearings	Comments
Bentonite	Polyalphaolefin	Without	–60	180	++	...	+	++	+	Low Temperature Grease
Bentonite	Ester	Without	–40	180	++	...	+	++	+	Low Temperature Grease
Aerosil	Silicon Oil	Without	–40	200	++	...	...	++	+	Broad Temperature Grease
Polyurea	Silicon Oil	>250	–40	200	+++	+	...	++	+	High Temperature Grease
Polyurea	Poly(phenyl ether)	>250	>0	220	+++	+	+	++	+	High Temperature Grease
PFFE	Silicon Oil	Without	–40	250	+++	+	...	++	++	High Temperature
PTFE	Perfluoropolyether	Without	–60	250	+++	+	+++	+++	++	Grease Broad Temperature Grease
FEP	...	Without	–60	220	+++	+	+++	+++	++	Broad Temperature Grease

TABLE 10.30—Desired attributes of multiple use greases. Reprinted with permission from the Lubrizol Corporation.

Service	Requirements
High Temperature	• High-temperature thickener • Oil of high viscosity • Oil of high flash point • Higher NLGI grade • Oxidation resistance
Low Temperature	• Low percentage of thickener • Lower NLGI grade • Oxidation resistance
Wide Operating Temperature Range	• High-temperature thickener • Good low-temperature torque • Good pumpability • Low evaporation • Oxidation resistance • Rust resistance
Subjection to Water	• Low washout • Low spray-off • Firm consistency • Rust resistance
Extreme Pressure	• Low wear test values • High EP test values • Solid additives, if indicated • Oil of high viscosity (preferred)
Multi-purpose	• EP-antiwear usual • Oxidation resistance • Rust resistance • Acceptable pumpability • Water resistance

must meet minimum requirements with respect to the chemical, physical, and technological properties, as prescribed in DIN 51 528 and DIN 51 502. The strengths and weaknesses of the soap greases are shown in Table 10.28 and the nonsoap greases are provided in Table 10.29 [733].

Multiple Use Greases

Multiple use greases are those that are formulated for use both in industrial and automotive applications. Table 10.30 lists the desired properties of the multiple use greases. Industrial use of these greases is much larger than automotive use; the ratio is about 60:40. Generally, the automotive wheel bearing greases, due to their superior performance characteristics, can easily fulfill the needs of the most industrial applications. However, in certain special circumstances the automotive products do not suffice and there is a need for a customized product. For example, for use in steel rolling mills, a product with superior water resistance and higher thermal stability is needed.

Automotive Requirements

As stated earlier, SAE specifies greases for the automotive use and according to the importance of properties needed for specific applications. See Table 10.31, which lists the desired properties. As stated earlier, NLGI classifies greases (Table 10.13) according to NLGI Service Classification System, which primarily relates to automotive applications, that is, for chassis and wheel bearing lubrication.

Wheel Bearings Greases

Wheel bearings are the most critical grease-lubricated components of a motor vehicle. Rolling-element bearings, such as tapered roller bearings, are commonly used in the wheel assemblies. The bearings generally operate under severe conditions of speed and load, and in hostile environments

TABLE 10.31—Relative importance of grease properties for automotive use [744].

Property	Wheel Bearings	Universal Joints	Chassis	ELI Chassis	Multi-purpose
Mechanical and Structural Stability	H	M	L	H	H
Oxidation Resistance	H	M	L	H	H
High Temperature Service	H	M	L	M	H
Friction and Wear	M	H	M	H	H
Corrosion	M	M	M	H	M
Washout	M	M	M	H	M

Note: H=Highest; M=Moderate; L=Least (SAE Information Report J310).

TABLE 10.32—*DN* values for various NLGI grade greases[a] [729].

NLGI Grade	*DN* Value
3	0–50,000
2	50,000–125,000
1	125,000–250,000
0	250,000–350,000

[a]This general *DN* guide holds for multi-purpose greases whose oil viscosity is below 250 mm^2/s at 40°C (1200 Saybolt Second Universal). Lower *DN* values may have to be used with greases having either tacky properties or higher oil viscosities.

(mud, water, snow, dust, etc.). They also experience shock loading and high temperatures during the use of brakes. Therefore, wheel bearing grease must maintain its consistency as well as provide the EP protection. The consistency change will result in softening or the excessive loss of oil, which will lead to leakage and malfunction of the brakes. In addition to the normal modes of failure typical of the rolling element bearings, wheel bearings are exposed to oscillatory motion during transport and under certain operating conditions. This will cause fretting wear. This underscores the need to use the lubricating grease with good EP properties.

Universal Joint Grease

Grease for universal joints must have high load-carrying capacity and also possesses properties listed in Table 10.31 [744]. Some wheel bearing greases can be used in this application.

Chassis Grease

Chassis grease is applied with grease guns through grease fittings at the equipment manufacturers' suggested intervals. For this application, the grease with high apparent viscosity at high shear rates is usually used.

Extended Lubrication Interval (ELI) Chassis Grease

This type of grease is used in suspensions, drivelines, and steering systems having sealed joints that are prepacked during manufacture or assembly. These parts do not require relubrication for long periods of time.

Multi-purpose Grease

Multi-purpose greases combine the properties of two or more specialized greases. This allows the use of a single type of grease for many different applications. A good multipurpose grease must have a high melting point, and operate well at continuous temperatures of 120°C (250°F), or more. In addition, it must be water resistant, possess good corrosion properties, and have excellent mechanical and oxidation stabilities. Most multi-purpose greases are based upon barium, lithium, and calcium complex soaps A good multipurpose grease is suitable for lubricating chassis, wheel bearings, universal joints, and miscellaneous other automotive parts.

While using conventional greases, the speed (r/min) of a bearing or a gear is usually not a problem since most lubricating greases can easily handle speeds of up to 3500 r/min. The normal speed of the electric motors is 1750 r/min and the normal industrial equipment runs at less than 1750 r/min. However, some large grease manufacturing companies [729] use ASTM D3336 Standard to test their greases using lab equipment that operates at speeds reaching 10,000 r/min. While at these higher speeds, most lubricating greases still perform, but the equipment can get noisy. The choice of a proper grease for such high speed applications can be aided by the use of *DN* value, which takes into account the equipment speed. The *DN* value, also called the speed factor, is the product obtained by multiplying the diameter of a bearing by the number of rev/min, that is, *DN* = diameter (in mm) × N (r/min). For journal (sleeve) bearings, the shaft diameter must be used and for antifriction bearings, the pitch diameter must be used. For a diameter given in inches, multiplying by 25.4 will yield the value in mm. Table 10.32 identifies the grease selection based on *DN* values [729].

Track Roller Lubricants

For this application, a soft NLGI Grade 0 or a semi-fluid grease with low thickener content is used. Track roller assemblies are used on crawler tractors or other track-laying vehicles. Lubricating greases used in this application must act as a good seal against abrasive materials, have good load-carrying capacity, offer protection against wear, at both fast and slow speeds and under high-load and shock load operation, and resist water washing. They are used extensively in front idlers, track-support idlers, and track rollers in all older model tractors where the seal leakage can be a problem. However, in the newer equipment with improved seals and bearings, these lubricants are being replaced by the engine oils, gear oils, and conventional greases.

Mineral Oils Mixed with Solids

These types of greases are heavy lubricants for specialized applications. They are usually used to lubricate the rough-fitting machine parts, operating under heavy pressures or loads and relatively slow speeds. Examples of such equipment include concrete mixers, bearings and rollers on conveyors, and heavy construction equipment.

Heavy Asphaltic-type Oils Blended with Lighter Oils

These lubricants, although classified as greases, are essentially very viscous, heavy oils that are used to lubricate open-type gearing and wire rope. These oils form a heavy protective film when applied to a surface and heated. The reason for the use of the light oil is that it not only improves the low-temperature fluidity of the asphaltic oil, but it also makes it easier to handle and use.

Extreme-Pressure Greases

These greases are formulated by the use of the film-strength additives. The function of these additives is to improve a lubricant's film-forming ability, or in other words facilitate the formation of a durable lubricant film. Additives used to formulate the EP greases include compounds of chlorine, phosphorus, and sulfur. Chlorine compounds include organic halides and chlorinated waxes; phosphorus compounds include phosphites, phosphates, and dithiophosphates; and sulfur compounds include organic polysulfides and dithiocarbamates. A number of factors affect the quality of the surface film. These include the types of the additive and the equipment's operating conditions, such as load, speed, surface condition, and design. Load-carrying capacity in a grease is determined by tests such as Timken or Four-Ball EP. These are commented on in the tests section.

Roll Neck Greases

Roll neck greases are specialized products that are used almost exclusively for lubricating plain bearings in rolling equipment. This grease is very hard, is of NLGI Consistency Grade 6, which is carved to a shape suited for application to the bearing of heavily loaded equipment.

Application and Industry Trends

As mentioned earlier, the lubricating grease market is ever changing. The user-related factors that are influencing the worldwide development of the future greases include the following:

- Improved performance
  - Longer life
  - Extreme pressure performance
  - Water resistance
  - Oxidation stability
  - Thermal stability
  - Improved compatibility with seals, bearing housings, and soft metals
- New component design
  - Smaller and lighter in weight
  - Higher operating speeds and heavier loads
  - New fabrication materials, such as ceramics and plastics
- Greater safety and environmental acceptability
  - "Cradle to Grave" responsibility
  - Removal of heavy metals, such as barium, zinc, and lead
  - No chlorine
  - Biodegradability
  - Low eco-toxicity
- Global product acceptance
- Cost-effectiveness

It is important that we consider the lubricating grease a value-added product and not a commodity, paying little attention to its cost but more to its performance. This is because the greases are the only lubricants that are effective in specialty applications, such as bearing lubrication, where the liquid lubricants will not be adequate. The complex nature of the lubricating greases and the intricate manufacturing process further justifies this consideration.

Formulation Examples

EP Grease: 5.0–10.0 % Olefin copolymer viscosity modifier/thickener, 0.5–2.0 % zinc dialkyl dithiophosphate or zinc dialkyldithiocarbamate antiwear agent, 1.0–4.0 % sulfur-phosphorus EP agent, 0.5–1.0 % arylamine oxidation inhibitor, 0.5–0.8 % metal sulfonate corrosion inhibitor, 1.0–5.0 % molybdenum disulfide or graphite antiseize agent (friction modifier). The balance is naphthenic oil.

EP Grease: 3–15 % Lithium 12-hydroxystearate soap thickener (depending upon grease consistency grade desired), 4 % alkali metal borate EP/AW agent, 1 % antimony dipentyldithiocarbamate load-carrying additive, 1–2 % amine salts of dibutyl phosphate and dibutyl thiophosphate EP agent, and 1–2 % sulfurized isobutylene EP agent/oxidation inhibitor. The balance is mineral oil, formulation extracted from Ref [750].

Grease for Constant Velocity Joints: 6.5 % Lithium simple soap thickener, 2 % sulfur-free molybdenum-amine complex friction modifier, 1 % molybdenum dialkyl dithiophosphate EP agent, 2 % primary/secondary mixed alcohol derived zinc dialkyl dithiophosphate antiwear agent/oxidation inhibitor, 1.3 % calcium sulfonate rust inhibitor, 0.5 % sulfur-phosphorus EP additive system, 0.3 % arylamine oxidation inhibitors. The balance is semi-synthetic base fluid comprising naphthenic oil, paraffinic oil, PAO, and dioctyl sebacate in a 25:50:20:5 weight ratio (formulation extracted from Ref [751]).

Grease with improved rust prevention and abrasion resistance: 10 % Diurea thickener, 3 % sodium thiosulfate and hydrated sodium thiosulfate, 2 % calcium salicylate and magnesium salicylate mixture, and 2 % benzotriazole yellow metal corrosion inhibitor. The balance is purified mineral oil (formulation extracted from Ref [752]). Reference also provides analogous formulations using lithium soap, lithium complex soap, and aluminum complex soap.

General Purpose Grease: 8 % Lithium 12-hydroxystearate soap thickener, 1 % hindered phenol (BHT) oxidation inhibitor, and 2 % calcium sulfonate rust inhibitor. The balance is 59 % paraffinic-naphthenic oil blend and 30 % dioctyl sebacate (formulation extracted from Ref [753]).

Biodegradable Grease: 1.5 % Sodium alginate thickener, 30 % glycerol emulsion stabilizer/compatibilizer, 2 % emulsifier, 36.5 % water, and 30 % beef tallow film-forming agent after emulsification (formulation extracted from Ref. [754]).

Biodegradable Grease: 5 % Ammonium alginate thickener, 50 % greater than C_{18} alcohols emulsion stabilizers/compatibilizers, 10 % graphite friction modifier, 10 % water, 25 % rapeseed oil film-forming agent after emulsification (formulation extracted from Ref [754]).

11

Metalworking and Machining Fluids

DISCUSSION IN THIS CHAPTER PERTAINS TO fluids that facilitate metalworking and machining operations. The nature of the various working operations is examined, along with the properties of the lubricants needed for each. Discussion includes metalworking fluid classifications, composition, formulating, and testing. Representative examples of the various metalworking fluid formulations are also provided at the end of the chapter.

Metalworking is the process of converting the bulk metal into *a component*, or *a part*, and primarily involves two types of operations: Those that produce metal debris and those that produce no debris. The former type is classified as the metal removal operations and the latter type is classified as the metal forming operations. Cutting and grinding are examples of the first type and drawing, stamping, and bending are examples of the second type. All metalworking operations involve bringing two solids, a tool and a work-piece, together to create a new part or a shape. The process involves high friction, high pressures, high temperatures, and tool wear, and it is the job of the lubricant, or the metalworking fluid, to control them. Metalworking fluids accomplish this by providing cooling, lubrication, and protection against corrosion. They, therefore, improve the efficiency of the operation, and hence increase productivity.

Irrespective of the type of the metal cutting operation, whether it is turning, milling, drilling, planing, shaping, broaching, or sawing, the mechanism of action of all cutting tools is the same. That is, the cutting is performed by the tool either as it moves across the metal surface being machined, or the tool is stationary and the metal surface moves against it. In either case, the process is accompanied by plastic deformation of the metal surface at the front of the cutting edge of the tool and the rubbing of the formed chip with the tool surface, as shown in Fig. 11.1. The temperature estimates in the cutting zone are 900°C on the tool's cutting edge, 500°C on the chip, and 200°C on the work-piece. As we move away from this zone, the temperature of the tool drops to 400°C on the tool's outer edge and to 200°C on the chip. About 75% of the heat generated is due to the deformation of the metal and the other 25% results from the friction due to sliding of the chip on the tool face. Metal deformation occurs due to shear or plastic flow along the shear plane that extends from the edge of

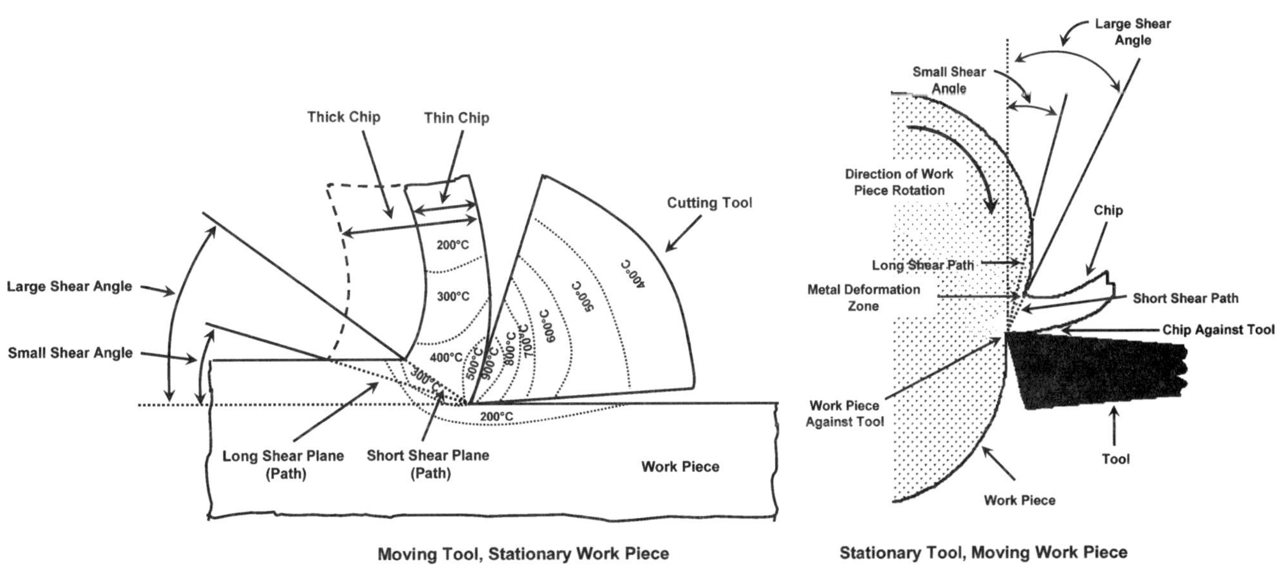

Large Shear Angle	Small Shear Angle
1. Shallow Cut	1. Deep Cut
2. Thin Chip	2. Thick Chip
3. Low Energy Requirement	3. High Energy Requirement
4. Low Coefficient of Friction	4. High Coefficient of Friction
5. Low Heat Generation	5. High Heat Generation

Fig. 11.1—Metal cutting process [755].

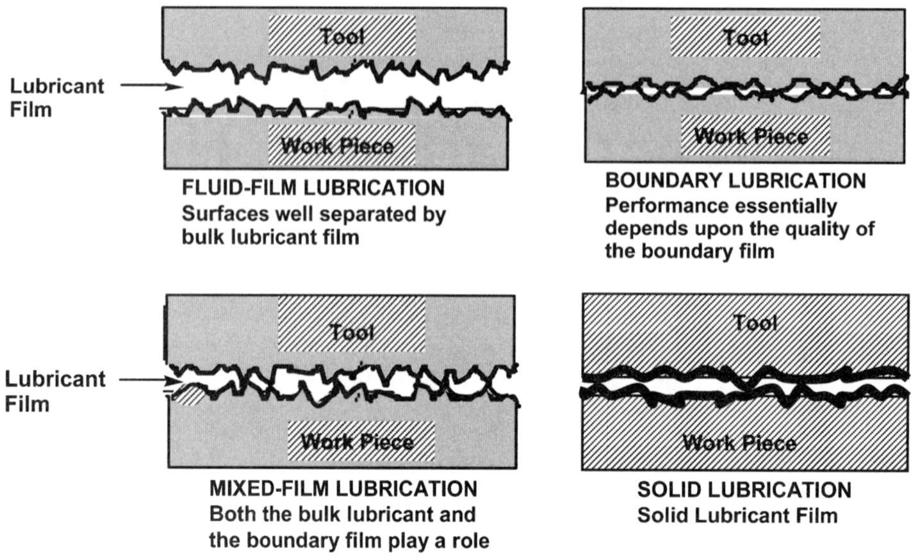

Fig. 11.2—Lubrication regimes in mechanical equipment [12].

the tool to the surface of the work-piece metal, see Fig. 11.1. Below the shear plane the metal is undisturbed; above it the metal is deformed and ultimately results in the formation of a chip. It is critical that the high temperatures in the cutting zone are decreased, otherwise extensive tool wear and rough finish of the work-piece will occur. In the case of the brittle materials, high temperatures in addition may cause the fracture of the metal along the shear plane, which will form a discontinuous or segmental-type chip. This is undesired since it will interrupt the continuous cutting action. In the case of the ductile metals, high temperatures will also cause a built-up-edge on the nose of the tool, which will result in a severe plowing action and again poor finish of the work-piece surface will occur. The built-up edge is the stagnant mass of the metal that sheared away from the body of the chip by the high tool-face friction.

The efficiency of the cutting operation may be improved by reducing tool-face friction in two ways. One way is to increase the shear angle shown in Fig. 11.1 and the second is to use a lubricant. An increase in the shear angle produces a thinner chip and deforms less metal. The result is lower friction and heat generation and reduced energy consumption. However, controlling friction by the use of a lubricant with cooling ability or the ability to form a low-friction, low shear lubricating film at the chip-tool interface is a more effective strategy. This is because a lubricant will reduce extensive metal-to-metal contact and facilitate sliding of the chip on the tool face; thereby reducing tool wear, increasing tool life, reducing the built-up edge, providing smooth finish, and permitting fast-machining speeds.

Grinding is a special type of cutting where extremely small chips are removed by a large number of irregularly shaped abrasive particles bonded in a wheel. Excessive heat is a problem in grinding operations as well. The function of a grinding fluid is to improve the grinding action as well as act as a coolant. The problem is that a liquid cannot be present at the point of grinding contact and hence cannot cool the surface. However, the fluids that contain extreme-pressure additives mediate the heat generation by forming a low-friction EP film.

Unlike the metal removal operations that are mechanistically quite similar to one another, metal forming operations are quite diverse. However, they do have common-

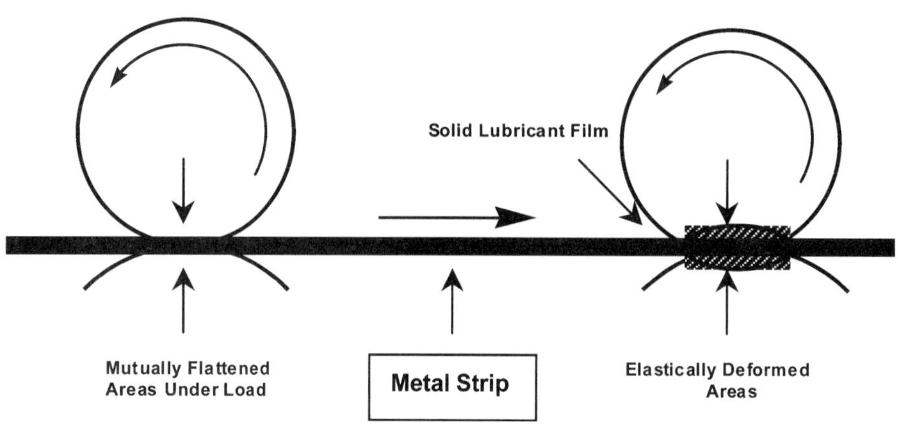

Fig. 11.3—Elasto-hydrodynamic lubrication.

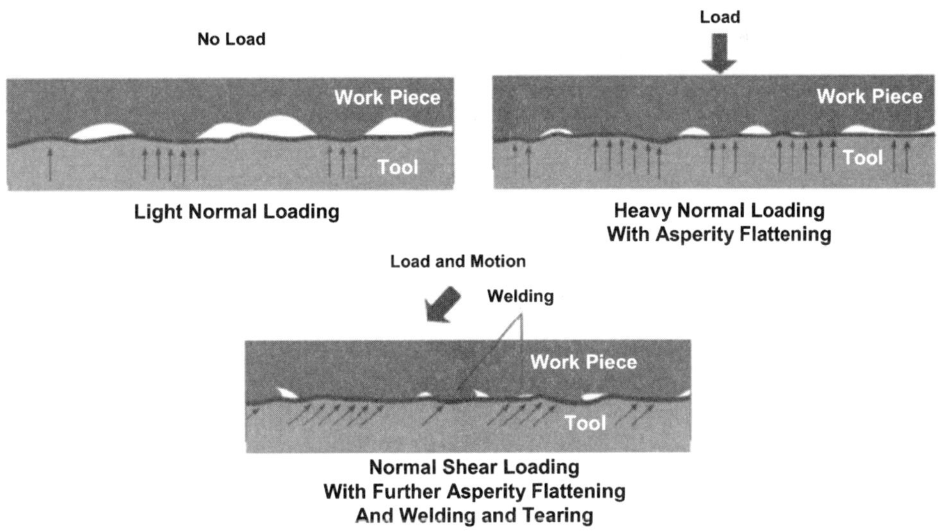

Fig. 11.4—Surface changes in metal forming operations [756].

alities among them. They all involve high pressures and temperatures, wear, and friction. Two of the latter factors definitely need to be controlled.

Lubrication

Mechanical equipment primarily experiences four types of lubrication environments, which are hydrodynamic or full-film lubrication, elasto-hydrodynamic lubrication, mixed-film lubrication, and boundary lubrication [11,13–15]. These are depicted in Figs. 11.2 [11] and 11.3. The cross-hatched area in Fig. 11.3 indicates a solid lubricant film that causes elastic deformation of the metal surfaces.

Metalworking operations are designed to have at least some metal-to-metal contact. This is because total separation of the surfaces will make the process inefficient. For example, in the metal removal operations, separation will lead to poor tool work-piece contact, which will interfere in the chip formation. In metal forming operations, this will lead to low friction and hence loss of control. This implies that in metalworking processes, the lubricant must be designed to perform in elasto-hydrodynamic, mixed-film, and boundary lubrication regimes. In general, metal removal operations require boundary lubrication and metal forming operations require elasto-hydrodynamic and mixed-film lubrication. The metal removal operations therefore involve little or no lubricant film and the metal forming operations involve a fairly thin lubricant film. As mentioned earlier, the metal forming operations primarily involve high pressures and high temperatures. Figure 11.4 depicts the effect of the load and movement during metal forming operations. The asperities start to flatten as the load on the work-piece increases. This increases contact and the heat generated due to loading causes high spots to weld [756]. The use of an extreme pressure lubricant helps avoid this. However, in certain situations, some lubricants cause metal movement when the metal is being pressed against the die, especially if they are not shear stable. The result is tearing of the welds, as shown in the figure. Metal movement can only be controlled by altering the lubricant properties.

Lubricant film in the metal forming operations is of two types: Wedge and squeeze. *Wedge* type films occur when two nonparallel surfaces in motion converge and *squeeze* type films occur when the parallel surfaces in motion come together [757]. In metal forming operations, the wedge type films are more common than the squeeze type. As the lubricant enters the converging zone, there is an increase in pressure that causes an increase in lubricant viscosity. This is a consequence of the lubricant molecules coming closer. The increase in viscosity changes the nature of the lubrication in the contact zone from being hydrodynamic to elasto-hydrodynamic. Based upon the pressure and speed considerations, we believe that we have moved from the right end of the curves, shown in Fig. 1.3, closer to the border of the mixed-film lubrication regime. As the lubricant exits, the gap widens and there is a decrease in viscosity and an increase in velocity. This causes a reversal to hydrodynamic lubrication, that is, a movement away from the mixed-film regime. The result is the separation of the surfaces and hence minimal metal-to-metal contact. A drop in viscosity is a consequence of the widening gap that permits the lubricant molecules to move away from one another.

The quality of the lubricant film determines the metal forming efficiency: The higher the operation speed, the higher the film thickness. Similarly, the higher the viscosity, the higher is the film thickness. Therefore, both these conditions force the converging surfaces to move away from each other (the gap widens), which improves the operation efficiency. For example, in the rolling operations, the speed of the strip determines the film thickness. Since the temperature and pressure also impact viscosity, hence the lubricant film thickness; these factors also affect the efficiency of the metal forming operations. Different base fluids have differing response to temperature and pressure, and therefore their properties can influence the properties of the metal forming lubricants. Naphthenic base stocks are usually preferred over paraffinic base stocks. This is because they not only experience a faster viscosity decrease with an increase in temperature (have a lower viscosity index); they also undergo a greater viscosity increase with an increase in pressure. These factors make lubricants derived from these base

TABLE 11.1—Classification of metalworking fluids according to DIN 51 385 [4,759].

No.	Fluid Class	Description
0	*Metal working fluid*	Fluid used as coolant and lubricant in the cutting and sometimes in the forming of materials
1	*Fluid not miscible with water*	Fluid not mixed with water before use
2	*Fluid miscible with water*	Fluid mixed with water before use
	2.1 Emulsifiable fluid	Water-miscible fluid which can form the dispersed phase of an oil-in-water emulsion
	2.2 Emulsion-forming fluid	Water-miscible fluid which can form the dispersing phase of a water-in-oil emulsion
	2.3 Water-soluble fluid	Fluid which forms solutions with water. These include true solutions as well as solutions of association colloids, e.g. "soap solutions"
3	*Fluid mixed with water*	Fluid mixed with water (water-miscible fluid ready for use)
	3.1 Metal working emulsion (oil-in-water)	Emulsifiable fluid mixed with water (mixture ready for use)
	3.2 Metal working emulsion (water-in-oil)	Emulsion-forming fluid mixed with water (mixture ready for use)
	3.3 Metal working solution	Water-soluble fluid mixed with water (mixture ready for use)
4	*Metal working liquid*	Liquid metal working fluid, used as coolant and lubricant in the cutting and sometimes in the forming of materials. *Note:* This term can be used instead of "metal working fluid" when fluids with other aggregate states are to be excluded

stocks extremely effective in high-pressure applications, such as rolling.

As mentioned earlier, *squeeze* films form when two parallel surfaces converge, a situation that occurs in operations such as upsetting. The film thickness increases due to the fast approaching surfaces and the lubricant viscosity increases because of an increase in pressure. This hinders the lubricant flow out of the contact zone, which results in a squeeze film. Deformation of the work-piece during the operation results in a flatter surface, thereby making the squeeze film thinner. The thickness of the squeeze films depends on the lubricant viscosity, speed, and load (force), in the same way as that of the wedge films, but for different reasons.

Lubrication regime defines a lubricant's ability to support load (modify friction and reduce metal transfer) and can be determined by taking into account the film thickness and the combined roughness. The ratio of the two, represented by λ, equals film thickness (ρ) divided by combined roughness of the surfaces (σ), or ρ/σ. The combined roughness is the roughness amplitude of the two surfaces relative to their average levels. The value of λ equal to 0.5 implies boundary regime; the value of λ between 0.5 and 3.0 implies mixed film regime; and the value of λ greater than 3 implies fluid-film or hydrodynamic regime [757]. Unlike metal removal operations that primarily involve the boundary lubrication regime, metal forming operations start in the boundary regime but move first into mixed-film and then hydrodynamic regimes as the operation picks up speed. Lubricant properties, such as viscosity, and film-forming additives play a role in boundary and mixed-film regimes. Refer to Chapter 4 for the details on film-forming agents.

Viscosity

Viscosity is critical to determining the quality of the lubricant film. In metal-forming applications, it determines the effectiveness of the film in separating the tool from the work-piece and therefore controlling friction and wear. Metal removal operations, on the other hand, have diverse lubrication needs, and hence optimum lubricant viscosity must be estimated for each operation. This is accomplished by considering the ability of the lubricant to enter and remain in the contact zone, the durability of the lubricant film, the desired rate of spreading, and its cooling capability. These parameters are measured by the use of the film-strength tests, such as the Timken Test (ASTM D2782), the Four-Ball Tests (ASTM D2783 and D4172), and the SAE #1 Test.

Metalworking Fluid Classification

Classification Based on Base Fluid

Primary functions of a metalworking fluid include lubrication, cooling, and corrosion inhibition. Metalworking and machining fluids fall under three general classes: Oil-based, water-based, and the solid suspensions [758]. Oil-based fluids are mineral oils with or without additives. These fluids contain only one type of oil, or a mixture of two or more oils, and may contain one additive or a combination of additives, such as those containing sulfur, chlorine, phosphorus, or other elements. Also used are emulsifiable (miscible, soluble) oils and miscible fluids that contain little or no oil. These are based solely on chemicals and provide the industry a means to achieve even higher machining speeds. Water-based fluids, on the other hand, are micellar solutions of oil or additives or both in water. These fluids are of three types: *Soluble oils, semi-synthetic fluids,* and *synthetic fluids*.

The worldwide use of the oil-based fluids is estimated at 45 % and that of water-based fluids is estimated at 53 %; with *synthetics, semi-synthetics,* and *soluble oils,* having a share of 4, 16, and 33 %, respectively. Solid suspensions make up the rest. The ASTM Standard D2881 [758] and the European Standard DIN 51 385 [759] classify metalworking lubricants according to their nature and functions. Classes of metalworking fluids are described below and are also described in Table 11.1 [4]. The essential difference between the various classes is that as one moves from *straight oil* to *synthetic fluid,* the amount of water increases and the amount of oil decreases.

1. *Straight Oil* (not dilutable)—not supportive of the microbial growth.
2. *"Soluble" Oil* (oil-based emulsifiable concentrate, diluted with water at the point of use)—more likely to support bacterial growth. It primarily contains oil, emulsifiers, and other additives. When diluted, the typical dilution ratio is between 5 to 20 parts of water to 1 part of concentrate. This produces an emulsion. Oil provides the lubrication and the corrosion protection and the water provides cooling. In some grinding applications, the dilution ratio can be as high as 200 to 1.
3. *Semi-synthetic Fluid* (oil-in-water emulsion concentrate; diluted prior to use)—more likely to support bacterial growth. These contain oil and additives emulsified in water. These preformed emulsions when diluted further are used to provide lubrication, cooling, and corrosion protection.
4. *Synthetic Fluid* (water-based solution concentrate; diluted prior to use)—more likely to support fungal growth. It contains no oil but is a solution of the corrosion inhibitors and the friction reducing additives in water. These fluids provide cooling and corrosion protection but their lubrication properties arise from the synthetic lubricity components.

Oil-based Fluids

These lubricants, commonly referred to as *straight oils,* are mineral oil-based, vegetable oil-based, or synthetic fluid-based. Vegetable oil-based fluids are more expensive than the mineral oil-based fluids but have the advantage of a higher degree of biodegradability. This facilitates the deposal of the spent fluid. With respect to the thermo-oxidative stability, which is a concern in all vegetable oils, rapeseed, castor, or coconut oils are the best. This is because either their saturate content or the mono-unsaturate content is high. The oil-based fluids do not contain any water; hence they do not support microbial growth. Therefore, they do not need any biocides. In addition, these oils have good ability to wet surfaces, provide good rust protection, and are most trouble-free of all the metal removal fluids. In addition to the higher cost than the water-based fluids, these have the disadvantages of the mist formation and hence pose a potential fire hazard. They may or may not contain any surface reactive additives. Additive-free oils are used only in light-duty applications involving metals with high machinability, such as aluminum, magnesium, and brass. Additive-treated oils primarily utilize hydrotreated naphthenic base stocks and contain polyol esters and fatty oils as lubricity agents and sulfur, chlorine, and phosphorus-derived extreme pressure additives. However, there is a growing trend towards the use of the paraffinic base stocks because the derived fluids have a lower tendency to produce mist. The cooling ability of these fluids is not as good as that of the water-based systems, but it is respectable.

Straight oils are used in severe cutting operations involving screw machines, cold headers, and other machines having a common sump for both hydraulic and machining lubricants. Because of this and their excellent lubricating ability, these lubricants are often used in high-temperature operations that involve high unit pressures and high spot temperatures. That is, those involving slow machine speeds. They are also the fluids of choice in most heavy-duty operations and for the machining of extremely hard metals. Straight oils can be classified as either "active" or "inactive." These oils are labeled *active* if they contain additives capable of releasing active sulfur (see discussion under the extreme pressure agents in Chapter 4). Otherwise, they are classified as *inactive*. Sulfur-releasing ability is high when the sulfurized olefins and the sulfurized fats used as additives are of high sulfur content, that is, their sulfur to olefin ratio is greater than 1. *Active* fluids are commonly used in machining steel. However, some metal removal operations, such as blanking, do employ *inactive fluids*. *Inactive* fluids contain fatty oils, and fatty oil-mineral oil mixtures or inactive sulfur additives, or a combination thereof. The in-house use of these fluids is higher than the water-based chemistry and the residues cause some housekeeping problems.

Water-based Fluids

These fluids are emulsions consisting of an organic phase and an aqueous phase. In emulsions, one phase is considered a dispersed phase and the other a continuous phase, usually the larger of the two. When the organic material, that is the oil and or the additives, is the dispersed phase and water is the continuous phase, the emulsions are called normal or oil-in-water emulsions. If, however, water is the dispersed phase and the organic material is the continuous phase, the emulsions are called invert, or water-in-oil emulsions. The two types of emulsions are depicted in Fig. 11.5 [459].

Metalworking and machining fluids are usually of oil-in-water type emulsions. The emulsions are of transient stability (ASTM D3342, D3707, and D3709). Their stability depends upon a variety of factors. These include the nature of the oil phase, the amount and the type of the emulsifier/s, the *p*H, the operating temperatures (ASTM D3707), the nature and the amount of additives, and the impurities, either inherently present or externally introduced into the system. All these factors can lead to coalescence of the fine droplets into larger ones and lead to oil-water separation. In general, the emulsions that are used as metalworking fluids are kinetically stable but thermodynamically unstable [757]. This means that such emulsions maintain their integrity during use, but have the tendency to phase separate when not in use. While the phase separation in the bulk fluid is undesired, the emulsions must phase separate on surfaces to release the oil/additives that improve lubrication. Oil availability in these emulsions depends upon its amount being present in the formulation and the emulsion stability. As a rule, the higher the oil content and the lower the emulsion stability, the greater is the oil availability. The problem is that the two essential

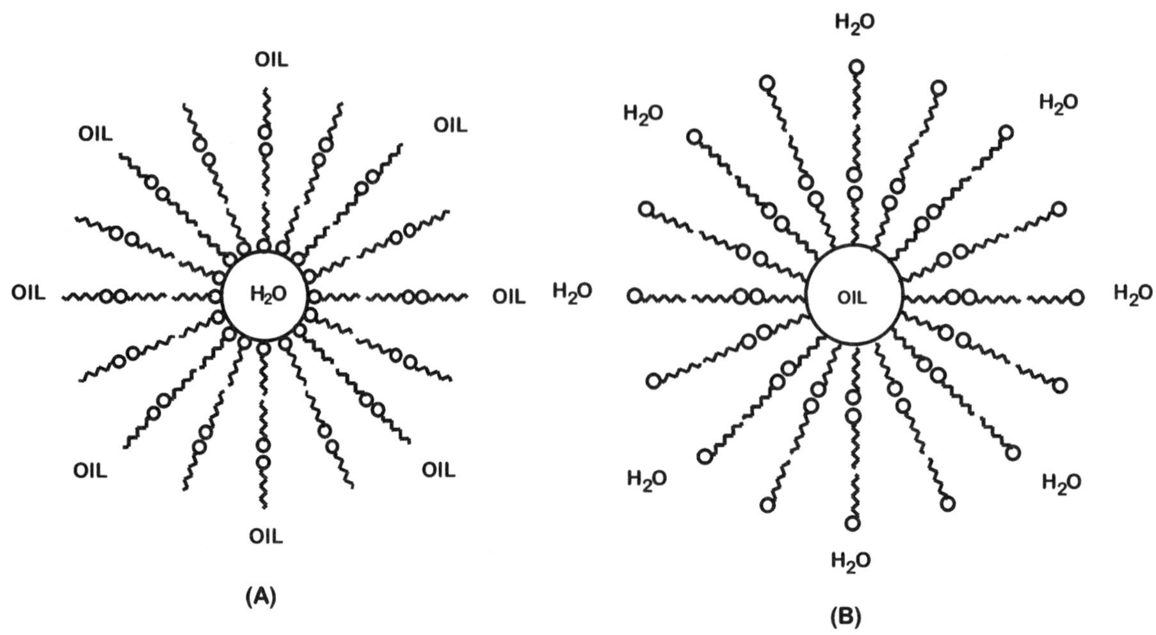

Fig. 11.5—A representation of (a) water-in-oil emulsion and (b) oil-in-water emulsion.

properties of the emulsions, namely the lubricating ability and the cooling capacity, are inversely related. Emulsions of high oil content (lower water content) not only have inadequate cooling ability, but they are also of lower stability, which decreases their useful life. To further complicate matters, some metalworking operations, such as rolling, tend to destabilize these high oil-based emulsions. In other words, there is lubrication-cooling tradeoff when the emulsion type metalworking fluids are used.

Emulsions are usually supplied as concentrates, which need to be diluted with water prior to use. In addition to cooling and lubricity (ASTM D2782, D2783, D4172, and D5619), these fluids possess inherent rust prevention properties (ASTM D4627), and detergency. Another attribute of these fluids is the ability to incorporate additional performance additives. The main disadvantages are their sensitivity towards hard water, susceptibility to microbial attack, and skin sensitivity.

Water-based fluids are susceptible to microbial attack, primarily because of their low organic (high water) content. Bacterial infestation of the fluid can lead to an objectionable odor due to the formation of the hydrogen sulfide, a change of color to gray, the formation of black stains on the workpiece, and foaming. Fungal growth can produce slime and is more difficult to control than the bacterial growth. The major side effect of the microbial attack is the emulsion's tendency to phase separate. Monitoring and control of the emulsion quality are therefore important. The common techniques used for this purpose are summarized in Table 11.2. The quality of the emulsions in metalworking fluids is maintained by taking the following steps.

1. The use of the biocides.
2. Maintaining the *p*H (bacteria are dormant at *p*H of greater than 8.8; 8.8–9.2 is a good range).
3. Controlling the dissolved oxygen levels.
4. Periodic removal of the contaminants, such as the tramp oil.
5. The use of the filtration equipment.

It is important to periodically check the organic (oil) content of the emulsion by breaking it. An acid or salt, such as sodium chloride, is used for this purpose. Alternatively, one can use a refractometer to determine the oil content. If the oil content is too low, it may be appropriate to top up the fluid with the fresh concentrate. Since the contaminants, such as the tramp oils, metal debris, and microbes, can destroy the integrity of emulsions, they should either be removed or controlled. Tramp oils are removed via skimming and the metal debris is removed via filtration. As mentioned above, the microbes are controlled by the periodic addition of the antimicrobial agents.

Foaming is another problem (ASTM D892) with emulsions that relates to the presence of the emulsifiers. Foaming not only leads to poor (spotty) lubrication, but it also impairs the fluid's heat transfer properties. While the new emulsions are more stable than the old emulsions, the lubricating ability (ASTM D2782, D2783, D4172, and D5619) of the emulsions improves with use. This is because the debris resulting from the metalworking operations and the decomposition

TABLE 11.2—Metalworking fluid monitoring techniques.

♦ Appearance and color	♦ Odor
♦ *p*H (ASTM D1293)	♦ Reserve alkalinity (ASTM D974)
♦ Emulsion stability (ASTM D1479)	♦ Low-shear foam (ASTM D3601)
♦ Corrosion inhibition (ASTM D4627)	♦ Microbial load (ASTM D3946)
♦ Relative bio-resistance (ASTM D3946)	

TABLE 11.3—Droplet size versus emulsion appearance.

Droplet size in microns	Appearance
>1	Milky white
0.1 to 1	Blue white
0.005 to 0.1	Translucent to semi-transparent
<0.005	Transparent to translucent

products from the additives, especially the EP agents, tend to destabilize them via nucleation. Ultimately, a complete breakdown of the emulsion occurs and a new batch must be used. The life of an emulsion during use is called its "batch life." In general, the less stable emulsions have a shorter batch life.

While the emulsions contain a variety of chemicals, discussed in the additive section, the chemical types that are critical to their formation and stability are emulsifiers and coupling agents. Emulsifiers are the key to the formation of emulsions, and coupling agents help improve their stability. Emulsifiers are surfactants that reduce the surface tension of the oil-water interface, thereby promoting miscibility and leading to colloidal dispersions. In many cases, a combination of emulsifiers is used and their selection depends upon the type of base oil to be emulsified and the nature of the other additives present in the formulation [757].

Emulsions are made by mechanically mixing the organic phase that comprises additives and emulsifier/s, called the *"concentrate,"* with water. Since the amount of water in the water-based fluids can be as high as 95 %, its quality not only determines the initial stability of the emulsion but also its "batch life." The presence of greater than 200 ppm calcium or magnesium carbonate (ASTM D511 and D513) can lead to the emulsion stability problems. And the presence of chloride and or sulfur levels (ASTM D512 and D516) of greater than 150 ppm will promote corrosion, instability, and rancidity. Soft water of less than 50 ppm hardness (ASTM D1126), on the other hand, leads to foaming in many products. While regular water is usually acceptable, the use of the distilled, deionized, or reverse osmosis water is recommended, if the problems are encountered.

Emulsion appearance can vary from being transparent to milky white, depending upon the droplet size. Table 11.3 describes the emulsion appearance as a function of the droplet size and Fig. 11.6 shows the appearance physically [757]. Oil-in-water emulsions with the droplet sizes of approximately 0.1–0.2 μm, sometimes referred to as micro-emulsions, are preferred in some metalworking applications. This is because their smaller droplet size makes them both kinetically and thermodynamically more stable. Macro-emulsions with the droplet sizes in the range of 0.2–10 μm are only kinetically stable. This means that the micro-emulsions prefer the dispersed form and are therefore less likely to phase separate. Of course, the droplet size depends upon factors such as the nature of the oil and the amount and type of the emulsifier used. Despite their greater kinetic and thermodynamic stability, the *p*H (ASTM D1293), oil-water ratio, and temperature have a profound effect on the stability of the micro-emulsions. Incidentally, the larger droplet size and the tendency to coalesce are not always undesired. Hence, macro-emulsions are often used in many once-through applications, where the immediate breakdown is desired to release oil for superior lubrication. The same properties are useful for the effluent treatment.

As mentioned earlier, the emulsions for metalworking use are usually produced prior to use by mechanically mix-

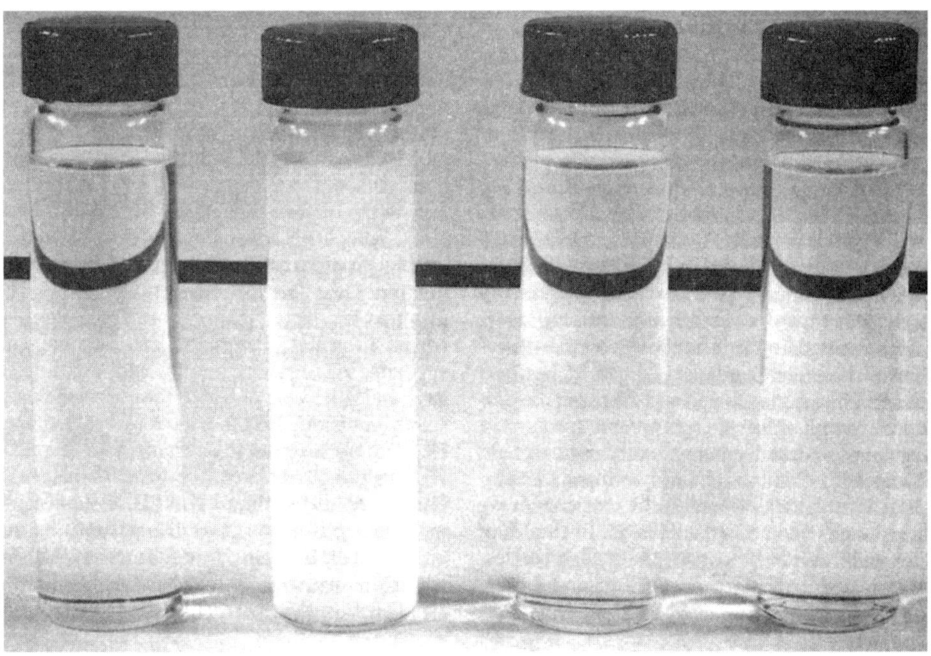

Straight Oil	Soluble Oil Macro-emulsion	Semi-synthetic Fluid Micro-emulsion	Synthetic Fluid Micellar Solution

Fig. 11.6—Physical appearance of commonly used metalworking lubricants [757].

ing the additive concentrate with water. The concentrate comprises oil, emulsifier/s, and a variety of additives. The concentrate-to-water ratio in such dilutions is between 1:10 (10 %) to 1:60 (1.5 %). However, this ratio may be changed if the operation requires a greater degree of cooling or more lubricity. Metalworking fluids for operations requiring more cooling contain a higher percentage of water than those requiring more lubrication. Micro-emulsions are stable (ASTM D3342, D3707, and D3709), are resistant to microbial attack (ASTM D3946), and are effective coolants. These attributes are primarily due to their low organic content and high emulsifier levels. Despite the listed advantages, the micro-emulsions suffer from the disadvantages of being expensive, are difficult to dispose of, and have an extensive tendency to foam (ASTM D892).

Composition of Water-based Fluids

As stated in the previous section, water-based fluids are classified as *soluble oils, synthetics,* and *semi-synthetics,* depending upon their oil content and the emulsion type. *Soluble oils* are macro-emulsions containing about 2–10 % naphthenic or paraffinic oil. *Semi-synthetic* fluids are micro-emulsions containing less than 2 % oil. *Synthetic fluids* contain no oil and are predominantly solutions of the water-soluble organic compounds or polymers in water. Please note that these oil concentrations pertain to ready-to-use finished fluids. The cooling ability of each type is related to its water content. Therefore, *synthetic fluids,* with the highest water content, are extremely effective in operations, such as grinding, that generate a lot of heat. As mentioned earlier, the appearance of an emulsion is related to its droplet size. When the droplet size is small, as in the case of the micro-emulsions, the fluids appear transparent, and when it is relatively large, as in the case of macro-emulsions, the fluids appear translucent. This is shown in Fig. 11.6. The formation of the micro-emulsions generally requires a higher amount of emulsifier.

Please note that the terms *synthetic* and *semi-synthetic* that are used to describe metalworking fluids have a different meaning than the same terms used to describe automotive lubricants. In the case of the metalworking fluids, these terms pertain to water-based systems, such as micellar solutions and emulsions. In the case of the automotive lubricants, however, the terms *synthetic* and *semi-synthetic* pertain to the type of base fluid used to formulate these lubricants. *Synthetic* implies the use of the synthetic base stocks, such as polyolefins, polyglycols, carboxylic acid esters, etc., and *semi-synthetic* implies a combination of the synthetic base stocks and petroleum base stocks. Despite this difference, both types of fluids meet the general definition of synthetic as being the *man-made* materials [226].

Soluble Oils (Macro-emulsions)

These lubricants are also known as emulsifiable oils or water miscible oils and are emulsions of the mineral oils, fatty oils, or both in water. The term soluble oil is inappropriate since these are not solutions but are emulsions. That is, they are blends of water, oil, and an emulsifying agent. The emulsifying agent, or surfactant, is necessary to form these emulsions. Commonly used oils include mineral oils, both paraffinic and naphthenic types, and natural or fatty oils. Fatty oil-based lubricants have excellent lubricating properties. This is due to the presence of the linear hydrocarbon group

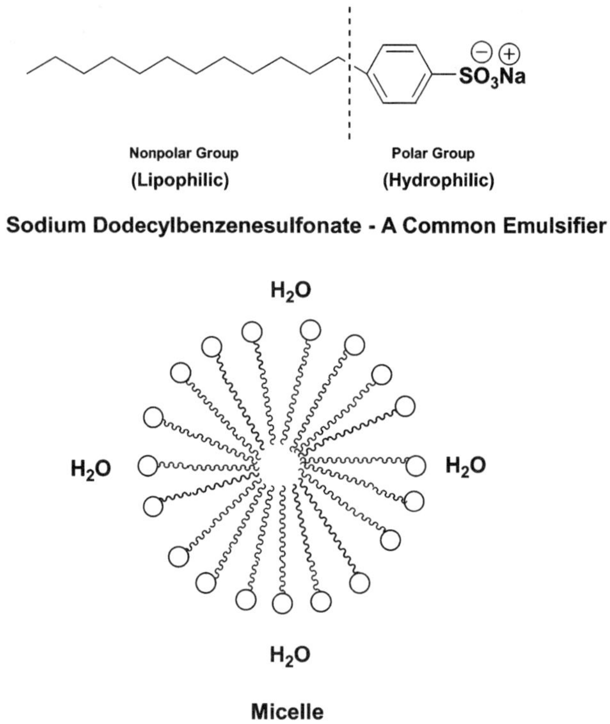

Fig. 11.7—Micellar structure of an emulsifier.

or groups that have friction reducing characteristics (ASTM D2670, D5183, and D5619). Because of the high oil content (>2.0 %), these emulsions are milky in appearance and are easily destabilized by salts, bacteria, and other materials, such as the tramp oil, that can deactivate the emulsifiers. Sulfur, chlorine, and phosphorus-containing additives are used to make these oils suitable for more severe metalworking operations, such as broaching, tapping, and threading. They may also contain rust and foam inhibitors and biocides. In view of their excellent cooling and lubricating properties, *soluble oils* are used for metal cutting operations that involve high speeds, low pressures, and generate high temperatures. In addition, these oils leave an oily film on the metal, giving it the corrosion protection.

Synthetic Fluids (Micellar Solutions)

Micelles are aggregates of emulsifier molecules that occur in water. These result from the tendency of the oleophilic portion of the molecule to avoid the highly polar water solvent and of the hydrophilic portion of the molecule to associate with the water, as is depicted in Fig. 11.7. Synthetic metalworking fluids do not contain any oil, petroleum, or synthetic, and are simply the solutions of the organic additives in water. That is why they are sometimes called chemical fluids. Since most organic materials are hard to dissolve in water, the high polarity of the additives is necessary for solubility. Soaps and other surfactants are often added to facilitate solubilization. Because the synthetic fluids are oil-free, they have poor lubricating ability but have excellent cooling ability. These fluids are therefore ideal for high-speed machining (metal removal) operations that generate a substantial amount of heat. In these operations, these fluids are more effective than the straight oils. Synthetics contain fatty amines, fatty amides, or fatty carboxylic acid salts for rust

TABLE 11.4—Characteristics of common metalworking fluids.

Fluid	Mineral Oil Content	Emulsifier	Contains Water	Comments
Mineral Oil Based				
Straight Oil	100%	None	No	None
Soluble Oil	30%	Yes	Yes	Macro-emulsion, needs to be diluted with water prior to use
Semi-synthetic	≤ 30%	Yes	Yes	Micro-emulsion, needs to be diluted with water prior to use
Mineral Oil Free				
Solution - Synthetic	0%	No	Yes	Nonmicellar, needs to be diluted with water prior to use
Emulsion - Synthetic	0%	Yes	Yes	Needs to be diluted with water prior to use
Straight Oil - Synthetic	-	No	No	These are based upon synthetic base stocks, such as PAOs, PAGs, and synthetic esters

inhibition (ASTM D665) and TEFLON® (PTFE, polytetrafluoroethylene), glycols and poly(alkylene glycol) for lubricity. Since these fluids are solutions and not emulsions, they do not suffer from destabilization problems and are useful for high-speed machining applications. Their higher stability is due to their small micelle size, which also makes them appear clear, and the low organic content. The micelle size in these fluids typically ranges between 0.005 and 0.015 μm. The fluids that do not contain any metal or amine carboxylates are quite stable in hard water. They sometimes contain borates and phosphates for water softening/friction modification; soaps and wetting agents for lubrication and the reduction of the surface tension; phosphorus, chlorine, and sulfur compounds for the extreme pressure properties; and germicides and biocides to reduce biological degradation. Despite their cleanliness, excellent hard water stability, long service life, and effectiveness under high machining speeds, they have the disadvantages of being expensive, have the tendency to wash away greases and paints from the machinery, and leave behind hard crystalline residues.

Synthetics can be formulated to shed tramp oil, that is, the undesirable contaminant oil, for easy skim-off. These fluids are resistant to bacterial degradation (ASTM D3946) because of the low organic content and have good workpiece visibility because of the clarity. The disadvantages include reduced lubricity due to the absence of the petroleum oils; tendency to leave hard crystalline residues; and high alkalinity, higher cost, and a greater tendency to foam. The higher cost and the greater tendency to foam relate to the higher amount of the emulsifiers and the coupling agents that are necessary to formulate them. Foam is usually controlled by the use of the foam inhibitors. Synthetics, in addition, are difficult to dispose of because it is not easy to separate the organics from water without sophisticated separation techniques. Sodium nitrite ($NaNO_2$) corrosion inhibitor in water is a true solution. It was previously used for applications where effective cooling was the only consideration. The use of the sodium nitrite is being discontinued because of the carcinogenic nitrosamine formation, resulting from its interaction with proteins. Because the additives used to formulate these fluids are synthetic in origin, these fluids are appropriately called synthetic lubricants, or the chemical coolants. Synthetics, like emulsions and microemulsions, are obtained by diluting the additive concentrate with water, typically in 1 : 10 to 1 : 50 ratio.

Semi-synthetic Fluids (Micro-emulsions)

These fluids are in between the *soluble oils* and the *synthetic fluids* in terms of their oil content. In every other aspect they resemble the *soluble oils*. Because of the low oil content, these fluids also appear clear. However, some semi-synthetic fluids of the high oil content contain greater than 3.0 % oil and are translucent. As mentioned earlier, the emulsion appearance is a function of the droplet size: The larger the droplet size, the milkier the appearance. These fluids consist of fine colloidal dispersions of organic and inorganic materials in water. The water content in these fluids can vary between 10 and 60 %. Organic materials commonly used in these fluids include fat, sulfur, and chlorine-containing chemicals, which impart extreme-pressure properties. Biocides and corrosion inhibitors can also be added to obtain the biological stability and rust inhibition. The advantages of these fluids include the ability to incorporate both the oil-soluble and the water-soluble chemicals, good lubricity, improved wettability, and easy waste treatment and disposal. However, the high levels of the tramp oil can destabilize these fluids.

Table 11.4 summarizes some of the characteristics of the liquid metalworking fluids that were discussed in the preceding sections.

Solid Dispersions

These lubricants contain solids suspended in water or in oil. Most solids are inorganic in origin, although at times organic polymers are also used. Dispersions are made by mechanically agitating the finely divided solids in the presence of a high molecular weight dispersant. Common solids used to formulate these lubricants include graphite, molybdenum disulfide (MoS_2), metal powders, metal oxides, metal halides, mica, and polytetrafluoroethylene or TEFLON®. Dispersions are hard to maintain because the solid particles are quite large and hence have an increased tendency to settle. During use, these lubricants form low-shear solid films, films that are easily removed when the two contacting surfaces slide, at the tool work-piece interface, which protect surfaces against metal-to-metal contact. The use of these fluids is limited to certain metal forming operations, such as extrusion and forging. They are rarely used in the metal removal lubricants because of their propensity to settle in the presence of debris. Commercial passive EP (PEP) agents, which are colloidal dispersions, are used in these lubricants.

Classification Based on End-use

Metalworking operations generate a significant amount of heat, which if not controlled or quickly dissipated, will lead to tool damage and inefficiencies in the metalworking operations. Metalworking fluids generally cool as well as lubricate; hence, they are sometimes called metalworking coolants, or lubricant coolants. These fluids are formulated to fulfill the cooling and wear control needs of the specific operations because each one differs in these requirements. This makes formulation or selection of a modern metalworking lubricant that meets all the process and product requirements a complicated task. In addition to cooling and lubrication, other properties sought in a metalworking fluid include appropriate viscosity (ASTM D445), ability to wet and adhere to surfaces (ASTM D2782, ASTM D5183, D2783, and D4172), and noncorrosivity to ferrous and nonferrous metals (ASTM D665 and D130). Metalworking fluids are sprayed or poured at the metal-tool interface to dissipate heat, lubricate, protect the freshly exposed metal surfaces against corrosion, and remove debris away from the critical areas.

Metalworking lubricants, based on the mode of their operation, can be classified as metal forming fluids, metal removal fluids, and miscellaneous others. Others include metal protecting fluids, metal treating fluids, and slide way lubricants. Because these fluids are not *directly* involved in metalworking, the present discussion will largely deal with the metal forming and metal removal fluids, which are of the most predominant type. Metal removal fluids are lubricants that are used in applications where the metal is removed from the work-piece in order to obtain the desired shape. Such applications include cutting, drilling, broaching, turning, grinding, milling, threading, reaming, boring, and sawing. The primary functions of the lubricant are cooling, facilitate debris removal, and minimize tool wear. Metal forming fluids are lubricants that are used in operations where the metal in the work-piece is plastically deformed to obtain the desired shape. Such operations involve molding of the metal by the processes of bending, stretching, e.g., wire drawing, and pounding. The primary function of a lubricant in these operations is to reduce friction. Lower friction helps in increasing the tool life and in lowering the energy usage. Metal forming operations include hot rolling, cold rolling, foil rolling, forging, wire drawing, tube drawing, deep drawing, ironing, extrusion, and spinning.

Heat removal has an effect on the surface finish and cold welding of the tool and the work-piece, which affects the tool life. The frictional heat generated during some metalworking operations, depending upon the metal hardness, can reach temperatures of 1000 °C or higher. In addition, extensive cutting pressures at points of contact can lead to specific surface loads of up to 5000 N/mm^2. Both these factors can cause local welding during cutting. Shearing of the welded spots will not only increase the roughness of the surfaces, but it will also expose fresh surfaces that are more prone to welding. Consequently, additional frictional heat will be generated. The function of the fluid is to dissipate this heat and reduce the number and the size of the welded spots. Lubrication effectiveness depends upon the properties of the oil and the presence or the absence of the friction reducing additives. Cooling ability, on the other hand, is a function of the amount of water and the thermal conductivity and specific heat of all of the components present in the formulation. See Table 7.18 for values of these thermodynamic parameters for oil and water.

Metal Removal Fluids

As stated earlier, the metal removing operations are of two types: those where the work-piece is moved against the stationary tool and those where the tool is moved against the stationary work-piece. In both cases, the tool cuts into the work-piece, resulting in the chip formation. These fluids, also called the metal cutting fluids, are utilized in operations that are used to remove the excess metal in the process of manufacturing a new part. Both oil and water-based fluids are used for these operations. Oil-based fluids can be of petroleum, synthetic, or biological (vegetable and animal) origin. These fluids are designed to perform the following key functions:

1. Cooling to prolong the tool life.
2. Lubrication to minimize friction, and hence improve the surface finish.
3. Facilitate removal of the chips and the metal debris.
4. Protect the freshly exposed surfaces against rust and corrosion.
5. Increase productivity and reduce cost through faster material removal rates and lower power consumption.

Heat produced during metal removal is primarily frictional and the most is generated during the chip formation. Additional heat results from deformation of the metal and during the travel of the chip across the tool surface [4,757]. The primary function of the lubricant in metal removal operations is to reduce friction as well as remove heat quickly. It must also remove the metal debris, resulting from the cutting and grinding operations, away from the work-piece. Otherwise, extensive tool wear will occur. While water is an excellent coolant, it lacks the ability to reduce friction and wear. Therefore, these fluids contain the friction reducing and wear control additives. Friction reducing additives primarily generate protective surface films via physical interaction. Wear control additives, on the other hand, generate such films via a chemical reaction. The former type includes fatty materials, such as vegetable oils and animal fats, and the latter type includes sulfur, chlorine, and phosphorus derivatives. The presence of these additives also minimizes welding of the generated metal debris onto the tool edge to form the "built-up" edge. The friction and wear control not only reduces blemishing, but it also improves the surface finish of the work-piece.

The material used for the cutting tools is selected to facilitate metal (chip) removal. Vibration, the metal feed rate, cutting speeds, and the lubricant availability in the cutting zone also play a roll in this process. These factors must therefore be taken into consideration when selecting proper tool

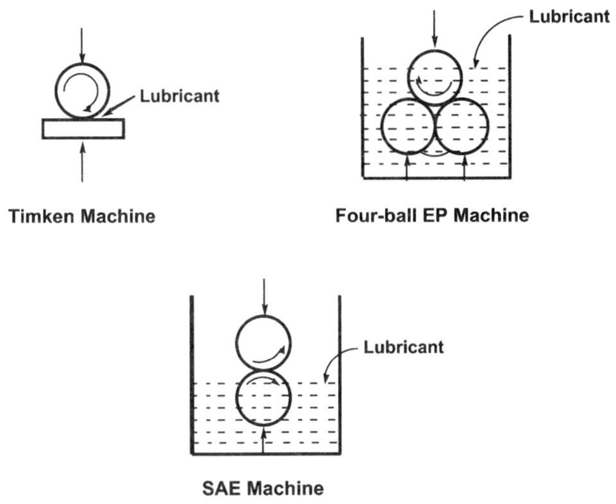

Fig. 11.8—Film-forming test machines [4].

material. For high-speed cutting operations, steel, special cutting alloys, ceramics, and cermets are commonly used. Cermets are composites that contain ceramics and metals bonded together. The material hardness increases in the order listed and so does the generated temperature, from approximately 800 °C to 1200°C. The geometrical shape of the cutting edge also determines the ease of the chip formation, which in turn affects the deformation and the friction zones. That is where the cutting fluid plays a roll, by providing the lubrication and cooling, hence controlling wear. The selection of a proper lubricant is therefore important because it will affect the cutting speed, tool life, surface finish, and the precision of the work-piece. Tool shape, the depth of the cut, and the temperature are also important considerations.

A cutting fluid's ability to reduce friction can be tested by the use of the traditional film-strength test machines. These include the Almen-Wieland rig, Falex Pin-and-vee Block tester, Timken Wear and Lubricant Testing machine (ASTM D2782), Four-Ball Test rig (ASTM D5183, D2783, and D4172), and SAE #1 Test rig [27,761]. However, a correlation with the field performance is lacking. Thread tapping tests (ASTM D5619), used by some additive suppliers, may also be meaningful in assessing the efficiency of the cutting oils. For cutting operations, one can use either straight cutting oils; which are mineral oils containing active chemical ingredients; oil-water emulsions with or without these ingredients; or aqueous solutions.

Test elements of three film strength machines are shown in Fig. 11.8 [4]. These machines are used to measure the effect of load and temperature on the film-forming ability of the lubricant, as reflected by the friction coefficient and wear. Results for the newly formulated oils are compared with those of the reference oils to measure their relative effectiveness. Because of the differences in the test conditions, the results from different machines are not directly comparable. The Almen-Wieland, Falex, Timken, and Four-Ball test machines examine the effects of sliding friction only. The SAE #1 tester, on the other hand, evaluates a lubricant's performance both under sliding and under rolling conditions.

The Almen-Wieland rig employs a soft steel pin that rotates in a soft steel split bushing at a constant speed. The load is increased in stages until either seizure occurs or the maximum load is reached. The Falex tester also uses a soft steel pin rotating at a constant speed but differs from the Almen-Wieland machine in that the pin rotates between the two hard V-shaped bearing blocks, instead of in a soft steel bushing. The test includes a run-in stage and two stages with the increasing loads.

The Timken machine consists of a rotating cylindrical steel ring pressed against a steel block. The machine is operated for ten minutes to determine the highest pressure at which there is no scoring of the rotating ring or the block.

In the Four-Ball tester, a one-half inch rotating steel ball is pressed against three stationary balls of the same size and quality. The system of balls is in a holder that contains enough lubricant to cover the three stationary balls. This setup provides three small areas of initial point contact. Because of this, high specific pressures are created at low loads, which lead to the deformation of the contact surfaces. The load is increased until the frictional heat welds the balls together. Weld load and wear scar diameter as a function of load are measured. Wear-reducing additives tend to form scars of smaller diameter, and welding occurs at higher pressures.

The SAE #1 tester consists of the two cylindrical rollers that are rotated against each other at different relative speeds. The test oil lubricates the lower roller, which is mechanically driven. The load is increased progressively until the failure occurs. The SAE machine simulates both the rolling friction and the sliding friction. The ratio of the two types of friction can be changed by changing the rotating speeds of the cylinders.

These tests are the only way to assess the suitability of lubricants for equipment where mixed and boundary lubrication conditions predominate. However, the data obtained from these tests are subjective. For slide way lubricants, the data from these tests can be substantiated by testing in the FZG (ASTM D5182), Gleason, and the Ryder Gear Tests [4].

Metal removal operations by their very nature, i.e., due to high pressures and high temperatures, do not permit hydrodynamic lubrication. Hence, lubricants containing film-forming additives are usually required. For materials of moderate hardness and for operations involving high speeds, friction reducing agents provide the necessary performance. However, for difficult to cut materials and operations involving high pressures and slow speeds, the extreme pressure agents are needed.

The selection of the proper lubricant depends upon the nature of the metal and the severity of the cutting operation. In general, the operations that employ slow cutting speeds place a higher demand on the lubricant than those that employ high cutting speeds. Figure 11.9 shows the relative severity of different cutting operations in terms of friction and heat generation and the demand they place on the lubricant with respect to the need for friction reduction and cooling [690]. Straight oils are used in operations where the speeds are slow, pressures are high, and the metal is very hard. This is because straight oils have high lubricity and good antiweld properties. Conversely, the water-based fluids are used when the machining speeds are slow, pressures are lower, and the stock removal is lower. This is because the water is

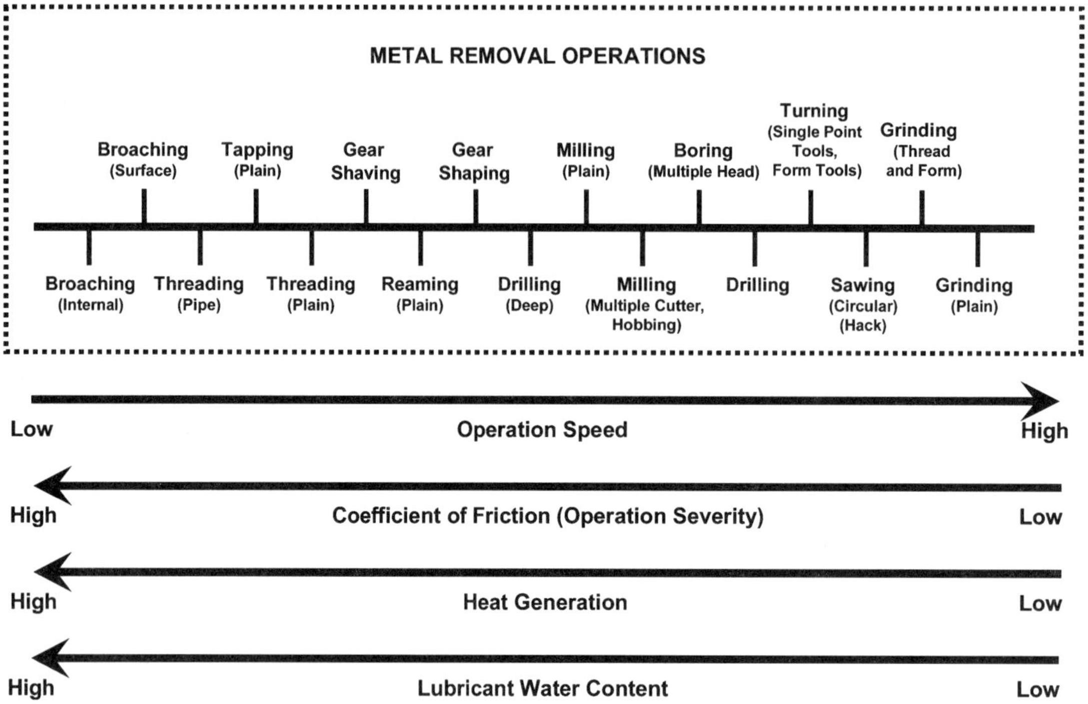

Fig. 11.9—Relative severity of some metal-removal operations [690,757].

better coolant than the oil. More specifically, straight oils and emulsifiable oils that contain combinations of sulfur, fat, phosphorus, overbased sulfonates, or chlorine are the lubricants of choice for severe operations, such as broaching, tapping, deep hole drilling, heavy forming (e.g., drawing, cold heading), form grinding and honing. Semi-synthetics are useful for the moderate severity operations, such as milling, turning, grinding, and stamping; and synthetics are ideal for most high-speed operations such as drilling. Please note that these uses are not absolute and an operator has the option of using a fluid other than that suggested if the operation has severity different than that suggested here.

Tool Wear

Besides the severity of the machining operation, tool wear is also a function of the quality of the cutting fluid and the machinability of the metal. Cutting fluid is essential to minimizing wear and equipment damage. Common types of wear that can occur in cutting tools due to the lubricant failure are as follows:

1. *Adhesive wear*—Welding between the tool surface and the work-piece surface can occur due to the high local temperatures. On shear, these spots can either produce the metal debris or result in the metal transfer from the tool to the work-piece. This kind of damage is more likely to occur at slow cutting speeds.
2. *Abrasive wear* occurs when the fragments of cutting or the wear debris embedded in the chip abrade the rake face of the tool.
3. *Diffusive wear* occurs when the metal atoms from one crystal lattice diffuse into the crystal lattice with the lower atomic density (concentration). This type of wear is likely when the tools made of hard materials are used

TABLE 11.5—Attributes of tool materials.

Tool Material	Attribute
Carbon Steel	Short service, limited effectiveness, loses hardness at relatively low temperatures
High-speed Steel (HSS)	Maintains high cutting edge at high temperatures, contains tungsten (W), molybdenum (Mo), chromium (Cr), cobalt (Co), vanadium (V)
Cast Alloys	Cobalt-based, more brittle than HSS, fills the gap between HSS and carbide tools
Cermets	Carbide-based, extremely hard, lack strength and toughness (need good support), not used where there is heavy vibration or shock
Ceramics	Cemented metallic oxides, hardness close to diamond, do not pick up heat, are brittle and therefore chip easily
Diamonds	Hardest of all abrasives, large stones are used in dressing tools, small stones are used in grinding wheels

TABLE 11.6—Machining characteristics of machined metals.

Machined Metal	Characteristics
Steels	
Free Machining	Modified for machinability with added phosphorus or sulfur
Alloy steels	Contain chromium (Cr), cobalt (Co), nickel (Ni), molybdenum (Mo), tungsten (W), vanadium (V), singly or in combination; more difficult to machine
Stainless steel	Difficult to machine because of high chromium (Cr) and or nickel (Ni) that work harden
Cast Iron	
Gray cast iron	Hard to machine
Malleable cast iron	Hard to machine
Nodular iron	Hard to machine
White cast iron	Hard to machine
Wrought iron	Hard to machine
Aluminum and its Alloys	Machine easily relative to carbon steel; pure aluminum may be gummy
Magnesium and its Alloys	Excellent machinability; chemically reactive hence use of water or acidic materials in the fluid must be avoided
Copper and its Alloys	
Single phase alloys, such as bronze, copper-nickel, copper-silicon	Hard to machine
Alloyed copper, lead-bronze, tin-bronze, naval brass, aluminum-bronze	Medium machinability
Alloys that contain substantial amounts of alloy materials	Free machining
Nickel and Its Alloys	
Pure nickel Incone (nickel-chromium-iron alloy) Hast alloy (nickel-copper alloy containing small amounts of iron, molybdenum and chromium) Monel (nickel-copper alloy)	Draggy and difficult to machine
Nickel-silver alloy	Good machinability
Heat Resistant Alloys	
Contain nickel, chromium, molybdenum, and titanium	Hard to machine and hence need heavily additized cutting fluids
Plastics	Thermoplastics melt and thermoset plastics char at low temperatures, hence cooling is most important. Soluble oils are often used

to machine softer metals, for example, during the use of the tungsten carbide tools.

4. *Corrosive wear* results from the attack of the atmospheric oxygen, moisture, or the decomposition products from the antiwear and extreme-pressure additives on the tool surface. Such additives include chlorine, sulfur, and phosphorus-containing chemicals.
5. *Premature wear* is the progressive loss of tool material at the tool work-piece interface and occurs due to the plastic flow of the metal at high temperatures.

Machinability primarily relates to metal hardness, although other factors, such as cutting speed, tool strength, and power consumption, also play a role. In general, the harder the metal, the greater is the tool wear. It is therefore important to select proper tool material for the job at hand. Commonly machined materials include steels, both alloy and stainless, cast iron, aluminum and its alloys, magnesium and its alloys, copper and its alloys, nickel and its alloys, heat resistant alloys that contain nickel, chromium, molybdenum, tungsten, and titanium, and plastics. Tool materials include carbon steel, high-speed steel, cast alloys, ceramics, cermets, and diamonds. The attributes of the various tool materials are provided in Table 11.5 and the characteristics of the metals machined are provided in Table 11.6.

Of the metals listed, ferrous metals such as carbon steel, low alloy steel, stainless steel, and aluminum are the largest volume metals machined. Copper, brass, and titanium are the next group, followed by the nickel-based alloys, cobalt-based alloys, magnesium, zinc, tin, beryllium, zirconium, tungsten, molybdenum, tantalum, uranium, and vanadium. The order with respect to the ease of machinability, from the easiest to the most difficult, is as follows:

1. Magnesium (Mg) and its alloys
2. Zinc (Zn) and its alloys
3. Brass
4. Aluminum (Al) and its alloys
5. Cast iron (Fe)
6. Bronze
7. Copper (Cu)
8. Carbon steel
9. Alloy steels
10. Stainless steel

TABLE 11.7—Cutting fluid recommendations by machinability groups [762].

Operation Type	Ferrous (>70 %)	Ferrous (50–70 %)	Ferrous (>40–50 %)	Ferrous (<40 %)	Nonferrous[a] (100 %)	Nonferrous[a] (≤100 %)
Broaching						
Internal	4,7	4,7	4,5,7	4,1,7	1,7	6,7
Surface	4,7,8	4,7	4,7	4,1,7	1,7,8	6,7,8
Tapping	4,1,7	4,1,7	4,1,7	4,1,7	3,1,7	3,1,7
Threading	4,7,8	4,7,8	4,7	4,7	3,7	2,7
Gear Shaving	6	6	6	4	3	3
Reaming	6,7,8	6,7,8	4,7,8	4,7,8	3,7,8	3,7,8
Gear Cutting	6	6	6	4	3	3
Drilling	10,8,7,4	10,8,7,4	10,8,7,4	10,8,7,4	10,8,7,3	10,8,7,3
Deep Hole Drilling	4,6,7,8	4,6,7,8	4,6,7,8	4,6,7,8	3,7,8	3,7,8
Milling	8,7,6	8,7,6	8,7,6	8,7,6	8,7,3	8,7,3
Boring	4,7	4,7	4,7	4,7	3,7	3,7
Planing	7,8,10	7,8,10	7,8,10	7,8,10	7,8,10	7,8,10
Turning	10,7,8	10,7,8	10,7,8	10,8,7	10,8,7	10,8,7
Sawing	4,7,8	4,7,8	4,7,8	4,7,8	3,7,8	3,7,8
Surface Grinding	10,9,8	10,9	10,9	10,9	9,10	9,10
Thread Grinding	10,8,7	10,8	4,5,8	4,7,8	1,7	1,7,8
Form Grinding	10,8,7	10,8	4,5,8	4,7,8	1,7	1,7,8

1. Straight Mineral Oil
2. Straight Fatty Oil
3. Blend of Mineral and Fatty Oil
4. Sulfurized Mineral-Fatty Oil Blend
5. Sulfurized Mineral Oil
6. Sulfo-chlorinated Mineral Oil
7. Soluble Oil
8. Chemical Emulsion
9. Simple Chemical Solution
10. Complex Chemical Solution

[a]Water-based fluids must not be used on magnesium grinding and great care must be taken in fluid selection for magnesium machining.

11. Nickel (Ni) alloys
12. Titanium (Ti) and its alloys

Because the metalworking operations involve both the pure metals and the alloys, the selection and use of an appropriate lubricant is a complex process. Table 11.7 lists cutting fluid recommendations based upon machinability groups [762] and Table 11.8 recommends the cutting and grinding fluids for specific metals.

Operation severity is a function of the cutting speed. In general, the operations involving slow cutting speeds, such as broaching and tapping, are more severe than those that have high cutting speeds, such as turning and milling. Drilling and reaming fall in between the two extremes in terms of severity. High severity at slow speeds is because at these speeds the metal-to-metal contact is more extensive (boundary situation) and of longer duration than at high speeds which promote the hydrodynamic lubrication. Consequently, there is a greater need for extreme pressure agents and anti-weld additives in the former case than in the latter case, where either the lubricant viscosity or the friction-reducing additives suffice. Since good cooling and good lubrication are not easy to obtain in the same fluid, water or micro-emulsions (semi-synthetic fluids) are used for operations that require more cooling. And, the use of the straight oils is preferred for operations that require better lubrication.

The majority of operations require both cooling and lubrication. Hence, the macro-emulsions (soluble oils) that contain substantial amounts of both organic and aqueous components are often employed. The operating range of these fluids is enhanced by the supplemental use of the fatty additives. For a cutting fluid to perform effectively, a careful consideration of its cooling and lubricating requirements is essential. High-speed cutting operations, such as turning, milling, and drilling, result in high temperatures; hence the cooling ability of the fluid is critical. Water is an excellent coolant but has little or no lubricating ability. However, this deficiency is overcome by the use of the friction reducers and the EP agents, which are entrained into water by the use of the wetting agents and the emulsifiers. Slow-speed operations, such as broaching and tapping, on the other hand, experience high friction and consequently lead to heavy tool wear. The lubricants for these operations therefore require the use of the extreme pressure additives [691]. A new technique, called dry machining, is being explored. The purpose is to eliminate the purchase, handling, use, and disposal costs. Since this technique does not use any lubricant, it involves high temperatures. Thus far, the technique has been applied only to the boring operations. Efforts are underway to extend it to drilling operations [692,693].

The depth of cut and the surface finish are additional

TABLE 11.8—Cutting and grinding fluid recommendations by metal types.

Removal Operation	Cast Iron	Carbon Steel	Stainless and High Alloy Steel	Nickel and Monel	Copper Brass and Bronze	Aluminum	Magnesium
Broaching	Soluble oil or synthetic fluid	High activity sulfur-containing oil	High activity sulfur- and chlorine-containing oil	High activity sulfur-containing oil	Mineral oil-fatty oil blend or low activity sulfur-containing oil	Low activity sulfur-containing oil	Mineral oil-fatty oil blend
Threading	Soluble oil or synthetic fluid	High activity sulfur-containing oil	High activity sulfur- and chlorine-containing oil	High activity sulfur-containing oil	Mineral oil-fatty oil blend or low activity sulfur-containing oil	Low activity sulfur-containing oil	Mineral oil-fatty oil blend
Gear Cutting	Soluble oil or synthetic fluid	High activity sulfur-containing oil	High activity sulfur- and chlorine-containing oil	...	Mineral oil-fatty oil blend	...	...
Drilling and Reaming	Soluble oil or synthetic fluid	High activity sulfur-containing oil	High activity sulfur- and chlorine-containing oil	High activity sulfur-containing oil	Mineral oil-fatty oil blend	Mineral oil-fatty oil blend	Mineral oil-fatty oil blend
Boring and Turning	...	Soluble oil or synthetic fluid Soluble oil	Soluble oil	High activity sulfur-containing oil	Soluble oil or synthetic fluid	Soluble oil or synthetic fluid	Mineral oil-fatty oil blend
Automatic Screw Machining	...	High activity sulfur- and chlorine-containing oil	High activity sulfur- and chlorine-containing oil	High activity sulfur- and chlorine-containing oil	Mineral oil or mineral oil-fatty oil blend	Low activity sulfur-containing oil	Mineral oil-fatty oil blend
Milling	Soluble oil or synthetic fluid	Soluble oil or synthetic fluid	Soluble oil	High activity sulfur-containing oil	Mineral oil-fatty oil blend, soluble oil, or synthetic fluid	Soluble oil or synthetic fluid	Mineral oil-fatty oil blend
Thread Rolling	...	High activity sulfur-containing oil	High activity sulfur-containing oil	...	Mineral oil or a low activity sulfur-containing oil	Mineral oil-fatty oil blend	Low/medium viscosity mineral oil
Sawing	Soluble oil or synthetic fluid	Soluble oil or synthetic fluid	Soluble oil or synthetic fluid	Soluble oil or synthetic fluid	Soluble oil or synthetic fluid	Soluble oil or synthetic fluid	Mineral oil-fatty oil blend
Thread Grinding	...	High activity sulfur- and chlorine-containing oil	High activity sulfur-containing oil	—	Low activity sulfur-containing oil	Low activity sulfur-containing oil	Low activity sulfur-containing oil
Other Grinding	Soluble oil or synthetic fluid	Soluble oil or synthetic fluid	Soluble oil or synthetic fluid	Soluble oil or synthetic fluid	Soluble oil or synthetic fluid	Soluble oil or synthetic fluid	Mineral oil-fatty oil blend

considerations during metalworking fluid selection. Slow-speed operations where the cuts are deep and the good work-piece finish is important, the lubricity of the fluid is critical. The use of the straight oils is therefore appropriate. For high-speed operations that involve shallow cuts, quick heat dissipation is desired. This makes micro-emulsions or micellar solutions the lubricants of choice. In most cases, the fluid is applied under pressure into the cutting zone, which is to reduce friction, minimize metal transfer, and maximize cooling.

Metal Forming Fluids

These fluids are used for operations that depend on the plastic flow of the metal. Such operations include rolling, extrusion, drawing (drawing tubes, tube bending, deep drawing, and wire forming), forging, and sheet metal forming. Some of these operations involve both ambient (cold working) and high temperatures (hot working). Metal forming processes can be distinguished as being steady-state or nonsteady-state [757], each type with different lubrication requirements. In the steady-state processes, such as rolling, it is possible to lubricate the surface of the work-piece during its approach to the deformation zone. However, in the nonsteady-state processes, such as sheet metal forming, the lubrication is not usually possible because of the nature of the operation, and one must depend upon the preapplied lubricant film. Some processes, such as long billet extrusion, have characteristics of both; that is no lubricant application during certain parts of the operation and lubrication during the other parts of the operation.

Metal forming fluids are designed to perform a number of functions, which include the following:

1. Friction control between the work-piece and the die surfaces.
2. Minimizing the tool and die wear.
3. Efficient dissipation of heat.
4. Work-piece and the die surface protection, for example, against the metal pick-up and corrosion.

The nature of the metal forming operations is much different than the metal cutting or the metal removal operations. The challenge in the metal forming operations is not only to deliver the lubricant where needed but also to stay during the period of the operation. This is because a liquid lubricant will have the tendency to squeeze out, when pressure is applied to conform the sheet metal to the die to make the part. However, it is important to note that the use of the high viscosity oils is not a universal solution since if the operation temperature is too high; the viscosity advantage of using the thicker oils will be lost. The outflow in such operations is further facilitated under heavy loads when the surfaces in contact smooth out due to flattening of the asperities, as shown in Fig. 11.4. Under these circumstances, the solid lubricants and the chemically bonded boundary coatings, such as the phosphated surfaces, may be more suitable. If the liquid lubricant must be used because of the inherent advantages, it is customary to apply the lubricant manually

TABLE 11.9—Operation severity of metal drawing operations [756].

Severity	Operation
Low	1. Drawing a cup with 10% reduction in diameter.
	2. Forming with a maximum localized stretch of 10% in a 25 mm length.
	3. Bending 90° around a radius equal to the blank thickness with a flange equal to or greater than 25 times the blank thickness.
	4. Blanking and piercing sheet up to 2.5 mm thick with clearances of 10% of sheet thickness.
Moderate	1. Drawing a cylindrical shell with 20% reduction in diameter.
	2. Forming with a maximum localized stretch of 20% in a 25 mm length.
	3. 90° bend around a radius less than the blank thickness with a flange width of 15 times the blank thickness.
	4. Blanking and piercing sheet thicker than 2.5 mm.
High	1. Drawing a cylindrical shell with 30% reduction in diameter.
	2. Forming with maximum localized stretch of 30% in a 25 mm diameter.
	3. Bending sheet less than 3 mm thick 90° with a flange width of equal to, or less than 10 times the blank thickness.
Maximum	1. Deep drawing a cylindrical shell with 40% or greater reduction in diameter.
	2. Forming with localized stretch of about 40% in a 25 mm length.

on the selected areas of the work-piece. Such application methods are expensive, but because of the equipment design this may be the only practical way to lubricate effectively. Liquid lubricants used in metal forming operations considered here must contain chlorinated or sulfochlorinated additives that under heat will form effective low friction EP films between surfaces.

While such films are an advantage with respect to preventing tool wear because they minimize direct metal-to-metal contact between the tool and the work metal (blank) or the strip and the die. However, this can be a disadvantage in many metal-forming operations, such as forming sheet metal and deep drawing, because the controlled metal-to-metal contact (i.e., controlled friction) is used to manage the flow of the work-piece over the die, roll, or the tool surface. Obviously, in such operations some tool wear will occur. The wear rate depends upon the die material as well as the effectiveness of the lubricant. For example, the use of the solid lubricants or the chemically bonded work metal is likely to result in greater die wear and in the former case in part due to abrasion. With extreme conditions of pressure and temperature, welding will occur between the metal asperities, resulting in pick-up on either the work metal or the die surface. This pick-up will cause damage to both the die and the work-piece. Hence, it is important to select the lubricant by considering these potential problems. The use of the uncoated low-carbon steel is the best in such metal forming operations as press forming, bending, spinning, roll-forming, and punching.

Temperature in these operations is primarily controlled by reducing the friction. It can also be controlled by applying a greater amount of lubricant or water, if possible. Water has a greater specific heat and acts as a better heat sink than mineral oil. Obviously, in most cases oil-in-water emulsions have a real advantage since they contain oil that will reduce friction through lubrication and have higher viscosity and are therefore more likely to stay in place.

The newly formed work-piece has freshly exposed metal which needs to be protected against atmospheric corrosion, especially if the part is going to be stored for extended periods of time. That is where the metal treating fluids, such as the rust preventives, to be discussed later, play a role. These fluids usually contain soaps (metal carboxylates), gelled metal sulfonates, and waxes which provide a physical barrier between the metal and the environment.

As mentioned in the above discussion, the primary functions of a lubricant during metal forming operations are to reduce friction, which lowers the energy consumption, and to minimize wear, which increases the tool life. It was also mentioned that the friction reduction is both advantageous and disadvantageous, depending upon the nature of the operation. Advantageous because it will facilitate the release of the forged part from the die and improve the surface finish of the work-piece, and disadvantageous because it can lead to slippage and hence make the operation inefficient. Common metal forming lubricants include the following types. Additives for these lubricants include fatty acids and fatty compounds, extreme pressure agents (sulfurized and sulfochlorinated fats and oils), emulsifiers, coupling agents, inorganic solids, and dispersants.

1. Mineral oils
2. Compounded oils, blends of mineral oils and fatty oils
3. Synthetic oils and esters
4. Fatty acids and their derivatives
5. EP oils
6. Aqueous solutions and emulsions
7. Dry film coatings
8. Polymer solutions
9. Dispersions containing graphite, molybdenum disulfide (MoS_2), salts, glass, bentonite, lime, mica, and talc

Mineral oils are not very effective lubricants in these applications. They are acceptable only if their viscosity is not too high. However, they are easy to apply and to remove from the work-piece by the use of a solvent or the detergent cleaners. *Compounded oils,* which are mixtures of the mineral oils and the animal fat or vegetable oil, such as palm oil, are much superior. These are extremely efficient lubricants but are relatively expensive. *EP oils* are analogous to the compounded oils, except that they are blends of the mineral oils and chlorine, sulfur, nitrogen and phosphorus-containing

TABLE 11.10—Lubricants commonly used in press forming uncoated low carbon steel sheet [756].

Operation Severity	Lubricant Type/Composition	Ease of Removal		Rust Protection
		Water-based Cleaners	Degreasers/Solvents	
Water-based Lubricants				
Low	Water emulsion of 5-20% – General purpose soluble oil or wax	Very good	Good	Fair
Moderate	1. Water solution 5-20% soap	Very good	Very good	Fair
	2. Water emulsion of heavy-duty soluble oil (contains sulfurized or chlorinated additives)	Very good	Good	Fair
High	1. Soap-fat paste, diluted with water (may contain wax)	Fair	Poor	Fair
	2. Water-emulsion of heavy-duty soluble oil (contains a high concentration of sulfurized or chlorinated additives)	Very good	Good	Fair to good
Maximum	1. Pigmented soap-fat paste, diluted with water	Poor	Very poor	Good
	2. Dry soap or wax, applied from water solution or dispersion (may contain soluble filler such as borax)	Good	Very poor	Good
Oil-based Lubricants				
Low	1. Mill oil, residual (viscosity usually about 20 mm^2/s at 40°C)	Good	Very good	Fair
	2. Mineral oil (viscosity 5-65 mm^2/s at 40°C)	Good	Very good	Fair
	3. Evaporating oil (Refined paraffinic oil)	Removal not required	Removal not required	Nil
Moderate	1. Mineral oil (viscosity 20-65 mm^2/s at 40°C) plus 10-30% fatty oil	Good	Very good	Fair
	2. Mineral oil plus 2-20% sulfurized or chlorinated oil (EP oil)	Good to fair	Good	Fair to poor
High	1. Fatty oil	Fair	Fair	Fair
	2. Mineral oil (viscosity 20-65 mm^2/s at 40°C) plus 5-50% of:			
	(a) Non-emulsifiable chlorinated oil	Poor	Good	Very poor
	(b) Emulsifiable chlorinated oil	Good	Good	Very poor
	3. Concentrated phosphated oil	Fair	Fair	Fair
Maximum	1. Blend of pigmented soap-fat paste with mineral oil	Poor	Poor	Fair
	2. Concentrated sulfochlorinated oil (may contain some fatty oil)			
	(a) Non-emulsifiable	Very poor	Fair	Very poor
	(b) Emulsifiable	Good	Fair	Vary poor
	3. Concentrated chlorinated oil			
	(a) Non-emulsifiable	Very poor	Fair	Very poor
	(b) Emulsifiable	Good	Fair	Very poor

EP additives. These additives thermally react with metals to form low friction, low shear, sacrificial metal halide, sulfide, and or phosphide protective film on the work-piece and the tool surface.

Soap solutions are generally potassium or sodium stearate or oleate diluted with water. For high severity drawing operations, the soap concentration needs to be increased. These solutions are excellent lubricants in high speed operations since they provide cooling in addition to the friction reduction. In addition, like the mineral and compounded oils, these fluids are also easy to apply and to remove by the use of the water-based cleaners. *Emulsions* used in metal forming are of the oil-in-water type and are useful in moderate speed forming operations. They may contain EP additives for use in operations that involve boundary conditions. Another type of emulsion that is used in forming operations is the soap-fat emulsions. These emulsions are thick paste-like systems that contain 35–40 % fat and can be used as such for operations such as drawing and stamping or diluted with water prior to use; the concentration depending upon the operation severity. Application is easy but removal is difficult since the use of the water-based or solvent cleaners leaves a residue.

Dry film coatings include any material that will form a solid film. Commonly employed materials include solid soap, fats, such as lanolin and tallow, paraffin waxes, and organic polymers. Soap-based films can be continuously applied to the strip while removing the water by the use of the warm air. Film thickness can be controlled within the desired limits. Fats and waxes can be employed as solutions

TABLE 11.11—Lubricant effectiveness in metal drawing operations [4].

Lubricant	Ferrous Metals	Copper and Brass	Bronze	Light Alloy
Powdered Compounds	Most Preferred	Limited Use	Most Preferred	Limited Use
Drawing Grease	Most Preferred	Most Preferred	Most Preferred	Most Preferred
Straight Oil	Most Preferred	Most Preferred	Most Preferred	Most Preferred
Emulsion	Most Preferred	Most Preferred	Limited Use	Most Preferred
Soap Solution	Most Preferred	Most Preferred	Not Used	Not Used

TABLE 11.12—Drawing compound properties [4].

Property	Powdered Compounds	Drawing Grease	Straight Oil	Emulsion	Soap Solution
Lubricity	+	+	+	(+)	(+)
Cooling Effect	−	(+)	+	+	+
Filterability	−	−	+	(+)	+
Annealing	(+)	(+)	+	(+)	(+)
Corrosion Protection	−	(+)	+	+	−
Adhesion	(+)	+	+	(+)	−

Note: + = Pronounced; (+) = Limited; − None.

containing the low volatility solvents or as emulsions. Acrylic polymer films are produced by the use of the warm trichloroethylene solutions and polyethylene films are produced by placing the low-melting, low density polyethylene over the blanks, prior to their entry into the press. These coatings provide excellent dry film lubrication.

Graphite and molybdenum disulfide are two common *solid lubricants* that are used in the metal forming operations. They can be used as such, as dispersions, or as additives in other lubricant systems. Their ability to lubricate and the way they perform this function were discussed in Chapter 4 on Additives. Other solids that are used include clay, talc, and chalk. They are usually used in combination with pastes made from soap and fat. While these materials protect sliding surfaces from damage under extreme pressure by physical separation, they may or may not decrease friction, which is a function of their crystal structure.

Metal forming operations vary widely with respect to severity. Therefore, it is important that the selected lubricant meets the lubrication requirements of the intended operation. While low severity operations are not highly dependent

TABLE 11.13—Drawing lubricant recommendations.

Operation	Carbon and Alloy Steels	Stainless Steel	Aluminum	Copper and Its Alloys
Wire drawing	Fat-based soluble oil	Straight oil containing sulfur and chlorine-based EP additives	Straight mineral oil-fatty oil blend	Fat-based soluble oil
Bar drawing	Fat-based semi-synthetic fluid	Straight oil containing sulfur and chlorine-based EP additives or a fine soft metal (such as lead) dispersion	Straight mineral oil-fatty oil blend	Fat-based semi-synthetic fluid
Tube drawing	Fat-based semi-synthetic fluid	Straight oil containing sulfur and chlorine-based EP additives or a fine soft metal (such as lead) dispersion	Straight mineral oil-fatty oil blend	Fat-based soluble oil or semi-synthetic fluid
Rod breakdown	Solid Lubricant containing soap and dry lime	Straight oil containing sulfur and chlorine-based EP additives	Straight mineral oil-fatty oil blend	Fat-based soluble oil
Press drawing	Fat-based soluble oil or semi-synthetic fluid	Straight oil containing sulfur and chlorine-based EP additives	Fat-based soluble oil	Straight mineral oil-fatty oil blend or fat-based soluble oil or semi-synthetic fluid
Blanking and stamping	Low to medium viscosity straight mineral oil, mineral oil-fatty oil blend, or fat-based soluble oil	Straight oil containing sulfur and chlorine-based EP additives	Fat-based soluble oil	Low to medium viscosity straight mineral oil or Fat-based soluble oil or semi-synthetic fluid
Heavy cupping	Fat-based semi-synthetic fluid	Straight oil containing sulfur and chlorine-based EP additives	Straight fatty oil or fat-based soluble oil	Fat-based semi-synthetic fluid
Deep drawing (cold)	Fat-based semi-synthetic fluid	Straight oil containing sulfur and chlorine-based EP additives	Straight fatty oil or fat-based soluble oil	Straight fatty oil or fat-based semi-synthetic Fluid
Deep drawing (hot)	---	---	Solid lubricant, such as graphite, as an aqueous or oil dispersion	---

TABLE 11.14—Operation severity of metal rolling operations [756].

Steel Surface	Lubricant Properties
Hot-rolled with scale	Penetration, wetting and flushing are essential. Usually soluble oils with proper balance of EP additives and wetting agents.
Hot-rolled, pickled	Soluble or light oils with good pickled wetting. Rust preventative oils.
Cold-rolled	Heavy duty EP type water soluble oils with EP additives and good wetting and minimum drag out.
Alloy Steel, Metal-coated and Prepainted Alloy Steels	Light paraffinic oils and evaporating compounds, if steelcoating by itself is not sufficient.

TABLE 11.15—Rolling oil properties versus rolling operation parameters [4].

Operation Parameter	High Friction Coefficient	Low Friction Coefficient	Strong Cooling
Energy Consumption	−	+	−
Heat Evolution	−	+	−
Roll Wear	−	+	−
Roll Imprint	+	−	−
Slipping	+	−	−
Roll Pick−up	−	+	−
Rolling Speed	−	−	+
Planeness	−	−	+

Note: − = No Effect; + = Positive Effect.

upon lubricant properties, with increased operation severity proper lubricant properties become more critical. In these cases, a lubricant with greater friction control or EP performance may be required. Such lubricants contain friction-reducing additives and or the extreme-pressure agents. Since the friction modifiers are quite effective in the mixed-film lubrication regime, they lose their effectiveness with a significant rise in pressure, temperature, and the shear rate, three of the boundary lubrication conditions. That is when there is a need for the extreme-pressure additives. Operations where the operation severity changes during the process need a balanced combination of both types of additives. In addition, the EP agents are more effective than friction reducers for maintaining a boundary film between the tool and the work-piece under conditions of severe deformation, which extends the tool life. Table 11.9 provides some of the criteria used to identify operation severity in metal forming operations, as exemplified by metal drawing, and Table 11.10 provides a list of lubricants suitable for each severity class [756]. Lubricant selection for metal drawing operation is not easy but the information in Tables 11.11–11.13 attempts to facilitate the process. Table 11.11 suggests the lubricant preference by metal class [4], Table 11.12 lists the lubricant properties necessary [4], and Table 11.13 provides the actual lubricant recommendations.

For metal rolling operation, the type of lubricant selected depends upon the surface characteristics of the steel strip, roll speed, roll material, roll condition, and the severity of forming. Table 11.14 lists severity of various metal rolling operations and the desirable characteristics of a suitable lubricant [756]. Table 11.15 lists specific lubricant properties that affect various operation parameters [4]. For actual lubricant recommendations see Ref [639].

For bending and spinning operations, the use of the mill oil or light mineral oil is usually sufficient. But for prepainted steel, light oil or evaporating oil formulations are necessary. For spinning operations, such as manual spinning or when the prepainted steel is used, little or no lubricant is necessary since heat and friction can be controlled by the use of the special tools, such as roller heads. Only if necessary, one needs to use the mill oil or the light mineral oil. For punching coated steel, usually no lubricant is required; however, in some situations, the use of the punching oil is beneficial.

In many cases, the residual lubricant from the formed part must be removed to facilitate further processing operations, such as spot welding, coating, and enameling. Solvent degreasing or the use of the alkali cleaners are the common methods. These methods have the issues of toxicity, flammability, and the disposal of the spent solvent and cleaners. Obviously, the materials of low toxicity, low flammability, and easy disposal are highly desirable. Some of these can be used many times over before disposal. Another option is recycling, which will be discussed in Chapter 13 on Lubricant Recycling.

Miscellaneous Fluids

Metal Protecting Fluids

These fluids, also called preservative oils, help protect the freshly exposed metal surfaces against air, water, and corrosive materials. While most of these are oil-based, the use of the water-based fluids is gaining popularity because of the lower cost, ease of disposal, environmental compatibility, and the reduced volatile carbon content.

Metal Treating Fluids

These fluids are used for heat-treating operations, such as quenching and tempering. Machining a work-piece from hard metals is difficult. Hence, it is often desirable to use soft materials, where possible, and then impart to them the required hardness by an after-treatment. Work-pieces from the ferrous alloys are heated to 750–1100 °C, depending on their composition, and are quenched by immersing them in a quenching fluid. Quenching is the process of controlled cooling of the steel components by the use of a fluid to obtain certain metallurgical properties, such as increased hardness, strength, and wear resistance. The process involves heating and holding the metal, such as steel, at temperatures above 1200°C to disperse carbon and alloying elements throughout the metal mass and then cooling it rapidly. In iron, this produces the crystal structure, called martensite, which imparts hardness. Slow cooling, called annealing, alters the iron's crystal structure to pearlite, a soft ductile mechanical mixture of ferrite and cementite, which is softer. Maximum hardness can be imparted to all common tool and machinery steels by cooling them rapidly. The hardness achieved varies directly with the carbon content and the quenching speed. While steel's alloy content has little effect on maximum attainable hardness, higher alloy steels attain maximum hardness at the slower cooling rates. The cooling rate depends on the mass, geometry, and the surface condition of the work-piece, as well as the quenching medium or the fluid.

Common quenching fluids include water, brine, and other aqueous salt solutions, salt melts, quenching oils, fatty oils, and emulsions. Most industrial quenching is carried out by the use of the refined petroleum base stocks because they

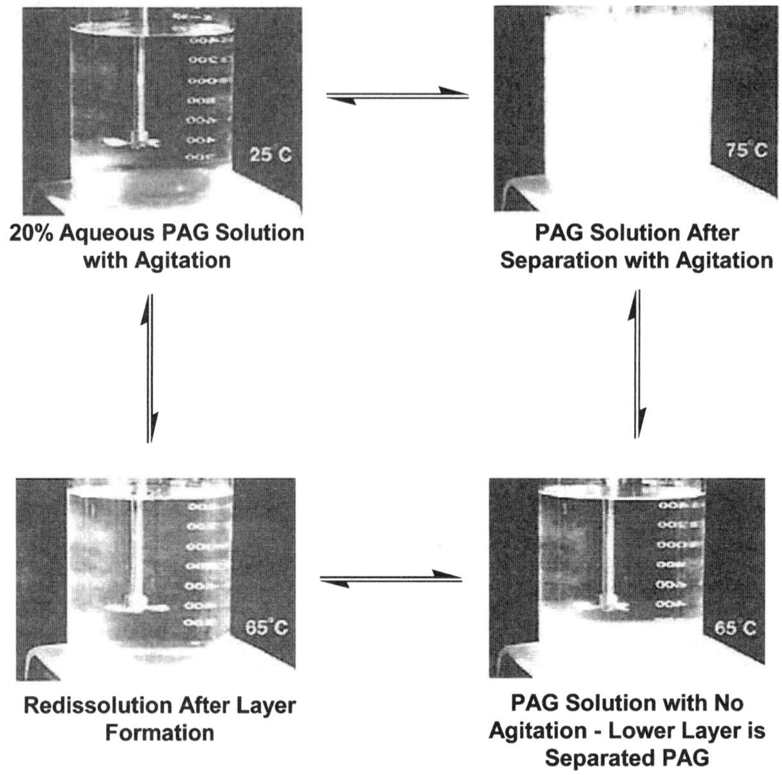

Fig. 11.10—Solubility reversal of aqueous PAG solution at high temperature [763].

are easy to handle, require little or no specialized equipment, are noncorrosive and nontoxic, and the quenched parts are easy to clean. These fluids must be nonirritating to the skin, provide corrosion protection, be chlorine and sulfur-free, and have proper viscosity, low volatility, high flash and fire points, and extended service life. These fluids act as heat transfer agents and can be oil-based or water-based.

Tempering is the heat treatment used to strengthen the metals and alloys. Usually the process involves two steps. In the first step, the metal and carbon, as in the production of steel, or the alloy metals, as in the case of hardened alloys, are heated to create a solid solution by heating them to the melting temperature and quenching. In the second step, tempering is carried out by holding it at temperatures lower than those necessary to produce solutions. The composition of the tempering fluids is analogous to that of the quenching fluids.

The base oils to formulate the oil-based quenching fluids can be synthetic or petroleum in origin. The performance specifications of these fluids are established by the OEMs and the end-users. Quench oil temperatures between 50°C and 90°C are the best for most applications. However, the higher temperatures increase the quenching speed. Because of the temperatures involved, the water content in the oil must be low, otherwise serious foaming will occur. The upper temperature limit for these fluids is determined by their potential flammability and, as a general rule, it should be at least 50°C below the flash point of the oil. Paraffinic oils, with good oxidation stability, are generally preferred. Marquenching is the process where the steel is quenched in the hot oil at 150–175°C and holding it until the temperature of the work-piece is equal to the temperature of the oil. This is followed by a prolonged period of slow cooling, which allows maximum hardness while minimizing work-piece distortion. Work-piece distortion commonly occurs in conventional quenching because of the nonuniform cooling rates of the different zones. Sometimes additives, such as oxidation inhibitors, are added to minimize acid and sludge-forming tendency of the fluid which extends its service life. Sludge, if formed, will mar the work-piece surface through deposition.

Oil-based formulations are being replaced by synthetic fluids due to their superior fire resistance and fewer disposal concerns. Common synthetic fluids are the nonflammable blends of water and water-soluble or water miscible polymers, such as poly(alkylene glycol)s (PAGs), poly(vinylpyrolidinone)s, and sodium acrylate polymers. While PAGs are completely soluble in water at room temperature, they have inverse solubility at higher temperatures. That is, the PAG comes out of solution and coats the metal surface or that of the part, thereby controlling the rate of cooling, which reduces quench cracking and distortion that occur in many other fluids. When the temperature drops below the threshold level, the PAG goes back into solution. This is shown in Fig. 11.10 [763]. Please note that in the figure, all transitions are reversible which is indicated by half arrows. With constant use, the PAG-based coolants will ultimately lose their viscosity due to shear. Since PAGs degrade slowly, it is easy to monitor and replenish the polymer in the coolant when necessary. PAGs are nontoxic and pose minimal environmental and health risks compared to the oil-based quenching fluids. They are also easy to remove from the work-piece during

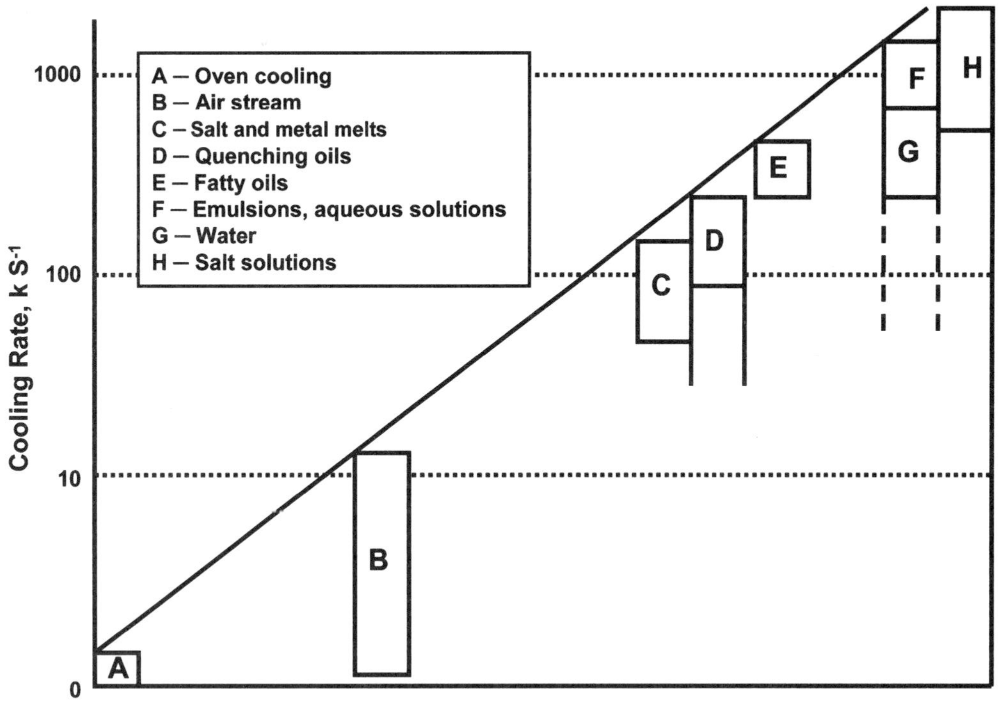

Fig. 11.11—Effectiveness of the various cooling media [4].

cleanup and are moderately biodegradable in the environment.

Because of the presence of water, these fluids can cause corrosion. Hence, they contain corrosion inhibitors. They are useful in quenching both ferrous and nonferrous metals. They have faster cooling ability than the oil-based quenchants because of the presence of water and unlike aqueous salt containing quenchants they do not leave crystalline residues on the work-piece surface. In addition, the quench severity can be controlled by altering the water to poly(alkylene glycol) ratio. The presence of water in these fluids makes them susceptible to microbial attack, which can be averted by the use of the biocides.

Figure 11.11 compares the effectiveness of the various cooling media [4]. As one can see that with respect to the cooling rate, water and water-based quenchants are extremely effective. However, they have a number of limitations, because of which their use is less desirable in some applications, such as steel hardening. The limitations include water loss due to boiling and the formation of the air pockets. The latter inhibits uniform cooling, which can cause distortion of the work-piece. These problems are overcome by constant stirring of the coolant. However, this decreases the water's cooling efficiency. Water as a coolant is applied for low alloy steel objects that are thin and are therefore easier to harden. When choosing water, it is important that it is devoid of salts, which can also cause inconsistent quenching. Brine (5–7 % sodium chloride solution) cools faster than the pure water (compare Items H and G in Fig. 11.11), but again it suffers from the same disadvantages. In addition, brine separates a layer of salt on the metal surface, which interferes with uniform cooling; hence it results in metal distortion. Brine is commonly used in quenching high carbon steels or parts requiring high hardness. Its most serious disadvantage is that it is extremely corrosive, both to the metal components and to the quenching equipment. Caustic solutions containing 5–10 % sodium hydroxide are also used, although, they need careful handling, storage, and use.

Slide Way Lubricants

These lubricants are used to lubricate slide ways and the accompanying pneumatic equipment. Slide ways are sliding surfaces on the bed of a machine along which a table or a carriage moves. Since the surfaces that slide over each other are flat, the area of contact is large. This leads to increased adhesive wear. In addition, these devices experience motion involving varying speeds, which causes sticking and slipping of the sliding surfaces. After the wear-in, the opposing surfaces form an even closer fit, which squeezes out and wipes away any lubricant that is in the path of motion. Essentially, a boundary lubrication condition exists. This type of lubrication to a degree is by design; otherwise, the excess lubricant would form a hydrodynamic wedge that will interfere with the smooth motion of the plane.

Boundary lubricated surfaces will adhere to each other, especially during slow speed operations. Adhesion occurs when the static friction either equals or exceeds the force of motion. If the adhesion is followed by movement due to the applied force, the phenomenon is called stick-slip. Regular occurrence of stick-slip causes not only vibration and noise but also damage to the work piece, tool, and the rider and way.

Slide way lubricants perform at extreme temperatures, high loads, moisture, and poor ambient air quality. They must therefore possess both the EP activity and the rust and corrosion-inhibiting properties. These lubricants are formulated with the friction reducers, primarily fatty carboxylic acid derivatives, which minimize stick-slip and the extreme pressure agents, which control the wear damage resulting

TABLE 11.16—Metalworking fluid additives.

Additive Type	Function	Typical Compounds
Water-based Fluids		
Alkalinity Buffers/pH	Control bacteria-derived acidity	Amines and inorganic bases
Antifoam Agents	Prevent foam	Silicones, silicates, and stearates
Antifungal Agents	Control fungal growth	Sodium omadine
Antimicrobials/Preservatives	Control microbial growth	Triazines, omadine, phenol, oxazolidines, and imidazolines
Metal Passivators	Prevent staining and leaching of yellow metals	Triazoles
Couplers	Emulsion Stability	Glycols, glycol esters, and alcohols
Emulsifiers	Emulsify organics in water	Sodium and lithium sulfonates; poly-ethoxylated phenols, alcohols and acids; metal and amine carboxylates (soaps of fatty acids)
Extreme Pressure/Lubricity Agents	Form heat-resistant chemical films that reduce friction	Fatty acids, fatty amides, fatty esters, and fatty acid salts (soaps); phosphates; sulfur and chlorine containing compounds
Odor Masks	Suppress odor	Natural and synthetic aromatic compounds
Thickeners	Improve viscosity	Polyacrylate esters and glycol esters
Corrosion and Rust Inhibitors	Control oxidative and corrosive damage to metal surfaces and tools	Carboxylic acid salts, amides, and amines
Oil-based Fluids		
Antifoam Agents	Prevent foam	Silicones, polymethacrylates, and stearates
Anti-misting Agents	Suppress mist formation	Polyisobutenes and other natural and synthetic polymers
Corrosion and Rust Inhibitors	Control oxidative and corrosive damage to metal surfaces and tools	Metal sulfonates and phosphates; organic acids and esters; and triazoles
Demulsifiers	Facilitate water separation and facilitate removal of tramp oil via skimming	Cationic and non-ionic polymeric surfactants
Dispersants	Prevent agglomeration of material particles that result during cutting and grinding	Polyamides and metal sulfonates
Extreme Pressure/Lubricity Agents	Form heat-resistant chemical films that reduce friction	Fatty acids, fatty amides, fatty esters, and fatty acid salts (soaps); phosphates; sulfur and chlorine containing compounds; basic sulfonates
Odor Masks	Suppress odor	Natural and synthetic aromatic compounds
Thickeners	Improve viscosity	Polyacrylate esters
Solid Lubricants	Improve film strength	Graphite, molybdenum disulfide, Poly(tetrafluoroethylene) – PTFE (TEFLON®)

from the boundary lubrication conditions. The operation at extreme temperatures leads to oxidative breakdown of the lubricant. This must be avoided, otherwise poor lubrication will result. This is accomplished by the use of the oxidation inhibitors. It is important to note that the stick-slip control requires combining proper machine design and superior lubricant quality.

Since the oil is removed due to the wiping action of the slide, lubricant is supplied at different points along the slide route. Suspended metallic fines or chips in the lubricant can lead to scratching, gouging, or abrading of the ways. The use of a properly formulated lubricant can minimize this type of damage, as well as control friction, chatter (noise), and stick-slip. Despite the fact that the lubrication is often once through, an oil circulating system may also be used to deliver the lubricant at different points along the slide. Please consult Chapter 9 on Miscellaneous Industrial Fluids for additional information on this topic.

Fluid Composition

Metalworking fluids are composed of a base fluid and a collection of chemicals, the same as most other lubricants. However, because of the diversity of the lubrication requirements in metalworking operations, the type and the quantity of the additives differ from operation to operation. The quality and quantity of the additives also depend upon the type and the inherent properties of the base fluid.

Base Fluid

The base fluid in metalworking lubricants can be vegetable oil, mineral oil, synthetic-based, or water. Straight oils use severely refined and hydrotreated mineral oils and synthetics, such as polyalphaolefins (PAOs), polybutenes, and poly(alkylene glycol)s. However, paraffinics are often the base oils of choice, primarily due to their lower cost. Nevertheless, the use of synthetic base stocks is increasing in fluids where tailored properties are desired and the petroleum base stocks are not effective.

Polyisobutylenes have the tendency to depolymerize at high temperatures, which makes them useful in rolling and drawing oils for ferrous and nonferrous metals, where mineral oil-derived lubricants cause staining during the subsequent annealing process. Polyalphaolefins, although used rarely in metalworking fluids, possess certain attributes that makes their use in metalworking fluids highly desirable. These include effectiveness over a broader temperature range and the lower hydrocarbon (HC) emissions than the petroleum oils of similar viscosities. Polyalphaolefins are also highly resistant to oxidative and thermal degradation. Polyalkylated, or multiply alkylated cyclopentanes (MACs) are a new class of synthetic hydrocarbons that are promising in terms of the future formulations. They have low pour points, high viscosity indices, and exceptionally low volatility. Other types of synthetic base stocks find limited use in metalworking applications.

For water-based fluids, naphthenic oils are preferred because of the ease of emulsibility and superior viscosity—pressure relationship. Poly(alkylene glycol)s are extensively used in metal removal fluid formulations, where their inverse solubility is an advantage. Inverse solubility is a unique property of PAGs which makes them *water-soluble* at low temperatures but *water-insoluble* at high temperatures. At high temperatures, they form a persistent lubricating film that contains additives at the hot tool-work piece interface, which can be easily cleaned off with water at low temperatures. PAG solution's solubility reversal was discussed earlier and is depicted in Fig. 11.10.

Additives

Additives used in metalworking fluids include emulsifiers, coupling agents, friction reducers, extreme pressure/antiwear agents, oxidation inhibitors, rust and corrosion inhibitors, foam inhibitors, and antimicrobial agents. Common types of additives that are typically used in the metalworking fluids are provided in Table 11.16, along with their chemistry and the mode of action. Discussion below pertains to the specific functions that these additives perform in metalworking fluids.

Emulsion Promoters

These additives facilitate the formation of emulsions and impart to their stability. These include emulsifiers and coupling agents. *Emulsifiers* are chemicals that are used to emulsify (solubilize) organic additives and or mineral and synthetic oils in water. These are used to create the soluble oils, synthetics, and semi-synthetics, all of which are water-based lubricants. Emulsifiers contain the functional groups that have the capability to associate with water as well as oil. These additives tend to be situated at the boundary between the oil and water (the oil-water interface), where they help reduce the interfacial tension and make the two phases miscible, to form a stable emulsion. Emulsion formation requires high-speed mixing or stirring.

Emulsifiers are classified as nonionic or ionic, depending upon whether the polar part is uncharged or charged. Ionic compounds can be subdivided further into cationic, if the charge is positive and anionic, if the charge is negative [458]. Nonionic emulsifiers that are often used in metalworking fluids include carboxylic acid amides and esters, polymeric ethers [poly(glycol ether)s], esters of polyhydric alcohols, and alkoxylated alkylphenols. Anionic emulsifiers include inorganic salts, primarily sodium, of the carboxylic acids, alkyl phosphoric acids, and aromatic sulfonic acids—both natural and synthetic. Because of the lower cost, these emulsifiers are used as the general-purpose additives. Cationic emulsifiers include mineral acid salts of amines and imidazolines. Compounds that find extensive use as emulsifiers include the following:

1. Alkali metal (primarily sodium) arylsulfonates, both petroleum and synthetic
2. Alkali metal carboxylates
3. Alkoxylated alcohols, phenols, fatty amines, fatty acids, and fatty amides
4. Sulfated fatty oils

The structures of some of the emulsifiers are shown in Fig. 4.157.

As mentioned in Chapter 4 on Additives, the efficiency of an emulsifier depends upon its molecular weight which is usually less than 2000 g/mol, its HLB (hydrophile-lipophile balance) value, water pH and hardness, the nature of the oil, and the operating conditions, such as temperature. HLB scale spans zero (0) to greater than 30. The higher numbers indicate oil compatibility decreasing and water compatibility increasing. Emulsifiers of HLB of greater than 13 generally lead to clear water solutions. Emulsifiers with an HLB of 3 to 6 are suitable for water-in-oil emulsions and those with an HLB of 8 to 18 are suitable for oil-in-water emulsions. The manner in which these additives form emulsions is shown in Fig. 4.156.

As a general rule, the nonionic emulsifiers are used in metalworking fluids based on naphthenic stocks, and the fatty acid carboxylates are used in those based on paraffinic stocks. Poly(alkylene glycol)s, also called hydroxyalkyl ethers, are sometimes avoided because their enhanced solubility in water does not allow clean separation for disposal. As mentioned earlier, the alkali metal salts (soaps) of carboxylic and sulfonic acids are among the most commonly used emulsifiers. However, the metal ion exchange with calcium and magnesium, whose salts are present in hard water, degrade emulsions by removing these soaps as insoluble calcium and magnesium salts. Nonionic emulsifiers lead to emulsions that are less sensitive to hard water.

Some emulsifiers show mild anti-rust performance (ASTM D665). Since emulsification is a liquid phase phenomenon and the rust inhibition occurs at the liquid-solid interface, optimizing both these properties is not easy. One can obtain both the emulsification and the anti-rust performance by altering the hydrocarbon chain length in the emulsifier molecule. However, in many cases, this strategy is not very effective and it is necessary to improve the anti-rust performance of the fluid by the use of supplemental additives. The common ones include sodium nitrite, borax, boric acid amides and esters, alkanolamines, and alkanolamides.

Coupling Agents (Couplers)

Some additives that need to be emulsified have a greater solubility in water than in oil. Consequently, they are not easy to formulate into the emulsifiable oil concentrate (package) without the presence of a large amount of water (as much as 50 %). Coupling agents facilitate the emulsification of water into the base oil, emulsifier system, and other additives. In the long term, these additives maintain the emulsion stability. Common coupling agents are low molecular weight alcohols, glycols (diols), and glycerols (triols), which become part of the non-oil portion of the package. This decreases the need for large amount of water necessary to solubilize additives thereby improving emulsification. It is important to note that although these components are organic in nature, they impart little or no lubricity to the final fluid, because of their relatively short hydrocarbon chains. Structures of some of these additives are given in Fig. 11.12. Common examples of these additives include the following:

1. Fatty alcohols, such as tridecyl alcohol
2. Glycols, such as ethylene glycol, diethylene glycol, and propylene glycol
3. Glycol ethers, such as propylene glycol monomethyl ether and hexylene glycol monomethyl ether
4. Fatty acids, such as caprylic acid, iso-nonanoic acid, and neo-decanoic acid
5. Nonionic surfactants, ethoxylated alcohol, nonylphenol

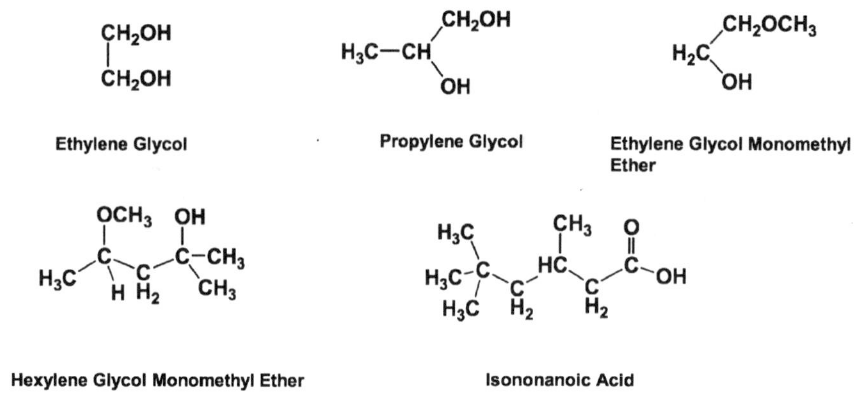

Fig. 11.12—Commonly used coupling agents.

ethoxylates, and poly(ethylene glycol) esters

Film-forming Agents

This additive class includes friction modifiers, also called the lubricity agents, film-strength additives and oiliness additives; anti-wear and extreme pressure agents, also called boundary additives and load-bearing additives; and corrosion inhibitors.

These additives are surface active because of their high polar to nonpolar ratio, and hence they have the tendency to separate on surfaces where they interact by a physical or a chemical mechanism to form protective surface films. See Chapter 4 on Additives.

One common structural feature of the friction modifiers is the presence of a fatty hydrocarbon group, a longer than C_{12} linear chain. In addition to fatty alcohols and acids, this class of additives includes fatty esters, natural and synthetic, fatty amides, fatty amines, and fatty alcohol-derived phosphites and alkyl acid phosphates. Those used most often in metalworking fluids include animal and vegetable oils, commonly called the triglycerides, alkyl and polyol esters of fatty acids, ethylene oxide/propylene oxide polymers, TEFLON®,

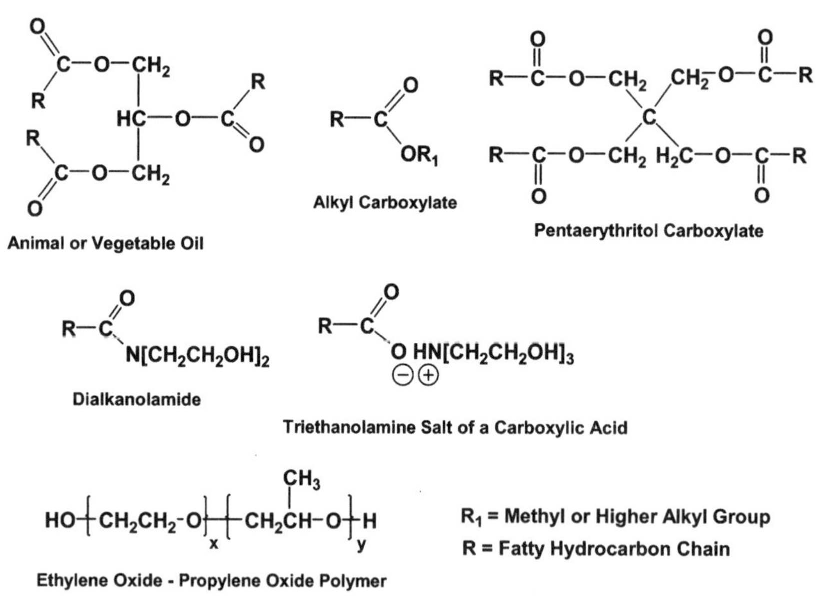

Fig. 11.13—Commonly used friction modifiers.

TABLE 11.17—Effective temperature range of lubricity agents and EP/antiwear agents [4].

Additive Type	Effective Temperature Range
Carboxylic acids, esters, and metal salts (soaps)	<200°C
Chloroparaffins and other chlorinated derivatives	180–450°C
Organophosphorus compounds, such as phosphoric acids and their derivatives	200–700°C
Organo-sulfur compounds, such as polysulfides	600–1000°C

TABLE 11.18—Tests used to determine EP/antiwear properties.

Chemical Class	Analyses	Performance Tests
Amine phosphates	% Nitrogen, phosphorus, and TAN/TBN	Four-ball wear, four-ball EP, FZG, and rust and oxidation
Methylene bis(dialkyl-dithiocarbamates)	% Sulfur, nitrogen, and residual chlorine and amine	Four-ball EP, FZG, Falex EP, and oxidation and corrosion
Sulfurized lard, esters, and fatty acids	% Total sulfur and active sulfur	Four-ball wear, four-ball EP, stick-slip, and copper corrosion
Triphenyl phosphorothioate	% Sulfur, phosphorus, and melting point	Four-ball EP, FZG, Falex EP, and oxidation and corrosion
Chlorinated paraffins and fatty acids	% Chlorine and acid value	Four-ball wear, Falex EP, Timken, and copper corrosion

alkanolamides, and mono, di, and tri-ethanolamine salts of carboxylic acids. The structures of some of these additives are shown in Fig. 11.13.

A greater degree of protection is needed in some applications, such as metal removal operations, than that provided by the friction modifiers. This is because these applications generate high temperatures which make physically and chemically adsorbed additive films easy to remove. That is where extreme pressure/antiwear agents become important. These additives chemically react with metal surfaces to form more tenacious protective films. The degree of EP protection in the equipment depends upon the conjunction temperature of the two metal surfaces in contact [6]; the higher the temperature, the greater the need for EP/antiwear protection. To be effective, it is important for the activation temperature of the EP additive to match the conjunction temperature. The effective temperature ranges of different types of additives when used in metalworking applications are provided in Table 11.17 [4]. Tests that are commonly used to determine the effectiveness of the various classes of the EP/antiwear agents are listed in Table 11.18.

Extreme pressure agents are primarily organic compounds of chlorine, sulfur, and phosphorus; although sometimes boron and sulfur-nitrogen compounds are also used. Please refer to Chapter 4 for their structure and the mechanism of performance. In all cases, the active elements chlorine, sulfur, and phosphorus react with the metal to form the metal halide, sulfide, and phosphite, phosphate, and phosphide protective films. These films are effective only below their eutectic point, or the decomposition temperature, of these salts. Chlorine compounds can lead to the metal corrosion by hydrogen chloride, which results from their hydrolysis. Hence, this must be taken into account prior to their use in cutting oils. In view of their effective range, sulfur compounds are used in heavy-duty and extra heavy-duty cutting operations. Metal sulfide films appear to have a higher load-carrying capacity and shear strength than the carboxylic acid soaps and metal phosphates.

As stated earlier, the film formation by these additives occurs by a two-step mechanism: Adsorption on the metal surface and thermal decomposition and reaction with the metal surface, due to frictional heat. The resulting metal salt films have low coefficients of friction and also demonstrate anti-weld properties, both of which minimize the tool wear. Metal chloride films have an approximate transition temperature of 600°C and metal sulfide films have a transition temperature of 1000°C. This makes the organic sulfides better extreme pressure additives than the organic chlorides, because the sulfide films can endure higher temperatures before becoming soft and getting removed. Figure 11.14 shows the temperature ranges over which the EP films from the metal cutting fluids will be effective [4]. Chlorinated fats,

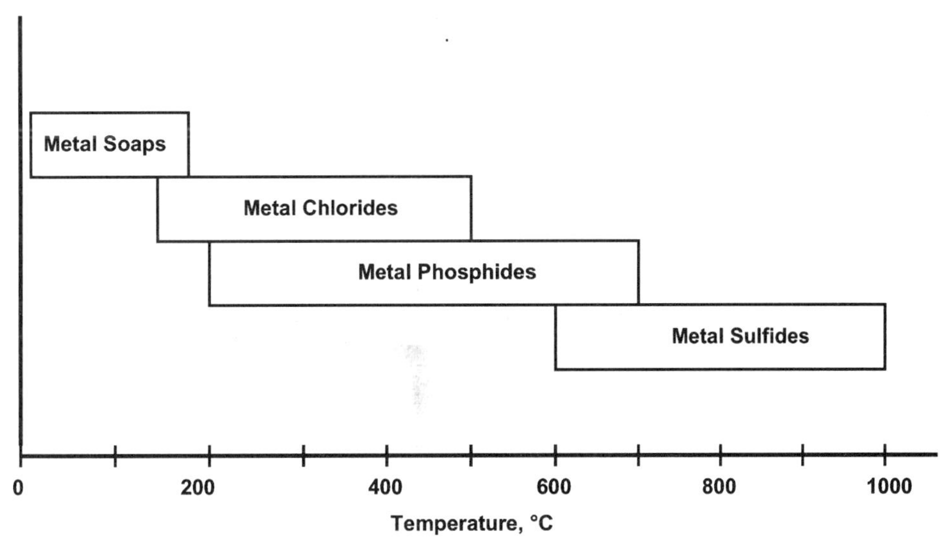

Fig. 11.14—Durability temperatures of boundary films resulting from cutting oil additives [4].

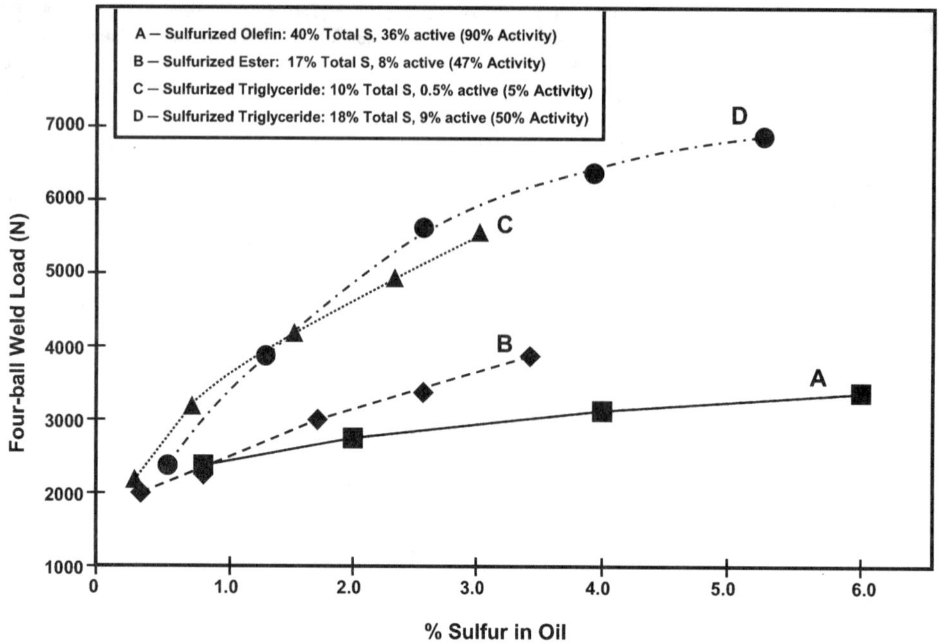

Fig. 11.15—EP performance versus surface activity and active sulfur [425].

chlorinated esters, and the chlorinated paraffins are the common organic chlorides used in formulating metalworking fluids.

Sulfurized fats, sulfurized oils, sulfurized paraffins, dissolved sulfur, and sulfochlorinated products are the common organic sulfides used. Typical sulfur level in these additives is between 10 and 40 %. The sulfur in these additives is of the two types: Active and reactive. Active sulfur is the dissolved or the easily releasable form of sulfur, which has the tendency to corrode yellow metals—a low temperature reaction. Because of this, the presence of the active sulfur in the metalworking fluid formulations is of concern. Experimentally, active sulfur is determined by the use of the ASTM D1662 test. Reactive sulfur, sometimes referred to as inactive sulfur, on the other hand, is the bound form of sulfur, for example as a sulfide or disulfide, and is released or reacts with the metal only at high temperatures. Inactive sulfur additives are commonly used to machine brass and copper alloys. Incidentally, active sulfurized mineral oils are excellent lubricants for machining the hard high carbon and alloy steels and are the most widely used fluids in the industry. These oils minimize tearing and rough finish, often encountered in machining such metals. The "active" sulfur contained in the oil tends to form an iron sulfide film on the steel surface, which because of lower shear strength facilitates the cutting action. Such oils may be dark or light-colored and odorous straight sulfurized or sulfochlorinated mineral or fatty oils. Excellent EP properties of these oils are depicted

TABLE 11.19—Performance properties of sulfurized products [425].

	Olefins		Triglycerides		Esters	
Property	*Inactive*	*Active*	*Inactive*	*Active*	*Inactive*	*Active*
Extreme Pressure	Low	Fair	Good	Very Good	Fair	Good
Antiwear	Good	Poor	Very Good	Low	Good	Low
Reactivity	Low	Very High	Low	High	Low	High
Copper Corrosion	Low	High	Low	High	Low	High
Oxidation Inhibition	Good	Poor	Good	Poor	Good	Low
Lubricity	Poor	Poor	Very High	Very High	Fair	Fair

TABLE 11.20—Synergy between sulfurized products and zinc dialkyl dithiophosphates [425].

Chemical Type	% Total Sulfur	% Active Sulfur	% Treatment	Copper Corrosion, 3 h at 100 °C	
				Chemical	Chemical+1.5 % ZnDTP[a]
Sulfurized Triglyceride	18	10.5	5	4c	3b
Sulfurized Triglyceride	15	5.0	5	3a	1b
Sulfurized Ester	17	8.5	5	3b	1b

[a]ZnDTP=Stabilized zinc di(2-ethylhexyl) dithiophosphate.

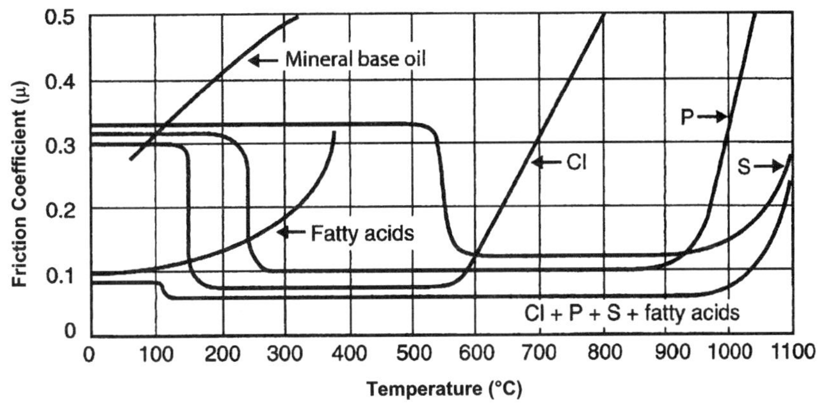

Fig. 11.16—EP additives versus temperature [764].

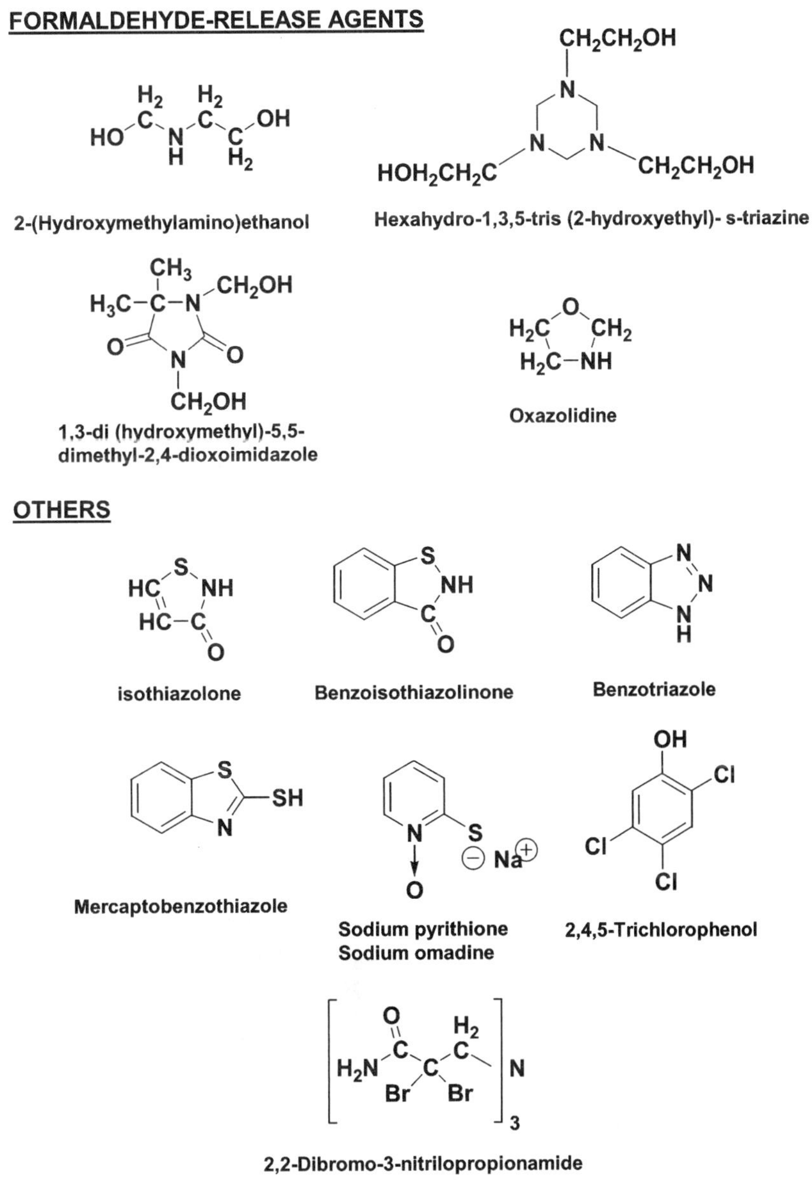

Fig. 11.17—Structures of some biocides.

TABLE 11.21—General composition of metal removal fluids.

Metal Removal Fluid	Composition[a]
Straight Cutting Oil	Mineral Oil; Friction Reducer (fatty acids, their metal salts and esters); Extreme Pressure Agent (chlorinated isobutylene, dibenzyl sulfide, organic phosphates).
Emulsifiable Cutting Oil	Water; Mineral Oil (primarily naphthenic); Emulsifier; Coupling Agent; Corrosion Inhibitor (alkanolamides); Foam Inhibitor (ethoxylated fatty alcohols); Antimicrobial Agent (triazines); Friction Reducer (fatty acid derivatives); Extreme Pressure Agent (chlorinated paraffins, sulfurized olefins, and organophosphorus compounds).
Water-soluble Cutting Oil	Water; Corrosion Inhibitor (sodium or potassium nitrite; mono-, di-, or tri-ethanolamine); Emulsifier (non-ionic surfactants; natural and synthetic fatty acid soaps; polyalkylene glycols)
Grinding Oil	Water; Mineral Oil; Friction Modifier (fatty acid derivatives); Extreme Pressure/Antiwear Agent (organic compounds of sulfur, phosphorus, and chlorine); Anti-misting Additive (polyacrylates and polyacrylamides).
Lapping Oil	Mineral Oil; Dispersant (alkenylsuccinimides); Lapping Agent.
Honing Oil (Precision Grinding)	Mineral Oil; Friction Reducer (fatty acids and triglycerides); Extreme Pressure Agent (chlorinated isobutylene, dibenzyl sulfide, organic phosphates).
Punching Lubricants (Emulsifiable and non-emulsifiable mineral, and solid dispersion type)	Mineral Oil; Friction Reducer (fatty oils, graphite, MoS_2, mica, or talc); Extreme Pressure Agent (organo-phosphorus, organo-sulfur, and organo-halogen compounds).

[a]Not all formulations contain all the listed additives.

in Fig. 11.15 [425]. As indicated by the four-ball weld load (DIN 51350 Part 2), both the sulfur activity and the substrate's affinity for the surface are important. The weld load of the more surface-active sulfurized esters and the fatty oils B, C, and D is higher than that of the less polar sulfurized olefin A, which is despite its very high sulfur activity. The effect of the sulfur activity is demonstrated by the weld load of the additive D relative to that of the additive C, which is negative or minor until the oil sulfur level becomes high. All in all, the surface activity of the additive is more important than the active sulfur level. It is important to note that the active sulfur-reactive sulfur ratio is a function of many factors, the sulfurization temperature being one. Typically, the fast high-temperature sulfurization by the use of the elemental sulfur yields products with high active sulfur [425]. Since all sulfurized products contain both types of sulfur and each type affects the lubricant properties differently, it is important to develop products with a proper balance of the two types of sulfur to meet the lubrication needs of a particular operation or the application. Table 11.19 shows the effects of the two types of sulfur on the various lubricant parameters. As one can see, a higher amount of active sulfur has a positive effect on reactivity and hence provides superior EP performance. However, with respect to all other parameters, it has either no effect or an undesirable effect. For all other lubricant parameters, a higher amount of inactive sulfur is more desirable. Copper activity of the active sulfur compounds can be reduced without diminishing their EP/antiwear performance. This may be achieved by the use of the zinc dialkyl dithiophosphates. Table 11.20 demonstrates the improving effect of the zinc dialkyl dithiophosphate on copper activity of the sulfurized products [425].

We commented earlier that the various types of film-forming additives have a specific temperature range, outside which they are either ineffective or lose their effectiveness. Based upon the data presented in Fig. 11.14 on the thermal durability of the metal films resulting from the thermal reaction of the film-forming additives; we surmise that if one is interested in maintaining the EP protection over a broader temperature range, one must combine various classes of film-forming additives. Figure 11.16 shows the effect of such a venture [440,764]. The figure shows that with respect to the friction control, the fatty additives although better than the straight mineral oil lose their effectiveness first. They are followed by the chlorine-containing materials, phosphorus-containing materials, and sulfur-containing materials. However, a combination of all four classes of additives is effective in keeping the friction at a very low level, i.e., at less than 0.1 μ, over the 0 to 1000°C range.

Because of the reactivity of the sulfurized hydrocarbons towards copper, bronze, and the other nonferrous metals, and the tendency of the chloro-paraffins to corrode metals via hydrolysis or thermolysis, the need for new extreme pressure agents exists. New extreme pressure additive technology based on overbased alkylbenzenesulfonates and carboxylates, which does not suffer from these disadvantages, has recently become available. Such additives, called the passive EP agents, are believed to function by forming a metal carbonate film at the tool work-piece interface. The high effectiveness of these additives in cutting, tapping, and threading operations suggests an alternative mechanism. That is, the high pressures at the tool-metal interface convert the amorphous calcium carbonate, present in the basic sulfonates and carboxylates, into crystalline salts that facilitate the metal removal. The films formed are of low shear strengths and high melting points. These additives do not contain phosphorus, sulfur, or chlorine in an active form but are synergistic with the sulfur-containing EP additives. In addition, they are less corrosive, easier to dispose of after use, cause little or no foaming, and are easily removed from the work piece surface. They can be used both for ferrous and nonferrous metals, which is an added benefit.

TABLE 11.22—General composition of metal forming and other fluids.

Lubricant	Composition[a]
METAL FORMING FLUIDS	
Hot Rolling Oil (emulsion type)	Water; Mineral Oil (primarily naphthenic); Emulsifier; Coupling Agent; Corrosion Inhibitor (alkanolamides); Antimicrobial Agent (triazines); Friction Reducer (fatty acid derivatives); Anti-misting additive (polyisobutylenes); and Extreme Pressure Agent (organo-phosphorus compounds).
Cold Rolling Oil (straight oil type)	Mineral oil (kerosine cut); Friction Reducer (dodecanol, fatty acids and esters); and Extreme Pressure Agent (organo-phosphorus, organo-sulfur, and organo-halogen compounds).
Cold Rolling Oil (emulsifiable type)	Water; Mineral or fatty oil; Friction Reducer (fatty alcohols, acids and esters); Emulsifier; Coupling Agent; Corrosion Inhibitor (alkanolamides, sulfonates); Foam Inhibitor (ethoxylated fatty alcohols); Antimicrobial Agent (triazines); and Extreme Pressure Agent (organo-phosphorus, organo-sulfur, and organo-halogen compounds).
Cold Rolling Oil (aqueous solution or synthetic type)	Water; Friction Reducer/EP Agent (organo-phosphorus compounds, poly(alkylene glycols); Corrosion Inhibitor (fatty amines).
Wire Drawing Lubricants (Straight oils and emulsion types)	Mineral Oil; Friction Reducer (animal or vegetable oils); Lubricant Carriers[b] (phosphoric acid, oxalic acid, lime); Extreme Pressure Agent (organic chlorides)
Tube Drawing Fluids (Straight oils and emulsion types)	Mineral Oil; Friction Reducer (animal or vegetable oil); Lubricant Carrier (phosphates, oxalates, lime); Extreme Pressure Agent (organic chlorides);
Deep Drawing Fluids (Straight oils and emulsion types)	Mineral Oil; Friction Reducer (animal or vegetable oil); Corrosion Inhibitor; Adhesion Improver (polymethacrylates); Extreme Pressure Agent (organo-phosphorus, organo-sulfur, and organo-halogen compounds).
Extrusion Lubricants (Powder, grease, solid, and straight oil types)	Mineral Oil; Friction Reducer (alkali metal soaps of fatty acids, fatty animal and vegetable oils); Lubricant Carrier (phosphoric acid, oxalic acid, lime); Extreme Pressure Agent (Organo-phosphorus, organo-sulfur, and organo-halogen compounds).
Cold Forging Lubricants	Mineral Oil; Lubricant Carrier (zinc phosphate); and Extreme Pressure Agent (organo-sulfur and organo-halogen compounds)
METAL TREATING FLUIDS	
Quenching Oils (straight oil type)	Mineral Oil; Oxidation Inhibitor (Alkylphenol); Quenching Promoter (high molecular weight hydrocarbons)
Quenching Oils (washable)	Mineral Oil; Emulsifier; Oxidation Inhibitor; Quenching Promoter (high molecular weight hydrocarbons)
Quenching Oils (emulsifiable type)	Mineral Oil; Emulsifier; Oxidation Inhibitor; Quenching Promoter (high molecular weight hydrocarbons); Antimicrobial Agent; Corrosion Inhibitor.
METAL PROTECTING FLUIDS	
Rust Preventive Oils (straight oil type)	Mineral Oil; Corrosion Inhibitor
Rust Preventive Oils (Emulsifiable type)	Mineral Oil; Emulsifier; Water; plasticizer (ethyl cellulose or cellulose acetate); micronized waxes; corrosion inhibitor (alkali and alkaline earth natural metal sulfonates; fatty carboxylates; naphthenates; and extensively oxidized hydrocarbons, both neutral and overbased, amino alcohols and triazoles)
SLIDE WAY LUBRICANTS	Mineral Oil; Friction Modifier (fatty acid derivatives); Antiwear Agent; Oxidation Inhibitor

[a]Not all formulations contain all the listed additives.
[b]These additives, applied through surface bonding, increase surface roughness which improves adhesion of the lubricant/EP additives.

Dispersants

These additives are mainly used to suspend inorganic solids, such as graphite and molybdenum disulfide, in specialty lubricants and include high molecular weight alkenylsuccinimides, alkylsalicylates, and alkylbenzenesulfonates.

Anti-misting Agents

Some metalworking operations generate mist, which is not only of concern because of the worker safety, but also because it leaves an undesirable oily residue on parts and equipment. Anti-misting agents are polymeric thickeners that are used to minimize mist generation in the mineral oil-derived fluids during use. These additives control the mist formation by modifying the droplet size and include ethylene-propylene copolymers, polyacrylates and polymethacrylates, polyacrylamides, polybutenes and polyisobutenes, polystyrenes, and styrene-butadiene copolymers.

Alkalinity Agents/Buffering Additives

These additives are used to impart reserve alkalinity (ASTM D974 and D4739) to the fluid and include alkanolamines, complex amines, and metal hydroxides and carbonates. These not only stabilize emulsions through a buffering action, but they also control corrosion by neutralizing the corrosive acids.

Corrosion Inhibitors

These additives help protect metals against corrosion by neutralizing acidic and other oxidizing species either present in the lubricant or produced during use. The selection of these additives depends upon the fluid type and the

TABLE 11.23—List of tests for metalworking fluids.

OIL-BASED FLUIDS	WATER-BASED FLUIDS
Corrosion	**Corrosion**
Ferrous metals (ASTM D665)	Ferrous metals (ASTM D4627, IP 287, and IP 125)
Cu and Cu alloys (ASTM D130)	Cu and Cu alloys (ASTM D130)
EP - Load Carrying	**EP - Load Carrying**
Falex EP Wear (Pin and Vee) (ASTM D3233)	Falex EP Wear (Pin and Vee) (ASTM D3233)
Four Ball Wear (ASTM D4172)	4-Ball Wear (ASTM D4172)
Four Ball Wear (Shell) (ASTM D2266)	4-Ball Wear (Shell) (ASTM D2266)
Four Ball EP Weld (ASTM D2783)	4-Ball EP Weld (ASTM D2783)
Timken Test (ASTM D2782)	Timken Test (ASTM D2782)
Metal Removal	**Metal Removal**
Falex #8 Tapping Torque	Falex #8 Tapping Torque
Lubrizol CNC Tapping Torque	Lubrizol CNC Tapping Torque
Lubrizol CNC Drill Wear/Life	Lubrizol CNC Drill Wear/Life
Lubrizol CNC Milling/Surface Finish	Lubrizol CNC Milling/Surface Finish
Lubrizol Pipe Threading Test	Lubrizol Pipe Threading Test
Other Tests	**Emulsion Stability**
GM Quenchometer (ASTM D3520)	Characteristics (IP 263)
Wolfson Probe Quenchalyzer	Moderate Timken aquarium Test
Stamping Lubricant Tester (ASTM D4173) (Draw Bead Simulator)	Lubrizol Centrifuge Procedure
Color (ASTM D1500)	**Other Tests**
Humidity Cabinet Rust (ASTM D1748)	Color (ASTM D 1500)
Turbine Oil Rust (ASTM D665)	Aqueous Cutting Fluid Corrosion (IP 125)
Panel Coker (Lubrizol Procedure)	Filter Paper Chip (IP 287)
Stick-Slip Test (Cincinnati Milacron)	Foam (IP 312)

metals to be protected. For straight oils, the neutral and overbased metal sulfonates are used to protect against iron corrosion (ASTM D665 and D4627). For the other types of metalworking fluids, these include fatty amines, neutral and basic barium and calcium alkylaromatic sulfonates, and metal and amine salts of carboxylic acids, boric acid, and the organic acid phosphates. Metal sulfonates are either alkylbenzene derived or alkylnaphthalene derived. Both fatty acids and high molecular weight oxidates (oxidized hydrocarbons) are used to manufacture soaps (metal carboxylates). Alkanolamides, imidazolines, and sarcosines, often used as rust inhibitors, are also the alkaline derivatives of the carboxylic acids. Of these, amines and overbased sulfonates, which are basic, perform by neutralizing acids and others perform by forming films that act a barrier against the environment. For yellow metal protection (ASTM D130), benzotriazole and dimercaptothiadiazole (DMTD) derivatives are used. Structures of these inhibitors are provided in Figs. 4.151 and 4.152.

Antimicrobial Agents

These additives, also known as biocides, are control specific. A fluid can therefore have a combination of these: One acts as an antibacterial agent and the other as an antifungal and yeast control agent. Microbial attack on the fluid is undesired because it causes a buildup of the acidic materials, corrosion of machinery and tools, destruction of the additives, objectionable odors, and produces materials that destabilize emulsions. Fungal attack can lead to a slimy material that can coat the machinery and tools, as well as clog the pumps and filters. This makes monitoring the microbe level in a fluid a necessity. Monitoring is carried out through commercial culture techniques. Standard practice to correct a problem is to use biocides. Often two different biocides are used in an alternating fashion to guard against microbes developing immunity [757].

Commonly used biocides include formaldehyde-release agents and others. Others can be further divided into the heterocyclics and the organo-halogen compounds. The structures of some of the biocides are shown in Fig. 11.17. 1,3-Di(hydroxymethyl)-5,5-dimethyl-2,4-dioxoimidazole, 2-hydroxymethyl-aminoethanol, hexahydro-1,3,5-tris(2-hydroxyethyl)-s-triazine, and oxazolidine are examples of the formaldehyde-release agents. These additives control bacterial growth by releasing formaldehyde, an antibacte-

TABLE 11.24—Parameters measured by various metalworking fluid tests.

TEST AND PROCEDURE	PURPOSE
Corrosion	
Copper (ASTM D130) 3 h at 100°C	Measures fluid's non-ferrous compatibility.
Turbine Oil Rust (ASTM D665) A-Distilled Water B-Synthetic Sea Water	Measures the ability of inhibited mineral oils to aid in preventing the rusting of ferrous metals in the presence of water.
Aqueous Cutting Fluid (IP125)	Measures corrosion protection of aqueous cutting fluids.
Filter Paper Chip Breakpoint (IP287)	Evaluates rust inhibition properties of aqueous cutting fluids.
Aluminum Cup Stain 20 min at 350°C	Cup appearance indicates non-ferrous corrosion properties.
Humidity Cabinet Rust (ASTM D1748)	Measures ability of preservative oils to protect metal parts from rusting under conditions of high humidity.
Salt Spray (MIL-B-117-64)	Steel part corrosion protection measured after exposure to 5% salt spray for 24 hours.
Cleveland Condensing Humidity Cabinet (ASTM D2247)	Measures anti-rust properties of metal preservative fluids on steel panels. Considered more severe than ASTM D1748 humidity test.
Extreme Pressure	
4-Ball Wear (ASTM D2266) 40 kg, 1200 rpm, 75°C,1 h Average Coefficient of Friction; Maximum Scar Diameter (mm)	Evaluates antiwear and anti-weld properties of lubricants.
Timken (ASTM D2782) OK Load (lb)	Measures abrasion resistance and load carrying capacity of lubricants.
Four Ball EP (ASTM D2783) Seizure (kg); Weld (kg); LWI (kg)	Evaluates extreme pressure and anti-weld properties of lubricants.
Falex EP (ASTM D3233)	Measures load carrying capacity and wear properties of lubricants.
Stability	
Foam (ASTM D892, IP312) Tendency/Stability (ml)	Determines foaming characteristics of lubricating oils at specific temperatures.
Panel Coker 4 h at 260°C, continuous splash	Determines relative stability of lubricants in contact with hot metal surfaces.
Demulsibility (ASTM D1401) (ml oil/ml water/ml emulsion)	Measures separation of oil and water from emulsion over time.
Emulsion Stability (IP263)	Measures emulsion stability in water.
Aquarium Biostability	Measures foam, bacteria, fungus, and odor over time in controlled aqueous environment.
Hydrolytic Stability (ASTM D2619)	Measures the ability of a fluid to resist hydrolysis and is determined by increasing acid number of the water phase.
Miscellaneous	
Color (ASTM D1500)	Visual determination of fluid color based on colorimetric readings.
GM Quenchometer (ASTM D3520) R. C. Index	Determines heat removal speed of quench oil.
Thread Tapping (Lubrizol test)	Evaluates fluid efficiency by measuring torque required during tapping operation in steel.
Pipe Threading (Lubrizol test)	Evaluates fluid efficiency by measuring torque required during threading operation on cast iron or stainless steel.
Stick-slip (Cincinnati Milacron test)	Measures static and dynamic coefficients of friction in slideway lubricants.
Bijur Filtration	Determines compatibility of lubricants with Bijur setup (specific to Bijur filter design)
Falex #8	Evaluates fluid efficiency by measuring torque required during tapping operation in steel.
SLT (Draw Bead Simulator)	Evaluates friction generated in a drawing process.
Reichert	Measures load-carrying properties of lubricants.

TABLE 11.25—Comparative Test Standards [27,759,765].

Test	ASTM Designation	ISO Designation	IP Designation	DIN Designation
Standard Test Method for Pour Point of Petroleum Products	D97-96a	3016-1994	15/95	DIN- ISO 3016:1982-10
Standard Test Method for Detection of Copper Corrosion from Petroleum Products by the Copper Strip Tarnish Test	D130-94	2160-1998	154/95	51 759
Standard Test Method for Kinematic Viscosity of Transparent and Opaque Liquids (the Calculation of Dynamic Viscosity)	D445-97	3104-1994	71/97	51 550
Standard Test Method for Rust-Preventing Characteristics of Inhibited Mineral Oil in the Presence of Water	D665-98	7120-1987	135/93	51585 A + B
Standard Test Method for Foaming Characteristics of Lubricating Oils	D892-98	6247-1998	146/82 312/74	51 566 E Edition 12/83
Standard Test Method for Acid and Base Number by Color-Indicator Titration	D974-97	6618-1997	139/98	
Standard Test Method for Water Separability of Petroleum Oils and Synthetic Fluids	D1401-98	6614-1994	-	51 599
Standard Test Method for Emulsion Stability of Soluble Cutting Oils (Discontinued 1988)	D1479	-	290/84	-
Standard Test Method for ASTM Color of Petroleum Products (ASTM Color Scale)	D1500-98	2049-1996	196/97	-
Standard Test Method for Rust Protection by Metal Preservatives in the Humidity Cabinet	D1748-83	-	366/84	51 359
Standard Test Method for Wear Preventive Characteristics of Lubricating Grease (Four-Ball Method)	D2266-99	-	239/97	51 350 Part 5
Standard Practice for Calculating Viscosity Index From Kinematic Viscosity at 40 and 100°C	D2270-93	2909-1981	226/91	DIN-ISO 2909:1997-10
Standard Test Method for Measurement of Extreme-Pressure Properties of Lubricating Fluids (Timken Method)	D2782-94	-	-	51 434
Standard Test Method for Measurement of Extreme-Pressure Properties of Lubricating Fluids (Four-Ball Method)	D2783-88	-	293/97	51 350 Part 2
Standard Test Method for Low-Temperature Viscosity of Automotive Fluid Lubricants Measured by Brookfield Viscometer	D2983-87	-	267/84	-
Standard Test Method for Foam in Aqueous Media (Bottle Test)	D3601-88	-	312/74	51 395-1 Dynamic Test
Standard Test Method for Wear Preventive Characteristics of Lubricating Fluid (Four-Ball Method)	D4172-94	-	293/97	51 350 Part 3
Standard Test Method for Iron Chip Corrosion for Water-Dilutable Metalworking Fluids	D4627-92	-	287/94 125/82	51 360 Part 2
Standard Test Method for Base Number Determination by Potentiometric Titration	D4739-96	-	276/95 417/96	-
Standard Test Method for Evaluating the Scuffing (Scoring) Load Capacity of Oils	D5182-97	-	334/93	-

rial agent. Formaldehyde results from the hydrolysis of these additives in water-based fluids. Heterocyclics include isothiazolone, benzisothiazolinones, morpholine, sodium pyrithione (sodium omadine), benzotriazole, and dimercaptothiadiazole (DMTD). Organo-halogen compounds include 2,4,5-trichlorophenol, bis-(2-chloroethyl) ethertetramethylenediamine copolymer, and 2,2-dibromo-3-nitrilopropionamide. A variety of other compounds that are outside these general classes are also used to control the microbial infestation and include *o*-phenylphenol, carbamates, dithiocarbamates, glutaraldehyde, and nitro alcohols.

Most of these compounds destroy bacteria directly. Sodium omadine is an effective antifungal agent. Materials, such as 2,2-dibromo-3-nitrilopropionamide, are useful in controlling bacteria and fungi, including yeast. These additives, by virtue of protecting against infestation from the all three types of microbes, minimize worker exposure and prolong emulsion batch life. Boron containing formulations usually do not experience bacterial growth. While choosing an antimicrobial agent, one must consider its toxicity, its effect on the emulsion stability, and regulations considering its discharge into the waste streams.

Foam Inhibitors (Antifoam Agents, Defoamers)

These additives are used to control foam (ASTM D892 and D3601, IP312) in metalworking fluids. Foam in lubricants occurs because of air or gas entrainment and impairs their flow and lubricating ability. Air entrainment is common during spraying and pumping of the fluids. Foam is particularly common in water-based fluids because they usually contain surfactant type water-soluble additives. Foam is undesired because it interferes with proper lubrication, inhibits cooling, and creates a problem if it overflows sumps and tanks. Foam inhibitors are readily dispersible additives, which are added to the fluid in a low concentration, i.e., in parts per million or parts per billion. They perform by reducing the surface tension of the bubbles (ASTM D1590), thereby resulting in their coalescence. Common types include amide and ester waxes, silicones and modified silicones, long-chain saturated alcohols, certain triglycerides, water-insoluble polyglycols, ethylene glycol-propylene glycol copolymers, and polyacrylates.

Silicone type inhibitors can adsorb on the metal surface and prevent paint or coating to adhere to the metal part. Their use is also undesired because of the environmental and waste treatment concerns, plating out effect or the "fisheyes," and the effects on cleaning and finishing. They are hard to remove by washing, which may be necessary for finishing the part via painting, varnishing, and enameling. The structures of some of the foam inhibitors are shown in Fig. 4.190.

Oxidation Inhibitors

These additives are used in lubricants to minimize deterioration of their organic component due to the oxidative attack. Metalworking fluids of high organic content such as straight oils are more susceptible to oxidation than fluids of low organic content, such as synthetic fluids. Other types fall in between the two with respect to oxidation susceptibility. Metalworking fluids employ all three types of oxidation inhibitors, viz., hydroperoxide decomposers such as zinc dialkyl dithiophosphates, dialkyldithiocarbamates, and dialkyl polysulfides; radical scavengers, such as hindered phenols and arylamines; and metal passivators, or chelators, such as thiadiazole derivatives, organic diamines, ethylenediaminetetraacetic acid (EDTA), and nitrilotriacetic acid. The use of the first two types predominates in fluids of high organic content and of the third type predominates in fluids of low organic content. Metal chelators, which control oxidation by complexing with the metal ions and making them innocuous, are of great importance in metalworking fluids because of the high probability of contamination of these fluids by metals and metal ions. Structures of the commonly used oxidation inhibitors are provided in Fig. 4.21.

Dyes

These additives are used to color code some lubricants and include anthraquinones, azo compounds, and triphenylmethane.

Odor Control Agents

These agents are used to mask undesirable odors that are either present in metalworking fluids because of the additives or are produced during use because of the microbial attack. Odor control agents include synthetic sassafras, pine oil, terpenes, such as terpinol and *d*-limonene, and methyl salicylate.

Inorganic/Organic Solids

These additives include graphite, MoS_2, metal powders, metal oxides, metal halides, mica, and tetrafluoroethylene polymer. Tetrafluoroethylene polymer, or TEFLON®, is one of the few organic materials that are used to formulate metalworking fluids. These additives are generally not used in metal removal lubricants, although some passive EP agents that are colloidal dispersions are used. However, the use of the inorganic solids is more common in formulating fluids for certain metal forming operations, such as extrusion and forging. The suspension process generally involves mechanically agitating the finely divided powders in oil or water, in the presence of an emulsifier or a dispersant.

Metalworking Fluid Formulations and Testing

Each metalworking operation places a different demand on the lubricant. A lubricant must therefore possess the specific properties to perform effectively. The properties that fulfill the common needs of most working operations include cooling, lubrication, wear control, and protection against corrosion. These fluids must meet performance specifications, which for these fluids are primarily established by the OEMs and the end-users. Table 11.21 provides the additive composition of the metal removal fluids and Table 11.22 provides the additive composition of the metal forming fluids and the miscellaneous others.

Test methods to evaluate performance of these fluids are not well standardized. Some are standardized tests, such as the ASTM, IP, and DIN tests, while the others are additive supplier or the end-user required tests. The standardized tests that are either presently used or can be used to judge the suitability of the metalworking fluids are listed in Table 11.23, and the parameters they evaluate are briefly described in Table 11.24. Many standards contain these tests, as summarized in Table 11.25. Details of these tests are available in the books on the ASTM, IP, and DIN Standards [27,759,765]. It is important to note that the tests across standards do not always match. They may differ because of the disparity in the hardware, the test method, or the way they are performed.

Formulation Examples

Fluid-based Formulations

Straight Oil: *Mineral Oil* (60 %), usually solvent-refined blends to achieve ISO 10, 15, 22, 32, 46, 68 grades; *Lubricity Agent* (15.3 %)—natural fats, synthetic esters; EP Additives (20 %) chlorinated paraffins, overbased sulfonates, phosphorus, sulfur, sulfurized fats, and sulfurized olefins; *Corrosion Inhibitor* (4.7 %).

Soluble Oil: *Mineral Oil* of 40 °C viscosity of 15–20 cSt (53.9 %); *Emulsifier/Corrosion Inhibitor* (30.3 %)—*anionic*—sulfonates, alkanolamine salts—*amphoteric*—amino acid soaps—*nonionic*—ethoxylates, alcohols, amines; *Other additives* (10.3 %)—foam inhibitor, bactericide/fungicide, extreme pressure agent/s; *Water* (5.5 %).

This additive package is diluted with water in 1:10 to 1:50 ratio to yield the finished fluid, which appears opaque.

Semi-synthetic Fluid: *Mineral Oil* (14.7 %); *Emulsifier/Corrosion Inhibitor* (50 %)—*anionic*—sulfonates, alkanolamine soaps, boron amides; *nonionic*—ethoxylates, alcohols, and amines; *Other Additives* (10.3 %)—foam inhibitor, oiliness agent, extreme pressure agent/s; *water* (25 %).

This additive package is diluted with water in 1:20 to 1:60 ratio to give the finished fluid that appears almost clear.

Synthetic Fluid: *Corrosion Inhibitor* (39.7 %)—alkanolamines, water soluble soaps, boron soaps/salts, fatty acids/esters; *Wetting/lubricity Agents* (7.2 %)—polyglycols and esters; *Other Additives* (3.1 %)—bactericide/fungicide, foam inhibitor, extreme pressure agent; *water* (50 %).

This additive package is diluted with water in 1:10 to 1:100 ratio to give the finished fluid that appears completely clear.

Application-based Formulations

Mineral Forming Fluid: *Extreme Pressure Agent* (20.0–35 %)—Chlorinated wax and sulfurized fat mixture; *Corrosion Inhibitor* (1.0–2.0 %)—alkanolamine; *Lubricity Agent/Friction Modifier* (10.0–20.0 %)—soap or lard oil. The balance is mineral oil.

Soluble-oil Forming Fluid: *Extreme Pressure Agent* (25.0–30.0 %)—Chlorinated wax and sulfurized fat mixture; *Corro-*

sion Inhibitor (4.0–6.0 %)—alkanolamine; *Lubricity Agent/Friction Modifier* (10.0–20.0 %)—soap or lard oil; *Foam Inhibitor* (0.5–1.0 %)—polyacrylate; *Buffer* (4.0–6.0 %)—diethanolamine or triethanolamine; *Biocide* (1.0–2.0 %)—triazine; *Emulsifier* (4.0–6.0 %)—soap. The balance is water.

Straight Oil Removal Fluid: *Extreme Pressure Agent* (10.0–20.0 %)—Chlorinated wax or sulfurized olefins; *Corrosion Inhibitor* (0.5–1.0 %)—alkanolamine or triazole; *Friction Modifier* (5.0–10.0 %)—carboxylic acid derivative. The balance is mineral oil.

Soluble-oil Removal Fluid: *Extreme Pressure Agent* (10.0–15.0 %)—Chlorinated wax or sulfurized olefins; *Corrosion Inhibitor/Buffer* (4.0–6.0 %)—alkanolamine; *Lubricity Agent/Friction Modifier* (5.0–10.0 %)—carboxylic acid derivative; *Foam Inhibitor* (1.0–2.0 %)—polymethacrylate; *Biocide* (2.0–3.0 %)—triazine; *Emulsifier* (5.0–10.0 %)—soap. The balance in mineral oil.

Semi-synthetic Removal Fluid: *Extreme Pressure Agent* (4.0–6.0 %)—Chlorinated wax or sulfurized olefins; *Corrosion Inhibitor/Buffer* (5.0–10.0 %)—alkanolamine; *Lubricity Agent/Friction Modifier* (5.0–8.0 %)—carboxylic acid derivative; *Foam Inhibitor* (1.0–2.0 %)—polymethacrylate; *Biocide* (2.0–3.0 %)—triazine; *Emulsifier* (5.0–10.0 %)—soap or petroleum sulfonate. The balance is water.

Synthetic Removal Fluid: *Extreme Pressure Agent* (2.0–3.0 %)—Sulfurized olefins or zinc dialkyl dithiophosphate; *Corrosion Inhibitor/Buffer* (5.0–10.0 %)—alkanolamine; *Lubricity Agent/Friction Modifier* (4.0–6.0 %)—carboxylic acid derivative; *Foam Inhibitor* (1.0–2.0 %)—polymethacrylate; *Biocide* (2.0–3.0 %)—triazine; *Emulsifier* (5.0–10.0 %)—soap or petroleum sulfonate. The balance is water.

Mineral Protecting Fluid: *Corrosion Inhibitor* (2.0–3.0 %)—Neutral metal sulfonate mixture. The balance is mineral oil.

Synthetic Protecting Fluid: *Corrosion Inhibitor* (0.5–1.0 %)—Amine carboxylate and amine borate mixture; *Lubricity Agent/Friction Modifier* (0.5–1.0 %)—phosphate ester; *Buffer* (5.0–10.0 %)—alkanolamine; *Biocide* (1.0–5.0 %)—triazine. The balance is water.

Mineral Treating Fluid: *Oxidation Inhibitor* (0.1–0.5 %)—Phenol and arylamine mixture; *Speed Improver* (1.0–10.0 %)—calcium sulfonate.The balance is mineral oils.

Synthetic Treating Fluid: *Corrosion Inhibitor* (1.0–2.5 %)—Alkanolamine; *Foam Inhibitor* (0.5–1.0 %)—polyacrylate; *Buffer* (5.0–10.0 %)—alkanolamine; *Biocide* (2.0–3.0 %)—triazine. The balance is water.

Aqueous Quenching Fluid: *Thickener* (10%)—Water-soluble ethylene oxide/propylene oxide coploymer; Corrision Inhibitor (0.03%)—nitrobenzoic acid or its alkali metal salts; *Corrosion Inhibitor Synergist* (0.03%)—salicylic acid or its alkali metal salts; *Buffer* (0.1%)—diethanolamone. The balance is water. Formulation extracted from Ref [763b].

12

Lubricant Testing

IN THIS CHAPTER WE DESCRIBE ANALYTICAL AND spectroscopic techniques and physical and chemical tests that are used to establish structural/compositional identity and physical and chemical properties of the additives, base stocks, and finished lubricants. Tribological (mechanical) tests are also included and so are the performance requirements of the finished lubricants and their in-service condition monitoring. The chapter also contains discussion pertaining to the new lubricant approval process and a list of ASTM and other standardized tests that are commonly used to assess lubricant quality.

Lubricant additives are either supplied individually or as a performance package for the user to blend in the base fluid of his or her choice to make a finished lubricant. The former is the case for nonautomotive lubricants and the latter is the case for automotive lubricants. In either case, the finished lubricant must meet the performance requirements established by a variety of technical organizations, OEMs, and the end-users. At high concentrations, the additive molecules exist as association structures, called micelles, due to intermolecular association via their polar ends. Upon dilution, deaggregation occurs and the additive molecules attain a more active form. Because the additives are reactive chemicals, they can interact with one another, when in a package, either synergistically or antagonistically [765]. The formulator's challenge is to deliver the intended performance by minimizing the antagonistic effects and maximizing the synergistic effects through careful balancing.

For automotive use, the viscosity modifier and the performance package are usually sold separately [308] and for the applications needing a viscosity modifier, it is blended in the base fluid along with the performance package to obtain the finished lubricant. Typically, the lubricant additive suppliers develop general-purpose performance packages, which when blended in widely available base stocks in a predetermined amount meet industry specifications. Additive suppliers may fine tune their packages for an individual company's use in its base stocks. Table 4.33 shows the classes of additives that are used to formulate engine lubricants and Table 4.34 contains the classes of additives that are used to formulate nonengine lubricants. Formulation examples were provided in previous chapters dealing with the various fluid types.

Introduction of a New Additive or a Product

The development of a new additive/formulation is initiated after a new product (lubricant) need is identified. The need for the new product is usually expressed by the OEMs and the end-users and relates either to the inadequate performance of the existing products in the current equipment or the perceived needs of the equipment under development. To fulfill this need, various organizations, such as SAE, API, ASTM, AGMA, ACEA, JASO, and OEMs, initiate the development of new performance specifications and the test methods.

Additive companies, either alone or in collaboration with a lubricant supplier, attempt to satisfy the performance requirements established for the new product. If the additive company is unable to develop the additive system using their existing technology base, they initiate a project to develop and test a new additive or additives. The newly developed additive or additives are blended with other additives in a customer's base oil and the lubricants thus obtained are screened in a number of physical, analytical, and mechanical tests, called the preliminary tests [4]. *Physical tests* deal with colligative molecular properties of the lubricant, such as density, viscosity, shear stability, emulsion-forming tendency, foaming characteristics, refractive index, flash point, cloud point, pour point, and evaporation loss. *Chemical tests*, in conjunction with spectroscopic methods, are used to characterize the lubricant. Important analytical tests pertain to structural analysis, hydrolytic stability, carbon residue, water, sulfur, ash content, acidity, alkalinity, alkaline residue, corrosion, corrosion protection, seal compatibility, and aging characteristics to predict a lubricant's service life. Some of these tests are standardized ASTM tests while others are proprietary bench tests. Bench tests, also called the screen tests, are accelerated tests that are devised to closely simulate conditions the lubricant is likely to experience in actual service. This kind of testing is quite common because it allows the evaluation of a large number of additives and formulations quickly and inexpensively. Candidates that demonstrate good performance in these tests are taken to the next stage to test performance, which is to test in full-scale tests.

The Approval Process

The purpose of the approval process is to ensure that the new lubricant meets the established performance criteria. Since the quality of the lubricant depends upon the quality of its components, preliminary tests are also performed on the base fluid, additives, and of course the lubricant itself to ensure quality and performance, prior to subjecting the lubricant to more expensive full-scale testing. Typically, additive quality is determined by establishing their identity and purity and the lubricant quality is assessed by its performance in the established tests. Full-scale testing is carried out using actual equipment either in a laboratory or in collaboration

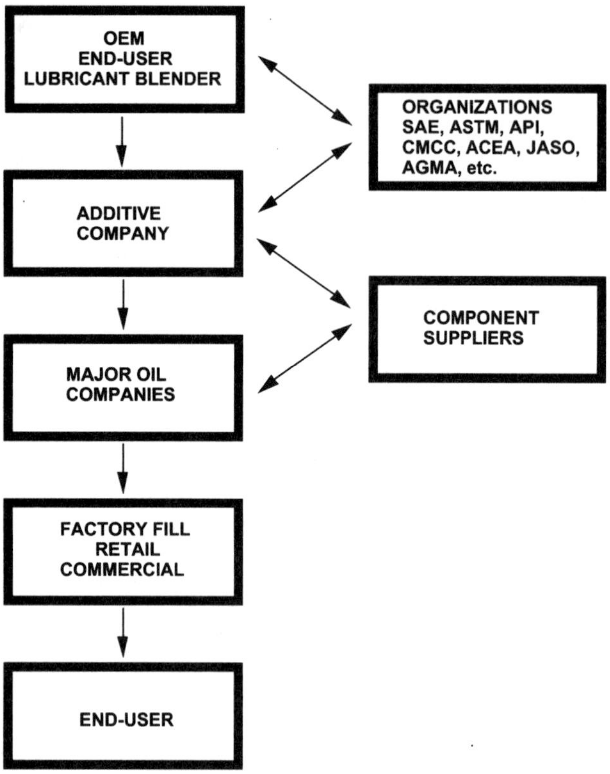

Fig. 12.1—New product development—route to the end-user.

with an end-user. For automotive products, field trials may also be necessary. The cost associated with the development and testing of the new product can be phenomenal. The lubricant that has successfully met all the performance requirements is ready to be marketed either through factory-fill or service-fill lubricant blenders. A flowchart of the process is shown in Fig. 12.1.

Physical and Analytical Tests

Additives

The first and foremost concern after an additive is manufactured is to establish its structural identity and purity. This is essential both from the perspective of conserving testing resources and for developing chemicals with optimal performance. Structural identity may be established by the use of analytical techniques available to a chemist. These include elemental analysis, wet chemical methods, functional group determination, molecular weight determination, and spectroscopic techniques. Some of the methods are provided in the ASTM standards while others are not. Elemental analyses are used to determine the amount of certain elements in the additive, the finished lubricant, or the used lubricant. Such elements include chlorine (Cl), bromine (Br), nitrogen (N), sulfur (S), phosphorus (P), and boron (B), and metals such as sodium (Na), potassium (K), calcium (Ca), magnesium (Mg), barium (Ba), zinc (Zn), copper (Cu), and molybdenum (Mo). Of these, some elements such as sulfur, calcium, magnesium, phosphorus, zinc, nitrogen, and chlorine originate from the additives. Others, such as copper, tin, iron, and lead are primarily wear elements. The methods to analyze these are provided in the ASTM Standards D4951, D5185, D4927, D5291, D4047, D1091, D874, D2622, D808, D4951, D4927, D4628, and D5185. Functional group analysis involves the use of both the wet chemical methods and the spectroscopic techniques. Commonly used wet chemical methods include neutralization number determination (ASTM D974, D664, D3339, D4739, D2896), saponification number (ASTM D94), bromine number (ASTM D1159), iodine value, acetylation reaction, reaction with Grignard reagent, and the Kjeldahl method. The neutralization numbers and saponification number provide information on the acid functional groups, such as sulfonic acid, carboxylic acid, phosphoric acid, and their derivatives. The neutralization numbers also provide information on the basic (alkaline) functional groups, such as amino groups and inorganic bases, such as metal oxides, hydroxides, and carbonates. Bromine number and iodine value are used to assess unsaturation in additives, base fluids, and lubricants. Acetylation and the reaction with Grignard reagent are used to determine the hydroxyl content of the additives. Grignard reagent is primarily used for situations, for example in the case of hindered phenols, where steric crowding does not allow a complete reaction with the acetylation reagent. The Kjeldahl method (ASTM D3228) is used to determine the elemental nitrogen. Spectroscopic methods include infrared (IR) and nuclear magnetic resonance (NMR) spectroscopy, and mass spectrometry (MS). These techniques are well established and are aptly covered in many publications [767–770]. The purity is determined by gas-liquid chromatography (GLC or GC), thin layer chromatography (TLC), liquid chromatography (LC), and gel permeation chromatography (GPC). These techniques are discussed in some detail in the latter part of this chapter. Additional tests include the molecular weight determination by vapor phase osmometry (VPO) and mass spectrometry (MS), oil solubility and package compatibility (ASTM D501), copper strip activity (ASTM D130), viscosity (ASTM D445), color (ASTM D1500), flash point (ASTM D92 and D93), volatility (ASTM D1078), melting point, boiling point, odor, clarity, and water content (ASTM D1744, D4928, and D4377). While some ASTM standards are designed to analyze only low quantities of the described elements, in many instances these methods can be used and *are* used to analyze larger concentrations as well, via dilution.

Base Fluids

Liquid lubricants are formulated by the use of a variety of base stocks, but mineral oils are by far the largest with respect to use. This is primarily because of their balanced properties, good additives response, ready availability, and reasonable cost. Mineral oil lubricants are therefore the first choice of the formulator for most applications. However, in other applications special properties are required that cannot be attained by the use of the mineral oils, even when formulated with the best additive technology. That is where the synthetic base stocks come into play. Synthetics offer good low-temperature fluidity, superior viscosity-temperature properties, low volatility and flammability, and excellent thermo-oxidative stability. For further details on these base stocks, please refer to Chapter 3.

Physical properties of the finished lubricants are primarily attributable to the structure and the properties of the lubricant base stocks, because they make up the bulk of the lubricant. Chemical properties, on the other hand, are due to

TABLE 12.1—Viscosity data on some lubricants [154].

Lubricant Type[a]	Viscosity (cSt) 100 °C	Viscosity (cSt) −40 °C	cSt Viscosity (40 °C) at Pressure 138 MPa	275.9 MPa	551.7 MPa	Viscosity Index
Fluoro Lubricant	2.9	500,000	2,700	200,000	>1,000,000	−132
Hydrocarbon (Aromatic)	3.4	50,000	800	36,000	>1,000,000	0
Hydrocarbon (Paraffinic)	3.9	14,000	340	12,000	27,000	100
Ester	4.4	3,600	110	500	4,900	151
Poly(glycol ether)	4.6	7,000	120	570	8,800	164
Phosphate[b]	4.6	8,000	...	...	...	164
Ester-base[b]	6.3	1,000	...	...	...	197
Silicone	9.5	150	160	700	48,000	195

[a]All Fluids—20 cSt at 40 °C and 0.1 MPa.
[b]Contains polymeric additive.

the presence of the additives used to formulate them. The properties that are critical to lubricant performance are listed below. These are the properties that are evaluated by the various physical and analytical tests:

1. Viscosity
   - Viscosity-temperature Relationship
   - Viscosity-pressure Relationship
   - Viscosity-shear rate Correlation
   - Viscosity-volatility Relationship
2. Vapor Pressure
3. Density
4. Bulk Modulus
5. Thermal Properties
6. Surface Tension
7. Gas Solubility
8. Foaming Tendency
9. Electrical Properties
10. Thermal Stability
11. Oxidation Stability

Viscosity

The viscosity values most frequently reported for a lubricant are at 40 °C and 100 °C (previously 100 °F and 210 °C) at atmospheric pressure and low-shear rates. Viscosity is a measure of a fluid's resistance to flow. The basic unit for absolute or dynamic viscosity is the Pascal-second (10 Poise). The common unit of absolute viscosity is centiPoise, cP (1 mPa·s). The most common method of viscosity measurement is described in the ASTM D445 standard. Viscometers are devices that are used to measure viscosity. Most depend on the force of gravity to drive the fluid through a capillary. The viscosity value thus obtained is referred to as kinematic viscosity. The unit of kinematic viscosity is Stoke (St) or centi-Stokes (cSt = 0.01 St). One centiStoke equals 1 mm^2/s. Absolute viscosity in centiPoise (cP) is equal to kinematic viscosity in centiStokes multiplied by the density of the fluid in kg/m^3. Two types of rotational viscometers are also used to measure viscosity. These are Brookfield (ASTM D2669 and ASTM D2983) and the cone and plate type. Brookfield viscometer is often used to measure low temperature viscosity since at low temperatures some fluids do not flow under the force of gravity. Viscosity index (VI), which is a measure of a lubricant's viscosity-temperature relationship, is based on 40 °C and 100 °C viscosity values (ASTM D2270).

Mineral oils have good viscosity indices (VIs) but synthetics have generally higher VIs and lower low-temperature viscosity, as shown in Table 12.1 [154]. It is important to note that the modern hydrocracking technology significantly improves these properties of the mineral base stocks (API Group II, Group III, and VHVI stocks) as well, and of course the use of the polymeric viscosity modifiers and pour point depressants further boosts them.

Consideration of the lubricant's viscosity-temperature and viscosity-pressure relationship is critical to applications that involve high temperatures and high pressures, such as the hydraulic systems. Bulk pressures in these systems range from atmospheric to 10,000 psi (0.1 and 69 MPa) and temperatures from −65 to 300 °F (−54 to 149 °C). Hydrodynamic bearings experience temperature of around 100 °F (55 °C) above these temperatures and the pressures of 10,000 psi (69 MPa) over the bulk system values. For elastohydrodynamic (EHD) contacts in gears, cams, and roller and ball bearings, the temperature may be 100 to 300 °F (55 to 167 °C) over the bulk value and the pressures may be in the 50,000 to 500,000 psi (345 to 3450 MPa) range. Boundary lubrication implies temperatures in the order of 650 °F (343 °C), or higher, and pressures in the same range as in the EHD contacts [154]. Viscosity-temperature and viscosity-pressure properties of the synthetics provide a much broader range, as shown in Table 12.1. The viscosity-pressure coefficient determines the quality of the elastohydrodynamic film in bearings and gears.

Many premium hydraulic fluids and lubricants are based on mineral oils, to which the polymeric viscosity modifiers are added to improve their high temperature performance. Some synthetic base stocks are also polymeric, for example, polysiloxanes (silicones), poly(glycol ether)s, polyesters, and poly(perfluoro ether)s. Polymers are susceptible to shear, which can cause a reduction in viscosity, reversibly or irreversibly, depending upon the magnitude and the duration of the shear forces. Such shear forces result from turbulent flow and in high contact applications, such as gears and bearings. These forces either deform the polymer, causing a temporary viscosity loss, or mechanically degrade it, causing a permanent viscosity loss. Shear rates in most lubricant applications range from very low values to 10^6 s^{-1}. Shear also occurs during cold starting of an automotive engine. During cold starts, the oils of 3000 to 50,000 cP (3 to 50 Pa·s) are subjected to shear rates of 10^3 to 10^4 s^{-1}.

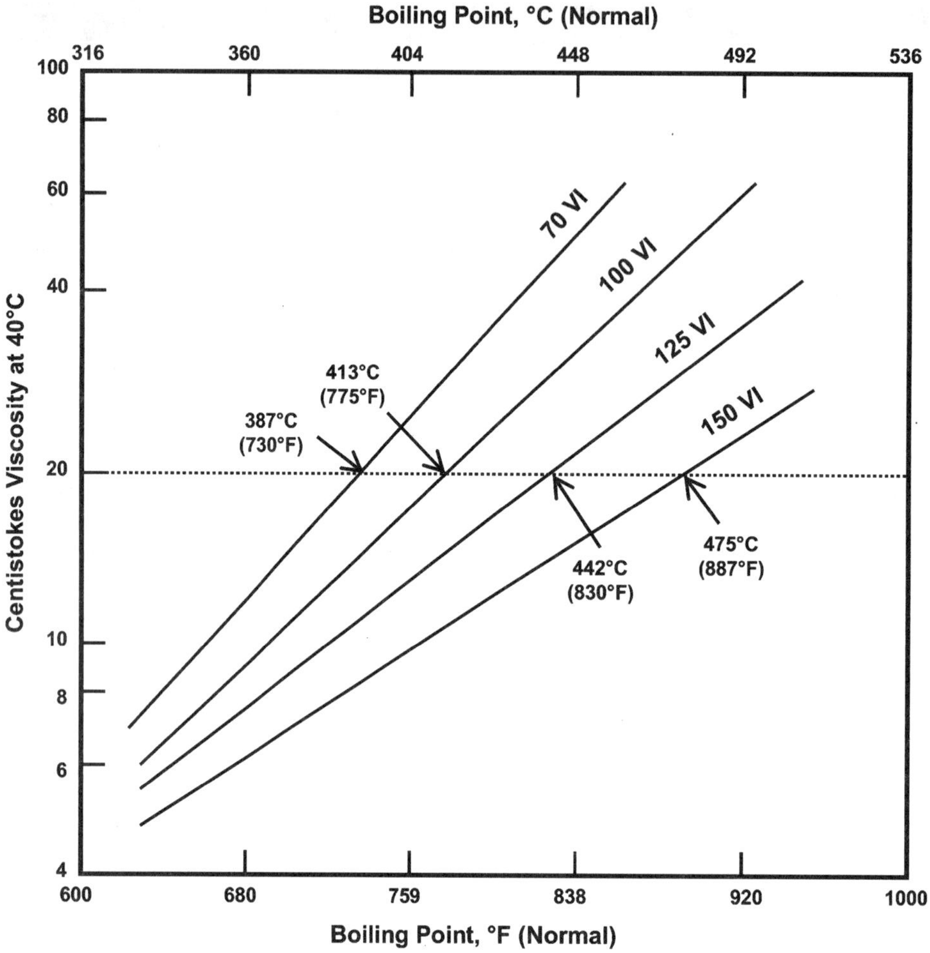

Fig. 12.2—Boiling point versus viscosity index [154].

Volatility and Flash Point

In automotive crankcase oils, vapor pressure, or volatility, is important in determining the rate of the oil consumption and the quality of the exhaust emissions. In other applications, such as steam turbines, high volatility can create a fire hazard. The volatility of an oil is tested according to the ASTM Test Method D92 (Cleveland Open Cup flash and fire points). High vapor pressure of the mineral oil at elevated temperatures is due to the presence of the highly volatile, low boiling components; hence the flash point of a well-distilled mineral oil can be higher by about 10 °F (5 °C) and fire point by about 100 °F (55.5 °C) than the normally obtained mineral oil. The flash point of 400 °F (204 °C) and a fire point of 440 °F (227 °C) are expected of a typical lubricating oil fraction. Gas chromatographic analysis of a typical mineral oil indicates that 5 to 95 % of the oil has a boiling range of 150 to 170 °C (302 to 338 °F). In general, the high VI oils have higher boiling points. Figure 12.2 shows viscosity-boiling point plots of the synthetic fluids and the narrow boiling range (30 °C) mineral oil of the different VIs [154]. For the oils of the same viscosity, the boiling point of the high VI oils is higher than that of the low VI oils. This is shown in the figure for oils that have a 40 °C viscosity of 20 cSt. Since the boiling points of the oil or its components are quite high, simple distillation cannot be used to determine them. However, one can either use vacuum distillation or gas chromatography (ASTM D2887). The latter procedure only works for oils that have a boiling point of less than 1000 °F (538 °C). Distillation (boiling) temperatures at reduced pressures (vacuum) can be converted into those at atmospheric pressures and *vice versa* by the use of a vapor pressure chart [154] or on line conversion [771]. Since the mineral oil contains components of different volatilities, vapor pressure of the mineral oil based lubricant is influenced by its more volatile components. In addition to oil consumption, evaporation, and safety (flammability), the volatility also affects the boundary lubrication. High volatility lubricants are known to cause more wear than the lubricants of low volatility [772,773].

Density and Specific Gravity

Density of a substance is defined by mass per unit volume and in liquids, such as lubricants, is expressed as gram/millilitre (g/mL). Relative density, also known as specific gravity, is a measure of the density of a material relative to another material. Specific gravity of the liquids is equal to the density of the liquid divided by the density of water, and in gases, it is the density of the gas divided by the density of air. Specific gravity has no units. For liquids, density, hence specific gravity, is typically measured at 60°F or 15.6°C. In petroleum products, the API gravity is used more often. As stated earlier, it is related to specific gravity at 60 °F. API gravity and specific gravity are inversely related. Table 12.2

TABLE 12.2—Coefficient of expansion for mineral oil lubricants [154].

Specific Gravity at 60°F (15.6°C)	API Gravity at 60°F (15.6°C)	Coefficient of Volumetric Expansion	
		Per °F	Per °C
1.076–0.967	0–14.9	0.00035	0.00063
0.966–0.850	15–34.9	0.00040	0.00072
0.850–0.776	35–50.9	0.00050	0.00090
0.775–0.742	51–63.9	0.00060	0.00108

shows the relationship between the two and the coefficient of volumetric expansion (estimated from ASTM Tables) [154]. A high API gravity value matches a low specific gravity and *vice versa*. Density of a material depends upon both pressure and temperature. Density change with temperature is called coefficient of thermal expansion and for liquids the more appropriate term is volumetric thermal expansion coefficient. This coefficient in liquids affects volume and is more sensitive to the boiling point of the hydrocarbon material or the component than to its density. Specific gravity is often used to identify specific lubricants, for example to distinguish between primarily paraffinic, naphthenic, and aromatic-based stocks (ASTM D3238).

Bulk Modulus

This fluid parameter expresses the resistance of a fluid to compression and is the reciprocal of compressibility. Compressibility varies with pressure, temperature, and molecular structure and is highly significant in fluids, such as hydraulic lubricants. It also plays a role in the viscosity-pressure, hydrodynamic, and EHD behavior of the lubricants. There are several methods to measure bulk modulus, including the ultrasonic methods [774]. As stated while discussing hydraulic fluids, entrained air, or another gas, in a hydraulic system under high pressure shows slow system response. This is due to greater volume reduction of the air containing oil under pressure than the compressibility of the original fluid [154].

Gas Solubility

Gas solubility in a lubricant affects many of its properties, such as viscosity, foaming tendency, bulk modulus, possibility of cavitation, heat transfer, oxidation, and boundary lubrication. In many cases, the gas is entrained at low pressures, which dissolves in the high-pressure portion of the lubrication and hydraulic systems. Hence, when the pressure is reduced, the gas comes out of the solution to produce foam or entrained gas bubbles. The dissolved oxygen from the air can also cause lubricant oxidation at high temperatures in bearings and other hot parts of the system. Gas solubility can be measured at temperatures up to 260 °C by using a gas chromatograph (GC) [154]. The gas solubility increases with an increase in pressure and decreases with an increase in temperature, approaching zero at the normal boiling point of the fluid. High molecular weight gases, such as CO_2, are less soluble in fluids than the low molecular weight gases, such as hydrogen and nitrogen. At the same time, an increase in temperature causes a smaller drop in the solubility of the lower molecular weight gases than of the higher molecular weight gases. The low molecular weight volatile components from the lubricant can have a similar effect on lubricant properties. However, because of their larger molecular size and weight, the adverse effects are quite small. The effect of gas solubility on the lubricant viscosity is shown in Table 12.3 [154]. These data substantiate our previous statement regarding the gases of higher molecular weight being less soluble than those of the lower molecular weight. Compare the gas solubilities in paraffinic neutral and paraffinic resin. The data also show that the fairly nonpolar gases, such as nitrogen, are more soluble in nonpolar hydrocarbon fluids than in polar esters. Examine the % viscosity change column of the table.

Foaming and Air Entrainment

The tendency of a lubricant to form and sustain foam generally increases as the fluid's molecular weight and the viscosity increases, or the temperature decreases. Foaming is caused by the escaping insoluble gases or the physical mixing of the excess gas with the fluid. Foaming can cause insufficient oil delivery to parts; for example bearings, causing premature failure; interfere with the proper operation of the equipment, such as the lubricating pumps; and may even show false oil level readings. Under some circumstances, foam may even overflow from the oil reservoirs. Foam formation can be minimized by altering the mechanical design where possible or by the use of the additives called the foam inhibitors. Common additives used to control foam include polysiloxanes (silicones) and polyacrylates. These additives perform by lowering the surface tension at the gas-liquid interface. Air entrainment causes similar problems as foam. In hydraulic systems, it in addition causes reduced response and in gear systems reduced heat transfer, hence the higher operating temperatures. Foam inhibitors are not very helpful in controlling the air entrainment. However, some addi-

TABLE 12.3—Effect of dissolved gases on lubricant viscosity at 1000 psi (7.0 MPa) pressure [154].

Fluid	Gas-free Viscosity at 100 °F (38 °C) cSt (mm²/s)	Gas-saturated Viscosity at 100 °F (38 °C) cSt (mm²/s)			% Change in Viscosity Due to N_2
		He	N_2	CO_2[a]	
Diester	14.9	...	12.9	...	−13
Paraffinic Neutral	15.4	14.0	13.1	4.0	−15
Paraffinic Bright Stock	625	...	470	...	−25
Polyester	630	...	515	...	−18
Naphthenic oil	3980	...	2620	...	−34
Paraffinic resin	5400	4600	3800	1500	−30

[a]Saturated at 500 psi (3.55 MPa).

tive suppliers market air-release agents, which are modified siloxanes, fatty acid esters, fatty alcohols, alkoxylates, and polyalcohols based on ethylene-propylene block copolymer.

Thermal Properties

Thermal properties of lubricants that are of interest are thermal conductivity, specific heat, and heat of vaporization. The first two are concerned primarily with the heat transfer and the last with the ease of evaporation. Thermal conductivity is the ability of a material to conduct heat. In liquid hydrocarbons it ranges from 0.14 W/m·K at 0 °C (273 K) to 0.11 W/m·K at 400 °C (673 K) [154]. These values should essentially hold for all mineral oil and synthetic hydrocarbon-based lubricants. Polarity and hydrogen bonding affect thermal conductivity; for example, ethylene glycol has a much higher thermal conductivity, of 0.31 W/m·K at 100 °C. Specific heat is the heat required to raise the temperature of a specific quantity of a substance, in our case a fluid, by one degree kelvin. The same as thermal conductivity, specific heat also varies linearly with temperature and increases with increased polarity or the hydrogen bonding of the molecules. Specific heat of water at 100 °C is about twice that of the oil [154] and for mineral oil and synthetic hydrocarbon lubricants, the values range from 0.45 Btu/lb-F (1882 J/kg·K) at 0 °C (273 K) to 0.78 Btu/lb-F (3263 J/kg·K) at 400 °C (673 K). The higher values for these parameters for a lubricant indicate a better heat dissipation ability. Heat of vaporization is the energy required to transform a given quantity of a substance into its vapor. Heat of vaporization depends on the pressure and the molecular weight or the boiling point of the oil. Most conventional mineral oil-based lubricants have latent heats of vaporization between 60 and 90 Btu/lb (140 to 209 kJ/kg) at atmospheric pressure. By comparison, the heat of vaporization of water is 969.7 Btu/lb (2255 kJ/kg) at atmospheric pressure. This implies that much less heat is required for evaporating oil than water, if the conditions are such that the two liquids have the same vapor pressure.

Electrical Conductivity

Electrical conductivity is a measure of a material's ability to conduct an electric current. If the electrical conductivity of a lubricant is high, it indicates the presence of ions and ion-forming materials, such as metals and metal-containing additives, water, and thermo-oxidative degradation products. In water-based lubricants, low electrical conductivity is desired since it is associated with metallic corrosion and the corrosion products promote oxidative degradation of the lubricant. Electrical conductivity of a well-refined and dry mineral oil and most synthetic lubricant base stocks is extremely low, in the order of 10^{-14} mho/cm^2. Electrical conductivity of the unused lubricant is of primary concern if the lubricant is to be used in an electrical environment, such as for transformers and some aircraft and industrial control systems.

Surface Tension

Surface tension is the force that makes the surface layer of a liquid behave like an elastic sheet and is the result of the intermolecular forces in a liquid that keep its molecules together. Surface tension involves a liquid and a gas or a vapor, and the related force interfacial tension involves two immiscible liquids. Two methods that can be used to measure surface tension/interfacial tension are ASTM D971 Rev A and D1331. Surface tension data for several base fluids are shown in Table 12.4 [154]. Interestingly, additives have a profound effect on the surface tension of the finished lubricant. For example, the addition of only 0.1 wt % silicone, commonly used to suppress foam in a mineral oil lubricant, causes a reduction in its surface tension to almost that of the silicone fluid itself and hence improve surface wetting of the bearings. Wetting is due to the preferential interaction of the fluid molecules with the molecules of the solid surface than with each other, that is, the adhesive forces are stronger than the cohesive forces. In general, the fluids of low surface tension have better wetting characteristics. The surface tension values of the fluids, listed in Table 12.4, suggest that the wetting tendency increases as we go down the table. However, as mentioned above, the presence of additives in the lubricant can easily change that.

TABLE 12.4—Surface tension of base fluids [154].

Fluid	Surface Tension dyne/cm (N/m)
Water	72($\times 10^{-3}$)
Mineral Oils	30–35($\times 10^{-3}$)
Esters	30–35($\times 10^{-3}$)
Methylsilicone	20–22($\times 10^{-3}$)
Fluorochloro Compounds	15–18($\times 10^{-3}$)

Interfacial tension between two immiscible liquids is approximately the difference between the surface tensions of the two liquids. Additives that create stable emulsions and micro-emulsions are capable of reducing the interfacial tension between the two phases to very low values, almost approaching zero, which results in thorough mixing of the two phases. While the emulsions in metalworking and hydraulic fluids are useful, in other applications such as automotive engine oils they are not. Detergents and dispersants, which are added to the automotive lubricants to neutralize and suspend combustion and oxidation products, reduce interfacial tension of the oil contaminated with 10 to 15 % water to such an extant that the stable emulsions that form cannot be easily separated. As mentioned earlier, the use of the additives, called demulsifiers, sometimes help. Both surface tension and interfacial tension are altered by additives and by the lubricant degradation products.

Thermal Stability

Thermal stability is the resistance of a lubricant to break down or to structurally change under the influence of heat and in the absence of oxygen. One of the methods used to measure this lubricant parameter is described in the ASTM Standard D2879. The method uses isoteniscope, a closed vessel with a manometer for measuring the rate of pressure increase at a specified heating rate. Thermo-gravimetric analysis (ASTM E1131, ISO 11358) and differential thermal analysis (ASTM E1782, ASTM E537) can also be used to evaluate thermal stability. Several thermal stability tests are described in the Federal Standards (FED-STD-791/2503.2 and FED-STD-791/2508.1) [775]. Thermal stability tests must allow for the decomposition of a significant portion of the test sample and provide an analysis of the liquid and solid decomposition products, as well as the gases formed. Mineral oils show a thermal stability of 650 to 700 °F

(343 to 371 °C). Synthetic hydrocarbons that are primarily aliphatic in nature, such as PAOs that are prepared by the polymerization of olefins followed by hydrogenation, show a thermal stability of 600 to 650 °F (316 to 343 °C). This is 50 °F (28 °C) or more below that of the mineral oil. Decomposition of the mineral oil results primarily in methane but also ethane and ethylene. A synthetic hydrocarbon on decomposition will produce a major amount of the monomer from which it was made [154]. Materials that contain aromatic rings in their structure are somewhat more stable and have a decomposition temperature of 850 to 900 °F (454 to 482 °C). This class includes poly(phenyl ether)s, chlorinated biphenyls, and condensed ring aromatic hydrocarbons, such as alkylated naphthalenes. Organic esters have thermal stability in 500 to 600 °F (260 to 316 °C) range, the ester functional group being the primary site for decomposition. Methyl esters have thermal stability similar to that of the mineral oil. Polymeric viscosity modifiers have lower thermal stability than their nonpolymeric analogues. Polymethacrylate have a decomposition temperature of 450 °F (232 °C) and polybutenes have a decomposition temperature of 550 °F (288 °C). Most additives used to formulate lubricants have thermal stability lower than that of the base fluids. For example, zinc dialkyl dithiophosphates that are used as oxidation inhibitors and EP/antiwear agents, degrade at 400 to 500 °F (204 to 260 °C). More active EP additives have even lower thermal stability [154].

Oxidation Stability

Most lubricant applications are in the presence of air or oxygen; hence a lubricant to have good oxidation stability is highly desirable. All hydrocarbon materials undergo oxidative degradation, which was described in the Additives chapter, Chapter 4. Unlike thermal stability which is inherent to the base stock, oxidation stability can be greatly improved by the use of the oxidation inhibitors. The consequences of oxidation are a lubricant's viscosity increase and the formation of acids and deposits, such as varnish and sludge. A wide variety of tests are available to assess a lubricant's oxidation stability. These include tests that are described in the ASTM Standards D2272 and D1313, and in the Federal Test Standard 791 (FED-STD-791), methods 2504, 3405.2, and 3407.2. These tests are suitable for measuring a lubricant's stable life and the effectiveness of the oxidation inhibitors. To monitor the oxidation process, a micro-oxidation test, such as the Penn State micro-oxidation test, has been developed along with the analytical procedures based on gel permeation chromatography (GPC) and atomic absorption spectroscopy (AAS) [776].

Oxidation life of a lubricant depends upon the service conditions. Khonsari and Booser estimated a mineral lubricant's life solely based on application's temperature, assuming no contamination [777]. The estimates are shown in Table 12.5. Obviously, in real life there is contamination, adverse catalytic effects from metals, such as copper and iron, and oxidation-inhibitor loss due to evaporation; all of which will affect an oil's expected life. By taking into account these factors, the researchers estimated oil life to be much shorter than that predicted. A real life temperature effect on the stability of a lubricant is provided in Fig. 12.3 [154]. In both cases, the lubricant life is longer at lower temperatures. Khonsari and Booser also estimated the service life of the synthetic lubricants as a function of temperature, see Fig. 12.4 [777]. Polyol esters are proposed to be thermally and oxidatively most stable and mineral oils to be the least stable. The others fall in between the two extremes with respect to the oxidative stability.

TABLE 12.5—Oxidation life of mineral oils under ideal conditions [777].

Fluid Type	Maximum Temperature for 1000-h Life
Uninhibited (used in once-through systems)	75 °C
Extreme-pressure Gear Lubricant	84 °C
Hydraulic Fluid	99 °C
Turbine Lubricants	106 °C
Heavily Refines, Hydrocracked	121 °C

Elemental and Structural Analysis

Petroleum, or crude oil, contains a wide variety of elements, some of which are present at percent levels and others at parts per million levels. However, refining processes used to manufacture fuels and mineral base oils remove most elements other than carbon, hydrogen, oxygen, nitrogen, and perhaps sulfur. Additives used to formulate lubricants contain elements that are used either to facilitate their solubility in base fluids or impart special properties. Common elements include nitrogen, sulfur, phosphorus, alkaline earth metals, zinc, copper, and molybdenum. A list of elements that are generally used in lubricants is provided in Table 12.6, along with their role [778].

A variety of analytical techniques are used for elemental analysis of the petroleum products and includes atomic spectroscopy (atomic absorption spectroscopy, AAS), and inductively coupled plasma atomic emission spectroscopy, ICP-AES), X-ray fluorescence (XRF), and micro-elemental techniques [779]. These analytical techniques are not only used for new oils but also for used oils, at the end of their service life. The objective in the latter case is to analyze wear metals, find their source, and devise ways to minimize wear in those machine parts. Wear metals originate primarily from the mechanical wear of the various parts of the equipment. Wear metals in the lubricant usually exist in a suspended form, which can settle on their own or after dilution of the lubricant by a low viscosity hydrocarbon solvent, such as pentane or hexane. Sometimes centrifugation is employed to separate them from the lubricant. Some wear metals are specific enough to be able to identify the machine part or parts from which the metal originated. A list of wear metals found in used combustion engine lubricant is provided in Table 12.7, along with their source [778]. Normal wear rate is slow; hence a sudden increase suggests a serious problem.

Elemental analysis of the used oils is generally performed by the use of the atomic spectroscopy techniques, particularly ICP-AES, because of its ability to analyze several elements at the same time. Rotating disk electrode emission spectrometry and ICP-mass spectrometry are also capable of analyzing multiple elements at the same time [779]. In addition, the ICP-mass spectrometry has the ability to detect elements even if their concentration is in parts per billion. However, this technique lacks the precision and the accuracy of the other atomic spectroscopy techniques [780]. ASTM methods for wear metal analysis include D5185,

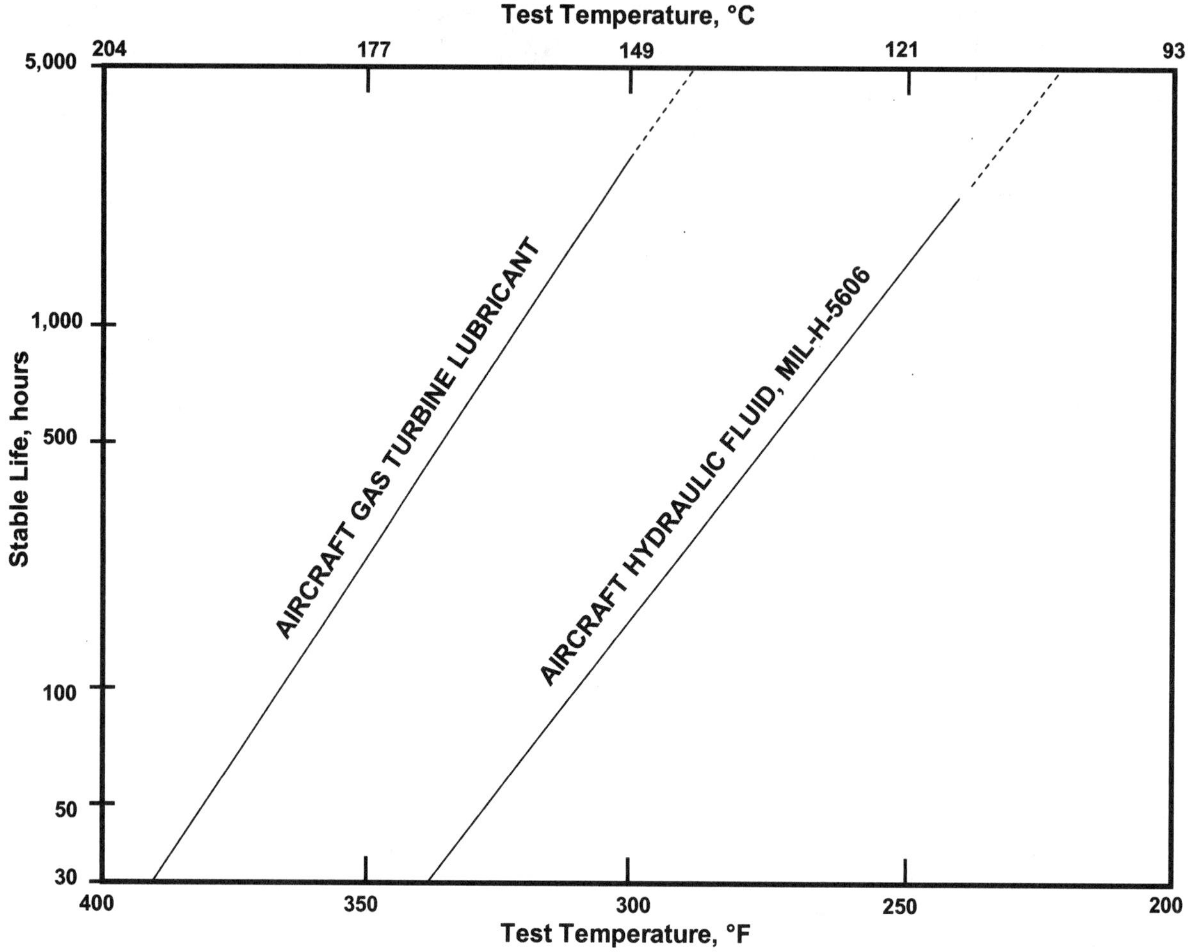

Fig. 12.3—Oxidation stability as a function of temperature [154].

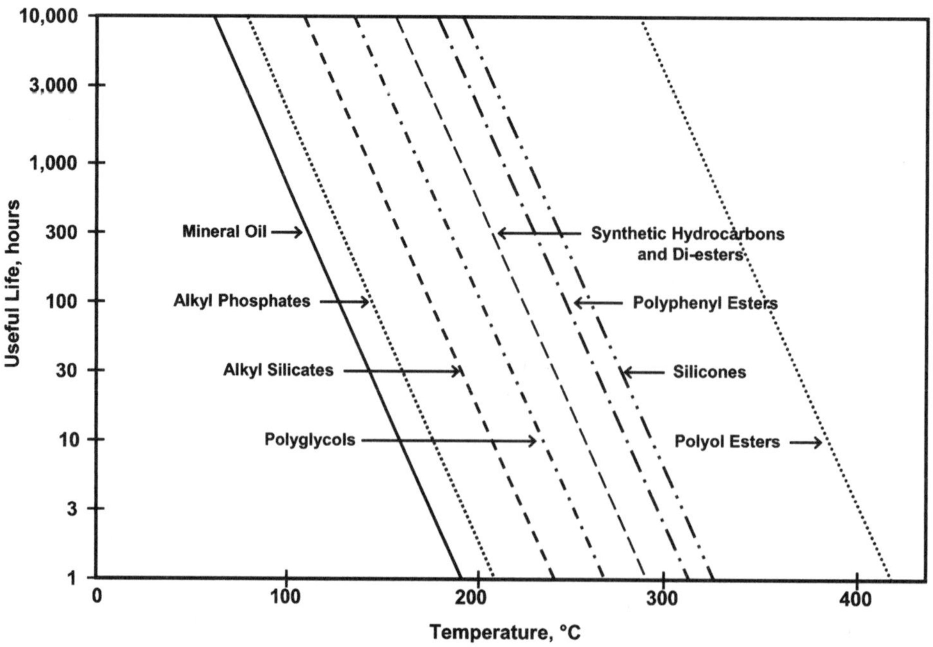

Fig. 12.4—Life expectancy of inhibited synthetic lubricating oils in air [777].

TABLE 12.6—Chemical elements present in lubricants and their role [778].

Element	Compounds	Performance/Application
Boron (B)	Borax and esters	Anti-wear agents, oxidation inhibitors, deodorant cutting oils, greases, brake fluids
Barium (Ba)	Sulfonates, phenates, dialkyl phosphates, phosphonates, and thiophosphonates	Detergent inhibitors, corrosion inhibitors, detergents, rust inhibitors, ATFs, greases
Calcium (Ca)	Sulfonates, phenates, carboxylates, and salicylates	Detergents, detergent inhibitors, dispersants
Cadmium (Cd)	Dialkyl dithiophosphates	Steam turbine oils
Chromium (Cr)	Salts	Grease additives
Copper (Cu)	Carboxylates, dithiocarbamates, and dialkyl dithiophosphates	Oxidation inhibitors, engine oils
Fluorine (F)	Fluorocarbons, fluoro-esters	Base fluids
Mercury (Hg)	Organic compounds	Bactericide in cutting oil emulsion
Magnesium (Mg)	Sulfonates and phenates	Detergents inhibitors
Molybdenum (Mo)	Molybdenum disulfide (MoS_2), dibutyldithiocarbamate, and alkyl phosphate	Greases, extreme-pressure additives
Nickel (Ni)	Cyclopentadienyl complexes	Antiwear agents, carbon deposit control, improved lubrication and combustion
Phosphorus (P)	Metal dialkyl dithiophosphates	Oxidation inhibitors, anti-wear agents, rust inhibitors
Lead (Pb)	Naphthenate	Extreme-pressure agent, greases, gear oils
Antimony (Sb)	Dialkyldithiocarbamates. Dialkyl phosphorodithionates	Antiwear/extreme-pressure agents, oxidation inhibitors
Selenium (Se)	Selenides	Oxidation inhibitors, bear corrosion control
Silicon (Si)	Silicone polymers	Foam inhibitors, base fluids
Tin (Sn)	Organic compounds	Anti-scuffing additives, metal deactivators
Zinc (Zn)	Dialkyl dithiophosphates, dithiocarbamates, and phenolates	Oxidation inhibitors, corrosion inhibitors, ant-wear additives, detergents, extreme-pressure additives, engine oils, hypoid gear lubricants, greases, turbine oils, ATFs, brake fluids

which is an ICP-AES method, and D6595 and D6728, both of which are based on the rotating disk electrode technology.

Wet chemistry methods, not very popular today, are combination methods involving gravimetry to convert the metal into its derivative, which is then analyzed by techniques such as photometry, titrimetry, or other wet analysis techniques. Bomb combustion methods also involve converting the metal into a water-soluble derivative and analyzing it using gravimetry or titrimetry. Most of these methods are being replaced by less tedious and more facile modern instrumental methods. For analyzing nonmetallic elements that are present in petroleum products in very small quantities, micro-elemental methods were devised. These methods involve burning the material to convert the element to be analyzed into a gaseous product, which is then analyzed by the use of a variety of techniques [779]. These ASTM methods are listed in Table 12.8.

Hydrocarbon Analysis

The next step after the elemental analysis is to identify the discrete chemical structures that make up the base oils and the additives; hence the lubricants. With respect to the base oils, the structural determination of the mineral base stocks is the most challenging since they are complex mixtures of the hydrocarbons of various types and various sizes. And, each component influences the base oil properties differently. Structure determination of the synthetic base stocks, on the other hand, is relatively straight forward since their manufacture employs reagents of well-defined structures and purity and in known proportions. Polymeric (or oligomeric) synthetics, such as polyalphaolefins, poly(alkylene glycol)s, and others, although structurally homogeneous, have a molecular weight distribution (polydispersity), which is sometimes beneficial to know to understand their behavior in certain operating environments. Polymeric additives, such as dispersants, viscosity modifiers, and pour point depressants require similar knowledge to understand their behavior. The diversity of the chemical structures present in the mineral oil requires sophisticated separation and analytical techniques to identify them. The methods used include gas chromatography, liquid chromatography, size exclusion chromatography or gel-permeation chromatography (GPC), ultraviolet (UV) spectroscopy, infrared (IR) spectroscopy, mass spectrometry (MS), and nuclear magnetic resonance (NMR) spectrometry. There are numerous ASTM standards that describe the use of these techniques. Of the spectroscopic techniques, the first three are the separation techniques and the last four are the identification techniques.

Gas Chromatography (GC)

This is one of the most common techniques used for separating hydrocarbons for identification by other techniques. In some cases, GC can also be used for identification, if the model compounds exist or are known. GC uses volatility as the primary property for separation. The first step is to convert the oil components into vapor, which is passed through a packed column along with an inert carrier gas, such as nitrogen or helium, to elute them. The carrier gas helps these components to travel through the column to a detector, which generates signals to be recorded on a chromatogram [781]. Two most common types of detectors used are flame ionization and thermal conductivity [782]. In flame ionization type detectors, the column effluent is mixed with hydrogen and air and ignited. The organic material such as hydrocarbons produces, among other species, ions and electrons, which create a signal. Thermal conductivity detectors contain a tungsten filament that is heated using a constant current [781]. The eluting hydrocarbon compounds increase the filament temperature, since they do not equal the cooling ef-

TABLE 12.7—Wear metals in used engine oil and their source [778].

Element	Possible Wear Source
Silver (Ag)	Wrist pin bearings in railroad and auto engines, silver plotted spline lubricating pump
Aluminum (Al)	Piston and bearings wear, push rods, air cooler, pump hosings, oil pumps, gear castings, box castings
Boron (B)	Coolant leakage in system
Cadmium (Cd)	Bearings
Chromium (Cr)	Ring wear, cooling system leakage, Cr-plated parts in aircraft engines, cylinder liners, seal rings
Copper (Cu)	Wear in bushings, injector shields, coolant core tubes, thrust washers, valve guides, connecting rods, piston rings, bearings, sleeves, bearing cages
Iron (Fe)	Wear from engine block, cylinder, gears, cylinder liners, valve guides, wrist pins, rings, camshaft, oil pump, crankshaft, ball and roller bearings, rust
Magnesium (Mg)	Cylinder liner, gear box housings in aircraft engines
Molybdenum (Mo)	Wear in bearing alloys and in oil coolers; various Mo-alloyed components in aircraft engines, piston rings
Sodium (Na)	Antifreeze leakage
Nickel (Ni)	Bearings, valves, gear platings
Lead (Pb)	Bearings, fuel blow-by, thrust bearings, bearing cages, bearing retainers
Antimony (Sb)	Crankshaft and camshaft bearings
Silicon (Si)	Dirt intrusion from improper air cleaner, seal materials
Tin (Sn)	Bearings and coatings of connecting rods and iron pistons
Titanium (Ti)	Various Ti-alloyed components in aircraft engines
Tungsten (W)	Bearings
Zinc (Zn)	Neoprene seals, galvanized piping

fect of the pure carrier gas and generate a signal. Other detection systems, such as atomic emission spectroscopy (AES) and chemiluminescence are used in specialized cases, for example for the detection of the sulfur compounds in light petroleum liquids (ASTM D5623). In GC, high volatility samples usually come out (elute) first. This method works well for high to medium volatility samples. Low volatility samples either take a long time to elute or stay on the column. Incidentally, volatility is a function of a chemical's molecular weight and the polar functional group, if present. For low volatility samples, a related method, called liquid chromatography, is employed. If the presence of a particular component is suspected in the sample and that component is otherwise available, it can be used to confirm its presence in the original sample. If not, GC can be used in combination with identification techniques, such as MS, IR, and UV. The use of the GC-MS and GC-FTIR is quite common and is extremely beneficial. GC is used for many other purposes besides hydrocarbon separation with the intention of identification [54]. Other lubricant-related uses include identifying contaminants, such as ethylene glycol which is used as a coolant in combustion engines, and the fuel dilution in diesel engine lubricant samples. ASTM standards that relate to the use of the GC for hydrocarbon analysis of mineral oils and the derived lubricants include the following:

1. ASTM D6352—Standard Test Method for Boiling Range Distribution of Petroleum Distillates in Boiling Range 174 to 700 °C by Gas Chromatography.
2. ASTM D2887 REV A—Standard Test Method for Boiling Range Distribution of Petroleum Fractions by Gas Chromatography.
3. ASTM D3524—Standard Test Method for Diesel Fuel Diluent in Used Diesel Engine Oils by Gas Chromatography.
4. ASTM D6417—Standard Test Method for Estimation of Engine Oil Volatility by Capillary Gas Chromatography.
5. ASTM D4291—Standard Test Method for Ethylene Glycol in Used Engine Oil.
6. ASTM D5480—Standard Test Method for Engine Oil Volatility by GC.
7. ASTM D3525—Standard Test Method for Gasoline Diluent in Used Engine Oils by Gas Chromatography.
8. ASTM D5623—Standard Test Method for Sulfur Compounds in Light Petroleum Liquids by Gas Chromatography and Sulfur Selective Detection.

Liquid Chromatography (LC)

As mentioned while discussing GC, LC is a closely related technique that also separates materials based on their physical and chemical properties. However, this technique employs a liquid instead of a gas to separate and analyze samples. Liquid chromatography techniques that are used for the analysis of the hydrocarbon petroleum products include absorption chromatography, high performance liquid chromatography (HPLC), size exclusion chromatography (SEC or GPC), fluorescent indicator absorption (FIA) and super critical fluid chromatography (SFC). In conventional LC, a liquid sample is introduced either neat or diluted with an appropriate solvent into a glass column prepacked with an appropriate solid material, such as silica, alumina, or some other solid medium. The sample is then washed down the column using a flowing stream of a solvent, starting with a low polarity solvent and progressing to higher polarity solvents until the sample is completely eluted from the column. The sequence of elution depends upon the strength of the affinity between the components of the sample towards the stationary phase (column packing). Those with least affinity (low polarity) will come out first, followed by the components with greater affinity (higher polarity). Absorption liquid chromatography is the method used most often for hydrocarbon analysis. In this method, the advantage is taken from the greater affinity (absorption) of certain components towards the stationary phase, which slows down their elution, thereby leading to their separation. In some cases, absorption by the stationary phase is so effective that the absorbed components can only be removed by washing with a very highly polar solvent [783]. Of the various LC methods,

TABLE 12.8—Wet analytical techniques [779].

Element	ASTM Method	Analytical Technique
Wet Methods		
Phosphorus	D1091	Photometry and Gravimetry
	D3231	Photometry
	D4047	Titrimetry
Vanadium	D1548	Photometry
Nitrogen	D3228	Kjeldahl Titration
Lithium/Sodium	D3340	Flame Photometry
Chlorine	D4929A	Potentiometry
	D5384	Titrimetry
Lead	D3341	Titrimetry
	D3348	Colorimetry
Sulfur, active	D4952	Colorimetry
Sulfur, mercaptan	D3227	Potentiometry
Bomb Combustion Methods		
Chlorine	D808	Gravimetry
Hydrogen	D1018	Gravimetry
Sulfur	D129	Gravimetry
	D1266	Gravimetry or Titrimetry
	D1552	Titrimetry or IR Detection
	D2784	Titrimetry or Turbidimetry
Micro-elemental Methods		
Carbon-Hydrogen, Nitrogen	D5291	IR or GC
Chlorine, Organic	D4929B	Microcoulometry
Nitrogen	D4629	Chemiluminescence
	D5762	Chemiluminescence
Oxygen	D5622	Reductive Pyrolysis
Sulfur	D3120	Oxidative Microcoulometry
	D4045	Hydrogenolysis and Rateometric Colorimetry
	D5453	UV Fluorescence
	D6667	UV Fluorescence

two that are used most often are high-performance liquid chromatography (HPLC) and gel-permeation chromatography (GPC).

In HPLC, the material to be analyzed is forced through a column with the stationary phase by a liquid at high pressure. The use of pressure provides the components less time to diffuse within the column, leading to improved resolution in the resulting chromatogram. Solvents that are commonly used include any miscible combination of water and an organic liquid, such as methanol and acetonitrile. Water may contain buffers or salts to assist in the separation of the sample components. HPLC chromatography is of four types [784]. These are normal phase chromatography, reverse phase chromatography, size exclusion chromatography, and ion-exchange chromatography. Normal phase HPLC (NP-HPLC) retains the analysis sample based on polarity. It uses a polar stationary phase and a nonpolar mobile phase, an eluent or solvent, and is used for analyzing polar materials. Conversely, reversed phase HPLC (RP-HPLC) consists of a nonpolar stationary phase and a polar mobile phase. This method is especially suitable for analyzing large nonpolar hydrocarbons and biomolecules. In this method, the polar molecules elute first and nonpolar molecules elute later. The eluted materials are detected by the use of either a UV or an RI (refractive index) detector, which generates signals to be recorded on a chromatogram [781]. The UV detector works well with compounds that contain multiple bonds or aromatic rings and the RI detector works well with compounds that are not UV active.

Size exclusion chromatography (SEC), also known as gel permeation chromatography (GPC) or gel filtration chromatography, separates materials based on their molecular size, or their hydrodynamic volume. SEC/GPC is the primary technique for determining the average molecular weights of natural and synthetic polymers. GPC was discussed in Chapter 4 under the viscosity modifiers section. Ion-exchange chromatography allows the separation of ions and polar molecules based on the charge properties of the molecules. It is not often used in the analysis of the petroleum products. ASTM test methods relating to the use of the liquid chromatography in the analysis of the petroleum products are listed below:

1. ASTM D2007—Test Method for Characteristic Groups in Rubber Extender and Processing Oils and Other Petroleum-Derived Oils by the Clay-Gel Absorption Chromatographic Method
2. ASTM D2549—Test Method for Separation of Representative Aromatics and Nonaromatics Fractions of High-Boiling Oils by Elution Chromatography

Other Chromatographic Techniques

Fluorescent Indicator Adsorption (FIA) is based on conventional open column liquid chromatography, with the difference that the column in addition to containing the normal silica gel is topped with a small portion of silica gel treated with a mixture of dyes. The dye mixture passes through the column with the sample and marks the boundaries (interfaces) of the adsorbent (silica gel) sections containing saturates and olefins, olefins and aromatics, and aromatics and isopropanol: the alcohol is used as an eluent. By measuring the separation of each dye, the relative proportions of the saturates, olefins, and aromatics can be determined fairly accurately. Super Critical Fluid Chromatography (SFC) uses a neither a liquid nor a gas, but uses a supercritical fluid, most commonly CO_2 [781]. A supercritical fluid is a fluid that has been heated above its critical temperature and pressure and has the properties between that of a gas and a liquid. For hydrocarbon analysis, SFC has an advantage over GC, because it uses lower temperatures, and over HPLC, because it offers a higher resolution. ASTM Standard D1319 (Universal Oil Products UOP Standard 311-02) describe the use of FIA to determine hydrocarbon types in liquid petroleum products.

Ultraviolet Spectroscopy (UV)

Ultraviolet (UV) spectroscopy, also known as UV spectrometry, is used to identify characteristic molecular fingerprints, enabling quantitative detection of specific molecules or molecular species in complex mixtures. UV spectrophotometer in association with high performance liquid chromatography (HPLC) is the most commonly used detection tool in hydrocarbon analysis. UV spectroscopy works only with specific organic structural types, those that have a high degree of conjugation, such as aromatics. UV absorption maxima occur at 190, 230, and 260 nm and correspond to mono-, di-, and poly-aromatics. The amounts of these components in the analysis sample can be quantified by the use of a simple formula based on molecular mass, and the average molar absorptive of each species at the these wavelengths [785]. One of the most common uses of the UV spectrometry is to determine the amount of the toxic impurities, particularly polyaromatic hydrocarbons, in lubricants and fuels. EPA has established limits on these compounds because of their carcinogenicity. Since UV spectroscopy is incapable of distinguishing between two structures with absorptions on the same wavelength, its real value in hydrocarbon analysis lies in its use in combination with separation techniques, such as liquid chromatography (LC) or gas chromatography (GC). Another closely relates technique, spectrofluorometry, is also used for hydrocarbon analysis [785].

Mass Spectrometry (MS)

This is a structure identifying technique which involves gas phase analysis of the samples. Those that are not gases are converted into gases through vaporization. The gas sample is then ionized either through electron ionization or chemical ionization to form primary and fragmented ions, which are separated according to masses by the use of a magnetic field and their mass-to-charge ratio and the relative abundance is recorded. Each molecule has a unique fragmentation pattern and the relative abundance and its structure can be reverse-synthesized from its fragments. The major limitations of the MS is that the complexity of the mass spectrum increases with increasing molecular size (molecular mass) and the number of components in the analysis sample, as is the case in mineral oil. These make analysis of the MS data a challenge. One way to overcome complexity and facilitate interpretation is to separate the mixture into components and then subjecting the components to the MS analysis. This is precisely what is done in GC-MS and LC-MS combination techniques. The most common use of mass spectrometry in hydrocarbon analysis is to determine the composition of the different process streams and the boiling fractions during the refining process. The ASTM standards pertaining to this analytical technique are listed below:

1. ASTM D2786—Standard Test Method for Hydrocarbon Type Analysis of Gas-Oil Saturates in Gas Oil Fractions by High Ionizing Voltage Mass Spectrometery.
2. ASTM D2789—Standard Test Method for Hydrocarbon Types in Low Olefinic Gasoline by Mass Spectrometery.
3. ASTM D3239—Standard Test Method for Aromatic Types Analysis of Gas-Oil Fractions by High Ionizing Voltage Mass Spectrometery.

Infrared Spectroscopy (IR)

Infrared spectroscopy is a widely applied nondestructive test method that is used to obtain an insight into a material's structure and or composition. Each material has a number of IR absorptions, which occur at specific frequencies (wave numbers). Most lubricants and mineral oils are mixtures and so are additives. Despite this, their IR spectra are sufficiently unique to help identify structural features of many of the individual components. However, in certain cases, a physical or chemical separation of the target components, for example by the use of gas or liquid chromatography, may facilitate structural identification of the components. The most important feature of the IR analysis is the ease to identify functional groups in organic compounds, most of which contain hetero atoms, such as nitrogen, oxygen, sulfur, and phosphorus. While the identification of these groups is nonequivocal, that of the purely hydrocarbon groups, such as multiple bonds and aromatic rings, is not always easy. This makes IR an invaluable tool in structure elucidation of the additives but less useful in identifying the structural features in purely hydrocarbon materials, such as mineral base oils. IR frequencies of the various organic groups are provided in Table 12.9 and an actual IR spectrum of ethanol is depicted in Fig. 12.5 [786]. Figure 12.6 identifies various regions in an infrared spectrum [787]. Table 12.10 provides infrared absorptions of some of the lubricant components [647].

Nuclear Magnetic Resonance (NMR) Spectroscopy

Nuclear magnetic resonance spectroscopy is a technique that exploits the magnetic properties of the atomic nuclei. Two important techniques are proton (^1H) NMR and carbon-13 (^{13}C) NMR spectroscopy, although some other magnetic nuclei can be measured as well [788]. NMR spectroscopy allows the identification of the individual atoms in a pure molecule. This technique is one of the most powerful techniques for determining structure of the organic compounds. Like infrared (IR) and ultraviolet (UV) spectroscopy, NMR is a form of absorption spectrometry where the amount of the absorbed electromagnetic radiation at a specific frequency can be related to certain chemical species which absorb at that frequency. The portion of the molecule that absorbs the energy transitions from the lower energy level to a higher energy level and the absorbed energy appears on the appropriate spectrum as a signal. Unlike IR spectroscopy, which examines the functional groups within

ABLE 12.9— (*Continued.*)

Bond	Type of Bond	Specific Type of Bond	Absorption Range and Intensity
-X =F, Cl, Br, I)	Fluoroalkanes	Ordinary	1000-1100 cm^{-1}
		Trifluoro-methyl	two strong, broad bands between 1100-1200 cm^{-1}
	Chloroalkanes		540-760 cm^{-1} (medium to weak)
	Bromoalkanes		below 600 cm^{-1}
	Iodoalkanes		below 600 cm^{-1}
-O	Nitro compounds	Aliphatic	1540 cm^{-1} (stronger band) and 1380 cm^{-1} (weaker band) - ALWAYS BOTH!
		Aromatic	1520, 1350 cm^{-1} (conjugation usually lowers the wave number)

olecule, or UV spectroscopy, which examines the moles as a whole, the NMR examines specific atoms and r environment in the sample. Proton and carbon-13 R are ideal for obtaining structural information of oric compounds since they are largely composed of carbon hydrogen atoms. Additional attribute of the NMR speccopy is that it can quantify the number of atoms in each ctural environment, which helps define the molecular cture of the compound accurately.

When the NMR active nuclei, such as ^{1}H or ^{13}C are ed in a magnetic field, they resonate at a specific frency, hence the name magnetic resonance for the techie, depending upon the strength of the magnetic field. al chemical environment affects the resonance frency of these atoms, altering their resonance frequency ntly. Since there are many chemical environments within molecule, each atom or group of atoms in the molecule nates at slightly different frequencies. These frequen-, which are magnetic field dependent, are converted into -independent variable by dividing the difference been the frequency of the signal and the frequency of the rence (tetramethylsilane) by the frequency of the magc field, as shown below:

$$\delta(\text{ppm}) = \frac{\text{Signal Frequency} - \text{Reference Frequency}}{\text{Operating Frequency of the Magnet}}$$

The value obtained is called chemical shift, which is represented by the Greek letter δ and is expressed in ppm. The reason for using the ppm unit for the chemical shift is because the signal frequency is in hertz (Hz) and magnetic frequency is in megahertz (MHz). By understanding the different chemical environments, the chemical shift can be used to obtain some structural information about the molecule in question, to assign signals (peaks) to an atom or a group of atoms. Modern analysis software allows the analysis of the area under each peak, to deduce the number of protons or carbons that are responsible for that peak by a mathematical process, called integration. The chemical shifts for peaks due to atoms or groups of atoms are known in the scientific literature. By comparing these values with the observed values for the sample, one can accurately identify various structural entities that exist in the molecule.

The chemical shifts ranges for protons in various electronic environments are shown in Fig. 12.7 [789] and their typical values are provided in Table 12.11 [789]. These values are not precise, and deviations of ±0.2 ppm, or more, are ex-

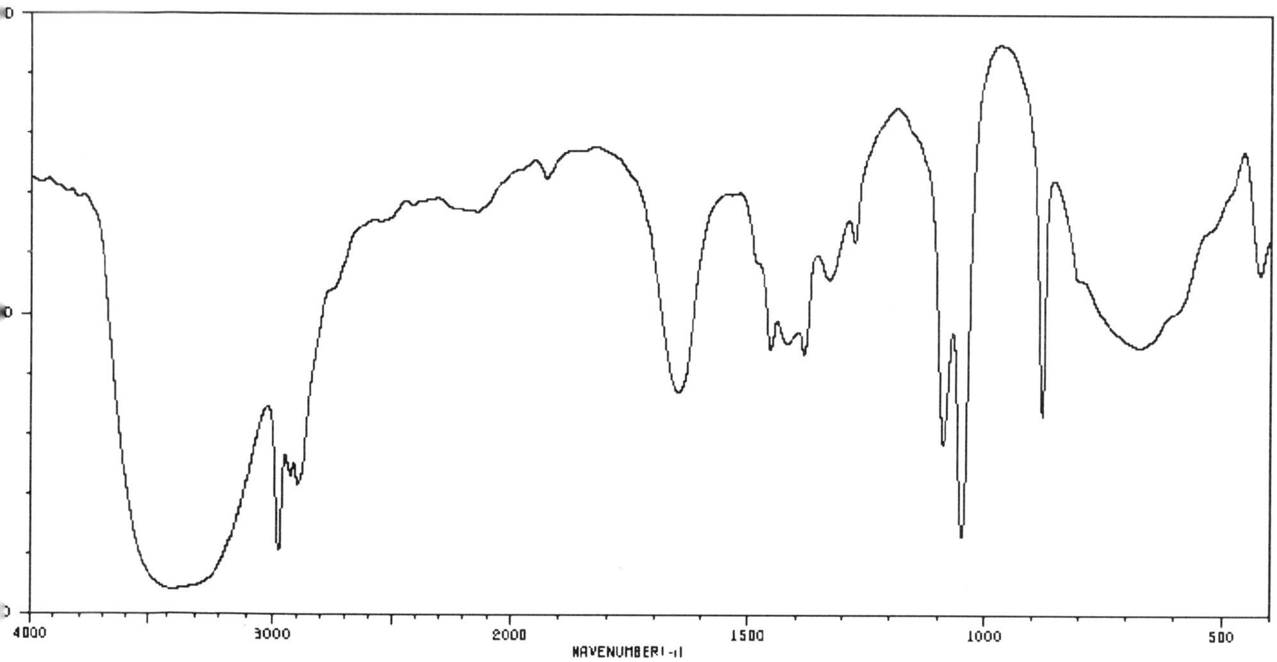

Fig. 12.5—Infrared spectrum of ethanol [786].

TABLE 12.9—Infrared spectroscopy correlation table [786].

Bond	Type of Bond	Specific Type of Bond	Absorption Range and Intensity
C-H	Alkyl	Methyl	1380 cm^{-1} (weak), 1260 cm^{-1} (strong) and 2870, 2960 cm^{-1} to medium)
		Methylene	1470 cm^{-1} (strong) and 2850, 2925 cm^{-1} (both strong to me
		Methine	2890 cm^{-1} (weak)
	Vinyl	$C=CH_2$	900 cm^{-1} (strong) and 2975, 3080 cm^{-1} (medium)
		C=CH	3020 cm^{-1} (medium)
		Mono-substituted alkenes	900, 990 cm^{-1} (both strong)
		Cis-di-substituted alkenes	670-700 cm^{-1} (strong)
		Trans-di-substituted alkenes	965 cm^{-1} (strong)
		Tri-substituted alkenes	800-840 cm^{-1} (strong to medium)
	Aromatic	Benzene/substituted benzene	3070 cm^{-1} (weak)
		Mono-substituted benzene	700-750 cm^{-1} (strong) and 700±10 cm^{-1} (strong)
		Ortho-di-substituted benzene	750 cm^{-1} (strong)
		Meta-di-substituted benzene	750-800 cm^{-1} (strong) and 860-900 cm^{-1} (strong)
		Para-di-substituted benzene	800-860 cm^{-1} (strong)
	Alkynes		3300 cm^{-1} (medium)
	Aldehydes		2720, 2820 cm^{-1} (medium)
C-C	Acyclic C-C	Mono-substituted alkenes	1645 cm^{-1} (medium)
		1,1-Di-substituted alkenes	1655 cm^{-1} (medium)
		Cis-1,2-di-substituted alkenes	1660 cm^{-1} (medium)
		Trans-1,2-di-substituted alkenes	1675 cm^{-1} (medium)
		Tri-substituted, tetra-substituted alkenes	1670 cm-$^{-1}$ (weak)
	Conjugated C-C	Dienes	1600, 1650 cm^{-1} (strong)
		With benzene ring	1625 cm^{-1} (strong)
		With C=O	1600 cm^{-1} (strong)
	Aromatic C=C		1450, 1500, 1580, 1600 cm^{-1} (strong to weak) - always A
	Triple C-C	Terminal alkynes	2100-2140 cm^{-1} (weak)
		Di-substituted alkynes	2190-2260 cm^{-1} (very weak, sometimes not visible)
C=O	Aldehyde/ketone	Saturated aliphatic/cyclic 6-membered	1720 cm^{-1}
		α,β-Unsaturated	1685 cm^{-1} (goes for aromatic ketones as well)
		Cyclic 5-membered	1750 cm^{-1}
		Cyclic 4-membered	1775 cm^{-1}
		Aldehydes	1725 cm^{-1} (influence of conjugation like with ketones)
	Carboxylic Acids and their derivates	Saturated carboxylic acids	1710 cm^{-1}
		Unsaturated/aromatic carboxylic acids	1680-1690 cm^{-1}
		Esters and lactones	1735 cm^{-1} (influence of conjugation and ring size like wit
		Anhydrides	1760 and 1820 cm^{-1} (both!)
		Halogenides	1800 cm^{-1}
		Amides	1650 cm^{-1} (associated amides)
		Carboxylates (salts)	1550-1610 cm^{-1} (goes for amino acid zwitterions as well
O-H	Alcohols, phenols		3610-3670 cm^{-1} (concentrating samples broadens the ba it to 3200-3400 cm^{-1})
	Carboxylic acids		3500-3560 cm^{-1} (concentrating samples broadens the ba it to 3000 cm^{-1})
N-H	Primary amines		doublet between 3400-3500 cm^{-1} and 1560-1640 cm^{-1} (s
	Secondary amines		above 3000 cm^{-1} (medium to weak)
	Ammonium ions		broad bands with multiple peaks between 2400-3200 cn
C-O	Alcohols	Primary	1050±10 cm^{-1}
		Secondary	around 1100 cm^{-1}
		Tertiary	1150-1200 cm^{-1}
	Phenols		1200 cm^{-1}
	Ethers	Aliphatic	1120 cm^{-1}
		Aromatic	1220-1260 cm^{-1}
	Carboxylic acids		1250-1300 cm^{-1}
	Esters		1100-1300 cm^{-1} (two bands - distinction to ketones, whi possess C-O!)
C-N	Aliphatic amines		1020-1220 cm^{-1} (often overlapped)
	C=N		1615-1700 cm^{-1} (similar conjugation effects to C=O)
	nitriles (triple C-N bond)		2210-2260 cm^{-1} (un-conjugated 2250, conjugated 2230
	Iso-nitriles (R-N-C bond)		2165-2110 cm^{-1} (2140 - 1990 cm^{-1} for R-N=C=S)

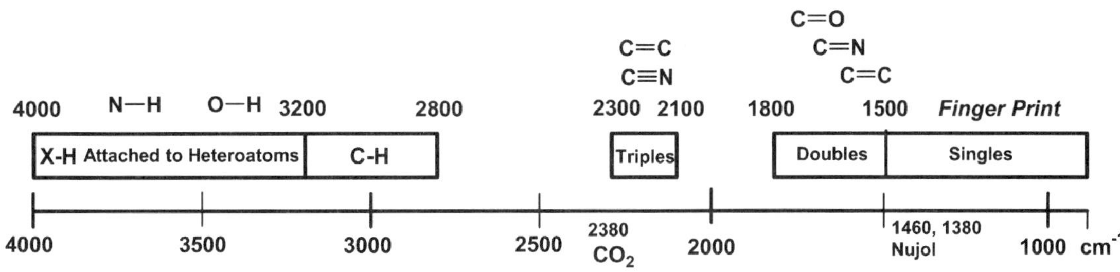

Fig. 12.6—Summary of infrared absorptions of bonds in organic molecules [787].

pected. The exact value of the chemical shift depends on the molecular structure and the solvent in which the spectrum is being recorded. Chemical shifts ranges for carbons in various electronic environments are shown in Fig. 12.8 and typical values are provided in Table 12.12 [54,790]. Figures 12.9 and 12.10 depict ^{1}H and ^{13}C NMR spectra for vanillin as an example [789,790]. Despite the versatility of the NMR spectroscopy to determine the chemical structures easily and accurately, its value in mineral oil analysis can only be realized by using it in conjunction with a separation technique, for example, chromatography. The use of the NMR for analyzing mineral oils and or its components is described in the ASTM Standard ASTM D5292, Test Method for Aromatic Hydrogen and Aromatic Carbon Contents of Hydrocarbon Oils by High Resolution Nuclear Magnetic Resonance Spectroscopy and the references [54,791–794].

By coordinating all the information gathered from the above-described separation and spectroscopic techniques, a formulator will have a good idea about the composition of the mineral base oil with respect to the amount of the paraffinic, naphthenic, and aromatic components. By combining this knowledge with the abundant data that are available on physical and chemical properties of the three types of hydrocarbons, a competent formulator will know which additives are necessary to design a lubricant that meets the performance needs of a specific application.

Lubricants

Tests that are used to assess the properties and the suitability of the lubricants for use in the intended application are listed in Tables 12.13–12.15. It is important to note that while the list is exhaustive, it is by no means complete. The detailed descriptions of these tests are available in publications by various standards-establishing organizations, such as ASTM, ISO, DIN, and U.S. Military. Here, we provide a short description of the most widely used physical and analytical tests.

TABLE 12.10—Infrared absorptions of lubricant components [647].

Contaminant/Lubricant Component	Adsorption, cm^{-1}	Type of vibration	Interferences
Water	3300-3500, 1600	O-H Stretch	Glycol, oxidants
Fuel	500-1000, 1600	C-H Out of plane deformations of aromatics	Unsaturation in lubricants
Oxidized Lubricant	1700-1750,1100-1200 1630,1270	C=O Stretch N=O Stretch (Oxidized from blow-by)	Viscosity index improvers Alkenes
Glycol Coolant	1040,3300-3500 1080, 3300-3500	C=O Stretch C-O Stretch	Oxidation Oxidation Inhibitors Aromatics
Zinc Dialkyl Dithiophosphate (ZDDP)	950-1050	P-O-C asymmetric stretch	Oxidation
Metal Sulfonate Detergent	1100-1200	SO_3 asymmetric	Oxidation
Polymethacrylate Pour-point Depressant	1732	C=O Stretch	
Dispersant			
Succinimide	1707	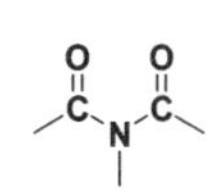	Oxidation
Mannich Base	1650	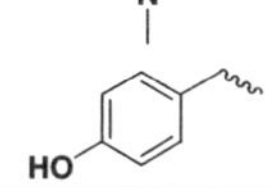	
Viscosity Index Improver Polymethacrylate	1732	C=O	
Styrene/Isoprene Polymer	700	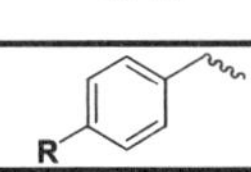	Oxidation, Fuel

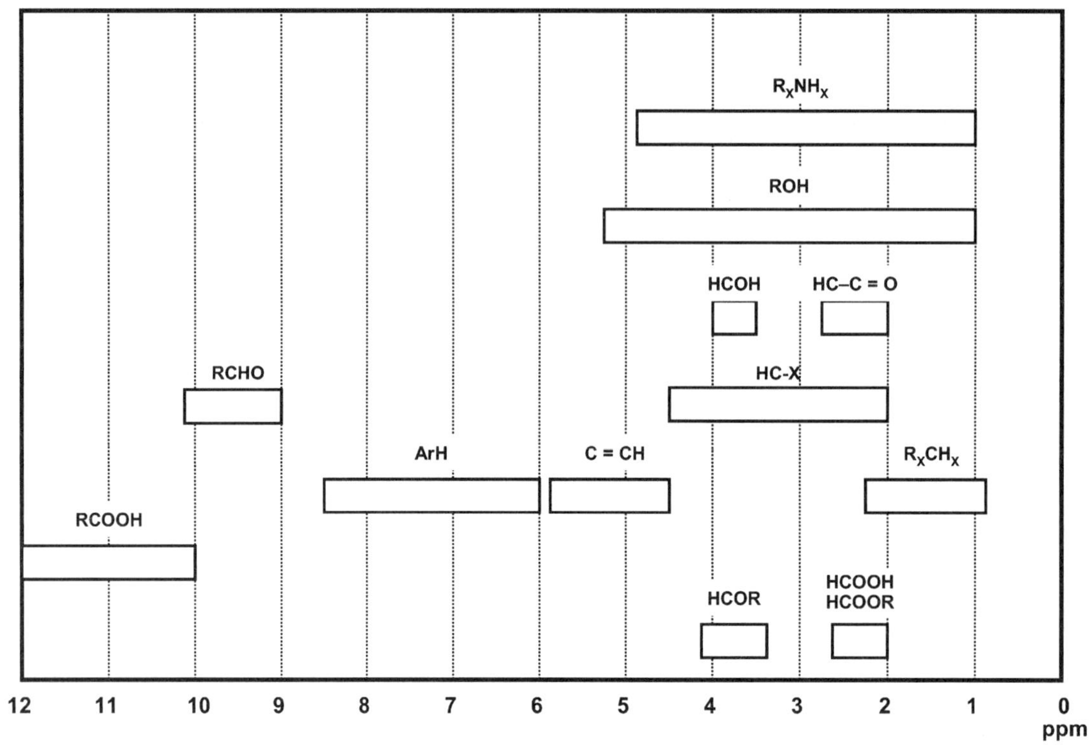

Fig. 12.7—Chemical shift ranges for protons in different electronic environments [789].

Neutralization Number (Acidic or Alkaline Condition) [ASTM D974, D664]

This test determines the quantity of the base in milligram of potassium hydroxide per gram or the acid also expressed as milligram of potassium hydroxide needed to neutralize acidic or alkaline compounds present in the new, in service, or the used lubricant. The end point of the neutralization reaction is determined either potentiometrically or colorimetrically. Of the two methods, the potentiometric method is superior since it works equally well with light-colored and dark-colored oils. In addition, it allows the measurement of the strong acid and the weak acid, the strong base and the weak base simultaneously. This method can also be used for lubricant production control and to determine a lubricant's remaining useful life.

Aniline Point (ASTM D611)

This test defines the temperature at which equal volumes of the oil and aniline, an aromatic compound, are miscible. This temperature is indicative of the solvency characteristics of the oil. In general, the lower the aniline point, the greater is the solvency of the oil.

Flash Point and Fire Point (ASTM D92, D93, D56, D1310)

These tests are used to judge the ignitability of an oil. Three types of equipment are commonly used for determining these lubricant parameters; these are the Tag Closed Tester, Pensky-Martens Closed Tester, and Cleveland Open Cup. Although flash and fire points provide a rough estimate of an oil's flammability, there are a number of other tests that are also used. In general, within a chemical type, flash point and fire point increase with an increase in the molecular weight [795].

Ash Content (ASTM D482, D874)

Ash is the amount of the incombustible material present in a lubricant. It is measured by burning the oil under prescribed conditions. The ASTM D482 Method is used to determine the amount of ash in petroleum products which do not contain ash-forming additives and the ASTM D874 is used for petroleum products that have metal-containing additives, such as zinc dialkyl dithiophosphates and metal sulfonates. The latter method measures sulfated ash, which is the ash that results from mixing the oil with sulfuric acid and then burning it in a high-temperature furnace. Sulfated ash is especially useful for determining the metal additive content of the new oils, which is related to an oil's detergency characteristics and deposit-forming tendency in the combustion engine lubricants.

Pour Point and Cloud Point (ASTM D97 and D2500)

These parameters primarily relate to the low-temperature properties of the fluids of high paraffinic content. The test methods used to assess an oil's response to cooling are pour point, cloud point, and flocculation point. Cloud point is the temperature at which the oil loses its clarity, i.e., develops haze or cloudiness due to the start of the crystal formation. Pour point is the lowest temperature at which the oil is observed to flow under specified conditions. These properties are important in the low-temperature operation of the automotive equipment. Flocculation point applies only to refrigeration lubricants and correlates with the oil's response to the presence of a refrigerant, such as Freon®. Flocculation point is the temperature at which the oil separates as a fine precipitate from a 90:10 mixture of the refrigerant and oil.

TABLE 12.11—Proton (^1H) chemical shifts [789].

Functional Group/Structure	CH_3	CH_2	CH	H
CH_2R	0.8	1.3	1.6	...
C=C	1.6	2.0	2.6	...
C≡C	1.7	2.2	2.8	...
C_6H_5	2.3	2.6	2.9	...
C_6H_6 (Monocyclic aromatic)	...	...	...	6.0–7.2
$C_{10}H_8$ (Polycyclic aromatic)	...	...	...	7.2–9.0
F	4.3	4.4	4.8	...
Cl	3.0	3.4	4.0	...
Br	2.7	3.4	4.1	...
I	2.2	3.2	4.2	...
OH	3.3	3.5	3.8	...
OR	3.3	3.4	3.7	...
OC_6H_5	3.8	4.0	4.3	...
OCOR	3.6	4.1	5.0	...
$OCOC_6H_5$	3.9	4.2	5.1	...
$OCOCF_3$	4.0	4.4		...
CHO	2.2	2.4	2.5	...
COR	2.1	2.2	2.6	...
COOH	2.1	2.3	2.6	...
COOR	2.0	2.3	2.5	...
$CONR_2$	2.0	2.1	2.4	...
CN	2.1	2.5	3.0	...
NH_2	2.5	2.7	3.0	...
NR_2	2.2	2.4	2.8	...
NRC_6H_5	2.6	3.0	3.6	...
NR_3^+	3.0	3.1	3.6	...
NHCOR	2.9	3.3	3.7	...
NO_2	4.1	4.2	4.4	...
SR	2.1	2.5	3.1	...
SOR	2.6	3.1		...
=O (Aliphatic aldehyde)	...	...	9.5	...
=O (Aromatic aldehyde)	...	...	10	...
M-H (Metal hydride)	...	...	−5 to −15	...

Any oil which prematurely produces a flocculent may hinder effective heat transmission.

Water Content (ASTM D95, D1744, D1533, and D96)

Lubricants used in many applications require a low water content. The applications include hydraulic, industrial gear, pneumatic tool, automotive, steam cylinder, turbine, transformer, and industrial circulating oil systems. Water content is the amount of moisture, expressed in weight %, volume %, or ppm. Three basic methods used to determine this lubricant parameter include distillation, centrifuging, and by the use of the coulometric Karl Fischer titration [795]. The centrifuge method is the least accurate but simple to perform. It measures not only the water content but also other hydrocarbon insoluble materials. The distillation, or the azoetrope, method is a little more accurate and can be used on all petroleum liquids. Coulometric, or electrometric, titration is the most accurate and measures water content even at parts per million levels. Concern for the lubricant's water content is because the presence of water can lead to corrosion, emulsion formation, and reduce a lubricant's oxidation resistance. In dielectric fluids, the presence of greater than 35 ppm water in the presence of suspended solids can adversely affect dielectric breakdown voltage.

TABLE 12.12—Typical ^{13}C chemical shifts [54].

Carbon Nuclear Environment	Chemical Shift, ppm
C (Alkane)	~0–30
Aliphatic and naphthenic carbon atoms	10.0–60.0
Carbon atoms in terminal CH_3 groups	14.1
Carbon atoms in CH_2 groups in the middle of a chain	29.2
C (Alkene)	~110–150
C—N	~50
C—O	~60
C—F	~70
Aromatic	~110–160
Ester, amide, acid	~160–170
Ketone, aldehyde	~200–220

Sulfur Content (ASTM D1266, D129, and D1662)

For a variety of lubricants, such as cutting fluids, gear oils, and slide way lubricants, the knowledge of the amount of sulfur present is important. These lubricants employ sulfur or sulfur compounds for EP/antiwear performance. Sulfur can be active or reactive (See Chapter 4 on Additives). The presence of chemically active sulfur leads to undesirable corrosion of yellow metals, such as copper and bronze. Conversely, its presence has the very desirable effect on lubricant's extreme pressure properties and load-carrying capacity. The amounts of active, reactive or chemically combined, and total sulfur in an oil is determined by the lamp combustion test (ASTM D1266), copper powder reactivity (ASTM D1662), the bomb oxidation test (ASTM D129), and X-ray spectrography (ASTM D2622). The bomb technique is universally applicable for measuring total sulfur content of a lubricant.

Copper Corrosion (ASTM D130, D1275, and ASTM D1261)

One of the requirements of a good lubricant is its compatibility with easily corrodible metals, such as copper. ASTM D130, the copper strip test, which is carried out at 50, 100, or 140 °C by immersing a polished copper strip in oil for a specified period of time and examining it for color and appearance change. Corrosivity of an oil is determined by comparing the tested copper strip to arbitrary standards ranging from a bright, relatively shiny appearance, to that of a carbon black or graphite tinted surface. See Table 7.14 for copper strip rating system.

Chemically combined sulfur is usually inert in this test. ASTM D1275 is the test method used to determine corrosive sulfur, both inorganic and organic, in electrical insulating oils of petroleum origin. D1275 test is run at 140 °C for 19 hours using a copper strip. The ASTM D1261 test is used to assess copper activity of a lubricating grease. ASTM D1402 is a similar test that uses the same equipment but assesses the effect of copper on the oxidation stability of grease, instead of its copper activity.

Carbon Residue (ASTM D189 and D524)

When an oil is subjected to evaporation and pyrolysis, it leaves a carbonaceous residue. Ramsbottom and Conradson methods are used to determine this lubricant property. This lubricant parameter is only of value while selecting hydrocarbon oils for very high-temperature applications, such as heat treating, air compressors, and high-temperature bearings.

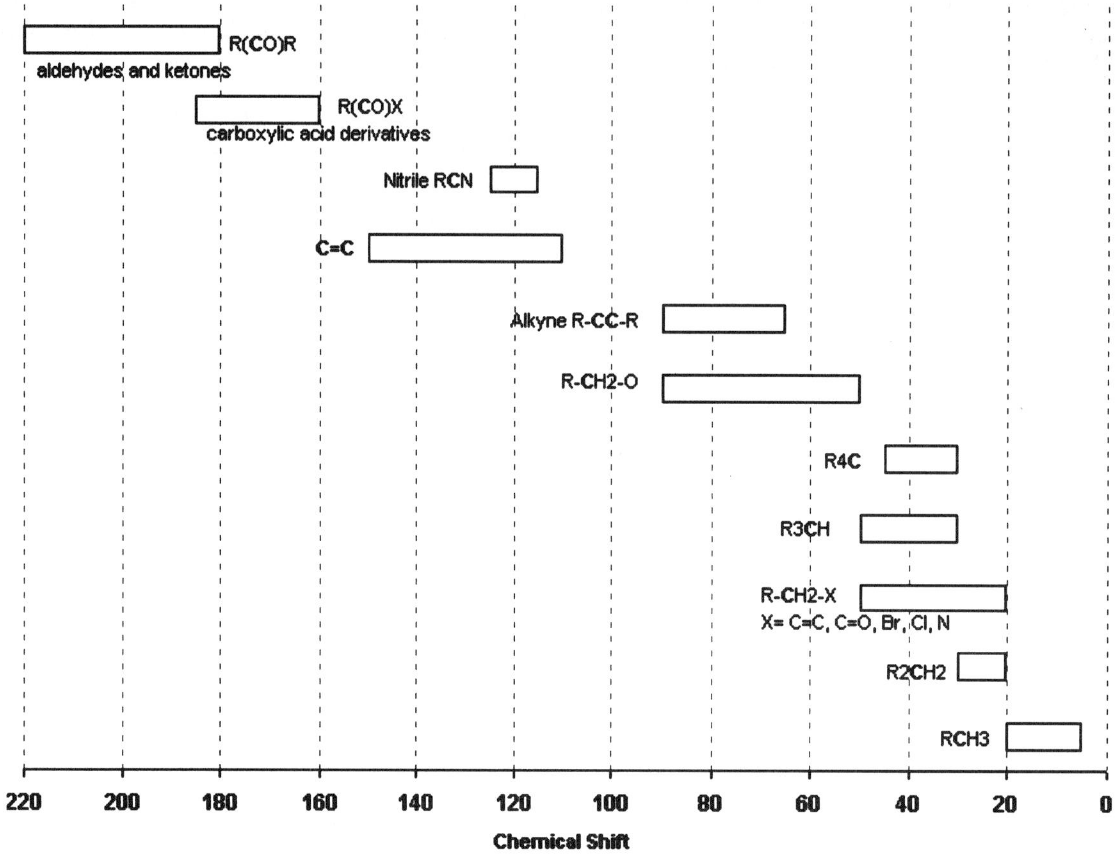

Fig. 12.8—Chemical shift ranges for carbons in different electronic environments [790].

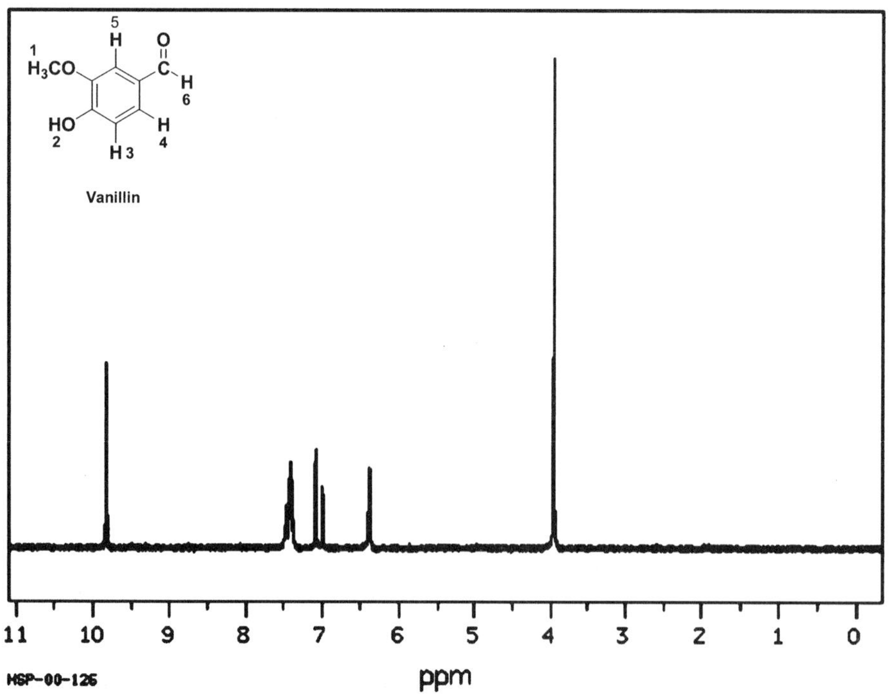

Position	Chemical Shift (ppm)
1	3.959
2	6.39
3	7.047
4	7.42
5	7.42
6	9.823

Fig. 12.9—^{1}H NMR of vanillin [789].

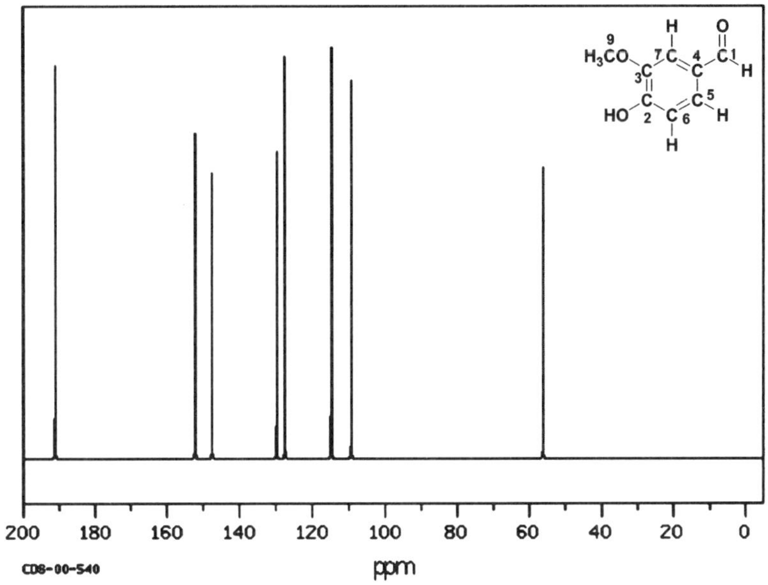

Carbon Position	Chemical Shift (ppm)
1	191.21
2	152.18
3	147.50
4	129.77
5	127.49
6	114.75
7	109.34
8	56.10

Fig. 12.10—^{13}C NMR of vanillin [790].

Specific Gravity (ASTM D941, D1217, and D1298)

Specific gravity is the ratio of the weight of a given volume of oil and an equal volume of water at a specified temperature. While specific gravity has some use, for example in estimating volume if the weight is known and the force required to pump a fluid, the API gravity is of greater value in lubricants. To determine API gravity, specific gravity measured at the desired temperature is converted into API gravity at 60 °F by the use of the Petroleum Measurement Tables, issued jointly by ASTM and IP (ASTM D1298). The API gravity scale measures the relative density of the various petroleum liquids. Refer to Chapter 2 for details. In general, the oils of higher API gravity, such as paraffinics and light crude oils, are commercially more important than those with lower API gravity, such as naphthenic and heavy crude oils.

Viscosity (ASTM D445, D2161, D2983)

Viscosity is considered the most significant and fundamental property of the lubricants. The importance of viscosity is demonstrated by the fact that it is one of the major factors in selecting a lubricant. Viscosity is equally important in new and is in-service oils. In new oils, it assures proper lubrication across a broad temperature range and in-service oils, it indicates the lubricant condition, degradation due to oxidation, and useful life. For a detailed discussion on viscosity, refer to Chapter 1 on Lubrication Fundamentals.

Chlorine Content (ASTM D808 and D1317)

Two methods for determining the chlorine content are the bomb method, which involves a gravimetric determination, and the volumetric or the titration method, which determines the amount of chlorine by converting it into sodium

TABLE 12.13—Examples of basic elements of tribological systems [799].

Tribological System/Process	Elements of the System: Tribo-element (1)[a]	Tribo-element (2)[a]	Interfacial Medium (3)	Surrounding Medium (4)
Sliding Bearing	Shaft	Bushing	Lubricant	Air
Band Clutch	Shaft	Band	...	Air
Disc Brake	Disc	Pad	Contaminant	Air
Worm Gear Set	Worm	Gear	Gear Oil	Air
Cam and Follower	Cam	Follower	Lubricant	Air
Printing Unit	Print-Head	Paper	Dye	Air
Audio Pick-up	Record	Sapphire Tip	...	Air
Electrical Contact	Ring	Brush	Spray	Cover Gas
Locomotion	Wheel	Rail	Contaminant	Air
Pipeline	Fluid	Pipeline	...	...
Wiredrawing	Wire	Die	Borax	Air
Hot Extrusion	Billet	Die	Glass	Air
Turning	Work Piece	Cutting Tool	Cutting Fluid	Air

[a]Moving or stationary.

TABLE 12.14—Physical properties of some lubricating oils [795].

	Viscosity (cSt)				Flammability (°C)				
Type of Oil	40°C	100°C	Viscosity index	Pour Point (°C)	Flash Point	Fire Point	Carbon Residue (%)	Neutralization Number (mg KOH/g of Oil)	Sulfated Ash %
Air Compressor, Reciprocating									
ISO Grade 32	30	5	60	−40	185	215	0.01–0.10	0.03–0.06	Trace
ISO Grade 150	141	11.5	61	−20	230	265	0.01	0.06	...
Air Compressor, Vane	47	7.0	60	−15	218	249	0.07	0.10	Trace
Aircraft Engine, Reciprocating									
ISO VG 65	110	11	100	−23	235	272	0.08	0.02	0.001
ISO VG 120	360	25	115	−10	230	327	0.20	0.03	0.001
Gas Turbine (Synthetic)									
Light	13	2.5	122	−60	235	255	...	...	...
Heavy	26	5.7	135	−50	265	295	0.20	0.21–0.40	0.001
Automatic Transmission	37	7.3	160	−45	190	220	0.26	0.48	0.07
Automotive Engine									
SAE 10W	42	7	100	−30	210	238	1.43	0.5	1.3
SAE 20W-20	75	9	100	−24	221	246	1.4	0.6	1.3
SAE 30	119	12	100	−18	227	260	1.2	0.6	1.1
SAE 40	180	16	100	−14	232	268	1.3	0.6	1.1
SAE 50	225	19	100	−14	238	282	1.4	0.7	1.1
Ball and Roller Bearing	70	8.5	80–100	−15	232	265	0.08	0.10	Trace
Diesel Engine									
SAE 10W	37.5	6	110	−29	221	249	0.83	1.0	0.78
SAE 20W	78	9.6	108	−23	227	266	0.86	...	0.75
SAE 30	123	12.7	105	−23	249	293	0.95	...	0.80
SAE 40	160	15.0	103	−15	249	293	1.07	...	0.75
SAE 50	241	19.6	101	−12	254	299	1.41	1.5	0.77
Electric Motor Bearings									
Light	32	5.5	>90	−23	193	237	0.01	0.02	None
Medium	65	6.4	90	−30	204	243	0.01	0.02	None
Heavy	130	12.1	95	−20	215	250	0.02	0.02	None
Gas Engine									
	111	12	90	−20	240	271	1.01	3.0	...
	185	13.5	66	−20	246	280	...	...	0.65
Gas Compressor	111	12	80	−20	240	271	1.01	3.0	...
Gear, General Purpose									
	68	8.8	95	−23	171	193	...	...	...
	172	13.2	80	−23	176	210	...	...	...
	291	19.2	80	−18	182	218	...	...	...
	399	23	85	−12	196	225	...	...	...
	508	27	80	-6	204	230	...	...	...
Gear, Automotive Grade									
SAE 90	130	14	100	−30	232	270	...	...	...
SAE 140	305	25	100	−18	238	280	...	...	...
Gear Exposed Chains									
	5,000	65	...	10	220	255	...	...	...
	26,000	175	...	18	225	265	...	...	...
	78,000	325	...	32	270	310	...	...	...
Aircraft									
Light	25	5	141	−42	152	177	0.01	0.25	0.01
Medium	69	8.8	106	−37	166	191	0.01	0.12	0.01
Hydraulic									
Light	31	5.3	100	−29	210	238	0.01	0.2–0.5	None
Medium	65	8.1	99	−20	230	260	0.02	0.2–0.5	None
Heavy	129	12.7	95	−20	245	275	0.2	0.10–0.5	None
Heat transfer	30	5.5	115	−10	205	232	<0.01	...	...
General Purpose (Naphthenic)									
ISO VG 15	15	3.2	40	−50	167	186	0.01	0.01	None
ISO VG 22	22	3.6	40	−45	168	188	0.01	0.02	None
ISO VG 32	32	4.8	35	−35	170	195	0.02	0.02	None
ISO VG 46	45	5.4	40	−30	182	207	0.02	0.02	None
ISO VG 100	97	8.8	40	−25	207	232	0.03	0.02	None
ISO VG 150	143	10.9	37	−20	204	230	0.03	0.02	None

TABLE 12.14— (Continued.)

Type of Oil	Viscosity (cSt) 40°C	Viscosity (cSt) 100°C	Viscosity index	Pour Point (°C)	Flammability (°C) Flash Point	Flammability (°C) Fire Point	Carbon Residue (%)	Neutralization Number (mg KOH/g of Oil)	Sulfated Ash %
ISO VG 220	228	15.0	33	−12	240	265	0.04	0.02	None
Machine Tool Ways									
Light	67	7.8	85	−30	190	214	...	0.2	0.001
Heavy	230	16.3	75	−22	214	245	...	0.2	0.005
General Purpose (Paraffinic)									
ISO VG 15	14	3.3	80	−4	171	190	0.01	0.01	None
ISO VG 22	22.4	4.1	80	−4	182	201	0.01	0.01	None
ISO VG 32	32	5.1	80	−10	196	226	0.02	0.02	None
ISO VG 46	46.5	6.4	90	−12	210	232	0.04	0.02	None
ISO VG 60	64.5	7.3	85	−12	226	248	0.10	0.03	None
ISO VG 100	108	9.6	70	−5	223	252	0.2	0.05	None
Paper Machine									
ISO VG 150	158	14.9	100	−10	261	295	0.4–0.80	0.25–1.0	0.3–0.4
ISO VG 220	218	20.0	100	−10	274	310	0.4–0.80	0.25–1.0	0.3–0.4
Pneumatic Tools									
Light—ISO VG 46	44	6.2	...	−34	174	204	0.4–0.5	0.32–0.60	0.01
Medium—ISO VG 100	116	9.6	...	−31	201	238	0.26	0.32	0.01
Heavy—ISO VG 150	143	11.9	...	−29	204	235	0.30	0.32	0.01
Refrigeration									
Light—ISO VG 22	21.7	3.9	60–70	−45	174	192	0.01	0.02	None
Medium—ISO VG 46	43	5.1	65	−40	190	212	0.02	0.02	None
Heavy—ISO VG 68	64	7.5	69	−37	216	230	0.02	0.02	None
Rolling Oil									
	4.6	...	...	−12	127	142	...	...	...
	32.3	4.8	...	−35	180	200	0.07	0.76	...
Railcar Journals	76	9.1	100	−29	180	210	...	0.01	...
Steam Cylinder (Usually contains an oiliness agent, such as tallow oil)									
ISO VG 460	475	30.5	100	−10	271	316	0.3–0.75	0.15	...
ISO VG 600	640	37	95	−3	289	332	...	0.10	...
ISO VG 1000	930	47	90	−1	288	339	...	...	...
ISO VG 1500	1,600	61	91	−1	313	351	...	...	...
Steam Turbine									
Light—ISO VG 32	32	5.5	112	−23	208	237	0.01	0.07	None
Medium—ISO VG 68	65	8.2	105	−17	212	260	0.02	0.08	None
Heavy—ISO VG 100	103	12.0	108	−15	248	286	0.03	0.08	None
Steel Mill Bearing	333	23.4	100	−15	249	287	0.26	0.03	
Textile Spindle									
ISO VG 10	9.5	2.6	85	−12	171	196	0.01	0.01	Trace
ISO VG 22	20.5	4.2	90	−12	189	213	0.06	0.02	None
Vacuum Pump (Low Vapor Pressure, <0.01 □m)	71	8.8	100	−12	240	282	0.08	0.03	...
Wire Rope									
	80 000	330	...	...	274	...	11.6	7.0	...
	270 000	650	...	...	274	...	12.0	7.0	...

chloride and titrating it with silver nitrate. Both methods are suitable for either new or the used oils, as well as greases. While the compounds containing chlorine and chlorine plus sulfur are often used in extreme pressure lubricants, such as cutting fluids, their use is on the decline, especially in Europe, because of its role in producing carcinogenic polychlorobiphenyls (PCBs) and dioxins. When used in a lubricant, it is important to determine the optimal amount to deliver the desired performance. Too small an amount will not provide appropriate friction and wear control and too large an amount will lead to increased corrosion.

Oxidation Stability (ASTM D943, D2272, D2893, D1313, D2446, D2893, and ASTM D942)

Oxidation resistance of a lubricant is a critical property, particularly in applications that involve high temperatures and exposure to air, such as internal combustion engines, steam turbines, and transformers. The most widely used method to assess oxidation resistance is ASTM D943, which is applicable to most lubricating oils, especially to those containing the oxidation inhibitors. Consequences of the poor oxidation stability are sludge, acids, foaming, emulsification, corrosion, and restricted oil flow in the operating units. Oxida-

TABLE 12.15—Typical bench test results for some lubricating oils [795].

Type of Oil	Aniline Point °C	Rust Prevention D 665	Copper Strip D 130	Sulfur/Chlorine Content % S	Sulfur/Chlorine Content % Cl	Metal Content %	Foaming Volume mL	Emulsion Volume mL	Oxidation Stability h to 2 mg KOG/g	Wear/EP Properties Timken OK Load, kg	Wear/EP Properties Four-Ball Wear Scar	Wear/EP Properties Falex, kg
Air Compressor												
Reciprocating	104	Pass A	1	...	...	...	0/0/0	40-4-0 (10 min)	1200	...	...	...
Vane	106	Pass B	1	...	...	...	0/0/0	40-40-0 (4 min)	1500+	...	...	...
Aircraft Engine												
Reciprocating	120	...	1	0.6	...	...	...	...	...	...	0.63	...
Gas Turbine	114	...	1	0.05	...	...	5/5/5	...	19 (204 °C)	2600	0.51	680
Automatic Transmission	102	Pass B	1b	0.17	...	0.04 (Zn)	10/20/10	...	...	...	...	...
Automotive Engine	110	...	1	0.32	0.05	0.14 (Zn)	0/0/0	...	...	...	...	...
Ball and Roller Bearing	111	Pass B	1	0.02	...	...	0/0/0	40-40-0 (10 min)	1500+	25	...	...
Diesel Cylinder	103	...	1	0.47	...	0.11 (Zn)	5/5/5	...	...	...	...	...
Electric Motor Bearings	106	Pass B	1	...	...	...	0/0/0	40-40-0 (4 min)	1500+	...	...	...
Gas Engine	112	...	1	0.32	0.05	0.135 (Zn)	0/0/0	...	...	...	...	...
Gas Compressor	112	...	1	0.32	0.05	0.135 (Zn)	0/0/0	...	...	...	...	...
Gear Oils												
General Purpose	...	Pass B	1	0.78	...	...	5/15/5	...	3.3	25	0.3 (42/250)	2040
Hypoid	102	Pass A	1	1.26	1.1	0.04 (P)	15/45/15	...	5.0	25	0.3 (42/250)	1700
Aircraft												
Light	...	...	...	...	...	...	...	...	...	16	...	1360
Medium	...	...	...	...	...	...	...	...	...	18	...	1475
Heat transfer	110	Pass A	1	...	...	...	...	...	115	...	...	...
Hydraulic	106	Pass B	1	...	...	...	0/0/0	40-40-0 (4 min)	1500+	>2.3	0.28	380
Machine Tool Ways	75	Pass A	1	0.73	...	...	5/30/25	...	...	11.4	...	...
Paper Machine	119	Pass B	1	0.12	...	...	10/50/20	40-40-0 (20 min)	1500+	...	0.55	...
Pneumatic Tools	...	...	...	0.31	3.2	...	5/10/5	Complete	...	...	...	...
Refrigeration	74	...	1	0.06	...	...	...	...	...	...	...	...
Steam Cylinder	120	...	...	...	...	...	...	...	...	...	0.27	400
Steam Turbine												
Light	105	Pass B	1	0.02	...	...	5/5/5	40-40-0 (8 min)	2500+ (250)	>2.3	0.28	380
Medium	112	Pass B	...	0.02	...	...	5/5/5	40-40-0 (8 min)	2500+	>2.3	0.25	500
Heavy	116	Pass B	1	0.03	...	...	5/5/5	40-40-0 (10 min)	1500+	2.4	0.35	510
EP Gear	116	Pass B	1	0.03	...	...	5/5/5	40-40-0 (10 min)	1500+	6.8	0.30	680
Steel Mill Bearing	122	...	...	...	...	...	5/15/5	40-40-0 (10 min)	...	...	...	...
Textile Spindle	87	Pass B	1	0.23	...	...	5/20/20	...	1000	...	...	...

tion resistance of the steam turbine oils is especially critical because of the danger of the turbine bearing failure. Other applications, such as gear, transformer, hydraulic, heat transfer, and gas turbine systems also require lubricants with excellent oxidation stability.

ASTM D2893 is used for evaluating the oxidation properties of the EP lubricants. However, the observed viscosity change and change in precipitation number may be partly due to the thermal effects. While the ASTM procedure requires a temperature of 95 °C (203 °F), some laboratories conduct the test at 121 °C (255 °F) to increase severity. ASTM D942 is used to determine the oxidation resistance of greases during storage in mechanical parts. In this test, a pressure vessel containing the grease at 99 °C is pressurized with oxygen and the vessel is held for a specified time. Little or no drop in pressure indicates oxidation stability. Grease must show good resistance to oxidation to yield a long expected service life.

Interfacial Tension (*ASTM D971, D2285*)

Lubricants differ in their wetting, spreading, and boundary lubrication ability due to the presence of additives or the oxidation degradation products. These differences are indicated by their interfacial tension (IFT) values. This lubricant parameter is measured by the use of a tensiometer. A drop in IFT indicates oil deterioration; hence it can be used to detect an increase in the rate of oxidation of the oil [796]. This information can be used to determine the need for oil maintenance or, if permissible, to replenish the consumed oxidation inhibitors in transformer oils, hydraulic fluids, and steam and marine turbine oils. IFT measurement on the new oils is significant only in transformer oils [795].

Demulsibility (*ASTM D1401 and D2711*)

These methods measure a lubricant's tendency to quickly separate from water. Good demulsibility is a requirement for most nonaqueous lubricants; a lubricant of poor demulsibility will form persistent emulsion, which can plug oil lines, insulate coolers, and accelerate failure of the bearings and other equipment parts due to inadequate lubricant film formation. Steam cylinder oils, used in closed systems, also require good separation from water to prevent the entrance of oil into the steam generating system [795]. The time for oil separation from an emulsion depends upon the presence of the foreign matter and the build-up of the oxidation products; both of which will decrease the emulsion's ability to break. For water-based hydraulic and metalworking fluids, emulsion stability is important, which in certain oil-in-water emulsions may be determined by the use of the ASTM Methods D1479 and D3342.

Load-carrying Capacity (*ASTM D1947, obsolete*)

A number of devices are used to evaluate the load-carrying capacity of the gear lubricants, Ryder gear test machine being one. Gear tests are useful in assessing the load-carrying capacity of the lubricants that contain antiwear, anti-scuff, and extreme pressure additives. These tests are intended for low-viscosity oils for circulating systems; hence they are not generally suitable for lubricants formulated with fatty materials for use in open or hypoid gears [795].

Extreme Pressure Properties (*ASTM D3233, D2782, and D2783*)

The ability to sustain load without seizure or scoring is an important property for applications such as gears that involve mixed film and boundary lubrication. One of the test machines that are used extensively to measure the EP properties of the lubricants is the Timken machine, which employs a hardened steel ring rotating against a flat steel block. The test can be used either for oils (ASTM D2782) or greases (ASTM D2509). Two other ASTM methods, Falex (D3233) and Four-ball (D2783) testers are also used. Please note that these tests are only screening tools and cannot be used to determine the true extreme pressure properties of the lubricants. Hence, it is recommended that the oils to be used in gear systems must be tested employing actual gears and operating parameters, such as load, speed, and temperatures, that are representative of the intended application [795].

Wear Properties (*ASTM D2670*)

The Falex tester is one of many that are used to test the wear properties of the lubricants. In the ASTM D2670 test procedure, the wear performance of the lubricants is compared on the basis of the final load before seizure of the pin and block assembly. Results from this test machine do not provide absolute values of the film strength or the extreme-pressure properties of the lubricant, but this test method is quite effective for comparing the wear properties of different lubricants.

Hydrolytic Stability (*ASTM D2619*)

This test method evaluates a lubricant's ability to resist hydrolysis. However, this test has limited use for screening and evaluating ester-based hydraulic fluids. This test is useful in monitoring the lubricant condition and assessing the effectiveness of hydrolysis inhibitors that are often used to protect phosphate and silicate ester fluids against hydrolysis.

Thermal Stability (*ASTM D2160*)

This test is used primarily for hydraulic oils to evaluate their resistance to property changes due to thermal stresses. This information is useful in selecting a lubricant for sealed hydraulic applications and heat transfer systems [795]. This test is not suitable for water-based fluids.

Low-Temperature Viscosity (*ASTM D2602*)

This test is used solely for automotive engine oil to measure the apparent viscosity and the response of the oil to a rotor in a device called Cold-Cranking Simulator. The device replicates cranking of an engine and its effect on the lubricant during cold climatic conditions.

Grease Penetration (*ASTM D217 and D1403*)

These tests assess consistency or hardness of grease via cone penetration. ASTM D217 uses a standard cone and ASTM D1403 employs fractional scale cones, which permit measurements on small volumes of grease. While the measurement can be carried out on an unworked grease, worked penetration simulates real-life use conditions and hence is more beneficial. A worked penetration value of about 265 to 295 is considered desirable for general purpose greases and a penetration of about 200 to 250 is considered appropriate for grease use in small double-sealed bearings since good "channeling" is more important than "feedability" of the grease [795].

Grease Dropping Point (*ASTM D566 and D2265*)

Dropping point is the temperature at which a grease exhibits a change from being a semisolid to a liquid. Of these tests, ASTM D2265 is used for high temperature greases. Please note that the dropping point only approximates the melting point of lubricating grease since greases lack a well-defined melting point. The measurement has little significance with respect to a grease's in-service performance.

Apparent Viscosity of a Grease (ASTM D1092 and D3232)

Rheological properties of a grease are determined by the use of pressure to make it flow through a cylinder into an attached capillary flow tube. The results are indicative of any flow problems that may occur during distribution and dispensing of grease from a central distribution system.

Evaporation Loss in Lubricating Greases (ASTM D972 and D2595)

These test methods can be used both for oils and greases. The tests may be carried out at any temperature between 99 and 150 °C. A significant loss of oil will result an increase in viscosity and in the worst case in a loss of the grease structure.

Wear Prevention and Extreme Pressure in Greases (ASTM D2266 and D2596)

Both these methods are Four-ball methods and are used to assess the wear prevention characteristic of a grease.

Leakage of Wheel Bearing Greases (ASTM D1263)

Greases used in automotive wheel bearings may leak due to thermal softening resulting from high speeds, loads, and excessive braking. The ASTM D1263 test determines the leakage tendency of a grease.

Water Washout (ASTM D1264)

This test measures washout of the greases at either 38 or 79 °C when sprayed with a water jet. Grease loss in one hour is a measure of its resistance to washout by water. Some greases reject water or incorporate it as droplets, others may absorb water forming emulsions or may even liquefy.

Grease Performance in Ball Bearings (ASTM Dl741, D3336, and D3337)

ASTM D1741, now obsolete, measured performance life and leakage tendency of a grease. ASTM D3336 is a similar test that is conducted at a high temperature of up to 371 °C. The ASTM D3337 method, analogous to D3336, employs small ball bearings. All three test methods are good for screening greases to identify those that are the most effective.

Lubricating Properties of Solids (ASTM D1367, D2510, D2511, D2625, D2649)

These tests are designed to evaluate lubricating properties of solids alone and in combination with oils and greases. The lubricating parameters they measure are listed below:

1. ASTM D1367 evaluates the lubricating qualities of graphite and other similar solids. The test can be used for comparative evaluation of the various solid lubricants in combination with other lubricating fluids.
2. ASTM D2510 determines the adhesion properties of solids by assessing the effectiveness and continuity of the solid-film formed by the lubricant.
3. ASTM D2511 evaluates thermal shock sensitivity of the solid lubricant film. Applications that involve extreme temperature variations may cause damage to the coated lubricant film. The evaluation is more qualitative than quantitative.
4. ASTM D2625 measures wear and load-carrying characteristics of the solid lubricants.
5. ASTM D2649 is used to evaluate corrosiveness of the solid lubricants, many of which contain chlorine, sulfur, and other elements. For this test, aluminum test panels are normally used. However, other metals may also be used, if so desired.

Mechanical or Tribological Tests

These tests help assess a lubricant's performance in laboratory tests that are designed to simulate the actual service conditions. Machine tests, such as SAE #1, Four-ball, Falex, FZG, and Ryder gear tests, evaluate a lubricant's effectiveness under mixed-film and boundary lubrication conditions. The outcome of the tests depends upon the tribological system as a whole—that is, the equipment, the operating conditions, the base oil quality, and the nature of the additives [4]. The type and the condition of materials that form the contact surfaces, surface load, equipment speed, nature of the contact (whether rolling or sliding), temperature, contaminants, and moisture are all important in this regard.

The base oil quality affects the lubricant's film-forming ability and the durability of such films under a variety of temperatures and pressures. Paraffinic base oils, for example, form stronger and more durable films than naphthenic or aromatic base oils. Additives, such as viscosity modifiers, friction modifiers, and extreme-pressure agents all enhance lubricant performance. Viscosity modifiers improve the lubricant's film-forming ability under elasto-hydrodynamic conditions by influencing its viscosity. Friction modifiers improve the nature of the lubricant film through physisorption or chemisorption on the surfaces and at the same time associating with the bulk lubricant. See the discussion in Chapter 4 on Friction Modifiers. Such films are less durable than those formed by the extreme-pressure agents. Extreme-pressure agents form irreversible films by chemically reacting with the surfaces.

Machines used for these tests measure the effect of load and temperature on the film-forming ability, as reflected by the friction coefficients and wear. Results for newly formulated oils are compared with those of the reference oils to measure their relative effectiveness. Commonly used test machines include the Almen-Wieland rig, the Falex tester, the Timken Wear and Lubricant testing machine, the Four-ball Test rig, and the SAE #1 tester, although there are many others. The test elements of the three commonly used machines are shown in Fig. 11.8.

These film strength tests are the only way to assess the suitability of the lubricants for driveline applications, such as gears, where mixed and boundary lubrication conditions predominate. However, the data obtained from these tests are subjective. To draw meaningful conclusions, the data must be substantiated by testing in real gear rig tests, such as FZG, Gleason, and Ryder tests, that simulate the operating conditions somewhat more closely. The configuration of Ryder, Gleason, and FZG machines is shown in Fig. 12.11.

The objective of mechanical and tribology testing is not only to ascertain that the lubricant provides adequate lubrication and wear control in the intended application, but also that it does it quickly and inexpensively. This is because the full-scale laboratory and field tests are long and expensive, as was stated earlier. Another objective of the tribology testing is to screen a large number of candidates to identify those that are the best with respect to both cost and performance.

The major challenge is to develop a test that is meaning-

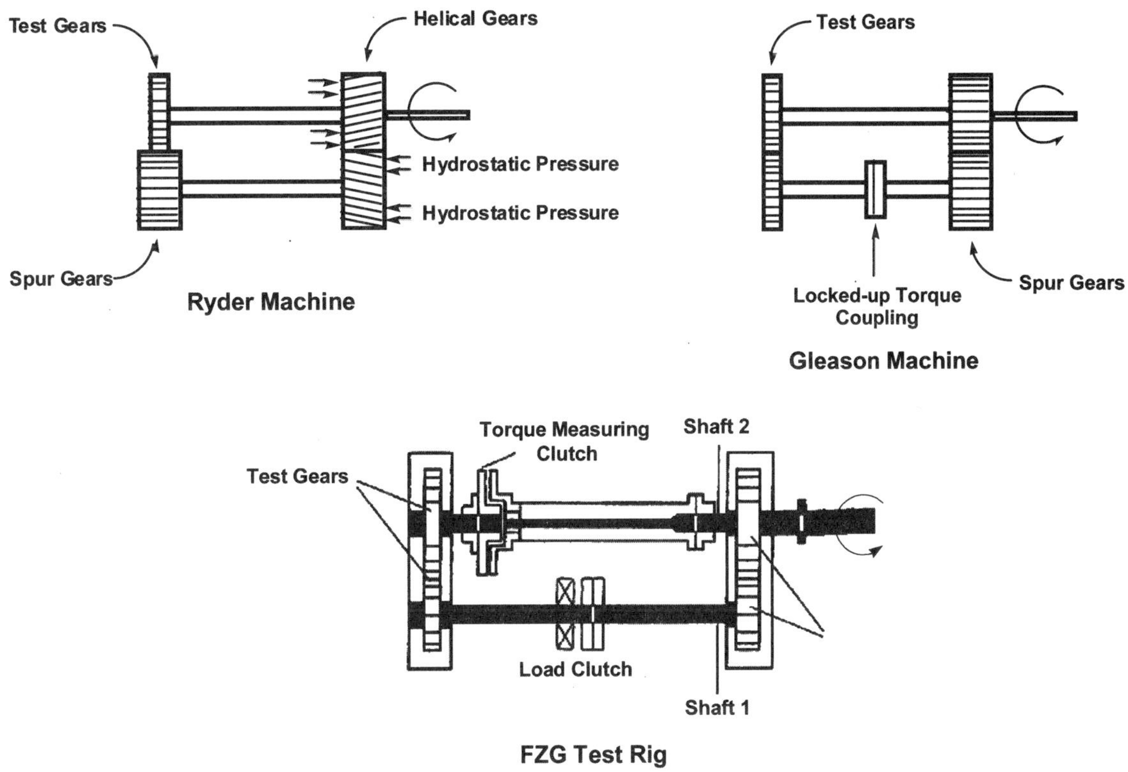

Fig. 12.11—Test elements of Ryder, Gleason, and FZG test machines.

ful; that is, it predicts a lubricant's performance in service. For a tribology test to be able to meet this goal, one must understand the lubrication needs of the intended application and simulate them in the laboratory. It is also important to recognize that while the tribology mechanism or mechanisms simulated in the laboratory may work effectively on a micro-scale, they may or may not indicate adequate performance on a macro-scale. The key to developing a meaningful test is to incorporate all of the important variables into the test so that a close conformity to the lubrication needs of the final application is achieved.

The performance of a tribo-system contact primarily depends upon how well its lubrication needs are met, either by the base oil itself or in combination with the additives that are added to enhance its lubrication ability. This translates into a lubricant's ability to form a suitable lubrication film under hydrodynamic, elasto-hydrodynamic, and boundary conditions. Hydrodynamic films generated within the conformal contacts may be tens of microns (μm) thick and hydrodynamic or elasto-hydrodynamic (EHD) films generated in nonconforming contacts may be around 1 μm thick, see Table 4.20. Hydrodynamic films are largely a consequence of the lubricant's bulk viscosity and in the case of the mineral oil lubricants depend primarily upon its paraffinic content. Elasto-hydrodynamic films, on the other hand, depend upon the lubricant viscosity in the contact zone and are mainly influenced by a mineral oil's naphthenic and aromatic content. Boundary films, which are present on the thin outer layers of the surface, may consist of surface oxides, adsorbed films, and chemical reaction films derived from the EP/antiwear additives present in the lubricant. These films are almost always less than 1 μm thick [797]. Chemically-generated surface films play a major role in applications involving extensive metal-to-metal contact, such as gears and bearings, and the highly stressed long-life mechanical systems.

As mentioned earlier, for a tribology test to be of practical value, it must have a clear link to the real life performance. For it to do so, the test must duplicate the performance mechanisms operating in the tribological system of the intended application. The major objective of the lubrication is to avoid or reduce the effects of friction and wear upon a mechanical system. A lubricant may also act as a cooling agent and remove heat from the location of the friction process, which will also help control wear. A schematic of a tribological system is provided in Fig. 12.12. [798]. A typical lubricated mechanical system has inputs that it transforms into outputs. The inputs are external to the system and comprise the operating variables. Those of importance are the type of motion, load, velocity, temperature, and duration. The outputs are the friction-related parameters, such as the frictional force and friction coefficient; vibration or noise, or both; and an increase in temperature, wear, and the contact conditions. These provide information as to how efficiently the tribological system is working. The manner in which the inputs are transformed into outputs is the technical function of the system and in turn depends upon the characteristics of the system, which include tribo-elements or metal pairs, lubricant, and the lubricating environment. Other factors that impact performance of the tribological system include surface metallurgy, topography, and the surface composition [798].

The first input for the tribological system is the type of motion. It is important to know if the motion involves rolling, sliding, spinning, or is of the impact type. Bearings in-

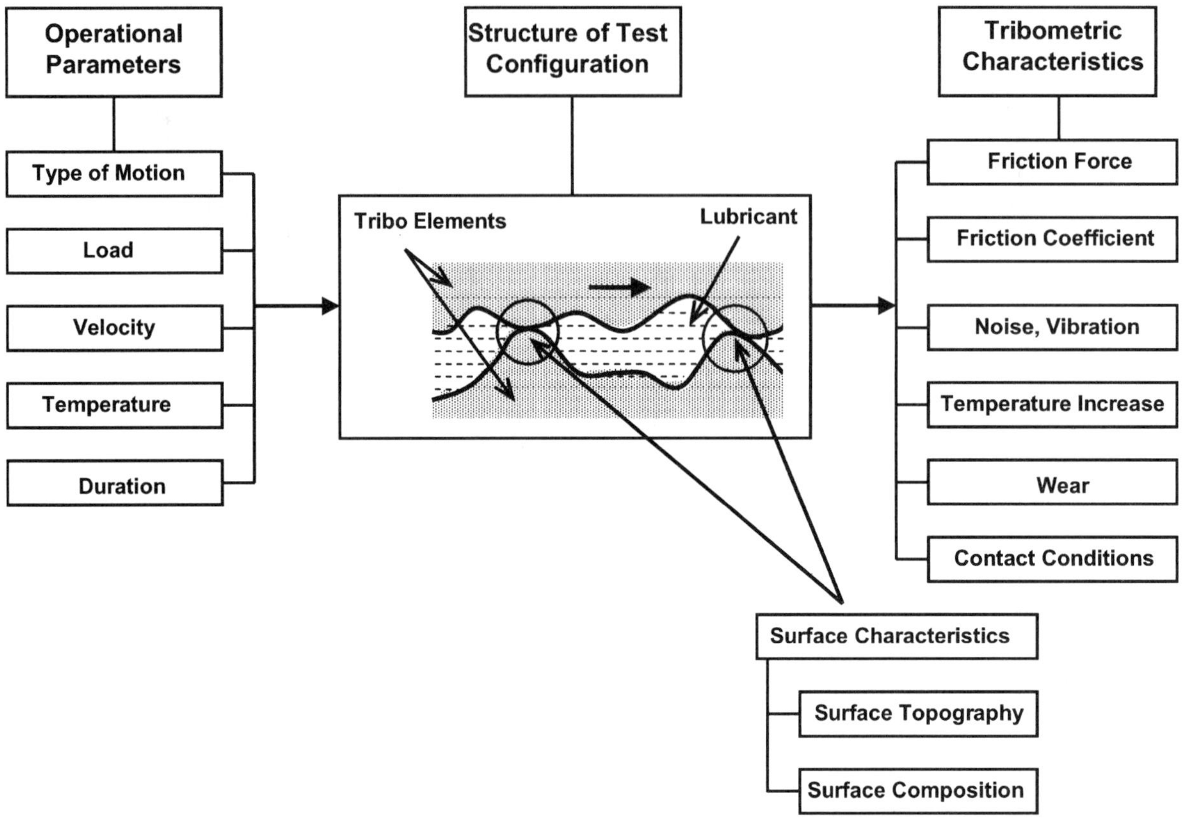

Fig. 12.12—Elements of a tribological system [798].

volve spinning motion or rolling motion, depending on their type; straight bevel gears have rolling motion, and hypoid gears have a combination of both rolling and sliding motion. Sliding motion is more strenuous with respect to contact and wear than rolling motion. In addition to characterizing the type of motion, its dependence on time is also important, such as being continuous, oscillating, reciprocating, or intermittent. Duration of the operation affects the lubricant viscosity due to the increased temperature, which will have an influence on lubrication quality in the tribological system. The load is the next significant input. Light (low) load implies less metal-to-metal contact than heavy (high) load and will affect frictional parameters of the system and the degree of wear. Velocity or speed will affect the amount of the lubricant in the contact zone, hence affect the nature of the lubrication. High speeds promote hydrodynamic lubrication and slow speeds promote mixed film lubrication. Temperature affects the lubricant viscosity, as well as promotes adhesion in the absence of the lubricant in the contact zone. For some tribological systems, other inputs, such as flow rate of the lubricant, vibration, and radiation must also be taken into consideration [798].

The first parameter for the tribological system is the properties of the tribological elements. These include geometry, metallurgy, chemical composition, thermal conductivity, elastic modulus, hardness, density, and other properties. Metallurgy defines the degree of hardness, adhesion tendency, propensity to wear, and the lubricant's effectiveness to form a lubricating film. Softer metals, for example, are more likely to have extensive content under load due to easy flattening and increased adhesion tendency than the harder metals, hence increased wear. Harder metals may be more prone to abrasive wear by the wear debris and the environment-related particulate matter. Chemical elements are usually added to improve mechanical properties of the bulk metal, such as hardness and ductility. Chemical treatment, such as phosphating, galvanizing, and case hardening, is carried out to improve appearance, corrosion protection, wear resistance, and friction control. Thermal conductivity is the ability of the metal to conduct heat. Metals differ in their ability to conduct heat, which is a function of their structure and the temperature. Thermal conductivity of some of the metals that are used in tribological elements is listed in a decreasing order below. Metals with high thermal conductivity will effectively take the heat away from the contact zone.

1. Silver (Ag), pure
2. Copper (Cu), pure
3. Gold (Au), pure
4. Aluminum (Al), pure
5. Brass (Cu plus 35–15 % Zn)
6. Iron (Fe), pure
7. Cast iron (Fe plus 2–3.5 % C plus 1–3 % Si)
8. Bronze (Cu plus 11 % Sn)
9. Carbon Steel (Fe plus 1.5–0.5 % C)

Elastic modulus, or modulus of elasticity, is the mathematical description of an object's or a substance's tendency to deform under the influence of force. Metal stiffness, as measured by Young's modulus, for some of the metals follows the order given below:

TABLE 12.16—Physical properties of some lubricating greases [795].

Grease Type	Soap Type	Color	Texture	Penetration (mm/10) at 25°C	Dropping Pt. (°C)	Base oil Viscosity (cSt) at 40°C	Bomb Oxidation Pressure Drop (500 hr, Kpa)	Effect on Copper	Leakage (g)	Evaporation (22 Hr), %
Automotive General Purpose										
Wheel and Chassis etc.	Calcium, lithium complex, sodium	Green, tan	Smooth, buttery	260-300	250+	150	60- 70	None	0-1.5	-
Ball and Roller Bearing										
Low Temperature	Lithium	Tan	Smooth	300	190	12-16	35-60	None	None	0.8 at 120°C
Medium Temperature	Sodium, sodium, calcium	Tan to brown	Buttery	280	200+	64-70	42-86	None	<1.0	0.8 at 125°C
High Temperature	Sodium	Tan	Buttery	260-300	195	107-150	28	None	<1.0	0.6 at 150°C
High Speed	Lithium, sodium	Tan	Buttery	280	260	100-130	14-90	None		0.5 at 100°C
High Load	Sodium	Brown	Fibrous	300	190	20 at 100°C	-	None	-	1.3 at 100°C
Extreme Pressure	Calcium, lithium, lead	Tan	Smooth and Buttery	280-300	230-260	108	56-120	None	None	-
Food Machines	Calcium	Tan, amber	Buttery	300+	265	71	40 (100 hr)	-	-	-
General Purpose	Calcium	Amber, tan	Smooth	280	90	60	-	None	None	-
Water Resistant	Calcium	Brown & black	Tacky	300	85-95	190	-	-	-	-
Open Bearings	Sodium-calcium	Tan	Solid	60	200	21	-	-	-	-
Railroad Bearing	Lithium-calcium	Amber	Fibrous	300+	170	160	50	-	-	-
Multipurpose	Lithium	Amber/green	Smooth, buttery	270	300+	100+	40-100	-	0.4	-
Steel Mill	Calcium, lithium, aluminum complex	Amber	Smooth	310	180	-	150	None	2.3	-
Textile	Calcium	Light brown	Tacky	320	200	-	42 30 (100 hr)	None	None	-
Wide Temperature Range	Non-soap	Brown-red	Smooth, buttery	300	260+	-	-	None	Pass	7.0 at 177°C

Wrought iron and steel > Titanium > Brass and bronze > Aluminum Alloy > Magnesium metal

Density of some of the metals used in the tribological elements follow the order:

Platinum > Gold > Tungsten > Lead > Silver > Copper > Iron > Steel > Tin > Titanium > Aluminum > Magnesium

Again, as with thermal conductivity, the effect of the elastic modulus and density on lubrication and wear cannot be predicted without considering the other properties of the metals.

Surface topography is also important. Finely finished surfaces or those with too rough a finish may experience more wear than those that have intermediate finish. This is because the former type of surface finish will not be able to hold the lubricant film effectively, thereby resulting in extensive surface contact; and the latter type will have extensive metal contact because of the large surface asperities. Surface composition effect is related to metal's ductility, which will affect its wear characteristics. In systems, such as worm gears, and other systems where the contacting parts are made of different (dissimilar) metals or of different materials, the degree of wear may be more or less than if both tribological elements were made of the same material, depending upon the material.

The lubricant and the environment are the remaining two factors of the tribological system. Properties of the lubricant may be classified into system-dependent properties and system-independent characteristics. System-dependent properties essentially depend upon the specifications of the whole tribological system and include viscosity, viscosity-pressure, viscosity-temperature, and EP/antiwear properties. System-independent properties include other physical and chemical properties of the lubricants [799]. It is important to note that it is critical to match the system-dependent characteristics of the lubricant with the technical function of the actual tribo-engineering system in which the lubricant is to be used. Environmental factors include chemical composition of the environment and the effect of its components, especially of the water vapor. If the inputs, or the operational parameters, are taken into account, along with the elements of the tribological system, it is possible to devise a quality tribological test. While it is impractical or possible to include all the system characteristics into a test, it is important to ensure that none of the essential operational aspects or the influencing parameters is overlooked.

A well designed test will provide the desired outputs listed in Fig. 12.12 under the tribometric characteristics. Table 12.13 shows examples of a number of tribological systems and their elements [799]. In ultra-high vacuum applications, the tribological system diminishes to only the interacting tribological elements since there is no lubricant and there exists almost no environment. The main interactions involve contact deformation, surface fatigue, abrasion, and adhesion. In air, these processes are supplemented by interactions with the atmosphere [800]. Finally, in a lubricated system, direct (contact) interactions between moving and stationary elements are prevented or influenced by the different mechanisms of lubrication.

The tribological tests predominantly assess the overall ability of a lubricant to permit the rubbing surfaces to operate without scuffing, seizing, or other signs of material destruction. The tests can be broadly classified into three groups. These are simple bench tests, tribo-technical tests, and full-scale tests, and in terms of usefulness they go from being the least effective to being the most effective [801].

Simple bench tests use simplified geometries that involve point, line, or flat contact, and are designed to differentiate between EP and nonEP oils; hence their accuracy is just adequate. Predicting the performance of the lubricants in real life applications purely on the basis of these tests is al-

TABLE 12.17—Typical bench test results for some lubricating greases [795].

Grease Type	Water Washout at 80°C, %	Low Temperature Torque (g cm)		% Oil Separation	Copper Strip Rating	Roll Stability	Wear Prevention	EP Property 4-Ball EP, kg	High Temperature Performance	Functional Bearing Life
		Start	*Run*				*Scar Diameter, mm*	*Load Wear Index/Weld Point*	*(hr)*	*(hr)*
Automotive General Purpose										
Wheel and Chassis etc.	3.0	13,000 (-40°C)	5,000	2.6	1	+8	0.38	45/275	580+ (150°C)	–
Ball and Roller Bearing										
Low Temperature	1.2	500 (-54°C)	200	4.5	1	–	0.36	41/250	1,000+ (125°C)	1,900 (100°C), 3,600 rpm 1,500 (120°C), 10,000 rpm
Medium Temperature	–	4,300 (-34°C)	1,800	2.0	1	–	–	–	400 (150°C)	2,500 (100°C), 3,600 rpm
High Temperature	7.0	1,000 (-43°C)	–	1.9	1	–	0.33	1/357	1,000 (150°C)	8,500+ (125°C), 3,600 rpm
High Speed	1.0	2,500 (-34°C)	–	–	1	+7	0.32	1/400	1,000+ (150°C)	2,000 (100°C), 3,600 rpm
High Load	–	–	–	1.0	1	–	–	–	–	–
Extreme Pressure	2.0	–	–	2.0	1	–	0.30	60/315	–	1,000 (120°C), 10,000 rpm
Food Machines	1.5	–	–	–	1	–	0.25 (10 kg, 1800 rpm)	25 (Timken, kg OK Load)	–	–
General Purpose	1.0	–	–	–	1	–	–	–	–	–
Water Resistant	1.0	–	–	–	1	–	–	–	–	–
Open Bearings	–	–	–	–	–	–	–	–	–	–
Railroad Bearing	–	–	–	–	1	–	–	–	–	–
Multi-purpose	0.10	3,000 (-30°C)	1,100	2 to 4	1	+10	0.4	38/250	1,200 (150°C)	10,000 (100°C), 3,600 rpm
Steel Mill	8	–	–	16	1	+3	0.49	35/225	–	–
Textile	–	–	–	0.2	1	–	–	–	–	–
Broad Temperature Range	5	3,390 (-65°C)	890	3.9	1	–	0.60	–	–	100-200 at R.T., 40,000 rpm, size 36 Brg

most impossible. However, these are useful for screening candidates to identify the best ones for further testing. Because of the limitations of the simple tests, more advanced tribo-technical tests must be performed. Such tests use as many variables as possible when simulating the actual performance. These tests use a complete tribo-engineering unit and are best performed in the laboratory, where it is possible to control the operating conditions better [799]. These tests use a number of test element geometries to simulate various contact situations that occur in real-life applications. These test geometries are shown in Figure 4.147. As stated earlier, the lubricant candidates that meet the preliminary tests must be subjected to full-scale tests to ascertain satisfactory performance in actual tribo-engineering systems. Since the cost of running such tests is quite high, they are only used to provide the final proof of a lubricant's performance. A systematic lubricant selection procedure is described by Horst Czichos [799].

Table 12.14 lists the typical physical properties and Table 12.15 lists the bench test results of some of the lubricants [795]. Tables 12.16 and 12.17 list the same for greases [795]. Please note that Table 12.17 is a continuation of Table 12.16, but it lists other grease properties. Tables 12.18–12.23 list the tests used to analyze and assess lubricant quality. Table 12.18 lists ASTM standards and the corresponding DIN, ISO, and IP standards. Table 12.19 lists DIN and DIN-ISO combined standards and Table 12.20 provides AFNOR standards and specifications [802]. Tables 12.21 and 12.22 comprise non-ASTM Standards other than DIN and AFNOR, and Table 12.23 lists standards and specifications that pertain to the military lubricants [803]. As one can see there is significant overlap across tables and some tests are included in more than one list. This is despite the fact that we attempted not to include the same tests in more than one table. Absolute classification in some cases was not possible because of the slightly different description for the test methods that may have been the same. We decided to exercise prudence and included both in our lists.

In-service Lubricant Analysis

The analysis of a lubricant in service is at least as important as that of a new lubricant or the used lubricant. This is because it can help identify problems in their infancy. Lubricant monitoring is routine in many industries and involves assessing physical appearance, viscosity, metals content, and the rate of the lubricant degradation. Once the problems are identified, they can be corrected either by replenishing the consumed additives or by the used lubricant replacement.

Oil and wear particle analyses are also valuable failure

TABLE 12.18—ASTM and related standards used in lubricant analysis and testing.

ASTM Standard	ISO Standard	Other Standards	Test Description
A1415	...	...	Test Method for Rubber Property-International Hardness
D56	...	...	Test Method for Flash Point by Tag Closed Cup Tester
D86	3405	IP 123	Test Method for Distillation of Petroleum Products at Atmospheric Pressure
D91	...	...	Test Method for Precipitation Number of Lubricating Oils
D92	2592	IP 36 DIN 51 376 FED-STD-791-1103	Test Method for Flash and Fire Points by Cleveland Open Cup
D93	2719	IP 34 NF EN 22719 DIN 51 758	Test Method for Flash Point by Pensky-Martens Closed Cup Tester
D94	6293	IP 136 DIN 51 559	Test Method for Saponification Number of Petroleum Products
D95	3733	IP 74 DIN 51 582	Test Method for Water in Petroleum Products and Bituminous Materials by Distillation
D96	...	...	Test Method for Water and Sediment Determination in Crude Oil by Centrifuge Method
D97	219/3016	IP 15 DIN 51 568 DIN 51 597 NFT 60105 FED-STD-791-201	Test Method for Pour Point of Petroleum Products
D117	...	...	Guide to Sampling Test Methods, Standard Practices, and Guides for Electrical Insulating Oils of Petroleum Origin
D128	...	...	Test Methods for Analysis of Lubricating Grease
D129		IP 1/IP 336 DIN 51 400 NF EN 24260	Test Method for Sulfur in Petroleum Products (General Bomb Method)
D130	2160	IP 154 DIN 51 759 FED-STD-791-5325	Test Method for Detection of Copper Corrosion from Petroleum Products by the Copper Strip Tarnish Test
D156	...	...	Test Method for Saybolt Color of Petroleum Products (Saybolt Chromometer Method)
D189	6615	IP 13 DIN 51 551	Test Method for Conradson Carbon Residue of Petroleum Products
D217	2137	IP 50	Test Method for Cone Penetration of Lubricating Grease
D287	...	...	Test Method for API Gravity of Crude Petroleum and Petroleum Products (Hydrometer Method)
D341	...	DIN 51 563	Standard Test Method for Viscosity-Temperature Charts for Liquid Petroleum Products
D445	3104	IP 71 DIN 51 550 DIN 51 372 DIN 53 015 FED-STD-791-305	Test Method for Kinematic Viscosity of Transparent and Opaque Liquids (the Calculation of Dynamic Viscosity)
D446	3105	IP 71 DIN 51 366 DIN 51562 Part 1	Specifications and Operating Instructions for Glass Capillary Kinematic Viscometers
D471	...	...	Test Method for Rubber Property Effect of Liquids
D482/ D874	3987/6245	IP 4/IP 223 DIN 52 005	Test Method for Ash from Petroleum Products
D524	4262	IP 14	Test Method for Ramsbottom Carbon Residue of Petroleum Products
D566	2176	IP 132 DIN 51 806	Test Method for Dropping Point of Lubricating Grease
D611	2977	IP 2	Test Methods for Aniline Point and Mixed Aniline Point of Petroleum Products and Hydrocarbon Solvents
D664	6619	IP 177 DIN EN 12634 FED-STD-791-5106 SAE ARP 5088	Test Method for Acid Number of Petroleum Products by Potentiometric Titration Method
D665/ D3603 D1748	7120	IP 135/I 287	Test Method for Rust-Preventing Characteristics of Inhibited Mineral Oil in the Presence of Water
D808	...	DIN 51 408	Test Method for Chlorine in New and Used Petroleum Products (Bomb Method)
D848	...	...	Test Method for Acid Wash Color of Industrial Aromatic Hydrocarbons
D874	3987	IP 163 DIN 51 575	Test Method for Sulfated Ash from Lubricating Oils and Additives

TABLE 12.18— *(Continued.)*

ASTM Standard	ISO Standard	Other Standards	Test Description
D877	...	...	Test Method for Dielectric Breakdown Voltage of Insulating Liquids Using Disk Electrodes
D891	...	IP 365/P 160 DIN 51 757	Standard Test Method for Specific Gravity, Apparent, of Liquid Industrial Chemicals
D892	6247	IP 146 FED-STD-3213	Test Method for Foaming Characteristics of Lubricating Oils
D893	...	...	Test Method for Insolubles In Used Lubricating Oils
D924	...	...	Test Method for Dissipation Factor (or Power Factor) and Relative Permittivity (Dielectric Constant) of Electrical Insulating Liquids
D942	...	IP 142	Test Method for Oxidation Stability of Lubricating Greases by the Oxygen Bomb Method
D943	4263	IP 280 BS 4388	Test Method for Oxidation Characteristics of Inhibited Mineral Oils
D971	...	...	Test Method for Interfacial Tension of Oil Against Water by the Ring Method
D972	...	IP 183 D51 581 FED-STD-791-351	Test Method for Evaporation Loss of Lubricating Greases and Oils
D974	6618	IP 139/354 DIN 51 558Part 1 FED-STD-791-5105	Test Method for Acid and Base Number by Color-Indicator Titration
D1078	...	IP 195	Test Method for the Determination of Distillation Characteristics of Volatile Organic Liquids (ASTM Procedure Now Obsolete)
D1091	...	IP 245 DIN 51 363	Test Method for Phosphorus in Lubricating Oils and Additives
D1092	...	...	Test Method for Measuring Apparent Viscosity of Lubricating Greases
D1093	...	...	Acidity of Hydrocarbon Liquids and Their *Distillation Residues*
D1159	3839	IP 130	Test Method for Bromine Numbers of Petroleum Distillates and Commercial Aliphatic Olefins by Electrometric Titration
D1160	...	...	Test Method for Distillation of Petroleum Products at Reduced Pressure
D1169	...	IEC* 60247	Test Method for Specific Resistance (Resistivity) of Electrical Insulating Liquids
D1193	...	...	Specification for Reagent Water
D1209	...	...	Test Method for Color of Clear Liquids (Platinum Cobalt Scale) (APHA Color)
D1217/ D1480	3838	IP 190	Test Method for Density and Relative Density (Specific Gravity) of Liquids by Bingham Pycnometer
D1218	...	...	Test Method for Refractive Index and Refractive Dispersion of Hydrocarbon Liquids
D1263	...	...	Test Method for Leakage Tendencies of Automotive Wheel Bearing Greases
D1264	...	...	Test Method for Determining the Water Washout Characteristics of Lubricating Greases
D1266	...	...	Test Method for Sulfur in Petroleum Products (Lamp Method)
D1275	...	IP 315	Test Method for Corrosive Sulfur in Electrical Insulating Oils
D1298/ D287	375/3675	IP 160 DIN 51 757	Practice for Density, Relative Density (Specific Gravity), or API Gravity of Crude Petroleum and Liquid Petroleum Products
D1310	...	...	Test Method for Flash Point and Fire Point of Liquids by Tag Open-Cup Apparatus
D1319	...	IP 156	Test Method for Hydrocarbon Types in Liquid Petroleum Products by Fluorescent Indicator Adsorption
D1401	6614	IP 412 DIN 51599	Test Method for Water Separability of Petroleum Oils and Synthetic Fluids
D1403	...	IP 310	Test Methods for Cone Penetration of Lubricating Grease Using One-Quarter and One-Half Scale Cone Equipment
D1404	...	...	Test Method for Estimation of Deleterious Particles in Lubricating Grease
D1478	...	...	Standard Test Method for Low-Temperature Torque of Ball Bearing Grease
D1480	...	...	Test Method for Density and Relative Density Specific Gravity of Viscous Materials by Bingham Pycnometer
D1481	3675	DIN 71 757 T 60 101	Test Method for Density and Relative Density Specific Gravity of Viscous Materials by Lipkin Bicapillary Pycnometer
D1492	...	...	Test Method for Bromine Index of Aromatic Hydrocarbons by Coulometric Titration
D1500/ D2392	2049	IP 17/IP 196 FED-STD-791-102	Test Method for ASTM Color of Petroleum Products (ASTM Color Scale)
D1533	...	...	Standard Test Method for Water in Insulating Liquids by Coulometric Karl Fischer Titration
D1552	...	...	Test Method for Sulfur in Petroleum Products (High-Temperature Method)
D1662	...	...	Standard Test Method for Active Sulfur in Cutting Oils

TABLE 12.18— (*Continued.*)

ASTM Standard	ISO Standard	Other Standards	Test Description
D1681	...	...	Standard Practice for Aquatic Toxicity Testing of Lubricants: Sample Preparation and Results Interpretation
D1742	DIN 51817	IP 121	Test Method for Oil Separation from Lubricating Grease During Storage
D1743	...	...	Test Method for Determining Corrosion Preventive Properties of Lubricating Greases
D1744	760/12937	DIN 51 777 FED-STD-791-3253	Test Method for Determination of Water in Liquid Petroleum Products by Karl Fischer Reagent (Discontinued 2000)
D1747	...	...	Test Method for Refractive Index of Viscous Materials
D1748	...	IP 366	Standard Test Method for Rust Protection by Metal Preservatives in the Humidity Cabinet
D1816	...	...	Test Method for Dielectric Breakdown Voltage of Insulating Oils of Petroleum Origin Using VDE Electrodes
D1831	...	...	Test Method for Roll Stability of Lubricating Grease
D1947 (Obsolete)	...	FED-STD-791-6508	Gear Test-Ryder
D1959	...	...	Test Method for Iodine Value of Drying Oils and Fatty Acids
D2007	...	...	Paraffins, aromatics, polar compounds and asphaltenes in rubber extender and processing oils
D2008	...	...	Test Method for Ultraviolet Absorbance and Absorptivity of Petroleum Products
D2070	...	...	Test Method for Thermal Stability of Hydraulic Oils
D2112	...	...	Test Method for Oxidation Stability of Inhibited Mineral Insulating Oil by Pressure Vessel
D2140	...	...	Standard Test Method for Separation of Representative Aromatic and Nonaromatic Fractions of High-Boiling Oils by Elution Chromatography
D2155/ E659	3988	FED-STD-791-1152	Auto-ignition Temperature
D2161	...	...	Practice for Conversion of Kinematic Viscosity to Saybolt Universal Viscosity or to Saybolt Universal Viscosity
D2240	...	...	Test Method for Rubber Property-Durometer Hardness
D2265	...	...	Test Method for Dropping Point of Lubricating Grease Over Wide Temperature Range
D2266	...	IP 239	Test Method for Wear Preventive Characteristics of Lubricating Grease (Four-Ball Method)
D2270	2909	IP 226 FED-STD-791-9111 FED-STD-791-2508 FED-STD-791-3411	Practice for Calculating Viscosity Index From Kinematic Viscosity at 40 and 100 °C
D2271	...	...	Test Method for Preliminary Examination of Hydraulic Fluids (Wear Test)
D2272	...	...	Test Method for Oxidation Stability of Steam Turbine Oils by Rotating Pressure Vessel
D2273	...	FED-STD-791-3004/3010	Test Method for Trace Sediment in Lubricating Oils
D2422	3448	...	Standard Classification of Industrial Fluid Lubricants by Viscosity System
D2440	...	...	Test Method for Oxidation Stability of Mineral Insulating Oil
D2500/ D5771	3015	IP 219 EN 23015	Test Method for Cloud Point of Petroleum Products
D2501	...	...	Test Method for Calculation of Viscosity-Gravity Constant (VGC) of Petroleum Oils
D2502	...	...	Test Method for Estimation of Mean Relative Molecular Mass of Petroleum Oils from Viscosity Measurements
D2503	...	...	Molecular weight relative of hydrocarbons compounds—thermoelectric measurement of vapor pressure
D2509	...	IP 326	Test Method for Measurement of Load-Carrying Capacity of Lubricating Grease (Timken Method)
D2510	...	...	Test Method for Adhesion of Solid Film Lubricants
D2532	...	FED-STD-791-307	Test Method for Viscosity and Viscosity Change After Standing at Low Temperature of Aircraft Turbine Lubricants
D2549	...	...	Test Method for Separation of Representative Aromatics and Nonaromatics Fractions of High-Boiling Oils by Elution Chromatography
D2595	...	...	Evaporation Loss of Lubricating Grease Over Wide Temperature Range
D2596	...	...	Test Method for Measurement of Extreme-Pressure Properties of Lubricating Grease (Four-Ball Method)
D2602	...	DIN 51 377	Test Method for Apparent Viscosity of Engine Oils at Low Temperature Using Cold -cranking Simulator Replaced in 1993 with D5293)
D2603/ D5621	...	...	Test Method for Sonic Shear Stability of Polymer-Containing Oils
D2619	15596	...	Test Method for Hydrolytic Stability of Hydraulic Fluids (Beverage Bottle Method)

TABLE 12.18— *(Continued.)*

ASTM Standard	ISO Standard	Other Standards	Test Description
D2622	8754	DIN EN ISO 14596	Test Method for Sulfur in Petroleum Products by Wavelength Dispersive X-Ray Fluorescence Spectrometry
D2625	...	...	Test Method for Endurance (Wear) Life and Load-Carrying Capacity of Solid Film Lubricants (Falex Pin and Vee Method)
D2649	...	...	Test Method for Corrosion Characteristics of Solid Film Lubricants
D2670	...	...	Test Method for Measuring Wear Properties of Fluid Lubricants (Falex Pin and Vee Block Method)
D2710	...	IP 299	Test Method for Bromine Index of Petroleum Hydrocarbons by Electrometric Titration
D2711	...	IP 19	Test Method for Demulsibility Characteristics of Lubricating Oils
D2714	...	...	Test Method for Calibration and Operation of the Falex Block—On-Ring Friction and Wear Testing Machine
D2717	...	...	Test Method for Thermal Conductivity of Liquids
D2766	...	...	Test method for specific heats of liquids and solids
D2779	...	...	Test Method for Estimation of Solubility of Gases in Petroleum Liquids
D2780	...	...	Standard Test Method for Solubility of Fixed Gases in Liquids
D2782	...	...	Test Method for Measurement of Extreme-Pressure Properties of Lubricating Fluids (Timken Method)
D2783	...	IP 293	Test Method for Measurement of Extreme-Pressure Properties of Lubricating Fluids (Four-Ball Method)
D2878	...	...	Test Method for Estimating Apparent Vapor Pressures and Molecular Weights of Lubricating Oils
D2879	...	...	Test Method for Vapor Pressure-Temperature Relationship and Initial Decomposition Temperature of Liquids by Isoteniscope
D2882	20763	...	Test Method for Indicating Wear Characteristics of Petroleum and Non-Petroleum Hydraulic Fluids in Constant Volume Vane Pump
D2887 D6352	3924	...	Test Method for Boiling Range Distribution of Petroleum Fractions by Gas Chromatography
D2893	...	IP 48	Test Method for Oxidation Characteristics of Extreme-Pressure Lubricating Oils
D2896	3771	IP 276	Test Method for Base Number of Petroleum Products by Potentiometric Perchloric Acid Titration
D2982	...	...	Glycol Base Anti-Freeze in Used Oil
D2983	...	IP 267 DIN 53 018	Test Method for Low-Temperature Viscosity of Automotive Fluid Lubricants Measured by Brookfield Viscometer
D3120	...	...	Trace Quantities of Sulfur in Light Liquid Petroleum Hydrocarbons by Oxidative Microcoulometry
D3228	...	...	Test Method for Total Nitrogen in Lubricating Oils and Fuel Oils by Modified Kjeldahl Method
D3232	...	...	Test Method for Measurement of Consistency of Lubricating Greases at High Temperatures
D3233	...	...	Test Method for Measurement of Extreme Pressure Properties of Fluid Lubricants (Falex Pin and Vee Block Methods)
D3238	...	...	Test Method for Calculation of Carbon Distribution and Structural Group Analysis of Petroleum Oils by the ndM Method
D3244	...	...	Standard Practice for Utilization of Test Data to Determine Conformance with Specifications
D3278	...	...	Test Methods for Flash Point of Liquids by Small Scale Closed-Cup Apparatus
D3300	...	...	Test Method for Dielectric Breakdown Voltage of Insulating Oils of Petroleum Origin Under Impulse Conditions
D3336	...	...	Test Method for Life of Lubricating Greases in Ball Bearings at Elevated Temperatures
D3337	...	...	Test Method for Determining Life and Torque of Lubricating Greases in Small Ball Bearings
D3339	7537	BS[x] 7393	Test Method for Acid Number of Petroleum Products by Semi-Micro Color Indicator Titration
D3427	9120	IP 313 DIN 51381	Test Method for Air Release Properties of Petroleum Oils
D3519	...	...	Test Method for Foam in Aqueous Media (Blender Test)
D3520	...	...	Test Method for Quenching Time of Heat-Treating Fluids (Magnetic Quenchometer Method)
D3524	...	...	Diesel Fuel Diluent in Used Engine Oils
D3525	...	...	Gasoline Diluent in Used Engine Oils by Gas Chromatography
D3527	...	...	Test Method for Life Performance of Automotive Wheel Bearing Grease

TABLE 12.18— *(Continued.)*

ASTM Standard	ISO Standard	Other Standards	Test Description
D3603	...	...	Rust-Preventing Characteristics of Steam Turbine Oil in the Presence of Water (Horizontal Disk Method)
D3702	...	...	Test Method for Wear Rate and Coefficient of Friction of Materials in Self Lubricated Rubbing Contact Using a Thrust Washer Testing Machine
D3704	...	...	Test Method for Wear Preventive Properties of Lubricating Greases Using the (Falex) Block-On-Ring Machine in Oscillating Motion
D3705	...	...	Test Method for Misting Properties of Lubricating Fluids
D3707	...	...	Test Method for Storage Stability of Water-in-Oil Emulsions under by the Oven Test Method
D3709	...	...	Test Method for Stability of Water-in-Oil Emulsions Under Low to Ambient Temperature Cycling Conditions
D3711	...	...	Test Method for Deposition Tendencies of Liquids in Thin Films
D3712	...	...	Test Method for Analysis of Oil-Soluble Petroleum Sulfonates by Liquid Chromatography
D3825	...	...	Test Method for Dynamic Surface Tension by the Fast Bubble Technique
D3827	...	...	Test Method for Estimation of Solubility of Gases in Petroleum and Other Organic Liquids
D3828	...	IP 303 DIN 51 755 NF EN ISO 7536	Test Method for Flash Point by Small Scale Closed Tester
D3829	...	...	Test Method for Predicting the Borderline Pumping Temperature of Engine Oil
D3934	1516	...	Test Method for Flash/No Flash Test-Equilibrium Method by a Closed-Cup Apparatus
D3941	1523	...	Test Method for Flash Point by the Equilibrium Method with a Closed-Cup Apparatus
D4006/ 95	3733/9029	IP 74/358	Water content-distillation method
D4007	...	...	Test Method for Water and Sediment in Crude Oil by the Centrifuge Method of(Laboratory Procedure)
D4047	4265	IP 149	Test Method for Phosphorus in Lubricating Oils and Additives by QuinolinePhosphomolybdate Method
D4048	...	IP 112	Test Method for Detection of Copper Corrosion from Lubricating Grease
D4049	...	...	Test Method for Determining the Resistance of Lubricating Grease to Water Spray
D4052/ D5002	12185	IP 365	Test Method for Density and Relative Density of Liquids by Digital Density Meter
D4059	...	DIN 51527-1 DIN EN 12766 IEC 60997	Polychlorinated biphenyls (PCB) content-GC method
D4170	...	...	Test Method for Fretting Wear Protection by Lubricating Greases
D4172	...	IP 293 DIN 51 350-5 CETOP TP 67H	Test Method for Wear Preventive Characteristics of Lubricating Fluid (Four-Ball Method)
D4175	...	...	Standard Terminology Relating to Petroleum, Petroleum Products, and Lubricants
D4289	...	...	Test Method for Elastomer Compatibility of Lubricating Greases and Fluids
D4290	...	...	Test Method for Determining the Leaking Tendencies of Automotive Wheel BearingGrease Under Accelerated Conditions
D4291	...	DIN 51 375	Trace Ethylene Glycol in Used Engine Oil
D4294	8754	IP 336	Test Method for Sulfur in Petroleum and Petroleum Products by Energy-Dispersive X-RayFluorescence Spectrometry
D4307	...	...	Determination of Additive Elements in Lubricating Oils by Inductively Coupled ofPlasma Atomic Emission Spectrometry Test
D4308	...	DIN 51 412	Standard Test Method for Electrical Conductivity of Liquid Hydrocarbons by PrecisionMeter
D4310	...	...	Test Method for Determination of the Sludging and Corrosion Tendencies of InhibitedMineral Oils
D4377	760	IP 356 DIN 51 777-1	Test Method for Water in Crude Oils by Potentiometric Karl Fischer Titration
D4378	...	...	Standard Practice for In-Service Monitoring of Mineral Turbine Oils for Steam and GasTurbines
D4425	...	...	Test Method for Oil Separation from Lubricating Grease by Centrifuging (Kopper's Method)
D4485	...	...	Specification for Performance of Engine Oils
D4530	10370	IP 398	Test Method for Determination of Carbon Residue (Micro Method)
D4624	...	IP 189 DIN 51757	Test Method for Measuring Apparent Viscosity Bi-capillary Viscometer at High-Temperature and High-Shear Rates
D4627	3838	IP 287 IP 125	Test Method for Iron Chip Corrosion for Water-Dilutable Metalworking Fluids

TABLE 12.18— *(Continued.)*

ASTM Standard	ISO Standard	Other Standards	Test Description
D4628	...	DIN 51 391 DIN 51 431	Test Method for Analysis of Barium, Cadmium, Magnesium, and Zinc in UnusedLubricating Oils by Atomic Absorption Spectrometry
D4629	...	IP 379	Test Method for Trace Nitrogen in Liquid Petroleum Hydrocarbons by Syringe/ InletOxidative Combustion and Chemiluminescence Detection
D4636	...	...	Test Method for Corrosiveness and Oxidation Stability of Hydraulic Oils, Aircraft TurbineEngine Lubricants, and Other Highly Refined Oils
D4682	...	...	Specification for Miscibility with Gasoline and Fluidity of Two-Stroke Cycle Gasoline EngineLubricants
D4683	...	...	Test Method for Measuring Viscosity at High Shear Rate and High Temperature byTapered Bearing Simulator
D4684	...	...	Test Method for Determination of Yield Stress and Apparent Viscosity of Engine Oils atLow Temperature
D4693	...	...	Test Method for Low-Temperature Torque of Grease-Lubricated Wheel Bearings
D4739	...	IP 276/417	Standard Test Method for Base Number Determination by Potentiometric Titration
D4741	...	...	Test Method for Measuring Viscosity at High Temperature and High Shear Rate byTapered-Plug Viscometer
D4742	...	...	Test for Oxidation Stability of Gasoline Engine Oils-Thin-Film Oxygen Uptake Method(TFOUT)
D4857	...	...	Test Method for Determination of the Ability of Lubricants to Minimize Ring Sticking andPiston Deposits in Two-Stroke-Cycle Gasoline Engines Other Than Outboards
D4858	...	...	Test Method for Determination of the Tendency of Lubricants to Promote Pre-ignition inTwo-Stroke-Cycle Gasoline Engines
D4859	...	...	Specification for Lubricants for Two-Stroke-Cycle Spark-Ignition Gasoline Engines-TC
D4863	...	...	Test Method for Determination of Lubricity of Two-Stroke-Cycle Gasoline EngineLubricants
D4898	4405/ 4406	...	Standard Test Method for Insoluble Contamination of Hydraulic Fluids by GravimetricAnalysis
D4927	...	...	Test Methods for Elemental Analysis of Lubricant and Additive Components—Barium,Calcium, Phosphorus, Sulfur, and Zinc by Wavelength-Dispersive X-Ray FluorescenceSpectroscopy
D4928	...	DIN EN ISO 12937	Test Methods for Water in Crude Oils by Coulometric Karl Fischer Titration
D4950	...	...	Classification and Specification for Automotive Service Greases
D4951	...	IP 288a	Test Method for Determination of Additive Elements in Lubricating Oils by InductivelyCoupled Plasma Atomic Emission Spectroscopy
D4998	...	...	Test Method for Evaluating Wear Characteristics of Tractor Hydraulic Fluids
D5001	...	...	Standard Test Method for Measurement of Lubricity of Aviation Turbine Fuels by the Ball-On-Cylinder Lubricity Evaluator (BOCLE)
D5119	...	...	Method for Evaluation of Automotive Engine Oils in the CRC L-38 Spark-Ignition ofEngine
D5133	...	...	Test Method for Low Temperature, Low Shear Rate, Viscosity/Temperature Dependenceof Lubricating Oils Using a Temperature Scanning Technique
D5182	14635	IP 334 DIN 51 354	Standard Test Method for Evaluating the Scuffing Load Capacity of Oils (FZG VisualMethod)
D5185	...	...	Test Method for Determination of Additive Elements, Wear Metals, and Contaminants inUsed Lubricating Oils and Determination of Selected Elements in Base Oils by InductivelyCoupled Plasma Emission Spectroscopy (ICP-AES)
D5191	...	...	Test Method for Vapor Pressure of Petroleum Products (Mini Method)
D5290	...	...	Measurement of Oil Consumption, Piston Deposits, and Wear in a Heavy-Duty ofHigh-Speed Diesel Engine-NTC-400 Procedure
D5291	...	...	Test Methods for Instrumental Determination of Carbon, Hydrogen, and Nitrogen inPetroleum Products and Lubricants
D5292	...	...	Test Method for Aromatic Hydrogen and Aromatic Carbon Contents of Hydrocarbon Oils byHigh Resolution. Nuclear Magnetic Resonance Spectroscopy
D5293	...	IP 350 IP 383	Test Method for Apparent Viscosity of Engine Oils Between −5 and −35 °C Using the Cold-Cranking Simulator
D5302	...	...	Test Method for Evaluation of Automotive Engine Oils for Inhibition of Deposit Formationand Wear in a Spark-Ignition Internal Combustion Engine Fueled with Gasoline andOperated Under Low-Temperature, Light-Duty Conditions (Sequence VE)
D5306	14935	...	Wick Ignition
D5453	...	...	Test Method for Determination of Total Sulfur in Light Hydrocarbons, Spark Ignition EngineFuel, Diesel Engine Fuel, and Engine Oil by Ultraviolet Fluorescence

TABLE 12.18— *(Continued.)*

ASTM Standard	ISO Standard	Other Standards	Test Description
D5480	...	DIN 51581.T2	Test Method for Engine Oil Volatility by Gas Chromatography
D5481	...	...	Test Method for Measuring Apparent Viscosity at High-Temperature and High-Shear Rateby Multi-cell Capillary Viscometer
D5482	...	...	Test Method for Vapor Pressure of Petroleum Products (Mini Method-Atmospheric)
D5483	...	...	Test Method for Oxidation Induction Time of Lubricating Greases by Pressure DifferentialScanning Calorimetry
D5533	...	...	Test Method for Evaluation of Automotive Engine Oils in Sequence IIIE, Spark-IgnitionEngine
D5534	...	...	Test Method for Vapor-Phase Rust-Preventing Characteristics of Hydraulic Fluids
D5570	...	...	Test Method for Evaluating Thermal Stability of Manual Transmission Lubricants in a CycleDurability Test
D5579	...	...	Test Method for Evaluating the Thermal Stability of Manual Transmission Lubricants in aCyclic Durability Test
D5619	...	...	Test Method for Comparing Metal Removal Fluids Using the Tapping Torque Test Machine
D5621	...	...	Standard Test Method for Sonic Shear Stability of Hydraulic Fluid of
D5662	...	...	Test Method for Determining Automotive Gear Oil Compatibility with Typical Oil SealElastomers
D5704	...	...	Test Method for Evaluation of the Thermal and Oxidative Stability of Lubricating Oils Usedfor Manual Transmissions and Final Drive Axles
D5706	...	...	Test Method for Determining Extreme Pressure Properties of Lubricating Greases Using aHigh-Frequency, Linear-Oscillation (SRV) Test Machine
D5707	...	...	Test Method for Measuring Friction and Wear Properties of Lubricating Greases Using aHigh-Frequency, Linear-Oscillation (SRV) Test Machine
D5760	...	...	Specification for Performance of Manual Transmission Gear Lubricants
D5762	...	...	Test Method for Nitrogen in Petroleum and Petroleum Products BY Boat-InletChemiluminescence
D5763	...	...	Test Method for Oxidation and Thermal Stability Characteristics of Gear Oils UsingUniversal Glassware
D5770	...	...	Test Method for Semi-quantitative Micro Determination of Acid Number of Lubricating OilsDuring Oxidation Testing
D5771	...	...	Test Method for Cloud Point of Petroleum Products (Optical Detection Stepped CoolingMethod)
D5772	...	IP 445	Test Method for Cloud Point of Petroleum Products (Linear Cooling Rate Method)
D5773	...	IP 446	Standard Test Method for Cloud Point of Petroleum Products (Constant Cooling RateMethod)
D5776	...	...	Standard Test Method for Bromine Index of Aromatic Hydrocarbons by ElectrometricTitration
D5800	...	CEC L-40-A-93 DIN 51 581	Test Method for Evaporation Loss of Lubricating Oils by the Noack Method
D5844	...	...	Test Method for Evaluation of Automotive Engine Oils for Inhibition of Rusting (SequenceIID)
D5846	...	...	Test Method for Universal Oxidation Test for Hydraulic and Turbine Oils Using theUniversal Oxidation Test Apparatus
D5862	...	...	Test Method for Evaluation of Engine Oils in Two-Stroke Cycle Turbo-Supercharged6V92TA Diesel Engine
D5864	...	...	Standard Test Method for Determining Aerobic Aquatic Biodegradation of Lubricants orTheir Components
D5949	...	...	Test Method for Pour Point of Petroleum Products (Automatic Pressure Pulsing Method)
D5950	...	...	Test Method for Pour Point of Petroleum Products Automatic Tilt Method)
D5966	...	...	Test Method for Evaluation of Engine Oils for Roller Follower Wear in Light-Duty DieselEngine
D5967	...	...	The Method of Evaluation of Diesel Engine Oils in T-8 Engine
D5968	...	...	Test Method for the Corrosiveness of Diesel Engine Oil
D5969	...	...	Test Method for Corrosion-Preventive Properties of Lubricating Greases in Presence ofDilute Synthetic Sea Water Environments
D5985	...	...	Test Method for Pour Point of Petroleum Products (Rotational Method)
D6006	...	...	Guide for Assessing Biodegradability of Hydraulic Fluids
D6022	...	...	Practice for Calculation of Permanent Shear Stability Index
D6045	...	...	Test Method for Color of Petroleum Products by the Automatic Tristimulus Method
D6046	...	...	Classification Of Hydraulic Fluids For Environmental Impact
D6074	...	...	Standard Guide for Characterizing Hydrocarbon Lubricant Base Oils
D6080	...	...	Practice for Defining the Viscosity Characteristics of Hydraulic Fluids

TABLE 12.18— *(Continued.)*

ASTM Standard	ISO Standard	Other Standards	Test Description
D6081	...	...	Practice for Aquatic Toxicity Testing of Lubricants: Sample Preparation and ResultsInterpretation
D6082	...	...	Test Method for High Temperature Foaming Characteristics of Lubricating Oils
D6121	...	...	Test Method for Evaluation of the Load Carrying Capacity of Lubricants Under Conditionsof Low Speed and High Torque Used for Final Hypoid Drive Axles
D6138	...	DIN 51 802	Test Method for Corrosion-Preventive Properties of Lubricating Greases Under ofDynamic Wet Conditions (EMCOR Test)
D6158	...	...	Specification for Mineral Hydraulic Oils
D6185	...	...	Practice for Evaluating Compatibility of Binary Mixtures of Lubricating Greases
D6186	...	...	Test Method for Oxidation Induction Time of Lubricating Oils by Pressure DifferentialScanning Calorimetry (PDSC)
D6184	...	...	Test Method for Oil Separation from Lubricating Grease (Conical Sieve Method)
D6202	...	...	Test Method for Automotive Engine Oils on the Fuel Economy of Passenger Cars andLight Duty Trucks in the Sequence VIA Spark Ignition Engine
D6224	...	...	Standard Practice for In-Service Monitoring of Lubricating Oil for Auxiliary Power PlantEquipment
D6278	20844	IP 294 DIN 51 382	Test Method for Shear Stability of Polymer Containing Fluids Using a European DieselInjector Apparatus
D6299	...	...	Applying Statistical Quality Assurance Techniques to Evaluate Analytical MeasurementSystem Performance
D6300	...	...	Determination of Precision and Bias Data for Use in Test Methods for Petroleum Productsand Lubricants
D6304	...	...	Test Method for Determination of Water in Petroleum Products, Lubricating Oils, andAdditives by Coulometric Karl Fischer Titration
D6335	...	...	Test Method for Determination of High Temperature Deposits by Thermo-Oxidation EngineOil Simulation Test
D6351	...	...	Test Method for Determination of Low Temperature Fluidity and Appearance of HydraulicFluids
D6352	...	...	Test Method for Boiling Range Distribution of Petroleum Distillates in Boiling Range from174 to 700° C by Gas Chromatography
D6375	...	...	Test Method for Evaporation Loss of Lubricating Oils by Thermo-gravimetric Analyzer(TGA) NOACK Method
D6384	...	...	Terminology Relating to Biodegradability and Eco-toxicity Of Lubricants
D6417	...	...	Test Method for Estimation of Engine Oil Volatility by Capillary Gas Chromatography
D6425	...	...	Test Method for Measuring Friction and Wear Properties of Extreme-Pressure LubricatingOils Using SRV Test Machine
D6443	...	IP 118/IP 244 DIN 51 431-2	Test Method for Determination of Calcium, Chlorine, Copper, Magnesium, Phosphorus,Sulfur, and Zinc in Unused Oils and Additives by Wavelength Dispersive X-rayFluorescence Spectrometery (Mathematical Correction Method)
D6448	...	...	Specification for Industrial Burner Fuels From Used Lubricating Oils
D6450	...	...	Standard Test Method for Flash Point by Continuously Closed Cup (CCCFP) Tester
D6481	...	...	Test Method for Determination of Phosphorus, Sulfur, Calcium, and Zinc in Lubrication Oilsby Energy Dispersive X-ray Fluorescence Spectrometery
D6483	...	...	The Method for Evaluation of Diesel Engine Oils in T-9 Engine
D6514	...	...	Test Method for High Temperature Universal Oxidation Test for Turbine Oils
D6546	...	...	Test Method for and Suggested Limits for Determining Compatibility of Elastomer Seals forIndustrial Hydraulic Applications
D6547	...	...	Test Method for Corrosiveness of a Lubricating Fluid to a Bi-metallic Couple
D6557	...	...	Test Method for Evaluation of Rust Preventive Characteristics of Automotive Engine Oils(Ball Rust Test)
D6560	...	IP 143 DIN 51 595	Test Method For Determination of Asphaltenes (Heptane Insolubles) in Crude Petroleumand Petroleum Products
D6593	...	...	Test Method for Evaluation of Automotive Engine Oils for Inhibition of Deposit Formation ina Spark-Ignition Internal Combustion Engine Fueled with Gasoline and Operated UnderLow-Temperature, Light Duty Conditions (Sequence VG)
D6594	...	...	Test Method for Evaluation of Corrosiveness of Diesel Engine Oil at 135 °C
D6595	...	...	Test Method for Determination of Wear Metals and Contaminants in Used Lubricating Oilsor Used Hydraulic Fluids by Rotating Disk Electrode Atomic Emission Spectrometry
D6616	...	...	Test Method for Measuring Viscosity at High Shear Rate by Tapered Bearing SimulatorViscometer at 100 °C
D6618	...	...	Test Method for Evaluation of Engine Oils in Diesel Four-Stroke-Cycle Supercharged 1M-PC Single Cylinder Oil Test Engine

TABLE 12.18— (*Continued.*)

ASTM Standard	ISO Standard	Other Standards	Test Description
D6681	...	...	Test Method for Evaluation of Engine Oils in a High Speed, Single-Cylinder Diesel Engine—Caterpillar 1P Test Procedure
D6709	...	...	Test Method for Evaluation of Automotive Engine Oils in the Sequence VIII Spark-IgnitionEngine (CLR Oil Test Engine)
D6750	...	...	Test Methods for Evaluation of Engine Oils in a High-Speed, Single-Cylinder DieselEngine—1K Procedure (0.4 % Fuel Sulfur) and 1N Procedure (0.04 % Fuel Sulfur)
D6837	...	...	Test Method for Measurement of Effects of Automotive Engine Oils on Fuel Economy ofPassenger Cars and Light-duty Trucks in Sequence VIB Spark Ignition Engine
D6743	...	DIN 51 528	Test Method for Thermal Stability of Organic Heat Transfer Fluids
D6795	...	...	Test Method for Measuring the Effect on Filterability of Engine Oils After Treatment withWater and Dry Ice and a Short 30 min Heating Time
D7042	...	...	Standard Test Method for Dynamic Viscosity and Density of Liquids by StabingerViscometer (and the Calculation of Kinematic Viscosity)
E29	...	...	Standard Practice for Using Significant Digits in Test Data to Determine Conformance withSpecifications
E135	...	...	Terminology Relating to Analytical Chemistry for Metals, Ores, and Related
E168	...	...	Standard Practices for General Techniques of Infrared Quantitative Analysis
E178	...	...	Practice for Dealing with Outlying Observations
E203	DIN 51 777	IP 356	Test Method for Water Using Volumetric Karl Fischer Titration
E234	...	IP 129/IP 230	Test Method for Bromine Index of Petroleum Hydrocarbons by Electrometric Titration
E326	...	...	Test Method for Hydroxyl Groups by Phthalic Anhydride Esterification
E502	...	...	Selection and Use of ASTM Standards for the Determination of Flash Point of Chemicalsby the Closed Cup Method
E659	...	DIN 51 794	Test Method for Auto-ignition Temperature of Liquid Chemicals
E691	...	...	Conducting an Inter-Laboratory Study to Determine the Precision of a Test Method
E729	...	...	Guide for Conducting Acute Toxicity Tests on Test Materials with Fishes, Macro-invertebrates, and Amphibians
E1022	...	...	Guide for Conducting Bio-concentration Tests with Fishes and Saltwater Bivalve Mollusks
E1023	...	...	Guide for Assessing the Hazard of a Material to Aquatic Organisms and Their Uses
E1064	...	IP 386	Test Method for Water in Organic Liquids by Coulometric Karl Fischer Titration
E1147	...	...	Test Method for Partition Coefficient (N-Octanol/Water) Estimation by LiquidChromatography
E1148	...	...	Test Methods for Measurement of Aqueous Solubility
E1195	...	...	Test Method for Determining a Sorption Constant (K_{oc}) for an Organic Chemical in Soil andSediments
E1242	...	...	Practice for Using Octanol-Water Partition Coefficient to Estimate Median LethalConcentrations for Fish Due to Narcosis
E1269	...	...	Standard Test Method for Determining Specific Heat Capacity by Differential ScanningCalorimetry
E1668	...	...	Guide for Determination of the Bio-accumulation of Sediment-associated Contaminants byBenthic Invertebrates
E1676	...	...	Guide for Conducting Laboratory Soil Toxicity or Bi-accumulation Tests with the LumbricidEarthworm Eisenia (elida)
E1687	...	...	Test Method for Determining Carcinogenic Potential of Virgin Base Oils in MetalworkingFluids
E1719	...	...	Vapor Pressure of Liquids by Ebulliometry
E1847	...	...	Practice for Statistical Analysis of Toxicity Tests Conducted under ASTM Guidelines
F313	4407/ 4408	FED-STD-791-3009	Particle Contamination-Microscopic
F661	4402	...	Particle Contamination-Optical
F2161	...	...	Standard Guide for Instrument and Precision Bearing Lubricants-Part 1 Oils
G40	...	...	Terminology Relating to Wear and Erosion
G125	...	...	Test Method for Measuring Liquid and Solid Material Fire Limits in Gaseous Oxidants
G 133/95	...	...	Test Method for Linearly Reciprocating Ball-on-Flat Sliding Wear
STP 315H	...	...	Multicylinder Test Sequence for Evaluating Automotive Engine Oils
STP 509A	...	...	Single Cylinder Engine Test for Evaluating the Performance of Crankcase Lubricants

TABLE 12.19—DIN standards.

Standard	Description
DIN 51 502	Lubricants and related materials; designation of lubricants and marking of lubricant containers, lubricationequipment and lubrication points
DIN 51 522	Heat transfer fluids–Requirements and testing
DIN EN 24260	Determination of sulfur content in petroleum products and hydrocarbons by the Wickbold combustionmethod (ISO 4260:1987)
DIN EN ISO 9227	Corrosion tests in artificial atmospheres–Salt spray tests (ISO 9227:2006)
DIN 51 524-1 Berichtigung 1	Pressure fluids-Hydraulic oils–Part 1: HL hydraulic oils; Minimum requirements, Corrigendum 1 to DIN 51524-1:2006-04
DIN 51 524-2 Berichtigung 1	Pressure fluids-Hydraulic oils–Part 2: HLP hydraulic oils; Minimum requirements, Corrigendum 1 to DIN 51524-2:2006-04
DIN 51 524-3 Berichtigung 1	Pressure fluids-Hydraulic oils–Part 3: HVLP hydraulic oils; Minimum requirements, Corrigendum 1 to DIN 51524-3:2006-04
DIN 51 345	Testing the compatibility of fire-resistant hydraulic fluids with metals
DIN 51 351	Testing of Lubricants–Determination of Flocculation Point of Refrigerator Oils by Pressure Tube Method
DIN 51 352-1	Testing of lubricants; determination of aging characteristics of lubricating oils; increase in Conradson carbon residue after aging by passing air through the lubricating oil
DIN 51352-2	Testing of lubricants; determination of aging characteristics of lubricating oils; Conradson carbon residue after aging by passing air through the lubricating oil in the presence of iron(III)oxide
DIN 51 355	Testing of Lubricants; Testing of the Characteristics of Gear Oils Preventing Corrosion on Steel in the Presence of Water; Stirring Methods A and B
DIN 51 358	Testing of corrosion preventing properties of lubricating oils for internal combustion engines by the seawater immersion test
DIN 51 360-1	Testing of cooling lubricants; determination of corrosion preventing characteristics of cooling lubricantsmixed with water; Herbert corrosion test
DIN 51 360-2	Testing of cooling lubricants; determination of corrosion preventing characteristics of cooling lubricantsmixed with water; chip/filter paper method
DIN 51 368	Determination of fraction separated by hydrochloric acid from water mix metal working fluids
DIN 51 369	Testing of cooling lubricants; determination of the pH value of water-mixed cooling lubricants
DIN 51 373	Testing of Fire Resistant Heat Transfer Fluids; Determination of Resistance to Oxidation Including anAssessment of the Catalyst Plates
DIN 51 378	Determination of carbon-type composition of mineral oils
DIN 51 379-2	Determination of molybdenum content of lubricating oils by wavelength-dispersive X-ray spectrometry
DIN 51 380	Determination of readily volatile components in used automotive engine oils by gas chromatography
DIN 51 381	Testing of Lubricating Oils, Governor Oils and Hydraulic Fluids; Determination of Air Release Properties
DIN 51 389 Part 2	Testing of lubricants; mechanical testing of hydraulic fluids by the vanepump method: method A for anhy-drous hydraulic fluids
DIN 51 390-2	Determination of the silicon content of petroleum products by wavelength-dispersive X-ray fluorescencespectrometry
DIN 51 391-2	Determination of additive elements content of lubricants by wavelength-dispersive X-ray fluorescencespectrometry (XRS)
DIN 51 394	Testing of lubricants; testing of low-viscosity lubricating oils for oxidation and corrosion inhibiting properties
DIN 51 396-1	Testing of lubricants-Determination of wear elements—Part 1: Direct determination by inductively coupledplasma optical emission spectroscopy (ICP-OES)
DIN 51 396-2	Determination of wear elements in lubricants by wavelength dispersive X-ray spectrometry
DIN 51 398	Testing of Lubricants; Procedure For Measurement of Low Temperature Apparent Viscosity by Means of theBrookfield Viscometer (Liquid Bath Method)
DIN 51 411	Testing of Liquid Mineral Oil Hydrocarbons; Determination of Saybolt Color
DIN 51 413-2	Analysis of liquid petroleum products by gas chromatography; determination of benzene content
DIN 51 417-2	Determination of mineral oil content of water-miscible metalworking lubricants by column chromatography
DIN 51 425	Testing of mineral oil hydrocarbons and similar materials-Analysis by high pressure liquid chromatography–General working principles
DIN 51 428	Cold Filter Plugging Point of Distillate Fuels
DIN 51 432	Determination of water and solvent in used oils and of their distillation residue by the distillation method
DIN 51 433	Determination of the mineral oil content of used oil by infrared spectrometry
DIN 51 443	Direct determination of the boron content of lubricating oils by atomic absorption spectrometry (AAS)
DIN 51 443-2	Direct determination of boron content of lubricants by inductively coupled plasma/optical emissionspectroscopy (ICP/OES)
DIN 51 444	Determination of bound nitrogen in petroleum products by combustion and chemiluminescence
DIN 51 448-1	Determination of certain liquid petroleum hydrocarbons by gas chromatography using a column switchingtechnique
DIN 51 502	Lubricants and Related Materials; Designation of Lubricants and Marking of Lubricant Containers,Lubrication Equipment and Lubrication Points
DIN 51 503-1	Lubricants-Refrigerator oils–Minimum requirements
DIN 51 503-2	Testing used lubricants for refrigeration compressors
DIN 51 506	Lubricants; VB and VC lubricating oils with and without additives and VDL lubricating oils; classification andrequirements

TABLE 12.19— *(Continued.)*

Standard	Description
DIN 51 515-1	Normal-duty L-TD type lubricants and governor fluids for use turbines
DIN 51 517-1	Lubricants-Lubricating oils–Type C lubricating oils-Part 1: Requirements and testing
DIN 51 517-2	Type CL lubricating oils–Part 2: Requirements and testing
DIN 51 517-2 Berichtigung 1	Lubricants-Lubricating oils–Part 2: Lubricating oils CL; Specifications, Corrigenda to DIN 51517-2:2004-01
DIN 51 517-3	Type CLP lubricating oils–Part 3: Requirements and testing
DIN 51 517-3 Berichtigung 1	Lubricants-Lubricating oils–Part 3: Lubricating oils CLP; Specifications, Corrigenda to DIN 51517-3:2004-01
DIN 51 519	ISO viscosity classification for industrial liquid lubricants
DIN 51 520	Non-aqueous metalworking fluids-Minimum requirements and testing
DIN 51 521	Type SE water-miscible metalworking fluids-Requirements and testing
DIN 51 524 Part 1	Pressure fluids; hydraulic oils; HL hydraulic oils: minimum requirements
DIN 51 535	Determination the tendency of petroleum products to form deposits in diesel engine turbochargers andintercoolers
DIN 51 538	Testing of lubricants for refrigeration compressors for resistance to ammonia
DIN 51 554-1	Testing of mineral oils; Test of susceptibility to aging according to Baader; Purpose, sampling, aging
DIN 51 554-2	Testing of mineral oils; test of susceptibility to aging according to Baader; testing at 110 Ã,Ã °C
DIN 51 554-3	Testing of mineral oils; Test of susceptibility to aging according to Baader; Testing at 95 Ã,Ã °C
DIN 51 561	Testing of mineral oils, liquid fuels and related liquids; measurement of viscosity using the Vogel-Ossag viscometer: Temperature range: approximately 10 to 150°C
DIN 51 562Part 1	Viscometry; determination of kinematic viscosity using the standard design Ubbelohde viscometer
DIN 51 563	Testing of Mineral Oils and Related Materials; Determination of Viscosity Temperature Relation; Slope m
DIN 51 566	(Draft) Testing of lubricating oils: determination of foaming characteristics
DIN 51 569	Testing of mineral oils, liquid fuels and related liquids; measurement of viscosity using the Vogel-Ossag viscometer:temperature range: −55 to approximately +10 °C
DIN 51 576	Determining the salt content of petoleum products
DIN 51 577-4	Determination of chlorine and bromine contents of petroleum products by energy-dispersive X-rayspectrometry using a miniature spectrometer
DIN 51 585	Testing of lubricants; testing of corrosive effect of steam turbine oils and hydraulic oils containing additives
DIN 51 586	Testing of Lubricants Determination of the Ageing Properties
DIN 51 587	Testing of lubricants: determination of aging behavior of steam turbine oils and hydraulic oils containingadditives
DIN 51 589-1	Determination of water separation ability of lubricating oils and low-flammability fluids after contact withsteam
DIN 51 592	(Draft) Testing of lubricants; determination of the content of undissolved matter in lubricating oils;membrane filter method
DIN 51 599	Testing of lubricating oils; determination of demulsification capacity by the stirring method
DIN 51 751	Distillation of Petroleum Products at Atmospheric Pressure
DIN 51 755	Testing of Mineral Oils and Other Combustible Liquids; Determination of Flash Point by the Closed TesterAccording to Abel-Pensky
DIN 51 775	Aniline Point and Mixed Aniline Point of Petroleum Products and Hydrocarbon Solvents
DIN 51 803	Base Number of Petroleum Products by Potentiometric Perchloric Acid Titration, Total Base Number, TBN
DIN 51 804	Cone Penetration of Lubricating Grease Using One-Quarter and One-Half Scale Cone Equipment
DIN 51 806-2	Oscillating test bench. Standard test method for fretting wear protection by lubricating greases
DIN 51 807-1	Testing of lubricants; Test of the behavior of lubricating greases in the presence of water; Static test
DIN 51 809	Acid Number of Petroleum Products by Potentiometric Titration, TAN
DIN 51 810	Determination of apparent viscosity of lubricating greases using the cone and plate viscometer
DIN 51 811	Testing of Lubricants; Testing of Corrosiveness to Copper of Greases; Copper Strip Tarnish Test
DIN 51 813	Determination of solid matter content of lubricating greases (particle sizes above 25 μm)
DIN 51 819-1	Testing rolling bearing lubricants using the FE 8 wear test machine–Principles
DIN 51 819-2	Testing rolling bearing lubricants using the FE 8 wear test machine–Test procedure
DIN 51821-1	Testing of lubricants; test using the FAG roller bearing grease testing apparatus FE9; general workingprinciples
DIN 51 825	Lubricants–Lubricating greases K-Classification and requirements
DIN 51 828-1	Determining the biodegradability of lubricants and related products–Part 1: General
DIN 51 848 Part 1	Testing of petroleum products; precision of test methods; general introduction; concepts and their applicationto petroleum standards specifying requirements
DIN 53 240-2	Catalyzed acylation and titrimetric method of determining hydroxyl value
DIN 53 241-1	Determination of iodine value by methods using Wijs solution
DIN 53 505	Testing of elastomers; Shore A and D hardness testing
DIN 53 521	Testing of rubber and elastomers: determination of their resistance to liquids, vapors and gases
DIN 53 538 Part 1	Standard reference elastomers; peroxide-cross-linked acrylonitrile/butadiene rubber (NBR) for characterizingservice fluids with respect to their action on NBR
DIN EN ISO 6743-4	Lubricants, industrial oils and related products (class L)–Classification–Part 4: Family H (Hydraulicsystems) (ISO 6743-4:1999)

TABLE 12.19— (*Continued.*)

Standard	Description
DIN EN 13016-1	Liquid petroleum products–Vapor pressure–Part 1: Determination of air saturated vapor pressure(ASVP); English version of DIN EN 13016-1
DIN EN 13131	Liquid petroleum products–Determination of nickel and vanadium content–Atomic absorption spectrometricmethod; English version of DIN EN 13131
DIN EN 14077	Petroleum products–Determination of organic halogen content–Oxidative microcoulometric method
DIN EN 14832	Petroleum and related products–Determination of the oxidation stability and corrosivity of fire-resistant phosphate esterfluids
DIN EN 14833	Petroleum and related products–Determination of the hydrolytic stability of fire-resistant phosphate esterfluids
DIN EN 14865-2	Railway applications–Axlebox lubricating greases–Part 2: Method to test the mechanical stability to covervehicle speeds up to 200 km/h
DIN EN ISO 12185	Crude petroleum and petroleum products–Determination of density using the oscillating U-tube method(ISO 12185:1996)
DIN EN ISO 12922	Lubricants, industrial oils and related products (class L)–Family H (Hydraulic systems)–Specifications forcategories HFAE, HFAS, HFB, HFC, HFDR and HFDU (ISO 12922:1999, including Technical Corrigendum1:2001)
DIN EN ISO 13736	Petroleum products and other liquids–Determination of flash point–Abel closed cup method (ISO13736:1997); English version of DIN EN ISO 13736
DIN EN ISO 15029-1	Petroleum and related products–Determination of spray ignition characteristics of fire-resistant fluids–Part1: Spray flame persistance; Hollow-cone nozzle method (ISO 15029-1:1999)
DIN EN ISO 20623	Petroleum and related products-Determination of the extreme-pressure and anti-wear properties of fluids–Four ball method (European conditions) (ISO 20623:2003)
DIN EN ISO 20763	Petroleum and related products–Determination of anti-wear properties of hydraulic fluids–Vane pumpmethod (ISO 20763:2004)
DIN EN ISO 20783-1	Petroleum and related products-Determination of emulsion stability of fire-resistant fluids–Part 1: Fluids incategory HFAE (ISO 20783-1:2003)
DIN EN ISO 20783-2	Petroleum and related products-Determination of emulsion stability of fire-resistant fluids–Part 2: Fluids incategory HFB (ISO 20783-2:2003)
DIN EN ISO 20823	Petroleum and related products–Determination of the flammability characteristics of fluids in contact withhot surfaces–Manifold ignition test (ISO 20823:2003)
DIN EN ISO 20843	Petroleum and related products-Determination of pH of fire-resistant fluids within categories HFAE, HFASand HFC (ISO 20843:2003)
DIN EN ISO 3675	Crude petroleum and liquid petroleum products–Laboratory determination of density–Hydrometer method(ISO 3675:1998)
DIN EN ISO 3679	Determination of flash point–Rapid equilibrium closed cup method (ISO 3679:2004)
DIN EN ISO 4263-1	Petroleum and related products–Determination of the aging behavior of inhibited oils and fluids–TOST test–Part 1: Procedure for mineral oils (ISO 4263-1:2003)
DIN EN ISO 4263-2	Determination of the aging behavior of inhibited oils and fluids-TOST test-Part 2: Procedure forcategory HFC hydraulic fluids (ISO 4263-2:2003)
DIN EN ISO 4263-3	Petroleum and related products-Determination of the aging behavior of inhibited oils and fluids-TOSTtest-Part 3: Anhydrous procedure for synthetic hydraulic fluids (ISO 4263-3:2006)
DIN EN ISO 4263-4	Petroleum and related products–Determination of the aging behavior of inhibited oils and fluids–TOST test–Part 4: Procedure for industrial gear oils (ISO 4263-4:2006)
DIN EN ISO 4267-2	Calculation of oil quantities for petroleum and liquid petroleum products–Part 2: Dynamic measurement(ISO 4267-2:1988)
DIN EN ISO 4404-1	Petroleum and related products–Determination of the corrosion resistance of fire-resistant hydraulic fluids–Part 1: Water-containing fluids (ISO 4404-1:2001)
DIN EN ISO 4404-2	Petroleum and related products–Determination of the corrosion resistance of fire-resistant hydraulic fluids–Part 2: Nonaqueous fluids (ISO 4404-2:2003)
DIN EN ISO 4630-1	Clear liquids-Estimation of color by the Gardner color scale–Part 1: Visual method (ISO 4630-1:2004)
DIN EN ISO 4630-2	Clear liquids-Estimation of color by the Gardner color scale–Part 2: Spectrophotometric method (ISO4630-2:2004)
DIN EN ISO 6271-1	Clear liquids–Estimation of color by the platinum-cobalt scale–Part 1: Visual method (ISO 6271-1:2004)
DIN EN ISO 6271-2	Clear liquids–Estimation of color by the platinum-cobalt scale–Part 2: Spectrophotometric method (ISO6271-2:2004)
DIN EN ISO 9038	Test for sustained combustibility of liquids (ISO 9038:2002)
DIN ISO 15380	Lubricants, industrial oils and related products (class L)–Family H (Hydraulic systems)–Specifications forcategories HETG, HEPG, HEES and HEPR (ISO 15380:2002)
DIN ISO 15597	Petroleum and related products–Determination of chlorine and bromine content–Wavelength-dispersive X-ray fluorescence spectrometry (ISO 15597:2001)
DIN-Fachbericht 143	Modern rheological test methods–Part 1: Determination of the yield point–Fundamentals and comparativetesting methods

TABLE 12.20—AFNOR standards and specifications used in lubricant analysis and testing [802].

AFNOR Standard	Test Description
NF E 48-614	Air Release Properties
NF E 48-690	Filterability of Hydraulic Fluids
NF E 48-691	Filterability of Hydraulic Fluids
NF EN 23015 (T60181)	Cloud Point Determination
NF EN ISO 3104 (T60100)	Kinematic Viscosity and Calculation of Dynamic Viscosity
NF EN ISO 3675 (T60101)	Density by the Use of Hydrometer
NF EN ISO 12185 (T60172)	Density by Oscillating U Tube Method
NF ISO 3733 (T60113)	Water Content by Distillation Method
NF ISO 6615 (T60116)	Carbon Residue by Conradson Method
NF ISO 6618 (T60112)	Acid and Base Number by Color Indicator
NF ISO 6296 (T60154)	Water Content by Karl Fischer Potentiometric Titration
NF ISO 8681/NF T60-500 (T60500)	Classification of Petroleum Products and Lubricants–Definition of Classes
NF ISO 14935 (T60177)	Determination of Wick Flame Persistence of Fire-resistant Fluids
NF EN 24260 (T60142)	Sulfur Content by Wickbold Combustion Method
NF M 07-003	Saybolt Color
NF M 07-015	Copper Corrosion
NF M 07-020	Centrifuge for Water and Sediment in Oils
NF M 07-024	Hydrocarbon Types by FIA
NF M 07-047	Oxidation Stability
NF T60-100	Kinematic Viscosity Determination and The Calculation of Dynamic Viscosity
NF T60-101	Laboratory Determination of Density
NF T60-102	Drop Point of Grease
NF T60-103	Closed Cup Flash Point of Lubricants and Fuel Oils
NF T60-104	Determination of Color (ASTM Scale)
NF T60-105	Pour Point Determination
NF T60-106	Determination of Additive Elements, Wear Metals and Contaminants in Unused and Used Lubricating Oils-Method by Inductively Coupled Plasma Atomic Emission Spectrometry (ICP/AES)
NF T60-107	Glass Capillary Kinematic Viscometers. Specifications and Operating Instructions
NF T60-110-1	Determination of Saponification Number–Part 1: Color-Indicator Titration Method
NF T60-110-2	Determination of Saponification Number-Part 2: Potentiometric Titration Method
NF T60-111	Aging Characteristics-Determination of Change in Conradson Carbon Residue after Oxidation
NF T60-112	Determination of Acid or Base Number
NF T60-114	Melting Point of Paraffins
NF T60-115	Determination of Asphaltene Content (Heptane Insoluble)
NF T60-116	Determination of Carbon Residue-Conradson Method
NF T60-117	Determination of The Ramsbottom Residue
NF T60-118	Determination of Flash and Fire Points-Cleveland Open Cup Method
NF T60-119	Determination of Cone Penetrability of Paraffinic Products
NF T60-123	Automated Penetrometer
NF T60-125	Determination of Water Separability of Petroleum Oils and Synthetic Fluids
NF T60-126	Mineral Oils-Determination of Interfacial Tension of Oil Against Water
NF T60-129	Determination of Foaming Characteristics of Lubricating Oils
NF T60-131	Electrical Insulating Oils. Detection of Corrosive Sulfur
NF T60-132	Cone Penetration and Water Resistance Properties of Lubricating Greases
NF T60-133	Lubricating Greases-Determination of Free Acidity and Alkalinity
NF T60-135	Determination Of Rust-Prevention Characteristics of Lubricating Greases
NF T60-136	Calculation of Viscosity Index From Kinematic Viscosity
NF T60-138	Determination of Flammability Characteristics of Fluids in Contact With Hot Surfaces-Manifold IgnitionTest
NF T60-140	Cone Penetration of Lubrication Grease Using One-Quarter or One-Half Scale Cone Equipment
NF T60-141	Industrial Liquid Lubricants. ISO Viscosity Classification
NF T60-142	Determination of Sulfur Content. Wickbold Combustion Method
NF T60-143	Lubricating Oils and Additives. Determination of Sulfated Ash
NF T60-144	Lubricating Greases-Determination of the Sulfated Ash
NF T60-148	Determination of the Kinematic Viscosity According to the Temperature
NF T60-149	Determination of Air-Release Properties of Steam Turbine and Other Oils-Impinger Method
9NF T60-150	Determination of the Aging Behavior of Inhibited Oils and Fluids-TOST Test-Part 1: Procedure forMineral Oils
NF T60-150-2	Determination of the Aging Behavior of Inhibited Oils and Fluids-TOST Test-Part 2: Method for HFCType Hydraulic Fluids
NF T60-151	Assessment of Rust-Preventing Characteristics of Turbine Oil in the Presence of Water
NF T60-152	Measurement of the Viscosity at Low Temperatures Using a Brookfield Viscosity Meter
NF T60-154	Determination of Water-Potentiometric Karl Fischer Titration Method
NF T60-156	De-Emulsification Number of New Non Inhibited Turbine Oils. Water Vapor Test
NF T60-159	Temporary Anticorrosion Products. Degreasing Ability by Dipping

TABLE 12.20— *(Continued.)*

AFNOR Standard	Test Description
NF T60-160-2	Corrosion Resistance of Non-aqueous Fire-Resistant Hydraulic Fluids
NF T60-162	Lubricants, Industrial Oils and Related Products (Class L) Classification. Part 0: General
NF T60-163	Lubricants, Industrial Oils and Related Products (Class L) Classification Part 1: Family A (Total Loss System)
NF T60-164	Lubricants, Industrial Oils and Related Products (Class L) Classification. Part 2: Family F (Spindle Bearings and Associated Clutches)
NF T60-167	Temporary Anti-Corrosion Products. Storage Stability (Cyclic Conditions)
NF T60-169	Oil and Related Products. HFAE, HFAS Category and HFC-Burning Hydraulic Oil pH Determination
NF T60-170	Temporary Anticorrosion Products. Assessment of the Covering Capacity by Dipping.
NF T60-171	Determination of Low-temperature Cone Penetration of Lubricating Greases.
NF T60-172	Crude Oil and Petroleum Products. Density Measurement. Oscillation U-Shaped Tube.
NF T60-173	Temporary Protection Against Corrosion. Evaluation of Rust Protection Properties. Humidity CabinetMethod
NF T60-174	Temporary Protection Against Corrosion. Evaluation of Corrosion Protection Properties. Cyclic HumidityCabinet Method
NF T60-177	Determination of Wick Flame Persistence of Fire-Resistant Fluids
NF T60-178	Petroleum Products. Quenching Oil. Silver Sensor Through the Static Test
NF T60-180	Crude Petroleum and Liquid or Solid Petroleum Products-Determination of Density or Relative Density-Capillary-Stoppered Pyknometer and Graduated Bicapillary Pyknometer Methods
NF T60-181	Determination of Cloud Point
NF T60-182	Test Method for Oxidation Stability of Gasoline Automotive Engine Oils by Thin-Film Oxygen Uptake(TFOUT)
NF T60-184-1	Determination of PCBs and Related Products-Part 1: Separation and Determination of Selected PCBCongeners by Gas Chromatography (GC) Using an Electron Capture Detector (ECD)
NF T60-184-2	Determination of PCBs and Related Products-Part 2: Calculation of Polychlorinated Biphenyl (PCB)Content
NF T60-185	Aqueous Machining Fluids. Foaming Tendency
NF T60-186	Aqueous Metal Working Fluids. Evaluation of Rust Prevention Properties on Contact with Ferrous Metals
NF T60-187	Aqueous Machining Fluids. Emulsifying And Solubilizing Suitability and Stability at Rest
NF T60-188	Aqueous Machining Fluids. Preparation of Synthetic Water for Testing. Preparation of Synthetic Water forTesting
NF T60-189	Lubricating Greases. Leakage Tendencies of Automotive Wheel Bearing Greases
NF T60-191	Petroleum Products and Lubricating Greases. Oil Separation on Storage of Grease. Static Conditionsunder Pressure
NF T60-192-1	Determination of Emulsion Stability of Fire-Resistant Fluids-Part 1: Fluids in Category HFAE
NF T60-192-2	Determination of Emulsion Stability of Fire-Resistant Fluids-Part 2: Fluids in Category HFB
NF T60-193	Aqueous Machining Fluids. Determination of the pH Value
NF T60-195	Aqueous Machining Fluids. Opacity under Diluted Form
NF T60-197	Aqueous and Nonaqueous Metal Working Fluids. Short and Long Term Storage Stability
NF T60-198	Lubricants. Evaluation of Primary Biodegradability. Method by Infrared Spectroscopy
NF T60-199	Lubricating Greases. Ability of Lubricating Greases to Resist to False Brinelling
NF T60-500	Petroleum Products and Lubricants. Method of Classification. Definition of Classes
NF T60-501	Lubricants, Industrial Oils and Related Products (Class L)-Classification-Part 3: Family D (Compressors)
NF T60-503	Lubricants, Industrial Oils and Related Products (Class L). Classification. Part 3B: Family D (Gas andRefrigeration Compressors)
NF T60-504	Lubricants, Industrial Oils and Related Products (Class L). Classification. Part 7: Family M (Metalworking)
NF T60-505	Lubricants, Industrial Oils and Related Products (Class L). Classification. Part 8: Family R (TemporaryProtection Against Corrosion)
NF T60-506	Lubricants, Industrial Oils and Related Products (Class L). Classification. Part 9: Family X (Greases)
NF T60-507	Lubricants, Industrial Oils and Related Products (Class L). Classification. Part 6: Family C (Gears)
NF T60-508	Lubricants, Industrial Oils and Related Products (Class L). Classification. Part 10: Family Y (Miscellaneous)
NF T60-509	Lubricants, Industrial Oils and Related Products (Class L). Classification. Part 11: Family P (PneumaticTools)
NF T60-510	Lubricants, Industrial Oils and Related Products (Class L). Classification. Part 12: Family Q (Heat TransferFluids)
NF T60-511	Lubricants, Industrial Oils and Related Products (Class L). Classification. Part 13: Family G (Slide ways)
NF T60-512	Lubricants, Industrial Oils and Related Products (Class L). Classification. Part 14: Family U (HeatTreatment)
NF T60-513	Lubricants, Industrial Oils and Related Products (Class L)-Classification-Part 4: Family H (HydraulicSystems)
NF T60-514	Lubricants, Industrial Oils and Related Products (Class L)-Classification-Part 15: E Family (InternalCombustion Engine Oils)
NF T60-521	Lubricants, Industrial Oils and Related Products (Class L)-Family T (Turbines)-Specifications forLubricating Oils for Turbines
NF T60-524	Lubricants, Industrial Oils and Related Products (Class L)-Family E (Internal Combustion Engine Oils)-Specifications for Two-stroke-cycle Gasoline Engine Oils (Categories EGB, EGC And EGD)
NF T60-525	Lubricants, Industrial Oils and Related Products (Class L). Family H (Hydraulic Systems). Specifications forCategories HH, HL, HM, HR, HV And HG

TABLE 12.20— (Continued.)

AFNOR Standard	Test Description
NF T60-526	Lubricants, Industrial Oils and Related Products (Class L)-Family H (Hydraulic Systems)-Specifications forCategories HFAE, HFAS, HFB, HFC, HFDR And HFDU
NF T60-530-1	Lubricants, Industrial Oils and Related Products (Class L)-Family C (Gears)-Part 1: Specifications forLubricants for Enclosed Gear Systems
NF T60-602	Extreme Pressure Lubricants for Industrial Gears. Evaluation of Oxidation Stability by Artificial Aging
NF T60-607	Determination of the UV Absorbance Index of DMSO Extract of Neat Cutting Oils
NF T60-608	Tapered Roller Bearing Lubricating Ability of a Grease Under High Load
NF T60-609	Determination of Acid Number
NF T60-611-1	Determination of Spray Ignition Characteristics of Fire-Resistant Fluids-Part 1: Spray Flame Persistence-Hollow-Cone Nozzle Method
NF T60-613	Aqueous Fluids For Metal Working-Evaluation of the Biological Resistance
NF T60-614	Determination of Flash/No Flash-Closed Cup Equilibrium Method
NF T60-615	Determination of Flash Point-Closed Cup Equilibrium Method
NF T60-616	Determination of Flash Point-Rapid Equilibrium Closed Cup Method
NF T60-617	Determination of Flash/No Flash-Rapid Equilibrium Closed Cup Method
NF T60-619	Determination of the Shear Stability of Polymer-Containing Oils Using a Diesel Injector Nozzle
NF T60-620	Determination of the Extreme Pressure and Antiwear Properties of Fluids-Four Ball Method (European Conditions)
NF T60-621	Determination of Antiwear Properties of Hydraulic Fluids-Vane Pump Method

analysis tools. The objective of these analyses is to determine the lubricant condition by physical inspection and the chemical analysis of the wear debris, contaminants, and reaction products from lubricants. Wear in mechanical equipment is inevitable and results from the surface contact between various machine parts, so is the lubricant breakdown due to oxidation and thermal degradation. Besides wear, the other types of metal damage also need to be assessed. This includes deposits and corrosion. Unfortunately, this type of damage is not as simple to assess as the lubricant sampling and analysis since it will require dismantling the equipment. The extent of the lubricant contamination or break down is determined by analyzing such lubricant parameters as color, viscosity, insolubles, etc.

Oil and wear particle analyses are noninterruptive diagnostic techniques that attempt to estimate the extent of wear in machine part/s and the lubricant condition by its reaction or decomposition products. Based upon these determinations, the equipment condition or the impending failure can be predicted [647]. Methods used for metal analysis include spectrometric methods, such as atomic absorption spectroscopy (AA), atomic emission spectroscopy (AES), inductively coupled plasma emission spectroscopy (ICPE) [804,805], and X-ray fluorescence (XRF) spectrometry [806]. Of these methods, AES and ICPE that rely on the detection of the light emitted by various elements are most popular. This is because of cost, speed, and other factors [647]. It is important to note that the metals in the oil come from both wear and those that are present in various additives. The latter group is represented by calcium, magnesium, barium, zinc, and molybdenum. Ferrography is another technique that is used for analyzing the particles resulting from mechanical wear that are present in the fluid. This technique uses microscopic examination and was developed in the 1970's for predictive maintenance, by analyzing ferrous particles in lubricating oils [807,808]. Ferrography allows potential determination of the amount and the type of wear, as well as the source of wear [647].

Full-scale Testing

As stated earlier, the quality of the finished lubricant is determined by its ability to meet the requirements of the established performance standards. Examples of such standards include the following:

- The SAE Viscosity Classification System
- API, ILSAC, and ACEA Engine Oil Standards for gasoline and diesel engine lubricants
- GM's DEXRON® Standard and Ford's MERCON® Standard for automatic transmission fluids
- API Standard for Gear Lubricants.
- U.S. Military Standards
- OEM Specifications
- End-user Requirements

Each of these standards accompanies a battery of performance tests in which proper lubricant performance must be demonstrated prior to use.

Performance Tests

The testing of the finished lubricant involves evaluating many of its properties and establishing the identity, the composition, and the properties of all of its components, i.e., additives and the base fluid. Testing involves both the preliminary testing and the full-scale testing. Full-scale testing entails testing a lubricant in a laboratory via accelerated tests in the real world equipment and by simulating the actual service conditions. That is, the tests are carried out using actual engines, transmissions, axles, hydraulic pumps, and so on. These tests usually evaluate more than one lubricant property at a time. At the end of each test, the equipment is disassembled and its parts rated based on different criteria. Standardization of the laboratory test conditions ascertains that the lubricants meet the performance requirements established by the various organizations. Standardized methods to test lubricants are published in the United States by ASTM and in Europe by CEC, DIN, IP, and NF. These organizations also ensure close adherence of the industry to these methods. Almost all lubricants are performance tested prior to their introduction into the market-

TABLE 12.21—Other Non-ASTM standards used in lubricant analysis and testing.

Standard	Test Description
SAE[n] Standards	SAE J183: Engine Oil Performance and Engine Service Classification (Other Than "Energy Conserving") SAE J254: Instrumentation and Techniques for Exhaust Gas Emissions Measurement
	SAE J300: Engine Oil Classification
	SAE J304: Engine Oil Tests
	SAE J308: Axle and Manual Transmission Lubricants
	SAE J726: Air Cleaner Test Code (Includes Piezometer Ring Specifications)
	SAE J1423: Passenger Car and Light-Duty Truck Energy-Conserving Engine Oil Classification
	SAE J1995: Engine Power Test Code-Spark Ignition and Compression Ignition-Gross Power Rating
IP 37	Acidity and Alkalinity of Lubricating Grease
IP 111	Calcium Content of Lubricating Oil
IP 225	Copper Content (Spectrophotometric)
IP 227	Corrosion-Silver
IP 306	Oxidation Stability of Straight Mineral Oils
IP 309	Cold Filter Plugging Point
IP 346	Polycyclic Aromatics in Petroleum Fractions-Dimethyl Sulfoxide Extraction Method
IP 436	Aromatics by HPLC
ISO 15595 FED-STD-791–5307/5308[e]	Shear Stability-Oxidation Stability
CETOP[i] RP 112H	Shear Stability-Diesel Injector
CEC[j] L-45-A-99	Shear Stability-Tapered Bearing
ISO 4406	Sizing Particles
ISO 4404 FED-STD-791-4011	Corrosion Prevention
ISO 6072 FED-STD-791-3604/3432	Seal Compatibility
ISO 6743/44	Lubricants, Industrial Oils, and Related Products (Class L); Classification. Family H (Hydraulic Systems)
ISO 6743-12	Lubricants, Industrial Oils and Related Products (Class L)-Classification-Part 12: Family Q (Heat Transfer Fluids)
FED-STD-791-3403	Fluid Compatibility
FED-STD-791-5003	Deposition Test
FED-STD-791-3410	Bearing Test
AFNOR[f] T2.13.7R1-1996	Hydraulic Fluid Power-Petroleum Fluids-Prediction of Bulk Moduli.
ISO 12922	Lubricants, Industrial Oils, and Related Products (Class L)-Family H (Hydraulic Systems)-Specificationsfor Categories HFAE, HFAS, HFB, HFC, HFDR and HFDU
ISO 15380	Lubricants, Industrial Fluids and Related Procedures (Class L), Family H (Hydraulic Systems)-Specifications forCategories HETG, HEPG, HEES and HEPR
ISO 5388	Stationary Air Compressors-Safety Rules and Code of Practice
JAS[g] JIS K 2514	Testing Method of Oxidation Characteristics of Engine Oils
ANSI[k]/ASHRAE[l] 34-1992	Determination of Refrigerant Lower Flammability Limit in Compliance with Proposed Addendum p toStandard 34.
ANSI/ASHRAE 97-1989	Sealed Glass Tube Method to Test the Chemical Stability of Materials for Use within Refrigerant Systems
ANSI/ASHRAE 34-2001	Designation and Safety Classification of Refrigerants
AGMA[m] 9005-D94	Industrial Gear Lubrication
Federal Test Method 3456	Channel Point of Federal Test Method Std. No. 791
FZG-PITS Test C i80 TS	Test Method for Evaluating the Influence of Oil Aging on the Pitting Load Capacity of Lubricants
GFC (Groupement Francais De Coordination) or CEC L-48-A00	Oxidative Stability of Lubricating Oils Used in Automotive Transmissions by Artificial Aging (Laboratory Test) MIL-PRF-2105E or SAE J2360-Lubricating Oil, Gear Multipurpose (Metric) Military Use
API 1509	Engine Service Classification and Guide to Crankcase Oil Selection

[a]BS—British Standards Institution.
[b]IEC—International Electrotechnical Commission.
[c]ISO—International Standardization Organization.
[d]DIN—Deutsches Institut für Normung.
[e]FED-STD—Federal Standard.
[f]AFNOR—Association Française de Normalisation.
[g]JSA—Japanese Standards Association.
[h]IP—Institute of Petroleum.
[i]CETOP—Comité Européen des Transmissions Oléohydrauliques et Pneumatiques.
[j]CEC—Coordinating European Council.
[k]ANSI—American National Standards Institute.
[l]ASHRAE—American Society of Heating, Refrigerating and Air-Conditioning Engineers.
[m]AGMA—American Gear Manufacturers Association.
[n]SAE—Society of Automotive Engineers.

TABLE 12.22—Non-ASTM tests used to evaluate lubricant properties.

ISO-International Standards Organization Test Methods	
ISO 2137: Automated Penetrometer	ISO 4263: Oxidation Stability
ISO 2137: Grease Worker	ISO 6614: Water Separability
ISO 2160: Copper and Silver Corrosion	ISO 6617: Oxidation Stability
ISO 3007: Vapor Pressure	ISO 7120: Rusting Properties of Oils
ISO 3015: Cloud Point	ISO 9120: Air release Properties
ISO 3016: Pour Point	ISO 9262: Low Temperature Brookfield Viscosity
ISO 3104: Kinematic Viscosity	ISO 12205: Oxidation Stability
ISO 3675: Specific Gravity	ISO 13357: Filterability of Hydraulic Fluids
ISO 3734: Centrifuge for Water and Sediment in Oils	ISO 13757: Hydrocarbon Types by FIA
ISO 4256: Copper Strip Corrosion	ISO 17025: Specific Gravity
IP-Institute of Petroleum Standards Test Methods	
IP 15: Cloud Point	IP 227: Silver Corrosion
IP 48: Oxidation Stability of Mineral Oils	IP229: Rotary Bomb Oxidation Test (RBOT)
IP 49: Automated Penetrometer	IP 267: Low Temperature Brookfield Viscosity
IP 50: Automated Penetrometer	IP 280: Oxidation Stability
IP 69: Vapor Pressure	IP 306: Oxidation Stability
IP 71: Kinematic Viscosity	IP 307: Oxidation Stability
IP 75: Centrifuge for Water and Sediment in Oils	IP 309: Cold Filter Plugging Point (CFPP)
IP 135: Rusting Properties of Oils	IP 310: Automated Penetrometer
IP 145: Centrifuge for Water and Sediment in Oils	IP 313: Air Release Properties
IP 146: Foaming Tendency of Oils	IP 319: Kinematic Viscosity
IP 154: Copper and Silver Corrosion	IP 331: Oxidation Stability
IP 156: Hydrocarbon Types by FIA	IP 335: Oxidation Stability
IP 160: Specific Gravity	IP 350: Cold-Crank Simulator
IP 161: Vapor Pressure	IP 359: Centrifuge for Water and Sediment in Oils
IP 179: Automated Penetrometer	IP 376: Automated Penetrometer
IP 219: Pour Point	IP 448: Filterability of Hydraulic Fluids
FTM-Federal Test Methods	
FTM 791-101: Saybolt Colorimeter	FTM 791-3201: Water Separability
FTM 791-201: Cloud and Pour Point	FTM 791-3211: Foaming Tendency of Oils
FTM 791-203: Pour Point Stability	FTM 791-3213: Foaming Tendency of Oils
FTM 791-305: Kinematic Viscosity	FTM 791-3470: Homogeneity and Miscibility of Oils
FTM 791-311: Automated Penetrometer	FTM 791-4011: Rusting Properties of Oils
FTM 791-311: Grease Worker	FTM 791-5307: Corrosiveness and Oxidation Stability
FTM 791-312: Automated Penetrometer	FTM 791-5308: Corrosiveness and Oxidation Stability
FTM 791-313: Grease Worker	FTM 791-5309: Copper Corrosion
FTM 791-313: Automated Penetrometer	FTM 791-5315: Rusting Properties of Oils
FTM 791-322: Oil Separation from Grease	FTM 791-5321: Corrosion of Lead by Lubricating Oils
FTM 791-334: Grease Torque Test	FTM 791-5325: Copper and Silver Corrosion
FTM 791-1303: Flock Point of Refrigeration Oils	FTM 791-5329: Rusting Properties of Oils
Miscellaneous Methods	
AACC 58-14: Automated Penetrometer	John Deere JDQ 71: Pour Point Stability
AN-G-15: Grease Worker	John Deere JDQ 74: Low Temperature Fluidity of Oils
AOCS Cs 16-60: Automated Penetrometer	JAN-H-792: Rust Protection in the Humidity Cabinet
AOCS CD 12-57: Oxidation Stability and Corrosion	NACE TM-01-72: Rusting Properties of Oils
EN 116: Cold Filter Plugging Point (CFPP)	UOT: Oxidation Stability and Corrosion
IHC BT-10: Oxidation and Corrosion	

place. The preliminary step to assess performance is the selection of the oils with demonstrated field performance for use as reference oils. These oils are run in the laboratory using production equipment from various manufacturers, simulating the actual service conditions. A good correlation between the field performance and the laboratory performance standardizes the test to evaluate the new lubricants.

The development of the laboratory tests is costly and time consuming. Hence, future performance requirements must be anticipated and defined as accurately as possible. This can only be achieved by the cooperation of the equipment manufacturers, lubricant and additive marketers, and the end-users. Laboratory tests, especially engine tests, are expensive to house, operate, and maintain. Despite continued efforts to devise tests to evaluate a lubricant's performance quickly and inexpensively [50], no good substitute for full-scale laboratory tests has yet been found, and the use of such tests is expected to continue in the foreseeable future.

Elements of a Quality Test

A quality test contains a number of key elements, which include the quality of the parts, the way in which the equipment is assembled, for example engine build, operating procedures, and the end of the test ratings. Low part-to-part variability is extremely critical to the proper evaluation of the lubricant. If parts vary in quality, it is difficult to assess whether the test results reflect the effectiveness of the lubricant, the part-to-part variance, or both. The way the equip-

TABLE 12.23—Military standards and specifications [803].

Specification	Description
MIL-L-2104	Lubricating Oil, Internal Combustion Engine, Combat/TacticalServices
MIL-L-2105 and 2105D	Lubricating Oil, Gear, Multi-purpose
MIL-PRF-23699F	Synthetic Aircraft Turbine Engine Oil
MIL-PRF-23827C	Aircraft and Instrument Grease
MIL-PRF-81322F	Aircraft Wide Temperature Range Grease
MIL-PRF-81329D	Solid Film Lubricant
MIL-PRF-83282D	Synthetic Fire Resistant Hydraulic Fluid
MIL-PRF-85336B	All Weather Lubricant for Weapons
MIL-L-1970IB	Semi-Fluid Lubricant for Weapons
MIL-G-21164D	Molybdenum Disulfide Grease
MIL-L-23398D	Solid Film Lubricant, Air-Cure
MIL-G-23549C	General Purpose Grease
MIL-G-25013E	Aircraft Bearing Grease
MIL-G-25537C	Aircraft Helicopter Bearing Grease
MIL-H-81019D	Hydraulic Fluid for Ultra Low Temperatures
MIL-S-81087C[a]	Silicone Fluid
MIL-G-81827 A	Aircraft High Loading and Antiwear Grease
MIL-L-81846	Instrument Ball Bearing Lubricating Oil
MIL-G-81937A	Ultra Clean Instrument Grease
DOD-L-85645A[a]	Dry Thin Film Lubricant
DOD-G-85733	High Temperature Catapult Grease
DOD-L-85734	Synthetic Helicopter Transmission Lubricant
VV-D-1078B	Silicone Fluid Damping Fluid
SAE J1899	Aircraft Piston Engine Oil, Ashless Dispersant
SAE J1966	Aircraft Piston Engine Oil, Non Dispersant
SAE AMS-G-4343	Pneumatic Systems Grease
SAE AMS-G-6032	Plug Valve Grease
MIL-H-22072C	Hydraulic Fluid for Catapults
A-A-59290	Arresting Gear Hydraulic Fluid
MIL-PRF-9000H	Diesel Engine Oil
MIL-PRF-17331H	Steam Turbine Lubricating Oil
MIL-PRF-17672D	Hydraulic Fluid
MIL-PRF-24139A	Multi-purpose Grease
DOD-PRF-24574	Lubricating Fluid for Oxidizing Mixtures
MIL-L-15719A	High Temperature Electrical Bearing Grease
MIL-T-17128C	Transducer Fluid
MIL-G-18458B	Exposed Gear and Rope Grease
MIL-H-19457D	Fire Resistant Hydraulic Fluid
MIL-L-24131B	Graphite and Alcohol Lubricant
MIL-L-24478C	Molybdenum Disulfide and Alcohol Lubricant
DOD-G-24508A	Multi-purpose Grease
DOD-G-24650	Food Processing Equipment Grease
DOD-G-24651	Food Processing Equipment Lubricating Oil
VV-L-825C	Lubricating Oil for Refrigerant Compressors
A-A-50433	Sea Water Resistant Grease
A-A-50634	Lubricating Oil for Compressors Using HFC-134A
A-A-59004A	Anti-Galling Compound
MIL-PRF-6081D	Jet Engine Lubricating Oil
MIL-PRF-608 5D	Aircraft Instrument Lubricating Oil
MIL-PRF-6086E	Aircraft Gear Petroleum Lubricating Oil
MIL-PRF-7808L	Aircraft Turbine Synthetic Engine Oil
MIL-PRF-7870C	Low Temperature Lubricating Oil
MIL-PRF-8188D	Corrosion Preventive Engine Oil [FSC 6850]
MIL-PRF-27601C	Hydraulic Fluid
MIL-PRF-27617F	Aircraft and Instrument Grease
MIL-PRF-32014	Aircraft and Missile High Speed Grease
MIL-PRF-83261 B	Aircraft Extreme Pressure Grease
MIL-PRF-83363C	Helicopter Transmission Grease
MIL-PRF-87100A	Aircraft Turbine SyntheticEngine Oil
MIL-PRF-87252C	Dielectric Coolant Fluid [FSC 9160]
MIL-PRF-87257A	Synthetic Fire Resistant Hydraulic Fluid
MIL-H-5606G1	Petroleum Hydraulic Fluid for Aircraft/Ordnance
DOD-L-25681D	Silicone Fluid with Molybdenum Disulfide
MIL-L-87177A	Synthetic Corrosion Preventive Lubricant

TABLE 12.23— *(Continued.)*

Specification	Description
MIL-PRF-2104G	Combat/Tactical Diesel Engine Oil
MIL-PRF-2105E	Multipurpose Gear Oil
MIL-PRF-3150D	Preservative Oil
MIL-PRF-6083F	Operational and Preservative Hydraulic Fluid
MIL-PRF-10924G	Automotive/Artillery Grease
MIL-PRF-12070E	Fog Oil
MIL-PRF-21260E	Preservative and Break-in Engine Oil
MIL-PRF-32033	Preservative and Water-Displacing Oil
MIL-PRF-46002C	Vapor Corrosion Inhibitor (VCI) Preservative Oil
MIL-PRF-46010F	Solid Film Lubricant
MIL-PRF-46147C	Solid Film Lubricant
MIL-PRF-46167C	Arctic Engine Oil
MIL-PRF-46170C	Synthetic Fire Resistant Hydraulic Fluid
MIL-PRF-46176B	Silicon Brake Fluid
MIL-PRF-53074A	Steam Cylinder Lubricating Oil
MIL-PRF-53131A	Precision Bearing Synthetic Lubricating Oil
VV-G-632B	General Purpose Industrial Grease
VV-G-671F	Graphite Grease
A-A-52039B	Automotive Engine Oil API Service SH
A-A-52036A	Commercial Heavy Duty Diesel Engine Oil
A-A-59354	Hydraulic Fluid for Machines
SAE J1703	Conventional Brake Fluid
MIL-PRF-63460D	Cleaner-Lubricant Preservative for Weapons
MIL-L-11734C	Synthetic Lubricant for Mechanical Fuse Systems
MIL-L-14107C	Low Temperature Weapons Lubricant
MIL-L-45983	Heat-Cured Solid Film Lubricant
MIL-L-46000C	Semi-Fluid Weapons Lubricant
MIL-G-46003A	Rifle Grease
MIL-L-46150	Semi-Fluid High Loading Weapons Lubricant
MIL-PRF-3572B	Colloidal Graphite in Oil
MIL-DTL-17111C	Power Transmission Fluid
MIL-PRF-26087C	Reciprocating Compressor Lubricating Oil
MIL-L-3918A[a]	Instrument Lubricating Oil for Jewel Bearings
MIL-L-46014[a]	Spindle Lubricating Oil
MIL-L-83767B[a]	Vacuum Pump Lubricating Oil
VV-C-846B	Emulsifiable Oil Type Cutting Fluids
A-A-50493A	Penetrating Oil
A-A-59113	Machine Tools/Slideways Lubricating Oil
A-A-59137	Breech Block Lubricating Oil (Naval Ordnance)
A-A-59173	Silicone Grease
A-A-59197	Fatty Oil for Metal Working Lubricants
SAE AS1241C	Fire Resistant Phosphate Ester Hydraulic Fluid

[a]Those specifications in bold italics had been designated as "Inactive for New Design" and no longer used.

ment is assembled may also have an impact on the laboratory test results. It is therefore important that the test equipment be built according to the established procedures and with parts of good and consistent quality.

As stated earlier, in the United States ASTM has the primary responsibility to ensure that the defined operating procedures for a particular test are followed. Most modern testing facilities use computer data acquisition and control systems to monitor the operation of a test. Ratings are conducted by qualified raters who are trained by a self-policing industry-supported group. Previously, much emphasis was placed on the operations and the ratings of a test. That emphasis is being replaced by the realization that the quality of parts, the manner in which the equipment is built, operating procedures, and the end of the test ratings all contribute toward a quality test.

Several controlling bodies ensure that the established test procedures are properly carried out and that the ratings are consistent among industry laboratories. For engine oils and gear oils, in the United States, these bodies include the Test Monitoring Center (TMC), ASTM surveillance panels, and SAE Lubricants Review Institute (LRI).

TMC, the key part of the ASTM monitoring system, is a neutral body that deals with the calibration and the monitoring of the test facilities to ensure that laboratory tests are conducted according to the procedures prescribed by ASTM [27,809]. TMC also supplies reference oils, validates stands for reference purposes, and approves laboratories for testing. If in TMC's opinion a laboratory is not operating according to established guidelines, TMC has the authority to invalidate such a laboratory. Since TMC oversees most test laboratories, it also provides statistics and identifies trends.

The surveillance panels are made up of experts, usually from the test laboratories, which monitor a specific test and ensure that the test is run as prescribed and is providing consistent and on-target data.

LRI ensures that the candidate products meet the U.S. Military performance requirements. To obtain approval, the additive or lubricant supplier submits the selected parts from the prescribed tests along with the test parameters and the test results to the LRI. If these are acceptable, a recommendation for approval is granted for the pertinent U.S. Military specification. Final approval is granted by the military upon review of the LRI's recommendations and supporting documentation. Since many of these candidate products also meet the performance requirements described in the SAE standards for lubricants marketed in the private sector, the LRI recommendations also impact commercial oils [810].

For automatic transmission fluids and tractor hydraulic fluids, the appropriate OEMs are responsible for developing the test procedures and monitoring the compliance of the various testing facilities to these procedures. For the industrial gear oils and the industrial hydraulic fluids, OEMs and AGMA are mainly responsible for these activities. For metalworking fluids, there are no established test procedures. Lubricant and additive marketers therefore select procedures that best describe the performance characteristics of their products.

Once the lubricants have met all the requirements established by SAE, API, ASTM, AGMA, ISO, OEMs, the U.S. Military, and the end-users, some lubricant and additive marketers choose to field test their products. This is the final test of a lubricant's performance and involves testing in equipment that is in actual service. Again, equipment parts are examined periodically to assess the lubricant's performance [811]. For automotive lubricants, fleets of commercial vehicles or vehicles which involve extensive use are employed. Such tests can take up to one year or longer. Because of the time and cost associated with such testing, ILSAC, a result of collaboration between AAMA and JAMA, has at present eliminated the fleet testing requirement in its passenger car engine oil standards. A lubricant with demonstrated good performance in the laboratory tests and field tests is ready for the marketplace.

13

Lubricants and the Environment

THIS CHAPTER DEALS WITH THE IMPACT OF lubricants on the environment. Primary emphasis is placed on the used lubricants since they may contain materials that are harmful to life or the environment, or both. Topics of lubricant conservation and the used oil reclamation, reprocessing, and disposal are also addressed and so are the concepts of the environmental compatibility, biodegradability, and toxicity of the lubricants.

Concern for the entry of used lubricant into the environment is on the rise, especially in industrialized countries. There are three main avenues to restrain the ever-increasing use of lubricants. These are to develop equipment, wherever and whenever possible, that does not require a lubricant, extend service intervals, and when possible recycle the used lubricant. In order to attain the extended service interval, one must use lubricants with extended useful life. Recycling is the option to minimize the used lubricant's entry into the environment. This translates into cost savings, with respect to buying a batch of a new lubricant as well as in disposal costs, and the potential damage to the environment, if the disposal method is inappropriate. Ways to minimize inadvertent entry of the lubricant into the environment is to use a closed system, where appropriate. A prime example is the modern automobile, where the automobile manufacturers have successfully minimized the loss of the lubricant or its volatile components into the environment through leakage and evaporation. They have achieved this by building closely fitting parts and recycling the volatiles into the engine by installing closed ventilation systems. Many industrial users of lubricants employ such self-contained systems to prevent the unintended lubricant loss into the environment.

Lubricant Deterioration In Service

Lubricants lose their effectiveness during use, due to degradation. The degradation rate depends upon operation severity, time in service, system temperatures, make-up rate, and the environmental conditions. Lubricant degradation may result from both physical and chemical factors, some of which are a result of extraneous contamination. A lubricant that has passed its effectiveness is either too contaminated or too viscous to use. Besides viscosity, other indications that suggest that the lubricant is ready to be supplemented with additives or replaced include increased corrosion of the system components and the formation of resin, varnish, sludge, and carbonaceous deposits. Degradation of the lubricant properties is a consequence of the additives being depleted. The function of the additives is to increase a lubricant's stability, hence improve its useful life, and to impart specific properties that are useful for its use in the intended application. Particulate contaminants, such as dirt and wear debris, cause abrasive wear; and water and acidic species from the lubricant oxidation and decomposition act as catalysts in further degradation of the lubricant. Estimates of the percent drop in additive concentration in an engine oil at the end of a drain period are provided below [812]:

1. Detergent (46 %)
2. Ashless dispersant (16 %)
3. Zinc dialkyl dithiophosphate (45 %)
4. Oxidation inhibitor package (45 %)
5. Viscosity modifier (9 %)

Oxidation is the major reason for the degradation of the most lubricants. Those based on carboxylate and phosphate esters in addition decompose thermally and hydrolytically. Oxidation products from the lubricants are either soluble and stay dissolved in the lubricant causing a viscosity increase, or are oil insoluble solids such as resins that form deposits when they separate on hot surfaces. Oxidation rate of the properly inhibited mineral oil is quite low at temperatures below 60 °C. However, as the temperature increases, so does the oxidation rate. As a rough approximation, the rate doubles for every 10 °C rise in temperature. In addition, certain contaminants, such as water and metals, can act as catalysts and accelerate the oxidation rate. The effect of the various contaminants on oxidation rate is shown in Fig. 13.1 [813]. In the figure, the total acid number (TAN) is correlated with the lubricant's useful life. Turbine oil is considered to have useful life until TAN reaches 2.0 mg KOH/g. As the figure shows, the wet lubricant, in the presence of either the iron or the copper, has a short life of less than 1000 hours, while in all other cases, the lubricant meets the test requirements even after 3500 hours. With respect to the acid number, dry and metal-free lubricant has the lowest acid number, which is followed by the lubricants that contain iron, copper, and water as contaminants. The presence of a significant amount of water in the lubricant has other repercussions as well, such as rusting. Besides temperature and the metal and water contaminants, other factors that increase the rate of oxidation include the amount of air in the lubricant and turbulence. Depending upon the lubricant, the amount of air dissolved in a lubricant can be as high as 10 % [814]. Besides the ASTM D943 test, another test that is often used to assess the oxidation stability of the turbine oils and a variety of other industrial fluids is ASTM D2272. Even the use of the simple physical parameters, such as color, odor, and the amount of insolubles, can be used to judge the health of a lubricant. Oxidation-related breakdown in most organic lubricants is similar, whether they are hydraulic fluids, gear oils, or turbine oils. Tables 13.1–13.5 compare new and used oils from various applications and Table 13.6 compares new

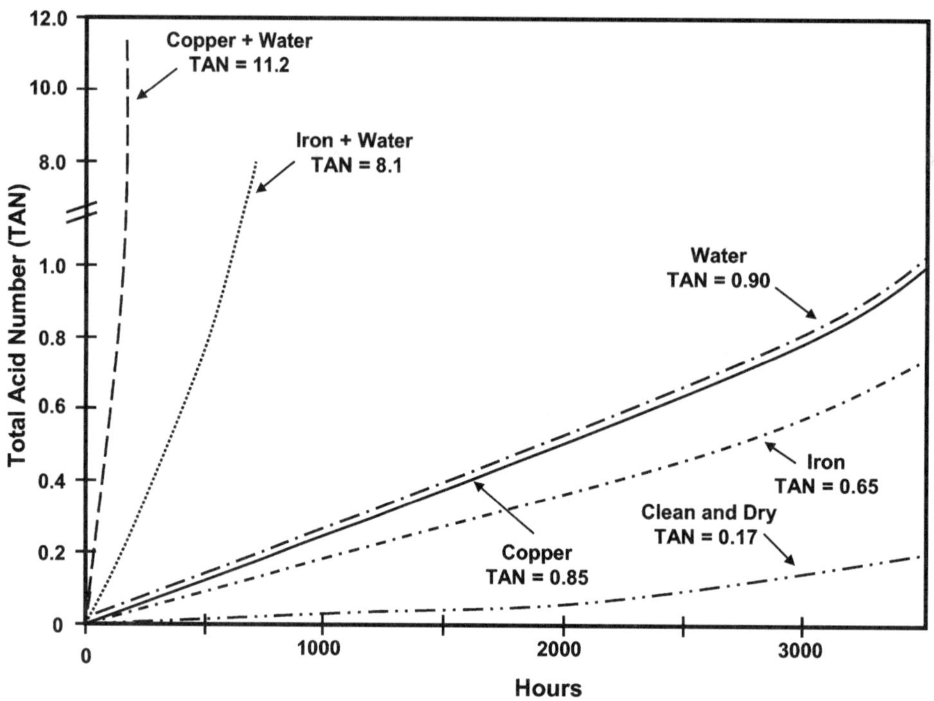

Fig. 13.1—Catalytic effects of contaminants on oil oxidation (modified ASTM D943) [813].

TABLE 13.1—Analysis of used turbine oil [813].

Tests	New oil	Used oil	% Change
Gravity (ASTM D287, °API)	29.8	28.9	-3.0
Viscosity (ASTM D445, mm^2/sec)			
40°C	62.16	67.61	+8.8
100°C	8.18	8.57	+4.8
Viscosity (SSU)			
100°F	322	351	+9.0
210°F	53.7	55.5	+3.4
Viscosity Index (ASTM D2270)	99	97	-2.0
Flash Point (ASTM D92, °C (°F)	230 (446)	225 (437)	-2.2
Color (ASTM D1500)	L 1.5	L 6.5	+333
Appearance	Bright	Hazy	Significant Decrease
Neutralization No. (ASTM D974)	0.13	4.10	+162.3
Water (ASTM D1744, ppm)	24	97	+304.0
Trace Sediment (ASTM D2273, % vol.)	< 0.001	0.02	Significant Increase
Emission Spectroscopy, ppm			
Aluminum	< 1	< 1	-
Barium	2	< 1	-
Boron	< 1	< 1	-
Calcium	< 5	< 5	-
Chromium	< 1	< 1	-
Copper	< 1	< 1	-
Iron	< 1	1	-
Lead	< 10	< 10	-
Silicon	< 1	< 1	-
Silver	< 1	< 1	-
Sodium	< 1	< 1	-
Tin	< 1	< 1	-
Zinc	< 10	< 10	-

TABLE 13.2—Analysis of used hydraulic oil [813].

Tests	New Oil	Used Oil	% Change
Gravity (ASTM D287, °API)	29.0	28.8	-0.7
Viscosity (ASTM D445, mm^2/sec)			
40°C	57.75	69.94	+21.1
100°C	7.62	8.75	+14.8
Viscosity (SSU)			
100°F	299	363	+21.4
210°F	51.8	55.7	+7.5
Viscosity Index (ASTM D2270)	93	96	+3.2
Flash Point (ASTM D92, °C (°F)	205 (401)	203 (397)	-1.0
Color (ASTM D1500)	3.5	Dil. 8	Significant Increase
Appearance	Bright	Dark	Significant Increase
Neutralization No. (ASTM D664)	1.0	1.92	+92
Insoluble Matter (ASTM D893, Wt %), Procedure A			
n-Pentane	Nil	2.64	Significant Increase
Toluene	Nil	0.46	Significant Increase
Resins	Nil	2.18	Significant Increase
Water	Absent	Absent	-
Additive Elements, Wt.			
Barium	0.12	0.10	-17.0
Phosphorus	0.06	0.04	-33.3
Zinc	0.06	0.04	-33.3
Emission Spectroscopy, ppm			
Aluminum	< 1	< 1	-
Boron	< 1	< 1	-
Chromium	< 1	2	-
Copper	< 1	14	Significant Increase
Iron	< 1	48	Significant Increase
Lead	< 10	< 10	-
Silicon	< 1	21	Significant Increase
Sodium	< 1	18	Significant Increase
Tin	< 1	3	-
Particulate Matter, Gravimetric Analysis (mg/100 mL)	2.0	25.8	+1190

and used press roll bearing grease [813]. In each case, the rightmost column shows a change in the lubricant parameter. In almost all cases, a significant negative change occurs in viscosity (ASTM D445), color (ASTM D1500), appearance, water content (ASTM D95 and D1744), acid number (ASTM D664), and the amount of insolubles (ASTM D893). This is indicated in the tables by a plus (+) sign. This suggests that the analyses of these lubricant parameters will provide useful information on a lubricant's condition. Of course, there are other application-specific tests that must also be run. The data in the table also indicate that an analysis of the trace metals is also beneficial in identifying contamination, confirming additive content (additive depletion), and indicating system wear. This is indicated in the tables by a minus (–) sign. Two procedures that are suitable for determining the trace metals are emission spectroscopy and atomic absorption. It is important to note that many of these liquid lubricant parameters do not apply to the lubricating greases. Analysis of grease on a routine basis is seldom required, because greases are not reused. Most analyses are performed to understand bearing failures, malfunction of equipment, or in setting relubrication schedules. Flame Photometric analysis data in Table 13.6 show the grease to be lithium soap grease that contains a lead-based extreme-pressure additive system. NLGI consistency grade change and reduced dropping point show the used grease to be considerably softer. The high amount of the insoluble matter and the iron content shows high wear.

The methods described so far are useful in assessing a lubricant's reclamation and recyclability potential. They have limited use in assessing the condition of a lubricant in service. One method that is useful in evaluating a lubricant's life quickly is to determine the amount of the oxidation inhibitor left in the lubricant since oxidation is the primary reason for lubricant degradation [815]. Oxidation inhibitors, natural or added, increase the thermo-oxidative stability of the lubricants. Their depletion over time will result in the deterioration of the physical and chemical properties of the lubricant. The problem with the commonly used measures of a lubricant's useful life, such as viscosity and acidity, is that

TABLE 13.3—Analysis of used paper machine oil [813].

Tests	New Oil	Used Oil	% Change
Gravity (ASTM D287, °API)	27.3	28.1	+2.9
Viscosity (ASTM D445), mm²/sec			
40°C	199.9	216	+8.0
100°C	16.60	17.83	+7.4
Viscosity (SSU)			
100°F	1061	1146	+8.0
210°F	86.5	91.7	+6.0
Viscosity Index (ASTM D2270)	86	89	+3.5
Appearance	Bright	Dark	Significant Change
Neutralization No. (ASTM D664)	1.9	1.6	-15.8
Water	Absent	Absent	-
Additive Elements (Wt %)			
Barium	0.15	0.10	-
Phosphorus	0.07	0.05	-
Zinc	0.07	0.04	-
Emission Spectroscopy (ppm)			
Aluminum	< 1	< 1	
Boron	< 1	< 1	
Chromium	2	2	
Copper	< 1	99	Significant Increase
Iron	1	58	Significant Increase
Lead	< 10	21	Significant Increase
Silicon	5	< 1	
Silver	< 1	< 1	
Sodium	< 1	1	
Tin	< 1	< 1	
Emulsion Test (ASTM D1401)	40-40-0 (25 min)	30-30-20 (60 min)	Significant Increase
Particulate Matter (mg/100 mL)	0.30	9.6	+3100
Sulfated Ash (ASTM D874) (wt %)	0.40	0.56	+40
Insoluble Matter (ASTM D893), Wt % Procedure A			
n-Pentane	Nil	0.52	+52
Toluene	Nil	0.15	+15
Resins	Nil	0.37	+37
Rust Test (ASTM D665) , Procedure A	Pass	Pass	-

they are not very responsive to changes in the lubricant's residual life. In addition, they are affected by the extraneous factors, such as fuel dilution, shear-related viscosity change, and removal of the oxidation products, through evaporation or corrosion [816]. Hence, the condition monitoring techniques based on viscosity and acidity are inappropriate in predicting a lubricant's remaining life. Techniques that have been deemed capable of achieving this goal include thermo-oxidative stressing, chemical-oxidative stressing, and electrochemical and instrumental methods.

Thermo-oxidative stressing involves long-term stability tests, which employ specific operating parameters, such as air flow, metal catalysts, sample size, temperature, etc., to simulate the operating conditions of the specific equipment. Methods used to determine the oxidation onset time or onset temperature include chemiluminescence, inverse gas chromatography [816], weight loss [817], gas evolution rate [818], and differential scanning calorimetry (DSC). Standard Test Method E2009-02 for Oxidation Onset Temperature of Hydrocarbons by Differential Scanning Calorimetry is used to determine the oxidation induction times of the lubricants. The rotating pressure vessel oxidation test (RPVOT) is an accelerated technique used extensively in monitoring the remaining oxidation inhibitor capacity of steam and gas turbine oils that are supplemented with phenolic and amine type inhibitors.

Chemical-oxidative stressing involves the use of a free radical generating material, such as cumene hydroperoxide. In this method, a sample of the lubricant diluted with hexadecane and cyclohexene at 60 °C is titrated with the free radical generating reagent. When the oxidation inhibitor is depleted, cyclohexene oxidizes causing a rapid decrease in

TABLE 13.4—Analysis of used gas engine oil [813].

Tests	New Oil	Used Oil	% Change
Gravity (ASTM D287), °API	27.5	27.8	+ 1.1
Viscosity (ASTM D445), mm^2/sec			
40°C	114.3	137.0	+19.9
100°C	12.48	14.24	+14.1
Viscosity (SSU)			
100°F	597	717	+20.1
210°F	69.6	76.6	+10.1
Viscosity Index (ASTM D2270)	100	101	+1
Flash Point (ASTM D92), °C (°F)	243 (469)	240 (464)	-1.2
Appearance	Bright	Dark	Significant Increase
Neutralization No. (ASTM D664)	3.0	5.2	+73.3
Insoluble Matter (ASTM D893), Wt % Procedure B			
n-Pentane	Nil	0.16	+16
Toluene	Nil	0.15	+15
Resins	Nil	0.01	+1
Water (hot plate)	Absent	Absent	-
Additive Elements (wt %)			
Calcium	0.19	0.17	-
Phosphorus	0.13	0.13	-
Zinc	0.14	0.14	-
Emission Spectroscopy (ppm)			
Aluminum	< 1	2	
Boron	< 1	< 1	
Chromium	< 1	8	Significant Increase
Copper	< 1	8	Significant Increase
Iron	< 1	15	Significant Increase
Lead	< 10	< 10	
Silicon	< 10	< 10	
Sodium	< 1	21	Significant Increase
Tin	< 1	< 1	
Infrared analysis			
Oxidation	Nil	Slight	-
Nitration	Nil	Moderate	-

the oxygen pressure. The indicator used is a nickel complex laser dye, which changes color by reacting with cumene hydroperoxide at the end point. The color change is monitored using a visible spectrophotometer. The electrochemical technique that is commonly used for this purpose is cyclic voltammetric analysis [819,820]. This technique determines the individual oxidation inhibitor concentrations through current-voltage relationships at the electrodes. The lubricant is dissolved (ester oils) or suspended (hydrocarbon oils and greases) into a solvent containing an electrolyte prior to analysis. The voltage and the magnitude at which the current flow increases are used to identify and quantify the inhibitor or inhibitors present in the lubricant. Instrumental techniques, such as gas chromatography and liquid chromatography have also been used for inhibitor analysis, but with varying degrees of success.

Used Oil—Environmental Considerations

Toxicity, safety, and environmental compatibility of the chemicals are becoming a growing concern. *Toxicity* determines the ability of the materials to harm life. While harm to humans is a major concern, the effect of chemicals on the environment as a whole cannot be ignored. Unused lubricants are generally considered less toxic than the used lubricants. Despite extensive efforts to reclaim and recycle the used oil, sooner or later some ends up in the environment. The exposure to used oils primarily occurs through skin absorption. Over the short-term, they can lead to skin irritation; over the long-term, they can act as carcinogens. The Occupational Safety and Health Administration (OSHA) requires all lubricant manufacturers to provide material safety data sheets (MSDSs) on their products. While each MSDS contains a variety of information, its primary purpose is to provide physical and health hazard data on the products so as to facilitate safe handling. Human exposure can be minimized by avoiding contact with the lubricant, by using protective equipment, such as gloves, oil-impervious clothing and boots, and by adopting explosion and fire prevention measures. Environmental protection requires that neither new nor used lubricant is released into the air, water, or soil. This can be

TABLE 13.5—Analysis of used EP gear oil [813].

Tests	New Oil	Used Oil	% Change
Gravity (ASTM D287), °API	26.0	25.9	-0.4
Viscosity (ASTM D445), mm^2/sec			
40°C	293	304	+3.8
100°C	23.42	24.15	+3.1
Viscosity (SSU)			
100°F	1556	1612	+3.6
210°F	116.6	119.9	+2.8
Viscosity Index (ASTM D2270)	99	101	+2.0
Flash Point (ASTM D92), °C (°F)	252 (485)	221 (430)	-12.3
Color (ASTM D1500)	5.0	L8.0 dil	Significant Increase
Appearance	Bright	Dark	Significant Increase
Neutralization No. (ASTM D664)	0.45	2.3	+411
Insoluble Matter (ASTM D893), Wt % by Procedure A			
n-Pentane	Nil	0.03	+3
Toluene	Nil	0.00	-
Resins	Nil	0.03	+3
Water (hotplate)	Absent	Absent	-
Chlorine (ASTM D808) (wt %)	Nil	0.13	+0.13
Sulfated Ash (ASTM D874) (wt %)	0.02	0.03	+50
Additive Elements (wt %)			
Sulfur (ASTM D1552)	0.79	0.38	-51.9
Phosphorus	0.024	< 0.01	-58
Emission Spectroscopy (ppm)			
Aluminum	< 1	1	-
Barium	< 1	< 1	-
Boron	< 1	< 1	-
Calcium	< 5	< 5	-
Copper	< 1	2	-
Iron	< 1	2	-
Lead	< 10	< 10	-
Silicon	< 1	< 1	-
Sodium	< 1	< 1	-
Tin	< 1	< 1	-
Zinc	< 10	10	-
Foam Test (ASTM D892) 5 Min blowing (mL)			
Sequence I	10	330	Significant Increase
Sequence II	25	340	Significant Increase
Sequence III	10	600	Significant Increase
10 Min settling (mL)			
Sequence I	0	25	Significant Increase
Sequence II	0	0	
Sequence III	0	300	Significant Increase

TABLE 13.6—Analysis of used press roll bearing grease [813].

Tests	New Grease	Used Grease	% Change
Dropping Point (ASTM D566), °F (°C)	358 (181)	290 (143)	-20.8
Penetration (ASTM D217)			
Un-worked	317	444	+40.0
Worked	323	448	+38.7
NLGI Consistency	1	000	Loss of consistency
Appearance	Smooth bright	Smooth bright	-
Color, Visual	Dark Brown	Black	Significant change
Texture	Smooth	Smooth	-
Odor	Natural	Oxidized	Significant change
Ash, Modified (ASTM D128), Wt %	2.9	4.29	+47.9
Spectrographic Analysis			
Major	Pb	Fe	Significant change
Minor	Ca, Li, Na	Al, Ca, Pb, Li, Mo, Na	
Trace	Cr, Cu, Fe, Mg	Cr, Cu, Mg, Mn, Si	
Water (ASTM D95), Wt %	Nil	0.2	+20
Insoluble Matter (ASTM D128)			
Wt %	0.06	2.47	+4017
Nature	No wear metals or dirt	Wear metals and dirt	Significant change
Infrared Analysis	Petroleum oil and lithium soap	Petroleum oil, lithium and calcium soap, and oxidation products	Significant change
Flame Photometric Analysis, Wt %			
Calcium	0.30	0.64	+113
Lithium	1.29	0.64	-50.4
Potassium	Trace	Trace	-
Sodium	0.01	0.04	-
Lead or PbO (ASTM D1262, Wt %)	1.78	0.79	-55.6

achieved only if proper procedures pertaining to collection, storage, and handling of the lubricants, lubrication, equipment maintenance, and the disposal of the used lubricants are in place. Because the lubricant disposal is costly and is subject to a number of federal, state, and local regulations, minimizing the volume of the used or the leaked lubricant is highly desired.

The interest in lubricant conservation, used oil recycling, and nonpolluting disposal is because of many reasons. These include progressively increasing crude oil prices, which affect the price of all petrochemicals including those of the lubricant base stocks, both mineral and synthetic, and a concern for the environment. It is estimated that the United States generates around 2.4 to 2.5 billion gallons of used oil each year [821]. Worldwide, this number soars to about 5.3 billion gallons. The U.S. Environmental Protection Agency (EPA) considers that out of 2.4 billion gallons of used oil 1.5 billion gallons is recoverable. Approximately 12 % of this volume is re-refined for re-use as lubricating oils, 12 % is reprocessed for use as metalworking fluids and other industrial oils, and 15 % is used for spraying on the roads and for other uses. And almost half is used as fuel, after mixing it with fresh residual and distillate fuels. In 1976, the U.S. Congress through Resource Conservation and Recovery Act of 1976 encouraged recycling of the used oil to produce recycled oils that are "substantially equivalent" to the new oils. It is estimated that if all the recoverable oil is recycled, ~30 % of the United States lubricant demand will be met [52].

Lubricant Conservation

Conserving oil should be the primary goal of every user/business because it minimizes the amount of the used oil to be recycled or disposed of, thereby conserving valuable resources. Today, controlling the generation of the used oil and its recycling is even more important due to the steadily increasing oil prices, supply shortages, environmentally damaging illegal disposal, and the cost of the legal disposal.

Conservation of oils can be achieved by taking the following steps:

1. In-use condition monitoring
2. Regular vehicle maintenance
3. Reduction in use
4. In-plant recycling and re-use
5. Efficient and economical retrieval from the waste streams
6. Utilizing out-plant commercial recycling processing
7. Re-use of the nonrecoverable oily waste as fuel

In the previous chapters, we discussed extended drain (service) intervals for automotive lubricants and pointed out the lubricant attributes and the equipment operating parameters that affect an oil's service duration. Factors and conditions that affect the oil service interval in an automobile include the following:

1. High-mileage engines
2. Diesel engines
3. Vehicles that use alcohol-gasoline blends
4. Short trip intermittent (stop and go) driving
5. Hot running conditions
6. Turbo-charged engines
7. Towing/heavy loads
8. Synthetic lubricants
9. High-capture efficiency oil filter
10. Predominant highway driving
11. Low-mileage engines (less than 50,000 miles)
12. On-board diagnostics

The first seven shorten the service interval and the last five extend it [822]. The primary reasons that negatively impact the drain interval are the increased entry of the blow-by gases from the combustion chamber into the crankcase, water accumulation in the oil, and the elevated operating temperatures; all of which lead to accelerated lubricant degradation due to oxidation. For the factors that have a positive impact on the service interval, the converse is true. Although prolonging the life of the oil in this manner will help the conservation effort, yet judging the optimal drain interval is what will have a direct impact on conservation. While the original equipment manufacturers publish guidelines that define realistic service interval for their equipment, consumers in the United States tend to go for a shorter drain interval. This is because of the marketing efforts of the service providers and the consumers' perceptions that changing the oil more often is good for the longevity of their automobile. While there is some merit to this philosophy, in most cases the oil is changed before its useful life is over. While more and more oil analysis laboratories are targeting passenger car owners to grow their market, it is not convenient for the consumers to take the in-use oil sample for analysis. In addition, the cost of such a service to a normal consumer is high.

To remove the uncertainty out of judging the proper drain interval, new on-board sensors and related technology is gaining popularity [822]. GM's Oil Life System (GMOLS) uses the engine's operating variables to suggest the best time for an oil change. The suggested drain interval is not only vehicle specific but also changes based on the driving conditions and the consumer's driving habits. For example, gentle highway driving in a warm climate will maximize the interval and short-trip driving in a cold climate or log duration fast speed highway driving in an extremely hot climate will shorten the interval between the oil changes. Depending on the vehicle, the former could be between 7000 and 12,000 miles and the latter could be 3000 miles, or less. Since most people's travel involves city and highway combination driving, the GMOLS is likely to suggest an oil change every 5000 to 6000 miles. GM data show that the OLS extends oil change intervals without risk to the engine. It does not use an oil sensor and maximizes engine oil performance by sensing engine speed (r/min), temperature, and other operating factors. Interestingly, the technology does not actually monitor even a single chemical or physical property of the oil. The sensor used in this technology is based on GM's determination that nearly all driving conditions can be grouped into one of four categories: easy freeway driving; high-temperature, high-load service; city driving; or extreme short-term, cold-start driving. GM discovered that the oil degradation in the first three categories was largely a function of the oil temperature. During extreme short-trip driving, the fourth category, the principle cause of oil degradation is water condensation and contaminants in the oil; the lower the oil temperature, the greater the contamination.

Daimler Chrysler's version of the oil monitor is called ASSYST in Europe and the Flexible Service System (FSS) in the United States. Like GM's sensor, the FSS uses a computerized system to track multiple engine operating conditions. In addition to the consumer driving habits, driving speeds, and temperatures, the system also monitors oil levels. It uses this information to determine the remaining time and the mileage before the next oil change and displays it. Daimler-Benz uses a digital oil quality sensor in its Mercedes-Benz vehicles, which measures changes in capacitance. Capacitance is a surrogate for the amount and the type of contaminants and oil degradation products present in the oil. An increase in dielectric constant (less resistance to electrical current flow) indicates oil contamination and degradation.

Delphi Corporation's INTELLEK® oil condition sensor uses both a computer algorithm and a sensor. The algorithm takes into account factors such as temperature, driving severity, oil level and oil type, all of which affect oil degradation. The sensing element measures various oil properties such as oil conductivity, water-glycol contamination, oil temperature, and determines the oil level.

Bosch is in the process of developing a multifunctional oil sensor that will determine oil level and oil condition in both spark-ignition and compression-ignition engines. The oil level sensor will allow the oil dipstick to be omitted from the automobile.

Eaton's fluid condition monitor (FCM) technology monitors multiple fluid properties by the use of impedance spectroscopy—a technology that measures multiple electrical properties of the fluid. It measures the surface properties of the fluid as well as the bulk properties and it can independently track multiple lubricant parameters. Measuring bulk properties provides information about changes in conductivity and dielectric constant, which indicate changes in the base fluid and the additives due to oxidation, water, acids, mixed fluids, and wear debris. Measuring surface properties provides a quantitative measure of the physical and chemical properties of the fluid at the fluid-metal interface.

While the primary function of in-use oil monitoring technologies is to optimize oil drain intervals, these can also provide information on viscosity, permittivity, conductivity, and temperature. These lubricant parameters provide an insight into the functioning of the engine or a change in lubricant quality. Viscosity and permittivity, or the dielectric constant, are the primary values that relate to the oil condition; viscosity increase is an indication of the oil oxidation and a viscosity decrease indicates mechanical wear (shear) and fuel dilution.

Conservation of the other lubricants, such as hydraulic oils, metalworking fluids, and greases require other strategies. These include preventing oil loss due to leakage; controlling oil degradation, filtering off wear-causing solids, and removing water and other contaminants to prolong a lubricant's useful life, thus avoiding its premature disposal. Leakage is prevented by ascertaining proper installation of the damage-resistant seals and protecting their integrity during

TABLE 13.7—Common analytical tests performed on used oil.

Test	Methodology
Contamination	
Moisture or water	ASTM D1744, D6304, and D4928 (Karl Fischer), ASTM E1252 (FT-IR), Distillation (ASTM E123, ASTM D1160), Crackle
Dirt or dust	Elemental Analysis (D6443), Multipurpose Filter
Soot	Thermo-gravimetric analysis (TGA) - (ASTM D5967-A4), FT-IR analysis, Pentane Insolubles (ASTM D893)
Hydrocarbons	GC (ASTM D3524), Flash Point (ASTM D56, D92, D93, and D3828)
Glycol coolant	FT-IR (ASTM E1252), Schiff's Reagent (ASTM D2982)
Particulates	Particle Count (Laser, Pore Blockage) (ISO 4406, ISO 11500 and ISO 11171), Gravimetric Analysis (ASTM D4898)
Physical Properties	
Viscosity and Viscosity Index	ASTM D445, D2983, and D2270
Neutralization Number (Acid Number, Base Number)	ASTM D664, D974, D2896, D4739
Oxidation	ASTM E1252 (FT-IR), RULER® (Remaining Useful Life Evaluation Routine, On-site Analysis Method)
Nitration	ASTM E1252 (FT-IR)
Rotary Pressure Vessel Oxidation Test (RPVOT)	ASTM D2272
Insolubles	ASTM D893
Demulsibility	ASTM D1401 and D2711
Foaming Characteristics	ASTM D892
Air-release Characteristics	ASTM D3427
Color	Gardner, ASTM D1500
Wear	
Spectrochemical	Atomic emission spectroscopy (AES) (ASTM D5185 and D6595), Inductively Coupled Plasma - Atomic Emission Spectrometer (ICP-AES), Atomic Absorption Spectroscopy (AA)
Ferrography	Microscopic Analysis
Filter Patch	Microscopic Analysis

use. Oil degradation either relates to thermal factors, oxidation, hydrolysis, and in the case of the water-based hydraulic and metalworking fluids to biodegradation. Temperature control to lower levels, where possible, minimizing water contamination, and controlling air entrainment can help slow down the oil degradation. Centrifugation, gravity settling, and distillation with or without vacuum can be used to remove solids and water; and the microbial degradation of the aqueous fluids can be minimized by the use of the biocides and sterilization. Other methods that can be used to remove solids and water include screening, paper, cloth, and diatomaceous earth (clay) and the membrane filtration.

Used Oil Recycling

EPA considers used oil to be a danger to the environment and has mandated its re-use and disposal. The EPA' s regulatory definition of the used oil is: *Any oil that has been refined from crude oil or any synthetic oil that has been used and as a result of such use is contaminated by physical or chemical impurities*. During normal use, contaminants such as oxidation and degradation products, dirt, metal scrapings, water, and chemicals enter the oil and over time the oil loses its ability to perform the intended functions. Such oil must be replaced with new or re-refined oil to do the job at hand. EPA uses three criteria to determine if the oil meets the definition of the used oil. These are summarized below:

1. Origin—Used oil must have been refined from the crude oil or made from synthetic materials. Animal and vegetable oils are not included in the EPA' s definition.
2. Use—Oils used as lubricants, hydraulic fluids, heat transfer fluids, buoyants, and for other similar purposes. Other fluids, for example, cleaning agents, antifreeze, and kerosine are excluded.
3. Contaminants—EPA' s definition requires the used oil to become contaminated as a result of being used. Physical contaminants include metal shavings, sawdust, or dirt, and chemical contaminants may include solvents, halogens, or salt water.

Used oils that meet EPA' s definition include the following:

1. Engine oils—typically includes gasoline and diesel engine crankcase oils and piston-engine oils for automobiles, trucks, boats, airplanes, locomotives and heavy equipment
2. Transmission fluids
3. Refrigeration oils

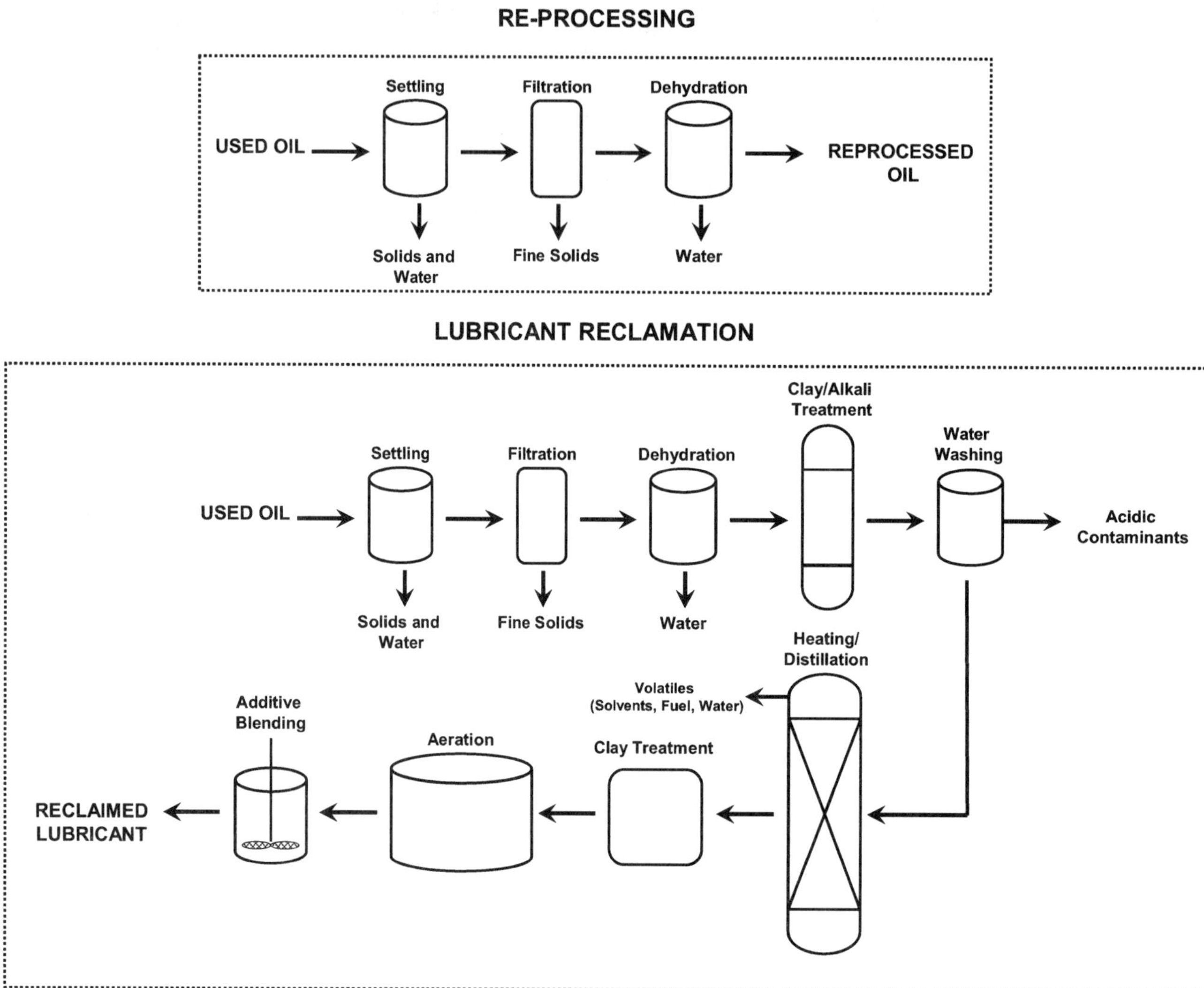

Fig. 13.2—Flow schemes for lubricant reprocessing and reclamation.

4. Compressor oils
5. Metalworking fluids and oils
6. Laminating oils
7. Industrial hydraulic fluids
8. Copper and aluminum wire drawing solutions
9. Electrical insulating oils
10. Industrial process oils
11. Oils used as buoyants
12. Synthetic oils—usually derived from coal, shale, or polymer-based starting materials

Oil Reconditioning

According to the EPA estimate, 380 million gallons of used oil are recycled each year. The volume equates to about half the annual engine oil use in the United States. Recycled used oil can either be used in the same application or in a different application. For use in the same application, the oil must be segregated, so as not to mix with the other types of oils. A prime example is used engine oil, which is obtained from automobile dealerships and quick oil change service organizations, such as Jiffy Lube, and is re-refined and sold in stores as engine oil. As mentioned earlier, out of 2.4 to 2.5 billion gallons, only 1.5 billion gallons of oil is considered recoverable by the EPA. This leaves behind ~1 billion gallons unaccounted for, which is somewhere in the environment. EPA notes that improperly disposed used oil is the largest single source of oil pollution in our nation's waters. The automotive gasoline engine oil, estimated at 757 million gallons, makes up almost one-third of the total lubricating oil sold in the United States. The engine oil service market comprises service providers and do-it-yourselfers. About 55 % of the oil change market is controlled by the first group and the remaining 45 % is controlled by the second group. The first group collects and recycles used oil efficiently and correctly. However, this is not the case for the do-it-yourselfers. While some take the used oil to collecting stations for recycle, others do not and the used oil ends up in storm drains, dumped on the ground, burned, or in some cases, used for other purposes, such as dust control. Industrial oils, even those used in automotive applications such as railroads and other transportation do not suffer from this dilemma. Industrial oil users have a greater degree of control on selection and use, which allows more effective conservation and recycling. Because of the

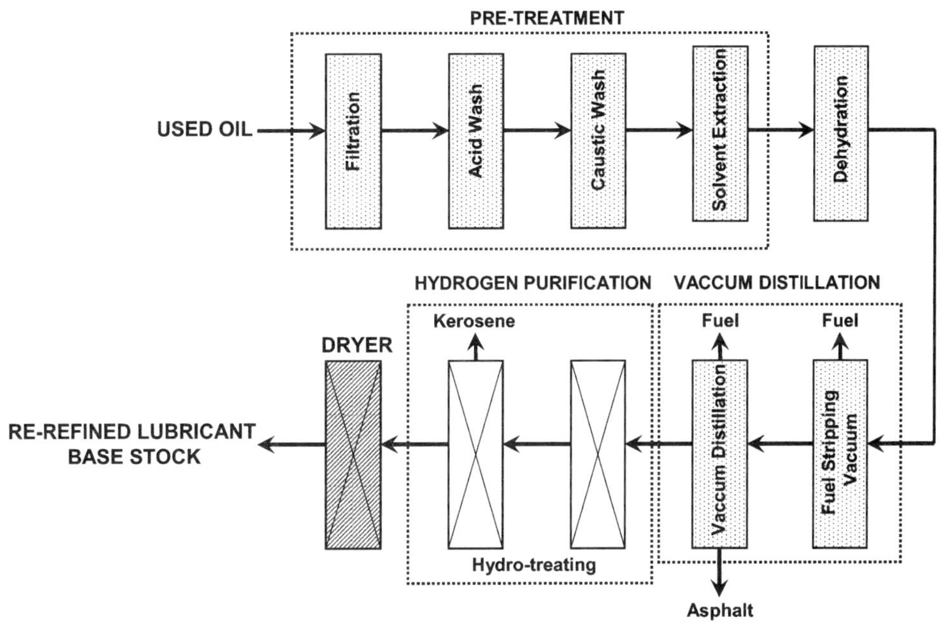

Fig. 13.3—A flow scheme for lubricant re-refining.

sheer volume, the used oils from the industrial applications are either reconditioned on-site and reused, or sent out for reclamation.

The recycling processes include mechanical purification, reconditioning or reprocessing, reclamation or re-refining, and incineration. In mechanical purification, there is no change in the chemical content of the oil. It involves just removing the particulate impurities.

In terms of simplicity, *reconditioning* (reprocessing) is the next. The process involves removing impurities from the used oil and using it again in the same application or in a different but related application. Typically, segregation of used oils at their source offers more of an opportunity for re-use in the original application. While this form of recycling might not restore the oil to its original condition, it does prolong its life. In some cases, it may be appropriate to replenish some of the spent additives, such as oxidation inhibitors, to improve oil's useful life and performance. The major advantage of this process over others, except mechanical purification, is that in most cases, it can be implemented on-site, which eliminates the cost of shipping. In-plant recycling, in addition, allows a secondary use for the used oil or the fluid. For example, a used hydraulic fluid with suitable additives can be used as a metalworking fluid.

Reclamation involves a higher degree of processing and is primarily used for industrial lubricants, such as hydraulic oils, gear oils, and metalworking fluids. It is more effective if the feed streams are unmixed, otherwise used oil may have to be processed for use as the furnace fuel oil. Reclamation involves the following steps:

- Settling, centrifuging, and filtering to remove solids
- Clay and alkali treatment to remove acidic contaminants, followed by washing to remove the resulting soaps
- Mild heating or distillation to remove volatile components
- Clay treatment to remove polar oxygenated materials or to improve color
- Aeration or biocide treatment to get rid of the bacterial contamination
- Additive blending to make up the depleted additives

Some feed streams need fewer steps than others. A commercial lubricant reclaiming company uses a host of analytical techniques to determine the degree of treatment required to operate cost effectively. The analyses that are generally performed prior to reclamation are listed in Table 13.7, along with the test methods. The steps listed for reprocessing and reclamation are shown diagrammatically in Fig. 13.2. It is important to point out that the quality of the reclaimed lubricant is either the same or lower than that of the lubricant prior to use. It is also important to make certain that none of the recoverable used oils, including those that are emulsifiable, are added to the waste water stream. This is because they can decrease the effectiveness and efficiency of water treatment/oil removal.

Re-refining is the most complex of the recycling processes and uses petroleum refining techniques, such as vacuum distillation and hydrotreating, both of which require specialized equipment. Because of this, only a limited number of companies are involved in this process. Re-refining, primarily used to recycle engine oils, results in clean high quality base stocks. The process involves pretreatment to reduce contaminant levels, multi-stage vacuum distillation, and catalytic hydrotreating. Pretreatment comprises application of heat, filtration, and treatment with acids, caustic, and other chemicals. The by-products of this process are asphalt, reclaimed fuel that finds entry into the crankcase during the low temperature operation, and kerosine resulting from hydro-treatment. For every gallon of the used oil that is re-refined, approximately 50 % is converted into re-refined base oil, 25 % distills as light ends (fuels), 15 % remains as bottoms (asphalt flux), and 10 % is water. The re-

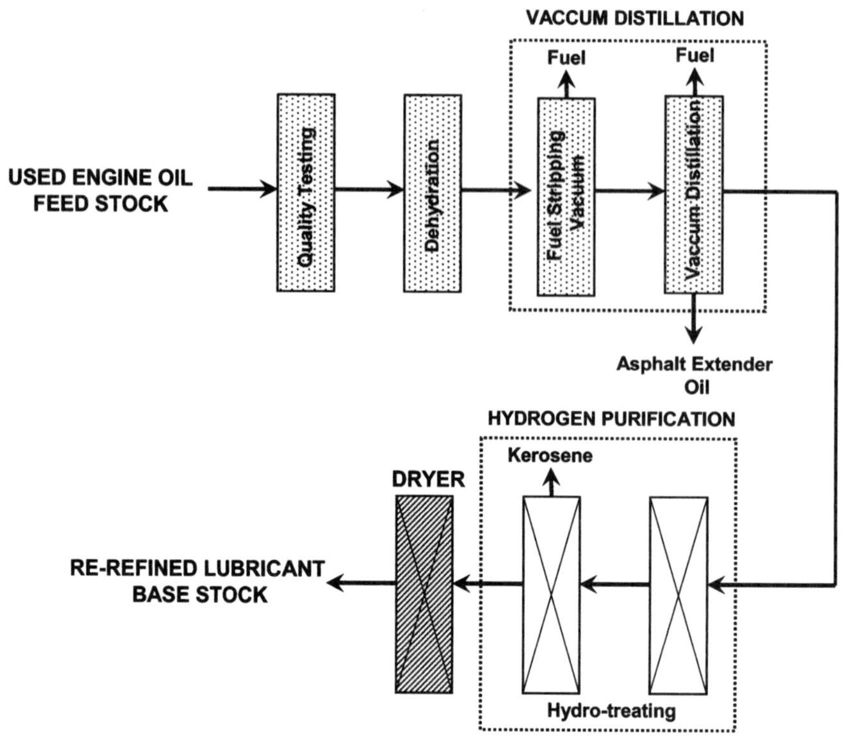

Fig. 13.4—Used engine oil recycling.

refined base oil produced is sold to the lubricant manufacturers that compound it with additives and then package the product for the industrial and automotive markets. The asphalt flux may be sold to roofing manufacturers, road asphalt suppliers for road building, or steel mills as a fuel. The water is treated and discharged.

A simplistic flow diagram of the re-refining operation is given in Fig. 13.3. Re-refining is usually used for large volume lubricant streams, such as engine oils. Currently, about 14 % of the used engine oil is re-refined. However, re-refining is not economically efficient for the new oil manufacturers and as a result the re-refined motor oil is more expensive than the new engine oil. Figure 13.4 depicts the process used by Safety Kleen to re-refine used engine oil [823]. Please note that this process is very similar to that used for re-refining of the other used oil streams, compare this figure with Fig. 13.3. The differences between the two processes pertains to quality control of the waste stream and the lack of pretreatment, presumably both because of the relatively clean nature of the used engine oil and its intended re-use as engine oil base stock after purification. Table 13.8 compares the properties of the re-refined oils produced by the IFP (Institut Francais Du Patrole) process with those of the fresh solvent neutral oils [4,824]. One can see that the properties of the re-refined base oils from the used oil compare well with those of the fresh solvent neutral oils of comparable viscosity. Hence, these oils are well suited to make finished engine oils, gear lubricants, hydraulic fluids, metalworking fluids, and greases. Properly re-refined oils not only meet the API Group I and Group II requirements but also have a comparable

TABLE 13.8—Quality comparison—Solvent refined fresh oils versus re-refined base oil [4].

	150 Neutral		350 Neutral		600 Neutral		Bright Stock	
Base Oil Parameter	**Fresh**	**Re-refined**	**Fresh**	**Re-refined**	**Fresh**	**Re-refined**	**Fresh**	**Re-refined**
Density, g/mL (ASTM D4052 and D1298)	0.875	0.874	0.835	0.882	0.895	0.888	0.910	0.903
Color (ASTM D1500)	<2	<1.5	<3	<2	<3.5	<2.5	<6.5	<5.5
Viscosity at 50 °C, mm²/s (ASTM D445/IP 71)	19–21	18–20	40–46	37–41	60–74	60–64	242–272	226–242
Viscosity Index (ASTM D2270)	97	95	95	95	95	95	95	95
Pour Point, °C (ASTM D97)	−15	−9	−9	−12	−9	−12	−9	−9
Flash Point, Open Cup, °C (ASTM D92)	200	215	215	245	240	255	290	275
Neutralization Number, mg KOH/g (ASTM D664 and D974)	0.05	0.03	0.03	0.05	0.05	0.05	0.05	0.03
Conradson Carbon, wt %	0.03	0.01	0.1	0.02	0.15	0.09	0.8	0.85
Ash, wt %	0	0	0	0	0	0	0	0

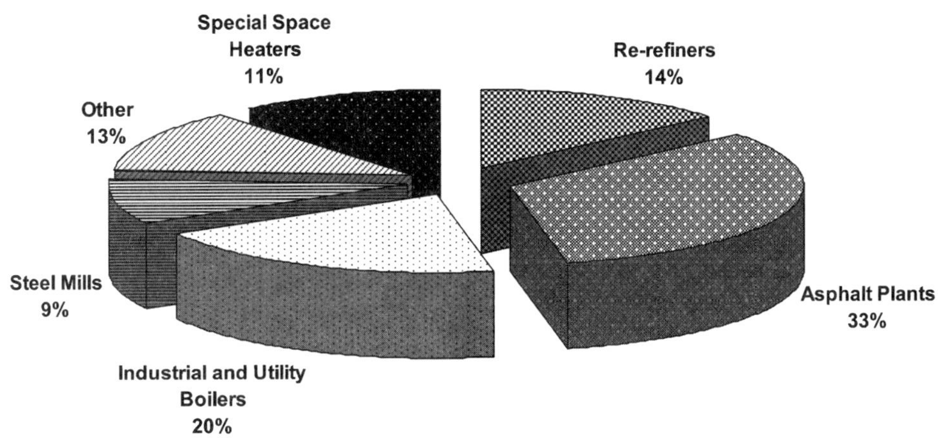

Fig. 13.5—Recycled oil applications.

quality. This is demonstrated by the use of these base oils in formulating lubricants that meet the newest standards for automotive lubricants listed below [823]:

1. API SM/ILSAC GF-4
2. API CI-4 Plus
3. DEXRON®-IIIG Automatic Transmission Fluid
4. U.S. Military Specification (e.g., MIL-PRF-2104G)

The cost of the re-refined base oils is comparable, although a little higher, to those obtained from the crude oil. Hopefully, there will be a cost advantage as the availability of the quality used oil increases [823]. Figure 13.4 shows various uses of the used oil. As one can see, almost 40 % is used in combustion-related applications and only 14 % is re-refined [825].

Lubricants and the Environment

Since there are limitations to the used lubricant recycling, both with respect to collection and indefinite reconditioning and reclamation, sooner or later some of the used lubricant is going to enter the environment. That is why the Federal and State governments have passed several laws and countless regulations covering the effect of the used oil on the environment and occupational health and safety. EPA has also established standards on emissions resulting from the burning of the waste oil as fuel. The emissions pertain to carbon monoxide (CO), sulfur oxides (SO_X), nitrogen oxides (NO_X), particulate matter (PM), particles less than 10 micrometres in size (PM_{10}), toxic metals, organic compounds, hydrogen chloride, and the global warming gases—carbon dioxide (CO_2) and methane (CH_4). Limits are also established on certain metals and organic pollutants. Metals include lead, antimony, arsenic, beryllium, cadmium, chromium, cobalt, manganese, nickel, selenium, and phosphorus. Organic materials that are regulated include phenol; chlorinated organics, such as PCBs, dichlorobenzene, and trichloroethylene; fused ring hydrocarbons, such as naphthalene, phenanthrene, anthracene, pyrene, benz(a)anthracene, chrysene, and benzo(a)pyrene; and aromatic esters, such as dibutyl phthalate, butyl benzyl phthalate, bis(2-ethylhexyl) phthalate.

Environmental Compatibility

As described in Chapter 7, a number of terms are used to describe a lubricant's compatibility with the environment. These include environmentally friendly, environmentally compatible, environmentally acceptable, environmentally responsible, environmentally aware, environmentally benign, environmentally harmless, environmentally safe, environmentally sensitive, and environmentally suitable. Two others that are used in the same context are green fluids, which are mostly manufactured from vegetable oils, and the food grade lubricants that are approved by the U.S. Department of Agriculture (USDA) through NSF. Food grade lubricants were described in Chapter 9 while discussing miscellaneous industrial lubricants. These lubricants are primarily used in the food industry where incidental food contact may occur. Although the food grade lubricants are made by the use of the U.S.P. grade white mineral oil, or its equivalent synthetic hydrocarbon oil, which is considered nontoxic, they do not have to meet the biodegradability criteria, commonly required of the environmentally compatible lubricants. Environmentally acceptable is the most widely used term which is also used by some ASTM committees to address the environmentally friendly lubricants. While presently there are no established standards for such lubricants, there is a consensus that such lubricants must be biodegradable and nontoxic. The problem is that the levels of these two parameters are subject to interpretation. Nonetheless, there is a general increase in the desirability of such lubricants, as indicated by their 5–10 % per year growth rate. Lubricants of high biodegradability are primarily based on vegetable oil, animal fat, or the derived ester base stocks. These base stocks are used in lubricants for applications, such as chain saws and hydraulic equipment, which are used in agriculture and forestry settings. This is to alleviate concerns for the lubricant's unintended release into the environment and causing harm.

One of the major challenges is to meaningfully evaluate a lubricant biodegradability. This requires the existence of suitable tests. An additional problem is defining the level of biodegradability that is considered acceptable. To exacerbate the problem further, of a dozen or so tests that are used worldwide to assess biodegradability, each uses different test criteria and conditions. These cause variances, making it difficult to develop a universal standard. Another complex issue that needs to be resolved is to distinguish between environmental friendliness and biodegradability. Unlike biodegrad-

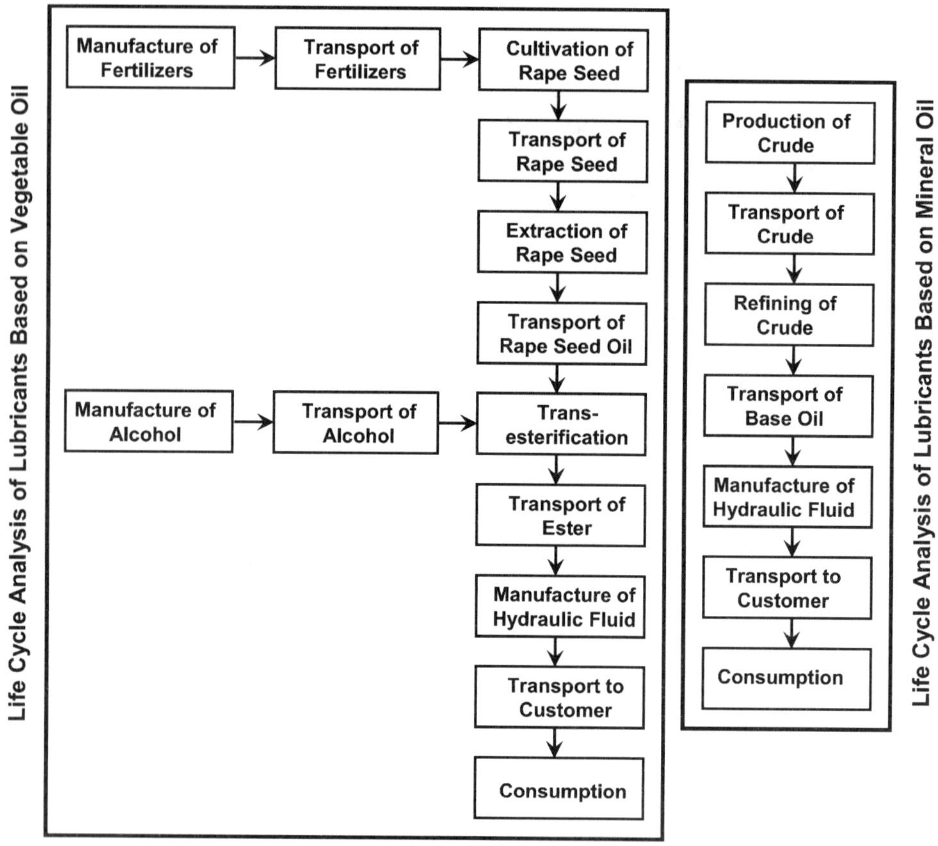

Fig. 13.6—Life cycle analyses of lubricants [828].

ability, which is based upon specific test data, the term environmental friendliness is purely subjective. According to one interpretation, environmentally friendly implies that the lubricant is not toxic and is biodegradable within a short time. There is no suggestion on how to test toxicity and which test organisms to use. According to another interpretation, the term environmentally friendly suggests the existence of some adverse environmental effects, but with minimal negative impact.

In Europe, there is consensus on which test methods and organisms to use and the duration of the tests. However, there is a need to agree on an acceptable test result for an environmentally acceptable lubricant. In addition to testing for short-term compatibility, it is also important to consider the long-term effects of the lubricants on the environment, which are a function of the nature of the base fluid, additive chemistry, and the ability to recycle. As mentioned earlier, environmental compatibility of the lubricant in off-road natural settings is absolutely critical. In these applications, up to 75 % of the lubricants used must be environmentally friendly. This is especially true in Europe, where strict national standards have led to wide adoption of the environmentally friendly fluids. European national standards that are presently in effect include the following. Further review of these standards is presented in Ref [829].

1. Germany (Blue Angel)
   - RAL UZ-79 Hydraulics
   - RAL UZ-48 Rapidly Biodegradable Chain Lubricants for Motor Saws
   - RALUZ-64 Lubricating oils and greases (also applicable to concrete mold release agents)
2. Sweden (Swedish Standard)
   - SS 155434 Hydraulics
   - SS 155470 Greases
3. Nordic Countries (Nordic Swan)—Lubricating oils (including chain oil, mold oil, hydraulic oil, two-stroke cycle-oil, lubricating grease, metal cutting fluid, and transmission/gear oil)
4. The Netherlands (VAMIL regulation)
   - Hydraulics
   - Greases
5. Austria
   UZ 14 Chain Saw oils
6. France (NF-Environment Mark)
   - Chain saw oils
   - ISO 15380
   - Hydraulics

The ecological requirements for these standards include biodegradability, aquatic toxicity, and renewability (recyclability). This suggests that to assess overall environmental effects of a lubricant beyond biodegradability may require more testing and understanding of how the lubricant components perform, functionally and ecologically, at all stages of their lifetime [826].

As stated in Chapter 3, Europe is considering Eco-labeling the lubricants. Eco-labeling is a voluntary scheme designed to encourage businesses to market products and services that are kinder to the environment and for Euro-

TABLE 13.9—Physical properties of environmental importance [831].

Property	Definition	Environmental Importance
Chemical Structure	♦ Molecular weight (range) ♦ Formula ♦ Functional groups ♦ Branching	♦ Determines other physical and chemical properties
Octanol/Water Partition Coefficien[a] (K_{ow})	♦ Equilibrium concentration ratio between octanol and water	♦ Characterizes partitioning to lipids in organisms or the organic carbon of sludge and soils ♦ QSAR estimates of toxicity, partitioning, and water solubility
Vapor Pressure[a] (VP)	♦ Partial pressure of vapor	♦ Characterizes tendency to leave liquid state and become a gas which can enter the air ♦ Allows estimation of evaporation and volatilization rates and extent
Water Solubility[a] (S_w)	♦ Aqueous concentration at saturation	♦ Characterizes extent to which it will dissolve in water ♦ Allow estimation of water concentration and volatility
Soil (Organic Carbon) Partition Coefficient (K_d or K_{oc})	♦ Equilibrium concentration between soil and water	♦ Characterizes the tendency to absorb to particles in the soil ♦ Allows estimation of mobility through soil and sediment
Dissociation Constant (K_a)	♦ Ionization of acids or bases	♦ Ionization extent at environmental *p*H ranges influences other properties
Melting[1] and Boiling Point[a]	♦ Critical temperature for phase changes: solid-liquid-gas	♦ Related to vapor pressure and solubility

[a]A range of values is appropriate for complex mixtures.

pean consumers, both public and private purchasers, to easily identify them. Flower is the symbol used for this purpose [827]. Eco-labeling is mainly driven by the wish of the European industry and the users to harmonize many of the existing national eco-labels, which are listed above. At this time, five classes of lubricants have been proposed to be included in the labeling scheme. These are hydraulic fluids, chain saw lubricants, gear oils, greases, and concrete mold release agents. The proposed criteria and the scientific evidence supporting the inclusion are based on the criteria described in the German Blue Angel and the Nordic Swan standards. Life cycle assessments (LCAs) are included while evaluating fluids. In the case of the lubricants, the biodegradability is not the main criterion, but all cradle-to-grave processes are. Figure 13.6 shows examples of the life cycle analyses of lubricants based on vegetable oil and mineral oil [828]. For details of the steps in these schemes and the typical findings and limitations of the past LCAs, see Ref [829]. Incidentally, the Council of European Union have recent adopted a new law, called the REACH legislation, which went into effect on June 1, 2007. The acronym stands for Registration, Evaluation, and Authorization of CHemicals. The objective of the new legislation is to avoid chemical contamination of air, water, soil, and the human environment [830].

The environmental compatibility requirements for off-road lubricants are changing, not only for base stocks but also for additives. For example, one of the most effective types of additives used in heavy-duty hydraulic fluids is zinc dialkyl dithiophosphates, potent antiwear and extreme-pressure additives. Zinc, being a heavy metal, is undesired and the new antiwear technology developed for use in biodegradable lubricants is metal-free, or ash-less. However, for the traditional hydraulic fluids, zinc dialkyl dithiophosphates are here to stay, primarily for economic reasons. As eco-friendliness moves beyond biodegradability, lubricant marketers may have to demonstrate an understanding of the lifecycle issues, sooner or later.

Need for Standardized Testing

The potential for materials to cause harm to the environment must be evaluated. The most obvious way is to examine the organisms in the areas where the material has been released or spilled. Although field studies are the most realistic way of evaluating all possible chemical effects, they are unable to quantify the level of harm caused by the release. This is because of the continuously changing exposure concentrations due to diffusion away from the point of release, movement to other media (i.e., transport between soil, water, sediment, and air), and degradation (biological and abiotic) of the material. A more reliable approach is to evaluate the inherent hazards of a material by testing its chemical properties in the environmental components under carefully controlled laboratory settings [831]. Toxicity to organisms may be determined in a single species test using suitable organisms under carefully controlled, constant environmental conditions. The advantages of using controlled conditions are that direct comparison among test samples is possible and the test results can be extrapolated to the environment. This is due to the development of the predictive models. Physical properties that are especially relevant to the environmental behavior of the substances include water solubility, vapor pressure, etc. While these properties are easy to measure in pure substances, they are not easy to measure in mixtures. Hence, for substances with multiple components, specific test methods must be applied that measure range rather than the average value of the property. It is also impor tant to know the chemical structure and composition of the test sample. Table 13.9 lists substance properties that are of importance in determining a chemical's entry into the environment [831,832].

The knowledge of the chemical structure and the composition of the substance, in this case the lubricant, are important because they impact many other properties. For example, hydrocarbon lubricants will have greater lipophilic character than the water-based lubricants or the extremely polar triaryl phosphate lubricants which have a considerable affinity towards water. Similarly, more polar additives, such as oleic acid and low molecular weight alkenylsuccinic acids, which are used in certain lubricants to reduce friction and control rust, will have significantly higher water solubility than their higher molecular weight analogues, such as alkenylsuccinimide dispersants. As the complexity of the formulation increases, as is the case in engine oils, the difficulty in predicting the environmental impact increases as well.

A partition coefficient is a measure of differential solubility of a compound in two solvents. The best known of these partition coefficients is the one based on the solvents octanol and water. *Octanol/water partition coefficient* (K_{ow} or P_{ow}) is a measure of the hydrophobic and the hydrophilic character of a substance. Hydrophobic character relates to absorption, bioavailability, metabolism, and toxicity in the living organisms. Octanol/water partition coefficient is also used to predict and model the migration of the dissolved hydrophobic organic compounds into the soil and the groundwater. It essentially indicates a substance's propensity towards bio-accumulation, which is one of the key parameter in QSAR (quantitative structure-activity relationship) calculations. This partition coefficient is measured by the use of the Shake Flask Method [833], HPLC (ASTM E1147), or the QSAR estimation method [834]. The values of K_{ow} and P_{ow}, expressed in logarithm, depend on the carbon number range and the chemical types present in the mixture. Within a chemical class, the value of K_{ow} increases with an increase in carbon number and indicates increasing solubility of the material in the organic phase (octanol). Across hydrocarbon classes, for the same carbon number, the aromatic compounds have a lower K_{ow} value than the aliphatic compounds; the value for linear paraffins being slightly higher than that for branched paraffins, which in turn is slightly higher than that for cyclic paraffins, or naphthenes.

Vapor pressure is the pressure of a vapor in equilibrium with its nonvapor phases. Most often the term is used to describe a liquid's tendency to evaporate. At any given temperature, for a particular substance, there is a pressure at which the vapor of that substance is in equilibrium with its liquid or the solid form. This is the equilibrium vapor pressure or the saturation vapor pressure of that substance at that temperature, and from the term vapor pressure it is understood to mean the saturation vapor pressure. A substance with a high vapor pressure at normal temperatures is often referred to as being volatile. The higher the vapor pressure of a material at a given temperature, the lower is the boiling point. The vapor pressure of a substance is important for determining both the rate of evaporation and the relative amount of the substance that will be in the air phase. Vapor pressure tends to decrease with increasing molecular weight and polarity. Thus, the materials that have low water solubility tend to have high vapor pressures and move from the water to the air fairly quickly [832,835].

For pure materials, direct pressure measurements may be used. For materials with very low vapor pressure values or for a mixture, a vapor saturation method is used. The stationary medium in a generator column is coated with the material and air is cycled over the medium until it is saturated [832]. Hydrocarbons, especially paraffinics, are likely to volatilize from water. The vapor pressure of the materials beyond a certain molecular weight becomes too low to be used to determine the material's fate in the environment.

Water solubility is the key property that determines the persistence, biodegradation, and toxicity of a material in the aquatic environment. Water solubility (S_w), also know as aqueous solubility, is the maximum amount of a substance that can dissolve in water at equilibrium at a given temperature and pressure. Water solubility has been correlated with the octanol-water partition coefficient (K_{ow}). In general, water solubility decreases with an increase in carbon number [835], but for the compounds with the same carbon number it also depends upon the structure, for example the presence of a polar group, and branching. Water solubility can be measured by progressively increasing the amount of material to water under prescribed conditions until a saturation point is reached. For complex mixtures, the individual components in the mixture may have different water solubility limits. Incidentally, the definition of solubility as it relates to pure substances does not apply to complex mixtures [832].

TABLE 13.10—Common biodegradability standards [747].

Standard	Country
ASTM D6046 (EPA)	USA
ISO/CD 15380	International
Umweltbundesamt (UBA) RAL-UZ Blue Angel	Germany
UBA WGK Water Hazard	Germany
VAMIL	Netherlands
City of Gothenburg "clean lubricant project"	Sweden
SS 15 54 34	Sweden

Adsorption is the retention of a gas, vapor, or dissolved materials on the surface of a solid and desorption is the release of the same from a solid under the same conditions. The adsorption of organic chemicals onto the organisms, sediment, or soil is an important transport mechanism that determines the fate of the chemicals in the environment. It also helps in explaining bioaccumulation of the chemicals in aquatic, sediment dwelling, and soil organisms and whether or not the chemicals will leach into the ground water. For organic materials, the important determinant of adsorption is the organic content of the soil or the sediment. Sorption constant (K_{oc}) for an organic chemical in the soil and the sediments may be determined by the use of the ASTM E1195 procedure [27].

Some organic compounds are ionic and they dissociate into ions in the aquatic environment. *Dissociation constant*, or the ionization constant (pKa), expresses a material's tendency to form ions. Dissociation greatly influences all other environmental processes and properties, such as sorption, bio-accumulation, and toxicity [836]. Also, the ionic species may have different aquatic toxicity than the unionized form, as a result of different mechanisms to transport the molecule across cell membranes [832].

Once a chemical enters into the environment, several things can happen. It can get transported within the environmental medium or across the environmental media, transform, or accumulate. Environmental media are air, water, and sediment (soil). Transport within the medium, for example, air and water, depends upon air and water currents. The material may follow the current or diffuse. Transport across media is exemplified by a material of high volatility that comes out of the water into the air. It is important to

TABLE 13.11—Environmental persistence classification—Aerobic fresh water (ASTM D6046 (also used by U.S. Military) [747].

Pw1	Greater than or equal to 60 % in 28 days =Ultimate (ASTM)/Readily (OECD) Biodegradable
Pw2	Greater than or equal to 60 % in 84 days (12 wks)
Pw3	Greater than or equal to 40 % in 84 days (12 wks)
Pw4	Less than 40 % in 84 days (12 wks)

note that the media are normally not homogeneous and contain other components. For example, water contains suspended solids and may be air bubbles, depending upon the turbulence; atmosphere may contain particulates and water vapor; and the soil may contain water vapor and air. An important driving force for diffusive transport is the tendency of the material to reach equilibrium between these phases and the partition coefficients quantify this equilibrium. Transformation of organic chemicals involves their breakdown, for example, to carbon dioxide and water. During the transformation, however, intermediate products may form. Hence the toxicity of a material is a combined result of the chemical itself, the intermediates, and the final products. Thereby, it is important to know the degradation products, if possible, to be able to evaluate their impact on the environment. Environment-related degradation of materials can be abiotic, i.e., hydrolysis, photolysis, and oxidation, or biotic (biodegradation). Biodegradation is a complex process and only certain microorganisms can accomplish it. Not all microorganisms can degrade all materials in all environments. For example, some microorganisms can degrade only the products from a previous biodegradation or modify the structure only slightly. Others can biodegrade materials to a greater degree. Hence, the biodegradation outcome depends both on the type and the number of organisms used to test biodegradability.

Abiotic degradation of lubricants in the environment is primarily due to photolysis, hydrolysis, and photo-oxidation. A variety of degradation tests are established by organizations, such as ASTM, ISO, OECD, CONCAWE, U.S. EPA, Environment Canada, and various regulatory bodies in the European Union.

Biodegradability

Biodegradation is defined as the chemical breakdown or transformation of a substance caused by organisms or their enzymes, and biodegradability is the ability of a material to degrade through biodegradation. Most organic compounds are thermodynamically unstable and in the environment are attacked by the microbes and get converted into carbon dioxide. Biodegradation is of two major types: Primary biodegradation and ultimate biodegradation. *Primary biodegradation* is defined as modification of a substance by microorganisms that results in a change in some measurable property of the substance. All in all, the molecule stays largely intact. Partial biodegradation is not necessarily a desirable property, since the intermediary metabolites formed may be more toxic than the original substrate. Therefore, ultimate biodegradation, or mineralization, is the preferred goal. *Ultimate biodegradation* is the degradation achieved when a substance is totally utilized by the microorganisms, resulting in the production of carbon dioxide, methane, water, mineral salts, and new microbial cellular constituents. The molecular breakdown in this process is extensive and results in the loss of biological, toxicological, chemical, and physical properties of the original product or substance. The Degradation/Accumulation Expert Group of the OECD Environment Committee has established a series of tests which classify compounds into the following three groups.

1. *Readily biodegradable*—implies rapid and complete mineralization of the chemical. It is an arbitrary classification of chemicals that have passed certain specified screening tests for ultimate biodegradability.
2. *Inherently biodegradable*—implies 20–70 % biodegradation in 28 days. This classification includes chemicals for which there is unequivocal evidence of biodegradation (primary or ultimate) in any biodegradability test.
3. *Nonbiodegradable*—negligible biotic removal of material under the test conditions.

Some of the worldwide biodegradability standards are listed in Table 13.10, the environmental persistence classification is shown in Table 13.11, and the common biodegradability tests are listed in Table 13.12 [747]. The most commonly used test to determine primary biodegradation is CEC-33-A-93 and the tests used for ultimate biodegradation are ASTM D5864 and EPA 560/6-82-003. OECD's test most recognized in the lubricant industry, and listed in the EPA

TABLE 13.12—Tests of biodegradability in aerobic aquatic environments [747].

Test Title	*Measurements*	*Sponsoring Organization*
Ultimate Biodegradation Tests:		
D5864, Test Method for Determining the Aerobic Aquatic Biodegradation of Lubricants	% Theoretical CO_2	ASTM
9429:1990, Technical Corrigendum 1, Water Quality-valuation in an Aqueous Medium of the "Ultimate" Biodegradability of Organic Compounds-Method by Analysis of Released Carbon Dioxide	% Theoretical CO_2	ISO
301B, CO_2 Evolution Test (Modified Sturm Test)	% Theoretical CO_2	OECD
Aerobic Aquatic Biodegradation Test	% Theoretical CO_2	US EPA
C.4-C: Carbon Dioxide (CO_2) Evolution	% Theoretical CO_2	EUC
L-33-A-934, Biodegradability of Two-Stroke Cycle Outboard engine Oils in Water (Formerly L-33-T-82)	% Loss of Extractable CH_2 Groups	CEC

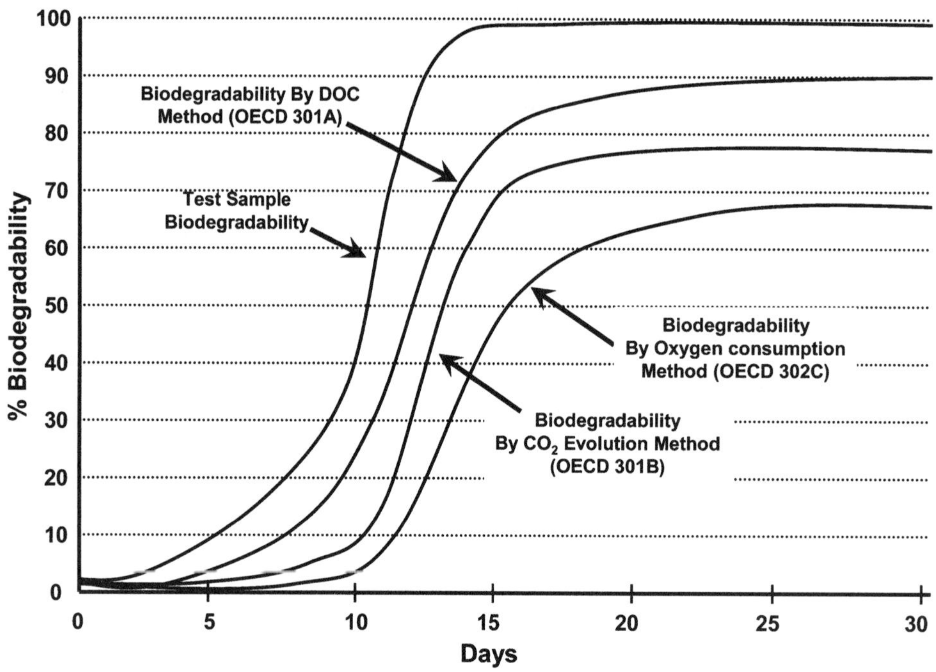

Fig. 13.7—Biodegradability methods comparison [831].

Guidelines, is OECD 301B CO_2 Evaluation Test (Modified Sturm): Biodegradability Assessment for Water Soluble and Insoluble Chemical Compounds and Formulations.

Eurospec distinguishes between ready and inherent biodegradability [747]. Inherent biodegradability is used to describe organic compounds, which are not readily biodegradable within 28 days. However, they must exhibit levels of degradation similar to those that are readily biodegradable over prolonged exposure periods. A chemical, which shows ultimate biodegradation in this situation, is likely to be considered as nonpersistent in the environment. A failure to show biodegradation after prolonged exposure renders the chemical as nonbiodegradable, or persistent. If a product does not pass the 28-day readily biodegradable part of the test, the chemical producer must consider if it wants to subject the material to inherently biodegradable portion of the test, which can take 84 days to complete. For a detailed discussion on biodegradability testing, see Ref [837].

Like abiotic degradation, biodegradation tests are many [838]. Certain standardized tests evaluate both the rate and the extent of biodegradation [831]. The rate is described by the terms inherent and ready. Inherently biodegradable means that the biodegradability is proven by any established test method. Degradation in the test must be at least 20 % but the duration is not specified, which may imply as long as needed. Tests used to determine inherent biodegradability include OECD 302A SCAS (ISO 9887), OECD 302B ZAHN-WELLENS (ISO 9888), and OECD 302C MITI (II). Readily biodegradable means that the substance (lubricant) degrades at least 80 % in salt water within 21 days. Most readily biodegradable fluids are based upon naturally occurring triglycerides (vegetable oils). Most common tests used to determine ready biodegradability are OECD 301A-F (ISO 7827, ISO 9439, MITI (I), ISO 10707, ISO 9408, and ISO 14593), OECD 310, and CEC L-33-A-93.

In these tests, the substance must demonstrate 60 % degradation to CO_2 or O_2 or 70 % removal of the dissolved organic carbon (DOC) within 28 days. Further, the pass criteria must be met within ten days after biodegradation exceeds 10 % of the mass loaded [839,840]. Ready biodegradation tests are so stringent that the substance that passes the test will rapidly degrade under most environmental conditions. ASTM Test Method D 5864, which includes the OECD 301B Modified Sturm procedures, evaluates biodegradation of the fully formulated lubricants or their components. It determines both the rate and the degree of the aerobic aquatic biodegradation, when exposed to an inoculum under laboratory conditions. The inoculum may be the activated sewage-sludge from a domestic sewage treatment plant, may be derived from soil or natural surface waters, or any combination of the three sources. The bacterial microorganisms used in the biodegradation tests are some of the simplest forms of life, and like all living organisms, are adversely affected by the presence of the chemical toxins. In this test, the low toxicity of the test sample is demonstrated by the microorganisms' ability to multiply and biodegrade the sample. The degree of biodegradability is measured by calculating the rate of conversion of the lubricant to CO_2. This test method is intended to specifically address the difficulties associated with testing the water insoluble materials and complex mixtures such as those that are found in many lubricants. This test method is designed to be applicable to all lubricants that are not volatile and are not inhibitory at the test concentration to the organisms present in the inoculum. For the water soluble test substances, the suggested reference substances are sodium benzoate or aniline. For the water insoluble test substances, the suggested reference substance is a low erucic acid rapeseed oil, also called LEAR, such as canola oil. Other organizations that have accepted the % theoretical CO_2 test method for determining the aerobic aquatic biodegradation of lubricants include ISO, OECD, U.S. EPA, and the EUC. A lubricant, hydraulic fluid, or lubricating grease is classified as readily biodegradable when 60 % or more of the carbon of the test material is converted into CO_2 in 28 days, as determined by the use of this test method. The test is considered complete when the CO_2 evolution reaches a plateau. The most established test methods used by the lubricant industry for evaluating biodegradability of their products are Method CEC-L-33-A-94 developed by the Coordinating European Council (CEC); Method OECD 301B, the Modified Sturm Test, developed by the Organization for Economic Cooperation and Development (OECD); and Method EPA 560/6-82-003, number CG-2000, the Shake Flask Test, adapted by the U.S. Environmental Protection Agency (EPA). These tests also determine the rate and the extent of aerobic aquatic biodegradation under laboratory conditions. The Modified Sturm Test and Shake Flask Test also calculate the rate of conversion of the lubricant into CO_2. The CEC test measures the consumption of the lubricant by analyzing the test material at various incubation times by the use of the infrared spectroscopy. Laboratory tests have shown that the degradation rates may vary widely among the various test methods indicated above and comparing the results from the different test methods for the same or similar materials is not easy. This is because % degradation is very dependent upon the type of biodegradability. A result of 100 % primary degradation in an inherent test cannot be compared with that of a ready test, where 100 % degradation is not possible and the conditions are more stringent. Also, there are many other variables to be concerned about [831].

OECD 301B modified Sturm test is adequate for soluble and insoluble organic, nonvolatile materials. This test measures the carbon dioxide evolved and therefore measures only the "complete" oxidation. Organic impurities will complicate the interpretation of the carbon dioxide production data. While the test is designed to last 28 days, it may end before if the degradation curve reaches a plateau for at least three determinations. The test may also extend beyond 28 days if the biodegradation has started but the plateau has not reached by Day 28. However, the test material will not be classed as readily biodegradable. The pass levels for ready biodegradability are 70 % removal of DOC (dissolved organic carbon) and 60 % of ThOD (theory oxygen depletion) or $ThCO_2$ (theory carbon dioxide) production for the respirometric methods. The lower limits are to take into account some of the carbon of the test chemical being retained in the new cells. These values must be attained within a 10-day window of the 28-day period of the test. The window starts when the degree of biodegradation reaches 10 % DOC and ThOD. Figure 13.7 compares biodegradability of the same material by various methods [831].

The CEC L-33-T-82, now listed as CEC L-33-A-934 test applies to most organic compounds, whether water soluble or insoluble and determines the overall biodegradability of the hydrocarbons, or similar compounds containing the methylene (CH_2) groups, measuring all transformations of the starting material including oxidation and hydrolysis. This CEC test is accepted in a number of Blue Angel Environmental Labels and requires 80 % or greater biodegradability. Despite being convenient and easy, the CEC test only measures the IR absorbance of the lipophilic molecules extractable into the chloroalkane solvent. It does not measure the water-soluble metabolites, which are difficult to extract and therefore cannot measure extensive degradation or mineralization. The CEC method has identified the following trends in lubricants [747].

1. Mineral oils, alkylated benzenes, PIB, PAOs, and poly(alkylene glycol)s have poor biodegradability, i.e., 0–40 %.
2. Vegetable oils (triglycerides), diesters, polyol esters show good biodegradability, (60–100 %). *n*-Alkyl monocarboxylic acids are common in nature and often appear as products from $\tilde{\beta}$ decarboxylation of the alkane moieties. Hence, they easily degrade as esters. The degradation products are also more soluble, although the effects on the ecosystem and groundwater contamination are not known.
3. Biodegradability of the aromatic polycarboxylate esters ranges from 5–80 %. Polyethers show poor biodegradability but have the advantage of being water-miscible. Because of which they do better in tests that are based on oxygen consumption, carbon dioxide evolution, or organic carbon removal. According to these tests, biodegradability is between 0 to 80 %. Polyethers of higher ethylene oxide content are more biodegradable than those with higher propylene oxide.
4. Biodegradability decreases if carbon number is less than C_4 or greater than C_{25} and as the chain branching increases.
5. Biodegradability depends on the available nitrogen and phosphorus in the environment and the inoculum size, if the lab test assessment is made after 21 to 28 days. Certain cultures of bacteria utilize tricresyl phosphate and zinc dialkyl dithiophosphate as carbon and phosphorus sources. Nitrogen heterocyclic may act as nitrogen sources and, if so, it may be possible to deliberately formulate blends in which the additives supply the necessary nitrogen, sulfur, and phosphorus for total base oil degradation.
6. Additives usually retard degradation in proportion to their concentration and are themselves poorly degraded, especially those with heterocyclic structures, such as triazine, and triazole.
7. Typical values in the CEC L-33-T-82 biodegradability test for common hydrocarbons are 15–35 % for mineral oil, 25–45 % for white oil, 70–100 % for natural and vegetable oil, 5–30 % for PAO, 0–25 % for polyether, 0–25 % for polyisobutylene, 5–80 % for phthalate and trimellitate esters, and 55–100 % for polyols and diesters.

Hydrocarbons are biodegradable to varying degrees. Part of the problem is their low water solubility, which decreases their availability to microorganisms. Aromatics, which are somewhat more soluble, give better results in biodegradability tests than paraffinics, and among paraffins the linear structures are considered more biodegradable than the branched structures. The outcome of the microbial oxidation is the formation of a carboxylic acid; hence the presence of the steric hindrance due to branching slows down the process somewhat.

Bioaccumulation is an increase in the concentration of a chemical in a biological organism over time. Materials that have the tendency to bioaccumulate are those that have high affinity towards lipids and are therefore resistant to meta-

TABLE 13.13—Commonly used additives in environmentally acceptable lubricants.

Additive	Compound	Water Poll. Class[a]
Oxidation Inhibitors	BHT and other phenols	1
	Alkylsubstituted diphenylamines	1
Deactivators for Cu, Zn, etc.	Benzotriazoles	2
Corrosion Inhibitors	Ester sulfonates	1
	Succinic acid esters	1
Antiwear Additives	Phosphoric esters	2
Pour Point Depressants	Styrene-maleic anhydride copolymers	Not identified
	Polymethacrylates	Not identified
Hydrolysis Protection	Carbodiimides	Not identified

[a]0=no danger; 1=little danger; 2=danger; 3=strongly endangering.

bolic breakdown by plants and animals. A number of standards, such as ASTM E1022, E1688, and E1676 are used to evaluate the tendency of a material to bioaccumulate [27,747].

Toxicity

Acute ecotoxicity of a substance is generally evaluated by conducting the acute toxicity tests. Such tests employ plants (algae), vertebrates (fish), and invertebrates (daphnia). Acute toxicity towards soil is determined by testing on invertebrates (earthworms) and plants (lettuce). Aquatic toxicity data on additives is rarely available. Interestingly, some additives, such as those used in food grade lubricants, are considered to be too toxic towards other life but are considered safe to humans. The most common test methods used for evaluating the acute toxicity of the lubricants are ASTM D6081 [27]; EPA 560/6-82-002, Sections EG-9 and ES-6; and OECD 203. These tests determine the concentration of a substance that produces a toxic effect on a specified percentage of test organisms in 96 hours. The acute toxicity test is normally conducted using rainbow trout and the results are expressed as concentration in parts per million (ppm) of the test material that results in a 50 % mortality rate after 96 hours (LC_{50}). A substance, such as a lubricant, is generally considered acceptable if aquatic toxicity (LC_{50}) exceeds 1000 ppm [841].

Standardization of toxicity testing is important because it takes variability out of the test, making test results more reproducible. In the standardized tests, biological variability is reduced by using the organisms of selected species that are of the same age range, similar in size, and in good health, with no observable abnormalities [838]. With quality data, it is possible to establish quantitative structure-activity relationships (QSARs), which can be used to predict toxicity to aquatic organisms based on structural or physical properties of the substance (ASTM E1242) [27]. QSAR is the process by which chemical structure is quantitatively correlated with a well defined process, such as biological activity or chemical reactivity [841]. As with the biodegradation test methods, each specific test provides data that are representative of a specific aspect of the natural environment. The dominating factor relating the toxicity test to the environment is the mode of exposure. As with the variability associated with the test organisms, each test material has unique characteristics that must be considered when conducting a standardized toxicity test. These include volatility, water solubility, and complexity/variability of the composition of the material. In order to relate the effects to exposure, one must define the duration of the exposure so that it is long enough to ensure maximum uptake of the material by the test organism. For testing low water solubility materials in acute toxicity tests, such periods range from two days for Daphnia tests to four days for most fish tests. Test guidelines usually require analytical conformation of the actual concentration over the course of the exposure period so that the collected data can be analyzed by the use of statistical techniques and to find the midpoint, or LC_{50} [841]. See the ASTM Standards E1847, E1023, and E729 [27]. The same approach is used to establish other effect endpoints, such as impact on growth or reproduction, for chronic testing. Typically, lubricants are considered "difficult materials" for aquatic toxicity testing, which is due to their low water solubility and being complex mixtures. Hence, designing, conducting, and interpreting aquatic toxicity studies on such materials involve special considerations [842].

There are a number of issues that need to be addressed because of the lubricants being of low water solubility. The acute LC_{50} values for a homologous series of chemicals are hard to distinguish because of the little differences in solubility across members. Also, as the hydrocarbon number increases there is a point where the material solubility in water is too low to show an effect on mortality. The presence of the water-soluble minor components can have a profound effect on the test results and the toxicity observed in such samples is not indicative of the toxicity of the bulk material. In some cases emulsifiers or solvents are used to improve the water solubility of the poorly soluble substances. If there is no toxicity associated with these dissolution aids themselves, there is no problem using this approach [841].

As mentioned earlier, the lubricants on account of being complex mixtures are difficult to test. Most lubricants comprise a base fluid as the major component, which either contains the same size molecules or is a mixture of homologues or isomers. These structural differences do not have much effect on aquatic toxicity. This complicates evaluating their toxicity. However, since most hydrocarbons are believed to show toxicity by the same mechanism and the effect to be additive, a toxic unit approach can be used to predict their acute toxicity [843]. The approach involves determining the threshold value (TV) of a component, which is the aqueous concentration of the component divided by the component's corresponding LC_{50}. Then the threshold values of all components in the mixture are added. If the result is greater than one, the composition is considered toxic, and if it is less than one, the composition is considered nontoxic. The procedure does not work well for substances of low water solubility and precise water solubility of the individual components is hard to determine [838]. Another approach, called "lethal loading" has been suggested to overcome these difficulties [842]. This methodology, specifically designed for complex mixtures of low water solubility, measures toxicity based on the amount of substance added to water and not the concentration of the dissolved components. The amount of substance added is termed as loading for simplicity. The toxicity values are expressed as LL_{50} to indicate that they are obtained from lethal loading and EL_{50} to indicate that they are obtained from effect loading. These designations were chosen to differentiate them from LC_{50} or EC_{50}, the toxicity results obtained from the solubility-based approach. Although for lubricants the lethal loading approach is preferred, it is not a good substitute for quantified exposure data. A list of standard toxicity methods is available in Ref [838]. As mentioned earlier, ecotoxicity employs plants, vertebrates, and invertebrates. The reference also lists plants and organisms that are used to determine the effect of chemicals on the environment. The basis of choosing these organisms is to test various members of the aquatic food web. Algae, invertebrates which graze on the algae, and fish which feed on the invertebrates are included in aquatic testing [841].

Environmentally Acceptable Lubricants

Earlier we talked about environmentally acceptable lubricants and mentioned that biodegradability and nontoxicity are two of the criteria for such fluids. Both these attributes depend on the base stocks and the additives used to formulate them. Base stocks differ in the degree of their biodegradability. Although there are conflicts with respect to the absolute degree of biodegradability, linear hydrocarbons and biological base stocks, such as vegetable oils, modified triglycerides, and synthetic esters—especially those made from the modified triglycerides—have the highest degree of biodegradability. Because of this, many environmentally acceptable lubricants use these materials as base fluids. Vegetable oils are finding increasing use as lubricant base stocks, not only because of their high degree of biodegradability but also due to their excellent lubricity, high viscosity index, high flash point; and above all because they originate from renewable sources and are non-toxic. High biodegradability makes them the ideal base stocks to formulate fluids to lubricate machinery operating in the natural environments. However, they have limitations with respect to hydrolysis, low-temperature properties, and oxidative stability. Many of these are being corrected by structural modifications and through genetic engineering to produce oils of even higher monounsaturate content. The most common sources of these base stocks include castor oil, rapeseed oil, olive oil, coconut oil, and palm oil. Out of these, the rapeseed and canola oils are preferred because of their high monounsaturate content, which makes them more oxidatively stable than those of the low monounsaturate (high polyunsaturate) content, such as soybean oil. These oils must be formulated with additives that either do not hurt the base stock's biodegradability and nontoxicity or are biodegradable and nontoxic themselves. Additives that are considered suitable for use in these lubricants are provided in Table 13.13.

TABLE 13.14—Constituent/property—Allowable levels.

Metal/Property	Limit
Arsenic	5 ppm maximum
Cadmium	2 ppm maximum
Chromium	10 ppm maximum
Lead	100 ppm maximum
Flash Point (Closed Cup)	100 °F (38 °C)
Total Halogens	4000 ppm maximum
PCBs	<2 ppm maximum

Disposal Issues

In the previous sections of this chapter we covered the topics of the lubricant conservation, recycling, and its entry into the environment and its impact. It was also mentioned that despite the preventive measures and regulatory control, a significant volume of the used lubricant is released into the environment (soil and water), either intentionally or unintentionally. The last topic that we need to address is the used lubricant disposal, which implies legal disposal. Burning and landfill are the focus of the discussion in this section.

Most countries have stringent regulations with respect to direct disposal in landfill and discharge into water. This is because even a small amount of lubricant can contaminate a large body of water, rendering it unsuitable for human consumption. However, most regulations permit entry of a threshold amount of oil into the nation's water supply that may escape treatment of the industrial waste water. Burning the lubricant that cannot be recycled as fuel, typically to generate electricity, is also governed by regulations, mainly because of the presence of the relatively high levels of additives. Burning generates airborne pollutants and ash, rich in toxic heavy metal compounds. The lubricant is therefore burned in facilities that are fitted with scrubbers to remove the airborne pollutants and particulates and have access to landfill sites to dispose of the resulting ash. While modern filtering, reconditioning, and re-refining technologies have facilitated recycling, there are situations where the used lubricant is not recyclable. For example, the used oil that does not meet EPA' s criteria of used oil, is heavily contaminated, or there are toxic components, such as PCBs, that are present. Under certain circumstances, the decision not to recycle may even be purely economics based. That is, burning the used oil allows the used oil generators to avoid the cost of collection and the cost of purchasing the clean fuels.

As mentioned earlier, direct disposal of the used lubricant in landfills is prohibited in many countries, which is to encourage recycling. Burning may be a better option, both because it provides high energy at a lower cost and because the used oil generates little revenue, which is especially the case if the oil's contamination levels are too high. Use as a burner fuel is a disposal option rather than a reprocessing option, as only the thermal (caloric) value of the oil is captured. It is estimated that 63 % of the total used oil is reprocessed for use as fuel, which is more profitable than re-refining it for use in lubricants. Prior to the use as fuel, used oil needs to be reprocessed. As a first step, the used oil is tested for certain types of contaminants, such as excess water, sediment, and polychlorinated biphenyls (PCBs). If the used oil fails this testing, it must be specially treated and managed. The uncontaminated used oil is slowly heated to separate the water from the oil. The water is sent to a wastewater treatment plant and the oil is filtered and blended with the crude oil or other potential fuels described below, and burned.

Used oil has a high caloric value of 10,000 Kcal/kg (39,700 BTU/lb); hence it is a good source of energy. There are three burning options for the used oil. These are incinerating at high temperatures, for example, in cement and lime kilns, burning untreated used oil, and blending it with the fuel oil, prior to burning. Incineration is the most effective way of destroying the used oil since it does not require any pretreatment. This is because the very high combustion temperatures and long retention times in the burning zone ensures the destruction of the toxic compounds. This minimizes the environmental disposal costs. However, it is a low value option. Burning the untreated used oil, especially in space heaters, is also inadvisable since it merely releases the harmful components into the environment. Used oil space heaters are essentially regular oil heaters that are slightly modified to burn less volatile and more viscous engine oils and other hydrocarbon lubricants. Though modified, these heaters operate without pollution controls. Burning used oil in space heaters is a small volume use and hence it is very difficult to get rid of the used oil in large volume by this method. Mixing the used oil with another fuel, such as gasoline, diesel, kerosene, Jet-A fuel, or fuel oil produces rich combustible mixtures and is a better option. Such mixtures can be used in industrial facilities if they are registered with the EPA to power various operations. Incidentally, such mixtures are not suitable for use in space heaters. In some countries, the used oil is sprayed on coal to improve its coking qualities.

Air emissions associated with burning the used oil result from the chemical constituents in the oil itself. Used oil contains residual additives or additive decomposition products. Additives that are of special concern are those that contain sulfur, phosphorus, halogens (chlorine and bromine), and transition and heavy metals, such as zinc, barium, lead, and molybdenum; as well as the aromatic organics, such as naphthalenes and phenols. In addition, many of the used oils contain wear metals, such as arsenic, chromium, cobalt, manganese, and nickel. While recycling can remove many of these products through filtration and re-refining, when burned in space heaters and in industrial furnaces, the contaminants in the used oil are discharged into the atmosphere or stay in the ash. Industrial facilities scrub emissions prior to their release into the air and dispose of ash by sending it to the landfills. Industrial users sometimes remove metal debris from the lubricant through filtration, prior to burning. It lowers the amount of ash that needs to be disposed of. In Fig. 13.5, we show various applications where the used oil is used. Each application has its own requirements and specifications, or both. Specifications for burning for energy recovery are summarized in Table 13.14. State-specific rules can vary, but the federal rules that must be adopted by each state as the minimum requirements are found under Title 40 of the Federal Regulations at Part 279. The EPA also maintains a Website for the used oil which can be accessed at www.epa.gov/epaoswer/haswaste/usedoil/.

Copyright Permissions

1. Title Cover Background — 2008 GMC Sierra Heavy-duty (GMT900) Allison 1000 Six-speed Automatic Transmission and 2005 GMC Canyon Double Overhead Cam In-Line 4-Cylinder Engine — Courtesy of General Motors.
2. Figure 4.140 and 11.16 — Mandakovic, R., "Assessment of EP Additives for Water Miscible Metalworking Fluids," J. Syn. Lubr, Vol. 16, No. 1, 1999, pp 13–26. © John Wiley & Sons Limited. Reproduced by permission.
3. Figures 3.14, 3.28, and 3.46 — van der Waal, G., "The relationship Between the Chemical Structure of Ester Base Fluids and Their Influence on Elastomer Seals and Wear Characteristics," J. Syn. Lubr. Vol. 1, No. 4, 1985, pp. 281–301. John Wiley & Sons Limited. Reproduced by permission.
4. Figures 4.30, 4.147, and 4.148 — Hamblin, P. C., Kristen, U., and Chasan, D., "A Review: Ashless Antioxidants, Copper Deactivators, and Corrosion Inhibitors, Their Use in Lubricating Oils," Lubrication Science, Vol. 2, 1990, pp. 287–318. John Wiley & Sons Limited. Reproduced by permission.
5. Tables 2.4, 3.17, 4.13, 5.25, 5.26, 5.28, 5.29, 5.37, 7.4, 7.5, 7.6, 7.7, 7.8, 7.9, 7.10, 7.40, 7.41, 7.42, 7.43, 8.10, 8.15, 9.3, 9.4, 10.10, and 10.30. © The Lubrizol Corporation. Reproduced by permission.
6. Figures 1.18, 1.20, 1.21, 1.22, 1.23, 1.24, 1.25, 1.26, 1.27, 4.64, 4.189, and 5.35. © The Lubrizol Corporation. Reproduced by permission.
7. Figures 4.160 — Rempp. P. and Herz, J. E. "Model Polymers," Concise Encyclopedia of Polymer Science and Engineering, J. I. Kroschwitz Editor, 1990, pp. 633–634, Wiley InterScience. John Wiley & Sons Limited. Reprinted with permission.
8. Tables 6.3 to 6.15 and 6.18 to 6.24. © DieselNet.com Reproduced by permission.
9. Figures 5.19 to 5.33. © Southwest Research Institute — SwRI® (http://www.swri.org). Reproduced by permission.
10. Figures 10.7, 10.14, 10.16, 10.17 and Tables 10.17, 10.19, 10.24 to 10.29. © National Lubricating Grease Institute, Missouri, USA. Reproduced by permission.
11. Figure 11.4 and Tables 11.9, 11.10, and 11.14. © BlueScope Steel Corporation. Reproduced by permission.
12. Figures 1.3, 4.8, 4.112, 4.115, 4.157, 4.158, 4.181, 4.182, 6.18, 6.25, 6.41, and 11.2; Table 2.1. Society of Automotive Engineers (SAE). Reproduced by permission.
13. Figure 8.9. Author Y. Kamura, "Some Problems in the Adhesion Theory of Friction," Fundamentals of Tribology, N. P. Suh and N. Saka Editors, Published by MIT Press, Cambridge, 1980, pp. 385-391. © MIT Press. Reproduced by permission.
14. Figures 4.145, 5.11, 5.13, 8.2, 8.10, 11.6, 12.12; Tables 10.5, 10.7, 10.31, and 12.12; and some items from Glossary of Terms. © ASM International. Reproduced by permission.
15. Figures 2.12, 4.120, 4.129, 4.130, 4.133, 4.134, 4.135, 4.139, 4.141; Tables 2.13 and 3.9. © Society of Tribologists and Lubrication Engineers (STLE). Reproduced by permission.
16. Figures 5.14 and 8.11. © Energy Institute UK. Reproduced with kind permission of the Energy Institute UK.
17. Tables 2.8, 4.24, 7.3, 7.24, 7.32, 7.35, 12.14, 12.15, 12.16, 12.17, 13.1, 13.2, 13.3, 13.4, 13.5, 13.6; Figures 3.49, 3.52, 4.154, 8.4, 13.1. Copyright 1983 — From CRC Handbook of Lubrication: Theory and Practice of Tribology, Vol. I — Application and Maintenance, E. Richard Booser Editor. Reproduced by permission of Taylor & Francis Group, LLC, a division of informa plc.
18. Tables 3.20, 3.21, 4.16, 4.17, 4.20, 7.18, 10.9, 12.1, 12.2, 12.3, 12.4, 12.13; Figures 4.102, 4.103, 4.126, 4.127, 4.136, 4.137, 4.145, 8.8, 12.2, and 12.3 Copyright 1988/1989 — From CRC Handbook of Lubrication: Theory and Practice of Tribology, Vol. II — Theory and Design, E. Richard Booser Editor. Reproduced by permission of Taylor & Francis Group, LLC, a division of informa plc.
19. Tables 3.1, 4.23, 7.44, 7.45, and 9.1; Figures 3.9, 3.10. 7.10, and 9.1 Copyright 1993/1994 — From CRC Handbook of Lubrication and Tribology, Vol. III — Monitoring, Materials, Synthetic Lubricants, and Applications, E. Richard Booser Editor. Reproduced by permission of Taylor & Francis Group, LLC, a division of informa plc.
20. Figures 4.35, 4.36, 4.37. 4.170, and 11.15; Tables 4.14, 4.15, 4.19, 4.21, 4.22, 4.29, 4.32, 10.8, 11.19, and 11.21 Copyright 2003 — From Lubricant Additives: Chemistry and Applications, Leslie R. Rudnick Ed., Marcel Dekker, Inc. Reproduced by permission of Taylor & Francis Group, LLC, a division of informa plc.
21. Figures 2.2, 3.15, 4.16, 4.19, 4.20, , 4.140, 4.142, 4.179, 4.183, 5.15, 4.40, 7.64; Tables 3.4 and 7.28. Springer. Reproduced by permission.
22. Figures 6.28, 6.29, 6.30, 6.36, 6.37, 6.38, 6.40, 6.46, and 6.47. Chevron Corporation. Reproduced by permission.
23. Fig. 3.54 (Source: Cargill); Fig. 4.97–4.99 (Source: Ken Crosthwaite et al., ExxonMobil, NPRA Lubricants and Waxes Paper, November 1999); Fig. 4.100 (Source: G. Bleimschein, J. Fotherigham, and A.Plaomar, BP Chemicals); Fig. 4.192 (Source: ILSAC); Figs. 5.43, 5.44 (Source: Cummins); Figs 6.33 (Source: Argonne). © Lubes'N'Greases, Reproduced by permission.

Appendix

Acronyms for Organizations and the Terms Related to Lubricants, Lubrication, and Tribology [667]

AAM—Alliance of Automobile Manufacturers.

AAMA—American Automobile Manufacturer Association.

ACC—American Chemistry Council.

Abrasion (Abrasive Wear)—A process in which hard particles or surface roughness (protuberances) are forced against and moving along a solid surface causing displacement of material. Also, see Abrasive Erosion.

Abrasive Erosion—Erosive wear caused by the relative motion of solid particles which are entrained in a fluid, moving nearly parallel to a solid surface.

ACEA—Association des Constructeurs Européens de l'Automobile (Association of European Automotive Manufacturers.

Acidity—The amount of free acid in any substance.

Acid Number—A measure of the amount of potassium hydroxide (KOH) needed to neutralize all or part of the acidity in a petroleum product.

ACS—American Chemical Society.

Additive—In lubrication, a material added to a base stock for the purpose of imparting new properties or of enhancing its existing properties. Major classes of additives include dispersants, detergents, oxidation inhibitors, viscosity modifiers, and antiwear/extreme pressure agents.

Additive Treat Level—The total percentage of all additives in a lubricating oil or grease.

Adherence—Physical attachment, either by adhesion or by other means, of a material to a surface.

Adhesion—Attraction or joining of two materials. The strength of adhesion is reflected by the force or forces necessary to separate them. In frictional contacts, adhesion is the attractive force between adjacent surfaces. In physical chemistry, it is the attraction between a solid surface, such as metal and a second liquid or solid phase, such as lubricating grease. In some instances, adhesion is reversible but in others, it is irreversible. When adhesion is solely due to interlocking of the asperities, it is called mechanical adhesion.

Adhesive Wear—Wear caused by the transfer of material from one surface to another during movement (tearing) of the solid-phase welded surfaces. Welding occurs due to frictional heat resulting from metal-to-metal contact. The particles removed from one surface are either permanently or temporarily attached to the other surface.

AF,AFR—Air/Fuel Ratio.

AFNOR—Association Française de Normalisation.

AFV—Alternative Fuel Vehicle.

Age Hardening—An increase in consistency (hardening) of greases over time. Also see Thixotropy.

AGELFI—Co-operative Research Organisation of the oil companies AGIP, ELF and FINA.

AGMA—American Gear Manufacturers' Association.

AGO—Automotive Gas Oil.

AHEM—Association of Hydraulic Equipment Manufacturers.

Air Entrainment—Incorporation of air in the form of bubbles as a dispersed phase in the bulk liquid. Air entrainment may be because of the mechanics of the system or by the release of the dissolved air due to a sudden change in environment, or both. Appearance of a liquid, such as being bubbly, opaque, etc., is an indication of air entrainment. It is important to note that the dissolved air can only be determined by analysis.

Alkali—While chemically any substance with basic properties (alkaline) is considered an alkali, the term usually implies hydroxides of ammonium, lithium, potassium, and sodium. The last three are members of the class of alkali metals. These substances are water soluble and have the ability to neutralize acids. The term is also extended to the hydroxides of the alkaline earth metals, such as magnesium, calcium, barium, and strontium.

Alkylaromatics (Alkylbenzene and Alkylnaphthalene)—These additives, made by the alkylation of benzene and naphthalene, are used as base fluids as well as sulfonated to sulfonic acids to make detergent additives.

Alkylbenzenesulfonic and Alkylnaphthalene-sulfonic Acids—These are used to make detergent additives by their reaction with an alkali metal or an alkaline earth metal base, with or without carbon dioxide.

Alkylphenol—Product of alkylation of phenol. Some alkylphenols, such as 2,6-di-*t*-butylphenol and its derivatives are used as oxidation inhibitors in lubricants. Others, such as dodecylphenol, are converted into metal salts, called phenates, for use as detergent additives.

Alkylsalicylic Acid—Made from alkylphenol by base catalyzed carbonation. Used as a starting material for making alkaline earth metal salicylate detergents for use in engine oils.

Almen EP Lubricant Tester—A journal bearing machine used for determining load-carrying capacity or extreme pressure properties of the gear lubricants, or both.

ALTENER—Alternative Energy Programme of the European Commission.

AMT—Automated Manual Transmission.

ANFAVEA—Auto Manufacturers Association (Brazil).

Anhydrous—Free of water. Term is often used to describe lubricating grease.

Aniline Point—The lowest temperature at which equal volumes of aniline and a hydrocarbon liquid (fuel or lubricant base stock) are completely miscible. A lower aniline point is a measure of high aromatic content of the liquid. Aniline point is used to assess solvency of a base stock or the cetane number of distillate diesel fuel. A product of high aniline point will be low in aromatics and naphthenes and, therefore, high in paraffins. In conjunction with API gravity, the aniline point may also be used to calculate the net heat of combustion of aviation fuels.

ANSI—American National Standards Institute.

Anticorrosion Additive—See Corrosion Inhibitor.

Antifoam Agent/Additive—See Foam Inhibitor.

Antifriction Bearing—A type of bearing that employs

rollers or balls. They are also called rolling element bearings. These may or may not contain a solid lubricant.

Antifriction Material—Material that exhibits low-friction or self-lubricating properties.

Antiknock—Resistance to detonation or pinging in spark-ignition engines.

Antioxidant—See Oxidation Inhibitor.

Antiscuffing Lubricant—Lubricant that is formulated to avoid scuffing, a type of wear damage. A more descriptive name is extreme-pressure lubricant.

Antiseizure Property—Ability of a bearing material to resist seizure during momentary lubrication failure.

Antistatic Additive—Additive that increases the conductivity of a hydrocarbon fuel to quickly dissipate electrostatic charge during fast dispensing, thereby reducing the fire/explosion hazard.

Antiwear Additive—Lubricant additive used to reduce wear. Additives or their reaction products form thin, tenacious chemical films on highly loaded parts to separate the potentially contacting surfaces.

Antiwear Agents—See Antiwear Additive.

Antiwear Number (AWN)—A variant of the wear coefficient (k). $AWN = -\log_{10} k$

Anti-weld Characteristic—See Antiseizure Property.

APE—Association of Petroleum Engineers (USA).

API—American Petroleum Institute.

API Engine Service Classification System—Classifications and designations for automotive engine oils developed by API in conjunction with SAE and ASTM.

API Gravity (API Degree)—Measure of density used in the U.S. petroleum industry where: $API° = [(141.5/\text{Specific Gravity at } 60\ °F)] - 131.5$.

API Oil Group—Classification of base fluids, both petroleum-derived and synthetic, based on VI and sulfur content, for use in formulating lubricants.

Apparent Area of Contact—Area of contact between two solid surfaces defined by the boundaries of their macroscopic interface.

Apparent Viscosity—Ratio of shear stress to shear rate of a non-Newtonian fluid, for example, thickened lubricants, such as viscosity modified oils and greases. In greases, the apparent viscosity depends on shear rate and temperature. Hence, both these parameters must be specified while reporting apparent viscosity.

Appearance/Bulk Appearance—General term relating to characteristics that are observable by visual inspection. It is usually reported for greases as bloom, color, luster, and texture. Bulk appearance determination involves an examination of the grease surface in an undisturbed state and is expressed as *bleeding, cracked, grainy, rough, smooth, bright, dull, etc*.

1. Bleeding, if there is free oil on the surface.
2. Cracked, if there are surface cracks.
3. Grainy, if the surface contains granules or lumps.
4. Rough, if the surface shows irregularities.
5. Smooth, if surface is free of irregularities.
6. Bright, if it reflects light with a relatively strong intensity.
7. Dull, if it reflects light with a relatively weak intensity. A high water content or certain thickeners and fillers may impart lubricating grease dull luster.

Aromatic—Derived from, or characterized by, the presence of a benzene ring.

Arylsulfonic Acid—Also see Alkylbenzenesulfonic Acid and Alkylnaphthalenesulfonic Acid. Alkylbenzene sulfonic acids can be natural or synthetic in origin. Natural sulfonic acids are obtained as by-products during base oil manufacture. Synthetic sulfonic acids are obtained from alkylation of benzene.

ASEAN—Association of South-East Asian Nations.

Ash—Metal-derived deposits formed in the combustion chamber and other engine parts during high-temperature operation.

Ash Content—Residue in weight percent left after combustion of a sample of a petroleum product.

ASLE—American Society of Lubrication Engineers, Now called Society of Tribologist and Lubrication Engineers (STLE).

ASM—American Society of Metals.

ASME—American Society of Mechanical Engineers.

Asperities—Small irregularities on a solid surface. Shape, size, and mechanical properties of asperities are the basis for developing many theoretical models for friction, lubrication, and wear.

Asphalt—Black to dark-brown solid or semi-solid material left during petroleum refining after removal of fuels and lubricant base stocks. Bitumens are the primary constituents of asphalt.

Asphaltic—Materials that are largely composed of or are similar to asphalt. Term is often used to describe lubricating oils derived from crude oils which contain asphalt.

ASTM—International (formerly American Society for Testing and Materials), whose D2 Committee deals with Petroleum Products and Lubricants.

ASTM Colorimeter—Apparatus widely used for determining the color of lubricating oil (ASTM Method D1500). The color so determined is known as ASTM Color.

ASTM Viscosity-Temperature Equation—Relates kinematic viscosity (v) with temperature.

ATC—Technical Committee of Petroleum Additives Manufacturers (Europe).

ATEIL—Association Technique de l'Industrie Européennes des Lubrifiants.

ATF—Automatic Transmission Fluid.

Auto-ignition—Spontaneous ignition and the resulting rapid reaction of a portion or all of the fuel-air mixture in an engine. The flame speed is many times greater than that of normal spark ignition. The noise associated with it is called knock.

Automatic Transmission Fluid (ATF)—Lubricant/hydraulic fluid used in automotive transmissions.

Bactericide—See Biocide.

Base—Compound that reacts with acids to form salts and water. Alkalis are water-soluble bases that are used in petroleum refining to remove acidic impurities. Oil soluble bases are used as lubricant additives to neutralize acids formed during the combustion and oxidation of the lubricant.

Base Number—Amount of acid (perchloric or hydrochloric) needed to neutralize part or all of a lubricant's basicity. When the acid in a sample is completely neutralized, the base number is called the total base number (TBN). Base number is usually expressed as mg KOH/g of sample, the same as the acid number.

Base Stock—Base fluid, usually a refined petroleum fraction or a selected synthetic material, into which additives are blended to produce finished lubricants. Base stock comprises major portion of the lubricant.

Basic Sulfonate—See Overbased Sulfonate.

Batch—Quantity of material handled or considered as a unit in processing.

Bath Lubrication—See flood lubrication.

Bentonite—Mineral montmorillonite, a magnesium-aluminum silicate, which is used as a component of drilling mud and a thickener in greases.

BHRA—British Hydromechanics Research Association.

Bingham Solid—An idealized solid that begins to flow appreciably only when the amount of applied stress exceeds its yield stress. After which the flow rate is proportional to the difference between the applied stress and the yield stress. Many lubricating greases can be considered as Bingham solids.

Biocide—Additive used to inhibit bacterial growth in the water-based hydraulic and metalworking fluids.

Biodegradation—Process of biochemical breakdown or transformation of a substance caused by microorganisms and their enzymes.

BIS—Bureau of Indian Standards.

Bitumen—Also called asphalt or tar, bitumen is the brown or black viscous residue left after the vacuum distillation of crude petroleum. It consists of high molecular weight hydrocarbons and minor amounts of sulfur and nitrogen compounds.

Black Oils—Lubricants that contain asphaltic materials, which improve adhesion and hence they are used to lubricate open gear systems and steel cables.

Bleeding—Oil released on the surface of a lubricating grease when it is not in service.

Blending—Process of mixing various lubricant base stocks to obtain base oils/fluids of desired physical properties.

BLF—British Lubricants Federation.

Block Grease—Lubricating grease that is sufficiently hard to maintain its shape in block or stick form.

Bloom—Typical blue or green surface color of a grease when viewed by reflected daylight. The color is associated with the absorption of the ultraviolet light in the oil and is not always visible when the sample is viewed by artificial light.

Bloom Fluorescence—Different color of an oil in reflected light than in transmitted light. For some applications, oils with yellowish-green bloom are preferred over those with bluish-green bloom. This demand is met either by special processing or by the use of fluorescent additives.

Blow-by—Unburned and partially burned fuel and combustion gases that travel past the piston rings into the internal combustion engines. This not only results in fuel dilution of the lubricant and its contamination, but also results in extensive oxidation of the oil by the chemically reactive species and free radicals that are present in the blow-by.

Blown Oil—Fatty oil artificially thickened by blown air.

BNP—Bureau de Normalisation des Petroles.

BOCLE—Ball on Cylinder Lubricity Evaluator.

Body—Nonscientific term used to designate viscosity or consistency.

BOI—Base Oil Interchanger.

Boiling Point—Temperature at which a substance experiences a phase change from being liquid to vapor. Boiling point is directly affected by pressure: It increases with an increase in pressure and decreases with a decrease in pressure.

Bonded Film Lubricant—See Bonded Solid Lubricant.

Bonded Solid Lubricant—Solid lubricant dispersed in a continuous matrix of a binder, or attached to a surface by the use of an adhesive material.

Boundary Lubricant—Lubricant that protects metal surfaces against wear under boundary conditions. This type of lubricating environment occurs when the loads are high and speeds are low. Boundary lubricants contain sulfur and phosphorus containing antiwear or extreme pressure (EP) additives that prevent surface wear damage by making resilient chemical surface films. Boundary conditions exist in bearings, gears, and some metalworking operations. Also see Elasto-hydrodynamic Lubrication, Extreme-pressure Lubrication, and Thin-film Lubrication.

Boundary Lubrication—Lubrication environment in which the lubricant film is too thin to effectively separate mutually rubbing surfaces and as a result, contact of surface asperities occurs. Consequently, friction and wear protection are determined by the chemical nature of the lubricant rather than its bulk properties.

Break-in—See running-in.

Break In—Operate a newly installed bearing, seal, or other tribo-component to condition its surface(s) for improved functional operation. The term run in is also used to describe this process.

Bright Stock—Refined, high viscosity lubricating base stock obtained from residual stocks by the use of processes such as acid treatment, solvent extraction, dewaxing, and or clay finishing. These stocks have low pour points, excellent solvency, and superb viscosity-pressure relationship.

Brookfield Viscosity—Apparent viscosity of a non-Newtonian fluid at controlled temperature and shear rate as measured by Brookfield viscometer.

BRT—Ball Rust Test.

BSI—British Standards Institution.

BSN—Bacharah Smoke Number.

BTC—British Technical Council of the Motor & Petroleum Industries (UK National Body in CEC).

BTU—British Thermal Units.

Built-up Edge—Material collected from the machining of a work piece that is collected at the edge of the cutting tool. Also see wedge formation.

Burning—Oxidation of a surface due to local heating in an oxidizing environment that commonly occurs in sliding contacts.

Burnish—Alteration of an originally machined surface to a more polished condition as a consequence of sliding or rolling action.

CAFE—Corporate Average Fuel Economy.

Cam—Eccentric shaft that is used in most internal combustion engines to coordinate opening and closing of the valves.

CARB—California Air Resources Board.

Carbon Residue—Coked material remaining after an oil has been exposed to high temperatures under controlled conditions.

Carboxylate Esters—See Diesters and Polyol Esters.

Catalytic Converter—An integral part of a vehicle's emissions control system. Oxidizing converters remove hydrocarbons and carbon monoxide (CO) from exhaust gases by oxidizing them to carbon dioxide (CO_2) and water, and reducing converters remove nitrogen oxide (NOx) emissions by chemically converting it into nitrogen (N_2) and water. Both use noble metal (platinum, palladium or rhodium) catalysts that can be "poisoned" by the phosphorus compounds present in the lubricant.

Catastrophic Wear—Sudden surface damage, deterioration, or a change of shape caused by wear, to an extent that the life of the part is appreciably shortened or its function is impaired.

Cavitation—Formation of vapor cavities within a liquid and their subsequent collapse caused by movement or vibration within the liquid film. Commonly occurs in greases and hydraulic fluids.

Cavitation Cloud—A collection of a large number of small cavitation bubbles.

Cavitation Damage—See Cavitation Erosion.

Cavitation Erosion—Progressive loss of material from a solid surface due to continuous exposure to cavitation; it is the collapse of the vapor bubbles that lead to high temperatures and local high-impact pressures which cause erosion.

CCD—Combustion Chamber Deposit.

CCMC—Comite des Constructeurs d'Autonobile de Marche Commun (Common Market Automobile Manufactures Association.

CCS—Cold Cranking Simulator.

CEC—Conseil Européens de Coordination pour les Développements des Essais de Performance des Lubrifiants et des Combustibles pour Moteurs (Coordinating European Council), whose test method CEC L-33-A93 (formerly CEC L-33-T82) is widely used to determine primary biodegradation of lubricants.

CEC-EFTC-CEC—Engine Fuels Technical Committee.

CEC-ELTC-CEC—Engine Lubricants Technical Committee.

CEC-TLTC-CEC—Transmission Lubricants Technical Committee.

CEFIC—European Chemical Indusrty Council.

CEN—Conseil Européens de Normalisation.

CentiStoke or centistokes(cSt)—Unit of kinematic viscosity. 1 cSt = 1 mm^2/s.

Cetane Index—Approximate cetane value of a diesel fuel calculated from physical properties, such as API gravity and mid-boiling point.

Cetane Number—Measure of ignition quality of a diesel fuel, as determined in a standard single cylinder test engine which measures ignition delay relative to primary reference fuels. High Cetane Number is associated with easy starting of a high-speed, direct-injection diesel engine and the less "white smoking" and the less "diesel knock" after the start-up. Cetane Number is determined by comparing the candidate diesel performance in the test engine with that of the various composition blends of normal cetane (*n*-cetane) and heptamethylnonane (HMN). Percentage by volume of *n*-cetane in a blend with HMN, which matches ignition quality of the test fuel, is the cetane number. See ASTM Method D613. For reference, normal cetane, a high-ignition fuel, has been assigned an arbitrary number of 100.

Cetane Number Improver—Substance which, when added to a diesel fuel, has the effect of increasing its cetane number. Common types of cetane improvers include nitroalkanes, nitrates, nitrocarbonates, and peroxides. 2-Ethylhexyl nitrate is one of the most often used cetane improver.

CFPP—Cold Filter Pugging Point.

CFR—Coordinating Fuel and Equipment Research Committee, composed of engine manufacturers, petroleum-refiners, petroleum-consumers, universities, government, and other technical people who supervise cooperative testing and study engine fuels for the Coordinating Research Council, Inc.

CGO—Cracked Gas Oil.

Chafing—Repeated rubbing of two solid bodies that can result in surface damage or wear, or both.

Channeling—Term is used in a variety of contexts.

1. Tendency of a grease or viscous oil to form air channels in a bearing or gear system, resulting in an intermittent lubricant film.
2. Grooves in a lubricating grease formed by passing rolling elements. Some gear lubricants and greases when they thicken, due to cold weather or other causes, form a groove or a channel through which the part to be lubricated moves without actually coming in full contact with the lubricant.
3. Low-temperature thickening of a flow-type lubricating grease to prevent uniform flow due to gravity.
4. In percolation filtration, channeling is when flow predominantly occurs only through certain portions of the clay bed.

Chatter—Elastic vibrations resulting from frictional or other instability.

Chemical Conversion Coating—Coating produced by a chemical or an electrochemical treatment of a metallic surface resulting in a superficial layer of compounds of the metal. Such coatings are usually pretreatments onto which fluid or solid lubricants may be applied. However, the coatings may themselves have lubricating properties.

Chemical Wear—See Corrosive Wear.

Chlorinated Lubricant—Lubricant containing a chlorine compound that reacts with the rubbing surfaces at elevated temperatures to form a protective chemical film; thereby preventing wear damage to the surface due to sliding. Also see extreme-pressure lubricant, sulfochlorinated lubricant, and sulfurized lubricant.

CIMAC—International Council on Combustion Engines.

Cloud Point—Temperature at which petroleum oil separates paraffin wax or other solid substances as crystals, imparting a cloudy appearance to the oil when it is chilled under prescribed conditions. Cloud point indicates the tendency of a material to plug filters or small orifices under cold weather conditions.

CIS—Commonwealth of Independent States.

CL (Cleanliness level)—A measure of relative freedom from contaminates.

CLR—Cooperative Lubrication Research.

CMA—Chemical Manufacturers Association.

CNG—Compressed Natural Gas.

CO—Carbon Monoxide.

CO_2—Carbon Dioxide.

Cocoa—Powdery debris resulting from fretting wear, primarily consisting of iron oxides, that is expelled from a ferrous metal joint near the location where fretting wear is occurring.

Coefficient of Adhesion—Ratio of the *normal force* required to separate two bodies to the *normal load* with which they were previously placed together.

Coefficient of Friction—Coefficient of *static* friction is the ratio of the tangential force initiating sliding motion to the load perpendicular to that motion. Coefficient of *kinetic friction* (usually called coefficient of friction) is the ratio of the tangential force sustaining sliding motion at constant velocity to the load perpendicular to that motion. In tribology, the dimensionless ratio of the friction force (F) between two bodies to the normal force (N) pressing these bodies together (Ref. 2): μ or $f = (F/N)$.

Cohesion—Molecular attraction causing particles of a substance to attract each other and stick together. Cohesion contributes to the resistance of a lubricating grease to flow.

Cold Cranking Simulator (CCS)—Intermediate shear rate viscometer that predicts the ability of an oil to assure satisfactory cranking speed in a cold engine.

Colloid—Substances of particle size larger than molecules, but small enough to possess reasonable dispersion stability in two phase systems. For example, lubricating grease is a colloidal system. See also Thickener.

Color—Predominant hue of lubricating grease, could be natural or due to an added dye. Some lubricants are color coded to facilitate identification or recommended use. Examples include transmission fluids, brake fluids, and two-stroke cycle engine oils. Color rating of dye-free oils is carried out by the use of many methods, including the ASTM D1500 procedure, under specified conditions and using specialized equipment. Some oils that are too dark for direct rating are diluted to report color.

Compatibility (Frictional)—Materials that exhibit good sliding behavior, including resistance to adhesive wear. Some materials that are considered metallurgically incompatible, for example, silver and iron, may be frictionally very compatible.

Compatibility (Lubricant)—Measure of the degree to which lubricants or lubricant components can be mixed without phase separation or deterioration in performance.

Compatibility (Metallurgical)—Measure of the extent to which materials are mutually soluble in the solid state.

Complex Grease or Complex Soap Grease—Lubricating grease thickened by a complex soap consisting of a normal soap and a complexing agent. The use of soap complexes yields products which have higher softening points than greases made from normal soaps.

Complex Soap—Soap formed by co-crystallization of two or more compounds. Complex soaps are mixtures of two or more components with structure or structures that improve their oil-retaining ability. More specifically, the dropping point of the resulting greases is effectively increased. Complexing agent used in such soaps is a low molecular weight carboxylic acid, such as acetic acid or lactic acid, or inorganic salts, such as metal carbonate or metal chloride.

Compounding—Process of mixing various additives in base oils/fluids to develop a finished lubricant of desired performance. Certain industrial oils, called compounded oils, contain fatty oils and similar materials which impart these lubricants special properties. While compounding is different from blending, it is important to note that most formulators do not recognize the subtle distinction and use the terms compounding and blending interchangeably.

Compression Ratio—In an internal combustion engine, the ratio of the volume of combustion space at bottom dead center to that at top dead center.

CONCAWE—Conservation of Clean Air and Water (Europe).

Cone Resistance Value (CRY)—A measure of the yield stress of a grease, obtained by static indentation with a cone. Also see Penetration.

Conformability—Quality of plain bearing material to adjust itself to shaft deflections and minor misalignments by deformation or by wearing away, without creating operating difficulties.

Consistency—Imprecise measure of the degree of a plastic material, such as a lubricating grease, to resist deformation under applied force (load).

Contact Lubrication (Obsolete Term)—Lubrication obtained when solid lubricant powders get rubbed into a surface.

Conversion Coating—See Chemical Conversion Coating.

Copper Strip Corrosion—Measure of the tendency of a petroleum product to corrode pure copper (ASTM D130).

Corrosion—Chemical related surface damage of the metal. Corrosion involves the attack of the metals by oxygen, sulfur, and acidic and basic chemicals. Chemically, corrosion is the oxidation of the metal to its oxides or salts. Rusting is the air-mediated formation of iron oxide and yellow metal corrosion is sulfur-mediated formation of copper sulfides.

Corrosion Inhibitor—An additive used to reduce corrosion. Metals whose corrosion is of primary concern in lubricant-related applications are iron, a copper, and lead.

Corrosive Wear—Wear caused by a chemical or an electrochemical reaction. See also Oxidative Wear.

cP—centiPoise, the unit of dynamic viscosity.

CR—Common Rail (Diesel Injection).

Cracking—Process of converting unwanted long-chain hydrocarbons into short-chain molecules by thermal or catalytic action. The latter is called catalytic cracking.

Crater Wear—Wear that occurs on the rake face of a cutting tool due to the contact with the material in the chip that is sliding along that face.

CRC—Coordinating Research Council. Organization supported by American Petroleum Institute and Society of Automotive Engineers, which administers the work of various committees pertaining to test work on fuels, lubricants, engines, etc.

Crown—Top of the piston in an internal combustion engine above the fire ring, which is exposed to direct flame impingement.

CRTV—Commercial Road Transport Vehicle.

cSt—CentiStoke, the unit of kinematic viscosity.

CSTCC—Continuously Slipping Torque Converter Clutch.

CUNA—Commissione Tecnica de Unificazione dell' Autoveicolo.

Cut—Product or fraction obtained by distillation within a specified temperature range.

Cutting Wear—See Abrasive Wear.

CVT—Continuosly Variable Transmission.

Cyclohexane Derivatives—These fluids were developed for use in automotive transmission applications requiring high traction.

DAP—Detroit Advisory Panel.

DASMIN—Deutsche Akkreditierungsstelle Mineralöl (Germany).

Debris—See Wear Debris.

Deformability—See Conformability.

Deformation Wear—Sliding wear involving plastic deformation of the wearing surface.

Delamination Wear—Wear process in which thin layers of material are removed from the wear surface.

Demulsibility—Measure of an organic water-immiscible fluid's ability to separate from water.

Density—Mass per unit volume. Useful in formulating lubricants with additives by volume and to calculate the amount of energy required to pump a fluid.

DEO—Diesel Engine Oil.

DEOP—Diesel Engine Oil Advisory Panel (API/EMA).

Detergent—Substance added to fuel or lubricant to keep engine parts clean. In engine oil formulations, the most commonly used detergents are basic metal sulfonates that have reserve basicity to neutralize acids formed from combustion and lubricant oxidation. They also possess the ability to suspend ordinarily oil-insoluble materials in the bulk lubricant.

Detergent/Dispersant—Additive package that combines a detergent with a dispersant. Both in combination are extremely effective in keeping harmful lubricant decomposition products away from surfaces.

Detergent Oil—Lubricating oil possessing sludge-dispersing properties for use in internal-combustion engines. Such oils hold sludge particles in suspension and thus promote engine cleanliness.

Detonation—Uncontrolled burning of the last portion (end gas) of the air-fuel mixture in the cylinder of a spark-ignition engine. Also known as "knock" or "ping."

DGMK—Deutsche Gesellschaft für Mineralölwissenschaft und Kohlechemie (Germany).

DI—Detergent Inhibitor.

DI—Direct Injection.

Dicarboxylic Acid Ester—Made from a dicarboxylic acid and an alcohol. Such esters are used as synthetic base fluids.

Dielectric Strength—Measure of the adequacy of electrical insulating oils of petroleum origin to be used in electrical devices, such as cables, transformers, oil circuit breakers, and similar apparatus.

Diester—See Dicarboxylic Acid Ester

Dilatant—Reversible increase in viscosity with an increase in shear stress.

Dilution of Engine Oil (Oil Dilution)—Contamination of the crankcase oil by the unburned fuel, leading to reduced viscosity and flash point. Reduced oil viscosity may lead to poor lubrication; hence increased wear.

DIN—Deutsches Institut für Normung.

Dirt Content—Measure of the size and concentration of foreign particles present in a lubricant. Lubricant decomposition products are also sometimes referred to as dirt.

Dispensability—Property of a grease which reflects the ease of a grease being dispensed from a container. Dispensability is related to pumpability and feedability.

Dispersant—Additive that is capable of dispersing oil-insoluble lubricant decomposition products, such as varnish, lacquer and sludge precursors.

Dispersant Oil—Heavy-duty oil that contains dispersant additive.

Distillate—Any of a wide range of products produced by distillation of crude petroleum.

Distillation—Process involving converting a volatile liquid into its vapor and then collecting it as a liquid. Distillation is also used to characterize volatility of a gasoline or distillate fuel.

***DN* Value**—Product of bearing bore diameter in millimetres and speed in revolutions per minute. *DN* value is used to select proper grease for bearings used in high speed applications.

DOHC—Double Overhead Cam (Twin Cam).

DOT—Department of Transportation.

Drawing Compound—Extreme-pressure lubricant used for such metalworking operations as wire drawing.

Drip Feed (Drop Feed) Lubrication—Lubrication system in which the lubricant is supplied to a device in the form of drops at regular intervals.

Drop (Dropping) Point—Temperature at which a drop falls from a grease through a specified orifice. It indicates grease transition from being a semisolid to a liquid state. In view of the test being a static test, dropping point does not necessarily coincide with maximum operating temperature of a grease.

Dry Film Lubricant—Solid material placed between two moving surfaces to prevent metal-to-metal contact, thus reducing friction and wear. Lubricating properties arise from sliding of the crystal lattice layers under low shear. Materials that are often used as lubricants include graphite and molybdenum disulfide. They are especially useful in mixed film lubrication regimes. They may be applied in the form of a paste, dispersion, or by spraying, dipping, and brushing in an air-drying carrier, which evaporates leaving a dry film.

Dry Sliding Wear—Sliding wear in which there is no intentional lubricant or moisture introduced into the contact area. Adsorbed species or oxides and other metal films that are either present or form during sliding may act as a lubricant. The proper term is unlubricated since solid lubricants, when used, will also form dry films.

Durometer Reading—Durometer is a device that is used to test the hardness of elastomers. Durometer reading is an index used to rank relative hardness of these materials.

Dynamic Viscosity—See Viscosity.

Elasto-hydrodynamic Lubrication (EHD)—

Lubrication regime which is characterized by high loads and high speeds in rolling elements where the mating parts deform elastically due to the incompressibility of the lubricant film under very high pressure.

ECCC—Electronically Controlled Converter Clutch.

EEB—European Environmental Bureau.

EELQMS—European Engine Lubricants Quality Management System.

EEV—Enhanced Environmental Friendly Vehicles and Engines.

EFI—Electronic Fuel Injection.

EFTC—Engine Fuels Technical Committee (of CEC).

EGR—Exhaust Gas Recycling.

EHD—Elastohdrodynamic Lubrication.

EKS—Electronic Knock Sensor.

ELTC—Engine Lubricants Technical Committee (of CEC).

EMA—Engine Manufacturer's Association.

EMD—General Motors Electromotive Division.

Emissions (Mobile Sources)—While emissions from an internal combustion engine include water, CO_2, CO, NOx and hydrocarbons (HC), only the last three are subject to stringent legislative control. Diesel engines in addition emit particulates (PM) which are also controlled. Efforts are being made to decrease the amount of the greenhouse gases, such as CO_2, in the environment due to a concern for global warming.

Emissions (Stationary Sources)—Emissions from stationary sources are hydrocarbons from evaporation of fuel and sulfur-derived emissions. These are in addition to CO and PM. Since fuel composition can influence emission of some of these pollutants, efforts are on the way to control sulfur content of heavy fuel oils used in such facilities.

Emulsifier—Additive that promotes the formation of a stable oil-water systems, called emulsions. Emulsifiers are bipolar additives with solubility in both phases. They are nonionic, if none of the atoms in their structure contain any electrical charges; anionic, if the charge on the organic portion is negative; and cationic, if the charge on the organic portion is positive.

Emulsion—Dispersion of fine droplets of one liquid (dispersed phase) in another liquid (continuous phase). Emulsions used as lubricants are oil-water systems that maintain their integrity with the help of a chemical compound called the emulsifier. Oil-in-water emulsions are called normal emulsions and water-in-oil emulsions are called invert emulsions.

ENGVA—European Natural Gas Association.

Engine Deposits—These include sludge, varnish, lacquer, and carbonaceous residues. They are derived both from the fuel combustion and the lubricant oxidation. Fuel combustion products contain unburned and partially burned fuel fragments which enter the crankcase as blow-by and cause break down of the lubricant. Water from the condensation of the combustion products, carbon, residues from fuel and lubricating oil additives, dust, and metal particles also contribute towards deposits.

Engine Oil—Oil used to lubricate internal combustion engine.

EOLCS—Engine Oil Lubricant Classification System (of API).

EP—Abbreviation for extreme pressure.

EPA—Environmental Protection Agency.

EPA Complex Model Gasoline—Gasoline designed to improve emissions quality. Limits were implemented on Reid Vapor Pressure (RVP), T-50 and T-90 (50 % and 90 % evaporated gasoline temperatures), oxygen content, aromatics content, and amounts of olefins, benzene, and sulfur. Effective date January 1, 1997.

EPA Simple Model Gasoline—Gasoline that was formulated to minimize volatile organic compound (VOC) emissions. Parameters manipulated included RVP and oxygen content and limits were placed on oxygen, benzene, sulfur, olefins, and T-90 content at levels equal to or lower than those of a refiner's 1990 baseline. Effective date January 1, 1995.

EP (Extreme Pressure) Agent—Additive used in a lubricant to impart or improve load-carrying capacity or antiweld quality. It prevents sliding metal surfaces from adhering to each other under extreme pressure conditions.

EPEFE—European Programme on Emissions, Fuels & Engine Technologies.

EP Lubricant—See Extreme-pressure Lubricant.

Erosion—Surface wear of a solid due to the motion of a fluid in contact that contains solid particles.

ESCS—Engine Service Classification System (API).

ESI—Extended Service Interval.

ETC—European Transient Test Cycle.

ETLP—Engine Test of Lubricants Panel (IP).

EU—European Union.

EUC—European Union Commission.

EUI—Electronic Unit Injector.

Evaporation Loss—Lubricant loss due to volatilization during use or storage.

Exhaust Gas Recirculation (EGR)—Process involving recycling of the exhaust gases into the carburetor or intake manifold to dilute the combustion mixture and reduce peak combustion temperatures. The result is the lowering of the amount of NOx formation.

Extreme Pressure Lubricant—Lubricating oil or grease which is supplemented with an EP agent to prevent metal-to-metal contact in operations involving high loads. Such lubricants are typically used in gear and bearing applications. These lubricants usually contain sulfur, halogens, or phosphorus. Another name for extreme pressure lubricant is anti-scuffing lubricant.

FA,FAR—Fuel Air Ratio.

Fat—Naturally-occurring ester which is a reaction product of fatty acid and glycerol. It is also called a triglyceride.

Fatigue—Cracking, flaking or spalling of a surface due to stresses beyond the endurance limit of the material.

Fatigue Wear—Removal of particles as a consequence of fatigue resulting from repeated localized stress. Fatigue wear in a part is usually indicated in the form of fracture or spall.

Fatty Acid—Organic acid derived from fats and natural oils.

Fatty (Natural) Oil—Fat (glycerol ester) that is liquid at room temperature.

FEO—Fuel Economy Oil.

Feedability—Ease with which a lubricating grease flows through a dispensing pump.

Ferrograph/Ferrography—An instrument used to determine the size distribution of the wear particles in lubricating oils used in mechanical systems. Ferrography, also called Magnetic Particle Analysis, is the technique used to analyze debris that is attracted to a magnet in the ferrograph.

Fiber—Grease soap thickeners occur in fibrous form. If fibers are very small (microscopic), grease appears smooth. If they are large or exist in bundles, the grease appears fibrous. Sodium soap thickened greases are usually fibrous, other soap greases are normally smooth.

Fiber Grease—Grease with a pronounced fibrous structure.

Fibril—Tiny fiber, barely visible even with an electron microscope. Fibers may be made of bundles of fibrils.

Filler—Materials, such as talc, pigments, and carbon black, that increase a lubricant's bulk or density. Fillers, commonly used in lubricating greases, may have no effects on the lubricating properties of a grease, except to increase its apparent viscosity.

Film Strength—Durability of the lubricant film against the effects of load, speed, and temperature without breaking down or rupturing. Rupture of such films occurs due to rubbing of the film by the asperities during sliding or rolling. High film strength implies high load-carrying capacity, which may or may not be true until proven through measuring due to a lubricating film.

Film Thickness—Distance between two surfaces due to a lubricant film that otherwise will be in contact.

Fire Point—Lowest temperature at which a petroleum product vaporizes to form air-vapor mixture above the liquid surface that burns continuously, when ignited by a small flame.

Flaking—Removal of material from a surface in the form of flakes or scale-like particles (pitting) as a result of fatigue.

Flash Point—Lowest temperature at which a fluid will support instantaneous combustion (a flash). Flash point is an important indicator of fire and explosion hazards associated with a petroleum product. It is determined by the use of Cleveland Open-cup (COC) Tester, Penskey-Martens Tester, and Tag Closed-cup Tester.

Flash Temperature—Maximum local temperature generated at some point in a sliding contact.

Flocculation (Floc) Point—Temperature at which wax or solids separate in an oil.

Flood Lubrication—System of lubrication in which a lubricant is supplied in a continuous stream at low pressure and subsequently drains away. Also known as bath lubrication.

Fluid Friction—Friction between the molecules of a gas or liquid in motion, expressed as shear stress. Unlike solid friction, fluid friction varies with speed and area.

Fluidity—Reciprocal of viscosity.

Fluoroester Fluids—These are the reaction products of perflluorocarboxylic acids with fluorinated alcohols. These are used as base fluids in specialty lubricants.

Foam—Collection of gas bubbles separated from one another by a thin liquid film.

Foaming—Process of formation of gas bubbles.

Foam Inhibitor—An additive used either to suppress foam formation of the petroleum products in service or to collapse it quickly, if it forms. Common foam inhibitors include silicone oil or polymers. These are used in extremely small quantities.

Four-Ball Tester—Of two types: Four-Ball Wear Tester and the Four-Ball EP Tester. These machines are used to evaluate a lubricant's antiwear qualities, frictional characteristics or the load-carrying capability. Its name is derived from four 1/2 in. steel balls used as test elements.

Fretting—Wear resulting from small amplitude (vibratory or oscillatory) motion between two surfaces; may produce red or black oxide. Also see False Brinelling.

Friction—Resistance to motion of one object over another. Friction depends on the degree of smoothness of the contacting surfaces, as well as the force (load) with which they are pressed together.

Friction Coefficient—See Coefficient of Friction.

Friction Force—Resisting force when one body moves or tends to move relative to the other under the action of an external force.

Friction Polymer—Amorphous, organic deposit that is produced when certain metals are rubbed together in the presence of certain liquids or gases.

Frosting—A form of ball bearing groove damage, appearing as a frosted area, which suggests that surface distress has occurred.

FTP—Federal Test Procedure (USA).

Full-film Lubrication—Type of lubrication where the solid surfaces are separated completely by a hydrodynamic fluid film.

FZG—Foschungstelle für Zhanräder und Getriebau: Technische Hochnische München.

FZG Test—A German gear test for evaluating the EP properties.

Galling—Severe form of scuffing associated with gross damage to the surfaces or failure. Also see adhesive wear.

Gaseous Fuels—Liquefied or compressed hydrocarbon gases (propane, butane or natural gas), which are being used in motor vehicles as replacements for gasoline and diesel fuel.

Gasket—A device that is usually made of a deformable material, which is used between two relatively static surfaces to prevent leakage.

Gasohol—Fuel mixture of 90 % gasoline and 10 % ethyl alcohol (ethanol).

Gasoline—Volatile mixture of liquid hydrocarbons, containing small amounts of additives and suitable for use as a fuel in spark-ignition, internal combustion engines.

GATT—General Agreement on Tariffs and Trade.

GDI—Gasoline (Engine) Direct Injection.

GE—General Electric Company.

Gel—An elastic mixture of a colloid and a liquid possessing a yield point and a jelly-like texture.

Gelling Agent—See Thickener.

GEPE—Groupe des Experts pour la et l'Engergie (Group of Experts for Pollution & Energy).

GM—General Motors Corporation.

Gravity—Expressed as mass per unit volume at a specified temperature, usually 60 °F (15 °C). Also see API Gravity.

Grease—See Lubricating Grease.

Grease Texture—Texture of a grease is determined by pressing a small portion of a grease together and then by slowly drawn apart. Texture is described as:

1. *Brittle,* if it ruptures or crumbles when compressed.
2. *Buttery,* if it separates in short peaks with no visible fibers.
3. *Long Fibers,* if it stretches or strings out into a single bundle of fibers.

GTL (Gas-to-Liquid) Base Stocks—Made by the conversion of the natural gas, primarily methane, into a synthetic hydrocarbon base fluid.

Gum—Rubber-like, sticky deposit, black or dark brown in color, which results from the oxidation and polymerization of the fuels and lubricating oils, or both. Harder deposits are described as lacquers or varnishes.

Hardness—Property of a solid material that indicates its resistance to penetration or abrasion by other bodies. Hardness of a solid depends on the method of its measurement.

HC—Hydrocarbons.

HD—Heavy Duty (vehicle or lubricant).

HDDO,HDDEO—Heavy-duty Diesel Engine Oil.

Heat Transfer Oil—Medium used to transfer heat away from a part or the equipment. Such oils are usually based on high-boiling petroleum fractions.

Heavy-duty Oils—Lubricants used in gasoline and diesel engines. These oils are stable against oxidation, protect bearings from corrosion, and have detergency and dispersancy.

Hershey Number—Dimensionless number used to evaluate performance of bearings. It is normally written as ZN/P, where Z denotes viscosity, N the frequency of rotation, and P the pressure. Often used in Stribeck curve to indicate various lubrication regimes.

HFO—Heavy Fuel Oil.

HFRR—High Frequency Reciprocating Rig.

Homogenization—Intimate mixing of a grease to produce a uniform dispersion of components.

HSDI—High Speed Direct Injection (diesel engine).

HT/HS—High Temperature/High Shear (viscosity).

Humidity Cabinet Test—Test used to evaluate the rust-preventing properties of metal protecting oils (metal preservatives) under conditions of high humidity.

Hydrated Soap—Soap with water associated with its structure, for example, that present in water-stabilized calcium soap grease.

Hydraulic Fluid—Fluids used as a medium to transfer hydraulic pressure to operate devices. They may or may not possess lubricant properties and may be water-based, mineral oil-based, or synthetic fluid-based.

Hydrocarbon—Compound that is made of carbon and hydrogen atoms only. The simplest hydrocarbons are gases at ordinary temperatures, but as their molecular weight increases, they become liquid, and ultimately solids (waxes). Hydrocarbons are the major constituents of petroleum.

Hydrodynamic Lubrication—Formation of a continuous lubricating fluid film of appropriate thickness between surfaces to prevent them from touching each other, thereby minimizing friction and wear. High lubricant viscosity, high equipment speed, and low load promote this type of lubrication. It is also called fluid film lubrication.

Hydrofinishing (Hydrogen Finishing)—Process for treating extracted base stocks with hydrogen to remove unsaturation; hence improve stability.

Hydrogenation—Chemical reaction between hydrogen and a material that has unsaturation or aromatic rings.

Hydrolytic Stability—Resistance of additives and certain synthetic lubricants against decomposition in the presence of water.

Hydrophilic—Affinity of a substance towards water.

Hydrophobic—Lack of affinity towards water. Related term is lipophilic—an affinity for organic materials. Usually hydrophobic materials are lipophilic.

Hydroprocessing (Hydrogen Processing)—Treating petroleum fractions with hydrogen, typically in the presence of a noble metal catalyst. Simple addition of hydrogen simply removes unsaturation. Hydrogenation under more severe conditions leads to bond breaking and bond forming, thereby altering chemical structures of the original hydrocarbons. When bond breaking is the primary outcome, the process is called hydrocracking, and when hydrocracking is followed by bond formation, it is called hydroisomerization.

Hydrostatic Lubrication—Lubrication in which the lubricant is supplied to a plain bearing under sufficient external pressure to separate the opposing surfaces by a continuous lubricant film.

Hypoid Gear Lubricant (Hypoid Oil)—Lubricant with extreme-pressure characteristics suitable for use in hypoid gears, which are often used in automobiles.

IChemE—Institute of Chemical Engineers (UK).

ICOMIA—International Council of Marine Industries Associations.

IDI—Indirect Injection.

IEA—International Energy Agency.

IFP—Institut Français de Pétrole.

ILSAC—International Lubricant Standardization Committee.

Impingement Corrosion—Form of erosion/corrosion generally associated with the impingement of a high-velocity flowing liquid containing air bubbles against a solid surface.

ILMA—Independent Lubricant Manufacturers Association.

Incompatibility—Fluids that do not mix well. Incompatibility may be reflected by phase separation, or loss of activity and or performance. This commonly occurs in lubricating greases and synthetic fluids.

Induction Period—Time during which oxidation of a material proceeds at a constant but a relatively slow rate. This period ends at the point where the oxidation rate increases sharply.

Inhibitor—Chemical whose presence in a petroleum product prevents or retards undesirable chemical changes from taking place, or in the condition of the equipment in which the product is being used. Common inhibitors include oxidation inhibitors and rust and corrosion inhbitors. These retard the rate of oxidation or corrosion.

Inorganic Thickener—See Nonsoap Thickener.

Insolubles—Contaminants in used oils that include dust, dirt, wear particles, and or oxidation products. In oils, their amount is often measured as pentane or benzene insolubles. In greases, they are measured by the use of prescribed reagents according to special analytical procedures, such as ASTM D128. Additional identifying analytical tests are required to determine the nature and composition of the isolated materials that comprise fillers, additives, and certain types of thickeners.

Insulating Oil—Oil used in circuit breakers, switches, transformers, and other electrical apparatus for insulating, cooling, or both. In general, such oils are highly-refined petroleum distillates of low volatility, with high resistance to oxidation and sludging.

IP—Institute of Petroleum, now called Energy Institute

ISO—International Standards Organization.

Iso-dewaxing—Hydroisomerization of the linear hydrocarbons (waxes) into branched hydrocarbons to improve low-temperature properties, such as pour point of the mineral base stocks. Also see Hydroprocessing.

IVD—Inlet Valve Deposit.

JALOS—Japanese Lubricating Oil Society.

JAMA—Japan Automobile Manufacturers Association.

JARI—Japan Automobile Research Institute.

JAST—Japan Society of Tribologists.

JASO—Japan Automobile Standards Organization.

JIS—Japan Industrial Standards.

JSAE—Society of Automotive Engineers, Japan.

Kinematic Viscosity—Measure of a fluid's resistance to flow under gravity at a specific temperature, usually 40 °C and 100 °C. Kinematic viscosity is easier to measure than dynamic viscosity. Kinematic viscosity can be converted into dynamic viscosity by dividing its value by fluid density, and vice versa. The units of kinematic viscosity are Stoke (St) and centiStoke (cSt), which correspond to cm^2/s and mm^2/s in Metric System. The units for dynamic or absolute viscosity are Poise (P) and centipoise (cP), which correspond to 0.1 and 0.001 Pa·s, respectively.

Kinetic Coefficient of Friction—Coefficient of friction under conditions of macroscopic relative motion between two bodies.

Knock—Audible hammering noise resulting from violent, explosive combustion of air-fuel mixture in a spark ignition engine.

KV—Kinematic Viscosity.

Lands—Circumferential areas between the grooves of a piston.

Lard Oil—Animal oil prepared from the fat of pig. Such oils are compounded with mineral oils to yield lubricants with special properties, especially cutting oils.

LCV—Light Commercial Vehicle.

LEV—Low Emissions Vehicles.

Liquefied Natural Gas (LNG)—Similar to LPG but consisting of lighter hydrocarbons, such as methane and ethane.

Liquefied Petroleum Gas (LPG)—Light hydrocarbon material, gaseous at atmospheric temperature and pressure, held in the liquid state by pressure to facilitate storage, transport and handling. Commercial liquefied gas consists essentially of propane, butane, or their mixture.

LMAO—Locomotive Maintenance Officer's Association.

Load—Force applied normal to the surface of one body by another contacting body or bodies. The term normal force is more precise and therefore preferred; however, the term normal load is also used. If applied vertically, the load can be expressed in mass units, but it is preferable to express force in as Newton units (N).

Load-carrying Capacity—Maximum load that a sliding or rolling system can support without failure; or maximum load or pressure that can be sustained by a lubricant, when used in a system under specific conditions, without failure.

Load Wear Index (LWI)—Measure of the relative ability of a lubricant to prevent wear under applied loads. See Four-Ball Test. It is calculated from the loads applied and corrected for elastic deformation of the balls under static loading and for the size of the wear scar. Formerly, it was called Mean Hertz Load.

LRI—Lubricants Review Institute (USA).

LSADO—Low Sulfer Automotive Diesel Oil.

Lubricant—Any substance interposed between two surfaces in relative motion for the purpose of reducing friction or wear that may occur. The term implies externally added substance. However, species such as oxide layers and tarnishes on certain metals can also prevent friction and wear.

Lubricant Compatibility—See Compatibility (Lubricant).

Lubricating Grease—A lubricant in which a lubricating fluid is thickened with soap or another material to a solid or semisolid consistency. Common soaps include calcium soap, sodium soap, barium soap, lithium soap, and aluminum soaps, both simple and complex. The liquid phase may be mineral oil or a synthetic base fluid.

Lubrication—Control of friction and wear by introducing a friction-reducing film between moving surfaces in contact. Such a film may be due to the presence of a fluid, solid, or a plastic substance. See Lubrication Regimes.

Lubrication Regimes—Lubrication environments defined on the basis of operating conditions, such as speed, load, etc., in lubricated tribo-systems. Different regimes have different amount of separation of surfaces and surface interaction; hence different frictional characteristics. Also see Boundary Lubrication, Elasto-hydrodynamic Lubrication, Full-film Lubrication, Hydrodynamic Lubrication, and Quasi-hydrodynamic Lubrication.

Lubricious—Relating to a substance or surface condition that tends to produce relatively low friction.

Lubricity—Ability of a lubricant to reduce friction and wear other than by its viscosity.

Luster—Reflected light intensity from a grease, its sheen or brilliance. See Appearance/Bulk Appearance.

Main Bearing—Bearing supporting the main power-transmitting shaft.

MDCL—Medium Diesel Cylinder Lubricant.

MDO—Marine Diesel Oil.

MGO—Marine Gas Oil.

Mechanical Activation—Acceleration or initiation of a chemical reaction by mechanical exposure of a nascent solid surface, for example obtained by a metal cutting operation.

Mechanical Stability (Grease)—Shear stability as tested in a standard rolling tester.

Mechanical Wear—Removal of material due to mechanical processes involving sliding, rolling, or repeated impact. Mechanical wear includes adhesive wear, abrasive wear, and fatigue wear, but not corrosive wear and thermal wear.

Melt Lubrication—Lubrication provided by steady melting of a lubricating species.

Metal Deactivator—Fuel or lubricant additive, usually a chelating agent, which through complexation deactivates metal surfaces and metal ions that will ordinarily catalyze fuel and lubricant breakdown; thereby inhibiting the formation of resin and other oxidation products. Copper in fuel lines and iron are of concern both in metallic form and in oxidized form (ions).

Metallic Wear—Wear due to rubbing or sliding contact between metallic materials. Its cause is cold welding of asperities, followed by separation and inter-surface metal transfer. Also see Adhesive Wear and Severe Wear.

Metallurgical Burn—Modification of the microstructure near the contact zone due to frictional temperature increase.

Metallurgical Compatibility—See Compatibility (Metallurgical).

Methylene-bridged Alkylphenol—Obtained by the reaction of alkylphenol with formaldehyde. Used as starting materials to make detergent additives for heavy-duty diesel engine oils.

Mineral Oil—Refined hydrocarbon oil obtained directly from petroleum refining.

MIRA—Motor Industry Research Association.

Mist Lubrication—Lubrication by an oil mist produced by injecting oil into a gas stream.

MITI—Ministry of International Trade and Industry (Japan).

MON—Motor Octane Number.

Morphology—Pertaining to structure and form.

MRV—Mini Rotary Viscometer.

MTAC—Multiple Test Acceptance Criteria.

MTBE—Methyl Tertiary Butyl Ether.

MTF—Manual Transmission Fluid.

Multi-grade Oil—Engine or gear oil that meets the requirements of more than one SAE viscosity grade, for example 10W-30, which meets the requirements of both SAE 10W and SAE 30. Such multi-viscosity oils provide broad-temperature performance and are usually made by adding a viscosity modifier to a low viscosity oil. In some case, such as narrow viscosity range gear oils (80W-90), multi-grades are made by blending base oils of two different viscosities.

Multiply-alkylated cyclopentanes, or MACs—These are made by the reaction of cyclopentadiene with an alcohol or an alkyl halide. They are used as base fluids in formulating high-temperature lubricants.

Multi-purpose Grease—Lubricating grease that meets the requirements of various applications. For example, it is suitable for use as a chassis lubricant, bearing lubricant, joint lubricant, water-pump lubricant, and cup grease.

MVEG—Motor Vehicle Emmisions Group (of the European Commision).

MVMA—Motor Vehicle Manufactures Association.

Naphthene—A group of cyclic hydrocarbons, which is also termed cycloparaffins or cycloalkanes. Polycyclic members are also found in the higher boiling fractions. The general formula for naphthenes is C_nH_{2n}, where n is a whole number.

Naphthenic—Petroleum-derived fluid, usually obtained from naphthenic crude oil, containing a high proportion of saturated cyclic structures.

Natural Sulfonate—Metal salt of the petroleum-derived (natural) sulfonic acid.

NEMA—National Electrical Manufacturers Association.

Neutral Oil—Petroleum-derived lubricant base stock that is devoid of acidity or alkalinity. Neutral oils are the basis of most commonly used automotive and diesel lubricants. They are light overhead cuts from vacuum distillation.

Neutralization Number—Measure of acidity or alkalinity of an oil. Neutralization number is the amount of milligrams of acid (HCI) or base (KOH) required to neutralize one gram of oil. It is expressed as mg of KOH/g of oil. Also see Acid Number and Base Number.

Newtonian Behavior—Lubricant exhibits Newtonian behavior if its shear rate is directly proportional to the shear stress. This constant proportion is the viscosity of the liquid.

Newtonian Flow—Fluid exhibits Newtonian flow if its viscosity is independent of the shear stress.

Newtonian Fluid—Fluid that exhibits Newtonian viscosity, that is, shear stress is proportional to the rate of shear.

Nitration—Process where nitrogen oxides react with hydrocarbon (petroleum) fluids at high temperatures, resulting in oxidation products that cause viscosity increase and deposit formation.

NLGI—National Lubricating Grease Institute.

NLGI Consistency Number—Scale comparing consistency (hardness) of greases. Scale range is from 000 to 6; higher consistency numbers indicate firmer (harder) greases. Greases both softer and harder than indicated by these consistency numbers are well known in the industry. However, such greases do not have an NLGI Number.

NMMA—National Marine Manufacturers Association.

NMOG—Non-Methane Organic Gases.

NOx—Nitrogen Oxides, primarily Nitrous Oxides.

Non-Newtonian Behavior—Fluids exhibit this type of behavior if they have variable relationship between shear stress and shear rate. This implies that they do not possess real viscosity but apparent viscosity. Such fluids, exemplified by greases and polymeric additive-containing lubricants, are called **Non-Newtonian fluids**.

Nonsoap Thickener—Lubricating grease thickener that is not derived from metal carboxylates (soaps). Such thickeners can be organic or inorganic. Former type includes polyurea and cellulosic materials and the latter type includes silica and modified clays.

NPRA—National Petroleum Refiners Association.

NVMA—National Vehicle Manufacturers Association.

OBD—On-board diagnostics.

OCP—Olefin Co-polymer.

Octane Number—Measure of a fuel's ability to prevent detonation in a spark ignition engine. Measured in a standard single cylinder, variable-compression-ratio engine by comparison with primary reference fuels which are blends of *iso*-octane (Octane Number 100) and *n*-heptane (Octane Number 0). Under mild conditions, the engine measures Research Octane Number (RON); under severe conditions it measure Motor Octane Number (MON). Posted Octane

Number of commercial gasolines, called Antiknock Index (AK), is the average of the two values.

Octane Requirement (OR) or Octane Number Requirement (ONR)—Lowest octane number reference fuel that will allow an engine to run knock-free under standard conditions of service.

Octane Requirement Increase (ORI)—As deposits in automobiles combustion system increase, the ORI of the engine increases. That is, higher octane number fuel is required for trouble-free operation. Addition of deposit control/cleanliness additives to gasoline help prevent this from happening.

ODI—Oil Drain Interval.

OECD—Organization Economic Cooperation and Development.

OEM—Original Equipment Manufacturer.

OICA—Organisation Internationale des Constructeurs d'Automobiles.

Oil—A liquid of vegetable, animal, mineral, or synthetic origin that feels slippery to the touch.

Oil Fog Lubrication—See Mist Lubrication.

Oil Groove—A channel or interconnected channels cut in a part to improve oil flow. When applied to a bearing, this strategy facilitates distribution of oil over the shaft as well.

Oiliness—Capability of a fluid to orient itself on metal surfaces and lower friction. Also see lubricity.

Oiliness Agent—Chemical with the ability to form an adsorbed film on metal that reduces friction.

Oil Pocket—A depression designed to retain oil on a sliding surface.

Oil Ring—A mobile ring whose inner surface rides a shaft or journal, causing the ring to rotate and pick up lubricant by dipping into the lubricant reservoir and delivering it to the top of the shaft for distribution to a bearing. An internal-combustion engine piston also contains an oil ring which controls the lubrication of the piston and cylinder walls.

Oil Ring Lubrication—See Oil Ring.

Oil Starvation—Situation in which a bearing, or another tribo-component, receives an inadequate supply of lubricant.

OPEST—Oil Protection and Emission System Test.

Organic Thickener—See Nonsoap Thickener.

ORI—Octane Requirement Increase.

Overbasing—Technique to impart base reserve in additives, called detergents. Such base exists in the additive in a colloidal form. Addition of overbased additives to a lubricant imparts ability to neutralize acidic oxidation products.

Oxidation—Reaction of oxygen with organic and inorganic materials. Hydrocarbon (petroleum) fluids oxidize to yield highly polar oxygenated materials of polymeric structure and carboxylic acids. These species lead to lubricant viscosity increase, deposit formation, and metal corrosion. Oxidation rate is accelerated by heat, light, metal catalysts and the presence of water, acids, or solid contaminants.

Oxidation Inhibitor—Additive used for the purpose of reducing the rate of oxidation and subsequent deterioration of oil or grease.

Oxidation Stability—Resistance of a petroleum product to oxidation and, therefore, it is a measure of a product's potential service or storage life.

Oxidative Wear—Corrosive wear in which a chemical reaction with oxygen or oxidizing environment predominates. This type of wear is usually a mild form of wear, but it can be severe in highly oxidative environment, such as those involving liquid oxygen or liquid fluorine.

Oxygenate—Oxygen-containing, metal-free organic compound, such as alcohol or ether, that can be used as a fuel or fuel supplement.

Oxygenated Fuels—Fuels for internal combustion engines that contain oxygenates, such as alcohol and methyl-*t*-butyl ether (MTBE), which increase gasoline's octane value.

Oxygenated Gasoline—Required annually from September 15 to March 15 for use in carbon monoxide (CO) nonattainment areas. Oxygenated gasoline contains a minimum of 2.0 % mass oxygen and a maximum of 1.0 % volume benzene.

Ozone and CO Nonattainment Areas—Any area in the continental United States that does not meet the 1990 Clean Air Act requirements for carbon monoxide or ground-level ozone pollutants.

PAG—See Poly(alkylene glycol).

PAH—Polyaromatic Hydrocarbon.

PAJ—Petroleum Association of Japan.

Pale Oil—Petroleum lubricating oil or process oil refined until its color by transmitted light is straw to pale yellow. Typically, it is a naphthenic oil.

Paraffinic—Type of petroleum fluid derived from paraffinic crude oil and containing a high proportion of straight chain saturated hydrocarbons, because of which it suffers from cold flow problems.

PCB,PCBs—Polychlorinated Biphenyls.

PCD—Passenger Car Diesel.

PCMO—Passenger Car Motor Oil.

PCV—Passenger Car Vehicle.

Penetration—Measure of consistency (hardness) based on an inverse penetration measurement (the softer the consistency, the higher the penetration number). Penetration is measured as the distance in millimetres that a standard needle or cone penetrates vertically into a sample of the material, under known conditions of loading, time, and temperature. Penetration may be measured without disturbing the grease (Undisturbed penetration), with minimum disturbance (Unworked penetration), and with substantial disturbance (Worked Penetration).

Pentane Insolubles—See Insolubles.

Percent Permanent Viscosity Loss (PPVL)—Measure of the PVL related to the viscosities of the fresh oil—equals PVL divided by fresh oil viscosity, multiplied by 100.

Percent Temporary Viscosity Loss (PTVL)—Difference between the viscosity of an oil measured at low and high shear stresses, divided by viscosity measured at low shear stress, multiplied by 100.

Perfluoropolyethers (PFPEs)—These are poly(alkylene glycol) analogues, where all hydrogen atoms in their structure are replaced by fluorine atoms. They are used in specialty applications due to unique properties.

Permanent Viscosity Loss (PVL)—Difference between the viscosity of fresh oil and that of the same oil after engine operation or special test conditions to determine polymer degradation.

Petroleum Oil—See Mineral Oil.

PFI—Port Fuel Injection.

Phosphate Esters—Made from phosphorus oxychloride and a hydroxy compound, such as an alcohol or a phenol. Phenol and alkylphenol esters are used as base fluids for fire-resistant lubricants and hydraulic fluids.

PIB—See Polyisobutylene.

Pitting—Surface damage in the form of cavities, which may be related to fatigue, overload, or corrosion.

Plastic Flow—Surface deformation of metal as a result of yielding under heavy load.

Plowing (ploughing)—Formation of grooves by plastic deformation of the softer of the two surfaces in relative motion.

PM—Particulate Matter.

PM2.5—Particulate matter of the diameter 2.5 μ or less.

PM10—Particulate matter with particle size below 10 microns.

Poise (P)—Metric unit of dynamic viscosity. Also see Kinematic Viscosity.

Polishing—A surface-finishing process utilizing successive grades of abrasive.

Polishing (Bore)—Excessive smoothening of the surface finish of the cylinder bore or cylinder liner in an engine to a mirror-like appearance, resulting in depreciation of ring sealing and oil consumption performance.

Polishing Wear—Mild form of wear that may involve fine-scale abrasion, plastic smearing of micro-asperities, tribochemical material removal, or a combination thereof.

Poly(alkylene glycol)—Made by acid or base-catalyzed polymerization of alkylene oxides, such as ethylene oxide and propylene oxide. Used as synthetic base fluids.

Polyisobutylene—Made by the polymerization of isobutene (isobutylene). Used as a starting material for making additives, such as dispersants, and after hydrogenation used as a viscosity modifier in lubricants.

Polyol Esters—These are the reaction products of polyhydric alcohols (polyols) and carboxylic acids. These are used as base fluids to design high temperature lubricants.

Poly(phenyl Ether)s, Polyphenyl Ethers, or PPEs—These materials are obtained by the reaction of an alkali metal phenate with a halogenated benzene. These are used in lubricants for high-temperature and radiation environments.

Positive Crankcase Ventilation (PCV)—System involves recycling the blow-by gases, which contain unburned hydrocarbons, from the crankcase through the carburetor intake manifold into the combustion chamber to burn them. The objective is to reduce hydrocarbon emissions.

Pour Point—The lowest temperature at which the fluid will flow when cooled under prescribed conditions. Pour point is an indicator of the ability of an oil or the distillate fuel to flow at cold operating temperatures.

Pour Point Depressant, PPD—Additive which lowers the pour point of a paraffinic oil which contains wax.

Pour Stability—Ability of a pour point depressant-treated oil to maintain its original ASTM pour point when subjected to storage at low temperature approximating winter conditions.

Precipitation Number—Number of millimetres (mL) of precipitate formed when 10 mL of lubricating oil is mixed with 90 mL of petroleum naphtha of defined quality and centrifuged under prescribed conditions. Precipitation number indicates the amount of asphaltic components in the lubricating oil, although a certain amount of paraffinic components may also separate.

Preignition—Ignition of the fuel-air mixture in a gasoline engine before the spark plug fires. Often caused by incandescent fuel- or lubricant-derived deposits in the combustion chamber, it wastes power and may damage the engine.

Pressure Lubrication—System of lubrication in which the lubricant is supplied under pressure to various parts of the equipment or the device.

Pressure-Viscosity Coefficient—Slope of a graph that shows variation of viscosity with pressure on a logarithmic scale. The term pressure-viscosity coefficient assumes a linear relationship between these two parameters.

Process Oil—Oil used as a component in another material or as a carrier for other products.

Protuberances—See Asperities.

PTF—Power Transmission Fluid.

Pumpability—Flow properties of a grease that permits proper delivery through lines, nozzles, and fittings of the grease dispensing systems and subsequent lubrication of the moving components. Also see Dispensability and Feedability.

Quasi-hydrodynamic Lubrication—A loosely defined regime of lubrication, especially in metalworking, where thin-film lubrication predominates.

Quenching Oil—Oil used for quenching metals during a heat-treating operation.

R&O Oil—Rust and oxidation inhibited oil. Such oils are highly-refined lubricating oils, which are formulated for long service in circulating systems, compressors, hydraulic systems, bearing housings, gear cases, etc.

REACH—Acronym for European environmental legislation pertaining to chemical contamination of the environment—**R**egistration, **E**valuation, and **A**uthorization of **CH**emicals.

Reclaimed Oil—Lubricating oil which, after undergoing a period of service, is collected, reprocessed, and sold for reuse.

Refining—Series of processes to convert crude oil and its fractions into finished petroleum products. Processes include thermal cracking, catalytic cracking, polymerization, alkylation, reforming, hydrocracking, hydroforming, hydrogenation, hydrogen treating, Hydrofining®, solvent extraction, dewaxing, de-oiling, acid treating, clay filtration, and de-asphalting.

Re-refining—Process of reclaiming used lubricant oils and restoring them to a condition similar to that of virgin stocks by filtration, clay adsorption, or more elaborate refining methods.

Reversibility—Ability of a grease to return to its normal grease-like consistency after temporary exposure to temperatures near or above its dropping point. Only a few types of greases possess this property.

Reynold's Equation—A basic equation of hydrodynamic lubrication.

Rheology—Deformation or flow characteristics, or both, of grease in terms of stress, strain, temperature, and time—commonly measured by Penetration and Apparent Viscosity.

Rheopectic Material—A material that shows an increase in viscosity with time under constant shear stress. After the removal of the shear stress, the viscosity slowly returns to the original level.

Rheopectic Grease—Lubricating grease which increases its consistency number (hardens) upon being subjected to shear.

Ridging—Form of plastic flow on the surface of the gear teeth that is characterized by a rippled appearance. Damage appears in the form of deep scratches that run as parallel ridges and is a consequence of plastic flow of the subsurface layer.

Rings—Circular metallic elements that ride in the grooves of a piston and provide compression sealing during combustion and to spread oil for lubrication.

Ring Sticking—Freezing of a piston ring in its groove in a piston engine or reciprocating compressor due to heavy deposits in the piston ring zone.

Ripple Formation (rippling)—Formation of periodic ridges and valleys transverse to the direction of motion.

Roller Bearing—See Anti-friction Bearing.

Rolling—Motion of two moving bodies of opposite curvature whose surface velocities in the common contact area are identical in magnitude and direction.

Rolling and Peening—Forms of plastic flow on the surface of the gear teeth that gives the appearance of being hammered. Metal may even be rolled over the teeth tips.

Rolling-contact Fatigue—Repeated stressing of a solid surface due to rolling contact between it and another solid surface or surfaces.

Rolling-contact Wear—Wear to a solid surface that results from rolling contact between that surface and another solid surface or surfaces.

Rolling-element Bearing—Bearing in which the moving parts are separated by balls, rollers, or needles.

RON—Research Octane Number.

RPM—Revolutions Per Minute.

Run In—Applying specified initial operating conditions to a tribological system to improve its long-term frictional or wear behavior, or both. It may involve conditions either more severe or less severe than the normal operating conditions and may also involve the use of special lubricants or surface chemical treatments, or both. Also see Break-in.

Running-in—Process by which machine parts improve in conformity, surface topography, and frictional compatibility during the initial stage of use. Chemical processes include formation of surface an oxide film and metallurgical processes include strain hardening.

Rust Preventative—Compound used to coat metal surfaces with a film that protects against rust, primarily in stored equipment.

RVP—Reid Vapor Pressure.

SAE—Society of Automotive Engineers.

SAE Viscosity Grade—An arbitrary number used for classifying engine oils, automotive transmission fluids, and gear oils, according to their viscosities, established by the Society of Automotive Engineers. SAE Grades are selected on the basis of the operating temperature and tribological requirements of the equipment.

Saponification—Making of a metal salt (soap) from a carboxylic acid ester, such as natural oils and fats, by the reaction of an alkali metal base, such as sodium hydroxide and potassium hydroxide.

Saponification Number—Measure of the amount of material in petroleum products that will saponify under test conditions. Saponification number, like the acid and base numbers, are expressed as number of milligrams of base, in this case sodium hydroxide, consumed by 1 g of oil. Saponifiable (fatty) material is not naturally present in petroleum products but is added to improve their lubricity.

Saybolt Color—Color standard for petroleum products (ASTM Method D156).

Saybolt Universal Viscosity (SUS or SSU)—It is based on time, in seconds, taken for 60 mL of fluid to flow through a capillary tube in a Saybolt Universal Viscometer at a given temperature (ASTM Method D88).

Scoring—Formation of significant scratches in the direction of sliding. Scoring may be due to local solid-phase welding or to abrasion. In the United States, the term scuffing is sometimes used as a synonym for scoring. It can be prevented by the use of antiwear, extreme pressure, and friction reducing additives.

SCR—Selective Catalyst Reduction.

Scratching—Fine abrasive furrows in the direction of sliding, which are produced by the cutting or plowing action, or both, of a sharp particle or protuberance moving along that surface.

Scuffing—See Scoring.

Seal—Device used to prevent leakage between moving parts, movement of the fluid from one chamber to another, or to exclude contaminants.

Seizure—Stopping of relative motion due to interfacial friction. Seizure may be accompanied by gross surface welding.

Selective Transfer—Process involving transfer and attachment of a specific species from one contacting surface to the other contacting surface during sliding.

Self-lubricating Bearing—See Anti-friction Bearing.

Self-lubricating Material—Solid material that shows low friction without the application of a lubricant. Examples include graphite, molybdenum disulfide, and polytetrafluoroethylene, and even composite materials with inherently low friction. Also see Solid Lubricant.

SEM—Scanning electron microscope/microscopy.

Separation Test—Test to determine the tendency of the oil to separate from a lubricating grease under conditions prescribed in ASTM Method D1742.

SHC—See Synthetic Hydrocarbons.

Shearing—Relative slipping or sliding between one part of a substance and an adjacent part. In solids, shearing involves breaking or altering of the crystal structure; in a fluid or plastic, shearing may or may not destroy the continuous nature of the substance.

Shear Rate—Rate of sliding within a flowing substance. See ASTM D1092.

Shear Stability—Resistance of a grease to changes in consistency (hardness) during mechanical working. In a lubricant, it is the ability of a lubricant to withstand the shearing action without degradation.

Shear Stability Index (SSI)—Measure of a viscosity

modifier's contribution towards percent loss in viscosity of an oil when subjected to engine operation or in specialized tests.

Shear Stress—Stress (force per unit area) that can cause shearing. Ratio of shear stress to shear rate in a fluid is its viscosity.

Shear-thickening—Increase in viscosity with an increase in shear rate or time.

Shear-thinning—Decrease in viscosity with an increase in shear stress or time. Decrease in viscosity may be temporary or permanent. The latter happens when the shear stress is sufficiently high to cause structural breakdown of the polymeric additives. Permanently sheared fluid has a lower viscosity than it had prior to shearing.

SHPD—Super High Performance Diesel (oil).

Silicate Esters—These are organic esters of orthosilicic acid, H_4SiO_4, which are made by the hydrolysis of alkoxysilicon chlorides with an alcohol. Used in lubricants for specialty application.

Silicones/Polysiloxanes—Made by water-catalyzed polymerization of dialkyl/diaryl silicon halides. Used as base fluids for specialty lubricants and as foam-inhibiting additives.

SIP—Styrene Isoprene Copolymers.

Slide-roll Ratio—Ratio of sliding velocity to rolling velocity, for example, in gear sets.

Sliding—Motion of two relatively moving bodies, in which their surface velocities in the common contact area are different with regard to magnitude or direction, or both.

Sliding Velocity—Difference between the velocities of each of the two surfaces relative to the point of contact.

Sludge—Semisolid to solid thick, dark residue containing oil, oxidation products, carbon, and water, which results from the low temperature operation of a combustion engine. Sludge is normally of mayonnaise consistency that accumulates on nonmoving interior surfaces of the engine and is usually removable by wiping, unless baked to a carbonaceous consistency.

Slumpability—Flow of a grease in a container under gravity, thereby allowing it to be fed into a pump or can. Also see Feedability.

Slurry—Mixture of solid particles in a liquid, of a consistency that is capable of being pumped as a liquid.

Slushing Oil—Mineral oil containing additives that enable it to protect the parts of a machine against rusting.

Soap—Alkali or alkaline metal salt of a carboxylic acid, especially a fatty acid. Also see Thickener, Complex Soap, and Saponification. While soaps are of importance as thickeners in greases, metal soaps formed in boundary lubrication regime by the in situ reaction of fatty additives and metal play an important role in reducing friction and wear.

Solid-film Lubrication—Lubrication by the application of solid lubricants.

Solid Lubricant—Solid used as a powder or a thin film on a surface to provide reduced friction and wear protection during the relative movement of the surfaces in contact. These lubricants have a layered (lamellar) structure, for example, graphite and molybdenum disulfide. Certain solids, such as glass, lubricate only in the melted state at the interface. These are not considered solid lubricants.

Soluble Oil—Mineral oil containing additives that enable it to form a stable emulsion with water.

Solvent Extraction—Refining process used to separate reactive components (unsaturated hydrocarbons) from the lubricant distillates in order to improve their oxidation stability, viscosity index, and additive response.

Solvent Refining—Process of extracting lubricant base stocks from heavy gas oil or another heavy, stripped crude stream by the use of selective solvents such as furfural or phenol.

Spalling—Separation of particles from a surface in the form of flakes, leaving behind a surface that consists of large pits, cavities, and related cracks. The term spalling is commonly associated with rolling-element bearings and with gear teeth. Spalling is usually a result of subsurface fatigue due to overloading and is more extensive than pitting.

Specific Gravity—Ratio of the weight (in air) of a given volume of a material to the weight (in air) of an equal volume of water at the stated temperature.

Spindle Oil—Oil of low viscosity used to lubricate light spindles operating at high speeds.

Splash Lubrication—System of lubrication in which the lubricant is splashed onto the moving parts.

Squeeze Film Lubrication—Development of fluid pressure sufficient to support a load between surfaces thickly coated or flooded with lubricant and rapidly moving toward each other. Because of viscosity (or apparent viscosity), the lubricant cannot immediately flow away from the area of contact. Squeeze-film lubrication occurs between gear teeth.

SSI—Shear Stability Index.

SSU (or SUS)—Saybolt Seconds Universal (or Saybolt Universal Seconds) a unit of vicosity. Also see Saybolt Universal Viscosity.

Starting Torque—Torque required for initiating rotary motion.

Static Coefficient of Friction—Coefficient of friction corresponding to the maximum frictional force that must be overcome to initiate motion of the two bodies in contact.

Stick-Slip—Transition from high coefficient of friction (stick) to low coefficient of friction (slip), as the relative velocity increases. This phenomenon can be explained in terms of the formation of the interfacial junctions on a microscopic scale first and then their destruction. Time duration between stick and slip will depend upon the velocity and the elastic characteristics of the system. Stick-slip in a system is indicated by a jerky motion.

Stoke (St)—Nonmetric unit of Kinematic viscosity.

STLE (Formerly ASLE)—Society of Tribologists and Lubrication Engineers.

Storage Stability—Measure of the ability of a lubricant to undergo prolonged periods of storage without showing any adverse conditions due to oxidation, oil separation, contamination, or any type of deterioration. Storage stability tests are often run on lubricants to determine their shelf life.

STOU—Super Tractor oil Universal.

Stribeck Curve—Graph showing the relationship between coefficient of friction and the dimensionless number (ZN/P), where Z is the dynamic viscosity, N is the equipment speed (revolutions per minute), and P is the load per unit of projected area.

Strong Acid Number—Quantity of base, expressed in terms of the equivalent number of mg of KOH, which is required to titrate the strong acid constituents present in 1 g of sample (ASTM Method D664 or D974).

Strong Base Number—Quantity of acid, expressed in terms of the equivalent number of mg of KOH, which is required to titrate the strong base constituents present in 1 g of sample (ASTM Method D664 or D974).

Structure—Physical arrangement of the thickener particles of the lubricating grease, additives, if any, and liquid lubricant. It is the nature and stability of this arrangement which determine the appearance, texture, and physical properties of the grease.

Sulfated Ash—Ash content (expressed as wt %) of oil determined by burning the oil, treating the residue with sulfuric acid, and evaporating to dryness.

Sulfochlorinated Lubricant—Lubricant containing chlorine and sulfur compounds, which react with the rubbing surfaces at elevated temperatures to form chemical protective films.

Sulfur-bridged Alkylphenol—Obtained by the reaction of an alkylphenol with elemental sulfur or sulfur chlorides. Such phenols are used as starting materials to make detergent additives for heavy-duty diesel engine oils.

Sulfurized Lubricant—Lubricant containing sulfur or a sulfur compound that reacts with the rubbing surfaces at elevated temperatures to form chemical protective films.

Surface Damage—Damage to a solid surface resulting from the mechanical contact with another substance, surface, or surfaces moving relative to it. Damage involves displacement or removal of the material. Wear is a form of surface damage in which the material is progressively removed. In another context, surface damage involves deterioration of function although no material loss from the surface has occurred.

Surface Distress—Distress occurs to the contacting surfaces in bearings and gears through intermittent contact as a consequence of sliding or surface fatigue, or both.

Surfactant—Chemical substance characterized by a strong tendency to form adsorbed interfacial films when in solution, emulsion, or suspension, thus producing effects such as low surface tension, penetration, boundary lubrication, wetting, and dispersing (Ref 1).

Syneresis (grease)—See Bleeding.

Synthetic Grease—A grease that contains a liquid lubricant that is not a mineral oil, but is synthetic.

Synthetic Hydrocarbons/Hydrocarbon Polymers—Base fluids and starting materials that contain carbon and hydrogen atoms only. See Polyolefins (PAOs, PIOs, and PIBs) and Alkylaromatics.

Synthetic Lubricant—See Lubricating Oil.

Synthetic Lubricating Oil—Lubricating fluid made by chemically reacting materials of a specific chemical composition to produce compound/s with desired and predictable properties. Also see Synthetic Oil/Fluid.

Synthetic Oil/Fluid—Oil produced from petroleum-derived raw materials by chemical synthesis, instead of obtaining it from petroleum refining. Examples include esters, ethers, silicones, glycols, and halogenated hydrocarbons.

Synthetic Sulfonate–A metal salt of a synthetic sulfonic acid.

Synthetic Thickener—See Nonsoap Thickener.

Tackiness—Grease description indicating its stickiness or adhesion capability.

Tag Closed-cup Tester—Instrument used to determine the flash point of flammable volatile materials flashing below 175 °F (80 °C), as described in ASTM Method D56.

TAN—Total Acid Number. See Acid Number.

TBN—Total Base Number. See Base Number.

Temporary Shear Stability Index (TSSI)—Measure of the viscosity modifier's contribution to an oil's temporary viscosity loss under high shear conditions. Temporary shear loss results from reversible deformation of the shape of the viscosity modifier in high shear regions of the engine.

Temporary Viscosity Loss (TVL)—Measure of decrease in dynamic viscosity under high shear rates compared to dynamic viscosity under low shear.

Texture—Texture of a grease, observed when a small portion of it is pressed together and then slowly separated. Also see Grease Texture.

Thermal Wear—Removal of material due to softening, melting, or evaporation during sliding or rolling.

Thickener—Solid material that is dispersed in a liquid lubricant to produce a grease. Common thickeners include soaps (metal carboxylates), silica, clays, and polyurea.

Thick-film Lubrication—State of lubrication in which the lubricant film thickness is much greater than the size of the surface asperities.

Thin-film Lubrication—Lubrication state where the lubricant film thickness is such that the friction and wear between the surfaces is determined by the properties of the surfaces as well as the viscosity of the lubricant.

Thixotropy—Property of a material to recover its consistency after a decrease due to shear. Lubricating grease exhibits this property in that the shear forces cause a decrease in consistency (softening), which reverts when the shear forces are removed.

Timken EP Test—This test is used in determine the load carrying capacity of an oil or a grease.

TISI—Thai Industrial Standards Institute

TLEV—Transitional Low Emissions Vehicle.

TPEO—Trunk Piston Engine Lubricant.

Transfer—Process in tribology by which material from one sliding surface becomes attached to another surface, possibly as the result of interfacial adhesion. Material may also back transfer to the surface from which it came. Also see Selective Transfer.

Transmission Oil—Oil used for transmission of hydraulic power and or to lubricate automobile transmission systems.

Tribo-—Prefix indicating a relationship to interacting surfaces in relative motion.

Tribo-chemistry—Chemistry dealing with the interacting surfaces in relative motion.

Tribo-element—Solid body that resides within a tribo-system and is bound by one or more tribo-surfaces.

Tribology—Science and technology of interacting surfaces in relative motion and the practices related to them. Tribology includes the study of friction, lubrication, and wear by erosion or by cavitation.

Tribometer—Instrument or test rig used to measure normal and frictional forces of moving surfaces and lubrication and wear behavior of materials or components.

Tribosystem—Functional combination of tribo-elements, including thermal and chemical surroundings.

Tribotechnology—Aspect of tribology that involves engineering applications of tribo-science and the design, development, analysis, and repair of the components for tribological applications.

U.S. Military Specification—Guide in determining the quality requirements of products used by the military services, published by the U.S. Department of Defense.

UEIL—European Union of Independent Lubricant Manufacturers.

UFIP—Union Française des Industries Pétrolières.

UHC—Unburned Hydrocarbon.

UK-PIA—United Kingdom Petroleum Industry Association.

ULEV—Ultra-low Emissions Vehicle.

UTAC—Union Technique de l'Automobile, du Motocycle et du Cycle.

UTTO—Universal Tractor Transmission Oil.

Vacuum Residue—Residue from vacuum distillation of crude oil.

Vapor-phase Lubrication—Type of lubrication in which one or more gaseous reactants are supplied to the vicinity of the surface to be lubricated and which subsequently react with the surface to form a chemical lubricating film.

Vapor Pressure Reid (RVP)—Pressure of the vapor accumulated above a sample of gasoline or other volatile fuel in a standard closed vessel (bomb) at 100 °F (37.8 °C). RVP is used to predict the vapor locking tendencies of the fuel in a vehicle's fuel system.

Varnish—Thin, insoluble, nonwipeable film occurring on some interior engine parts. It can cause sticking and malfunction of the moving parts with close clearances. Such film in a diesel engine is called lacquer. Varnish and lacquer deposits result from the oxidation or polymerization, or both, of fuels and lubricants, or both.

VI (Viscosity Index) Improver—Also called viscosity improver. It is usually a polymeric additive which is added to an oil to increase its viscosity index, that is, to reduce variation in oil viscosity with temperature. VI improvers are the additives that are used to make multi-grades.

Viscoelasticity—Combined property where the fluid is both viscous and elastic. Viscoelastic fluid deformation is dependent both upon temperature and strain rate.

Viscosity—Measure of a fluid's resistance to flow, which is due to friction that exists between fluid layers, called internal friction. Also see Kinematic Viscosity.

Viscosity Conversion Table—Table or chart which is used to convert kinematic viscosity, in centistokes, into Saybolt viscosity in seconds at the same temperature. Conversion into Saybolt Universal viscosities may be carried out by the procedures and tables described in ASTM Method D446, and into Saybolt Furol viscosities by those described in ASTM Method D666. Augmented tables are published in ASTM Special Technical Publication 43B.

Viscosity Index (VI)—Relationship of viscosity of a fluid to temperature in 40° and 100 °C range. High VI fluids lose viscosity with temperature at a slower rate than the low VI oils.

Viscosity Modifier—High molecular weight polymer that is used to maintain the viscosity of an oil in the desired viscosity range. Also see VI Improver.

VOC—Volatile Organic Compound.

Water Resistance—Resistance of an oil or the lubricating grease to be adversely effected by water. Water can cause hydrolysis of the additives or the base fluids such as carboxylate esters, phosphate esters, and poly(alkylene glycol)s, thereby diminishing lubricant performance. Some greases can experience soap washout, depending on the amount of water.

Water Spray Resistance—Ability of a grease to resist displacement from a surface by the impact of water spray.

Wear Debris—Particles that become detached in the wear process.

Wear Resistance—Resistance of a body against wear-related material removal.

Wear Scar—Portion of a solid surface that indicates removal of material due to wear.

Wear Scar Diameter—Dimensions of the wear scar.

White Oil—Highly refined lubricant stock used for specialty applications such as cosmetics and medicines.

Wick Lubrication—System in which the lubricant is delivered to the metal surface by means of a wick.

Wiping—Spreading or removal of material from one point and deposition at another point on the surface of the two bodies in sliding contact. Smeared metal is usually in a softened or a melted form.

Working—Subjecting the lubricating grease to intentional agitating or shearing action.

Worked Penetration—Penetration of a sample of lubricating grease immediately after it has been worked 60 strokes at 77 °F in a standard grease worker.

Yield Point (Yield Value or Yield Stress)—Minimum shear stress that produces flow in a plastic material.

ZDDP—Zinc Dialkyl Dithiophosphates.

ZEV—Zero Emissions Vehicle.

The Systeme International (SI) Units and Conversion Factors

This section of the Appendix provides a summary of the units of measurement that apply to the material included in this book, together with the appropriate conversion factors needed to change them into a "standard" unit of the SI. (See Tables A.1 and A.2.)

The SI is founded on seven SI base *units* for seven *base quantities* assumed to be mutually independent, as given in Table A.3. In addition, there are quantities, called *derived quantities,* which are defined in terms of the seven base quantities via a system of quantity equations [844]. Examples of the SI derived units are given in Tables A.4–A.6. SI also employs 20 prefixes to form decimal multiples and submultiples of SI unit, which are provided in Table A.7. Certain units are not part of the International System of Units per se, but are important and widely used. These units are accepted for use with the SI and are given in Table A.8. Table A.9 lists some of the commonly used Metric System units and symbols and Table A.10 provides the conversion factors.

TABLE A.1—European Lubricants Coding—ISO 6743.

LUBRICANT FAMILY	ISO STANDARD	SYMBOLS
Total loss system	**(ISO 6743/1)**	**A**
Open gears, wire ropes		AB
Undemanding applications		
Refined mineral oils		AN
Unrefined mineral oils		AY
Mold release		**B**
Gears	**(ISO 6743/6)**	**C**
Enclosed		
Mineral oils, R&O		CKB
EP/AW		CKC
EP/AW thermally stable		CKD
R&O with low friction		CKE
Splash lubrication (EP,AW)		CKG
Extreme temperature lubricants (very low or high temperatures)		CKS
CKS lubricants under high load		CKT
Open gears		
Asphaltic, R&O		CKH
Asphaltic, EP		CKJ
Grease		CKL
Intermittent application		CKM
Compressors	**(ISO 6743/3)**	**D**
Air compressors		
Reciprocating (oil lubricated compression chambers)		
Light duty (stage pressure ration < 3:1, discharge temp <160 C)		DAA
Medium duty		DAB
Heavy duty (where coke formation may occur with DAB)		DAC
Rotary vane or screw compressors		
Light duty (Discharge <800 Kpa , air temp <90 C)		DAG
Medium duty		DAH
Heavy duty (Discharge pressure >1500 Kpa or air > 110 C)		DAJ
Vacuum pumps		
Reciprocating, rotary drip-feed, rotary oil flooded (vane/screw)		
Low vacuum for non-aggressive gas (1000 to 1 mbar)		DVA
Low vacuum for aggressive gas (1000 to 1 mbar)		DVB
Oil sealed vacuum pumps (sliding, rotary vane or plunger)		
Medium vacuum for non-aggressive gas (1 to 0.001 m bar)		DVC
Medium vacuum for aggressive gas (1 to 0.001 m bar)		DVD
High vacuum for non-aggressive gas (.001 to .0000001 mbar)		DVE
High vacuum for aggressive gas (0.001 to 0.0000001 mbar)		DVF
Gas compressors		
Non-reactive gas (use oil)		
Gases which do not contain condensable materials		DGA
DGA gases containing moisture or condensable materials		DGB
Reactive gases (use synthetics)		
Gases with high solubility in mineral oil that reduce viscosity		DGC
Gases that chemically react with mineral oil		DGD
Very dry inert or reducing gases with dew point –40 C		DGE
Refrigeration compressors		
Ammonia evaporator above –40 C		DRA
Ammonia evaporator below –40 C (synthetic hydrocarbon)		DRB
Halocarbon refrigerant		DRC
Hydrocarbon refrigerant		DRD
Internal combustion engines	**(ISO 6743/15)**	**E**
Two-stroke engines		EG
Mineral base stocks-lubricity and detergency		EGB
Synthetic fluid to reduce smoke and exhaust deposits		EGC
With Synthetic fluid and added detergency, severe service		EGD
Spindle bearings, clutches	**(ISO 6743/2)**	**F**
Clutches, spindle bearings, bearings		FC
Spindle bearings, bearings		FD
Slide ways	**(ISO 6743/13)**	**G**
Hydraulic systems	**(ISO 6743/4)**	**H**
Automatic transmissions		HA
Fire resistant fluids		HF
Oil in water emulsions (soluble oils)		HFAE
Chemical solutions in water		HFAS
Water in oil emulsions (invert emulsions)		HFB
Water polymer solutions (water glycol)		HFC

TABLE A.1— *(Continued.)*

LUBRICANT FAMILY	ISO STANDARD	SYMBOLS
Synthetic fluids containing no water		HFD
Phosphate ester		HFDR
Chlorinated hydrocarbons		HFDS
Mixtures of HFDR and HFDS fluids		HFDT
Other compositions		HFDU
Hydraulic-slide way		HG
Refined mineral oil		HH
R&O		HL
Anti-wear		HM
Couplers & converters		HN
High VI R&O		HR
High VI AW		HV
Synthetic w/o fire resist		HS
Metal working	**(ISO 6743/7)**	**M**
Operations needing lubrication (oil)		
Fluid with antirust		MHA
Fluid with fat (friction reducing)		MHB
Fluid with inactive S, EP without fat		MHC
Fluid with active S, EP without fat		MHD
Fluid with inactive S with fat		MHE
Fluid with active S with fat		MHF
Greases, pastes, wax diluted with MHA		MHG
Soaps, powders, solids and blends thereof		MHH
Operations needing cooling (water)		
Soluble oils with rust prevention		MAA
Solubles with friction reduction		MAB
Solubles with EP		MAC
Solubles with friction reduction & EP		MAD
Semi-synthetics with rust prevention		MAE
Semi-synthetics with rust, AW &/or EP		MAF
Synthetics with rust preventions		MAG
Synthetics with friction and/or EP		MAH
Greases and pastes applied blended with water		MAI
Electrical insulation		**N**
Pneumatic tools	**(ISO 6743/11)**	**P**
Percussive pneumatic tools		
Non-inhibited straight mineral oil		PAA
Mineral oil with AW or corrosion prevention (CP)		PAB
Mineral oil with AW, CP, emulsifying and foam inhibition		PAC
Synthetic base fluid lubricants		PAD
Greasing (semi-fluid grease)		PAF
Rotary pneumatic tools		
Non-inhibited straight mineral oil		PBA
Mineral oil with CP		PBB
Mineral oil with AW, CP, emulsifying and foam inhibition		PBC
Synthetic base fluid lubricants		PBD
Heat transfer	**(ISO 6743/12)**	**Q**
Corrosion preventives	**(ISO 6743/8)**	**R**
Light duty		
Water-displacing fluid leaving a thin film		RA
Water-based fluid leaving an oily film		RB
RB products having water-displacing properties		RCC
More severe duty		
Undiluted fluid		RD
RD products having water-displacing properties		RDD
Solvent-based fluid leaving oily or greasy film		RE
RE products having water-displacing properties		REE
Solvent-based fluid leaving a waxy to dry film		RF
RF products having water-displacing properties		RFF
Solvent-based fluid leaving a bituminous film		RG
Water-based fluid leaving a waxy to greasy film		RH
Soft or thick grease		RK
Undiluted fluid (for protection coated materials)		RL
Fluid leaving waxy to dry film (to protect coated materials)		RM
Solvent-or water-based fluid leaving peelable film		RP
Plastic compound applied when melted		RT
Turbines	**(ISO 6743/5)**	**T**
Aircraft turbines		TA
Hydraulic		TH
Steam turbine		
Normal service, R&O		TSA
Synthetic fluids w/o fire-resistant properties		TSC

TABLE A.1— *(Continued.)*

LUBRICANT FAMILY	ISO STANDARD	SYMBOLS
Turbines	**(ISO 6743/5)**	T
Fire resistant phosphate		TSD
High load-carrying ability		TSE
Gas turbine		
Normal service with R&O		TGA
High temperature service with R&O		TGB
Synthetic fluids w/o fire-resistant properties		TGC
Fire-resistant (phosphate ester) for control system		TGD
High load-carrying ability		TGE
Control system (phosphate)		TCD
Heat treatment	**(ISO 6743/14)**	U
Grease applications	**(ISO 6743/9)**	X
Use 5 symbols after X		
Lower temperature (0 to <-40 C)		A through G
Upper temperature (+60 to >180 C)		A through G
Rust protection (A = bad through I = best)		A through I
Load (A = no EP, B = with EP)		A or B
NLGI Consistency		000 to 6
Example X-A-A-G-B-2 (is a EP, rust inhibited #2 EP grease, 0 to +60 C range)		
(**Note differences from DIN 51512 are base fluid code, EP code and code order**)		
Other applications	**(ISO 6743/10)**	Y
Process oils (dust, ink, etc)		
Benzene and naphthalene removal from coal gas		YA
Batching of fibers		YB
Caking prevention		YC
Dust suppression		YD
Extending rubber or plasticizing plastics (aromatic extracts)		YEA
Extending rubber or plasticizing plastics (naphthenics)		YEB
Extending rubber or plasticizing plastics (paraffinic)		YEC
Extending rubber or plasticizing plastics (white oil)		YED
Air filters for air conditioning systems		YF
Impregnation		YG
Cleaning and household		YH
Conditioning of leather		YL
Plant spraying		YM
Printing inks (heavy inks/absorption type printing)		YPA
Printing inks (light oils-heat-set drying or radiation set ink)		YPB
Crack detection (non-fluorescent mineral oils)		YR
Foam suppression		YS
Cosmetics (white oils)		YT
Pharmaceuticals (white oil)		YW
Electrolysis and metal plating		YX
Testing (calibration and standardizing)		YZ
Steam cylinders		Z

TABLE A.2—Viscosity conversion table—cSt versus SSU (SUS).

Viscosity (cSt)	Viscosity (SSU)	Viscosity (cSt)	Viscosity (SSU)	Viscosity (cSt)	Viscosity (SSU)
2	32.6	46	214	240	1112
3	36.0	50	233	250	1159
4	39.1	55	256	260	1205
5	42.5	60	279	270	1251
6	45.7	65	302	280	1297
7	49.0	70	325	290	1344
8	52.0	75	349	300	1390
9	55.7	80	372	315	1460
10	59.0	85	395	330	1529
11	62.5	90	418	350	1622
12	66.2	95	442	370	1715
13	70.0	100	465	390	1807
15	77.5	110	511	410	1900
17	85.5	120	558	430	1990
19	94	130	605	450	2090
21	100	140	649	470	2180
23	111	150	695	490	2270
25	120	160	742	500	2320
28	133	170	788	550	2540
30	142	180	834	600	2780
33	155	190	881	650	3010
35	164	200	927	700	3240
38	178	210	973	750	3470
40	187	220	1020	800	3700
43	200	230	1058		

TABLE A.3—SI base units [844].

	SI Base Unit	
Base Quantity	Name	Symbol
length	metre	m
mass	kilogram	kg
time	second	s
electric current	ampere	A
thermodynamic temperature	Kelvin	K
amount of substance	mole	mol
luminous intensity	candela	cd

TABLE A.4—Examples of SI derived units [844].

	SI Derived Unit	
Derived Quantity	Name	Symbol
area	square metre	m^2
volume	cubic metre	m^3
speed, velocity	metre per second	m/s
acceleration	metre per second squared	m/s^2
wave number	reciprocal metre	m^{-1}
mass density	kilogram per cubic metre	kg/m^3
specific volume	cubic metre per kilogram	m^3/kg
current density	ampere per square metre	A/m^2
magnetic field strength	ampere per metre	A/m
amount-of-substance concentration	mole per cubic metre	mol/m^3
luminance	candela per square metre	cd/m^2
mass fraction	kilogram per kilogram, which may be represented by the number 1	kg/kg=1

TABLE A.5—SI derived units with special names and symbols [844].

Derived Quantity	Name	Symbol	Expression in Terms of Other SI Units	Expression in Terms of SI Base Units
plane angle	radian	rad	...	$m \cdot m^{-1} = 1$ [(b)]
frequency	hertz	Hz	...	s^{-1}
force	Newton	N	...	$m \cdot kg \cdot s^{-2}$
pressure, stress	Pascal	Pa	N/m^2	$m^{-1} \cdot kg \cdot s^{-2}$
energy, work, quantity of heat	Joule	J	N·m	$m^2 \cdot kg \cdot s^{-2}$

TABLE A.6—Examples of SI derived units whose names and symbols include SI derived units with special names and symbols [844].

Derived Quantity	Name	Symbol
dynamic viscosity	Pascal second	Pa·s
moment of force	Newton metre	N·m
surface tension	Newton per metre	N/m
angular velocity	radian per second	rad/s
heat capacity, entropy	Joule per Kelvin	J/K
specific heat capacity, specific entropy	Joule per kilogram Kelvin	J/(kg·K)
specific energy	Joule per kilogram	J/kg
thermal conductivity	watt per metre Kelvin	W/(m·K)
permittivity	farad per metre	F/m
molar entropy, molar heat capacity	Joule per mole Kelvin	J/(mol·K)
exposure (x and γ rays)	coulomb per kilogram	C/kg
absorbed dose rate	gray per second	Gy/s

TABLE A.7—SI Prefixes [844].

Factor	Name	Symbol	Factor	Name	Symbol
10^{24}	yotta	Y	10^{-1}	deci	d
10^{21}	zetta	Z	10^{-2}	centi	c
10^{18}	exa	E	10^{-3}	milli	m
10^{15}	peta	P	10^{-6}	micro	μ
10^{12}	tera	T	10^{-9}	nano	n
10^{9}	giga	G	10^{-12}	pico	p
10^{6}	mega	M	10^{-15}	femto	f
10^{3}	kilo	k	10^{-18}	atto	a
10^{2}	hecto	h	10^{-21}	zepto	z
10^{1}	deka	da	10^{-24}	yocto	y

TABLE A.8—Non-SI units accepted for use with the SI units [844].

Name	Symbol	Value in SI units
minute	min	1 min=60 s
hour	h	1 h=60 min=3600 s
litre	L	$1\ L = 1\ dm^3 = 10^{-3}\ m^3$
metric ton or tonne	t	$1\ t = 10^3\ kg$
bar	bar	$1\ bar = 0.1\ MPa = 100\ kPa = 1000\ hPa = 10^5\ Pa$
ångström	Å	$1\ Å = 0.1\ nm = 10^{-10}\ m$
rad	rad	$1\ rad = 1\ cGy = 10^{-2}\ Gy$

TABLE A.9—Commonly used metric system units and symbols [845].

Quantity Measured	Unit	Symbol	Relationship
Length, width, distance, thickness, girth, etc.	millimetre	mm	10 mm = 1 cm
	centimetre	cm	100 cm = 1 m
	metre	m	
	kilometre	km	1 km = 1000 m
Mass ("weight")*	milligram	mg	1000 mg = 1 g
	gram	g	
	kilogram	kg	1 kg = 1000 g
	metric ton	t	1 t = 1000 kg
Time	second	s	
Temperature	degree Celsius	°C	
Area	square metre	m^2	
	hectare	ha	1 ha = 10 000 m^2
	square kilometre	km^2	1 km^2 = 100 ha
Volume	millilitre	mL	1000 mL = 1 L
	cubic centimetre	cm^3	1 cm^3 = 1 mL
	litre	L	1000 L = 1 m^3
	cubic metre	m^3	
Speed, velocity	metre per second	m/s	
	kilometre per hour	km/h	1 km/h = 0.278 m/s
Density	kilogram per cubic metre	kg/m^3	
Force	Newton	N	
Pressure, stress	kiloPascal	kPa	
Power	Watt	W	
	kilowatt	kW	1 kW = 1000 W
Energy	kiloJoule	kJ	
	MegaJoule	MJ	1 MJ = 1000 kJ
	kiloWatt hour	kW·h	1 kW·h = 3.6 MJ
Electric current	ampere	A	

TABLE A.10—Unit conversion factors [846].

Units	Multiplied by Factor	Converted Units
ampere-hours	3600	Coulombs (C)
ampere-hours	0.03731	Faradays
amps	amp × volt	Watts
angstrom unit	3.94E−06	inch
angstrom unit	0.0001	micron
atmospheres	76	cm of mercury
atmospheres	101.325	kilo Pascal
atmospheres	14.7	pounds/sq in.
barrels (oil)	158.9873	litre (L or l)
barrels (oil)	42	gallons (oil)
bars	0.9869	atmospheres
bars	100	kiloPascals (kPa)
bars	14.5	pounds/sq in.
Btu	1.06E+10	ergs
Btu (39 °F)	1059.67	Joules (J)
Btu/hr	0.2930711	Watts
calorie, gram (mean)	0.00396832	Btu (mean)
calorie	1.56E−06	horsepower-hour
calorie (mean)	4.19002	Joule
calorie	1.16E−06	kiloWatt-hour (kWh)
Celsius (degree C)	1.8+32	degree Fahrenheit
centimetres (cm)	0.3937	inches
centimetres	393.7	mils
centimetres of mercury (0 °C)	1333.2239	Pascal
centimetres of mercury	0.01316	atmospheres
centimetres of mercury	0.1934	pounds/sq in. (psi)
centiPoise	1.00E−03	Pascal second (Pa-s)
centiStokes	1.00E−06	metre sq per second
Coulombs	1.04E−05	Faradays
cubic centimetre	0.06102374	cu. inch
cubic centimetre	0.002113	pint (US liq.)
cubic centimetre	0.001057	quart (US liq.)
cycle per second (cps)	1	hertz
degrees Fahrenheit	−32 × 0.55555	degrees Celsius
degrees (angle)	0.01745329	radians
degrees/sec	0.1667	revolutions/min
dyne/sq cm	9.87E−07	atmospheres
dyne/sq cm	2.95E−05	inch of mercury at 0 °C
dynes	0.00102	grams
dynes	0.0000001	Joules/cm
dynes	0.00001	Joules/metre (Newton)
dynes	1.02E−06	kilograms
dynes	2.25E−06	pounds
dyne cm	1.00E−07	Newton metre
dyne per sq cm	0.1	Pascal (Pa)
dynes/sq cm	0.000001	bars
ergs	9.48E−11	Btu
ergs	7.37E−08	foot-pounds
ergs	2.78E−14	kiloWatt-h
ergs	2.78E−11	Watt-h
ergs/s	5.69E−06	Btu/min
ergs/s	4.43E−06	ft-lb/min
ergs/s	7.38E−08	ft-lb/s
ergs/s	1.34E−10	horsepower
ergs/s	1.43E−09	kg-calories/min
ergs/s	1.00E−07	Watts
Fahrenheit (°F)	−32 × 0.55555	Celsius (°C)
Faradays	26.8	ampere-hours
Faraday (chemical)	96495.7	coulomb per mole
Faraday (physical)	96521.9	coulomb per mole
Feet (ft)	30.48	centimetres

TABLE A.10— *(Continued.)*

Units	Multiplied by Factor	Converted Units
foot-pounds	0.001286	Btu
foot-pounds	1.36E+07	ergs
foot-pounds	0.3238	gram-calories
foot-pounds	5.05E−07	hp-h
foot-pounds	1.355818	Joules
foot-pounds	0.000324	kg-calories
foot-pounds	3.77E−07	kiloWatt-h
foot-pounds/min	0.001286	Btu/min
foot-pounds/min	0.01667	foot-pounds/s
foot-pounds/min	0.0000303	horsepower
foot-pounds/min	0.000324	kg-calories/min
foot-pounds/min	0.0226	Watts
foot-pounds/s	4.6263	Btu/h
foot-pounds/s	0.07717	Btu/min
foot-pounds/s	8.18E−04	horsepower
foot-pounds/s	1.01945	kg-calories/min
foot-pounds/s	1.355818	watts
gallons (Imperial)	4.54609	litres
gallons (US)	3785.412	cubic cm (cc)
gallons (US)	3.785412	litres
gallons (liq. British imp.)	1.20094	gallons (US liq.)
gallons (US)	0.83267	gallons (imp.)
gauss	1.00E−04	Tesla
grams	0.001	kilograms
grams	1000	milligrams
grams	0.035273962	ounces(avoirdupois)
grams	0.002204623	pounds
grams/cm	0.0056	pounds/inch
grams/cu. cm	0.03613	pounds/cu in
grams/litre	1000	parts/million
grams/sq cm	2.0481	pounds/sq ft
grams force/sq cm	98.0665	Pascal
gram-calories	0.0039683	Btu
gram-calories	4.19E−07	ergs
gram-calories	3.088	foot-pounds
gram-calories	1.56E−06	horsepower-h
gram-calories	0.001163	watt-h
gram-calories/s	14.286	Btu/h
gram-calories	9.30E−08	Btu
gram-centimetres	980.7	ergs
gram-centimetres	980700	Joules
horsepower	42.44	Btu/min
horsepower	33000	foot-lb/min
horsepower	550	foot-lb/s
horsepower (metric)	0.9863	horsepower
horsepower	10.68	kg-calories/min
horsepower	0.7457	kilowatts
horsepower (metric)	735.4988	watts
horsepower-h	2547	Btu
horsepower-h	2.68E+13	ergs
horsepower-h	1980000	foot-lb
horsepower-h	2684000	Joules
horsepower-h	641.1	kg-calories
horsepower-h	273700	kg-metres
horsepower-h	0.7457	kilowatt-h
inches	2.54	centimetres
inches	0.0254	metres
inches	25.4	millimetres
inches	1000	mils
inches of mercury (at 32 °F)	3386.389	Pascal
inches of mercury	0.03342	atmospheres

TABLE A.10— *(Continued.)*

Units	Multiplied by Factor	Converted Units
inches of mercury	0.03453	kgs/sq cm
inches of mercury	345.3	kgs/sq metre
inches of mercury	70.73	pounds/sq ft
inches of mercury	0.4912	pounds/sq in.
Joules	9.48E−04	Btu
Joules	10000000	ergs
Joules	0.7376	foot-pounds
Joules	0.0002389	kg-calories
Joules	0.102	kg-metres
Joules	0.0002778	watt-h
Joules/cm	10200	grams
Joules/cm	10000000	dynes
Joules/cm	100	Joules/metre (Newton)
Joules	22.48	pounds
Kelvin	t/°C=T/K −273.15	degree Celsius
kilos (slang)		kilograms
kilograms (kg)	980665	dynes
kilograms	1000	grams
kilograms	0.09807	Joules/cm
kilograms	9.807	Joules/metre (Newton)
kilograms	2.2046	pounds
kilograms	9.84E−04	tons (long)
kilograms	1.10E−03	tons (short)
kilograms/cu metre	0.001	grams/cu cm
kilograms/cu metre	0.06243	pound/cu ft
kilograms/cu metre	3.61E−05	pounds/cu in
kilograms/cu metre	3.41E−10	pounds/mil-foot
kilograms/metre	0.672	pounds/ft
kilograms/sq cm	980665	dynes
kilograms/sq cm	0.9678	atmospheres
kilograms/sq cm	28.96	inches of mercury
kilograms/sq cm	2048	pounds/sq ft
kilograms/sq cm	14.22	pounds/sq in.
kilograms/sq metre	9.68E−05	atmospheres
kilograms/sq metre	9.81E−05	bars
kilograms/cu metre	2.90E−03	inches of mercury
kilograms/cu metre	0.2048	pounds/sq ft
kilograms/cu metre	1.42E−03	pounds/sq in.
kilograms/sq mm	1.00E+06	kgs/sq metre
kilogram-calories	3.968	Btu
kilogram-calories	3088	foot-pounds
kilogram-calories	1.56E−03	hp-h
kilogram-calories	4186	Joules
kilogram-calories	426.9	kg-metres
kilogram-calories	4.186	kiloJoules
kilogram-calories	1.16E−03	kilowatt-h
kilogram-metres	9.29E−03	Btu
kilogram-metres	9.80E+07	ergs
kilogram-metres	7.233	foot-pounds
kilogram-metres	9.804	Joules
kilogram-metres	2.34E−03	kg-calories
kilogram-metres	2.72E−06	kilowatt-h
kilogram-force metres	9.80665	Newton metres
kilolitres	1000	litres
kilometres	3280.84	feet
kilometres	1000	metres
kilometres	0.6214	miles
kilometres	1094	yards
kilometres/h	0.6214	miles/h

TABLE A.10— *(Continued.)*

Units	Multiplied by Factor	Converted Units
kilowatts	56.92	Btu/min
kilowatts	737.6	foot-lb/s
kilowatts	1.341	horsepower
kilowatts	14.34	kg-calories/min
kilowatts	1000	watts
kilowatt-h	3413	Btu
kilowatt-h	3.60E+13	ergs
kilowatt-h	2.66E+06	foot-lb
kilowatt-h	1.341	horsepower-h
kilowatt-h	3.60059E +06	Joules
kilowatt-h	860.5	kg-calories
kilowatt-h	3.67E+05	kg-metres
kilowatt-h	4448.222	Newton
knots	1.8532	kilometres/h
knots	1	nautical miles/h
knots	1.151	statute miles/h (mph)
lb (short for 'pounds')	0.4536	kilograms
litres	1000	cu cm
litres	0.2642	gallons (U.S. liq.)
litres	2.113	pints (U.S. liq.)
litres	1.057	quarts (U.S. liq.)
litres/min	5.89E−04	cu ft/s
litres/min	4.40E−03	gals/s
metres	100	centimetres
metres	3.281	feet
metres	39.37	inches
metres	0.001	kilometres
metres	5.40E−04	miles (nautical)
metres	6.21E−04	miles (statute)
metres	1000	millimetres
metres	1.094	yards
metres/s	196.8	feet/min
metres/s	3.281	feet/s
metres/s	2.237	miles/h
micrograms	1.00E−06	grams
microns	1.00E−06	metres
miles (nautical)	1.852	kilometres
miles/h	1.609344	kms/h
miles/h	0.8684	knots
millibars	100	Pascals
millibars	0.0295299	inches of mercury
milligrams	0.001	grams
milligrams/litre	1	parts/million
millilitres	0.001	litres
millimetres	0.1	centimetres
millimetres	3.28E−03	feet
millimetres	0.03937	inches
millimetres	1.00E−06	kilometres
millimetres	0.001	metres
millimetres	39.37	mils
mils	2.54E−03	centimetres
mils	0.001	inches
mils	2.54E−02	millimetres
Newton	1.00E+05	dynes
Newton	0.2248	pound force (lbf)
ounces	28.349523	grams
ounces	0.0625	pounds
ounces (fluid)	0.02957	litres
oz or ozs (short for ounce)	28.349523	grams
pints (US liquid)	473.1765	cu cm

TABLE A.10— *(Continued.)*

Units	Multiplied by Factor	Converted Units
pints (US liquid)	0.125	gallons (US)
pints (US liquid)	0.4731765	litres
pints (US liquid)	473.1765	millilitres
pints (US liquid)	16	ounce (US fluid)
pints (US liquid)	0.5	quarts (liquid)
Poise (P)	0.1	Pascal s
Poiseuille	1	Pascal s
pounds	453.5924	grams
pounds	0.04448	Joules/cm
pounds	4.448	Joules/metre (Newton)
pound (lb)	0.4536	kilograms
pounds	16	ounces
pounds	0.0005	tons (short)
pound force (lbf)	4.448	Newton
pounds/ft	1.488	kgs/metre
pounds/in.	178.6	gms/cm
pounds/sq in.	0.06804	atmospheres
pounds/sq in.	2.036	inches of mercury
pounds/sq in.	6894.757	Pascal (Pa)
psi (pounds-force/sq in.)	6.894757	kiloPascal (kPa)
quarts (liquid)	946.4	cu cms
quarts (liquid)	0.25	gallons
quarts (liquid)	0.9463	litres
rad (ionizing radiation)	0.01	gray (Gy)(Joule/kg)
revolutions/min (r/min)	6	degrees/s
revolutions/min	0.1047198	radians/s
square centimetres	0.155	square inches
stokes	1.00E−04	metre sq per s
quarts (liquid)	0.25	gallons
quarts (liquid)	0.9463	litres
rad (ionizing radiation)	0.01	gray (Gy)(Joule/kg)
revolutions/s	60	revs/min
square centimetres	0.155	square inches
square inches	6.452	sq cm
square millimetres	1.55E−03	sq inches
stokes	1.00E−04	metre sq per s
tons (long)	1016.047	kilograms
tons (tonne)(metric)	1000	kilograms
tons (metric)	2205	pounds
tons (short)	2000	pounds
tons (short)	0.9071847	tonnes (metric)
torr	133.32237	pascal
watts	3.4129	Btu/h
watts	44.27	foot-lb/min
watts	1.34E−03	horsepower
watts	1.36E−03	horsepower (metric)
watts	0.01433	kg-calories/min
watt hour	3600	Joule
watt-hours	3.413	Btu

Index

S

T

U

V

W

References

[1] Sieloff, F. X. and Musser, J. L., "What Does the Engine Designer Need to Know About Engine Oils?" *Presented to Detroit Section of the Society of Automotive Engineers*, March 16, 1982.

[2] (a) "Kline Study Shows Shell Retakes Top Spot in Global Lubricants Market in 2006," July 30, 2007. http://www.klinegroup.com/news/lubricant_ci_073107.asp. (b) Competitive Intelligence for the Global Lubricants Industry, Year Four, 2005–2015, An Overview on Key Regions, Trends, and Leading Suppliers, Kline Report (http://www.klinegroup.com/news/speeches/ICIS9-12-13-06.pdf). (c) Lube Report. February 1, 2006; Vol. 6, Issue 5 and "Asia Overtakes North America in Lubricants Consumption According to Study by Kline & Company," Auto News from July 7, 2005. (d) Gill, G., "Lubes: A 62 Billion-dollar Baby by 2015," *Lube Report* (May 1, 2007).

[3] "Tribology," http://en.wikipedia.org/wiki/tribology

[4] Klamann, D., *Lubricants and Related Products—Synthesis Properties Applications International Standards*, Verlag Chemie, Hamburg, 1984, and reference therein.

[5] Bartenev, G. M. and Lavrentev, V. V., *Friction and Wear of Polymers*, Elsevier Publishing, NY, 1981, p. 203.

[6] Fein, R. S., "Boundary Lubrication," *CRC Handbook of Lubrication, Theory and Practice in Tribology, Vol. II, Theory and Design*, R. E. Booser, Ed., CRC Press, Boca Raton, 1988, pp. 49–68.

[7] Samuels, L. E., Doyle, E. D., and Turley, D. M., "Sliding Wear Mechanisms," *Fundamentals of Friction and Wear of Materials, Papers presented at the 1980 ASM Materials Science Seminar*, Pittsburgh, PA, October 4–5, 1980, D. A. Rigney, Ed., ASM International, 1981.

[8] Sargent, Jr., L. B., "On the Fundamental Nature of Metal-Metal Adhesion," *ASLE Transactions*, Vol. 21, 1978, pp. 285–290.

[9] Larsen-Basse, J., "Basic Theory of Solid Friction," *American Society of Metals Handbook, Friction, Lubrication, and Wear Technology*, ASM International, S. D. Henry, Ed., Vol. 18, 1992, pp. 27–38.

[10] Rabinowicz, E., "Dry Friction," *Standard Handbook of Lubrication Engineering*, J. J. O'Connor, Ed., McGraw-Hill Book Company, NY, Chapter 1, 1968, pp. 1–3 to 1–26.

[11] Schiemann, L. F. and Schwind, J. J., "Fundamentals of Automotive Gear Lubrication," SAE Paper 841213, Fuels and Lubricants Technology: An Overview, SP603, Society of Automotive Engineers, Oct. 1984, pp. 107–115.

[12] Pilon, E. L., "Gear Lubrication-I," *Lubrication*, Vol. 66, 1980, pp. 1–12.

[13] *Fundamentals of Tribology*, N. P. Suh and N. Saka, Eds., MIT Press, Cambridge, MA, 1980.

[14] Wedeven, L. D., "What is EHD?" *Journal of the American Society of Lubrication Engineers*, Vol. 31, pp. 291–296, June 1975.

[15] *Elasto-hydrodynamic Lubrication: The Fundamentals of Roller and Gear Lubrication*, D. Dowson and G. R. Higginson, Eds., Pergamon Press, Oxford, England, 1966.

[16] *Hydrostatic Lubrication* by R. Bassani and B. Piccigallo, Elsevier Publications, 1992.

[17] Davis, J. R., "Friction and Wear of Internal Combustion Engine Parts," *American Society of Metals Handbook, Friction, Lubrication, and Wear Technology*, ASM International, S. D. Henry, Ed., Vol. 18, 1992, pp. 553–562.

[18] Rosenberg, R. C., "General Friction Considerations for Engine Design," SAE Paper 821576, Society of Automotive Engineers, 1982.

[19] Brewer, A. F., "Lubricant Application Systems: Lubricant Dispensing Equipment," *Standard Handbook of Lubrication Engineering*, J. J. O'Connor, Ed., McGraw-Hill Book Company, NY, Chapter 25, 1968, pp. 25–1 to 25–44.

[20] Rizvi, S. Q. A., "Additives: Chemistry and Testing," *Tribology Data Handbook—An Excellent Friction, Lubrication, and Wear Resource*, E. Richard Booser, Ed., CRC Press, Boca Raton, 1997, pp. 117–137.

[21] (a) Chamberlin, W. B. and Saunders, J. D., "Automobile Engines," *CRC Handbook of Lubrication, Vol. I, Theory and Practice of Tribology: Applications and Maintenance*, Richard E. Booser, Ed., CRC Press, Boca Raton, *Figure courtesy of the Oldsmobile Division of the General Motors Corporation*, 1983, pp. 3–44. (b) Figure courtesy of the Oldsmobile Division of the General Motors Corporation.

[22] Greene, A. B., "Automotive Lubrication," *Lubrication and Lubricants*, E. R. Braithwaite, Ed., Elsevier Publishing Company, Amsterdam, Chapter 9, 1967, p. 428.

[23] Obert, E. F., *Internal Combustion Engines and Air Pollution*, Intext Educational Publishing, NY, 1968.

[24] (a)Rein, S. W., "Viscosity-I," *Lubrication*, Vol. 64, No. 1, 1978, pp. 1–12. (b) "Viscosity-II," *Lubrication*, Vol. 64, No. 1, 1978, pp. 13–32. (c) "Viscosity," *Lubrication*, Vol. 52, No. 3, 1966, pp. 21–48.

[25] Watermann, H., *Correlations between Physical Constants and Chemical Structure*, Elsevier, Amsterdam, 1958.

[26] Eirich, F., *Rheology: Theory and Applications*, Vol. 4, Academic Press, New York, 1967.

[27] *ASTM Annual Book of Standards*, Section 05, "Petroleum Products, Lubricants, and Fossil Fuels," ASTM International, West Conshohocken, PA.

[28] Manning, R. E. and Hoover, M. R., "Flow Properties and Shear Stability," *Fuels and Lubricants Handbook: Technology, Properties, and Testing*, MNL37WCD, ASTM International, West Conshohocken, PA, 2003, pp. 833–878.

[29] Willcock, D., *Mech. Eng.*, Vol. 66, 1944, p. 739; Vol. 67, 1944, p. 201.

[30] Klaus, E., *Mech. Eng.*, Vol. 85, No. 10, 1963, pp. 42–45.

[31] Czichos, H., "Basic Tribological Parameters," *American Society of Metals Handbook, Friction, Lubrication, and Wear Technology*, ASM International, S. D. Henry, Ed., Vol. 18, 1992, pp. 81–88.

[32] Anonymous Author, "Engineering and Design—Lubricants and Hydraulic Fluids," http://www.usace.army.mil/inet/usace-docs/eng-manuals/em1110-2-1424/.

[33] Rizvi, S. Q. A., "History of Automotive Lubrication," *SAE Technical Paper 961949, Presented at Fuels and Lubricant Meeting*, San Antonio, Tx, October 14–17, 1996.

[34] Robbins, W. K. and Hsu, C. S., "Petroleum (Composition)," *Kirk-Othmer Encyclopedia of Chemical Technology, Fourth Edition*, M. Grayson, Ed., John Wiley and Son, NY, Vol. 18, 1996, pp. 352–370.

[35] Schmidt, G. K. and Forster, E. J., "Modern Refining for Today's Fuels and Lubricants," SAE Paper 841206, Fuels and

Lubricants Technology: An Overview, SP.603, Society of Automotive Engineers, Oct. 1984, pp. 1–10.

[36] "Water-based Drilling Fluids Using Latex Additives," U.S. Patent 6,703,351 (March 9, 2004). Inventors: D. K. Clapper, C. Joseph Stowe II, R. G. Bland, T. Xiang, T., and S. Benaissa.

[37] Clark, R. K., "Petroleum (Drilling Fluids)," *Kirk-Othmer Encyclopedia of Chemical Technology, Fourth Edition,* M. Grayson, Ed., John Wiley and Son, NY, Vol. 18, 1996, pp. 370–405.

[38] "Drilling Mud Viscosifier," U.S. Patent 4,240,915 (December 23, 1980). Inventor: J. Block.

[39] Speight, J., "Petroleum (Refinery Processes)," *Kirk-Othmer Encyclopedia of Chemical Technology, Fourth Edition,* M. Grayson, Ed., John Wiley and Son, NY, Vol. 18, 1996, pp. 433–469.

[40] Rakow, M. S., "Petroleum Oil Refining." *Fuels and Lubricants Handbook: Technology, Properties, Performance, and Testing,* MNL37WCD, Ed., ASTM International, West Conshohocken, PA, Chapter 1, 2003, pp. 3–30.

[41] Prince, R. J., "Base Oils from Petroleum," *Chemistry and Technology of Lubricants,* R. M. Mortier and S. T. Orszulik, Eds., Blackie Academic and Professional-Chapman & Hall, London, UK, 1996, Chapter 1, pp. 1–33.

[42] Stipanovic, A. J., "Hydrocarbon Base Oil Chemistry," *Fuels and Lubricants Handbook: Technology, Properties, Performance, and Testing,* MNL37WCD, Ed., ASTM International, West Conshohocken, PA, Chapter 7, 2003, pp. 169–184.

[43] Congram, G. E., "Chemical Additives Aid Processing Efficiency," *The Oil and Gas Journal,* Feb. 9, 1976, pp. 102–105.

[44] Lubrication Theory and Practice, www.lubrizol.com

[45] Gedeon, B. J. and Yenni, N. L., "Use of "Clean" Paraffinic Processing Oils to Improve TPE Properties," *Presented at TPEs 2000,* Philadelphia, PA, Sept. 27–28, 1999.

[46] Sequeira, A., "Lubricant Base Oil Processing," *Lubrication,* Vol. 75, No. 1, 1989, pp. 1–12.

[47] Sequeira, A., *Lubricant Base Oil and Wax Processing,* Marcel Dekker Inc., NY, 1994.

[48] "1999 Lubricating Oil and Wax Capacities of Refiners and Refiners in the Western Hemisphere," *National Petroleum and Refiners Association (NPRA),* 1999.

[49] "Viscosity Index Tables for Celsius Temperatures," ASTM Data Series Publication-DS-39B, ASTM International, West Conshohocken, PA, Sept. 1965.

[50] Schilling, A., *Motor Oils and Engine Lubrication,* Scientific Publications, Great Britain, 1968.

[51] http://www.prod.exxonmobil.com/refiningtechnologies/pdf/refin_EMICT_santiesteban_japan050701.pdf. The link is no longer active.

[52] Johnson, R. L., "Conservation of Lubricants, Materials and Energy," *CRC Handbook of Lubrication, Vol. I, Theory and Practice of Tribology: Applications and Maintenance,* R. E. Booser, Ed., CRC Press, Boca Raton, 1983, pp. 551–565.

[53] Cornitius, T., "Base Stock Market Outlook," *Lubricants World,* 1998, pp. 16–23.

[54] Fitch, J. C. and Barnes, M., "Hydrocarbon Analysis," *Fuels and Lubricants Handbook: Technology, Properties, Performance, and Testing,* MNL37WCD, Ed., ASTM International, West Conshohocken, PA, Chapter 24, 2003, pp. 649–673.

[55] Cochrac, G. J. and Rizvi, S. Q. A., "Oxidation of Lubricants and Fuels," *Fuels and Lubricants Handbook: Technology, Properties, Performance, and Testing,* MNL37WCD, Ed., ASTM International, West Conshohocken, PA, Chapter 30, 2003, pp. 787–824.

[56] Fresco, G. B., Klaus, E. E., and Tewksbury, E. J., "Measurement and Prediction of Viscosity-Pressure Characteristics of Liquids," *ASME J. Lubr. Technol.,* Paper No. 68-Lub-8, 1968, pp. 1–7.

[57] Roelands, C. J. A., Vlugter, J. C., and Waterman, H. I., "The Viscosity-Temperature-Pressure Relationship of Lubricating Oils and Its Correlation with Chemical Constitution," *J. Basic Eng.,* Dec. 1963, pp. 601–610.

[58] Errichello, R., "Selecting Oils with High Pressure-Viscosity Coefficient-Increase Bearing Life by More Than Four Times," *Machinery Lubrication Magazine,* March 2004. "http://www.machinerylubrication.com/article_detail.asp?articleid=586&relatedbookgroup=Lubrication

[59] Blair, S., Klaus, E., and Bala, V., "An Experimental Verification of the Significance of the Reciprocal Asymptotic Isoviscous Pressure for EHD Lubricants," *ASLE Tribology Transactions,* Vol. 36, No. 2, 1993, pp. 153–162.

[60] Murray, D. W., MacDonald, J. M., White, A. M., and Wright, P. W., "The Effect of Basestock Composition on Lubricant Oxidation Performance," *Petroleum Review,* 1982, pp. 36–40.

[61] "Base Oils for Automotive Lubricants," Special Publication SP 526, Society of Automotive Engineers, Warrendale, PA, 1982, and references therein.

[62] Stipanovic, A. J., Smith, M. P., Firmstone, G. P., and Patel, J. A., "Compositional Analysis of Re-refined and Non-conventional Lubricant Base Oils: Correlations to Sequence VE and IIIE Engine Tests," *SAE Publication* 941978, Society of Automotive Engineers, Warrendale, PA, 1994.

[63] Firmstone, G. P., Smith, M. P., and Stipanovic, A. J., "A Comparison of Neural Network and Partial Least Squares Approaches in Correlating Base Oil Composition to Lubricant Performance in Gasoline and Industrial Oil Applications," *Special Publication SP 952534,* Society of Automotive Engineers, Warrendale, PA, 1995.

[64] Roby, S. H., Supp, J. A., Barrer, D. E., and Hogue, C. H., "Base Oil Effects in the Sequence IIID and Sequence VD Engine Tests," *SAE Publication 892108,* Society of Automotive Engineers, Warrendale, PA, 1989.

[65] "The Increasing Significance of Base Oils in the Evolution of Heavy-Duty Diesel Engine Oils," 1999 National Petroleum Refiners Association (NPRA), LW-99–130, Houston, TX, 1999.

[66] Yoshida, T., Igarashi, J., and Watanabe, H., "Pro-Oxidation Properties of Basic Nitrogen Components in Base Oil," *Proceedings of the 11th International Colloquium Technology,* Esslingen, Germany, Jan. 1998.

[67] Kramer, D. C., Ziemer, J. N., Cheng, M. T., Fry, C. E., Reynolds, R. N., and Lok, B. K., et al., "Influence of Group II and Group III Base Oil Composition on VI and Oxidation Stability," National Lubricating Grease Institute (NLGI) Paper 9907, Tucson AZ, 1999.

[68] Mangone, C., "Gas to Liquids—Conversions Produce Extremely Pure Base Oils," in*Machinery Lubrication Magazine,* Nov. 2002.

[69] "Physical and Chemical Properties of Engine Oils" *SAE J357, SAE Information Report, Completely Revised May 1984,* Society of Automotive Engineers.

[70] Asseff, P. A., "Lubrication Theory and Practice," Publication 183-320-59, The Lubrizol Corporation, Wickliffe, OH.

[71] Ripple, D. E., and Fuhrmann, J. F., "Performance Comparisons of Synthetic and Mineral Oil Crankcase Lubricant Base Stocks," *Presented at the 4th International Colloquium*, Esslingen, Germany, Jan. 1984.

[72] Smith, T. G., "Neopentyl Polyol Esters," *Synthetic Lubricants*, R. C. Gunderson and A. W. Hart, Eds., Reinhold Publishing Corporation, NY, Chapter 10, 1962.

[73] Goddard, D. R., "Synthetic Fluids," *Lubrication and Lubricants*, E. R. Braithwaite, Ed., Elsevier Publishing, Chapter 4, 1967, pp. 166–196.

[74] *Synthetic Lubricants and High-performance Functional Fluids*, 2nd Ed., L. R. Rudnick and R. L. Shubkin, Eds., Marcel Dekker, Inc., NY, 1999. Also see the 1993 Edition, R. L. Shubkin, Ed.

[75] Toxics Release Inventory: List of Toxic Chemicals Within the Polychlorinated Alkanes Category and Guidance for Reporting United States Office of Pollution, June 1999, Environmental Protection Prevention and Toxics Agency Washington, DC 20460, http://www.epa. gov/tri/guide_docs/1999/ polychloroalkanes1999. pdf

[76] Adrianus, H., and van Dijk, J., "The Fischer-Tropsch Synthesis: A Mechanistic Study Using Transient Isotopic Tracing," 2001, [http://alexandria. tue. nl/extra2/200111083.pdf].

[77] (a) http://en. wikipedia.org/wiki/Linear_alpha_olefins (b) http://nexant.ecnext.com/com2/gi_0255-148/Alpha-Olefins.html.

[78] Mazzone, L. C. A., and Fernandes, F. A. N., "Modeling of Fischer-Tropsch Synthesis in a Tubular Reactor," *Lat. Am. Appl. Res.* Vol. 36, 2006, pp. 141–148.

[79] Buenemann, T. F., Boyde, S., Randles, S., and Thompson, I., "Synthetic Lubricants-Non Aqueous," *Fuels and Lubricants Handbook: Technology, Properties, Performance, and Testing*, MNL37WCD, Ed., ASTM International, West Conshohocken, PA, Chapter 10, 2003, pp. 249–266.

[80] "Verfahren zur Herstellung von Polyethylen, Polypropylen und Copolymeren," Patent DE3007725, 1980.

[81] "Metallocene Catalysis Polymerization," http://www.pslc.ws/mactest/mcene.htm.

[82] Patent WO 96/23010, 01 August 1996 - Alpha-olefins and olefin polymers and processes therefore.

[83] "Post-metallocene Catalyst," http://en. wikipedia.org/wiki/post-metallocene_catalyst

[84] "Synthetic Lubricants by Oligomerization and Hydrogenation," U.S. Patent 3,780,128 (1973). Inventor: R. L. Shubkin.

[85] "Polymerization of Olefins with BF_3," U.S. Patent 3,382,291 (1968). Inventor: J. A. Brennan.

[86] "Process of Producing a C_6–C_{16} Normal Alpha-olefin Oligomer Having a Pour Point Below about –50 °F," U.S. Patent 3,763,244 (1973). Inventor: R. L. Shubkin.

[87] Shubkin, R. L., "Polyalphaolefins," *CRC Handbook of Lubrication, Theory and Practice in Tribology, Vol. III, Monitoring, Materials, Synthetic Lubricants, and Applications*, R. E. Booser, Ed., CRC Press, Boca Raton, 1993, pp. 219–235.

[88] Moore, L., Bagheri, V., Barrett, C., and Palm, J., "Next Generation Polyalphaolefins—The Next Step in the Evolution of Synthetic Hydrocarbon Fluids," Innovene USA LLC 11/22/05 revision; posted 11/22/05. http://www.innovene.com/ liveassets/bp_internet/innovene/STAGING/ home_assets/assets/products/polyalphaolefins/pdfs/ Next_Generation_PAOs.

[89] "Biodegradable Polyalphaolefin Fluids and Formulations Containing the Fluids," U.S. Patent 6,071,863 (2000). Inventors: R. Benda, J. V. Bullen, and A. J. Plomar.

[90] http://www.atiel.org/downloads/code_practice/ AppendixBissue13.pdf#search=/22polyinternalolefins/ 20pios/22

[91] "Europeans Adopt 6th Base Oil Group," by Tim Sullivan, Lube Report, Sept. 23, 2003.

[92] "Alkylation of Aromatic Hydrocarbons," U.S. Patent 2,939,890 (June 7, 1960). Inventor: G. L. Hervert.

[93] "Aromatic Lubricants and Their Method of Preparation," U.S. Patent 3,173,965 (March 16, 1965). Inventors: J. J. Pappas and F. H. Kant.

[94] "Preparation of Synthetic Hydrocarbon Lubricants," U.S. Patent 4,148,834 (April 10, 1979). Inventors: C. D. Kennedy and G. E. Nicks.

[95] "Alkylation of Aromatic Compounds," U.S. Patent 4,551,573 (Nov. 5, 1985). Inventor: R. L. Cobb.

[96] Aromatics Alkylation—Towards Cleaner, More Selective Processes. [http://www.prod.exxonmobil.com/ refiningtechnologies/pdf/ aromatics_alkylation_catalyst_gen_shape_sel_france.pdf].

[97] "Synthetic oil With a High Viscosity Number and a Low Pour Point (Zeolite/alumina)," U.S. Patent 6,491,809 (Dec. 10, 2002). Inventors: A. Forestiere, E. Benazzi, P. Briot, L. Lew, and S. Gautier.

[98] "Hydrocarbon Hydraulic Oils," U.S. Patent 3,478,113 (Nov. 11, 1965). Inventors: U. B. Bray and M. Fainman.

[99] "Preparation of Synthetic Hydrocarbon Lubricants," U.S. Patent 3,909,432 (Sept. 30, 1975). Inventors: S. McGuire, J. L. Riddle, G. E. Nicks, O. C. Kerfoot, and C. D. Kennedy.

[100] "Emulsifier/Detergent Compounds for Metalworking Lubricants," Thomas, S. C., Jr., Application 20050075252 April 7, 2005.

[101] Suva® Refrigerants—Suva® 134-a in Chillers, http:// refrigerants.dupont.com/Suva/en_US/pdf/h42444.pdf

[102] http://www.exxonmobilchemical.com/Public_Files/ Synthetics/Synthetic_ Lubricants_and_Fluids/Worldwide/ Data_Shcct_Syncsstic_v5.pdf

[103] "Condensation Product from Olefins and Hydrocarbons of Naphthalene Series and Process of Making the Same," U.S. Patent 1,667,214 (April 24, 1928). Inventor: R. Michel.

[104] "Hydrocarbon Oil and Process for Manufacture of the Same," U.S. Patent 1,815,022 (July 14, 1931). Inventor: G. H. B. Davis.

[105] "Methylene Linked Aromatic Pour Point Depressant," U.S. Patent 4,753,745 (June 28, 1988). Inventors: J. L. Kostusyk and S. Q. A. Rizvi.

[106] Randles, S. J., Stroud, P. M., Mortier, R. M., Orszulik, S. T., Hoyes, T. J., and Brown, M., "Synthetic Base Fluids," *Chemistry and Technology of Lubricants*, R. M. Mortier and S. T. Orszulik, Eds., Blackie Academic and Professional-Chapman & Hall, London, UK, 1996, Chapter 2, pp. 34–74.

[107] van der Waal, G., "The Relationship Between the Chemical Structure of Ester Base Fluids and Their Influence on Elas-

tomer Seals and Wear Characteristics," *J. Syn. Lub.*, Vol. 1, No. 4, 1985, pp. 281–301.

[108] Alber, F., Kloppenburg, D., Kloppenburg, H., Schär, D., and Utz, R., "Citric Acid Esters Based on Mixtures of Alcohols—New Kind of Polymer Additives and Plasticizers for a Variety of Polymers," (http://www.priorartdatabase.com/IPCOM/000029795/).

[109] *Kirk-Othmer Encyclopedia of Chemical Technology*, 4th ed., John Wiley & Sons, Inc., NY, Vol. 18, 1991, p. 1991.

[110] "Process for the Production of Trimellitic Anhydride," U.S. Patent 4,587,350 (May 6, 1986). Inventors: P. H. Kilner, J. P. Egan, Jr., S. G. Ceisel, and W. P. Schammel.

[111] "Preparation of Aliphatic Alpha, Omega-dicarboxylic Acids," U.S. Patent 6,437,180 (August 20, 2002). Inventors: M. Schelhaas, H. Greiving, and M. Jautelat.

[112] "Manufacturing Method of Neopentyl Glycol," U.S. Patent 6,268,539 (July 31, 2001). Inventors: H. Sen-Huang, T. Chia-Ruey, and C. Jung-Jen.

[113] "Process for the Preparation of Pentaerythritol," U.S. Patent 5,741,956 (April 21, 1998). Inventor: L. Eek.

[114] "Method for Purifying Trimethylolpropane, Which is Produced by Hydrogenation, by Means of Continuous Distillation," U.S. Patent 6,692,616 (February 17, 2004). Inventors: D. Dernbach, D. Kratz, A. Stammer, H. Rust, and G. Schulz.

[115] Murrenhoff, H. and Remmelmann, A., "Environmentally Friendly Oils," *Fuels and Lubricants Handbook: Technology, Properties, Performance, and Testing*, G. E. Totten Ed., ASTM International, West Conshohocken, PA, Chapter 11, 2003, pp. 267–295.

[116] Cerny, J., "Using Calorimetry to Measure Motor Oil Remaining Useful Life," *Practical Oil Analysis Magazine*, Sept. 2004. http://www.practicingoilanalysis.com/article_detail.asp?articleid=649

[117] Sanvordenker, K. S., "Material Compatibility of R134a in Refrigerant Systems," *Presented at The American Society of Heating, Refrigeration and Air-Conditioning Engineers, Winter Meeting*, Jan. 1989.

[118] Quinn, C., Traver, F., and Murthy, K., "Silicones," *Synthetic Lubricants and High Performance Functional Fluids*, R. L. Rudnick and R. L. Shubkin, Eds., Marcel Dekker, Inc., NY, 1999, pp. 267–286.

[119] *Coolanol® Silicate Ester Dielectric Heat Transfer Fluids*, http://www.exxonmobil.com/lubes/exxonmobil/emal/Coolanol_Silicate_Ester_Dielectric_Heat_Transfer_Fluids.html

[120] Jones, Jr., W. R., "The Properties of Perfluoroethers Used for Space Applications," NASA Technical Memorandum 106275, July 1998. http://gltrs.grc.nasa.gov/reports/1993/TM-106275.pdf

[121] Gumprecht, W. H., "PR-143—A New Class of High-temperature Fluids," *ASLE Transactions*, Vol. 9, 1966 pp. 24–30.

[122] Sianesi, D., Zamboni, V., Fontanelli, R., and Binaghi, M., "Perfluoropolyethers. Their Physical Properties and Behavior at High and Low Temperatures," *Wear*, Vol. 18, 1971, pp. 85–100.

[123] Sianesi, D., Pasetti, A., Zamboni, V., Fontanelli, R., and Bernardi, G. C., "Perfluoropolyethers by Photo-oxidation of Fluoro-olfins," *Chim. Ind.*, (*Milan*), Vol. 55, 1978, pp. 208–221.

[124] Ohsaka, Y., "Recent Advances in Synthetic Lubricating Oils., (8). Perfluorinated Polyethers," *Petrotech* (*Tokyo*), Vol. 8, No. 9, 1985, pp. 840–843.

[125] Kinzig, B. J., and Ravner, H., "Factors Contributing to the Properties of Fluoropolymer Barrier Films," *ASLE Trans.*, Vol. 21, 1978, pp. 291–298.

[126] Hilton, M. R., and Fleischaur, P. D., "Lubricants for High-vacuum Application," Aerospace Report No. TR-0091 (6945-03)-6, March 1993.

[127] Bell, G. A., Howell, J., and DelPasco, T. W., "Perfluoroalkyl Polyethers," *Synthetic Lubricants and High Performance Functional Fluids*, R. L. Rudnick, and R. L. Shubkin, Eds., Marcel Dekker, Inc., NY, 1999, p. 215.

[128] Gulino, D. A., and Coles, C. E., "Oxygen Plasma Effects on Several Liquid Droplet Radiator Fluids," *J. Spacecr. Rockets*, Vol. 25, No. 2, 1988, pp. 99–101.

[129] Jones, W. R., Jr., Johnson, R. L., Winer, W. O., and Sanborn, D. M., "Pressure-Viscosity Measurements for Several Lubricants to 5.5×10^8 per Square Metre (8×10^4 psi) and 149 °C (300 °F)," *ASLE Trans.*, Vol. 18, No. 8, 1975, pp. 249–262.

[130] Quinn, C., Traver, F., and Murthy, K., "Silicones," *Lubrication at the Frontier*, D. Dowson et al., Eds., *Proceedings of the 25th Leeds-Lyons Symposium on Tribology, 1998*, Elsevier, NY, 1999, p. 267.

[131] Aderin, M., Johnston, G. J., Spikes, H. A., and Caporiccio, G., "The Elasto-hydrodynamic Properties of Some Advanced Non Hydrocarbon-Based Lubricants," *Lubr. Eng.*, Vol. 48, No. 8, 1992, pp. 633–638.

[132] Lagow, R. J., and Margrave, J. L., "Direct Fluorination: A New Approach to Fluorine Chemistry," *In Progress in Inorganic Chemistry*, Vol. 26, 1979, pp. 162–210. S. J. Llppard, Ed., Wiley, NY.

[133] Jones, W. R., Bierschenk, T. R., Juhlke, T. J., Kawa, H., and Lagow, R. J., "Improved Perfluoroalkyl Ether Fluid Development," *Ind. Eng. Chem. Res.*, Vol. 26, 1987 pp. 1930–1935.

[134] Blerschenk, T. R., Kawa, H., Juhlke, T. J., and Lagow, R. J., "The Development of New, Low-Cost Perfluoroalkylether Fluids with Excellent Low and High-Temperature Properties," *NASA CR 182155*, May 1988.

[135] Conley, P. L., and Bohnerj, J. J., "Experience with Synthetic Fluorinated Fluid Lubricants," *24th Aerospace Mechanisms Symposium*, NASA CP 3062, April 18–20, 1990, pp. 213–230.

[136] "Method of Quenching Metal," U.S. Patent 5,490,886 (February 13, 1996). Inventor: D. G. Placek.

[137] "Military Fluids—MIL-H-19457 D (SH)," http://www.radcoind.com/MIL-H-19457D. htm.

[138] *Ullmann's Encyclopedia of Industrial Chemistry*, 5th ed., W. Gerhartz, Exec. Ed., Deerfield Beach, FL: VCH Publishers, Vol A1, 1985 to Present, VA11 364.

[139] Purdy, E. M., "Use of CTFE as an Additive to Enhance Fire Resistance of Single Hydraulic Fluid," TARDEC Report 13610, Nov. 1994.

[140] *Chemical and Process Technology Encyclopedia*, D. Considine, McGraw-Hill Education, 1974, p. 286.

[141] "Halovac Inert Pump Fluids," http://www.inlandvacuum.com/halovac.html.

[142] *Hawley's Condensed Chemical Dictionary*, 12th ed., R. J. Lewis, Sr., Ed., Van Nostrand Rheinhold Co., NY, 1993.

[143] (a) Santolubes Product Brochure. (b) Joaquim, M., "Polyphenyl Ether Lubricants," *Synthetic Lubricants and High-*

performance Functional Fluids, R. L. Rudnick, and R. L. Shubkin, Eds., Marcel Dekker, Inc., NY, 1999, p. 239. Swedbers, S., "What is up next?" Lubes 'n' greases. Vol. 13, July 18, 2008, pp. 674.

[144] Herber, Sr., J. F. and Joaquim, M., "Polyphenyl Ethers: Lubrication in Extreme Environments," http://www.chemassociates.com/products/findett/PPEs_LubEng.pdf.

[145] OS 124, Polyphenyl Ether Fluid by Santovac®, Commercial Literature, Santovac® Fluids, Saint Charles, MO, 1999.

[146] Venier, C. G. and Casserly, E. W., "Cycloaliphatics," in *Synthetic Lubricants and High Performance Functional Fluids*, R. Shubkin, Ed., 1993. pp. 241–269.

[147] Venier, C. G. and Casserly, E. W., "Cycloaliphatics," *Synthetic Lubricants and High-performance Functional Fluids*, R. L. Rudnick and R. L. Shubkin, Eds., Marcel Dekker, Inc., NY, 1999, p. 325.

[148] "Lubricants Comprising Novel Cyclopentanes, Cyclopentadienes, Cyclopentenes, and Mixtures Thereof and Methods of Manufacture," U.S. Patent 4,721,823 (January 6, 1988). Inventors: C. G. Venier and E. W. Casserly.

[149] "Control of the Reaction Parameters Changes the Alkylate Composition [Substituted Cyclopentadienes, Methods of Manufacture and Uses," U.S. Patent 5,144,095 (September 1, 1992). Inventors: C. G. Venier and E. W. Casserly,

[150] "MACS vs. PFPEs: Choosing between Multiply Alkylated Cycolpentanes and Perfluoropolyethers for Vacuum Applications," http://www.nyelubricants.com/pdf/literature/Lube-Insights-Pennzane. pdf.

[151] Venier, C. G. and Casserly, E. W., "Multiply-Alkylated Cyclopentanes (MACs): A New Class of Synthesized Hydrocarbon Fluids," *Lubr. Eng.* Vol. 47, No. 7, 1991, pp. 586–591.

[152] Brown, P. L., Miller, J. B., Jones, Jr., W. R., Rasmussen, K., Wheeler, D. W., Rana, M., and Peri, F., "The Clouds and the Earth's Radiant Energy System Elevation Bearing Assembly Life Test," *33rd Aerospace Mechanisms Symposium*, Pasadena, CA, NASACP-1999-209259, 1999, pp. 197–212.

[153] Jones, Jr., W. R. and Jansen, M. J., "Space Tribology," NASA/TM—2000-209924, March 2000. http://gltrs.grc.nasa.gov/reports/2000/TM-2000-209924. pdf

[154] Klaus, E. E. and Tewksbury, E. J., "Liquid Lubricants," *Handbook of Lubrication: Theory and Practice of Tribology, Volume II, Theory and Design*, R. Booser, Ed., CRC Press, Inc., Boca Raton, 1983, pp. 229–253.

[155] "Metric Conversion Chart—Pressure," http://www.engnetglobal.com/tips/convert.asp?catid=17

[156] Errichello, R., "Selecting Oils with High Pressure-Viscosity Coefficient—Increase Bearing Life by More Than Four Times," *Machinery Lubrication Magazine*, March 2004.

[157] Lehmus, M. and Holmqvist, H. G., "Top Tier Base Oil Trends Reflect European Engine Oil Demands," *Lubricants World*, Nov. 2000. pp. 25–26.

[158] DePuy, C. H. and King, R. W., "Pyrolytic Cis Elimination," *Chemical Reviews*, Vol. 60, 1960, pp. 431–445.

[159] *High Temperature Lubrication*, A. R. Lansdown, Mechanical Engineering Publications Ltd., London, 1988.

[160] Szeri, A. Z., *Tribology—Friction, Lubrication, and Wear*, Hemisphere Publications, Washington, DC, 1980.

[161] Mock, III G. B., "Checking Plastic and Lubricant Compatibility Helps Avoid Warranty Problems," www.nyelubricants.com/pdf/1nyeplasticcomp.pdf.

[162] Fuhr, J., Oetinger, J., Totten, G. E., and Webster, G. M., "Mineral Oil Heat Transfer Fluids," *Fuels and Lubricants Handbook: Technology, Properties, Performance, and Testing*, MNL37WCD, Ed., ASTM International, West Conshohocken, PA, Chapter 21, 2003, pp. 573–586.

[163] Reddy, R. G., Wu, B., and Roger, R. D., "Ionic Liquids as Thermal Energy Storage Media for Solar Thermal Electric Power Systems," http://www.eere.doe.gov/troughnet/pdfs/rreddy_ionicfluids.pdf.

[164] Joaquim, M. E., and Herber, J. F., "Lubrication of Electronic Connectors and Equipment in Radiation Environments," http://www.chemassociates.com/products/findett/PPEs_Radiation2.pdf.

[165] Bolt, R. O., "Radiation Effects on Lubricants," *CRC Handbook of Lubrication, Vol. I, Theory and Practice of Tribology: Applications and Maintenance*, R. E. Booser, Ed., CRC Press, Boca Raton, 1983, pp. 209–233.

[166] Carroll, J. G., and Bolt, R. O., *Radiation Effects on Organic Materials*, R. O. Bolt and J. G. Carroll, Eds., Academic Press, NY, 1963.

[167] Harrington, R., "Elastomers for Use in Radiation Fields IV. Effects of Gamma Radiation on Miscellaneous Elastomers and Rubber-like Materials," *Rubber Age*, Vol. 83, No. 1, 1958, p. 472.

[168] "Eco-Label," http://europa.eu.int/scadplus/leg/en/lvb/l28020.htm

[169] "Concept Background Document: Development of Criteria for the Award of the European Eco-label to Lubricants," by D. Theodori, R. J. Saft, H. Krop, P. van Broekhuizen http://europa.eu.int/comm/environment/ecolabel/pdf/lubricants/backgrounddoc_111203.pdf

[170] "Lubricating Oil Basestocks," Summary prepared by American Petroleum Institute, July 24, 2001 http://www.petroleumhpv.org/Product_Categories/Lube_Oils/2004_03_03_Lubes_Revised_Robust_Summary.pdf

[171] "Biodegradability and Alternatives," http://mbm.net.au/b100/biodegrade.html

[172] "A Comparison of the Environmental Performance of Olefin and Paraffin Synthetic Base Fluids (SBF)," Nov. 2006. http://www.americanchemistry.com/s_acc/bin.asp?CID=1075&DID=4210&DOC=FILE.PDF

[173] Randles, S. J., and Wright, M., "Environmentally Friendly Ester Lubricants for the Automotive and Engineering Industries—Esters, Base Oils for Synthetic Lubricants," *International Congress about Lubricants, Lubricants for Future and the Environment*, Brussels, Sept. 1991, p. 197.

[174] "Gearing Up Performance—An Introduction to Synthetic Lubricants by Jeffrey Lay," [http://www.nyelubricants.com/pdf/nye_gear_tech_v2.pdf]

[175] Applewhite, T. H., "Fats and Fatty Oils," *Kirk-Othmer Encyclopedia of Chemical Technology*, Third Edition, M. Grayson, Ed., John Wiley and Son, NY, Vol. 9, 1981, pp. 795–831.

[176] Krizek, F. Ch., and Coffin, P. S., "Vegetable Oil Based Lubricants and Their Applications," *Presented at the 13th World Petroleum Congress*, Buenos Aires, 1991.

[177] (a) Honary, L. A. T., University of Northern Iowa, "Biodegradable/Bio-based Lubricants and Greases," *Machinery Lubrication Magazine*, Sept. 2001. (b) Erhan, S. Z. and Sharma, B. K., "Environmentally Friendly Biobased Lubricants," http://www.apfct.com/download/report/PM%20Session/877.pdf.

[178] Begstra, R., "Green means Go," *Lubes-N-Greases*, November 2004, pp. 40–41.

[179] "Fats, Oils, Fatty Acids, Triglycerides," http://www.scientificpsychic. com/fitness/fattyacids.html.

[180] (a) Erhan, S. Z. and Perez, J. M., *Biobased Industrial Fluids and Lubricants,* L. R. Rudnick, Ed., AOCS Press, Champaign, IL, 2002. (b) Swedberg, S., "What is Bio—and What is Next?" *Lube & Greases*, Vol. 14, No. 7, July 18, 2008, pp. 6–14.

[181] Barwell, F. T., "Lubricant Additives—The History of Their Introduction and Assessment," *Erdol-Erdgas-Kohle*, Vol. 102, No. 6, 1986, pp. 308–314.

[182] Barwell, F. T., "Lubricant Additives—The History of Their Introduction and Assessment," *Erdol-Erdgas-Kohle*, Vol. 102, No. 12, 1986, pp. 561–567.

[183] Rizvi, S. Q. A., "Additives for Automotive Fuels and Lubricants," *Journal of the Society of Tribologists and Lubrication Engineers*, Vol. 34, 1999, p. 55–.

[184] "Lubricant Additive Chemistry," W. C. Gergel, *Presented at the International Symposium on Technical Organic Additives and Environment,* Interlaken, Switzerland, May 24–25, 1984.

[185] "Lubricating Oil Additives-A Chemist's Eye View," J. F. Ford, *J. Inst. Pet.*, Vol. 54, 1968, pp. 188–210.

[186] "Additive Expansion Unlikely, Says Afton," by Nancy DeMarco, Lube Report, March 7, 2006.

[187] Rizvi, S. Q. A., "Dispersants," *Lubricant Additives: Chemistry and Applications,* Chapter Leslie R. Rudnick, Ed., Marcel Dekker, Inc., NY, Chapter 5, 2003, pp. 137–170.

[188] Rizvi, S. Q. A., "Detergents," *Lubricant Additives: Chemistry and Applications,* Chapter Leslie R. Rudnick, Ed., Marcel Dekker, Inc., NY, Chapter 4, 2003, pp. 113–136.

[189] "Lubrication Seminar,"—Lubrizol Report No. C-10337, published July 1991.

[190] Bouman, C. A., *Properties of Lubricating Oils and Engine Deposits,* McMillan and Co., London, 1950, pp. 69–92.

[191] Kreuz, K. L., "Diesel Engine Chemistry as Applied to Lubricant Problems," *Lubrication*, Vol. 56, 1970, pp. 77–88.

[192] Kreuz, K. L., "Gasoline Engine Chemistry as Applied to Lubricant Problems," *Lubrication*, Vol. 55, 1969, pp. 53–64.

[193] Nahumck, W. M., Hyndman, C. W., and Cryvoff, S. A., "Development of the PV-2 Engine Deposit and Wear Test," *An ASTM Task Force Progress Report, SAE Publication 872123, Presented at International Fuels and Lubricants Meeting and Exposition,* Toronto, Canada, November 2–5, 1987.

[194] Covitch, M. J., Humphrey, B. K., and Ripple, D. E., "Oil Thickening in the Mack T-7 Engine Test—Fuel Effects and the Influence of Lubricant Additives on Soot Aggregation," *Presented at SAE Fuels and Lubricants Meeting,* Tulsa, Oklahoma, October 23, 1985.

[195] Lewinsky, P. M., Cooke, V. B., and Andrews, C. A., "Lubrication Oil Requirements for Low Emission Diesel Engines," *Presented at AVL Conference, Engine and Environment,* Graz, Austria, August 1–12, 1989.

[196] Ingold, K. U., "Inhibition of Autoxidation of Organic Substances in Liquid Phase," *Chemical Reviews*, Vol. 61, 1961, pp. 563–589.

[197] "Electromagnetic Spectrum," http://en.wikipedia.org/wiki/Electromagnetic_spectrum.

[198] Rasberger, M., "Oxidative Degradation and Stabilization of Mineral Oil based Lubricants," *Chemistry and Technology of Lubricants,* R. M. Mortier and S. T. Orszulik, Eds., Blackie Academic and Professional-Chapman & Hall, London, UK, 1996, Chapter 4, pp. 98–143.

[199] Abou El Naga, H. H. and Salem, A. E. M., "Effect of Worn Metals on the Oxidation of Lubricating Oils," *Wear*, Vol. 96, 1984, pp. 267–283.

[200] Al-Malaika, S., Marogi, A. and Scott, G., "Mechanisms of Antioxidant Action: Transformations Involved in the Antioxidant Function of Metal Dialkyl Dithiocarbamates, III," *J. Appl. Polym. Sci.*, Vol. 33, 1987, pp. 1455–71.

[201] "Molybdenum Sulfur Antioxidant Lube Additive," U.S. Patent 4,995,996 (February 26, 1991). Inventors: C. L. Coyle, M. A. Greaney, E. L. Stiefel, and J. N. Francis.

[202] Johnson, M. D., Korcek, S. and Zinbo, M., "Inhibition of Oxidation by ZDTP and Ashless Antioxidants in the Presence of Hydroperoxides at 160 °C," Lubricant and Additive Effects on Engine Wear, SP-558, Fuels and Lubricants Meeting, San Francisco, CA, October 31–November 3, 1983, pp. 71–81.

[203] "2,6-Di-tertiarybutyl-4-methylphenol," U.S. Patent 2,265, 582 (December 9, 1941). Inventors: D. R. Stevens and W. A. Gruse.

[204] "Process for the Production of Hydroxyalkylphenyl Derivatives," U.S. Patent 4,228,297 (October 14, 1980). Inventors: J. Haeberli, K. P. Park, A. F. Vellturo, and G. F. Nurnberger.

[205] "Procedure for Producing Stearyl-beta.-(3,5-dibutyl-4-hydroxyphenyl) propionate and Bis-(beta., (3,5-dibutyl-4-hydroxybenzyl)-methylcarboxyethyl)sulphide," U.S. Patent 4,694,099 (September 15, 1987). Inventors: M. Ahlfors, S. Koskimies, L. Lahtinen, and P. Idelman.

[206] "Liquid Phenolic Sulphur-containing Antioxidants," U.S. Patent 7,026,438 (April 11, 2006). Inventors: H. Camenzind, P. Dubs, R. Martin, D. E. Chasan, G. Demme, and J. Robbins.

[207] Hamblin, P. C., Kristen, U. and Chasan, D., "A Review: Ashless Antioxidants, Copper Deactivators, and Corrosion Inhibitors, Their Use in Lubricating Oils," *Lubr. Sci.*, 2, 1990, pp. 287–318.

[208] *Developments in Polymer Stabilization,* G. Scott, Ed., Elsevier Applied Science Publishers, 1984, pp. 65–104.

[209] Vijh, A. K., "Electrochemical Mechanisms of the Dissolution of Metals and the Contaminants Oxidation of Lubricating Oils under High-temperature Friction Conditions," *Wear*, Vol. 104, 1985. pp. 151–158.

[210] Bartz, W. J., "Lubrication of Industrial and Automotive Gears: A Review," *Performance and Testing of Gear Oils and Transmission Fluids,* R. Tourret and E. P. Wright, Eds., Heyden and Son, Chapter 16, 1981, pp. 247–287.

[211] Scott, G., "New Developments in the Mechanistic Understanding of Antioxidant Behavior," *J. Appl. Polym. Sci.*, Applied Polymer Symposium, 35, 1979, pp. 123–149.

[212] Migdal, C., "Antioxidants," *Lubricant Additives: Chemistry and Applications,* Leslie R. Rudnick, Ed., Marcel Dekker, Inc., NY, Chapter 7, 2003, pp. 1–27.

[213] Gatto, V. J., Moehle, W. E., Cobb, T. W., and Schneller, E. R., "Oxidation Fundamentals and Its Application to Turbine Oil Testing," *J. ASTM Int.*, Vol. 3, No. 4, 2006.

[214] Phillips, W. D., "Turbine Lubricating Oils and Hydraulic Fluids," *Fuels and Lubricants Handbook: Technology, Properties, Performance, and Testing,* MNL37WCD, Ed., ASTM International, West Conshohocken, PA, Chapter 12, 2003, pp. 297–351.

[215] Lachowicz, D. R. and Kreuz, K. L., "Peroxynitrates: the Unstable Products of Olefin Nitration with Dinitrogen Tetroxide in the Presence of Oxygen. A New Route to α-Nitroketones," *J. Org. Chem.*, Vol. 32, 1967, pp. 3885–3888.

[216] Covitch, M. J., Graf, R. T., and Gundic, D. T., "Microstructure of Carbonaceous Diesel Engine Piston Deposits," *Lubricant Engineering*, Vol. 44, 1988, p. 128.

[217] Covitch, M. J., Richardson, J. P., and Graf, R. T., "Structural Aspects of European and American Diesel Engine Piston Deposits," *Lubr. Sci.*, Vol. 2, 1990, pp. 231–251.

[218] Personal Communication, R. E. Kornbrekke, Lubrizol Research and Development.

[219] Kornbrekke, R. E., Patrzyk-Semanik, P., Kirchner-Jean, T., Raguz, M. G., and Bardasz, E., "Understanding Soot-mediated Oil Thickening—Part 6: Base Oil Effects," SAE Technical Paper 982665, Society of Automotive Engineers, October 1, 1998. Also see Parts 1–5 by E. Bardasz et al., *SAE Papers 952527*, October 1995; 961915, October 1, 1996; 971692, May 1, 1997; 976193, May 1, 1997; and 972952, October 1, 1997.

[220] "Lubricating Oil Compositions Containing Alkenyl Succinimides of Tetraethylene Pentamine," U.S. Patent 3,361,673 (January 2, 1968). Inventors: F. A. Stuart, R. G., anderson, and A. Y. Drummond.

[221] Cooper, A. R., "Molecular Weight Determination," *Concise Encyclopedia of Polymer Science and Engineering*, J. I. Kroschwitz, Ed., Wiley InterScience, 1990, pp. 638–639.

[222] Ravve, A., "Molecular Weights of Polymers," *Organic Chemistry of Macromolecules*, Marcel Dekker, NY, Chapter 3, 1967, pp. 39–54.

[223] Meunier, D. M., "Molecular Weight Determination," Chapter 46, [http://prenhall.com/settle/chapters/ch46.pdf].

[224] Deanin, R. D., *Polymer Structure, Properties, and Applications*, Cahner Books, NY, Chapter 3, 1972, p. 53.

[225] Randall, J. C., "Microstructure," *Concise Encyclopedia of Polymer Science and Engineering*, J. I. Kroschwitz, Ed., Wiley InterScience, 1990, p. 625.

[226] Rizvi, S. Q. A., "Lubricant Additives and Their Functions," *American Society of Metals Handbook*, 10th ed., S. D. Henry, Ed., ASM International, Vol. 18, 1992, pp. 98–112.

[227] Rizvi, S. Q. A., "Additives and Additive Chemistry," *Fuels and Lubricants Handbook: Technology, Properties, Performance, and Testing*, MNL37WCD, Ed., ASTM International, West Conshohocken, PA, Chapter 9, 2003, pp. 199–248.

[228] Fotheringham, J. D., "Polybutenes," *Synthetic Lubricants and High-performance Functional Fluids*, R. L. Shubkin, Ed., Marcel Dekker, Inc., NY, 1993.

[229] "Ethylene Alpha-olefin Polymer Substituted Mono- and Dicarboxylic Acid Dispersant Additives," U.S. Patent 5,435,926 (July 25, 1995). Inventors: A. Gutierrez, R. A. Kleist, W. R. Song, A. Rossi, H. W. Turner, H. C. Welborn, and R. D. Lundberg.

[230] "Novel Ethylene Alpha-olefin Copolymer Substituted Mannich Base Lubricant Dispersant Additives," U.S. Patent 5,017,299 (May 21, 1991). Inventors: A. Gutierrez, W. R. Song, R. D. Lundberg, and R. A. Kleist.

[231] "Lubricant Additives," U.S. Patent 5,384,055 (January 24, 1995). Inventor: H. Ashijan.

[232] "Production of Borated Ashless Dispersants," U.S. Patent 4,985,156 (January 15, 1991). Inventors: H. Ashijan and H. A Gawel.

[233] "Acylating Agents and Dispersants for Lubricating Oil and Fuels," U.S. Patent 6,846,783 (January 25, 2005). Inventors: S. Bartley, J. D. Burrington, and R. Yodice.

[234] "Dispersant and Dispersant Viscosity Index Improvers from Selectively Hydrogenated Aryl-substituted Olefin Containing Diene Copolymers," U.S. Patent 6,248,702 (June 19, 2001). Inventors: T. S. Coolbaugh, F. C. Loveless, J. E. Marlin, II, and D. N. Matthews.

[235] Stambaugh, R. L., "Viscosity Index Improvers and Thickeners," *Chemistry and Technology of Lubricants*, R. M. Mortier and S. T. Orszulik, Eds., Blackie Academic and Professional-Chapman & Hall, London, UK, 1996, pp. 144–180.

[236] "Nitrogen-containing Esters and Lubricants Containing Them," U.S. Patent 4,604,221 (August 5, 1986). Inventors: C. P. Bryant and H. M. Gerdes.

[237] "Nitrogen-free Ester of Carboxy Containing Interpolymers," U.S. Patent 5,256,752 (October 26, 1993). Inventors: D. M. Dishong and R. M. Lange.

[238] "Lubricating Oil Additives," U.S. Patent 6,071,862 (June 6, 2000). Inventors: J. R. Shanklin, Jr. and N. C. Mathur.

[239] Colyer, C. C. and Gergel, W. C., "Detergents/Dispersants," *Chemistry and Technology of Lubricants*, R. M. Mortier and S. T. Orszulik, Eds., CH Publishers, Inc., NY, Chapter 3, 1992, pp. 62–82.

[240] "Lubricating Oils Containing Metal Derivatives," U.S. Patent 2,628,942 (February 17, 1953). Inventors: J. R. Morris and J. R. Roach.

[241] "Modified Olefin-diolefin Resin," U.S. Patent 2,634,256 (April 7, 1953). Inventors: W. J. Sparks, D. W. Young, Roselle, and J. D. Garber.

[242] "Reaction Product of High Molecular Weight Succinic Acids and Succinic Anhydrides with an Ethylene Polyamine," U.S. Patent 3,172,892 (March 9, 1965). Inventors: W. M. Le Suer and G. R. Norman.

[243] "Novel Carboxylic Acid Acylating Agents, Derivatives Thereof, Concentrate and Lubricant Compositions Containing the Same, and Processes for Their Preparation," U.S. Patent 4,234,435 (November 18, 1980). Inventors: N. A. Meinhardt and K. E. Davis.

[244] "Lubricant," U.S. Patent 3,215,707 (November 2, 1965). Inventor: R. J. Rense.

[245] Weill, J. and Sillion, B., "Reaction of Chlorinated Polyisobutene with Maleic Anhydride: Mechanism of Catalysis by Dichloromaleic Anhydride," *Revue de l'Institut Francais du Petrole*, Vol. 40, No. 1, 1985, pp. 77–89, Also see J. Weill's Ph.D. Dissertation, 1982.

[246] "Process for Manufacturing Anhydrides of Alkenyl Dicarboxylic Acids," U.S. Patent 4,433,157 (February 21, 1984). Inventors: J. Weill, J. Garapon, and B. Sillion.

[247] "Treatment of Organic Compounds to Reduce Chlorine Level," U.S. Patent 5,708,097 (January 13, 1998). Inventors: C. K. Baumanis, M. M. Maynard, A. C. Clark, M. R. Sivic, C. P. Kovall, and D. L. Westfall.

[248] "Low Chlorine Polyalkylene Substituted Carboxylic Acylating Agent Compositions and Compounds Derived Therefrom," U.S. Patent 5,885,944 (March 23, 1999). Inventors: J. K. Pudelski, M. R. Sivik, K. F. Wollenberg, R. Yodice, J. Rutter, and J. G. Dietz.

[249] "Low Chlorine Content Composition for Use in Lubricants and Fuels," U.S. Patent 6,077,909 (June 20, 2000). Inventors: J. K. Pudelski, C. J. Kolp, J. G. Dietz, C. K. Baumanis, S. L. Bartley, and J. D. Burrington.

[250] Wollenberg, K. F. and Pudelski, J. K., "Preparation, NMR Characterization and Lubricant Additive Application of Novel Polyisobutenyl Phthalic Anhydrides," *Symposium on Recent Advances in the Chemistry of Lubricant Additives*, Paper Presented before the Division of Petroleum Chemistry, Inc., 218th National Meeting of the American Chemical Society, New Orleans, LA, August 22–26, 1999.

[251] "One-step Process for the Preparation of Alkenyl Succinic Anhydride," U.S. Patent 5,319,030 (June 7, 1994). Inventors: J. J. Harrison, R. Ruhe, Jr., and R. William.

[252] "Two-step Thermal Process for the Preparation of Alkenylsuccinic Anhydride," U.S. Patent 5,625,004 (April 29, 1997). Inventors: J. J. Harrison, R. Ruhe, Jr., and R. William.

[253] Morrison, R. T. and Boyd, R. N., "The Diels-Alder Reaction," *Organic Chemistry*, 3rd ed., Section 27.8, Allyn and Bacon, Inc., Boston, 1976, pp. 876–878.

[254] "Hydroxy-group Containing Acylated Nitrogen Compounds Useful as Additives for Lubricating Oil and Fuel Compositions," U.S. Patent 5,696,067 (December 9, 1997). Inventors: P. E. Adams, M. R. Baker, and J. G. Dietz.

[255] "Mixed Carboxylic Compositions and Derivatives and Use as Lubricating Oil and Fuel," U.S. Patent 6,030,929 (February 29, 2000). Inventor: J. K. Pudelski.

[256] "Substituted Carboxylic Acylating Agent Compositions and Derivatives Thereof for Use in Lubricants and Fuels," U.S. Patent 5,912,213 (June 15, 1999). Inventors: M. R. Baker, J. G. Dietz, and R. Yodice.

[257] "Acylated Nitrogen Compounds Useful as Additives for Lubricating Oil and Fuel Compositions," U.S. Patent 5,856,279 (January 5, 1999). Inventor: M. R. Baker.

[258] "Process for Preparing Condensation Product of Hydroxy-substituted Aromatic Compounds and Glyoxylic Reactants," U.S. Patent 6,001,781 (December 14, 1999). Inventors: M. R. Baker, K. M. Hull, and D. L. Westfall.

[259] "Intermediates Useful for Preparing Dispersant-viscosity Improvers for Lubricating Oils," U.S. Patent 6,117,941 (September 12, 2000). Inventors: P. E. Adams, R. M. Lange, R. Yodice, M. Baker, and J. Dietz.

[260] "Nitrogen Containing Dispersant-viscosity Improvers," U.S. Patent 6,492,306 (December 10, 2002). Inventors: M. R. Sivik, R. M. Lange, D. M. Vargo, and P. E. Adams.

[261] "Alkylation of Phenols," *Kirk-Othmer Encyclopedia of Chemical Technology*, Vol. 1, 1963, pp. 894–895 and "Ion Exchange," *Kirk-Othmer Encyclopedia of Chemical Technology*, Vol. 2, 1967 pp. 871–899, Interscience Publishers, NY.

[262] "Aromatic Mannich Compound-containing Composition and Process for Making Same," U.S. Patent 6,179,885 (January 30, 2001). Inventor: J. R. McAtee.

[263] "Manufacture of Alkylphenol Compounds," U.S. Patent 4,202,199 (May 13, 1980). Inventors: F. Merger, and G. Nestler.

[264] "Methods for Preparing Alkylated Hydroxyaromatics," U.S. Patent 5,663,457 (September 2, 1997). Inventor: C. J. Kolp.

[265] "Ethylene Amines," *Kirk Othmer's Encyclopedia of Chemical Technology*, 2nd ed., Interscience Publishers, NY, Vol. 7, 1965, pp. 22–37.

[266] "Amines II. Reactions," *Organic Chemistry*, R. T. Morrison, and R. N. Boyd, Eds., 3rd ed., Allyn and Bacon, Inc., Boston, Section 23.1, 1976 pp. 745–748.

[267] "High Molecular Weight Nitrogen-containing Condensates and Fuels and Lubricants Containing Same," U.S. Patent 5,053,152 (October 1, 1991). Inventor: T. F. Steckel.

[268] "Preparation of Tris(hydroxymethyl)aminomethane," U.S. Patent 4,233,245 (November 11, 1980), Inventors: J. Bourguignon, M-X Sion, and M. Moreau.

[269] "Dispersants and Process for Their Preparation," U.S. Patent 3,980,569 (September 14, 1976). Inventors: J. F. Pindar, J. M. Cohen, and C. P. Bryant.

[270] "Artificial Resins and Method of Making," U.S. Patent 2,098,869 (November 9, 1937). Inventors: J. Harmon, and F. M. Meigs.

[271] "Aminoalkylation and Amidoalkylation," *Advanced Organic Chemistry, Reactions, Mechanisms, and Structure* by J. March, 4th ed., Chapter 11 on Aromatic Electrophilic Substitution, Section 1–25, Wiley-Interscience Publication, 1992, pp. 550–551.

[272] "Ethylenediamine Salts of Thiophosphonic Acids," U.S. Patent 3,185,728 (May 25, 1965). Inventors: E. E. Schallenberg and R. G. Lacoste.

[273] "Friction Reducing Agents for Lubricants," U.S. Patent 3,702,824 (November 14, 1972). Inventor: R. C. Slicht.

[274] "Olefin-thionophosphine Sulfide Reaction Products, Their Derivatives, and Use Thereof as Oil and Fuel Additives," U.S. Patent 4,042,523 (August 16, 1977). Inventor: S. J. Brois.

[275] "Olefin-thionophosphine Sulfide Reaction Products, Their Derivatives, and Use Thereof as Oil and Fuel Additives," U.S. Patent 4,100,187 (July 11, 1978). Inventor: S. J. Brois.

[276] "Oil Containing Dispersant VII Olefin Copolymer," U.S. Patent 4,715,975 (December 29, 1987). Inventors: M. M. Kapusciniski, B. J. Kaufman, and C. S. Liu.

[277] "Dispersant-antioxidant Multifunction Viscosity Index Improver," U.S. Patent 4,699,723 (October 13, 1987). Inventors: M. M. Kapuscinski and R. E. Jones.

[278] "Dispersant Anti-oxidant VI Improver and Lubricating Oil Composition Containing Same," U.S. Patent 4,948,524 (August 14, 1990). Inventors: M. M. Kapuscinski, T. E. Nalesnik, R. T. Biggs, H. Chafetz, and C. S. Liu.

[279] "Lubricant for Use in Diesel Engines," U.S. Patent 6,187,721 (February 13, 2001). Inventors: I. Goldblatt, M. McHenry, K. Henderson, D. Carlisle, N. Ainscough, M. Brown, and R. Tittel.

[280] "Graft Copolymers and Lubricants Containing Such as Dispersant-viscosity Improvers," U.S. Patent 5,298,565 (March 29, 1994). Inventors: R. M. Lange and C. V. Luciani.

[281] "Dispersant Viscosity Index Improvers," U.S. Patent 6,083,888 (July 4, 2000). Inventor: R. J. Sutherland.

[282] "Lubricating Oils and Fuels Containing Graft Copolymers," U.S. Patent 3,506,574 (April 14, 1970). Inventors: R. L. Stambaugh and R. D. Bakule.

[283] "Viscosity Index Improvers with Dispersant Properties Prepared by Reaction of Lithiated Hydrogenated Copolymers with 4-Substituted Aminopyridines," U.S. Patent 4,402,843 (September 6, 1983). Inventor: W. J. Trepka.

[284] "Viscosity Index Improvers with Dispersant Properties Prepared by Reaction of Lithiated Hydrogenated Copolymers with Substituted Aminolactams," U.S. Patent 4,402,844

(September 6, 1983). Inventor: W. J. Trepka.

[285] "Viscosity Improvers for Lubricating Oil Compositions," U.S. Patent 6,124,249 (September 26, 2000). Inventors: J. G. Seebauer and C. P. Bryant.

[286] "Dispersant-viscosity Improvers for Lubricating Oil Compositions," U.S. Patent 5,969,068 (October 19, 1999). Inventors: C. P. Bryant, B. A. Grisso, and R. Cantiani.

[287] "Dispersant-viscosity Improvers for Lubricating Oil Compositions," U.S. Patent 5,821,313 (October 13, 1998). Inventors: M. R. Sivic and C. P. Bryant.

[288] "Dispersant-viscosity Improvers for Lubricating Oil Compositions," U.S. Patent 5,512,192 (April 30, 1996). Inventors: R. M. Lange and C. V. Luciani.

[289] "Dispersant-viscosity Improvers for Lubricating Oil Compositions," U.S. Patent 5,540,851 (July 30, 1996). Inventor: R. M. Lange.

[290] "Graft Copolymers from Solvent-free Reactions and Dispersant Derivatives Thereof," U.S. Patent 5,429,758 (July 4, 1995). Inventors: K. Hayashi, T. R. Hopkins, and C. R. Scharf.

[291] "Novel VI Improver, Dispersant, and Anti-oxidant Additive and Lubricating Oil Composition Containing Same," U.S. Patent 4,863,623 (September 5, 1989). Inventor: T. E. Nalesnik.

[292] "Functionalized Graft Co-polymer as a Viscosity Index Improver, Dispersant, and Anti-oxidant Additive and Lubricating Oil Composition Containing Same," U.S. Patent 5,409,623 (April 25, 1995). Inventors: M. K. Mishra and I. D. Rubin.

[293] "Lubricating Oil Containing Dispersant Viscosity Index Improver," U.S. Patent 5,520,829 (May 28, 1996). Inventors: M. K. Kapuscinski, C. S. Liu, L. D. Grina, and R. E. Jones.

[294] "Process for Making Dispersant Viscosity Index Improvers," U.S. Patent 5,486,563 (January 23, 1996). Inventor: R. J. Sutherland.

[295] "Star-shaped Dispersant Viscosity Index Improver," U.S. Patent 4,077,893 (March 7, 1978). Inventor: T. E. Kiovsky.

[296] "Multifunctional Viscosity Index Improver-dispersant Antioxidant," U.S. Patent 5,439,607 (August 8, 1995). Inventor: A. O. Patil.

[297] "Lubricants Containing Ashless Antiwear-dispersant Additive Having Viscosity Index Improver Credit," U.S. Patent 5,698,500 (December 16, 1997). Inventors: J. R. Baranski and C. A. Migdal.

[298] "Dispersant Viscosity Index Improvers," U.S. Patent 5,360,564 (November 1, 1994). Inventors: R. J. Sutherland and R. B. Rhodes.

[299] Brady, J. E. and Humiston, G. E., "Chemical Bonding: General Concepts—Polar Molecules and Electronegativity," *General Chemistry: Principles and Structure*, 2nd ed., Section 4.8, John Wiley and Son, NY, 1978, pp. 114–117.

[300] "Dispersant Additives and Process," U.S. Patent 5,804,667 (September 8, 1998), and U.S. Patent 5,936,041 (August 10, 1999). Inventors: W. B. Diana, J. V. Cusumano, K. R. Gorda, J. Emert, W. B. Eckstrom, D. C. Dankworth, J. E. Stanat, and J. P. Stokes.

[301] "Ashless Dispersants Formed from Substituted Acylating Agents and Their Production and Use," U.S. Patent 5,241,003 (August 31, 1993). Inventors: D. J. Degonia and P. G. Griffin.

[302] "Oil Soluble Dispersant Additives Useful in Oleaginous Compositions," U.S. Patent 5,026,495 (June 25, 1991). Inventors: J. Emert, R. D. Lundberg, and A. Gutierrez.

[303] "Middle Distillate Containing Storage Stability Additive," U.S. Patent 4,948,386 (August 14, 1990). Inventors: R. L. Sung, B. J. Kaufman, and K. J. Thomas.

[304] "Non-ash Containing Lubricant Oil Composition," U.S. Patent 3,189,544 (June 15, 1965). Inventors: H. Ratner and R. F. Bergstrom.

[305] "Reaction Products of Hydrocarbon-substituted Succinic Acid-producing Compound, an Amine, and an Alkenyl Cyanide," U.S. Patents 3,278,550 (October 11, 1966) and 3,366,569 (June 30, 1968). Inventors: G. R. Norman and W. M. Le Suer.

[306] "Cleavage of Amine Oxides," *Advanced Organic Chemistry*, J. March, Ed., 4th ed., Chapter 17 on Eliminations, Section 7–8, Wiley-Interscience Publication, 1992, pp. 1018–1019.

[307] "Hydrolysis of Amides," Section 20–13, pp. 671–672 and "Alkaline and Acidic Hydrolysis of Esters," Sections, 20.17 and 20.18, pp. 677–681, *Organic Chemistry*, R. T. Morrison and R. N. Boyd, 3rd ed., Allyn and Bacon, Inc., Boston, 1976.

[308] Baczek, S. K., and Chamberlin, W. B., "Petroleum Additives," *Encyclopedia of Polymer Science and Engineering, Second Edition* John Wiley and Sons, Vol. 11, 1988, p. 22.

[309] "Engine Oil Viscosity Classification," *SAE J300—Revised May 2004*, SAE Standard, Society of Automotive Engineers.

[310] "Multigrade Lubricating Compositions Containing No Viscosity Modifier," U.S. Patent 5,965,497 (October 12, 1999). Inventors: D. R. Adams and P. Brice.

[311] "High Functionality Low Molecular Weight Oil Soluble Dispersant Additives Useful in Oleaginous Compositions," U.S. Patent 5,788,722 (August 4, 1998). Inventors: J. Emert and R. D. Lundberg.

[312] "Polymers Derived From Ethylene and 1—Butene for the Use in the Preparation of Lubricant Dispersant Additives," U.S. Patent 6,030,930 (February 29, 2000). Inventors: J. Emert, A. Rossi, S. Rea, J. W. Frederick, and M. W. Kim.

[313] "Lubricating Oil Containing Dispersant VII and Pour Depressant," U.S. Patent 4,668,412 (May 26, 1987). Inventors: W. P. Hart and C. S. Liu.

[314] "Ethylene Alpha-olefin/diene Interpolymer-substituted Carboxylic Acid Dispersant Additives," U.S. Patents 5,759,967 (June 2, 1998) and 5,681,799 (October 28, 1997). Inventors: W. R. Song, A. Rossi, H. W. Turner, H. W. Welborn, R. D. Lundberg, A. Gutierrez, and R. A. Kleist.

[315] "Ethylene Alpha-olefin Polymer Substituted Mono- and Dicarboxylic Acid Dispersant Additives," U.S. Patent 5,433,757 (July 18, 1995). Inventors: W. R. Song, A. Rossi, H. W. Turner, H. C. Welborn, and R. D. Lundberg.

[316] "Borated Ethylene Alpha-olefin Copolymer Substituted Mannich Base Lubricant Dispersant Additives," U.S. Patent 5,382,698 (January 17, 1995). Inventors: W. R. Song, R. D. Lundberg, A. Gutierrez, and R. A. Kleist.

[317] "Polyalkylene Polysuccinimides and Post-treated Derivatives Thereof," U.S. Patent 6,146,431 (November 14, 2000). Inventors: J. J. Harrison and W. A. Ruhe, Jr.

[318] "Ready Reference for Lubricant and Fuel Performance," *Lubrizol Publication*; 2004.

[319] "Engine Oil Having Dithiocarbamate and Aldehyde/epoxide for Improved Seal Performance, Sludge and Deposit Performance," U.S. Patent 6,121,211 (September 9, 2000). Inventors: C. F. Stachew, W. D. Abraham, J. A. Supp, J. R. Shanklin, and G. D. Lamb.

[320] "Viton Seal Compatible Dispersant and Lubricating Oil Composition Containing Same," U.S. Patent 5,188,745 (February 23, 1993). Inventors: C. A. Migdal and T. E. Nalesnick.

[321] "Dibasic Acid Lubricating Oil Dispersant and Viton Seal Additives," U.S. Patent 4,663,064 (May 5, 1987). Inventors: T. E. Nalesnik and C. M. Cusano.

[322] "Lubricating Oil Dispersant and Viton Seal Additives," U.S. Patent 4,636,322 (January 13, 1987). Inventor: T. E. Nalesnik.

[323] "Dispersant Additives," U.S. Patent 6,127,322 (October 3, 2000). Inventors: R. M. Scott and R. W. Shaw.

[324] "Method for Preparing Engine Seal Compatible Dispersant for Lubricating Oils Comprising Reacting Hydrocarbyl Substituted Dicarboxylic Compound with Aminoguanidine or Basic Salt Thereof," U.S. Patent 5,080,815 (January 14, 1992). Inventors: D. J. Fenoglio, P. R. Vettel, and D. W. Eggerding.

[325] "Compositions for Control of Induction System Deposits," U.S. Patent 5,679,116 (October 21, 1997). Inventors: L. J. Cunningham, D. P. Hollrah, and A. M. Kulinowski.

[326] "Deposit Control Additives and Fuel Compositions Containing the Same," U.S. Patent 5,334,228 (August 2, 1994). Inventors: H. Ashjian, M. P. Miller, D.-M. Shen, and M. M. Wu.

[327] "Additive Formulation for Fuels Incorporating Ester Function Products and a Detergent-Dispersant," U.S. Patent 5,433,755 (July 18, 1995). Inventors: P. Mulard, Y. Labruyere, A. Forestiere, and R. Bregent.

[328] "Gear and Transmission Lubricant Compositions of Improved Sludge-Dispersibility, Fluids Comprising the Same," U.S. Patent 5,665,685 (September 9, 1997). Inventor: S. Takigawa.

[329] "Automatic Transmission Fluids and Additives Thereof," U.S. Patent 5,441,656 (August 15, 1995). Inventors: H. Otani and R. J. Hartley.

[330] "Hydraulic Automatic Transmission Fluid with Superior Friction Performance," U.S. Patent 4,253,977 (March 3, 1981). Inventor: R. O'Halloran.

[331] "Lubricating Oil Composition for Automatic Transmission," U.S. Patent 5,972,854 (October 26, 1999). Inventors: T. Ichihashi, H. Igarashi, J. Deshimaru, and T. Ikeda.

[332] "Automatic Transmission Fluid Composition," U.S. Patent 5,958,286 (September 28, 1999). Inventor: M. Kitanaka.

[333] "Automatic Transmission Fluids Having Enhanced Performance Capabilities," U.S. Patent 5,972,851 (October 26, 1999). Inventors: S. Srinivasan, D. W. Smith, and J. P. Sunne.

[334] "Gear Oil Additive Concentrates and Lubricants Containing them," U.S. Patent 6,096,691 (August 1, 2000). Inventors: G. S. Conary and R. J. Hartley.

[335] "Lubricating Oil Composition for High-speed Gears," U.S. Patent 5,756,429 (May 26, 1998). Inventor: T. Ichihashi.

[336] "Hydraulic Fluids," U.S. Patent 5,767,045 (June 16, 1988). Inventor: H. T. Ryan.

[337] "Use of Dispersant Additives as Process Antifoulants," U.S. Patent 5,368,777 (November 29, 1994). Inventor: D. R. Forester.

[338] "Multifunctional Antifoulant Compositions," U.S. Patent 4,927,561 (May 22, 1990). Inventor: D. R. Forester.

[339] "Multifunctional Antifoulant Compositions and Methods of Use Thereof," U.S. Patent 4,775,458 (October 4, 1988). Inventor: D. R. Forester.

[340] "Process For Making Overbased Calcium Sulfonate Detergents Using Calcium Oxide and a Less Than Stoichiometric Amount of Water," U.S. Patent 6,015,778 (January 18, 2000). Inventors: A. J. Rolfes and S. E. Jaynes.

[341] "Magnesium Low Rate Number Sulphonates," U.S. Patent 5,922,655 (July 13, 1999). Inventors: D. Moulin, J. A. Cleverley, and C. H. Bovington.

[342] "Method of Preparing Overbased Barium Sulfonates," U.S. Patent 3,959,164 (May 25, 1976). Inventor: A. R. Sabol.

[343] "Overbased Alkali Metal Sulfonates," U.S. Patent 4,867,891 (September 19, 1989). Inventor: M. W. Hunt.

[344] "Metal Phenates," U.S. Patent 4,221,673 (September 9, 1980). Inventors: R. Robson, B. Swinney, and R. D. Tack.

[345] "Preparation of Overbased Magnesium Phenates," U.S. Patent 4,435,301 (March 6, 1984). Inventors: C. G. Brannen and M. W. Hunt.

[346] "Process for the Preparation of Sulfurized Overbased Phenate Detergents," U.S. Patent 4,973,411 (November 27, 1990). Inventors: T. C. Jao, C. K. Esche, E. D. Black, and R. H. Jenkins Jr.

[347] "Methods for Preparing Group II Metal Overbased Sulfurized Alkylphenols," U. S. 4,971,710 (November 20, 1990). Inventor: T. V. Liston.

[348] "Production of Overbased Metal Phenates," U.S. Patent 4,104,180 (August 1, 1978). Inventors: V. C. E. Burnop.

[349] "Method of Preparing Overbased Lubricating Oil Additives," U.S. Patent 4,057,504 (November 8, 1977). Inventors: M. Shiga, K. Hirano, and M. Matsushita.

[350] "Recovery of Overbased Alkaline Earth Metal Additives from Centrifugates," U.S. Patent 4,910,334 (March 20, 1990). Inventors: F. A. Stuart and W. A. Tyson, Jr.

[351] "Process for Preparing Overbased Naphthenic Micronutrient Compositions," U.S. Patent 4,243,676 (January 6, 1981). Inventor: W. R. Ali.

[352] "Process for Overbased Petroleum Oxidate," U.S. Patent 5,013,463 (May 7, 1991). Inventor: F. J. Slama.

[353] "Preparation of Basic Salt," U.S. Patent 4,810,398 (March 7, 1989). Inventors E. M. G. A. Van Kruchten and R. R. Van Well.

[354] Weamer, G. L., "Additives—A Way to Quality in Motor Oils," Petroleum Refiner, 38, 1959, pp. 215–219.

[355] "Metal Salts of Organic Acids of Phosphorus," U.S. Patent 2,785,128 (March 12, 1957). Inventor: A. H. Popkin.

[356] "Preparation of Overbased Magnesium Sulfonates," U.S. Patent 4,148,740 (April 10, 1979). Inventors: V. J. Cease and G. R. Kirk.

[357] Price, C. C., "The Alkylation of Aromatic Compounds by the Friedel-Crafts Method," *Organic Reactions,* John Wiley and Sons, Inc., Vol. III, Chapter 1, 1946, pp. 1–82.

[358] "Alkylation of Aromatics with Olefins in the Presence of an Alumina Catalyst," U.S. Patent 4,219,690 (August 26, 1980). Inventor: S. M. Kovach.

[359] "Process and System for Alkylation of Aromatic Compounds," U.S. Patent 5,770,782 (June 23, 1998). Inventors: J. F. Knifton, P. R. Anantaneni, and M. Stockton.

[360] Olah, G. A., *Friedel-Crafts and Related Reactions,* Interscience-Wiley Publishers, Vol. 2, 1964.

[361] "Aromatic Alkylation," U.S. Patent 2,754,341 (July 10, 1956). Inventor: E. V. Kirkland.

[362] "Process for the Production of Light Alpha Olefins by Oligomerization of Ethylene," U.S. Patent 5,345,023 (September 6, 1994). Inventors: Y. Chauvin, D. Commereuc, F. Hugues, H. Olivier, and L. Saussine.

[363] "Process for the Conversion of Ethylene into Light Alpha Olefins with the Use of Additives Based on Quaternary Ammonium Salts," U.S. Patent 5,817,905 (October 6, 1998). Inventors: D. Commereuc, Y. Glaize, F. Hugues, P. Panzarella, and L. Saussine.

[364] "Oligomerization Reaction System," U.S. Patent 4,020,121 (April 26, 1977). Inventors: A. T. Kister, and E. Lutz.

[365] "Continuous Manufacturing Process for Alpha-olefins," U.S. Patent 6,555,723, April 29, 2003. Inventor: R. S. Schiffino.

[366] "Manufacture of Alpha-olefins," U.S. Patent 6,103,946 (August 15, 2000). Inventors: M. S. Brookhart III and B. L. Small.

[367] "Process for Producing High Vinylidene Polyisobutylene," U.S. Patent 6,683,138 (January 27, 2004). Inventors: E. C. Baxter III, G. Valdez, C. Lobue, and T. Lowry.

[368] "Cationic Polymerization of 1-olefins," U.S. Patent 4,605,808 (August 12, 1986). Inventor: J. R. Samson.

[369] "Aromatics Alkylation Process," U.S. Patent 5,191,135 (March 2, 1993). Inventors: F. G. Dwyer and Q. N. Lee.

[370] Suter, C. M., and Weston, A. W., "Direct Sulfonation of Aromatic Hydrocarbons and Their Halogen Derivatives," *Organic Reactions*, John Wiley and Sons, Inc., Vol. 3, Chapter 4, 1946, pp. 141–197.

[371] "Conversion of Sulfonic Acids into a Hydrocarbon Oil of Superior Oxidation Stability," U.S. Patent 4,188,500 (February 12, 1980). Inventors: D. S. Bosniack and P. F. Korbach. See the "Background of Invention" portion of the patent.

[372] "Verfahren zur Herstellung von Sulfonsauren," DRP 735096, 1940. Inventors: C. Platz and K. Schimmelsschmidt.

[373] "Reaction of Petroleum Oils with Sulfur Dioxide and Chlorine," U.S. Patent 2.263.312 (November 18, 1941). Inventor: C. F. Reed.

[374] "Fatty Acids," http://www.cyberlipid.org/fa/acid0001.htm#3.

[375] "Lubricating Oil Compositions Containing Saligenin Derivatives," U.S. Patent 6,310,011 (October 30, 2001). Inventors: J. L. Karn, J. A. Kocsis, L. E. King, P. E. Adams, and V. A. Notaro.

[376] "Oil Formulations Containing Overbased Multi-functional Additive," U.S. Patent 4,954,273 (September 4, 1990). Inventors: R. A. Denis and F. W. Koch.

[377] "Lubricating Oil Compositions Containing a Polyoxyalkylene Carboxylic Acid Salt Additive," U.S. Patent 5,250,203 (October 5, 1993). Inventors: R. A. Denis and F. W. Koch.

[378] "Carboxylation with Carbon Dioxide. Kolbe-Schmitt Reaction," *Advanced Organic Chemistry, Reactions, Mechanisms, and Structure* by J. March, 4th ed., Chapter 11 on Aromatic Electrophilic Substitution, Section 1-20, Wiley-Interscience Publication, 1992, pp. 546–547.

[379] "Additives for Lubricating Oils and Processes for Producing Them," U.S. Patent 5,089,158 (February 18, 1992). Inventors: E. M. G. A. Van Kruchten and G. W. J. Heimerikx.

[380] "Phosphorus Pentasulfide," http://en.wikipedia.org/wiki/Phosphorus_pentasulfide

[381] Huang, N. Z., "Overbased Products Using Non-ionic Substrates—Chemistry and Properties," *Presented at the Symposium on Recent Advances in the Chemistry of Lubricant Additives, Division of Petroleum Chemistry, 218 National Meeting of the American Chemical Society*, New Orleans, LA, August 22–26, 1999.

[382] "Non-conventional Overbased Materials," U.S. Patent 5,556,569 (September 17, 1996). Inventor: N. Z. Huang.

[383] "Metal Complexes and Methods of Making Same," U.S. Patent 2,777,874 (January 15, 1957). Inventors: P. A. Asseff, T. W. Mastin, and A. Rhodes.

[384] "Metal-containing Lubricant Compositions," U.S. Patent 4,767,551 (August 30, 1988). Inventors: M. W. Hunt and S. Kennedy.

[385] "Process for Overbasing via Metal Borate Formation," U.S. Patent 6,090,757 (July 18, 2000). Inventor: T. F. Steckel.

[386] "Method of Preparation of Superbased Salts," U.S. Patent 2,695,910 (November 30, 1954). Inventors: P. A. Asseff, T. W. Mastin, and A. Rhodes.

[387] "Sulfurized Overbased Compositions," U.S. Patent 5,484,542 (January 16, 1996). Inventors: J. M. Cahoon, J. L. Karn, N. Z. Haung, and J. P. Roski.

[388] "Process for Overbasing via Metal Borate Formation," U.S. Patent 5,064,545 (November 12, 1991). Inventor: T. F. Steckel.

[389] "Borated Basic Metal Salt and Lubricating Oil Composition," U.S. Patent 4,539,126 (September 3, 1985). Inventors: J. J. Bleeker, M. Booth, M. G. F. M. van Grieken, W. J. Krijnen, and G. D. van Wijngaarden.

[390] "Overbased Carboxylate Gels," U.S. Patent 5,919,741 (July 6, 1999). Inventors: S. E. Jaynes and W. R. Sweet.

[391] "Mixed Carboxylate Overbased Gels," U.S. Patent 5,401,424 (March 28, 1995). Inventors: J. N. Vinci and W. R. Sweet.

[392] "Calcium Containing Micellar Complexes," U.S. Patent 3,766,067 (October 16, 1973). Inventor: R. L. McMillen.

[393] "Magnesium Overbasing Process," U.S. Patent 5,173,203 (December 22, 1992). Inventors: W. P. Nichols and J. L. Karn.

[394] "Basic Alkali Metal Sulfonate Dispersions, Process for Their Preparation, and Lubricants Containing Same," U.S. Patent 5,037,565 (August 6, 1991). Inventor: L. E. King.

[395] "Compounds Useful as Detergent Additives for Lubricants and Lubricating Compositions," U.S. Patent 5,021,174 (June 4, 1991). Inventors: P. Koch and A. Di Serio.

[396] Hunter, M. E., and Baker, R. F., "Corrosion," *Fuels and Lubricants Handbook: Technology, Properties, Performance, and Testing*, G. E. Totten, Ed., ASTM International, West Conshohocken, PA., Chapter 31, 2003, pp. 825–832.

[397] Boner, C. J., "Theory of Action and Performance," *Gear and Transmission Lubricants*, Reinhold Publishing Company, NY, Chapter 8, 1964.

[398] Bhushan, B., and Gupta, B. K., "Physics of Tribological Materials," *Handbook of Tribology; Materials, Coatings, and Surface Treatments*, McGraw-Hill Inc., NY, Chapter 3, 1991.

[399] Buckley, D. H., "Properties of Surfaces," *CRC Handbook of Lubrication, Theory and Practice of Tribology, Vol. II: Theory and Design*, R. E. Booser Ed., CRC Press, Boca Raton, 1988, pp. 17–30.

[400] "Improved Fuel Economy," http://www.fueleconomy.gov/feg/why.shtml

[401] http://www.dieselnet.com

[402] Oliver, C. R., Reuter, R. M., and Sendra, J. C., "Fuel Efficient Gasoline-Engine Oils," *Lubrication*, Vol. 67, 1981, pp. 1–12.

[403] McFall, D., "Road Map for Engine Oils: 'Six D' Fuel Economy Test Points the Way," *Lubes'N'Greases*, April 2006, p. 1.

[404] Kick, G. K., "Automotive Engine Lubrication," *Standard Handbook of Lubrication Engineering*, J. J. O'Connor,

Ed., McGraw-Hill Book Company, NY, Chapter 32, 1968, p. 32–37.

[405] McFall, D., "Sequence VI Fuel Economy Test," *Lubes'n'Greases*, March 8, 2000, pp. 6–12.

[406] McFall, D., "Light-Duty Diesel: What Are We Missing?" *Lubes'N'Greases*, November 2002, pp. 23–31.

[407] O'Brien, J. A., "Lubricant Additives," *CRC Handbook of Lubrication, Theory and Practice of Tribology, Vol. II, Theory and Design*, R. E. Booser, Ed., CRC Press, Boca Raton, 1988, pp. 301–315.

[408] Lansdown, A. R., "Extreme Pressure and Anti-wear Additives," *Chemistry and Technology of Lubricants*, R. M. Mortier and S. T. Orszulik, Eds., VCH Publishers, Inc., NY, Chapter 12, 1992, pp. 269–281.

[409] Crawford, J., and Psaila, A., "Miscellaneous Additives and Vegetable Oils," *Chemistry and Technology of Lubricants*, R. M. Mortier and S. T. Orszulik, Eds., Blackie Academic and Professional-Chapman & Hall, London, UK, 1996, Chapter 6, pp. 181–202.

[410] (a) McFall, D., "On the Road to New Diesel Regs," *Lube & Greases*, November 2002, pp. 23–27. (b) Hoocendoohn, R., and Kenbeek, D., "Friction Modifiers to the Rescue," *Lube & Greases*, November 2003, pp. 14–20. (c) *ibid.*, *Lubricant's World*, November 2003, pp. 16–18.

[411] Kenbeek, D., and Benemann, T. F., "Organic Friction Modifiers," *Lubricant Additives: Chemistry and Applications*, Leslie R. Rudnick, Ed., Marcel Dekker, Inc., NY, Chapter 7, 2003, pp. 203–221.*Lubricant's World*, November 2003, pp. 16–18.

[412] "Friction," http://www.synlube.com/howsyn.htm

[413] Taylor, R. I., "Car Lubricants: Fact & Fiction," *Tribology and Lubrication Technology*, October 2004.

[414] Stewart, R. M., Spohn, C. R., and Meyer, W. A. P., "Influence of Lubricating Oil Formulation on Winter Hot-starting Engine Torque Requirements," SAE Paper No. 690157, *Paper presented at SAE Automotive Engineering Congress*, Detroit, Jan. 1969.

[415] Jayne, G. J., Matthews, B. M., and Thomas, A. S., "Hypoid Gear Oils for the 1980s," *Performance and Testing of Gear Oils and Transmission Fluids*, R. Tourret and E. P. Wright, Ed., Heyden and Son, Chapter 19, 1981, pp. 307–319.

[416] Feng, I. M., Perilstein, W. L., and Adams, M. R., "Solid Film Deposition and Non-sacrificial Boundary Lubrication," *ASLE Transactions*, Vol. 6, 1963, pp. 60–66.

[417] "DMTD Derivatives as Antiwear/antioxidant Additives," U.S. Patent 5,318,712 (June 17, 1994). Inventors: S. Q. A. Rizvi, R. M. Lange, and R. E. Gapinski.

[418] "Dynamic Light Scattering Studies of Neutral Di-isobutyl Zinc Dithiophosphate," E. S. Yamaguchi et al., *Tribol. Trans.*, Vol. 40, No. 2, 1997, pp. 330–337.

[419] Jones, R. B., and Coy, R. C., "The Thermal Degradation and EP Performance of Zinc Dialkyl Dithiophosphate Additives in White Oil," *ASLE Transactions*, Vol. 24, 1981, pp. 77–90.

[420] Brazier, A. D., and Elliot, J. S, "The Thermal Stability of Zinc Dithiophosphates," *J. Institute of Petroleum*, Vol. 53, 1967, pp. 63–76.

[421] Bateman, L., and Moore, C. G., "Reaction of Sulfur with Olefins," *Organic Sulfur Compounds*, N. Kharasch, Ed., Pergamon Press, Vol. 1, Chapter 20, 1961, pp. 210–228.

[422] "Sulfur-containing Lubricant and Functional Fluid Compositions," U.S. Patent 4,800,031 (January 24, 1989). Inventors: S. A. Di Biase and J. W. Pialet.

[423] "Sulfur-containing Compositions, and Additive Concentrates, Lubricating Oils, Metal Working Lubricants and Asphalt Compositions Containing Same," U.S. Patent 4,740,322 (April 26, 1988). Inventors: S. A. Di Biase and R. L. Sowerby.

[424] "Composition of and Method for Polysulfides Having a Reduced odor level," U.S. Patent 6,362,374 (March 26, 2002). Inventors: D. R. Forester, R. L. Sowerby, B. B. Malik, and V. A. Gober.

[425] Rossrucker, T., and Fessenbecker, A., "Sulfur Carriers," *Lubricant Additives: Chemistry and Applications*, Leslie R. Rudnick, Ed., Marcel Dekker, Inc., NY, Chapter 9, 2003, pp. 259–291.

[426] Thorn, G. D., and Ludwig, R. A., *The Dithiocarbamates and Related Compounds*, Elsevier Publishing Company, 1962.

[427] Gerrard, W., and Hudson, H. R., "Organic Derivatives of Phosphorous and Thiophosphorous Acids," *Organic Phosphorus Compounds*, G. M. Kosolapoff and L. Maier, Eds., Wiley InterScience, Vol. 5, Chapter 13, 1973, p. 21.

[428] Cherbuliez, E., "Organic Derivatives of Phosphoric Acid," *Organic Phosphorus Compounds*, G. M. Kosolapoff and L. Maier, Eds., Wiley InterScience, Vol. 6, Chapter 15, 1973, p. 211.

[429] "Introduction to Wear," http://ocw.mit.edu/NR/rdonlyres/Mechanical-Engineering/2-800Fall-2004/3EB7EE5A-43BF-4B4B-860B-1FD57DD8207F/0/ch4a_wear_intro.pdf

[430] Nieman, G., Rettig, H., and Lechner, G., "Scuffing Tests on Gear Oils in FZG Apparatus," *ASLE Transactions*, Vol. 4, 1961, p. 71.

[431] Forbes, E. S., and Battersby, J., "The Effect of Chemical Structure on the Load-carrying and Adsorption Properties of Dialkyl Phosphites," *ASLE Trans.*, Vol. 17, No. 4, 1974, pp. 263–270.

[432] Sakurai, T., and Sato, K., "Chemical Reactivity and Load-carrying Capacity of Lubricating Oils Containing Organic Phosphorus Compounds," *ASLE Trans.*, Vol. 13, 1970, pp. 252–261.

[433] McFadden, C., Soto, C., and Spencer, N. D., "Adsorption and Surface Chemistry in Tribology," *Tribology International*, Vol. 30, No. 12, 1997, pp. 881–888.

[434] Rowe, C. N., "Lubricated Wear," *CRC Handbook of Lubrication, Theory and Practice in Tribology, Vol. II, Theory and Design*, R. E. Booser, Ed., CRC Press, Boca Raton, 1988, pp. 209–225.

[435] Barcroft, F. T., and Daniel, S. G., "The Action of Neutral Organic Phosphates as EP Additive," *ASME J. Basic Eng.*, 64-Lub-22, 1964.

[436] Dorinson, A., "The Additive Action of Some Organic Chlorides and Sulfides in the Four-Ball Lubricant Test," *ASLE Transactions*, Vol. 16, No. 1, 1973, pp. 22–31.

[437] Forbes, E. S., and Reid, A. J. D., "Liquid Phase Adsorption/Reaction Studies of Organo-sulfur Compounds and Their Load-carrying Mechanism," *ASLE Transactions*, Vol. 16, No. 1, 1973, pp. 50–60.

[438] Plaza, S., "Some Chemical Reactions of Organic Disulfides in Boundary Lubrication," *ASLE Transactions*, Vol. 30, No. 4, 1987, pp. 493–500.

[439] Kawamura, M., Moritani, H., Esaki, Y., and Fujita, K., "The Mechanism of Synergism Between Sulfur- and Phosphorus-type EP Additives," *ASLE Transactions*, Vol. 29, No. 4,

1986, pp. 451–456.

[440] Phillips, W. D., "Ashless Phosphorus-containing Lubricating Oil Additives," *Lubricant Additives: Chemistry and Applications*, Leslie R. Rudnick, Ed., Marcel Dekker, Inc., NY, Chapter 3, 2003, pp. 45–108.

[441] Anonymous Author, "Fundamentals of Wear," *Lubrication*, Vol. 12, No. 6, 1957, pp. 61–78.

[442] Anonymous Author, "Lubricant Additives," *Ind. Lubr., and Trib.*, Vol. 49, No. 1, 1997, pp. 15–30.

[443] Mandakovic, R., "Assessment of EP Additives for Water Miscible Metalworking Fluids," *J. Syn. Lubr*, Vol. 16, No. 1, 1999, pp. 13–26.

[444] Rounds, F. G., "Additive Interactions and Their Effect on the Performance of a Zinc Dialkyl Dithiophosphate," *ASLE Transactions*, Vol. 21, No. 2, 1978, pp. 91–101.

[445] (a) Rounds, F. G., "Some Effects of Amines on Zinc Dialkyl dithiophosphate Antiwear Performance as Measured in 4-Ball Wear Tests," *ASLE Transactions*, Vol. 24, No. 4, 1981, pp. 431–440. (b) Bovington, C. H., "Friction, Wear, and Role of Additives," *Chemistry and Technology of Lubricants*, R. M. Mortier and S. T. Orszulik, Eds., Blackie Academic and Professional-Chamber & Hall, London, UK, 1996, Chapter 12, pp. 320-370.

[446] Liang, H., Totten, G. E., and Webster, G. M., "Lubrication and Tribology Fundamentals," *Fuels and Lubricants Handbook: Technology, Properties, Performance, and Testing*, MNL37WCD, Ed., ASTM International, West Conshohocken, PA, Chapter 35, 2003, p. 961.

[447] Lauer, D., "Industrial Applications," *CRC Handbook of Lubrication, Theory and Practice in Tribology, Vol. III, Monitoring, Materials, Synthetic Lubricants, and Applications*, R. E. Booser, Ed., CRC Press, Boca Raton, 1993, pp. 195–205.

[448] (a) Lancaster, J. K., "Solid Lubricants," *CRC Handbook of Lubrication, Theory and Practice in Tribology, Vol. II, Theory and Design*, R. E. Booser, Ed., CRC Press, Boca Raton, 1988, pp. 269–290. (b) Mariani, G., "Selection and Applications of Solid Lubricants as Friction Modifiers," *Lubricant Additives: Chemistry and Applications*, L. R. Rudnick, Ed., Marcel Dekker, Inc., New York, 2003, Chapter 6, pp. 171–201.

[449] (a) Peterson, M. B., and Johnson, R. L., "Friction Studies of Graphite and Mixtures of Graphite with Several Metallic Oxides and Salts at temperatures to 1000 °F," TN-3657, National Aeronautics and Space Administration, Washington, DC, 1956. (b) Sliney, H. E., "Solid Lubricants," *American Society of Metals Handbook, Friction, Lubrication, and Wear Technology*, S. D. Henry, ed., ASM International, Vol. 18, 1992, pp. 113–122. (c) http://www.a-m.de/englisch/lexikon/molybdaendisulfid-bild1.htm.

[450] Haltner, A., and Flom, D. G., "Lubrication in Space," *CRC Handbook of Lubrication, Vol. I, Theory and Practice of Tribology: Applications and Maintenance*, R. E. Booser, Ed., CRC Press, Boca Raton, 1983, pp. 389–404.

[451] Fusaro, R. L., "Friction and Wear Life Properties of Polyimide Thin Films," TN-D-6914, National Aeronautics and Space Administration, Washington, DC, 1972.

[452] Brydson, J. A., Platics *Materials*, 3rd ed., Butterworth, London, 1975.

[453] Theberge, J. E., "Properties of the Internally Lubricated, Glass-fortifies Thermoplastics for Gears and Bearings," *Proc. ASLE Int. Con. Solid Lubr.*, SP-3, American Society of Lubrication Engineers, Park Ridge, IL, 1976.

[454] Braithewaite, E. R., and Greene, A. B., "A Critical Analysis of the Performance of Molybdenum Compounds in Motor Vehicles," *Wear*, Vol. 46, 1978, p. 405.

[455] Messina, J., "Rust-inhibited, Non-reactive, Pefluorinated Polymer Greases," *Proc. ASLE Conf. Solid Lubr. SP-3*, American Society of Lubrication Engineers, Park Ridge, IL, 1976.

[456] *Corrosion—Causes and Prevention*, F. N. Speller, McGraw-Hill Publishing, 1935.

[457] "Standard Electrode Reduction and Oxidation Potential," http://www.siliconfareast.com/ox_potential.htm.

[458] Becher, P., *Emulsions: Theory and Practice, American Chemical Society Monograph Series*, Reinhold Publishing Corporation, Chapter 6, 1957, pp. 209–231.

[459] Karsa, D. R., "Industrial Applications of Surfactants," *Industrial Applications of Surfactants—An Overview*, D. R. Karsa, Ed., Published by Royal Society of Chemistry, Cambridge, England, 1987.

[460] Hancock, R. I., "Macromolecular Surfactants," *Surfactants*, Th. F. Tadros, Ed., Academic Press, Chapter 12, 1984, pp. 287–321.

[461] Salomon, M. F., Galic, M., and Jaynes, S. E., "Application of a New Fleet Test Procedure to Investigate Emulsion Formation Tendencies of Motor Oils," SAE Technical Paper 912398, 1991.

[462] Rempp., P., and Herz, J. E., "Model Polymers," *Concise Encyclopedia of Polymer Science and Engineering*, J. I. Kroschwitz, Ed., Wiley InterScience, 1990, pp. 633–634.

[463] Van Krevelen, D. W., and Hoftyzer, P. J., "Typology of Polymers," *Properties of Polymers: Correlations with Chemical Structure*, Elsevier Publishing Company, Amsterdam, Chapter 2, 1972.

[464] Freeman, W. J., "Characterization of Polymers," *Concise Encyclopedia of Polymer Science and Engineering*, J. I. Kroschwitz, Ed., Wiley InterScience, 1990, pp. 139–142.

[465] Tung, L. H., "Fractionation," *Concise Encyclopedia of Polymer Science and Engineering*, J. I. Kroschwitz, Ed., Wiley InterScience, 1990, p. 417.

[466] Muller, H. G., "Mechanism of Action of Viscosity Index Improvers," *Tribol. Int.*, June 1978, pp. 189–192.

[467] Covitch, M., "Olefin Copolymer Viscosity Modifiers," *Lubricant Additives: Chemistry and Applications*, L. R. Rudnick, Ed., Marcel Dekker, Inc., NY, Chapter 10, 2003, pp. 293–327.

[468] Kleiser, W. M., Walker, H. M., and Rutherford, J. A., "Determination of Lubricating Oil Additive Effects in Taxicab Service," Soc. Automotive Engr. Tech. Paper Ser. No. 912386, 1991.

[469] Kinker, B. G., "Polymethacrylate Viscosity Modifiers," *Lubricant Additives: Chemistry and Applications*, L. R. Rudnick, Ed., Marcel Dekker, Inc., NY, Chapter 11, 2003, pp. 329–353.

[470] Arlie, J. P., Dennis, J., and Parc, G., 2003, "Viscosity Index Improvers I, Mechanical and Thermal Stabilities of Polymethacrylates and Polyolefins," IP Paper 75–005, Inst. Petr., London, 1975.

[471] *Polymer Handbook*, J. Branderup, and E. H. Immergut Eds., Interscience, NY, V-5 to V-11, 1966.

[472] Kinker, B. G., "Automatic Transmission Fluid Shear Stability Trends and Viscosity Improver Trend," *Proc. 98 Intl. Symp.*

Tribology of Vehicle Trans., Yokohama, 1998, pp. 5–10.

[473] Watson, R. W., and McDonnell, Jr., T. F., "Additives—The Right Stuff for Automotive Engine Oils," Fuels and Lubricants Technology: *An Overview SP.603*, Oct. 1984, Society of Automotive Engineers, pp. 17–28.

[474] "Lubricant Additives and the Environment," CEFIC, Brussels, Belgium (1993).

[475] "Fuel Additives and the Environment," CEFIC, Brussels, Belgium (1994).

[476] McFall, D., "Is Scrutiny of Engine Oils Slipping?," *Lubes'N'Greases*, Jan. 8, 2001 p. 6.

[477] *Internal Combustion Engines*, V. Ganesan, Mcgraw-Hill Inc., NY, 1995.

[478] Norbye, J. P., *The Wankel Engine: Design, Development, Applications*, Chilton Book Company, Philadelphia, 1971.

[479] "Gas and Expansion Turbines," *Van Nostrand's Scientific Encyclopedia*, Fifth Edition, D. M. Considine, Ed., Van Nostrand Reinhold Company, NY, 1976, pp. 1138–1148.

[480] "Engine Service Classification System and Guide to Crankcase Oil Selection," *API Publication 1509*, American Petroleum Institute, Sept. 2002, www.api.org/certifications/engineoil/pubs/index. cfm

[481] "The ILSAC GF-4 Standard for Passenger Car Engine Oils," *International Lubricant Standardization and Approval Committee*, January 14, 2004, www.ilma.org/resources/ilsac_finalstd011404.pdf

[482] McFall, D., "Automakers Push 'Start' Button on GF-5 Oils," *Lube Report*, Vol. 4, No. 52, Dec. 29, 2004.

[483] Taylor, R. I., Mainwaring, R. and Mortier, R. M., "Engine Lubricant Trends Since 1990," [http://www.iantaylor.org.uk/papers/IMechEReview.pdf#search=%22%E2%80%9CEngine%20Lubricant%20Trends%20Since%201990%22%22].

[484] "Performance Specification MIL-PRF-2104H: Lubricating Oil, Internal Combustion Engine, Combat Tactical Service," Published 12 July 2004 [http://www.wbdg.org/ccb/FEDMIL/prf2104h.pdf].

[485] "Two-stroke Cycle Gasoline Engine Lubricant, Recertified," SAE J1536—Revised November 2005, SAE Standard, Society of Automotive Engineers.

[486] "Engine Oil Performance and Engine Service Classification (Other Than 'Energy Conserving')," SAE J183—Revised March 1, 2006, SAE Standard, Society of Automotive Engineers.

[487] "ACEA and API Tables," http://www.infineum.com/information/tables.html

[488] Chitty, D. B., "A Rig Test Approach to the Investigation of Valve Train Failures in Automotive Engines," *Performance Testing of Lubricants for Automotive Engines and Transmissions*, C. F. McCue, J. C. Cree, and R. Tourret, Eds., Applied Science Publishers, Barking, England, 1973.

[489] McFall, D., and Tocci, L., "Diesel Oils Trend Toward OEM Specs," *Lube Report*, January 3, 2006, http://www.imakenews.com/lng/e_article000508398.cfm?x=b11,0,w.

[490] "ACEA 2007 Sequences," http://www.acea.be/images/uploads/pub/070308_ACEA_sequences_2007_LD_and_HD.pdf

[491] "International Tests and Specifications for Automotive Oils," SAE J2227, *Surface Vehicles Information Report*, Issued April 2002.

[492] "Global Performance Specifications for Light Duty Diesel Engine Oil (Global DLD-1, DLD-2 and DLD-3)," http://www.enginemanufacturers.org/admin/library/upload/287.pdf

[493] "Recommended Guideline: Global Performance Specification for Diesel Engine Oil (Global DHD-1)," http://www.jama-english.jp/release/release/2001/dhd-1.pdf

[494] McFall, D., "Diesel Oil Specs: Converging or Diverging?," *Lubes-N-Greases*, Aug. 2002, pp. 6–13.

[495] Sullivan, T., "Heavy-duty Oil Specs: Convergence on Horizon?" *Lube Report*, Vol. 6, No. 45, Nov. 8, 2006.

[496] Carter, B. H., "Marine Lubricants," *Chemistry and Technology of Lubricants*, R. M. Mortier and S. T. Orszulik, Eds., Blackie Academic and Professional-Chapman & Hall, London, UK, 1996, Chapter 10, pp. 287–305.

[497] "Shipboard Investigations with Selected Fuels of Tomorrow," G. J. Hellingman and S. Barrow, *Presented at 14th International Congress on Combustion Engines*, Helsinki, 1981.

[498] Scott, R., "Stationary Natural Gas Engine Lubrication," *Machinery Lubrication*, Noria Corporation, Sept. 2004, http://www.machinerylubrication.com/article_detail.asp?articleid=663&pagetitle=Stationary%20Natural%20Gas%20Engine%20Lubrication

[499] Scott, R., "Reciprocating Natural Gas Compressors," *Machinery Lubrication*, Noria Corporation, Nov. 2003.

[500] Lansdown, A. R., "Aviation Lubricants," *Chemistry and Technology of Lubricants*, R. M. Mortier and S. T. Orszulik, Eds., Blackie Academic and Professional-Chapman & Hall, London, UK, 1996, Chapter 9, pp. 264–305.

[501] "TURBO OIL 274—Turbine Engine & Accessory," *BP Lubricant Brochure*, http://www.bp.com/liveassets/bp_internet/aviation/air_bp/STAGING/local_assets/downloads_pdfs/b/brochure_bp_turbo_oil_274_sept_02.pdf

[502] "TURBO OIL 25—Turbine Engine & Accessory," *BP Lubricant Brochure*, http://www.bp.com/liveassets/bp_internet/aviation/air_bp/STAGING/local_assets/downloads_pdfs/b/brochure_bp_turbo_oil_25_sept_02.pdf

[503] "TURBO OIL 2389—Turbine Engine & Accessory," *BP Lubricant Brochure*, http://www.bp.com/liveassets/bp_internet/aviation/air_bp/STAGING/local_assets/downloads_pdfs/b/brochure_bp_turbo_oil_2389_march_03.pdf

[504] "TURBO OIL 2380—Turbine & Accessory," *BP Lubricant Brochure*, http://www.bp.com/liveassets/bp_internet/aviation/air_bp/STAGING/local_assets/downloads_pdfs/b/brochure_bp_turbo_oil_2380_sept_02.pdf

[505] Kudolis, C., "Choosing an Oil for Two-Cycle Engines," *E.A.A. Experimenter™ Magazine*, April, 1994, http://www.theultralightplace.com/choosing_oil.htm

[506] "Two-stroke Cycle Gasoline Engine Lubricants: Performance and Service Classification," SAE J2116—July 2003, SAE Standard, Society of Automotive Engineers.

[507] "FC-W™: Four-Stroke Cycle, Water-Cooled Gasoline Engine Lubricant Certification Procedure Manual," http://www.nmma.org/certification/local/downloads/documents/FCW_Certification_Procedure_Manual.pdf

[508] Courtesy of the Suzuki Motor Corporation.

[509] *CRC Diesel Engine Rating Manual* (*CRC Manual No. 15*), 1986.

[510] *Sludge Rating Manual* (*CRC Manual 12*), 1976.

[511] *CRC Varnish Rating Manual* (*CRC Manual No. 9*), 1971.

[512] *CRC Rust Rating Manual* (*CRC Manual No. 7*), *1962*; Pub-

lished by Automotive Vehicle Fuel, Lubricant, and Equipment Research Committee of the Coordinating Research Council, Inc., 1986.

[513] SWRI. http://www.swri.org.

[514] Figure Courtesy of Petro-Canada.

[515] API Base Oil Interchangeability Guidelines for Passenger Car Motor Oils and Diesel Engine Oils, Paragraph E.1.3.

[516] "Synthetic Lubricating Oils," U.S. Patent 4,737,297 (April 12, 1988). Inventors: T. Yoshida and H. Watanabe.

[517] Shugarman, A. L., "Selecting Synthetics for Off-Highway Crankcase Applications," *Machinery Lubrication Magazine*, March 2002.

[518] Culley, S. A., McDonnell, T. F., Ball, D. J., Kirby, C. W., and Hawes, S. W., "Emission Control Systems and Phosphorus in Oils," *Automot. Eng.*, Nov. 1996, pp. 35–44.

[519] Kayama, F., Horiguchi, H., and Hamamatsu, A., "Potential Health Effects of Alkylphenols in Japan," *Japanese Medical Association Journal {JMAJ}*, Vol. 3, No. 3, 2003, pp. 108–114.

[520] Schettler, T., "Prenatal Exposure to Endocrine Disrupting Chemicals: Effects on Children's Development and Health," http://www.epa.gov/Region8/humanhealth/children/SchettlerPrenatalExposures-toEndocrineDisruptors.pdf

[521] Miller, D., Wheals, B. B., Beresford, N., and Sumpter, J. P., "Estrogenic Activity of Phenolic Additives Determined by an In Vitro Yeast Bioassay," *Environ. Health Perspect.*, Vol. 109, No. 2, 2001, pp. 133–138.

[522] (a) McGeehan, J. of Chevron, "Uncovering the Problems with Extended Oil Drains," *Machinery Lubrication Magazine*, September 2001. (b) DeMarco, N., "GM, California to Bust 3,000 Mile Myth," *Lube Report*, Vol. 8, No. 24, June 11, 2008.

[523] McFall, D., "Extended of Extreme," *Lubes & Greases*, March 2000, pp. 16–20.

[524] "Cummins Centinel Available for Aftermarket Purchase," www.cummins.com (Search for Cummins Centinel).

[525] Gatliff, B., "Cummins Centinel," www.embeded.com/story/OEG2001061850078.

[526] "Fuel economy additive and lubricant composition containing same," U.S. Patent 5,863,873 (January 26, 1999). Inventor: C. H. Bovington.

[527] "Engine lubricant using molybdenum dithiocarbamate as an antioxidant top treatment in high sulfur base stocks," U.S. Patent 6,706,672 (March 16, 2004). Inventor: J. Kelley.

[528] "Low ash lubricant composition," U.S. Patent 5,259,967 (November 9, 1993). Inventor: D. E. Ripple.

[529] "Diesel crankcase lubricant composition," U.S. Patent 4,282,107 (August 4, 1981). Inventors: B. H. Zoleski, R. L. Sung, G. R. Schierberg, and G. P. Hebert.

[530] "Oxidation and corrosion resistant diesel engine lubricant," U.S. Patent 4,758,363 (July 19, 1988). Inventors: R. L. Sung, B. H. Zoleski, and R. L. O'Rourke.

[531] "Lubricating oil compositions," U.S. Patent 7,053,027 (May 30, 2006). Inventors: L. Chambard and T. Garner.

[532] Rose, M., "Gas Engine Oils—A Question of Balance," Infineum International Limited, *Machinery Lubrication Magazine* March 2004.

[533] E. A. A. Experimenter™ Magazine, April, 1994. Article submitted by C. Kudolis (EAA 299878) http://www.ultralighthomepage.com/OIL/oil.html

[534] "Low sulfur lubricant composition for two-stroke engines," U.S. Patent 6,391,833 (May 21, 2002). Inventors: F. Ilmain, S. Addagarla, and J. Cazin.

[535] "Air Pollution Emissions Overview," http://www.epa.gov/oar/oaqps/emissns.html

[536] "Pollutants," http://www.epa.gov/oms/invntory/overview/pollutants/index.htm

[537] "Integrated Risk Information System," http://www.epa.gov/iris/

[538] "Control of Hazardous Air Pollutants from Mobile Sources, 40 CFR Parts 59, 80, 85 and 86," http://www.epa.gov/oms/regs/toxics/proposedrule.pdf

[539] "Technology Transfer Network—1999 National Scale Air Toxics Assessment," www.epa.gov/ttn/atw/nata1999

[540] "Ambient Air Quality Standards," http://www.arb.ca.gov/research/aaqs/aaqs2.pdf

[541] "Ambient Air Quality Standards (NAAQS)," www.epa.gov/air/criteria.html

[542] Eyzat, P. and Guibet, J. C., "A New Look at Nitrogen Oxides Formation in Internal Combustion Engines," *SAE Paper 680124*, 1968.

[543] Zeldovich, Y. B., Sadovnikov, P. Y., and Frank-Kamenetskii, D. A., *Oxidation of Nitrogen in Combustion*, Academy of Sciences, U.S.S.R., Moscow-Leningrad, 1947.

[544] Lavoie, G. A., Heywood, J. B., and Keck, J. C., "Experimental and Theoretical Study of Nitric Oxide Formation in Internal Combustion engines," *Combust. Sci. Technol.* Vol. 1, 1970, 313.

[545] Newhall, H. K., and Shahed, S. M., "Kinetics of Nitric Oxide Formation in High Pressure Flames," *13th Symposium (International) on Combustion*, Salt Lake City, Aug. 1970.

[546] "Knock and Antiknock Action," M. M. Sturgis, *SAE Trans.* Vol. 63, 1955, 253.

[547] Patterson, D. J., and Henein, N. A., *Emissions from Combustion Engines and Their Control*, Ann Arbor Science Publishers, 1972.

[548] Data Source—Chevron's website. "Fuels/Motor Oil," http://www.chevron.com/products/prodserv/fuels/

[549] "Heavy Duty On-Highway Diesel Engine Certification Test Procedure Issues at the 2007—2010 Emissions Levels," by Shirish A. Shimpi of Cummins Inc., Chairman EMA Calibration Standards Task Force, http://www.epa.gov/oms/regs/hd2007/wrkshop2/shimpi.ppt

[550] "Advanced Petroleum Based Fuels—Diesel Emission Control Project," http://www.nrel.gov/vehiclesandfuels/apbf/apbf_dec.html

[551] *Automotive Fuels Handbook*, K. Owen and T. Coley, Society of Automotive Engineers, Warrendale, PA, 1990.

[552] Brinkman, N. D., et al., "Effect of Fuel Volatility on Drivability at Low and Intermediate Ambient Temperatures," SAE Technical Paper 830593, 1983.

[553] Barker, D. A., and Dunn, M. R., "Drivability Number—A Gasoline Volatility Parameter Related to Cold Start Passenger Car Performance," SAE Technical Paper 831756, 1983.

[554] Bitting, B., et al., "Intake Valve Deposits—Fuel Detergency Requirements Revisited," SAE Technical Paper 872117, 1987.

[555] Taniguchi, B. Y., et al., "Injector Deposits—The Tip of Intake System Deposit Problem," SAE Technical Paper 861534, 1986.

[556] Courtesy of Chevron-http://www.chevron.com/products/prodserv/fuels/bulletin/motorgas/

[557] Hagen, D. F. and Holiday, G. W., "The Effects of Engine Operating and Design Variables on Exhaust Emissions," SAE Publication TP-6, 1964, p. 206.

[558] Heisler, H., *Advanced Engine Technology*, Society of Automotive Engineers, 1997.

[559] "Proton Exchange Membrane Fuel Cell," http://en.wikipedia.org/wiki/Proton_exchange_membrane_fuel_cell

[560] "Diesel Retrofit Technology Verification," http://www.epa.gov/otaq/retrofit/index.htm

[561] "Recent Developments in Integrated Exhaust Control Technologies for Diesel Engines," ftp://ftp.arb.ca.gov/carbis/ds/archpres/meca/diesel.ppt

[562] "How BlueTec® Works," http://merceds-benz.com/specials/scr/en/index_nocom_en. htm

[563] The Effect of Cetane Number Increases Due to Additives on NOx Emissions from Heavy-Duty Highway Engines, EPA420-S-02–012, June 2002. http://www.epa.gov/otaq/models/analysis/s02012.pdf

[564] "Biodiesel," National Biodiesel Board, www.biodiesel.org

[565] Henein, N. A., and Bolt, J. A., "The Effect of Some Fuel and Engine Factors on Diesel Smoke," SAE Technical Paper 690557, 1969.

[566] Stone, R., *Introduction to Internal Combustion Engines*, Society of Automotive Engineers, 1993.

[567] "Exhaust Emissions, Spark-ignition (SI) Engines," Grözlinger, W., *Automotive Handbook*, R. Bosch, GmbH, 1993, pp. 478–482.

[568] Courtesy of Toyota Motor Corporation.

[569] Stone, R., and Kwan, E., "Variable Valve Actuation Mechanisms and the Potential for Their Application," SAE Technical Paper 890673, SAE Transactions, Vol. 98, 1989.

[570] "Diesel Fuel Technical Review," www.chevron.com/products/prodserv/fuels/diesel/documents/Diesel_Fuel_TechnicalReview.pdf

[571] Courtesy of R. Bosch GMBH.

[572] Newton, K., Steeds, W., and Garret, T. K., "Potential for Future Development," *The Motor Vehicle*, Chapter 17, 1996, pp. 389–395.

[573] Courtesy of Cummins Inc., Moore, W., "Living with Cool-EGR," www.contructionequipment.com/prevention/cd031a004.asp

[574] Sullivan, T., "Charting a Safe Course on EGR," The Effect of EGR on Various Lubricant Parameters [Argonne Data], *Lubes'N'Greases*, Sept. 2001.

[575] Wagner, R. M., Green Jr., J. B., Storey, J. M., and Daw, C. S., "Extending Exhaust Gas Recirculation Limits in Diesel Engines," Oakridge National Laboratory Research Article.

[576] "SAE Fuels and Lubricant Standards," *2007 SAE Handbook*, Society of Automotive Engineers, Inc., Warrendale, PA.

[577] Hauser, K. R., and Crossby, T. A., "The Impact of Intake Valve Deposits on Exhaust Emissions," SAE Technical Paper 922259, 1992.

[578] Tupa, R. C., "Port Fuel Injector Deposits—Causes/Consequences/Cures," SAE Technical Paper 872113, 1987.

[579] Esaki, Y., Ishiguro, T., Suzuki, N., and Nakada, M., "Mechanism of Intake Deposit Formation—Part I—Characterization of Deposits," SAE Technical Paper 900151, 1990.

[580] Price, R. J., Wilkinson, P. T., Jones, D. A., and Morley, C., "A Laboratory Simulation and Mechanism for the Fuel Dependence of SI Combustion Chamber Deposit Formation," *SAE Technical Paper 952445, Presented at Fuels and Lubricant Meeting*, Toronto, Ontario, Oct. 16–19, 1995.

[581] Daly, D. T., Bannon, S. A., Fog, D. A., and Harold, S. M., "Mechanism of Combustion Chamber Formation," *SAE Technical Paper 941889, Presented at Fuels and Lubricant Meeting*, Baltimore, MD, Oct. 17–20, 1994.

[582] Kalghatagi, G. T., "Combustion Chamber Deposits in Spark-ignition Engines: A Literature Review," SAE Technical Paper 952443, 1995.

[583] "Drinking Water," http://www.epa.gov/mtbe/water.htm

[584] Guidance on Quantifying NOx Benefits for Cetane Improvement Programs for Use in SIPs and Transportation Conformity, EPA420-B-04-005, June 2004, http://www.epa.gov/OMS/guidance/420b04005.pdf

[585] NFPA Standard T 2.13.7R1-1997: Hydraulic Fluid Power-Petroleum Fluids-Prediction of Bulk Moduli, National Fluid Power Association, Milwaukee, WI 1997.

[586] Totten, G. E., Webster, G. M., and Yeapple, F. D., "Physical Properties and Their Determination," *Handbook of Hydraulic Fluid Technology*, G. E. Totten, Ed., Marcel Dekker, NY, 2000 p. 253.

[587] Esposito, A., *Fluid Power with Applications*, Prentice Hall, Englewood Cliffs, NJ, 1988.

[588] SAE J754A Lubricant Types—Construction and Industrial Machinery, *SAE Handbook*, Part 3, Society of Automotive Engineers, Warrendale, PA, 2004, 40.88.

[589] Kushwaha, R. L. and Liu, J., "Farm and Construction Equipment," *Handbook of Lubrication and Tribology, Vol. I, Applications and Maintenance*, 2nd ed., G. E. Totten, Ed., CRC Press, Chapter 21, 2006, pp. 21-1–21-18.

[590] Cleveland, W. S., "Tractor Wet Brake and Wet Clutch Friction Properties," *NLGI Spokesman*, July 1987 pp. 135–138.

[591] Hanley, W. J., "Farm and Construction Equipment," *CRC Handbook of Lubrication, Vol. I, Theory and Practice of Tribology: Applications and Maintenance*, R. E. Booser, Ed., CRC Press, Boca Raton, 1983, pp. 249–267.

[592] "Ready Reference for Hydraulic and Industrial Fluids," *Lubrizol Publication*, May 2005.

[593] Michael, P. W., Herzog, S. N., and Marougy, T. E., "Fluid Viscosity Selection Criteria for Hydraulic Pumps and Motors," *Proceedings of the 48th National Conference on Fluid Power*, Chicago, National Fluid Power Association, Milwaukee, WI, 2000, pp. 313–325.

[594] Givens, W. A., and Michael, P. W., "Hydraulic Fluids," *Fuels and Lubricants Handbook: Technology, Properties, Performance, and Testing*, G. E. Totten Ed., ASTM International, West Conshohocken, PA, Chapter 13, 2003, pp. 353–381.

[595] Lloyd, B. J., "Water Water Everywhere," *Lubes-N-Greases*, Vol. 11, No. 3, 1997, pp. 44–53.

[596] Brew, D., Khonsari, M., and Ball, J. H., "Current Research in Cavitating Fluid Films," STLE Special Publication SP-28, Society of Tribologists and Lubrication Engineers, Park Ridge, IL, 1990, pp. 63–65.

[597] Barnes, M., "Particle Counting—Oil Analysis 101," *Practicing Oil Analysis Magazine*, July 2002.

[598] "Ultra-clean Super Hydraulic Oil," http://

tds.econocophillips.com/catalogs/Conoco/Industrial%20Oils/Con%20UC%20Super%20Hydraulic%20Oil%20TDS%20Web.pdf

[599] "Utilizing ASTM Specification Tests to Predict Hydraulic Fluid Performance," http://www.lubricantsuniversity.com/index.php?option=com_content&task=view&id=229&Itemid=144

[600] "Project Book Hydraulic Bosch Rexroth," http://www.boschrexroth.com/business_units/bri/de&/branchen/awz/utomotive/fiat-gm_m40.pdf.

[601] Day, M., "What Happened to NAS 1638?" *Practicing Oil Analysis Magazine*, Nov. 2002, [http://www.practicingoilanalysis.com/article_detail.asp?articleid=409&relatedbookgroup=ContaminationControl].

[602] Ashby, D. M., "O-ring Specification—Some Important Considerations," *Hydraul. Pneum.*, Aug. 1999, pp. 53, 54, 90.

[603] Carpsjo, C., *Proceedings of International Symposium on Performance Testing of Hydraulic Fluids*, Institute of Petroleum, London, 3–6 Oct. 1978.

[604] "Hydraulic Fluids: 3. Chemical and Physical Information," www.atsdr.cdc.gov/toxprofiles/tp99-c3.pdf.

[605] *Vickers Guide to Alternative Fluids*, Eaton Corp., Southfield MI, Nov. 1992.

[606] Totten, G. E., and Sun, Y., "Water-Glycol Hydraulic Fluids," *Handbook of Hydraulic Fluids Technology*, G. E. Totten, Ed., Marcel Dekker, NY, 2000, pp. 917–982.

[607] NFPA/T2.13.1:R3-1997, Recommended Practice—Hydraulic Fluid Power—Use of Fire Resistant Fluids in Industrial Systems, NFPA, Milwaukee WI, 1997.

[608] MIL-H-5606G Military Specification: Hydraulic Fluid, Petroleum Base, Aircraft, Missile and Ordinance, NATO code number H-515, U.S. Department of Defense, Fort Belvoir, VA, 1994.

[609] Hohn, B. R., Michaelis, K., and Dobereiner, R., "Load-carrying Capacity of Fast Biodegradable Gear Lubricants," *Lubr. Eng.*, Vol. 55, 1999, p. 37.

[610] *Hydraulic Fluid Information*, Denison Corp., Marysville, OH, 1995, p. 2.

[611] Sharma, S. K., Snyder, Jr., C., E., Gschwender, L. J., Lang, J. C., and Schreiber, B. F., "Stuck Servovalves in Aircraft Hydraulic Systems," *Lubr. Eng.*, Vol. 55, No. 7, July 1999.

[612] Michael, P. W., and Webb, S., "Future Fluids—What is Coming Down the Hydraulic Line," *OEM Off-Highway*, Jan. 1998, pp. 34–36.

[613] Wambach, Jr., W. E., "Hydraulic Systems and Fluids," *CRC Handbook of Lubrication, Vol. I, Theory and Practice of Tribology: Applications and Maintenance*, R. E. Booser, Ed., CRC Press, Boca Raton, 1983, pp. 149–166.

[614] Clason, D., "Zinc-based Additives Continue to Deliver," *Machinery Lubrication Magazine*, Nov. 2003.

[615] Bartz, W. J., "Environmentally Acceptable Hydraulic Fluids," Technische Akademie, Esslingen, Germany.

[616] (a) Hope, K., "Balancing Lubricant Properties with Vegetable Oil and PAO Blends," STLE Philadelphia, PA, May 8, 2007, http://www.teraspeed.com/RLI/STLE%20philadelphia.ppt (b) "Government-Industry Forum on Non-Food Uses of Crops (Gifnfc 7/7) Case Study: Plant Oil Based Lubricants in Total Loss & Potential Loss Applications," Final Report, 16 May 2002, Prepared by: Prof. Dr. M. Schneider and Dr. Paul Smith, http://www.nnfcc.co.uk/nnfcclibrary/productreport/download.cfm?id=36#search=%22environmentally%20friendly%20hydraulic%20oils%20murrenhoff%22

[617] "Lubricating Oil Composition Useful in Hydraulic Fluids," U.S. Patent 6,756,346 (June 29, 2004). Inventors: Y. Baba and K. Hanyuda.

[618] "Water-Based Hydraulic Fluid," U.S. Patent 4,481,125 (November 6, 1984). Inventor: R. V. Holgado.

[619] Brown, K., "Condition-Monitoring of Phosphate Ester Hydraulic Fluids," *Machinery Lubrication Magazine*, Nov. 2002.

[620] Deen, H. E., and Ryer, J., "Automatic Transmission Fluids—Properties and Performance," H. E. Deen, and J. Ryer in *Fuels and Lubricants Technology: An Overview, SP.603*, Society of Automotive Engineers, Oct. 1984, pp. 117–127.

[621] Based upon information on Automatic Transmission Fluid by L. Carley C2007, [http://www.aa1car.com/library/atf.htm].

[622] Artman, D. M., and Copes, R. G., "ATF From Performance Challenges to Market Opportunities," *Presented at the 1994 NPRA National Fuels and Lubricants Meeting*, Nov. 3–4, 1994.

[623] (a) Sullivan, T., "GM Rolls Out DEXRON®-VI," *LUBE REPORT*, April 12, 2005; http://www.imakenews.com/lng/e_article000384801.cfm?x=b11,0,w Tocci, L., "Ford Unveils New Transmission Fluid," Lube Report, Vol. 7, No. 18, April 25, 2007.

[624] (a) Tocci, L., "Ford Pulls a Plug on MERCON®," *Lube Report*, April 12, 2006, Vol. 6 No. 15. (b) Tocci, L., "Ford Unveils New Transmission Fluid," *Lube Report*, Vol. 7, No. 18, April 15, 2007.

[625] Based on the information taken from *Driveline Abstracts*, June 2007, Southwest Research Institute, San Antonio, TX [www.swri.org/d06/Driveline/DrivelineAbstratcs.pdf].

[626] Graham, R., and Oviatt, W. R., "Automatic Transmission Fluids—Developments Toward Rationalization," *Presented at CEC 1985 International Symposium*, Wolfsburg, Germany, June 7, 1985.

[627] Beimesch, B. J., "Automotive Applications," *CRC Handbook of Lubrication, Theory and Practice in Tribology, Vol. III, Monitoring, Materials, Synthetic Lubricants, and Applications*, R. E. Booser, Ed., CRC Press, Boca Raton, 1993.

[628] Perez, J. M., and Klaus, E. E., "Dibasic Acid and Polyol Esters," *CRC Handbook of Lubrication, Theory and Practice in Tribology, Vol. III, Monitoring, Materials, Synthetic Lubricants, and Applications*, R. E. Booser, Ed., CRC Press, Boca Raton, 1993, pp. 237–252.

[629] "Continuously Variable Transmission," http://en.wikipedia.org/wiki/Continuously_variable_transmission Marsz, S. J., "A CVT in a Compact Efficient Package," Machine Design, June 2005.

[630] Hamid, S., "Maximizing Bearing Life with EHD Lubricants," *Machinery Lubrication Magazine*, Sep. 2004.

[631] "Traction Fluid with Alkane Bridged Dimer," U.S. Patent 6,828,283 (December 7, 2004). Inventors: T. J. Chapaton, T. W., and C. and J. L. Linden.

[632] "Tractions Fluids Having Excellent Low Temperature Properties," U.S. Patent 6,602,830 (August 5, 2003). Inventors: K. C. Fey, G. F. Viaplana Gordon, T. K. Hilty, R. K. King, T. M. Leaym, G. M. Wieber, and E. D. Groenhof.

[633] "Tractions Fluids," U.S. Patent 6,623,399 (September 23, 2003). Inventors: K. C. Fey, G. F. Viaplana Gordon, T. K. Hilty, R. K. King, T. M. Leaym, L. A. Stark-Kasley, G. M. Wieber, and E. D. Groenhof.
[634] "Method for Lubricating A Continuously Variable Transmission," U.S. Patent 6,730,640 (May 4, 2004). Inventors: R. L. Sowerby, C. D. Tipton, and J. L. Sumiejski.
[635] "Traction Fluids Formulation," U.S. Patent 6,372,696 (April 16, 2002). Inventor: C. D. Tipton.
[636] "Cyclic Oligomer Traction Fluid," U.S. Patent 7,045,488 (May 16, 2006). Inventors: S. J. Bartley, D. M. Barr, R. M. Lange, and C. D. Tipton.
[637] "High Boron Fluids for Continuously Variable Transmissions," U.S. Patent 6,451,745 (September 17, 2002). Inventor: W. C. Ward.
[638] "Lubricant composition," U.S. Patent 6,465,399 (October 15, 2002). Inventors: N. Koishikawa and Y. Tatsumi.
[639] Bastian, E. L. H., "Metalworking," *Standard Handbook of Lubrication Engineering*, J. J. O'Connor, Ed., McGraw-Hill Book Company, NY, Chapter 23, 1968, pp. 23-1 to 23-25.
[640] Piermaria, D., and Michaelis, K., "Recognizing Gear Failures," published 6/21/2007, (http://machinedesign.com/ContentItem/67968/Recognizinggearfailures.aspx)]
[641] Volin, R. H., "Literature Review: Techniques and Applications of Mechanical Signature,"
[642] Martin, H. R., "Vibration Analysis," *American Society of Metals Handbook, Friction, Lubrication, and Wear Technology*, S. D. Henry, Ed., ASM International, Vol. 18, 1992 pp. 293–298.
[643] "Gear Diagnostic Parameters (Vibration Based)," http://mfpt.org/Archive%20OLD%20Pages/mfpt_geardiagnosticparam.html.
[644] "A Model-Based Gear Diagnostic Technique," W. Wang and A. K. Wong, [http://dspace.dsto.defence.gov.au/dspace/bitstream/1947/4127/1/DSTO-TR-1079%20PR.pdf]
[645] "Diagnostics of Gearing and Gear Couplings Using Envelop Spectrum Methods," A. Barkov and N. A. Barkova; http://www.geartools.co.uk/mambo/index.php?option=com_content&task=view&id=54&Itemid=1
[646] Kennedy, Jr., F. E., "Surface Temperature Measurement," *American Society of Metals Handbook, Friction, Lubrication, and Wear Technology*, S. D. Henry Ed., ASM International, Vol. 18, 1992, pp. 438–44.
[647] Lockwood, F. E. and Dalley, R., "Lubricant Analysis," *American Society of Metals Handbook, Friction, Lubrication, and Wear Technology*, S. D. Henry, Ed., ASM International, Vol. 18, 1992, pp. 299–312.
[648] Ives, L. K., "Electron Microscopy," *American Society of Metals Handbook, Friction, Lubrication, and Wear Technology*, S. D. Henry, Ed., ASM International, Vol. 18, 1992 pp. 376–392.
[649] Bala, V., "Gear Lubricants," *Fuels and Lubricants Handbook: Technology, Properties, Performance, and Testing*, G. E. Totten, Ed., ASTM International, West Conshohocken, PA, Chapter 16, 2003, pp. 431–463.
[650] Herring, D. H., and G. D. Lindell, "Heat Treating Heavy Duty Gears," GEARSolutionsonline, www.gearsolutionsonline.com
[651] Errichello, R. "Friction, Lubrication, and Wear of Gears," *American Society of Metals Handbook, Friction, Lubrication, and Wear Technology*, S. D. Henry, Ed., *ASM International*, Vol. 18, 1992, pp. 535–545.
[652] Dekock, J., "Laser Heat Treating Advances for the Gear Industry," GEARSolutionsonline, www.gearsolutionsonline.com
[653] Herring, D. H., and Balme, S. D., "Oil Quenching Technologies for Gears," GEARSolutionsonline, www.gearsolutionsonline.com.
[654] "New Standards Released: ANSI/AGMA 1003-H07 and ANSI/AGMA 1103-H07; ANSI/AGMA 9005-E-02," www.agma.org//AM/Template.cfm?Section=Home
[655] Benedict, G. H., "Gears," *Standard Handbook of Lubrication Engineering*, J. J. O'Connor, Ed., McGraw-Hill Book Company, NY, Chapter 20, 1968, pp. 20-1 to 20-30.
[656] Stucker, J. B., "Rear Axle and Gear Box Lubricants," *CRC Handbook of Lubrication, Vol. I, Theory and Practice of Tribology: Applications and Maintenance*, R. E. Booser, Ed., CRC Press, Boca Raton, 1983 pp. 55–69.
[657] Pilon, E. L., "Gear Lubrication-II," *Lubrication*, Vol. 66, 1980, pp. 13–24.
[658] Johnson, W. A., "Laboratory and Field Findings Pertaining to Axle Lubricants," *Presented at SAE Fuels and Lubricants Meeting*, Chicago, June 5–10, 1960.
[659] Kamura, Y., "Some Problems in the Adhesion Theory of Friction," *Fundamentals of Tribology*, N. P. Suh and N. Saka, Eds., Published by MIT Press, Cambridge, 1980 pp. 385–391.
[660] Roux, F., "Some New Oil Testing Devices, in the Development Stage, for European Engines," *Automotive Engines and Transmissions*, C. F. McCue, J. C. Cree, and R. Tourrets Eds., Applied Science Publishers, Barking, England, 1973.
[661] Radovich, J. L., "Gears," *CRC Handbook of Lubrication, Vol. II, Theory and Practice of Tribology: Theory and Design*, R. E. Booser, Ed., CRC Press, Boca Raton, 1988, pp. 539–564.
[662] "Lubrication of Gears," http://www.qtcgears.com/Q410/QTC/Q410P440.htm
[663] "Gears Lubrication," http://www.agroengineers.com/gears/gears-lubrication.shtml
[664] "Axle and Manual Transmission Lubricant Viscosity Classification," SAE J306-SAE Standard, Society of Automotive Engineers.
[665] "Lubricant Service Designations for Automotive Manual Transmissions and Axles," API Publication No. 1560, American Petroleum Institute.
[666] Sutherland, J. M., "Proposed Automotive Gear Lubricant Categories: Their Impact on the Industry," *Presented at NPRA National Fuels and Lubricants Meeting*, Houston, TX, Nov. 2–3, 1989.
[667] Blau, P. J., "Glossary of Terms," *American Society of Metals Handbook, Friction, Lubrication, and Wear Technology*, S. D. Henry, Ed., ASM International, Vol. 18, 1992, pp. 1–21.
[668] "Lubricating Automotive Gears," Report No. C-7626, The Lubrizol Corporation, May 1983.
[669] Olszewski, W. F., Nieswender, D. D., Stanek, J. M., and Taylor, D. J., "Performance Characteristics of Hypoid Gear Lubricants," *Performance and Testing of Gear Oils and Transmission Fluids*, R. Tourret and E. P Wright, Eds., Heyden and Son, Chapter 17, 1981, pp. 293–305.
[670] "Lubricant Service Designations for Automotive Manual Transmissions, Manual Transaxles, and Axles," *API Publica-*

tion 1560, 7th ed., July 1995, http://www.upmpg.com/pdf/gear_lube_designations.pdf

[671] "Limited Slip Differentials," *Lubrication*, 564, 1970, pp. 49–60.

[672] Chocholek, S. E., "The Development of Differential for the Improvement of Traction Control," www.torsen.com/files/traction_control_article.pdf.

[673] Nice, K., "How Differentials Work," http://auto.howstuffworks.com/differential.htm

[674] "Progress is Slow on PG-1 and PG-2," *Lubrizol NewsLine*, Vol. 111, Jan. 1993.

[675] Root, D. C., "Selecting the Right Gear Oil," , Vol. 32, 1976, p. 8.

[676] "Sulfurized Polyisobutylene Based Wear and Oxidation Inhibitors," U.S. Patent 6,884,855 (April 16, 2005). Inventors: K. D. Nelson and F. Plavac.

[677] "Sulfide- and Polysulfide-containing Lubricating Oil Additive Compositions and Lubricating Compositions Containing the Same," U.S. Patent 6,689,723 (February 10, 2004). Inventors: W. T. Sullivan, P. R. Kanga, T. S. Coolbaugh.

[678] "Manual Transmission Lubricants with Improved Synchromesh Performance," U.S. Patent 6,617,287 (September 9, 2003). Inventors: M. P. Gahagan and B. M. O'Connor.

[679] "Oil Additive Concentrates and Lubricants of Enhanced Performance Capabilities," U.S. Patent 5,328,619 (July 12 1994). Inventor: G. S. Conary.

[680] "Turbine," http://en.wikipedia.org/wiki/Turbines

[681] "Gas Compressor," http://en.wikipedia.org/wiki/Gas_compressor

[682] Garg, D., Totten, G. E., and Webster, G. M., "Compressor Lubricants," *Fuels and Lubricants Handbook: Technology, Properties, Performance, and Testing*, G. E. Totten, Ed., ASTM International, West Conshohocken, PA, Chapter 14, 2003, pp. 383–412.

[683] Kist, K. R. and Doperalski, E. J., "Brief Introduction to the Screw Compressor," Paper 68E, *AIChE 86th National Meeting*, 1979.

[684] Cohen, S. C., "Development of a 'Synthetic' Compressor Oil Based on Two-stage Hydrotreated Petroleum Basestocks," *Lubr. Eng.*, Vol. 44, No. 3, 1987, pp. 230–238.

[685] Matthews, P. H. D., "Lubrication of Reciprocating Compressors," *Journal of Synthetic Lubricants*, Vol. 6, No. 4, 1990, pp. 292–317.

[686] Short, G. D., "Development of Synthetic Lubricants for Extended Life in Rotary Screw Compressors," *Lubr. Eng.*, Vol. 40, No. 8, 1983, pp. 463–470.

[687] Van Ormer, H. P., "Trim Compressed Air Cost with Synthetic Lubricants," *Power*, 1987, pp. 43–45.

[688] Matthews, P. H. D., Walkden, R. H., and Williams, G. C., "Current and Future Developments of Lubricants for Industrial Air Compressors," Paper Number C477/023/94, IMechE Conference, *Proceedings of the Institute of Mechanical Engineers*, 1994, pp. 175–183.

[689] Takashima, H., "Gas Compressor Oils," *Toraiborojisuto*, Vol. 38, 1993, pp. 135–139.

[690] Kajdas, C., "Industrial Lubricants," *Chemistry and Technology of Lubricants*, R. M. Mortier and S. T. Orszulik, Eds., Blackie Academic and Professional-Chapman & Hall, London, UK, 1996, Chapter 8, pp. 228–265.

[691] Kajdas, C., "Additives for Metalworking Lubricants. A Review," *Sci. Lubr.*, Vol. 1, pp. 385–409.

[692] Sutherland, J. W., Olson, W. W., and Bergstrom, A., "Dry Machining," Summary of Manufacturing Research 1996–97, Department of Mechanical Engineering and Engineering Mechanics, Michigan Technological University, Houghton, MI.

[693] Cantor, N. M., "Dry or Near-dry Machining Won't Replace Traditional Metalworking Fluid Technology Anytime Soon, But They Do Offer Advantages for Some Niche Markets," *Tribology and Lubrication Technology*, Nov. 2003, p. 31.

[694] Tolfa, J. C., "Synthetic Lubricants Suitable for Use in Process and Hydrocarbon Gas Compressors," *Lubr. Eng.*, Vol. 47, No. 4, 1990, pp. 289–295.

[695] Miller, J. W., "Synthetic and HVI Compressor Lubricants," *Journal of Synthetic Lubricants*, Vol. 6, No. 2, 1989, pp. 107–122.

[696] Miller, J. W., "Synthetic Lubricants and Their Industrial Applications," *Journal of Synthetic Lubricants*, Vol. 1, No. 2, 1984, pp. 136–152.

[697] Garg, D. R., "Polyalkylene Glycol-Based Compressor Lubricants," Presented at the *6th Annual Reciprocating Compressor Conference*, Salt Lake City, UT, 23–26 Sept. 1993.

[698] Patzau, S. and Szchawnicka, E., "Oils for Air and Technical Gas Compressors," *Trybologia*, Vol. 20, No. 4, 1989, pp. 18–21.

[699] Randles, S. J., "Esters," *Synthetic Lubricants and High Performance Functional Fluids*, R. L. Rudnick and R. L. Shubkin, Eds., Marcel Dekker, Inc., NY, 1999, pp. 63–102.

[700] Miller, J. W., "New Synthetic Food Grade Rotary Screw Compressor Lubricants," *Lubr. Eng.*, Vol. 40, No. 7, 1983, pp. 433–436.

[701] Oberle, J. and Rajewski, T., "The Development of Lubricants for Ammonia Refrigeration Systems," *Presented at the HAR 19th Annual Meeting*, 23–26 March 1997, New Orleans, LA.

[702] Mills, A. J., Tempest, M. A., and Thomas, A. S., "Performance Testing of Rotary Screw Compressor Fluids in Europe," *Lubr. Eng.*, Vol., No. 5, 1986, pp. 278–286.

[703] Henderson, D. R., Rajewski, T. E., Speich, C. F., and Grieg, B., "Lubricants in Refrigerant Systems," *ASHRAE Handbook-Fundamentals*, ASHRAE Publications, Atlanta, Chapter 7, 1997.

[704] (a) Devos, R., Davis, TH. 1997, "HFC-134a Qualification, Industry Refrigerator Capillary Data," *Proceedings of International Conference on Ozone Protection Technologies*; 1997, November 12–13. Baltimore, MD, Arlington, VA: Alliance for Responsible Atmospheric Policy, p. 10. (b) DeVos, R., "R-134a Qualification-Industry Refrigerator Capillary Data," *ASHRAE Transactions*, BN-97-7-4, 1997, p. 1.

[705] Michels, H. H. and Sienel, T. H., "Refrigeration Lubricants," *Fuels and Lubricants Handbook: Technology, Properties, and Testing*, G. E. Totten, Ed., ASTM International, West Conshohocken, PA, Chapter 15, 2003, pp. 413–430.

[706] Sunami, M., Takidawa, K., Suda, S., and Sasoki, U., "New Immiscible Refrigeration Lubricants for HFCs," *ASHRAE Transactions*, SD-95-83, 1995, p. 940.

[707] Sunami, M., Takidawa, K., and Suda, S., "New Immiscible Refrigeration Lubricants for HFCs," *Proceedings of ASHRAE Purdue CFC Conference*, Purdue University, West Lafayette, IN, 1994, p. 129.

[708] Gunderson, R. C., and Hart, A. W., *Synthetic Lubricants*, Reinhold, NY, 1962.

[709] Glova, D. J., "High Temperature Stability of Refrigerants in

Lubricating Oils," *ASHRAE Transactions*, Vol. 90, No. 4, 1984, p. 806.

[710] Takagi, M., "New Type Refrigeration Lubricants for Alternative Refrigerants," *Energy Resources*, Vol. 16, 1995, p. 497.

[711] "Food Grade Lubricants Reduce Contamination Threats for Food and Beverage Manufacturers," by D. Hodson, Shell Cassida, *Machinery Lubrication Magazine*, Jan. 2004, http://www.machinerylubrication. com/ article_detail.asp?articleid=572&relatedbookgroup =Lubrication2

[712] "Understanding Food-Grade Lubricants," by M. Williamson, *Machinery Lubrication Magazine*, Jan. 2003.

[713] "Title 21: Food and Drugs," www.accessdata. fda.gov/ scripts/cdrh/cfdocs/cfCFR/CFRSearch.cfm.

[714] "Inventory of Effective Food Contact Substance Notifications," http://www.cfsan.fda.gov/~dms/opa-fcn.html

[715] "Transformer," http://en.wikipedia.org/wiki/ Transformer

[716] "Transformer Oil Purchase Specification—Test Limits," http://www.doble.com/services/ lab_services_transformer_oil_specs.html

[717] "Concrete Form Release Compositions," U.S. Patent 6,811,810 (November 2, 2004). Inventors: V. S. LaFay and S. L. Neltner.

[718] Vacuum Pump Fluids, http://www.2spi.com/catalog/vac/vac-fluids.html

[719] Burke, L. J., "Compressors and Vacuum Pumps," *CRC Handbook of Lubrication, Vol. I, Theory and Practice of Tribology: Applications and Maintenance*, R. E. Booser, Ed., CRC Press, Boca Raton, 1983, pp. 135–147.

[720] Cichelli, E., Unangst, W. C., and Stanzione, L. A., "Wire Rope and Chain," *CRC Handbook of Lubrication, Vol. I, Theory and Practice of Tribology: Applications and Maintenance*, R. E. Booser, Ed., CRC Press, Boca Raton, 1983, pp. 193–208.

[721] Turner, J. E. and Barnes, C., "Lubrication Basics for Wire Ropes," *Machinery Lubrication Magazine*, July 2002.

[722] "Turbine Oils with Excellent High Temperature Oxidative Stability," U.S. Patent 6,326,336 (December 4, 2001). Inventors: V. J. Gatto, H. Ryan, and R. N. Iyer.

[723] "Additive Composition for Industrial Fluid," U.S. Patent 6,645,920 (November 11, 2003). Inventor: B. J. Butke.

[724] "Turbine Lubricant," U.S. Patent 4,101,431 (July 18, 1978). Inventor: L. F. Kuntschik.

[725] "High Temperature Compressor Oil," U.S. Patent 5,156,759 (October 20, 1992). Inventor: D. H. Culpon, Jr.

[726] (a) "Refrigerant Lubricant Compositions," U.S. Patent 5,152,926 (October 6, 1992). Inventor: W. L. Brown. (b) "Pneumatic Tool Lubricants," U.S. Patent 2,918,429 (December 22, 1959). Inventors: Melvin E. Gililland and Richard C. Givens.

[727] *NLGI Lubricating Grease Guide*, National Grease Institute, Missouri, 1987. Revised 1996, 2006

[728] Wilson, T. C., "Modern Automotive Greases," SAE Paper 841212, Fuels and Lubricants Technology: An Overview, SP.603, Society of Automotive Engineers, Oct. 1984, pp. 101–106.

[729] Boehringer, R. H., "Grease," *American Society of Metals Handbook, Friction, Lubrication, and Wear Technology*, Scott D. Henry, Ed., ASM International, Vol. 18, 1992, pp. 123–131.

[730] (a) Gow, G., "Lubricating Grease," *Chemistry and Technology of Lubricants*, R. M. Mortier and S. T. Orszulik, Eds., Blackie Academic and Professional-Chapman & Hall, London, UK, Chapter 11, 1992, pp. 306–319. (b) Figure courtesy of SKF.

[731] Sinistyn, V. V., *The Choice and Application of Plastic Greases*, Khimiya, Moscow.

[732] Khan, K., "Global Grease Production Rises Sharply," *Lube Report*, Vol. 8, No. 25, June 18, 2008.

[733] Bartz, W. J., "Lubricating Greases Properties and Their Classification Especially Synthetic Greases," *NLGI Spokesman*, July 2004, Vol. 68, No. 4, pp. 15–24.

[734] "Synthetic Lubricating Greases Containing Metal Chelates of Schiff Bases," U.S. Patent 5,147,567 (September 15, 1992). Inventors: V. S. Agarwala, A. A. Conte, Jr., K. S. Rajan, and P. K. Sen.

[735] "Polyurea Grease with Reduced Oil Separation," U.S. Patent 4,759,859 (July 26, 1988). Inventor: J. A. Waynick.

[736] "Polyurea-based Extreme Pressure Greases," U.S. Patent 4,100,081 (July 11, 1978). Inventor: J. L. Dreher.

[737] Rumierz, J., "Lubricants for Rolling Element Bearings," *American Society of Metals Handbook, Friction, Lubrication, and Wear Technology*, S. D. Henry, Ed., ASM International, Vol. 18, 1992, pp. 132–138.

[738] Silverstein, R. and Rudnick, L., "Additives for Grease Applications," *Lubricant Additives: Chemistry and Applications*, L. R. Rudnick, Ed., Marcel Dekker, Inc., NY, Chapter 19, 2003, pp. 513–539.

[739] King, J. P., *Presented at 1998 NLGI Annual Grease Meeting*, Dublin, Ireland.

[740] Tocci, L., "Transparent Grease Bears a Heavy Load," *Lubes-N-Greases*, April 1997, Vol. 3, pp. 52–54.

[741] Ogawa, T., Kimura, H., and Hayama, M., *NLGI Spokesman*, Vol. 62, No. 6, 1998, pp. 28–36.

[742] Ruge, I. W., "Lubricating Greases—Characteristics and Selection," *CRC Handbook of Lubrication, Theory and Practice in Tribology, Vol. II, Theory and Design*, R. E. Booser, Ed., CRC Press, Boca Raton, 1988, pp. 255–267.

[743] "Dielectric Constants," http://www.orioninstruments. com/html/dielectric_constants.asp

[744] "Ready Reference for Grease," a *Lubrizol Publication*.

[745] "Automotive Lubrication Greases," SAE J310-Revised June 9, SAE Standard, SAE Handbook, Society of Automotive Engineers.

[746] Bartz, W. J., "Lubricating Greases Properties and their Classification Especially Synthetic Greases," *Presentation in 2003 NLGI Back to Basics Course*.

[747] "Understanding Biobased/Biodegradable and the Industry's Standardized Tests and Definitions," http://www. renewablelube.com/products/pdf/standardized. pdf

[748] Verdura, T. M., Brunette, G., and Shah, R., "Lubricating Greases," *Fuels and Lubricants Handbook: Technology, Properties, Performance, and Testing*, MNL37WCD, Ed., ASTM International, West Conshohocken, PA, Chapter 20, 2003, pp. 557–572.

[749] Hunter, M. E., and Baker, R. F., "Corrosion," *Fuels and Lubricants Handbook: Technology, Properties, Performance, and Testing*, MNL37WCD, Ed., ASTM International, West Conshohocken, PA, Chapter 31, 2003, pp. 825–832.

[750] "Grease Composition with Improved Extreme Pressure and

Antiwear Properties," U.S. Patent 5,246,604 (September 21, 1993). Inventor: P. F. Vartanian.

[751] "Grease Composition for Constant Velocity Joints," U.S. Patent 6,656,890 (December 2, 2003). Inventors: G. Fish, J. S. Cooper, J. E., T. D. Davies, and G. R. Isaac.

[752] "Grease Composition with Improved Rust Prevention and Abrasion Resistance Properties," U.S. Patent 6,800,595 (October 5, 2004). Inventors: T. Ozaki, N. Shinoda, and K. Nakazawa.

[753] "General-purpose Grease Composition," U.S. Patent 4,582,616 (April 15, 1986). Inventors: T. Kita, M. Ikeda, and A. Aseya.

[754] "Biodegradable Grease Composition," U.S. Patent 4,115,282 (September 19, 1978). Inventor: H. D. Grasshoff.

[755] "Merchant, M. E., "Deformation," *Standard Handbook of Lubrication Engineering*, J. J. O'Connor, Ed., McGraw-Hill Book Company, NY, Chapter 8, 1968, pp. 8-1 to 8-3.

[756] Forming: Lubrication of Steel Sheet and Strip for Forming, Technical Bulletin Tb-F1 Rev 2, November 2003 [www.bluescopesteel.com.au/go/tools-and-resources/technical-library/forming-bulletins]

[757] Laemmle, J. T., "Metalworking Fluids," *American Society of Metals Handbook, Friction, Lubrication, and Wear Technology*, S. D. Henry, Ed., ASM International, Vol. 18, 1992, pp. 139–149.

[758] "Standard Classification of Metalworking Fluids and Related Materials," ASTM Standard D 2881, *Annual Book of ASTM Standards*, ASTM International, West Conshohocken, PA.

[759] *DIN Handbuch*—Mineralöle und Brennstoffe, Beuth Verlag GmbH, Berlin, May 2001.

[760] http://www.aral-lubricants.com/lubricants/com/technical/lubricants_abc/metalfluid.content.html

[761] Sargent, L. B., Jr., "Lubricant Performance," *Standard Handbook of Lubrication Engineering*, J. J. O'Connor, Ed., McGraw-Hill Book Company, NY, Chapter 27, 1968, pp. 27-1 to 27-21.

[762] Kelley, R., and Foltz, G., "Cutting Fluids," *CRC Handbook of Lubrication, Theory and Practice in Tribology, Vol. II, Theory and Design*, R. E. Booser, Ed., CRC Press, Boca Raton, 1988, pp. 357–369.

[763] (a) Liscic, B., Tensi, H. M., Totten, G. E., and Webster, G. M., "Non-lubricating Process Fluids: Steel Quenching Technology," *Fuels and Lubricants Handbook: Technology, Properties, Performance, and Testing*, George E. Totten, Ed., Chapter 22, pp. 587–634, ASTM International, West Conshohocken, Pennsylvania (2003). (b) "Corrosion Inhibiting Quenchant Compositions," U.S. Patent 4,595,425 (June 17, 1986). Inventor Ronald H. Harding.

[764] Mandakovic, R., "Assessment of EP Additives for Water Miscible Metalworking Fluids," *J. Syn. Lubr.*, Vol. 16, No. 1, 1999, pp. 13–26.

[765] *Standard Methods for Analysis and Testing of Petroleum and Related Products and British Standard 2000 Parts 1997*, John Wiley and Sons, N.Y., 1997.

[766] "Interactions of Additives and Lubricating Base Stocks," S. M. Hsu and A. L. Cummings, Lubricant and Additive Effects on Engine Wear, SP-558, Fuels and Lubricants Meeting, San Francisco, CA, Oct. 31–Nov. 3, 1983, pp. 61–70.

[767] De Hoffmann, E., Charette, J., and Stroobant, V., *Mass Spectrometry: Principles and Applications*, John Wiley & Sons, Dec. 1996.

[768] Colthup, N. B., Daly, L. H., and Wiberley, S. E., *Introduction to Infrared and Raman Spectroscopy*, 3rd ed., Academic Press, Sept. 1990.

[769] Macomber, R. S., *A Complete Introduction to Modern NMR Spectroscopy*, John Wiley & Sons, Dec. 1997.

[770] Derome, A. E., "Modern NMR Techniques for Chemistry Research," *Tetrahedron Organic Chemistry Series*, J. E. Baldwin and P. D. Magnus, Senior Eds., Pergamon Press, Oxford, 1993.

[771] "Reduced Pressure Boiling Point Calculation by JAVA Applet," http://www.pirika.com/chem/TCPEE/BP/RPBP.htm

[772] Klaus, E. E., and Bieber, H. E., "Effects of Some Physical and Chemical Properties of Lubricants on Boundary Lubrication," *ASLE Trans.*, Vol. 7, No. 1, 1964.

[773] Fein, R. S., "Chemistry in Concentrated Conjunction Lubrication," Interdisciplinary Approach to the Lubrication of Concentrated Contacts, National Aeronautics and Space Administration, Washington, D.C., Chapter 12, 1970.

[774] Balasubramaniam, K., Sankaran, S., and Rajwade, A. V., "Ultrasonic NDE Rheological Measurement Tools For Industrial Process Control," (http://www.ndt.net/article/wcndt2004/pdf/in-process_ndt-nde/738_balasubramaniam. pdf).

[775] "Alphabetical List of Federal Standards," http://apps.fss.gsa.gov/pub/fedspecs/fedspecs.cfm?sort=8&firstTime=Y&RequestTimeout=500

[776] Lockwood, F. E., and Klaus, E. E., "Ester Oxidation Under Simulated Boundary Lubrication Conditions," *ASLE Trans.*, Vol. 24, 1981, p. 276.

[777] Khonsari, M., and Booser, E. R., "Predicting Lube Life—Heat and Contaminants Are the Biggest Enemies of Bearing Grease and Oil," *Machinery Lubrication Magazine*, Sep. 2003.

[778] Nadkarni, R. A., "Overview," *Modern Instrumental Methods of Elemental Analysis of Petroleum Products and Lubricants, ASTM STP 1109*, R. A. Nadkarni, Ed., ASTM International, West Conshohocken, PA, 1991, p. 1.

[779] Nadkarni, R. K., "Elemental Analysis," *Fuels and Lubricants Handbook: Technology, Properties, Performance, and Testing*, MNL37WCD, Ed., ASTM International, West Conshohocken, PA, Chapter 26, 2003, pp. 707–716.

[780] Williams, M. C., *Modern Instrumental Methods of Elemental Analysis of Petroleum Products and Lubricants, ASTM STP 1109*, R. A. Nadkarni, Ed., ASTM International, West Conshohocken, PA, 1991, p. 96.

[781] Shugar, G., and Ballinger, J., *Chemical Technicians' Ready Reference Handbook*, 4th Ed., McGraw-Hill, Inc., NY, 1996, pp. 831–896.

[782] Altgelt, K., and Gouw, T., *Chromatography in Petroleum Analysis*, MarcellDekker, Inc., NY, 1979.

[783] Norris, T. A., "Chromatography II," *Lubrication*, Vol. 65, No. 2, 1979.

[784] "High Performance Liquid Chromatography," http://en.wikipedia.org/wiki/High_performance_liquid_chromatography

[785] Denis, J., Briant, J., and Hipeaux, J., "Lubricant Properties—Analysis and Testing," Editions Technip, Paris 2000.

[786] "Infrared Spectroscopy Correlation Table," http://en.

wikipedia.org/wiki/Infrared_spectroscopy_correlation_table

[787] "Infrared Spectroscopy," http://en.wikipedia.org/wiki/Infrared_spectroscopy

[788] "Nuclear Magnetic Resonance Spectroscopy," http://en.wikipedia.org/wiki/Nuclear_magnetic_resonance_spectroscopy

[789] "Proton NMR," http://en.wikipedia.org/wiki/Proton_NMR.

[790] "Carbon-13 NMR," http://en.wikipedia.org/wiki/Carbon-13_NMR

[791] Sahoo, S. K., Sharma, B. K., and Singh, I. D., "NMR Studies of Lube Distillates for Making Quality Base Oils," Preprints (ACS Division of Petroleum Chemistry), Vol. 45, 1999.

[792] Sarpal, A. S., Kapur, G. S., Mukherjee, S., and Jain, S. K., "Characterization by 13C-NMR Spectroscopy of Base Oils Produced by Different Processes," *Fuel*, Vol. 76, No. 10, 1997, pp. 931–937.

[793] Nessel, C. S., Coker, D. T., King, A. G., and Mumford, D. L., "Carcinogenic Assessment of Lubricant Base Oils by Proton NMR," Preprints (ACS Division of Petroleum Chemistry), Vol. 42, No. 1, 1997, pp. 251–254.

[794] Adhvaryu, A., Pandey, D. C., and Singh, I. D., "Effect of Composition on the Degradation Behavior of Base Oil," Preprints (ACS Division of Petroleum Chemistry, Inc.), Vol. 42, No. 1, 1997, pp. 225–228.

[795] Baker, A. E., "Lubricant Properties and Test Methods," *CRC Handbook of Lubrication, Vol. I, Theory and Practice of Tribology: Applications and Maintenance*, R. E. Booser Ed., CRC Press, Boca Raton, 1983, pp. 481–515.

[796] Moules, C., "The Role of Interfacial Tension Measurement in the Oil Industry," http://nttworld.com/docs/camtel.pdf

[797] Wedeven, L. D., "Bench Test Modeling," *Fuels and Lubricants Handbook: Technology, Properties, Performance, and Testing*, G. E. Totten Ed., ASTM International, West Conshohocken, PA, Chapter 36, 2003, pp. 963–1015.

[798] Czichos, H., "Design of Friction and Wear Experiments," *American Society of Metals Handbook*, 10th Ed., S. D. Henry Ed., ASM International, Vol. 18, 1992, pp. 480–488.

[799] Czichos, H., "Systems Analysis," *CRC Handbook of Lubrication, Theory and Practice in Tribology, Vol. II, Theory and Design*, R. E. Booser, Ed., CRC Press, Boca Raton, 1988, pp. 645–661.

[800] John, P. W. M., *Statistical Design and Analysis of Experiments*, Macmillan, London, 1971.

[801] Junemann, H., "Mechanical Tests for Lubricants," *Erdol Köhle*, Vol. 29, 1976, p. 259.

[802] List of Standards, http://64.233.179.104/translate_c?hl=en&u=http://www.bz01.com/stdfile/NF1.html&prev=/search%3Fq%3Dafnor%2Bt60-133%26hl%3Den]

[803] Rudnick, L., "Summary of Lubricant Standard Test Methods and Some Product Specifications," *Lubricant Additives: Chemistry and Applications*, L. R. Rudnick, Ed., Marcel Dekker, Inc., New York, Chapter 23, 2003, pp. 599–616.

[804] Granchi, M. P., Biggerstaff, J. A., Hilliard, L. J., and Grey, P., "Use of a Robot and Flow Injection for Automated Sample Preparation and Analysis of Used Oils by ICP Emission Spectrometry," *Spectrochim. Acta*, Vol. 42B (No. 1/2), 1987, pp. 169–180.

[805] Evans, S. J., and Klueppel, R. J., "Analysis of Oil With Inductively Coupled Plasma: Total Process Automation," *Spectrochim. Acta*, Vol. 40B (No. 1/2), 1985, pp. 49–55.

[806] Liu, Y., Harding, A. R., and Leyden, D. E., "Determination of Wear Metals in Oil by X-Ray Spectrometry," *Anal. Chim. Acta*, Vol. 180, 1986, pp. 349–355.

[807] Bowen, E. R., and Seifert, W. "Ferrography—A New Tool for Analyzing Wear Conditions," *Presented at Fluid Power Testing Symposium*, Milwaukee, WI, 16–18 August 1976.

[808] Anderson, D. P., and Driver, R. D., "Equilibrium Particle Concentration in Engine Oil," *Wear*, Vol. 56, 1979, pp. 415–419.

[809] Colyer, C. C., "The ASTM Test Monitoring System—Purpose and Operation," *ASTM Stand. News*, Oct. 1981, pp. 8–12.

[810] Colyer, C. C., "The SAE Lubricants Review Institute—Purpose and Operation," *Presented at the International Congress and Exposition*, Detroit, MI, Feb. 28—March 4, 1983.

[811] Colyer, C. C., Davis, K. E., Desing, M. E., Jaynes, S. E., and Copan, W. G., "Field Testing: The Final Proof of Performance," C. C. Colyer, *Presented at the XXII FISITA Congress*, Dearborn, MI, Sept. 25–29, 1988.

[812] Environmental Health and Safety Publications Series on Emission Scenario Documents No. 10 EMISSION SCENARIO DOCUMENT ON LUBRICANTS AND LUBRICANT ADDITIVES Environment Directorate Organisation for Economic Co-operation and Development Dec. 2004.

[813] Abner, E., Jr., "Lubricant Deterioration in Service," *CRC Handbook of Lubrication, Vol. I, Theory and Practice of Tribology: Applications and Maintenance*, R. E Booser Ed., CRC Press, Boca Raton, 1983, pp. 517–532.

[814] Keller, G. R., *Hydraulic System Analysis, Hydraulic & Pneumatics*, Cleveland, 1969.

[815] Kaufman, R. E., "Rapid Determination of Remaining Useful Lubricant Life," *CRC Handbook of Lubrication, Theory and Practice in Tribology, Vol. III, Monitoring, Materials, Synthetic Lubricants, and Applications*, R. E. Booser, Ed., CRC Press, Boca Raton, 1993, pp. 89–100.

[816] Sen, A. K., and Kumar, R., "Oxidation Stability of Lubricants by Inverse Phase Chromatography," *Lubr. Eng.*, Vol. 47, No. 3, 1991, p. 211.

[817] Naga, H. H., and Samen, A. E., "The Effect of Base Stock Volatility on Lubricating Oil Oxidation Stability," *Lubr. Eng.*, Vol. 44, No. 11, 1988, p. 931.

[818] Ravner, H., and Wohltjen, H., "The Determination of the Oxidative Stability of Several Deuterated Lubricants by an Electronic Gas Sensor," *Lubr. Eng.*, Vol. 39, No. 11, 1983, p. 701.

[819] Herguth, W. R., and Phillips, S., "Comparison of Common Analytical Techniques to Voltammetric Analysis of Antioxidants in Industrial Lubricating oils," *STLE 1994-Annual Meeting*, Kansas City. [http://www.herguth.com/technical/cyclic_voltametry.pdf#search=%22astm%20cyclic%20voltammetric%20%22].

[820] Kaufman, R. E., "Remaining Useful Life Measurements of Gas Turbine Engine Oils, Diesel Engine Oils, Automotive Oils, Hydraulic Fluids, and Greases Using Cyclic Voltammetric Methods," in *Condition Monitoring; International Conference Proceedings*, Joint Oil Analysis Program, Technical Support Center, Naval Air Station, Pensacola, FL, Nov. 1992.

[821] Becker, D. A., and Brinkman, D. W., "Recycling (Oil)," *Kirk-Othmer Encyclopedia of Chemical Technology, Fourth Edition*, M. Grayson Ed., John Wiley and Son, NY, Vol. 21, 1996, pp. 1–10.

[822] Gebarin, S. K., and Fitch, J., "Determining Proper Oil and Fil-

ter Change Intervals: Can Onboard Automotive Sensors Help," *Practicing Oil Analysis Magazine*, Jan. 2004, [http://www.practicingoilanalysis.com/article_detail.asp?articleid=562&relatedbookgroup=OilAnalysis].

[823] Bachelder, D. L., "Recycling Used Engine Oil by Re-refining," *Presented at 2005 Used Oil/HHW Conference "Building Hope for Healthy Communities, One Action at a Time,"* San Diego, CA, April 28, 2005, http://www.ciwmb.ca.gov/HHW/Events/AnnualConf/2005/April28/Session4/DIYers/ReRefine.pdf.

[824] Quang, D. V. et al., "Hydrocarbon Process," Vol. 53 No. 4, 1974, p. 129.

[825] Blankenship, F., "Do You Know Where Your Used Oil Goes?," *Machinery Lubrication Magazine*, Jan. 2005.

[826] Tocci, L., "What Does Biodegradable Really Mean?," *Lubes'n'Greases*, Vol. 10, No. 5, 2004, pp. 40–43.

[827] "European Eco-Label Home Page," http://ec.europa.eu/environment/ecolabel/index_en.htm

[828] Carnes, K., "Off-road Hydraulic Fluids: Beyond Biodegradability," *Tribology & Lubrication Technology*, Sept. 2004, pp. 33–40.

[829] Theodori, D., Saft, R. I., Krop, H., and Van Broekhuizen, P. V., "Concept Background Document: Development of Criteria for the Award of the European Eco-label to Lubricants," Presented Nov. 27, 2003, by the IVAM Research and Consultancy on Sustainability, Amsterdam, Netherlands. [http://ec.europa.eu/environment/ecolabel/pdf/lubricants/backgrounddoc_111203.pdf].

[830] Gill, G., "Blenders Brace as EU Adopts REACH," *Lube Report*, Vol. 6, No. 51, Dec. 20, 2006.

[831] Hinman, M. L., "Environmental Characteristics of Fuels and Lubricants," *Fuels and Lubricants Handbook: Technology, Properties, Performance, and Testing*, MNL37WCD, Ed., ASTM International, West Conshohocken, PA, Chapter 34, 2003, pp. 885–908.

[832] "A Practical Guide to Protecting Man and the Environment, Machelen," Exxon Chemical Europe, Inc., Machelen, Belgium, 1999.

[833] Lyman, W. J., "Octanol/Water Partition Coefficient," *Handbook of Chemical Property Estimation Methods: Environmental Behavior of Organic Compounds*, W. J. Lyman, W. F. Reehl, and D. H. Rosenblatt, Eds., McGraw-Hill, NY, 1982, pp. 1–54.

[834] Lipnick, R. L., "Structure-Activity Relationships, in Fundamentals of Aquatic Toxicology: Effects," *Environmental Fate and Risk Assessment*, G. M. Rand, Ed., Taylor & Francis, Washington, DC, 1995, pp. 609–656.

[835] *A Critical Review of Toxicity Values and an Evaluation of the Persistence of Petroleum Products for Use in Natural Resources Damage Assessments*, American Petroleum Institute, Washington, DC, 1995, p. 196.

[836] Lyman, W. J., "Transport and Transformation Processes, in Fundamentals of Aquatic Toxicology: Effects," *Environmental Fate and Risk Assessment*, G. M. Rand, Ed., Taylor and Francis, Washington DC, 1995, pp. 449–492.

[837] "OECD Series On the Test Guidelines Programme, Number 2—Detailed Review Paper on Biodegradability Testing Environment Monograph No. 98," Published by Organisation for Economic Co-Operation and Development, Paris 1995, OCDE/GD (95) 43, www.olis.oecd.org/...dd3491a17c199db2c12569fa005d35af/e846f5796657ed89c12561d40042a27b/$FILE/ENE51276.ENG

[838] ISO 10708: "Water Quality—Evaluation in an Aqueous Medium of the Ultimate Aerobic Biodegradability of Organic Compounds—Determination of Biochemical Oxygen Demand in a Two-phase Bottle Test," ISO, Geneva, 1997, p. 17.

[839] OECD Guideline for Testing of Chemicals: 301-Ready Biodegradability, Organization for Economic Cooperation and Development, Paris, 1992, p. 62.

[840] Harmonized Test Guidelines: 835-Fate, Transport and Transformation Test Guidelines—OPPTS 835.3110 Ready Biodegradability, U.S. Environmental Protection Agency, Office of Prevention, Pesticides, and Toxic Substances, Washington, DC, 1998, p. 45.

[841] "Lubricating Oil Base Stocks," API 2004 Publication, http://www.petroleumhpv.org/Product_Categories/Lube_Oils/2004-03-03_Lubes_Revised_Robust_Summary.pdf

[842] Guidance Document on Aquatic Toxicity Testing of Difficult Substances and Mixtures, OECD, Paris, 2000, p. 53.

[843] Guidance Document on Application and Interpretation of Single Species Tests in Environmental Toxicology, Environment Canada, Environment Technology Centre, Ottawa, Ontario, Canada, 1999, p. 203.

[844] "The NIST Reference on Constants, Units, and Uncertainty: International System of Units," http://www.physics.nist.gov/cuu/Units/index.html

[845] "Commonly Used Metric System Units, Symbols, and Prefixes," http://lamar.colostate.edu/~hillger/common.html

[846] http://www.simetric.co.uk/si_tz.htm